MINI Cooper

(R55, R56, R57)

Service Manual

Cooper, Cooper S, John Cooper Works (JCW)
including Clubman, Convertible
2007, 2008, 2009, 2010, 2011, 2012, 2013

BentleyPublishers.com

BENTLEY PUBLISHERS® | Automotive Reference™

Bentley Publishers, a division of Robert Bentley, Inc.
1734 Massachusetts Avenue
Cambridge, MA 02138 USA
800-423-4595 / 617-547-4170

Information that makes the difference®

BentleyPublishers®.com

Technical contact information

We welcome your feedback. Please submit corrections and additions to our MINI technical discussion forum at:

`http://www.BentleyPublishers.com`

Updates and corrections

We will evaluate submissions and post appropriate editorial changes online as updates or tech discussion. Appropriate updates and corrections will be added to the book in future printings. Check for updates and corrections for this manual before beginning work on your vehicle. See the following web address for additional information:

`http://www.BentleyPublishers.com/updates/`

WARNING—Important Safety Notice

Do not use this manual for repairs unless you are familiar with automotive repair procedures and safe workshop practices. This manual illustrates the workshop procedures for some maintenance and service work. It is not a substitute for full and up-to-date information from the vehicle manufacturer or for proper training as an automotive technician. Note that it is not possible for us to anticipate all of the ways or conditions under which vehicles may be serviced or to provide cautions as to all of the possible hazards that may result.

We have endeavored to ensure the accuracy of the information in this manual. Please note, however, that considering the vast quantity and the complexity of the service information involved, we cannot warrant the accuracy or completeness of the information contained in this manual.

FOR THESE REASONS, NEITHER THE PUBLISHER NOR THE AUTHOR MAKES ANY WARRANTIES, EXPRESS OR IMPLIED, THAT THE INFORMATION IN THIS MANUAL IS FREE OF ERRORS OR OMISSIONS, AND WE EXPRESSLY DISCLAIM THE IMPLIED WARRANTIES OF MERCHANTABILITY AND OF FITNESS FOR A PARTICULAR PURPOSE, EVEN IF THE PUBLISHER OR AUTHOR HAVE BEEN ADVISED OF A PARTICULAR PURPOSE, AND EVEN IF A PARTICULAR PURPOSE IS INDICATED IN THE MANUAL. THE PUBLISHER AND AUTHOR ALSO DISCLAIM ALL LIABILITY FOR DIRECT, INDIRECT, INCIDENTAL OR CONSEQUENTIAL DAMAGES THAT RESULT FROM ANY USE OF THE EXAMPLES, INSTRUCTIONS OR OTHER INFORMATION IN THIS MANUAL. IN NO EVENT SHALL OUR LIABILITY, WHETHER IN TORT, CONTRACT OR OTHERWISE, EXCEED THE COST OF THIS MANUAL.

Before attempting any work on your MINI, read **001 Warnings and Cautions** and any **WARNING** or **CAUTION** that accompanies a procedure in the manual. Review the **WARNINGS** and **CAUTIONS** each time you prepare to work on your MINI.

Your common sense and good judgment are crucial to safe and successful service work. Read procedures through before starting them. Think about whether the condition of your car, your level of mechanical skill or your level of reading comprehension might result in or contribute in some way to an occurrence which might cause you injury, damage your car or result in an unsafe repair. If you have doubts for these or other reasons about your ability to perform safe repair work on your car, have the work done at an authorized MINI dealer or other qualified shop.

Part numbers listed in this manual are for identification purposes only, not for ordering. Always check with your authorized MINI dealer to verify part numbers and availability before beginning service work that may require new parts.

Special tools required to perform certain service operations are identified in the manual and are recommended for use. Use of improper tools may be detrimental to the car's safe operation as well as the safety of the person servicing the car.

The vehicle manufacturer will continue to issue service information updates and parts retrofits after the editorial closing of this manual. Some of these updates and retrofits will apply to procedures and specifications in this manual. We regret that we cannot supply updates to purchasers of this manual.

ISBN 978-0-8376-1730-5 *Editorial closing 04 / 2014 Job code: BM13-02

Library of Congress Cataloging-in-Publication Data for the 2011 edition

Mini Cooper (R55, R56, R57) service manual : Cooper, Cooper S, John Cooper Works (JCW), including Clubman, Convertible : 2007, 2008, 2009, 2010, 2011.

p. cm.

Includes index.

ISBN 978-0-8376-1671-1 (alk. paper)

1. Mini-Cooper automobiles--Maintenance and repair--Handbooks, manuals, etc. I. Robert Bentley, inc.

TL215.M465M54 2011

629.28'722--dc23

2011022900

Manufactured in the United States of America

202509R001

Foreword

For the MINI owner with basic mechanical skills and for independent auto service professionals, this manual includes many of the specifications and procedures that were available in an authorized MINI dealer service department as this manual went to press. The MINI owner with no intention of working on his or her car will find that owning and referring to this manual will make it possible to be better informed and to more knowledgeably discuss repairs with a professional automotive technician.

For those intending to do maintenance and repair on their MINI, it is essential that safety equipment be used and safety precautions observed when working on the vehicle. A minimum safety equipment list includes hand protection, eye protection and a fire extinguisher. A selection of good quality hand tools is also needed. This includes a torque wrench to ensure that fasteners are tightened in accordance with specifications. In some cases, the text refers to special tools that are recommended or required to accomplish adjustments or repairs. These tools are often identified by their MINI / BMW special tool number and illustrated.

Disclaimer

We have endeavored to ensure the accuracy of the information in this manual. When the vast array of data presented in the manual is taken into account, however, no claim to infallibility can be made. We therefore cannot be responsible for the result of any errors that may have crept into the text. Please also read the **WARNING—Important Safety Notice** on the copyright page at the beginning of this book.

Prior to starting a repair procedure, read the procedure, **001 Warnings and Cautions** and the warnings and cautions that accompany the procedure. Reading a procedure before beginning work helps you determine in advance the need for specific skills, identify hazards, prepare for appropriate capture and handling of hazardous materials, and the need for particular tools and replacement parts such as gaskets.

Bentley Publishers encourages comments from the readers of this manual with regard to errors, and suggestions for improvement of our product. These communications have been and will be carefully considered in the preparation of this and other manuals. If you identify inconsistencies in the manual, you may have found an error. Please contact the publisher and we will endeavor to post applicable corrections on our website. Posted updates and corrections should be reviewed before beginning work. Please see the following web address:

`http://www.BentleyPublishers.com/updates/`

MINI will continue to issue service information and parts retrofits after the editorial closing of this manual. Some of this updated information may apply to procedures and specifications in this manual. For the latest information, please see the following web address:

`http://www.mintechinfo.com/`

MINI USA offers extensive warranties, especially on components of the fuel delivery and emission control systems. Therefore, before deciding to repair a MINI that may be covered wholly or in part by any warranties issued by MINI USA, consult your authorized MINI dealer. You may find that the dealer can make the repair either free or at minimum cost. Regardless of its age, or whether it is under warranty, your MINI is both an easy car to service and an easy car to get serviced. So if at any time a repair is needed that you feel is too difficult to do yourself, a trained MINI technician is ready to do the job for you.

Bentley Publishers

001 Warnings and Cautions

PLEASE READ THESE WARNINGS AND CAUTIONS BEFORE PROCEEDING WITH MAINTENANCE AND REPAIR WORK.

WARNINGS—
See also CAUTIONS

- Read the important safety notice on the copyright page at the beginning of the book.
- Some repairs may be beyond your capability. If you lack the skills, tools and equipment or a suitable workplace for any procedure described in this manual, we suggest you leave such repairs to an authorized MINI dealer service department or other qualified shop.
- Thoroughly read each procedure and the **WARNINGS** and **CAUTIONS** that accompany the procedure. Also review posted corrections at `www.BentleyPublishers.com/updates/` before beginning work.
- If any procedure, tightening torque, wear limit, specification or data presented in this manual does not appear to be appropriate for a specific application, contact the publisher or the vehicle manufacturer for clarification before using the information in question.
- Do not reuse fasteners that are worn or deformed. Many fasteners are designed to be used only once and become unreliable and may fail when used a second time. This includes, but is not limited to, nuts, bolts, washers, self-locking nuts or bolts, circlips and cotter pins. Replace these fasteners with new parts.
- Do not work under a lifted car unless it is solidly supported on stands designed for the purpose. Do not support a car on cinder blocks, hollow tiles or other props that may crumble under continuous load. Do not work under a car that is supported solely by a jack. Do not work under the car while the engine is running.
- If you are going to work under a car on the ground, make sure that the ground is level. Block the wheels to keep the car from rolling. Disconnect the battery negative (–) terminal to prevent others from starting the car while you are under it.
- Do not run the engine unless the work area is well ventilated. Carbon monoxide kills.
- Remove rings, bracelets and other jewelry so that they cannot cause electrical shorts, get caught in running machinery, or be crushed by heavy parts.
- Tie long hair behind your head. Do not wear a necktie, a scarf, loose clothing, or a necklace when you work near machine tools or running engines. If your hair, clothing, or jewelry were to get caught in the machinery, severe injury could result.
- Do not attempt to work on your car if you do not feel well. You increase the danger of injury to yourself and others if you are tired, upset or have taken medication or any other substance that may keep you from being fully alert.
- Illuminate your work area adequately but safely. Use a portable safety light for working inside or under the car. Make sure the bulb is enclosed by a wire cage. The hot filament of an accidentally broken bulb can ignite spilled fuel, vapors or oil.
- Catch draining fuel, oil, or brake fluid in suitable containers. Do not use food or beverage containers that might mislead someone into drinking from them. Store flammable fluids away from fire hazards. Wipe up spills at once, but do not store the oily rags, which can ignite and burn spontaneously.
- Observe good workshop practices. Wear goggles when you operate machine tools or work with battery acid. Wear gloves or other protective clothing whenever the job requires working with harmful substances.
- Greases, lubricants and other automotive chemicals contain toxic substances, many of which are absorbed directly through the skin. Read the manufacturer's instructions and warnings carefully. Use hand and eye protection. Avoid direct skin contact.
- Disconnect the battery negative (–) terminal whenever you work on the fuel system or the electrical system. Do not smoke or work near heaters or other fire hazards. Keep an approved fire extinguisher handy.
- Friction materials (such as brake pads and shoes or clutch discs) contain asbestos fibers or other friction materials. Do not create dust by grinding, sanding, or by cleaning with compressed air. Avoid breathing dust. Breathing any friction material dust can lead to serious diseases and may result in death.
- Batteries give off explosive hydrogen gas during charging. Keep sparks, lighted matches and open flame away from the top of the battery. If hydrogen gas escaping from the cap vents is ignited, it may ignite gas trapped in the cells and cause the battery to explode.
- The air-conditioning system is filled with chemical refrigerant, which is hazardous. Make sure the system is serviced only by a trained technician using approved refrigerant recovery/recycling equipment, trained in related safety precautions, and familiar with regulations governing the discharge and disposal of automotive chemical refrigerants.

Continued on next page

WARNINGS (continued)

- Do not expose any part of the A/C system to high temperatures such as open flame. Excessive heat increases system pressure and may cause the system to burst.
- Some aerosol tire inflators are highly flammable. Be extremely cautious when repairing a tire that may have been inflated using an aerosol tire inflator. Keep sparks, open flame or other sources of ignition away from the tire repair area. Inflate and deflate the tire at least four times before breaking the bead from the rim. Completely remove the tire from the rim before attempting any repair.
- Cars covered by this manual are equipped with a supplemental restraint system (SRS) that automatically deploys airbags and pyrotechnic seat belt tensioners in case of a frontal or side impact. These are explosive devices. Handled improperly or without adequate safeguards, they can be accidently activated and cause serious injury.
- The ignition system produces high voltages that can be fatal. Avoid contact with exposed terminals and use extreme care when working on a car with the engine running or the ignition switched ON.
- Place jack stands only at locations specified by the manufacturer. The vehicle lifting jack supplied with the vehicle is intended for tire changes only. Use a heavy duty floor jack to lift vehicle before installing jack stands. See **020 Maintenance**.
- Battery acid (electrolyte) can cause severe burns. Flush contact area with water, then seek medical attention.
- Aerosol cleaners and solvents may contain hazardous or deadly vapors and are highly flammable. Use only in a well ventilated area. Do not use on hot surfaces (engines, brakes, etc.).
- Due to risk of personal injury, be sure the engine is cold before beginning work on the cooling system.

CAUTIONS—
See also WARNINGS

- If you lack the skills, tools and equipment, or a suitable workshop for any procedure described in this manual, leave such repairs to an authorized MINI dealer or other qualified shop.
- MINI is constantly improving its cars and sometimes these changes, both in parts and specifications, are made applicable to earlier models. Part numbers listed in this manual are for reference only. Check with your authorized MINI dealer parts department for the latest information.
- Before starting a job, make certain that you have the necessary tools and parts on hand. Read all the instructions thoroughly, and do not attempt shortcuts. Use tools appropriate to the work and use only replacement parts meeting MINI specifications.
- Use pneumatic and electric tools only to loosen threaded parts and fasteners. Do not use these tools to tighten fasteners, especially on light alloy parts. Use a torque wrench to tighten fasteners to the tightening torque specification listed.
- Be mindful of the environment and ecology. Before you drain the crankcase, find out the proper way to dispose of the oil. Do not pour oil onto the ground, down a drain, or into a stream, pond or lake. Dispose of waste in accordance with federal, state and local laws.
- Electronic control modules cannot withstand temperatures from a paint-drying booth or a heat lamp in excess of 203°F (95°C). Do not subject to temperatures in excess of 185°F (85°C) for more than two hours.
- Before doing any electrical welding on cars equipped with ABS, disconnect the battery negative (–) terminal (ground strap) and the ABS control module connector. See **340 Brakes**.
- Make sure the ignition is switched OFF before disconnecting the battery.
- Label battery cables before disconnecting. On some models, battery cables are not color coded.
- Disconnecting the battery may erase fault code(s) stored in control module memory. Using special MINI diagnostic equipment, check for fault codes prior to disconnecting the battery cables. If the malfunction indicator light (MIL) is illuminated, see **OBD On-Board Diagnostics** at the back of this manual. (This light may be identified as the Check Engine light or the Service Engine Soon light.) If any other system faults are detected (indicated by an illuminated warning light), see an authorized MINI dealer.
- If a normal or rapid charger is used to charge battery, disconnect the battery and remove it from the vehicle in order to avoid damaging the vehicle.
- Do not quick-charge the battery (for boost starting) for longer than one minute. Wait at least one minute before boosting the battery a second time.
- Connect and disconnect a battery charger only with the battery charger switched OFF.
- Slow-charge a sealed or "maintenance free" battery at an amperage rate that is approximately 10% of the battery's ampere-hour (Ah) rating.
- Do not allow battery charging voltage to exceed 16.5 volts. If the battery begins producing gas or boiling violently, reduce the charging rate. Boosting a sulfated battery at a high charging rate can cause an explosion.

002 Vehicle Identification and VIN Decoder

Vehicle Identification Number (VIN), decoding

Some of the information in this manual applies only to cars of a particular model year or range of years. For example, 2007 refers to the 2007 model year but does not necessarily match the calendar year in which the car was manufactured or sold. To be sure of the model year of a particular car, check the Vehicle Identification Number (VIN) on the car.

The VIN is a unique sequence of 17 characters assigned by MINI to identify each individual car. When decoded, the VIN tells the country and year of manufacture; make, model and serial number; assembly plant and even some equipment specifications.

The MINI VIN is located on the top of the dashboard, on the driver's side where the number can be seen through the windshield. The table below explains some of the codes in the VIN for 2007 - 2013 MINI models.

Three platform codes are used by the manufacturer to identify the various MINI models:

- R55: Cooper Clubman, Cooper Clubman S
- R56: Cooper, Cooper S
- R57: Cooper convertible, Cooper S convertible

Sample VIN: WMWRC33506TK12345
position: 1 2 3 4 5 6 7 8 9 10 11 12 - 17

VIN position	Description	Decoding information	
1	Manufacturing Country	W	Germany
2	Manufacturer	M	MINI
3	Vehicle Type	W	Passenger vehicle
4	Series	M S Z	MINI to model year 2010 Hardtop from model year 2011 Clubman and Convertible models from 2011
5	Model	F G L M N P R S U V	Hardtop through 2010, Clubman from 2011 Clubman S from 2011 Clubman through 2010 Clubman S through 2010 Convertible S from 2011 Convertible S from 2011 Convertible through 2010 Convertible S to 2010 Cooper from 2011 Cooper S from 2011
6	Engine type	3, 7, 9	1.6 liter 4 cylinder gasoline engine
7	Vehicle type	3 C	to model year 2009 from model year 2010
8	Restraint system	1, 5	
9	Check digit		0 - 9 or X, calculated by NHTSA
10	Model year	7 8 9 A B C D	2007 2008 2009 2010 2011 2012 2013
11	Assembly plant	T	Oxford, UK
12-17	Serial number		Sequential production number for specific vehicle

This page intentionally left blank

010 Product Familiarization

GENERAL

BM11010004

The information included in this section is based on introductory information for MINI vehicles sold in the USA and Canada. The content provided here is intended to serve as a product familiarization guide. Note that the information presented is subject to change and should be used as a general reference only.

> ***WARNING—***
> - *Always check the MINI factory repair information or the publisher's website at www.bentleypublishers.com for information that may supersede any information included in this section.*

PRODUCT OVERVIEW

BM11010005

Following on the success of the first generation MINI, the second generation (R56) MINI was introduced to the US market in 2007 with two updated models - the R56 Cooper and Cooper S.

The R56 models were completely enhanced and upgraded. Every area of the R56 was modified to give the best possible product. Handling, ride quality and overall fuel economy was greatly improved by the use of new materials, engine technology and vehicle electronics.

Externally, at first glance, the R56 does not differ much from the first generation. The R56 maintained the unique and unmistakable features that make a MINI - a MINI. However, under the skin, the complete drivetrain was updated. There were two new completely redesigned engines, which share almost nothing in common with the previous (W10/11) engines.

Later body styles included the R55 Clubman (2008 introduction) and R57 Convertible (2009 introduction). And for 2011, the model lineup received a minor facelift, including two new engines.

The R56 models entered the market first with the 175 hp MINI Cooper S and the 120 hp MINI Cooper.

The MINI Cooper S was powered by a 1.6 liter turbocharged engine with extremely high specific output of more than 100 hp per liter, allowing acceleration to 60 mph in 7.1 seconds and providing a top speed of 225 km/h or 140 mph.

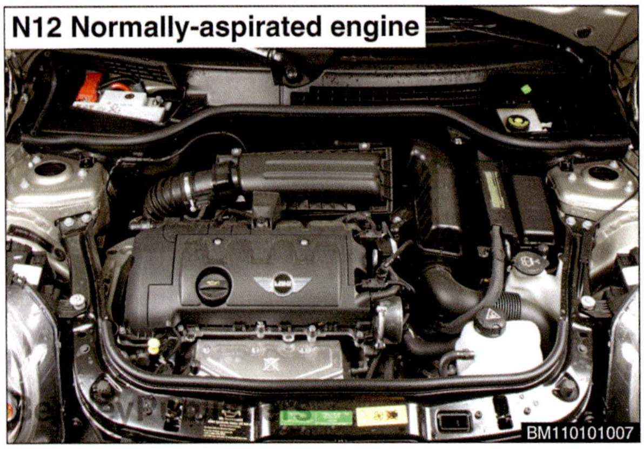

The innovative normally-aspirated engine in the MINI Cooper also displaced 1.6 liters and featured fully variable valve control with a 0 to 60 mph acceleration time of 9.1 seconds, and top speed of 203 km/h or 126 mph.

Technical Data - MINI Cooper MK II

		MINI Cooper	MINI Cooper S
Body			
No of doors/seats		3/4	3/4
Length/width/height (unladen))	mm	3,699 / 1,683 / 1,40	3,714 / 1,683 / 1,407
Wheelbase	mm	2,467	2,467
Track front/rear	mm	1,459/1,467	1,453/1,461
Turning circle	m	10.7	10.7
Fuel tank capacity (approx.)	ltr	40	50
Cooling system including heater	ltr	5.2	5.2
Engine oil	ltr	4.2	4.2
Manual transmission fluid		Lifetime fill	Lifetime fill
Weight, unladen, according to DIN / EU[1]	kg	1,065 / 1,140	1,130 / 1,205
Max permissible weight to DIN	kg	1,515	1,580
Max permissible axle load front/rear	kg	855 / 74	885 / 755
Max roof load	kg	75	75
Luggage compartment according to DIN	ltr	160–680	160–680
Air resistance	c_X x A	0.33 x 1.97	0.36 x 1.97
Engine			
Configuration/No of cylinders/valves		Inline4/4	Inline4/4
Capacity	cc	1,598	1,598
Bore/stroke	mm	77.0 / 85.8	77.0 / 85.8
Compression ratio	:1	11.0	10.5
Fuel	RON	from 91	from 91
Max output	kW/bhp	88 / 12	128 / 175
at	rpm	6,000	5,500
Max torque	Nm	160	240 (260 with overboost)
at	rpm	4,250	1,600 – 5,000 (1,700 – 4,500 with overboost)
Alternator	A/W	120 / 1,680	120 / 1,680
Chassis and suspension			
Front suspension	Singe-joint MacPherson spring strut axle with anti-dive		
Rear suspension	Longitudinal struts with centrally pivoted control arms, Z-axle		
Brake, front		Disc, vented	Disc, vented
Diameter	mm	280	294
Brake, rear		Disc brake	Disc brake
Diameter	mm	259	259
Driving stability systems	Hydraulic two-circuit brake system with anti-lock brakes (ABS), electronic brake force distribution (EBD), cornering brake control (CBC) and brake assistant. Optional ASC+T traction control and dynamic stability control (DSC) with hill start a ssistant. Mechanical parking brake acting on the rear wheels.		
Steering	Electrohydraulic power steering (EPS)		
Overall transmission ratio	:1	14.1	14.1
Tires		175/65 R15 H	195/55 R16 87V RSC
Wheels		5,5J x 15 Alloy	6,5J x 16 Alloy
Performance			
Power-to-weight ratio, DIN	kg/kW	12.1	8.8
Output per litre	kW/ltr	55.1	80.1
Acceleration 0–62 mph	s	9.1	7.1
0–1000 m	s	30.2	27.5
Ground clearance	mm	139	130

[1]Weight of the car in road trim (DIN) plus 75 kg for driver and luggage.

Powertrain

Engines

R55, R56, and R57 MINI Coopers are powered by 1.6 liter four cylinder engines, manufactured in joint cooperation with PSA Peugeot Citroën. The "new generation" of MINI engines are designated with a prefix of "N".

MINI Cooper engine applications (R55, R56, R57)	
Cooper 2007 - 2010 2011 - 2013	 N12, double VANOS, Valvetronic (118 hp @ 6000 rpm) N16, double VANOS, Valvetronic (121 hp @ 6000 rpm)
Cooper S 2007 - 2010 2011 - 2013	 N14, single VANOS (171 hp @ 5500 rpm) N18, double VANOS, Valvetronic (181 hp @ 5500 rpm)
John Cooper Works (JCW) 2009 - 2013	 N14 single VANOS (208 hp @ 6000 rpm)

BM11010009

The N12 engine is the early normally-aspirated MINI engine. Among many of the technical changes, the N12 engine features the following:

- All aluminum two-piece crankcase design with bedplate
- Infinitely variable double VANOS.
- Variable valve timing systems (Valvetronic).
- Lightweight alloy cylinder head.
- Weight-optimized roller cam followers.
- Volume controlled oil pump.
- Composite - instead of conventional cast - camshafts and weight-optimized crankshaft.
- Cast iron cylinder liners.
- Cast iron crankshaft.

BM010008

For the R56 Cooper S, the N14 engine was developed to meet the expected performance criteria of the "S". The N14 engine features many of the same technological advances (except Valvetronic) as the N12, with the following additional items:

- Infinitely variable single VANOS (Intake only).
- Direct fuel injection.
- Twin-scroll exhaust turbocharger.
- Auxiliary electric coolant pump for turbocharger cooling.

The main purpose of the direct fuel injection system is to combine high levels of specific power output with low fuel consumption figures. Furthermore, it improves smooth running and contributes to exemplary exhaust emissions. Low-volume, turbocharged engines with high power densities offer a considerably more favorable rate of fuel consumption by comparison with naturally aspirated engines with high cylinder capacities.

Direct injection system (N14 and N18 engines)

Map-controlled oil pump (N18, N16)

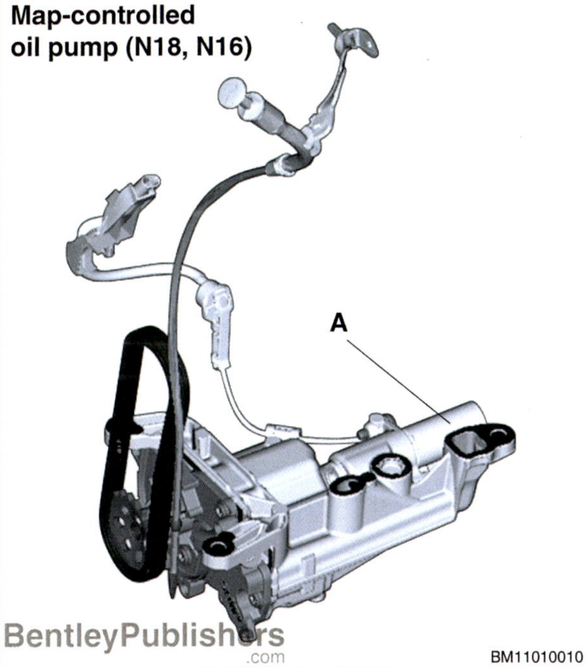

The N18 and N16 are the successors to the N14 and N12 engines. They will were introduced in the LCI (life cycle impulse) series in August 2010 for the 2011 model year. Both of engines share many technical features that were first introduced in the N14 and N12.

The N16, like the N12 engine, is equipped with manifold injection while the N18 engine is equipped with direct fuel injection similar to the N14 engine. The fuel injectors have been noise-optimized by adding a shim that absorbs noise.

1. High pressure pump
2. High pressure line
3. High pressure fuel injector
4. Fuel rail

The N18 engine is used in the Cooper S models. The N18 represents an advancement in terms of power, fuel consumption and technical innovations. The N18 is a TVDi (turbo-valvetronic direct injection) engine. The combination of these 3 technologies along with double vanos enables high power output (181hp) with low fuel consumption. The N18 engine is equipped with a dual scroll exhaust turbocharger, double VANOS with Valvetronic III and direct fuel injection.

The N16 replaced the N12 engine in the MINI Cooper models. Technically, the N16 engine shares the same features as the N12. These features include valvetronic, double vanos, and manifold injection.

Both engines feature an electronic volume-controlled oil pump. This is another measure for optimizing fuel consumption. The pump delivers only as much oil as is necessary under the respective engine operating conditions. No surplus oil quantity is delivered in low-load operating ranges. This operating mode reduces the pump work and therefore the fuel consumption of the engine while also slowing down the oil aging process.

The electrohydraulic pressure control valve mounted on the pump (**2**) controls the pump output. It is operated based on a characteristic map within the DME (ECM) which in turn is based on feedback from the oil pressure sensor on the oil filter housing.

Transmissions

The transmission is mounted in line with the engine on the driver side of the engine bay. The final drive/differential assembly is integral with the transmission housing and provides drive to the front wheels.

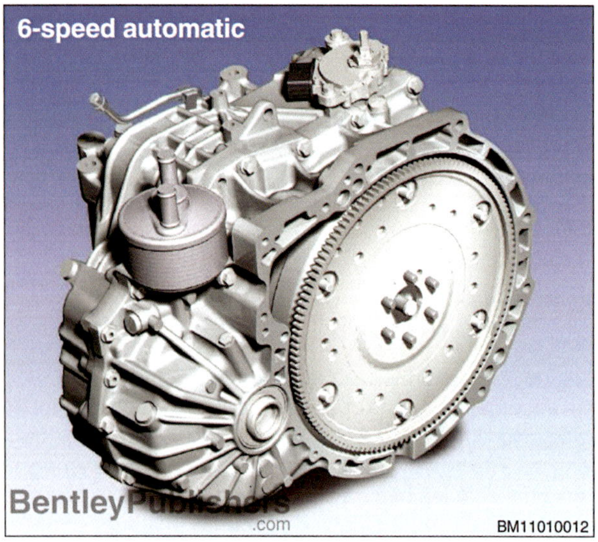

For automatic transmission models, a 6-speed GA6F21WA unit is used. The gearbox supplier is AISIN AW CO. LTD, Japan. In connection with Agitronic, the GA6F21WA transmission is characterized by variable gearshift times, ranging from comfortable up to sports-orientated. Agitronic enables individual gear selection by means of the shift paddles on the steering wheel even in drive range D without having to select manual mode (Steptronic) beforehand with the selector lever.

AISIN Automatic transmission	
Gear ratios 1st / 2nd / 3rd	4.044 / 2.371 / 1.556
Gear ratios 4th / 5th / 6th	1.159 / 0.852 / 0.672
Reverse gear ratio	3.193
Final drive ratio	3.683 (Cooper S) 4.103 (Cooper)

The sport button features three different gearshift speeds. This means, in addition to the comfortable and agile gearshift speed, there is now also a sports speed.

Preconditions for the various shift speeds:

Comfortable shift speed:

- Drive range D without using shift paddles
- Accelerator pedal pressed but without full load or kick-down

Agile shift speed:

- Drive range D while using shift paddles
- Drive range SD or manual
- Accelerator pedal pressed with full load or kick-down

The manual transmission fitted to the MINI Cooper S is a 6-speed unit manufactured by Getrag. Two different units were used based on model.

- Cooper - 6-speed GS6-55BG
- Cooper S, the 6-speed GS6-53BG

Differences between GS6-53BG and GS6-55BG:

GS6-53BG has a transmission shaft differential cover. Further distinguishing features of the transmission can be seen in the housing shape within the clutch housing.

In addition to the externally visible distinguishing features of the two transmissions, the GS6-53BG is approximately 24.5 mm longer than the GS6-55BG. Even the distance between the transmission input shaft and the differential output (output shafts connection) varies.

GS6-53BG/DG (Cooper S)

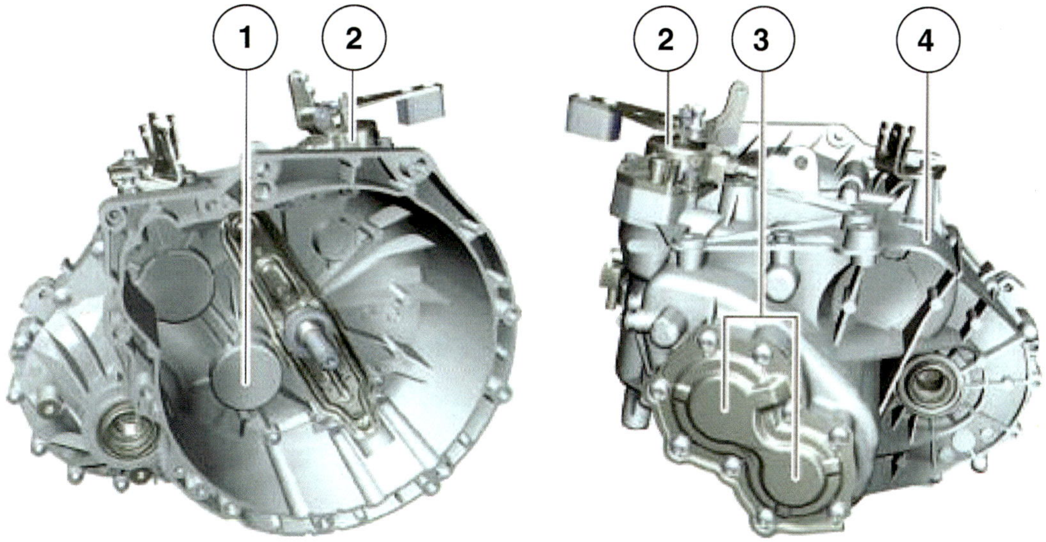

GS6-55BG (Cooper)

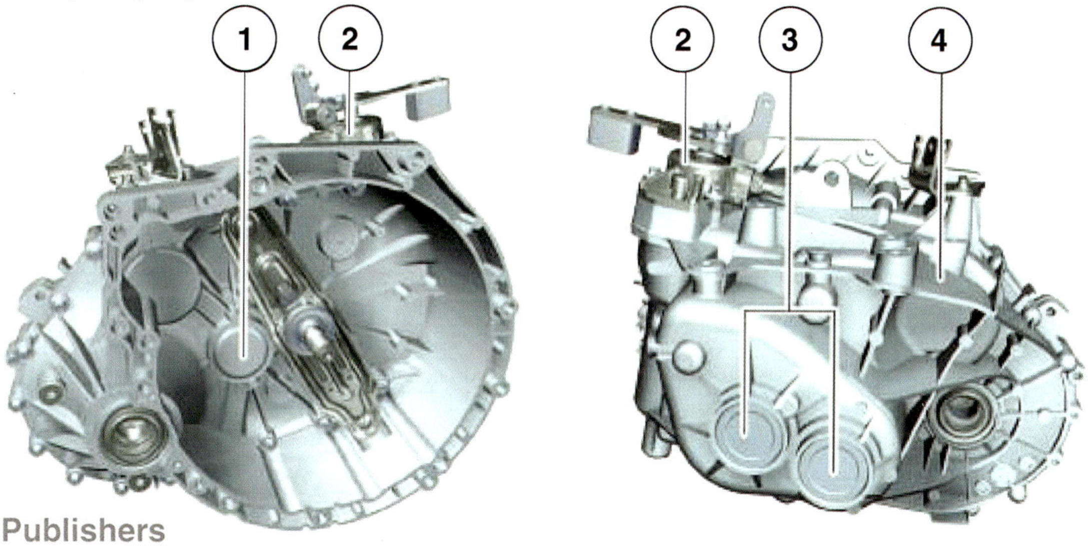

1. Differences in clutch housing
2. Shifting mechanism
3. Differences on shaft end
4. Transmission housing

BM11010014

Body Exterior

The R56 body shell structure is designed for a high level of corrosion endurance with 12 years corrosion protection. The body is 100% galvanized. Welded seams have been avoided and spotweld joints and adhesive bonds have been used where possible. All cavities are sealed with wax. The body is coated with anti-corrosion dipping paint, sealed with PVC and protected from corrosion by the main coat and final coat of paint.

Production of the MINI Convertible (R57).

Clubman (R55)

The R55 Clubman launched in the spring of 2008. It is longer and taller than the R56. The width remains the same. Compared to the R56, the wheelbase and vehicle interior of the R55 has increased by 80 mm. In comparison, the rear overhang on the R55 is 158 mm longer. The knee room available for the rear passengers has increased by 77.5 mm and the headroom by 16 mm.

MINI Clubman

The Clubman incorporates an ingenious door configuration on the right side of the car. The clubdoor swings open in the opposing direction relative to the main passenger door to dramatically increase the access to the rear seating area.

Convertible (R57)

The R57 Convertible was released in the first quarter of 2009. The R57 technology was primarily based on that of the R56. The convertible top technology is a further development of the top found on the previous generation convertible (R52). DSC is standard equipment and the engine and gearbox are carried over from the R56.

Technical highlights of the R57 in comparison to the R56:

- Fully automatic soft top
- Innovative, active overturn safety system

Body Interior

The car's interior has been engineered to provide outstanding ergonomics, comfort and safety. The interior of the R56 MINI is the result of a genuine revolution in design covering all areas and features within the car. Optimized ergonomics, greater space, more sophisticated and stylish materials, and quality of finish reflect the premium character of the MINI.

The car's developers attached utmost importance to simple and ergonomic control by the driver. Hence, all control elements are in arranged as a function of their frequency of use and need for operation.

- The interior highlight is the round, center speedometer Apart from the analog-face, the center display incorporates the audio controls as well as the optional entertainment and navigation functions. The most distinctive feature, however, is the speed indicator which surrounds all other functions like a 360-degree ring. And with the entire unit being placed high up within the instrument panel, all function buttons can be positioned for optimum ergonomic control.

- Other features typical of MINI are the vertical arrangement of the rotary air ventilation knobs, the two cup holders integrated in the base section of the center console, as well as the toggle switches carried over from decades of racing heritage.

- The conventional ignition key is replaced by a round, electronic signal transmitter. Once inserted into the slot next to the steering wheel, this signal transmitter enables the driver to start the engine by simply pressing the start/stop button. And with optional Comfort Access, the driver is even able to leave the signal transmitter in their pocket, the only requirement being that the transmitter is on their body or within the car in order to open the doors and start the engine.

The MINI Cooper door design is a major feature of the interior. The design of the armrest matches the design of the center console. The armrest and storage compartment in the door are stylishly framed in silver and the surface of the armrest is covered with black leather.

Front seats have integrated side airbags in the side cushions. Height adjustment is provided as standard. Sports seats are standard on the Cooper S. Lumbar support and two-level seat heating are seat options. Seats are available in leather, partial leather, leatherette or cloth.

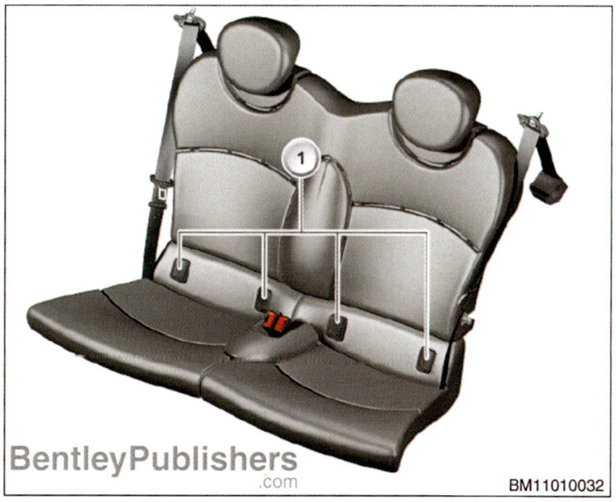

The rear seats can be folded down 50:50. The procedure for folding them down can be carried out very easily. A release lever on each seat allows the folding backrest to be locked and released The rear seats are equipped with ISOFIX (1) brackets.

The luggage compartment on the R56 has a volume of 160 liters (compared to 150 liters on the previous R53/R50). With the rear seats folded down, the R56 has a luggage volume of 680 liters. There is also an accessory outlet on the left of the luggage compartment (1 bottom).

Safety

The R56 MINI Cooper uses an extremely strong and rigid passenger safety cell, with front and rear crumple zones and side impact door beams.

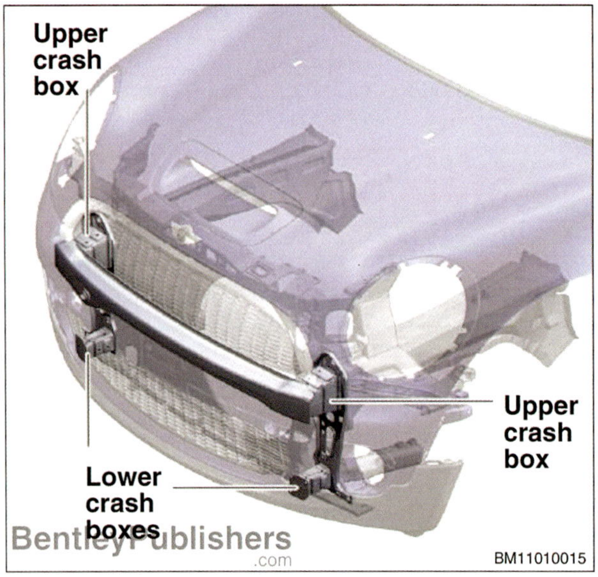

BM11010015

The outstanding crash protection can be attributed to the use of high-strength steels and aluminum components (in the door area). Although the vehicle has short overhangs and deformation zones, crash loads are well transferred. The front end features crash cans on the front axle carrier in addition to the bumper carrier and the crash boxes.

In accidents at low speeds, the crash boxes prevent damage to expensive components such as the body, lights and cooling system. Energy is dissipated by the deformation of the crash boxes.

The solid passenger cell is a prerequisite for putting the restraint systems to the best possible use. In addition to front airbags for the driver and front passenger, side airbags (pelvis-thorax airbags) in the front seats, belt force limiters and belt tensioners for the driver and front passenger, head-level airbags (curtain) for all seats and a steering column unit with crash elements ensure that occupants are restrained in a great variety of collision situations.

Active safety is enhanced by the standard anti-lock brakes (ABS), Electronic Brakeforce Distribution (EBD), Corner Brake Control (CBC) and Dynamic Stability Control (DSC). Additionally, Dynamic Traction Control (DTC), a sportier mode of DSC, was an available option on some later models.

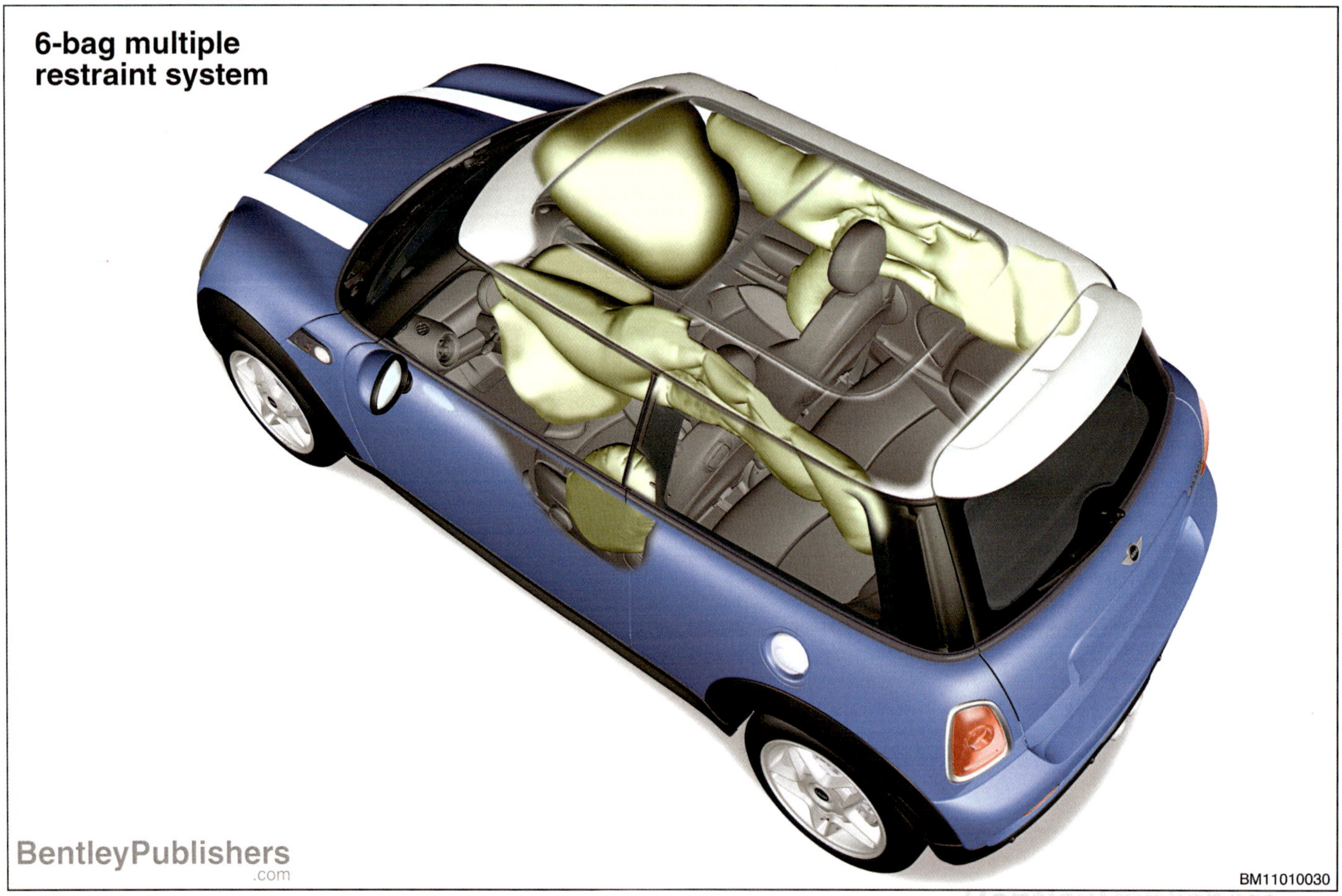

BM11010030

Suspension, Steering, Brakes

Innovations in suspension technology for the second generation MINI Cooper further enhance the go-kart feeling, benefiting from front-wheel drive, a low centre of gravity, and a wide track.

On the MINI Cooper S, the suspension is tuned to reflect the vehicle's additional power. The MINI Cooper comes standard with 15-inch wheels, while the MINI Cooper S features 16-inch rims with runflat tires allowing the driver to continue even after a complete loss of tire pressure.

Front suspension

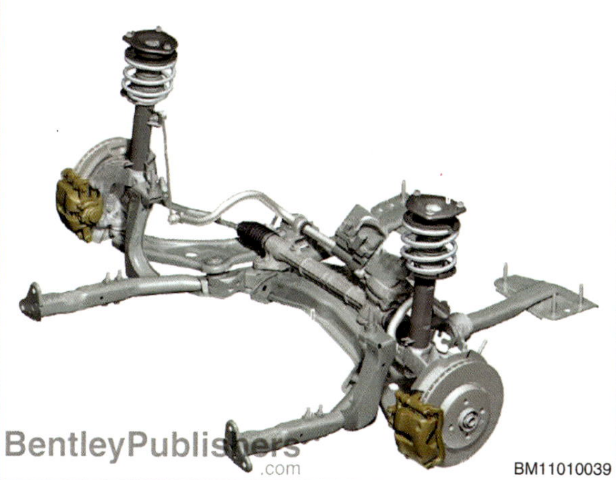

McPherson spring struts on the front axle ensure excellent wheel guidance, keeping the steering almost entirely free of drive forces both in fast bends, when accelerating and applying the brakes.

Rear suspension

The central arm rear axle, a concept quite unique in the MINI segment, ensures optimum road contact at all times through its elaborate kinematics. All the suspension know-how and competence of the BMW Group has gone into the development of the rear axle, use of longitudinal aluminium arms serving to reduce weight by 6 kilos or 13 lbs from the former rear axle.

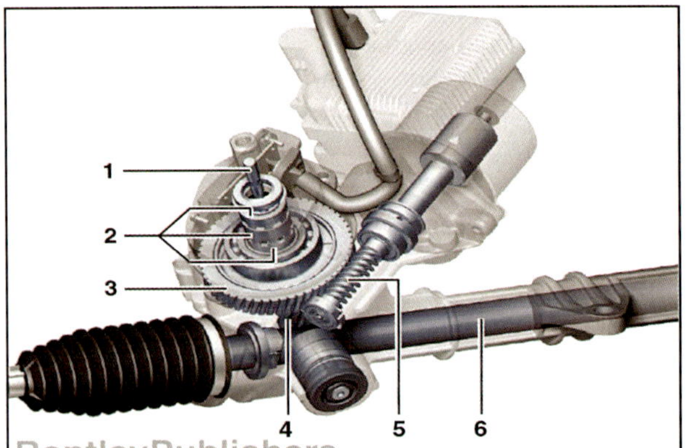

1 Torsion rod of steering torque sensor
2 Steering torque sensor
3 Worm gear
4 Rack and pinion
5 Worm shaft
6 Steering rack

BM11010001

Electromechanical power steering (EPS) fitted on the R56 replaces the electrohydraulic power steering (EHPS) used on the first generation MINI cars.

The use of the electromechanical power steering system provides many advantages. In comparison to conventional hydraulic systems, the EPS steering serves to reduce weight by a significant margin. Another advantage is that the electric servo motor is only activated when power assistance is required by the driver. So through its low weight and on-demand operation, the new steering is able to further reduce fuel consumption

With its electromechanical control, the steering system can be programmed to various settings. The "basic" setting provides well-balanced power assistance as a function of road speed, allowing the driver to maneuver and park the car with minimum effort. In the interest of safety and enhanced precision, steering forces have to be increased at higher speeds, and the integrated, active return motion moves the steering wheel back precisely to its central position in all situations.

In addition, an optional driving dynamics control button (SPORT button) makes it possible to switch between standard mode and sports mode. The sports mode, for example, activates a noticeably more dynamic control map characterized by higher steering forces and enhanced steering precision.

John Cooper Works

The John Cooper Works (JCW) model is based on the MINI Cooper S with the N14 engine.

The complete performance upgrade is only available from the factory. The entire engine has been redesigned and these modification cannot be easily performed at a MINI dealership.

Some JCW components such as springs, strut brace, aero and trim kit can be installed in the dealership.

The challenge for the JCW upgrade was to surpass the performance of the MINI Cooper S without making compromises in terms of fuel consumption and comfort. In other words: It was to be a high performance sports car.

Engine Technology:

- Turbocharged inline 4 cylinder with direct injection
- Displacement: 1.6 liters
- Top Speed: 147 mph
- Horsepower: 208 hp @ 6,000 rpm
- Torque: 191 lb-ft @ 1,850 - 6,500 rpm

Summary of Changes

Only components that were needed to achieve higher performance and a more sporty sound were modified.

- Cylinder head gasket revised in terms of heat distribution, resulting in improved coolant flow.
- Cylinder head wall thickness changed in some places in relation to water jacket to remove local peaks caused by increase in performance.
- Improved intake valve seat material and extra-hardened intake-valves.
- Strengthened piston side faces, and adapted to a reduced compression ratio of 10.0.
- Charge air pipe improved to reduce pressure drop.
- Intake air filter with integrated air mass sensor.
- Turbocharger adapted to changed conditions.
- Turbocharger and exhaust manifold materials adapted to higher exhaust gas temperature (about 30°C higher).
- Catalytic converter is a two stage design.
- Red Brembo® front brake calipers with 17 inch wheels and 12.4 inch brake rotors.
- Appearance and acoustics, plus specific sound impedance of the exhaust system were improved to create a special sound.

Dashboard and controls

1. Outside mirror adjustment switch
2. Cruise control
3. Light switch
   Head light flasher
   Turn signal indicators
   Onboard computer
4. Instrument lighting control
5. Wiper/washer switch
6. Ignition lock
7. START/STOP button
8. Hazard warning flasher switch
9. OBD socket
10. Hood release
11. Steering wheel adjustment control
12. Horn button
13. Driving stability control

1. "Always open" gauge (Convertible only)
2. Tachometer with indicator and warning lights
3. Display for clock, outside temperature, current vehicle speed
4. Speedometer with indicator and warning lights
5. Fuel gauge
6. Indicator for computer, odometer, trip odometer, service interval, program display for automatic transmission, initializing flat tire monitor
7. Trip odometer reset

020 Maintenance

TABLES

General

This repair group details service and maintenance procedures for the 2007 through 2013 MINI Cooper models.

MINI Cooper vehicles are equipped with condition based service (CBS) capabilities that monitors vehicle systems and components and suggests service based on the actual current condition of components. Following the CBS recommended service intervals helps ensure safe and dependable operation of the vehicle.

Aside from keeping your vehicle in the best possible condition, proper maintenance plays a role in maintaining full protection under MINI's new-vehicle warranty coverage. If in doubt about the terms and conditions of your vehicle's warranty, an authorized MINI dealer should be able to explain them.

MINI is constantly updating their recommended maintenance procedures and requirements. The information contained here may not include updates, revisions or part supersessions made by MINI since the publication of the documents supplied with the vehicle. If there is doubt about a specific model or model year, or what intervals to follow, consult an authorized MINI dealer.

Warnings, Cautions and Notes

Throughout this manual are many passages with the headings **WARNING**, **CAUTION**, or **NOTE**. These headings have different meanings.

WARNING—
- *Text under this heading warns of unsafe practices that are likely to cause injury, either by direct threat to the person(s) performing the work or by increased risk of accident or mechanical failure while driving.*

CAUTION—
- *Text under this heading calls attention to important precautions to be observed during the repair work that will help prevent accidentally damaging the vehicle or its parts.*

NOTE—
- *A note contains helpful information, tips or pointers which help in doing a better job and completing it more easily.*

Read **WARNING, CAUTION** and **NOTE** headings before you begin repair work. See also **001 Warnings and Cautions**.

Condition Based Service (CBS)

CBS basics

The MINI maintenance system, called condition based service (CBS) notifies the driver of needed service based on the condition of monitored components. CBS recommends maintenance only for components that are worn.

CBS measures, monitors and determines the required maintenance of several service items independently of each other. The instrument cluster prompts the driver whenever one of the CBS items requires maintenance or replacement. CBS also details the recommended, due and overdue required maintenance.

For many maintenance tasks, CBS uses sensors to monitor vehicle driving conditions, then uses this data to calculate current and future service requirements. CBS sorts service requirements according to date due and notifies the driver when the vehicle requires service.

CBS tasks that are monitored are displayed in the tachometer instrument and the car communication computer (CCC) (cars fitted with navigation). The display shows remaining times or distances for the following maintenance tasks.

- Engine oil and filter change
- Front brake pad replacement
- Rear brake pad replacement
- Cabin microfilter replacement
- Brake fluid change
- Vehicle check
- Safety and emissions inspections

Automatic transmission fluid, oxygen sensor and spark plug replacement intervals are not shown in the tachometer or the CCC (cars with navigation). These items still require periodic replacement. See **Maintenance tables** in this repair group.

Tachometer CBS display uses an upper and lower panels:

1. Colored vehicle symbol:
   orange = normal
   yellow = service due
   red = service overdue.
2. Remaining distance to next service.
3. Date for next service.

MINI remote key

BentleyPublishers.com BM11020003

Service requirement information is stored in the remote key while the vehicle is driven. This information can be accessed by an authorized MINI service center at the time of a service appointment.

For CBS to work correctly, make sure correct local date and time are set in the vehicle instrument cluster.

Service recognition

The CBS system recognizes when a service is required using one of three methods:

Adaptive. Sensor data and algorithms are used to determine the required maintenance interval for a service based on individual user driving style. Examples:

- Oil change interval is determined by fuel consumption.
- Front and rear brake pad replacement intervals are determined from two-stage brake lining wear sensor signals on left front and right rear brake pads; and from algorithms using these input parameters: travel distance, wheel speed, braking pressure, braking time, and braking frequency.

Fixed time (months) determines required maintenance for the following:

- Brake fluid change
- Mandated safety and emissions inspections

Fixed distance (miles) determines required maintenance for the following:

- Spark plug replacement (dependent on engine)
- Air filter element replacement
- Automatic transmission fluid change
- Cabin microfilter replacement
- Oxygen sensor replacement
- Vehicle check

See **Maintenance tables** in this repair group for details.

CBS items, viewing and resetting

The tachometer displays CBS data with graphics and text. When one or more service operations are carried out (example: front brake pads and rotors are changed), reset the service interval.

Vehicle with navigation: The car communication computer (CCC) in center dashboard also displays CBS information. CBS resetting functions can be found in the **Settings** menu.

BM11020010

- Enter vehicle and close doors. Switch ignition ON (KL15) but do not start engine.

◄ Press and hold trip reset button on tachometer **(arrow)** for 10 seconds.

BM11020006

◄ Upper panel in tachometer is illuminated with a service item. Lower panel indicates remaining time or mileage left for that service Item.

- Push BC (board computer) button on end of turn signal stalk to scroll display to next available CBS item.

BM11020025

◄ To reset, push and hold BC button again. Lower panel shows RESET.

- Repeat procedure for additional resets.

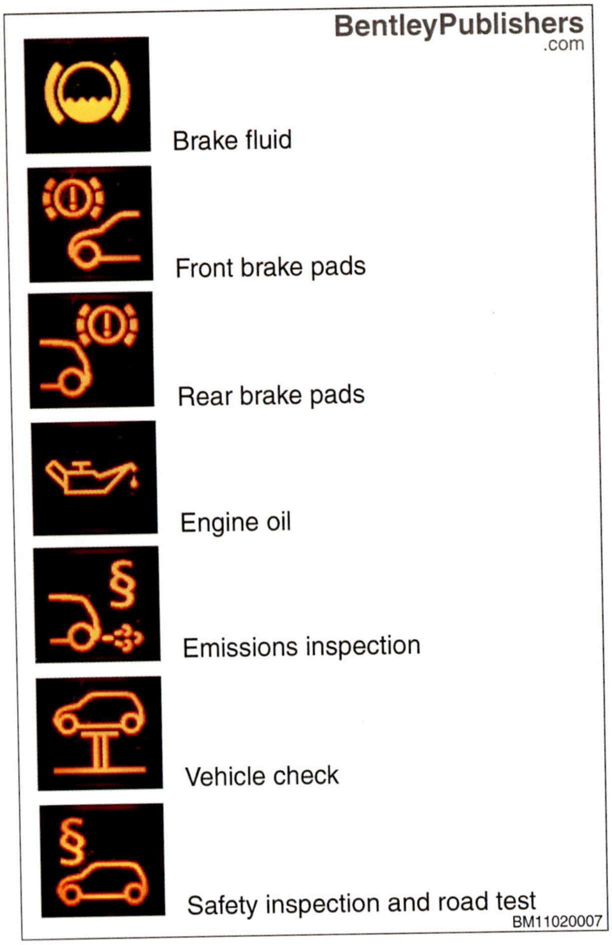

Maintenance tables

The service interval display in the tachometer signals the need for maintenance and inspection.

- Oil change tasks are in **Table a**.
- Condition based service (CBS) tasks are in **Table b**.
- Service and maintenance tasks required at set mileage intervals are in **Table c**.
- Other recommended periodic vehicle inspection tasks are in **Table d**.

MINI is constantly upgrading recommended maintenance procedures and requirements. The information contained here is as accurate as possible at the time of publication. If there is any doubt about what procedures apply to a specific model or year, or what intervals should be followed, remember that an authorized MINI dealer has the latest information on factory-recommended maintenance.

Except where noted, the maintenance items listed apply to all models and model years covered by this manual. The number in the **Additional repair information** column refers to the repair group in this manual where additional information can be found.

> ***CAUTION—***
> - *The publisher of this manual recommends that engine oil and filter be replaced a intervals not greater than 7500 miles or one year, whichever comes first.*

Table a. Engine oil service	**Additional repair information**
Engine oil and oil filter, change per CBS prompt based on fuel consumption or: • 15,000 miles or 12 months (vehicles manufactured to 02/2012) • 10,000 miles or 12 months (vehicles manufactured from 03/2012)	**119**
Brief diagnostic tests: • Check Control, check messages. • Indicator and warning lights, scan for fault codes using BMW scan tool or equivalent. • CBS display, reset. • Tires, check tread, adjust pressures. • Tire pressure monitor, reset.	**020, 300**
Parking brake, check operation, adjust as necessary.	**340**
Washer fluid, top off	

Table b. CBS prompted service	Additional repair information
Front brake pads, replace (CBS prompt based on pad thickness).	340
Rear brake pads, replace (CBS prompt based on pad thickness).	340
Brake fluid, change every 2 years (CBS prompt based on time elapsed since last change).	340
Cabin microfilter, change • Not in CBS menu • Replace every other oil change (normal driving) • Replace once every two years (low-mileage driving)	020

Table c. Additional services	Additional repair information
Every 30,000 miles	
JCW models: Spark plugs, replace.	020
Every 45,000 miles	
Air filter element, replace. Reduce change interval in dusty conditions.	020
Every 60,000 miles	
N14, N16, N18 engine except JCW: Spark plugs, replace.	020
Every 100,000 miles	
N12 engine: Spark plugs, replace.	020
Automatic transmission fluid (ATF), replace.	240
Every 120,000 miles	
Oxygen sensors, replace.	180
*MINI states that engine cooling system and manual transmission are filled with life-time fluid and do not require fluid replacement.	

Table d. Vehicle check (CBS prompt or every second oil change)	Additional repair information
Horn, headlight flasher, hazard warning flasher, check operation,	
Instrument cluster and control lighting, blower, check operation	
Lighting system (dome lights, glove compartment light, flashlight, trunk or cargo compartment lights, turn signals, hazard warning flasher, back-up lights, brake lights, license plate lights), check operation.	630
Seat belts, check condition and function.	720
Wipers and washers, check.	611
Body, check for corrosion.	
Tires, check tread depth, wear pattern, condition, inflation pressure, correct pressure as necessary.	
Run-flat indicator, initialize.	020
Battery, check state of charge, charge as required.	121
Airbags, inspect for torn covers, damage or attached stickers.	721

Table d. Vehicle check (CBS prompt or every second oil change) (continued)	Additional repair information
Mirrors, inspect.	510
Engine coolant, check level and concentration.	020, 170
Engine accessory belt, inspect	
Windshield washer, intensive cleaning system, check protection level, fluid level, top off if necessary.	
Brake system connections and lines, check for leaks, damage and correct positioning.	
Underbody (transmission, rear axle, fuel lines, exhaust system) check for damage, leaks and corrosion.	
Steering components, check for clearance, damage and wear.	
Convertible: Lower and raise convertible top to test. Actuate rollover protection using scan tool to test.	
Road test, check for performance, braking, steering, suspension, noises.	

Basic Service Information

Diagnostic trouble codes (DTCs), accessing

With a generic OBD II scan tool connected to the 16-pin OBD II plug, diagnostic trouble codes (DTCs) can be obtained, along with the conditions associated with the illumination of the malfunction indicator light (MIL) or Check Engine light. A more advanced or MINI-dedicated diagnostic scan tool accesses additional information.

◀ Connect diagnostic scan tool to OBD II connector (**arrow**) under left side of dashboard.

– Place transmission selector lever in PARK or NUETRAL. Engage parking brake. Make sure ignition is OFF.

– Start engine and let idle.

– Navigate scan tool menu on screen to display DTCs.

For additional DTC and OBD information, see **OBD On-Board Diagnostics**.

Tightening fasteners

When tightening bolts or nuts on a component, it is good practice to tighten the bolts gradually and evenly to avoid misalignment or over-stressing the component. For components sealed with gaskets, this method helps ensure that the gasket seals properly.

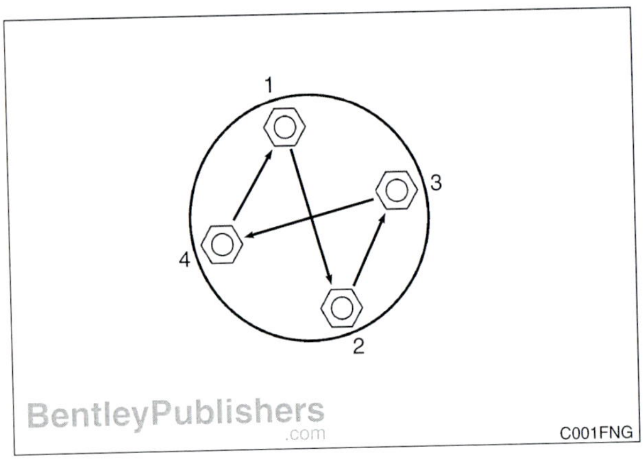

◀ Where there are several fasteners, tighten them gradually in a crisscross pattern. Repeat the sequence until all the bolts are evenly tightened to the proper specification.

For some repairs a specific tightening sequence is necessary, or particular order of assembly is required. Such special conditions are noted in the text. Where no specific torque is listed, the table **General bolt tightening torques in Nm (max. permissible)** can be used as a general guide for tightening fasteners.

NOTE—
- *Metric bolt classes or grades are marked on the bolt head.*
- *Do not confuse wrench size with bolt diameter.*
- *Values in table are for reference only.*

General bolt tightening torques in Nm (max. permissible)						
Bolt class (according to DIN 267)						
Bolt diameter	**5.6**	**5.8**	**6.8**	**8.8**	**10.9**	**12.9**
M5	2.5	3.5	4.5	6	8	10
M6	4.5	6	7.5	10	14	17
M8	11	15	18	24	34	40
M10	23	30	36	47	66	79
M12	39	52	62	82	115	140
M14	62	82	98	130	180	220
M16	94	126	150	200	280	340
M18	130	174	210	280	390	470

Non-reusable fasteners

Many fasteners must be replaced with new ones once they are removed. These include but are not limited to: bolts, nuts (self-locking, nylock, etc.), roll pins, clips and washers. Genuine MINI parts should be the only replacement parts used for this purpose.

Some bolts are designed to stretch during assembly and are permanently altered, rendering them unreliable once removed. These are known as torque-to-yield fasteners. Always replace fasteners where instructed to do so. Failure to replace these fasteners could cause personal injury or vehicle damage. See an authorized MINI dealer for applications and ordering information.

Parts

Genuine MINI parts

Replacement parts from an authorized MINI dealer are designed and manufactured to the same high standards as the original parts. They will be the correct material, manufactured to same specifications, and guaranteed to fit and work as intended. Most genuine MINI parts carry a limited warranty.

MINI is constantly updating and improving their vehicles, often making improvements during a given model year. MINI may recommend a newer, improved part as a replacement, and your authorized dealer parts department will know about it and provide it. The MINI parts organization is best equipped to deal with your parts needs.

Many of the maintenance and repair tasks in this manual call for the installation of new parts, or the use of new gaskets and other materials when reinstalling parts. Most often, the needed parts should be on hand before beginning the job. Read the complete procedure to determine which parts are needed. When ordering parts, be sure to have the model, model year, VIN and date of manufacture. The label on the driver's door pillar specifies the month and year that the vehicle was built.

For some bigger jobs, partial disassembly and inspection is required to determine a complete parts list. Read the procedure carefully and, if necessary, make arrangements to get the necessary parts while your vehicle is disassembled.

Non-returnable parts

Some parts cannot be returned, even for credit. The most common example is electrical parts, which are almost universally considered non-returnable. Buy electrical parts carefully, and be as sure as possible that a replacement is needed, especially for expensive parts such as electronic control modules. It may be wise to let an authorized MINI dealer or other qualified shop confirm your diagnosis before purchasing an expensive non-returnable part.

Tools

Many reputable tool manufacturers offer good quality, moderately priced tools with a lifetime guarantee. These are your best buy. They cost a little more, but they are good quality tools that will do what is expected of them. Sears' Craftsman® line is one such source of good quality tools.

Some of the repairs covered in this manual require the use of special tools, such as special tools for setting camshaft timing, pullers or specialized electrical test equipment. These special tools are called out in the text and can be purchased through an authorized MINI dealer. As an alternative, some special tools mentioned may be purchased from the following tool manufacturers or distributors:

Specialty tool suppliers

Assenmacher Specialty Tools, Inc.
6440 Odell Place
Boulder, CO 80301
800-525-2943
www.asttool.com

Baum Tools Unlimited, Inc.
PO Box 5867
Sarasota, FL 34277
800-848-6657
www.baumtools.com

Mac Tools
4635 Hilton Corporate Drive
Columbus, OH 43232
800-622-8665
www.mactools.com

Samstag Sales
115 Main St. N., Suite 216
Carthage, TN 37030
615-735-3388
www.samstagsales.com

Snap-On Technologies, Inc.
2801 80th St.
Kenosha, WI 53141-1410
262-656-5200
www.snapon.com

Identification Plates and Labels

Vehicle identification number (VIN)

This is a combination of letters and numbers that identify the particular vehicle. The VIN appears on the state registration document and on the vehicle itself. One location is on the right front strut tower in the engine compartment (**arrow**), another in the lower left corner of the windshield. The last 7 digits are usually all that is required for the MINI parts department to identify the vehicle.

Date of manufacture

Build date may be necessary when ordering replacement parts or determining if any of the warranty recalls are applicable to your vehicle. The label on the driver side B-pillar specifies the month and year (**arrow**) that the vehicle was built.

Raising Vehicle

> ***WARNING—***
> - *Do not work under a raised vehicle unless it is solidly supported on jack stands that are intended for that purpose.*
> - *When raising the vehicle using a hydraulic lift or jack, carefully position the lift arm pads or jack pad to prevent damaging the vehicle body.*
> - *Watch the lift arms or jack closely. Make sure the vehicle stays stable and does not shift or tilt.*
> - *Do not use wood, concrete blocks, or bricks to support a vehicle. Wood may split. Blocks and bricks, while strong, are not designed for that kind of load, and may break or collapse.*
> - *Use care when removing major (heavy) components from one end of the vehicle. The sudden change in weight and balance can cause vehicle to tip off lift or jack stands.*
> - *Do not support vehicle at engine oil pan, transmission or fuel tank. Serious damage may result.*

Lifting vehicle

Use proper lift points (**arrows**) to raise vehicle safely and avoid damage.

Raising vehicle safely using vehicle jack

- Park vehicle on flat, level surface.
- If changing a tire, loosen lug bolts before raising vehicle. See **Changing a tire (Hardtop or Convertible)** or **Changing a tire (Clubman)** in this repair group.
- Block wheel that is opposite and farthest from jack to prevent vehicle from unexpectedly rolling.

Place jack into position. Make sure jack is resting on flat, solid surface. Use a board or other support to provide a firm surface for jack, if necessary.

- Raise vehicle slowly while constantly checking position of jack and vehicle.

> ***WARNING—***
> - *Do not rely on the transmission or the parking brake to keep the vehicle from rolling. Always block the opposite wheel.*

Working under vehicle safely

- Disconnect negative (–) cable from battery so that vehicle cannot be started. Let others know what you are doing.

> ***CAUTION—***
> - *Prior to disconnecting the battery, read the battery disconnection cautions given in* **001 Warnings and Cautions**.

- Raise vehicle slowly. See **Lifting vehicle** or **Raising vehicle safely using vehicle jack** in this repair group.

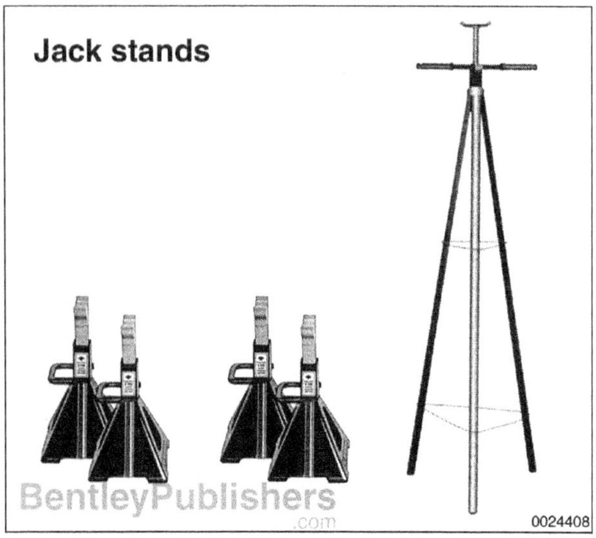

Use at least two jack stands to support vehicle. Use jack stands designed for the purpose of supporting a vehicle.

- Place jack stands on firm, solid surface. If necessary, use a flat board or similar solid object to provide a firm footing.
- Lower vehicle slowly until its weight is fully supported by jack stands. Watch to make sure that the jack stands do not tip or lean as the vehicle settles on them.
- Observe jacking precautions again when raising vehicle to remove jack stands.

Emergencies

See also:

- **540 Sunroof** for emergency sunroof closing.
- **541 Convertible Top** for emergency convertible top closing.

Flat tire indication

In case of a flat tire or extensive loss of air in a tire, instrument warning (!) illuminates in yellow or red. An acoustic signal also sounds.

- Reduce speed and stop vehicle with caution. Avoid sudden braking and steering maneuvers.
- Identify damaged tire and replace. See **Changing a tire (Hardtop or Convertible)** or **Changing a tire (Clubman)** in this repair group.

For flat tire monitor (FTM) and tire pressure monitor (TPM) information, see **300 Suspension, Steering and Brakes–General**.

Changing a tire (Hardtop or Convertible)

The space-saver spare tire is located under the car, directly underneath the cargo compartment.

Models equipped with safety (run-on-flat) tires installed at the factory are not fitted with a spare tire. See the owner's manual for information on using the MINI Mobility System (if equipped) with onboard tool kit and tire change set.

- Stop vehicle on as flat a surface as possible, in a place where you can be easily seen by other drivers. Avoid stopping just over the crest of a hill.

> ***WARNING—***
> - *If a tire goes flat while driving, pull well off the road. Changing a tire on a busy street or highway is extremely dangerous. If necessary, drive a short distance on the flat tire to get to a safe place. It is much better to ruin a tire or rim than to risk being injured.*

- Switch emergency flashers ON and set out flares or emergency markers well behind vehicle. Use wheel chock in vehicle tool kit to block wheel diagonally opposite to one being changed. Have passengers get out of vehicle and stand well away from road.
- Open hatch or tailgate and remove cargo floor liner.

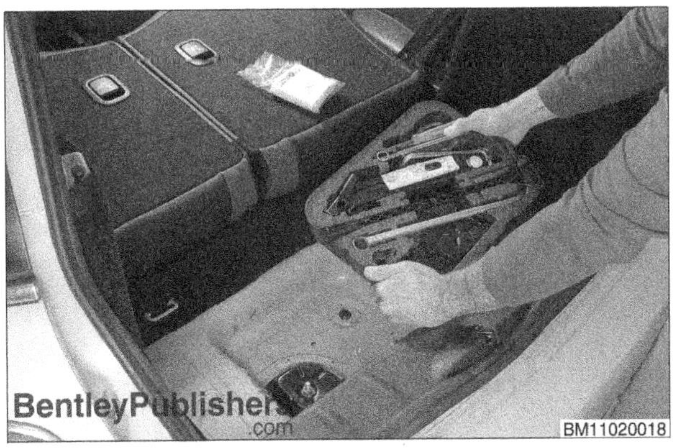

Remove tool kit from rear cargo area.

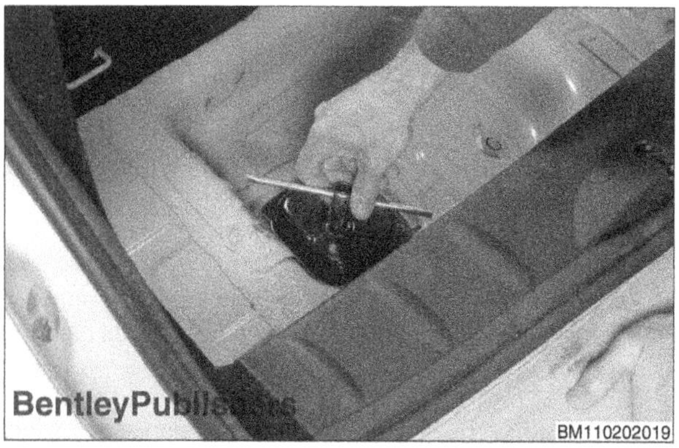

BM110202019

Working in space under tool kit, use special wrench tool to remove cover hold down nut and remove cover.

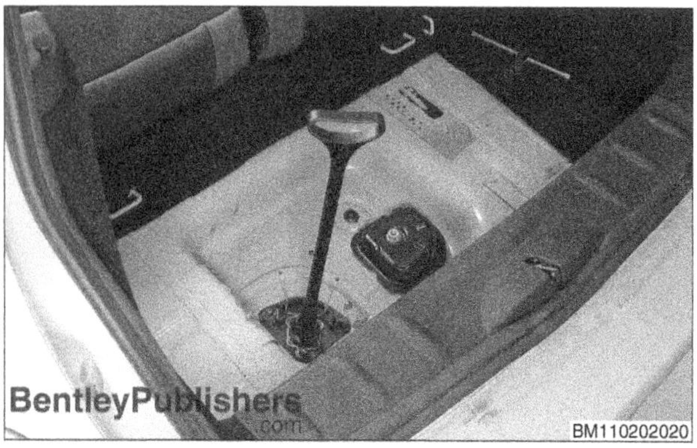

BM110202020

Thread spare tire lowering handle on spare tire mounting stud.

BM11020021

While holding weight of spare tire with lowering handle, pinch in on retaining clips (**arrows**) to release and lower spare tire to ground with handle. Unscrew lifting handle.

BM11020022

Disconnect air valve extension hose and remove spare tire from tray. Remove dust cap from extension hose and place on tire valve.

- Loosen wheel bolts while vehicle is on ground, but leave them a little snug.

BM11020017

- Place jack in lifting point (**inset**) nearest wheel being changed on a firm level surface. Raise vehicle only far enough so that the wheel is fully off ground and then remove wheel bolts and wheel.

- Remove air pressure valve extension from spare tire, if applicable.

- Install spare tire. Install wheel bolts and tighten them hand tight using lug wrench.

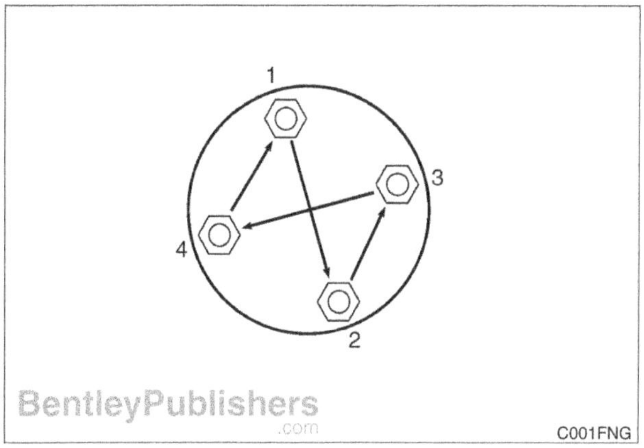

C001FNG

- Lower vehicle. With all wheels on ground, fully tighten nuts in a crisscross pattern. Check inflation pressure of spare tire.

Tightening torque	
Wheel to wheel hub	140 ± 10 Nm (103 ± 7 ft-lb)

- Replace tools in tool kit and store flat tire in cargo compartment. Plastic bag provided in tool kit can be used to store tire.

Changing a tire (Clubman)

The space-saver spare tire is located under the car, directly underneath the cargo compartment. Models equipped with safety (run-on-flat) tires installed at the factory are not fitted with a spare tire. See the owner's manual for information on using the MINI Mobility System (if equipped).

- Stop vehicle on as flat a surface as possible, in a place where you can be easily seen by other drivers. Avoid stopping just over the crest of a hill.

> ***WARNING —***
> - *If a tire goes flat while driving, pull well off the road. Changing a tire on a busy street or highway is extremely dangerous. If necessary, drive a short distance on the flat tire to get to a safe place. It is much better to ruin a tire or rim than to risk being injured.*

- Switch emergency flashers ON and set out flares or emergency markers well behind vehicle. Use wheel chock in vehicle tool kit to block wheel diagonally opposite to one being changed. Have passengers get out of vehicle and stand well away from road.

- Open rear split doors and fold up floor mat.

- Lift out tire change set. Unscrew tire retaining nut and lift out space-saver spare tire.

- Loosen wheel bolts while vehicle is on ground, but leave them a little snug.

◀ Place jack in lifting point (**inset**) nearest wheel being changed. Use a board to provide a firm footing for jack if ground is soft. Raise vehicle only far enough so that the wheel is fully off ground and then remove wheel bolts and wheel.

– Install spare tire. Install wheel bolts and tighten them hand tight using lug wrench.

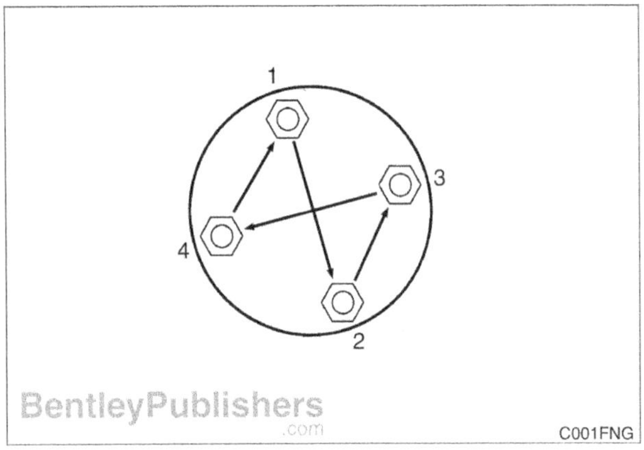

◀ Lower vehicle. With all wheels on ground, fully tighten nuts in a crisscross pattern. Check inflation pressure of spare tire.

Tightening torque	
Wheel to wheel hub	140 ± 10 Nm (103 ± 7 ft-lb)

– Replace tools in tool kit and store flat tire in cargo compartment. Plastic bag provided in tool kit can be used to store tire.

Jump starting

WARNING —
- *Batteries produce explosive and noxious gasses. Keeps sparks and flames away. Do not smoke near batteries.*
- *Do not jump-start the engine if you suspect that the battery is frozen. Trapped gas may explode. Allow the battery to thaw first.*
- *Do not quick-charge the battery (for boost starting) for longer than one minute, and do not exceed 16.5 volts at the battery with the boosting cables attached. Wait at least one minute before boosting the battery a second time.*

Vehicle with discharged or dead battery can be jump-started if the good battery is 12 volts and of the approximate same capacitance in Ah (ampere-hours). This information can be found on the battery label.

CAUTION—
- *Jump starting your MINI using a larger battery or a vehicle with a higher output charging system can cause permanent damage to vehicle electronics, including expensive control modules. If a jump start is absolutely necessary, be sure the battery in the donor vehicle is of the same charging capacity.*

See owner's manual for proper instruction on jump-starting. When jump-starting the engine, note the following warnings.

Indicator and warning lights

Many vehicle systems are self-monitored for faults while driving. Generally, a red warning light that comes on during driving should be considered serious. If you cannot immediately determine the seriousness of the warning light, stop the vehicle in a safe place and switch the engine OFF as soon as possible. Consult the owner's manual for additional information on the warning light and the recommended action.

If the malfunction indicator light (MIL) (Check Engine light) illuminates or flashes, it indicates that an emissions-related fault has occurred. Faults such as a bad oxygen sensor or a faulty fuel injector can cause the exhaust or evaporative emissions system to exceed specified limits. When these limits are exceeded, the MIL is switched ON. The vehicle can be safety driven with the light on, although the emission systems should be checked as soon as possible. See **OBD On Board Diagnostics** for more information on the MIL and the on-board diagnostic system.

Towing

> ***CAUTION—***
> - *Do not tow the vehicle on all four wheels except for very short distances to move it to a safe place.*
> - *Do not tow with sling-type equipment. The front spoilers and bumper covers may sustain damage.*

Front

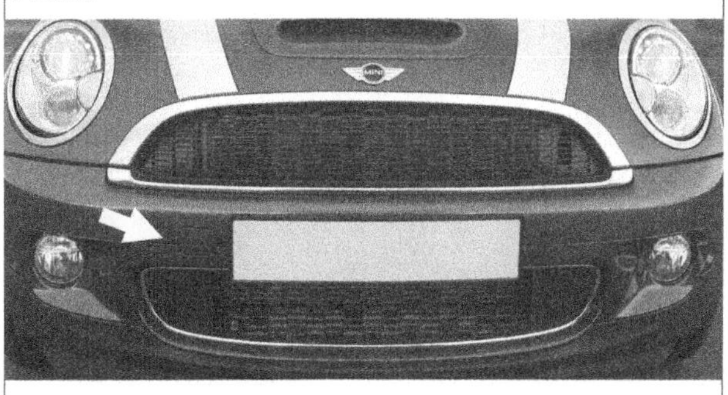

Rear

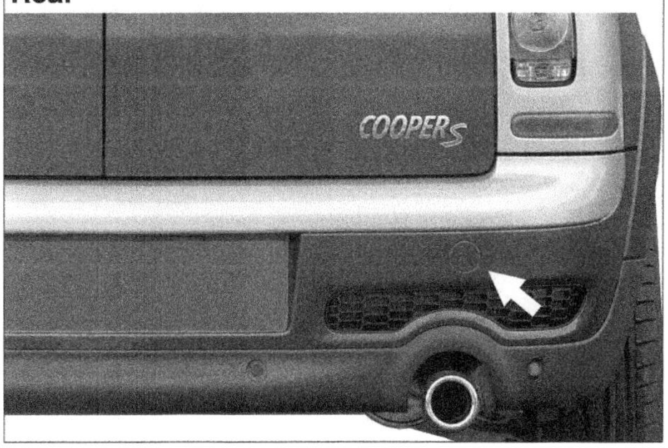

◄ To access threaded towing eye socket, pry open trim cover on front or rear bumper.

– Remove towing eye from cargo compartment tool kit and screw into front or rear towing socket threads until it bottoms firmly.

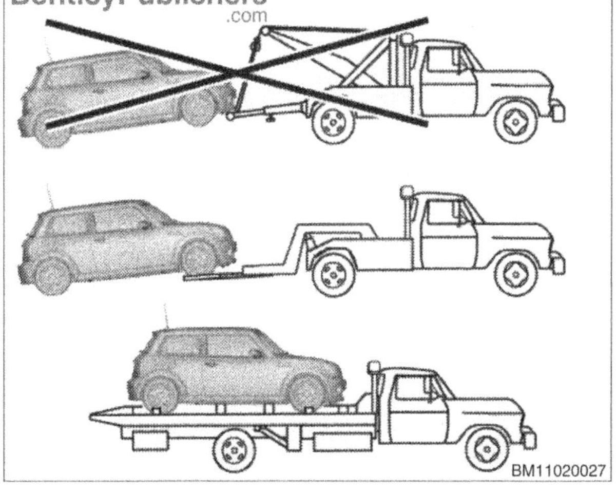

◄ Do not tow with front wheels on ground unless absolutely necessary. Have front wheels placed on dollies, or use flat bed truck. Use towing eye for winching vehicle on flat bed.

> ***CAUTION—***
> - *Tow vehicle with transmission in N (neutral).*

ENGINE COMPARTMENT

Engine compartment components

Cooper S with N14 engine shown. Other engine configurations are similar.

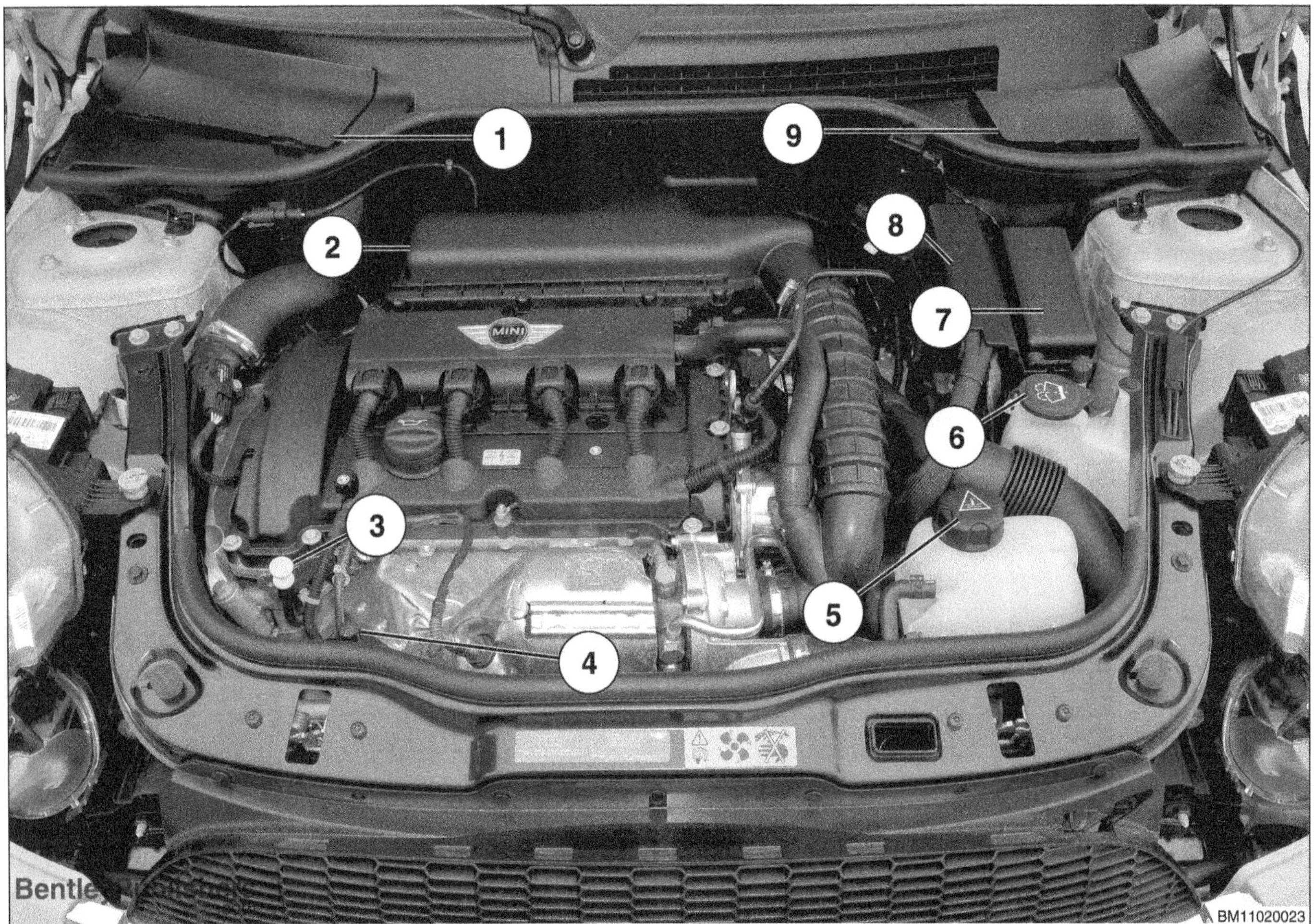

1. Battery (under cover)
2. Air filter housing
3. Engine oil dipstick
4. A/C refrigerant service port
5. Coolant reservoir
6. Windshield washer fluid reservoir
7. Engine compartment power distribution box (fuses and relays)
8. Engine control module (ECM) housing
9. Brake fluid reservoir (under cover)

Ignition coil cover, removing and installing

Cooper (non-turbo) engine ignition coils are hidden under the engine upper cover. Cooper S (turbo) coils are not equipped with a cover.

To remove Cooper upper engine cover (ignition coil cover):

< Remove screws (**arrows**).

- Tilt and pull cover forward to unclip from retaining clips at cylinder head cover.

> ***CAUTION—***
> • *Do not break plastic clips.*

- Installation is reverse of removal.

Air Filter

Air filter element, replacing (Cooper)

- Switch ignition OFF.

< Remove air filter cover fasteners (**A**, **B**).

- Tilt up air filter cover and remove filter element. Note orientation of filter element.
- Clean out air filter housing.
- Install new air filter element and upper air filter housing. Reinstall screws and bolt.

Air filter element, replacing (Cooper S, JCW)

- Switch ignition OFF.

< Disconnect harness connector from mass air flow sensor (**A**). Loosen air duct clamp (**B**).

- Detach air duct from air filter housing.
- Remove air filter cover fastener screws (**arrows**).
- Swing up air filter upper housing and remove filter element. Note orientation of filter element.
- Clean out air filter housing.
- Install new air filter element and close upper air filter housing. Reinstall screws. Reconnect harness connector and air duct.
- Reattach air duct. Use new clamp if necessary.

Battery

Battery maintenance

The original equipment battery in MINI models is maintenance free. The original electrolyte normally lasts the entire service life of the battery under moderate climate conditions.

Replace battery with one rated equal to the original battery. See **121 Battery, Starter, Alternator** for additional battery information.

Battery rating	
Cooper, Cooper S	12 volt, 55 Ah or 70 Ah

Be sure to review battery reconnection notes in **121 Battery, Starter, Alternator** for reintializing vehicle after a power interruption.

Intelligent battery sensor (IBS) notes

Not all MINI models are fitted with an intelligent battery sensor (IBS). An IBS-equipped car can be quickly identified by IBS electronics integrated in the terminal end of the negative battery cable.

When replacing the battery in an IBS-equipped car, register the new battery to the vehicle using MINI scan tool. The IBS measures the voltage, current and temperature of the battery and calculates the battery state of charge (SoC) to optimize vehicle power usage.

In simple terms, battery registration reinitializes power management based on the characteristics of the new battery and clears charging system voltage adaptations. This ensures correct operation of the vehicle electrical system.

Battery, checking and cleaning

The battery is located in right rear of engine compartment, under a plastic plenum door.

– Simple maintenance of the battery and its terminal connections ensures maximum starting performance, especially in winter when colder temperatures reduce battery power.

– Make sure that:
  - Battery cables are tight.
  - Battery terminals, cable clamps and battery case are free of white deposits that indicate corrosion and acid salts. Even a thin layer of dust containing conductive acid salts can cause battery discharge.

– Remove battery corrosion:
  - Disconnect battery cables. Disconnect negative (–) cable first.
  - Remove battery from vehicle.
  - Clean terminal posts and cable clamps with a wire brush.
  - Clean main chassis ground terminal next to battery. Wash away corrosion with a baking soda and water solution to neutralize battery acid.
  - Reinstall battery and reconnect cable clamps, positive (+) cable first.
  - Lightly coat outside of terminals, hold down screws and clamps with petroleum jelly, grease, or a commercial battery terminal corrosion inhibitor.

> ***WARNING —***
> - *Battery acid is extremely dangerous. Take care to keep it from contacting eyes, skin, or clothing. Wear eye protection. Extinguish smoking materials and do not work near open flames.*

> ***CAUTION—***
> - *Prior to disconnecting the battery, read the battery disconnection cautions given in* **001 Warnings and Cautions**.

BRAKES

Brake fluid, replacing

MINI recommends replacing the brake fluid every two years. Brake fluid absorbs moisture easily, and moisture in the fluid affects brake performance and reliability.

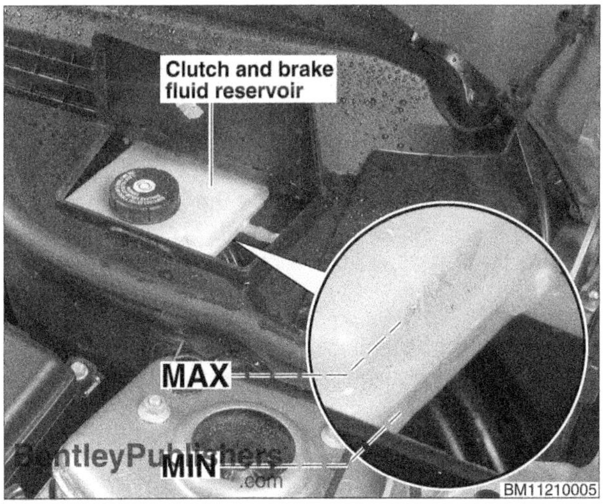

When replacing or adding brake fluid, use only new fluid from a previously unopened container. Do not use brake fluid that has been bled from the system, even if it is brand new. Use only DOT 4 brake fluid.

See **340 Brakes** for brake fluid flushing procedures and additional brake service information.

Brake pad and rotor wear, checking

Disc brakes are fitted at all four wheels.

– Check brake pad thickness whenever wheels are off or brake work is being done. Also inspect:
  - Brake rotors: Scoring, excessive wear indicated by large edge lip.
  - Condition of caliper seal: Check for leakage.
  - Condition of caliper slide bolts.

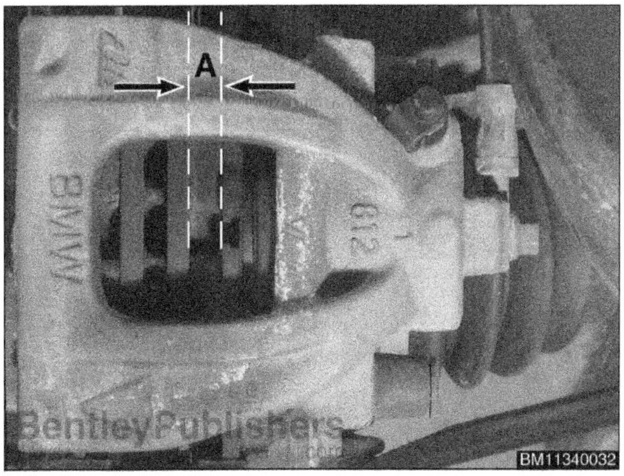

Thickness of inner brake pad can be estimated without removing brake caliper.

Brake pad lining minimum thickness (A)	
Front or rear pad	3.0 mm (0.12 in)

Brake system, inspecting

Routine maintenance of the brake system includes maintaining brake fluid in the reservoir, checking brake pads for wear, checking parking brake function and inspecting the system for fluid leaks or other damage.

– Inspect unions and brake calipers for signs of fluid leaks.

– Inspect rigid lines for corrosion, dents, or other damage. Inspect flexible hoses for cracking. Replace faulty hoses or lines.

> ***WARNING—***
> - *Incorrect installation or overtightened hoses, lines and unions may cause chafing or leakage. This can lead to partial or complete brake system failure.*

Parking brake, checking

The parking brake system may require periodic adjustment depending on use.

– Adjust parking brake if lever can be pulled up more than 6 notches.

– Check that parking brake cables move freely.

Parking brake description and adjustment procedure are in **340 Brakes**.

COOLING SYSTEM

Cooling system, checking

Antifreeze raises the boiling point and lowers the freezing point of engine coolant. The manufacturer recommends coolant that is a 50 / 50 mixture of distilled water and ethylene glycol-based antifreeze.

Routine cooling system maintenance consists of maintaining the coolant level and inspecting hoses. MINI states that the engine cooling system is filled with life-time fluid and does not require fluid replacement.

> ***WARNING—***
> - *Due to risk of personal injury, be sure the engine is cold before opening any part of the cooling system or removing the coolant reservoir cap.*

> ***CAUTION—***
> - *Use only MINI approved antifreeze when filling the cooling system. Use of antifreeze containing phosphates is harmful to the cooling system.*

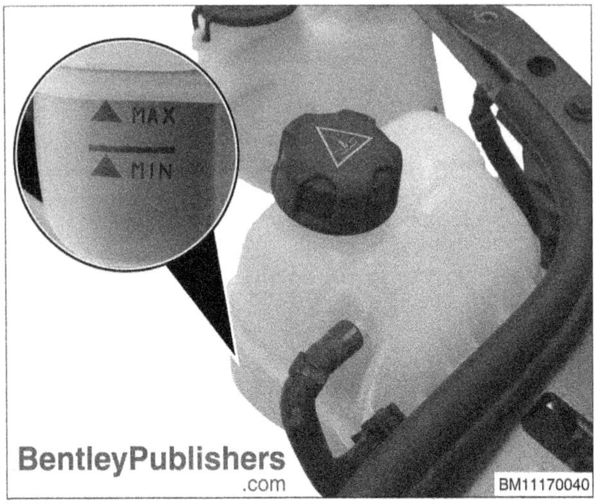

Maximum and minimum coolant levels are marked on coolant expansion tank (overflow reservoir).

– Add coolant to engine:

- Open reservoir cap by twisting slowly counterclockwise to allow any pressure to escape safely. Then open cap.
- If coolant is low, slowly add coolant to correct level. Do not overfill.

Engine coolant		
Model	**Approximate capacity: liter (US qt.)**	**Coolant specification**
Cooper	5.3 (5.6)	50% mixture of MINI phosphate / nitrite free antifreeze and water
Cooper S	6 (6.3)	

– Check for tight, dry hose connections. Coolant seepage at clamped connection indicates either that hose clamp is loose, hose is damaged, or connection is dirty or corroded. Dried coolant has a chalky appearance.

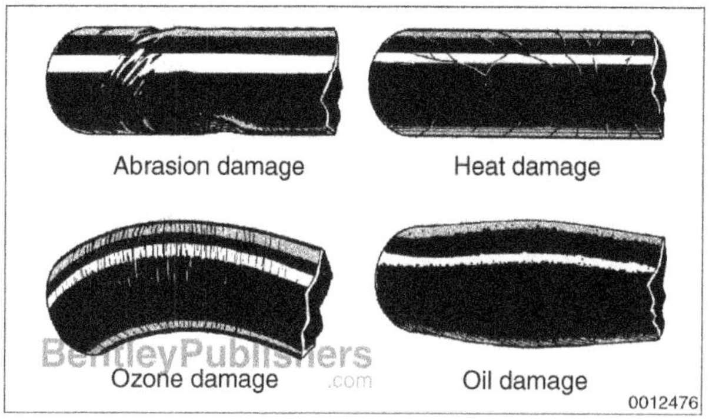

◄ Check for firm hoses. Replace any hose that is cracked, has become soft and limp, or has been contaminated by oil. The illustration shows examples of damage to coolant hoses. Any of these conditions is cause for hose replacement. *(Illustration courtesy of Gates Rubber Company, Inc.)*

As a preventive measure, replace cooling system hoses every four years.

Engine Accessory Belt

Accessory belt service

The accessory belt and pulleys transfer power from the engine crankshaft to various accessories. MINI models use one serpentine belt to operate belt-driven components.

Inspect accessory belt with engine OFF. If belt shows signs of wear, cracking, glazing, or missing sections, replace immediately.

> ***WARNING—***
> - *Observe care when replacing the belt. Personal injury could result if tensioner springs back into position uncontrolled.*

> ***CAUTION—***
> - *Mark belt rotation direction if planning to reuse an old belt.*

To reduce the chance of belt failure while driving, replace belt every four years.

Accessory belt, replacing

Layout of accessory belt.

1. Coolant pump pulley
2. Friction wheel assembly
3. Vibration damper (crankshaft pulley)
4. Accessory belt
5. Belt tensioner
6. Alternator
7. A/C compressor

Note that coolant pump is not driven by accessory belt. See **170 Radiator and Cooling System** for coolant pump and friction wheel disassembly.

– Remove right front wheel housing liner. See **410 Fenders, Engine Hood**.

– Remove right headlight assembly. See **630 Lights**.

– Remove engine hood lock carrier. See **410 Fenders, Engine Hood**.

Use box wrench to rotate (**curved arrow**) belt tensioner into service position. Slide in lock pin (**arrow**) to secure. (Radiator support removed for purpose of illustration.)

– Detach accessory belt from alternator pulley.

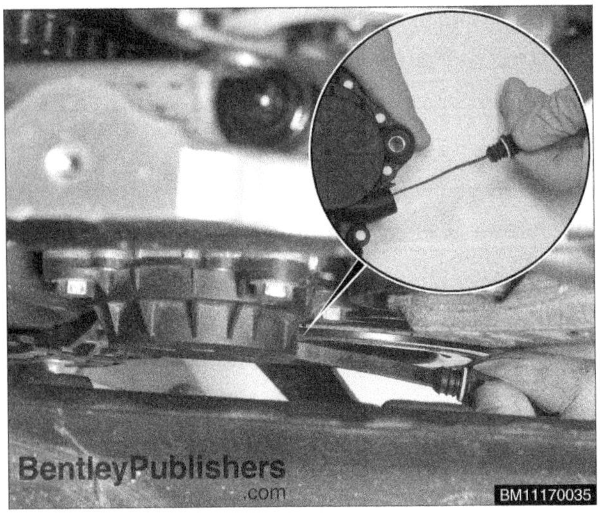

Place friction wheel in service position by pulling strap (**inset**) all the way out and holding.

– Detach accessory belt from crankshaft vibration damper and A/C compressor pulley. Check belt for cracks and coolant or oil residue. Replace if damaged or contaminated.

– When reinstalling, make sure belt is correctly seated in pulley grooves.

Engine Oil

Engine lubricant

Engine oil specification	
N12, N14, N16, N18 engine	SAE 5W - 30 MINI long-life rating (LL-01) synthetic oil for US market API rating SM or higher

Engine oil capacity	
Engine	Approximate capacity (Incl. oil filter)
N12, N14, N16, N18	4.2 liters (4.4 US qt)

Engine oil, checking level

Use MINI specified oil to top off engine oil level between oil changes. See **Engine lubricant** in this repair group.

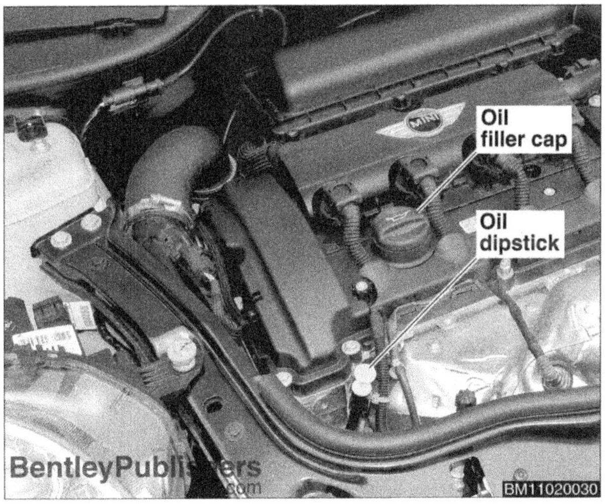

◄ Check engine oil level with dipstick located at right front of engine compartment.

- Check oil level with vehicle on level surface, after engine has been stopped for at least a few minutes.
- Check level by pulling out dipstick and wiping it clean. Reinsert all the way and withdraw again. Oil level is correct if it is between two marks near end of dipstick.
- If oil level is low, add oil by removing oil filler cap on top of cylinder head cover.
- Add only amount needed to bring oil level to upper mark on dipstick, using an oil of correct viscosity and grade.

> ***CAUTION—***
> - *Do not overfill engine oil. Overfilling may lead to engine damage.*

Engine oil and filter, changing

A complete oil change requires new oil, a new filter insert kit and a new drain plug sealing washer.

- Run engine for a few minutes to warm engine oil. Shut engine OFF.
- With vehicle on level ground, place 1½ gallon oil drain pan under engine sump.

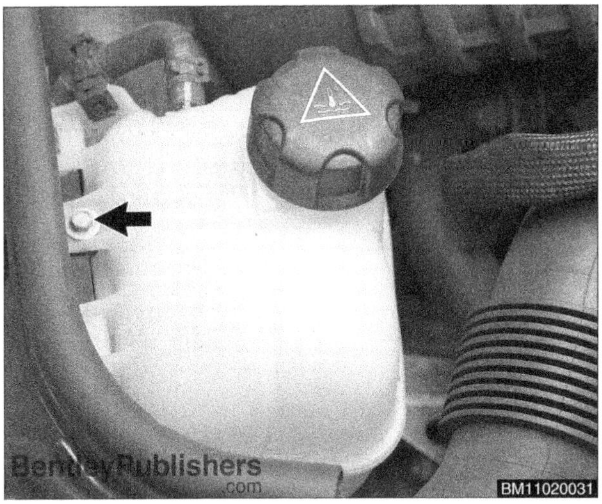

◄ Remove coolant reservoir mounting screw (**arrow**). Pull coolant reservoir toward right side of engine compartment and secure. Do not detach coolant hoses.

Using a 27 mm socket, loosen oil filter housing cover (**arrow**) slowly. Wait for oil to drain back into crankcase then remove oil filter cover with filter element.

> ***CAUTION—***
> - *Catch dripping oil in shop towels to prevent oil contamination of hoses and electrical harness insulation.*

Remove and discard filter element and O-ring seal.

- Install new O-ring and lubricate with engine oil.
- Insert new filter element in cover.

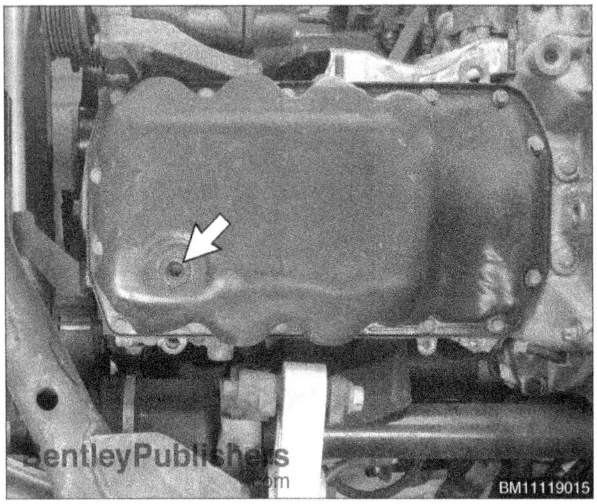

Use Allen key to loosen drain plug (**arrow**) at engine oil sump. Remove plug by hand and let oil flow into drain pan.

> ***WARNING—***
> - *Pull the loose plug away from the hole quickly to avoid being scalded by hot oil. It runs out quickly when the plug is removed. Use gloves to protect your hands.*

– When oil flow has diminished to an occasional drip, reinstall drain plug with a new sealing washer.

> ***CAUTION—***
> - *Be sure to tighten the engine oil drain plug using only hand tools. Power tools can strip the threads of the plug and oil pan.*

Tightening torque	
Engine oil drain plug to oil pan (M18 x 1.5)	30 Nm (22 ft-lb)

– Install and tighten oil filter housing cover.

Tightening torque	
Cover to oil filter housing	25 Nm (18 ft-lb)

– Refill crankcase with new oil. See **Engine lubricant** in this repair group for oil specification and approximate oil capacity. Use dipstick to check for correct oil level.

– Start engine and check that oil pressure warning light immediately goes out.

– Allow engine to run for a few minutes to circulate new oil, then check for leaks at drain plug and oil filter. Stop engine and recheck oil level.

FUEL SYSTEM

Fuel filter

The fuel filter, located in the fuel tank, is integrated with the right side fuel tank sensor assembly. No service or replacement interval is specified for the fuel filter. See **160 Fuel tank and Fuel Pump** for fuel filter replacing.

Fuel tank and fuel lines, inspecting

Inspect the fuel tank, fuel lines and fuel system for damage or leaks. Check for fuel leaks in the engine compartment or fuel odors in the passenger compartment. Inspect evaporative emissions hoses, checking carefully at charcoal canister and evaporative emissions purge system. See **130 Fuel Injection** and **160 Fuel Tank and Fuel Pump** for component locations and additional information.

INTERIOR VENTILATION MICROFILTER

Interior ventilation microfilter, replacing

> ***CAUTION—***
> *• Do not operate heating and cooling system without the microfilter in place.*

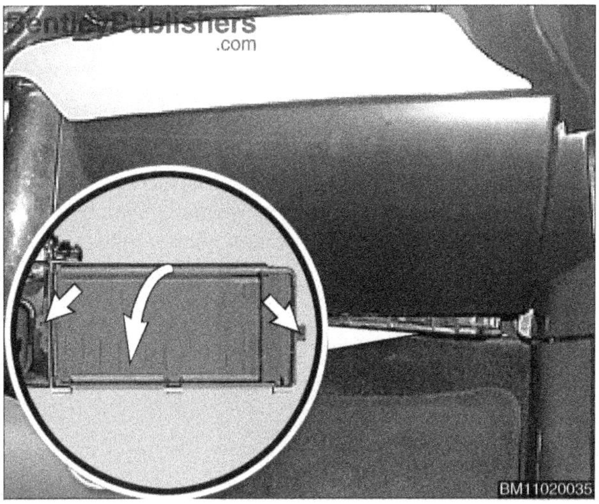

Working at microfilter housing below and behind glove compartment:

• If necessary, remove microfilter housing cover screws (**arrows**).

• Flip open catches and pull down housing cover (**flared arrow**).

- Slide microfilter out of housing (**arrow**).
- Using anti-fungal cleaning solvent (BMW part no 0 447 913 or equivalent), clean inside of housing and housing cover. Wipe away excess fluid and dry thoroughly.
- Microfilter installation is reverse of removal. Make sure filter is correctly seated on guides.

SPARK PLUGS

Spark plugs, replacing

> ***WARNING —***
> *• Remove sparks plugs only after the engine has cooled.*

For spark plug application information, consult a reputable MINI parts dealer or the web sites for Beru or NGK, the authorized spark plug suppliers for MINI. The recommended spark plug replacement intervals are in **Table e**. The spark plug electrode gap is not adjustable.

See also **120 Ignition System**.

Table e. Spark plug change intervals	
N12 engine	100,000 miles (160,000 km)
N14, N16, N18 engine (exc. JCW)	60,000 miles (100,00 km)
JCW	30,000 miles (50,000 km)

- Switch ignition OFF.
- **N12, N16 engine**: Remove ignition coil cover. See **Ignition coil cover, removing and installing** in this repair group.

- Unlock (**curved arrow**) each ignition coil connector. Pull (**straight arrow**) connector off coil. Pull coil straight out of spark plug hole.
- Use special 12-point spark plug socket to remove spark plug.

  Keep spark plugs in the order that they were removed from the cylinders. If diagnosing engine misfire or other malfunctions, knowing which spark plug came out of which cylinder may be helpful.

– Installation is reverse of removal.
  - Thread plugs into cylinder head by hand to prevent cross-threading.

Tightening torque	
Spark plug to cylinder head	23 ± 3 Nm (17 ± 2 ft-lb)

– When reinstalling ignition coils to spark plugs or ignition coil connectors to ignition coils, make sure that they snap securely into place.

TRANSMISSION

Transmission service

Automatic transmission. Automatic transmission fluid (ATF) change interval is 100,000 miles (160,000 km).

Manual transmission. There is no scheduled manual transmission fluid change interval.

During periodic underbody vehicle inspection, be sure to check the transmission for fluid leaks. Transmission fluid level checking and replacement procedure are covered in **230 Manual Transmission** and **240 Automatic Transmission**.

OTHER MECHANICAL MAINTENANCE

Oxygen sensors

OBD II enhanced emission standards require the engine control module (ECM) to monitor oxygen content in the exhaust with oxygen sensors placed before and after the catalytic converter. If the ECM detects that catalytic converter or oxygen sensor efficiency has degraded past a certain preprogrammed limit, it switches the malfunction indicator light (MIL) ON and stores a diagnostic trouble code (DTC) in the ECM. See **OBD On Board Diagnostics** for more information on OBD II systems. See **180 Exhaust System** for oxygen sensor replacement.

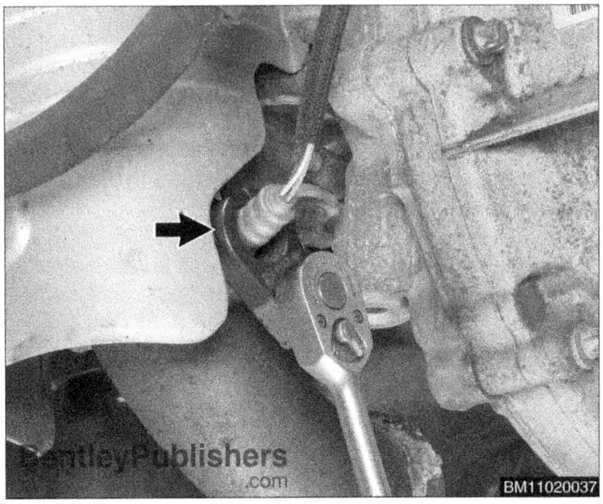

A special socket for replacing the oxygen sensor (**arrow**) is available from most automotive parts stores. The socket has a groove cut down one side to allow the sensor to be installed without damaging the wire harness.

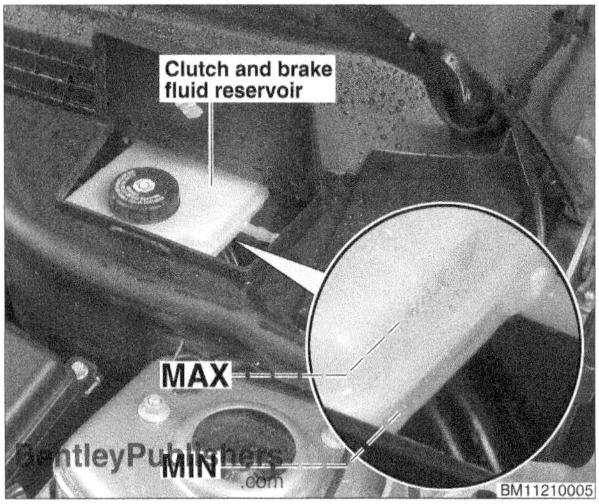

Clutch fluid, checking

In vehicle equipped with manual transmission and clutch, the hydraulic clutch and the brake system share the same reservoir and the same fluid. Clutch fluid level and brake fluid level are checked at the same time. Top up reservoir with DOT 4 low viscosity brake fluid.

Clutch fluid and clutch hydraulics service are covered in **210 CLutch**.

Drive axle joint (CV joint) boots, inspecting

Inspect CV joint protective boots closely for cracks and any other damage that allows contaminants to get into CV joints. If a rubber boot fails, water and dirt quickly damage the joint.

See **310 Front Suspension, Drive Axles** for CV joint boot service.

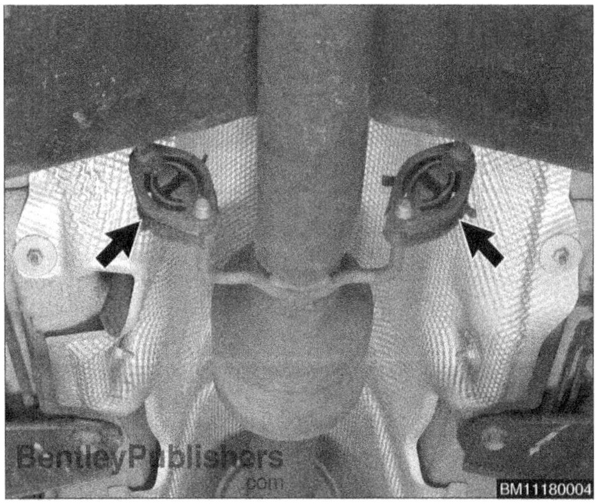

Exhaust system, inspecting

Exhaust system life varies widely according to driving habits and environmental conditions. If short-distance driving predominates, the moisture and condensation in the system do not fully dry out, causing early corrosion damage and more frequent replacement.

Scheduled maintenance of the exhaust system is limited to inspection:

- Check to see that exhaust hangers (**arrows**) are in place and properly supporting the system and that the system does not strike the body.
- Check for restrictions due to dents or kinks.
- Check for weakness or perforation due to rust.

See also **180 Exhaust System**.

Suspension, inspecting

Inspection of the front suspension and steering includes a check of moving parts for wear and excessive play. Inspect ball joint and tie-rod rubber seals and boots for cracks or tears that could allow the entry of dirt, water, and other contaminants. See **310 Front Suspension, Drive Axles**.

Inspection of the rear suspension includes a check of moving parts for wear and excessive play. Inspect suspension bushings for wear or deterioration. See **330 Rear Suspension**.

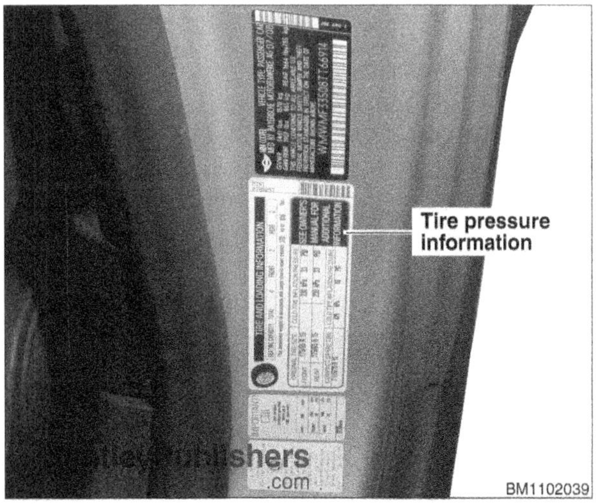

Tires, checking inflation pressure

Recommended tire pressures are on the driver side B-pillar sticker. Note that higher pressures are recommended when the vehicle is more heavily loaded.

Correct tire pressures are important to handling and stability, fuel economy and tire wear. Tire pressures change with temperature. Check tire pressures often during seasonal temperature changes.

> ***WARNING—***
> * *Do not inflate any tire to a pressure higher than the maximum inflation pressure listed on the sidewall. Use care when adding air to warm tires. Warm tire pressures can increase as much as 4 psi (0.3 bar) over their cold pressures.*

Tires, rotating

MINI does not recommend tire rotation. Rotating the tires may adversely affect road handling and tire grip.

Wheels, aligning

MINI recommends checking the front and rear alignment once a year and whenever new tires are installed.

See **320 Steering and Wheel Alignment** for alignment specifications.

BODY AND INTERIOR MAINTENANCE

Body and hinges, lubricating

- Lubricate door locks and lock cylinders with an oil that contains graphite.
- Lubricate body and door hinges, hood latch and door check rods with SAE 30 or SAE 40 engine oil.
- Lubricate seat runners with multipurpose grease.
- If door weather-strips are sticking, lubricate them with silicone spray or talcum powder. Do not apply any oil to rubber parts.
- Lubricate hood release cable.
- Use minimum amount of winter lock deicer. Alcohol in deicer washes grease out of lock assemblies and may cause locks to corrode internally or to become difficult to operate.

Exterior washing

The longer dirt is left on paint, the greater the risk of damaging the glossy finish, either by scratching or by the chemical effect dirt particles may have on the painted surface.

Do not wash the vehicle in direct sunlight. If the engine hood is warm, allow it to cool. Beads of water not only leave spots when dried rapidly by the sun or heat from the engine, but also can act as small magnifying glasses and burn spots into the finish.

- Wash vehicle with a mixture of lukewarm water and a car wash product.
- Rinse using plenty of clear water.

– Wipe body dry with soft cloth towel or chamois to prevent water spots.

Interior care

– Remove dirt spots with lukewarm soapy water or a dry foam cleaner.

– Use spot remover for grease and oil spots. Do not pour liquid directly on carpet or fabric, but dampen a clean cloth and rub carefully, starting at edge of spot and working inward. Do not use gasoline, naphtha, or other flammable substances.

Leather upholstery and trim

– Periodically clean leather upholstery and trim. Use a slightly damp cotton or wool cloth to get rid of dirt in creases and pores that can cause brittleness and premature aging.

– On heavily soiled areas, use mild detergent (such as Woolite®) or other specially formulated leather cleaners.

– Dry trim and upholstery completely using a soft cloth.

Regular use of a good quality leather conditioner reduces drying and cracking of the leather.

Seat belts

Dirt and other abrasive particles damage seat belt webbing.

– If it is necessary to clean seat belts, use a mild soap solution. Avoid bleach and other strong cleaning agents, which may weaken belt webbing.

> ***WARNING —***
> - *Do not clean the seat belt webbing using dry cleaning or other chemicals. Allow wet belts to dry before allowing them to retract.*

– Inspect condition of belt webbing and function of retractor mechanisms. See **720 Seat Belts** for seat belt troubleshooting.

Windshield wiper blade maintenance

Common problems with the windshield wipers include streaking or sheeting, water drops after wiping and blade chatter. Streaking is usually caused when wiper blades are coated with road film or car wash wax.

– Clean blades using soapy water. If cleaning blades does not cure problem, replace them.

– In general, replace wiper blades twice a year, before and after the cold season.

See **611 Wipers and Washers** for more information.

100 Engine General

GENERAL

This section covers general information on engines and engine components. See **Table a** for engine application information and specifications.

See also:

- **020 Maintenance** for engine compartment views and accessory belt service, oil change, air filter service and spark plug replacement
- **110 Engine Removal and Installation**
- **113 Cylinder Head Removal and Installation**
- **117 Camshaft Timing Chain**
- **119 Lubrication System** for lubrication system description and crankcase seal service
- **120 Ignition System** for ignition coil, camshaft sensor, crankshaft sensor and knock sensor service
- **130 Fuel Injection** for engine management system description and component replacement
- **170 Radiator and Cooling System** for radiator, coolant pump and coolant service
- **180 Exhaust System** for exhaust manifold, oxygen sensor and turbocharger service

Engine ID tag

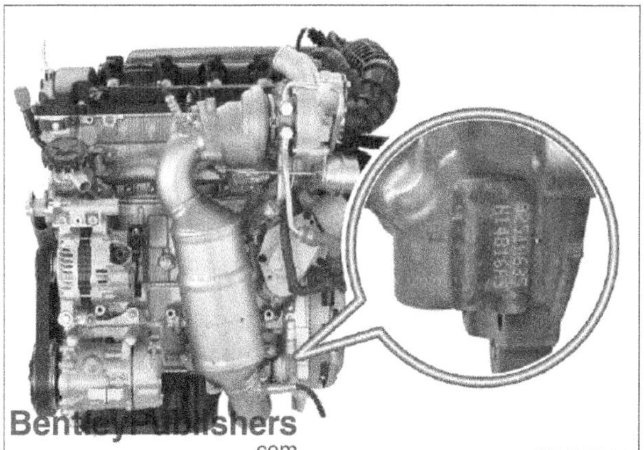
BM11100001

Engine ID tag with engine serial number is on left side of engine block, above engine oil pan.

Engine identifying features

The same engine, jointly developed by BMW Group and PSA Peugeot Citroën, is used in all MINI versions. The main visual difference between the engines is that the turbocharged engine (Cooper S) is equipped with a turbocharger on the top and front of the engine. Vehicles produced from 08 / 2010 were updated with new features. However, the main design of the engines remained similar.

BM11100005

N12 engine is the normally aspirated engine available in MINI Cooper models. A few notable features are:

- All aluminum two-piece crankcase design
- Infinitely variable valve timing (double VANOS)
- Variable valve lift (Valvetronic)
- Lightweight alloy cylinder head
- Weight-optimized roller cam followers
- Volume controlled oil pump
- Composite camshafts
- Cast iron cylinder liners
- Cast iron crankshaft
- Coolant pump with "switch-OFF" control
- Single serpentine belt drive

BM11100004

N16 engine features many of the same technological systems as the N12 engine, with the following additional features:

- Map-controlled oil pump
- Infinitely variable valve timing (double VANOS)
- Variable valve lift (Valvetronic) with 3-phase servomotor and integrated eccentric shaft sensor
- Composite camshafts

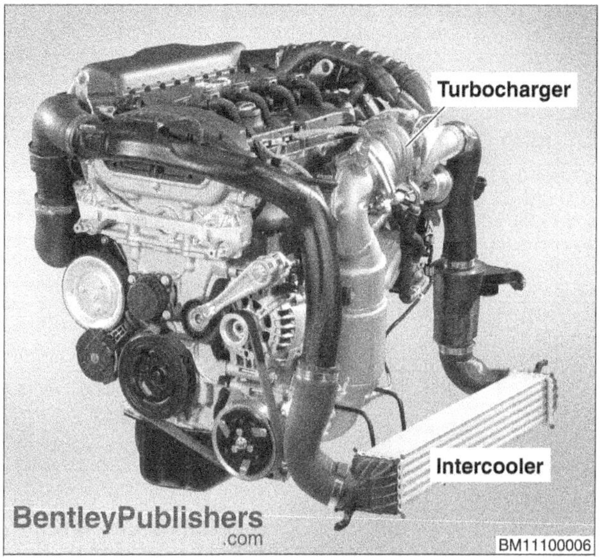

N14 engine features many of the same technological systems as the N12 engine, with the following additional features:

- Infinitely variable intake valve timing (single VANOS)
- Direct fuel injection
- Twin-scroll exhaust turbocharger
- Auxiliary electric coolant pump for turbocharger cooling

N18 engine features many of the same technological systems as the N14 engine, with the following additional features:

- Infinitely variable valve timing (double VANOS)
- Variable valve lift (Valvetronic) with 3-phase servomotor and integrated eccentric shaft sensor
- Map-controlled oil pump
- Composite camshafts
- New piston design

Engine applications

Table a. MINI Cooper engine applications and specifications							
Year, model	**Engine code, type**	**Displacement cc (cid)**	**Bore x stroke mm (in)**	**Comp. ratio**	**Torque lb-ft @ rpm**	**Horsepower hp @ rpm**	**Engine management**
1.6 liter 4-cylinder (normally aspirated)							
2007 - 2010 Cooper, Clubman	N12B16AB, double VANOS, Valvetronic	1598 (97.6)	85.7 x 77 (3.37 x 3.03)	11 : 1	118 @ 4250	118 @ 6000	Siemens DME MEV17.2
2009 - 2010 Convertible	N12B16AB, double VANOS, Valvetronic	1598 (97.6)	85.7 x 77 (3.37 x 3.03)	11 : 1	118 @ 4250	118 @ 6000	Siemens DME MEV17.2
2011 - 2013 Cooper, Clubman	N16B16M0, double VANOS, Valvetronic	1598 (97.6)	85.7 x 77 (3.37 x 3.03)	11 : 1	118 @ 4250	121 @ 6000	Siemens DME MEV17.2.2
1.6 liter 4-cylinder (turbocharged)							
2007 - 2010 Cooper S, Clubman S	N14B16AC, single VANOS	1598 (97.6)	85.8 x 77 (3.37 x 3.03)	10.5 : 1	177 @ 1600 - 5000	171 @ 5500	Siemens DME MED17.2
2009 - 2010 Convertible S	N14B16AC, single VANOS	1598 (97.6)	85.7 x 77 (3.37 x 3.03)	10.5 : 1	177 @ 1600 - 5000	171 @ 5500	Siemens DME MED17.2
2011 - 2013 Cooper S	N18B16M0, double VANOS, Valvetronic	1598 (97.6)	85.7 x 77 (3.37 x 3.03)	10.5 : 1	191 @ 1700 - 6500	181 @ 5550	Siemens DME MEVD17.2.2
1.6 liter 4-cylinder John Cooper Works (JCW) (turbocharged)							
2009 - 2013 JCW	N14B16TO, single VANOS	1598 (97.6)	85.7 x 77 (3.37 x 3.03)	10.0 : 1	207 @ 1850 - 5600	208 @ 6000	Siemens DME MED17.2

110 Engine Removal and Installation

GENERAL

This repair group includes an overview of engine removal and installation.

See also:

- **020 Maintenance** for engine cover removal
- **100 Engine–General** for engine codes and applications
- **121 Battery, Starter, Alternator**
- **170 Radiator and Cooling System**
- **180 Exhaust System**
- **230 Manual Transmission**
- **240 Automatic Transmission**
- **310 Front Suspension, Drive Axles**
- **410 Fenders, Engine Hood**

Warnings and Cautions

WARNING—

- *The fuel system is designed to retain pressure even when the ignition is OFF. When working with the fuel system, loosen fuel lines slowly to allow residual fuel pressure to dissipate. Avoid spraying fuel.*
- *Before beginning any work on the fuel system, place a fire extinguisher in the vicinity of the work area.*
- *Fuel is highly flammable. When working around fuel, do not disconnect any wires that could cause electrical sparks. Do not smoke or work near heaters or other fire hazards.*
- *Loosen the fuel tank cap to release pressure in the tank before working on the tank or lines.*
- *When disconnecting a fuel hose, wrap a shop towel around the end of the hose to prevent fuel spray.*
- *Due to risk of personal injury, be sure the engine is cold before beginning the removal procedure.*
- *Do not use a work light with an incandescent bulb near any fuel. Fuel may spray on the hot bulb causing a fire.*
- *Make sure the work area is properly ventilated.*

> *CAUTION—*
> - *If the MIL (malfunction indicator light, also called Check Engine or Service Engine Soon light) is illuminated, see* **OBD On-Board Diagnostics** *for DME fault code information.*
> - *If other system faults are indicated by an illuminated ABS, SRS or DSC warning light, see the appropriate repair group in this manual or a qualified MINI repair shop for more information on fault codes.*
> - *Cover painted surfaces before beginning engine removal.*
> - *As an aid to installation, label components, wires and hoses before removing them.*
> - *Do not reuse gaskets, O-rings or seals during reassembly.*

ENGINE REMOVAL AND INSTALLATION

Removing and installing N12, N16 (naturally aspirated) and N14, N18 (turbocharged) engines is similar.

Engine, removing and installing

> *WARNING—*
> - *Due to risk of personal injury, be sure the engine is cold before beginning the removal procedure.*

– Remove engine hood. See **410 Fenders, Engine Hood**.

– Disconnect negative (–) cable from battery.

> *CAUTION—*
> - *Prior to disconnecting the battery, read the battery disconnection cautions in* **001 Warnings and Cautions**.

– Raise vehicle and support safely.

> *WARNING—*
> - *Make sure the vehicle is stable and well supported at all times. Use a professional automotive lift or jack stands designed for the purpose. A floor jack is not adequate support.*

– Remove exhaust system. See **180 Exhaust System**.

– Drain engine oil.

– Drain engine coolant and detach coolant and heater hoses from engine. See **170 Radiator and Cooling System**.

– Remove air cleaner housing and intake air ducts. See **130 Fuel Injection**.

– Using suitable equipment, evacuate refrigerant from A/C system.

– Detach main engine fuel hose at left rear of engine compartment.

> *WARNING—*
> - *When disconnecting the fuel hose, wrap a shop towel around the end of the hose to prevent fuel spray.*

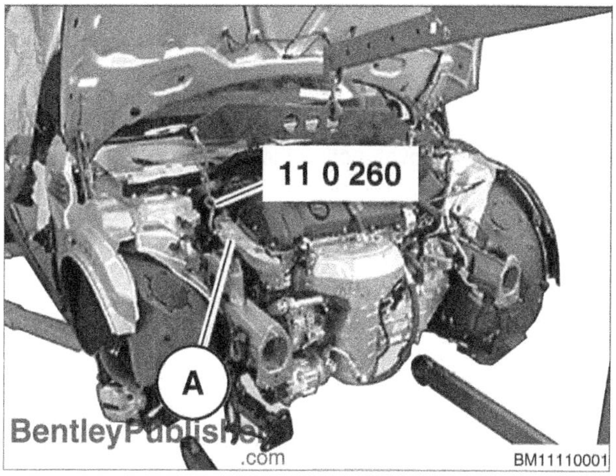

- Remove drive axles. See **310 Front Suspension, Drive Axles**.
- Detach vacuum line from brake booster.
- Disconnect engine wiring harness and lay aside.
- Remove radiator support complete with radiator and intercooler (if applicable). See **410 Fenders, Engine Hood**.
  - Detach intercooler air ducts as you pull radiator support forward.

◄ Attach MINI special tool **11 0 260** or equivalent to lifting eye (**A**) on engine.

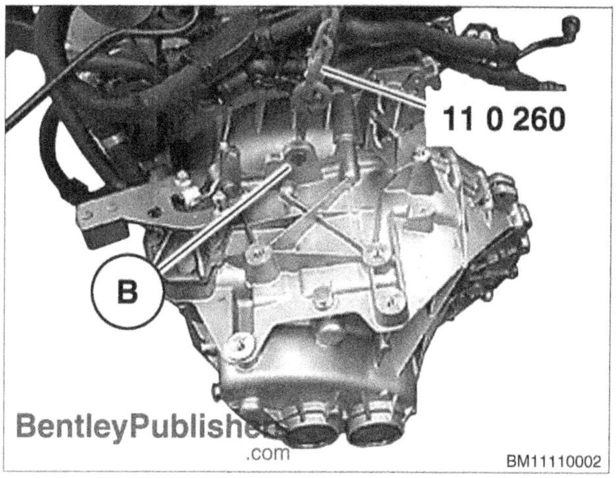

◄ Attach MINI special tool **11 0 260** or equivalent to lifting eye (**B**) on transmission.

- Raise engine - transmission unit slightly and remove engine - transmission mounts.

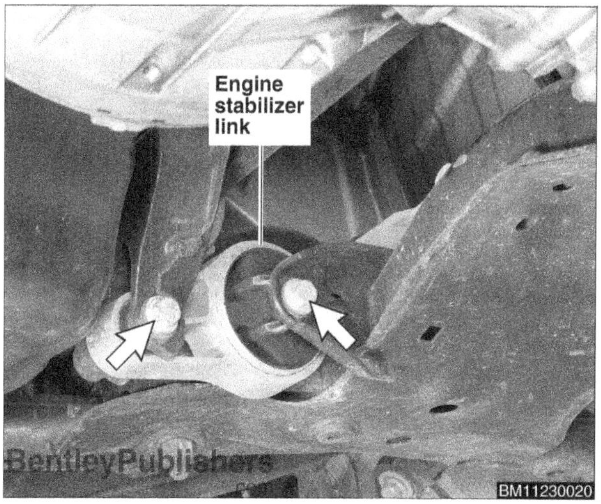

◄ Working underneath engine, remove engine stabilizer mounting bolts (**arrows**) and remove engine stabilizer.

Carefully remove engine - transmission assembly from engine compartment toward front, checking for any wiring, fuel lines, or mechanical parts that might become snagged as assembly is removed.

– Installation is reverse of removal, noting the following:
  - Replace gaskets, O-rings and seals.
  - Change engine oil and filter and check all other fluid levels. See **020 Maintenance**.
  - Recharge refrigerant.
  - Refill and bleed cooling system. See **170 Radiator and Cooling System**.

113 Cylinder Head Removal and Installation

Cylinder head and cylinder head gasket

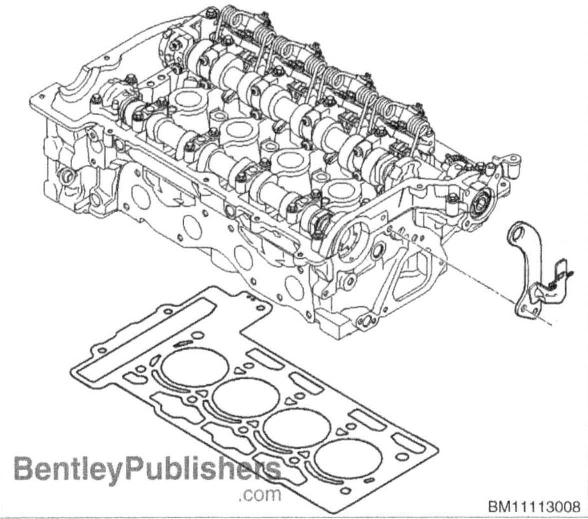

GENERAL

This repair group provides removal and installation procedures for the cylinder head and associated components.

The cylinder head is equipped with variable camshaft timing (sometimes referred to as VANOS, from the German words *VAriable NOckenwellen Steuerung*). Special tools and procedures are required to remove and install camshaft adjustment units and to time camshafts. Be sure to read each procedure through before starting work.

See also:

- **020 Maintenance**
- **100 Engine–General** for engine code and application information.
- **117 Camshaft Timing Chain** for camshaft timing chain and camshaft adjuster (VANOS) service
- **130 Fuel Injection** for air filter housing, intake manifold removal and Valvetronic motor
- **170 Radiator and Cooling System**
- **180 Exhaust System** for exhaust manifold and turbocharger

Engine identification table

Table a. Engine identification	
Year, model	**Engine designation**
2007 - 2010 Cooper, Clubman, Convertible	N12
2007 - 2010 Cooper S, Clubman S, Convertible S	N14
2011 - 2013 Cooper, Clubman, Convertible	N16
2011 - 2013 Cooper S, Clubman S, Convertible S	N18

Warnings and Cautions

WARNING—
- *To avoid personal injury, be sure the engine is cold before beginning work on engine components.*
- *Use extreme caution when draining and disposing of engine coolant. Coolant is poisonous and lethal to humans and pets. Pets are attracted to coolant because of its sweet smell and taste. Seek medical attention immediately if coolant is ingested.*
- *The fuel system is designed to retain pressure even when the ignition is OFF. When working with the fuel system, loosen fuel lines slowly to allow residual fuel pressure to dissipate. Avoid spraying fuel. Use shop towels to capture leaking fuel.*
- *Before beginning work on the fuel system, place a fire extinguisher in the vicinity of the work area.*
- *Fuel is highly flammable. When working around fuel, do not disconnect wires that could cause electrical sparks. Do not smoke or work near heaters or other fire hazards.*
- *Wear eye protection and protective clothing to avoid injuries from contact with fuel.*
- *Unscrew the fuel tank cap to release pressure in the tank before working on fuel lines.*
- *Do not use a work light with an incandescent bulb near fuel. Fuel may spray on the hot bulb causing a fire.*
- *Make sure the work area is properly ventilated.*

CAUTION—
- *Follow torque instructions, including angle of rotation specifications, when installing fasteners.*

CYLINDER HEAD

Cylinder head cover, removing and installing

Cylinder head cover service is similar for N12, N16 (naturally aspirated) and N14, N18 (turbocharged) models. The procedures are combined with differences noted.

– Make sure ignition is OFF. Remove ignition key.

– Remove intake air housing. See **130 Fuel Injection.**

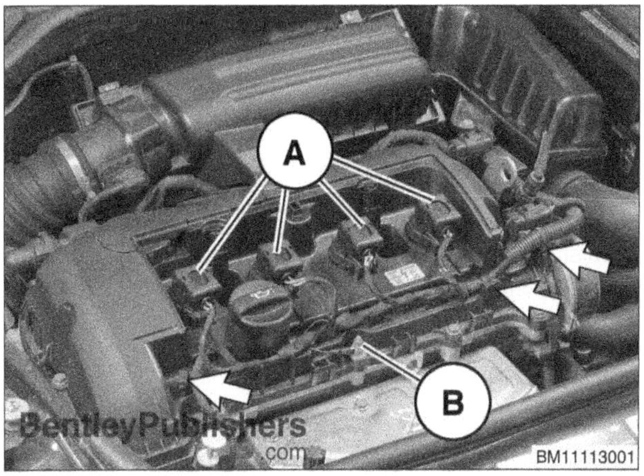

◄ Remove ignition coils (**A**). See **120 Ignition System**.

– Unclip engine wiring harness (**arrows**) and remove ground fastener (**B**). Lay wiring harness aside.

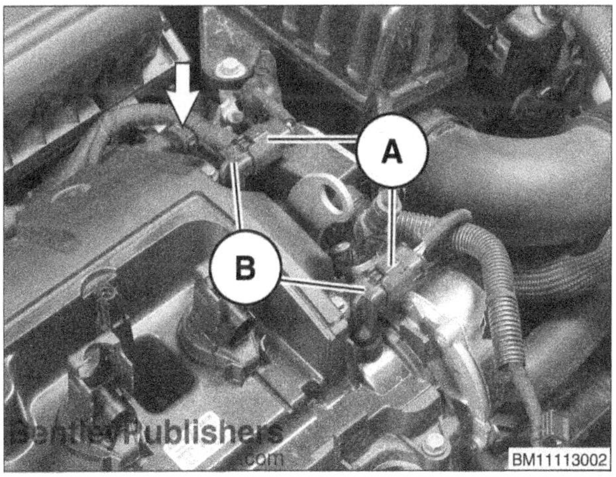

◄ **N12, N16 engine**: Disconnect electrical connectors (**A**) and unclip engine wiring harness (**arrow**).

– **N12, N16 engine**: Remove sensors (**B**), if necessary.

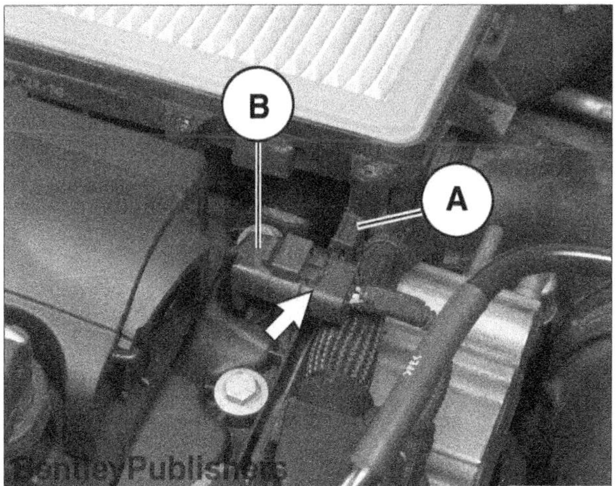

◄ **N14, N18 engine**: Disconnect electrical connector (**arrow**) and unclip engine wiring harness (**A**).

– **N14, N18 engine**: Remove sensor (**B**), if necessary.

Cylinder head cover, removing and installing

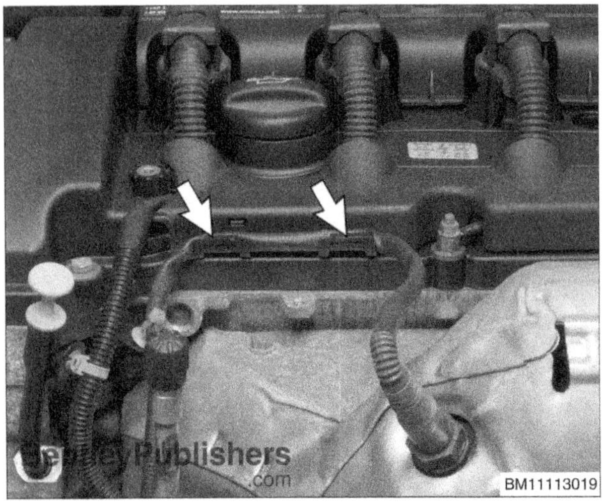

< **N14, N18 engine** Detach oxygen sensor harness (**arrows**) from cylinder head cover.

< **N12, N16 engine**: Working at right front of cylinder head cover, use small flat blade screwdriver to release crankcase breather hose clip (**inset**). Detach hose from cover.

< **N12, N16 engine**: Working at rear of cylinder head cover, unclip wiring harness (**arrows**).

◄ Unclip wiring harness loom (**A**) by sliding up and out of bracket in direction of **arrow**.

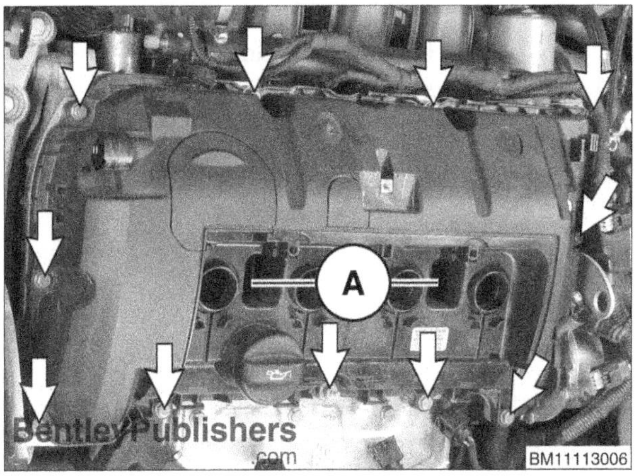

◄ Remove cylinder head cover fasteners (**arrows**) then fasteners (**A**) in center of cylinder head cover.

– Lift cylinder head cover off cylinder head.

◄ Reinstall cylinder head cover.

- Coat cylinder head contact surfaces with thin layer of anti-friction compound such as glycerine.
- Place small quantity of silicone sealant at half moon gasket area.
- Tighten gradually using numbered sequence, starting with inside fasteners.

Tightening torque	
Cylinder head cover to cylinder head	10 Nm (7.3 ft-lb)

– Remainder of assembly is reverse of removal.

Tightening torque	
Ground connection to cylinder head	10 Nm (7.3 ft-lb)

113

Cylinder head and timing chain components (N12, N16 engines)

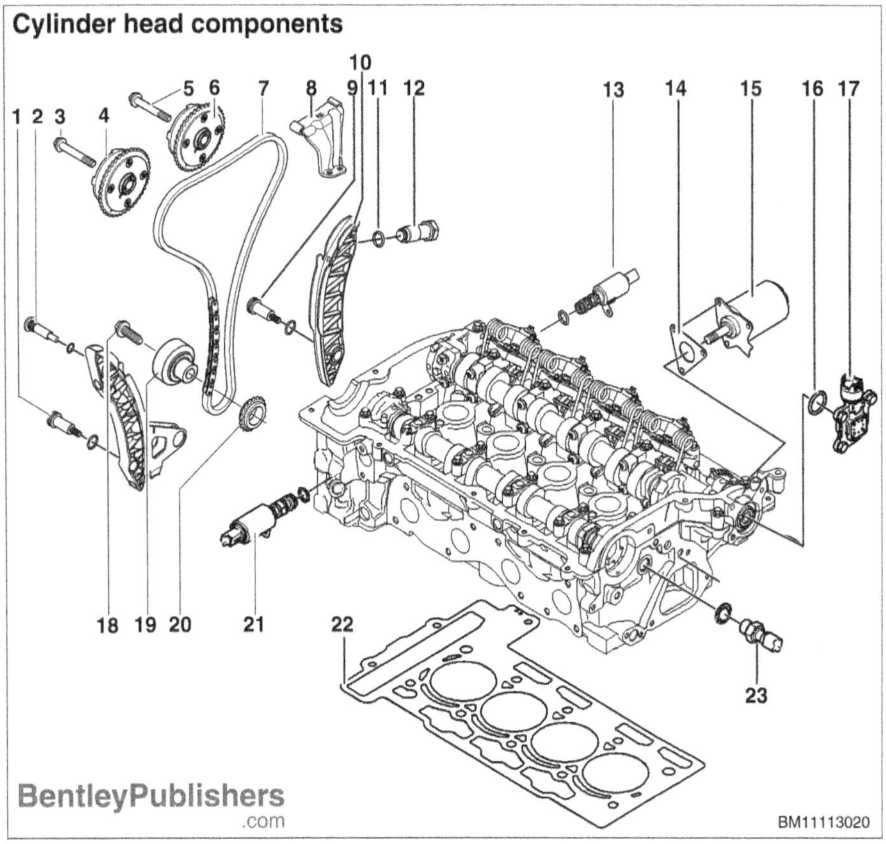

1. Timing chain guide bearing bolt
2. Timing chain guide bearing bolt
3. VANOS Torx fastener
   - Always replace
   - Tightening torque
     Stage 1: 20 Nm (15 ft-lb)
     Stage 2: additional 180°
4. Exhaust VANOS adjuster
5. VANOS Torx fastener
   - Always replace
   - Tightening torque
     Stage 1: 20 Nm (15 ft-lb)
     Stage 2: additional 180°
6. Intake VANOS adjuster
7. Timing chain
8. Slide rail
9. Timing chain guide bearing bolt
10. Timing chain tensioner guide
11. O-ring
12. Timing chain tensioner
    - Tightening torque: 65 Nm (48-ft-lb)
13. Intake VANOS solenoid
    - Tightening torque 9 Nm (7 ft-lb)
14. Valvetronic motor gasket
15. Valvetronic motor
16. O-ring
17. Eccentric shaft sensor
18. Bolt (damper hub to crankshaft)
    - Always replace
    - Use holding tool (MINI special tool **11 9 280)** to loosen / tighten bolt
    - Tightening torque
      Stage 1: 50 Nm (36.8 ft-lb)
      Stage 2: additional 180°
19. Damper hub
20. Crankshaft timing chain sprocket
21. Exhaust VANOS solenoid
    - Tightening torque 9 Nm (7 ft-lb)
22. Cylinder head gasket
23. Oil pressure switch

Cylinder head and timing chain components (N14 engine)

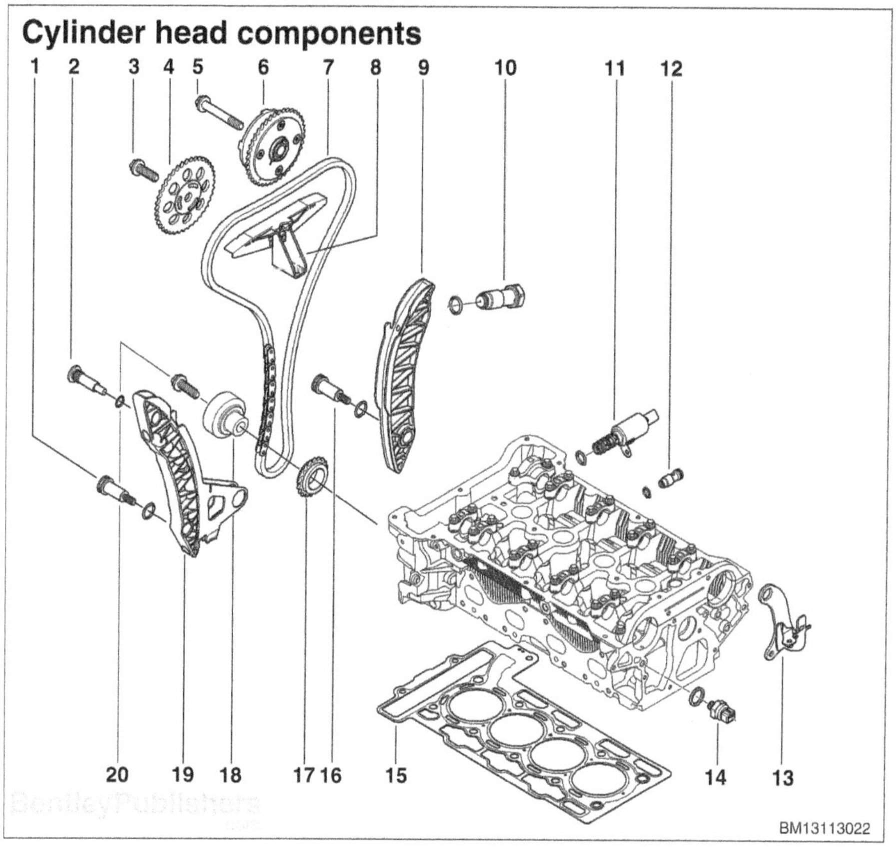

1. Timing chain guide bearing bolt
2. Timing chain guide bearing bolt
3. Exhaust camshaft sprocket fastener
   - Always replace
   - Tightening torque
     Stage 1: 20 Nm (15 ft-lb)
     Stage 2: additional 90°
4. Exhaust camshaft sprocket
5. VANOS Torx fastener
   - Always replace
   - Tightening torque
     Stage 1: 20 Nm (15 ft-lb)
     Stage 2: additional 180°
6. Intake VANOS adjuster
7. Timing chain
8. Slide rail
9. Timing chain tensioner guide
10. Timing chain tensioner
    - Tightening torque: 65 Nm (48-ft-lb)
11. Intake VANOS solenoid
    - Tightening torque 9 Nm (7 ft-lb)
12. Non return valve
13. Lifting eyelet
14. Oil pressure switch
15. Cylinder head gasket
16. Timing chain guide bearing bolt
17. Crankshaft timing chain sprocket
18. Damper hub
19. Timing chain guide
20. Bolt (damper hub to crankshaft)
    - Always replace
    - Use holding tool (MINI special tool **11 9 280)** to loosen / tighten bolt
    - Tightening torque
      Stage 1: 50 Nm (36.8 ft-lb)
      Stage 2: additional 180°

Cylinder head and timing chain components (N18 engine)

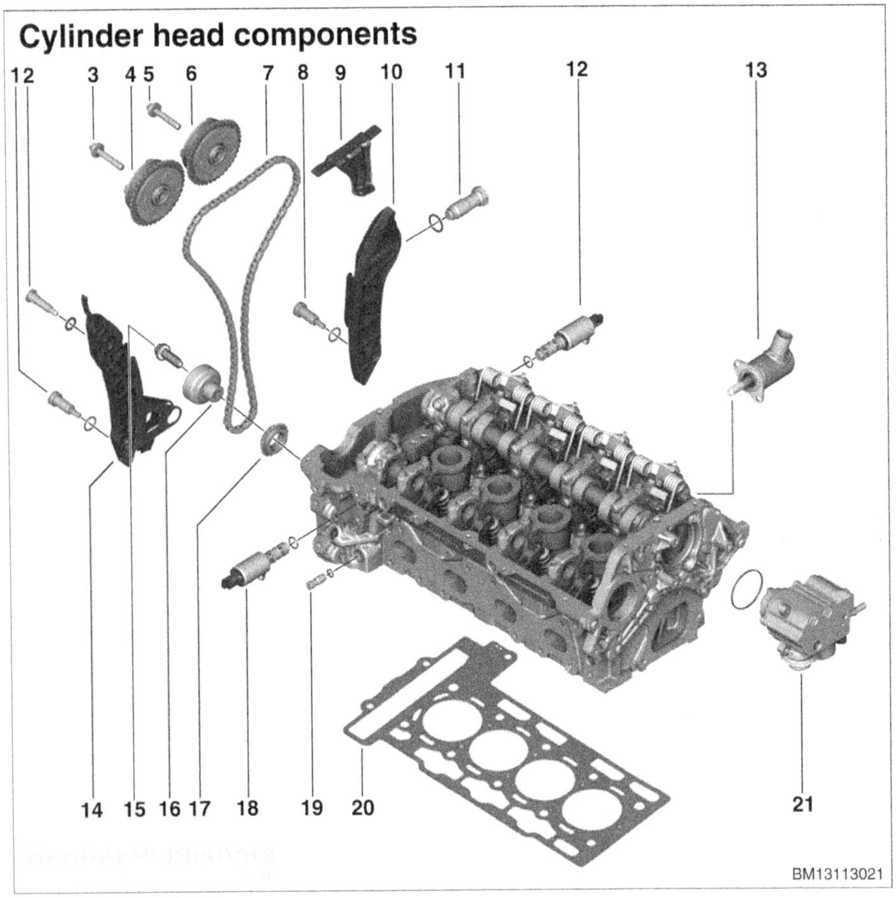

1. Timing chain guide bearing bolt
2. Timing chain guide bearing bolt
3. VANOS Torx fastener
   - Always replace
   - Tightening torque
     Stage 1: 20 Nm (15 ft-lb)
     Stage 2: additional 180°
4. Exhaust VANOS adjuster
5. VANOS Torx fastener
   - Always replace
   - Tightening torque
     Stage 1: 20 Nm (15 ft-lb)
     Stage 2: additional 180°
6. Intake VANOS adjuster
7. Timing chain
8. Timing chain guide bearing bolt
9. Slide rail
10. Timing chain tensioner guide
11. Timing chain tensioner
    - Tightening torque: 65 Nm (48-ft-lb)
12. Intake VANOS solenoid
    - Tightening torque 9 Nm (7 ft-lb)
13. Valvetronic motor
14. Timing chain guide
15. Bolt (damper hub to crankshaft)
    - Always replace
    - Use holding tool (MINI special tool **11 9 280)** to loosen / tighten bolt
    - Tightening torque
      Stage 1: 50 Nm (36.8 ft-lb)
      Stage 2: additional 180°
16. Damper hub
17. Crankshaft timing chain sprocket
18. Exhaust VANOS solenoid
    - Tightening torque 9 Nm (7 ft-lb)
19. Non return valve
20. Cylinder head gasket
21. High pressure fuel pump

Cylinder head service

When removing and installing the cylinder head, special tools and procedures are required to remove and install camshaft adjustment units and Valvetronic eccentric shaft and to time camshafts. Read the entire procedure before beginning repairs.

> ***CAUTION—***
> - *Disassembly, removal and assembly of camshafts, camshaft adjustment units or cylinder head without special tools poses the risk of damage or breakage: Valves may be bent by contact with the piston crowns.*

Camshaft timing is covered in **117 Camshaft Timing Chain**.

If you remove a cylinder head and determine that it requires significant reconditioning work, a remanufactured cylinder head is available from an authorized MINI dealer.

Cylinder head, removing and installing

The cylinder head bolts stretch when tightened. Replace them any time they are removed. Use angle protractor to torque.

> ***WARNING—***
> - *To avoid personal injury, be sure the engine is cold before beginning the removal procedure.*

– Disconnect negative (–) cable from battery.

> ***CAUTION—***
> - *Prior to disconnecting the battery, read the battery disconnection cautions in* **001 Warnings and Cautions**.

– Raise vehicle and support safely.

> ***WARNING—***
> - *Make sure the vehicle is stable and well supported at all times. Use a professional automotive lift or jack stands designed for the purpose. A floor jack is not adequate support.*

– With exhaust system fully cooled off, remove exhaust system and exhaust manifold. See **180 Exhaust System**.

– Drain engine coolant. See **170 Radiator and Cooling System**.

– Drain engine oil. See **020 Maintenance**.

– Remove cylinder head cover. See **Cylinder head cover removing and installing** in this repair group.

– Remove intake manifold. See **130 Fuel Injection**.

Remove engine oil dipstick tube fastener (**arrow**) then remove dipstick tube from engine.

– Detach coolant hoses from cylinder head.

– Remove VANOS unit(s). See **117 Camshaft Timing Chain**.

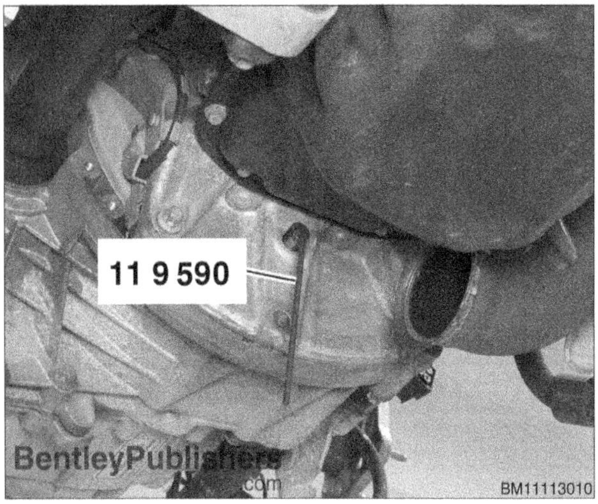

Using MINI special tool **11 9 590**, secure crankshaft in 90° piston position.

– Place radiator support in service position. See **410 Fenders, Engine Hood**.

– Support engine from above.

– Remove upper alternator fasteners. Do not remove alternator.

– Remove right side engine mount.

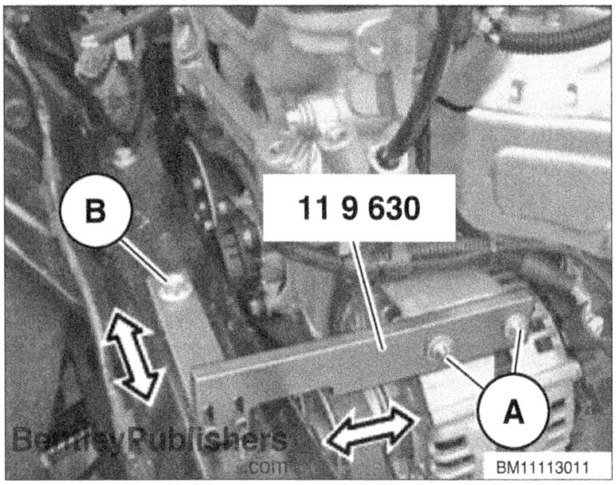

Install MINI special tool **11 9 630** using alternator (**A**) and engine mount (**B**) fasteners.

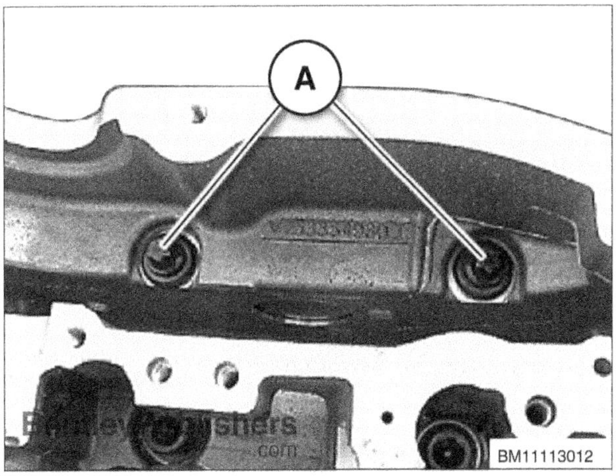

< Remove cylinder head fasteners (**A**).

< Remove cylinder head fastener (**A**).

< Using MINI special tool **11 2 250**, remove cylinder head fasteners starting in center of cylinder head and working outwards.

– With help from an assistant, lift cylinder head off engine.

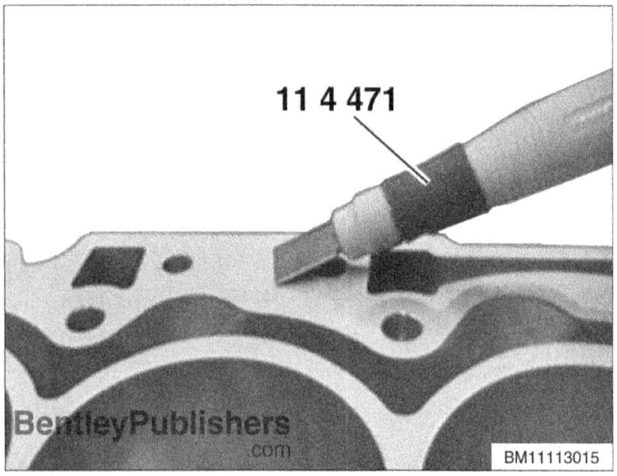

Using MINI special tool **11 4 471**, remove coarse gasket remnants from engine block and cylinder head sealing surfaces.

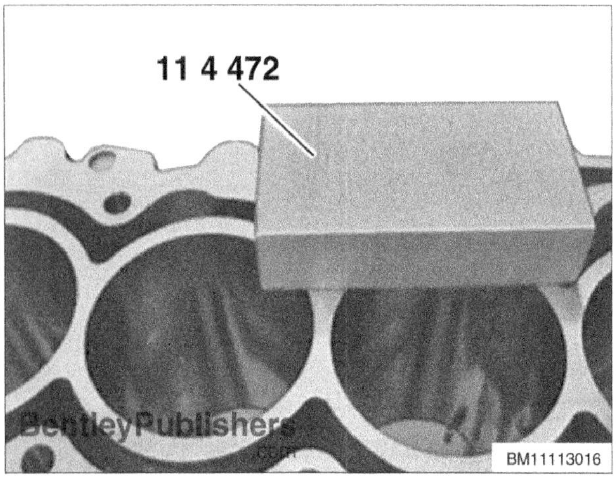

using MINI special tool **11 4 472**, remove fine gasket remnants from engine block and cylinder head sealing surfaces.

– Check cylinder head for straightness and warpage.

Cylinder head specification	
Longitudinal maximum warpage	0.10 mm (0.003 in)
Transversal maximum warpage	0.05 mm (0.001 in)

When installing cylinder head:

- Check alignment dowels in engine block for damage.
- Use new cylinder head fasteners. Do not clean coating from new bolts.
- Transfer washers from old cylinder head fasteners to new.
- Torque center bolts first then working outwards following numbered sequence.

Tightening torques	
Cylinder head to engine block M8 x 35 mm (replace fasteners)	30 Nm (22.1 ft-lb)
Cylinder head to engine block M8 x 95 mm (replace fasteners) • Stage 1 • Stage 2 • Stage 3	 15 Nm (11 ft-lb) Additional 90° Additional 90°
Cylinder head to engine block M10 x 145 mm (replace fasteners) • Stage 1 • Stage 2 • Stage 3	 30 Nm (22.1 ft-lb) Additional 90° Additional 90°

– Remainder of installation is reverse of removal.

Tightening torques	
Cylinder head cover to cylinder head	10 Nm (7.3 ft-lb)
Ground connection to cylinder head	10 Nm (7.3 ft-lb)
Lifting eyelet to cylinder head M8 x 20 mm, M8 x 25 mm	20 Nm (14.7 ft-lb)
Eccentric shaft sensor to cylinder head M6 x 40 mm	10 Nm (7.3 ft-lb)
Intake camshaft bearing cap to cylinder head M8 x 30.5 mm	10 Nm (7.3 ft-lb)
Exhaust camshaft bearing cap to cylinder head M6 x 40 mm	10 Nm (7.3 ft-lb)
Eccentric shaft bearing block to cylinder head M6 x 40 mm, M6 x 70 mm	10 Nm (7.3 ft-lb)

117 Camshaft Timing Chain

GENERAL

This repair group provides removal and repair information for crankshaft vibration damper, engine timing chain and camshaft adjustment (VANOS) units. Camshaft timing procedure is also included. Repair procedures in this repair group assume that the engine is installed in the engine bay.

Special tools and procedures are required to remove and install camshaft adjustment units and to time camshafts. Be sure to read each procedure through before starting work.

See also:

- **020 Maintenance**
- **100 Engine–General** for engine code and application information
- **113 Cylinder Head Removal and Installation** for cylinder head cover removal
- **130 Fuel Injection** for air filter housing and intake manifold removal
- **170 Radiator and Cooling System** for radiator drain and refill

Engine identification table

Table a. Engine identification	
Year, model	**Engine designation**
2007 - 2010 Cooper, Clubman, Convertible	N12
2007 - 2010 Cooper S, Clubman S, Convertible S	N14
2011- 2013 Cooper, Clubman, Convertible	N16
2011- 2013 Cooper S, Clubman S, Convertible S	N18

Warnings

> ***WARNING—***
> - *Due to risk of personal injury, be sure the engine is cold before beginning work on engine components.*
> - *Use extreme caution when draining and disposing of engine coolant. Coolant is poisonous and lethal to humans and pets. Pets are attracted to coolant because of its sweet smell and taste. Seek medical attention immediately if coolant is ingested.*
> - *Fuel is highly flammable. When working around fuel, do not disconnect wires that could cause electrical sparks. Do not smoke or work near heaters or other fire hazards.*

VIBRATION DAMPER

Vibration damper, removing and installing

The N12, N16 (normally aspirated) and N14, N18 (turbocharged) engines use a 2-piece crankshaft vibration damper. The hub is bolted to the crankshaft through its end face and the belt pulley is bolted to the hub.

– Raise front of vehicle and support in a safe manner.

> ***CAUTION—***
> - *Make sure the vehicle is stable and well supported at all times. Use a professional automotive lift or jack stands designed for the purpose. A floor jack is not adequate support.*

– Remove right front wheel well liner. See **410 Fenders, Engine Hood**.

– With engine fully cooled off, mark direction of rotation on engine accessory belt. Remove belt. See **020 Maintenance**.

◄ Remove vibration damper fasteners (**arrows**) and remove vibration damper.

> ***CAUTION—***
> - *Do not remove vibration damper hub bolt (**A**). If this bolt is removed, the timing chain drive sprocket becomes free to rotate, resulting in valve damage.*

– Installation is reverse of removal.

Tightening torque	
Vibration damper to hub M8 x 16 mm	28 Nm (20.6 ft-lb)

TIMING CHAIN SERVICE

Timing chain components (N12, N16 engines)

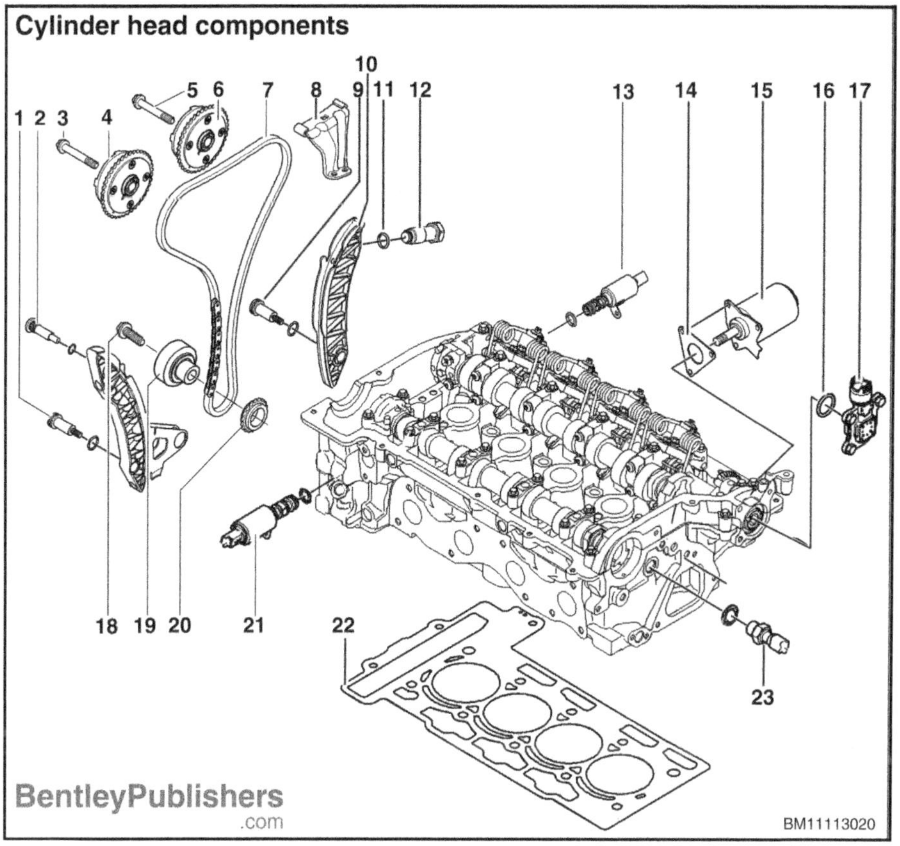

1. Timing chain guide bearing bolt
2. Timing chain guide bearing bolt
3. VANOS Torx fastener
   - Always replace
   - Tightening torque
     Stage 1: 20 Nm (15 ft-lb)
     Stage 2: additional 180°
4. Exhaust VANOS adjuster
5. VANOS Torx fastener
   - Always replace
   - Tightening torque
     Stage 1: 20 Nm (15 ft-lb)
     Stage 2: additional 180°
6. Intake VANOS adjuster
7. Timing chain
8. Slide rail
9. Timing chain guide bearing bolt
10. Timing chain tensioner guide
11. O-ring
12. Timing chain tensioner
    - Tightening torque: 65 Nm (48-ft-lb)
13. Intake VANOS solenoid
    - Tightening torque 9 Nm (7 ft-lb)
14. Valvetronic motor gasket
15. Valvetronic motor
16. O-ring
17. Eccentric shaft sensor
18. Bolt (damper hub to crankshaft)
    - Always replace
    - Use holding tool (MINI special tool **11 9 280)** to loosen / tighten bolt
    - Tightening torque
      Stage 1: 50 Nm (36.8 ft-lb)
      Stage 2: additional 180°
19. Damper hub
20. Crankshaft timing chain sprocket
21. Exhaust VANOS solenoid
    - Tightening torque 9 Nm (7 ft-lb)
22. Cylinder head gasket
23. Oil pressure switch

Timing chain components (N14 engine)

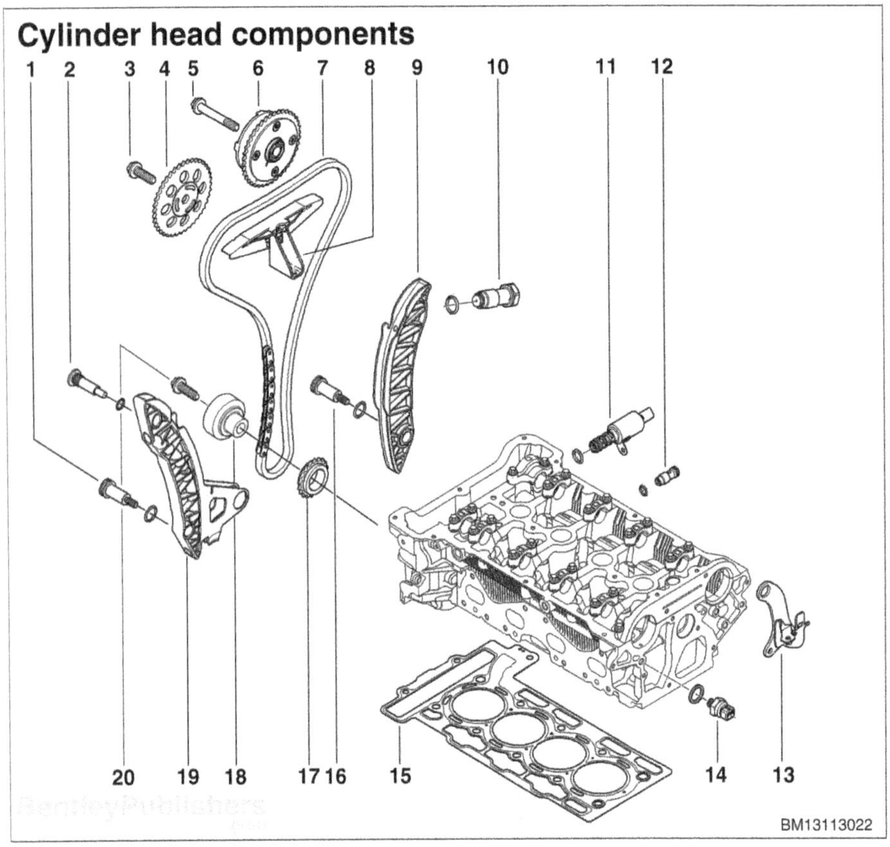

1. Timing chain guide bearing bolt
2. Timing chain guide bearing bolt
3. Exhaust camshaft sprocket fastener
   - Always replace
   - Tightening torque
     Stage 1: 20 Nm (15 ft-lb)
     Stage 2: additional 90°
4. Exhaust camshaft sprocket
5. VANOS Torx fastener
   - Always replace
   - Tightening torque
     Stage 1: 20 Nm (15 ft-lb)
     Stage 2: additional 180°
6. Intake VANOS adjuster
7. Timing chain
8. Slide rail
9. Timing chain tensioner guide
10. Timing chain tensioner
    - Tightening torque: 65 Nm (48-ft-lb)
11. Intake VANOS solenoid
    - Tightening torque 9 Nm (7 ft-lb)
12. Non return valve
13. Lifting eyelet
14. Oil pressure switch
15. Cylinder head gasket
16. Timing chain guide bearing bolt
17. Crankshaft timing chain sprocket
18. Damper hub
19. Timing chain guide
20. Bolt (damper hub to crankshaft)
    - Always replace
    - Use holding tool (MINI special tool **11 9 280)** to loosen / tighten bolt
    - Tightening torque
      Stage 1: 50 Nm (36.8 ft-lb)
      Stage 2: additional 180°

Timing chain components (N18 engine)

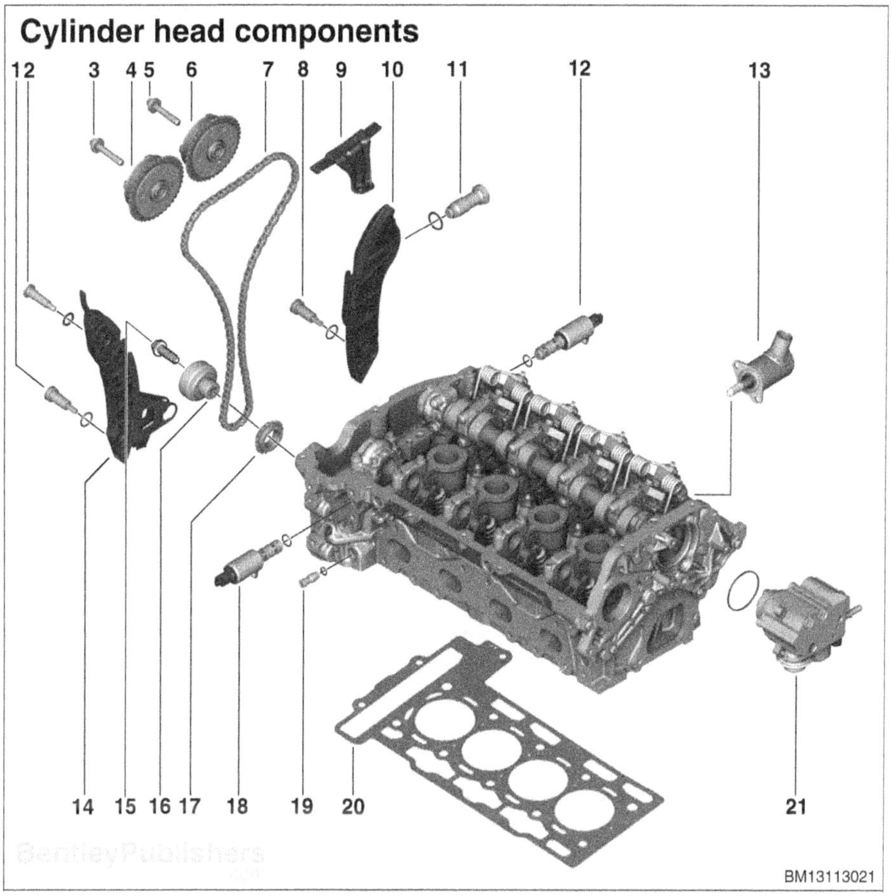

1. Timing chain guide bearing bolt
2. Timing chain guide bearing bolt
3. VANOS Torx fastener
    - Always replace
    - Tightening torque
      Stage 1: 20 Nm (15 ft-lb)
      Stage 2: additional 180°
4. Exhaust VANOS adjuster
5. VANOS Torx fastener
    - Always replace
    - Tightening torque
      Stage 1: 20 Nm (15 ft-lb)
      Stage 2: additional 180°
6. Intake VANOS adjuster
7. Timing chain
8. Timing chain guide bearing bolt
9. Slide rail
10. Timing chain tensioner guide
11. Timing chain tensioner
    - Tightening torque: 65 Nm (48-ft-lb)
12. Intake VANOS solenoid
    - Tightening torque 9 Nm (7 ft-lb)
13. Valvetronic motor
14. Timing chain guide
15. Bolt (damper hub to crankshaft)
    - Always replace
    - Use holding tool (MINI special tool **11 9 280)** to loosen / tighten bolt
    - Tightening torque
      Stage 1: 50 Nm (36.8 ft-lb)
      Stage 2: additional 180°
16. Damper hub
17. Crankshaft timing chain sprocket
18. Exhaust VANOS solenoid
    - Tightening torque 9 Nm (7 ft-lb)
19. Non return valve
20. Cylinder head gasket
21. High pressure fuel pump

Timing chain, removing and installing

Camshaft timing chain service is similar for all engines. The procedures have been combined, with differences noted.

When removing and installing the camshaft timing chain, special tools and procedures are required to remove and install VANOS units and to time camshafts. Read the entire procedure before beginning repairs.

> ***WARNING—***
> - *To avoid personal injury, be sure the engine is cold before beginning the procedure.*

> ***CAUTION—***
> - *Disassembly, removal and assembly of camshafts or VANOS units without special tools poses the risk of damage or breakage: Valves may be bent by contact with the piston crowns.*
> - *Throughout this procedure, unless otherwise specified, crankshaft and camshafts remain locked against rotation using MINI special tools.*

– Switch ignition OFF and remove key.

– Disconnect negative (–) cable from battery.

> ***CAUTION—***
> - *Prior to disconnecting the battery, read the battery disconnection cautions in* **001 Warnings and Cautions**.

– Raise vehicle and support safely.

> ***WARNING—***
> - *Make sure vehicle is stable and well supported at all times. Use a professional automotive lift or jack stands. A floor jack is not adequate support.*

– Remove spark plugs. See **020 Maintenance**.

– Remove cylinder head cover. See **113 Cylinder Head Removal and Installation**.

– Remove vibration damper. See **Vibration damper, removing and installing** in this repair group.

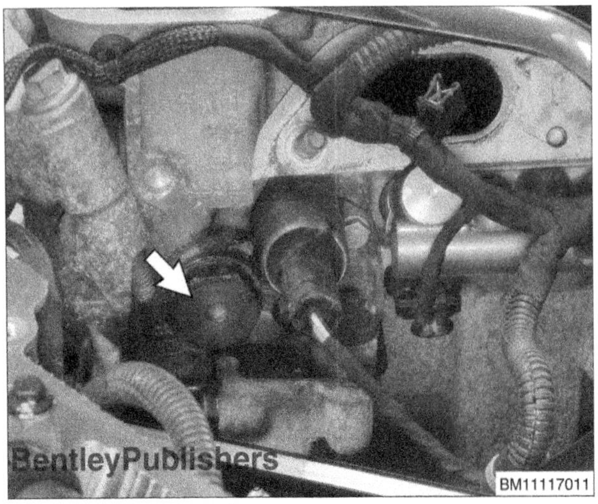

◄ Remove timing chain tensioner (**arrow**). Be prepared to catch any dripping oil in a shop towel.

– **N12, N16, N18 engine**: Remove intake and exhaust VANOS units. See **VANOS units removing and installing (N12, N16, N18 engine)** in this repair group.

– **N14 engine**: Remove intake VANOS unit. See **VANOS unit removing and installing (N14 engine)** in this repair group.

◄ **N14 engine**: Remove exhaust camshaft sprocket fastener (**arrow**).

– Disengage timing chain from camshaft sprocket and secure from falling into timing cover.

– Remove sprocket from camshaft.

– Remove crankshaft seal. See **119 Lubrication System**.

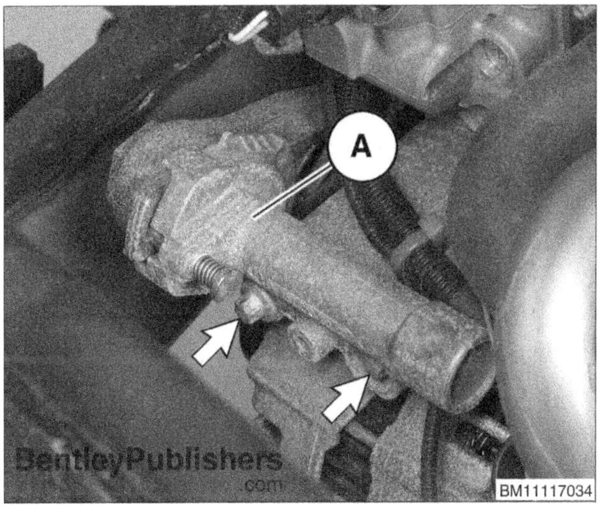

◄ Remove drive belt tensioner fasteners (**arrows**) and remove tensioner (**A**).

– **N14 engine**: Remove coolant pump friction gear. See **170 Radiator and Cooling System**.

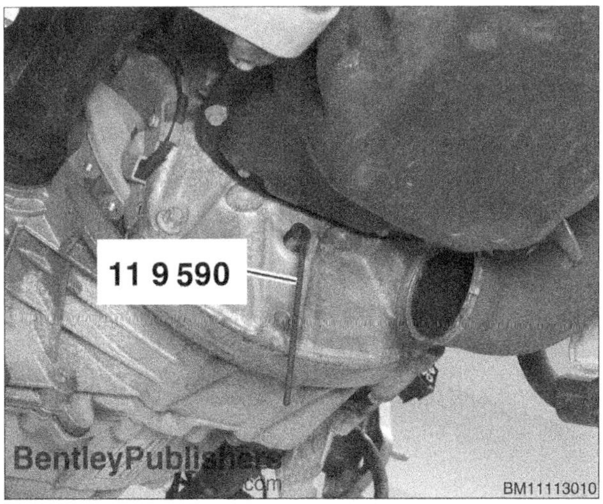

◄ Using MINI special tool **11 9 590**, secure crankshaft in 90° piston position. Tool remains in place during entire repair process.

> ***CAUTION—***
> *• Holes used to balance engine and for aligning engine can be mixed up. Verify engine is in 90° position by viewing piston through spark plug hole.*

– Place radiator support in service position. See **410 Fenders, Engine Hood**.

– Support engine from above.

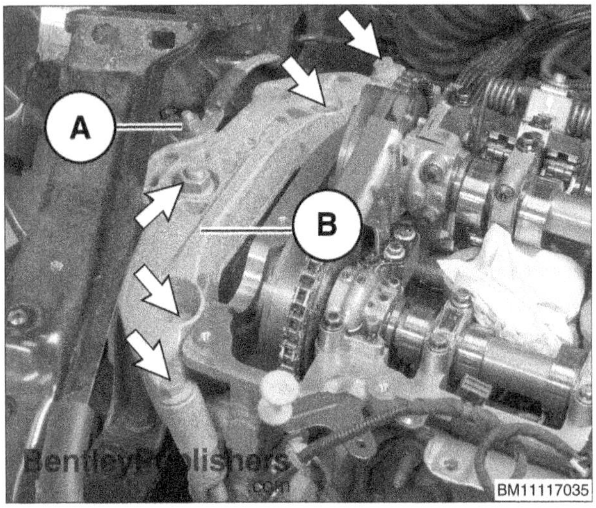

⋞ Remove engine ground strap nut (**A**), then remove right side engine mount bracket fasteners (**arrows**) and remove mount bracket (**B**).

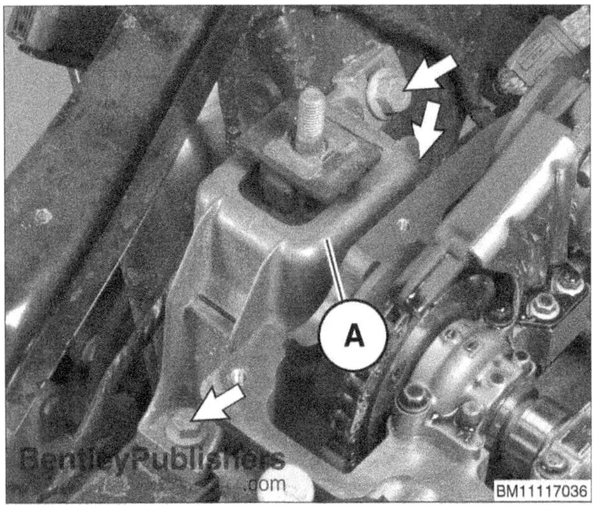

⋞ Remove engine mount fasteners (**arrows**) then remove engine mount (**A**).

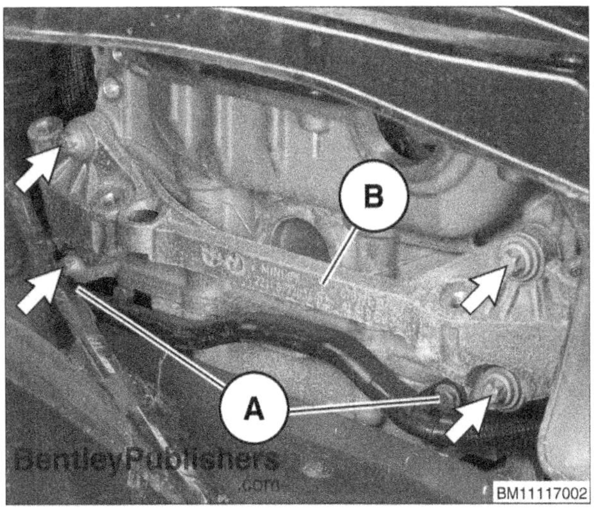

⋞ Remove wiring harness loom fasteners (**A**).

– Remove engine mount adapter plate fasteners (**arrows**) then remove engine mount adapter plate (**B**).

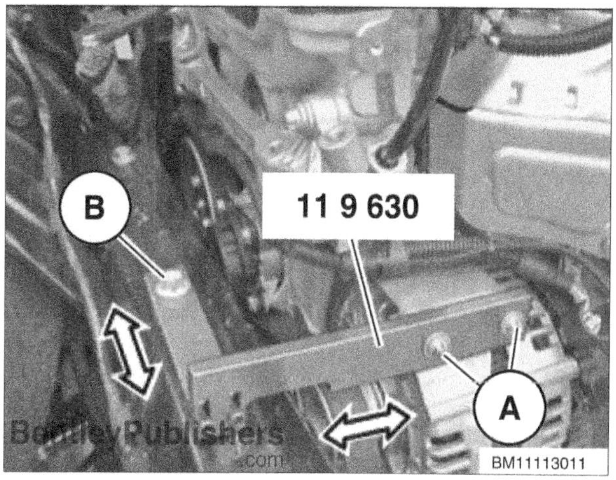

Install MINI special tool **11 9 630** using alternator (**A**) and engine mount (**B**) fasteners. If MINI special tool is unavailable, support engine from below using a suitable jack stand.

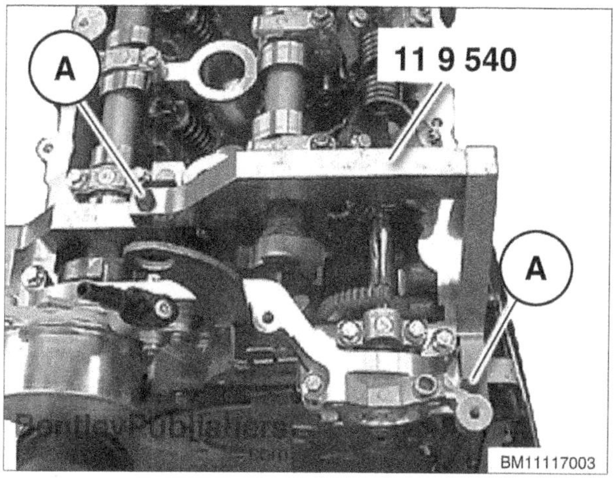

N12, N16 engine: Using MINI special tool **11 9 540**, lock camshafts in 90° piston position. Tighten down tool with supplied fasteners (**A**). Tool remains in place during entire repair process.

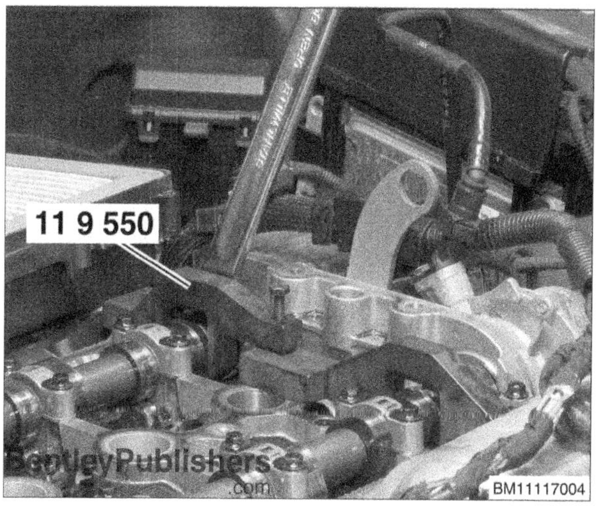

N14 engine: Using MINI special tool **11 9 550**, lock camshafts in 90° piston position. Tool remains in place during entire repair process

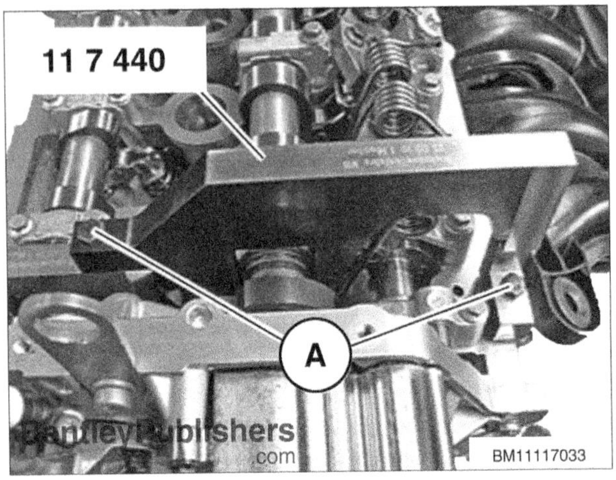

N18 engine: Using MINI special tool **11 7 440**, lock camshafts in 90° piston position. Tool remains in place during entire repair process.

Install MINI special tool **11 9 280** using crankshaft vibration damper fasteners (**arrows**).

With an assistant holding MINI special tool **11 9 280**, remove crankshaft damper hub fastener.

< Remove timing chain guide rail and tensioner guide rail bearing bolts (**A**).

– Carefully remove timing chain with cassette and lower sprocket from engine as one unit.

Camshaft timing chain, installing

< Orientation of timing chain sprockets (viewed through timing cover):

- **A**: Camshaft timing chain
- **B**: Oil pump drive chain

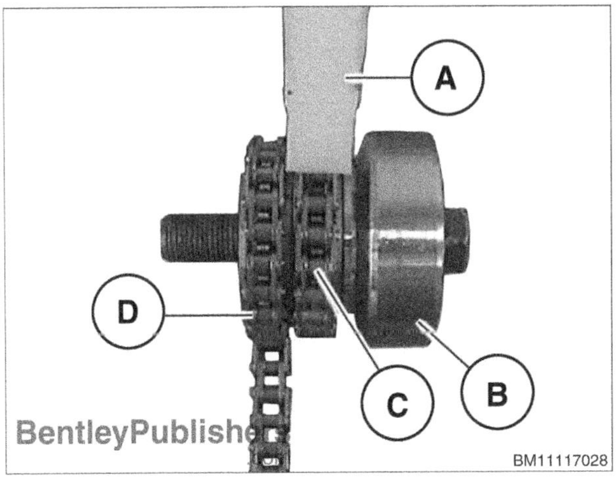

< Orientation of timing chain sprockets:

- **A**: Timing chain cassette
- **B**: Crankshaft hub
- **C**: Camshaft timing chain
- **D**: Oil pump drive chain

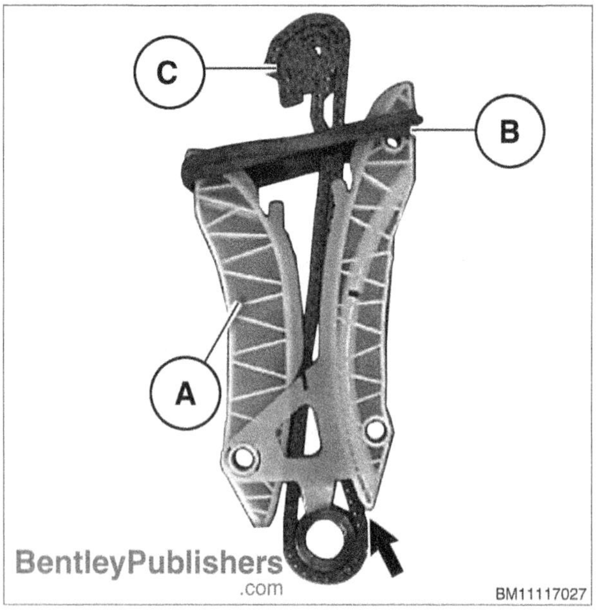

Prior to camshaft timing chain installation:

- Secure timing chain cassette (**A**) using large rubber band (**B**).
- Hold timing chain (**C**) tight so that lower sprocket is resting against guide (**arrow**).

Attach oil pump sprocket (**B**) to crankshaft (**A**).

– Install timing chain cassette and secure in place.

Tightening torque	
Timing chain guide bearing bolts to block M16x1.5 (replace seal)	30 Nm (22.1 ft-lb)

– Install crankshaft damper hub using new fastener.

Lightly oil threads of new hub-to-crankshaft center bolt. Using MINI special tool **11 9 280** and a torque wrench with angle protractor, torque crankshaft damper hub fastener (**A**).

Tightening torque	
Damper hub to crankshaft (use new fastener) • Stage 1 • Stage 2	 50 Nm (36.8 ft-lb) additional 180°

– **N12, N16, N18 engine**: Install VANOS units. See **VANOS units, removing and installing (N12, N16, N18 engine)** in this repair group.

– **N14 engine**: Install exhaust camshaft sprocket

Tightening torque	
Exhaust sprocket to camshaft M10 x 30 mm (use new fastener) • Stage 1 • Stage 2	 20 Nm (15 ft-lb) additional 90°

– **N14 engine**: Install intake VANOS unit. See **VANOS unit, removing and installing (N14 engine)** in this repair group.

– Remove crankshaft and camshaft locking tools.

– Using a ratchet on crankshaft damper center bolt, turn engine clockwise two full rotations.

– Check engine timing. See **Camshaft timing, checking and adjusting** in this repair group.

– Install crankshaft seal. See **119 Lubrication**.

– Remainder of installation is reverse of removal.

Camshaft timing, checking and adjusting

Engine timing is adjusted when pistons are locked in 90° position. This is an updated procedure that eliminates using cylinder 1 top dead center (TDC) when setting camshaft and crankshaft correlation.

Special tools and procedures are required to remove and install VANOS units and to time camshafts. Read the entire procedure before beginning repairs.

> ***WARNING—***
> - *To avoid personal injury, be sure the engine is cold before beginning the procedure.*

> ***CAUTION—***
> - *Disassembly, removal and assembly of camshafts or VANOS units without special tools poses the risk of damage or breakage: Valves may be bent by contact with the piston crowns.*
> - *Throughout this procedure, unless otherwise specified, crankshaft and camshafts remain locked against rotation using MINI special tools.*

– Switch ignition OFF and remove key.

– Disconnect negative (–) cable from battery.

> ***CAUTION—***
> - *Prior to disconnecting the battery, read the battery disconnection cautions in* **001 Warnings and Cautions**.

– Raise vehicle and support safely.

> ***WARNING—***
> - *Make sure the vehicle is stable and well supported at all times. Use a professional automotive lift or jack stands designed for the purpose. A floor jack is not adequate support.*

– Remove ignition coil cover. See **020 Maintenance**.

– Remove spark plugs. See **020 Maintenance**.

– Remove cylinder head cover. See **113 Cylinder Head Removal and Installation**.

◄ Using crankshaft damper center bolt, turn engine over by hand in clockwise direction until **IN** and **EX (inset)** can be seen on top of camshafts.

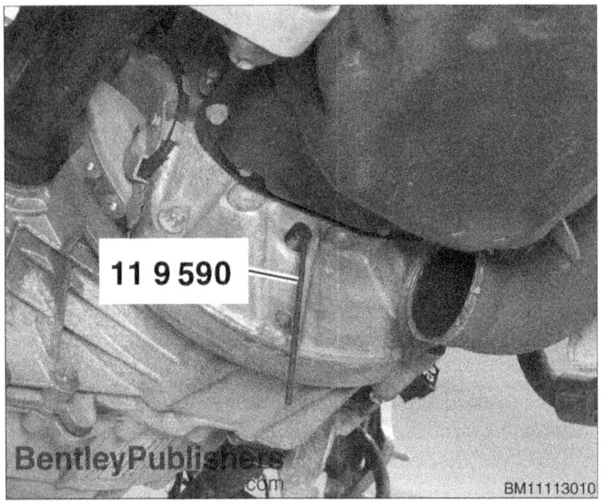

◄ Using MINI special tool **11 9 590**, secure crankshaft in 90° piston position. Tool remains in place during entire repair process.

> ***CAUTION—***
> - *Holes used to balance engine and hole for aligning engine can be mixed up. Verify engine is in 90° position by viewing piston position through spark plug hole.*

◄ Cylinder 1 exhaust camshaft lobe (**arrow**) points at an angle to the right in a lower inward direction (**lines**).

◄ Cylinder 1 intake camshaft lobes (**arrow**) points at an angle to the left in an upper inward direction (**lines**).

If intake and exhaust camshafts are properly aligned, reassemble engine.

If intake or exhaust camshafts are not properly aligned, use following steps to align.

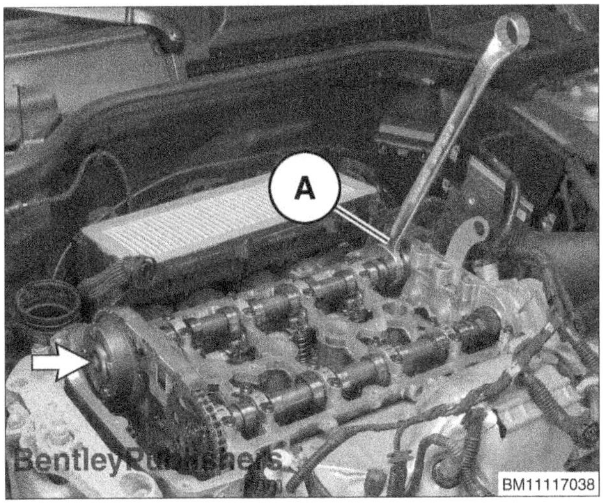

◄ Counterhold intake camshaft at hexagon boss (**A**) and loosen VANOS unit fastener (**arrow**).

– Repeat step for exhaust camshaft.

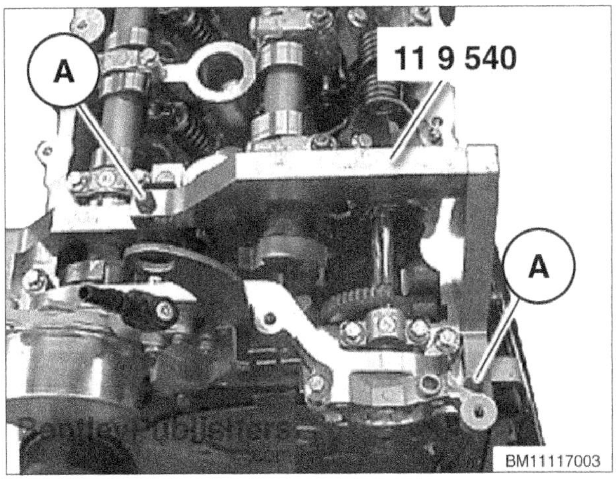

◄ **N12, N16 engine**: Using MINI special tool **11 9 540**, lock camshafts in place.

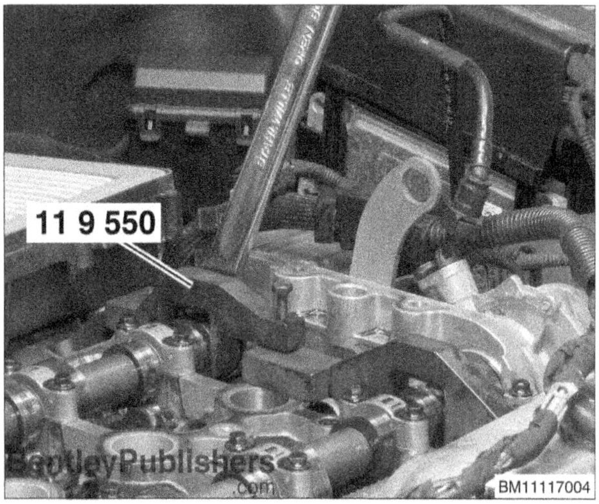

N14 engine: Using MINI special tools **11 9 551** and **11 9 552**, lock camshafts in place.

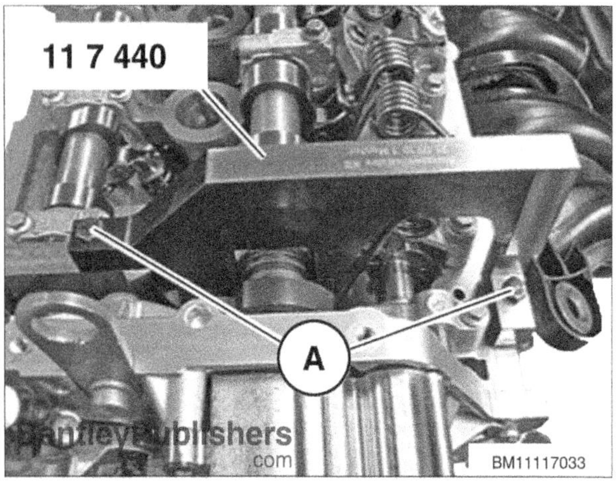

N18 engine: Using MINI special tool **11 7 440**, lock camshafts in 90° piston position. Tool remains in place during entire repair process

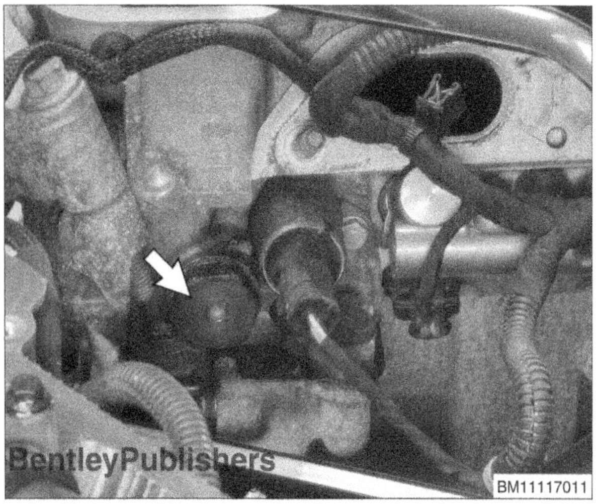

Remove timing chain tensioner (**arrow**). Be prepared to catch any dripping oil in a shop towel.

Install MINI special tool **11 9 340** in timing chain tensioner mounting hole.

- Using MINI special tool **11 9 340** (dummy tensioner) pretension timing chain to 0.6 Nm (5.3 in-lb).

N12, N16, N18 engine: Using a torque wrench and angle protractor (**A**), torque intake VANOS unit fastener.

Tightening torque	
VANOS unit to camshaft (use new fastener) • Stage 1 • Stage 2	 20 Nm (15 ft-lb) additional 180°

N12, N16, N18 engine: Using a torque wrench and angle protractor, torque exhaust VANOS unit fastener (**A**).

Tightening torque	
VANOS unit to camshaft (use new fastener) • Stage 1 • Stage 2	 20 Nm (15 ft-lb) additional 180°

N14 engine: Using a torque wrench and angle protractor, torque intake VANOS unit fastener (**A**).

Tightening torque	
VANOS unit to camshaft (use new fastener) • Stage 1 • Stage 2	20 Nm (15 ft-lb) additional 180°

– **N14 engine**: Torque exhaust camshaft sprocket

Tightening torque	
Exhaust sprocket to camshaft M10 x 30 mm (use new fastener) • Stage 1 • Stage 2	20 Nm (15 ft-lb) additional 90°

– Remove dummy tensioner and install timing chain tensioner. Use new seal.

Tightening torque	
Timing chain tensioner to cylinder head M22 x 1.5 (replace seal)	65 Nm (48 ft-lb)

– Recheck camshaft timing.

– Reassemble engine.

> *CAUTION—*
> • *Remove crankshaft and camshaft locking tools before rotating or starting engine.*

CAMSHAFT ADJUSTMENT (VANOS) UNITS

The MINI engine cylinder head is equipped with variable camshaft timing or VANOS (from the German words *VAriable NOckenwellen Steuerung*). Special tools and procedures are required to remove and install camshaft adjustment units and to time camshafts. Be sure to read each procedure through before starting work.

VANOS unit solenoid valves, removing and installing

All MINI Cooper engines are equipped with intake camshaft VANOS solenoid valves located on the right side of the cylinder head. Engines equipped with dual VANOS have an additional exhaust camshaft solenoid valve located on the left side of the cylinder head.

NOTE—
- *If VANOS solenoid and related fault codes are stored, remove the solenoids and inspecting for sludge. If dirty, clean and reinstall. Resetting fault memory to see if the fault(s) returns.*
- *On cars with dual VANOS, the solenoids can be swapped from side to side to help determine if the problem is caused by the solenoid.*

– Remove ignition coil cover. See **020 Maintenance**.

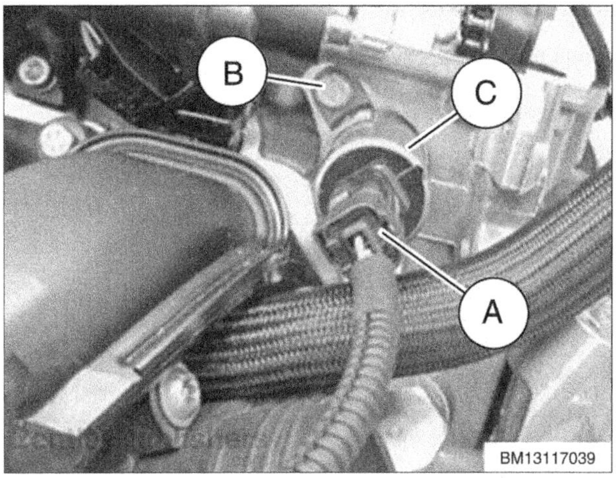

◄ Intake camshaft VANOS solenoid valve:
- Disconnect electrical connector (**A**).
- Remove fastener (**B**).
- Remove solenoid valve (**C**). If necessary, valves can be rotated in cylinder head to aid in removal and installation.

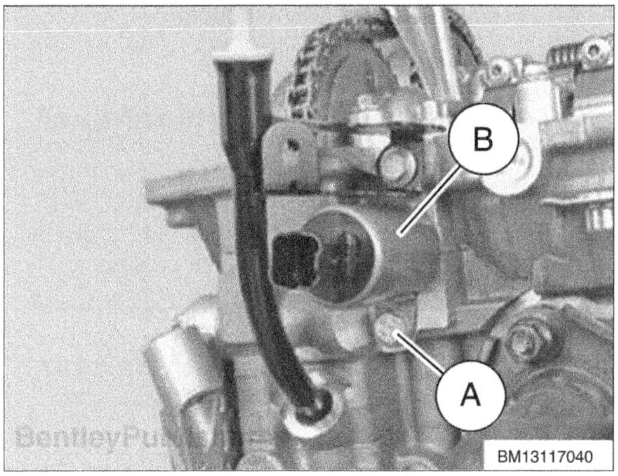

◄ Exhaust camshaft VANOS solenoid valve:
- Disconnect electrical connector.
- Remove fastener (**A**).
- Remove solenoid valve (**B**). If necessary, valves can be rotated in cylinder head to aid in removal and installation.

– Installation is the reverse of removal, making sure of the following:
- Replace O-ring and coat lightly with oil.
- Ensure valves are clean and free of dirt.

Tightening torque	
VANOS solenoid to cylinder head (replace sealing O-ring)	9 Nm (7 ft-lb)

VANOS units, removing and installing (N12, N16, N18 engine)

Engine timing is adjusted when pistons are locked in 90° position. This is an updated procedure that eliminates using top dead center (TDC) when setting camshaft and crankshaft correlation.

Special tools and procedures are required to remove and install VANOS units and to time camshafts.

> ***WARNING—***
> - *To avoid personal injury, be sure engine is cold before beginning the procedure.*

> ***CAUTION—***
> - *Disassembly, removal and assembly of camshafts or VANOS units without special tools poses the risk of damage: Valves may be bent by contact with the piston crowns.*
> - *Throughout this procedure, unless otherwise specified, crankshaft and camshafts remain locked against rotation using MINI special tools.*

– Check camshaft timing. See **Camshaft timing, adjusting** in this repair group.

VANOS units, removing

- Switch ignition OFF and remove key.
- Disconnect negative (–) cable from battery.

> *CAUTION—*
> - *Prior to disconnecting the battery, read the battery disconnection cautions in* **001 Warnings and Cautions**.

- Raise vehicle and support safely.

> *WARNING—*
> - *Make sure vehicle is stable and well supported at all times. Use a professional automotive lift or jack stands. A floor jack is not adequate support.*

- Remove ignition coil cover. See **020 Maintenance**.
- Remove spark plugs. See **020 Maintenance**.
- Remove cylinder head cover. See **113 Cylinder Head Removal and Installation**.

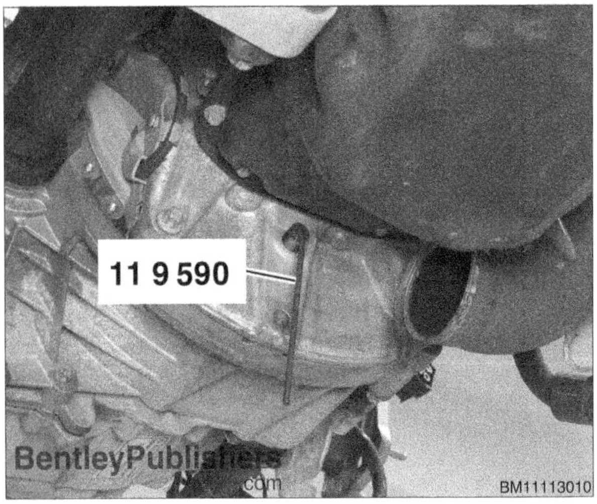

Using MINI special tool **11 9 590**, secure crankshaft in 90° piston position. Tool remains in place during entire repair process.

> *CAUTION—*
> - *Holes used to balance engine and for aligning engine can be mixed up. Verify engine is in 90° position by viewing piston position through spark plug hole.*

Remove fasteners (**A**), then remove timing chain slide rail (**arrow**).

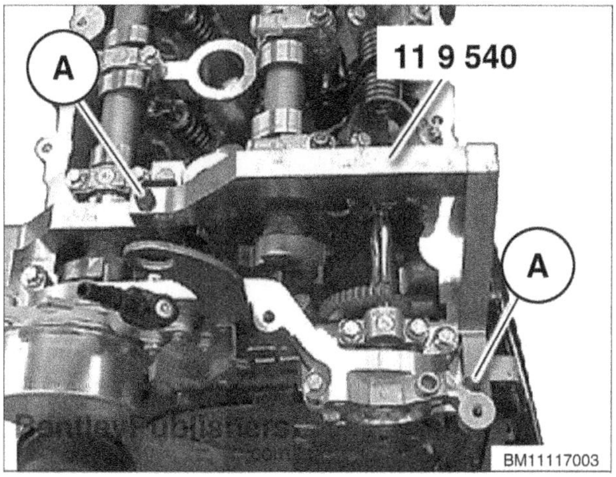

N12, N16 engine: Using MINI special tool **11 9 540**, lock camshafts in 90° piston position. Tighten down tool with supplied fasteners (**A**). Tool remains in place during entire repair process.

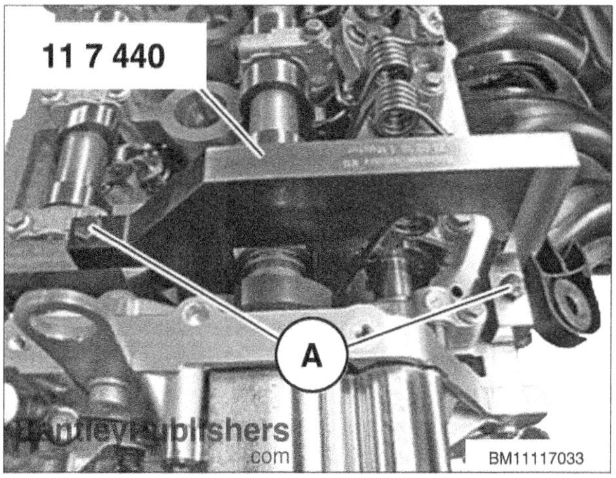

N18 engine: Using MINI special tool **11 7 440**, lock camshafts in 90° piston position. Secure with supplied fasteners (**A**). Tool remains in place during entire repair process

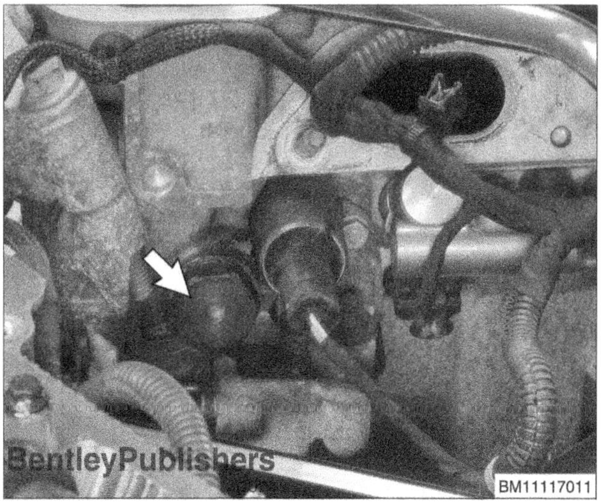

Remove timing chain tensioner (**A**). Be prepared to catch any dripping oil in a shop towel.

VANOS units, removing and installing (N12, N16, N18 engine)

Working at right side front of cylinder head, remove timing chain guide rail bearing bolt (**arrow**).

Working at front of camshafts:

- Loosen and remove exhaust and intake VANOS unit mounting fasteners (**arrows**). Discard fasteners.
- Disengage timing chain from camshaft adjustment unit sprockets.

– Remove intake VANOS unit (**A**).

– Remove exhaust VANOS unit (**B**).

VANOS units can be mixed up. Check markings prior to installation.

- Intake VANOS unit marked **IN**.
- Exhaust VANOS unit marked **EX**.

VANOS units, installing

◀ Use new bolt (**A**) to reattach exhaust VANOS unit (**B**). VANOS unit may be installed in any position initially. Finger-tighten mounting bolts for now.

– Install timing chain on exhaust VANOS unit (**B**).

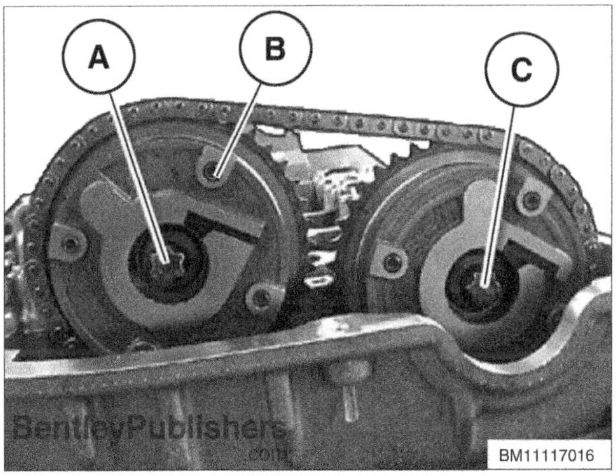

◀ Use new bolt (**A**) to reattach intake VANOS unit (**B**). VANOS unit may be installed in any position initially. Finger-tighten mounting bolts (**A**, **C**) for now.

– Install timing chain on intake VANOS unit

◀ Install timing chain slide rail (**arrow**) and torque fasteners (**A**).

Tightening torque	
Timing chain guide rail to cylinder head M8 (replace seal)	20 Nm (14.7 ft-lb)

VANOS units, removing and installing (N12, N16, N18 engine)

Install MINI special tool **11 9 340** in timing chain tensioner mounting hole and pretension timing chain to 0.6 Nm (5.3 in-lb).

Using a torque wrench and angle protractor, torque intake VANOS unit fastener (**A**).

Tightening torque	
VANOS unit to camshaft (use new fastener) • Stage 1 • Stage 2	20 Nm (15 ft-lb) additional 180°

Using a torque wrench and angle protractor, torque exhaust VANOS unit fastener (**A**).

Tightening torque	
VANOS unit to camshaft (use new fastener) • Stage 1 • Stage 2	20 Nm (15 ft-lb) additional 180°

– Remove dummy tensioner and install timing chain tensioner. Use new seal.

Tightening torque	
Timing chain tensioner to cylinder head M22 x 1.5 (replace seal)	65 Nm (48 ft-lb)

– Check camshaft timing. See **Camshaft timing, adjusting** in this repair group.

– Reassemble engine.

> ***CAUTION—***
> - *Be sure to remove crankshaft and camshaft locking tools before rotating or starting engine.*

VANOS unit, removing and installing (N14 engine)

Engine timing is adjusted when pistons are locked in 90° position. This is an updated procedure that eliminates using top dead center (TDC) when setting camshaft and crankshaft correlation.

Special tools and procedures are required to remove and install VANOS units and to time camshafts. Read the entire procedure before beginning repairs.

> ***WARNING—***
> - *To avoid personal injury, be sure the engine is cold before beginning the procedure.*

> ***CAUTION—***
> - *Disassembly, removal and assembly of camshafts or VANOS units without special tools poses the risk of damage or breakage: Valves may be bent by contact with the piston crowns.*
> - *Throughout this procedure, unless otherwise specified, crankshaft and camshafts remain locked against rotation using MINI special tools.*

– Check camshaft timing. See **Camshaft timing, adjusting** in this repair group.

VANOS unit, removing

– Switch ignition OFF and remove key.

– Disconnect negative (–) cable from battery.

> ***CAUTION—***
> - *Prior to disconnecting the battery, read the battery disconnection cautions in* **001 Warnings and Cautions**.

– Raise vehicle and support safely.

> ***WARNING—***
> - *Make sure the vehicle is stable and well supported at all times. Use a professional automotive lift or jack stands designed for the purpose. A floor jack is not adequate support.*

– Remove ignition coil cover. See **020 Maintenance**.

– Remove spark plugs. See **020 Maintenance**.

– Remove cylinder head cover. See **113 Cylinder Head Removal and Installation**.

Using crankshaft damper center bolt, turn engine over by hand in clockwise direction until **IN** and **EX** can be seen on top of camshafts.

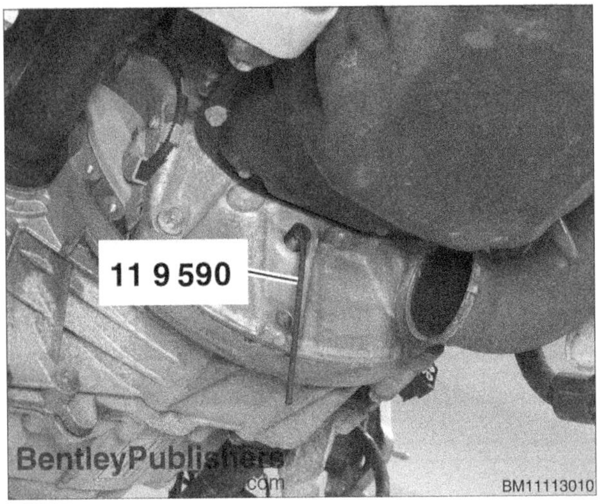

Using MINI special tool **11 9 590**, secure crankshaft in 90° piston position. Tool remains in place during entire repair process.

> ***CAUTION—***
>
> - *Holes used to balance engine and for aligning engine can be mixed up. Verify engine is in 90° position by viewing piston position through spark plug hole.*

Using MINI special tools **11 9 550**, lock camshafts in place.

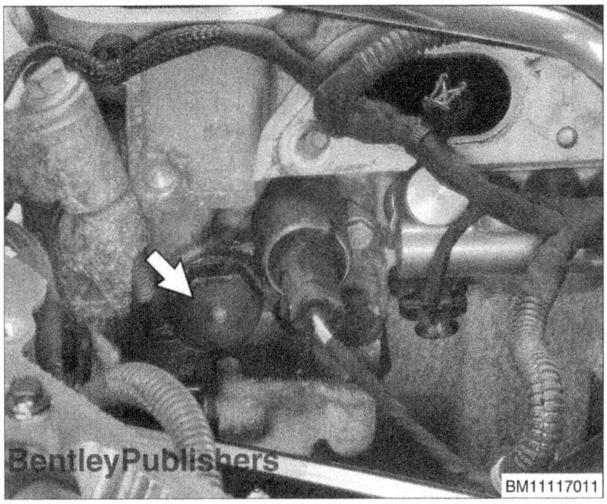

◄ Remove timing chain tensioner (**A**). Be prepared to catch any dripping oil in a shop towel.

◄ Remove timing chain guide rail bearing bolt (**A**).

◄ Remove fasteners (**A**) then remove timing chain slide rail (**arrow**).

Working at front of camshaft:

- Loosen and remove intake VANOS unit mounting fastener (**arrow**). Discard fastener.
- Disengage timing chain from camshaft adjustment unit sprocket.

– Remove intake VANOS unit.

VANOS unit, installing

Use new bolt to reattach intake VANOS unit. VANOS unit may be installed in any position initially. Finger-tighten mounting bolt (**arrow**) for now.

Install timing chain slide rail (**arrow**) and torque fasteners (**A**).

Tightening torque	
Timing chain guide rail to cylinder head M8 (replace seal)	20 Nm (14.7 ft-lb)

Install MINI special tool **11 9 340** in timing chain tensioner mounting hole and pretension timing chain to 0.4 Nm (3.5 in-lb).

Using a torque wrench and angle protractor, torque intake VANOS unit fastener (**A**).

Tightening torque	
VANOS unit to camshaft (use new fastener) • Stage 1 • Stage 2	 20 Nm (15 ft-lb) additional 180°

- Remove dummy tensioner and install timing chain tensioner. Use new seal.

Tightening torque	
Timing chain tensioner to cylinder head M22x1.5 (replace seal)	65 Nm (48 ft-lb)

- Check camshaft timing. See **Camshaft timing, adjusting** in this repair group.
- Reassemble engine.

> ***CAUTION—***
> • *Remove crankshaft and camshaft locking tools before rotating or starting engine.*

119 Lubrication System

General

This repair group covers lubrication system troubleshooting and crankshaft seal replacement. Oil pan removal and oil pump replacement are also covered.

See also:

- **020 Maintenance** for oil and oil filter change
- **100 Engine–General** for engine identification and application

Engine identification table

Table a. Engine identification	
Year, model	**Engine designation**
2007 - 2010 Cooper, Clubman	N12
2009 - 2010 Convertible	N12
2007 - 2010 Cooper S, Clubman S	N14
2009 - 2010 Convertible S	N14
2011 - 2013 Cooper, Clubman	N16
2011 - 2013 Cooper S, Clubman S	N18

Warnings and Cautions

WARNING—
- *Due to risk of personal injury, be sure the engine is cold before beginning work on engine components.*

CAUTION—
- *Prior to disconnecting the battery, read the battery disconnection cautions in* **001 Warnings and Cautions**.
- *Follow torque instructions, including angle of rotation specifications, when installing fasteners.*

ENGINE LUBRICATION

Engine lubricant

Engine oil specification	
N12, N14, N16, N18 engine: MINI long-life rating (LL-01) synthetic oil for US market, SAE 5W - 30	API rating SM or higher

Engine oil capacity	
Engine	**Approximate capacity (incl. oil filter)**
N12	4.2 liters (4.4 US qt)
N14	4.2 liters (4.4 US qt)
N16	4.2 liters (4.4 US qt)
N18	4.2 liters (4.4 US qt)

Engine lubrication (N12, N14 engine)

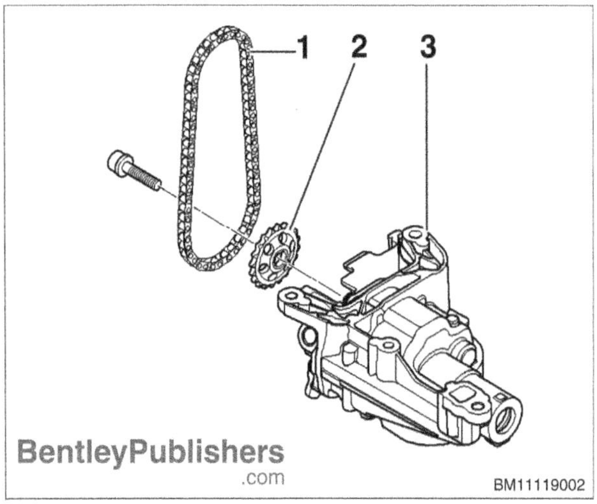

To meet the oiling requirements of N12 and N14 engines, a chain-driven volumetric-flow controlled oil pump is utilized. The VANOS system requires a large volume of oil, particularly at low engine speeds.

1. Oil pump drive chain
2. Oil pump sprocket
3. Engine oil pump

The volumetric-flow controlled pump design allows:

- Increased power output
- Reduced weight
- Optimized fuel consumption
- Reduced exhaust emissions

The volumetric-flow controlled oil pump delivers only as much oil as is necessary. No surplus quantities of oil are delivered in low load operating ranges. This reduces the fuel consumption of the engine and slows down the oil wear rate.

Engine lubrication (N16, N18 engine)

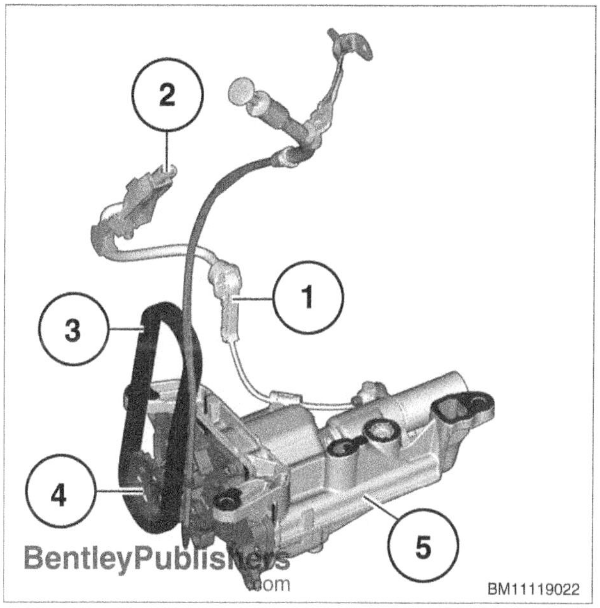

To meet the oiling and fuel efficiency requirements of N16 and N18 engines, a chain-driven electronically volumetric-flow controlled oil pump is utilized. Oil delivery volume is dependent on engine operation and controlled by the engine control module (ECM). The VANOS system requires a large volume of oil, particularly at low engine speeds.

1. Wiring harness engine block feed-through
2. Map-controlled oil pump electrical connector
3. Oil pump drive chain
4. Oil pump sprocket
5. Engine oil pump

The ECM activates an electrohydraulic pressure control valve which affects the oil pressure at the slide valve control mechanism within the oil pump, altering oil pump output.The electrohydraulic pressure control valve mounted on the pump controls pump output. Oil pump output is based on a characteristic map within the ECM and feedback from the oil pressure sensor.

OIL PRESSURE WARNING SYSTEM

> ***CAUTION—***
> - *If the red oil pressure warning light comes on or flashes on while driving, assume that the oil pressure is low. Stop the engine immediately and make arrangements to test oil pressure.*

Oil pressure, checking (N12, N14 engine)

Test oil pressure by removing oil pressure switch and installing oil pressure gauge in its place.

– Loosen oil filter housing to allow engine oil to drain back into oil pan. Tighten cover.

Detach harness connector (**arrow**) from oil pressure switch. If necessary, remove ignition coil cover. See **020 Maintenance**.

– Remove switch. Be prepared to catch leaking oil with a shop towel.

> *CAUTION—*
> - *Thoroughly clean around the oil pressure switch before removing it.*

– Install pressure gauge in place of oil pressure switch.

– With gauge installed, start engine and allow to reach operating temperature. Check oil pressure. See **Table b**.

NOTE—
- *For the most accurate test results, make sure the engine oil and filter are new and the oil of the correct grade.*

Table b. Engine oil pressure	
N12, N14 engine: At idle, engine at operating temperature	0.7 bar (10.1 psi)
N12 engine: Maximum regulated pressure (3000 rpm), engine at operating temperature	1.5 - 3.0 bar (21.7 - 43.5 psi)
N14 engine: Maximum regulated pressure (3000 rpm), engine at operating temperature	1.7 - 3.0 bar (24.6 - 43.5 psi)

– Remove pressure gauge and reinstall pressure switch with new sealing washer.

Tightening torque	
Oil pressure switch to cylinder head N12, N14 engine (M16 threads, lubricate O-ring with clean oil)	20 Nm (14.7 ft-lb)

If testing shows low oil pressure, one or more of the following conditions may be indicated:
- Worn or faulty oil pump or faulty pump pressure relief valve
- Worn or damaged engine bearings
- Severe engine wear

Any of these conditions indicate the need for major repairs.

Oil pressure, checking (N16, N18 engine)

Test oil pressure by removing oil filter housing and installing MINI special tool **11 7 460** in its place.

– Loosen oil filter housing to allow engine oil to drain back into oil pan. Tighten cover.

– Remove oil filter housing.

BM11119023

Remove oil filter element (**A**) from oil filter housing and insert in MINI special tool **11 7 460**.

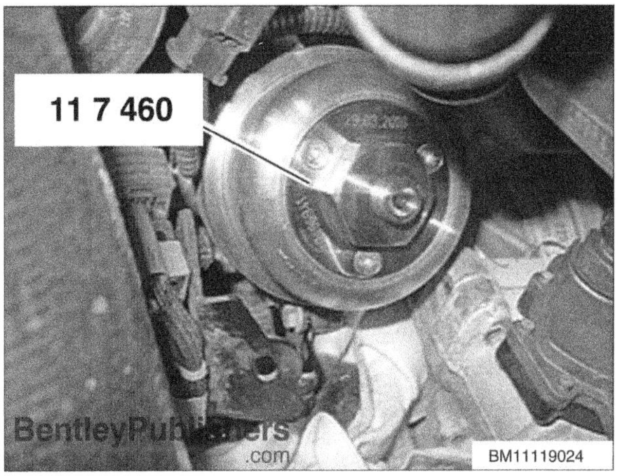

BM11119024

Install MINI special tool **11 7 460** in place of oil filter housing.

- Connect oil pressure gauge to MINI special tool **11 7 460**.
- With gauge installed, start engine and allow to reach operating temperature. Check oil pressure. See **Table c**.

NOTE—

- *For the most accurate test results, make sure the engine oil and filter are new and the oil of the correct grade.*

Table c. Engine oil pressure	
N16, N18 engine: At idle, engine at operating temperature	0.7 bar (10.1 psi)
N18 engine: Maximum regulated pressure (3000 rpm), engine at operating temperature	1.15 - 6.45 bar (16.6 - 93.5 psi)

- Remove pressure gauge and reinstall oil filter housing with new sealing O-ring.

Tightening torque	
Oil filter housing to housing replace O-ring	25 Nm (18.4 ft-lb)

If testing shows low oil pressure, one or more of the following conditions may be indicated:

- Worn or faulty oil pump or faulty pump pressure relief valve
- Worn or damaged engine bearings
- Severe engine wear

Any of these conditions indicate the need for major repairs.

Oil pressure warning switch, replacing (N12, N14 engine)

- Loosen oil filter housing (**arrow**) to allow engine oil to drain back into oil pan.

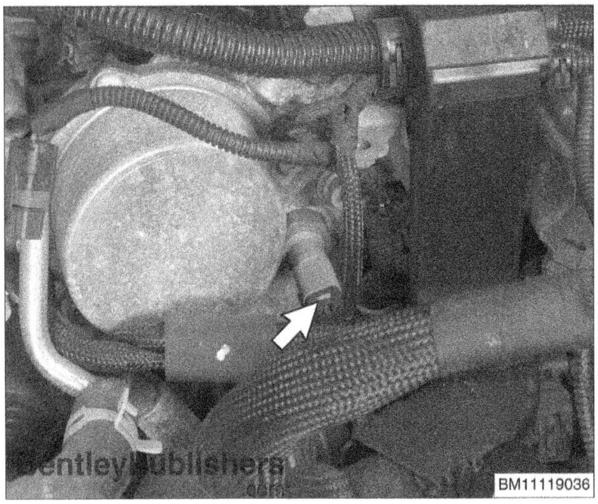

- Disconnect electrical connector (**A**) from oil pressure switch and remove switch (**B**). Be prepared to catch leaking oil with a shop towel.

> *CAUTION—*
> - *Thoroughly clean around the oil pressure switch before removing it.*

– Installation is reverse of removal. Tighten oil filter housing cap and top off oil, if necessary.

Tightening torque	
Oil pressure switch to cylinder head N12, N14 engine (M16 threads, lubricate O-ring with clean oil)	20 Nm (14.7 ft-lb)

Oil pressure warning switch, replacing (N16, N18 engine)

- Loosen oil filter housing to allow engine oil to drain back into oil pan.

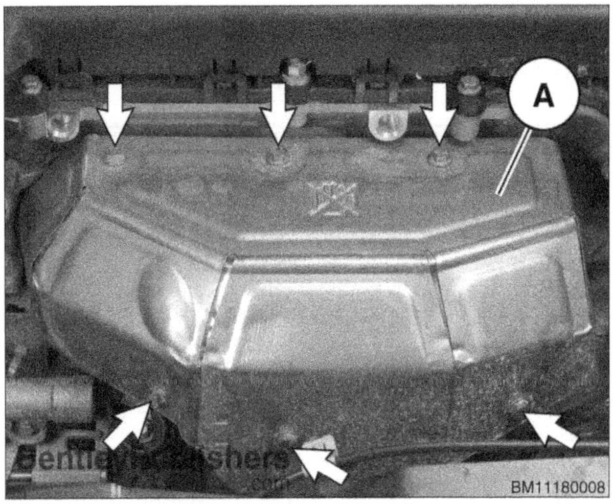

◄ **N16 engine**: Remove exhaust heat shield fasteners (**arrows**), then remove exhaust heat shield (**A**).

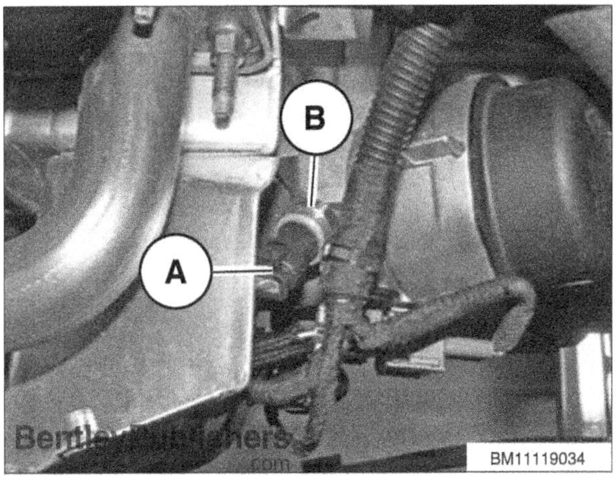

◄ **N16 engine**: Disconnect electrical connector (**A**) from oil pressure switch and remove switch (**B**). Be prepared to catch leaking oil with a shop towel.

> ***CAUTION—***
> - *Thoroughly clean around the oil pressure switch before removing it.*

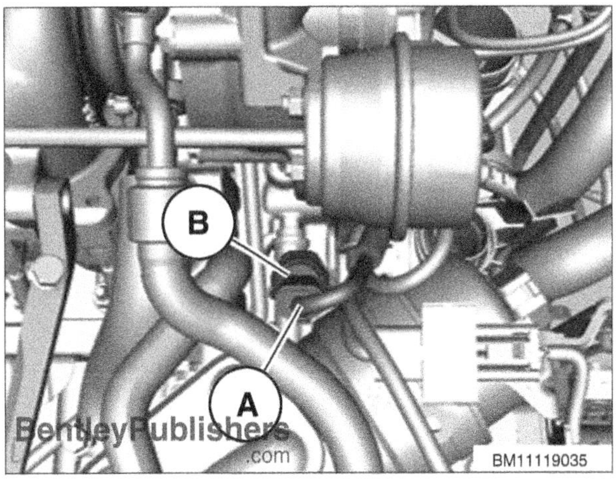

◄ **N18 engine**: Disconnect electrical connector (**A**) from oil pressure switch and remove switch (**B**). Be prepared to catch leaking oil with a shop towel.

> ***CAUTION—***
> - *Thoroughly clean around the oil pressure switch before removing it.*

– Installation is reverse of removal. Tighten oil filter housing cap and top off oil, if necessary.

Tightening torque	
Oil pressure switch to oil filter housing N16, N18 engine (M16 threads, lubricate O-ring with clean oil)	25 Nm (18.4 ft-lb)

CRANKSHAFT SEALS

Crankshaft front seal, removing and installing

> ***WARNING —***
> *• To avoid personal injury, be sure the engine is cold before beginning the procedure.*

– Raise vehicle and support safely.

> ***WARNING —***
> *• Make sure the vehicle is stable and well supported at all times. Use a professional automotive lift or jack stands designed for the purpose. A floor jack is not adequate support.*

– Remove engine accessory belt. See **020 Maintenance**.

◄ Remove vibration damper. See **117 Camshaft Timing Chain**.

> ***CAUTION—***
> *• Do not remove vibration damper hub bolt (**arrow**). If this bolt is removed, the timing chain drive sprocket becomes free to rotate, resulting in valve damage.*
> *• MINI recommends removing A/C line to gain access to crankshaft seal. Have A/C system professionally discharged and recharged.*

◄ Using a drift (**A**), drive crankshaft seal inwards at top (**arrow**) until bottom tilts outwards.

– Remove crankshaft seal from crankcase.

– Before installing new seal:
  - Clean seal seating surface.
  - Apply light coal of engine oil to seal lip running surface on crankshaft flange.

Place MINI special tool **11 9 601** against vibration damper hub and tighten down using damper bolts (**A**).

Tightening torque	
Special tool 11 9 601 to vibration damper hub	15 Nm (11 ft-lb)

– Apply a light coat of engine oil to MINI special tool **11 9 601**.

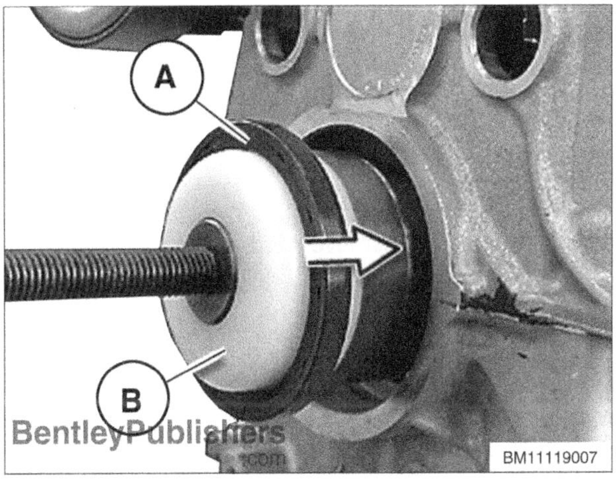

Place crankshaft seal (**A**) and installation sleeve (**B**) over tool and against crankcase opening.

- Plastic sleeve is supplied with seal and is used as installation guide sleeve.
- Push crankshaft seal (**A**) over support sleeve (**B**) toward crankcase (**arrow**).

Remove plastic sleeve from MINI special tool **11 9 601**.

- Use MINI special tools **11 9 602** and **11 9 603** to press in seal.
- Reinstall crankshaft vibration damper.

Tightening torque	
Vibration damper to vibration damper hub (M8 x 16 mm)	28 Nm (20.6 ft-lb)

- Remainder of assembly is reverse of disassembly.

Crankshaft rear main seal, removing and installing

Crankshaft rear main seal (flywheel seal) replacement requires removal of the transmission and flywheel.

- Remove transmission. See **230 Manual Transmission** or **240 Automatic Transmission**.
- Manual transmission vehicle: Remove clutch pressure plate and disc. See **210 Clutch**.
- Remove flywheel. See **210 Clutch** or **240 Automatic Transmission**.
- Drain engine oil.

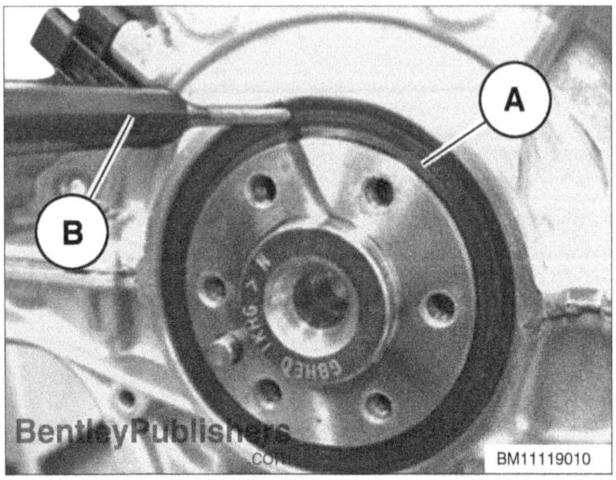

- Using a drift (**B**), drive rear main seal seal (**A**) inwards at top until bottom tilts outwards.
- Remove rear main seal from crankcase.
- Before installing new seal:
  - Clean seal seating surface.
  - Apply light coal of engine oil to seal lip running surface on crankshaft flange.

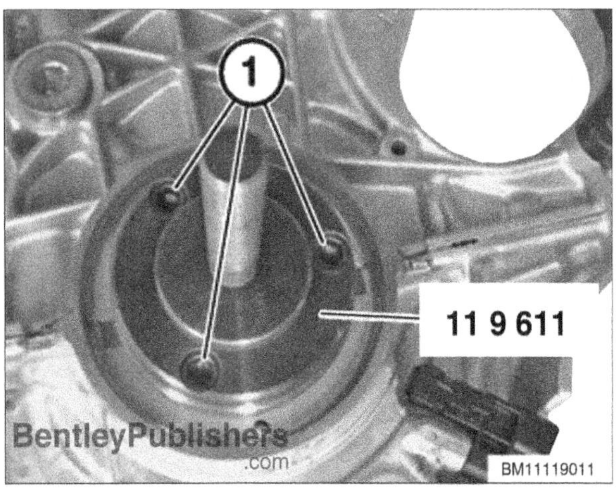

- Place MINI special tool **11 9 611** against vibration damper hub and tighten down using supplied bolts (**A**).

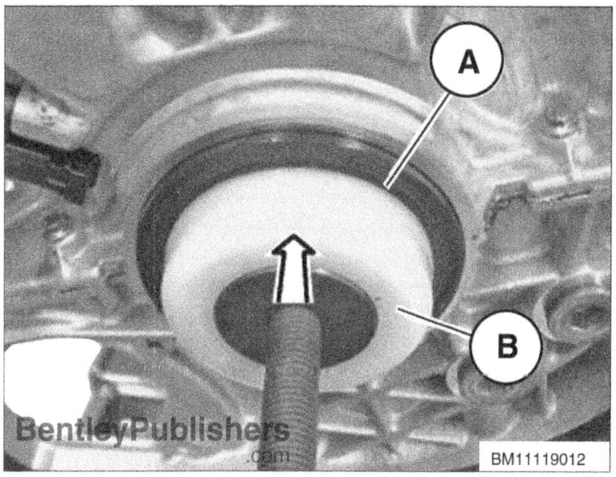

Place rear main seal (**A**) and installation sleeve (**B**) over tool and against crankcase opening.

- Plastic sleeve is supplied with seal and is used as installation guide sleeve.
- Push rear main seal (**A**) over support sleeve (**B**) toward crankcase (**arrow**).

Install seal using MINI special tools **1 19 612** and **11 9 613**.

- Press seal partially into crankcase and check to make sure seal is not misaligned.
- Continue pressing in until special tool **11 9 612** is flush with crankcase. Remove special tools.

– Reassemble engine.

OIL PAN AND OIL PUMP

Oil pan, removing and installing

– Disconnect negative (-) battery cable.

> ***CAUTION—***
> - *Prior to disconnecting the battery, read the battery disconnection cautions in* **001 Warnings and Cautions.**

– Raise car and support safely.

> ***WARNING—***
> - *Make sure the car is stable and well supported at all times. Use a professional automotive lift or jack stands designed for the purpose. A floor jack is not adequate support.*

– Remove exhaust system. See **180 Exhaust System**.

– Drain engine oil. See **020 Maintenance**.

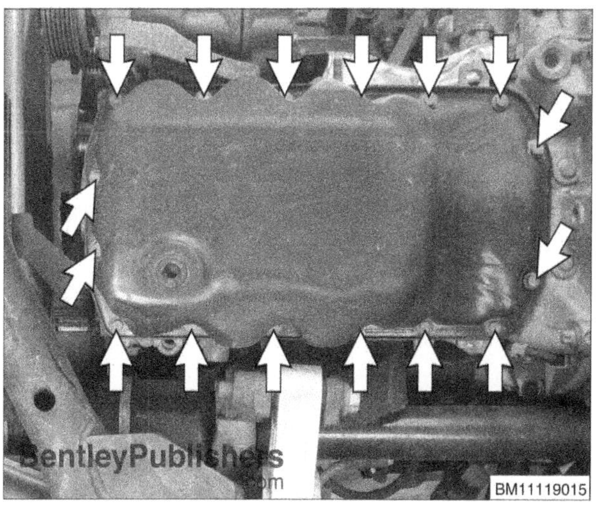

Remove oil pan fasteners (**arrows**).

- Using MINI special tools **11 9 581** and **11 9 582** to remove fastener (**A**) above exhaust manifold.

- Remove oil pan.

> ***CAUTION—***
> *• If the oil pan does not separate easily from the engine cylinder block, tap lightly with a rubber mallet to break it free. Do not use prying tools.*

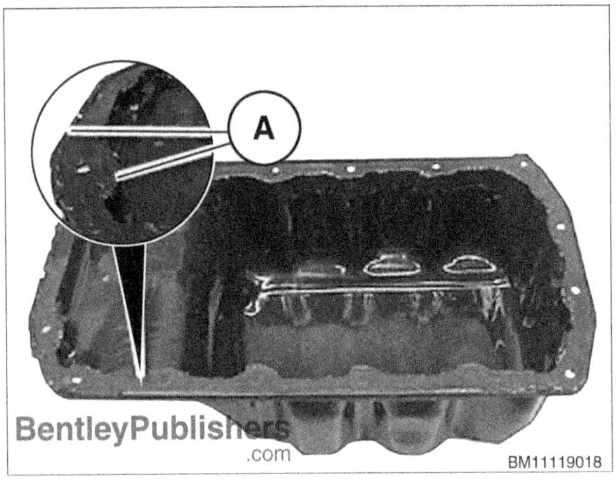

- Remove protruding or sticking-out sealing beads (**A**) from oil pan.

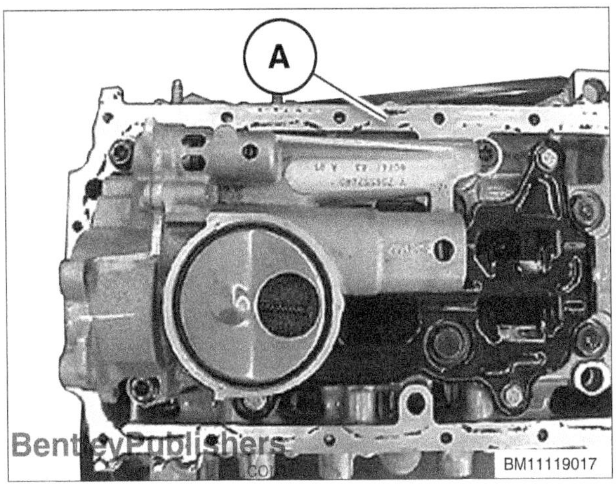

- When reinstalling oil pan:
  - Thoroughly clean old gasket material from engine block sealing surface and use a new metal substrate gasket.
  - Mount oil pan and insert bolts.
  - Tighten bolts in a crisscross pattern.

> ***CAUTION—***
> *• Do not use silicone sealant. Use a metal substrate gasket.*

Tightening torque	
Oil pan to crankcase	12 Nm (8.8 ft-lb)
Drain plug to oil pan	30 Nm (22.1 ft-lb)

- Remainder of installation is reverse of removal. Fill engine with oil, run engine and check for leaks.

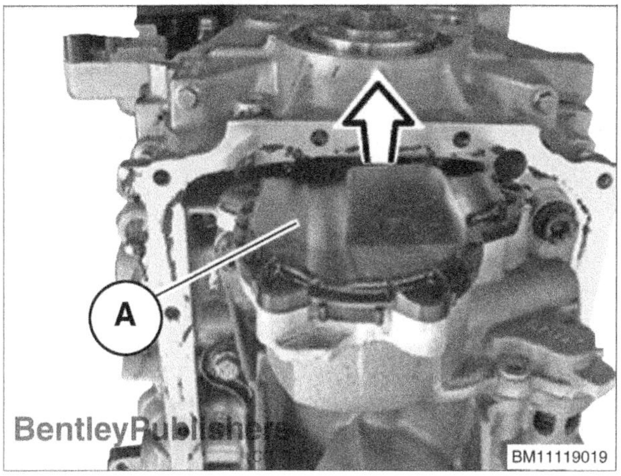

Oil pump, removing and installing (N12, N14 engines)

– Remove oil pan. See **Oil pan, removing and installing** in this repair group.

◄ Pull cover (**A**) off oil pump in direction of **arrow**.

◄ Counterhold crankshaft hub central bolt (**B**) and loosen oil pump sprocket fastener (**A**).

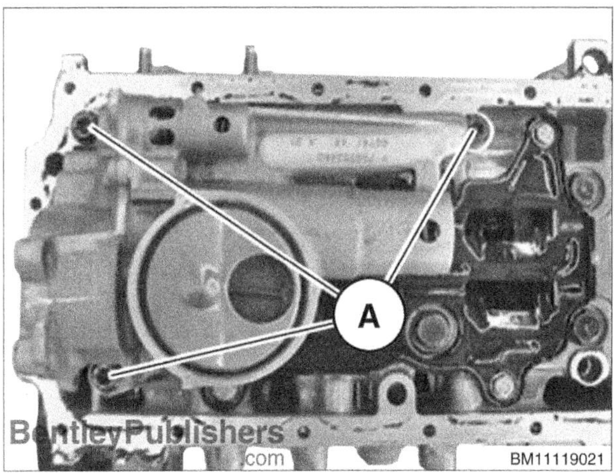

◄ Remove oil pump fasteners (**A**) and remove oil pump from crankcase.

– Installation is reverse of removal. Fill engine with oil, run engine and check for leaks.

Tightening torque	
Oil pump sprocket to oil pump (replace fastener) • Stage 1 • Stage 2	 5 Nm (44 in-lb) additional 90°
Oil pump to crankcase	25 Nm (18.4 ft-lb)

Oil pump, removing and installing (N16, N18 engine)

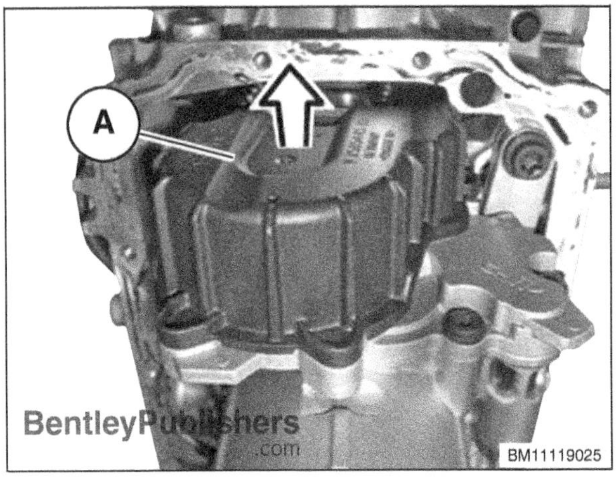

– Remove oil pan. See **Oil pan, removing and installing** in this repair group.

< Pull cover (**A**) off oil pump in direction of **arrow**.

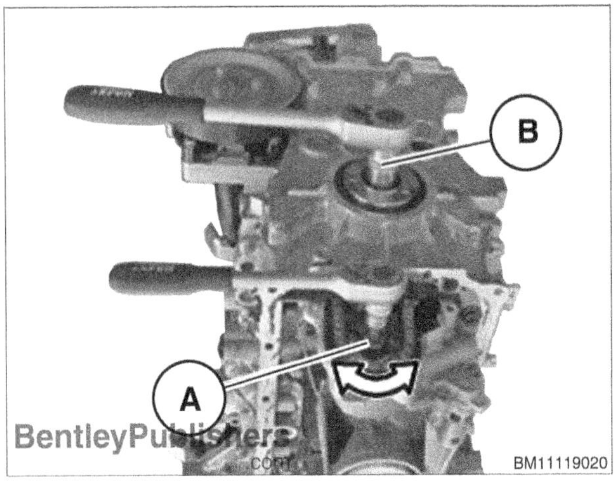

< Counterhold crankshaft hub central bolt (**A**) and loosen oil pump sprocket fastener (**B**).

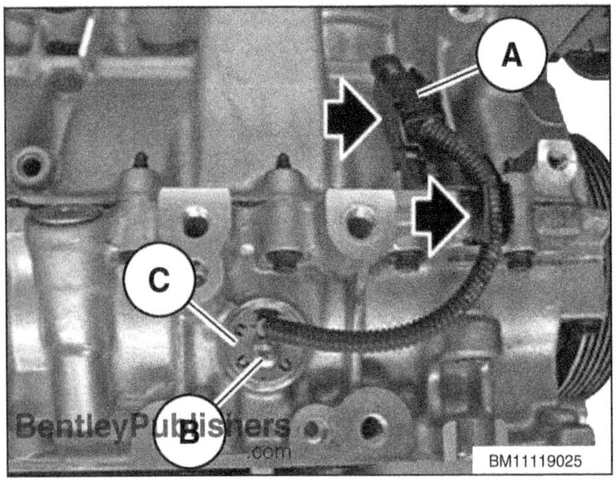

< Disconnect electrical connector (**A**). Remove fastener (**B**), then remove retaining plate (**C**).

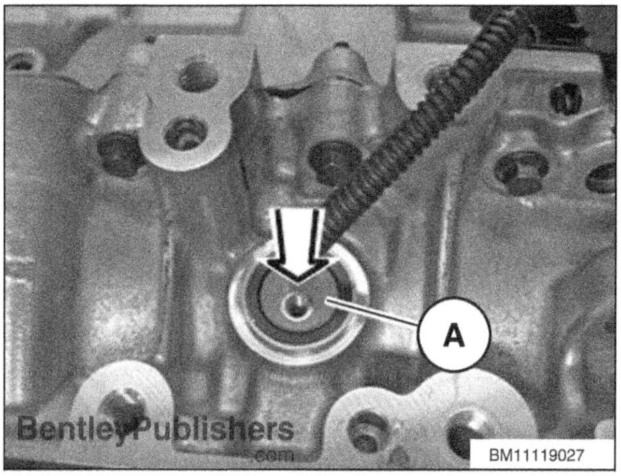

Press support (**A**) into engine block.

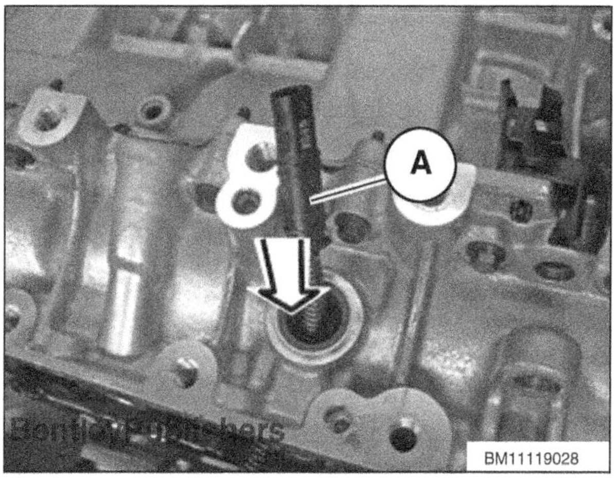

Feed wiring harness through hole in engine block.

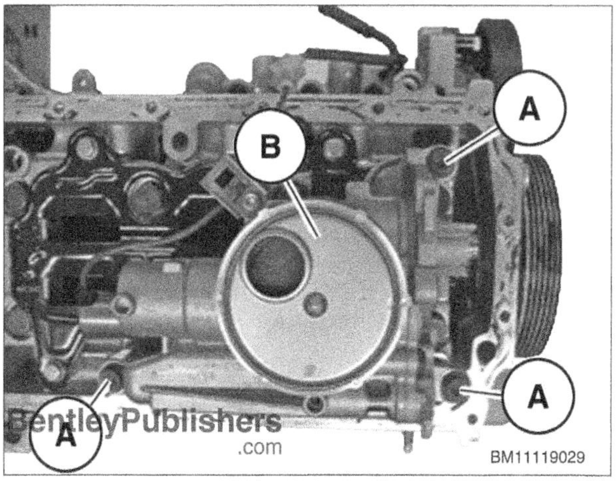

Remove oil pump fasteners (**A**) and remove oil pump (**B**) from crankcase.

Installation is reverse of removal. Remember to:

- Replace wiring harness O-ring (**A**).
- Fill engine with oil, run engine and check for leaks.

Tightening torque	
Oil pump sprocket to oil pump (replace fastener) • Stage 1 • Stage 2	 5 Nm (44 in-lb) additional 90°
Oil pump to crankcase	25 Nm (18.4 ft-lb)
Wiring harness clip to crankcase	4 Nm (35 in-lb)

Oil pump solenoid, removing and installing (N16, N18 engine)

– Remove oil pan. See **Oil pan, removing and installing** in this repair group.

– Remove oil pump. See **Oil pump (N16, N18 engine), removing and installing** in this repair group.

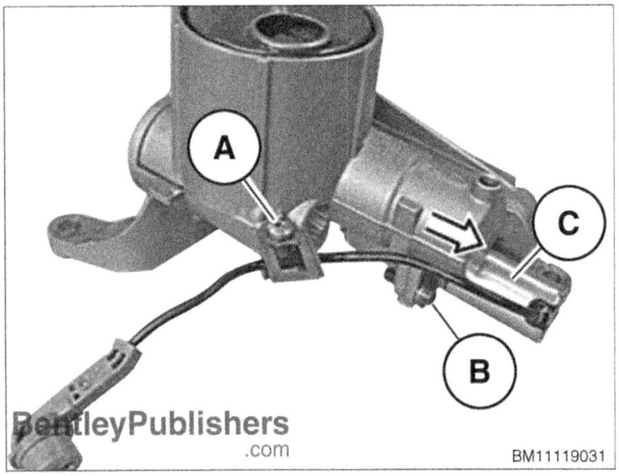

Remove wiring harness fastener (**A**).

– Remove oil pump solenoid fastener (**B**) then remove solenoid (**C**) from oil pump.

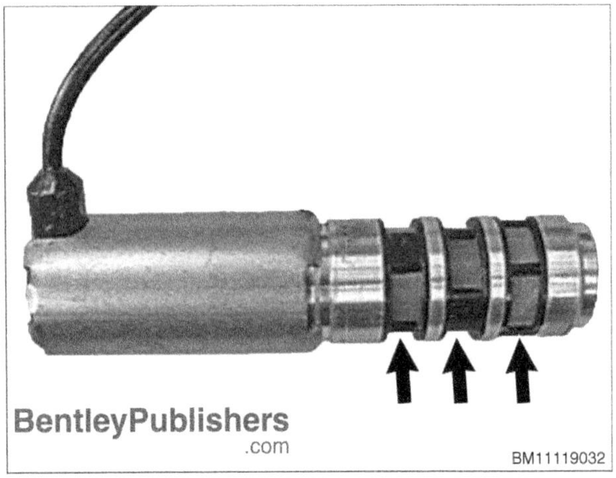

Installation is reverse of removal. Remember to:

- Check that oil pump solenoid strainer (**arrows**) is free from debris.
- Replace wiring harness O-ring (**A**).
- Fill engine with oil, run engine and check for leaks.

Tightening torque	
Oil pump sprocket to oil pump (replace fastener) • Stage 1 • Stage 2	 5 Nm (44 in-lb) additional 90°
Oil pump to crankcase	25 Nm (18.4 ft-lb)
Wiring harness clip to crankcase	4 Nm (35 in-lb)

120 Ignition System

General

This repair group covers ignition component troubleshooting and replacement.

See also:
- **020 Maintenance** for spark plug replacement
- **100 Engine–General** for engine applications
- **130 Fuel Injection** for DME system applications, description and power supply fuses
- **ECL Electrical Component Locations**
- **ELE Electrical Wiring Diagrams**
- **OBD On-Board Diagnostics**

Engine identification table

Table a. Engine identification

Year, model	Engine designation
2007 - 2010 Cooper, Clubman	N12
2009 - 2010 Convertible	N12
2007 - 2010 Cooper S, Clubman S	N14
2009 - 2010 Convertible S	N14
2011 - 2013 Cooper, Clubman	N16
2011 - 2013 Cooper S, Clubman S	N18

Ignition firing order

Table b. Firing order	
4-cylinder engine	1-3-4-2

◄ Cylinder 1 (**arrow**) is at the passenger side of the engine.

Disabling ignition system

The ignition system operates in a lethal voltage range. Disable the ignition system any time engine service or repair work is being done that requires the ignition to be switched ON.

One way of disabling the ignition is by removing the DME main relay. The relay is located in the front power distribution panel in the left side of the engine compartment. DME fuse, relay and power supply information is in **130 Fuel Injection**.

> ***CAUTION—***
> - *Relay locations vary. Use care when identifying relays. See* **ECL Electrical Component Locations**.

Ignition coils

◄ MINI engines use a distributorless ignition system with individual ignition coils (**numbered**) for each cylinder. There is no distributor cap or ignition rotor. (N12 engine shown other engines similar.)

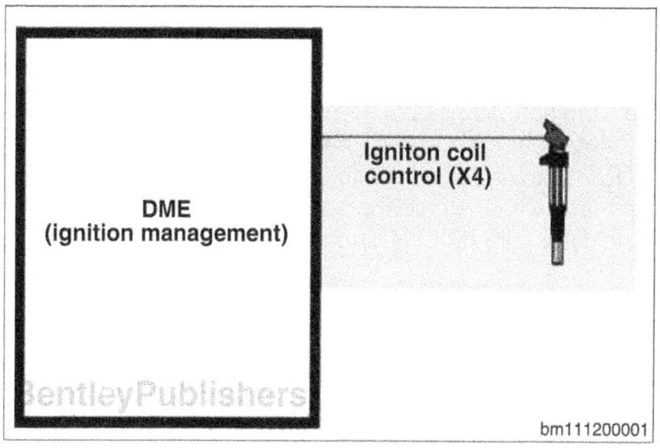

Each coil is separately controlled and monitored by the ECM.

When the ignition is switched ON, the ECM receives wake-up (KL 15w) via the bus network and a hardwire KL15 wake-up signal directly from the car access system (CAS). It then energizes the ECM main relay and the ignition unloader relay to supply power to the engine electronics fuse carrier and ignition coils.

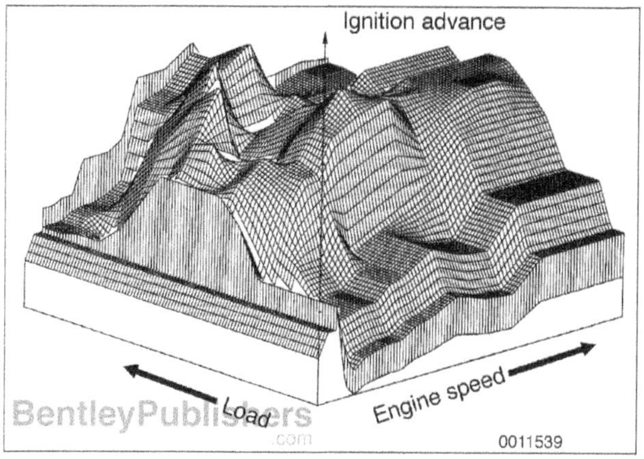

Ignition timing is electronically mapped and not adjustable. The ECM uses engine load, engine speed, coolant temperature, knock detection, and intake air temperature as the basic inputs for timing control.

A three-dimensional map similar to the one shown is digitally stored in the ECM. The initial ignition point is determined by the crankshaft sensor during cranking. Once the engine is running, the ECM refers to the stored map to continually adjust ignition timing based on operating conditions.

Crankshaft sensor

The crankshaft sensor detects crankshaft position and rotation speed via a toothed pulse wheel mounted on the crankshaft. If the ECM does not receive an impulse signal from the crankshaft sensor during cranking, the engine does not start. If the OBD II system misfire detection protocol detects a catalyst damaging fault due to a malfunction in crankshaft sensor components, the malfunction indicator light (MIL) is illuminated.

The crankshaft position sensor is supplied 12 volts from the engine electronics fuses and ground from the ECM. It produces a 5 volt square wave signal to the ECM when the engine rotates.

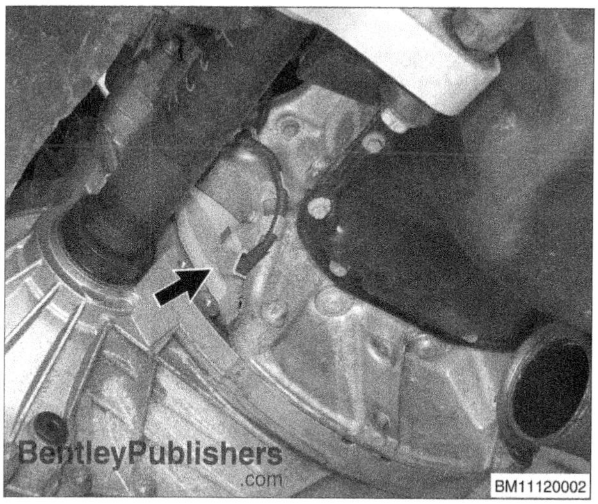

Crankshaft sensor (**arrow**) is at left rear of cylinder block behind drive axle. It is behind a protective cover.

The crankshaft sensor is also referred to as engine speed or reference sensor.

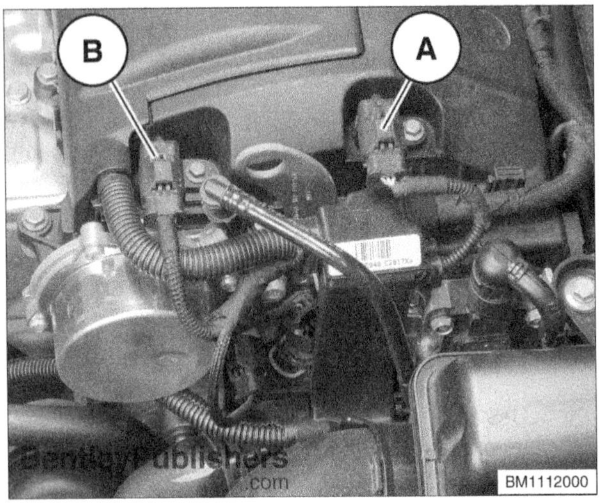

Camshaft sensors

Each camshaft sensor responds to an impulse wheel mounted at the end of the camshaft. The signal from camshaft sensors are used by the ECM for cylinder recognition, spark timing, sequential fuel injection, VANOS (camshaft timing) and Valvetronic (variable valve lift) control. If a fault with the camshaft sensor is detected the malfunction indicator light (MIL) is illuminated

The camshaft position sensor is supplied 12 volts from the engine electronics fuses and ground from the ECM. It produces a 5 volt square wave signal to the ECM when the engine rotates.

Intake camshaft sensor (**A**): Left front of cylinder head. (N12, N14, N16, N18 engine)

Exhaust camshaft sensor (**B**): Right front of cylinder head. (N12, N16, N18 engine)

The camshaft sensor is also called the cylinder identification sensor.

Knock sensor

The knock sensor monitors the combustion chambers for engine-damaging knock. A knock sensor is a piezoelectric microphone tuned to the frequencies of engine knock or detonation. If engine knock is detected, ignition spark is retarded by the ECM. If a fault with the knock sensor is detected the malfunction indicator light (MIL) is illuminated

Knock sensor (**arrow**) is bolted to rear of engine block next to starter motor.

Warnings and Cautions

WARNING—
- *Do not touch or disconnect any cables from the ignition coils while the engine is running or being cranked by the starter.*
- *The ignition system produces high voltages that can be fatal. Avoid contact with exposed terminals. Use extreme caution when working on a car with the ignition switched ON or the engine running.*
- *Connect and disconnect the DME system wiring and test equipment leads when the ignition is OFF.*
- *Before operating the starter without starting the engine (for example when testing compression), disable the ignition. See* **Disabling ignition system** *in this repair group.*

CAUTION—
- *Do not attempt to disable the ignition by removing the coils from the spark plugs.*
- *Maintain a high level of cleanliness when servicing ignition coils. Fuel or oil residue can cause a breakdown in the electrical resistance qualities of silicone used in production. This in turn can result in a failure of the ignition coil.*
- *Do not connect any test equipment that delivers a 12-volt power supply to terminal 15 (+) of the ignition coil. The current flow may damage the engine control module (ECM). Connect test equipment only as specified by MINI or the equipment maker.*
- *Connect or disconnect ignition system wires, multiple wire connectors and ignition test equipment leads only while the ignition is OFF. Switch multimeter functions or measurement ranges with test probes disconnected.*
- *Do not disconnect the battery while the engine is running.*
- *Prior to disconnecting the battery cables, read the battery disconnection cautions in* **001 Warnings and Cautions**.
- *Wait at least 1 minute after switching the ignition OFF before removing the ECM connector. If the connector is removed before this time, residual power in the system relay may damage the control module.*
- *Use a digital multimeter for electrical tests. Use an LED test light for quick tests.*
- *Follow torque instructions, including angle of rotation specifications.*

TROUBLESHOOTING

On-board diagnostics

If faults arise, or if the malfunction indicator light (MIL) is illuminated, begin troubleshooting by connecting MINI scan tool to the data link connector (OBD II plug) (**arrow**). For information on how to access diagnostic trouble codes (DTCs), see **020 Maintenance**.

Misfire detection

Engine misfire in one or more cylinders may be caused by malfunctions in various subsystems. The OBD II system incorporated into the engine management systems is designed to detect and warn of misfire faults during engine operation. See **OBD On-Board Diagnostics.**

WARNING—
- *Ignition misfires can cause high hydrocarbon exhaust emissions and catalytic converter damage. For this reason, if a severe misfire is detected, the fuel injector to the specific cylinder is switched OFF and the MIL is illuminated. A misfire may also overheat the catalytic converter(s), a fire hazard.*

Oscilloscope diagnostic diagrams

One way to diagnose faulty engine management components or functions is to use an oscilloscope to analyze spark quality with the engine running.

See **Table c** for a list of common ignition coil voltage faults and related causes.

Table c. Ignition secondary voltage diagnostics		
	Secondary voltage low	**Secondary voltage high**
Spark plug electrode gap	Too small	Too big
Spark plug electrode condition		Worn/burnt
Spark plug electrode temperature	Too high	Too low
Engine compression	Too low	Too high
Spark plug wires		Faulty
Fuel air mixture		Too lean

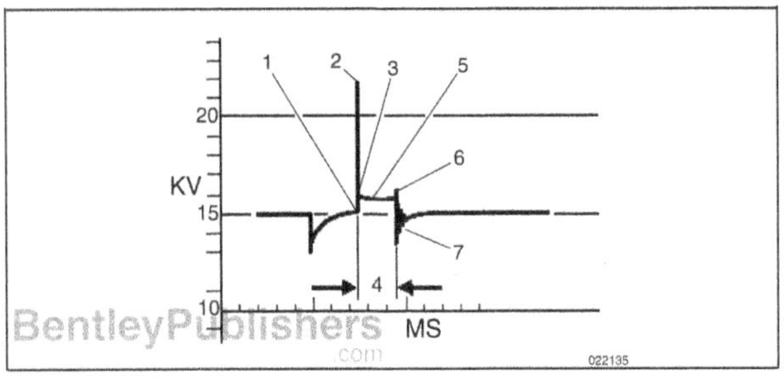

Normal scope trace of spark at idle

1. Start of ignition voltage peak
2. Level of ignition voltage
3. Level of combustion voltage
4. Period of combustion
5. Combustion curve characteristics
6. Start of spark decay
7. Termination oscillations

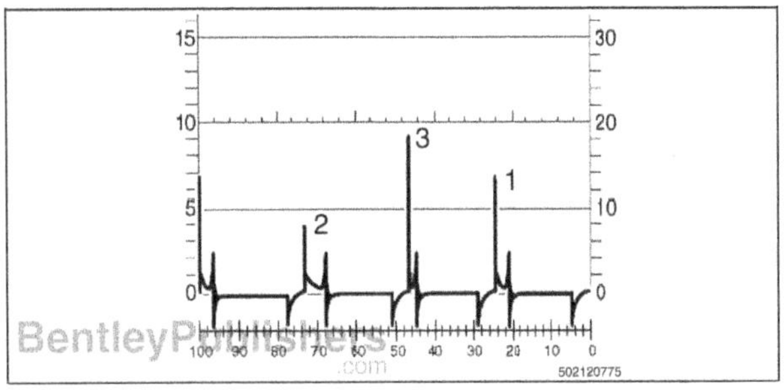

Ignition spark at idle speed

1. Normal ignition voltage peak: Good spark plug
2. Low voltage peak: Closed plug gap
3. High voltage peak: Open plug gap

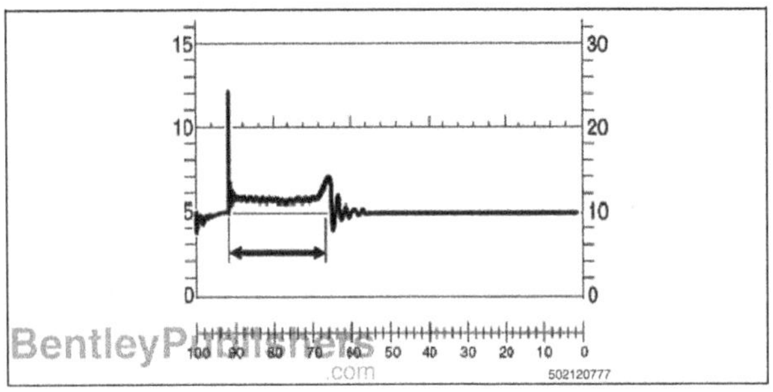

Long combustion period: Small spark plug gap

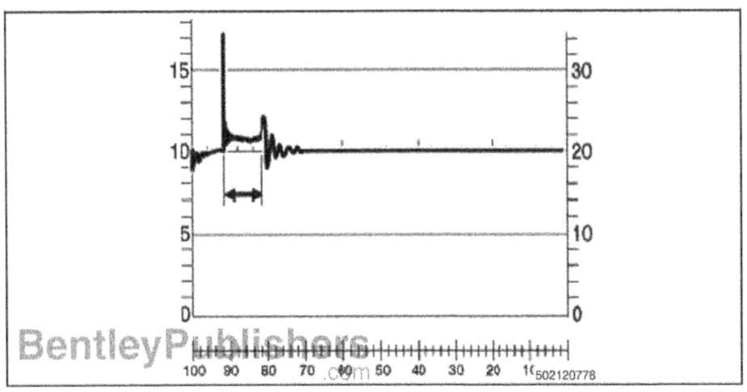

Short combustion period: Large spark plug gap

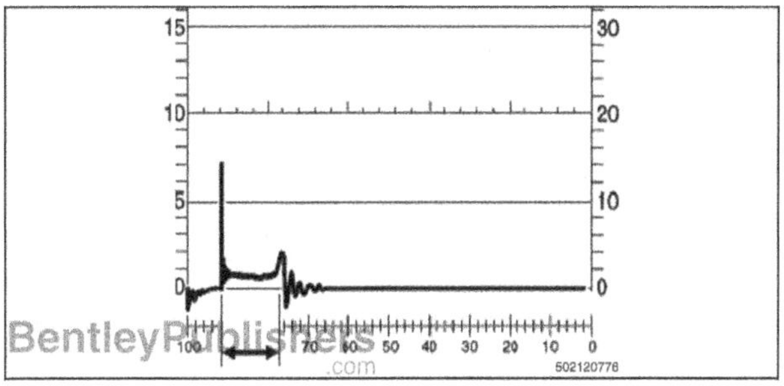

Normal combustion period at idle

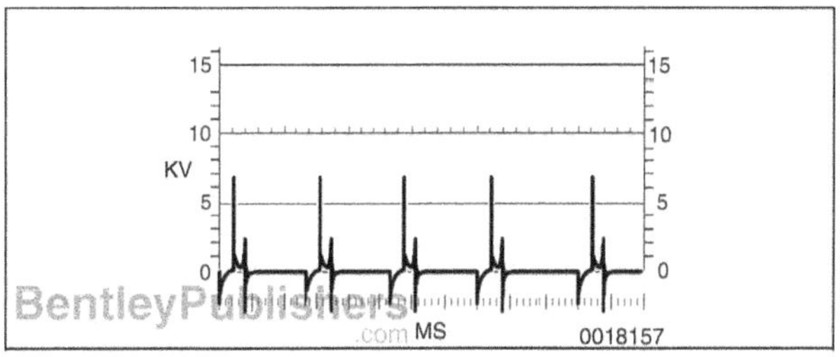

Normal oscilloscope pattern for ignition system at idle

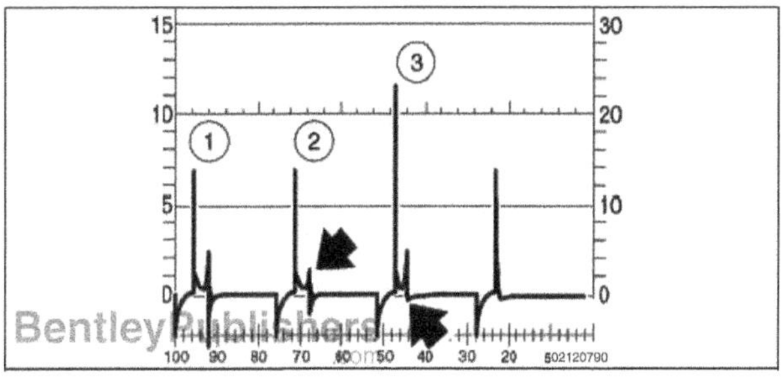

Ignition voltage peaks at idle

1. Normal ignition peaks
2. Downward peak (**arrow**) shortened: Ignition coil defective
3. Downward peak (**arrow**) missing completely: Ignition coil defective

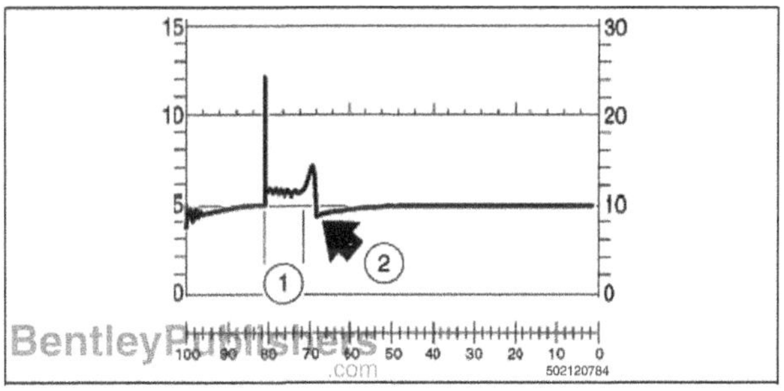

Defective ignition coil

1. Short spark period
2. Spark voltage line (**arrow**) with very slight drop

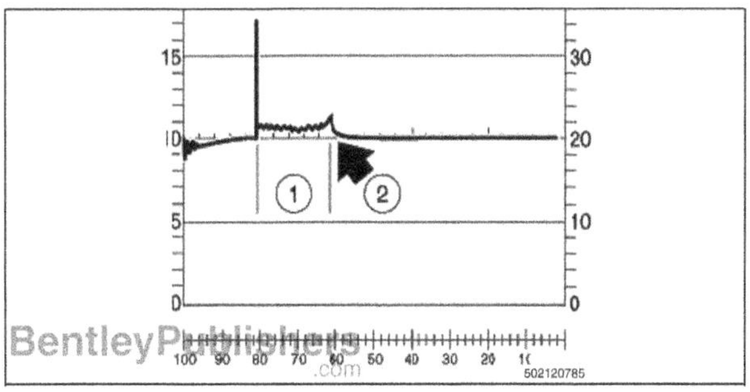

Defective ignition coil

1. Normal combustion period
2. Spark voltage line (**arrow**) absent

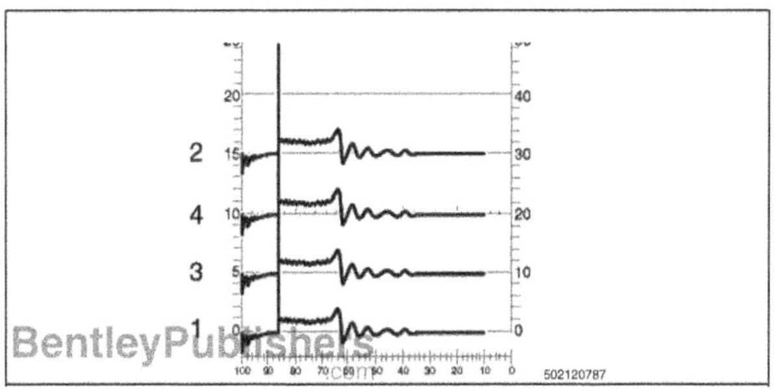

Normal secondary voltage patterns (4-cylinder pattern shown)

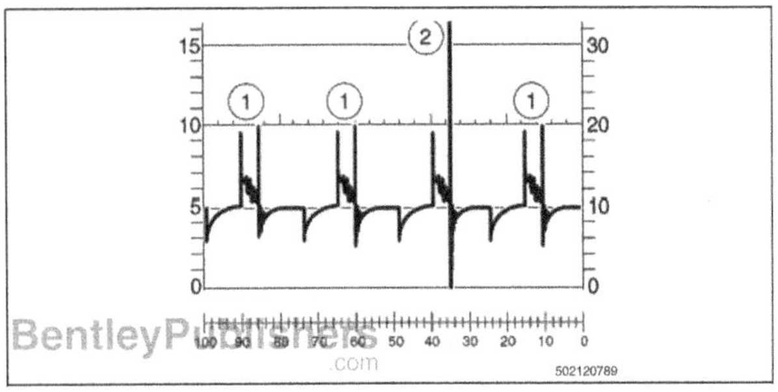

Ignition voltage peaks in response to sudden acceleration load:

1. Normal ignition pattern. Beginning of dying out pattern is not much higher than ignition voltage peak.
2. Beginning of dying out pattern considerably higher than ignition voltage peak. Fault in injection system:
   - Lean fuel mixture
   - Defective fuel injector
   - Low compression in cylinder

Ignition Coil Service

There is a separate ignition coil above each spark plug. Spark plug replacement is covered in **020 Maintenance**.

Ignition coil, replacing

– Make sure ignition is switched OFF.

– **N12, N16 engine:** Remove ignition coil cover. See **020 Maintenance**.

Unlock (**curved arrow**) ignition coil connector. Pull (**straight arrow**) connector off coil.

Slide rod-shaped ignition coil straight out of spark plug hole.

> ***CAUTION—***
> - *Maintain a high level of cleanliness when servicing ignition coils. Fuel or oil residue can cause a breakdown in the electrical resistance qualities of silicone used in production. This in turn can result in a failure of the ignition coil.*

– Installation is reverse of removal.

– Check for fault codes and reset ECM memory.

Crankshaft Sensor Service

Crankshaft sensor, removing and installing

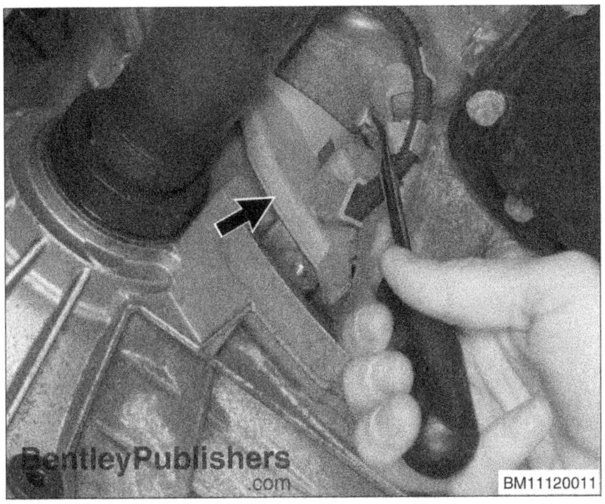

◀ Remove plastic rivet. Then remove sensor protective cover (**arrow**).

◀ Disconnect electrical connector (**A**). Remove fastener (**arrow**). Then remove crankshaft sensor from engine.

– Installation is reverse of removal.

Tightening torque	
Crankshaft sensor to engine	5 Nm (44 in-lb)

Camshaft Sensor Service

Camshaft sensor, removing and installing

Exhaust camshaft sensor is shown, intake camshaft sensor is similar.

◀ Remove ignition coil cover fasteners (**arrows**). Then remove ignition coil cover.

- Disconnect camshaft sensor electrical connector (**B**).
- Remove camshaft sensor fastener (**A**). Remove camshaft sensor from cylinder head.
- Installation is reverse of removal.

Tightening torque	
Crankshaft sensor to engine (replace O-ring)	8 Nm (70 in-lb)

Knock Sensor Service

Knock sensor, removing and installing

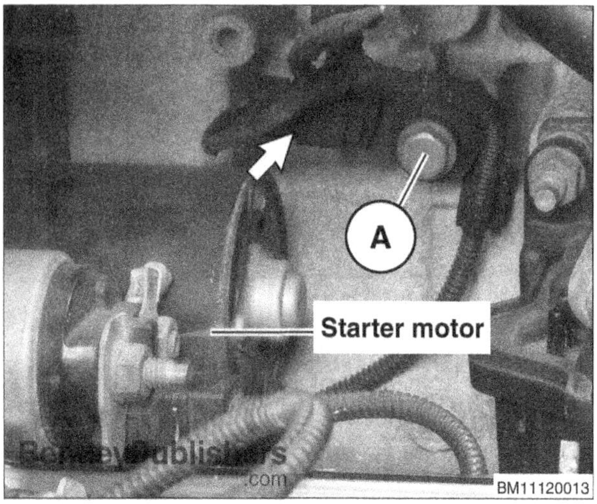

- Disconnect knock sensor electrical connector (**arrow**).
- Remove knock sensor fastener (**A**) and remove knock sensor from engine block.
- Installation is reverse of removal.

Tightening torque	
Knock sensor to engine (replace O-ring)	20 Nm (14.7 ft-lb)

121 Battery, Starter, Alternator

GENERAL

This repair group covers the battery, alternator, starter and associated components of the electrical system.

See also:

- **020 Maintenance** for accessory belt replacement and battery routine service.
- **600 Electrical System–General** for bus system information, power reset procedure and general electrical troubleshooting
- **ECL Electrical Component Locations**
- **ELE Electrical Wiring Diagrams**

Engine electrical system

The alternator and starter are wired directly to the battery. To prevent accidental shorts that might blow a fuse or damage wires and electrical components, disconnect the negative (–) battery cable before working on the electrical system.

Different versions of alternators, starters and batteries are used in MINI models. Replace components according to the original equipment specifications.

Troubleshooting

Basic battery, starter and charging system troubleshooting information is in **Table a**.

Table a. Battery, starter and charging system troubleshooting

Symptom	Probable cause	Corrective action
Engine does not crank	Fault in car access system (CAS).	Try another ignition key. If problem persists, contact your authorized MINI dealer.
Engine cranks slowly or not at all, solenoid clicks when starter is operated.	Battery cables loose, dirty or corroded.	Clean or replace cables. See **020 Maintenance**.
	Battery discharged.	Charge battery and test. Replace if necessary.
	Battery to body ground cable loose, dirty or corroded	Inspect ground cable. Clean, tighten or replace if necessary.
	Poor connection at starter motor terminal X6512.	Check connections at starter motor.
	Starter motor or solenoid faulty.	Test starter.
Battery does not stay charged more than a few days.	Short circuit draining battery.	Test for parasitic current draw.
	Short driving trips and high electrical drain on charging system does not allow battery to recharge.	Evaluate driving style. Where possible, reduce electrical consumption when making short trips.
	Engine accessory belt loose, worn or damaged.	Inspect or replace accessory belt. See **020 Maintenance**.
	Battery faulty.	Test battery and replace if necessary.
	Battery cables loose, dirty or corroded.	Clean or replace cables. See **020 Maintenance**.
	Alternator faulty.	Test alternator.

Warnings and Cautions

WARNING—
- *Wear goggles, rubber gloves, and a rubber apron when working around the battery or battery acid (electrolyte).*
- *Battery acid contains sulfuric acid and can cause skin irritation and burning. If acid is spilled on your skin or clothing, flush the area at once with large quantities of water. If electrolyte gets into your eyes, flush them with large quantities of clean water for several minutes and call a physician.*
- *A battery that is being charged or is fully charged gives off explosive hydrogen gas. Keep sparks and open flames away. Do not smoke.*
- *Battery lead and battery acid are both hazardous materials. When disposing of an old battery, follow federal, state and local regulations on hazardous waste disposal.*

CAUTION—

- *Use a digital multimeter when testing automotive electrical components.*
- *If a repair procedure specifies disconnecting the battery, follow the instruction for safety reasons.*
- *Prior to disconnecting the battery, read the battery disconnection cautions in* **001 Warnings and Cautions***.*
- *Disconnecting the battery cables may erase fault codes stored in ECM memory.*
- *Using a conventional battery charger to jump start the vehicle may damage electronic control modules. Use an electronic vehicle power supply or another 12-volt battery for jump-starting. Jump starting is covered in* **020 Maintenance***.*
- *Disconnect the negative (–) battery cable first and reconnect it last. Cover the battery post with an insulating material whenever the cable is removed.*
- *Do not disconnect battery, alternator or starter wires while the engine is running.*
- *Do not reverse the battery cables. Even a momentary wrong connection can damage the alternator or other electrical components.*
- *Do not depend on insulation color to tell battery positive and negative cables apart. Label cables before removing.*
- *Power windows and sunroof may fail to function properly after disconnecting and reconnecting the battery. Reinitialize windows and sunroof. See* **Battery reconnection notes** *in this repair group.*
- *Do not make any additional connections at the negative terminal of the battery.*
- *Do not modify the battery ground cable.*
- *Do not make any connections between the intelligent battery sensor (IBS) and the sensor screw.*
- *Do not use force when disconnecting the ground terminal from the battery.*
- *Do not pull at the ground cable.*
- *Do not use the IBS as a pivot point to lever off the ground terminal.*
- *Use a torque wrench when reinstalling the battery ground cable.*
- *On cars equipped with IBS on the negative battery terminal, do not release or tighten the IBS screw (screw with Torx head).*
- *Follow torque instructions, including angle of rotation specifications, when installing fasteners.*

Battery

Battery basic

The wet-cell lead-acid battery is rated by ampere-hours (Ah) and cold cranking amps (CCA) listed on the battery. Three Ah ratings are available depending on vehicle equipment.

- 40 Ah size-optimized
- 55 Ah size-optimized
- 70 Ah

The advantages of the size-optimized versions are lower weight and smaller dimensions.

Battery location and connections

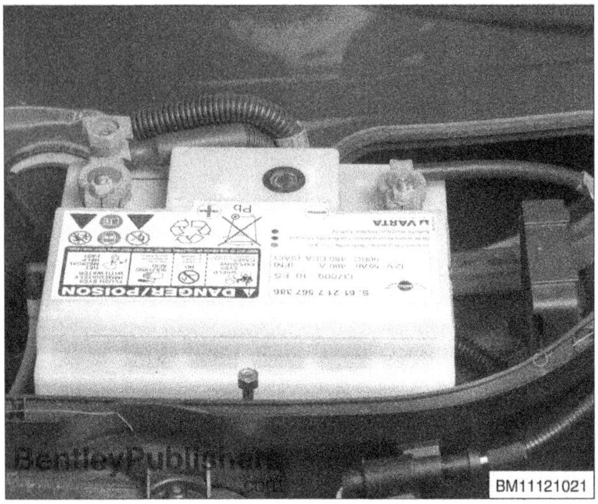

The battery is mounted in the right rear of the engine compartment, underneath the right side cowl cover.

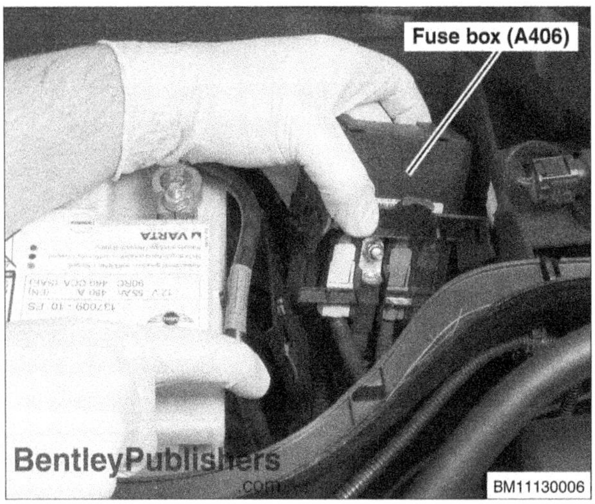

A power distribution box (A604) is mounted on the vehicle battery. This central power supply unit supplies power to the main electric loads.

The fuses are integrated in the housing of the power distribution box and cannot be serviced separately. When replacing a fuse, replace power distribution box as a complete unit.

The power distribution box on the battery powers the following loads:

- Power distribution box (A400a) in left side engine compartment
- Junction box (A4010) behind right kick panel trim.
- Valvetronic (via relay box)

For power distribution panel details, see **ECL Electrical Component Locations**.

Intelligent battery sensor (IBS)

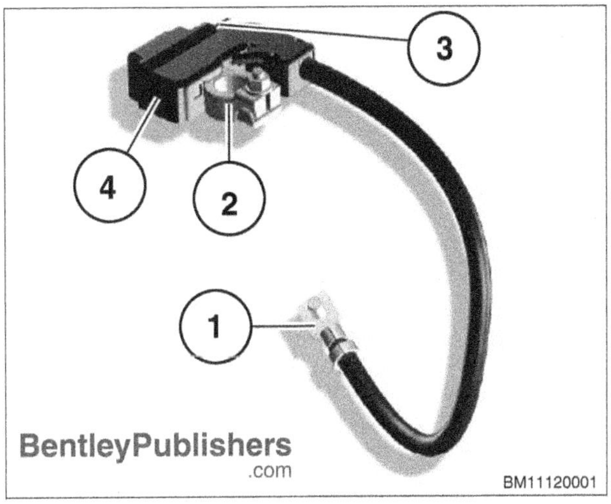

◄ Not all MINI models come equipped with IBS. An IBS-equipped car can be quickly identified by the electronics integrated in the terminal end of the negative battery cable:

1. Battery ground cable
2. B- connection to IBS
3. Bit-serial data (BSD) interface
4. Intelligent battery sensor (IBS)

The IBS is sensitive to mechanical stress and strain. See **Warnings and Cautions** in this repair group.

Replace IBS as a complete unit with the battery ground cable. The ground cable serves as a heat sink for the IBS.

If IBS is replaced, allow vehicle to rest unused at least 3 hours. Only then can the new IBS determine battery status.

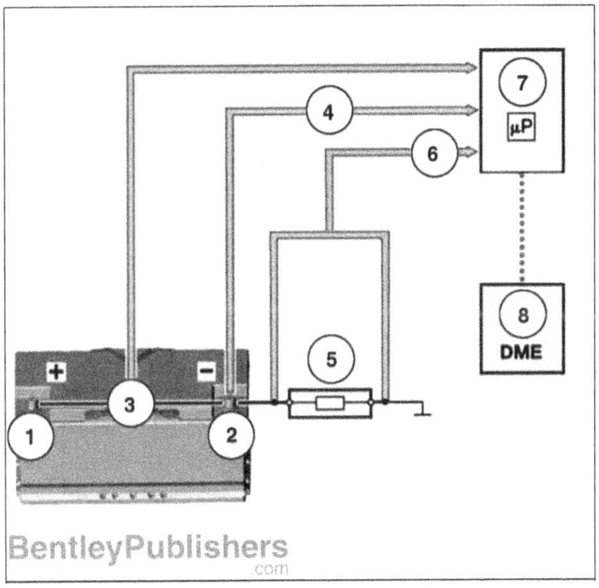

◄ The IBS microprocessor monitors and measures battery conditions such as:

- Terminal voltage via measurement from B+ to ground
- Charge / discharge current via integrated shunt resistor
- Temperature of battery acid via integrated temperature sensor

1. B+ battery positive terminal
2. Battery ground terminal
3. Voltage measurement
4. Temperature measurement
5. Integrated shunt resistor
6. Current measurement
7. IBS microprocessor
8. Engine control module (ECM)

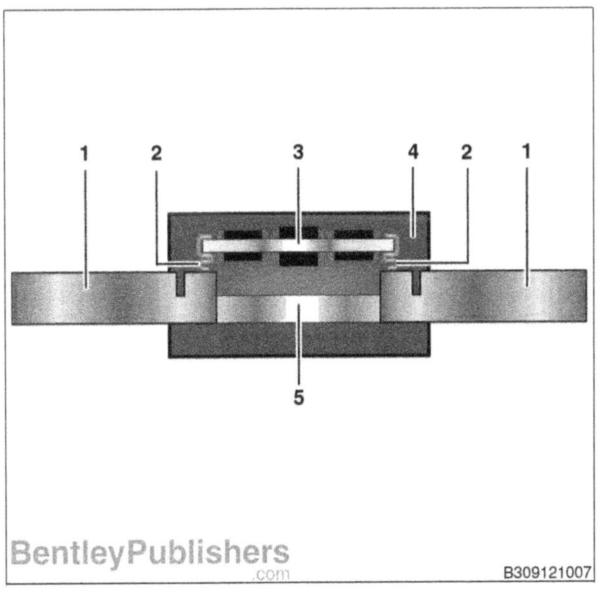

◄ IBS components:

1. Copper plate
2. Gull wing tabs
3. PC board with evaluation electronics
4. Housing
5. Manganin-copper alloy resistor maintains constant temperature regardless of current flow, used to measure current flow.

IBS evaluation function

IBS electronics continuously measure the following values under all vehicle operating conditions:

- Voltage (6 v to 16.5 v)
- Starting current (0 A to 1000 A)
- Charging current (20 A to 200 A)
- Temperature (-40°C to 105°C or -40°F to 221°F)
- Parasitic draw or closed circuit current (0 A to 10 A)

When the vehicle is stationary, the IBS is programmed to wake up every 40 seconds and measure battery condition within approx. 50 ms in order to save power. The measured values, communicated to the engine control module (ECM) via binary serial data (BSD) interface, are used to calculate the following:

- State of charge (SoC) shows the current charge of the battery. SoC is used during key OFF periods to insure that the battery maintains a sufficient charge to start the engine at least one more time.
- State of health (SoH) tracks the history of the battery in the vehicle. Charge / discharge cycles and times are monitored. SoH helps the ECM determine correct battery charging rates and anticipated battery life. Internal resistance of the battery is calculated by IBS from current and voltage drop during engine start. These values are used by the ECM to calculate the SoH of the battery.

Software contained in the IBS microprocessor utilizes measured values to calculate battery SoC during vehicle sleep mode and compares this information with that received from the ECM pertaining to the battery SoC / SoH during the period between engine OFF and deactivation of the DME main relay.

Battery data is stored every 2 hours over a 6 hour time frame, providing three snapshots of battery SoC information. The stored information / snapshot data is overwritten every 6 hours. Whenever the ignition is activated, IBS updates the ECM with the current closed circuit histogram / battery status information. The ECM evaluates the new data and, if a parasitic draw (closed-circuit current draw) is identified, a fault is stored in fault memory.

IBS diagnosis

A fault code is stored in the ECM when the IBS is defective. The ECM boosts idle speed in order to sufficiently charge the battery.

The IBS can only be diagnosed through the ECM. The self-diagnosis function checks the voltage, current, temperature, terminal 15 wake up signal as well as system errors in the IBS.

Voltage measurement

If IBS is shorted to ground, ECM fault code displays “Voltage Fault DME ON”. The IBS is unable to wake up the ECM.

If IBS is shorted to B+, ECM fault code displays “Voltage Fault, DME not ON”. The vehicle does NOT enter sleep mode.

Current measurement

Current measurement is a dynamic process. If an implausible current flow value is detected, ECM fault code displays “Current Fault”.

IBS wake-up

The IBS constantly monitors battery SoC, even when the vehicle is asleep. A wake-up signal is sent to the ECM if the SoC falls below a predetermined threshold. Upon wake-up, the ECM switches OFF auxiliary electrical loads in order to preserve battery charge.

IBS wakes up the vehicle once during a key-off cycle. Once awakened, the vehicle returns to sleep mode.

Wake-up signal faults

The IBS detects wake-up line faults under the following conditions:

- ECM switched ON.
- Ignition switched ON (voltage high at IBS).
- Terminal 15 running via alternator output at BSD.

If voltages at IBS terminal 15 and terminal 15 via the BSD are not equal, a fault is indicated either in the BSD line or the IBS.

Battery, testing

MINI batteries are rated as follows:

- **Ah rating** is determined by the average amount of current the battery can deliver over time without dropping below a specified voltage.
- **CCA rating** is determined by the battery's ability to deliver starting current at 0°F (-18°C) without dropping below a specified voltage.

Battery testing determines battery condition and state of charge. Before testing the battery, check that the battery cables are tight and free of corrosion.

Electronic battery testing

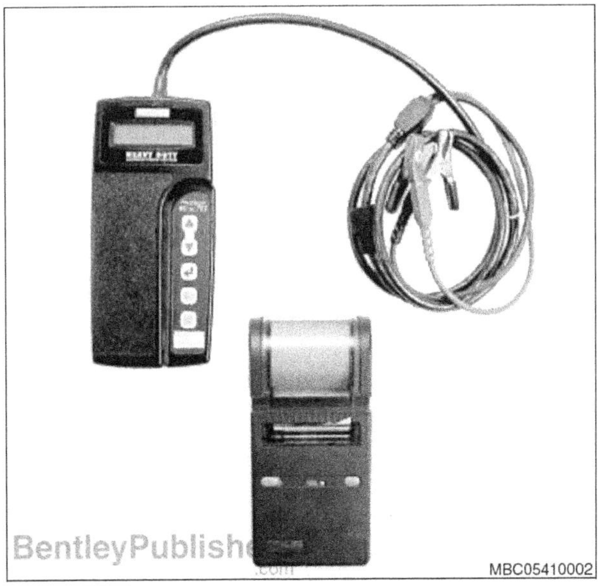

MBC05410002

Electronic battery testers such as Midtronics® MCR 717 are recommended for testing all battery types.

Hydrometer testing

On conventional or low-maintenance batteries, check the specific gravity of the electrolyte using a hydrometer. The more dense the concentration of sulfuric acid in the electrolyte, the higher the state of charge.

Before hydrometer testing, load the battery with 15 amperes for one minute. If the battery is installed in the vehicle, this can be done by switching the headlights ON for about one minute without the engine running. The state of battery charge based on specific gravity values are in **Table b**.

Note that electrolyte temperature affects hydrometer reading. Check the electrolyte temperature with a thermometer. Add 0.004 to the hydrometer reading for every 6°C (10°F) that the electrolyte is above 27°C (80°F). Subtract 0.004 from the reading for every 6°C (10°F) that the electrolyte is below 27°C (80°F).

Table b. Specific gravity of battery electrolyte at 27°C (80°F)	
Specific gravity	**State of charge**
1.265	Fully charged
1.225	75% charged
1.190	50% charged
1.155	25% charged
1.120	Fully discharged

If the specific gravity is at or above 1.225 but the battery lacks power for starting, determine the battery's service condition with a load voltage test. See **Battery load voltage test** in this repair group.

If the average specific gravity of the six cells is below 1.225, charge battery and retest. See **Battery charging notes** in this repair group.

Battery open-circuit voltage test

– Before testing, load battery with 15 amperes for one minute with battery load-tester or switch headlights ON for about one minute without engine running. Connect digital voltmeter across battery terminals. Open-circuit voltage levels are given in **Table c**.

Table c. Open-circuit voltage and battery charge	
Open-circuit voltage	**State of charge**
12.6 or more	Fully charged
12.4	75% charged
12.2	50% charged
12.0	25% charged
11.7 or less	Fully discharged

– If open-circuit voltage is OK but battery still lacks power for starting, perform a load voltage test. See **Battery load voltage test** in this repair group.

– If open-circuit voltage is below 12.4 volts, recharge battery and retest. See **Battery charging notes** in this repair group.

Battery load voltage test

A battery load tester is required for a load voltage test. The test is made by applying a high resistive load to the battery terminals and simultaneously measuring battery voltage. For the most accurate results, make sure the battery is fully charged. Disconnect battery cables before making the test. Replace the battery if the voltage is below that listed in **Table d**.

> ***WARNING—***
> • *Wear protective goggles and clothing when performing a load test.*

Table d. Battery load test–minimum voltage	
Ambient temperature	**Voltage***
27°C (80°F)	9.6
16°C (60°F)	9.5
4°C (40°F)	9.3
-7°C (20°F)	8.9
-18°C (0°F)	8.5
* Measure after applying a 200 amp load for 15 seconds.	

Battery parasitic draw, testing

If the vehicle battery is discharged for an unknown reason, perform the parasitic draw test designed to detect excessive current flow from the battery when everything in the vehicle is shut down. This is also known as a closed circuit current measurement.

– Check that battery voltage is 12 volts or higher. If lower, recharge battery.

B309121029

< Remove negative cable and install battery isolation switch (OTC tool 7645 or equivalent).

– Close switch. Connect digital multimeter in series across switch. Set meter to milliamp (mA) scale.

– Lock engine hood switch into hood closed position using screwdriver or similar. With exception of engine hood, keep all other doors and lids closed.

– To simulate normal parasitic draw conditions: Switch ignition ON and activate all electrical consumers, including any accessories. Switch ignition OFF. Open and close driver's door (simulates somebody getting out). Lock car, arming alarm if installed.

– Open battery isolation switch and monitor current draw. Wait 60 - 70 minutes.

– Maximum allowable parasitic draw after waiting:

Parasitic draw specification	
Base model	10 mA
Fully equipped model	18 mA
Approximate maximum	40 mA

– If draw exceeds maximum allowable, find consumer(s) at fault by removing individual fuses and noting change in parasitic draw. Repair or replace component(s) found at fault.

Fuse locations and applications are in **ECL Electrical Component Locations**.

430-watt Deutronic power supply

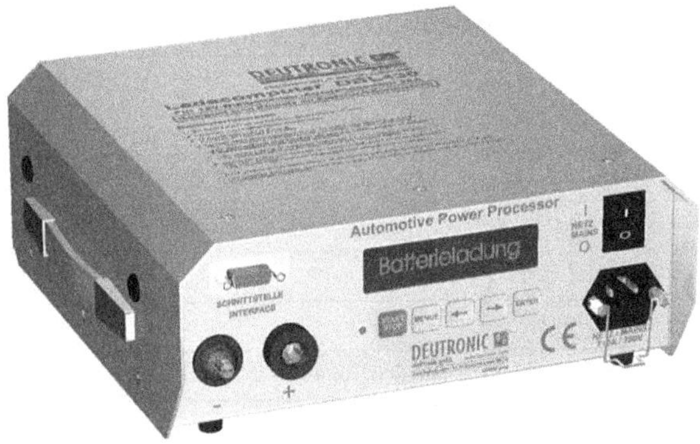

B309121013

70-amp ACCTIVA Professional power supply

B309121014

Battery charging notes

Consistent, clean voltage is essential when charging the battery or maintaining battery charge while coding or programming control modules. A transformer-based battery charger is inadequate for this task. MINI recommends the use of a vehicle power supply instead of a battery charger.

A power supply is a constant voltage, high-amperage power station with electronically controlled functions. It insures clean DC voltage and eliminates AC. It also provides fast, efficient and safe battery charging. Once proper battery voltage is obtained, the power supply floats the charge to maintain that voltage.

Battery charging voltage	
MINI recommendation	14.8 v

Power supply units from different manufacturers are illustrated on this page.

If using a power supply, recharge battery without removing it. Otherwise, remove the battery from the engine compartment during charging. See **Battery, removing and installing** in this repair group.

Prolonged charging with a conventional charger can cause battery electrolyte evaporation to a level that can damage. It is best to use a low-current charger (6 amperes or less) to prevent battery damage caused by overheating.

WARNING—
- *Hydrogen gas given off by the battery during charging is explosive. Do not smoke. Keep open flames away from the top of the battery, and prevent electrical sparks by turning off the battery charger before connecting or disconnecting it.*

CAUTION—
- *Battery electrolyte (sulfuric acid) can damage the car. If electrolyte is spilled, clean the area with a solution of baking soda and water.*
- *Allow a frozen battery to thaw before attempting to recharge it.*
- *Do not exceed 16.5 charging voltage at the battery.*

Battery, charging

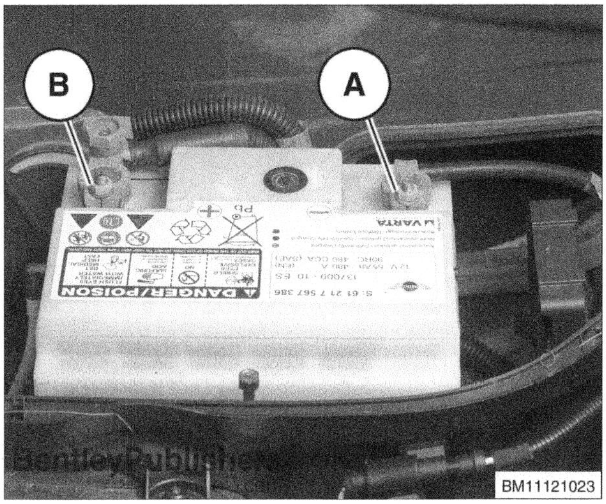

BM11121023

◄ Connect vehicle power supply charger to battery B+ (**B**) and ground (**A**). Follow manufacturer's instructions for setting charging voltage.

Battery, removing and installing

> *CAUTION—*
> - *Prior to disconnecting the battery, read the battery disconnection cautions given in* **001 Warnings and Cautions**.

– Remove right cowl cover. See **513 Interior Trim**.

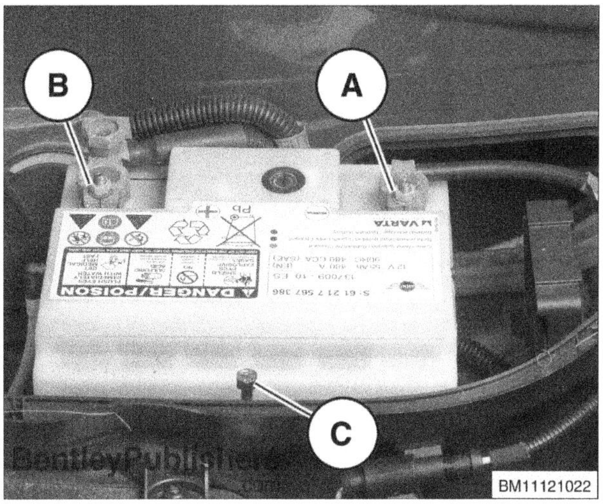

BM11121022

◄ Working at top of battery:

- Loosen fastener (**A**) and disconnect negative cable first.
- Loosen positive cable fastener (**B**) and disconnect cable.

> *CAUTION—*
> - *Do not damage intelligent battery sensor (IBS) if equipped, which is attached to ground cable. See* **Warnings and Cautions** *in this repair group.*

- Remove battery hold-down bolt (**C**).
- Disconnect battery vent hose at side of battery, if applicable.

– Lift battery out of cowl.

– Clean away corrosion in and around battery tray and on cables ends.

– When reinstalling battery, reconnect negative cable last.

Tightening torques	
Battery post pinch bolt (M6 bolt)	5 Nm (44 in-lb)
Battery hold down bolt	6 Nm (53 in-lb)

Once the battery is reinstalled, the following steps are necessary:

- If higher rating battery is installed, recode CAS with correct battery specifications using MINI scan tool.
- Register battery. See **Replacement battery registration** in this repair group.
- Follow **Battery reconnection notes** in this repair group.

Battery registration (vehicle with intelligent battery sensor)

On vehicle fitted with intelligent battery sensor (IBS), register replacement battery to vehicle using MINI scan tool. Without new battery registration, odd electrical system behavior and electrical system faults may result.

To register battery, run service function “Register battery replacement” to notify vehicle power management (software in the engine electronics and intelligent battery sensor) that a new battery is fitted in the vehicle. During battery registration, the following operations are performed:

- Current mileage is stored.
- Stored values from old battery (charge status, current, voltage, temperature, etc.) are deleted.
- Power management is initialized.

Battery reconnection notes

In addition to **Warnings and Cautions** in this repair group and in **001 Warnings and Cautions,** observe the following whenever the battery is disconnected or accidentally discharged and then reconnected.

Clock

– Reset.

Radio

– Reenter presets.

Steering angle sensor

– Electronically calibrate using MINI scan tool, if necessary.

Power window regulator motors

Initialize as follows:

– Switch ignition ON and close window completely.

– Release window button, then hold switch in Close direction again for approx. 1 second.

– Open window completely.

– Release window button, then hold switch in Open direction again for approx. 1 second.

– Close window completely again.

– Release window button, then hold switch in Close direction again for approx. 1 second.

– Check to make sure one-touch control, anti-trapping and comfort functions are activated.

See **512 Door Windows** for additional information.

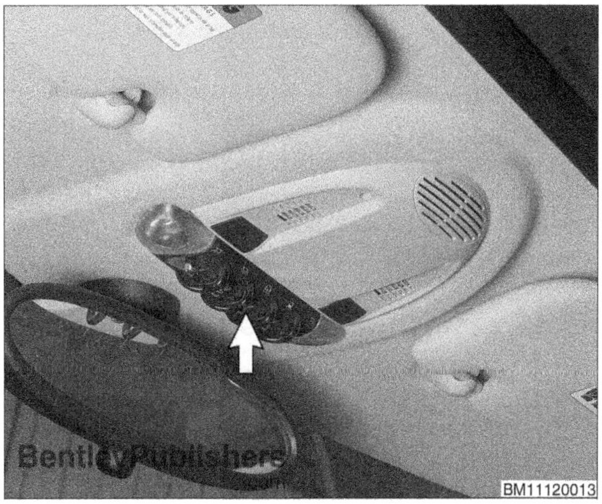

Sunroof

Initialize as follows (if applicable):

◄ Switch ignition ON and press and hold sunroof switch (**arrow**) in Raised direction.

– After sunroof reaches Raised position, keep switch pressed for approx. 15 seconds longer.

– Initialization is complete when sunroof attempts briefly to lift again.

STARTER

Starter troubleshooting

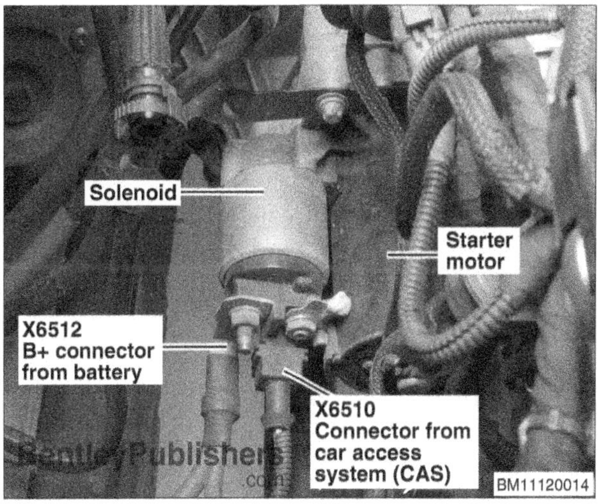

◄ The starter is bolted to the engine bell housing under the intake manifold. Starter wiring connects to the solenoid, above the starter, as follows:

- Large wire (X6512 or B+ connector) is direct battery voltage.
- Smaller wire (X6510) operates starter solenoid via ignition switch and car access system (CAS).

– If starter fails to operate, check the following:

- Car access system (CAS): Try another ignition key. If no faults can be found, have CAS checked using MINI scan tool. CAS replacement is covered in **CAS module, removing and installing** in this repair group.
- Manual transmission: Clutch switch module is connected to starting system. Make sure clutch is pressed all the way to the floor. See **210 Clutch**.
- Automatic transmission: PARK / NEUTRAL position signal for starter neutral safety function is communicated from transmission gear position switch to CAS module. Failure of this function causes a no-start condition.

– If starter and solenoid click repeatedly without starting vehicle:

- Check battery state of charge.
- Inspect battery and starter cables, terminals and ground connections for good contact. In particular, make sure ground connections between battery, body and engine are completely clean and tight.
- If no faults are found, starter may be faulty and should be replaced.

– Access to starter is difficult as it is under the intake manifold. If possible, Check for battery voltage at terminal X6510 of starter motor with ignition in START position.
  - If voltage is not present, check wiring between CAS and starter terminal.
  - Check CAS and other inputs that disrupt power to solenoid. See **ELE Electrical Wiring Diagrams**.
  - If voltage is present and no other visible wiring faults can be found, problem is most likely in starter motor.

CAS module, removing and installing

> *CAUTION—*
> - *CAS module is static sensitive. Static discharge may damage it permanently. Handle using static prevention equipment and techniques. See* **600 Electrical System–General**.

– Remove upper dashboard trim panel. See **513 Interior Trim**.

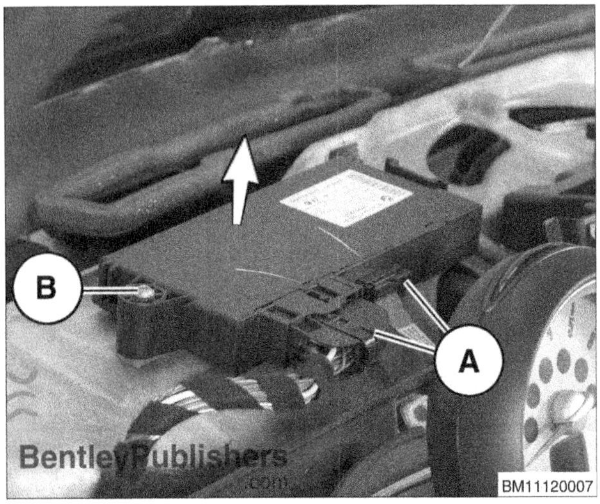

– Working at top of dashboard:
  - Detach ribbon harness and electrical connector (**A**) from CAS module.
  - Remove module mounting screw (**B**) and remove from dashboard.

– Inspect ribbon cable and replace if damaged. (Ribbon cable connects CAS to ignition key slot and START STOP button.)

– When installing, make sure CAS module guide fits correctly into mounting slot.

– After reassembly, carry out vehicle coding and programming. See **600 Electrical System–General**.

Starter, removing and installing

The starter is removed from below the vehicle.

– Switch ignition OFF. Disconnect negative (–) battery cable and cover battery terminal to keep cable from accidentally contacting terminal.

> *CAUTION—*
> - *Prior to disconnecting the battery, read the battery disconnection cautions in* **001 Warnings and Cautions**.

– Remove engine air filter housing and ducts. See **130 Fuel Injection**.

– Raise vehicle and support safely.

> *WARNING—*
> - *Make sure the vehicle is stable and well supported at all times. Use a professional automotive lift or jack stands designed for the purpose. A floor jack is not adequate support.*

– Remove right front wheel.

◀ **N12, N16 engine**: Working beneath intake manifold, remove fuel tank vent solenoid (**arrow**).

◀ **N14, N18 engine**: Loosen fasteners (**arrows**) and set vacuum canister (**A**) aside.

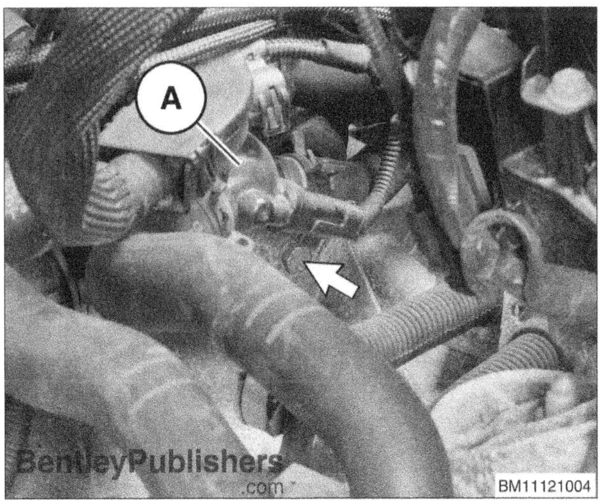

◀ Working below thermostat at transmission bell housing (**A**), remove starter fastener (**arrow**).

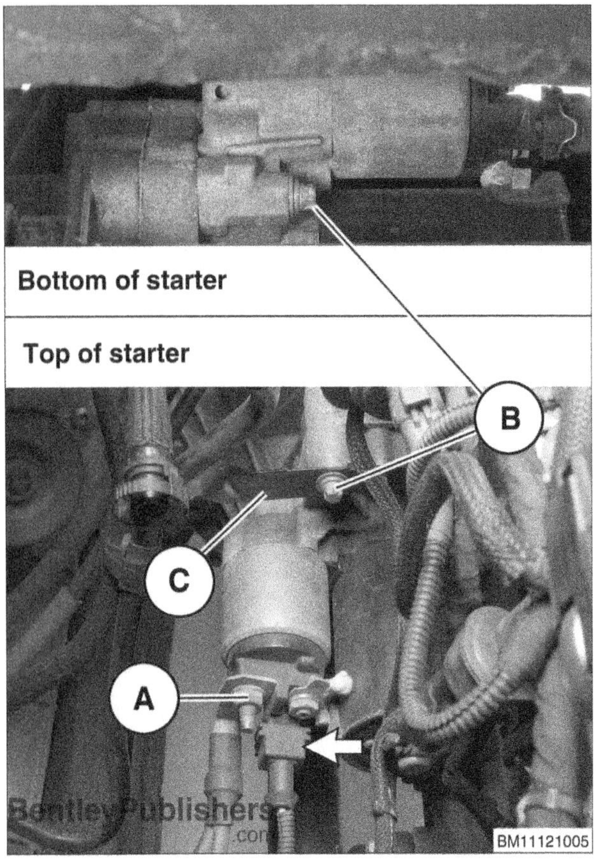

Working at starter solenoid electrical connections:

- Remove fastener (**A**) and detach B+ cable from solenoid.
- Press in wire clip to release solenoid electrical connector (**arrow**).
- Remove starter fasteners (**B**).
- Remove holder (**C**) and remove starter motor from engine.

– Before installing new starter, rotate engine and inspect flywheel or torque plate gear teeth through starter hole. Replace if damaged. See **210 Clutch** or **240 Automatic Transmission**.

– Rotate starter toward front of vehicle to install. Reattach electrical connectors.

Tightening torques	
B+ cable to starter (M8)	14 Nm (10.3 ft-lb)
Starter to bell housing	20 Nm (15 ft-lb)

ALTERNATOR AND CHARGING SYSTEM

Alternator basics

The alternator is bolted to the front of the engine block. It is driven by the engine accessory belt. A belt idler pulley is bolted to the bottom of the alternator housing. Accessory belt replacement is covered in **020 Maintenance**.

Several versions of alternators are used in MINI vehicles. Be sure that the replacement alternator is correct for the application.

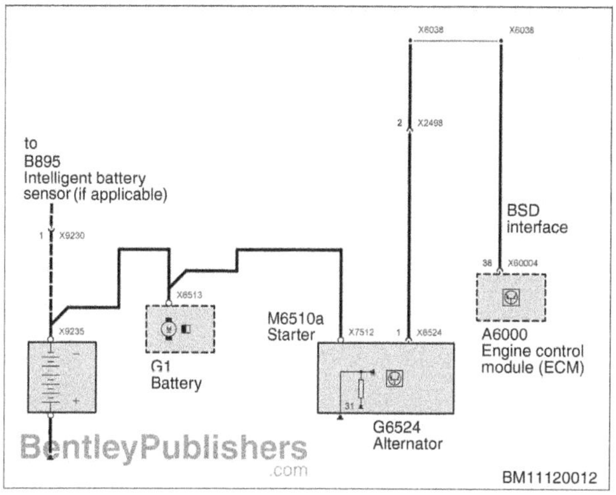

Alternator interface

The alternator communicates with the engine control module (ECM) via bit-serial data (BSD) interface. This single wire connection allows the ECM to adapt its calculations and control data to alternator output. The ECM controls the following functions:

- Alternator activation and deactivation.
- Informing the voltage regulator of the nominal voltage value to be set.
- Alternator load response control.
- BSD line diagnosis.
- Alternator fault code storage.
- Charge indicator light activation in instrument cluster.

Charging system troubleshooting

Some charging system tests require special test equipment. If the test equipment is not available, charging system fault diagnosis can be performed by an authorized MINI dealer or other qualified repair shop. See **Table a** for general electrical component troubleshooting.

Before checking the alternator, make sure the battery is fully charged and capable of holding a charge. Check that the battery terminals are clean and tight and the engine accessory drive belt is properly tensioned and not severely worn.

Charging system quick-check

– Use a digital multimeter to measure voltage across the battery terminals with ignition OFF and then again with engine running. Battery voltage should be about 12.6 volts with key OFF and between 13.5 and 14.5 volts with engine running.

– Run engine at about 2000 rpm and switch ON electrical loads (fans, lights, rear window defroster, and wipers). With all accessories ON, battery voltage should be above 12.6 volts.

The regulated voltage (engine running, battery charged, accessories and lights OFF) is usually between 13.2 and 14.5, depending on temperature and operating conditions. If the voltage is higher than 14.8, the voltage regulator is most likely faulty.

Charging system, checking

CAUTION—
- *Do not disconnect the battery while the engine is running. Damage to the alternator or engine electronic systems may result.*
- *Only use a digital multimeter when testing charging system components.*

– Switch ignition ON. Check that dashboard warning lights comes ON. If warning lights do not come ON, check for fault codes.

- Check for battery voltage at alternator terminal 30 (B+) (**arrow**):
  - Connect voltmeter between terminal 30 and ground and check for battery voltage. If voltage is not present, check wiring for faults.
- Connect MINI scan tool or oscilloscope to check alternator function.

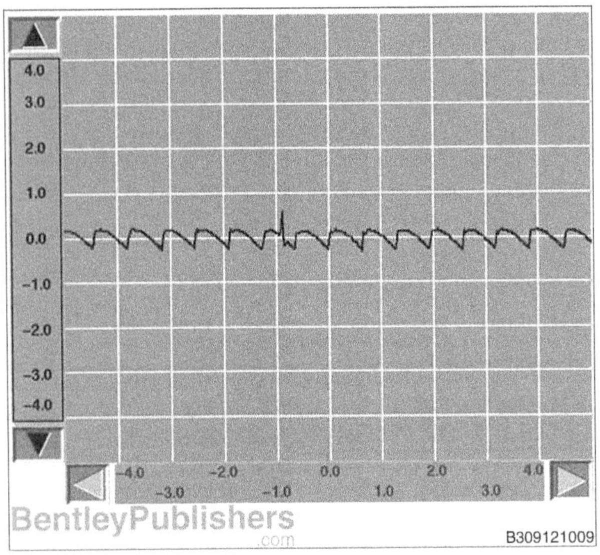

Oscilloscope test patterns

- Normal alternator pattern.

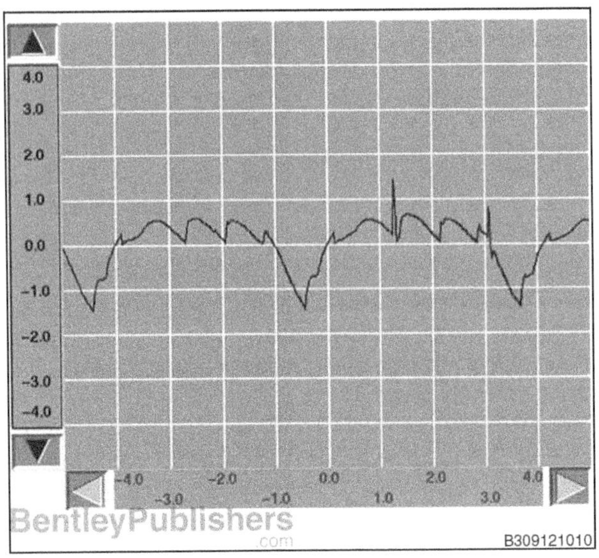

- Alternator with defective diode.

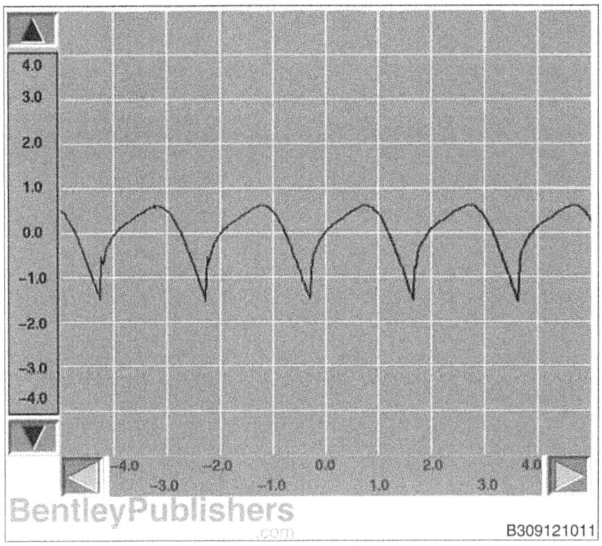

Alternator with broken winding.

- If test indicate a defective alternator, replace unit. Repair parts are not available for the alternator.

Alternator, removing and installing

- Switch ignition OFF. Disconnect negative (–) battery cable and cover battery terminal to keep cable from accidentally contacting terminal.

> *CAUTION—*
> - *Prior to disconnecting the battery, read the battery disconnection cautions in* **001 Warnings and Cautions**.

- Raise vehicle and support safely.

> *WARNING—*
> - *Make sure the vehicle is stable and well supported at all times. Use a professional automotive lift or jack stands designed for the purpose. A floor jack is not adequate support.*

- Place radiator support in service position. See **410 Engine, Fenders, Hood**.

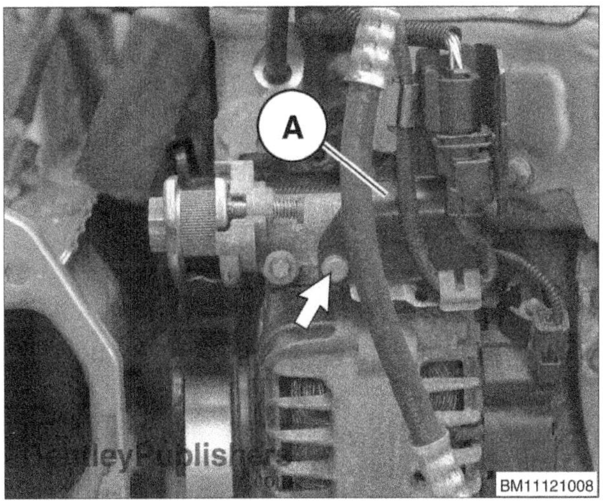

N14, N18 engine: Remove wiring harness bracket fasteners (**arrows**) and move bracket (**A**) aside.

Using a wrench, bring accessory belt tensioner in to service position (**curved arrow**). Secure position of tensioner using locking pin (**straight arrow**).

- Remove engine accessory belt. See **020 Maintenance**. Mark direction of rotation on belt before removing.

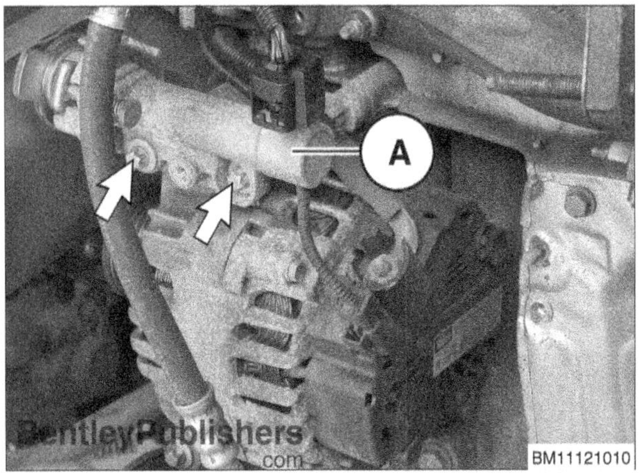

◄ Remove tensioner fasteners (**arrows**) and remove tensioner (**A**) from engine.

◄ Disconnect electrical connector (**arrow**). Remove nut (**A**) and remove positive cable (**B**) from alternator.

◄ Remove lower alternator fastener (**arrow**).

> ***CAUTION—***
> • *Be careful not to bend A/C line.*

– Remove alternator from vehicle.

– Installation is reverse of removal.

Tightening torques	
Alternator cable to alternator	13.5 Nm (10 ft-lb)
Drive belt tensioner to alternator	20 Nm (14.7 ft-lb)
Alternator to engine block (M8)	20 Nm (14.7 ft-lb)
Friction gear to engine block	8 Nm (6 ft-lb)

– Reassemble vehicle. Remember to:

- Reinstall accessory belt using previously made direction-of-rotation marks.
- Make sure accessory belt grooves engage pulleys correctly.
- After reconnecting battery, see **Battery reconnection notes** in this repair group.

130 Fuel Injection

TABLE

General

This repair group covers service and repair of engine management systems. Information in this repair group is organized by engine.

See also:

- **020 Maintenance** for air filter, fuel filter and spark plug replacement
- **100 Engine–General** for engine identification
- **120 Ignition System** for ignition coil, camshaft sensor, crankshaft sensor and knock sensor service
- **160 Fuel Tank and Fuel Pump** for fuel pump and fuel pump relay service, fuel pressure tests and fuel pressure regulator replacement
- **180 Exhaust System** for oxygen sensor service
- **ECL Electrical Component Locations**
- **ELE Electrical Wiring Diagrams**
- **OBD On-Board Diagnostics**

Engine identification table

Table a. Engine identification

Year, model	Engine designation
2007 - 2010 Cooper, Clubman, Convertible	N12 (non-turbo)
2007 - 2010 Cooper S, Clubman S, Convertible S	N14 (turbo)
2011 - 2013 Cooper, Clubman, Convertible	N16 non-turbo)
2011 - 2013 Cooper S, Clubman S, Convertible S	N18 (turbo)

Warnings and Cautions

WARNING—

- *The fuel system is designed to retain pressure even when the ignition is OFF. When working with the fuel system, loosen the fuel lines slowly to allow residual fuel pressure to dissipate. Avoid spraying fuel. Use shop towels to capture leaking fuel.*
- *Before beginning work on the fuel system, place a fire extinguisher in the vicinity of the work area.*
- *Work only on the fuel system when engine temperature is below 40°C (104°F).*
- *Fuel is highly flammable. When working around fuel, do not disconnect wires that could cause electrical sparks. Do not smoke or work near heaters or other fire hazards.*
- *Wear eye protection and protective clothing to avoid injuries from contact with fuel.*
- *When working on an open fuel system, wear suitable hand protection, as prolonged contact with fuel can cause illnesses and skin disorders.*
- *Unscrew the fuel tank cap to release pressure in the tank before working on fuel lines.*

> ***WARNING —***
> - *Do not use a work light with an incandescent bulb near fuel. Fuel may spray on the hot bulb causing a fire.*
> - *Make sure the work area is properly ventilated.*
> - *Due to risk of personal injury, be sure the engine is cold before beginning work on engine components.*
> - *The ignition system produces high voltages that can be fatal. Avoid contact with exposed terminals. Use extreme caution when working on a car with ignition switched ON or engine running.*

> ***CAUTION—***
> - *Renew fuel system hoses, clamps and O-rings any time they are removed.*
> - *Prior to disconnecting the battery, read the battery disconnection cautions in* **001 Warnings and Cautions**.
> - *Connect and disconnect the DME system wiring and test equipment leads only when the ignition is switched OFF.*
> - *Wait at least 1 minute after switching the ignition OFF before removing the engine control module (ECM) connector. If the connector is removed before this time, residual power in the system relay may damage the control module.*
> - *Tests or repair procedures in this section may set fault codes (DTCs) in the ECM and illuminate the malfunction indicator light (MIL). After repairs are completed, clear DTC memory using a MINI scan tool or equivalent.*
> - *Fuel system cleaners and other chemical additives other than those specifically recommended by MINI may damage catalytic converters, oxygen sensors or other fuel supply components.*
> - *Do not connect any test equipment that delivers a 12-volt power supply to terminal 15 (+) of the ignition coils. The current flow may damage the ECM. Connect test equipment only as specified by MINI or the equipment maker.*
> - *Relay positions can vary. Confirm relay location and function by identifying wiring in socket using wiring diagrams. See* **ELE Electrical Wiring Diagrams**.
> - *Use a digital multimeter for electrical tests. Use an LED test light for quick tests.*

Driveability Troubleshooting

The self-diagnostic DME engine management systems monitor and store diagnostic trouble codes (DTCs). If the malfunction indicator light (MIL) illuminates, it indicates that an emissions-related fault has occurred and that one or more DTCs are stored in the engine control module (ECM).

If faults arise, or if the MIL is illuminated, begin troubleshooting by connecting MINI scan tool to the OBD II plug under the dashboard (**arrow**).

The capabilities of OBD II software has the potential to save diagnostic time and avoid incorrect component replacement. See **On-board diagnostics** in this repair group.

Basic engine settings

Idle speed, idle mixture (%CO), and ignition timing are not adjustable. The DME system is adaptive and automatically compensates for changes in the engine due to age, minor wear or small problems such as a disconnected vacuum hose. However, the adaptive range is limited. Once the limits are exceeded, driveability problems become noticeable.

Poor initial driveability may be encountered when the battery is disconnected and reconnected. When the battery is disconnected, ECM adaptive memory may be reset. The system readapts after about ten minutes of driving.

System voltage

Digital motor electronics (DME) requires that the system (battery) voltage be maintained within a narrow range of DC voltage. DC voltage levels beyond or below the operating range, or any AC voltage in the electrical system can cause havoc.

– When troubleshooting an illuminated MIL, make sure the battery is fully charged and capable of delivering all its power to the electrical system. An undercharged battery can amplify AC alternator output ripple.

– To make a quick check of battery charge, measure voltage across battery terminals with all cables attached and ignition OFF. A fully charged battery measures 12.6 volts or slightly more, compared to 12.15 volts for a battery with a 25% charge.

The DME system operates at low voltage and current levels, making it sensitive to small increases in resistance. The electrical system is routinely subject to corrosion, vibration and wear, so faults or corrosion in the wiring harness and connectors are not uncommon.

– Check battery terminals for corrosion or loose cable connections. See **121 Battery, Starter, Alternator**.

– If a battery cable connection has no visible faults but is still suspect, measure voltage drop across the connection. A large drop indicates excessive resistance, meaning that the connection is corroded, dirty or damaged. Clean or repair and retest. See **600 Electrical System–General** for voltage drop test procedure.

– Visually inspect wiring, connectors, switches and fuses. Loose or damaged connectors can cause intermittent problems, especially small terminals in ECM connectors. Disconnect wiring harness connectors to check for corrosion, and use electrical cleaning spray to remove contaminants.

Main grounds

Good grounds are critical to proper DME operation. If a ground connection has no visible faults but is still suspect, measure the voltage drop across the connection. A large voltage drop means high resistance. Clean or repair the connection and retest.

For voltage drop testing, see **600 Electrical System-General**.

For ground locations, see **ECL Electrical Component Locations**.

On-board diagnostics

Second generation on-board diagnostics (OBD II) software and hardware is incorporated in the engine management systems. The OBD II system monitors components that influence exhaust and evaporative emissions. If a problem is detected, the OBD II system stores the associated diagnostic trouble code (DTC) and condition.

If vehicle emissions levels exceed 1.5 times Federally mandated criteria, the OBD II system illuminates the malfunction indicator light (MIL) in the instrument cluster.

Scan tool connected to OBD II plug (**arrow**) below left dashboard.

Professional diagnostic scan tools available at the time of this printing include factory MINI equipment, Autologic Diagnostics and a small number of aftermarket MINI-specific tools.

In addition to professional scan tools, there are inexpensive generic OBD II scan tool software programs and handheld units available. Although these have limited capabilities as compared to the dedicated tools, they are powerful diagnostic tools.

NOTE—
- *Fault memory (diagnostic trouble codes) can only be erased and reset using a scan tool. Removing the connector from the ECM or disconnecting the battery does not erase DTC memory.*

ENGINE MANAGEMENT POWER SUPPLY

Power supply fuses

See **ECL Electrical Component Locations** for fuse panel access information.

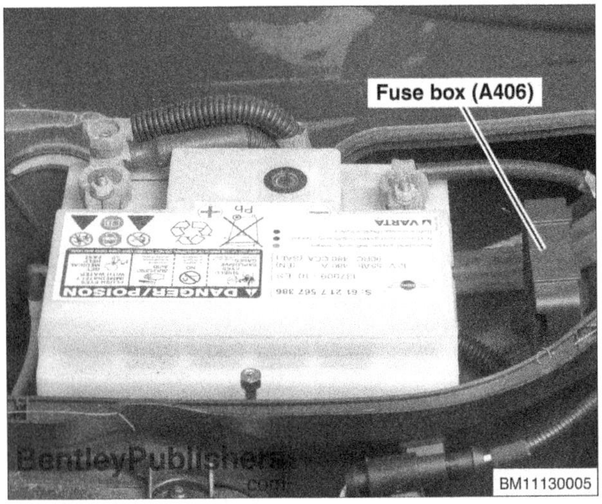

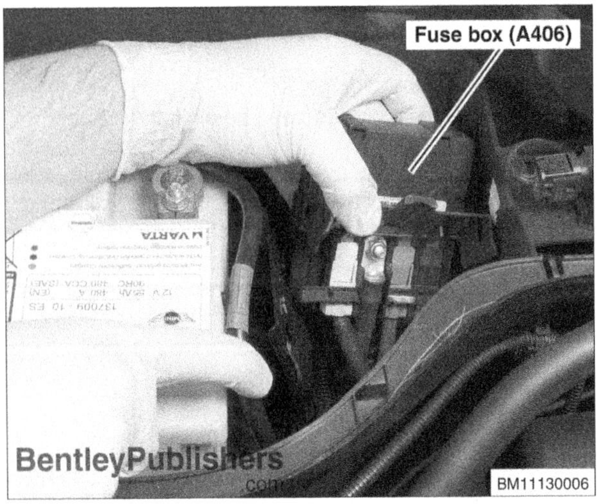

High amperage power supply fuse panel (**A406**), attached to side of vehicle battery.

- **F67** 150A: Power supply to fuses located in A400a (not replaceable separately).
- **F68** 40A: Power supply to Valvetronic relay (if equipped) and fuel injector relay (if equipped).

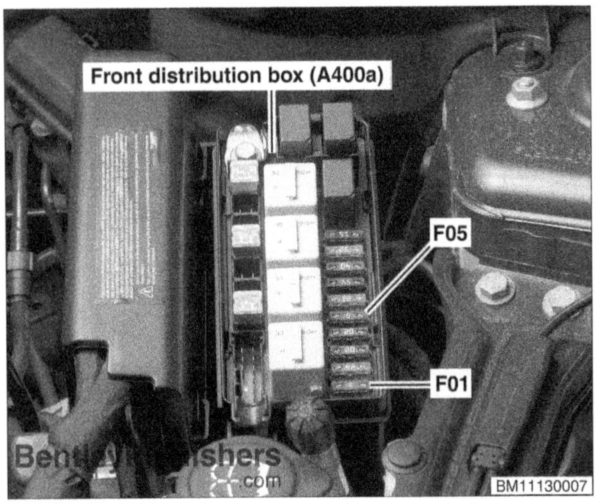

Front power distribution box (**A400a**), left side of engine compartment:

- **F01** 5A: Power supply to engine control module (ECM).
- **F02** 25A: Power supply to engine control module (ECM).
- **F03** 20A: Power supply to engine control module (ECM).
- **F04** 20A: Power supply to engine control module (ECM).
- **F05** 5A: Power supply to diagnosis module for fuel tank vapor leakage (DMTL) (Valvetronic relay (if equipped) and fuel injector relay (if equipped), Valvetronic relay (if equipped) and fuel injector relay (if equipped).

DME main relay

The DME main relay is energized via the ECM and supplies battery positive (B+) power to the following:

- ECM
- Fuel injectors
- Fuel tank leak detection pump (DMTL)
- Valvetronic relay

If the DME main relay is faulty, the engine does not start.

◄ DME main relay (**K6300**) located in left side engine compartment in front power distribution box (**A400a**).

NOTE—

- *Relay locations vary. Confirm relay identification by matching wiring colors and terminal numbers. See* ***ELE Electrical Wiring Diagrams****.*

DME power supply circuit

DME main relay

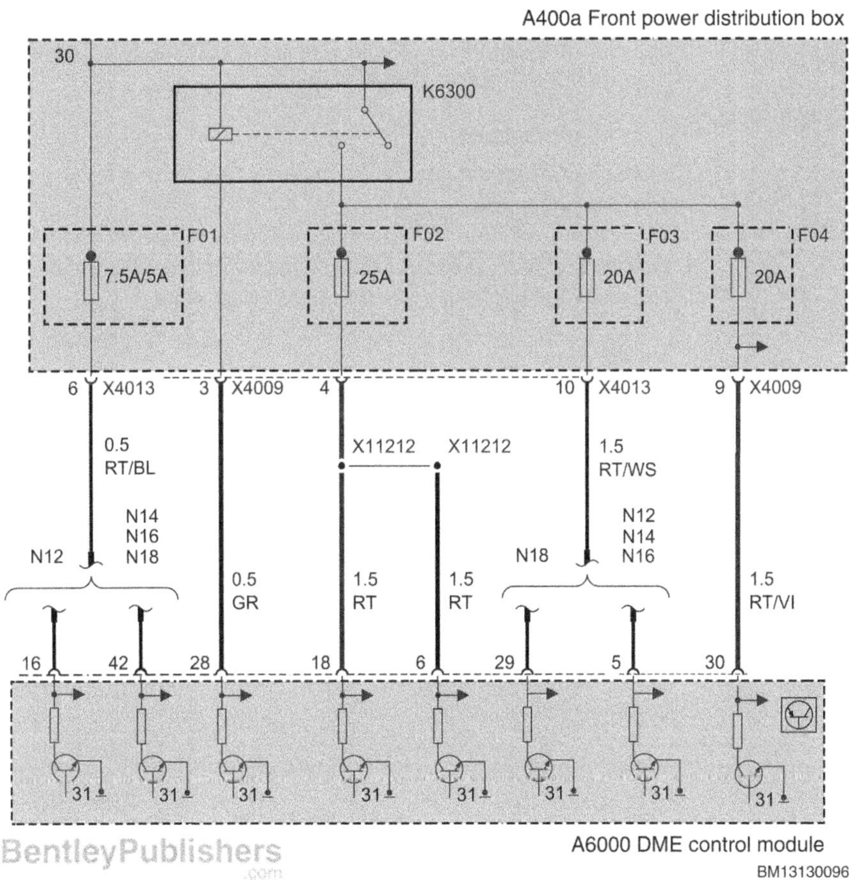

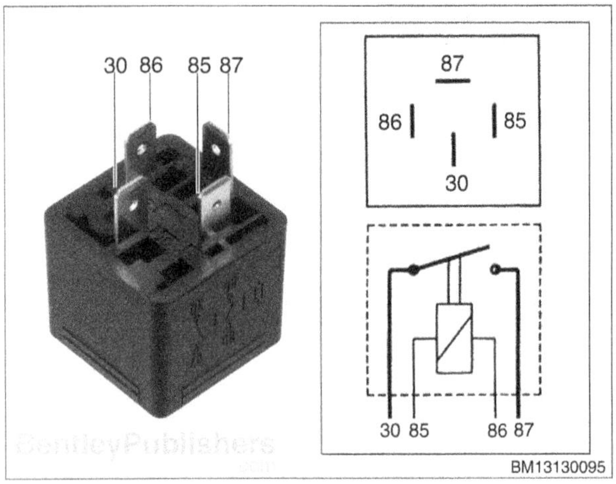

DME main relay testing

With ignition off, remove DME main relay (**K6300**) in left side engine compartment front power distribution box (**A400a**).

NOTE—

- *Relay locations vary. Confirm relay identification by matching wiring colors and terminal numbers. See* **ELE Electrical Wiring Diagrams**.

– Check for voltage at relay socket **30**.

– If battery voltage is present, continue testing.

– If battery voltage is not present, check the following:
  - Large red wire in relay socket.
  - Fuse **F67** in high amperage fuse holder (**A406**) next to battery.
  - See **ELE Electrical Wiring Diagrams** for more details.

– Check for ground at relay socket **85** with ignition ON.

– If ground is present, continue testing.

– If ground is not present, signal from ECM is missing. Check wire between relay and ECM 28-pin connector X60004. See **Table b** (non-turbo), or **Table c** (turbo) in this repair group.

– If no faults are found:
  - Check ECM grounding.
  - ECM may be defective.

– Reinstall relay. With ignition ON, check for battery voltage at relay socket **87** (check from underside of relay socket).

NOTE—

- *Some relays may have an additional terminal 87 (labeled terminal 87a).*

– If battery voltage is present, relay circuit is function correctly.

– If battery voltage is not present and all earlier tests are OK, relay is faulty. Replace.

– If no faults are found during relay testing but power is not reaching ECM or other components, check fuses **F01** - **F05** in front power distribution box (**A400a**). See **Power supply fuses** in this repair group.

– When finished testing, check for fault codes and reset ECM memory.

SIEMENS DME (NON-TURBO)

DME engine management system description

See **Table a** for help identifying your engine.

The DME manages and monitors the following functions:

Air

- Electronic throttle
- Mass air flow measurement
- Intake manifold temperature and differential pressure sensor
- Valvetronic II (N12 engine)
- Valvetronic III (N16 engine)

Fuel

- Fuel supply
- Fuel injection

Ignition

- Direct ignition
- Knock control
- Primary and secondary ignition monitoring

Emissions

- OBD II compliance
- Pre and post-catalyst oxygen sensors
- Electrically heated DME-map-controlled thermostat
- Electrically DME-map-controlled oil pump (N16 engine)
- Misfire detection
- Evaporative emission control and leak detection
- Malfunction indicator light (MIL)

Performance controls

- Double VANOS control
- Output of injection signal (TI) for fuel economy gauge
- Output of engine rpm (TD) for tachometer
- A/C compressor control
- Electric radiator cooling fan
- CAN-bus communication
- Dynamic stability control (DSC) interface
- Car access system (CAS)
- Cruise control
- ECM programming

Engine control module (ECM)

The engine control module (ECM) (**arrow**) is mounted next to the front distribution box (A400a) at the left side of the engine compartment.

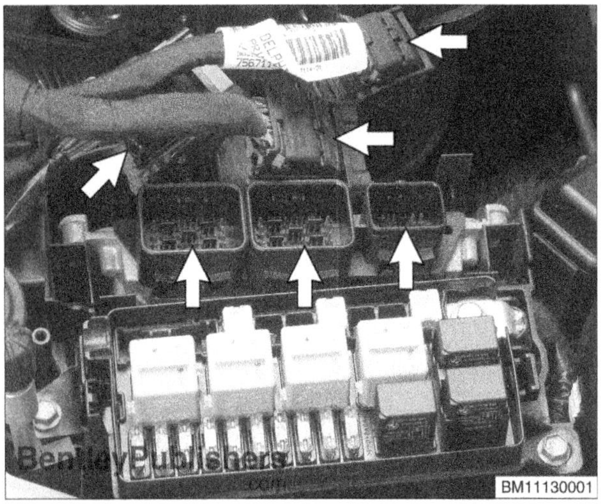

The ECM is flash-programmable and features 3 main electrical harness connectors (**arrows**), with a total of 138 pins.

The EEPROM (chip) in the ECM is coded to the vehicle. The ECM from another vehicle cannot be swapped for testing purposes.

Fuel metering

The ECM meters pressurized fuel by changing the opening time (pulse width) of the fuel injectors. The exact amount of fuel injected is determined by the amount of time the injectors are open. To ensure that injector pulse width is the only factor that determines fuel metering, fuel pump pressure is maintained by a pressure regulator. The injectors are mounted to a common fuel rail.

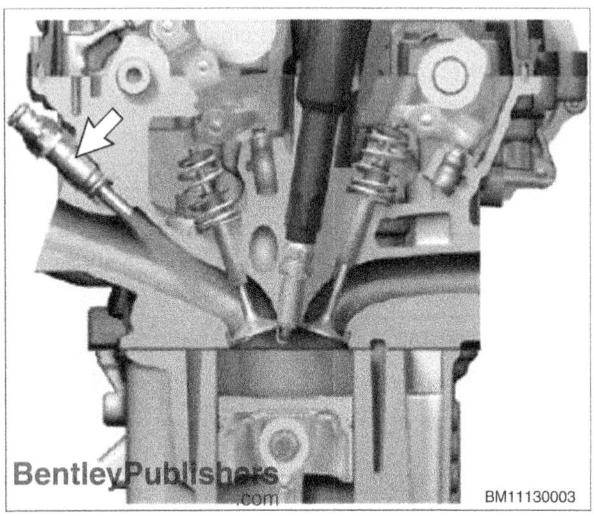

The fuel injectors are mounted into a machined bore in the cylinder head. This design allows the fuel injectors to be closely mounted to the intake valves.

The ECM monitors engine operating conditions to determine injector opening duration. Each injector can be individually controlled for cylinder selective fuel trim.

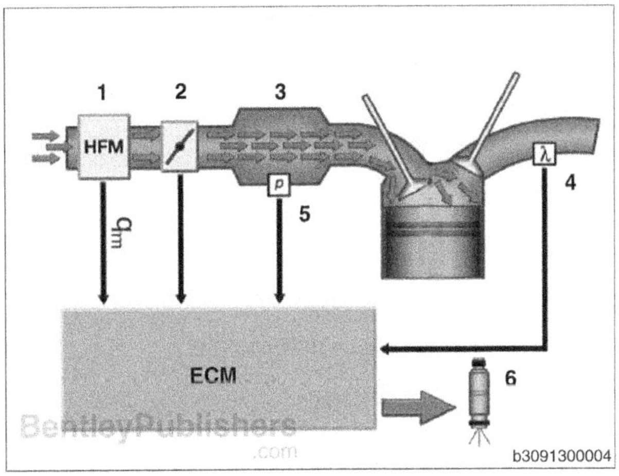

Air intake

Air entering the engine passes through a pleated paper element in the air filter housing. Intake air mass is then measured by a mass air flow sensor. A reference current is used to heat a thin film in the sensor when the engine is running. The current needed to hold the temperature of the film constant is used to calculate the mass of the intake air.

1. Air mass measurement (HFM)
2. Throttle valve
3. Intake manifold
4. Variable intake valve lift (Valvetronic)
5. Residual oxygen measurement in exhaust
6. Intake manifold vacuum
7. Engine speed
8. Injection timing

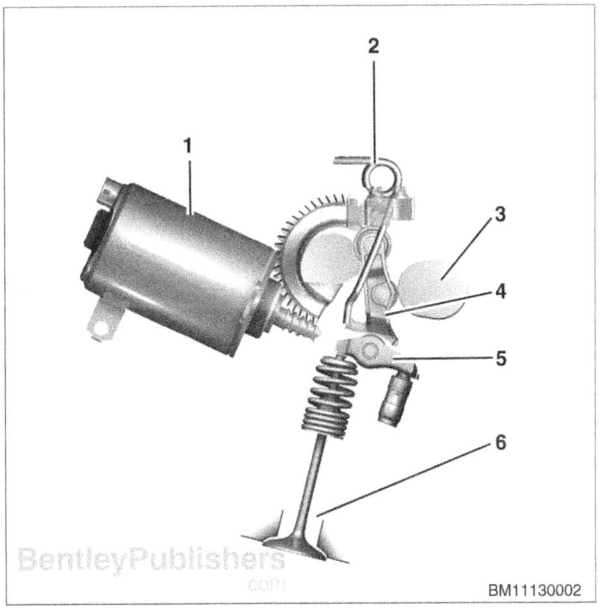

Idle speed control

The ECM controls Idle speed by varying intake valve lift via the Valvetronic system.

1. Valvetronic motor
2. Return spring
3. Intake camshaft
4. Intermediate lever
5. Intake rocker
6. Intake valve

Idle speed is not adjustable. The basic functions and parameters of idle speed control are as follows:

- Control of cold air intake volume.
- Smooth idle speeds regardless of load and inputs.
- Smooth transition from acceleration to deceleration.

Idle speed stabilization is active during the following conditions:

- Engine warm up
- Heating or A/C activation
- Drive gear selected (automatic transmission)
- Varying cooling fan speeds

Throttle control

Engine load control is not controlled by the throttle module (EDK) but is instead carried out by the Valvetronic function of the ECM. However, the throttle may be slightly closed to allow sufficient manifold vacuum for the crankcase ventilation and canister purge systems.

Knock (detonation) control

The knock sensor monitors and controls ignition knock through the ECM. The knock sensor, functioning like a microphone, converts mechanical vibration (knock) into electrical signals. The ECM is programmed to react to frequencies that are characteristic of engine knock and adapt the ignition timing point accordingly.

Knock sensor replacement is covered in **120 Ignition System**.

Air filter housing and ducts, removing and installing (non-turbo)

- Switch OFF ignition and remove remote key from key slot.

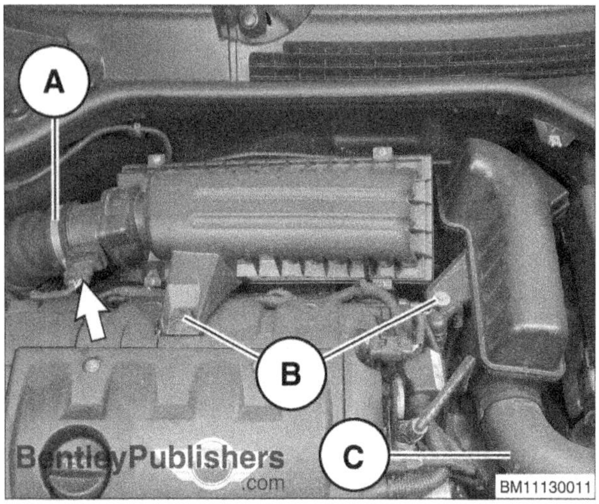

- Loosen hose clamp (**A**) and disconnect mass air flow sensor electrical connector (**arrow**).
- Remove intake silencer housing fasteners (**A**).
- Pull intake port (**C**) off intake silencer housing (**A**) and remove intake silencer housing.

- Loosen clamp (**arrow**) and detach air intake hose (**A**) from throttle housing.

> ***CAUTION—***
> • *Throttle housing in constructed of plastic. Be careful not to damage when disconnecting hose.*

- Installation is reverse of removal.

Tightening torques	
Intake filter housing lid fasteners	2.5 Nm (22.1 in-lb)
Intake filter housing to cylinder head cover	8 Nm (5.9 ft-lb)

Mass air flow sensor, replacing (non-turbo)

A faulty air flow sensor can produce the following problems:

- Difficult to restart when engine is hot.
- Engine starts then stalls.
- Engine starts and runs only with accelerator pedal depressed.

– Switch ignition OFF and remove key.

◄ Working at left front of engine compartment at air filter housing, disconnect mass air flow sensor electrical connector (**arrow**) and loosen intake air duct clamp (**A**).

– Remove intake air duct from mass air flow sensor.

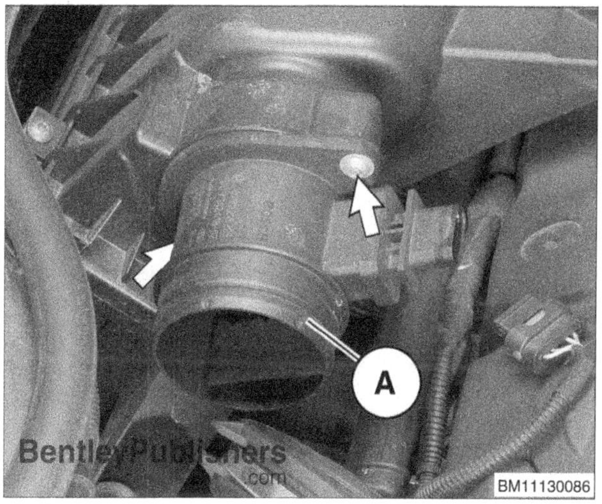

◄ Remove fasteners (**arrows**) then remove mass air flow sensor (**A**).

– Check intake ducts for cracks and vacuum leaks.

◄ Installation is reverse of removal.

- Inspect O-ring (**arrow**) and replace if necessary.

– After reinstalling, check for fault codes and reset ECM memory.

Tightening torques	
Mass air flow sensor to housing	3 Nm (26 in-lb)

Intake manifold differential pressure sensor, removing and installing (non-turbo)

– Switch ignition OFF and remove key.

– Remove air filter housing and ducts. See **Air filter housing and ducts, removing and installing (non-turbo)** in this repair group.

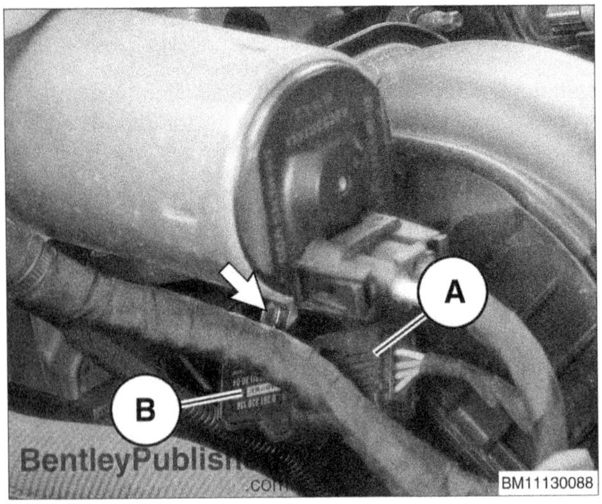

BM11130088

< Working at back of intake manifold near Vavletronic motor:

- Detach sensor electrical connector (**A**).
- Remove fastener (**arrows**).
- Remove sensor (**B**).

BM11130080

< Installation is reverse of removal. Inspect O-ring (**arrow**) and replace if necessary.

– After reinstalling, check for fault codes and reset ECM memory.

Tightening torques	
Sensor to intake manifold	5 Nm (44.2 in-lb)

Throttle housing, replacing (non-turbo)

The throttle housing (EDK) is not used for engine load control. Engine load control is carried out by the Valvetronic function of the ECM. However, the throttle may be slightly closed to allow sufficient manifold vacuum for the crankcase ventilation and canister purge systems. There is no mechanical (cable) link between the accelerator pedal and the throttle plate.

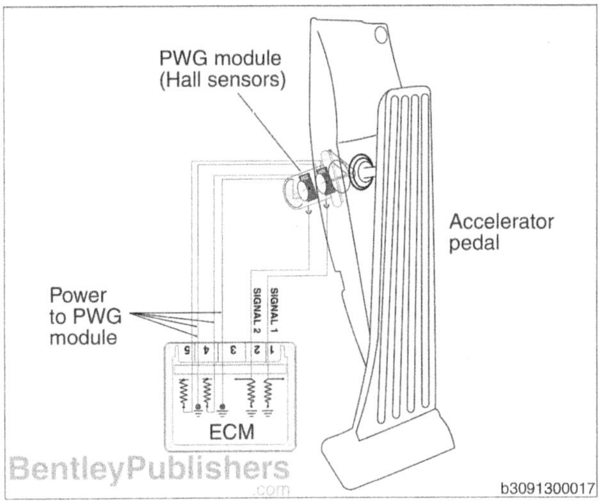

The accelerator pedal module (PWG) at the pedal assembly communicates pedal position directly to the ECM. The module provides two variable voltage signals (via two Hall sensors) to the ECM for pedal position and rate of movement.

The ECM provides an independent 5 v power supply and ground for each Hall sensor. As the accelerator pedal is moved from rest to full throttle, the sensors produce a variable voltage signal.

The output of the Hall sensors is checked for plausibility. The voltage range of Hall sensor 1 is approximately 0.5 to 4.5 v. Hall sensor 2 ranges from approximately 0.5 to 2.5 v.

The throttle housing unit is nonadjustable. If found to be faulty, replace as a complete unit.

- Switch ignition OFF and remove key.

- Remove air filter housing and air intake ducts. See **Air filter housing and ducts, removing and installing (non-turbo)** in this repair group.

Working at left side of intake manifold:

- Remove mounting fasteners (**arrows**).
- Tilt throttle housing up far enough to detach electrical connector.
- Remove throttle housing from vehicle.

- Installation is reverse of removal. Replace profile gasket (sealing O-ring) between throttle assembly and intake manifold.

Tightening torques	
Throttle housing to used intake manifold	7 Nm (61 in-lb)
Throttle housing to new intake manifold	10 Nm (88 in-lb)

- After reinstalling, use MINI scan tool to check for fault codes and reset ECM memory. Reset throttle plate adaptation values following on-screen directions. If the adaptation process is not completed correctly, the engine does not start.

Valvetronic motor, removing and installing (non-turbo)

– Switch ignition OFF and remove key.

– Remove ignition coil cover, if equipped. See **020 Maintenance**.

– Disconnect battery negative (-) cable.

> *CAUTION—*
> • *Prior to disconnecting the battery, read the battery disconnection cautions in* **001 Warnings and Cautions**.

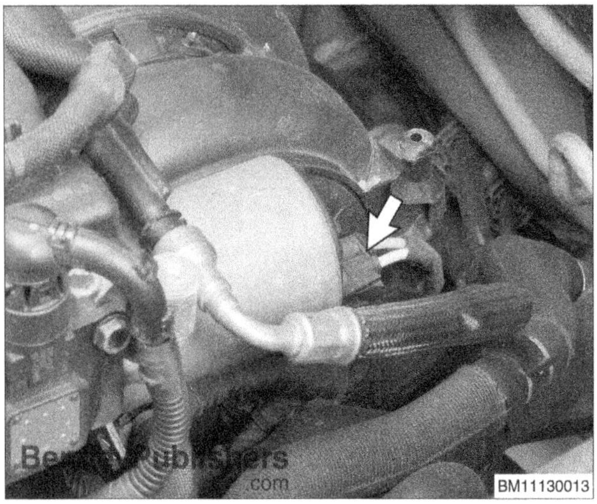

◄ Disconnect Valvetronic motor electrical connector (**A**).

– Detach wiring harness at retaining clip on Valvetronic motor.

◄ Remove Valvetronic motor fasteners (**arrows**).

BM11130015

Use Allen wrench to rotate Valvetronic motor worm shaft drive counterclockwise (**curved arrow**) to slide motor out of cylinder head.

> *CAUTION—*
> * *Do not rotate shaft to maximum to end stop.*

- Installation is reverse of removal.
- After reinstalling, check for fault codes and reset ECM memory.

Tightening torque	
Valvetronic motor to cylinder head	8.5 Nm (75.2 in-lb)

Fuel rail and injectors, removing and installing (non-turbo)

Fuel injectors are electrically controlled solenoid valves that provide precisely metered and atomized fuel into the engine intake ports.

Injectors receive voltage from the DME main relay. The engine control module (ECM) controls injector opening by activating the ground circuit for the injector valve solenoids. The ECM varies the duration (in milliseconds) of injector opening to regulate air / fuel ratio.

- Switch ignition OFF and remove key.
- Remove air filter housing. See **Air filter housing and ducts, removing and installing (non-turbo)** in this repair group.
- Disconnect battery negative (-) cable.

> *CAUTION—*
> * *Prior to disconnecting the battery, read the battery disconnection cautions in* **001 Warnings and Cautions**.

BM11130016

Unscrew schræder valve cap (**arrow**) from fuel rail. Connect air line adapter to fuel rail fitting.

* Unscrew fuel tank cap to release pressure.
* Using a brief burst of compressed air (maximum of 3 bar or 43.5 psi), blow fuel back into fuel tank.

> *WARNING—*
> * *Fuel in the fuel line is under pressure (approx. 3 - 5 bar or 45 - 75 psi) and may be expelled forcibly. Do not smoke or work near heaters or other fire hazards. Keep a fire extinguisher handy.*
> * *Unscrew the fuel tank cap to release pressure in the tank before working on the fuel line.*
> * *Plug open fuel lines and fittings.*

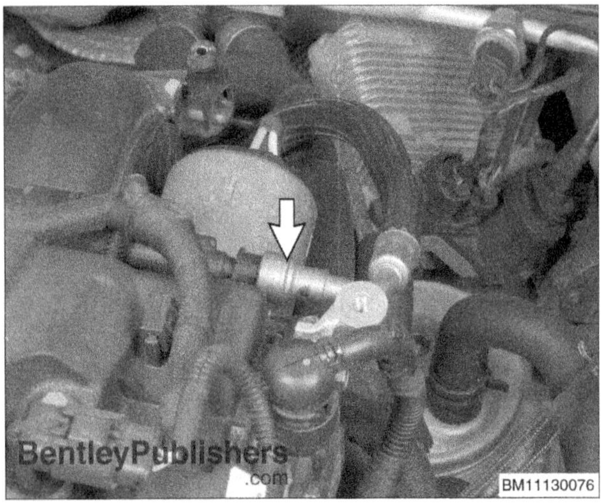

< Working at left rear of engine, separate fuel line connector (**arrow**).

> ***WARNING —***
> * *Wrap a clean shop towel around fitting before disconnecting. Residual fuel pressure is present in the fuel line.*

– Seal fuel line and fuel rail opening.

< Remove fuel rail fasteners (**arrows**).

< Raise fuel rail from engine slightly, then disconnect fuel injector electrical connectors (**arrows**).

Remove individual injectors:

- Pry retaining clip from injector (**A**).
- Pull injector from rail (**arrow**).

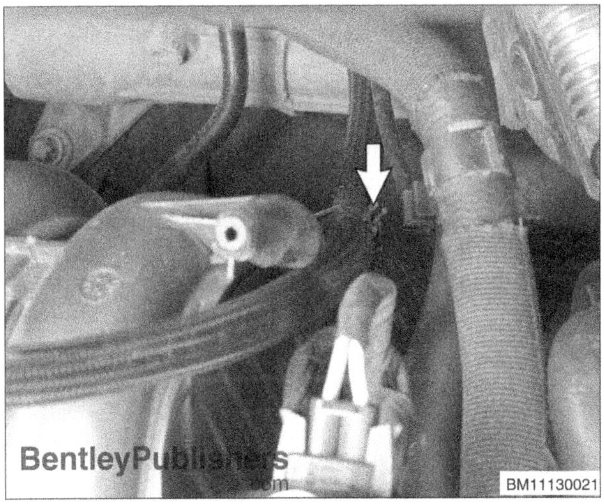

Inspect fuel supply line (**arrow**) for damage. If damaged, install new fuel line with anti-chafing protection. Also install a double C clip to ensure that clearance from other hoses is provided.

– Installation is reverse of removal. Remember to:

- Fit new sealing O-rings when installing injectors. For ease of installation, lightly lubricate O-rings with assembly lubricant.
- Check that injector electrical connections are correctly fitted and that injectors are fully seated prior to installing fuel rail mounting bolts.
- Replace wire ties.

Tightening torque	
Fuel rail to cylinder head	8 Nm (70.8 in-lb)

– After reinstalling, check for fault codes and reset ECM memory.

Intake manifold, removing and installing (non-turbo)

– Switch ignition OFF and remove key.

– Remove air filter housing. See **Air filter housing and ducts, removing and installing (non-turbo)** in this repair group.

– Disconnect battery negative (-) cable.

> ***CAUTION—***
> - *Prior to disconnecting the battery, read the battery disconnection cautions in* **001 Warnings and Cautions**.

Working above throttle housing, disconnect VANOS solenoid electrical connector (**arrow**) then detach harness from throttle housing (**A**).

Disconnect throttle housing electrical connector (**arrow**).

- Raise vehicle and support safely.

> ***WARNING—***
> - *Make sure the vehicle is stable and well supported at all times. Use a professional automotive lift or jack stands designed for the purpose. A floor jack is not adequate support.*

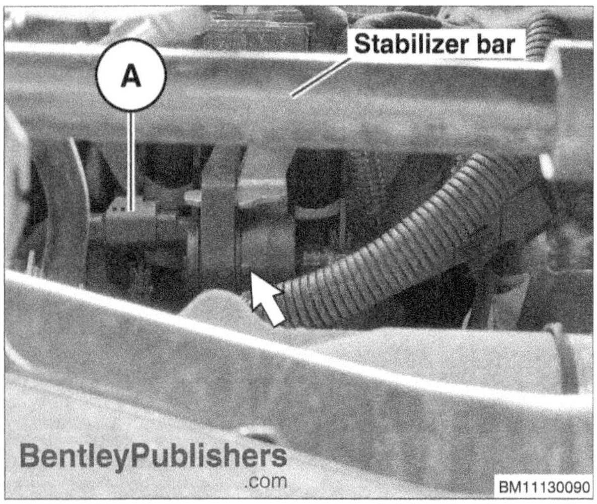

Working underneath vehicle, disconnect fuel tank vent valve electrical connector (**A**).

- Remove fuel tank vent valve (**arrow**) from engine.
- Disconnect engine breather lines and secure aside.

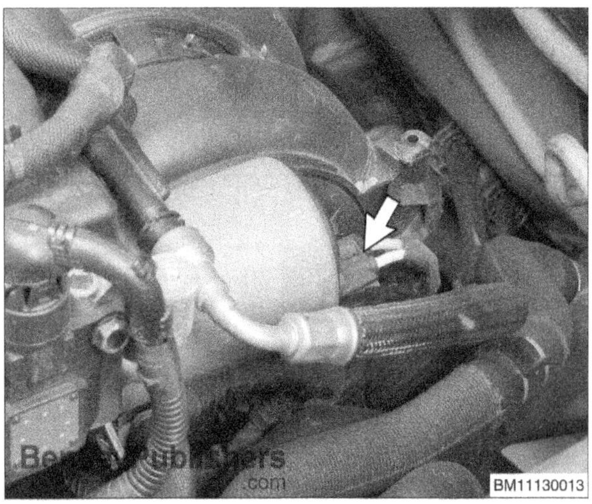

Disconnect Valvetronic motor electrical connector (**arrow**).

- Detach wiring harness at retaining clip on Valvetronic motor.

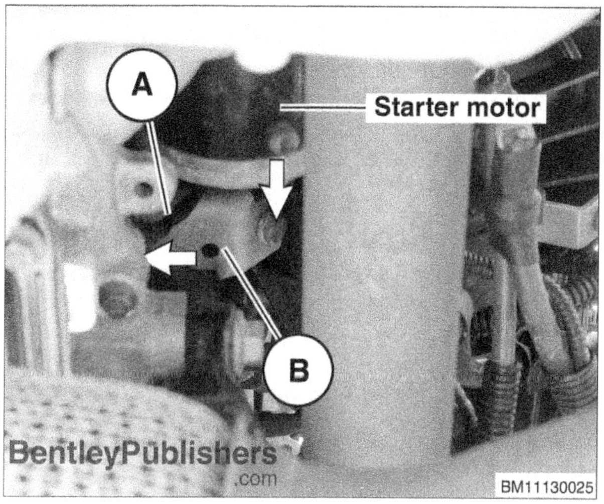

◀ Working beneath vehicle, detach wiring harness at retaining clip (**A**)

– Remove fasteners (**arrows**) and remove bracket (**B**).

◀ Remove remaining intake manifold fasteners (**arrows**), then remove intake manifold.

– Plug cylinder head intake ports to prevent debris from entering engine.

– Installation is reverse of removal. Remember to replace all seals.

Tightening torque	
Intake manifold to cylinder head (8 x 125 mm bolt)	15 Nm (11 ft-lb)
Intake manifold to cylinder head (hexagon nut)	20 Nm (14.7 ft-lb)
Intake manifold support bracket to crankcase	20 Nm (14.7 ft-lb)

Crankcase breather valve (non-turbo)

◀ The internal crankcase ventilation system is integrated into the plastic cylinder head cover. The crankcase breather valve (**arrow**) is not serviceable as a separate component.

The pressure control valve varies the vacuum applied to the crankcase breather depending on engine load. The valve is balanced between spring pressure and the amount of manifold vacuum.

At idle when the intake manifold vacuum is high, vacuum reduces the valve opening, allowing a small amount of crankcase vapors to be drawn into the intake manifold.

At part to full load conditions when intake manifold vacuum is lower, the spring opens the valve and additional crankcase vapors are drawn into the intake manifold.

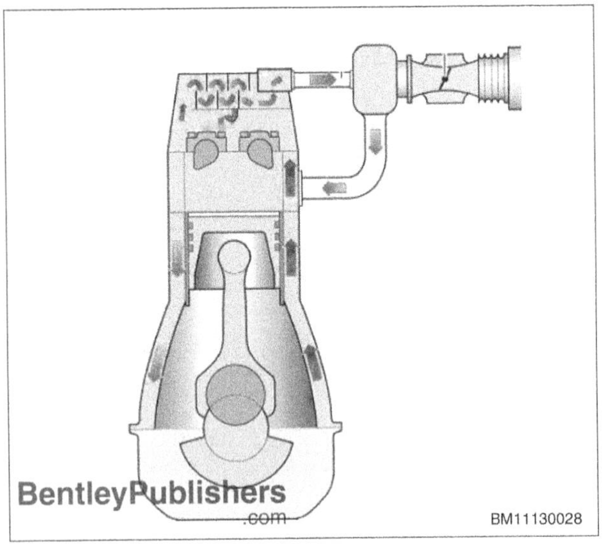

Oil separation is carried out via a labyrinth system and a cyclone separator incorporated into the cylinder head cover. By having the system components integrated into the cylinder head cover, the crankcase gases are heated by the engine rather than an electric heater as on the external system. Once the liquid oil is separated from the crankcase vapors, the oil is allowed to drain back through check valves back into the engine.

If the diaphragm valve in the breather housing leaks, full intake vacuum may be applied to the crankcase, resulting in excessive oil consumption, irregular idle, whistling or howling noises or oil smoke in the exhaust.

Specifications for non-turbo engines

The pressure control valve:

- opens when pressure is less than 30 mbar
- closes when pressure is greater than 30 mbar

A faulty crankcase breather valve causes significant deviations from the specified value:

- Crankcase ventilation clogged: Ambient pressure +100 mbar
- Crankcase ventilation internal leak: ambient pressure -170 mbar

Engine coolant temperature (ECT) sensor, replacing (non-turbo)

The engine coolant temperature (ECT) sensor is a negative temperature coefficient (NTC) sensor. As coolant temperature rises, resistance through the sensor decreases.

The ECM varies ignition timing and air / fuel mixture based on engine coolant temperature. The ECT sensor is supplied a 5 volt reference voltage. The voltage drop across the sensor varies as the coolant temperature (sensor resistance) changes.

If the ECT sensor input is faulty or not plausible, the MIL is illuminated when OBD II fault criteria are exceeded. The ECM assumes a substitute value (80°C / 176°F) to maintain engine operation. Ignition timing is set to a safely conservative basic setting.

The ECT sensor is located on the left side of the cylinder head in thermostat housing.

NOTE—

- *On N16 engines, the coolant temperature sensor may not be separately available from MINI (coolant temperature sensor integrated in thermostat housing). Consult an authorized MINI parts department for the latest parts supersession information.*

– Switch ignition OFF and remove key.

– Remove air filter housing. See **Air filter housing and ducts, removing and installing (non-turbo)** in this repair group.

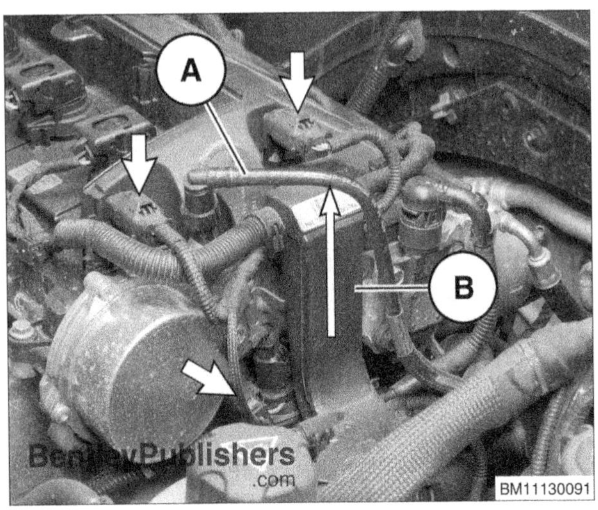

Unlock and detach electrical connectors (**arrows**).

– Unlock and disconnect vacuum line from vacuum pump (**A**).

– Carefully pull cable duct (**B**) upwards slightly.

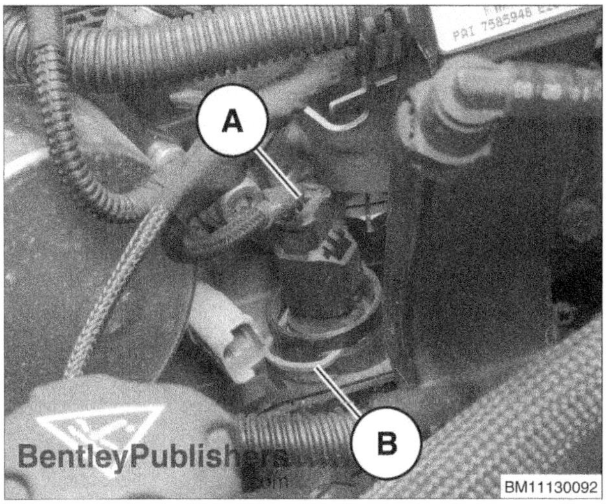

With engine fully cooled off, disconnect ECT sensor connector (**A**).

- Remove locking clip (**B**) and remove sensor from cylinder head. Be prepared to catch small amount of coolant.
- Installation is reverse of removal:
  - Use new sealing O-ring when installing sensor.
  - Replace lost coolant.
- After reinstalling, check for fault codes and reset ECM memory.

ECM pin assignments (non-turbo)

The engine control module (ECM) (**arrow**) is located in the right rear of the engine compartment in the E-box.

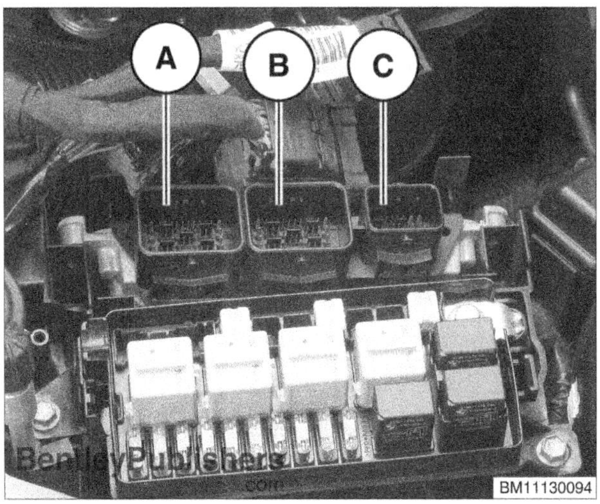

The ECM has 3 main electrical harness connectors with a total of 138 pins.

- **A:** X60004 (voltage, ground supply, signals)
- **B:** X60231 (signals)
- **C:** X60232 (signals)

ECM pin assignments for non-turbo engines are given in **Table a**. This information can be helpful when diagnosing faults to or from the ECM.

Generally, absence of voltage or continuity means there is a wiring or connector problem. Test results with incorrect values do not necessarily mean that a component is faulty. Check for loose, broken or corroded connections and wiring before replacing components. For engine management system electrical schematics, see **ELE Electrical Wiring Diagrams**.

When making checks at the ECM, use a break-out box to allow tests to be made with the connector attached to the ECM. This prevents damage to the small terminals in the connector. As an alternative, the connector housing can be separated so that electrical checks can be made from the back of the connector.

> ***CAUTION—***
> - *Wait at least one minute after switching the ignition OFF before removing the connector from the ECM. If the connector is removed before this time, residual power in the system relay may damage the ECM.*
> - *Connect or disconnect the control module connector and meter probes with the ignition OFF.*

Table b. ECM pin assignments (non-turbo)			
Connector X60004 53-pin black (voltage, ground supply, signals)			
Pin	**Type**	**Component or function**	**Note**
1	input	Ignition coil voltage supply	Power saving relay, terminal 15
2	input	Valvetronic voltage supply	Valvetronic relay
3	ground	Ground	
4	ground	Ground	
5	input	Voltage supply, terminal 87	Fuse F05
6	input	Voltage supply, terminal 87	Fuse F02
7	-	Not used	
8	output	Valvetronic relay control	Valvetronic relay
9	input	Neutral sensor signal	Neutral sensor
10	input/output	TD signal	OBD II socket
11	output	Fuel pump relay control	Fuel pump relay
12	-	Not used	
13	output	Cooling fan relay control	Cooling fan relay
14	output	DMTL valve signal	Diagnostic module for fuel tank leakage
15	output	DMTL valve heating diagnosis	Diagnostic module for fuel tank leakage
16	input	Voltage supply, terminal 30	Fuse F01
17	output	DMTL valve signal	Diagnostic module for fuel tank leakage
18	input	Voltage supply, terminal 87	Fuse F02
19	input	Wake up signal terminal 15	Terminal 15 wake up connector
20	output	Accelerator pedal module voltage supply	Accelerator pedal module
21	output	Coolant pressure sensor voltage supply	Coolant pressure sensor
22	input	Brake light test signal	Brake light switch
23	-	Not used	
24	-	Not used	
25	output	Engine breather heating relay control	Engine breather heating relay

Table b. ECM pin assignments (non-turbo) (continued)			
Connector X60004 53-pin black (voltage, ground supply, signals) (continued)			
Pin	**Type**	**Component or function**	**Note**
26	input	Brake light switch signal	Brake light switch
27	input	Start signal	Car access system
28	output	DME main relay activation	DME main relay
29	-	Not used	
30	input	Voltage supply, terminal 87	Fuse F04
31	-	Not used	
32	input	Brake vacuum sensor signal	Brake vacuum sensor
33	input	Coolant pressure sensor signal	Coolant pressure sensor
34	output	Accelerator pedal module signal	Accelerator pedal module
35	output	Accelerator pedal module signal	Accelerator pedal module
36	-	Not used	
37	-	Not used	
38	input/output	Bit serial data (BSD) signal	Intelligent battery sensor (IBS)
39	input	Clutch switch signal	Clutch switch
40	input/output	PT-CAN high signal	Powertrain bus
41	-	Not used	
42	-	Not used	
43	-	Not used	
44	output	Accelerator pedal module ground	Accelerator pedal module
45	input	Coolant pressure sensor ground	Coolant pressure sensor
46	output	Accelerator pedal module voltage supply	Accelerator pedal module
47	output	Accelerator pedal module ground	Accelerator pedal module
48	ground	Brake vacuum sensor ground	Brake vacuum sensor
49	ground	Neutral sensor ground	Neutral sensor
50	input	Wake up signal terminal 15	Car access system
51	input/output	PA bus signal	PA bus connector
52	input/output	PT-CAN low signal	Powertrain bus\
53	ground	ground	

Connector X60211 53-pin brown (signals)			
Pin	**Type**	**Component or function**	**Note**
1	output	Valvetronic actuator signal	Valvetronic actuator timing actuator
2	output	Valvetronic actuator signal	Valvetronic actuator timing actuator
3	-	Not used	
4	-	Not used	
5	output	Voltage supply	Terminal 15 connector

Connector X60211 53-pin brown (signals) (continued)			
Pin	**Type**	**Component or function**	**Note**
6	output	Fuel tank vent valve signal	Fuel tank vent valve
7	ground	Exhaust camshaft sensor ground	Exhaust camshaft sensor
8	ground	Crankshaft sensor ground	Crankshaft sensor
9	ground	Hot-film air mass meter ground	Hot-film air mass meter
10	input	Electric throttle valve actuator signal	Electric throttle valve actuator
11	input	Electric throttle valve actuator signal	Electric throttle valve actuator
12	-	Not used	
13	-	Not used	
14	ground	Engine temperature sensor ground	Engine temperature sensor
15	input	Oxygen sensor	Bank 1 sensor 1
16	input	Oxygen sensor	Bank 1 sensor 1
17	output	Voltage supply, terminal 15	Terminal 15 connector
18	input	Oxygen sensor signal	Bank 1 sensor 2
19	-	Not used	
20	-	Not used	
21	input	Intake temp / differential pressure sensor signal	Intake temp / differential pressure sensor
22	-	Not used	
23	input	Intake temp / differential pressure sensor signal	Intake temp / differential pressure sensor
24	input	Oxygen sensor	Bank 1 sensor 2
25	ground	Oxygen sensor	Bank 1 sensor 2
26	ground	Electric throttle valve actuator ground	Electric throttle valve actuator
27	ground	Oxygen sensor	Bank 1 sensor 1
28	input	Oxygen sensor	Bank 1 sensor 1
29	output	Voltage supply, terminal 15	Terminal 15 connector
30	input	Oxygen sensor heating	Bank 1 sensor 1
31	-	Not used	
32	input	Engine temperature sensor signal	Engine temperature sensor
33	input	Knock sensor signal	Knock sensor
34	input	Knock sensor signal	Knock sensor
35	-	Not used	
36	output	Electric throttle valve actuator voltage supply	Electric throttle valve actuator
37	output	Intake temp / differential pressure sensor supply	Intake temp / differential pressure sensor
38	output	Exhaust camshaft sensor voltage supply	Exhaust camshaft sensor
39	output	Crankshaft sensor voltage supply	Crankshaft sensor
40	input	Intake air temperature signal	Hot-film air mass meter
41	output	Voltage supply, terminal 15	Terminal 15 connector
42	-	Not used	

Connector X60211 53-pin brown (signals) (continued)			
Pin	**Type**	**Component or function**	**Note**
43	ground	Intake temp / differential pressure sensor ground	Intake temp / differential pressure sensor
44	-	Not used	
45	output	Fuel injector, cylinder 1 signal	Fuel injector, cylinder 1
46	-	Not used	
47	output	Fuel injector, cylinder 1 signal	Fuel injector, cylinder 2
48	output	Fuel injector, cylinder 1 signal	Fuel injector, cylinder 3
49	input	Characteristic map thermostat signal	Characteristic map thermostat
50	-	Not used	
51	-	Not used	
52	-	Not used	
53	-	Not used	

Connector X60212 32-pin gray (signals)			
Pin	**Type**	**Component or function**	**Note**
1	output	Intake VANOS solenoid signal	Intake VANOS solenoid
2	output	Exhaust VANOS solenoid signal	Exhaust VANOS solenoid
3	-	Not used	
4	-	Not used	
5	-	Not used	
6	-	Not used	
7	output	Fuel injector, cylinder 1 signal	Fuel injector, cylinder 4
8	output	Friction wheel signal	Friction wheel
9	output	Ignition coil signal	Ignition coil signal 3
10	output	Ignition coil signal	Ignition coil signal 1
11	input	Oil pressure switch signa;	Oil pressure switch
12	-	Not used	
13	-	Not used	
14	input	Hot-film air mass meter signal	Hot-film air mass meter
15	input	Exhaust camshaft sensor signal	Exhaust camshaft sensor
16	input	Intake camshaft sensor	Intake camshaft sensor
17	output	Ignition coil signal	Ignition coil signal 2
18	output	Ignition coil signal	Ignition coil signal 4
19	ground	Eccentric shaft sensor	Eccentric shaft sensor
20	input	Eccentric shaft sensor signal	Eccentric shaft sensor
21	input	Eccentric shaft sensor signal	Eccentric shaft sensor
22	input	Intake camshaft sensor voltage supply	Intake camshaft sensor
23	ground	Intake camshaft sensor ground	Intake camshaft sensor

Connector X60212 32-pin gray (signals) (continued)			
Pin	**Type**	**Component or function**	**Note**
24	input	Crankshaft sensor signal	Crankshaft sensor
25	input	Electric throttle valve actuator signal	Electric throttle valve actuator
26	input	Electric throttle valve actuator signal	Electric throttle valve actuator
27	-	Not used	
28	input	Eccentric shaft sensor signal	Eccentric shaft sensor
29	input	Eccentric shaft sensor signal	Eccentric shaft sensor
30	ground	Eccentric shaft sensor	Eccentric shaft sensor
31	input	Eccentric shaft sensor signal	Eccentric shaft sensor
32	input	Eccentric shaft sensor signal	Eccentric shaft sensor

Siemens DME (Turbo)

DME engine management system description

The DME manages and monitors the following functions:

Air

- Electronic throttle
- Mass air flow
- Intake manifold pressure sensor
- Intake temperature and pressure sensor
- Valvetronic III (N18 engine)

Fuel

- Fuel supply
- Fuel injection
- Fuel rail pressure sensor

Ignition

- Direct ignition
- Knock control
- Primary and secondary ignition monitoring

Emissions

- OBD II compliance
- Pre and post-catalyst oxygen sensors
- Electrically heated DME-map-controlled thermostat
- Electrically DME-map-controlled oil pump (N18 engine)
- Misfire detection
- Evaporative emission control and leak detection
- Malfunction indicator light (MIL)

Performance controls

- VANOS control (N14 engine)
- Double VANOS control (N18 engine)
- Output of injection signal (TI) for fuel economy gauge
- Output of engine rpm (TD) for tachometer
- A/C compressor control
- Electric radiator cooling fan
- CAN-bus communication
- Dynamic stability control (DSC) interface
- Car access system (CAS)
- Cruise control
- ECM programming

Engine control module (ECM)

The engine control module (ECM) (**arrow**) is mounted next to the front distribution box (A400a) at the left side of the engine compartment.

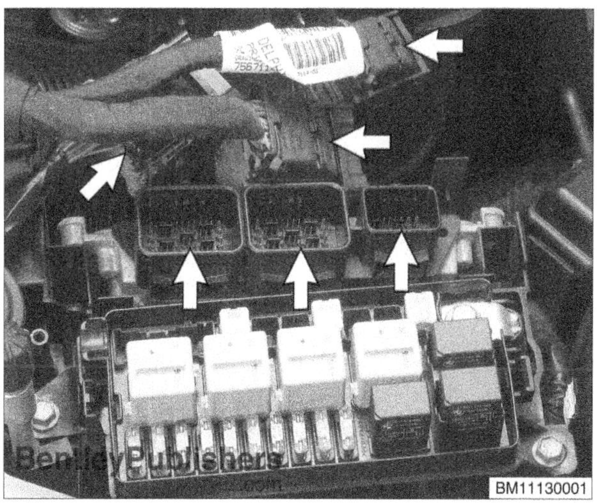

The ECM is flash-programmable and features 3 main electrical harness connectors, with a total of 138 pins.

The EEPROM (chip) in the ECM is coded to the vehicle. The ECM cannot be swapped from another vehicle for testing purposes.

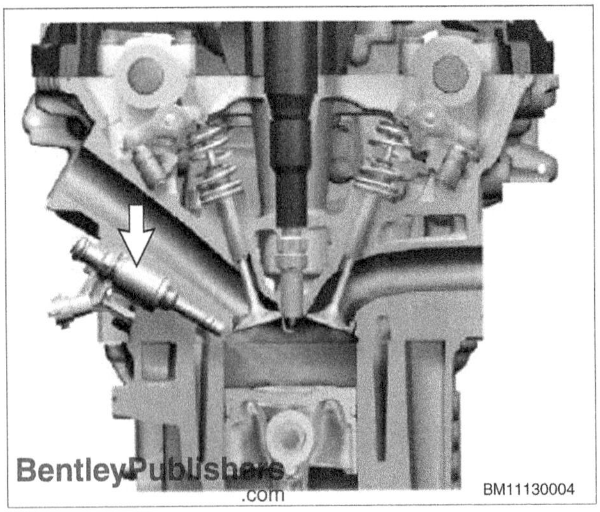

Fuel metering

The ECM meters pressurized fuel by changing the opening time (pulse width) of the fuel injectors. The exact amount of fuel injected is determined by the amount of time the injectors are open.

◄ MINI model turbo charged engines utilize direct fuel injection. The fuel injector (**arrow**) is mounted into a machined hole in the cylinder head near the spark plug.

A direct injection turbo charged engine achieves a higher compression ratio when compared to one utilizing manifold injection. At the same time, exhaust-gas temperature is reduced under full load. Another advantage of this injection system is the improved efficiency in part-load operation.

The ECM monitors engine operating conditions to determine injector opening duration. Each injector can be individually controlled for cylinder selective fuel trim.

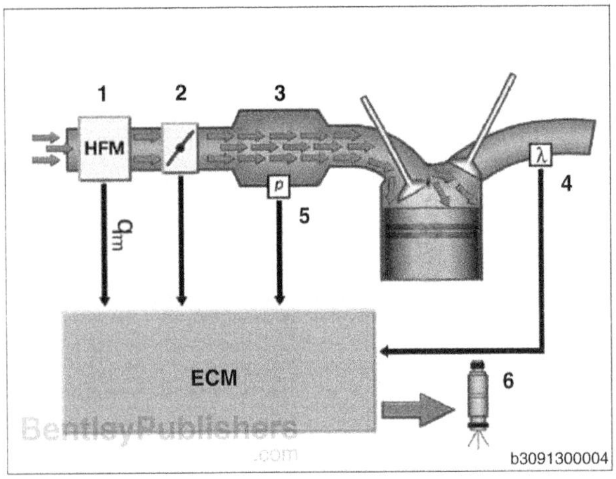

Air intake

◄ Air entering the engine passes through a pleated paper element in the air filter housing. Intake air mass is then measured by a mass air flow sensor. A reference current is used to heat a thin film in the sensor when the engine is running. The current needed to hold the temperature of the film constant is used to calculate the mass of the intake air.

1. Air mass measurement (HFM)
2. Throttle valve
3. Intake manifold
4. Residual oxygen measurement in exhaust
5. Intake manifold vacuum
6. Injection timing

Air intake ducting plays a significant role due to the requirements of a turbocharged engine. In principle, the energy of escaping exhaust gases is utilized to precompress the inducted fresh air and thus introduce a greater air mass into the combustion chamber. This is only possible if the air intake ducting is leak free.

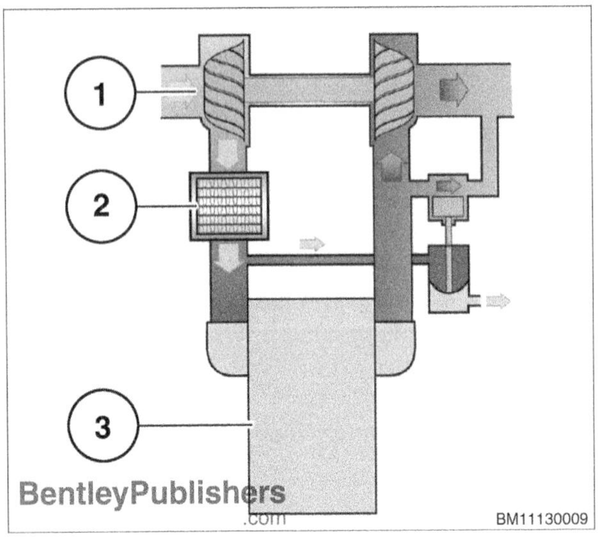

◄ Fresh air is drawn in via the air cleaner and charge-air suction lines to the turbochargers (**1**) and compressed. Because the turbochargers can get very hot during operation, they are connected to the engine coolant and engine oil circuits. The charge air is greatly heated when compressed in the turbocharger, making it necessary for the air to be cooled in an intercooler (**2**).

The compressed and cooled charge air is routed from the intercooler via the throttle valve and intake manifold into the engine (**3**). The system is equipped with several sensors and actuators in order to ensure that the volume of air is optimally adapted to engine operating conditions.

- Drive gear selected (automatic transmission)
- Varying cooling fan speeds

Idle speed control

The N14 ECM controls Idle speed by varying throttle valve actuation. Idle speed is not adjustable. The basic functions and parameters of idle speed control are as follows:

- Control of cold air intake volume.
- Smooth idle speeds regardless of load and inputs.
- Smooth transition from acceleration to deceleration.

Idle speed stabilization is active during the following conditions:

- Engine warm up
- Heating or A/C activation
- Drive gear selected (automatic transmission)
- Varying cooling fan speeds

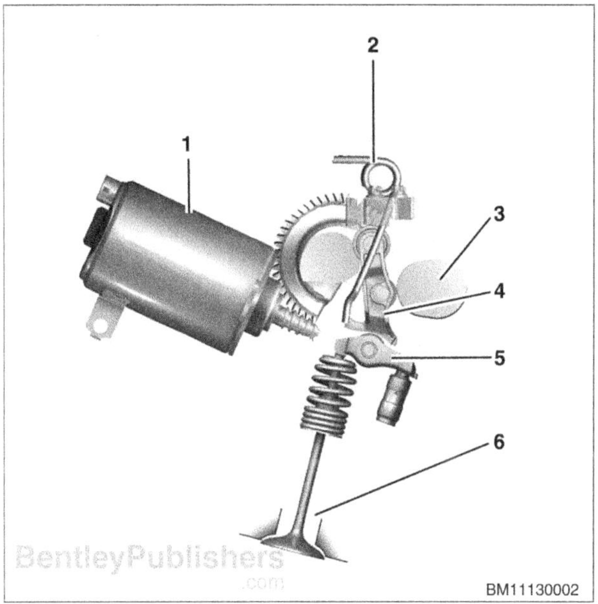

The N18 engines controls Idle speed by varying intake valve lift via the Valvetronic system.

1. Valvetronic motor
2. Return spring
3. Intake camshaft
4. Intermediate lever
5. Intake rocker
6. Intake valve

Idle speed is not adjustable. The basic functions and parameters of idle speed control are as follows:

- Control of cold air intake volume.
- Smooth idle speeds regardless of load and inputs.
- Smooth transition from acceleration to deceleration.

Idle speed stabilization is active during the following conditions:

- Engine warm up
- Heating or A/C activation

Throttle control

The turbocharged engines features electronic throttle control (EDK). There is no throttle cable between the accelerator pedal and the throttle housing. EDK integrates the driver command with the requirements of the traction control system (DSC) and cruise control.

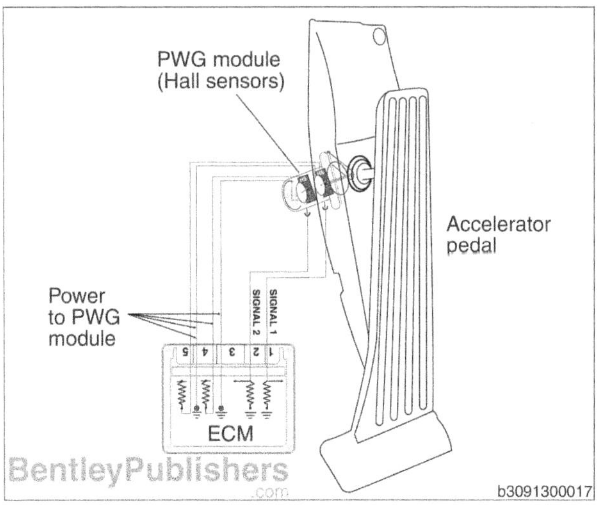

The accelerator pedal module (PWG) at the pedal assembly communicates pedal position directly to the ECM. The module provides two variable voltage signals (via two Hall sensors) to the ECM for pedal position and rate of movement.

The ECM provides an independent 5 v power supply and ground for each Hall sensor. As the accelerator pedal is moved from rest to full throttle, the sensors produce a variable voltage signal.

The output of the Hall sensors is checked for plausibility. The voltage range of Hall sensor 1 is approximately 0.5 to 4.5 v. Hall sensor 2 ranges from approximately 0.5 to 2.5 v.

The throttle housing unit is nonadjustable. If found to be faulty, replace as a complete unit.

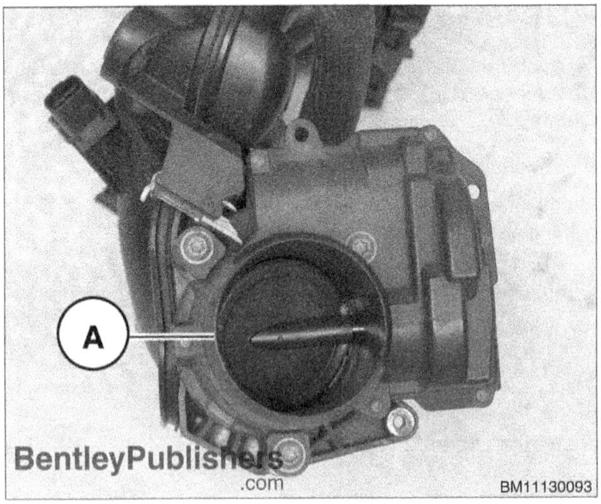

In the Valvetronic turbocharged motor (N18), the throttle module (EDK) (**A**) is not needed for engine load control. Engine load control is carried out by the Valvetronic function of the ECM. However, the throttle may be slightly closed to allow sufficient manifold vacuum for the crankcase ventilation and canister purge systems.

Knock (detonation) control

The knock sensor monitors and controls ignition knock through the ECM. The knock sensor functions like a microphone, converting mechanical vibration (knock) into electrical signals. The ECM is programmed to react to frequencies that are characteristic of engine knock and adapt the ignition timing point accordingly.

Knock sensor replacement is covered in **120 Ignition System**.

Air filter housing and ducts, removing and installing (turbo)

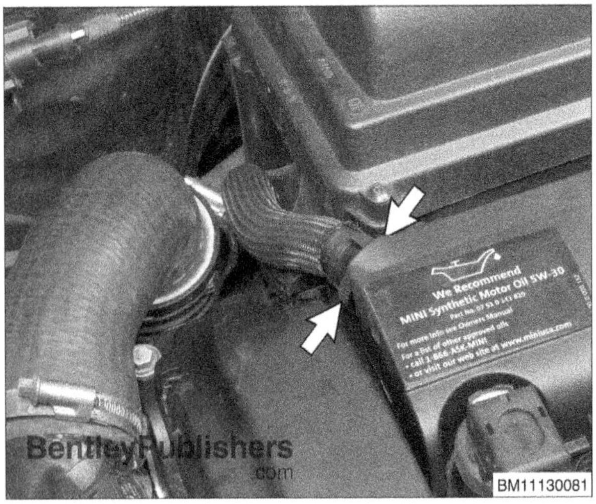

Disconnect hose from cylinder head cover by squeezing (**arrows**) and pulling off.

◀ Remove air filter housing mounting fastener (**arrow**).

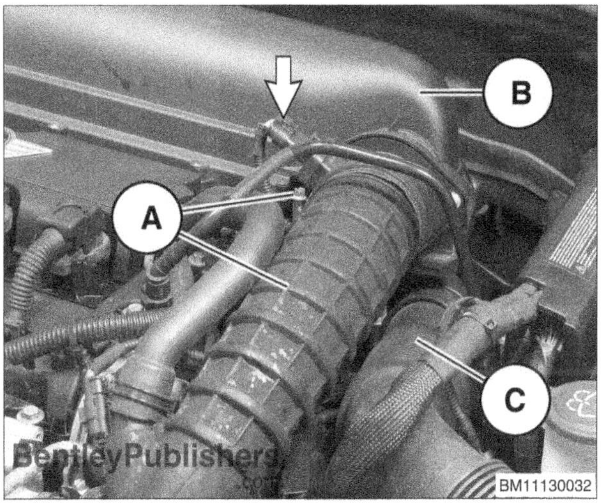

◀ Loosen clamp and detach duct (**A**) from intake silencer housing (**B**).

- Slightly push suction line (**C**) in at latches and pull off intake pipe at air filter housing.
- Disconnect mass air flow electrical connector (**arrow**).
- Remove air filter housing together with intake pipe.
- Remove coolant expansion tank fastener. Place coolant tank to side. Do not disconnect hoses.

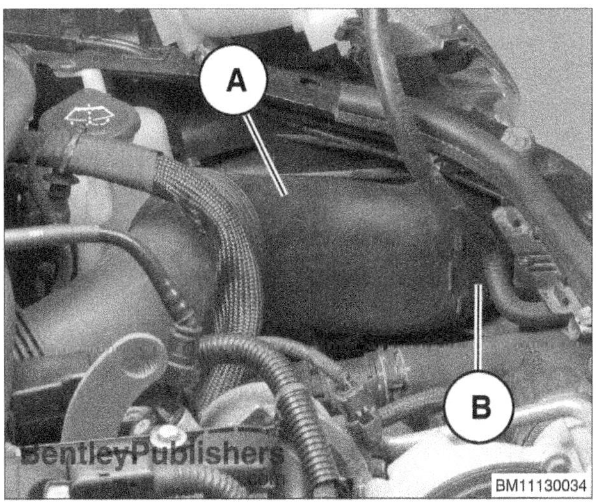

◀ Unclip air intake duct (**A**) from body panel (**B**) and remove.

- Installation is reverse of removal.

> ***CAUTION—***
> *• Turbocharger failure may result if charge air ducts are not installed dry and free of grease. Install using antiseize compound only.*

Tightening torque	
Air filter housing to intake manifold	6 Nm (53 in-lb)

- After reinstalling, check for fault codes and reset ECM memory.

Charge air ducts, left side, removing and installing (turbo)

– Release screw and lift coolant expansion tank out of rubber mount. Lay coolant expansion tank to one side. Do not detach coolant hoses. See **170 Radiator and Cooling System**.

◄ Working at turbocharger, loosen clamp (**arrow**) then disconnect charge air duct from turbocharger (**A**).

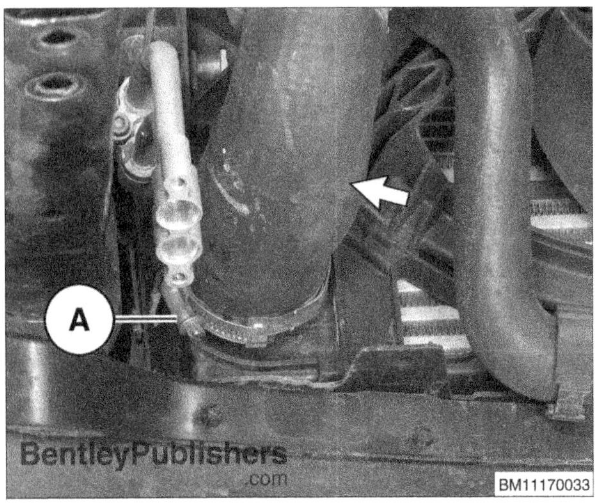

◄ Working at bottom of radiator support, loosen clamp (**A**) then disconnect charge air duct (**arrow**) from intercooler.

◄ Remove fastener (**arrow**) then remove charge air duct (**A**)

– Installation is reverse of removal.

> ***CAUTION—***
> - *Turbocharger failure may result if charge air ducts are not installed dry and free of grease. Install using antiseize compound only.*

– After reinstalling, check for fault codes and reset ECM memory.

Charge air ducts, right side, removing and installing (turbo)

– Remove right side headlight. See **630 Lights**.

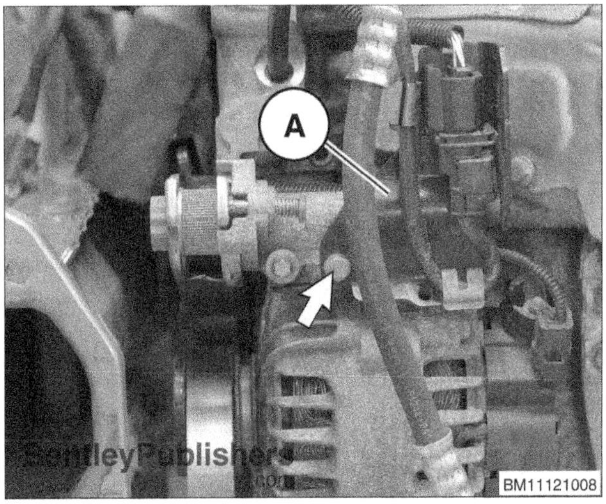

< Remove fastener (**arrow**). Then remove bracket (**A**).

– Raise vehicle and support safely.

> *WARNING—*
> • *Make sure the vehicle is stable and well supported at all times. Use a professional automotive lift or jack stands designed for the purpose. A floor jack is not adequate support.*

< Working at bottom of radiator support, remove charge air duct fastener (**arrow**).

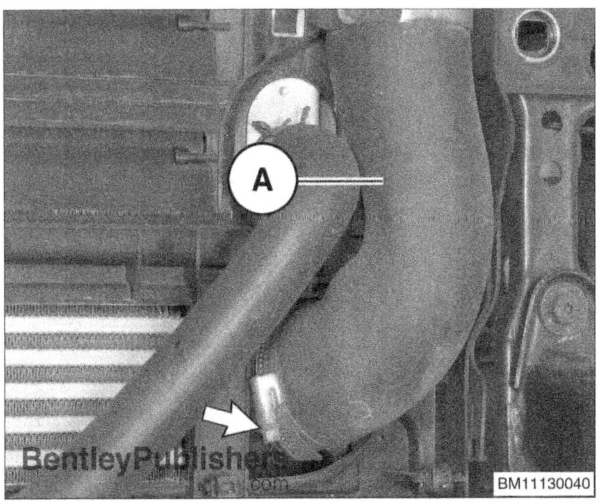

< Working at right side of intercooler, loosen clamp (**arrow**). Then disconnect charge air duct (**A**) from intercooler.

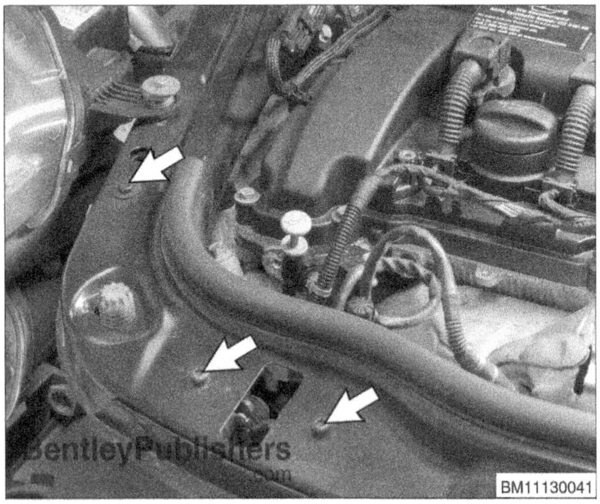

Remove fasteners (**arrows**).

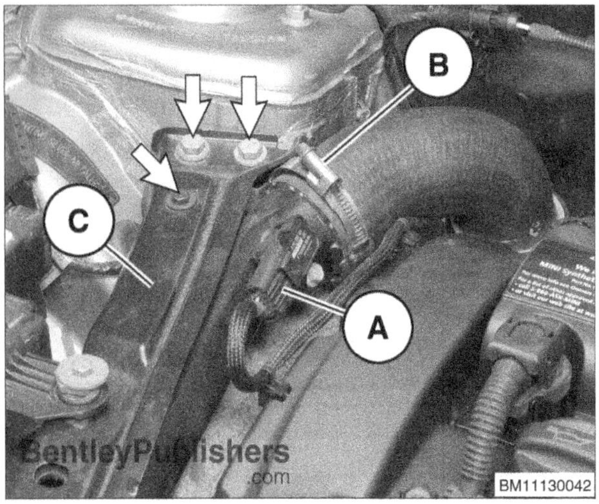

Remove fasteners (**arrows**).

- Disconnect mass air flow sensor connector (**A**).
- Release clamp (**B**).
- Raise lock carrier (**C**) slightly.
- Detach charge-air duct from hose and remove.
- Installation is reverse of removal.

> ***CAUTION—***
> - *Turbocharger failure may result if charge air ducts are not installed dry and free of grease. Install using antiseize compound only.*

Tightening torques	
Radiator support fasteners	10 Nm (7.3 ft-lb)
Radiator support brace to body	22 Nm (16.2 ft-lb)

- After reinstalling, check for fault codes and reset ECM memory.

Mass air flow sensor, removing and installing

A faulty air flow sensor can produce the following problems:
- Difficult to restart when engine is hot.
- Engine starts then stalls.
- Engine starts and runs only with accelerator pedal depressed.

- Switch ignition OFF and remove key.

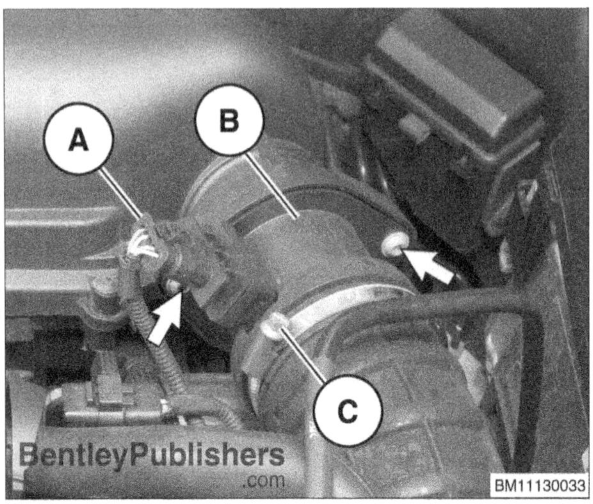

Working at air filter housing:
- Detach mass air flow sensor electrical connector (**A**).
- Loosen clamp (**C**) and remove air duct from sensor.
- Remove fasteners (**arrows**).
- Remove mass air flow sensor (**B**).

- Check intake ducts for cracks and vacuum leaks.

- Installation is reverse of removal.
- After reinstalling, check for fault codes and reset ECM memory.

Tightening torques	
Mass air flow sensor to housing	4.6 Nm (40.7 in-lb)

Intake manifold charge air pressure sensor, removing and installing

- Switch ignition OFF and remove key.
- Remove air filter housing and ducts. See **Air filter housing and ducts, removing and installing (turbo)** in this repair group.

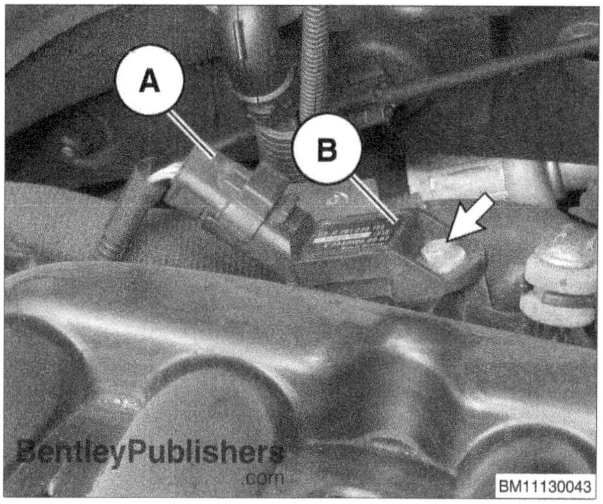

◄ Working at back of intake manifold:
- Detach sensor electrical connector (**A**).
- Remove fastener (**arrows**).
- Remove sensor (**B**).

◄ Installation is reverse of removal. Inspect O-ring (**arrow**) and replace in necessary.

- After reinstalling, check for fault codes and reset ECM memory.

Tightening torques	
Sensor to intake manifold	5 Nm (44.2 in-lb)

Charge air duct pressure sensor, replacing (turbo)

– Switch ignition OFF and remove key.

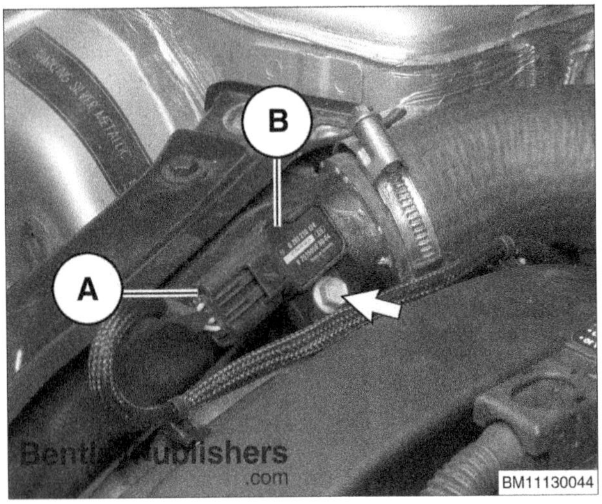

◄ Working at charge air duct:
- Detach sensor electrical connector (**A**).
- Remove fasteners (**arrows**).
- Remove sensor (**B**).

◄ Installation is reverse of removal. Inspect O-ring (**arrow**) and replace in necessary.

– After reinstalling, check for fault codes and reset ECM memory.

Tightening torques	
Sensor to charge air duct	5 Nm (44.2 in-lb)

Throttle housing, removing and installing (turbo)

N14 engine: The throttle housing is used for load control. There is no mechanical (cable) link between the accelerator pedal and the throttle plate.

N18 engine: The throttle housing (EDK) is not used for engine load control. Engine load control is carried out by the Valvetronic function of the ECM. However, the throttle may be slightly closed to allow sufficient manifold vacuum for the crankcase ventilation and canister purge systems.

NOTE —
- *A MINI scan tool is required to set throttle plate adaptation. If the adaptation process is not completed correctly, the engine does not start.*

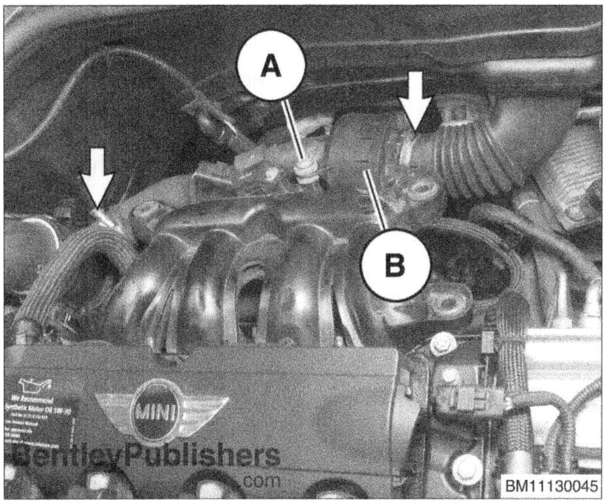

– Switch ignition OFF and remove key.

– Remove air filter housing and ducts. See **Air filter housing and ducts, removing and installing (turbo)** in this repair group.

◀ Loosen hose clamps (**arrows**).

– Remove fastener (**A**).

– Detach sound generator (**B**) from air duct and remove.

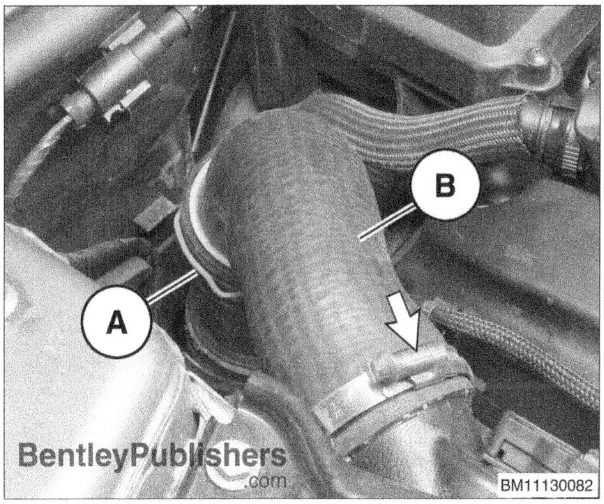

◀ Loosen hose clamp (**arrow**) and remove spring clamp (**A**). Then remove hose (**B**).

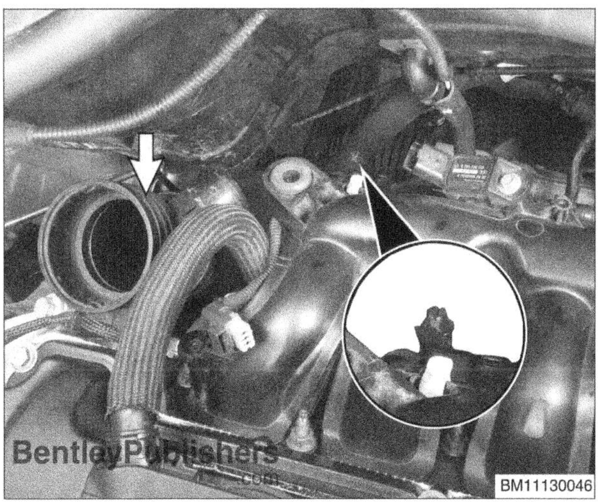

◀ Loosen clamp (**inset**) and detach charge air duct (**arrow**).

> ***CAUTION—***
> *• Use care not to damage plastic throttle housing when disconnecting hose.*

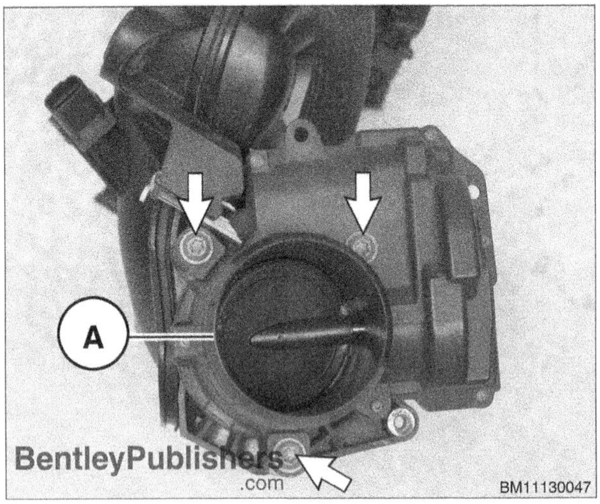

Working at left side of intake manifold:

- Remove fasteners (**arrows**).
- Tilt throttle housing (**A**) up far enough to access electrical connector.

– Disconnect electrical connector and remove throttle housing from vehicle.

– Installation is reverse of removal.

– Replace profile gasket (sealing O-ring) between throttle assembly and intake manifold.

Tightening torques	
Hose clamp to sound generator	3 Nm (26.5 in-lb)
Throttle housing to used intake manifold	7 Nm (61 in-lb)
Throttle housing to new intake manifold	10 Nm (88 in-lb)

After reinstalling, use MINI scan tool to check for fault codes and reset ECM memory. Reset throttle plate adaptation values following on-screen directions. If the adaptation process is not completed correctly, the engine does not start.

Valvetronic motor, removing and installing (N18 engine)

– Switch ignition OFF and remove key.

– Remove ignition coil cover, if equipped. See **020 Maintenance**.

– Disconnect battery negative (-) cable.

> ***CAUTION—***
> - *Prior to disconnecting the battery, read the battery disconnection cautions in* **001 Warnings and Cautions**.

– Disconnect Valvetronic motor electrical connector.

Remove Valvetronic motor fasteners (**arrows**).

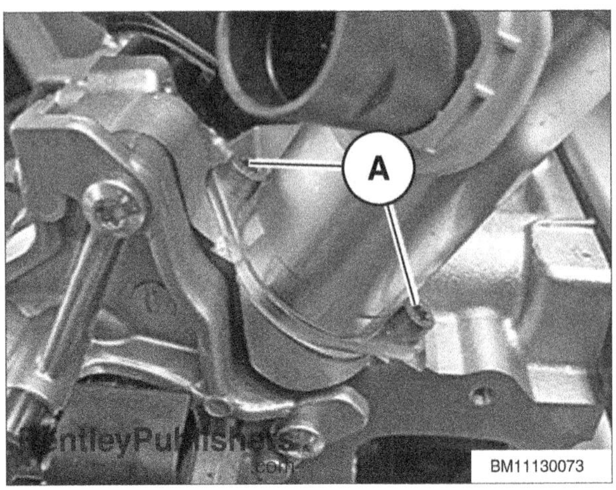

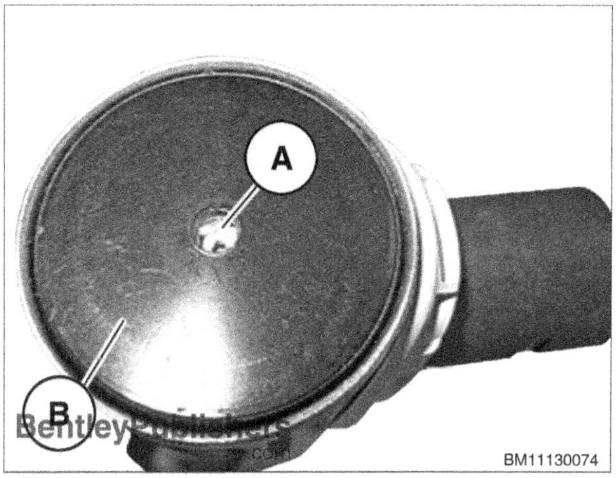

BM11130074

- Remove protective film from Valvetronic motor (**A**) adjusting screw (**B**).

NOTE—
- *Photo shows protective film removed.*

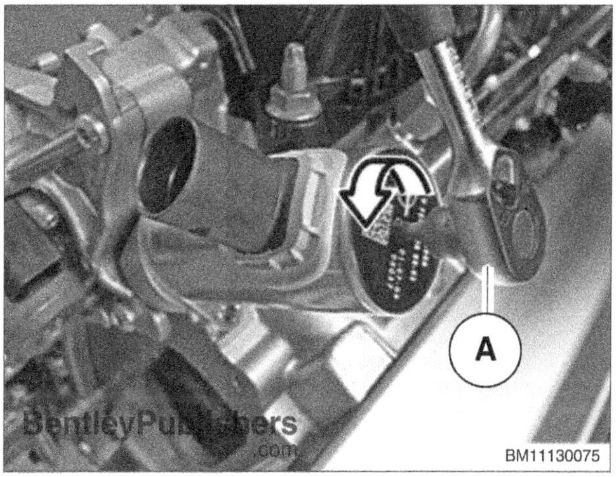

BM11130075

- Use Allen wrench (**A**) to rotate Valvetronic motor worm drive counterclockwise and slide motor out of cylinder head cover (**arrow**).

CAUTION—
- *Do not rotate shaft to maximum to end stop.*

– Remove Valvetronic motor from engine.

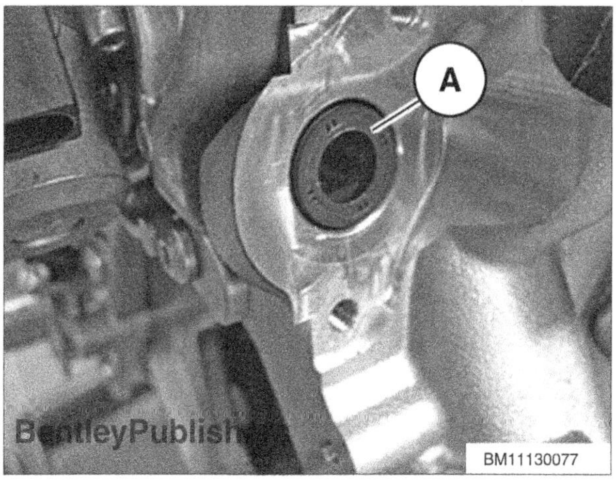

BM11130077

- If Valvetronic motor seal (**A**) remains in cylinder head after removing Valvetronic motor, remove seal (**A**) from cylinder head with plastic prying tool.

Valvetronic motor, removing and installing (N18 engine)

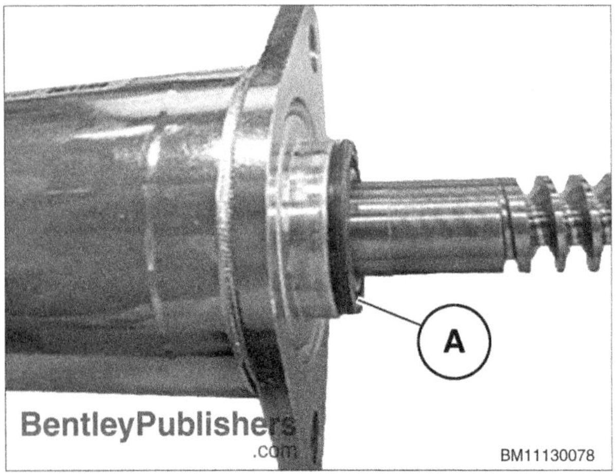

When installing, replace Valvetronic motor seal (**A**).

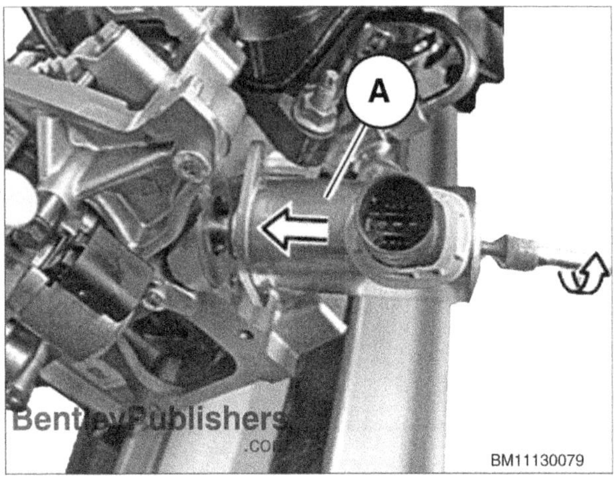

Screw Vavletronic motor (**A**) worm gear drive in direction of **arrow** evenly over intermediate shaft gear to limit position.

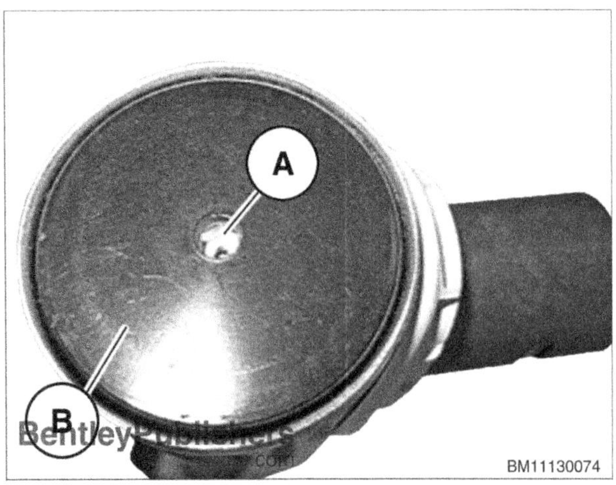

Seal Valvetronic motor adjustment screw hole with protective film.

- Remainder of installation is reverse of removal.
- After reinstalling, use MINI scan tool to check for fault codes and reset ECM memory.

Fuel injector service (turbo)

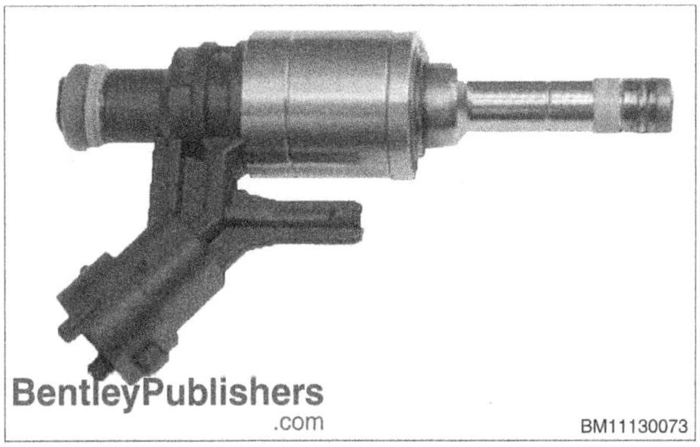

MINI turbo engine utilizes a high precision injection (HPI) direct fuel injection system. Fuel injectors are electrically controlled solenoid valves that meter atomized fuel directly into the engine combustion chamber. The ECM switches the high pressure fuel injectors using the ground side of the circuit.The ECM varies the duration (in milliseconds) of injector opening to regulate air / fuel ratio.

There are two versions of injectors used on the N14 engine.

- **Version 1**: Used on cylinder head marked with MCGU10xxxxxxx:
  Version 2: Used on cylinder head marked with MCGU15xxxxxxx:

NOTE—
- *Version 2 injector is used on the N18 engine.*

Locate and record cylinder head number to determine injector version.

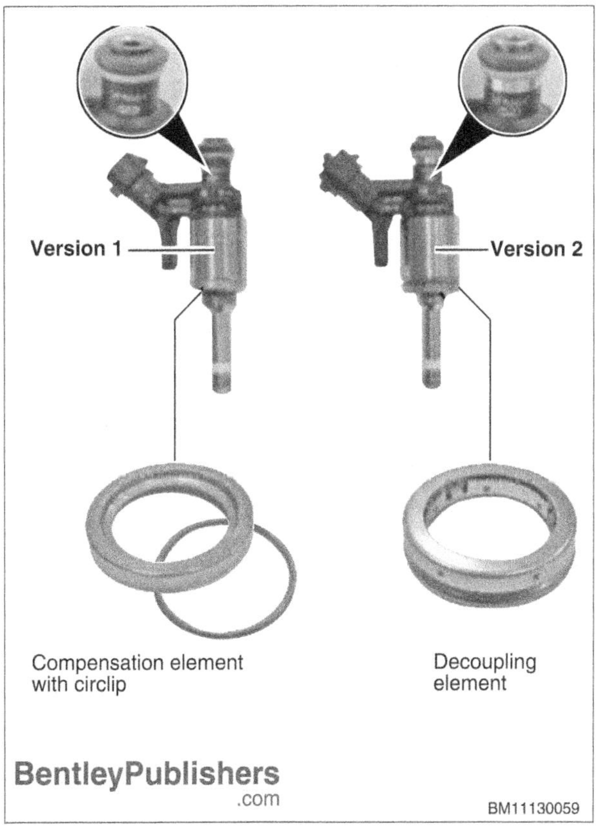

Injector differences:

Version 1 (first generation):
- injector connecting tube is rounded.
- First part of encapsulation made of plastic.
- Uses compensation element retained with circlip.

Version 2 (second generation):
- injector (**B**) connection tube is slightly thicker and not rounded.
- First part of encapsulation made of metal.
- Uses decoupling element (vibration isolator).

NOTE—
- *Decoupling element must be replaced with new part before reinstalling injector. A clicking sound should be heard when new decoupling element is fully seated.*
- *When replacing MCGU10 type cylinder head with MCGU15 type cylinder head, Version 1 injectors can be reused, but must be modified by removing compensation element and circlip and installing new decoupling element.*

Fuel injectors, removing (turbo)

– Disconnect battery negative (-) cable.

> ***CAUTION—***
> - *Prior to disconnecting the battery, read the battery disconnection cautions in* **001 Warnings and Cautions**.

– Remove intake manifold. See **Intake manifold, removing and installing (turbo)** in this repair group.

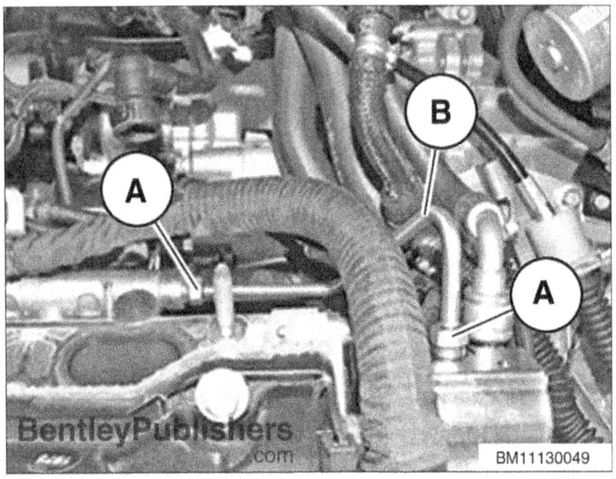

◄ Loosen fuel line nuts (**A**) then remove fuel line (**B**).

> ***WARNING—***
> - *Fuel in the fuel line is under pressure (up to 200 bar or 2900 psi) and may be expelled forcibly. Do not smoke or work near heaters or other fire hazards. Keep a fire extinguisher handy.*
> - *Wrap a clean shop towel around fitting before disconnecting. Residual fuel pressure is present in the fuel line.*
> - *Work only on fuel system when engine temperature is below 40°C (104°F).*
> - *Unscrew the fuel tank cap to release pressure in the tank before working on the fuel lines.*
> - *Plug open fuel lines and fittings.*
> - *To reduce fuel pressure, place towel over injector and carefully tap with a wrench.*

> ***CAUTION—***
> - *Maintain a clean work area when servicing high pressure fuel system. Contaminants can cause a system malfunction.*
> - *Seal all fuel system openings with protective caps.*
> - *Replace high pressure fuel lines if tightened more than 10 times.*

◄ Seal high-pressure rail and high-pressure pump with plugs, MINI special tools **32 1 270**.

– Disconnect electrical connectors from fuel injector.

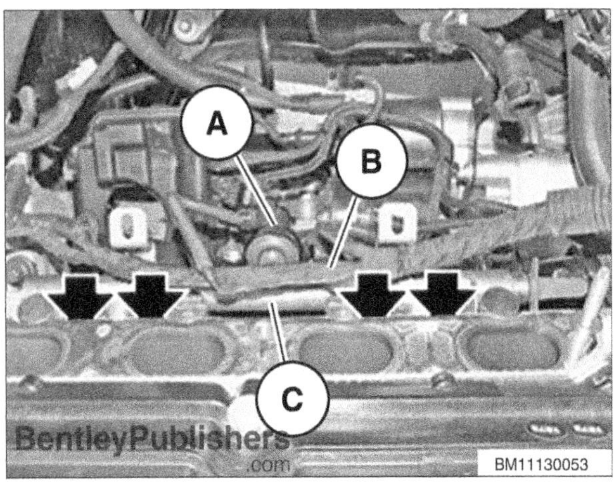

◄ Disconnect electrical connector for high-pressure fuel sensor (**A**).

– Detach wiring harness (**B**) from fuel rail (**C**).

– Release fasteners (**arrows**).

– Detach fuel rail (**C**) from cylinder head and remove.

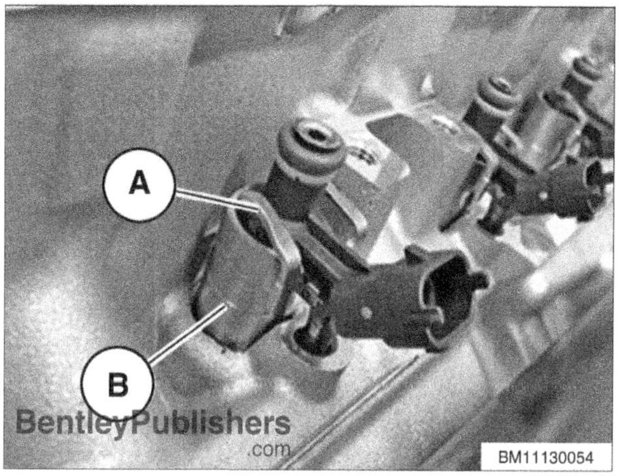

Remove fuel injector retainer (**A**) and discard.

- Pull fuel injector up (**B**) to remove.

Procedure for fuel injector stuck in cylinder head

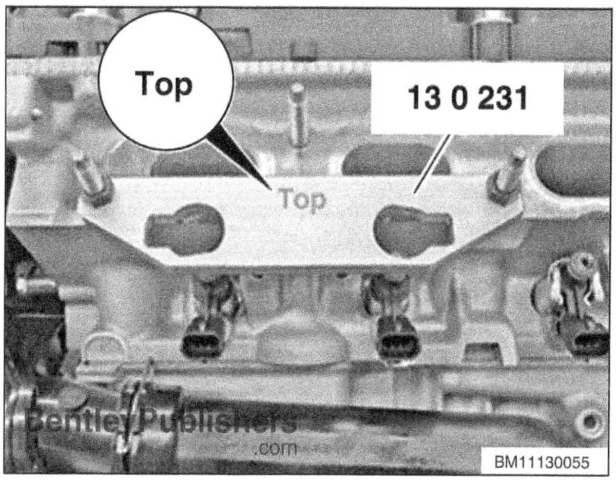

Install MINI special tool **13 0 231** on cylinder head. Note that "Top" faces up.

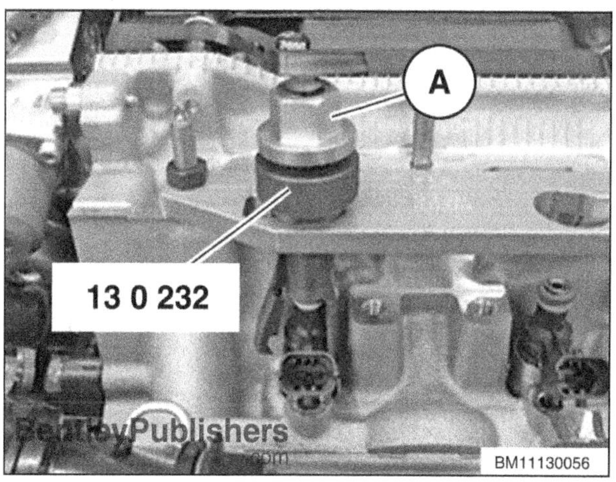

Mount MINI special tool **13 0 232** to MINI special tool **13 0 231**

- To secure injector, turn nut (**A**) by hand until noticeable resistance is felt.

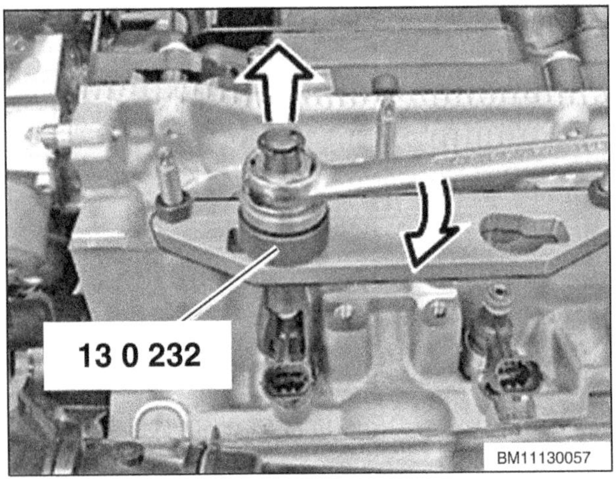

Turn nut on MINI special tool **13 0 232** several turns until injector is released from hole.

- Remove MNI special tool **13 0 231** from cylinder head.
- Remove injector from cylinder head.

Fuel injector, installing (turbo)

Installing

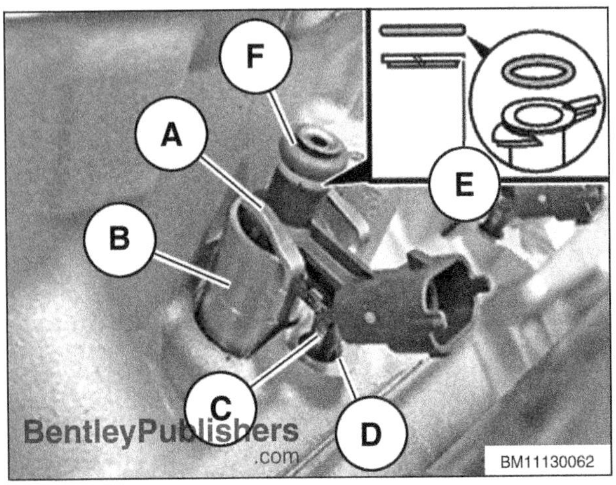

- When installing Version 2 injector: remove old decoupling element and install new decoupling element. Press decoupling element on to injector until audible click is heard.

Be sure guide pin (**C**) of fuel injector (**B**) is correctly positioned in designated hole (**D**) in cylinder head.

- Replace seal (**F**) and support ring (**E**), making sure support ring (**E**) is seated correctly.
- Use new retainer (**A**).
- Replace teflon seal. See **Fuel injector Teflon seal, replacing (turbo)** in this repair group.

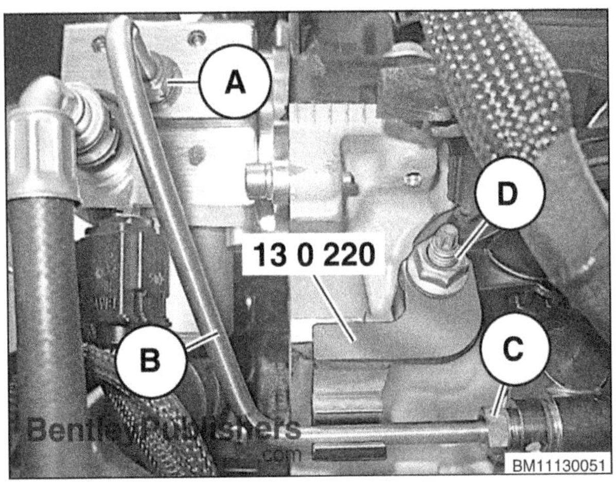

When tightening screw connections, first tighten nut (**C**) and then nut (**A**)

- Preinstall fuel line (**B**). Hand-tighten nuts (**C** and **A**).
- Install MINI special tool **13 0 220** for securing pressure line (**B**) to screw fastening point of intake manifold (**D**).

Tightening torque	
Bracket to intake manifold mounting stud	20 Nm (14.7 ft-lb)

- Tighten nuts (**C**, **A**) according to torque specifications.

Tightening torques	
N14 engine Fuel line nuts: M12 x 1 • Stage 1: Intial tightening • Stage 2 • Stage 3: After warm up for 5 minutes	 15 Nm (11 ft-lb) 33 Nm (24.3 ft-lb) 33 Nm (24.3 ft-lb)
N18 engine M12 x 1 nuts	 26 Nm (19.1 ft-lb)

– Remove MINI special tool **13 0 220** and reinstall intake manifold nut.

Tightening torques	
Fuel rail to cylinder head	19 Nm (14 ft-lb)
Bracket to intake manifold mounting stud	20 Nm (14.7 ft-lb)

– Remainder of installation is reverse of removal.

– Check fuel system for leaks.

Fuel injector teflon seal, replacing (turbo)

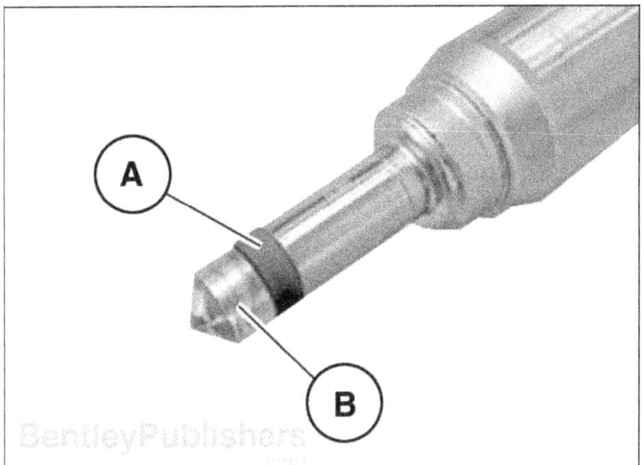

b3091300067

Before replacing PTFE sealing ring (**A**) make sure your hands and work surface are clean and free of oil. Avoid mechanical contact with injector tip (**B**).

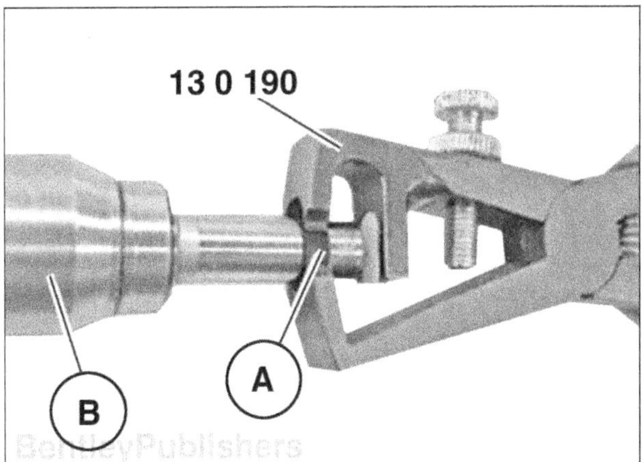

b3091300068

Using special tool 13 0 190, remove PTFE sealing ring (**A**) from injector (**B**). Use a lint-free cloth to remove combustion residue from cylindrical part of injector tip. Do not clean injector tip.

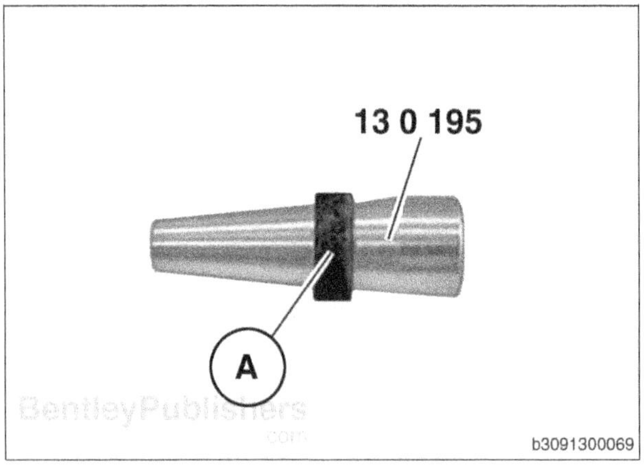

Slide new PTFE sealing ring (**A**) onto mounting taper **13 0 195**.

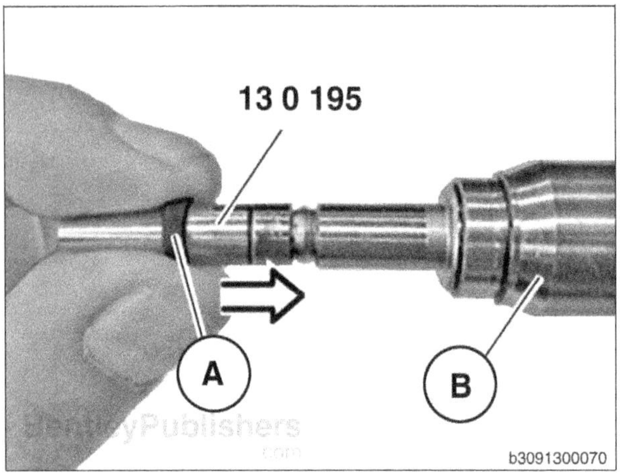

Using fingers and mounting taper (**13 0 195**) install PTFE sealing ring (**A**) on injector (**B**).

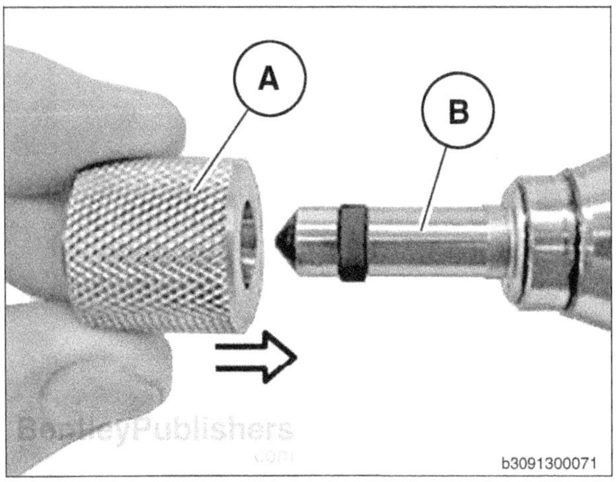

To bring PTFE sealing ring to its installation dimension, slide three mounting sleeves (**A**) with decreasing diameters on injector (**B**).

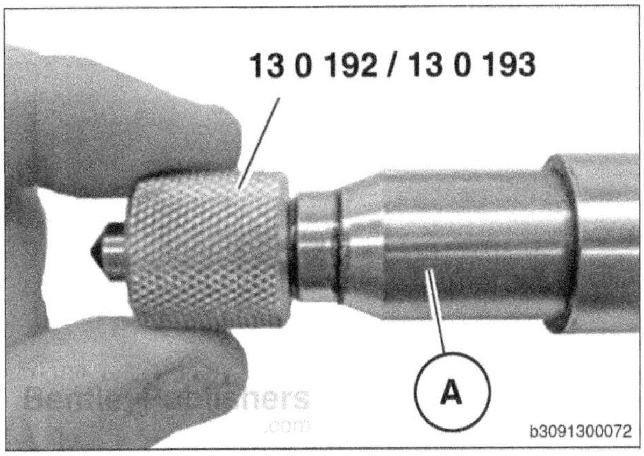

First slide **13 0 192** (large diameter sleeve) on injector (**A**). Then slide mounting sleeve **13 0 193** on injector (**A**).

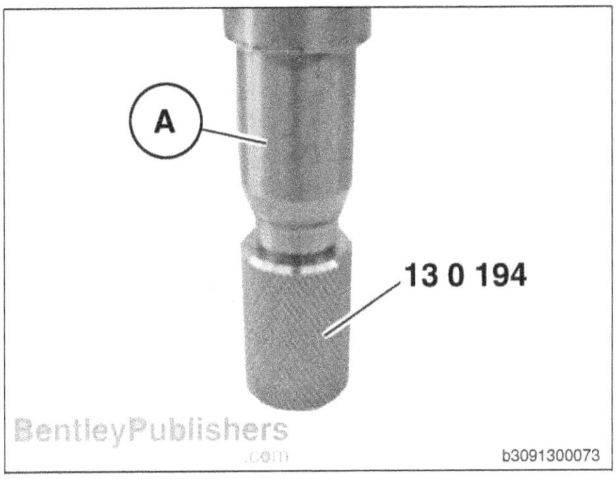

Next press injector (**A**) into mounting sleeve **13 0 194** (small diameter). Install injector in cylinder head within 10 minutes of removing mounting sleeve.

Intake manifold, removing and installing (turbo)

- Switch ignition OFF and remove key.
- Remove air filter housing. See **Air filter housing and ducts, removing and installing (turbo)** in this repair group.
- Disconnect battery negative (-) cable.

> *CAUTION—*
> - *Prior to disconnecting the battery, read the battery disconnection cautions in* **001 Warnings and Cautions**.

- Disconnect intake manifold pressure sensor electrical connector (**A**).

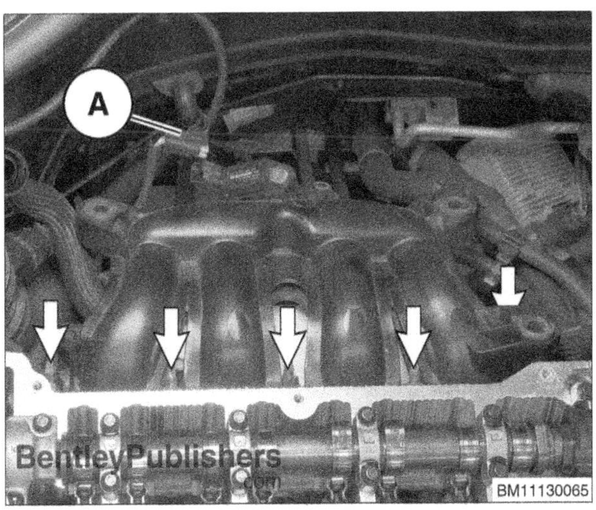

Remove remaining intake manifold fasteners (**arrows**), then move intake manifold away from cylinder head to gain access to components mounted underneath.

- Plug cylinder head intake ports to prevent debris from entering engine.

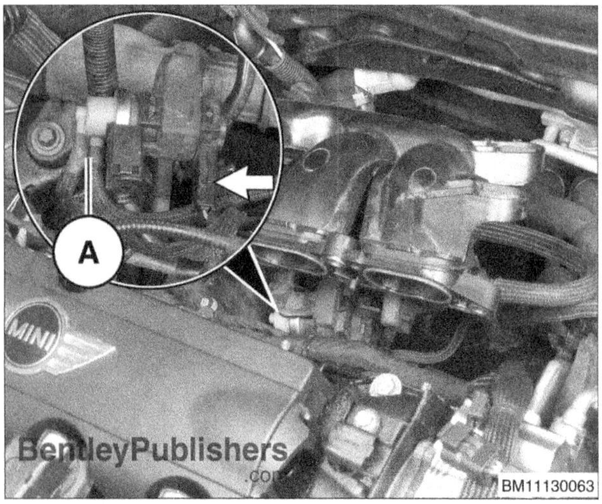

- Disconnect vacuum lines on vacuum connection (**A**).
- Disconnect electrical connecter (**arrow**) at EPPC (pressure converter) valve.

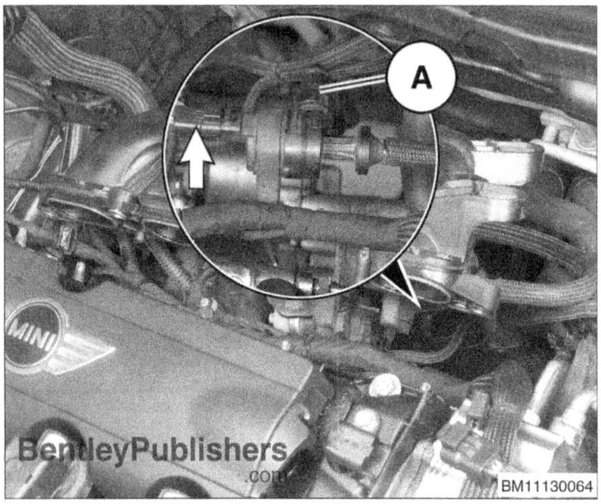

- Disconnect electrical connecter (**arrow**) on fuel tank vent valve.
- Detach hose from tank vent valve (**A**).
- Remove intake manifold from cylinder head.

- When reconnecting EPPC, note that OUT connector on EPPC is identified with a green ring.
  - Vacuum line (**A**) is fitted with a green ring (OUT).
  - Vacuum line (**B**) without green ring (VAC).
- Installation is reverse of removal. Remember to replace all seals.

Tightening torque	
Intake manifold to cylinder head (8 x 125 mm bolt)	15 Nm (11 ft-lb)
Intake manifold to cylinder head (hexagon nut)	20 Nm (14.7 ft-lb)
Intake manifold support bracket to crankcase	20 Nm (14.7 ft-lb)

Crankcase breather valve (turbo)

The crankcase ventilation system is integrated into the plastic cylinder head cover and is not serviceable as a separate component.

The pressure control valve varies the vacuum applied to the crankcase breather depending on engine load. The valve is balanced between spring pressure and the amount of manifold vacuum.

At idle when the intake manifold vacuum is high, vacuum reduces the valve opening, allowing a small amount of crankcase vapors to be drawn into the intake manifold.

At part to full load conditions, when intake manifold vacuum is lower, the spring opens the valve and additional crankcase vapors are drawn into the intake manifold.

Oil separation is carried out via a labyrinth system and a cyclone separator. Once the liquid oil is separated from the crankcase vapors, the oil is allowed to drain back through check valves into the engine.

The MINI turbocharged engine greatly influences intake manifold pressure and hence, the operation of the pressure control valve.

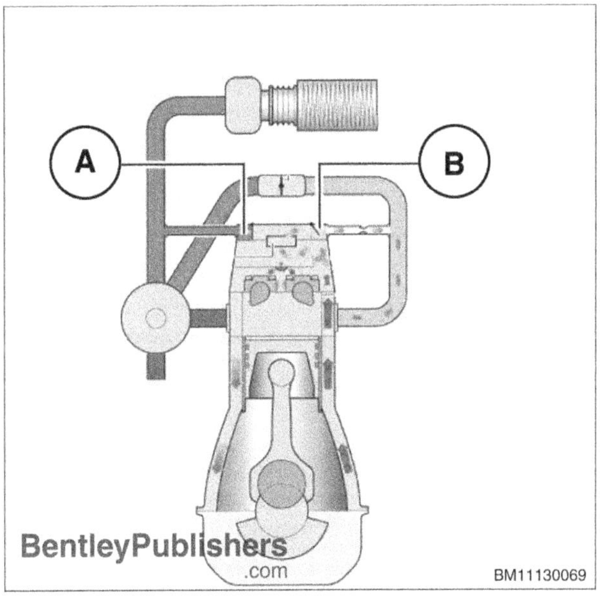

Turbo engine crankcase venting system is designed to function in two modes. One mode is under load and the other is during deceleration or idle. There is a check valve (**A**) in the cylinder head cover to prevent boost air from the turbocharger from entering into the crankcase.

When under deceleration or in idle mode, crankcase gasses are sent to the cyclone separator. The liquid oil is separated and allowed into the crankcase via the drainback valve.

The cleaned crankcase gas (less oil) is sent to intake manifold via the pressure control valve and non-return valve (**B**). The intake manifold is under low pressure (vacuum), which allows the gasses to enter the engine.

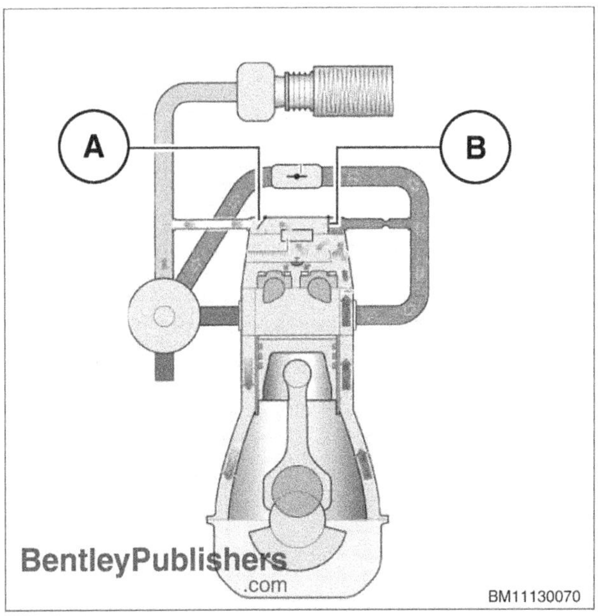

In acceleration mode (under load), the turbocharger supplies boost pressure which in turn increases the pressure in the intake manifold. The check valve (**A**) prevents boost pressure from entering the crankcase.

The crankcase gasses flow the same path through the cyclone separator and pressure control valve. However, it is at this point where the gasses are rerouted to the suction side pipe via the non-return valve (**B**). The gasses are then fed into the turbocharger and ultimately into the engine via the intake manifold.

If the diaphragm valve in the breather housing leaks, full intake vacuum may be applied to the crankcase, resulting in excessive oil consumption, irregular idle, whistling or howling noises or oil smoke in the exhaust.

Specifications for turbo engines

The pressure control valve:

- opens when pressure is less than 30 mbar
- closes when pressure is greater than 30 mbar

A faulty crankcase breather valve causes significant deviations from the specified value:

- Crankcase ventilation clogged: Ambient pressure +100 mbar
- Crankcase ventilation internal leak: ambient pressure -170 mbar.

Engine coolant temperature (ECT) sensor, replacing (turbo)

The engine coolant temperature (ECT) sensor is an negative temperature coefficient (NTC) sensor. As coolant temperature rises, resistance through the sensor decreases.

The ECM varies ignition timing and air / fuel mixture based on engine coolant temperature. The ECT sensor is supplied a 5 volt reference voltage. The voltage drop across the sensor varies as the coolant temperature (sensor resistance) changes.

If the ECT sensor input is faulty or not plausible, the MIL is illuminated when OBD II fault criteria are exceeded. The ECM assumes a substitute value (80°C / 176°F) to maintain engine operation. The ignition timing is set to a safely conservative basic setting.

The ECT sensor is located on the left side of the cylinder head in thermostat housing.

NOTE—

- *On N18 engines, the coolant temperature sensor may not be separately available from MINI. Consult an authorized MINI parts department for the latest parts supercession information.*

– Switch ignition OFF and remove key.

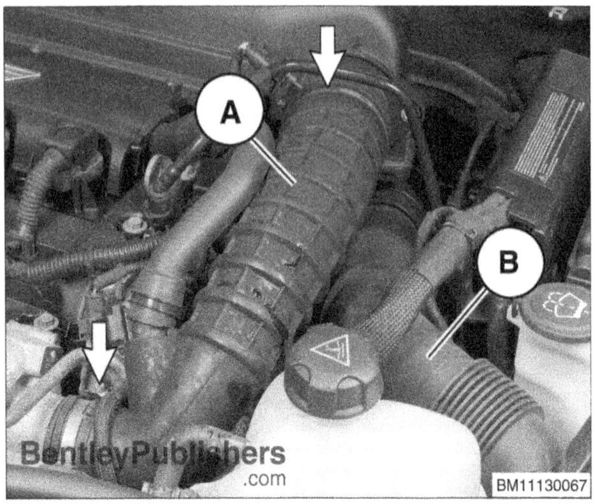

< Loosen air duct clamps (**arrows**) then lay air duct (**A**) aside.

– Remove intake air pipe (**B**).

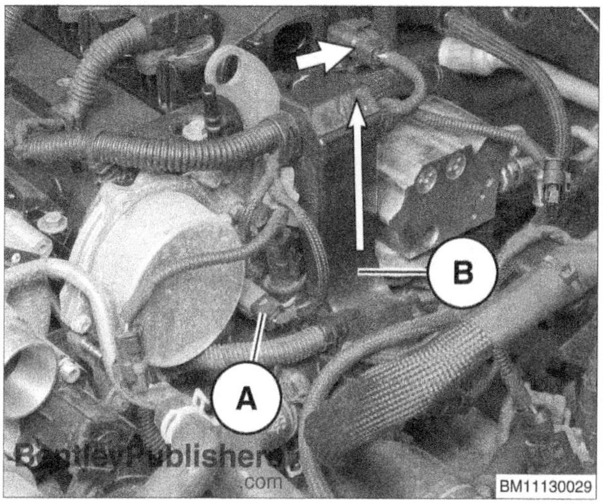

◀ Unlock and detach camshaft sensor electrical connectors (**arrow**).

– Disconnect electrical connector (**A**).

– Unlock and disconnect vacuum line at vacuum pump.

– Carefully pull cable duct (**B**) upwards slightly.

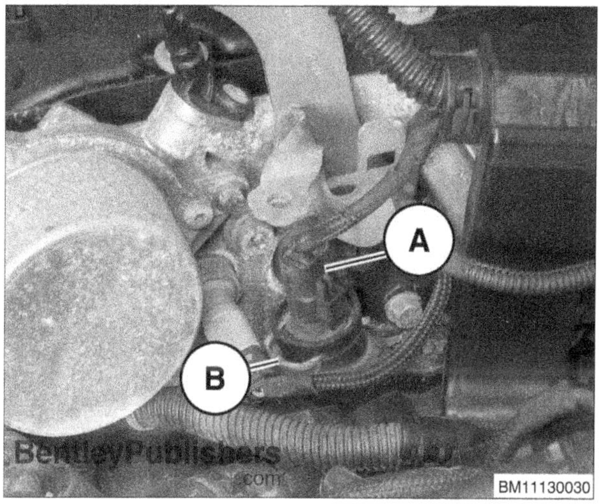

◀ With engine fully cooled off, disconnect ECT sensor connector (**A**).

– Remove locking clip (**B**) and remove sensor from cylinder head. Be prepared to catch small amount of coolant.

– Installation is reverse of removal:
  - Use new sealing O-ring when installing sensor.
  - Replace lost coolant.

– After reinstalling, check for fault codes and reset ECM memory.

ECM pin assignments (turbo)

The engine control module (ECM) (**arrow**) is located in the right rear of the engine compartment in the E-box.

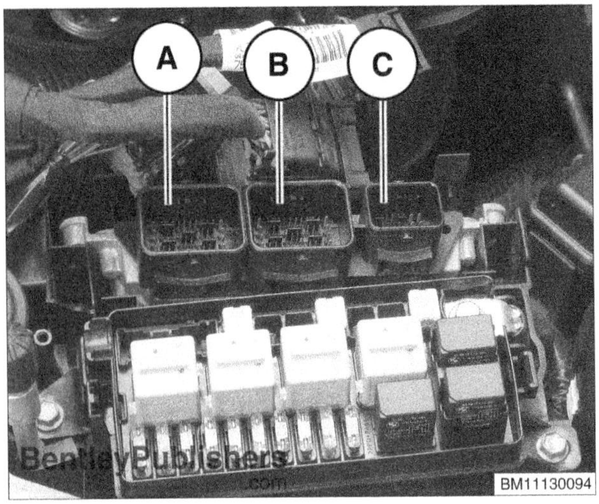

The ECM has 3 main electrical harness connectors with a total of 138 pins.

- **A:** X60004 (voltage, ground supply, signals)
- **B:** X60231 (signals)
- **C:** X60232 (signals)

ECM pin assignments for turbo engines are given in **Table c**. This information can be helpful when diagnosing faults to or from the ECM.

Generally, absence of voltage or continuity means there is a wiring or connector problem. Test results with incorrect values do not necessarily mean that a component is faulty. Check for loose, broken or corroded connections and wiring before replacing components. For engine management system electrical schematics, see **ELE Electrical Wiring Diagrams**.

When making checks at the ECM, use a break-out box to allow tests to be made with the connector attached to the ECM. This prevents damage to the small terminals in the connector. As an alternative, the connector housing can be separated so that electrical checks can be made from the back of the connector.

> ***CAUTION—***
> - *Wait at least one minute after switching the ignition OFF before removing the connector from the ECM. If the connector is removed before this time, residual power in the system relay may damage the ECM.*
> - *Connect or disconnect the control module connector and meter probes with the ignition OFF.*

Table c. ECM pin assignments (turbo)			
Connector X60004 53-pin black (voltage, ground supply, signals)			
Pin	**Type**	**Component or function**	**Note**
1	input	Ignition coil voltage supply	Power saving relay, terminal 15
2	output	Fuel injector relay signal	Fuel injector relay
3	ground	Ground	
4	ground	Ground	
5	input	Voltage supply, terminal 87	Fuse F05
6	input	Voltage supply, terminal 87	Fuse F02
7	-	Not used	
8	output	Fuel injector relay signal	Fuel injector relay
9	input	Neutral sensor signal	Neutral sensor
10	input/output	TD signal	OBD II socket
11	output	Fuel pump relay control	Fuel pump relay
12	output	Cooling fan relay control	Cooling fan relay
13	output	Cooling fan relay control	Cooling fan relay
14	output	DMTL valve signal	Diagnostic module for fuel tank leakage
15	output	DMTL valve heating diagnosis	Diagnostic module for fuel tank leakage
16	input	Voltage supply, terminal 30	Fuse F01
17	output	DMTL valve signal	Diagnostic module for fuel tank leakage
18	input	Voltage supply, terminal 87	Fuse F02
19	input	Wake up signal terminal 15	erminal 15 wake up connector
20	output	Accelerator pedal module voltage supply	Accelerator pedal module
21	output	Coolant pressure sensor voltage supply	Coolant pressure sensor
22	input	Brake light test signal	Brake light switch
23	-	Not used	
24	-	Not used	
25	output	Engine breather heating relay control	Engine breather heating relay
26	input	Brake light switch signal	Brake light switch
27	input	Start signal	Car access system
28	output	DME main relay activation	DME main relay
29	-	Not used	
30	input	Voltage supply, terminal 87	Fuse F04
31	-	Not used	
32	input	Brake vacuum sensor signal	Brake vacuum sensor
33	input	Coolant pressure sensor signal	Coolant pressure sensor
34	output	Accelerator pedal module signal	Accelerator pedal module
35	output	Accelerator pedal module signal	Accelerator pedal module
36	-	Not used	

Table c. ECM pin assignments (turbo) (continued)			
Connector X60004 53-pin black (voltage, ground supply, signals) (continued)			
Pin	**Type**	**Component or function**	**Note**
37	-	Not used	
38	input/output	Bit serial data (BSD) signal	Intelligent battery sensor (IBS)
39	input	Clutch switch signal	Clutch switch
40	input/output	PT-CAN high signal	Powertrain bus
41	-	Not used	
42	-	Not used	
43	-	Not used	
44	output	Accelerator pedal module ground	Accelerator pedal module
45	input	Coolant pressure sensor ground	Coolant pressure sensor
46	output	Accelerator pedal module voltage supply	Accelerator pedal module
47	output	Accelerator pedal module ground	Accelerator pedal module
48	ground	Brake vacuum sensor ground	Brake vacuum sensor
49	ground	Neutral sensor ground	Neutral sensor
50	input	Wake up signal terminal 15	Car access system
51	input/output	PA bus signal	PA bus connector
52	input/output	PT-CAN low signal	Powertrain bus\
53	ground	ground	

Connector X60211 53-pin brown (signals)			
Pin	**Type**	**Component or function**	**Note**
1	output	Valvetronic actuator signal (N18 engine only)	Valvetronic actuator timing actuator
2	output	Valvetronic actuator signal (N18 engine only)	Valvetronic actuator timing actuator
3	-	Not used	
4	-	Not used	
5	output	Voltage supply	Terminal 15 connector
6	output	Fuel tank vent valve signal	Fuel tank vent valve
7	ground	Exhaust camshaft sensor ground	Exhaust camshaft sensor
8	ground	Crankshaft sensor ground	Crankshaft sensor
9	ground	Hot-film air mass meter ground	Hot-film air mass meter
10	input	Electric throttle valve actuator signal	Electric throttle valve actuator
11	input	Electric throttle valve actuator signal	Electric throttle valve actuator
12	-	Not used	
13	input	Intake temp / pressure boost sensor signal	Intake temp / pressure boost sensor
14	ground	Engine temperature sensor ground	Engine temperature sensor
15	input	Oxygen sensor	Bank 1 sensor 1
16	input	Oxygen sensor	Bank 1 sensor 1

Connector X60211 53-pin brown (signals) (continued)			
Pin	**Type**	**Component or function**	**Note**
17	output	Voltage supply, terminal 15	Terminal 15 connector
18	input	Oxygen sensor signal	Bank 1 sensor 2
19	-	Not used	
20	-	Not used	
21	input	Intake pipe pressure sensor signal	Intake pipe pressure sensor
22	-	Not used	
23	input	Intake pipe pressure sensor signal	Intake pipe pressure sensor
24	input	Oxygen sensor	Bank 1 sensor 2
25	ground	Oxygen sensor	Bank 1 sensor 2
26	ground	Electric throttle valve actuator ground	Electric throttle valve actuator
27	ground	Oxygen sensor	Bank 1 sensor 1
28	input	Oxygen sensor	Bank 1 sensor 1
29	output	Voltage supply, terminal 15	Terminal 15 connector
30	input	Oxygen sensor heating	Bank 1 sensor 1
31	-	Not used	
32	input	Engine temperature sensor signal	Engine temperature sensor
33	input	Knock sensor signal	Knock sensor
34	input	Knock sensor signal	Knock sensor
35	output	Crankshaft sensor signa;	Crankshaft sensor
36	output	Electric throttle valve actuator voltage supply	Electric throttle valve actuator
37	output	Intake pipe pressure sensor supply	Intake pipe pressure sensor
38	output	Exhaust camshaft sensor voltage supply	Exhaust camshaft sensor
39	output	Crankshaft sensor voltage supply	Crankshaft sensor
40	input	Intake air temperature signal	Hot-film air mass meter
41	output	Voltage supply, terminal 15	Terminal 15 connector
42	-	Not used	
43	ground	Intake temp / differential pressure sensor ground	Intake temp / differential pressure sensor
44	-	Not used	
45	output	Fuel volume control valve signal	Fuel volume control valve
46	-	Not used	
47	output	Thrust air control valve signal	Thrust air control valve
48	output	Turbocharger coolant pump signal	Turbocharger coolant pipe
49	input	Characteristic map thermostat signal	Characteristic map thermostat
50	-	Not used	
51	-	Not used	
52	-	Not used	
53	-	Not used	

Connector X60212 32-pin gray (signals)			
Pin	**Type**	**Component or function**	**Note**
1	output	Intake VANOS solenoid signal	Intake VANOS solenoid
2	output	Exhaust VANOS solenoid signal (N18 engine only)	Exhaust VANOS solenoid
3	output	Fuel injector, cylinder 4 voltage supply	Fuel injector, cylinder 4
4	output	Fuel injector, cylinder 1 voltage supply	Fuel injector, cylinder 1
5	output	Fuel injector, cylinder 3 voltage supply	Fuel injector, cylinder 3
6	output	Fuel injector, cylinder 2 voltage supply	Fuel injector, cylinder 2
7	output	Wastgate valve signa;	Wastegate valve
8	-	Not used	
9	output	Ignition coil signal	Ignition coil signal 2
10	output	Ignition coil signal	Ignition coil signal 4
11	input	Oil pressure switch signa;	Oil pressure switch
12	-	Not used	
13	input	Crankshaft sensor signal	Crankshaft sensor
14	input	Hot-film air mass meter signal	Hot-film air mass meter
15	-	Not used	
16	input	Intake camshaft sensor	Intake camshaft sensor
17	output	Ignition coil signal	Ignition coil signal 3
18	output	Ignition coil signal	Ignition coil signal 1
19	-	Not used	
20	input	Fuel rail pressure sensor signal	Fuel rail pressure sensor
21	-	Not used	
22	input	Fuel rail pressure sensor supply	Fuel rail pressure sensor
23	output	Turbocharger coolant pump signal	Turbocharger coolant pump
24	input	Fuel rail pressure sensor supply	Fuel rail pressure sensor
25	input	Electric throttle valve actuator signal	Electric throttle valve actuator
26	input	Electric throttle valve actuator signal	Electric throttle valve actuator
27	output	Fuel injector, cylinder 2 signal	Fuel injector, cylinder 2
28	output	Fuel injector, cylinder 4 signal	Fuel injector, cylinder 4
29	output	Fuel injector, cylinder 1 signal	Fuel injector, cylinder 1
30	output	Fuel injector, cylinder 3 signal	Fuel injector, cylinder 3
31	input	Intake temp / differential pressure sensor signal	Intake temp / differential pressure sensor
32	-	Not used	

160 Fuel Tank and Fuel Pump

TABLE

GENERAL

This repair group covers service information for the fuel supply system.

See also:

- **100 Engine–General** for model year, engine code, and DME applications
- **130 Fuel Injection**
- **ECL Electrical Component Locations** for fuel pump fuse and relay access information

Fuel tank

The saddle type plastic fuel tank is mounted underneath the center of the car (under rear seat). Mounted in the fuel tank are the electric fuel pump, fuel filter, fuel level sending units and suction jet pump. Connecting lines for the evaporative emission control system are also attached to the tank.

Fuel tank capacity for MINI models is listed in **Table a**.

Table a. Fuel tank capacity for MINI models	
Normally aspirated models	40 liters (10.6 US gal)
Turbocharged models	50 liters (13.2 US gal)

Fuel delivery systems

MINI models use two main fuel delivery systems:

- Non-turbo engines use a multi-port injection system
- Turbo models use a direct injection system.

Non-turbo engine fuel injectors are installed in machine-bored holes in the cylinder head. An in-tank electric fuel pump delivers pressure-regulated fuel to the fuel rail. The fuel pump is DME controlled via the fuel pump relay (K96). It can operate at up to 5 bar (72.5 psi) and 95 liters (25 gallons) per hour.

Direct fuel injection on the turbo engine allows for precise metering of the mixture and higher compression. This is made possible by locating a fuel injector for each cylinder directly in the combustion chamber. The same low pressure fuel system used on non-turbo vehicles is used to supply fuel to a high pressure fuel pump that is mechanically driven off the back of the intake camshaft. The high pressure fuel system provides fuel to the injectors between 50 and 120 bar (725 and 1740 psi).

High pressure direct injection

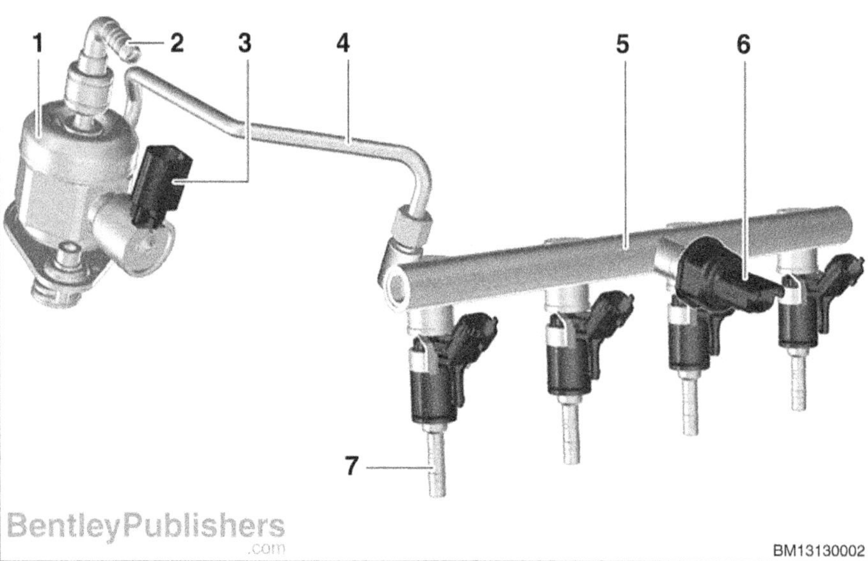

Fuel preparation (turbo models)

1. High pressure pump (Bosch pump shown)
2. Low pressure line
3. Fuel supply control valve
4. High pressure line
5. High pressure fuel rail
6. High pressure sensor
7. Solenoid valve injector

Electric fuel pump

An electric fuel pump is mounted in the fuel tank in tandem with the left side fuel level sender and suction jet pump. A suction jet pump is located in the left side of the tank to transfer fuel to the right side. The combined fuel filter / pressure regulator delivers pressurized fuel to the fuel injection system. The system is a non-return design; there is one fuel line to the fuel injectors and excess fuel flows directly from the fuel filter / pressure regulator back into the tank.

Warnings and Cautions

WARNING—

- *The fuel system is designed to retain pressure even when the ignition is OFF. When working with the fuel system, loosen the fuel lines slowly to allow residual fuel pressure to dissipate. Avoid spraying fuel. Use shop towels to capture leaking fuel.*
- *Before beginning work on the fuel system, place a fire extinguisher in the vicinity of the work area.*
- *Work only on the fuel system when engine temperature is below 40°C (104°F).*
- *Electric fuel pump starts each time a door is opened.*
- *Fuel is highly flammable. When working around fuel, do not disconnect wires that could cause electrical sparks. Do not smoke or work near heaters or other fire hazards.*
- *Wear eye protection, face shield and protective clothing to avoid injuries from contact with fuel.*
- *When working on an open fuel system, wear suitable hand protection, as prolonged contact with fuel can cause illnesses and skin disorders.*
- *Unscrew the fuel tank cap to release pressure in the tank before working on fuel lines.*
- *Do not use a work light with an incandescent bulb near fuel. Fuel may spray on the hot bulb causing a fire.*
- *Make sure the work area is properly ventilated.*
- *Due to risk of personal injury, be sure the engine is cold before beginning work on engine components.*

CAUTION—

- *Prior to disconnecting the battery, read the battery disconnection cautions given in* **001 Warnings and Cautions**.
- *Before making any electrical tests with ignition switched ON, disable the ignition system as described in* **120 Ignition System**. *Be sure the battery is disconnected when replacing components.*
- *To prevent damage to the ignition system or other DME components, including the engine control module (ECM), connect and disconnect wires and test equipment with ignition OFF.*
- *Cleanliness is essential when working with the fuel system. Thoroughly clean the fuel line unions before disconnecting any of the lines. Plug open fuel lines and ports.*
- *Use new seals, O-rings and hose clamps when replacing fuel system components.*

CAUTION—
- *Use only clean tools. Keep removed parts clean and sealed or covered with a clean, lint-free cloth, especially if completion of the repair is delayed.*
- *Do not move the car while the fuel system is open.*
- *Avoid using high pressure compressed air to blow out lines and components. High pressure can rupture internal seals and gaskets.*

FUEL SYSTEM TROUBLESHOOTING

This troubleshooting section covers the low pressure side of the fuel supply system on both normally-aspirated engine and turbo engines.

Fault diagnosis and fuel pressure tests on the high pressure side of the fuel injection system used on turbo engines is best carried out using a MINI scan tool.

Electric fuel pump fuse and relay

Begin troubleshooting fuel system faults by checking fuel pump fuse and relay. Also, check DME main relay. The fuel pump fuse and relay are in the junction box (JB)(**A4010**) behind the right kick panel (base of right A-pillar).

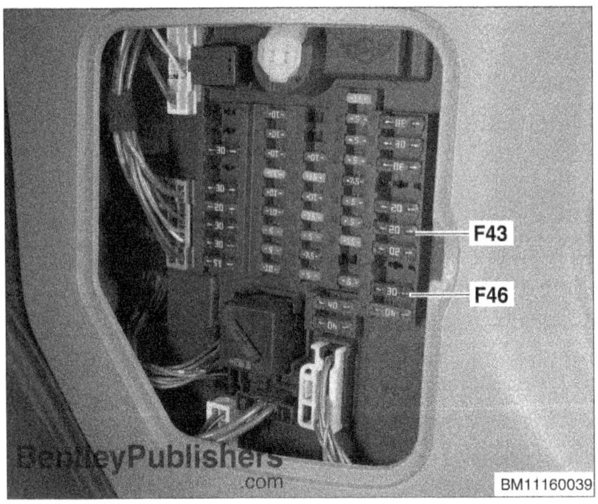

◄ Fuel pump supply circuit fuse in JB.

Electric fuel pump fuse applications	
2007 - 2010 models	Fuse **F43** (20A)
2011 - 2013 models	Fuse **F46** (20A)

– Engine control module (ECM) supplies switched ground to relay. During starting, fuel pump runs as long as ignition switch is in START position and continues to run once engine starts. If electrical system faults interrupt power to fuel pump, the engine does not run.

Operating electric fuel pump for tests

This procedure explains how to operate the fuel pump for testing purposes without having to run the engine.

NOTE—
- *On N14 and N18 engines, fuel pressure tests and fault diagnosis on the high pressure side of the fuel injection system is done using a BMW scan tool.*

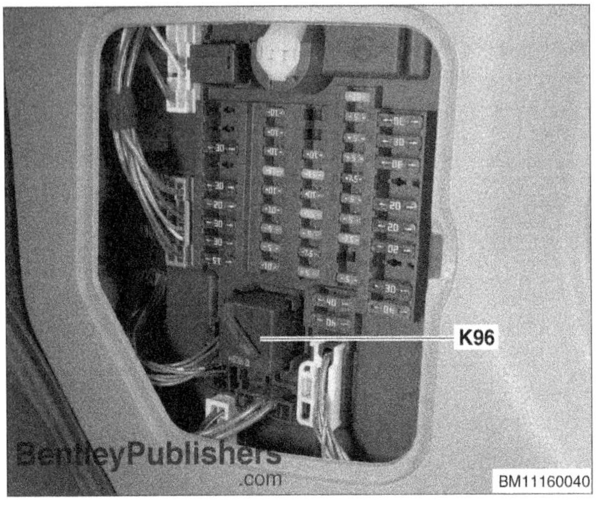

◄ Remove fuel pump relay (**K96**) in JB.

CAUTION—
- *Relay and fuse positions and wire colors vary. One way to verify a relay position or wire colors is to compare the wiring colors at the relay socket to the colors indicated on the wiring diagrams in* **ELE Electrical Wiring Diagrams**.

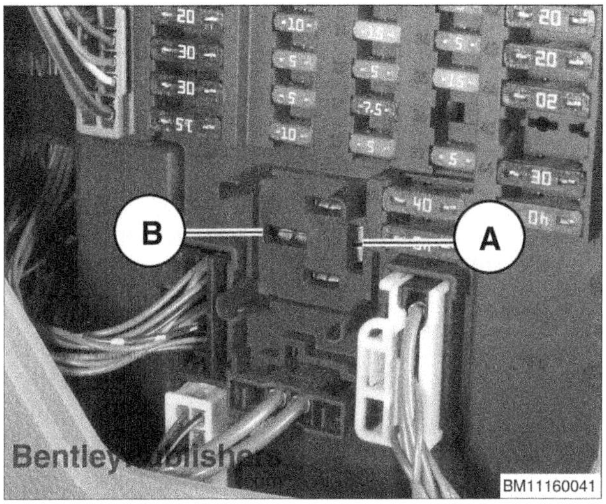

BM11160041

◄ Connect relay socket terminal 30 (**A**) to relay socket terminal 87 (**B**) with fused jumper wire.

- Use 14 ga wire to make jumper and include an in-line fuse holder with a 15A fuse. To avoid relay socket damage from repeated connecting and disconnecting, also include a toggle switch in jumper harness.
- Fuel pump operates as soon as jumper wire is attached.

— If pump does not run with jumper installed, fault could be in fuel pump, fuel pump fuse or wiring to pump. Check pump, fuse and wiring. See **Fuel pump electrical circuit, testing** in this repair group.

— After completing tests, remove jumper harness and install relay.

160

Fuel pump electrical circuit, testing

If fuel pump does not operate or there is no fuel pressure, first test fuel pump electrical circuit.

— Remove rear seat to access fuel pump and fuel level sender. See **Fuel tank component access** in this repair group.

BM11160042

◄ Disconnect electrical connector from fuel pump and fuel level sender by sliding lock (**arrow**) and pulling connector up.

BM11160043

◄ Connect voltmeter between fuel pump leads (**arrows**) on connector.

— Activate fuel pump. See **Operating electric fuel pump for tests** in this repair group.

— Use digital multimeter to test for battery voltage at fuel pump connector leads.

— If voltage and ground are present, fuel pump is probably faulty. If there is no voltage, check wiring between fuel pump and fuel pump relay.

— After completing tests, reconnect fuel pump harness connector.

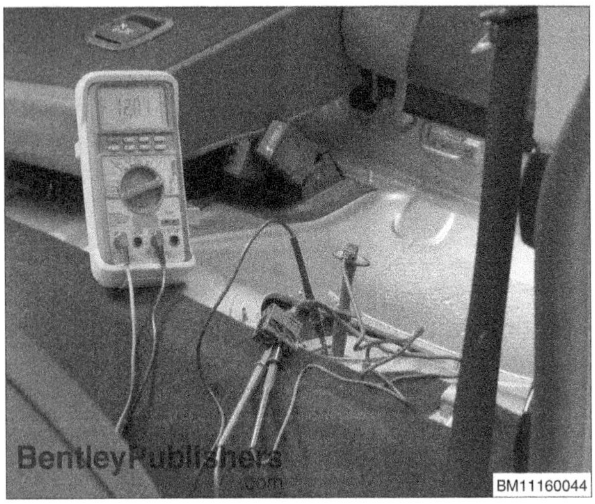

Fuel pump power consumption, testing

If fuel delivery is erratic or poor, or if fuel pump makes abnormally loud noises, test pump power consumption.

Install a digital ammeter in fuel pump circuit using the following steps.

– Remove rear seat to access fuel pump and fuel level sender. See **Fuel tank component access** in this repair group.

◄ Disconnect electrical connector from fuel pump and fuel level sender by sliding lock (**arrow**) and pulling connector up.

– Make sure battery voltage at fuel pump connector is at least 12.6 volts. Charge battery if necessary.

– Attach digital ammeter between pump lead and harness. Connect jumper wire between pump and harness.

> ***CAUTION—***
> *• Do not allow the test leads to short to ground.*

◄ Activate fuel pump. See **Operating electric fuel pump for tests** in this repair group.

– Compare ammeter reading with specification listed in table.

Fuel pump current	
Current consumption	8 - 13 amps

– Higher than normal power consumption by fuel pump may indicate a worn pump, causing intermittent fuel starvation due to pump overheating and seizure. Replace pump.

– Lower than normal power consumption may indicate blockage in a fuel line. Before replacing fuel pump, be sure to check that return line and pump pickup (inside fuel tank) are not obstructed.

Fuel pressure gauge, installing (non-turbo)

- Use a fuel pressure gauge with a minimum range of 0 to 7 bar (0 to 100 psi).

> ***WARNING—***
> - *The fuel system is designed to retain pressure even when the ignition is OFF. When working with the fuel system, loosen fuel lines slowly to allow residual fuel pressure to dissipate. Avoid spraying fuel. Use shop towels to capture leaking fuel.*
> - *Fuel pump starts each time door is opened.*
> - *Make sure the fuel pressure gauge is securely connected to the fuel rail to prevent it from coming loose under pressure.*

- Remove air filter housing and ducts. See **130 Fuel Injection**.

< Disconnect fuel line (**arrow**).

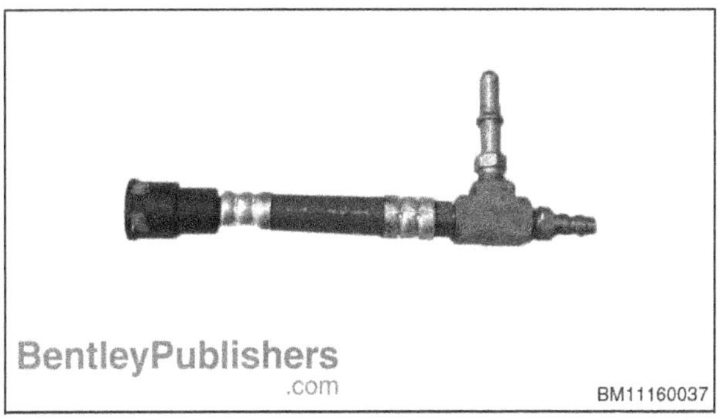

< Install compatible T-style test adapter and attach fuel pressure gauge.

Fuel pressure gauge, installing (turbo)

Use a fuel pressure gauge with a minimum range of 0 to 7 bar (0 to 100 psi).

> ***WARNING —***
> - *The fuel system is designed to retain pressure even when the ignition is OFF. When working with the fuel system, loosen fuel lines slowly to allow residual fuel pressure to dissipate. Avoid spraying fuel. Use shop towels to capture leaking fuel.*
> - *Fuel pump starts each time a door is opened.*
> - *Make sure the fuel pressure gauge is securely connected to the fuel rail to prevent it from coming loose under pressure.*

– Remove splash shield for transmission.

◄ Working underneath vehicle on driver side near subframe and power steering rack, disconnect fuel line (**arrow**).

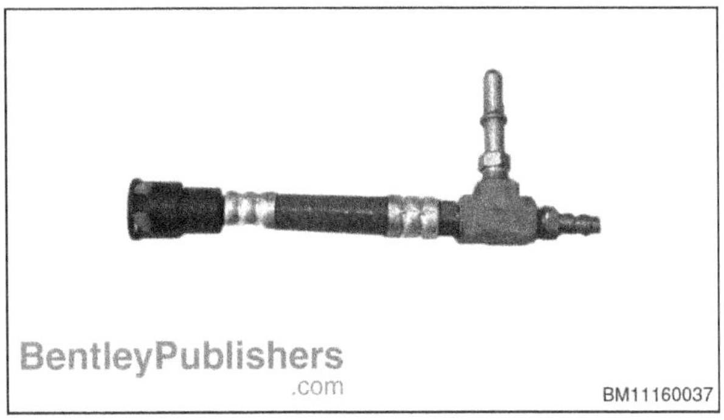

◄ Install compatible T-style test adapter and attach fuel pressure gauge.

Fuel delivery, testing

Checking electric fuel pump fuel delivery is a fundamental part of troubleshooting and diagnosing the DME system. Fuel pressure directly influences fuel delivery. On turbocharged engines, only the low pressure supply side of the direct fuel injection system can be measured. High side pressure of the fuel system can only be viewed using a MINI scan tool to display the high pressure fuel sensor measured value. There are two significant low pressure fuel delivery values to be measured:

- **System pressure** is created by fuel pump and maintained by pressure regulator. System pressure is adjustable only by the ECM.
- **Residual pressure** is pressure maintained in closed system after engine and electric fuel pump are shut off.

System pressure, engine not running

– Attach fuel pressure gauge. See **Fuel pressure gauge, installing (non-turbo)** or **Fuel pressure gauge, installing (turbo)** in this repair group.

– Activate fuel pump. See **Operating electric fuel pump for tests** in this repair group. Compare fuel pressure to specifications in accompanying table.

CAUTION—
- *The fuel pump is capable of developing a higher pressure than that regulated by the pressure regulator. Do not allow pressure to rise above 6.5 bar (94 psi). Damage to the fuel lines or fuel system components could result.*

Fuel pressure specification, engine not running	
System pressure	5 bar (72.5 psi)

– After completing tests without running engine, check for fault codes and reset ECM memory.

System pressure, engine running

If fuel pressure tests show normal pressure but engine lacks power under acceleration, test fuel pump pressure control regulation function.

– With fuel pressure gauge attached, start engine and check fuel pressure.

Fuel pressure specification, engine running	
System pressure	5 bar (72.5 psi)

Residual pressure

For quick restarts and to avoid vapor lock when the engine is hot, the fuel injection system retains fuel pressure after the engine is shut off. This residual pressure is primarily maintained by the fuel pressure regulator and a check valve at the fuel pump outlet.

– Attach fuel pressure gauge. See **Fuel pressure gauge, installing (non-turbo)** or **Fuel pressure gauge, installing (turbo)** in this repair group.

– Start engine and allow it to run for approximately one minute. Note fuel pressure reading. Shut engine OFF.

– Note fuel pressure after approximately 20 minutes. Make sure pressure does not drop more than 0.5 bar from nominal system pressure with engine not running.

Fuel pressure specification, engine not running	
System pressure	5 bar (72.5 psi)

– When finished, disconnect pressure gauge and fitting. Replace Schræder valve cap.

– If fuel system does not maintain pressure:
  - Visually check for leaks in fuel lines and unions.
  - Check for leaking injector(s).
  - Check for faulty fuel pump check valve.

LOW PRESSURE FUEL SYSTEM

Fuel tank component access

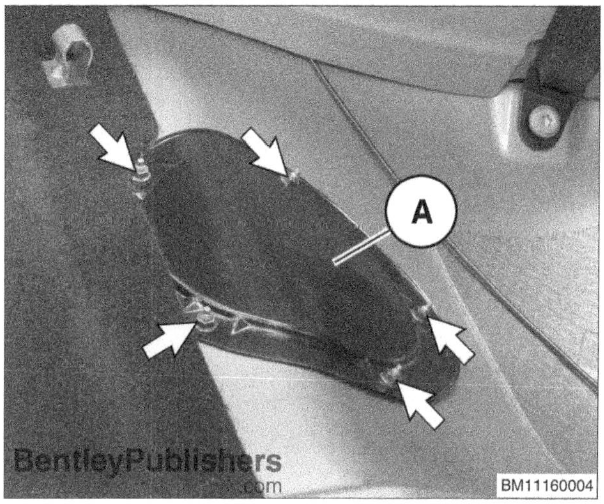

Each lobe of the fuel tank is equipped with a separate service cover. The right cover allows access to the right side fuel tank sender, fuel filter and fuel pressure regulator. The left cover allows access to the left side fuel sender and the fuel pump.

The photo illustrates right side cover. Left side is similar.

– Remove rear seat cushion. See **520 Seats**.

– Peel back carpet to expose fuel tank service cover.

< Working above fuel tank:

- Remove fasteners (**arrows**).
- Remove service cover (**A**).

– Installation is reverse of removal.

Fuel filter and fuel level sender (right side), removing

– Disconnect negative cable from battery.

> *CAUTION—*
> - *Prior to disconnecting the battery, read the battery disconnection cautions given in* **001 Warnings and Cautions**.

– Drain fuel tank. See **Fuel tank, draining** in this repair group.

– Remove left side fuel pump and level sender. See **Fuel pump and fuel level sender (left side), removing** in this repair group.

– Attach stiff wire to fuel sending unit hose pack through left side fuel tank service opening. This helps with installation.

– Remove right side fuel pump access cover. See **Fuel tank component access** in this repair group.

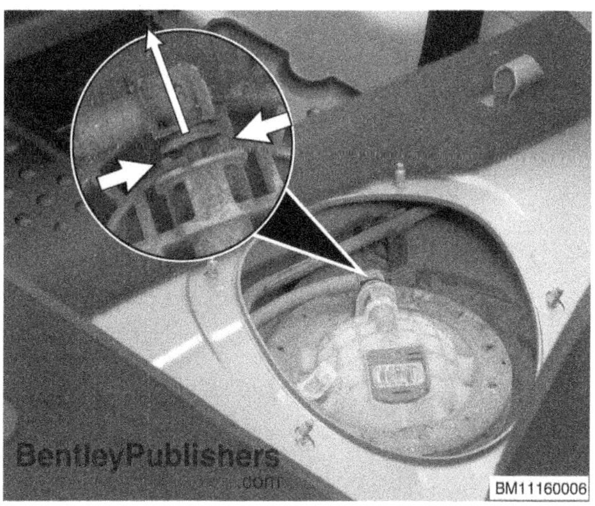

< Unlock quick release tabs by squeezing together (**arrows**) and disconnect fuel line.

> *WARNING—*
> - *When disconnecting fuel hose, wrap shop towel around end of hose to prevent spray of fuel under pressure.*

Fuel filter and fuel level sender (right side), removing

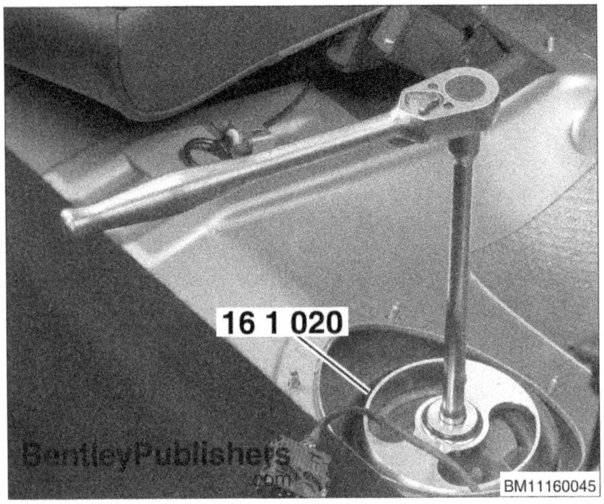

Using MINI special tool **16 1 020**, remove fuel pump service cap.

Remove fuel line holder from sender unit. Lift sender unit and lines out of fuel tank.

> ***CAUTION—***
> - *Do not bend fuel sender arm.*
> - *Be prepared to catch any dripping fuel.*

Fuel level sender, replacing

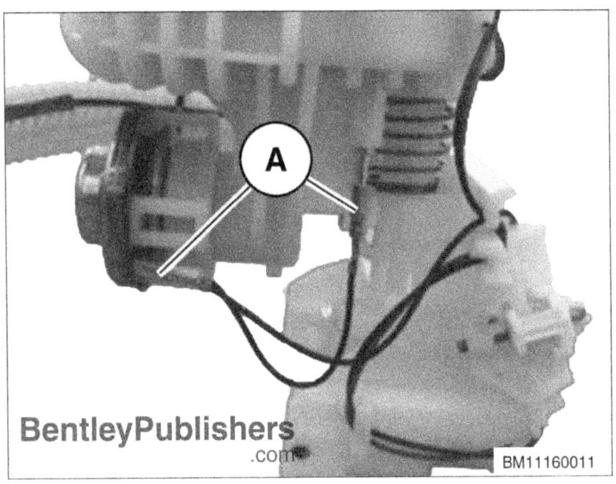

Disconnect electrical connector (**A**).

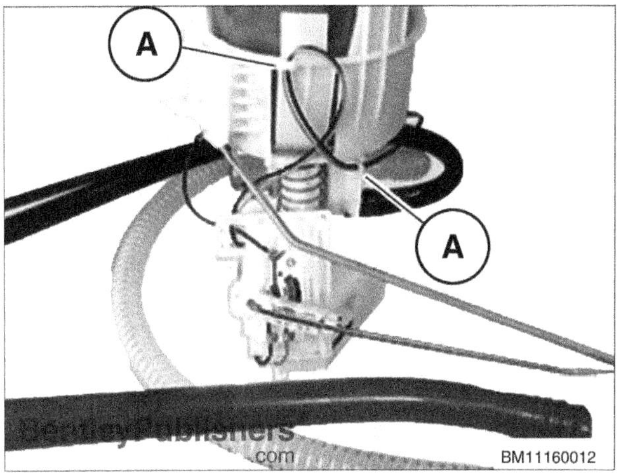

◀ Remove wire from holders (**A**).

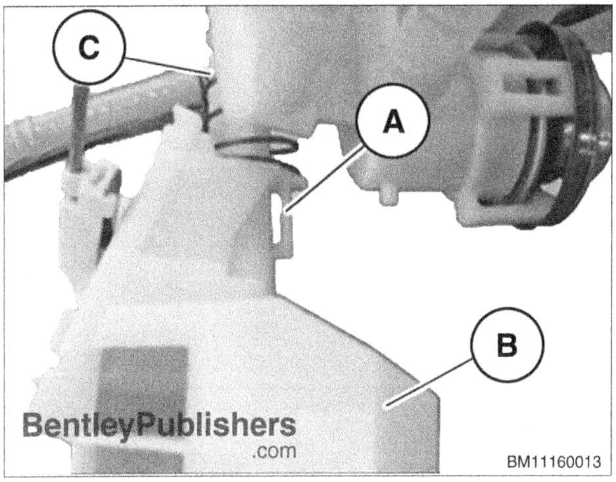

◀ Release retaining hook (**A**) and pull sensor (**B**) off in downward direction.

- Note installation of spring (**C**) for reassembly.

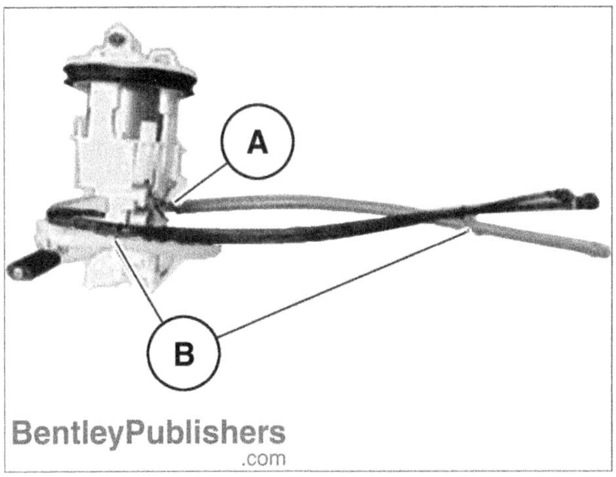

◀ Release line (**A**) from line clips (**B**).

- Installation of fuel level sender is reverse of removal. Check that electrical wires and fuel lines are properly routed.

- See **Fuel pump and fuel level sender (left side), installing** in this repair group for fuel pump installation.

Fuel filter and fuel level sender (right side), installing

– Working inside fuel tank, carefully install fuel pump and fuel level sender. If necessary rotate hose pack and tilt toward fuel level sender.

> ***CAUTION—***
> - *To avoid kinking hoses, be sure to rotate sensor unit and hoses in the same direction.*
> - *Do not bend fuel sender arm.*

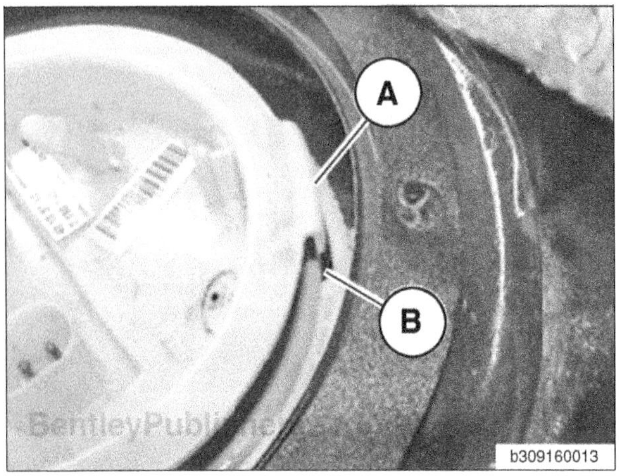

◀ When installing check to see that service cap lug (**A**) engages with fuel tank slot (**B**).

– Install service cap hand tight.

– Use MINI special tool 16 1 020 to tighten service cap.

Tightening torques	
Service cap to fuel tank metal lock ring	45 ± 5 Nm (33 ± 3.6 ft-lb)
Service cap to fuel tank plastic lock ring	85 ± 5 Nm (62.9 ± 3.6 ft-lb)

– Connect fuel line.

– Install fuel tank access cover.

> ***CAUTION—***
> - *After finishing repairs but before starting engine, be sure there is at least 5 liters (1.5 gallons) of fuel in the tank. The fuel pump is damaged if run without fuel.*

Fuel pump and fuel level sender (left side), removing

– Drain fuel tank. See **Fuel tank, draining** in this repair group.

– Disconnect negative cable from battery.

> ***CAUTION—***
> - *Prior to disconnecting the battery, read the battery disconnection cautions given in* **001 Warnings and Cautions**.

– Remove left side fuel sender access cover. See **Fuel tank component access** in this repair group.

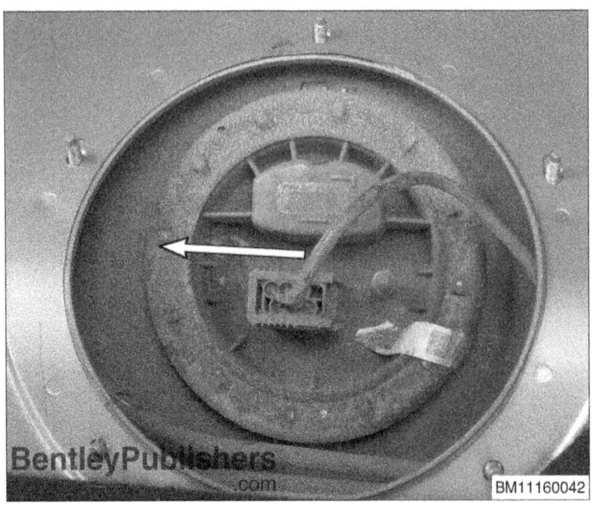

◀ Disconnect electrical connector by sliding lock in direction of **arrow**.

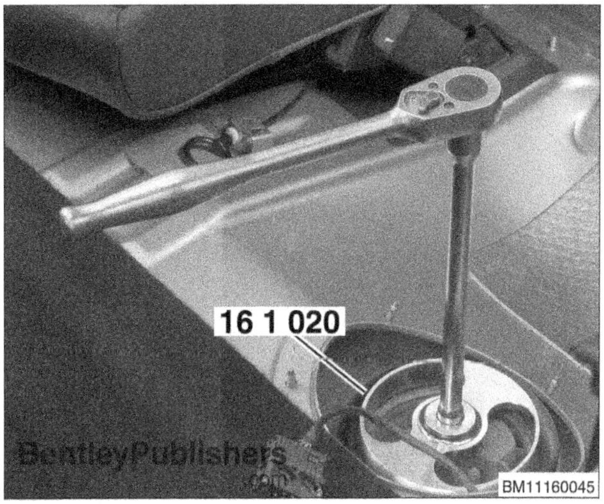

- Using MINI special tool **16 1 020,** remove fuel pump service cap.
- Carefully raise sender out of tank, but do not remove.

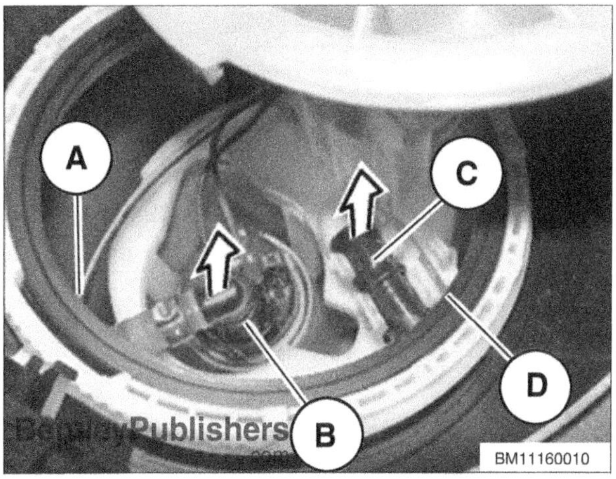

- Disconnect electrical connector (**A**) inside fuel tank.
- Unlock fuel line (**B**) at quick-release fastener and detach.
- Unclip fuel line (**C**).
- Carefully lift sender out of tank with hose pack. Remove and discard fuel tank seals (O-rings).

Fuel level sender, replacing

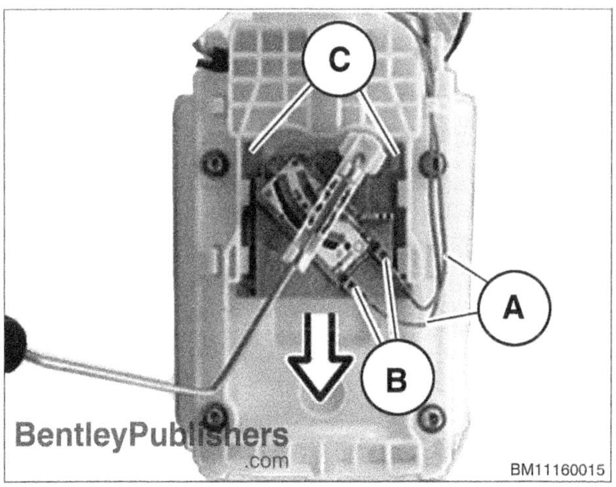

- Detach lines (**A**) at retaining lugs (**B**).
- Release fuel level sender from hooks (**C**) and pull off downwards.
- Installation of fuel level sender is reverse of removal. Check that electrical wires and fuel lines are properly routed.
- See **Fuel pump and fuel level sender (left side), installing** in this repair group for fuel pump installation.

Fuel pump and fuel level sender (left side), installing

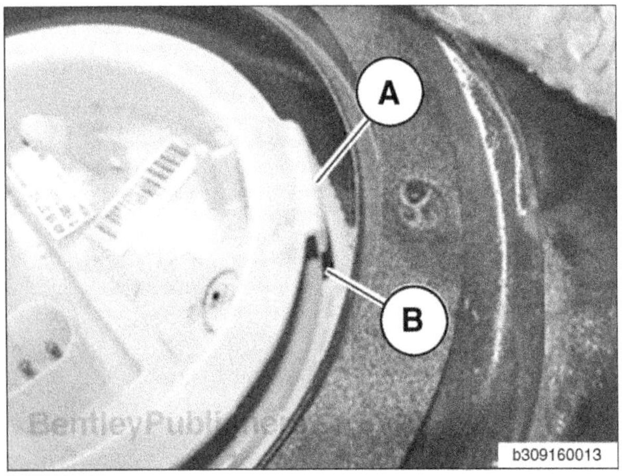

When installing check to see that service cap lug (**A**) engages with fuel tank slot (**B**).

– Install service cap hand tight.

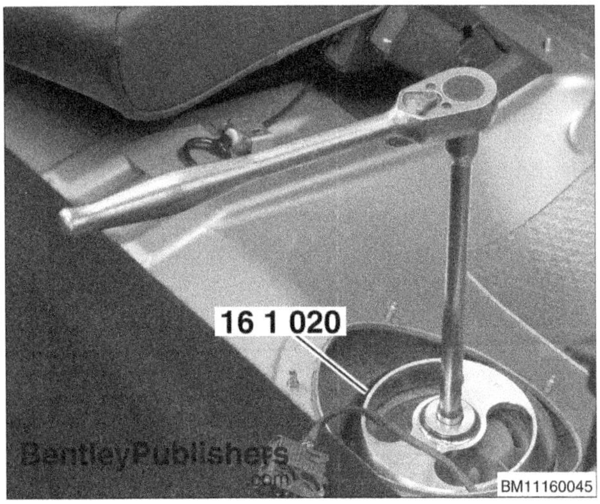

Using MINI special tool **16 1 020** tighten service cap.

Tightening torques	
Service cap to fuel tank metal lock ring	45 ± 5 Nm (33 ± 3.6 ft-lb)
Service cap to fuel tank plastic lock ring	85 ± 5 Nm (62.9 ± 3.6 ft-lb)

– Connect fuel line.

– Install fuel tank access cover.

> ***CAUTION—***
> • *After finishing repairs but before starting engine, be sure there is at least 5 liters (1.5 gallons) of fuel in the tank. The fuel pump is damaged if run without fuel.*

Fuel filter, removing and installing

– Remove right side fuel level sender and fuel level sensor. See **Fuel filter and fuel level sender (right side), removing** in this repair group.

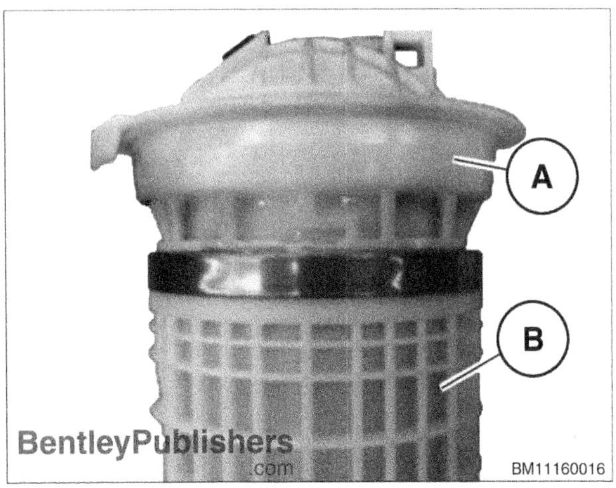

Release cap (**A**) on fuel filter housing (**B**) by one turn counterclockwise (bayonet lock) and pull off to remove.

◀ Pull fuel filter (**A**) out. If fuel filter is stuck, twist filter while pulling it out.

– Clean inside of fuel filter housing (**B**).

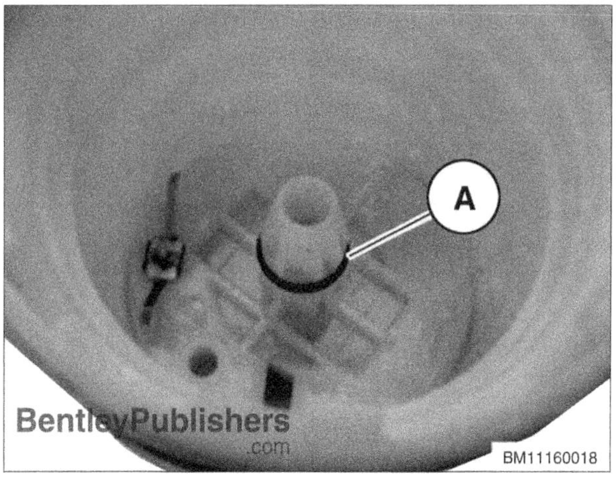

◀ Remove and discard O-ring (**A**). Replace with new.

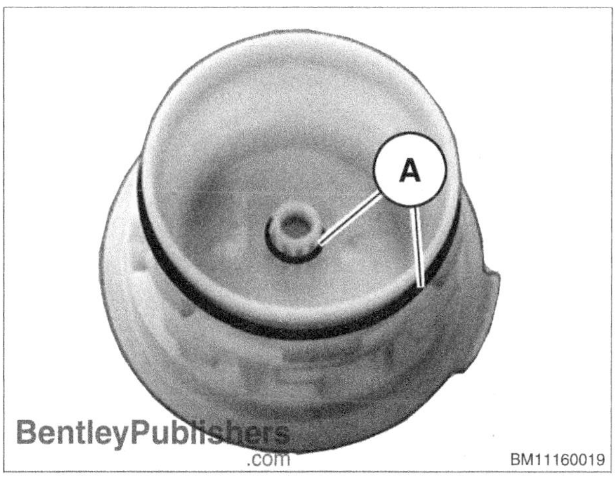

◀ Remove and discard O-rings (**A**). Replace with new.

– Installation is reverse of removal.

> ***CAUTION—***
> • *After finishing repairs but before starting engine, be sure there is at least 5 liters (1.5 gallons) of fuel in the tank. The fuel pump is damaged if run without fuel.*

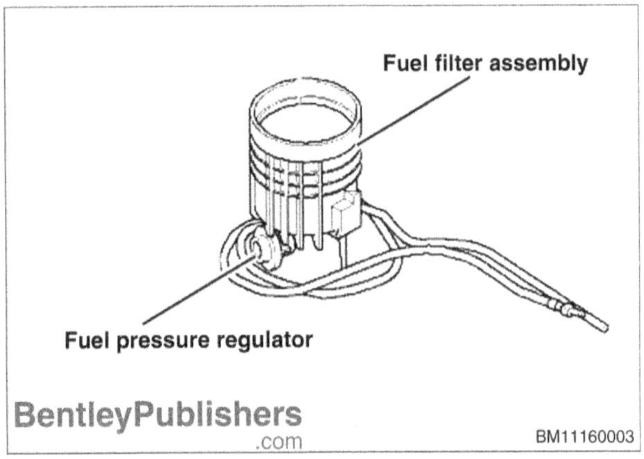

Fuel pressure regulator

System fuel pressure is created by the in-tank fuel pump and maintained by the fuel pressure regulator.

Fuel pressure regulator is integrated with the fuel filter assembly and right side fuel level sender inside the fuel tank. The fuel pressure regulator is not available separately from MINI.

Fuel tank, draining

When draining fuel tank, use a safe storage unit and an approved fuel pumping device.

> ***WARNING—***
> - *Before starting to work on tank removal, make sure hot components, such as the exhaust system, are completely cooled down.*
> - *Fuel may be spilled. Do not smoke or work near heaters or other fire hazards.*

– Start engine and allow to run 10 - 15 seconds to fill fuel compensating siphon assembly. This allows both lobes of fuel tank to be drawn off through fuel filler pipe.

– Disconnect negative cable from battery.

> ***CAUTION—***
> - *Prior to disconnecting the battery, read the battery disconnection cautions given in* **001 Warnings and Cautions**.

– Remove fuel tank filler cap.

– Slide suction hose into filler neck about 130 cm (51 in), twisting as necessary. Withdraw fuel into storage unit.

– Monitor fuel level reduction in both lobes:
  - Remove rear seat cushion and access both fuel tank sender harness connectors. See **Fuel tank component access** in this repair group.
  - Use multimeter to measure resistance at both senders. Resistance should drop as fuel level drops.

– If siphoning mechanism is faulty, drain left tank lobe separately by removing sender cover and pumping fuel directly out of left lobe.

– Remove suction hose from tank filler neck carefully to avoid damaging filler neck baffle plate.

> ***CAUTION—***
> - *After finishing repairs but before starting engine, be sure there is at least 5 liters (1.5 gallons) of fuel in tank. The fuel pump is damaged if run without fuel.*

Fuel tank, removing and installing

- Drain fuel tank. See **Fuel tank, draining** in this repair group.
- Remove both rear wheel housing liners. See **410 Fenders, Engine Hood**.
- Remove underbody panelling on left and right (if equipped).
- Remove exhaust system. See **180 Exhaust System**.
- Remove exhaust heat shields.

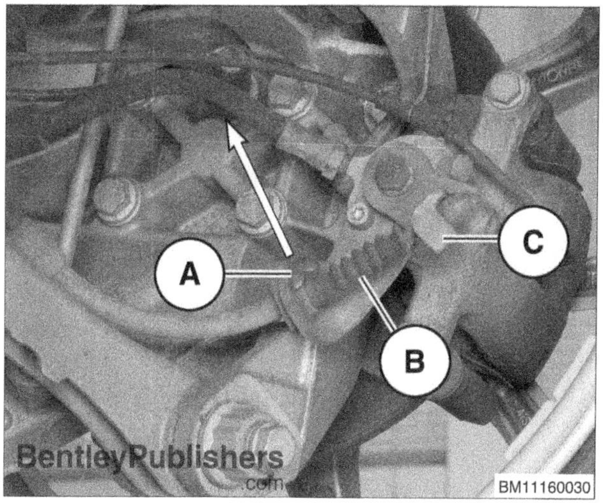

< Working at rear brake caliper:
- Remove parking brake cable locking clip (**A**) in direction of **arrow**.
- Disengage parking brake cable (**B**) from actuating lever (**C**).
- Feed out parking brake cable (**B**) downward.

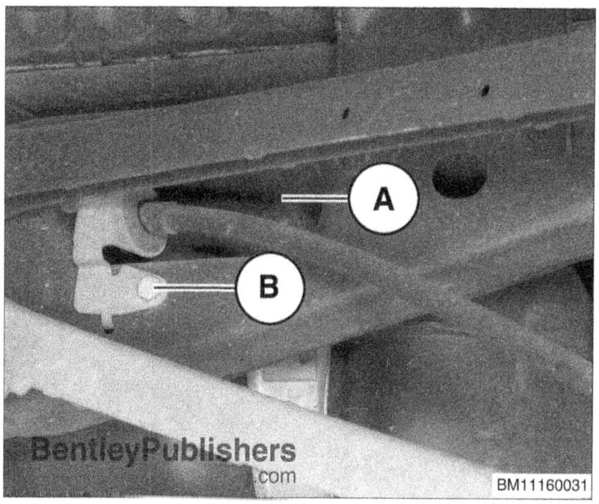

< Remove fasteners (**B**) from rear axle carrier.

- Disengage parking brake cables from holders and feed out through rear axle carrier (**A**). Right side shown. Left is similar.

NOTE—

- *Parking brake cables remain attached to parking brake handle.*

Unlock quick-release fasteners (**arrows**) on activated carbon canister and detach vapor lines.

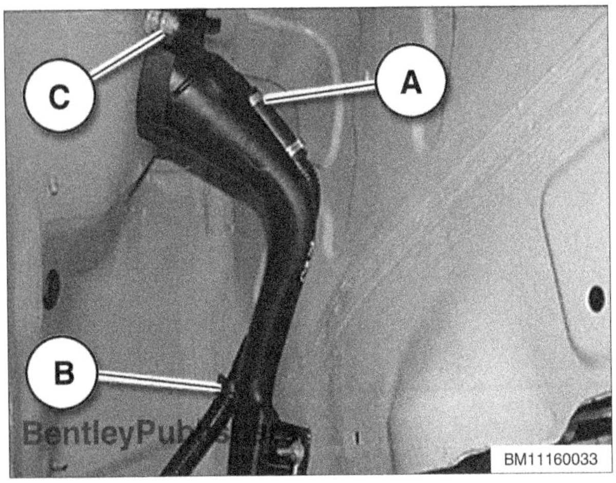

Loosen hose clamp (**A**) and disengage fuel filler neck vent line from holder (**B**).

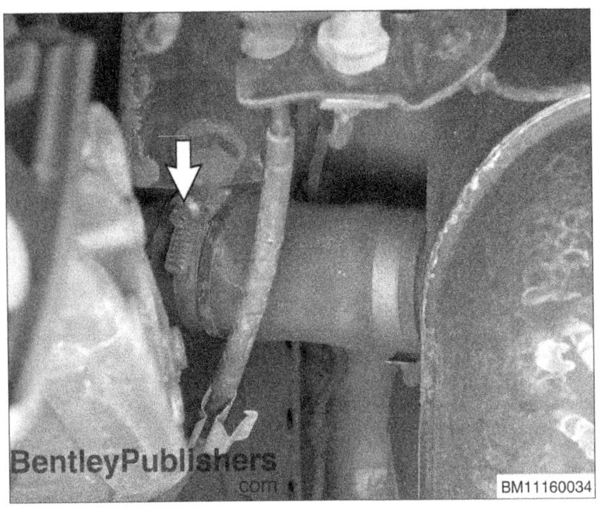

Loosen hose clamp (**arrow**) and detach fuel filler hose from fuel tank.

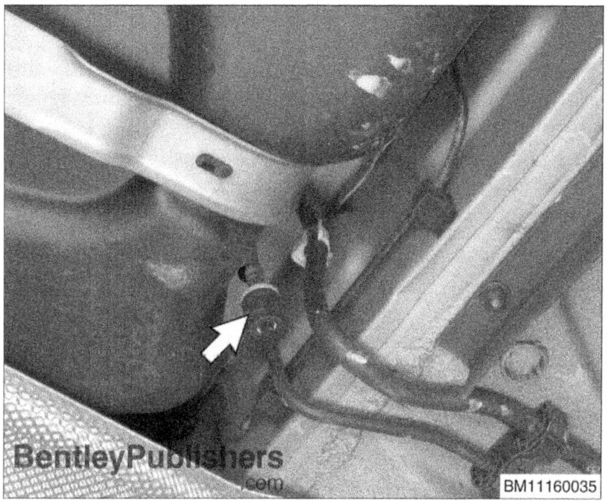

◀ Unlock and disconnect quick-release fastener (**arrow**) of fuel feed line.

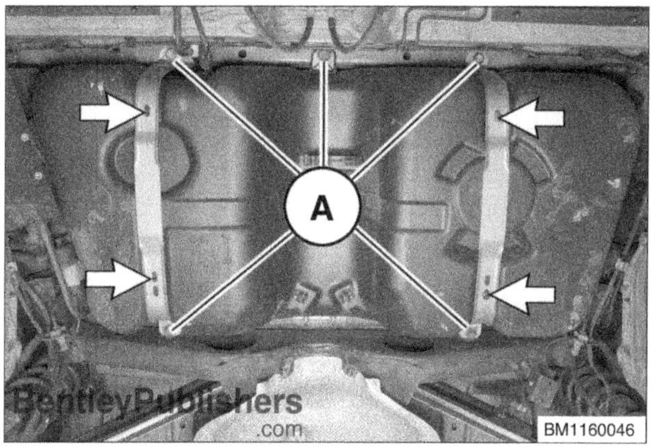

◀ Remove expander rivets (**arrows**).

- Release fasteners (**A**) and lower fuel tank enough to access top of fuel tank.
- Disconnect electrical connector on left sending unit.
- Carefully guide fuel tank vent lines through opening in body.
- Lower fuel tank.
- Installation is reverse of removal. Remember to:
  - Replace hose clamps and O-rings.
  - Properly route all lines and wires.
  - Check fuel system for leaks.

Tightening torques	
Parking brake cable to body	10 Nm (7.3 ft-lb)
Fuel filler neck hose clamp	3 Nm (26 in-lb)
Fuel tank strap to body	19 Nm (14 ft-lb)

HIGH PRESSURE FUEL SYSTEM

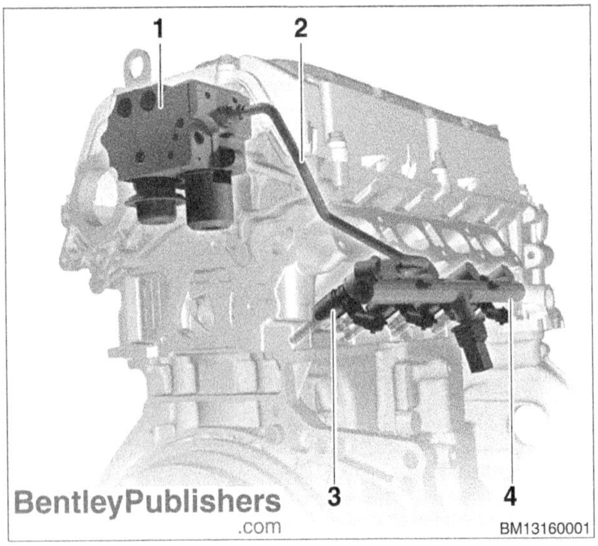

Mini Cooper S models with turbocharged engines are equipped with direct fuel injection utilizing a high pressure pump mechanically driven off the end of the intake camshaft. Fuel is supplied to the high pressure pump at a pressure of 5 bar by the electric fuel pump in the fuel tank. Fuel is delivered by the electric fuel pump corresponding to engine requirements.

1. High pressure camshaft-driven pump
2. High pressure supply line
3. High pressure fuel injector
4. Fuel rail

If a fault is diagnosed in the system, such as failure of the high pressure sensor, the fuel-supply control valve is de-energized; fuel then bypasses into the rail. In case of limp-home mode, turbocharging is deactivated by an opening of the wastegate valves.

Additional causes of limp-home mode can be:

- Implausible high pressure sensor values
- Failure of the fuel supply control valve
- Leakage in the high pressure system
- Failure of the high pressure pump
- Failure of the high pressure sensor

NOTE—

- *N14 and N18 engines produced to March 2012 are quipped with a Continental high pressure pump. N18 engines produced after March of 2012 are equipped with a Bosch high pressure pump.*

High pressure fuel pump, removing (turbo)

– Disconnect negative cable from battery.

CAUTION—

- *Prior to disconnecting the battery, read the battery disconnection cautions given in* **001 Warnings and Cautions**.

– Remove intake manifold. See **130 Fuel Injection**.

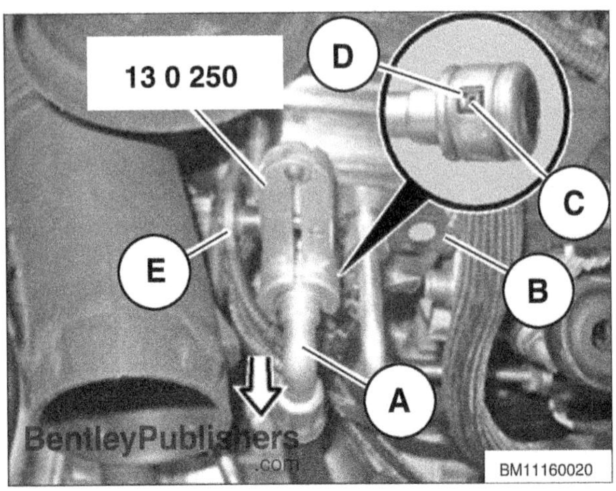

Working at high pressure fuel pump, place MINI special tool **13 0 250** on fuel delivery line (**A**), positioning retaining lugs of tool in recesses of quick-release coupling (**D**).

NOTE—

- *Continental high pressure pump shown, Bosch pump is similar.*

– Tighten knurled screw (**E**) to secure MINI special tool **13 0 250**.

– Place cloth over fuel delivery line (**A**) and high pressure pump (**B**).

– To unlock, push fuel delivery line (**A**) towards high pressure pump (**B**), then detach in direction of **arrow**.

– Remove MINI special tool **13 0 250**.

– Seal fuel delivery line and high pressure fuel pump openings to prevent contamination.

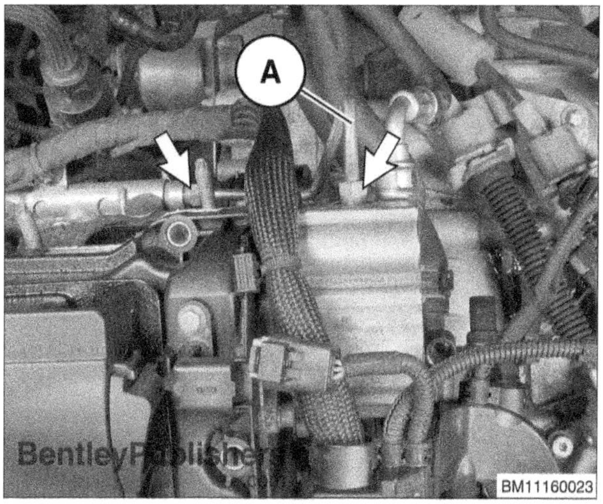

< Continental High pressure pump:

- Loosen fuel line nuts (**arrows**).
- Remove fuel delivery line (**A**).
- Seal fuel delivery line openings to prevent contamination.

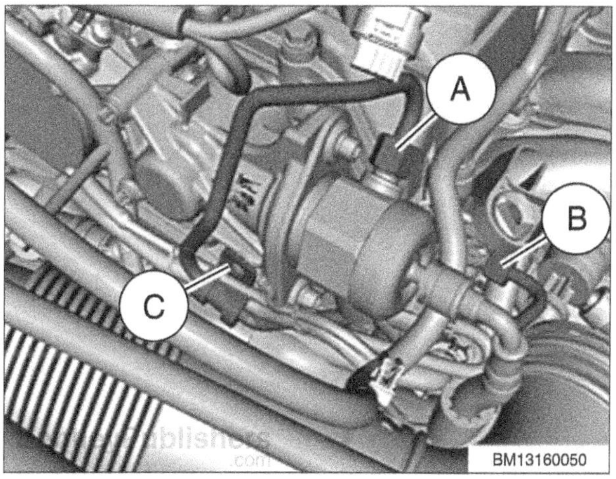

< Bosch High pressure pump:

- Loosen fuel line nuts (**A, B**).
- Remove support bracket fastener (**C**) and remove fuel supply line.
- Seal fuel delivery line openings to prevent contamination.

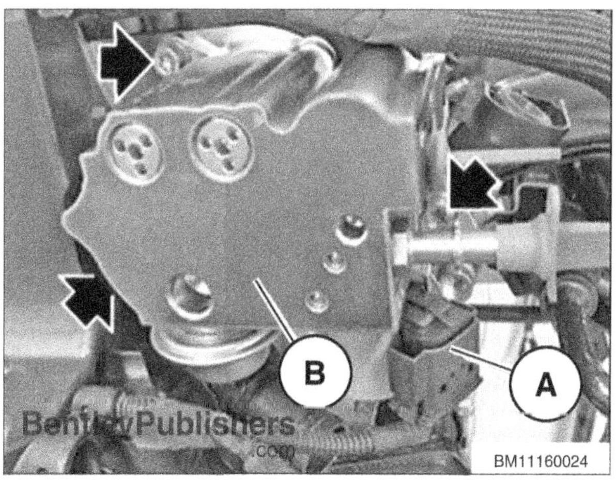

< Continental High pressure pump:

- Disconnect electrical connector (**A**).
- Remove fasteners (**arrows**) and discard.
- Detach high pressure fuel pump (**B**) from cylinder head and remove. Be prepared to catch dripping oil with a cloth.

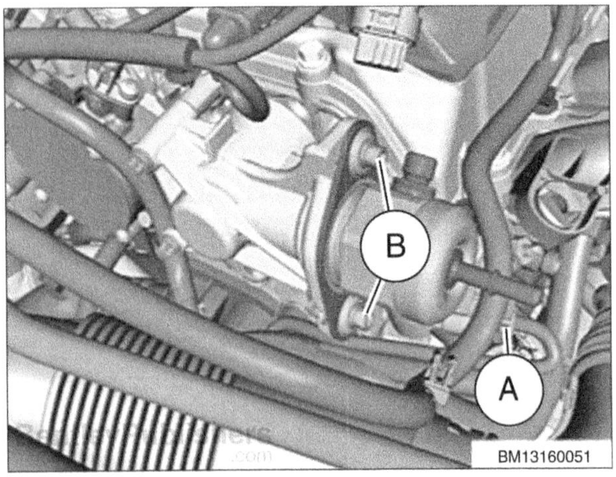

Bosch high pressure pump:

- Disconnect electrical connector (**A**).
- Remove fasteners (**B**) and discard.

NOTE—
- *Drive tappet behind pump may fall when removing pump from housing. Be prepared to catch tappet during pump removal.*

High pressure fuel pump, installing (turbo)

- Clean cylinder head and pump sealing surface.

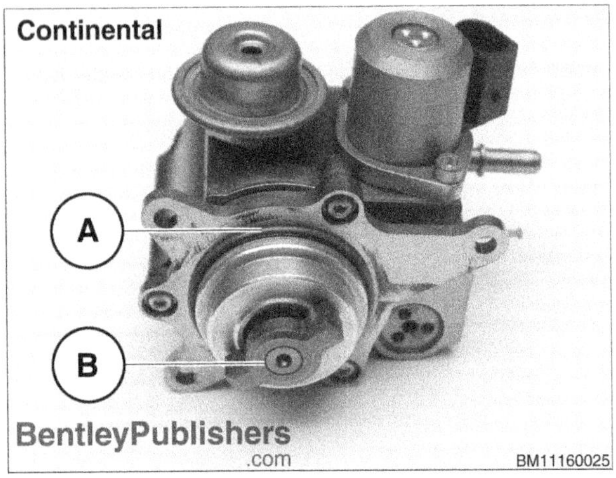

Continental high pressure pump:

- Replace O-ring (**A**).
- Rotate drive (**B**) of high pressure fuel pump to correct position prior to installation.

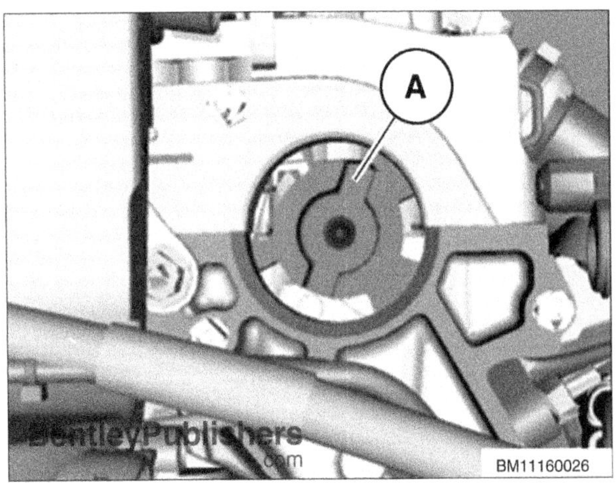

Continental high pressure pump:

- Align drive of pump with groove (**A**) on intake camshaft.

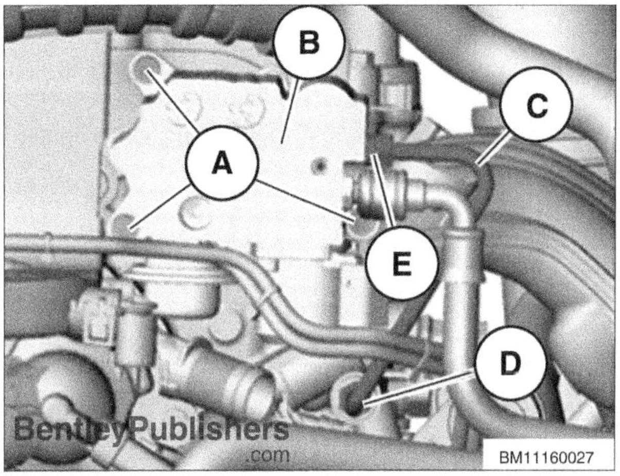

Continental high pressure pump:

– Use the following sequence when installing fuel line:
  - Lightly coat fuel line connections with transmission fluid.
  - Hand tighten high pressure fuel pump fasteners (**A**).
  - Preinstall fuel line (**C**).
  - Tighten nuts (**D**, **E**) finger tight.
  - Tighten high pressure fuel pump fasteners (**A**) to proper torque.
  - Tighten nuts (**D**, **E**) to proper torque.

Tightening torques (Continental high pressure pump)	
High pressure fuel pump to cylinder head • N14 engine (replace fasteners) • N18 engine (replace fasteners)	 11 Nm (97 in-lb) 10 Nm (89 in-lb)
Fuel line to high pressure fuel pump • N14 engine stage 1 stage 2 stage 3 (warm engine 5 minute, retorque) • N18 engine	 15 Nm (11 ft-lb) 33 Nm (24 ft-lb) 33 Nm (24 ft-lb) 26 Nm (19 ft-lb)

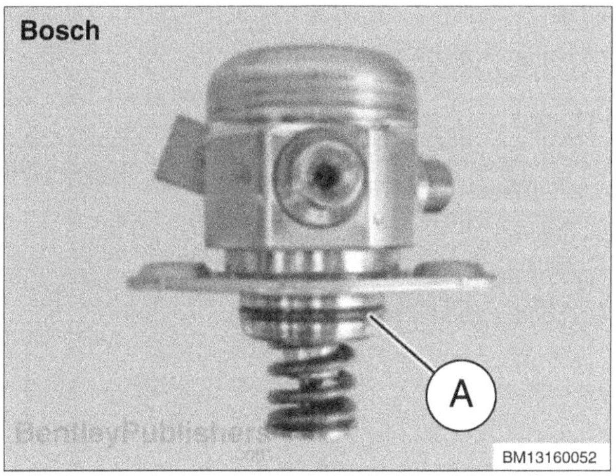

Bosch high pressure pump:

– Replace O-ring (**A**).

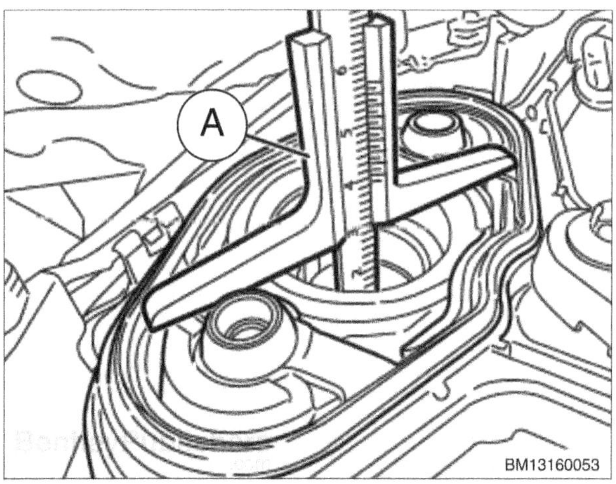

Bosch high pressure pump:

– Cam of high pressure pump drive must be set to bottom dead center before installing high pressure pump.
  - Rest depth gauge (**A**) flat on flange for high pressure pump.
  - Rotate engine at crankshaft belt drive in clockwise direction until depth gauge shows camshaft lobe has reached bottom most position.

High pressure fuel pump, installing (turbo)

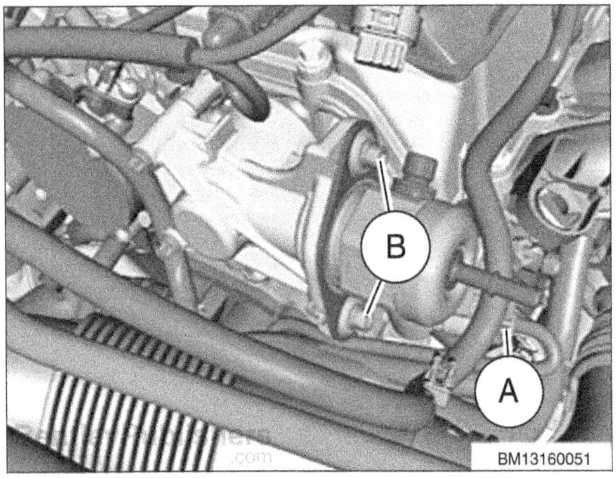

< Bosch high pressure pump:

– Install fasteners (**B**) and slowly draw pump down by tightening fasteners alternately in stages (90° increments).
  - When high pressure fuel pump is flush to mounting flange tighten fasteners to final torque.
  - Connect electrical connector (**A**)

> ***CAUTION—***
> *• Damage may occur if fasteners are not tightened in equal increments.*

Tightening torques (Bosch high pressure pump)	
High pressure fuel pump (replace fasteners)	12 ±1 Nm (106 in-lb)

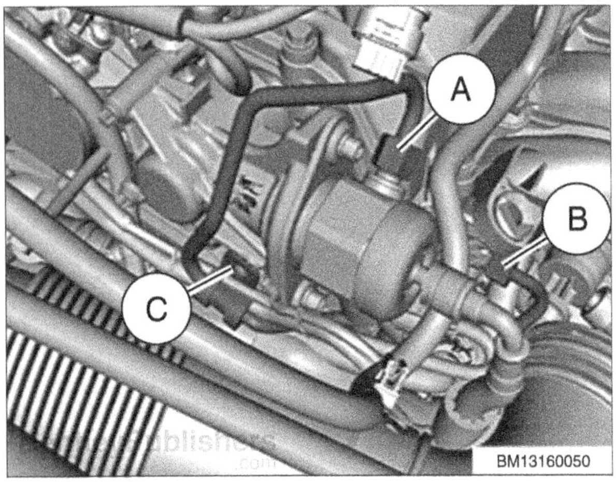

< Bosch high pressure pump:

– Use the following sequence when installing fuel line:
  - Lightly coat fuel line connections with transmission fluid.
  - Tighten nuts (**A, B**) finger tight.
  - Install and tighten support bracket fastener (**C**).
  - Tighten nut (**B**) on high pressure rail.
  - Tighten nut (**A**) on high pressure pump.

Tightening torque	
Fuel line to high pressure fuel pump	26 Nm (19 ft-lb)

– Reassemble engine.

– Check fuel system for leaks.

– Check for fault codes and reset ECM memory.

Fuel Tank Evaporative Control

Evaporative control, also referred to as running losses control, is designed to prevent fuel system evaporative losses from venting into the atmosphere. The components of this system allow control and monitoring of evaporative losses by on-board diagnostic (OBD II) software incorporated into the engine control module (ECM). Listed below are the main components of the evaporative control system and their functions:

- Carbon canister stores fuel vapors.
- Plumbing ducts vapors from fuel tank to canister and from canister to intake manifold.
- Carbon canister purge valve is controlled by engine control module (ECM).
- Pressure regulator shunts excess fuel volume directly back to fuel tank before it circulates through the fuel lines.
- Leak detection unit (DMTL) pressurizes fuel tank and evaporative system to monitor system leaks.

Evaporative system troubleshooting

< Start by accessing diagnostic trouble codes (DTCs) using a MINI scan tool.

- For purposes of OBD II emissions compliance, the DME system sets a diagnostic trouble code (DTC) when it detects a leak in the evaporative control system.
- Malfunction indicator light (MIL) is illuminated upon second recurrence of fault. See **OBD On-Board Diagnostics**.

– When leak testing, observe the following conditions to obtain plausible results:

- Fuel tank ¼ to ¾ full.
- Vehicle parked for at least 2 hours to allow fuel to reach ambient temperature. Ideal fuel temperature is 10°- 20°C (50° - 68°F).
- Do not refuel immediately before leak test.

– If a leak is detected, check the following areas:

- Fuel filler cap leaking or off.
- Fuel tank ventilation lines leaking at fuel tank or activated carbon canister.
- Tank ventilation valve leaking (in engine compartment).
- Fuel level sensor and fuel pump assembly cover leaking.

For evaporative system component replacement, see **Fuel Supply Components** in this repair group.

Evaporative system leak test

Use the following leak test when diagnosing evaporative system leak fault codes.

– Working underneath left rear bumper, gain access to activated carbon canister. See **Activated carbon canister, removing and installing** in this repair group.

◄ Remove fresh air line (**arrow**) from DMTL pump.

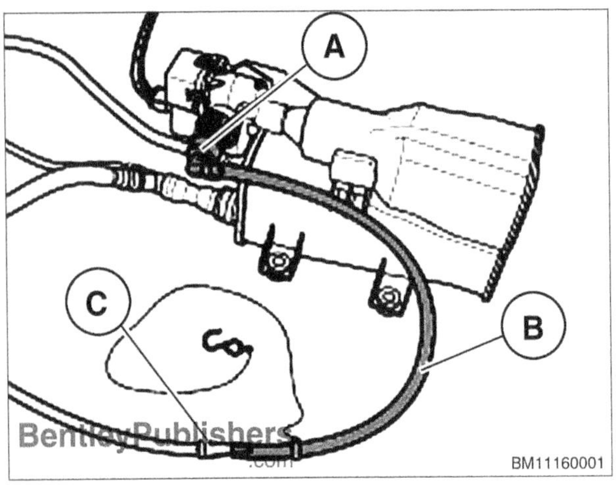

◄ Install MINI test adapter 83 30 0 433 207 (**B**) (or suitable substitute) at DMTL fresh air connection (**A**).

– Connect evaporative system smoke machine (**C**) to test adapter (**B**).

> ***WARNING—***
> - *Use only smoke machine that utilizes nitrogen to pressurize fuel system.*
> - *The fuel system may release a substantial amount of fuel-rich gasoline vapors during testing. Work in a well ventilated area.*

– Remove fuel filler cap.

– Activate smoke machine and fill fuel system with smoke until it escapes from fuel filler neck.

– Turn off smoke machine.

– Reinstall fuel filler cap.

– Reactivate smoke machine and fill fuel system with smoke.

> ***WARNING—***
> - *Do not exceed a maximum fuel system pressure of 0.3 bar (4 psi) when using smoke machine.*

– Examine fuel system components for signs of escaping smoke. Repair or replace faulty components.

– After completing tests:
- Attach DMTL fresh air line.
- Check for fault codes and reset ECM memory.

Activated carbon canister, removing and installing

The carbon canister is on the left side underneath the rear bumper.

– Raise rear end of car and support safely.

> ***CAUTION—***
> *• Make sure car is stable and well supported at all times. Use a professional automotive lift or jack stands. A floor jack is not adequate support.*

◄ Remove right side rear bumper splash shield fasteners (**arrows**).

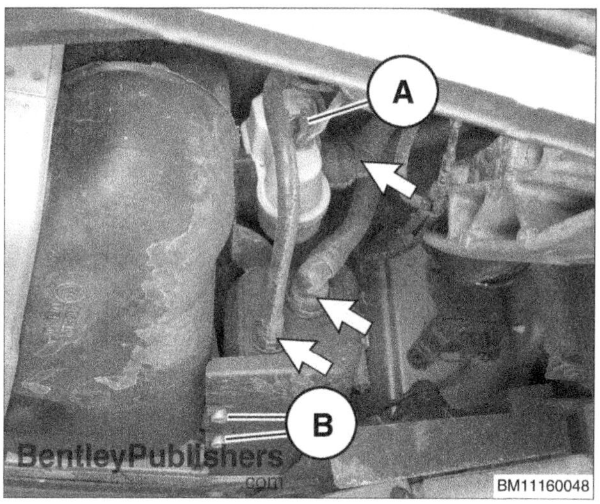

◄ Remove activated carbon canister:

- Remove line connections at leak detection pump (**arrow**).
- Disconnect electrical harness connector (**A**).
- Remove fasteners (**arrows**) and remove carbon canister from vehicle.

– Installation is reverse of removal.

Fuel tank leak detection unit (DMTL), removing and installing

The DMTL is mounted to the activated carbon canister.

– Raise car and support safely.

> ***WARNING—***
> - *Make sure the car is stable and well supported at all times. Use a professional automotive lift or jack stands designed for the purpose. A floor jack is not adequate support.*

– Remove activated carbon canister. See **Activated carbon canister, removing and installing** in this repair group.

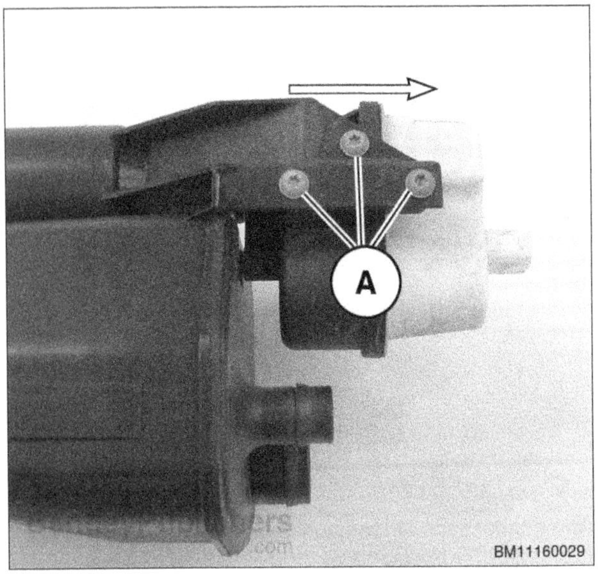

◄ Remove fasteners (**A**) and pull DMTL unit out of carbon canister in direction of **arrow**.

– Installation is reverse of removal.

170 Radiator and Cooling System

GENERAL

This repair group covers component repair information for the engine cooling system.

See also:

- **020 Maintenance** for coolant hose inspection
- **100 Engine–General** for engine code and application information
- **130 Fuel Injection** for engine coolant temperature (ECT) sensor service
- **640 Heating and Air-conditioning** for heater core replacement

Electrically heated thermostat

An electrically heated DME controlled thermostat is externally mounted above the coolant pump. The engine control module (ECM) activates the thermostat to maintain engine coolant temperature within a narrow range.

If the electronics fail, the mechanical function of the thermostat acts as a fail-sate.

Radiator and expansion tank

The radiator is a crossflow design. An expansion tank provides for coolant expansion at higher temperatures and easy monitoring of the coolant level.

Automatic transmission fluid (ATF) is circulated through an additional heat exchanger (ATF cooler).

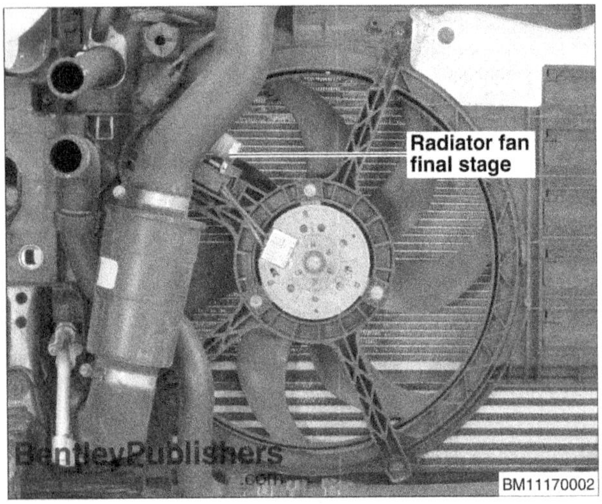

Electric cooling fan

Models covered by this manual use an electric cooling fan mounted on the engine side of the radiator.

The electric cooling fan is controlled by the engine control module (ECM) via the output final stage mounted on the fan housing next to the fan motor. The fan is operated using a pulse width modulated signal. Fan circuit wiring is protected by a high-amperage fuse. Electric fan activation is based on the following inputs to the ECM:

- Radiator outlet temperature
- Calculated catalytic converter temperature
- Vehicle speed
- Battery voltage
- Calculated refrigerant pressure

When the vehicle is first started, the ECM activates the electric fan briefly at 20% of its maximum speed, then switches OFF. This is for diagnostic monitoring. The voltage generated by the fan when it slows down (acting as a generator) must match the stored rpm values in the fan output stage to confirm that the fan is operating correctly.

If the ECM fault memory indicates a cooling fan fault, check that the fan is not seized and that it spins freely.

Note that when A/C is switched ON, the electric fan is not immediately turned on. After the engine is switched OFF, the fan may continue to run at varying speeds for up to 10 minutes, based on calculated catalyst temperature.

Automatic transmission fluid (ATF) cooler

ATF lines connect the transmission to the ATF cooler (heat exchanger) at top of transmission.

On cold engine start-up, the engine coolant is heated more quickly than the transmission fluid. Heat from engine coolant is used to warm up the ATF faster, reducing drag in the transmission and improving fuel mileage.

Once ATF reaches normal operating temperature, it is hotter than the engine coolant. The ATF heat exchanger then acts as a cooling device.

Warnings and Cautions

WARNING—

- *At normal operating temperature the cooling system is pressurized. Allow the system to cool as long as possible before opening (a minimum of one hour), then release the cap slowly to allow safe release of pressure.*
- *Releasing cooling system pressure lowers the coolant boiling point. The coolant may boil suddenly. Use heavy gloves and wear eye and face protection to guard against scalding.*
- *Use extreme care when draining and disposing of engine coolant. Coolant is poisonous and lethal to humans and pets. Pets are attracted to coolant because of its sweet smell and taste. Seek medical attention immediately if coolant is ingested.*

CAUTION—

- *Avoid adding cold water to the coolant while the engine is hot or overheated. If it is necessary to add coolant to a hot system, do so only with the engine running and coolant pump operating.*
- *To avoid excess silicate gel precipitation in the cooling system and loss of cooling capacity, use MINI coolant or equivalent low silicate antifreeze.*
- *If oil enters the cooling system, flush the radiator, expansion tank and heating circuit with cleaning agent. MINI recommends removal of the radiator and expansion tank to flush.*
- *When working on the cooling system, cover the alternator to protect it against coolant drips.*
- *Prior to disconnecting the battery, read the battery disconnection cautions given in* **001 Warnings and Cautions**.

170

TROUBLESHOOTING

Begin the diagnosis of cooling system problems with a thorough visual inspection. If no visual faults are found, check the engine control module (ECM) fault memory for stored diagnostic trouble codes (DTCs) using a MINI scan tool.

Common cooling system faults can be grouped into one of 4 categories:

- Cooling system leaks
- Poor coolant circulation
- Radiator cooling fan faults
- Electrical or electronic faults

Cooling system inspection

– Check operation of electric radiator fan:
  - Start engine, turn A/C on, MAX setting. Electric fan should run.

– Check coolant hoses for cracks or softness. Check clamps for looseness. Check coolant level and check for evidence of coolant leaks from engine.

– Check that radiator fins are not blocked with dirt or debris. Clean radiator using low-pressure water or compressed air. Blow outward from engine side out.

– At normal operating temperature, cooling system is pressurized. This raises boiling point of coolant. Leaks may prevent system from becoming pressurized. Pressure test cooling system to help pinpoint hard-to-find leaks. See **Cooling system pressure test** in this repair group.

– If cooling system is full of coolant and holds pressure:
  - Use an appropriate scan tool to interrogate engine control module (ECM) for radiator fan or DME control circuit faults.
  - Check for failed thermostat or coolant pump impeller.
  - Check for clogged or plugged radiator or coolant passages.

Cooling system pressure test

A cooling system pressure tester is used to test for coolant leaks, including internal ones. Common sources of internal coolant leaks are a faulty cylinder head gasket, a cracked cylinder head, or a cracked engine block.

> ***WARNING—***
> - *At normal operating temperature the cooling system is pressurized. Allow the system to cool before opening. Remove the cap slowly to allow safe release of pressure.*

- With engine cold, install pressure tester (MINI special tools 17 0 101 / 17 0 113 or equivalent) to expansion tank. Pressurize system to specification listed in **Table a**.
  - Pressure should not drop more than 0.1 bar (1.45 psi) for at least two minutes.
  - If pressure drops rapidly and there is no sign of external leakage, cylinder head gasket may be faulty. Perform compression and leak-down tests.
  - Test expansion tank cap using pressure tester with correct adapter (MINI special tool 17 0 114 or equivalent). Replace faulty cap or cap gasket.

> ***CAUTION—***
> - *Exceeding the specified test pressure could damage the radiator or other system components.*

Table a. Cooling system test pressures	
Component	**Test pressure**
Radiator	1.4 bar (20.3 psi)
Radiator cap	2 bar (29 psi)

Combustion chamber leak test

- If you suspect that combustion chamber pressure is leaking into the cooling system past the cylinder head gasket, use an exhaust gas analyzer to test the vapors rising from the coolant at the expansion tank.

> ***CAUTION—***
> - *Use an extension tube above the reservoir neck to maintain distance between the surface of the coolant and the gas analyzer nozzle. The gas analyzer is easily damaged if it is allowed to inhale liquid coolant.*
> - *While running engine to check for causes of overheating, watch the coolant temperature carefully to avoid engine damage.*

Thermostat

If the engine overheats or runs too cool and no other cooling system tests indicate trouble, the thermostat may be faulty.

The electrically heated thermostat is monitored by the OBD II diagnostic software. The fault may lie in the DME. Check engine control module (ECM) fault memory for stored diagnostic trouble codes (DTCs) using a MINI scan tool.

Cooling System Service

Coolant, draining and filling

> ***WARNING—***
> - *Allow cooling system to cool before opening or draining the cooling system.*

– Raise car and support safely.

> ***WARNING—***
> - *Make sure the car is stable and well supported at all times. Use a professional automotive lift or jack stands designed for the purpose. A floor jack is not adequate support.*

– Remove splash shield from under engine.

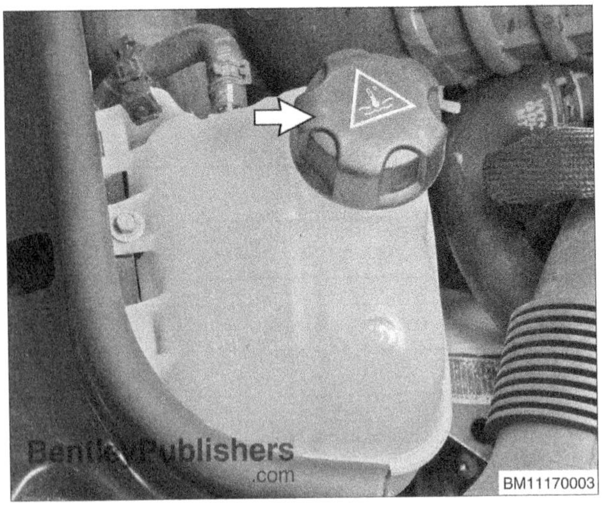

◄ Remove cap (**arrow**) from radiator expansion tank.

> ***CAUTION—***
> - *Cover alternator when working on oil, coolant or fuel circuits to protect from damage.*

– Place a 5-gallon pail underneath radiator.

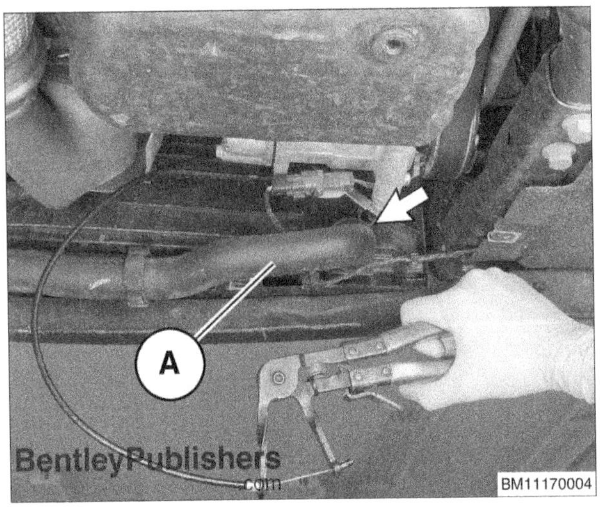

◄ **N12, N16 engine**: Using hose clamp pliers, release hose clamp on lower coolant hose (**arrow**).
- Remove coolant hose (**A**).

◄ **N14, N18 engine**: Release hose clamp on lower coolant hose (**arrow**) with hose clamp pliers.
- Remove coolant hose (**arrow**).

– Catch and dispose of coolant properly.

> ***WARNING—***
> - *Coolant is poisonous. It is especially lethal to pets. Clean up spills immediately and rinse the area with water. If coolant is ingested, immediately seek medical attention.*

– Before filling cooling system, reattach removed hoses.

– Using a coolant mixture of 50% antifreeze and 50% distilled water, fill expansion tank slowly to top.

> ***CAUTION—***
> - *Open radiator bleed screw when filling cooling system. See* **Cooling system, filling and bleeding** *in this repair group.*
> - *Tap water may cause corrosion of radiator and engine components.*
> - *Do not reuse coolant.*

170

– Bleed cooling system. See **Cooling system, filling and bleeding** in this repair group.

Cooling system, filling and bleeding

Trapped air in the cooling system can prevent proper coolant circulation and cause overheating. Whenever the coolant is drained and filled, bleed the cooling system.

> ***CAUTION—***
> - *Carry out this procedure when replacing cooling system components or refilling the cooling system. Failure to follow this procedure may result in incomplete filling of cooling system, overheating and engine damage.*

Bleeding without vacuum bleeder

– Before refilling radiator:
- Switch ignition ON. Do not start engine.
- Set heater temperature controls to full warm.
- Set blower control to low.
- Set seat heater control to maximum.

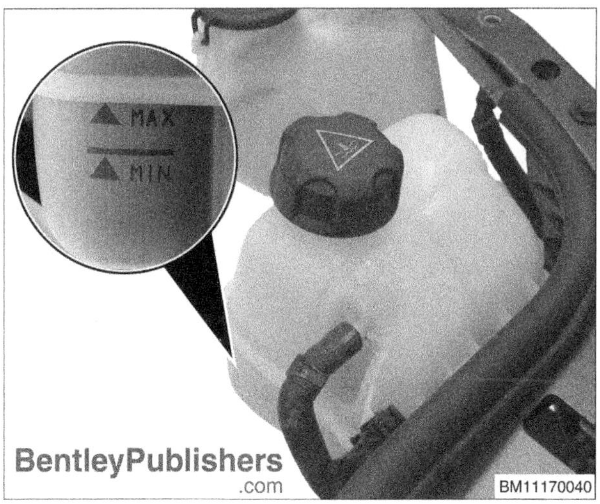

◄ Using a coolant mixture of 50% antifreeze and 50% distilled water, fill expansion tank slowly to MAX (**inset**).

> ***CAUTION—***
> - *Use genuine MINI coolant or its equivalent to avoid the formation of harmful, clogging deposits in the cooling system.*

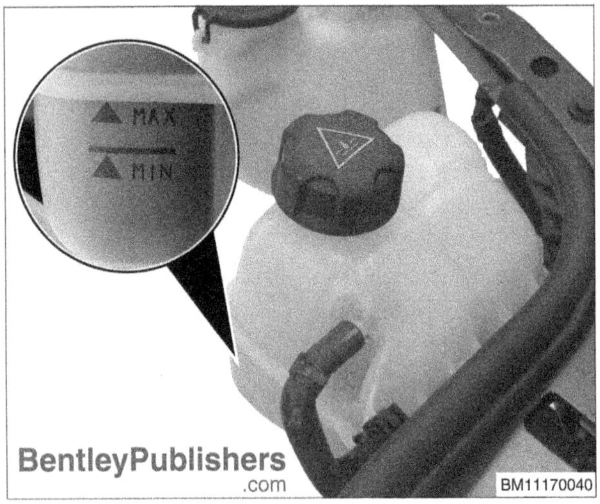

< Start engine and run at idle speed for one minute (coolant reservoir cap open). Adjust coolant level to MAX (**inset**). Close cap.

- Run engine until it reaches operating temperature.
- After engine has cooled, recheck coolant level.
- Top up so that coolant level is at MAX.
- If trapped air cannot be bled manually, pressure-bleed system using cooling system vacuum bleeder.

Table b. Cooling system capacities	
Capacity	5.2 L (1.4 gal)

Bleeding with vacuum bleeder

Use MINI special tools **17 00 039**, **17 0 107 and 17 0 102** to carry out this bleeding procedure.

– Before refilling radiator:

- Check that cooling system expansion tank is empty.
- Filling unit container should have 1 - 2 liters more than cooling system capacity.
- Position filling unit container at same height as coolant expansion tank.
- Set heater temperature controls to full warm.
- Set blower control to low.
- Set seat heater control to maximum.

Placing cooling system under vacuum

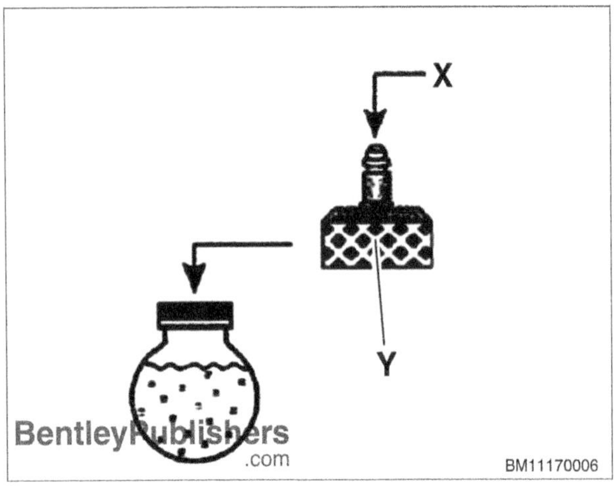

< Select suitable adapter (**Y**) and connect to coolant expansion tank.

– Connect filler unit to adapter connection (**X**).

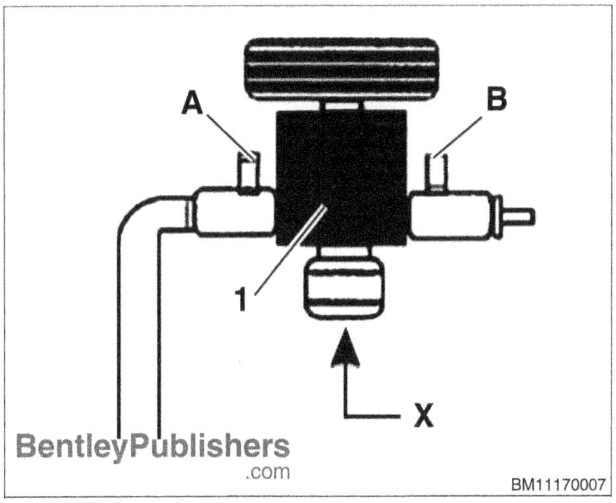

Close shut-off valves (**A**, **B**) on filling unit (**1**).

- **X** Expansion tank connection

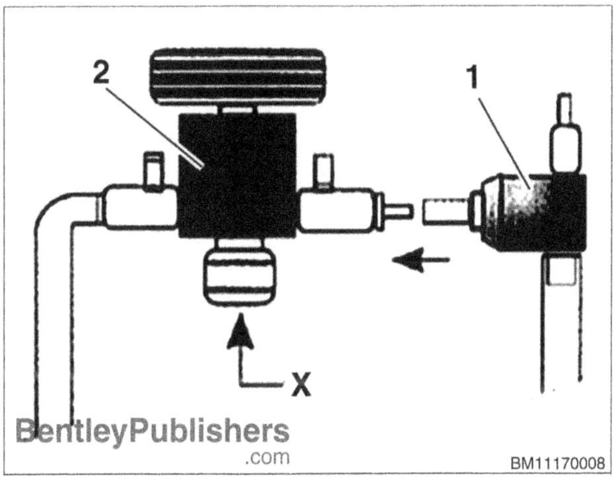

Connect venturi nozzle (**1**) to filling unit (**2**).

- **X** Expansion tank connection

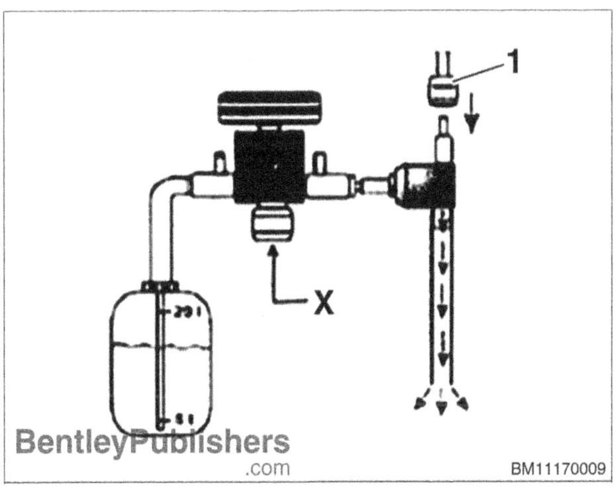

Connect compressed air (**1**).

- 6 - 10 bar (87 - 145 psi)
- **X** Expansion tank connection

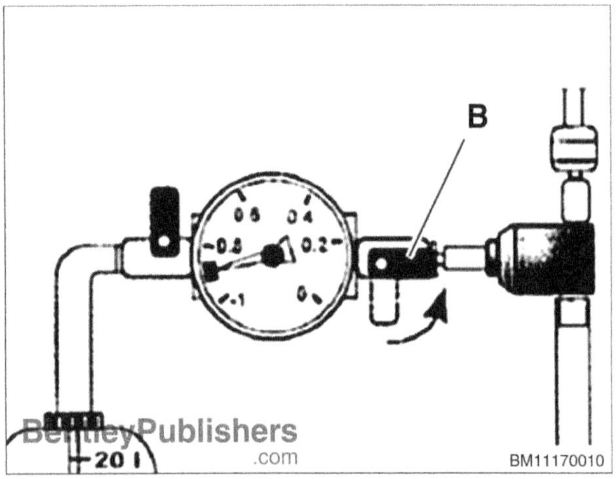

◀ Open shut-off valve (**B**).

– Venturi nozzle produces a flow noise.

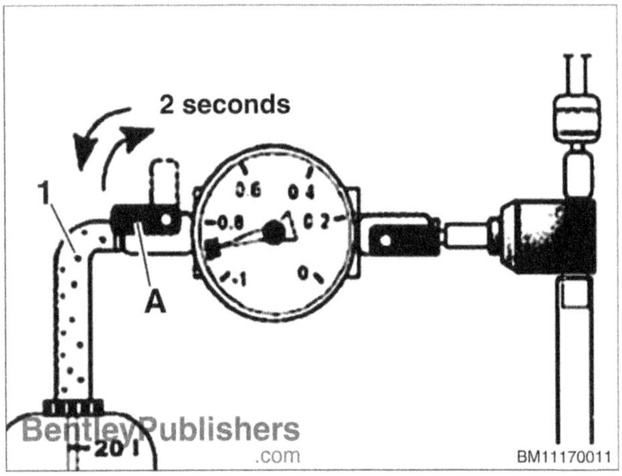

◀ Open shut-off valve (**A**) until filling hose (**1**) is free of bubbles.

– Close shut-off valve (**A**) again. Filling hose (**1**) is vented in this way.

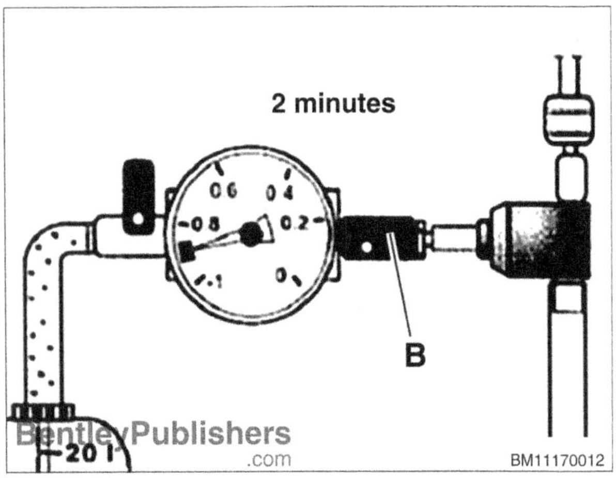

◀ Shut-off valve (**B**) remains open. Generate vacuum in coolant system for approx. 2 minutes. Final vacuum is reached at a vacuum of -0.7 to -0.95 bar. (Green scale on vacuum gauge).

NOTE—
- *The coolant hoses contract during vacuum build-up.*

– Close shut-off valve (**B**) again.

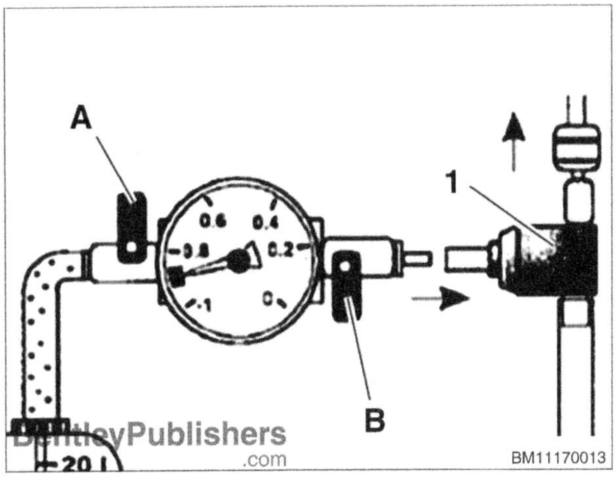

◀ Close both shut-off valves (**A**, **B**). Then seal venturi nozzle (**1**).

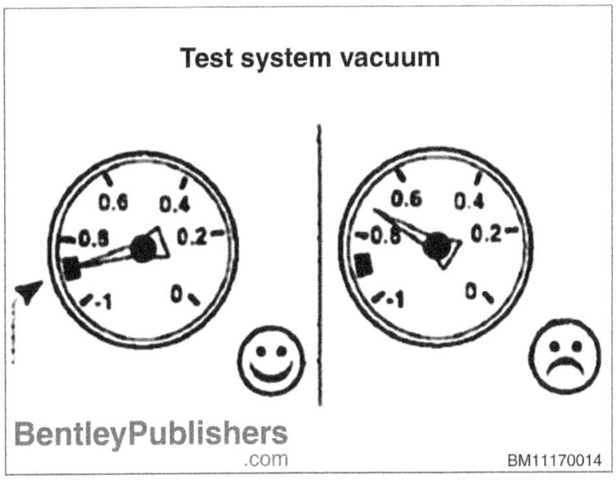

◀ Make sure cooling system holds vacuum for 30 seconds. If needle in vacuum gauge falls, this indicates a leak in cooling system.

- If the vacuum remains constant, proceed with filling.
- In case of leaks, check cooling system for leaks.

Filling cooling system under vacuum

– Position filling unit container at same height as coolant expansion tank.

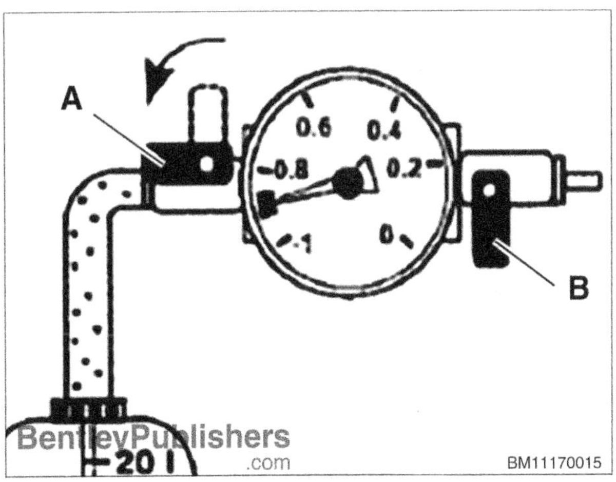

◀ Shut-off valve (**B**) remains closed during filling process.

– To fill cooling system, open shut-off valve (**A**) to filling unit container.

– Add coolant.

– Filling process is finished when needle in vacuum gauge is at 0 bar or no longer falls.

– If necessary, reduce remaining vacuum. Open shut-off valve (**B**) to do so.

– Remove filling unit with adapter from expansion tank.

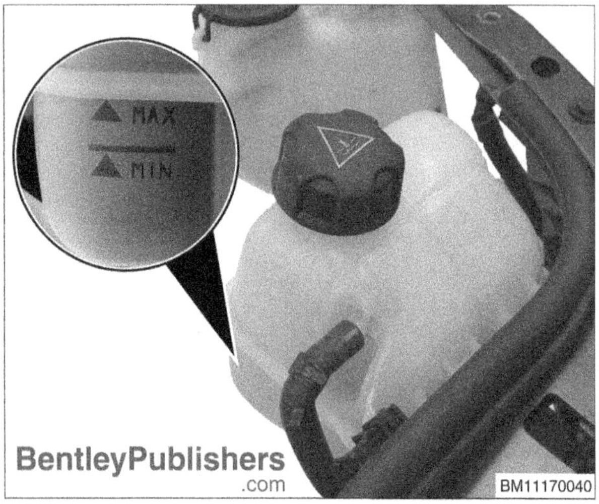

Start engine and run at idle speed for one minute (cap open). Adjust coolant level to MAX (**inset**). Close cap.

- Run engine until it reaches operating temperature.
- After engine has cooled, recheck coolant level.
- Top up so that coolant level indicator is at MAX.

Table c. Cooling system capacities	
Capacity	5.2L (2.17 gal)

Radiator, removing and installing

– Raise car and support safely

> ***WARNING —***
> - *Make sure the car is stable and well supported at all times. Use a professional automotive lift or jack stands designed for the purpose. A floor jack is not adequate support.*

– Drain radiator. See **Coolant draining and filling** in this repair group.

> ***WARNING —***
> - *Allow cooling system to cool before opening or draining system.*

– Remove front bumper. See **510 Exterior Trim, Bumpers**.

– Connect an approved refrigerant recovery unit to A/C system and follow manufacturer's instructions to discharge system.

> ***WARNING —***
> - *Do not attempt to discharge or charge the A/C system without proper equipment and training. Damage to the vehicle and personal injury may result.*

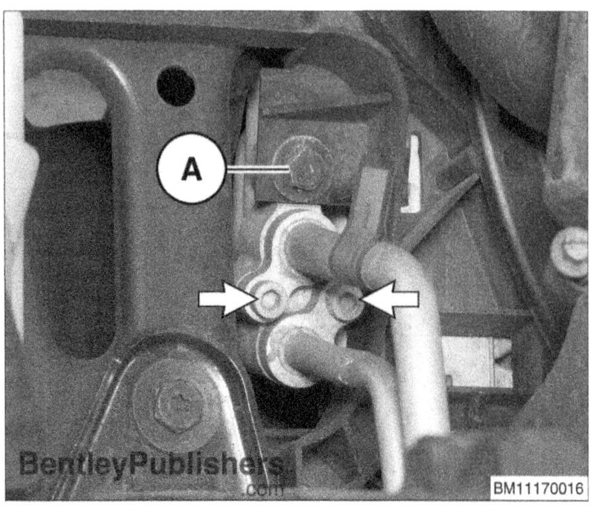

Working at rear of radiator support, remove refrigerant line fasteners (**arrows**) and mounting fastener (**A**).

N14, N18 engine: Remove A/C condenser (**A**) fasteners (**arrows**).

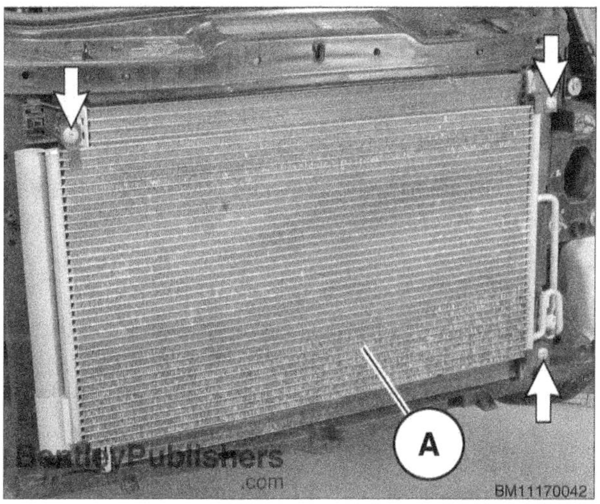

N12, N16 engine: Remove A/C condenser (**A**) fasteners (**arrows**).

- Carefully tilt A/C condenser forward until you can access radiator hoses.

> ***CAUTION—***
> - *Secure A/C condenser from falling to prevent A/C lines from bending.*

N12, N16 engine: Unclip (**arrows**) radiator (**A**) at top left and right and tilt forward.

Using hose clamp pliers, release upper radiator hose clamps (**arrows**).

Using hose clamp pliers, release lower radiator hose clamps (**arrow**).

- Carefully slide radiator up to remove.
- Installation is reverse of removal. Make sure:
  - Radiator seats firmly in lower mounts.
  - Electric cooling fan cowl clips in securely.
  - Refill and bleed cooling system. See **Cooling system, filling and bleeding** in this repair group.
  - Check cooling system for leaks.

Electric cooling fan, removing and installing

- Switch ignition OFF.

N12, **N16 engine**: Disconnect electrical connector (**A**).

- **N12**, **N16 engine**: Remove cooling fan fastener (**arrow**).
- **N14**, **N18 engine**: Remove left side charge air duct. See **130 Fuel Injection**.

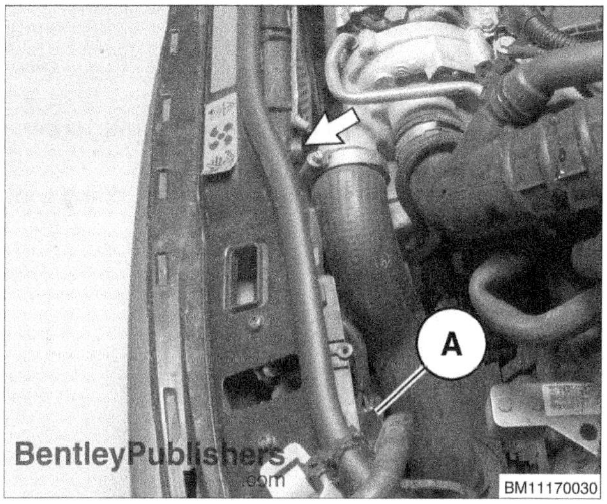

◄ **N14, N18 engine**: Disconnect electrical connector (**A**).

– **N14, N18 engine**: Remove cooling fan fastener (**arrow**).

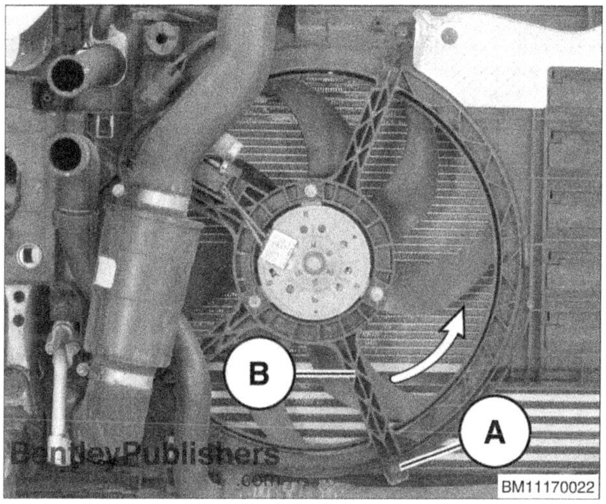

◄ Remove fastener (**A**), unlock fan cowl duct (**B**) in direction of **arrow** and remove.

– Installation is reverse of removal. Make sure:
- Fan cowl snaps securely into clips on lower edge.
- Fan electrical connector snaps firmly into place and wires and hose are routed as before.

Coolant expansion tank, removing and installing

– Drain cooling system. See **Coolant, draining and filling** in this repair group.

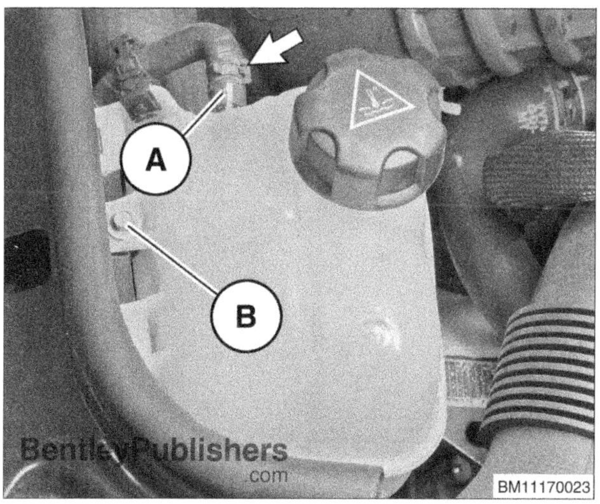

◄ Remove coolant hose spring clamp (**arrow**) with hose clamp pliers and remove hose (**A**) from expansion tank.

– Release expansion tank fastener (**B**).

– Remove expansion tank from rubber mount and raise slightly until lower coolant hose is accessible.

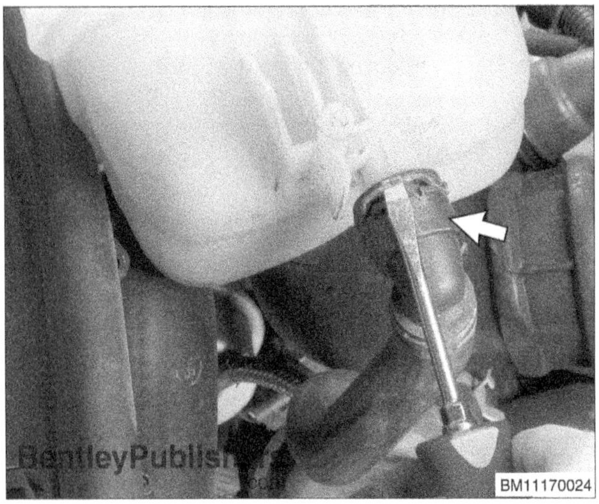
BM11170024

◀ Unlock and detach lower coolant hose (**arrow**) and remove expansion tank.

– Installation is reverse of removal. Remember to:
  - Make sure locating pin at bottom of tank fits securely into rubber grommet.
  - Refill and bleed cooling system. See **Cooling system, filling and bleeding** in this repair group.
  - Check cooling system for leaks.

Thermostat, removing and installing

– Drain cooling system. See **Coolant, draining and filling** in this repair group.

BM11170025

◀ Working at left rear of cylinder head, release lock (**A**) on coolant pipe in direction of **arrow**.

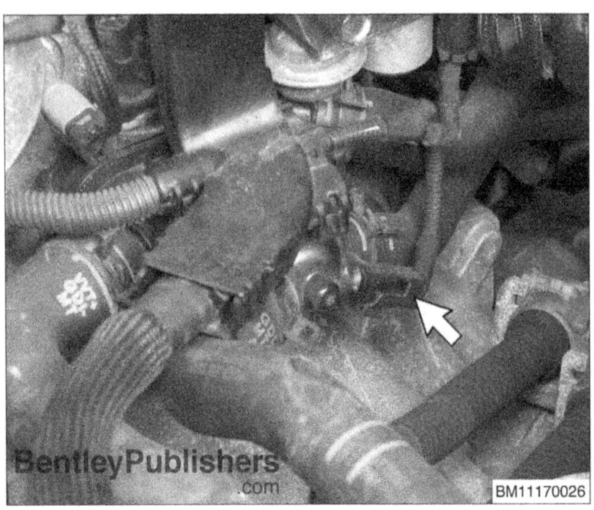
BM11170026

◀ Disconnect electrical connector (**arrow**).

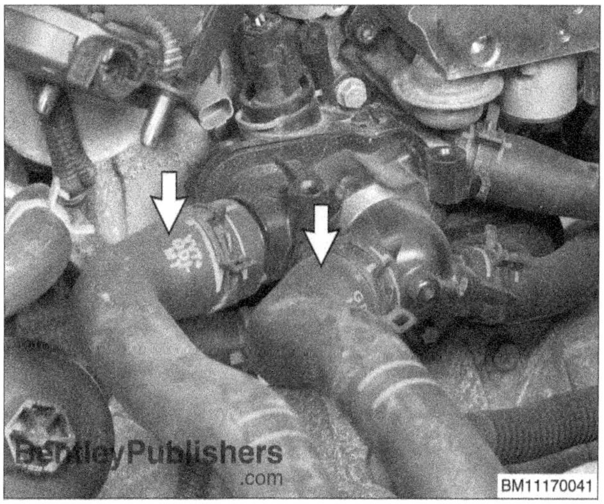

◄ Remove coolant hoses (**arrows**) from thermostat housing.

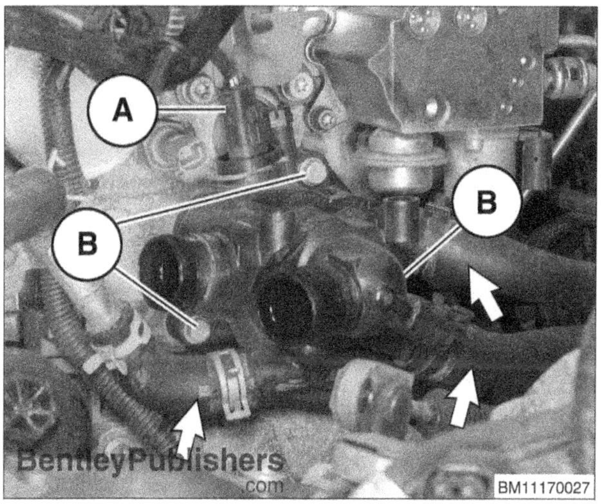

◄ Remove coolant hoses (**arrows**) from thermostat housing.

- Disconnect plug connection (**A**) on coolant temperature sensor.
- Release thermostat housing fasteners (**B**).

– Installation is reverse of removal. Remember to:

- Refill and bleed cooling system. See **Cooling system, filling and bleeding** in this repair group.
- Replace thermostat housing sealing O-ring.
- Check cooling system for leaks.

Tightening torque	
Thermostat housing to cylinder head	8 Nm (71 in-lb)

Coolant pump, removing and installing

– Drain cooling system. See **Coolant, draining and filling** in this repair group.

◄ Place friction wheel in service position by pulling strap (**inset**) all the way out and holding.

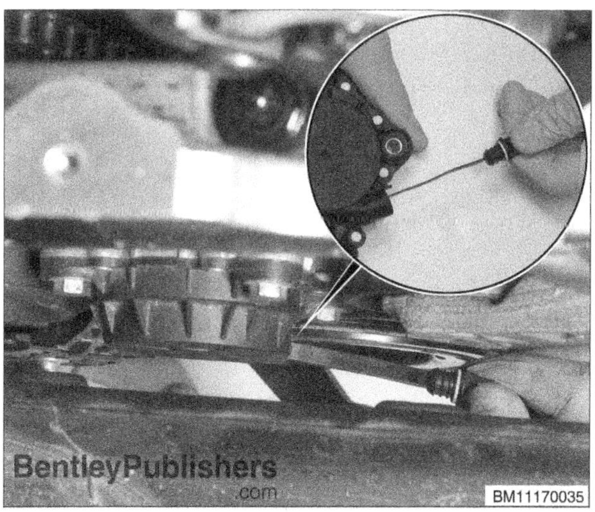

Coolant pump, removing and installing

Remove coolant pump pulley fasteners (**arrows**) then remove coolant pump pulley.

– Remove friction wheel. See **Friction wheel, removing and installing** in this repair group.

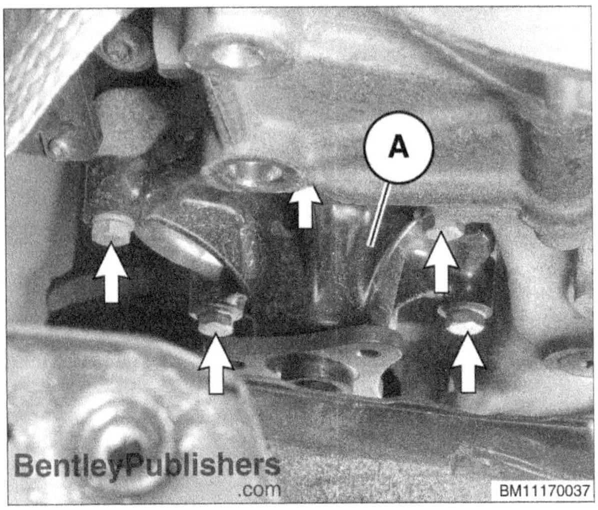

Remove fasteners (**arrows, inset**) and remove coolant pump (**A**) from engine block.

Installation is reverse of removal. Remember to:

- Inspect coolant pump pulley friction surface (**A**) for wear. Replace if necessary.
- Reinstall friction gear.
- Refill and bleed cooling system. See **Cooling system, filling and bleeding** in this repair group.
- Check cooling system for leaks.

Auxiliary coolant pump, removing and installing (N14, N18 engine)

– Drain cooling system. See **Coolant, draining and filling** in this repair group.

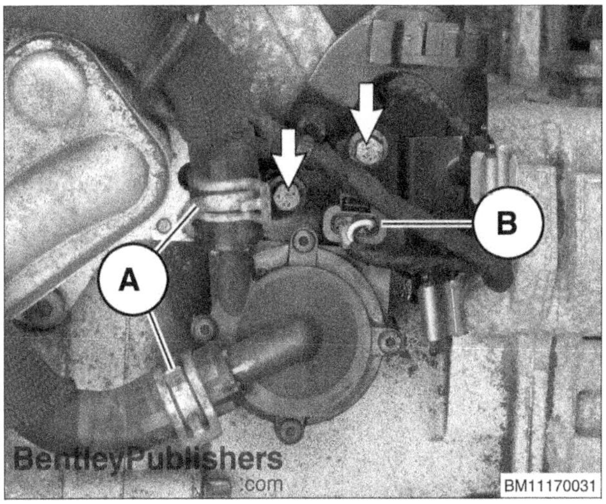

◄ Working above transmission:

- Release hose clamps (**A**).
- Detach coolant hoses from auxiliary coolant pump.
- Disconnect electrical connector (**B**).
- Remove fasteners (**arrows**).

– installation is reverse of removal. Remember to:

- Refill and bleed cooling system. See **Cooling system, filling and bleeding** in this repair group.
- Check cooling system for leaks.

Tightening torque	
Coolant pump to transmission	8 Nm (71 in-lb)

Friction wheel, removing and installing

– Remove right front wheel housing liner. See **410 Fenders, Engine Hood**.

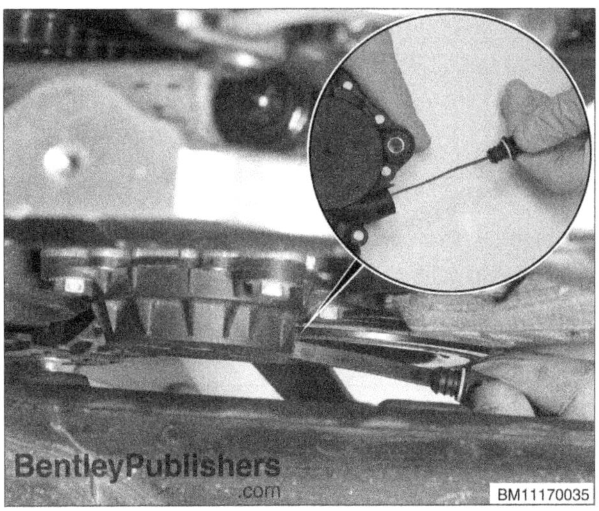

◄ Place friction wheel in service position by pulling strap (**inset**) all the way out and holding.

170

Friction wheel, removing and installing

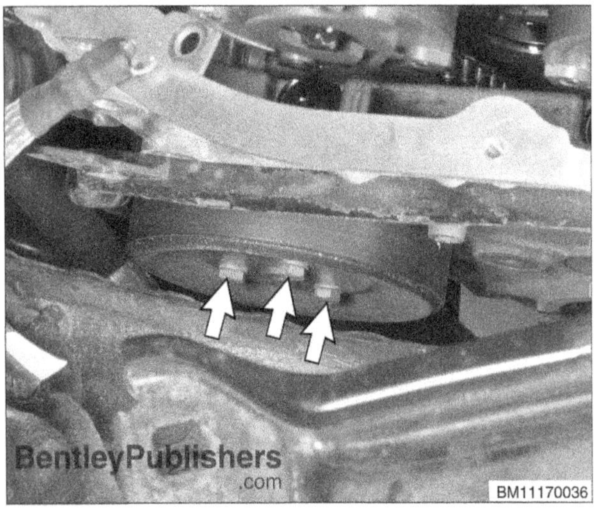

Remove coolant pump pulley fasteners (**arrows**) then remove coolant pump pulley.

Remove friction wheel fasteners (**arrows**) then remove friction wheel from engine.

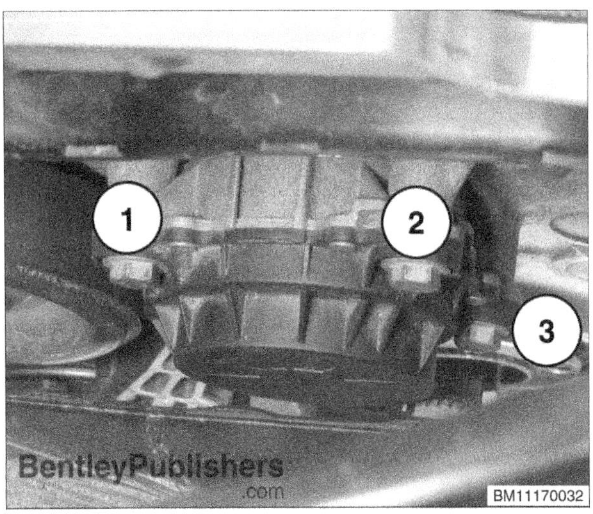

Installation is reverse of removal. Remember to:

- Check coolant pump pulley friction surface for wear. Replace if necessary.
- Torque friction wheel fasteners in sequence: **1** - **2** - **3**.

Tightening torque	
Friction wheel to engine block	9 Nm (79 in-lb)

Intercooler, removing and installing (N14, N18 engine)

– Raise car and support safely

> ***WARNING—***
> - *Make sure the car is stable and well supported at all times. Use a professional automotive lift or jack stands designed for the purpose. A floor jack is not adequate support.*

– Drain radiator. See **Coolant draining and filling** in this repair group.

> ***WARNING—***
> - *Allow cooling system to cool before opening or draining system.*

– Remove front bumper. See **510 Exterior Trim, Bumpers**.

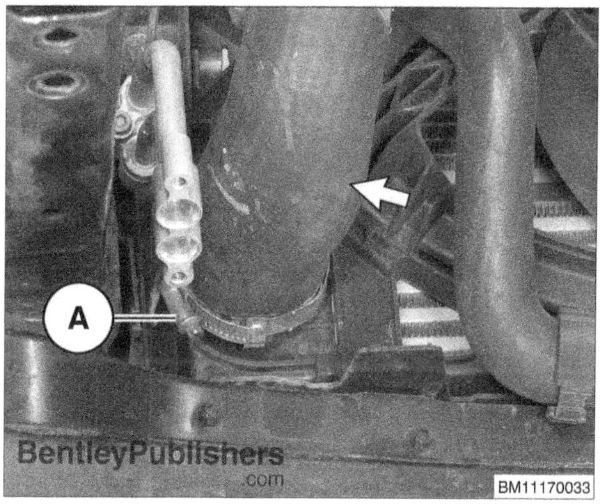

◀ Loosen hose clamps (**A**) then disconnect left and right charge air ducts (**arrow**) from intercooler. Left side shown, right is similar.

◀ Remove fasteners (**arrows**) at sides of intercooler (**A**) and remove from vehicle.

– Installation is reverse of removal.

180 Exhaust System

GENERAL

This repair group covers removal and replacement of the exhaust system including exhaust manifold and oxygen sensors.

Warnings

WARNING—
- *The exhaust system and catalytic converter operate at very high temperatures. Allow components to cool before servicing. Wear protective clothing to prevent burns. Do not use flammable chemicals near a hot catalytic converter.*
- *Corroded exhaust system components crumble easily and often have exposed sharp edges. To avoid injury, wear eye protection and heavy gloves when working with old exhaust parts.*

EXHAUST SYSTEM

The exhaust system is designed to be maintenance free, although regular inspection is warranted due to the harsh operating conditions. Under normal conditions, catalytic converters do not require replacement unless damaged

Use new fasteners, clamps, rubber mounts and gaskets when replacing exhaust components. A liberal application of penetrating oil to the exhaust system nuts and bolts in advance may make removal easier.

Exhaust system, removing and installing

The removal and installation procedures given here are general procedures for all models. Where model-specific differences exist, they are noted in the text. The removal of the exhaust system as a complete unit is recommended. Once the complete system is removed from the car, individual pipes and mufflers can be more easily replaced.

> ***WARNING—***
> - *Exhaust gases are colorless, odorless, and very toxic. Run the engine only in a well-ventilated area. Immediately repair any leaks in the exhaust system or structural damage to the car body that might allow exhaust gases to enter the passenger compartment.*

> ***CAUTION—***
> - *Use care not to drag or bang oxygen sensors. Oxygen sensors are easily damaged.*
> - *When detaching and reattaching oxygen sensor connectors, make sure front and rear connectors are not mixed up.*
> - *The exhaust system is heavy. Work with an assistant*

– Raise vehicle and support safely.

> ***WARNING—***
> - *Make sure the vehicle is stable and well supported at all times. Use a professional automotive lift or jack stands designed for the purpose. A floor jack is not adequate support.*

– **Convertible**: Remove under body cross brace.

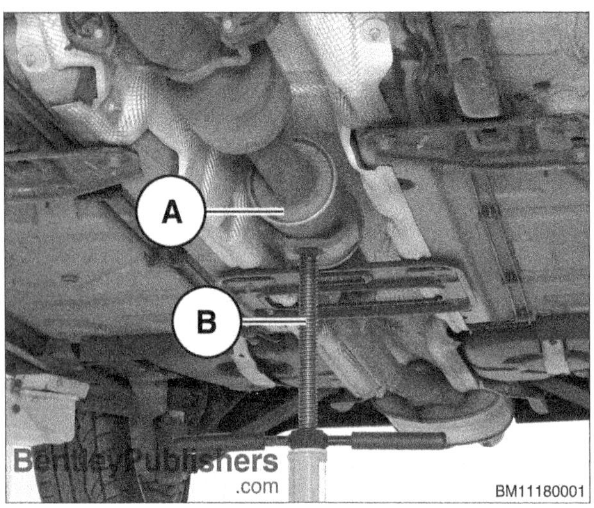

◄ Support exhaust system (**A**) with suitable jack (**B**).

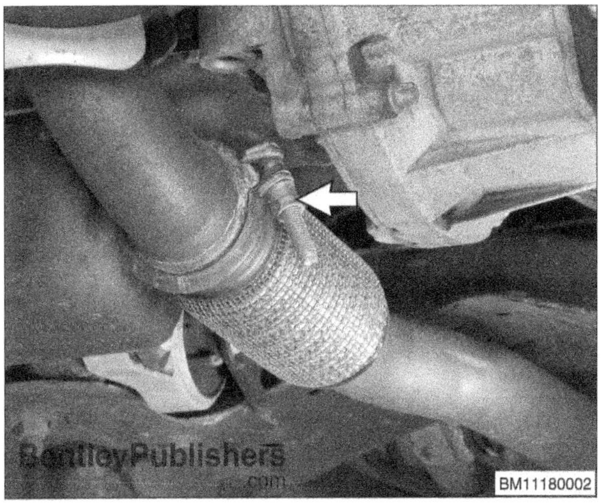

Loosen exhaust clamp (**arrow**).

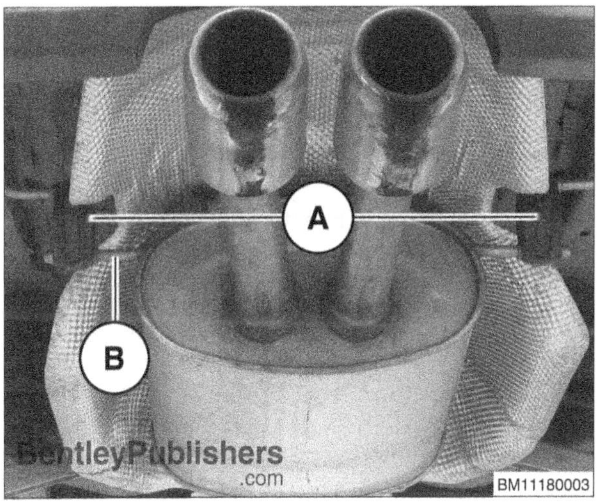

Spray rubber mount (**A**) with lubricant and detach from rear muffler (**B**).

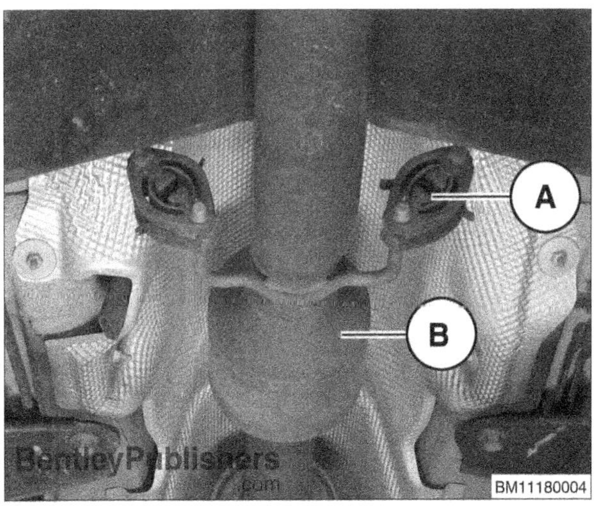

Spray rubber mount (**A**) with lubricant and detach from exhaust pipe (**B**).

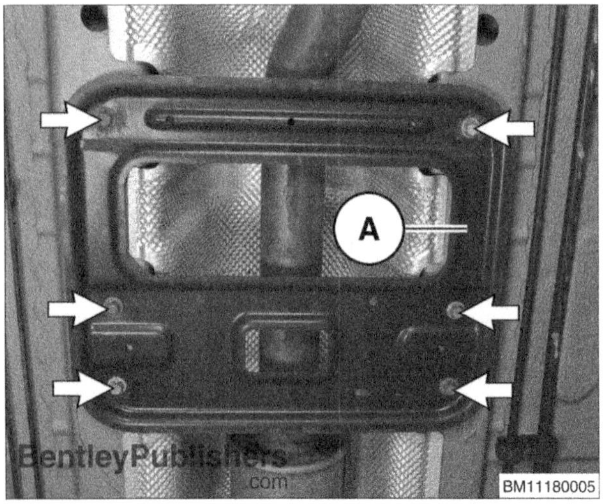

Remove fasteners (**arrows**) then remove reinforcement plate (**A**).

- Lower and remove exhaust system with assistance of a second person.
- Installation is reverse of removal. Remember to:
  - Use new gaskets and hardware. Coat hardware threads with anti-seize paste.
  - Loosely install exhaust system mounting hardware and hangers. Make sure there is sufficient clearance between exhaust system and vehicle underbody at every point before tightening fasteners to their final torque.
  - Check exhaust system for leaks.

Tightening torque	
V-band clamp to catalytic converter	25 Nm (18.4 ft-lb)

Exhaust system components, removing and installing

Exhaust components are welded together at the factory and must be cut off to be replaced. MINI offers individual exhaust components as replacement parts.

Component replacement is best accomplished by first removing the complete exhaust system. See **Exhaust system, removing and installing** in this repair group.

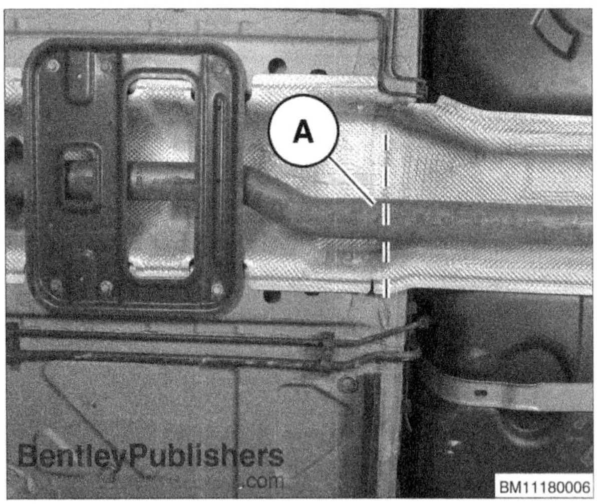

If the separation point is not marked by a punch mark in the pipe, determine the separation point by means of the pipe length on the new muffler.

- Use exhaust pipe cutter (MINI special tool 00 2 210 or equivalent) to cut pipe(s). Deburr cut pipe end(s).

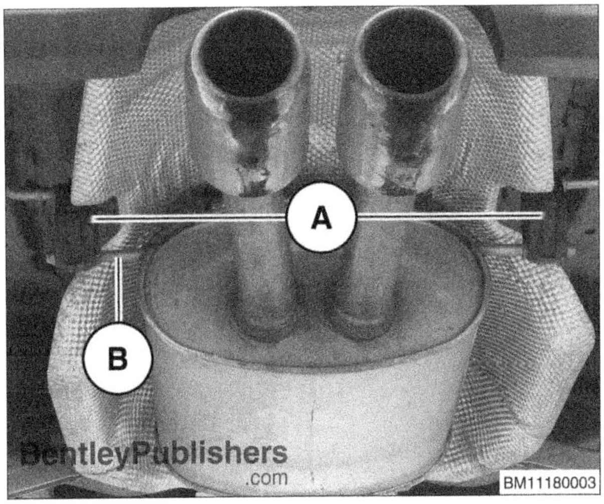

Muffler replacing

◀ Spray rubber mount (**A**) with lubricant and detach from rear muffler (**B**).

– Lower and remove muffler with assistance of a second person.

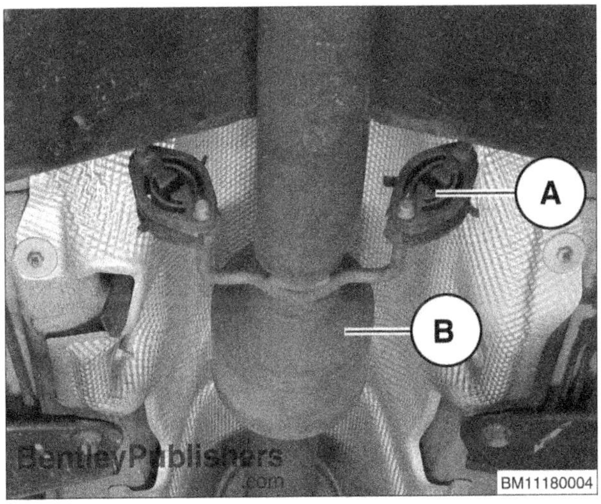

Front silencer, replacing

◀ Spray rubber mount (**A**) with lubricant and detach from exhaust pipe (**B**).

– Lower and remove front silencer with assistance of a second person.

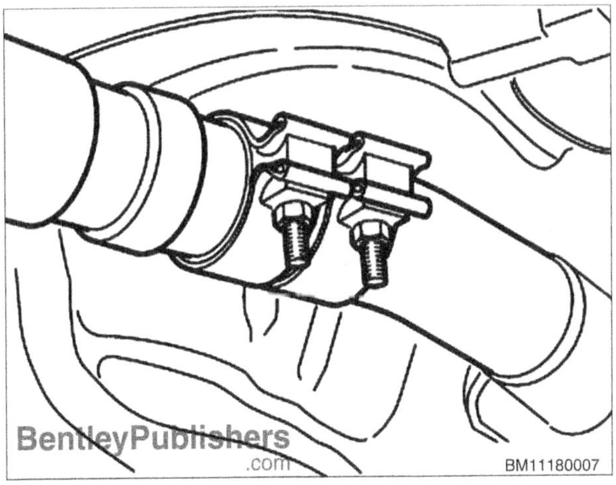

◀ Assemble pipes using band clamps.

– Installation is reverse of removal. Remember to:
  - Make sure there is sufficient clearance between exhaust system and vehicle underbody at every point.
  - Check exhaust system for leaks.

Tightening torque	
Band clamp to exhaust pipes	45 Nm (33 ft-lb)

Exhaust Manifold

Exhaust manifold, removing and installing (N12, N16 engine)

- Place radiator support in service position. See **410 Fenders, Engine Hood**.
- Remove both oxygen sensors. See **Oxygen sensors, removing and installing** in this repair group

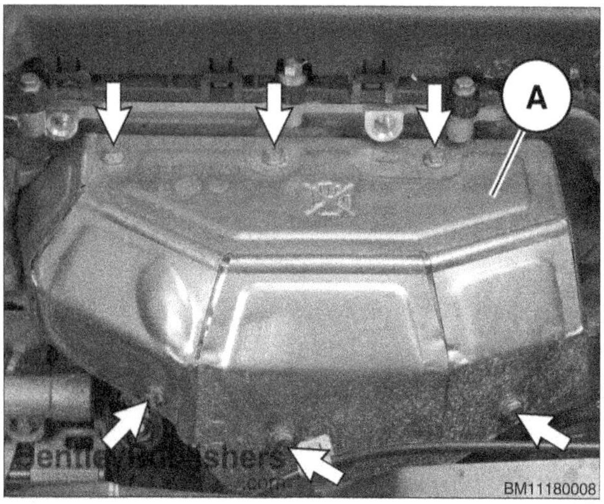

◄ Remove exhaust manifold heat shield fasteners (**arrows**), then remove heat shield (**A**).

◄ Remove catalytic converter heat shield fasteners (**arrows**), then remove heat shield.

◄ Working at catalytic converter, remove fasteners (**arrows**).

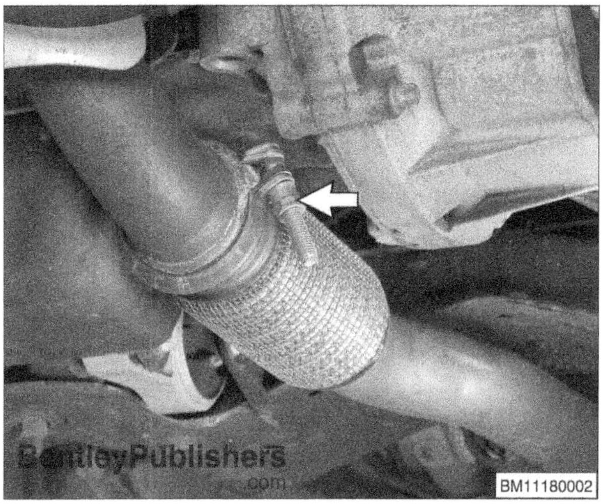

Loosen exhaust clamp (**arrow**).

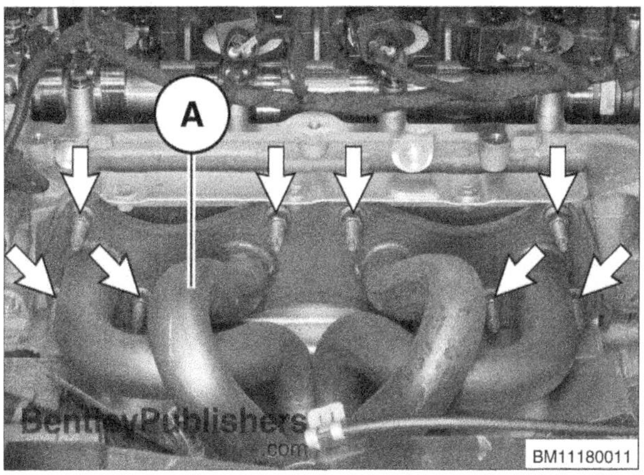

Remove nuts (**arrows**) and remove exhaust manifold (**A**).

- Installation is reverse of removal. Remember to:
  - Clean sealing surfaces of manifold and cylinder head.
  - Use new gaskets and hardware.
  - Check exhaust system for leaks.

Tightening torque	
Catalytic converter to engine block	25 Nm (18.4 ft-lb)
Exhaust heat shield to manifold	4 Nm (35.4 in-lb)
Exhaust manifold to cylinder head • Replace nuts • Coat threads with copper paste	25 Nm (18.4 ft-lb)
V-band clamp to catalytic converter	25 Nm (18.4 ft-lb)

Exhaust manifold and turbocharger, removing and installing (N14, N18 engine)

- Place radiator support in service position. See **410 Fenders, Engine Hood**.
- Remove both oxygen sensors. See **Oxygen sensors, removing and installing** in this repair group
- Drain engine coolant. See **170 Radiator and Cooling System**.
- Remove coolant expansion tank. See **170 Radiator and Cooling System**.
- Remove catalytic converter. See **Catalytic converter, removing and installing (N14, N18 engine)** in this repair group.
- Remove charge air ducts. See **130 Fuel Injection**.

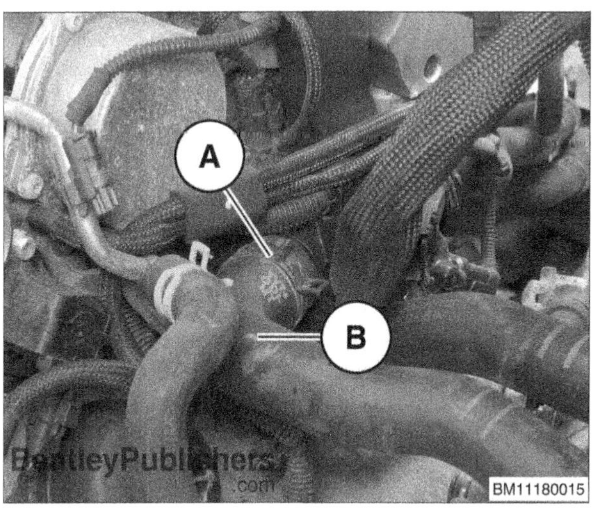

Using spring clamp pliers, release tension on spring clamp (**A**) then remove radiator hose (**B**).

Remove banjo bolts (**arrows**) then remove coolant lines (**A**).

Remove vacuum hose (**arrow**).

- Disconnect electrical connector (**A**).
- Release fastener (**B**).

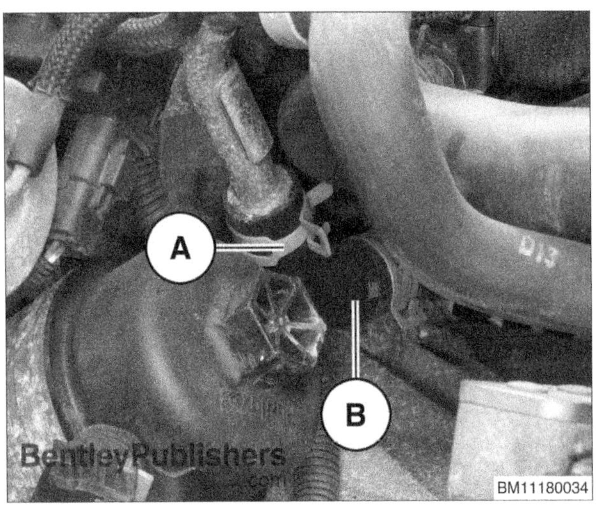

Using spring clamp pliers, release tension on spring clamp (**A**) then remove coolant hose (**B**).

– Remove upper turbo coolant line from turbo.

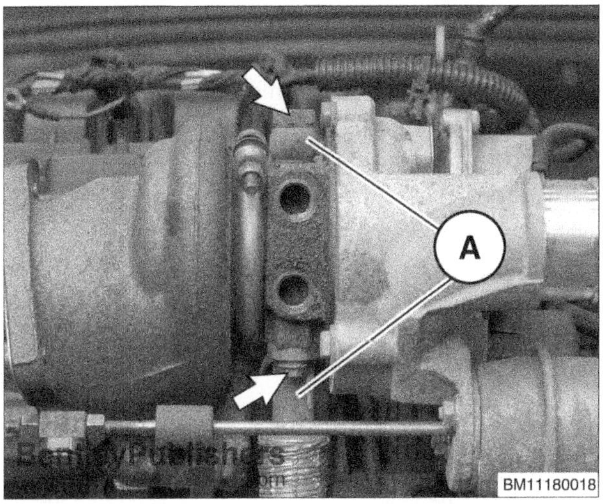

Remove oil line fasteners (**arrows**), then remove oil lines (**A**).

Remove exhaust manifold support bracket fasteners (**arrows**), then remove support bracket (**A**).

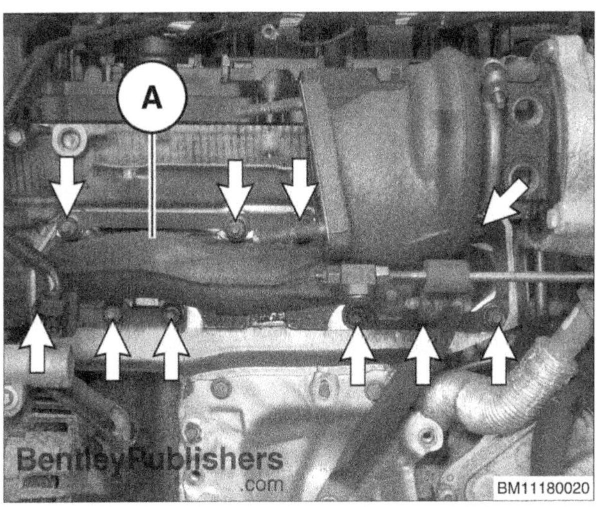

Remove nuts (**arrows**) and remove exhaust manifold (**A**).

< Remove nuts (**arrows**), then remove turbocharger (**A**) from exhaust manifold (**B**).

– Remainder of installation is reverse of removal. Remember to:
- Clean sealing surfaces of manifold and cylinder head.
- Use new O-rings, gaskets and hardware. Coat hardware threads with anti-seize paste.
- Check air intake system for leaks.
- Check exhaust system for leaks.

Tightening torque	
Catalytic converter to engine block	25 Nm (18.4 ft-lb)
Coolant line banjo bolts to turbocharger (M12 x 1.5)	35 Nm (25.8 ft-lb)
Exhaust heat shield to manifold	4 Nm (35.4 in-lb)
Exhaust manifold to cylinder head • Replace nuts • Coat threads with copper paste	25 Nm (18.4 ft-lb)
Oil feed line to turbocharger (M6 x 100 mm)	30 Nm (22 ft-lb)
Oil return line to turbocharger (M8 x 125 mm)	8 Nm (70.8 in-lb)
Stud to turbocharger	15 Nm (11 ft-lb)
Support bracket to engine block	19 Nm (14 ft-lb)
Support bracket to turbocharger	25 Nm (18.4 ft-lb)
Turbocharger to exhaust manifold	20 Nm (14.7 ft-lb)
V-band clamp to catalytic converter	25 Nm (18.4 ft-lb)

Oxygen Sensors

Oxygen sensors, removing and installing

Oxygen sensors are installed at the front and rear of the catalytic converters. Prior to sensor removal, use MINI scan tool to read out and clear ECM fault codes.

> ***CAUTION—***
> - *Do not drag or bang oxygen sensors. Oxygen sensors are easily damaged.*
> - *To avoid possible engine damage, do not mix up oxygen sensor connectors. Mark connectors before disassembling.*

– Raise vehicle and support safely.

> ***WARNING —***
> - *Make sure the vehicle is stable and well supported at all times. Use a professional automotive lift or jack stands designed for the purpose. A floor jack is not adequate support.*

– Disconnect oxygen sensor electrical connectors.

Precatalyst oxygen sensor (N12, N16 engine)

◀ Using oxygen sensor socket (**A**), loosen and remove oxygen sensor (**B**).

– Installation is reverse of removal. Remember to:

- Apply a small amount of anti-seize compound to sensor threads.
- Make sure sensor electrical harness connectors are routed as before and connected correctly.

> ***CAUTION—***
> *• Do not overuse anti-seize paste. Do not contaminate tip of sensor with paste or lubricants.*

Tightening torque	
Oxygen sensor to exhaust pipe (use MINI special tool 11 7 020)	50 Nm (37 ft-lb)

Post-catalyst oxygen sensor (N12, N16 engine)

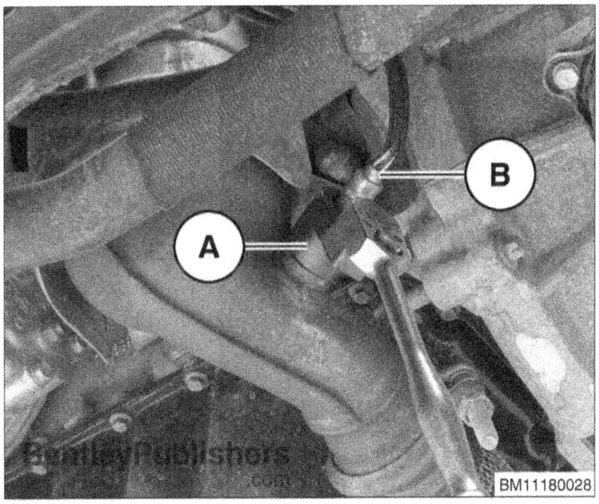

◀ Using oxygen sensor socket (**A**), loosen and remove oxygen sensor (**B**).

– Installation is reverse of removal. Remember to:

- Apply a small amount of anti-seize compound to sensor threads.
- Make sure sensor electrical harness connectors are routed as before and connected correctly.

> ***CAUTION—***
> *• Do not overuse anti-seize paste. Do not contaminate tip of sensor with paste or lubricants.*

Tightening torque	
Oxygen sensor to exhaust pipe (use MINI special tool 11 7 020)	50 Nm (37 ft-lb)

Precatalyst oxygen sensor (N14, N18 engine)

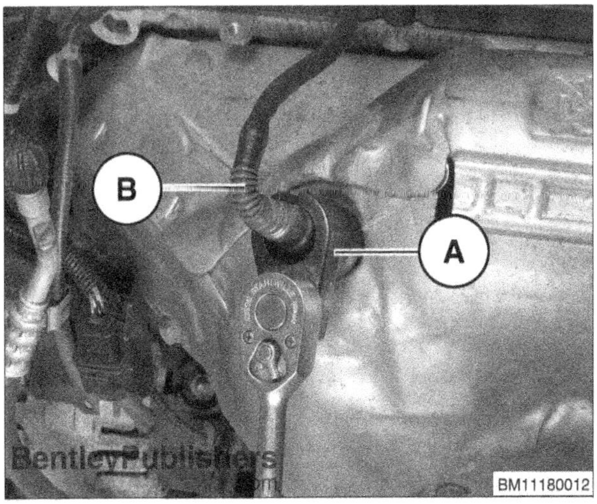

◀ Using oxygen sensor socket (**A**), loosen and remove oxygen sensor (**B**).

– Installation is reverse of removal. Remember to:

- Apply a small amount of anti-seize compound to sensor threads.
- Make sure sensor electrical harness connectors are routed as before and connected correctly.

> ***CAUTION—***
> *• Do not overuse anti-seize paste. Do not contaminate tip of sensor with paste or lubricants.*

Tightening torque	
Oxygen sensor to exhaust pipe (use MINI special tool 11 7 020)	50 Nm (37 ft-lb)

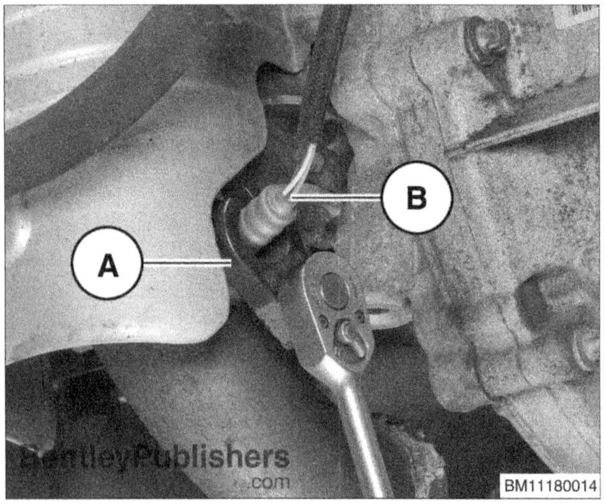

Post-catalyst oxygen sensor (N14, N18)

◄ Using oxygen sensor socket (**A**), loosen and remove oxygen sensor (**B**).

– Installation is reverse of removal. Remember to:
  - Apply a small amount of anti-seize compound to sensor threads.
  - Make sure sensor electrical harness connectors are routed as before and connected correctly.

> ***CAUTION—***
> - *Do not overuse anti-seize paste. Do not contaminate tip of sensor with paste or lubricants.*

Tightening torque	
Oxygen sensor to exhaust pipe (use MINI special tool 11 7 020)	50 Nm (37 ft-lb)

CATALYTIC CONVERTER

MINI models covered by this manual utilize two catalytic converters. One located near engine and one located in the center exhaust pipe.

Catalytic converter, removing and installing (N12, N16 engine)

N12 and N16 engine near-engine catalyst is integrated into the exhaust manifold and not replaceable separately. See **Exhaust manifold, removing and installing (N12, N16 engine)** in this repair group. To replace secondary converter see **Exhaust system components, removing and installing** in this repair group.

Catalytic converter, removing and installing (N14, N18 engine)

– Place radiator support in service position. See **410 Fenders, Engine Hood**.

– Remove both oxygen sensors. See **Oxygen sensors, removing and installing** in this repair group

◄ Remove fastener (**arrow**), then remove bracket (**A**).

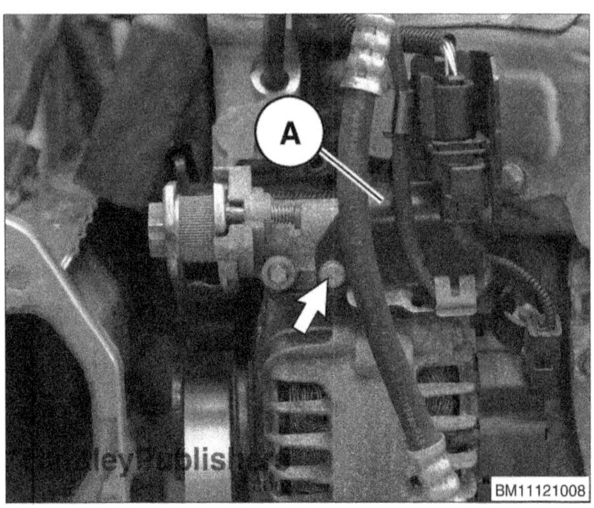

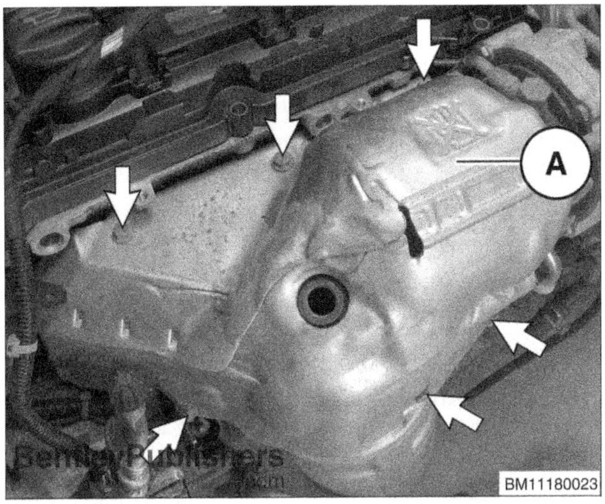

Remove exhaust manifold heat shield fasteners (**arrows**), then remove heat shield (**A**).

Remove catalytic converter heat shield fasteners (**arrows**), then remove heat shield.

Working at catalytic converter, remove fasteners (**arrows**).

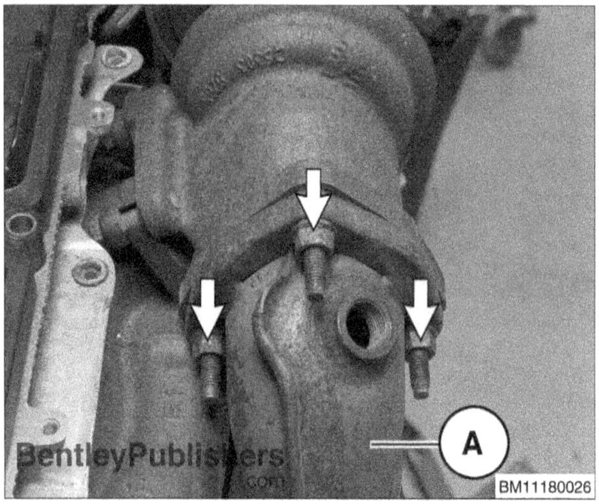

Remove fasteners (**arrows**) then remove catalytic converter (**A**).

– Installation is reverse of removal. Remember to:
  - Clean sealing surfaces of manifold and catalytic converter.
  - Use new O-rings, gaskets and hardware. Coat hardware threads with anti-seize paste.
  - Check air intake system for leaks.
  - Check exhaust system for leaks

Tightening torque	
Catalytic converter to engine block	25 Nm (18.4 ft-lb)
Catalytic converter to exhaust manifold • Stage 1 • Stage 2	 7 Nm (61.9 in-lb) 40 Nm (29.5 ft-lb)
Exhaust heat shield to manifold	4 Nm (35.4 in-lb)
Stud to turbocharger	15 Nm (11 ft-lb)
Support bracket to catalytic converter	20 Nm (14.7 ft-lb)
Support bracket to engine block	20 Nm (14.7 ft-lb)
V-band clamp to catalytic converter	25 Nm (18.4 ft-lb)

EXHAUST COMPONENTS

Exhaust components (N12, N16 engine))

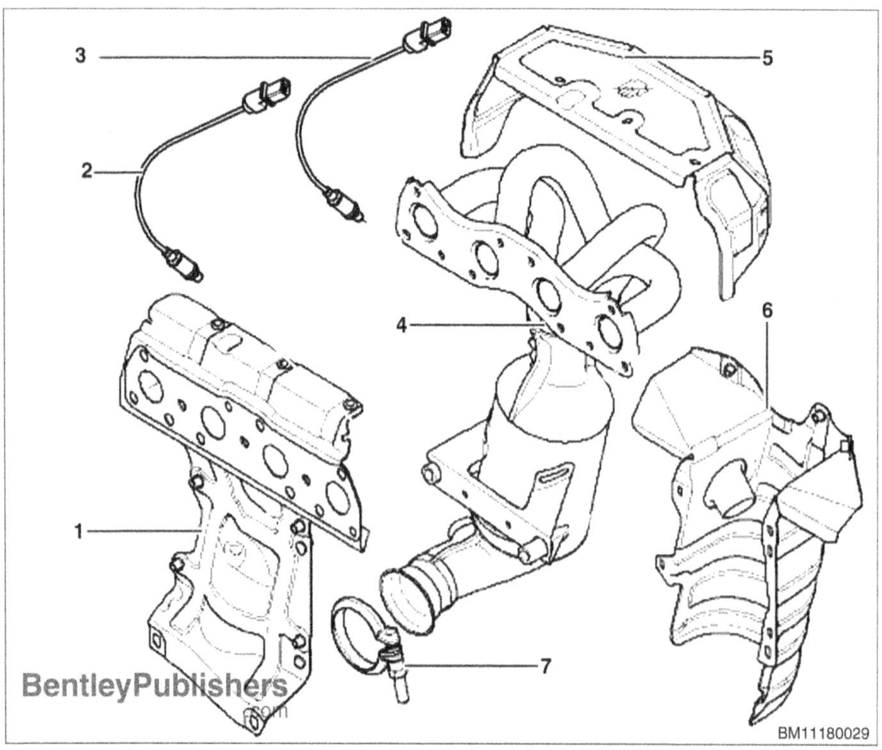

1. Exhaust manifold gasket with heat shield
2. Precatalyst oxygen sensor
3. Post-catalyst oxygen sensor, front
4. Exhaust manifold
5. Heat shield
6. Heat shield
7. V-band clamp
   25 Nm (18.4 ft-lb)

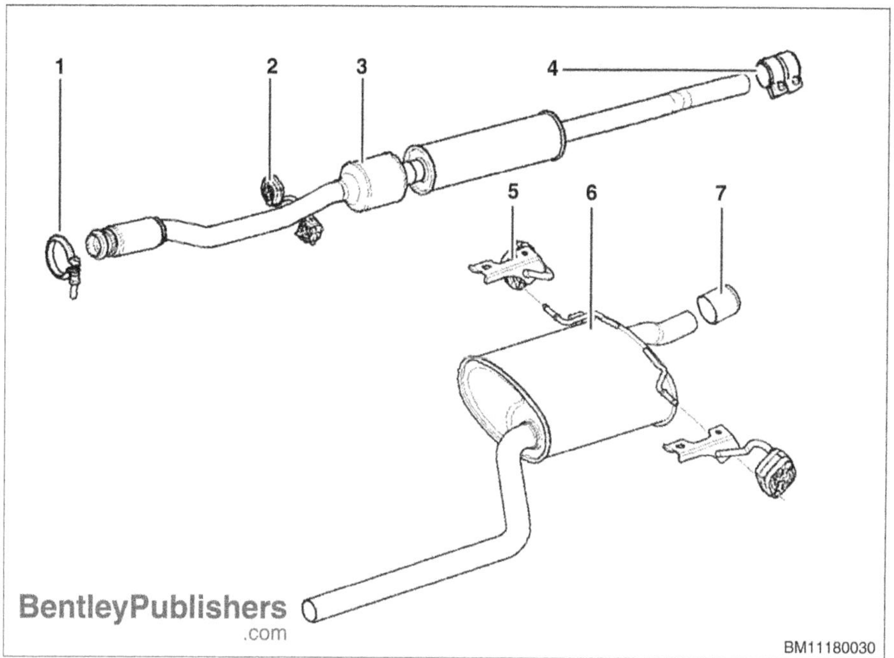

1. V-band clamp
   25 Nm (18.4 ft-lb)
2. Exhaust hanger
3. Front silencer pipe with catalytic converter
4. Exhaust band clamp
   45 Nm (33 ft-lb)
5. Exhaust mounting bracket with hanger
6. Muffler
7. Chrome muffler tip

Exhaust components (N14, N18 engine)

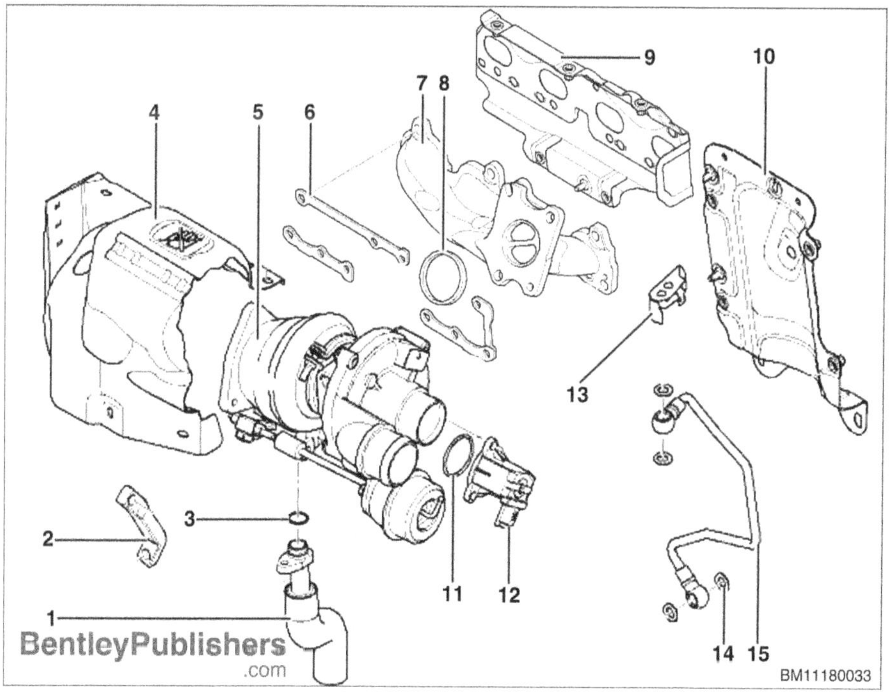

1. Oil return line
2. Support bracket
3. O-ring
4. Heat shield
5. Turbocharger
6. Mounting strip
7. Exhaust manifold
8. Gasket
9. Exhaust manifold gasket with heat shield
10. Heat shield
11. O-ring
12. Recirculation valve
13. Bracket
14. Gasket
15. Oil supply line

Exhaust components (N14, N18 engine)

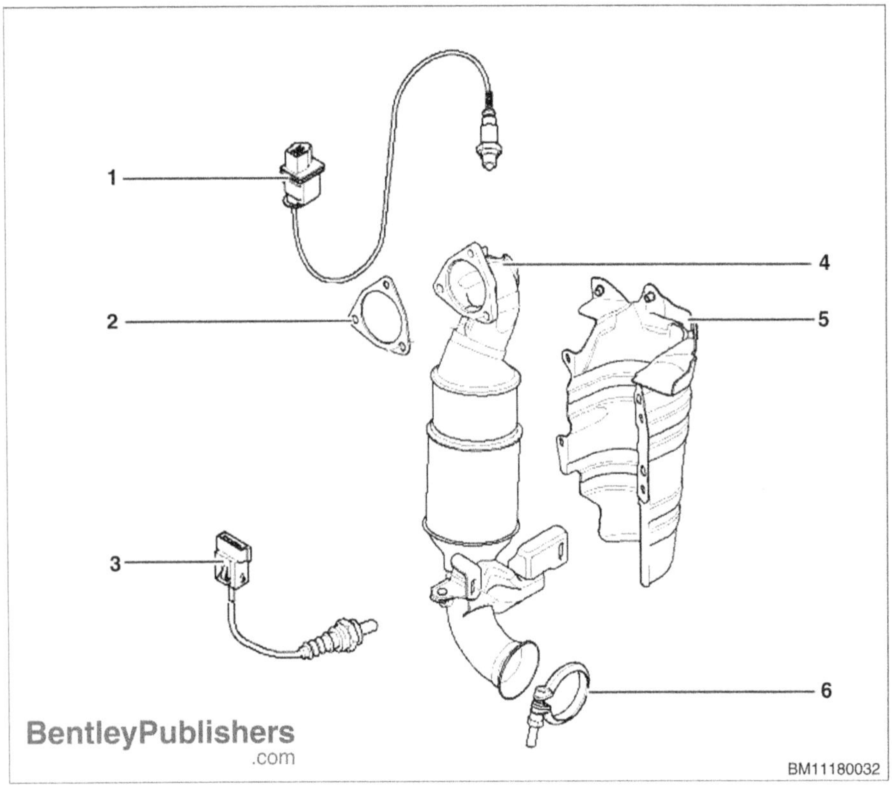

1. Exhaust manifold gasket with heat shield
2. Precatalyst oxygen sensor
3. Post-catalyst oxygen sensor, front
4. Exhaust manifold
5. Heat shield
6. Heat shield
7. V-band clamp
   25 Nm (18.4 ft-lb)

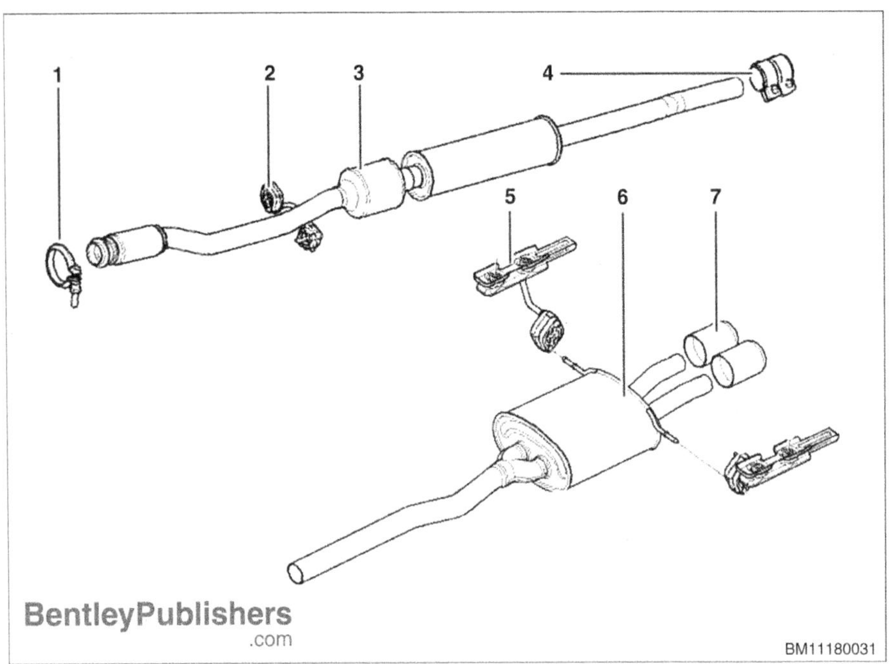

1. V-band clamp
   25 Nm (18.4 ft-lb)
2. Exhaust hanger
3. Front silencer pipe with catalytic converter
4. Exhaust band clamp
   45 Nm (33 ft-lb)
5. Exhaust mounting bracket with hanger
6. Muffler
7. Chrome muffler tip

200 Transmission–General

TABLES

General

This section provides general information for drivetrain components: clutch, clutch hydraulics and transmissions.

See also:

- **119 Lubrication System** for rear main seal replacement
- **210 Clutch** for clutch mechanical and hydraulic repairs and flywheel replacement
- **230 Manual Transmission** for manual transmission fluid service, seal replacement and transmission replacement
- **240 Automatic Transmission** for ATF change and transmission replacement
- **310 Front Suspension** for front differential, drive axles and CV joints

Drivetrain

Cooper models are equipped with either a 6-speed manual or 6-speed automatic transmission. The transmissions is mounted on the left side of the engine bay in line with the engine. The differential assembly is integral with the transmission housing and provides drive to the front wheels via jointed drive axles.

Manual Transmission

Manual transmission and clutch

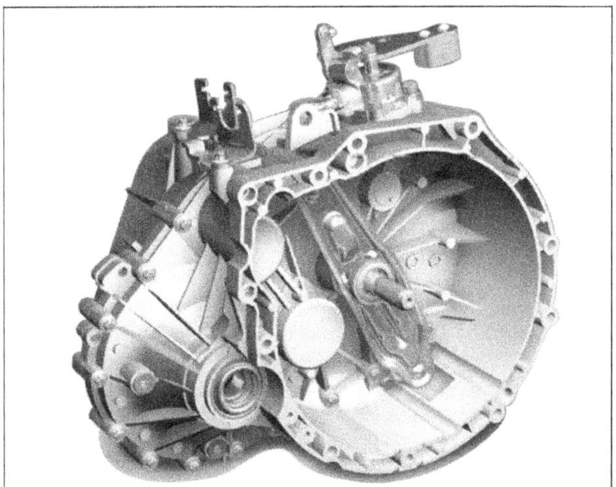

BM11200002

Two versions of the Getrag 6-speed manual transmissions are fitted in MINI Cooper models. Manual transmission application information, recommended fluids and fill capacities are in **Table a**.

See **230 Manual Transmission** for additional details.

A conventional clutch is used. For more information, see **210 Clutch**.

Table a. MINI Cooper manual transmission applications (6-speed)				
Model (engine)	**BMW designation**	**Manufacturer designation**	**Recommended fluid**	**Fill capacity (max)**
Cooper, Clubman, Convertible (N12, N16)	GS6-55BG	Getrag G255	MTF-LT-4 (yellow sticker) BMW part no. 83 22 2 344 589	1.7 liters (1.8 US qt)
Cooper S, Clubman S, Convertible S, John Cooper Works (N14, N18)	GS6-53BG / DG	Getrag G253		1.9 liters (2.0 US qt)

AUTOMATIC TRANSMISSION

6-speed automatic transmission

The automatic transmission manufactured by Aisin is a 6-speed with Agitronic. This feature enables individual gear selection in Drive (D) position by means of shift paddles on the steering wheel. The steering wheel shift paddles may also be used with the transmission in Sport (S) position.

The Aisin transmission is equipped with different gear ratios, depending on turbo or normally aspirated application. For more information, see **240 Automatic Transmission**.

Table b. MINI Cooper automatic transmission (6-speed)	
BMW designation	GA6F21WA
Manufacturer designation	Aisin F21
Recommended fluid	JWS-3309 ATF BMW part no. 83 22 0 402 413

210 Clutch

GENERAL

This repair group covers replacement of clutch hydraulic and mechanical components.

See also:

- **200 Transmission–General** for transmission application information
- **230 Manual Transmission** for transmission removal

Warnings and Cautions

WARNING—
- *Brake and clutch hydraulic fluid is poisonous, highly corrosive and dangerous to the environment. Wear safety glasses and rubber gloves when working with hydraulic fluid.*
- *Do not siphon with your mouth.*
- *Dispose of hydraulic fluid properly.*
- *Friction material in the clutch disc produces dangerous dust. Do not breathe in dust. Use water to wet down components and collect dripping mixture in a shop towel.*

CAUTION—
- *Do not exceed 2 bar (28 psi) pressure at the brake and clutch fluid reservoir when bleeding or flushing the hydraulic system.*
- *Use new hydraulic fluid from a fresh, unopened container. Hydraulic fluid absorbs moisture from the air. This can lead to corrosion problems in the brake and clutch hydraulic systems, and also lowers the fluid boiling point.*
- *Immediately clean away any fluid spilled on painted surfaces and wash with water, as hydraulic fluid removes paint.*

Clutch Hydraulics

Clutch hydraulic operation

The clutch pedal operates a hydraulic master cylinder supplied with fluid from the brake fluid reservoir. The clutch master cylinder fluid line connects to a slave cylinder which operates the clutch release mechanism.

A soft or spongy feel to the clutch pedal, long pedal free-play, or grinding noises from the gears while shifting may indicate problems with the clutch hydraulics. In these circumstances it is best to start with a clutch fluid flush, followed, if necessary, by replacement of hydraulic parts.

The clutch hydraulic system shares the fluid reservoir and fluid with the brake hydraulic system.

Hydraulic fluid application	
Brake and clutch fluid	DOT 4 low viscosity

Clutch hydraulic system, bleeding and flushing

If brake and clutch hydraulic fluid is murky or muddy, or has not been changed within two years, flush system and change fluid using a pressure bleeder.

– Working in engine compartment, flip open brake fluid reservoir cover in left side cowl panel.

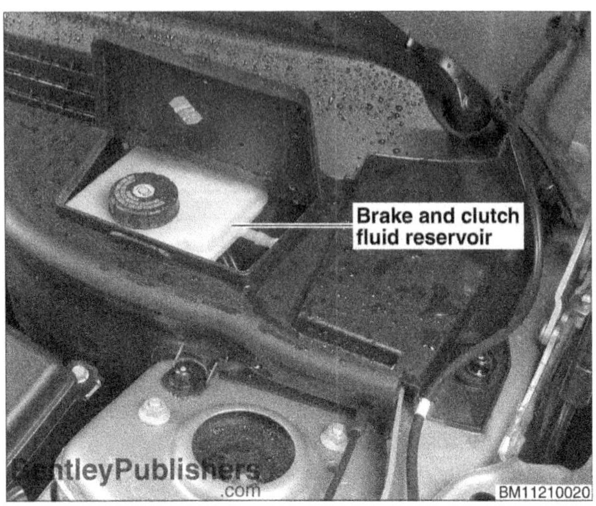

Remove hydraulic fluid reservoir cap. Using a clean syringe, remove fluid from reservoir. Refill reservoir with clean DOT 4 low viscosity brake fluid.

– Attach pressure bleeder to hydraulic fluid reservoir and pressurize.

> ***CAUTION—***
> - *Do not exceed 2 bar (28 psi) pressure at the fluid reservoir when bleeding or flushing the hydraulic system.*

– Raise car and remove splash shield to gain access to slave cylinder underneath transmission.

> ***WARNING —***
> - *Make sure the car is stable and well supported at all times. Use a professional automotive lift or jack stands designed for the purpose. A floor jack is not adequate support.*

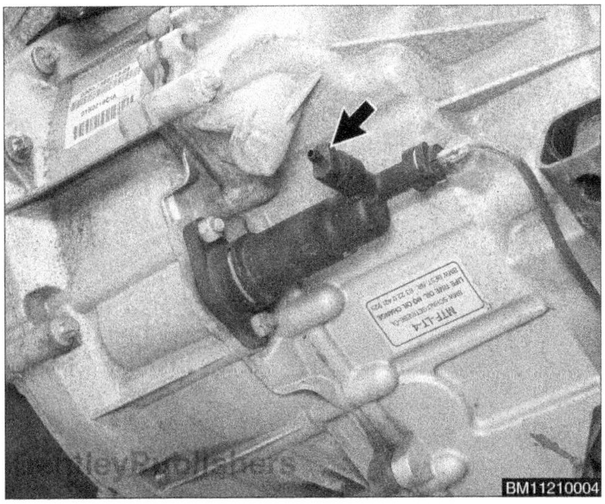

Connect hose from clutch slave cylinder bleeder valve (**arrow**) to container.

- Open bleeder valve and allow fluid to expel until clean fluid comes out, free of air bubbles.
- Close bleeder valve and disconnect pressure bleeding equipment from fluid reservoir.
- Slowly operate clutch pedal about 10 times.
- Fill reservoir with clean fluid as necessary.

Clutch master cylinder, replacing

– Working in engine compartment, flip open brake fluid reservoir cover in left side cowl panel.

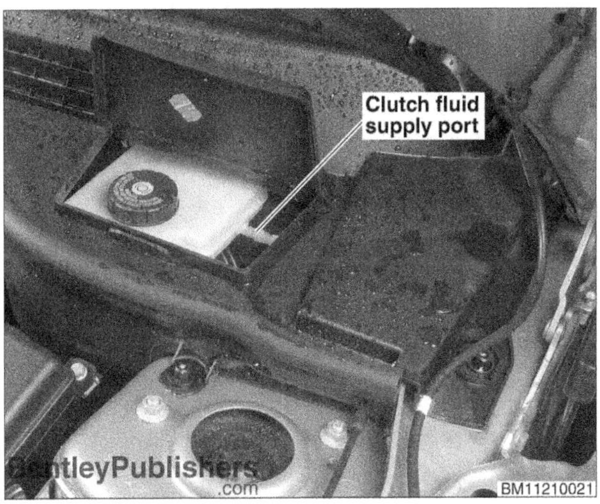

Remove hydraulic fluid reservoir cap.

- Using a clean syringe, remove fluid from reservoir until fluid level falls below clutch fluid supply port.
- Detach clutch master cylinder fluid supply hose from port.
- Plug open fluid ports.

– Working in passenger compartment under left side dashboard, remove lower dashboard trim panel (pedal cluster trim). See **513 Interior Trim**.

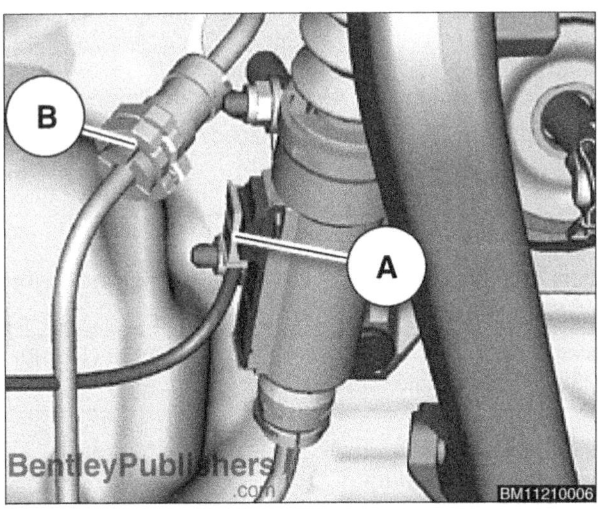

Working underneath left side of dashboard:

- Detach clutch master cylinder module electrical connector (**A**).
- Separate fluid line junction (**B**). Plug open fluid ports.

> ***CAUTION—***
> *• Place pan under fluid lines to catch dripping fluid.*

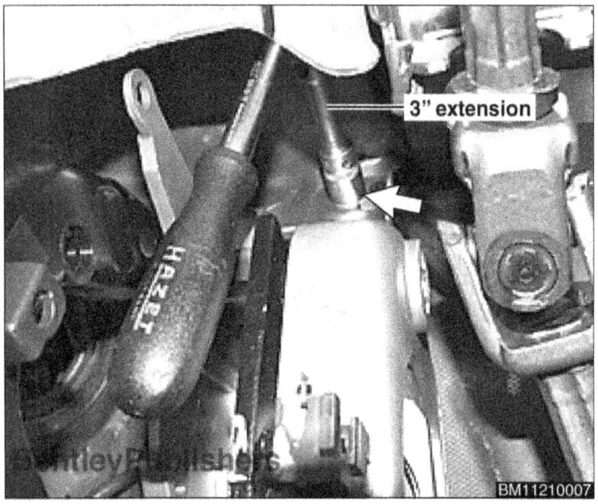

Working at pedal cluster, use ratchet with 3-inch extension to remove pedal cluster mounting bolt (**arrow**).

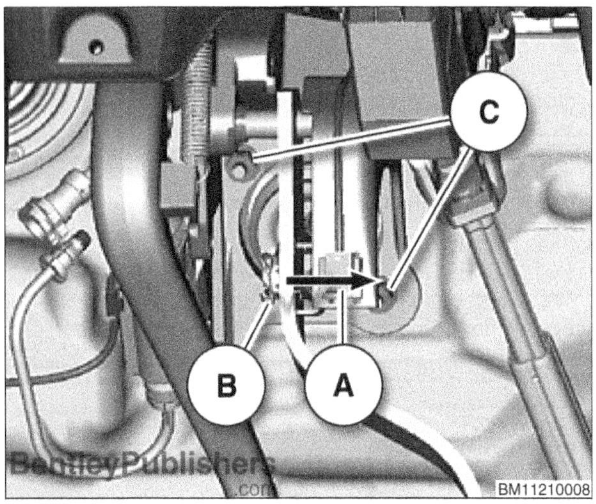

Working at pedals:
- Detach stop light switch electrical connector (**A**).
- Remove clip (**B**) and slide out brake pedal pin (**arrow**).
- Remove pedal cluster mounting nuts (**C**).
- Lower pedal cluster and remove from vehicle.

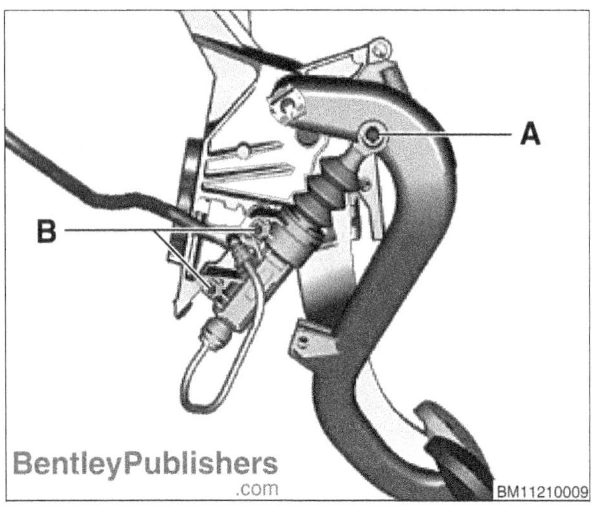

Working on clutch master cylinder at pedal cluster:
- Squeeze pin (**A**) and press out.
- Remove master cylinder mounting nuts (**B**) and remove cylinder from pedals.

– Installation is reverse of removal. Keep in mind the following:
- Replace pin retaining clips if bent or damaged.
- Replace self-locking nuts.

Tightening torques	
Clutch master cylinder to pedal cluster (replace M6 self-locking nuts with new)	9 Nm (7 ft-lb)
Pedal cluster to body (replace M8 self-locking nuts with new)	21 Nm (17 ft-lb)

– When finished, bleed clutch hydraulics. See **Clutch hydraulic system, bleeding and flushing**.

Clutch slave cylinder, removing and installing

– Working in engine compartment, flip open brake fluid reservoir cover in left side cowl panel.

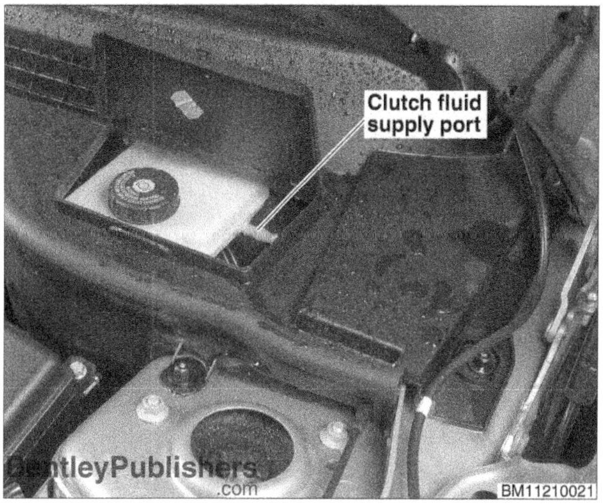

◀ Remove hydraulic fluid reservoir cap. Using a clean syringe, remove fluid from reservoir until fluid level falls below clutch fluid supply port.

– Raise car and remove splash shield to gain access to slave cylinder underneath transmission.

> ***WARNING—***
> - *Make sure the car is stable and well supported at all times. Use a professional automotive lift or jack stands designed for the purpose. A floor jack is not adequate support.*

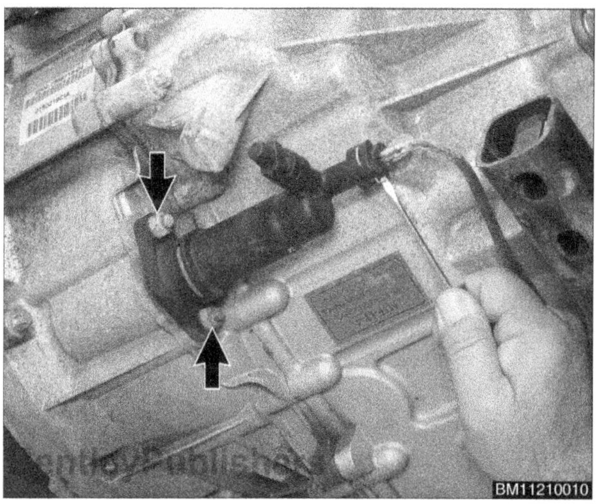

◀ Working at slave cylinder:

- Remove mounting nuts (**arrows**).
- Pry off circlip and separate hydraulic line at slave cylinder. Be prepared to catch dripping fluid. Plug open hydraulic ports.
- Remove slave cylinder from transmission.

> ***CAUTION—***
> - *If slave cylinder is to be reused, do not deflect push rod. Fluid can escape if push rod is deflected.*

– When reinstalling, clean and lightly grease slave cylinder push rod tip.

Tightening torque	
Clutch slave cylinder to transmission	21 Nm (17 ft-lb)

– When finished, bleed clutch hydraulics. See **Clutch hydraulic system, bleeding and flushing**.

Clutch Mechanical

Clutch disc

MINI models with manual transmissions use a single-disc hydraulically actuated clutch.

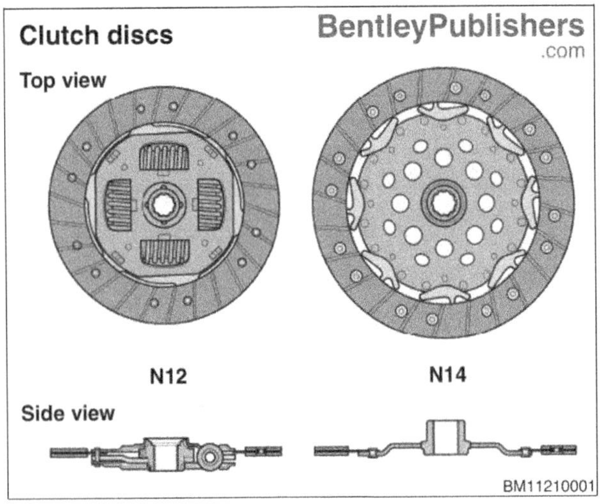

The clutch disc comes in 2 versions:

- **N12, N16 engine** (normally aspirated model): Traditional clutch disc with integral damper springs. Diameter = 200 mm (8 in).
- **N14, N18 engine** (turbocharged model): Larger diameter clutch disc without damper springs. Torsion forces are absorbed by dual mass flywheel. Diameter = 228 mm (9 in).

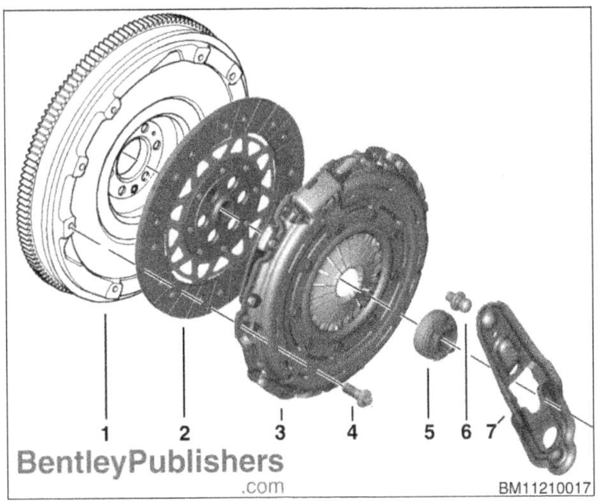

Clutch components (N14, N18 engines)

1. Dual-mass flywheel
2. Clutch disc (diameter = 228 mm or 9 in)
3. Pressure plate
4. M8 x 16 mm 8.8 grade Torx bolt
   - Replace with new
   - Tighten to 28 Nm (21 ft-lb)
5. Release bearing (throw-out bearing)
6. Ball pivot
7. Release lever

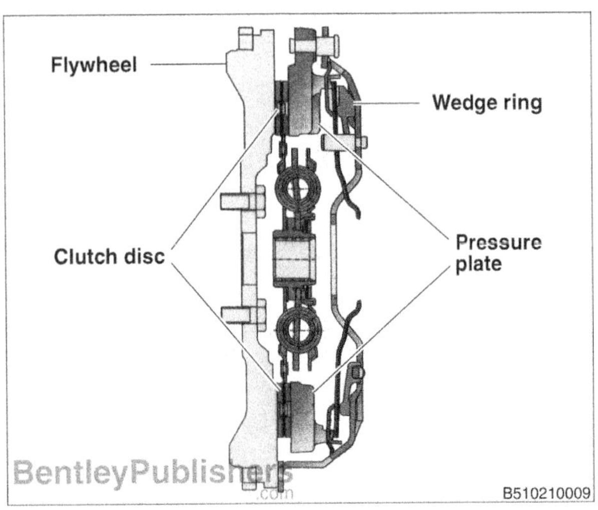

Self-adjusting clutch (SAC)

The SAC pressure plate uses a wedge ring which rotates against the diaphragm springs to accommodate for clutch disc wear. The wedge ring adjusts by means of spring tension. Special tools are needed to apply and relieve spring tension as the clutch pressure plate is removed and installed.

Flywheel

The flywheel comes in two versions:

- **N12 engine** (normally aspirated model): Conventional single mass flywheel.
- **N14 engine** (turbocharged model): Dual-mass flywheel with three-part mechanical torsional-vibration damper.

Clutch replacement

> *WARNING—*
> - *Friction material in the clutch disc produces dangerous dust. Do not breathe in dust. Use water to wet down components and collect dripping mixture in a shop towel.*

Remove the transmission from the engine to access clutch mechanical components. Replace the clutch disc, pressure plate and release bearing during a clutch overhaul.

Be sure to check the bottom of the bellhousing for oil. If engine oil is found, check for a faulty rear crankshaft main oil seal. See **119 Lubrication System** for rear main seal replacement.

Clutch, removing and installing

– Remove transmission. See **230 Manual Transmission**.

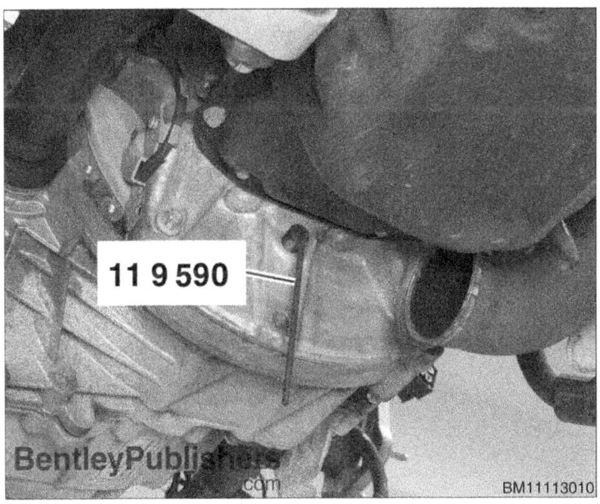

Fit special tool **11 9 590** to counterhold flywheel during pressure plate removal.

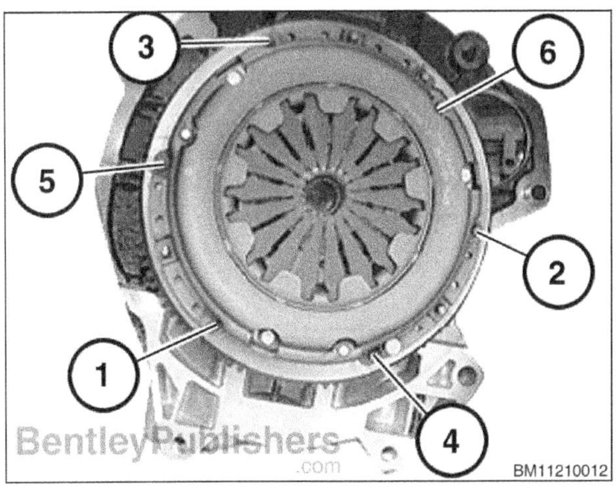

Gradually loosen pressure plate bolts in a crisscross pattern (**1** - **6**) and remove bolts.

– Remove pressure plate and clutch discdisc.

> *WARNING—*
> - *Friction material in the clutch disc produces dangerous dust. Do not breathe in dust. Use water to wet down components and collect dripping mixture in a shop towel.*

– Clean flywheel and check for wear and damage. If necessary, replace flywheel. See **Flywheel, removing and installing** in this repair group.

– Replace release bearing. See **Clutch release bearing, replacing** in this repair group.

– Check clutch disc for damage, rust and wear. Replace disc if friction material measures 1 mm (0.04 in) or less above rivet head on transmission side.

– Slide clutch disc on transmission input shaft and check that it moves freely along shaft before continuing with installation.

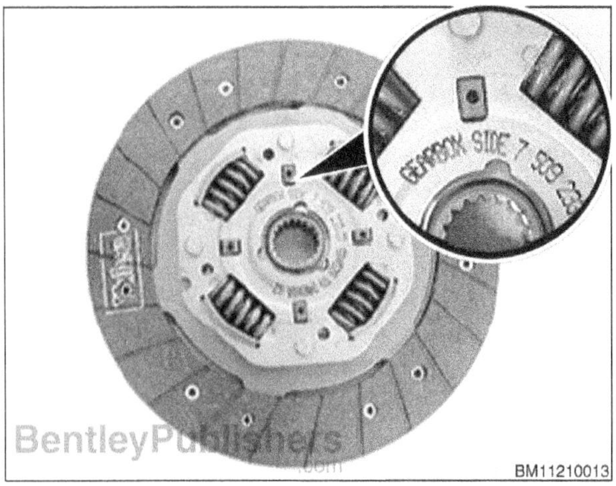

◄ Note clutch disc installation direction mark (**inset**). Face shown is transmission side. (Clutch disc with damper spring center for N12, N16 engine illustrated.)

◄ Install clutch disc and pressure plate on flywheel. Depending on engine, use the following clutch disc centering tool:

- **N12, N16 engine**: Special tool **21 2 290**
- **N14, N18 engine**: Special tool **21 6 110**

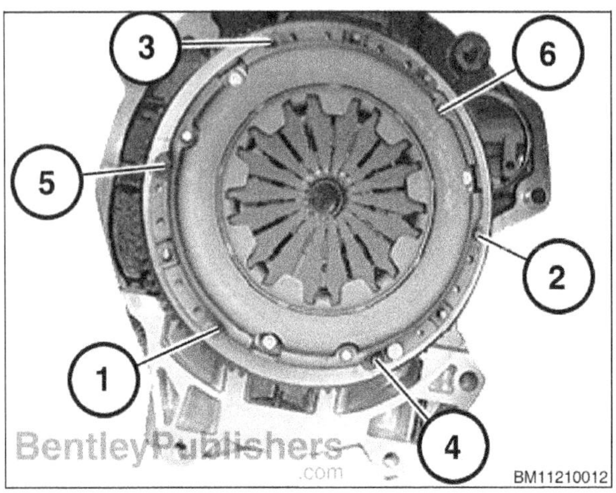

◄ Install pressure plate using new mounting bolts. Gradually tighten in a crisscross pattern (**1 - 6**).

- While tightening, periodically rotate clutch disc centering tool to help maintain clutch disc alignment.
- When finished, remove flywheel locking and clutch disc centering tools.

Tightening torque	
Clutch pressure plate to flywheel (replace M8 x 16 mm 8.8 grade Torx bolts with new)	28 Nm (21 ft-lb)

Self-adjusting clutch (SAC), removing and installing

Special tools are needed when removing and installing self-adjusting clutch (SAC). Be sure to read the procedure through before you start work.

– Remove transmission. See **230 Manual Transmission**.

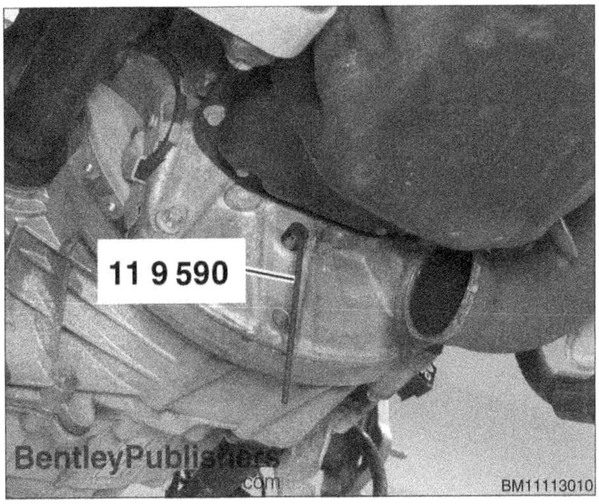

◄ Fit special tool **11 9 590** to counterhold flywheel during pressure plate removal.

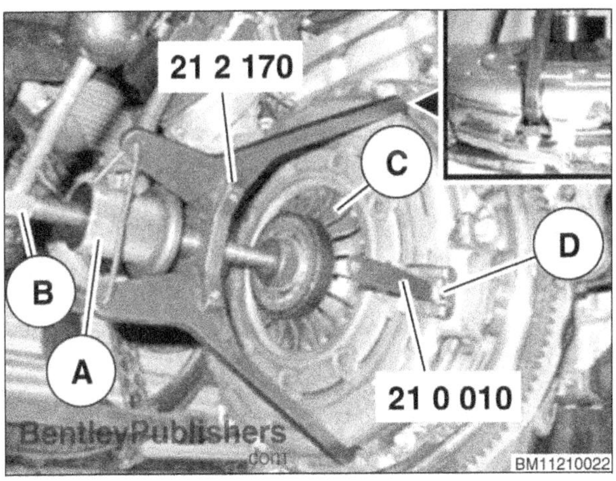

◄ If SAC is to be reused:

- Install MINI special tool **21 2 170** as shown. Make sure tool fingers are inserted in holes near where diaphragm spring is riveted to cover. See **insert**.
- Tighten knurled knob (**A**).
- Screw in spindle (**B**) approx. 6 - 8 mm (¼ - ⅓ in) until diaphragm spring (**C**) is pretensioned.
- Insert MINI special tool **21 0 010** into cutout (**D**) and deactivate clutch adjustment.
- Leave special tool in position until refitting takes place.

– Gradually loosen pressure plate bolts in a crisscross pattern and remove bolts.

– Remove pressure plate and clutch disc.

> ***WARNING —***
> - *Friction material in the clutch disc produces dangerous dust. Do not breathe in dust. Use water to wet down components and collect dripping mixture in a shop towel.*

– Clean flywheel and check for wear and damage. If necessary, replace flywheel. See **Flywheel, removing and installing** in this repair group.

– Replace release bearing. See **Clutch release bearing, replacing** in this repair group.

– Check clutch disc for damage, rust and wear. Replace disc if friction material measures 1 mm (0.04 in) or less above rivet head on transmission side.

– Slide clutch disc on transmission input shaft and check that it moves freely along shaft before continuing with installation.

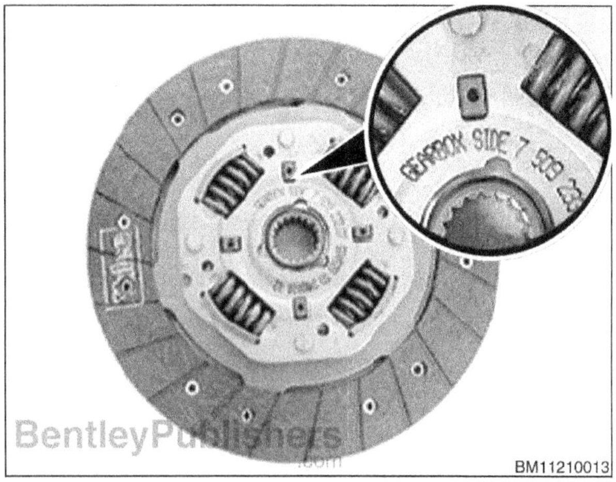

Note clutch disc installation direction mark (**inset**). Face shown is transmission side. (Clutch disc with damper spring center for N12 engine illustrated.)

Center clutch disc using special tool **21 0 000**.

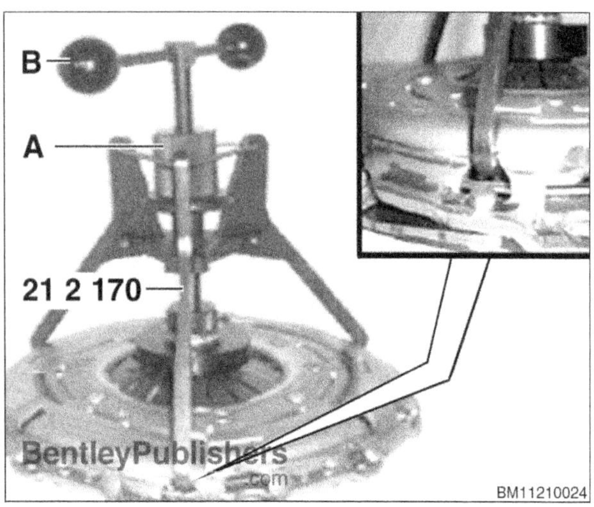

If installing new SAC:

- Install MINI special tool **21 2 170** on pressure plate as shown. Make sure tool fingers are inserted in holes near where diaphragm spring is riveted to cover. See **insert**.
- Tighten knurled knob (**A**).
- Screw spindle (**B**) approx. 10 - 12 mm (3/8 - 1/2 in) until diaphragm spring is pretensioned. Do not overstress spring.

– Install pressure plate to flywheel using new mounting bolts. Gradually tighten in crisscross pattern. Make sure special tool **21 2 170** remains undisturbed on pressure plate until mounting bolts are fully torqued.

Tightening torque	
Clutch pressure plate to flywheel (replace M8 bolts with new)	23 Nm (17 ft-lb)

– When finished, remove flywheel locking and clutch disc centering tools.

– Install transmission. See **230 Manual Transmission**.

Clutch release bearing, replacing

– Remove transmission. See **230 Manual Transmission**.

◀ Push spring clip (**arrow**) through clutch release lever and lift lever out of transmission bell housing.

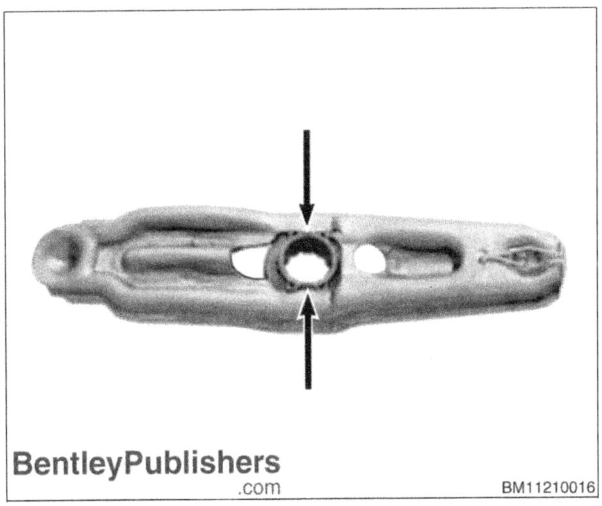

◀ With clutch release lever on bench and release bearing face down, press in retainers slightly (**arrows**) to detach release bearing.

– When reinstalling:
  - Check release lever for damage or cracks. Replace if necessary.
  - Clean release bearing and release lever sliding surfaces. Do not grease sliding parts.
  - Check spring clip and release lever ball pivot for damage. Replace if necessary.
  - Clean sliding surface of release bearing guide sleeve on transmission. Coat lightly with grease.
  - Lightly grease ball pivot before installing release lever.

– Reinstall transmission. See **230 Manual Transmission**.

Flywheel, removing and installing

– Remove transmission. See **230 Manual Transmission**.

– Remove clutch. See **Clutch replacement** in this repair group.

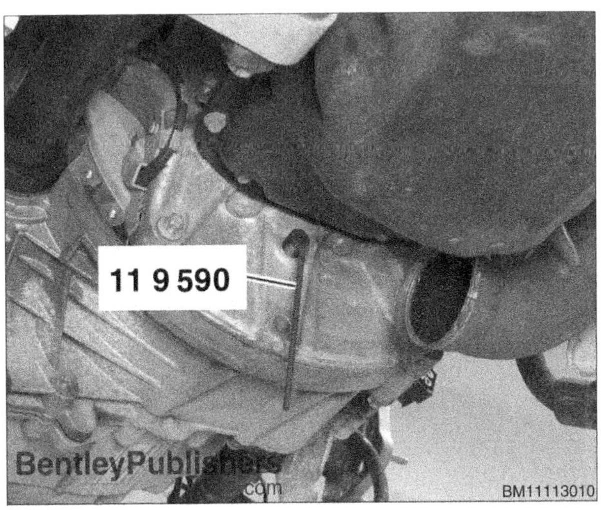

◀ Install special tool **11 9 590** to counterhold flywheel.

Remove flywheel mounting bolts (**arrows**) and discard. Remove flywheel.

- Clean crankshaft threaded holes.

When installing, note that crankshaft sensor gear is loose on crankshaft. Make sure flywheel bore is matched to dowel pin (**arrow**) on sensor gear when mounting flywheel to crankshaft.

- Reinstall flywheel using new Torx bolts.

Tightening torque	
Flywheel to crankshaft (replace M9 x 1.25 x 25 mm Torx T55 bolts with new) • Stage 1 • Stage 2 • Stage 3	 8 Nm (6 ft-lb) 30 Nm (22 ft-lb) additional 90°

- Reassemble clutch. See **Clutch replacement** in this repair group.
- When finished, remove flywheel locking tool.
- Reinstall transmission. See **230 Manual Transmission**.

230 Manual Transmission

GENERAL

This repair group covers manual transmission service, including transmission removal and installation. Internal transmission repair is not covered. Special tools and procedures are required to disassemble and service the internal geartrain.

See also:

- **210 Clutch**
- **250 Gearshift Linkage**

Manual transmission applications and gear ratios

The MINI front-wheel drive transmission is integral with the differential.

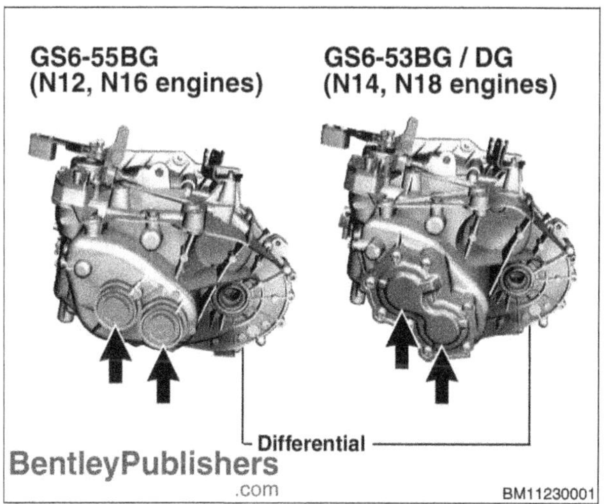

The two Getrag 6-speed manual transmissions in MINI models are visually similar, with slight differences in transmission shaft end castings (**arrows**). See **Table a**. Gear ratios are in **Table b**.

Table a. Manual transmission applications

Model (engine)	MINI designation	Manufacturer designation
Cooper, Clubman, Convertible (N12, N16)	GS6-55BG	Getrag G255
Cooper S, Clubman S, Convertible S (N14, N18)	GS6-53BG / DG	Getrag G253

A limited slip differential is optional on the GS6-53BG / DG transmission.

Table b. Manual transmission gear ratios		
Gear	GS6-55BG	GS6-53BG / DG
1st	3.214: 1	3.308: 1
2nd	1.792: 1	2.130: 1
3rd	1.194: 1	1.483: 1
4th	0.914: 1	1.139: 1
5th	0.784: 1	0.949: 1
6th	0.683: 1	0.816: 1
Reverse	3.143: 1	3.231: 1
Final drive	4.353: 1	3.647: 1

Warnings and Cautions

WARNING—
- *Do not reuse self-locking fasteners. They are designed to be used only once and may fail if reused. Replace with new.*

CAUTION—
- *After separating the engine and transmission, check that bell housing alignment dowels are undamaged and placed correctly before reassembly. Misalignment of engine flywheel to torque converter may result in clutch slipping, shudder, or poor disengagement; noise from transmission input shaft; transmission popping out of gear; difficulty changing gears; or internal damage to the transmission.*
- *Fill manual transmission only with MINI approved synthetic oil. Filling with incorrect lube could cause the following damage*
  - *— Premature synchromesh ring wear.*
  - *— Reduced service life of bearings.*
  - *— Tooth profile damage.*
  - *— Faster wear of selector forks.*
  - *— Damage to radial shaft oil seals.*

TRANSMISSION FLUID SERVICE

The manual transmission comes filled from the factory with "Lifetime" synthetic fluid and does not require periodic fluid changes.

Table c. MINI Cooper manual transmission applications (6-speed)				
Model (engine)	**BMW designation**	**Manufacturer designation**	**Recommended fluid**	**Fill capacity (max)**
Cooper, Clubman, Convertible (N12, N16)	GS6-55BG	Getrag G255	MTF-LT-4 (yellow sticker) BMW part no. 83 22 2 344 589	1.7 liters (1.8 US qt)
Cooper S, Clubman S, Convertible S, John Cooper Works (N14, N18)	GS6-53BG / DG	Getrag G253		1.9 liters (2.0 US qt)

Transmission fluid level, checking

- Drive vehicle for a few miles to warm transmission.
- Raise car and support safely.

> ***WARNING—***
> - *Make sure the car is stable and well supported at all times. Use a professional automotive lift or jack stands designed for the purpose. A floor jack is not adequate support.*

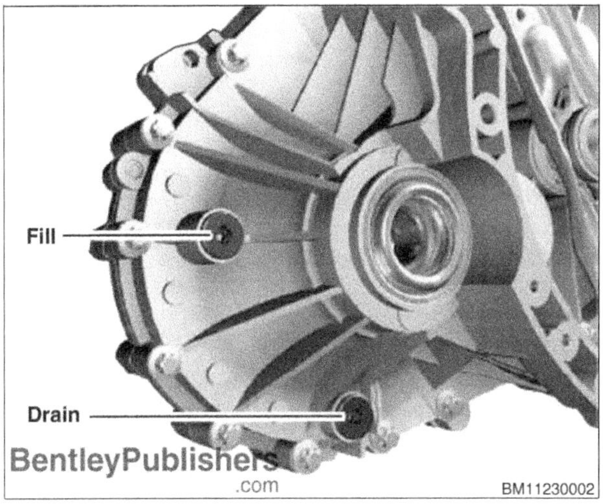

◄ Unscrew transmission fill plug. Make sure fluid is even with bottom of hole or drips slightly.

- Clean plug and reinstall.

Tightening torque	
Transmission drain or fill plug to transmission	43 ± 4 Nm (32 ± ft-lb)

Transmission fluid, replacing

- Drive vehicle for a few miles to warm transmission.
- Raise car and support safely.

> ***WARNING—***
> - *Make sure the car is stable and well supported at all times. Use a professional automotive lift or jack stands designed for the purpose. A floor jack is not adequate support.*

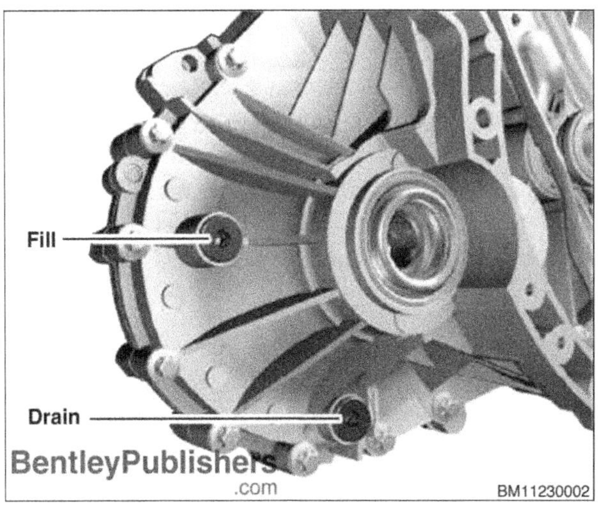

◄ Unscrew transmission drain plug and allow fluid to completely drain into a suitable container.

- Clean drain plug and reinstall.
- Unscrew transmission fill plug and fill transmission until fluid overflows slightly.
- Clean fill plug and reinstall.

Tightening torque	
Transmission drain or fill plug to transmission	43 ± 4 Nm (32 ± ft-lb)

TRANSMISSION SERVICE

Back-up light switch, replacing

– Remove air filter (N12, N16 engine) or intake air ducts (N14, N18 engine) to access top of transmission.

◄ Working between shifter cable boots, detach electrical harness connector (**arrow**) and unscrew back-up light switch.

– Replacement is reverse of removal.

Tightening torque	
Back-up light switch to transmission	16 Nm (12 ft-lb)

Drive axle seal, replacing

Drive axle flange seal replacement procedure is similar on left and right. When reinstalling a drive axle, be sure to replace flange seal first.

– Raise car and support safely.

> ***WARNING —***
> *• Make sure the car is stable and well supported at all times. Use a professional automotive lift or jack stands designed for the purpose. A floor jack is not adequate support.*

– Drain transmission fluid.

– Remove drive axle. See **310 Front Suspension, Drive Axles**. Be prepared to catch dripping oil.

◄ Insert seal puller (MINI special tool **23 0 120**) into seal to remove. Take care not to damage seal seating surface or differential case.

◄ When installing, coat sealing lip of new seal with gear oil. Drive seal with MINI special tool **23 0 370** until seal is flush with differential case.

– Remainder of installation is reverse of removal. Remember to:

- Use assembly guard, supplied with new seal, to protect seal lip against damage. See **310 Front Suspension, Drive Axles**.
- Push drive axle shaft into differential case past resistance of snap ring until an audible snap is heard. Remove assembly guard.
- Fill transmission with fluid and check for leaks.

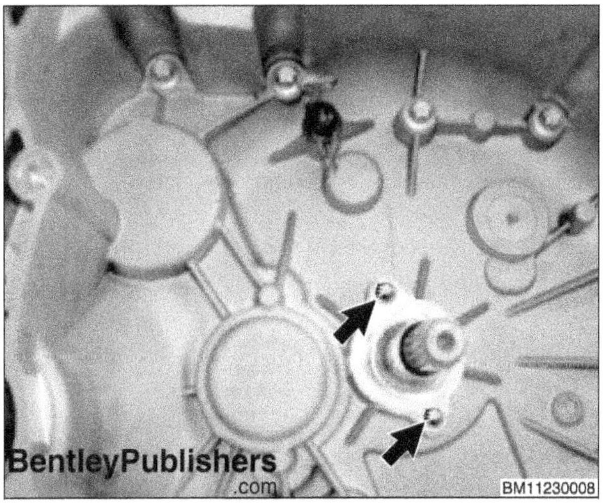

Input shaft seal, replacing

- Remove transmission. See **Transmission Removal and Installation** in this repair group.
- Remove clutch release lever and bearing. See **210 Clutch**.
- Remove clutch release bearing guide tube mounting bolts (**arrows**) and lift off guide tube.
- Drive a hole into input shaft seal using a centre punch.

> *CAUTION—*
> *• Do not make the hole using a drill. Drilling debris may fall into the transmission and cause damage.*

- Screw MINI special tool **23 0 490** into edge of seal, then use slide hammer to remove seal. Take care not to damage seal seating surface.

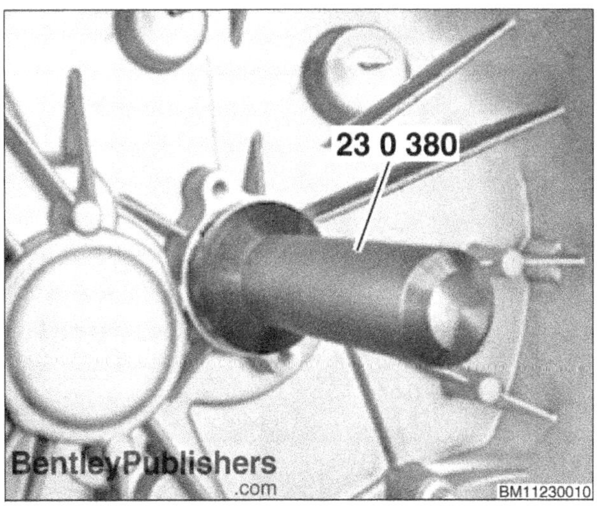

- When installing, coat sealing lip of new seal with gear oil. Use MINI special tool **23 0 380** to drive seal firmly home.
- Remainder of installation is reverse of removal. Remember to check and top off transmission with fluid.

Transmission Removal and Installation

Remove and install the transmission with the vehicle on a lift. Support engine using appropriate equipment. Use a transmission jack to lower the transmission.

Transmission removal procedure is a complicated job which includes the following steps:

- Support and raise engine; covered in this repair group.
- Remove transmission supports and mounts; covered in this repair group.
- Remove front wheels and drive axles. See **310 Front Suspension, Drive Axles**.
- Remove complete exhaust system. See **180 Exhaust System**.
- Remove wheel housing liners. See **410 Fenders, Engine Hood**.
- Detach front bumper. See **510 Exterior Trim, Bumpers**.
- Remove underbody panels. See **020 Maintenance**.
- Disconnect steering rack from steering shaft. See **320 Steering**.
- Remove front subframe. See **310 Front Suspension, Drive Axles**.

This section describes transmission removal and installation without engine removal. If additional engine repair is needed, the engine and transmission may be removed as a complete unit with little additional work. See **110 Engine Removal and Installation**.

> ***WARNING—***
> - *Front subframe and transmission removal may upset the balance of the vehicle on a lift.*

Transmission, removing and installing

– Disconnect negative (-) cable from battery.

> ***CAUTION—***
> - *Prior to disconnecting the battery, read the battery disconnection cautions given in* **001 Warnings and Cautions**.

– **N12 engine**: Remove engine air cleaner housing and air ducts. See **130 Fuel Injection**.

– **N14 engine**: Remove turbo air ducts above transmission. See **130 Fuel Injection**.

◄ Use engine support bar to support engine and transmission.

> ***CAUTION—***
> - *Do not allow engine support bar to crush hood release Bowden cable.*
> - *Do not raise engine until engine and transmission supports are loosened as described in the following steps.*

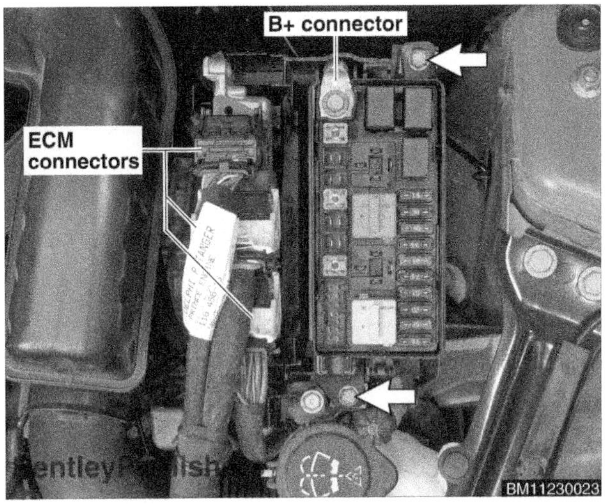

Working at left rear engine compartment:

- Remove front power distribution and fuse panel cover.
- Disconnect battery B+ connector from fuse panel.
- Remove distribution panel mounting bolts (**arrows**) and push panel aside.
- Detach engine control module (ECM) connectors and pull ECM out of holder. See **130 Fuel Injection**.

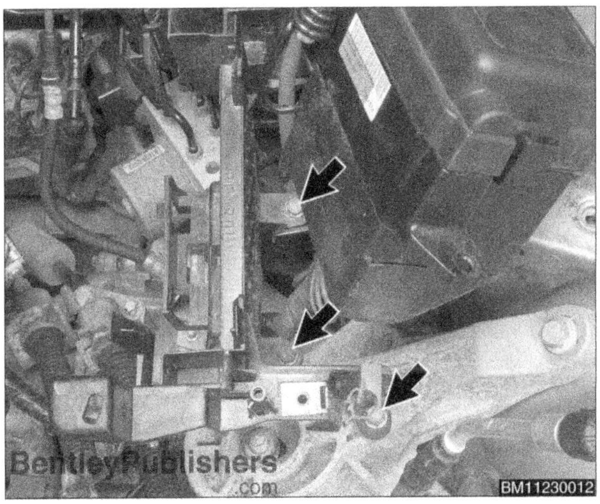

Remove fuse panel bracket mounting bolts (**arrows**) and lift off bracket.

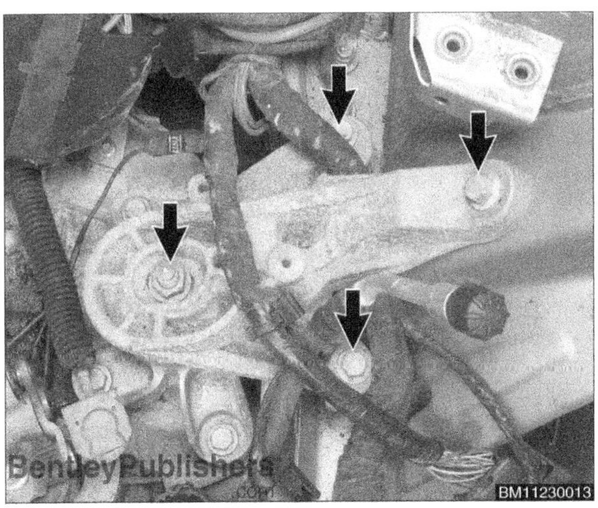

Working at transmission upper support bracket, left front inner fender wall, remove mounting fasteners (**arrows**) and lift off support bracket.

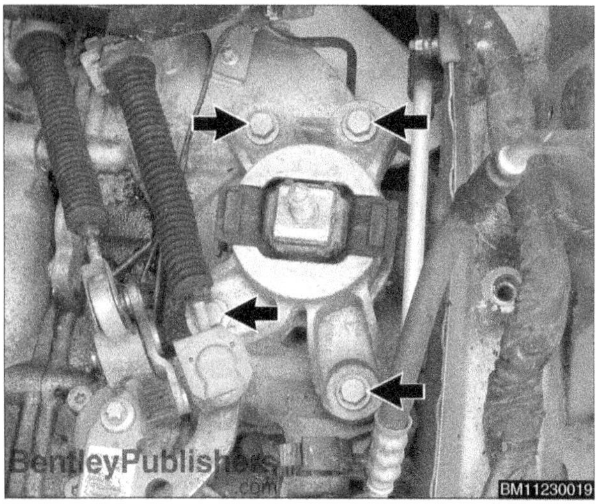

< Remove transmission upper mount bolts (**arrows**) and lift off mount.

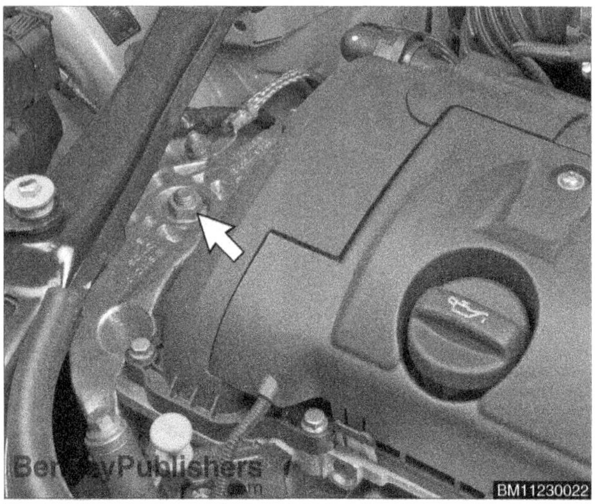

< Working at right side engine support bracket, loosen engine mounting nut (**arrow**).

– Raise vehicle and support safely.

> *WARNING—*
> - *Make sure the vehicle is stable and well supported at all times. Use a professional automotive lift or jack stands designed for the purpose. A floor jack is not adequate support.*

< Working underneath rear of engine, remove engine stabilizer link mounting bolts (**arrows**). Remove stabilizer link.

– Use engine support bar to raise engine approx 10 mm (3/8 in)

– Drain transmission fluid.

– Separate exhaust pipe from catalytic converter. Remove entire exhaust system. See **180 Exhaust System**.

– Remove front subframe and drive axles. See **310 Front Suspension, Drive Axles**.

> *CAUTION—*
> - *While steering spindle is detached from steering rack, make sure steering wheel is not rotated. Rotating the wheel may damage the airbag and horn contact spring.*

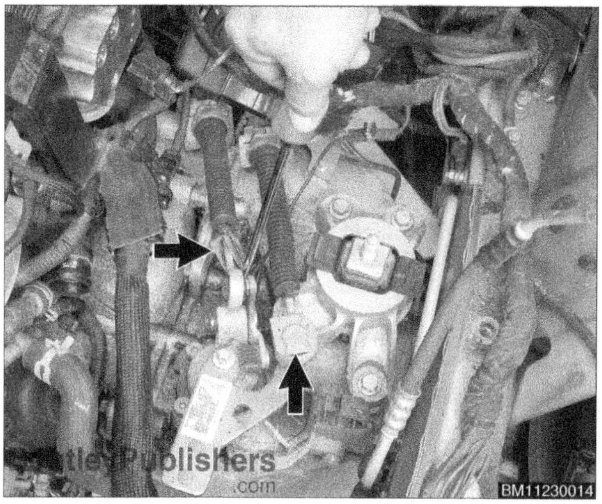

◀ Working at top of transmission housing:

- Slide selector cable ball joint locking clips (**arrows**) off and use screwdriver to pry cables off selector levers.
- Remove cable bracket mounting bolt and lift off bracket.

– Detach back-up light switch and neutral sensor switch connectors. Remove connector harnesses from brackets on transmission.

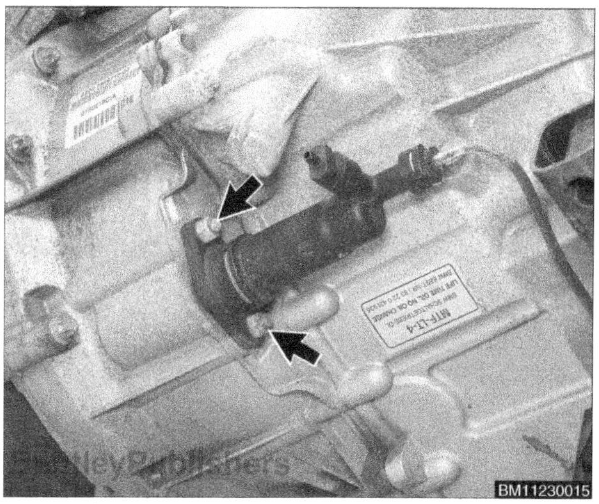

◀ Working underneath transmission at slave cylinder:

- Remove mounting nuts (**arrows**).
- Do not separate hydraulic line from cylinder.
- Remove slave cylinder from transmission and hang aside.

> ***CAUTION—***
> - *If slave cylinder is to be reused, do not deflect push rod. Fluid can escape if push rod is deflected.*

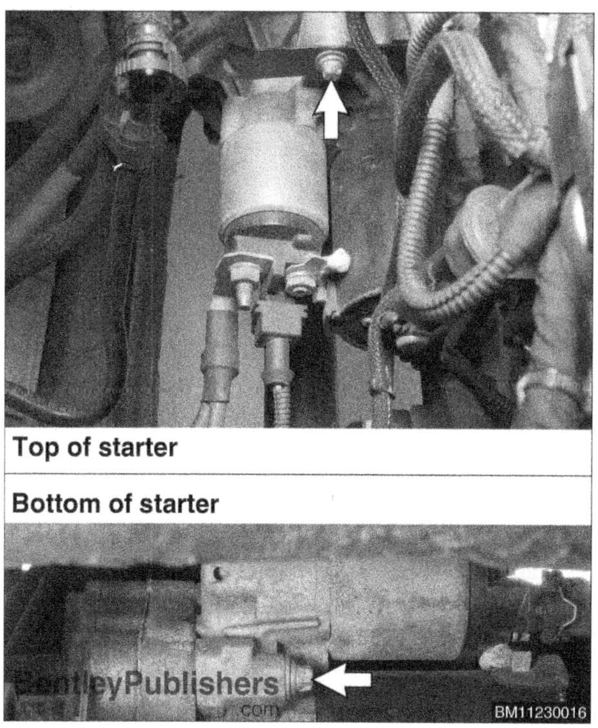

◀ Working at starter motor, remove upper and lower mounting bolts (**arrows**). Remove starter from transmission and hang aside.

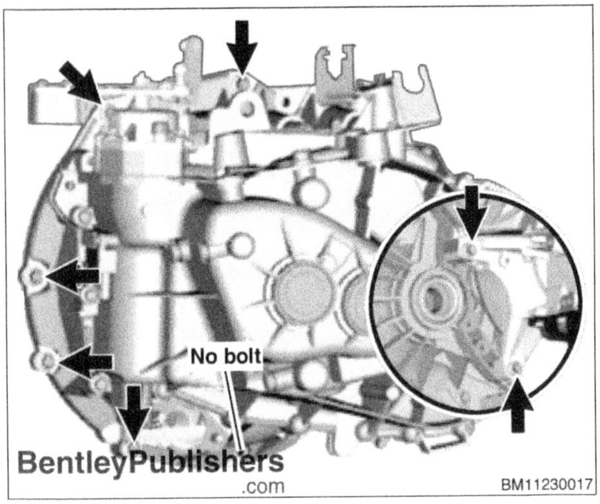

- Place transmission jack underneath transmission, then remove transmission bell housing bolts (**arrows**). Note that there is no bolt fitted to lowest point on transmission.

- Pull transmission straight back to disengage transmission input shaft from clutch, then tip down to remove.

> ***WARNING —***
> *• Be sure the vehicle is properly supported. The removal of the transmission may upset the balance of the vehicle on the lift.*

> ***CAUTION—***
> *• Do not allow transmission to hang on input shaft and clutch disc during removal and installation.*

- Remove grease and contaminants from transmission input shaft. Do not grease shaft.

- Clean clutch release bearing and lever.

- Inspect clutch disc for wear. Replace complete clutch if necessary. See **210 Clutch**.
  - If reusing clutch, lightly grease clutch disc hub splines.
  - On new clutch, splines are already greased.

- Before reinstalling transmission, check alignment dowels (**arrows**) in bell housing flange for correct seating. Replace if damaged.

- Installation is reverse of removal. Remember to:
  - Replace self-locking fasteners with new.
  - Replace drive axle flange seals. See **Drive axle seal, replacing** in this repair group.

Tightening torques	
Ball joint, inner, to subframe (replace M14 x 1.5 ZNS3 self-locking nut with new)	70 Nm (ft-lb) + additional 90°
Ball joint, outer, to steering knuckle (replace M12 x 1.5 ZNS3 self-locking nut with new	70 Nm (ft-lb) + additional 90°
Bumper to front subframe (M8 x 18 mm bolts)	19 Nm (14 ft-lb)
Clutch slave cylinder to transmission (M8 nuts)	21 Nm (15.4 ft-lb)
Control arm bracket and stabilizer anchor to subframe (M14 x 1.5 x 105 mm self-tapping bolts)	165 Nm (122 ft-lb)
Drive axle bearing block to engine block, right side (M10 bolts x 45 bolts)	38 Nm (28 ft-lb)
Drive axle collar nut (replace M22 x 1.5 nut with new, stake to axle after tightening)	182 Nm (134 ft-lb)
Engine right side support bracket to engine mount (replace M12 x 1.5 mm 10.9 grade self-locking nut with new)	100 Nm (74 ft-lb)

Tightening torques (continued)	
Engine stabilizer link to bracket (M12 x 90 mm bolts)	108 Nm (80 ft-lb)
Exhaust pipe to catalytic converter (V-band clamp)	25 ± 3 Nm (18 ± 2 ft-lb)
Front power distribution and fuse panel to bracket (M6 bolts)	5 Nm (44 in-lb)
Front subframe to chassis: • Front bolts M12 x 85 mm • Rear bolts M12 x 62 mm • Control arm bushing carrier bolts (replace M10 x 60 mm 10.9 grade bolts with new)	 100 Nm (74 ft-lb) 100 Nm (74 ft-lb) 59 Nm (44 ft-lb) + additional 90°
Road wheel to hub	140 ± 10 Nm (103 ± 7 ft-lb)
Stabilizer link to stabilizer bar (replace M10 self-locking nuts)	56 Nm (41 ft-lb)
Starter motor to transmission (M8 bolts)	20 Nm (14.9 ft-lb)
Steering shaft universal joint pinch bolt to steering rack (replace M8 x 1.25 x 41 mm bolt with new)	28 Nm (21 ft-lb)
Tie rod end to steering knuckle (replace M12 x 1.5 self-locking nut)	65 Nm (48 ft-lb)
Transmission to engine (M10 bolts)	38 Nm (28 ft-lb)
Transmission upper mount to transmission (M10 x 45 mm 8.8 grade bolts)	38 Nm (28 ft-lb)
Transmission upper support bracket to body (replace M10 bolts with new) • Stage 1 • Stage 2	 25 Nm (18 ft-lb) + additional 90°
Transmission upper support bracket to transmission mount (replace M12 x 1.5 mm 10.9 grade self-locking nut with new)	100 Nm (74 ft-lb)

– When finished, fill transmission with correct fluid. See **Transmission Fluid Service** in this repair group.

– Bleed clutch hydraulics if necessary. See **210 Clutch**.

– Have vehicle professionally aligned.

– Vehicle with DSC: Use MINI scan tool to carry out steering angle sensor adjustment.

240 Automatic Transmission

240

General

This repair group covers maintenance and replacement of the automatic transmission.

Automatic transmission internal repairs are not covered. Transmission repairs require special service equipment and knowledge. If transmission internal service is required, consult an authorized MINI dealer about a factory reconditioned unit or a transmission rebuild.

See also:

- **119 Lubrication System** for crankshaft rear main (flywheel) seal.
- **170 Radiator and Cooling System** for draining and filling engine coolant.
- **200 Transmission–General** for drivetrain information.
- **250 Gearshift Linkage** for manual override of gearshift lock

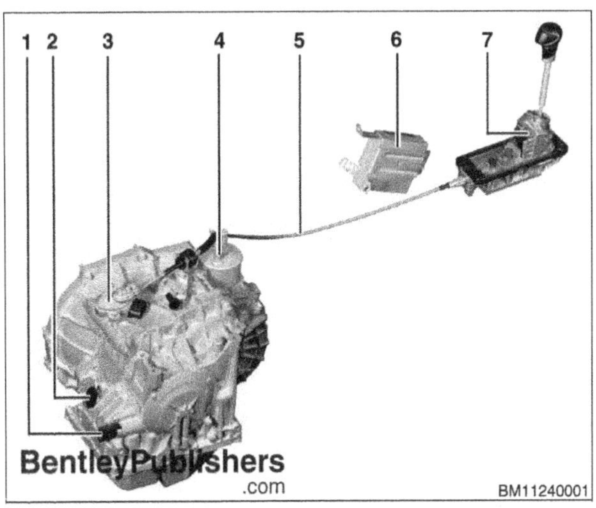

Automatic transmission and controls

1. Multiple pin connector 1 to EGS control module
2. Multiple pin connector 2 to EGS control module
3. Gear position switch
4. ATF cooler
5. Shift selector Bowden cable
6. EGS control module (under left side dashboard)
7. Selector lever unit

Automatic transmission overview

The Aisin 6-speed automatic transmission is electronically controlled by EGS control module. The vehicle may be driven in fully automatic mode (drive range D) or in Steptronic mode (M/S range), allowing manual shifting of gear range. This transmission is fitted to both normally aspirated (N12, N16 engine) and turbocharged (N14, N18 engine) MINI models.

Table a. Aisin GA6F21WA 6-speed automatic transmission	
Transmission output	Max. torque 240 Nm (177 lb-ft)
Gear ratios	1st gear: 4.044
	2nd gear: 2.371
	3rd gear: 1.556
	4th gear: 1.159
	5th gear: 0.852
	6th gear: 0.672
	Reverse: 3.193
Final drive	N12, N16 engine = 4.103 N14, N18 engine = 3.683
Gear positions	P-R-N-D-M/S
Control	Electrohydraulic
Weight	Including transmission oil and torque converter: approximately 82 kg (181 lb)
Overall length	348 mm (13.7 in)
Towability	50 km at max. 50 kph (30 mph at max. 30 mph)
Maintenance	Maintenance free with lifetime oil fill

Shift characteristics

The Agitronic feature enables gear selection by means of shift paddles on the steering wheel. Gears can be shifted without having to select Steptronic even if the shift selector is in drive range D.

The SPORT button enables three different gear shifting speeds: Comfortable, agile and sports speeds. This range is realized by 3-stage hydraulic gearshift pressure, providing short shift times.

Gear position switch

The gear position switch on the transmission housing is connected to the selector lever via Bowden cable. The switch sends signals for gear positions P, R, N, and D to the EGS control module. Gear position is displayed on the instrument cluster. The display is controlled by EGS control module via PT-CAN.

PARK / NEUTRAL position signal for starter neutral safety function is communicated from gear position switch to CAS module. See **Starter troubleshooting** in **121 Battery, Starter, Alternator**.

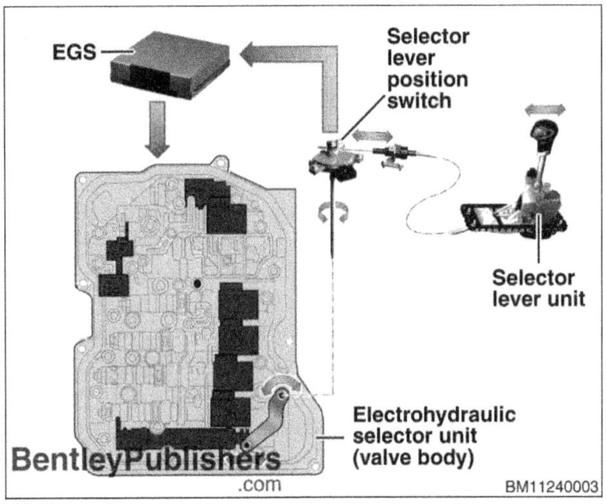

Electrohydraulic selector unit (valve body)

Electrohydraulic selector unit (valve body), above the transmission oil pan, sets hydraulic pressures for drive and brake clutches. The Bowden cable from the selector lever regulates the position switch which is connected directly to the valve body.

Electronic gearbox control (EGS)

EGS control module, underneath left side of dashboard above the pedal assembly, uses PT-CAN to exchange data with the following control modules:

- Engine control module (ECM)
- Dynamic stability control (DSC) module
- Car access system (CAS)
- Junction box electronics
- Instrument cluster

EGS functions

Pedal position sensor signal, sent via CAN-bus to EGS control module, triggers gear selection.

Injection signal (the engine load status signal) plus throttle-valve signal and pulling-away signal trigger gear change.

Engine Intervention. During a gear change, transmission electronics complement engine electronics by retarding ignition timing about 200 ms. This reduces engine torque momentarily and improves shift quality. Transmission load and shift time are reduced.

Kick-down. If the driver presses the accelerator pedal to the floor, the accelerator pedal module full-throttle voltage value of 4.3 V is exceeded. This is interpreted by the ECM as a transmission gear kick-down signal. The information is sent to EGS via PT-CAN.

Transmission input speed Hall sensor on the gearbox input shaft signals EGS the exact transmission input speed for the following functions:

- Control, adaptation and monitoring of gear changes.
- Control and monitoring of converter lockup clutch.
- Diagnosis of shift elements and plausibility checking of engine and gearbox output speeds.

Transmission output speed Hall sensor on the parking interlock gear signals EGS the exact transmission output speed for the following functions:

- Gear change point selection.
- Diagnosis of shift elements and plausibility checking of engine and turbo unit output speeds.

DSC. The engine limited to 4500 rpm during DSC intervention. The transmission shifts up automatically above 4500 rpm to reduce torque to the wheels. Thanks to DSC control and front-wheel drive, no special winter program is needed in EGS,

Transmission fluid temperature sensor is in the valve body. Transmission fluid temperature information is needed for the following functions:

- Matching shift pressure (system pressure) and pressure increase and decrease during gear changes.
- Activating and deactivating temperature-dependent functions such as engaging converter lockup clutch.

Brake light switch signal via PT-CAN is used to control selector lever interlock (shift lock) and other transmission control functions such as downhill detection function.

Engine speed data are needed for safety functions. For example, EGS prevents downshifting that could overrev the engine.

Gradient detection (in shift lever positions D and SD) detects whether the vehicle is travelling uphill or downhill by comparing engine torque and the selected gear to wheel speed.

- When travelling uphill, the EGS selects the gear which best provides the vehicle with sufficient traction.
- When travelling downhill, the gear is held after brake use or manual downshift. The transmission shifts down when brake use is detected.

Fast-off detection. Rapid throttle closure, due to the driver removing the foot rapidly from accelerator pedal, makes the transmission upshift. This is not desirable because it hinders engine braking when fast deceleration is needed. In addition, in many cases, using the brake pedal again shortly after shifting up or down considerably increases shift frequency. Fast-off detection function blocks upshifting in the circumstances described.

Cornering detection. Shifting up in corners is prevented above a defined lateral acceleration. The radius of the corner is detected from the differences in speed of the inside and outside wheels.

Shift characteristics. EGS determines shift up and shift down points as functions of:

- Driver's desired load (accelerator pedal and throttle-valve position).
- Transmission output speed.
- Selector lever setting.
- Kick-down signal.

Shift points vary, depending on fluid temperature sensor signal.

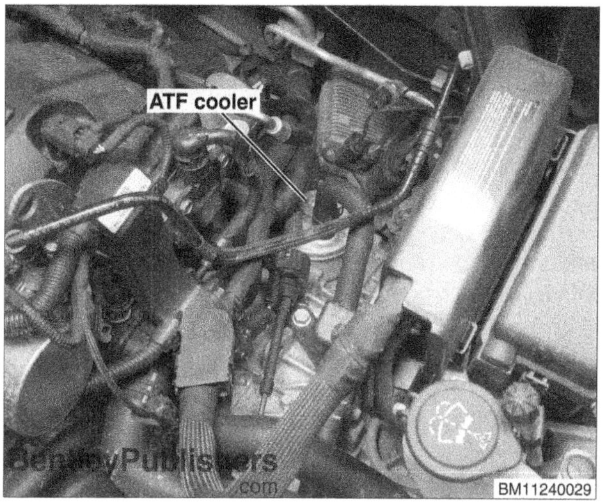

Transmission fluid cooler

The ATF cooler, bolted to the transmission housing, connects to the engine coolant circuit via two lines. The ATF cooler speeds up transmission fluid heating while the engine warms up. When the engine reaches operating temperature, coolant is used to reduce transmission fluid temperature.

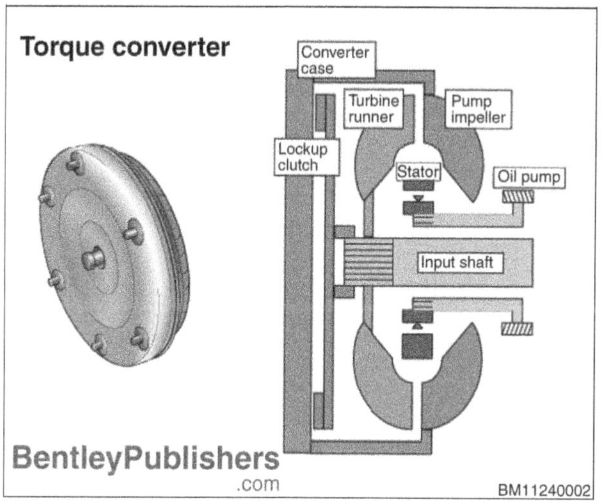

Torque converter with lockup clutch

Torque from the engine is transferred to the transmission by a torque converter fitted with a lockup clutch, ensuring a positive lock between the engine and automatic gearbox and reducing fuel consumption.

The lockup clutch is closed in 2nd to 6th gears depending on the following conditions:

- Accelerator pedal position.
- Transmission output speed: Locked from 1800 rpm up.
- Drive program.
- Transmission fluid temperature: Unlocked below 20°C / 68°F so that transmission oil heats up faster.
- Selected gear.

Diagnostics

Faults in sensors, actuators and EGS control module are stored in fault code memory and can be read using MINI scan tool.

Limp-home mode

If the EGS control module detects a fault, it activates an emergency operation program (limp-home mode). Third gear is selected, allowing the trip to be completed at a slower speed. The instrument cluster displays the letters EP (Emergency Program).

Warnings and Cautions

WARNING—

- *Do not reuse self-locking fasteners. They are designed to be used only once and may fail if reused. Replace with new.*

CAUTION—

- *When performing any repair which involves separating the engine and transmission, check that bell housing alignment dowels are undamaged and in place before reassembly. Misalignment of engine flywheel to torque converter may result in transmission internal damage.*
- *When removing interior trim clips, avoid damaging trim pieces by using a plastic prying tool or a screwdriver with the tip wrapped with masking tape.*
- *Fill transmission only with MINI approved synthetic fluid. Filling with incorrect lube could cause the following damage:*
  - *— Premature wear.*
  - *— Reduced service life of bearings.*
  - *— Damage to oil seals.*

AUTOMATIC TRANSMISSION FLUID (ATF) SERVICE

Check transmission periodically for fluid leaks.

Automatic transmission fluid (ATF)	
Fluid specification	JWS-3309 ATF MINI part no. 83 22 0 402 413

ATF level, checking

ATF level, which varies according to temperature, is checked by the overflow method.

- Place gear selector lever in PARK (P).
- Connect MINI scan tool to vehicle.
- Raise car and support in a safe manner. Make sure vehicle is level

> ***WARNING —***
> - *Make sure the car is stable and well supported at all times. Use a professional automotive lift or jack stands designed for the purpose. A floor jack is not adequate support.*

Use MINI special tool **24 4 240** (Torx T55) to remove ATF filler plug on transmission. Place a clean drip tray underneath transmission to catch fluid.

Remove ATF inspection plug (**arrow**) from transmission sump.

- Top off ATF through fill hole until it overflows at inspection hole.

Automatic transmission fluid (ATF)	
Fluid specification	JWS-3309 ATF MINI part no. 83 22 0 402 413

- Start engine and run at idle.
- Check and see if any more ATF runs out of inspection hole. If not, continue to add ATF
- Press brake pedal and slowly shift from P to D twice. Put gear selector back in P and release brake pedal.

– Check and see if any ATF runs out of inspection hole.

– Read ATF temperature with MINI scan tool.

ATF temperature specification for level check	
ATF operating temperature	35° - 45°C (95° - 113°F)

– With ATF at operating temperature, top off ATF until it runs out of inspection hole. A gravity fed ATF filler tool is best to use for filling transmission.

– Install inspection and filler plugs to transmission. Be sure to replace sealing gaskets.

Tightening torques	
ATF plugs to transmission: • M10 • M14 • M18	 27 Nm (20 ft-lb) 23.5 + 10 Nm (17 + 7 ft-lb) 23.5 + 10 Nm (17 + 7 ft-lb)

ATF, replacing

A good way to be sure the correct amount of fluid is refilled is to measure the amount of fluid that was initially drained out. Note that this assumes the correct amount of fluid was in the transmission prior to draining.

When refilling, pump in the same amount of fluid that was initially drained out. Then follow the procedure in the manual for checking the ATF level. See **ATF level, checking** in this repair group.Recheck fluid level after driving car and shifting through all gears.

– Raise car and support in a safe manner. Make sure vehicle is level

> ***WARNING—***
> *• Make sure the car is stable and well supported at all times. Use a professional automotive lift or jack stands designed for the purpose. A floor jack is not adequate support.*

◄ Remove ATF inspection plug (**arrow**) from transmission sump.

> ***WARNING—***
> *• Do not drain fluid before transmission is fully cooled off.*

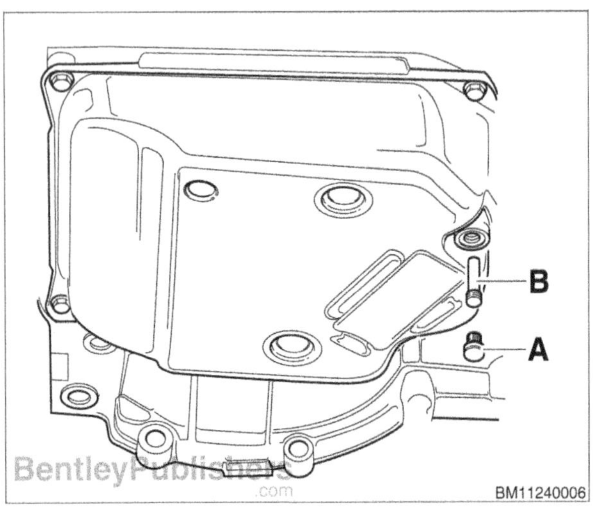

◄ With inspection plug (**A**) removed, unscrew overflow tube (**B**) from inside inspection hole.

– Drain ATF into suitable container. Note amount drained to help with refilling.

– When ATF is completely drained, reinstall overflow tube and tighten.

> ***CAUTION—***
> *• Do not overtighten overflow tube.*

– Hand tighten inspection plug using new seal.

– Fill transmission with ATF through filler hole. A gravity fed ATF filler tool is best to use for filling transmission.

Automatic transmission fluid (ATF)	
Approximate capacity (not including torque converter)	4.5 liters (4.8 US qt)
Lifetime fluid	JWS-3309 ATF MINI part no. 83 22 0 402 413

– Check ATF level. See **ATF level, checking** in this repair group.

– Install filler plugs to transmission. Be sure to replace sealing gasket.

Tightening torques	
ATF plugs to transmission: • M10 • M14 • M18	 27 Nm (20 ft-lb) 23.5 + 10 Nm (17 + 7 ft-lb) 23.5 + 10 Nm (17 + 7 ft-lb)

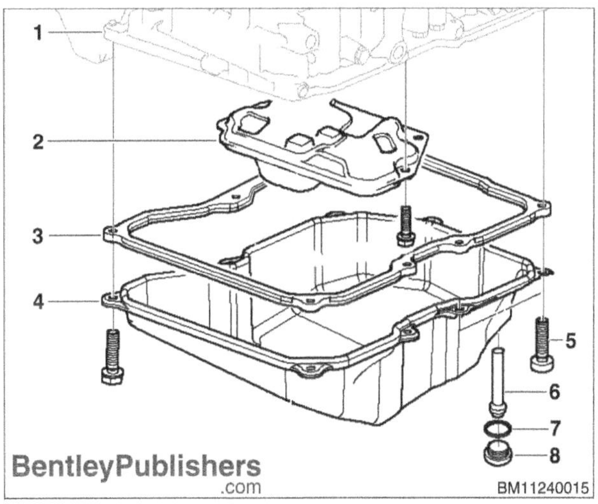

Fluid strainer and fluid sump components

1. Automatic transmission housing
2. Fluid strainer
3. Sump gasket
4. Sump
5. Torx-head bolt
6. Fluid overflow tube
7. Sealing O-ring
8. Fluid inspection plug

Fluid strainer, removing and installing

– Disconnect negative (-) cable from battery.

> ***CAUTION—***
> • *Prior to disconnecting the battery, read the battery disconnection cautions given in* **001 Warnings and Cautions**.

– **N12 engine**: Remove engine air cleaner housing and air ducts. See **130 Fuel Injection**.

– **N14 engine**: Remove turbo air ducts above transmission. See **130 Fuel Injection**.

< Use engine support bar to support engine and transmission.

> ***CAUTION—***
> • *Do not allow engine support bar to crush hood release Bowden cable.*
> • *Do not raise engine until engine and transmission supports are loosened as described in the following steps.*

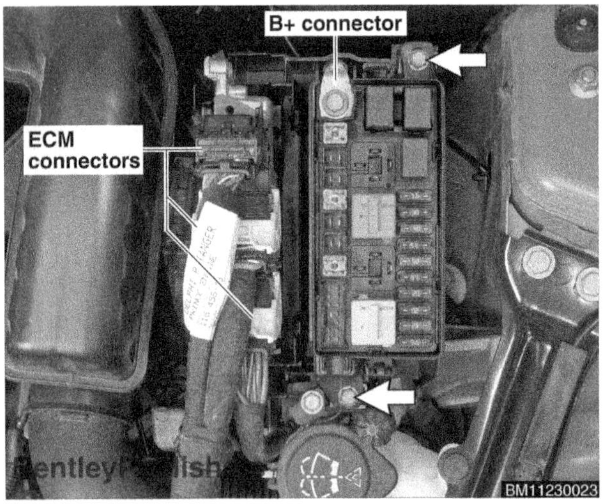

Working at left rear engine compartment:

- Remove front power distribution and fuse panel cover.
- Disconnect battery B+ connector from fuse panel.
- Remove distribution panel mounting bolts (**arrows**) and push panel aside.
- Detach engine control module (ECM) connectors and pull ECM out of holder. See **130 Fuel Injection**.

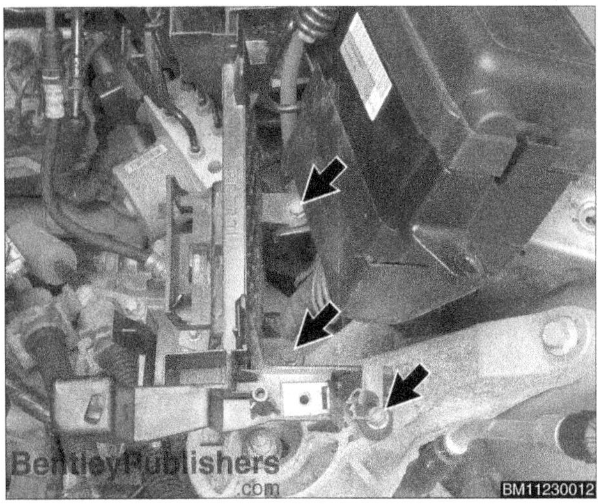

Remove fuse panel bracket mounting bolts (**arrows**) and lift off bracket.

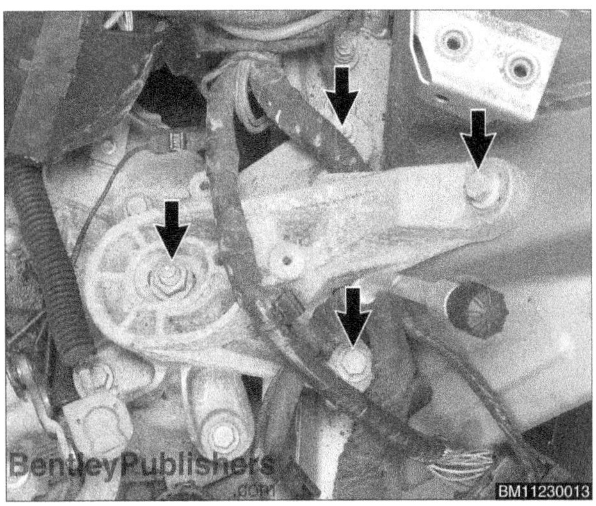

Working at transmission upper support bracket, left front inner fender wall, remove mounting fasteners (**arrows**) and lift off support bracket.

◄ Working at right side engine support bracket, loosen nut (**arrow**).

– Raise vehicle and support safely.

> ***WARNING —***
> *• Make sure the vehicle is stable and well supported at all times. Use a professional automotive lift or jack stands designed for the purpose. A floor jack is not adequate support.*

◄ Working underneath rear of engine, remove engine stabilizer link mounting bolts (**arrows**). Remove stabilizer link.

– Use engine support bar to raise engine approx 10 mm (3/8 in)

– Remove fluid inspection plug and fluid overflow tube and drain transmission fluid into clean container. See **ATF, replacing** in this repair group. Note quantity of drained fluid.

> ***WARNING —***
> *• Do not drain fluid before transmission is fully cooled off.*

– Remove 8 fluid sump bolts, including one Torx-head. See **Fluid strainer and fluid sump components** in this repair group. Lower sump.

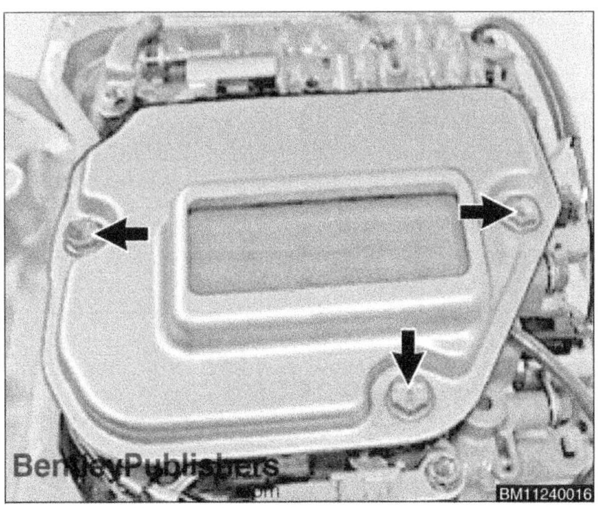

◄ Remove fluid strainer mounting bolts (**arrows**). Lower strainer and clean transmission valve body contact surfaces. Install new strainer.

Tightening torque	
Fluid strainer to transmission valve body (M6 bolts)	11 Nm (8 ft-lb)

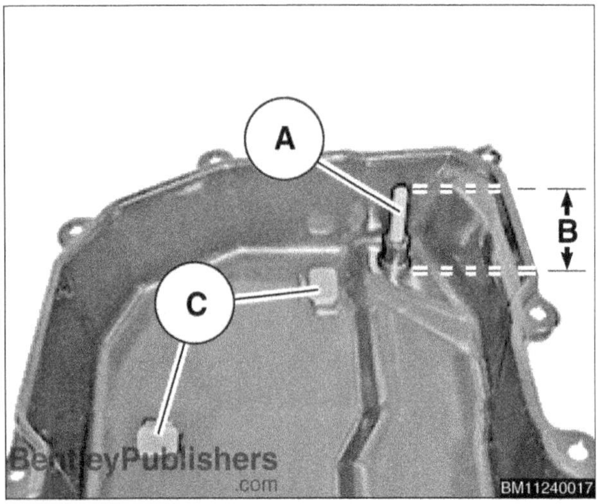

If new sump or new overflow tube (**A**) is to be fitted, install overflow tube and make sure overflow tube extension (**B**) inside sump pan matches specifications.

Overflow tube extension inside sump	
Measurement **B**	37.35 - 38.05 mm (1.47 -1.50 in)

> ***CAUTION—***
> *• Do not overtighten overflow tube.*

– Wipe metal flakes off sump magnets (**C**).

– Installation is reverse of removal. Remember to
  • Replace sump gasket and O-ring seals.
  • Replace self-locking fasteners with new.
  • Fill transmission with fluid when finished. See **ATF, replacing** in this repair group.
  • Run engine and check for fluid leaks.

Tightening torques	
ATF plugs to transmission: • M10 • M14 • M18	 27 Nm (20 ft-lb) 23.5 + 10 Nm (17 + 7 ft-lb) 23.5 + 10 Nm (17 + 7 ft-lb)
Engine right side support bracket to engine mount (replace M12 x 1.5 mm 10.9 grade self-locking nut with new)	100 Nm (74 ft-lb)
Engine stabilizer link to bracket (M12 x 90 mm bolts)	108 Nm (80 ft-lb)
Front power distribution and fuse panel to bracket (M6 bolts)	5 Nm (44 in-lb)
Transmission sump to transmission housing (M6 bolts)	8 Nm (6 ft-lb)
Transmission upper support bracket to body (replace M10 bolts with new) • Stage 1 • Stage 2	 25 Nm (18 ft-lb) + additional 90°
Transmission upper support bracket to transmission mount (replace M12 x 1.5 mm 10.9 grade self-locking nut with new)	100 Nm (74 ft-lb)

ATF cooler, removing and installing

> ***WARNING—***
> • *To avoid personal injury, be sure the engine is cold before beginning this procedure.*

– Drain engine coolant. See **170 Radiator and Cooling System**.

– **N12 engine**: Remove engine air cleaner housing and air ducts. See **130 Fuel Injection**.

– **N14 engine**: Remove turbo air ducts above transmission. See **130 Fuel Injection**.

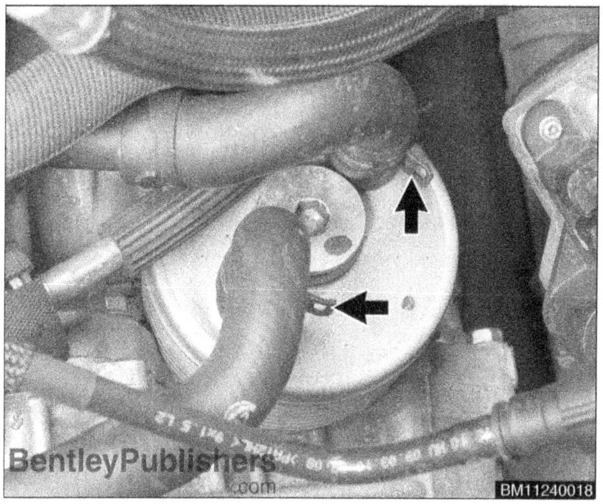

◄ Working at transmission fluid cooler, loosen hose clamps (**arrows**) and detach engine coolant lines. Be prepared to catch dripping antifreeze.

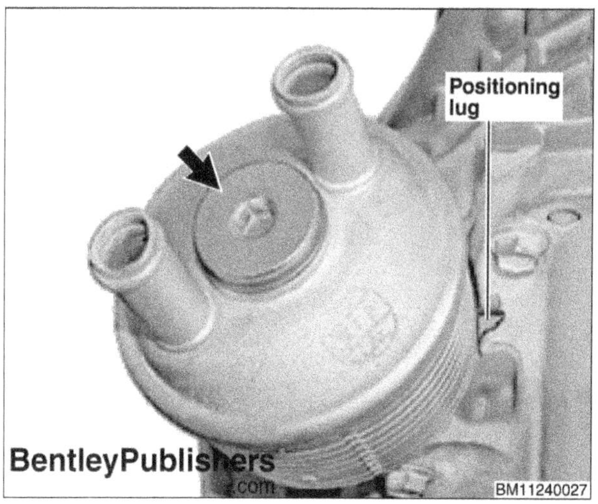

◄ Remove ATF cooler mounting bolt (**arrow**) and lift cooler off transmission housing.

• Note positioning lug on side of cooler.

• Be prepared to catch dripping ATF.

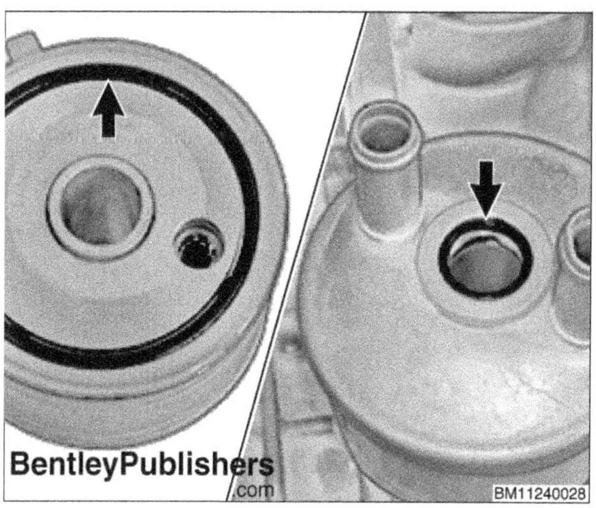

◄ Replace sealing O-rings (**arrows**). Make sure seal seating surfaces are clean.

– Installation is reverse of removal. Remember to:

• Moisten sealing O-rings with clean ATF.

• Use positioning lug on side of ATF cooler to align cooler to transmission.

• Fill cooling system and check for leaks. See **170 Radiator and Cooling System**.

• Warm up transmission, check ATF level and check for leaks. See **ATF level, checking** in this repair group.

Tightening torque	
ATF cooler to transmission housing (M12 bolt)	35 Nm (26 ft-lb)

Transmission Service

Back-up light switch function

The back-up lights are directly supplied with voltage (12 V) by the gear position switch.

Gear position switch, replacing and adjusting

Removing

– Remove air filter assembly and intake hoses. See **130 Fuel Injection**.

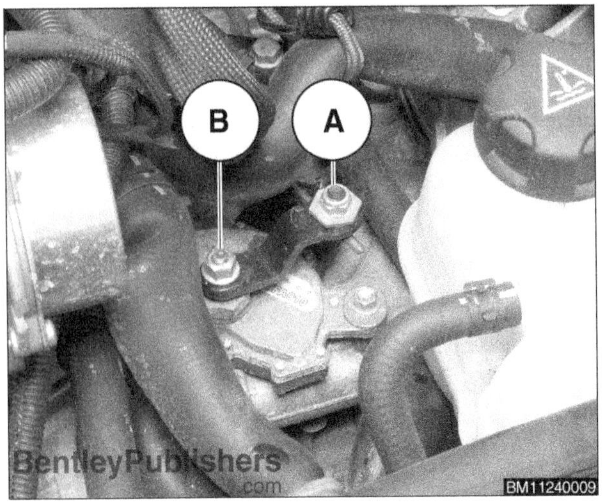

◄ Working at top of transmission at gear position switch:
- Loosen Bowden cable lock nut (**A**).
- Remove gear selector lever mounting nut (**B**).

– Detach Bowden cable housing from transmission case and lift cable end and gear position lever off gear selector shaft.

– Disconnect gear position switch electrical connector.

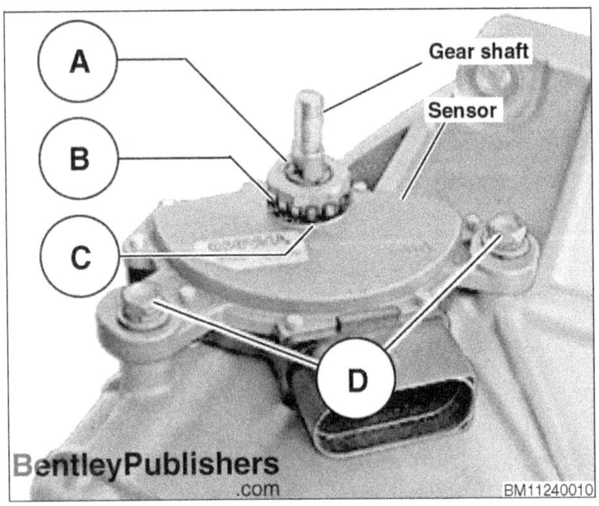

◄ Working at gear position switch:
- Remove mounting nut **A** and lift off retaining clip **B** and shim **C**.
- Remove mounting bolts **D** and lift switch off gear shaft.

Installing

– Installation is reverse of removal. Do not tighten switch mounting bolts (**D**) completely until switch is adjusted.

Tightening torque	
Gear position switch mounting nut (M14 nut)	7 Nm (62 in-lb)

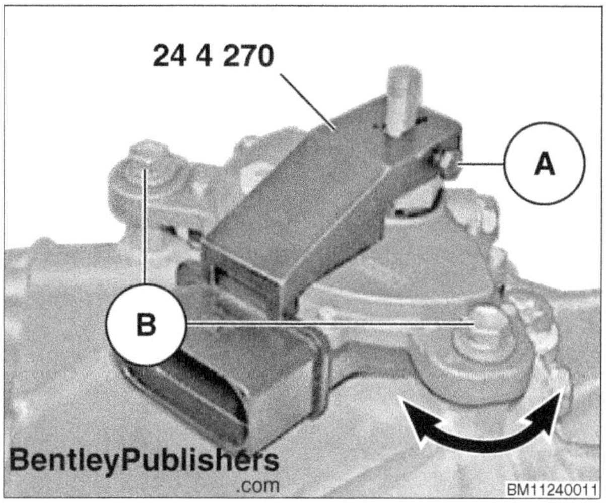

Adjusting

◀ Place MINI special tool **24 4 270** over gear position switch as shown.

- Tighten lock screw (**A**).
- Align gear position switch to N (neutral). Tighten mounting bolts (**B**), then remove special tool.

Tightening torque	
Gear position switch to transmission (M6 screws)	5.5 Nm (48 in-lb)

– Place gear position lever over gear shaft and tighten nut.

Tightening torque	
Gear position lever to gear shaft (M8 nut)	12 Nm (9 ft-lb)

– Make sure Bowden cable sheath is firmly anchored to transmission housing. Place selector lever in P (PARK).

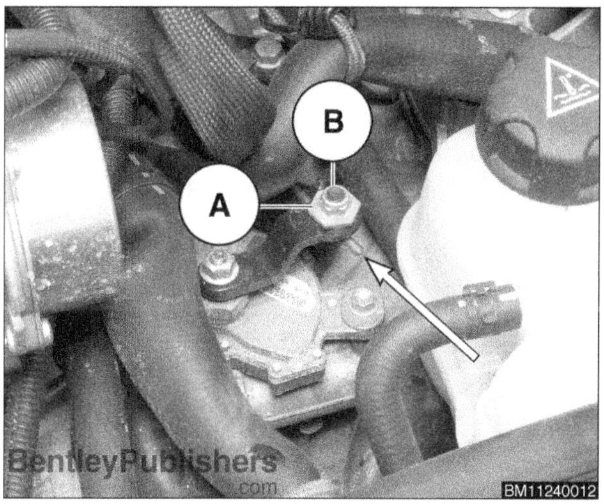

◀ Grip clamping sleeve (**A**) and loosen Bowden cable lock nut (**B**). Press Bowden cable firmly into sheath (**arrow**), then tighten lock nut.

Tightening torque	
Bowden cable lock nut (M8 nut)	12 Nm (9 ft-lb)

– Remaining installation is the reverse of removal. Be sure to reattach switch harness connector.

– When finished, road test vehicle and verify shifting.

Drive axle seal, replacing

Drive axle flange seal replacement procedure is similar on left and right. When reinstalling a drive axle, be sure to replace flange seal first.

– Raise car and support safely.

> ***WARNING—***
> - *Make sure the car is stable and well supported at all times. Use a professional automotive lift or jack stands designed for the purpose. A floor jack is not adequate support.*

– Drain ATF. See **ATF, replacing** in this repair group.

– Remove drive axle. See **310 Front Suspension, Drive Axles**. Be prepared to catch dripping oil.

240

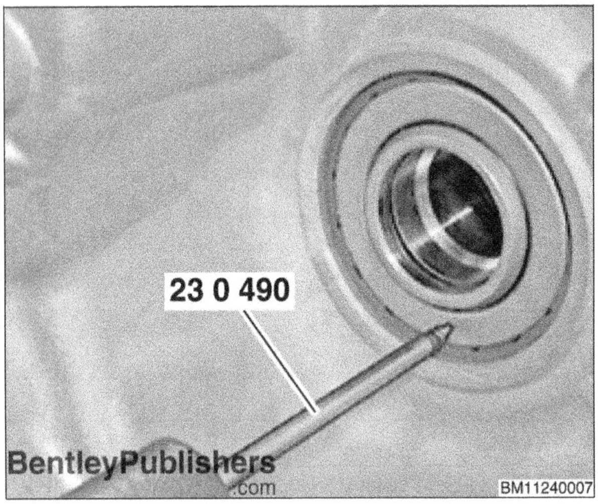

– Use a center punch to drive a small hole into drive axle seal.

> ***CAUTION—***
> *• Do not use a drill to create hole in drive axle seal. Metal shavings could result in transmission failure.*

< Screw the tip of MINI special tool **23 0 490** into seal and pull out using slide hammer.

< When installing, coat sealing surfaces of new seal with ATF. Drive seal firmly home with MINI special tool **24 4 300**.

– Remainder of installation is reverse of removal. Remember to:
  - Use assembly guard, supplied with new seal, to protect seal lip against damage. See **310 Front Suspension, Drive Axles**.
  - Push drive axle shaft into differential case past resistance of snap ring until an audible snap is heard. Remove assembly guard.
  - Fill transmission with fluid and check for leaks. See **ATF, replacing** in this repair group.

Torque converter seal, replacing

The automatic transmission input shaft seal is often referred to as the torque converter seal.

– Remove transmission. See **Transmission, removing and installing** in this repair group.

– Remove torque converter. See **Torque converter, removing and installing** in this repair group.

< Remove torque converter seal using universal puller (MINI special tool **00 5 010** shown).

– Clean sealing surfaces before installing new seal.

– Lubricate sealing lip of new seal with clean ATF.

Install new seal using installation tool (MINI special tool **24 4 250**). Place seal on installation tool and press seal fully into position.

Transmission Removal and Installation

Remove and install the transmission with the vehicle on a lift. Support engine using appropriate equipment. Use a transmission jack to lower the transmission.

Transmission removal procedure is a complicated job which includes the following steps:

- Drain engine coolant; covered in **170 Radiator and Cooling System**.
- Support and raise engine; covered in this repair group.
- Remove transmission supports and mounts; covered in this repair group.
- Remove front wheels and drive axles. See **310 Front Suspension, Drive Axles**.
- Remove complete exhaust system. See. See **180 Exhaust System**.
- Remove wheel housing liners. See **410 Fenders, Engine Hood**.
- Detach front bumper. See **510 Exterior Trim, Bumpers**.
- Remove underbody panels. See **020 Maintenance**.
- Disconnect steering rack from steering shaft. See **320 Steering**.
- Remove front subframe. See **310 Front Suspension, Drive Axles**.

This section describes transmission removal and installation without engine removal. If additional engine repair is needed, the engine and transmission may be removed as a complete unit with little additional work. See **110 Engine Removal and Installation**.

> ***WARNING—***
> - *Front subframe and transmission removal may upset the balance of the vehicle on a lift.*

Transmission, removing and installing

– Disconnect negative (-) cable from battery.

> ***CAUTION—***
> - *Prior to disconnecting the battery, read the battery disconnection cautions given in* **001 Warnings and Cautions**.

– **N12 engine**: Remove engine air cleaner housing and air ducts. See **130 Fuel Injection**.

– **N14 engine**: Remove turbo air ducts above transmission. See **130 Fuel Injection**.

– Drain engine coolant. See **170 Radiator and Cooling System**.

< Use engine support bar to support engine and transmission.

> ***CAUTION—***
> - *Do not allow engine support bar to crush hood release Bowden cable.*
> - *Do not raise engine until engine and transmission supports are loosened as described in the following steps.*

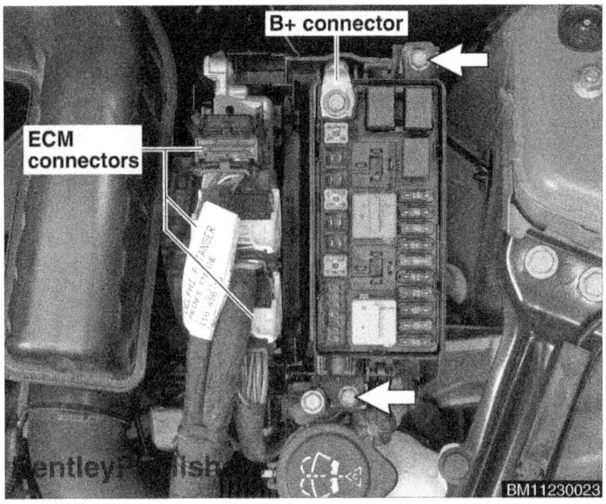

Working at left rear engine compartment:

- Remove front power distribution and fuse panel cover.
- Disconnect battery B+ connector from fuse panel.
- Remove distribution panel mounting bolts (**arrows**) and push panel aside.
- Detach engine control module (ECM) connectors and pull ECM out of holder. See **130 Fuel Injection**.

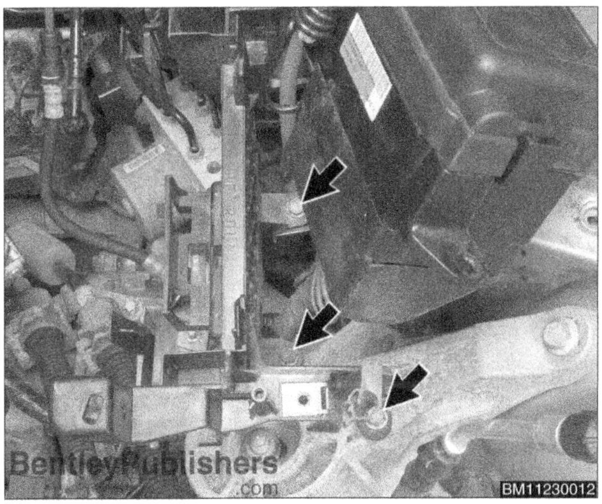

Remove fuse panel bracket mounting bolts (**arrows**) and lift off bracket.

Working at transmission upper support bracket, left front inner fender wall:

- Remove mounting bolts (**arrows**) then lift off support bracket.

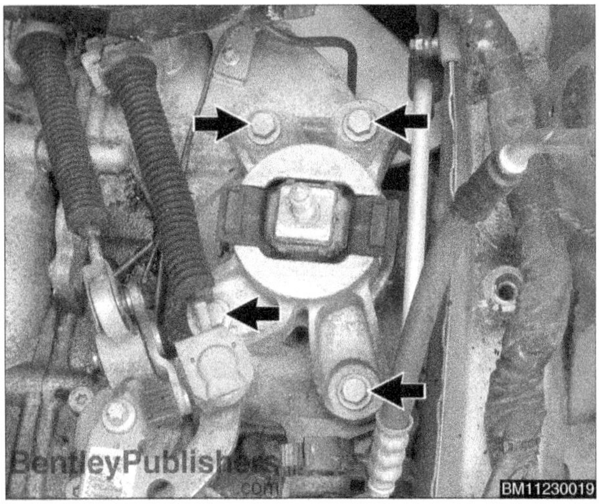

◄ Remove transmission upper mount bolts (**arrows**) and lift off mount.

◄ Working at right side engine support bracket, loosen nut (**arrow**).

– Raise vehicle and support safely.

> *WARNING —*
> • *Make sure the vehicle is stable and well supported at all times. Use a professional automotive lift or jack stands designed for the purpose. A floor jack is not adequate support.*

◄ Working underneath rear of engine, remove engine stabilizer link mounting bolts (**arrows**). Remove stabilizer link.

– Use engine support bar to raise engine approx 10 mm (3/8 in)

– Separate exhaust pipe from catalytic converter. Remove entire exhaust system. See **180 Exhaust System**.

– Remove front subframe and drive axles. See **310 Front Suspension, Drive Axles**.

> *CAUTION—*
> • *While steering spindle is detached from steering rack, make sure steering wheel is not rotated. Rotating the wheel may damage the airbag and horn contact spring.*

– Remove starter motor. See **121 Battery, Starter, Alternator**.

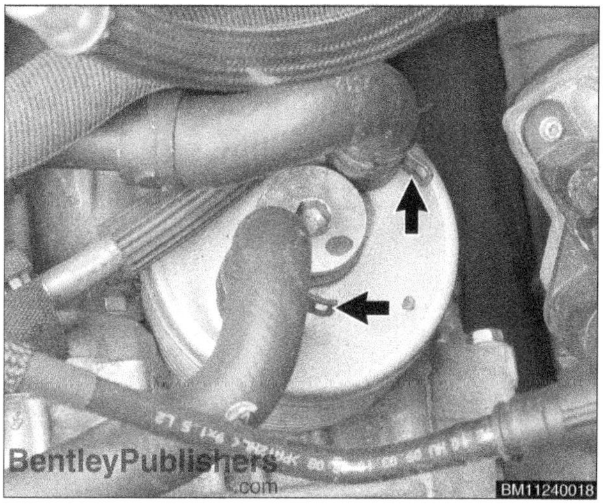

Working at transmission fluid cooler, loosen hose clamps (**arrows**) and detach engine coolant lines. Be prepared to catch dripping antifreeze.

Detach EGS electrical multiple connectors (**arrows**) at transmission. Remove connector harnesses from brackets on transmission. Note wiring harness routing to help in reassembly.

Working at top of transmission at gear position switch:

- Loosen Bowden cable lock nut (**arrow**).
- Detach gear position switch electrical connector.
- Slide Bowden cable sheath out of bracket on transmission and lay end of cable aside.

– Use transmission jack to support weight of transmission.

◀ Rotate engine until torque converter mounting nut (**arrow**) becomes visible through starter opening.

- Remove nut, then rotate engine to remove remaining 5 mounting nuts

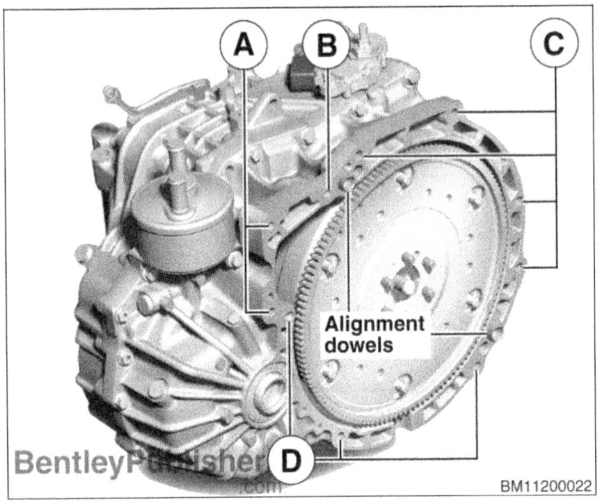

◀ Remove transmission to engine mounting bolts. Note that bolt sizes and lengths vary:

- **A** = M8 x 85 mm starter bolts
- **B** = M8 x 60 mm starter bolt
- **C** = M10 x 40 mm 8.8 grade ZNS3 bolt
- **D** = M10 x 60 mm 8.8 grade ZNS3 bolt

– Remove transmission by pulling back and down. Lower jack slowly while watching carefully to make sure no lines, hoses or wires become snagged.

> ***WARNING—***
> - *Be sure the vehicle is properly supported. The removal of the transmission may upset the balance of the vehicle on the lift.*

> ***CAUTION—***
> - *Do not allow the torque converter to fall off the transmission input shaft.*

– Inspect engine flywheel (torque plate) for cracks or elongated holes. Replace if necessary.

– If torque converter seal is leaky, or torque converter position on transmission input shaft was disturbed during removal, replace torque converter seal. See **Torque converter seal, replacing** in this repair group.

– Before reinstalling transmission, check alignment dowels in bell housing flange for correct seating. Replace if damaged.

– Installation is reverse of removal. Replace self-locking fasteners with new. Be sure to use correct length bolts at bell housing.

Tightening torques	
Ball joint, inner, to subframe (replace M14 x 1.5 ZNS3 self-locking nut with new)	70 Nm (ft-lb) + additional 90°
Ball joint, outer, to steering knuckle (replace M12 x 1.5 ZNS3 self-locking nut with new	70 Nm (ft-lb) + additional 90°

Tightening torques (continued)	
Bumper to front subframe (M8 x 18 mm bolts)	19 Nm (14 ft-lb)
Control arm bracket and stabilizer anchor to subframe (M14 x 1.5 x 105 mm self-tapping bolts)	165 Nm (122 ft-lb)
Drive axle bearing block to engine block, right side (M10 bolts x 45 bolts)	38 Nm (28 ft-lb)
Drive axle collar nut (replace M22 x 1.5 nut with new, stake to axle after tightening)	182 Nm (134 ft-lb)
Engine right side support bracket to engine mount (replace M12 x 1.5 mm 10.9 grade self-locking nut with new)	100 Nm (74 ft-lb)
Engine stabilizer link to bracket (M12 x 90 mm bolts)	108 Nm (80 ft-lb)
Exhaust pipe to catalytic converter (V-band clamp)	25 ± 3 Nm (18 ± 2 ft-lb)
Front power distribution and fuse panel to bracket (M6 bolts)	5 Nm (44 in-lb)
Front subframe to chassis: • Front bolts M12 x 85 mm • Rear bolts M12 x 62 mm • Control arm bushing carrier bolts (replace M10 x 60 mm 10.9 grade bolts with new)	 100 Nm (74 ft-lb) 100 Nm (74 ft-lb) 59 Nm (44 ft-lb) + additional 90°
Outer ball joint to control arm (replace M14 self-locking nuts with new)	175 Nm (129 ft-lb)
Road wheel to hub	140 ± 10 Nm (103 ± 7 ft-lb)
Stabilizer link to stabilizer bar (replace M10 self-locking nuts with new)	56 Nm (41 ft-lb)
Starter motor to transmission (M8 bolts)	20 Nm (14.9 ft-lb)
Steering shaft universal joint pinch bolt to steering rack (replace M8 x 1.25 x 41 mm bolt with new)	28 Nm (21 ft-lb)
Tie rod end to steering knuckle (replace M12 x 1.5 self-locking nut with new)	65 Nm (48 ft-lb)
Torque converter to flywheel (M10 x 1 nuts)	57 Nm (42 ft-lb)
Transmission to engine (M10 bolts)	38 Nm (28 ft-lb)
Transmission upper mount to transmission (M10 x 45 mm 8.8 grade bolts)	38 Nm (28 ft-lb)
Transmission upper support bracket to body (replace M10 bolts with new) • Stage 1 • Stage 2	 25 Nm (18 ft-lb) + additional 90°
Transmission upper support bracket to transmission mount (replace M12 x 1.5 mm 10.9 grade self-locking nut with new)	100 Nm (74 ft-lb)

– Adjust Bowden cable setting at gear position switch. See **Gear position switch, replacing and adjusting** in this repair group.

240

– When finished, fill transmission with correct fluid. See **ATF, replacing** in this repair group.

– Have vehicle professionally aligned.

– Vehicle with DSC: Use MINI scan tool to carry out steering angle sensor adjustment.

Torque converter, removing and installing

– Remove transmission. See **Transmission, removing and installing** in this repair group.

◄ Install special tools **24 4 260** on two torque converter mounting studs.

– Remove torque converter by pulling on handles of removal tools.

- Be prepared to catch ATF as it flows out of torque converter.
- Place torque converter on clean surface so that excess fluid does not run out.

– Torque converter seal replacement is covered in **Torque converter seal, replacing** in this repair group.

– Install torque converter to transmission by aligning opening on converter hub with oil pump driver. Press torque converter on transmission shaft as far as it will go. Converter slips inward and snaps in place in pump impeller.

> ***CAUTION—***
> - *If converter is not installed correctly, transmission oil pump driver may be damaged when transmission is installed to engine.*

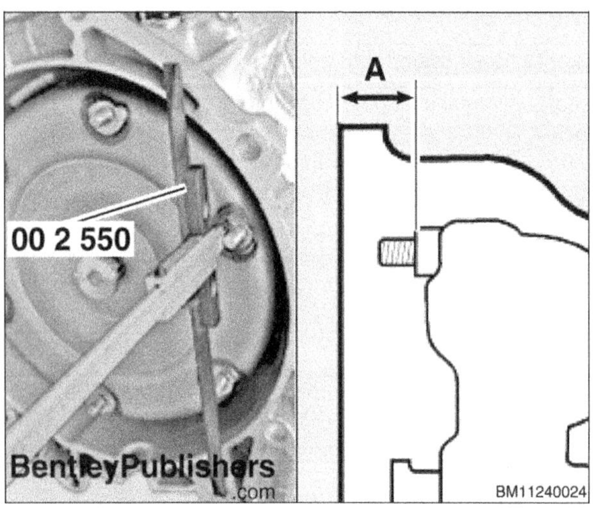

◄ Determine distance between bell housing flange opening and base of torque converter mounting stud. Use straight edge (MINI special tool **00 2 550**) and depth gauge to measure value.

Torque converter stud clearance specification	
Base of converter stud to bell housing flange opening (measurement **A**) • N12 engine • N14 engine	 11. 8 mm (0.465 in) 17.5 mm (0.689 in)

– Reinstall transmission and fill transmission with correct fluid. See **ATF, replacing** in this repair group.

Flywheel, removing and installing

The flywheel in automatic transmission models is also called torque plate.

- Remove transmission. See **Transmission, removing and installing** in this repair group.

◄ Install special tool **11 9 590** to counterhold flywheel.

◄ Remove flywheel mounting bolts (**arrows**) and discard. Remove flywheel.

- Clean crankshaft threaded holes.

◄ When installing, note that crankshaft sensor gear is loose on crankshaft. Make sure flywheel bore is matched to dowel pin (**arrow**) on sensor gear when mounting flywheel to crankshaft.

- Reinstall flywheel using new bolts.

Tightening torque	
Flywheel to crankshaft (replace M9 x 1.25 x 20 mm bolts with new) • Stage 1 • Stage 2 • Stage 3	 8 Nm (6 ft-lb) 30 Nm (22 ft-lb) additional 90°

- When finished, remove flywheel locking tool.

- Reinstall transmission. See **Transmission, removing and installing** in this repair group.

240

250 Gearshift Linkage

GENERAL

This repair group covers transmission gearshift service for manual and automatic transmissions.

To gain access to the complete gearshift mechanism, remove the exhaust system. See **180 Exhaust System**.

Automatic transmission shift lock

When the automatic transmission gear selector lever is in P (PARK) or N (NEUTRAL), the shift lock solenoid blocks the lever from moving. With the ignition ON and the brake applied, the engine control module (ECM) energizes the solenoid to release shift lock and allow gear selection.

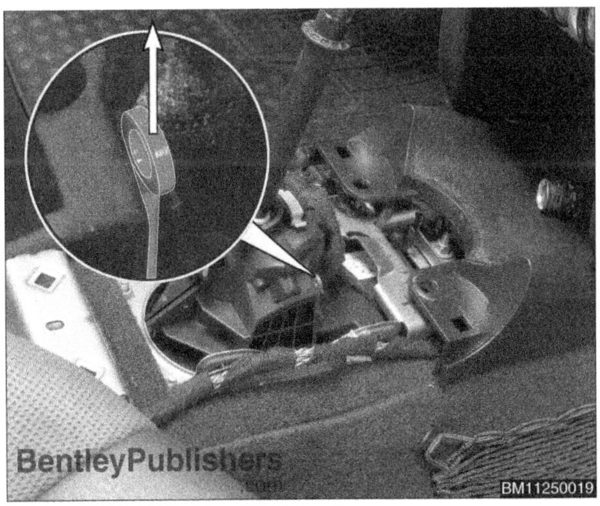

If the battery is dead or for any other reason the selector lever cannot be moved:

- Pull up selector lever bezel. See **Selector lever bezel, removing and installing** in this repair group.
- Use hook tool to pull (**arrow**) loop on right side of selector lever to release shift lock. (Center console removed for clarity.)

Gearshift knob, removing and installing

– Place manual gearshift in neutral. Position automatic selector lever in P (PARK).

Remove shift knob by pulling straight up and off shift lever. Automatic gear selector shown. Manual shift knob is removed the same way.

> *CAUTION—*
> - *Do not twist knob or locating key can be damaged.*

– Installation is reverse of removal. Remember to align gearshift knob with locating key on gearshift lever and push knob straight down on lever without twisting.

Caution

> ***CAUTION—***
> * *When removing interior trim clips, avoid damaging trim pieces by using a plastic prying tool or a screwdriver with the tip wrapped with masking tape.*

Manual Transmission Gearshift

Manual gearshift components

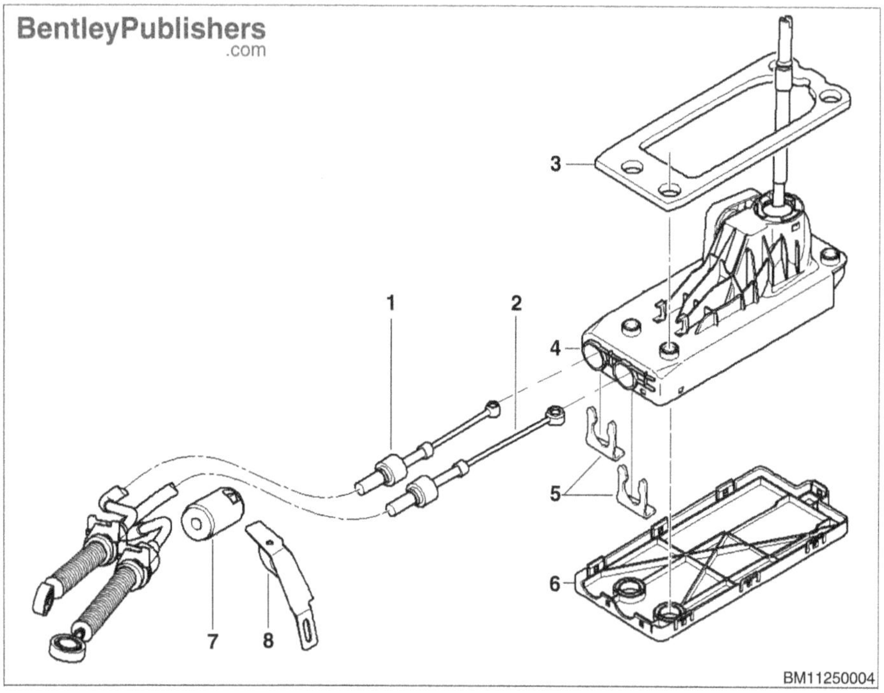

1. Selector shaft Bowden cable
2. Shifting shaft Bowden cable
3. Gasket
4. Shifter cassette
5. Bowden cable locking clips
6. Cover
7. Vibration damper
8. Shift cable bracket

Shift lever bezel, removing and installing

– Remove shift knob off by pulling straight up. Do not twist knob. See **Gearshift knob, removing and installing** in this repair group.

Use plastic prying tool to pry up (**arrows**) bezel ring at base of shift boot.

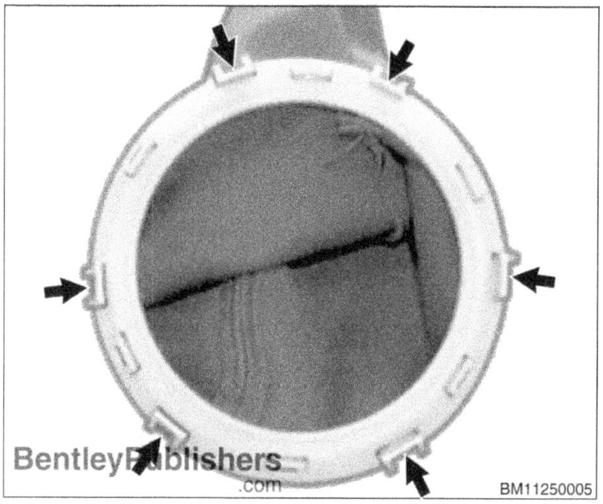

When reinstalling, note that spacing of retaining lugs (**arrows**) is uneven.

Shift lever mechanism, removing and installing

- Remove shift lever knob. See **Gearshift knob, removing and installing** in this repair group.
- Remove shift lever bezel and shift boot. See **Shift lever bezel, removing and installing** in this repair group.
- Remove center console. See **513 Interior Trim**.

Working at base of shifter mechanism, release locking clip (**arrow**).

- Raise vehicle and support safely.

> ***WARNING—***
> - *Make sure the vehicle is stable and well supported at all times. Use a professional automotive lift or jack stands designed for the purpose. A floor jack is not adequate support.*

- Remove exhaust system. See **180 Exhaust System**.
- Remove exhaust system heat shield.

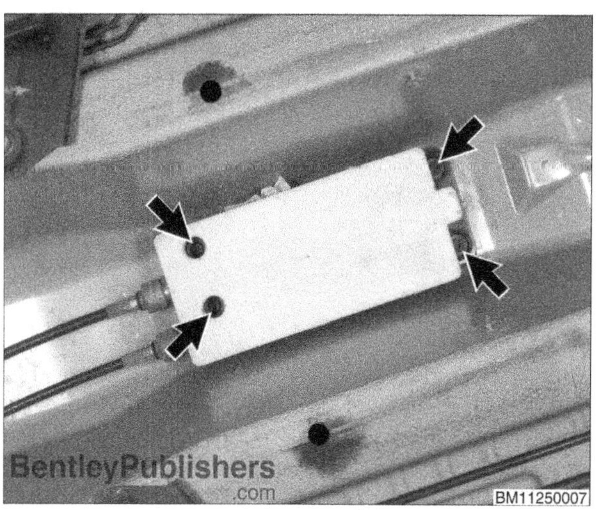

Working underneath vehicle below shifter mechanism, remove shifter cassette mounting bolts (**arrows**). Pull cassette backward to remove.

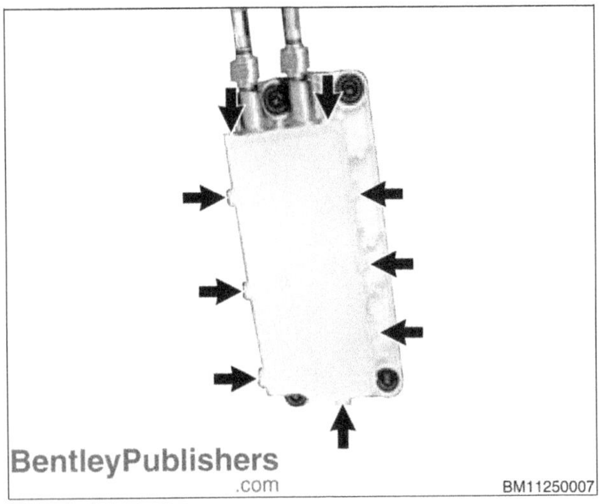

Release shifter cassette cover locking tabs (**arrows**) and lift off cover.

> **CAUTION—**
> - *Do not damage locking tabs during cover removal.*

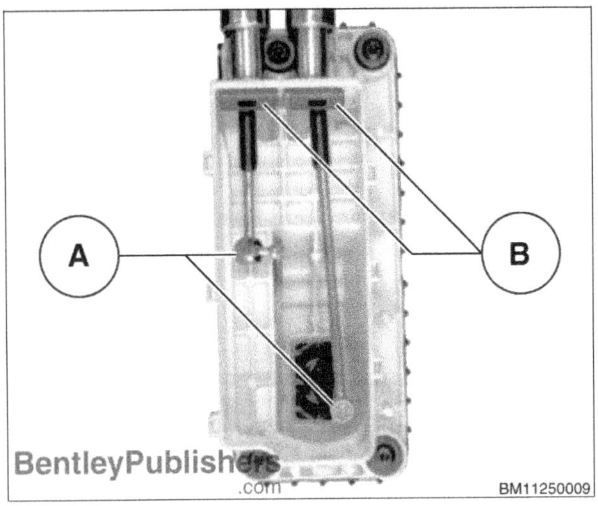

Working inside shifter cassette:

- Separate Bowden cable assembly ball sockets (**A**) from shift lever balls.
- Remove cable locking clips (**B**) and pull cables out of cassette.

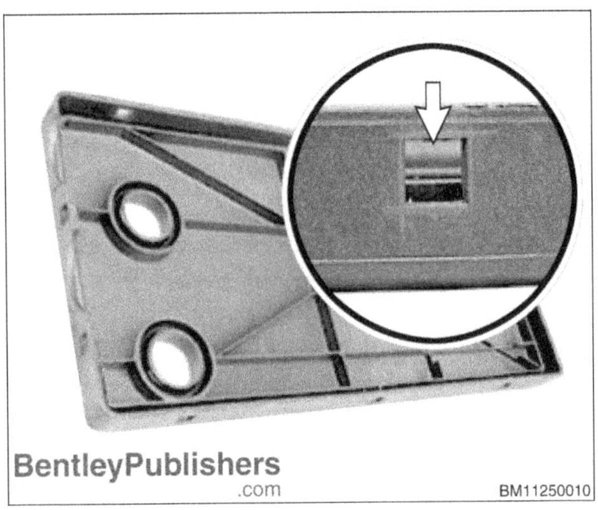

When reassembling cassette:

- Inspect cassette cover to make sure locking tabs (**arrow**) are not damaged.
- Replace moisture seals if defective.

◄ Lift cassette into position by clipping cassette front retaining lugs (**arrows**) to floor plate.

– Remainder of installation is reverse of removal.

Shift cable assembly, replacing

– Disconnect negative (-) cable from battery.

> ***CAUTION—***
> • *Prior to disconnecting the battery, read the battery disconnection cautions given in* **001 Warnings and Cautions**.

– **N12, N16 engine**: Remove engine air cleaner housing and air ducts. See **130 Fuel Injection**.

– **N14, N18 engine**: Remove turbo air ducts above transmission. See **130 Fuel Injection**.

– Remove shift lever knob. See **Gearshift knob, removing and installing** in this repair group.

– Remove shift lever bezel and shift boot. See **Shift lever bezel, removing and installing** in this repair group.

– Remove center console. See **513 Interior Trim**.

– Raise vehicle and support safely.

> ***WARNING—***
> • *Make sure the vehicle is stable and well supported at all times. Use a professional automotive lift or jack stands designed for the purpose. A floor jack is not adequate support.*

– Remove shifter cassette and detach Bowden shift cables. See **Shift lever mechanism, removing and installing** in this repair group.

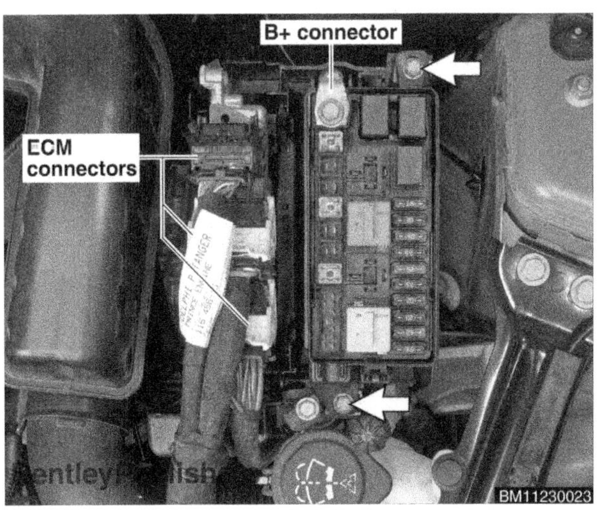

◄ Working at left rear engine compartment:

- Remove front power distribution and fuse panel cover.
- Disconnect battery B+ connector from fuse panel.
- Remove distribution panel mounting bolts (**arrows**) and push panel aside.
- Detach engine control module (ECM) connectors and pull ECM out of holder. See **130 Fuel Injection**.

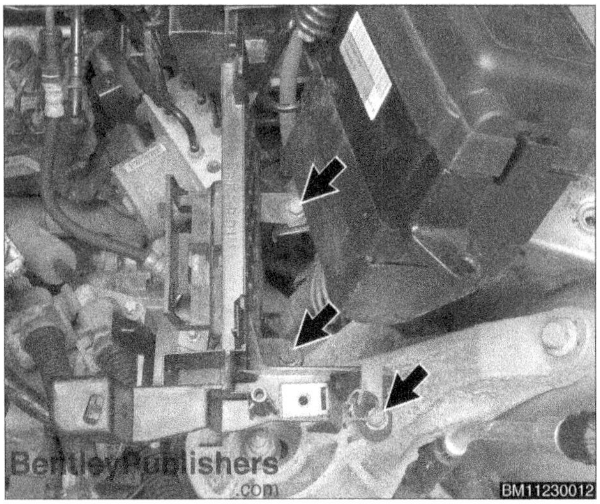

Remove fuse panel bracket mounting bolts (**arrows**) and lift off bracket.

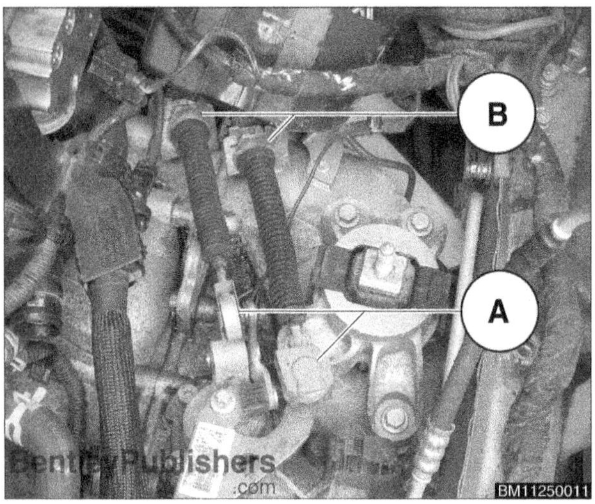

Working at top of transmission housing:

- Separate Bowden cable ball sockets (**A**) from transmission gear selector levers.
- Remove cable sheath anchor clips (**B**) and detach cables from transmission housing.

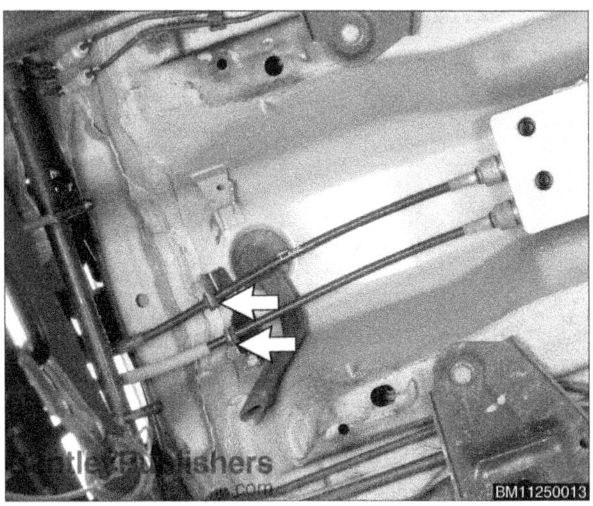

Working underneath vehicle, detach shift cables from brackets (**arrows**) on floor plate.

AUTOMATIC TRANSMISSION GEARSHIFT

Gear selector components

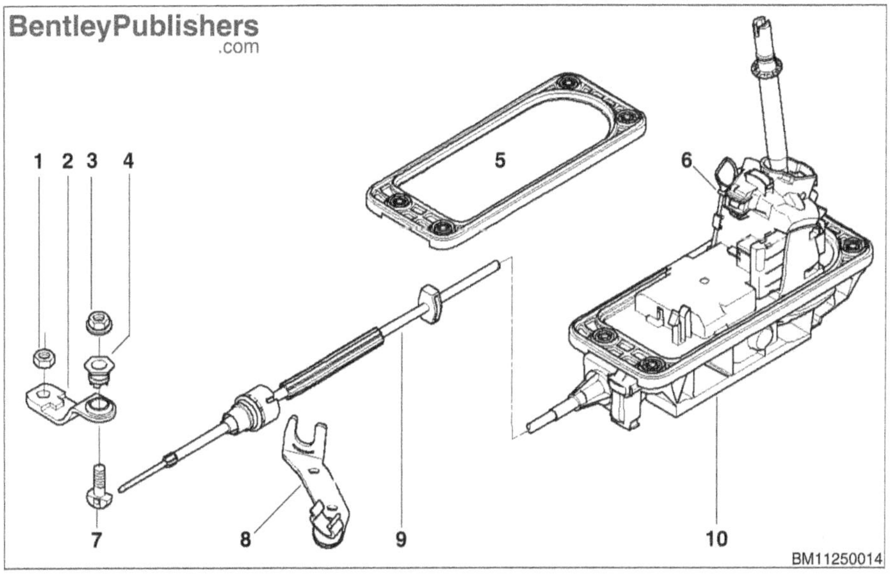

1. M8 self-locking collar nut
2. Gear selector lever
3. Lock nut
4. Clamping sleeve
5. Gasket
6. Shift lock emergency release cable
7. Cable lug
8. Cable bracket
9. Shift cable
10. Selector lever assembly

Selector lever bezel, removing and installing

- Remove gearshift knob. See **Gearshift knob, removing and installing** in this repair group.

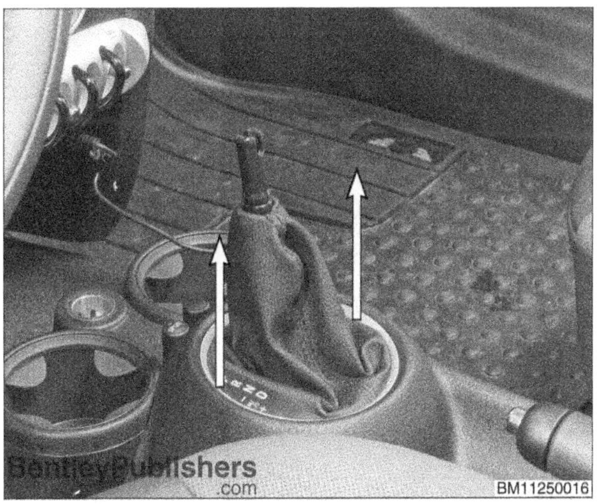

◄ Pull bezel up (**arrows**) to release from console.

- Lift bezel and slide boot over shaft.

◄ Separate electrical connector (**arrow**) and lift off bezel.

- Installation is reverse of removal.

Selector lever assembly, removing and installing

- Place selector lever in P(PARK) and remove gearshift knob. See **Gearshift knob, removing and installing** in this repair group. Remove selector lever bezel. See **Selector lever bezel, removing and installing** in this repair group.

- **N12, N16 engine**: Remove engine air cleaner housing and air ducts. See **130 Fuel Injection**.

- **N14, n18 engine**: Remove turbo air ducts above transmission. See **130 Fuel Injection**.

◄ Working in engine compartment above automatic transmission:
  - Counterhold shift cable clamping sleeve and loosen lock nut (**A**) to release cable end from transmission gear selector lever.
  - Rotate cable sheath locking sleeve (**arrow**) and detach shift cable from hold-down brackets.

- Raise vehicle and support safely.

> ***WARNING—***
> - *Make sure the vehicle is stable and well supported at all times. Use a professional automotive lift or jack stands designed for the purpose. A floor jack is not adequate support.*

- Remove exhaust system. See **180 Exhaust System**.

- Remove exhaust system heat shield.

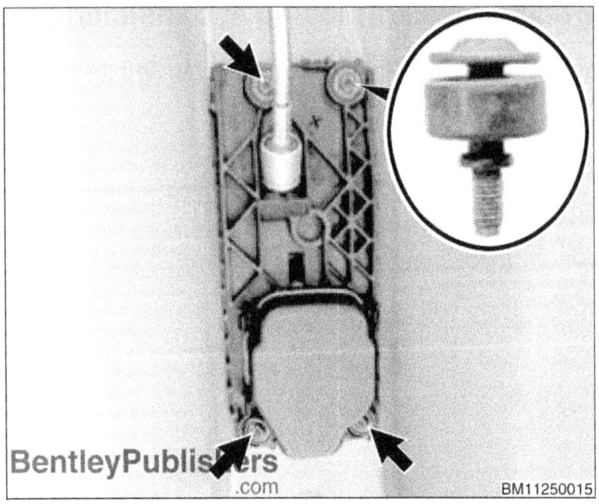

◄ Working underneath vehicle below gear selector mechanism, remove shifter cassette mounting bolts (**arrows, inset**).
  - Pull cassette backward to remove.
  - Detach electrical connector inside cassette to allow removal.

- Detach shift cable from bracket on floor plate and remove shift cassette with shift cable.

- Installation is reverse of removal.

Tightening torque	
Shifter assembly to floor pan (M6 bolts)	8 Nm (6 ft-lb)

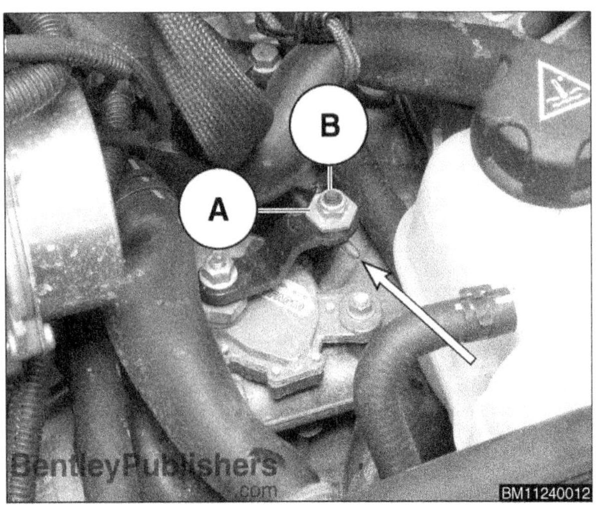

◄ When finished:
  - Make sure Bowden cable sheath is firmly anchored to transmission housing. Place selector lever in P (PARK).
  - Grip clamping sleeve (**A**) and loosen Bowden cable lock nut (**B**). Press Bowden cable firmly into sheath (**arrow**), then tighten lock nut.

Tightening torque	
Bowden cable lock nut (M8 nut)	12 Nm (9 ft-lb)

300 Suspension, Steering and Brakes–General

GENERAL

This section covers general information for front and rear suspension, braking, steering and integrated braking and suspension systems including antilock braking (ABS) and traction control (ASC + T, DSC).

Also included is information on wheels, tires and tire pressure monitoring systems (FTM, TPM).

SUSPENSION

The suspension system provides a firm and responsive ride with good levels of comfort and good acoustic properties. The weight distribution between the front and rear axles is biased to the front due the front wheel drive configuration.

Front suspension

The front suspension features antidive and antisquat geometry due to the 1: 1 ratio of stabilizer bar movement to strut movement. The layout of the front suspension minimizes camber loss due to side forces, thus improving handling and steering response.

The front suspension consists of the following main components:

- MacPherson strut assemblies, including coil springs and swiveling strut top mounts.
- Lower control arms.
- Front subframe.
- Stabilizer bar and links.
- Front wheel hubs and bearings.

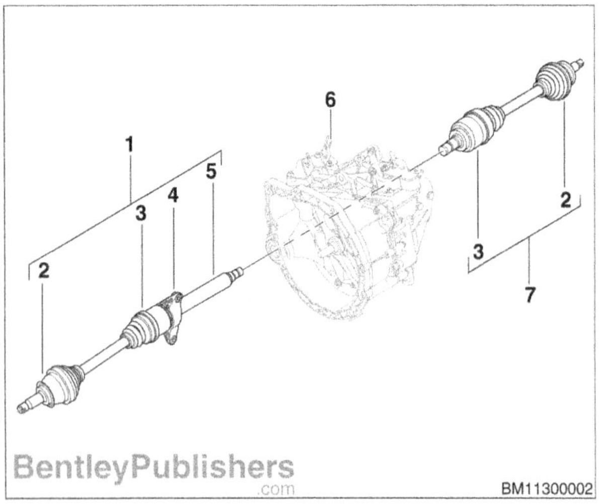

Front drive axles

Power to the front wheels is transmitted through two drive axles (half-shafts) with two constant velocity (CV) joints each. Due to the requirement to steer the wheels, the outer joints are designed with a large degree of articulation.

1. Right drive axle
2. Outer CV joint
3. Inner CV joint
4. Support bearing
5. Intermediate shaft
6. Transmission
7. Left drive axle

Rear suspension

The multilink rear suspension is used for all models of MINI. Benefits of this suspension are light weight, exceptional handling characteristics, ride comfort and good acoustic properties.

Main rear suspension components include:

- Rear springs
- Rear struts (shock absorbers)
- Semi-trailing arms
- Upper control arms
- Lower control arms
- Rear subframe
- Stabilizer bar
- Rear wheel hubs and bearings

STEERING

The MINI features a electromechanical power assisted steering system on all models. The steering gear is of conventional rack-and-pinion design.

The main components of the steering system are:

- Electromechanical steering (EPS) system with control module.
- Steering torque sensor.
- Electric motor with motor position sensor.
- Step-down gear (consisting of worm gear and worm shaft).
- Rack and pinion steering gear.

EPS control module is networked with DME, ASC + T or DSC, steering angle sensor (in SZL) and instrument cluster using the vehicle bus system.

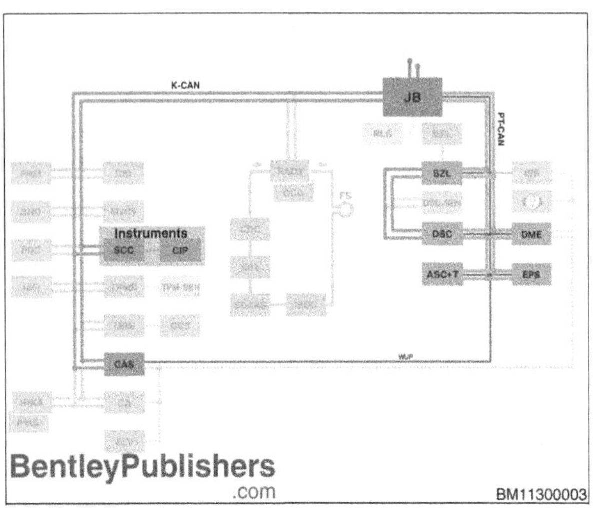

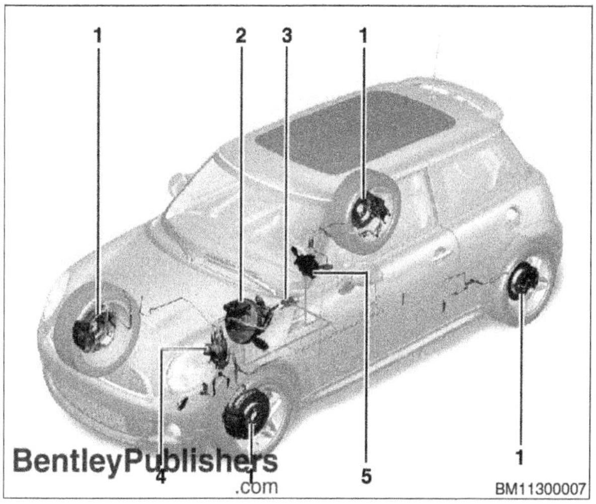

BRAKE SYSTEM

The brake system features discs and calipers at all four wheels. The hydraulic system has a dual circuit that is split diagonally.

MINI models are equipped with antilock braking (ABS) and standard automatic stability control + traction (ASC+T) or optional dynamic stability control (DSC).

Brake system components include:

1. Wheel brake assembly with ABS wheel speed sensor and brake pad sensor (left front and right rear wheels only).
2. Brake master cylinder, reservoir, vacuum operated brake booster.
3. DSC acceleration sensor (with optional DSC).
4. ABS hydraulic unit and control module.
5. Steering angle sensor (with optional DSC).

Brake booster

A mechanical vacuum pump, driven directly by the camshaft, supplies the brake booster with vacuum.

MINI models equipped with ASC+T are fitted with a brake booster with mechanical braking assistant (MBA) for emergency brake assistance.

Models with optional DSC are fitted with a conventional brake booster without MBA as the emergency braking assistance is provided by DSC.

Externally, both versions of the brake booster are identical, the only difference being the type identification plate.

300

Electronic brake and stability control

Safe vehicle handling and braking is primarily achieved by highly developed, state of the art chassis, suspension and brake design. However, even on a perfectly balanced vehicle, critical situations may occur during braking, accelerating and cornering which result in loss of directional stability. MINI is equipped with the following systems to assist the driver in these extreme situations:

Antilock braking system (ABS), standard on MINI models, prevents lock-up of road wheels during brake application, thus maintaining vehicle stability and steering control under emergency conditions. The system cannot be disabled by the driver. ABS includes the following functions:

- Electronic brake force distribution (EBV)
- Cornering brake control (CBC)
- Engine drag torque control (MSR)

Automatic stability control + traction (ASC + T), standard on MINI models, prevents the driven front wheels from spinning when engine torque is applied and maintains an equal torque distribution to both wheels. This ensures that optimum traction and directional stability are maintained.

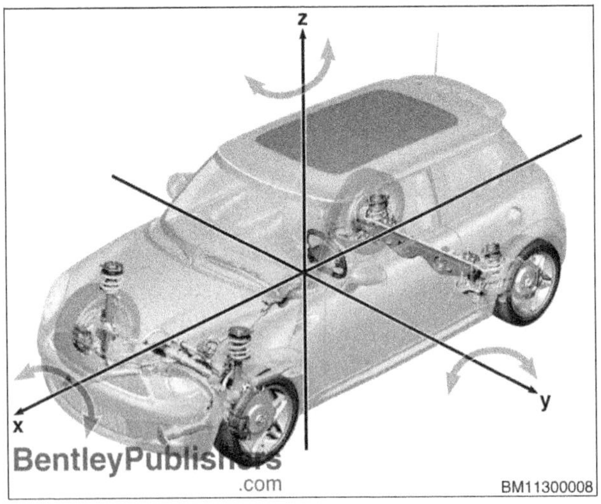

Dynamic stability control (DSC), optional on MINI models, interfaces with ABS, EBV, CBC, MSR and ASC + T but also monitors:

- Steering wheel movements via the steering angle sensor.
- Lateral (sideways) and vehicle rotational (yaw) speed via DSC acceleration sensor.

For additional details about electronic brake and suspension systems, see **340 Brakes**.

Wheels and Tires

> ***WARNING —***
> - *Replace tires in pairs and on the same axle. Preferably, replace all of the tires at the same time. Install tires of the same type and tread pattern.*
> - *New tires tend to be slippery. General industry guidelines suggest a break-in period of at least 350 miles (560 Km).*
> - *Do not use tires of a lesser speed rating and/or load range than originally installed.*

> ***CAUTION—***
> - *MINI specifies that only approved wheels be used. Consult an authorized MINI dealer or the vehicle owner's manual for additional information.*
> - *Make sure new tires are the same size as originally installed or of an approved alternative size. If you change sizes, be aware that gear ratios and speedometer readings may change.*

Tire and wheel information

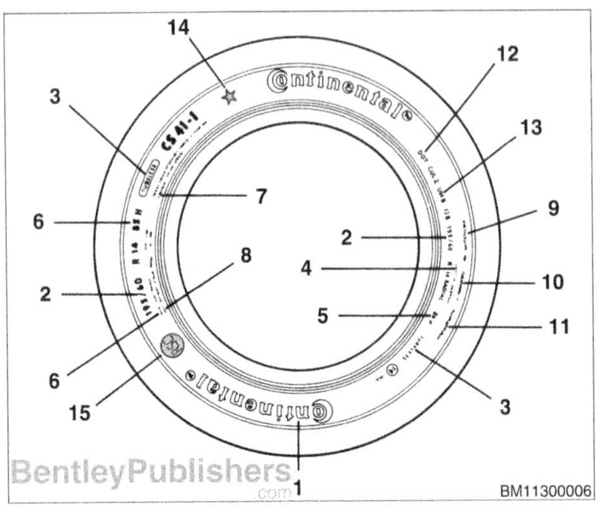

Overview of typical information found on tire sidewall that complies with United States Federal Motor Vehicle Safety Standard (FMVSS) 109 and MINI specifications. Information on USA Department of Transportation (DOT) required uniform tire quality grading is noted.

1. Brand name
2. Tire dimensions (size)
3. Type (tubeless)
4. Country of origin
5. Load index and speed rating
6. Load range (A, B, C)
7. Tread and sidewall ply and construction
8. Maximum cold inflation pressure
9. Tread wear index number (DOT)
10. Traction index number (AA, A, B, C) (DOT)

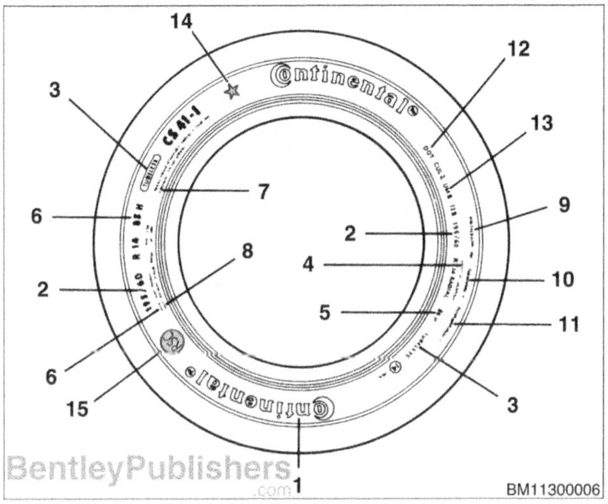

11. Temperature index number (A, B, C) (DOT)
12. USA DOT compliant
13. Manufacturers production and date coding
14. MINI recommendation designation (approval star)
15. Run-on flat designation (RSC) where applicable

When installing new tires or wheels, note the following:

- MINI specifies that only approved tires be used. Only radial tires are approved, and retreads are specifically not recommended.
- MINI specifies that only approved wheels be used. Consult an authorized MINI dealer or the owner's manual for additional information.
- Balance new tires and fit rims (wheels) with new valve stems.
- Wheels and tires may require rebalancing as they accumulate mileage due to settling and wear.
- Tire size is critical to the proper operation of the antilock brake system and traction control system.
- Numerous styles of wheels in 15, 16, 17 and 18 inch diameters are available from an authorized MINI dealer.

If in doubt about tire application or safety information, consult the owner's manual or your MINI dealer.

Tire pressure monitoring

Factory recommended cold tire inflation pressures vary based on numerous factors including tire size, type, load index, and speed rating.

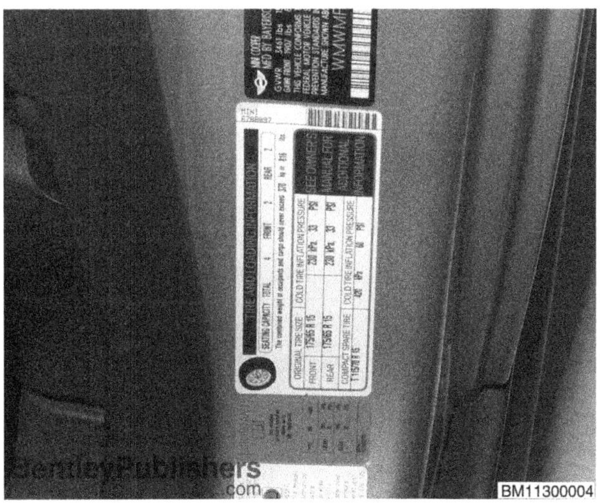

A driver side B-pillar sticker provides information on tire pressures for original equipment tires. Smaller additional stickers remind the driver to consult the owner's manual for information on resetting the flat tire monitor and to limit vehicle speed when pressures are lower.

MINI vehicles are equipped with one of two available technologies for automatic monitoring of tire pressures:

- **Flat tire monitor (FTM)** compares wheel speed information from the ABS wheel speed sensors to determine if one tire is losing air.
- **Tire pressure monitor (TPM)** uses a sensor inside each wheel and tire assembly to directly measure tire pressure and temperature.

TPM replaced FTM as of October 2007 production.

Note that tire pressure monitoring may be referred to as RDC, German acronym for *Reifendruck Control* (tire pressure control).

Flat tire indication

In case of a flat tire or extensive loss of air in a tire, instrument warning (!) illuminates in yellow or red. An acoustic signal also sounds.

– Reduce speed and stop vehicle with caution. Avoid sudden braking and steering maneuvers.

– Identify damaged tire and replace. See **020 Maintenance**.

Run-on-flat tires

In case of a flat tire warning in the instrument cluster:

– Cautiously reduce speed to below 50 mph (80 kph). Avoid sudden braking and steering maneuvers. Do not exceed a speed of 50 mph (80 kph).

> ***WARNING—***
> - *If the vehicle is not equipped with run-on-flat tires, do not continue to drive. Continuing to drive with a flat tire could cause a severe accident.*

At the next opportunity, check air pressure in all four tires. If tire pressure in all four tires is correct, tire monitoring may not have initialized correctly. Initialize system. See tire pressure monitor initializing procedures in this repair group.

In case of complete tire pressure loss, use **Table a** to estimate safe distance for continued driving.

Table a. Driving distance with flat run-on flat tire

Condition	Approx. safe distance to drive
1 - 2 persons, no luggage	155 miles (250 km)
2 persons, full luggage 4 persons, no luggage	94 miles (150 km)
Full load: 4 persons, full luggage	30 miles (50 km)

> ***WARNING—***
> - *In case of a completely flat run-on-flat tire, drive cautiously and do not exceed a speed of 50 mph (80 kph).*
> - *Vehicle handling changes with a flat tire. This can mean reduced tracking stability in braking, extended braking distance and altered steering characteristics.*
> - *If unusual vibration or loud noises occur while driving on a damaged tire, this may be an indication that the tire has failed completely. Reduce speed and pull over as soon as possible at a suitable location. Otherwise parts of the tire could come loose, resulting in an accident. Do not continue driving.*

Flat tire monitor (FTM)

Flat tire monitor (FTM) uses input from the ABS wheel speed sensors to keep track of tire inflation pressures as the vehicle is driven. In case of pressure loss in a tire, the rolling radius and therefore the rotation speed of the wheel changes. This change is detected and the system alerts the driver using warning lights in the instrument cluster.

> ***CAUTION—***
> - *Flat tire monitor does not warn the driver of sudden, severe tire damage caused by external factors, nor can it identify gradual loss of pressure that inevitably occurs in all four tires over time.*

Tire pressure monitor (TPM)

If the vehicle is equipped with tire pressure monitoring (TPM), tire pressures are individually and automatically monitored by the TPM system.

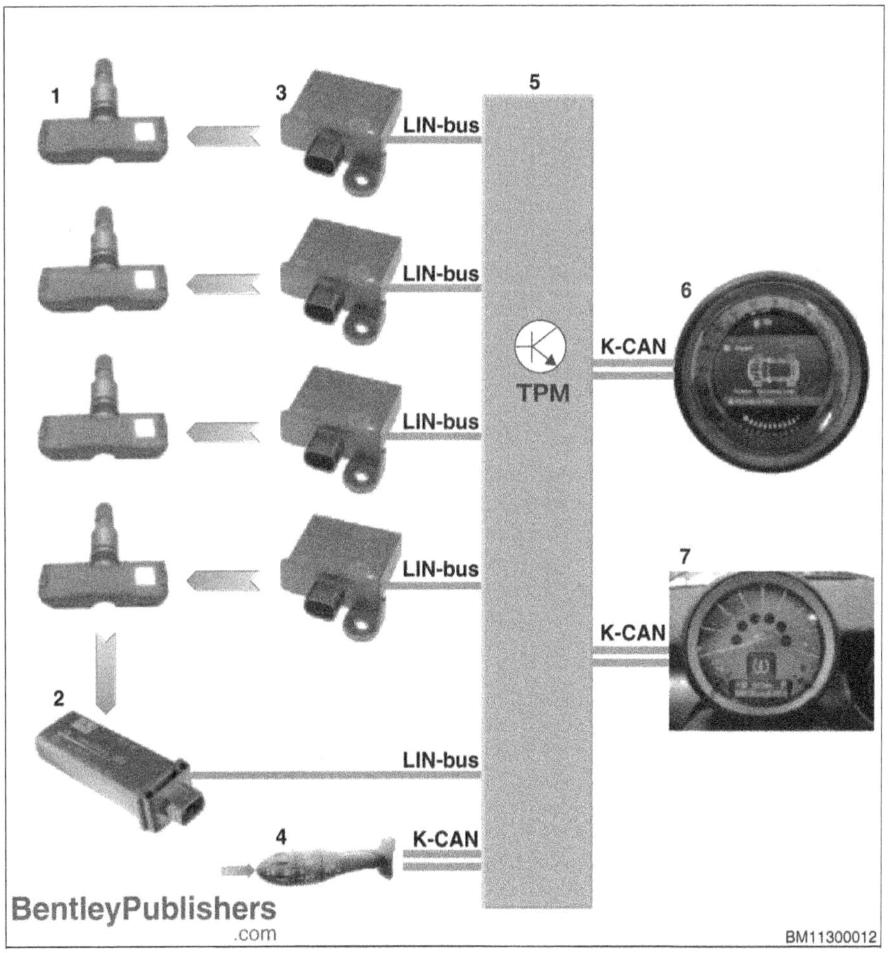

1. Wheel antenna modules
2. Central TPM antenna
3. Wheel well signal receiver modules
4. Reset button
5. TPM control module
6. Central car computer (CCC) display
7. Instrument cluster

300

A TPM antenna in each wheel communicates tire pressure data to a signal receiver in each wheel housing.

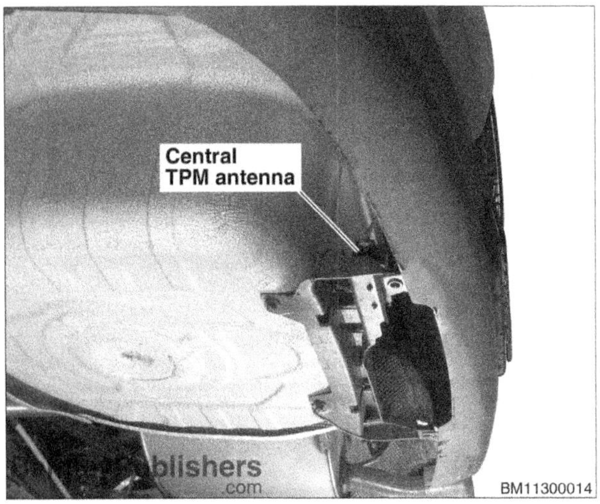

The signals are then sent to the central TPM antenna under the rear bumper.

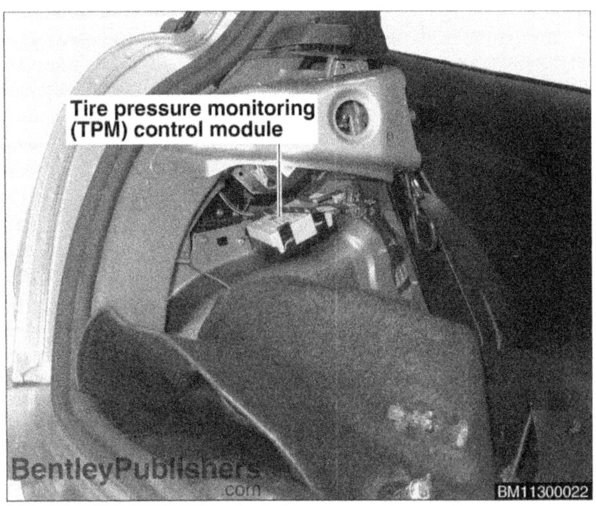

The TPM control module, in left rearbehind cargo compartment trim, processes information received from the central antenna and compares it to tire pressure data learned during initializing. When pressure in a particular tire deviates from the specified value, the control module sends a fault telegram to the instrument cluster and displays a Check Control message.

Initializing tire pressure monitoring (instrument cluster)

The procedures for initializing flat tire monitor (FTM) or tire pressure monitor (TPM) are similar. Initialize tire monitoring each time tire inflation pressures are corrected and after every tire or wheel change. Initialization is completed during driving and can be interrupted at any time. When driving resumes, initialization continues automatically.

Do not initialize the system if you are driving with snow chains or a space-saver spare tire.

Make sure tires are at the minimum allowable pressure. If the tire pressure is too low, the system does not complete the reset procedure.

Tire pressures	
Minimum TPM requirement	approx 1.4 - 1.6 bar (20 - 23 psi)

– Start engine but do not start driving.

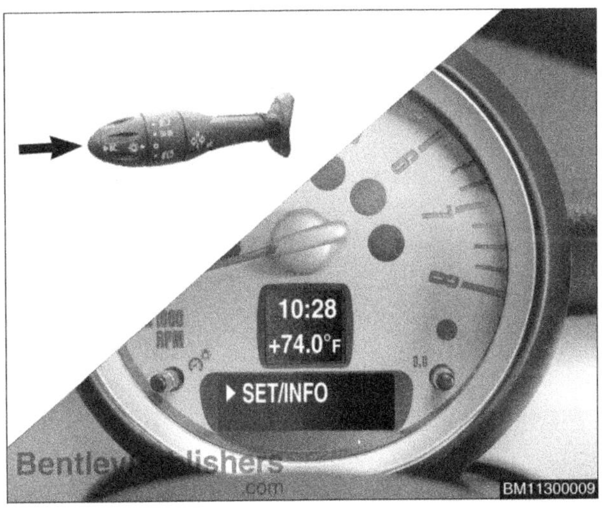

< Briefly press button in turn indicator lever repeatedly until **SET/INFO** is displayed.

– Press and hold button until display changes.

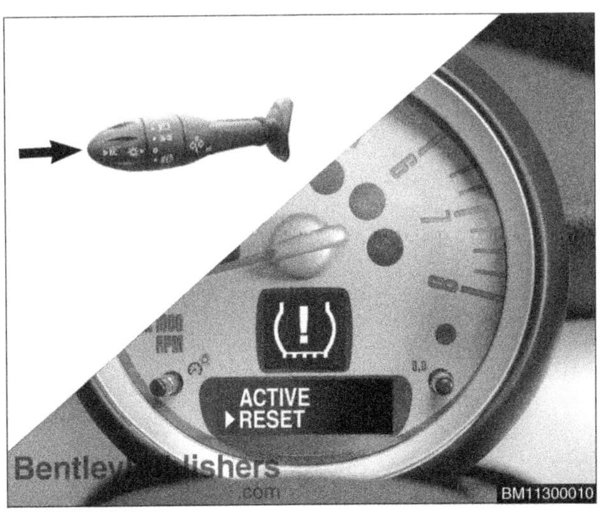

< Briefly press button repeatedly until (!) symbol and **RESET** or **ACTIVE RESET** are displayed.

– Press and hold button until display changes to **RESETTING**.

– Start driving. Initialization is completed while vehicle is moving.

300

Initializing tire pressure monitoring (joystick)

Initialize tire monitoring each time tire inflation pressures are corrected and after every tire or wheel change. Initialization is completed during driving and can be interrupted at any time. When driving resumes, initialization continues automatically.

Do not initialize the system if you are driving with snow chains or a space-saver spare tire.

Make sure tires are at the minimum allowable pressure. If the tire pressure is too low, the system does not complete the reset procedure.

Tire pressures	
Minimum TPM requirement	approx 1.4 - 1.6 bar (20 - 23 psi)

– Start engine but do not start driving.

– Press HOME button on center console.

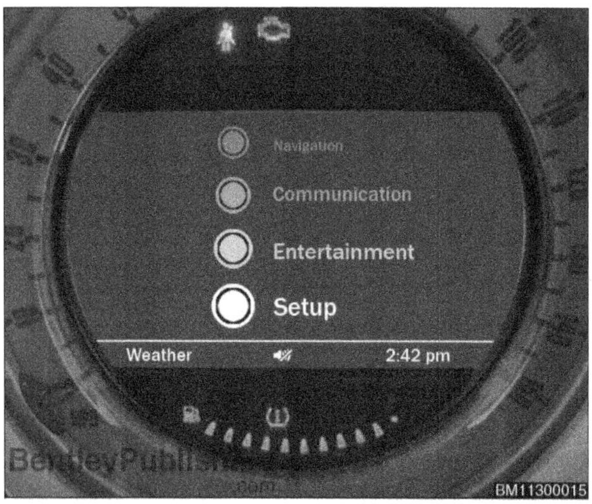

◄ Rotate MINI joystick to **Setup** and press.

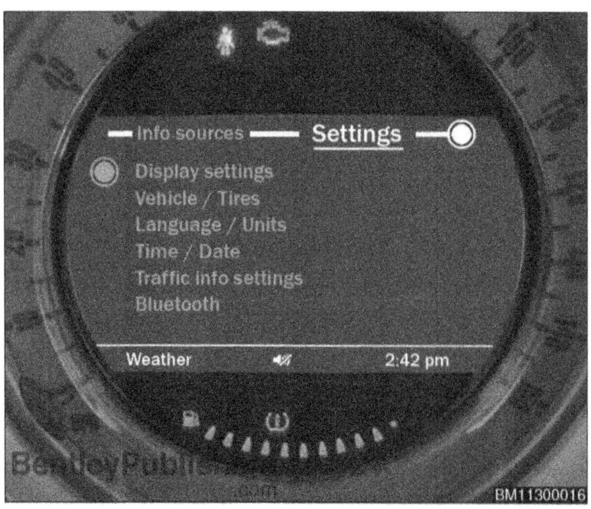

◄ Rotate joystick to **Settings** and press.

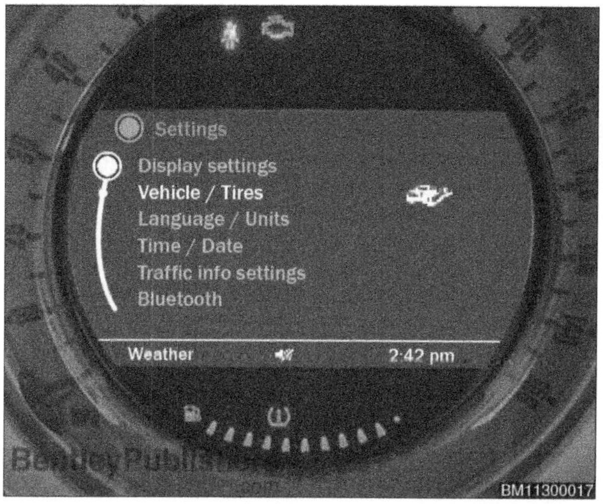

BM11300017

◄ Rotate joystick to **Vehicle / Tires** and press.

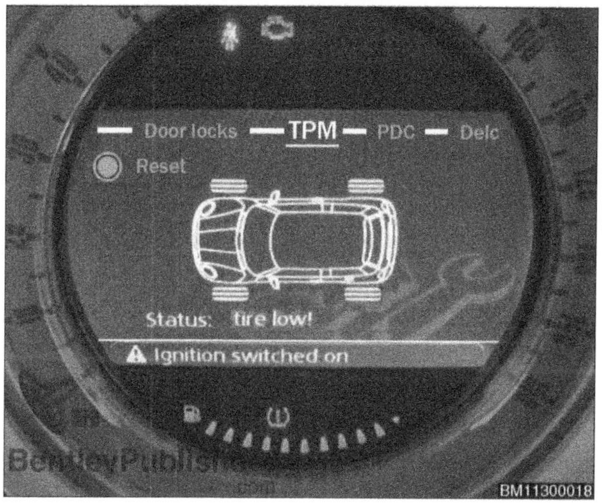

BM11300018

◄ Rotate joystick to **TPM** and press.

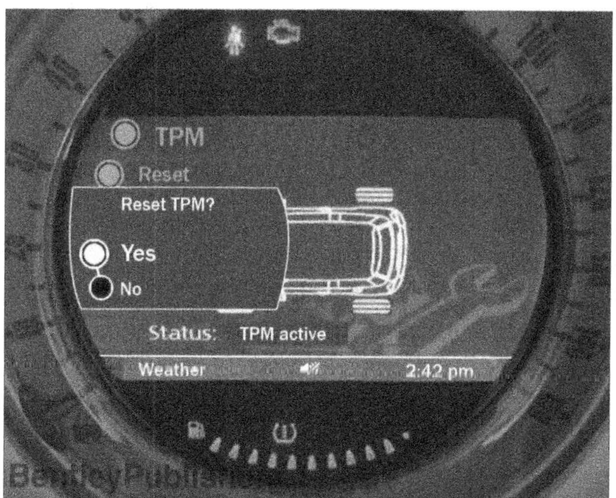

BM11300019

◄ Select **Reset** and press joystick. Select **Yes** and press joystick.

– Drive vehicle in order to complete TPM resetting procedure.

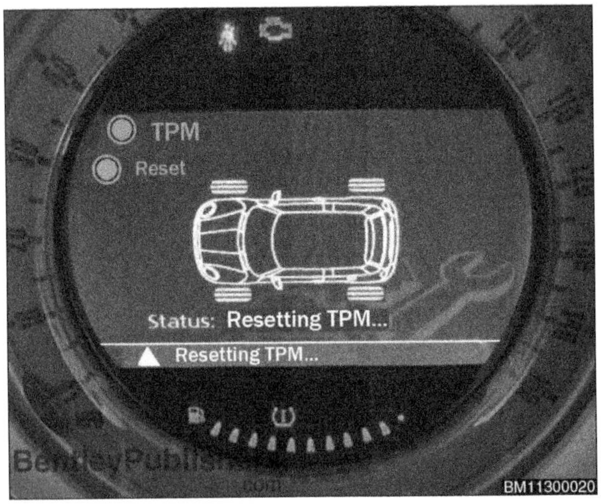

During this process, display shows **Status: Resetting TPM**. Tires display gray until process is complete. This should take only a few minutes.

When reset is complete, the yellow TPM indicator goes out. Correctly inflated tires show up as green in display. For each tire, the tire ID, pressure and temperature are stored in the TPM control module.

310 Front Suspension, Drive Axles

310

GENERAL

This repair group covers front suspension and drive axle repair or replacement.

See also:

- **300 Suspension, Steering and Brakes-General** for a description of the front suspension and components.
- **320 Steering and Wheel Alignment** for wheel alignment specifications, as well as specifications for setting ride height.

Front suspension components

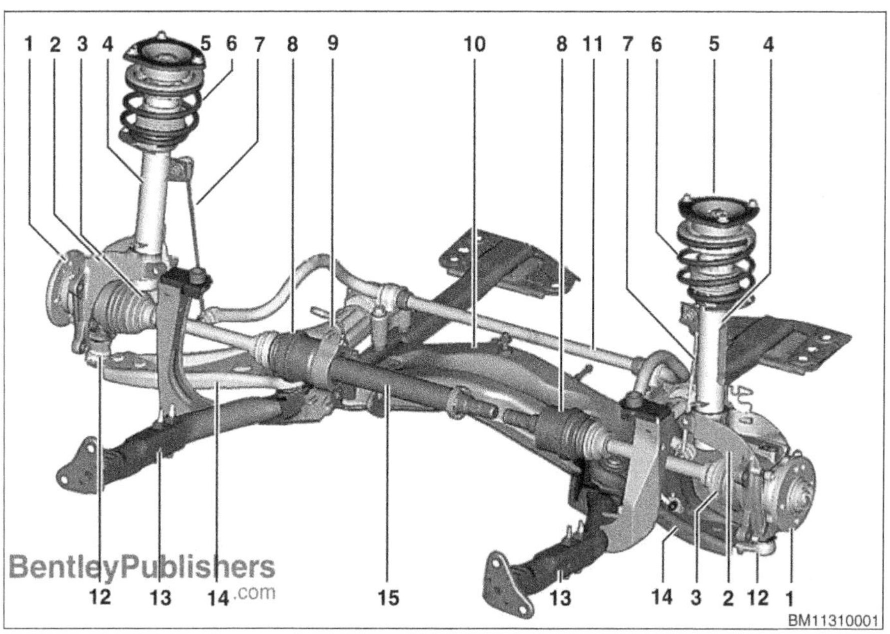

1. Wheel hub
2. Steering knuckle
3. Outer CV joint
4. Strut
5. Upper strut bearing
6. Coil spring
7. Stabilizer bar link
8. Inner CV joint
9. Right axle inner bearing block
10. Subframe
11. Stabilizer bar
12. Ball joint
13. Bumper bracket
14. Control arm
15. Right axle intermediate shaft

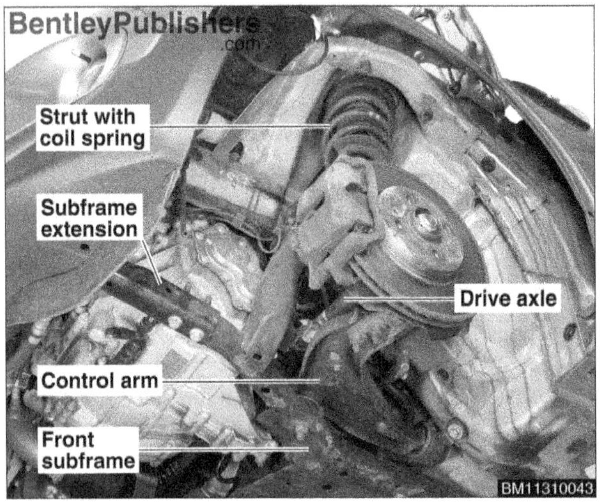

Front suspension design

Front suspension components are mounted to a rigid subframe bolted to the body at two mounting points at the rear and two mounting points at the front. The subframe is made from tubular steel, keeping its weight relatively low despite its sturdy structure.

Bumper brackets bolted to subframe extension rails are used as crash deformation elements in front of the subframe.

Springs and shock absorbers are designed as MacPherson strut assemblies. This provides anti-dive control when braking and starting off and further advantages in terms of weight and space savings.

Lower control arms are supported by ball joints and bushings. They are attached to the steering knuckles by outer ball joints.

A stabilizer bar is anchored to the subframe and attached at each end to the strut tube.

Wheel bearing are incorporated in the wheel hubs and are part of the steering knuckle assembly.

The transmission drives the left and right drive axles which, in turn, drive the front wheels.

During wheel alignment, adjust toe at the tie rods. Correct camber by removing a locating pin on the strut top bearing and rotating the strut top inside the strut tower. See **320 Steering and Wheel Alignment**.

Warnings and Cautions

WARNING—
- *Do not reuse self-locking fasteners. They are designed to be used only once and may fail if reused. Replace with new.*
- *Do not attempt to disassemble the strut assembly without a spring compressor designed specifically for the job.*
- *Front subframe removal may upset the balance of the vehicle on a lift.*

CAUTION—
- *Be sure to have wheel alignment carried out if fasteners for any of the following components are loosened and retorqued:*
  -Steering rack.
  -Ball joint.
  -Tie rod end to tie rod.
- *Have a wheel alignment performed if any of the following components are replaced:*
  -Front subframe.
  -Steering rack.
  -Steering rack boot.
  -Steering knuckle.
  -Ball joint.
  -Control arm.
  -Tie rod end.
  -Tie rod.

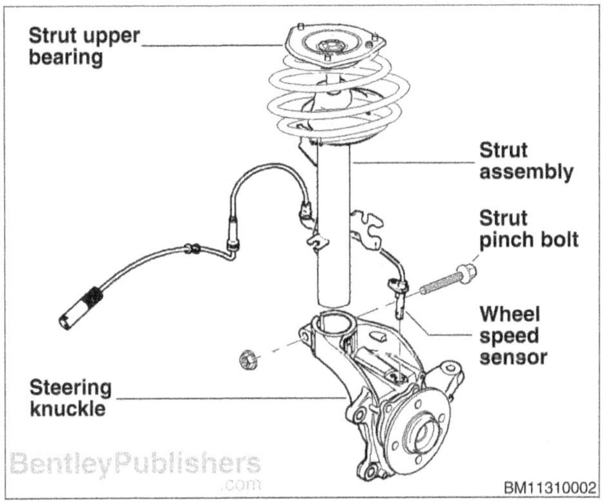

FRONT STRUT ASSEMBLY

The front suspension shock absorbers are MacPherson struts. The strut is a major component of the front suspension and supports the spring. Most strut assembly components are available as replacement parts.

If a shock absorber or strut is defective on one side of the vehicle, replace both shock absorbers on that axle. If a coil spring is broken due to corrosion, replace both coil springs on the axle.

Front strut, upper strut mount or spring replacement is a two-step procedure:

- Removal of strut assembly from the vehicle.
- Disassembly and replacement of components on a work bench.

Front strut assembly, removing and installing

– Raise front of car and support in a safe manner.

> ***WARNING—***
> - *Make sure the car is stable and well supported at all times. Use a professional automotive lift or jack stands designed for the purpose. A floor jack is not adequate support.*

310

– Remove front wheel.

– Detach tie rod end from steering knuckle. See **320 Steering and Wheel Alignment**.

– Detach stabilizer link from strut housing. See **Front stabilizer bar link, removing and installing** in this repair group.

Disengage brake hose from strut bracket (**arrow**).

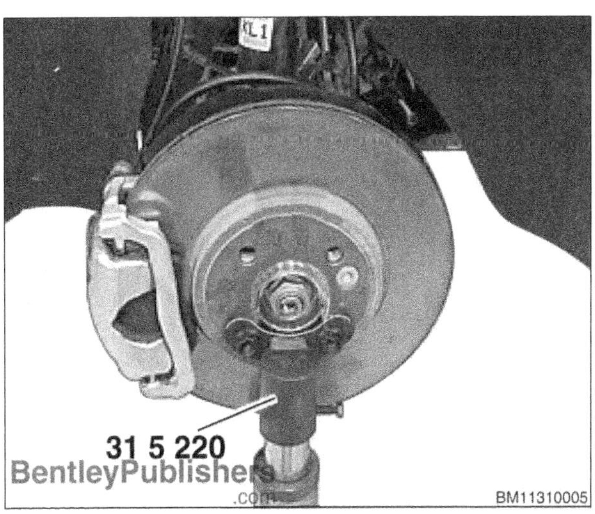

Use floor jack to support steering knuckle or brake rotor. MINI special tool **31 5 220** is shown supporting brake rotor in illustration.

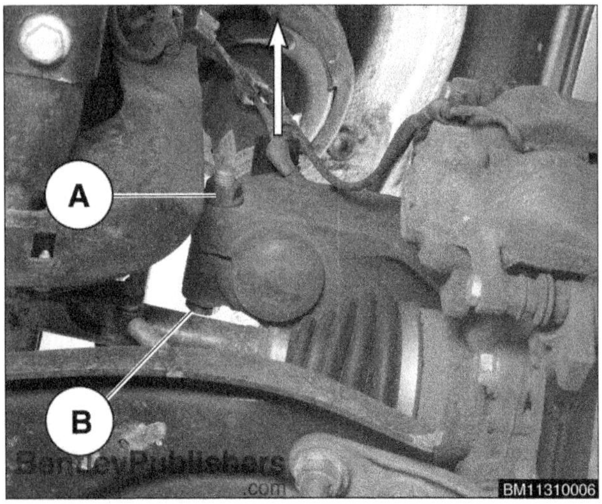

Working at base of strut:

- Disengage ABS wheel speed sensor electrical harness from strut bracket (**arrow**).
- Remove nut (**A**) and slide out strut pinch bolt (**B**).

– Carefully lower floor jack until strut disengages from steering knuckle.

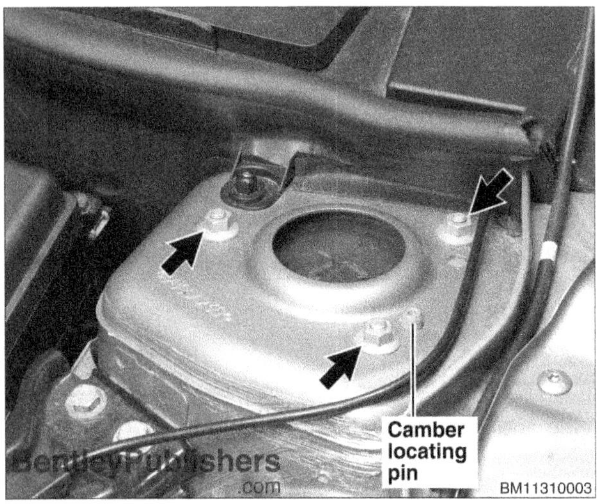

Working in engine compartment at strut upper bearing:

- Note position of camber locating pin. If pin is missing, remove one upper bearing fastener and mark position of strut upper bearing studs on strut tower to maintain original camber.
- Secure strut against falling out.
- Remove strut upper bearing fasteners (**arrows**).

– Lower strut out of wheel arch.

– If replacing strut, coil spring or strut upper bearing, disassemble and reassemble strut assembly on bench. See **Front strut, disassembling and assembling** in this repair group.

– Remainder of installation is reverse of removal.

- Clean strut upper bearing contact surface in strut tower.
- Replace self-locking fasteners.
- Strut top bearing without camber pin: Mount strut top to strut tower using marks made previously.
- Make sure locating boss on strut tube slides in steering knuckle pinch collar gap. Keep strut and steering knuckle contact area grease-free.
- Face strut pinch bolt head in direction of travel. Tighten bolt, not nut.
- Align wheels when job is complete.

Tightening torques	
Road wheel to hub	140 ± 10 Nm (103 ± 7 ft-lb)
Strut assembly pinch bolt (replace M12 x 1.5 x 80 mm ZNS3 bolt with new; tighten bolt only)	100 Nm (74 ft-lb)
Strut upper bearing to strut tower (replace M8 collar nut with new)	34 Nm (25 ft-lb)
Tie rod end to steering knuckle (replace M12 x 1.5 ZNS3 nut with new)	65 Nm (48 ft-lb)
Stabilizer link to strut housing (replace M10 ZNS3 nuts with new)	56 Nm (41 ft-lb)

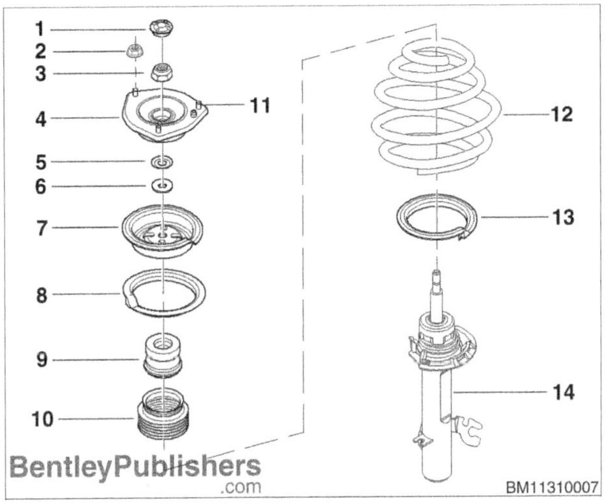

Front strut assembly components

1. Protective cap
2. Strut upper bearing mounting nut
   - Replace M8 x 1.25 collar nut with new
   - Tighten to 34 Nm (25 ft-lb)
3. Strut upper retaining nut
   - M12 x 1.5 ZNS3 nut
   - Tighten to 64 Nm (47 ft-lb)
4. Strut upper bearing
5. Upper bearing dust shield
6. Thrust washer
7. Upper spring seat
8. Spring pad
9. Rubber bump stop
10. Dust boot
11. Camber locating pin
12. Coil spring
13. Spring pad
14. Strut

Front strut, disassembling and assembling

If replacing strut, upper strut mount or coil spring, first remove the strut assembly from the vehicle. See **Front strut assembly, removing and installing** in this repair group.

MINI recommends the use of MINI spring compressor (special tool 31 3 340). A number of different spring compressor tools are available. Photos in this section illustrate the function of one such tool in use.

> ***WARNING—***
> - *Do not attempt to disassemble the strut assembly without a spring compressor designed specifically for this job.*
> - *Prior to each use, check special tool for functionality.*
> - *Do not use a damaged tool.*
> - *Do not make any modifications to tool.*
> - *Use correct size spring retainers when compressing coil spring.*
> - *When assembling MINI spring compressor (special tool 31 3 340), make sure spring retainer plates are felt and heard to snap into place. Check seating of spring retainers carefully.*

Clamp spring compressor in shop vise.

502310996

Position coil spring between spring retainers so that 3 coils lie between retainers (**arrows**). Compress spring.

> ***WARNING—***
> - *Make sure that when tensioned, the spring coil rests completely in the spring holder recess.*
> - *Do not tighten or loosen spring compressor with an impact tool.*
> - *Tighten down the coil spring until stress on the strut upper bearing is relieved.*
> - *Only loosen upper strut nut if spring coils are completely inserted in the spring holder grooves. If necessary, loosen compressor, reposition and recompress.*

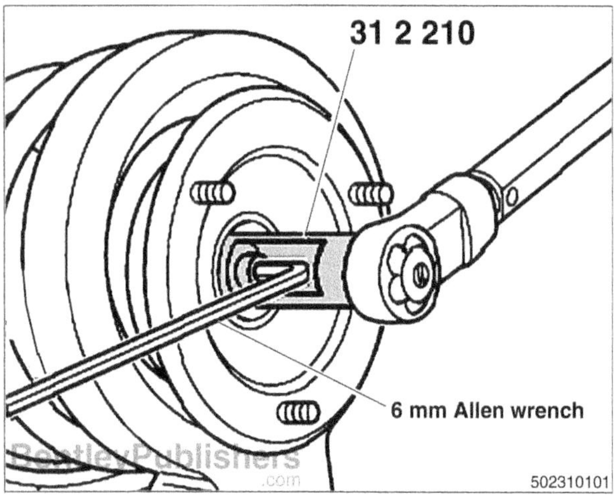

502310101

Use MINI special tool **31 2 210** or equivalent to remove upper strut nut. Counterhold strut shaft using 6 mm Allen wrench.

> ***CAUTION—***
> - *Do not remove strut nut with impact tool.*

- Remove upper strut bearing and related components.
- If a new coil spring is being installed, relieve tension on spring compressor and remove coil spring.
- Check strut dust boot, rubber bump stop and spring pads. Replace as necessary.
- Replace strut, upper strut mount or spring, as needed. If one spring is corroded or damaged, replace both springs as a matched pair.
- When reassembling, be sure components are installed in correct locations and order. See following illustrations.

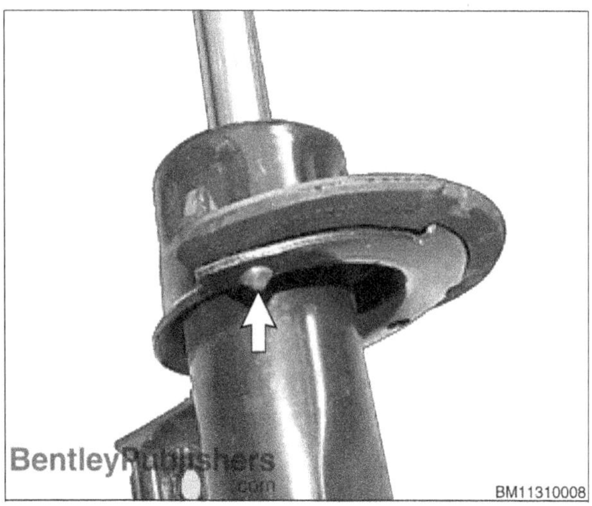

BM11310008

Position lower spring pad so that rubber pin (**arrow**) locks into hole in spring seat.

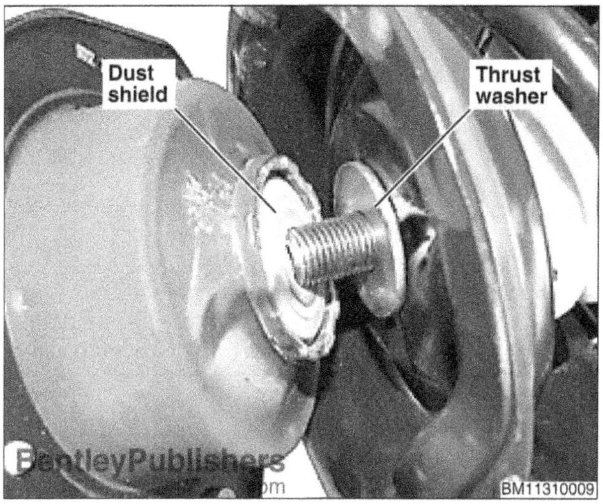

◀ Mount upper bearing dust shield and thrust washer in correct order.

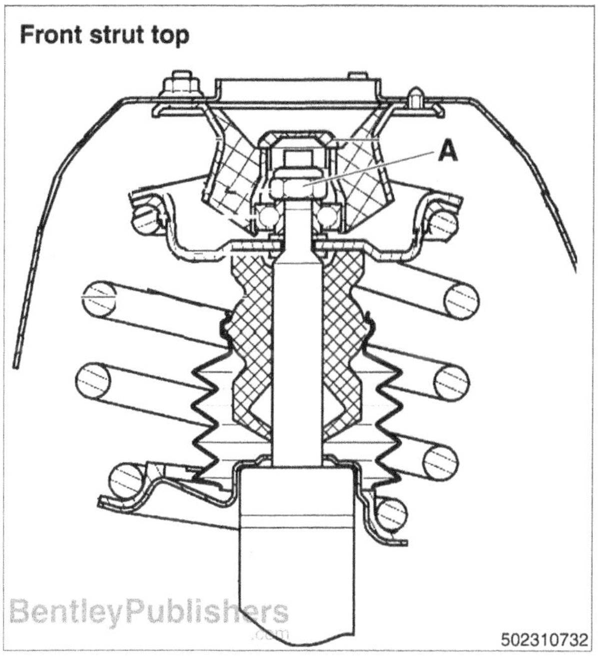

◀ With spring compressed, reassemble strut components

- Use a new upper strut self-locking nut (**A**). Counterhold strut shaft with 6 mm Allen wrench and tighten nut fully.
- Illustration shows a typical automotive strut assembly. MINI models are similar but not identical.

> ***CAUTION—***
> *• Do not tighten strut nut with impact tool.*

Tightening torques	
Strut shaft to strut top bearing (replace M12 x 1.5 self-locking nut with new)	64 Nm (47 ft-lb)

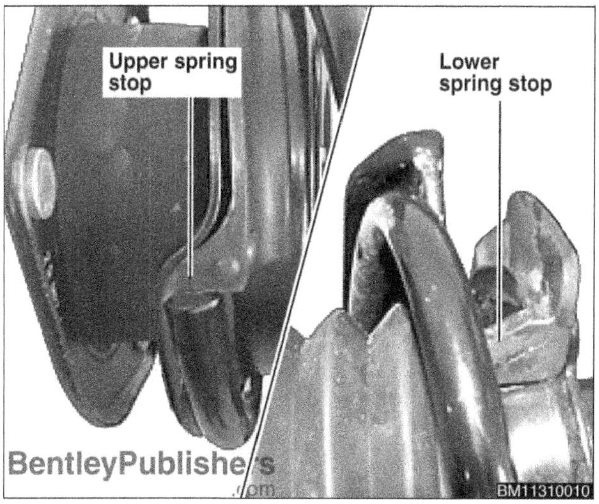

◀ Make sure upper and lower coil spring ends butt up to spring seat stops. Align components correctly to each other before releasing spring compressor.

– Check installation position of strut dust boot.

– Reinstall strut assembly in vehicle. See **Front strut assembly, removing and installing** in this repair group.

– Have vehicle professionally aligned.

Front Subframe

The front subframe provides rigid mounting points for suspension and steering components. The engine and transmission are suspended from the chassis above and are located and stabilized by the front subframe.

> ***CAUTION—***
> - *Removal or replacement of the subframe may affect suspension and steering geometry, including front wheel alignment. Make appropriate matching marks during removal and have the front end aligned once repairs are complete.*

Front subframe components

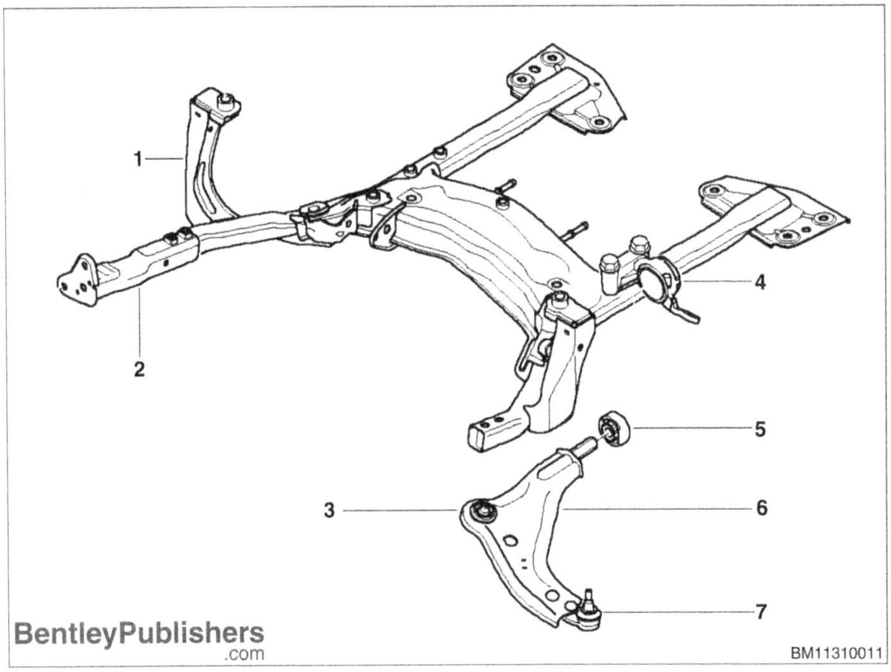

1. Subframe
2. Bumper bracket (crash deformation element)
3. Inner ball joint
4. Control arm bushing carrier
5. Control arm bushing
6. Control arm
7. Outer (steering knuckle) ball joint

Front subframe, removing and installing

Complete removal of the front subframe is necessary for some jobs, such as transmission removal or subframe replacement. In other cases, such as front control arm or stabilizer bar replacement, it is necessary to lower the subframe approx. 120 mm (5 inches). The procedures in either case is similar. After the job is completed, perform wheel alignment.

> ***WARNING—***
> - *Front subframe removal may upset the balance of the vehicle on a lift.*

– Raise front of car and support in a safe manner.

> ***WARNING—***
> - *Make sure the car is stable and well supported at all times. Use a professional automotive lift or jack stands designed for the purpose. A floor jack is not adequate support.*

- Remove front wheels.
- Separate exhaust pipe from catalytic converter. Remove entire exhaust system. See **180 Exhaust System**.
- Remove front bumper trim and front bumper. See **510 Exterior Trim, Bumpers**.
- Partially remove front wheel housing liners. See **410 Fenders. Engine Hood**.
- Partially remove underbody panelling on left and right.
- If applicable, detach ride-height sensor connector and expose line up to engine carrier.

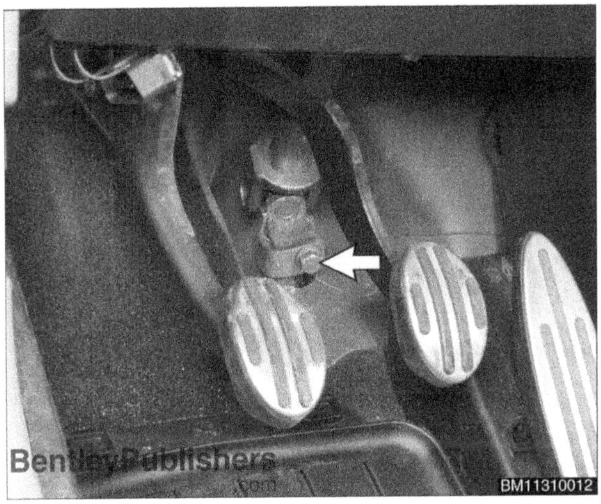

Working underneath left side dashboard, remove steering universal joint pinch bolt (**arrow**) and detach upper from lower steering shaft.

> *CAUTION—*
> - *While steering spindle is detached from steering rack, make sure steering wheel is not rotated. Rotating the wheel may damage the airbag and horn contact spring.*

- Detach left and right tie rod ends from steering knuckles. See **320 Steering and Wheel Alignment**.
- Detach both outer ball joints from steering knuckles. See **Outer ball joint, replacing** in this repair group.
- Detach both stabilizer links from stabilizer bar. See **Front stabilizer bar link, removing and installing** in this repair group.
- Remove ABS wheel speed sensor harness brackets from left and right upright corners of subframe.

Working underneath rear of engine, remove engine stabilizer link mounting bolts (**arrows**). Remove stabilizer link.

310

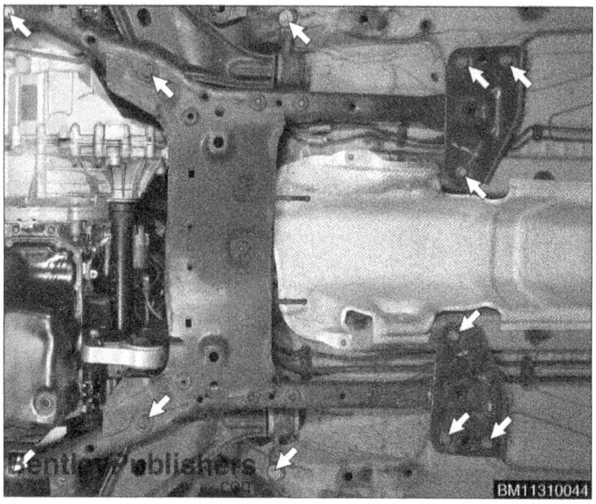

While supporting subframe with suitable jack, remove subframe mounting bolts (**arrows**).

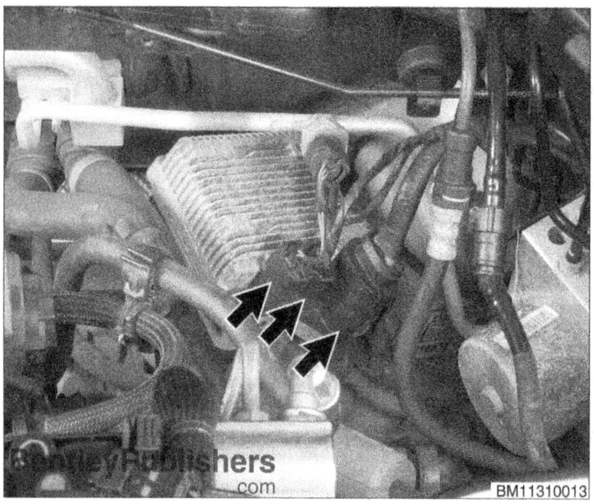

Lower subframe a few inches and detach electromechanical steering electrical connectors (**arrows**).

– Detach both control arms from subframe. See **Front control arm, removing and installing** in this repair group.

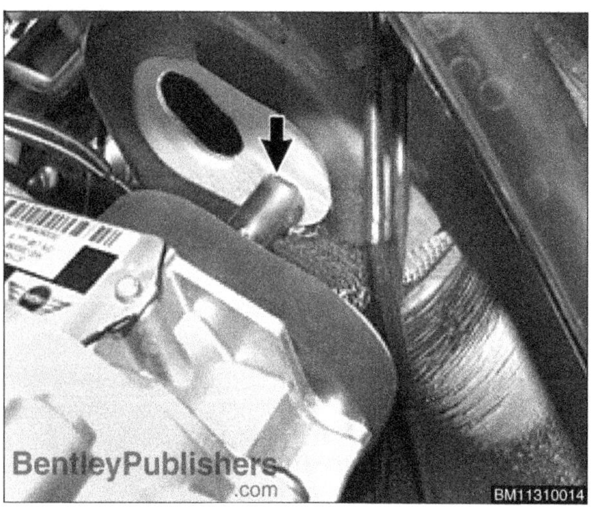

Lower subframe, watching carefully to make sure no electrical lines, hoses or components become snagged.

> ***CAUTION—***
> - *When lowering subframe, make sure steering rack stub shaft (**arrow**) does not collide with bulkhead.*

– If replacing subframe with new, transfer components from old subframe before reassembly.

Be sure to replace steering rack foam rubber seal (**arrow**) before reassembly.

– Check subframe mounting bolt hole threads. Repair with Helicoil if necessary.

– When reinstalling:
  - Raise subframe into place, watching carefully to make sure no electrical lines, hoses or components become snagged.
  - Attach engine stabilizer link loosely before reinstalling subframe mounting bolts.
  - Use correct length subframe bolts in different locations:
    -Front bolts = M12 x 85 mm
    -Control arm bushing carrier bolts = M10 x 60 mm
    -Rear bolts = M12 x 62 mm
  - Replace self-locking fasteners.

Tightening torques	
Ball joint, inner, to subframe (replace M14 x 1.5 ZNS3 self-locking nut with new)	70 Nm (ft-lb) + additional 90°
Ball joint, outer, to control arm (replace M14 x 1.5 self-locking nuts with new)	175 Nm (129 ft-lb)
Ball joint, outer, to steering knuckle (replace M12 x 1.5 ZNS3 self-locking nut with new)	70 Nm (ft-lb) + additional 90°
Bumper to front subframe (M8 x 18 mm bolts)	19 Nm (14 ft-lb)
Control arm bracket and stabilizer anchor to subframe (M14 x 1.5 x 105 mm self-tapping bolts)	165 Nm (122 ft-lb)
Engine stabilizer link to bracket (M12 x 90 mm bolts)	108 Nm (80 ft-lb)
Exhaust pipe to catalytic converter (V-band clamp)	25 ± 3 Nm (18 ± 2 ft-lb)
Front subframe to chassis: • Front bolts M12 x 85 mm • Rear bolts M12 x 62 mm • Control arm bushing carrier bolts (replace M10 x 60 mm 10.9 grade bolts with new)	 100 Nm (74 ft-lb) 100 Nm (74 ft-lb) 59 Nm (44 ft-lb) + additional 90°
Road wheel to hub	140 ± 10 Nm (103 ± 7 ft-lb)
Stabilizer link to stabilizer bar (replace M10 self-locking nuts)	56 Nm (41 ft-lb)
Tie rod end to steering knuckle (replace M12 x 1.5 self-locking nut)	65 Nm (48 ft-lb)
Universal joint pinch bolt to steering shaft (replace M8 x 1.25 x 41 mm bolt with new)	28 Nm (21 ft-lb)

– Have vehicle professionally aligned.

– Vehicle with DSC: Use MINI scan tool to adjust steering angle sensor.

310

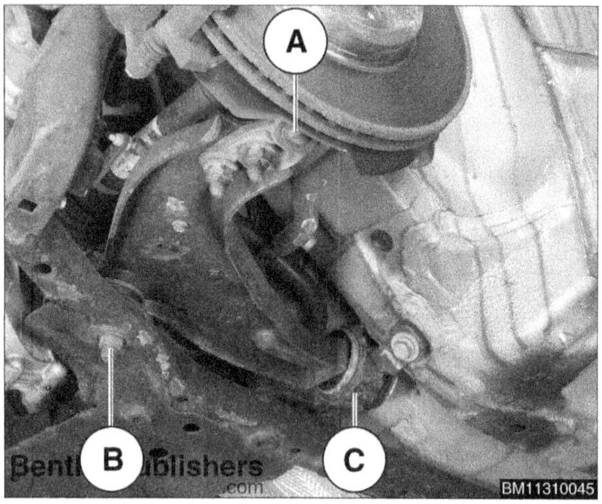

Front Control Arms

Each front control arm has three attachment points:

- **A** Outer ball joint attached to steering knuckle
- **B** Inner ball joint attached to front subframe
- **C** Bushing and bushing carrier attached to front subframe and body

The outer ball joint and rear bushing are available as replacement parts.

Front control arm, removing and installing

– Raise front of car and support in a safe manner.

> ***WARNING—***
> - *Make sure the car is stable and well supported at all times. Use a professional automotive lift or jack stands designed for the purpose. A floor jack is not adequate support.*

– Remove front wheels.

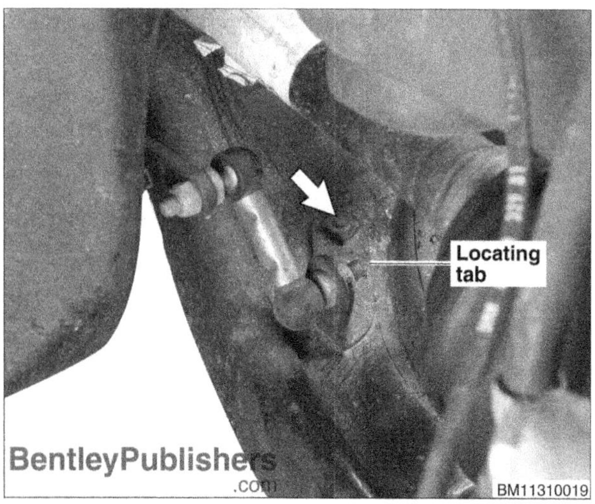

Vehicle with ride height sensor: Remove ride height sensor bracket bolt (**arrow**) at left control arm. Note position of sensor bracket locating tab.

Remove outer ball joint mounting nut. Use MINI special tool **31 2 310** or equivalent to press ball joint off steering knuckle.

◀ Remove inner ball joint mounting nut. Screw MINI special tool **31 1 040** to ball joint shaft, which is press-fit into subframe bore. Strike tool from below (**arrow**) to release ball joint from subframe.

– Lower front subframe approx. 120 mm (5 in). See **Front subframe, removing and installing** in this repair group.

– Detach stabilizer bar from subframe and tie up. See **Stabilizer bar, removing and installing** in this repair group. Note that stabilizer bar anchor bolts are also used to attach control arm bushing carrier to subframe.

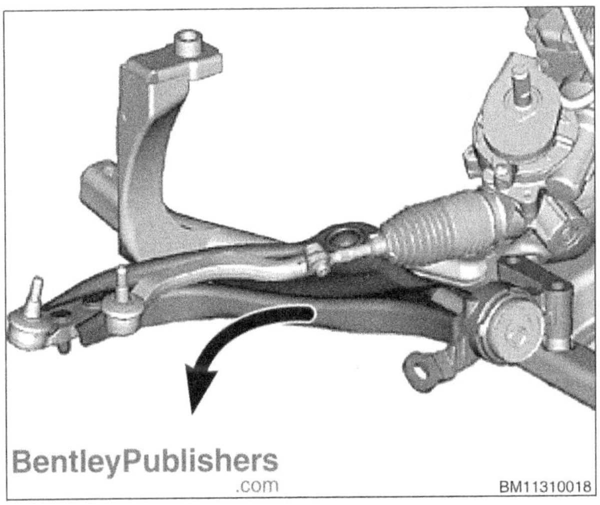

◀ Tilt control arm to remove.

– Installation is reverse of removal. Note the following:
  - Raise front subframe, watching carefully to make sure no lines, hoses or other components become snagged.
  - Use correct length subframe bolts in different locations. See **Front subframe, removing and installing** in this repair group.
  - Replace self-locking fasteners.
  - Keep ball joint shafts and corresponding bores clean and grease-free.
  - If applicable, position ride height sensor on left control arm using locating tab.

Tightening torques	
Ride height sensor bracket to control arm (M6 bolt)	6 Nm (53 in-lb)

– Have vehicle professionally aligned.

– Vehicle with DSC: Use MINI scan tool to adjust steering angle sensor.

Outer ball joint, replacing

– Raise front of car and support in a safe manner.

> ***WARNING—***
> - *Make sure the car is stable and well supported at all times. Use a professional automotive lift or jack stands designed for the purpose. A floor jack is not adequate support.*

– Remove front wheel.

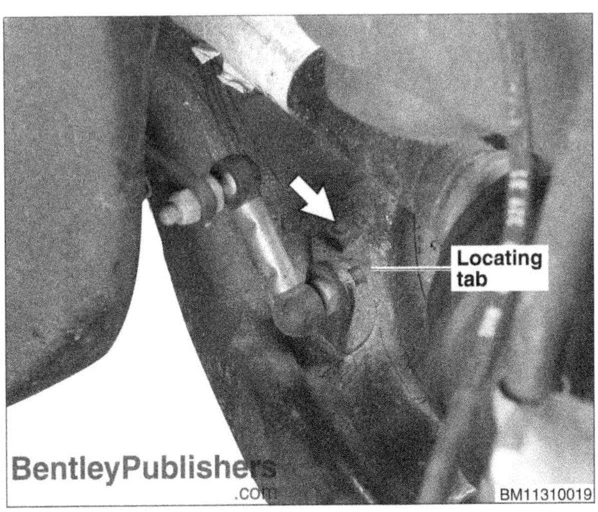

◀ Vehicle with ride height sensor: Remove ride height sensor bracket bolt (**arrow**) at left control arm. Note position of sensor bracket locating tab.

310

Remove outer ball joint mounting nut at steering knuckle. Use MINI special tool **31 2 310** or equivalent to press ball joint off steering knuckle.

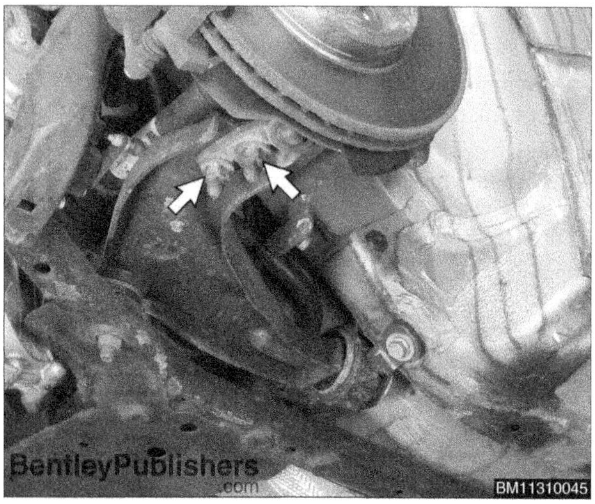

Remove outer ball joint mounting nuts (**arrows**) at control arm. Remove ball joint.

- Installation is reverse of removal. Replace self-locking fasteners.

Tightening torque	
Ball joint, outer, to control arm (replace M14 x 1.5 self-locking nuts with new)	175 Nm (129 ft-lb)
Ball joint, outer, to steering knuckle (replace M12 x 1.5 ZNS3 self-locking nut with new)	70 Nm (ft-lb) + additional 90°
Road wheel to hub	140 ± 10 Nm (103 ± 7 ft-lb)

Front control arm bushing, replacing

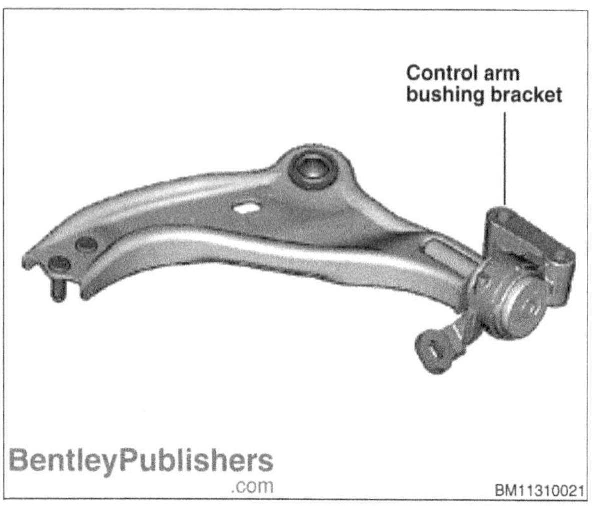

Lower front subframe to remove control arm and gain access to control arm bushing.

> ***CAUTION—***
> • *Replace control arm bushings in pairs.*

- Remove front control arm. See **Front control arm, removing and installing** in this repair group.

◄ Mount control arm in press and use heat gun to heat rubber bushing. Press out control arm pin, making sure that rubber mount inner sleeve does not stay on pin.

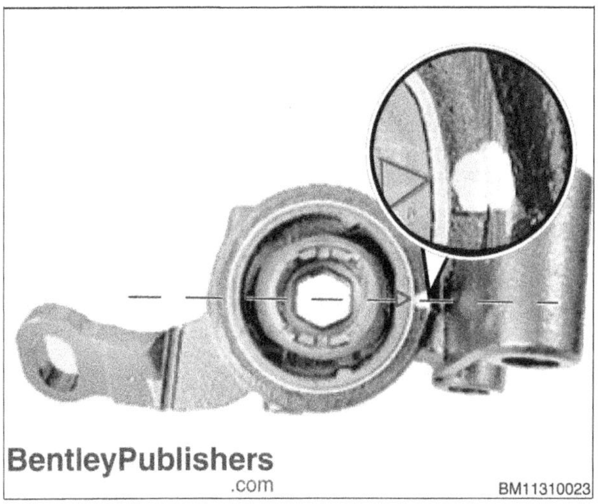

◄ Mark alignment of bushing on bracket (**inset**) before pressing out bushing.

◄ Use press tools to force bushing out of bracket. Reinstall new bushing, making sure arrowhead on bushing points toward previously made mark on bracket. See previous illustration.

When installing control arm to bushing:

- Coat control arm pin and inner sleeve of bushing with generous coating of CircoLight®.
- Align control arm to bushing and use press to mount control arm to bushing (**arrow**). Press arm in to stop. Check gap between control arm and rubber bushing after pressing.

Control arm bushing	
Gap between control arm and rubber bushing	6 ± 2 mm (0.24 ± 0.08 in)

– Immediately bolt bracket and control arm to subframe. See **Front control arm, removing and installing** in this repair group.

– Reassemble front suspension and place unladen vehicle on ground for 30 minutes. Avoid major suspension deflections.

After 30 minutes, CircoLight® evaporates fully, allowing control arm to seat firmly and correctly in rubber bushing.

FRONT STABILIZER BAR

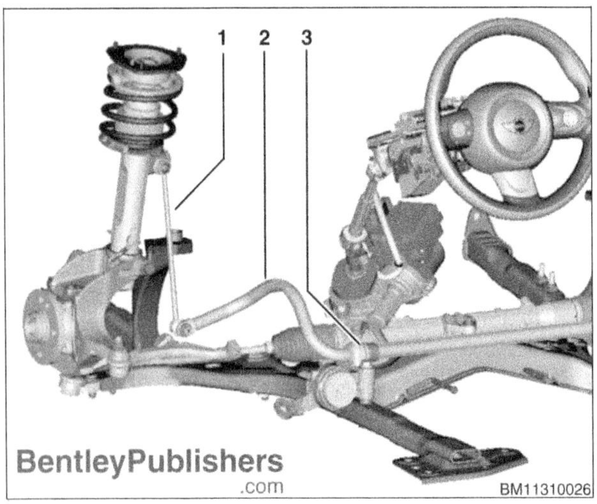

Lower subframe to remove front stabilizer.

Front stabilizer components:

1. Stabilizer link
2. Stabilizer bar
3. Stabilizer anchor

Front stabilizer bar link, removing and installing

– Raise front of car and support in a safe manner.

> ***WARNING—***
> - *Make sure the car is stable and well supported at all times. Use a professional automotive lift or jack stands designed for the purpose. A floor jack is not adequate support.*

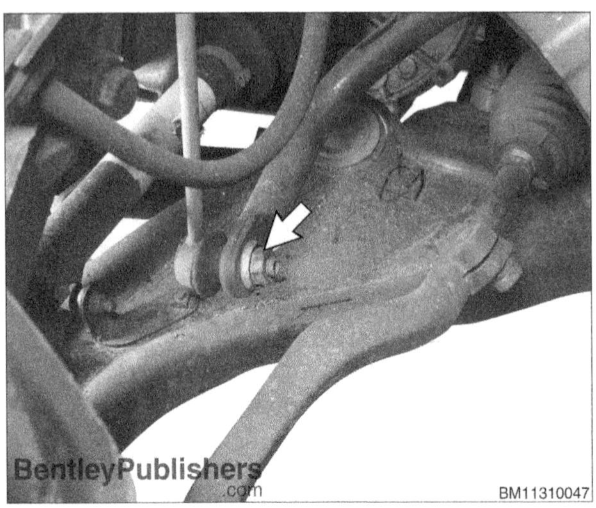

Detach stabilizer bar link from stabilizer bar. Depending on stabilizer link design, there are two methods of counter holding stabilizer link ball joint shaft:

- Model illustrated: Use Allen wrench to counterhold shaft while removing mounting nut (**arrow**).
- Alternate method: Use thin flat open-end wrench to counterhold.

◄ Detach upper end of stabilizer bar link. Counterholding technique is similar to previous step.

– Installation is reverse of removal. Replace self-locking nuts.

Tightening torque	
Stabilizer bar link to stabilizer bar or strut tube bracket (replace M10 ZNS3 self-locking collar nut with new)	56 Nm (41 ft-lb)

Front stabilizer bar, removing and installing

– Raise front of car and support in a safe manner.

> ***WARNING—***
> • *Make sure the car is stable and well supported at all times. Use a professional automotive lift or jack stands designed for the purpose. A floor jack is not adequate support.*

– Remove front wheels.

– Lower front subframe approx. 120 mm (5 in). See **Front subframe, removing and installing** in this repair group.

– Detach stabilizer links from stabilizer bar. See **Front stabilizer bar link, removing and installing** in this repair group.

◄ Remove stabilizer bar anchor bolts (**arrows**) on left and right and slide stabilizer to side to remove.

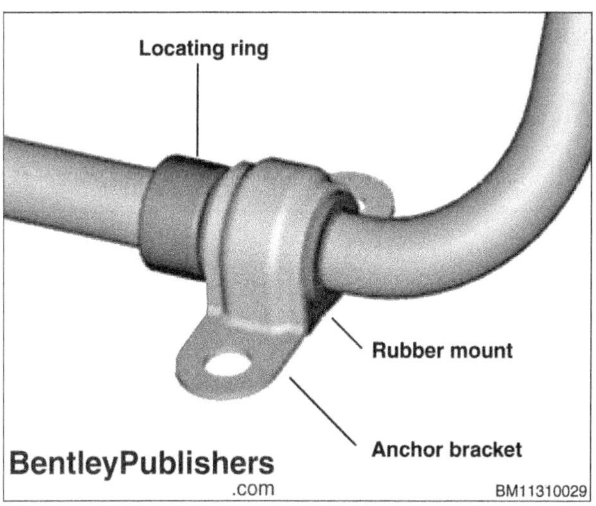

◄ Remove stabilizer bar anchor bracket and slide rubber mounts off stabilizer bar.

– When installing rubber mount:
  - Keep stabilizer bar clean and grease-free.
  - Make sure rubber mount is engaged correctly in locating ring. *Do not* use lubricant or adhesive such as CircoLight® between rubber mount and stabilizer bar.

– Place stabilizer bar in correct location above subframe. Coat outside of stabilizer rubber mounts with CircoLight® adhesive and antiseize, then place stabilizer anchor brackets over mounts and install bolts.

Tightening torque	
Control arm bracket and stabilizer anchor to subframe (M14 x 1.5 x 105 mm self-tapping bolts)	165 Nm (122 ft-lb)

– Remainder of installation is reverse of removal. Note the following:
  - Raise front subframe, watching carefully to make sure no lines, hoses or other components become snagged.
  - Use correct length subframe bolts in different locations. See **Front subframe, removing and installing** in this repair group.
  - Replace self-locking fasteners.
  - Keep ball joint shafts and corresponding bores clean and grease-free.
  - If applicable, position ride height sensor on left control arm using locating tab.

Tightening torques	
Ride height sensor bracket to control arm (M6 bolt)	6 Nm (53 in-lb)
Stabilizer bar link to stabilizer bar (replace M10 ZNS3 self-locking collar nut with new)	56 Nm (41 ft-lb)

– Have vehicle professionally aligned.

– Vehicle with DSC: Use MINI scan tool to adjust steering angle sensor.

FRONT WHEEL BEARINGS

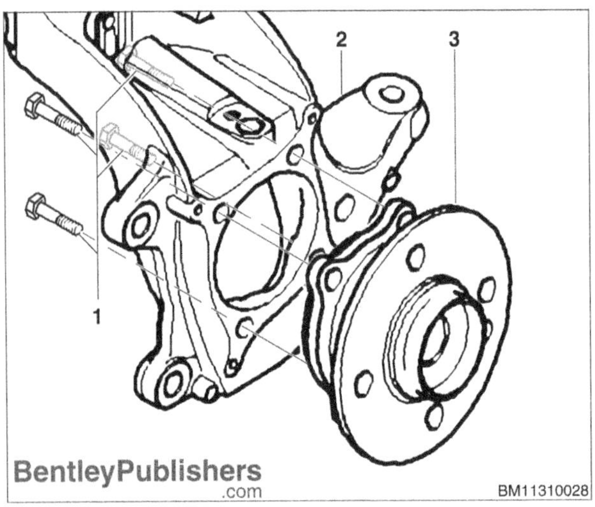

Front wheel bearings are permanently sealed and require no maintenance. The bearing is integral with the wheel hub. The wheel bearing and hub assembly is bolted to the steering knuckle.

1. M10 x 30 mm ZNS3 10.9 grade bolts
   - Replace with new
   - Tighten in 2 stages:
     Stage 1: 20 Nm (15 ft-lb)
     Stage 2: additional 90°
2. Steering knuckle
3. Wheel bearing and hub assembly

Front wheel bearing, removing and installing

– Raise front of car and support in a safe manner.

> ***WARNING —***
> - *Make sure the car is stable and well supported at all times. Use a professional automotive lift or jack stands designed for the purpose. A floor jack is not adequate support.*

– Remove front wheel.

◀ Have an assistant press down on brake pedal to keep wheels from turning. Loosen flanged axle nut (**arrow**) on wheel hub. Remove nut.

– Remove brake caliper and disc. See **340 Brakes**. Do not detach caliper brake fluid line.

> ***CAUTION—***
> - *Hang brake caliper aside using stiff wire to prevent the brake fluid hose from stretching.*

◀ Undo ABS wheel speed sensor mounting bolt (**arrow**). Remove sensor to avoid damaging it while removing wheel hub.

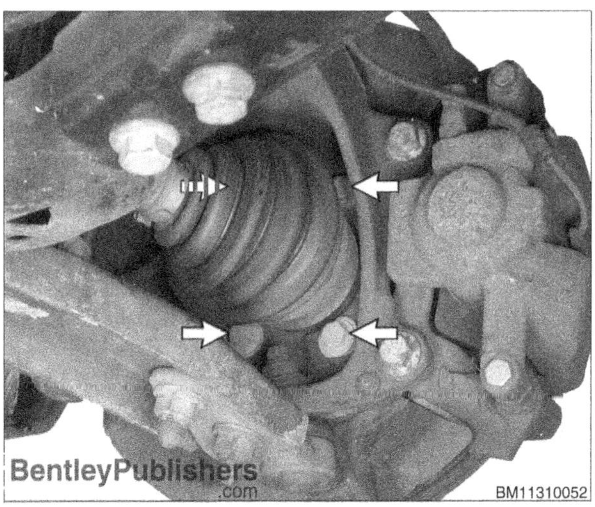

◀ Remove four bolts (**arrows**) securing bearing hub to steering knuckle.

– Detach wheel hub from steering knuckle and drive shaft.

– Installation is reverse of removal.
  - Replace self-locking wheel bearing hub mounting bolts with new.
  - Clean ABS wheel speed sensor bore and lubricate with Staburags® NBU 12/K or equivalent grease before installing sensor.
  - When installing self-locking flanged axle nut:
    -Use new nut.
    -Lightly oil contact surface of nut and wheel bearing hub. Do not allow oil to get on threads.
    -Final tightening with vehicle on ground and assistant with foot on brake pedal.
    -Stake nut to lock in place.

Tightening torques	
Bearing hub to steering knuckle (replace M10 x 30 mm ZNS3 10.9 grade bolts with new): • Stage 1 • Stage 2	 20 Nm (15 ft-lb) additional 90°
Brake caliper to steering knuckle (replace M12 x 1.5 self-locking bolts with new)	110 Nm (81 ft-lb)
Brake disc to bearing hub (replace Torx M10 x 13 with new)	27 Nm (20 ft-lb)
Drive axle outer CV joint to bearing hub (replace M22 x 1.5 flange nut with new)	182 Nm (134 ft-lb)
Road wheel to hub	140 ± 10 Nm (103 ± 7 ft-lb)
Wheel speed sensor to steering knuckle	8 Nm (6 ft-lb)

Drive Axles

Drive axles use constant-velocity (CV) joints on both ends. The outer joints are of the traditional CV joint variety, whereas the inners are referred to as tripod joints.

An intermediate shaft on the right insures equal length swinging drive axles, reducing the intensity of torque steer. The shaft is one piece with the inner tripod joint cup. It is supported at the right end by a bearing bolted to the engine oil pan.

For replacement parts, both inner and outer CV joints boots or complete axles are offered by MINI.

The inner tripod joint, intermediate shaft and intermediate shaft support bearing are not available separately.

Drive axle components

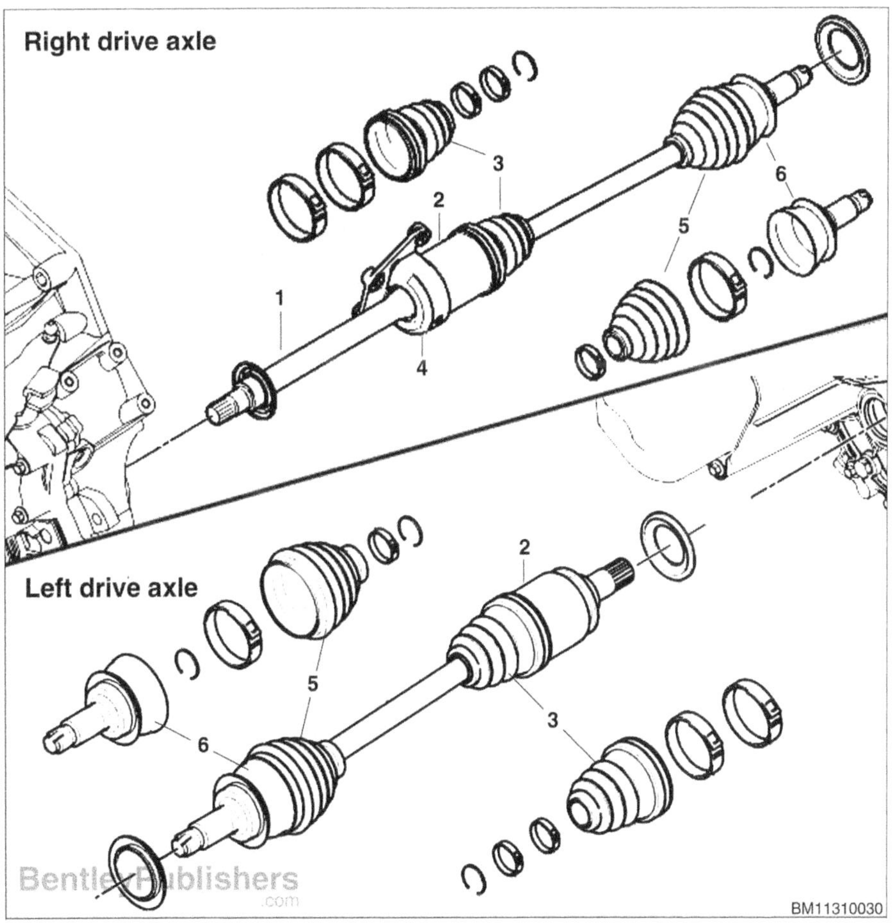

1. Intermediate shaft
2. Inner tripod joint
3. Inner CV joint boot
4. Intermediate shaft bearing housing
5. Outer CV joint boot
6. Outer CV joint

Drive axle, removing and installing

– Raise front of car and support in a safe manner.

> *WARNING—*
> *• Make sure the car is stable and well supported at all times. Use a professional automotive lift or jack stands designed for the purpose. A floor jack is not adequate support.*

– Manual transmission: Drain transmission oil. See **230 Manual Transmission**.

– Remove front wheel.

< Have an assistant press down on brake pedal to keep wheels from turning. Loosen flanged axle nut (**arrow**) on wheel hub. Remove nut.

– Detach tie rod end from steering knuckle. See **320 Steering and Wheel Alignment**.

– Detach outer ball joint from steering knuckle. See **Outer ball joint, replacing** in this repair group.

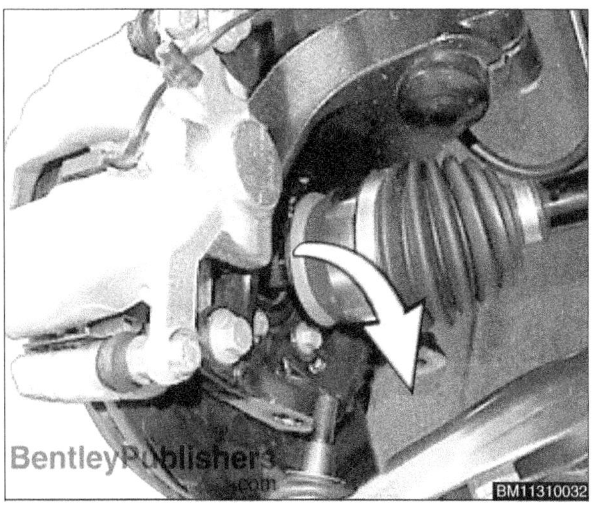

< Swing steering knuckle outward and press stub end of drive axle inward. If necessary, hit threaded end of drive axle with plastic hammer to drive out of wheel bearing hub. Swing end of drive axle out of back of bearing hub (**arrow**).

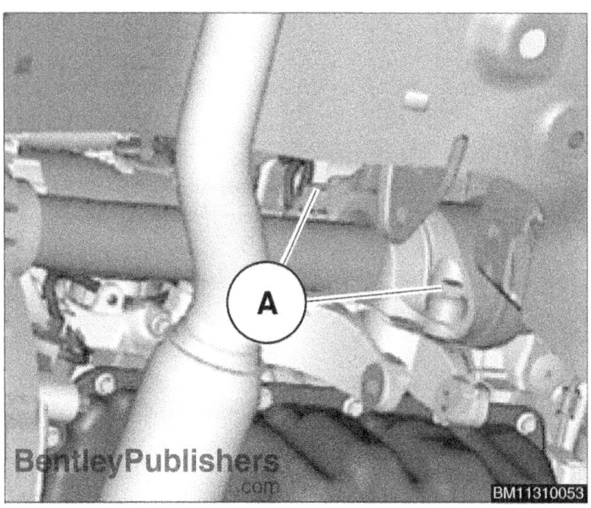

< Right drive axle: Remove intermediate shaft bearing housing mounting bolts (**arrows**).

– Use pry bar to lever drive axle inner end out of differential flange. Be prepared to catch dripping transmission fluid.

– Inspect drive axle. Replace CV joint boots or axle as necessary.

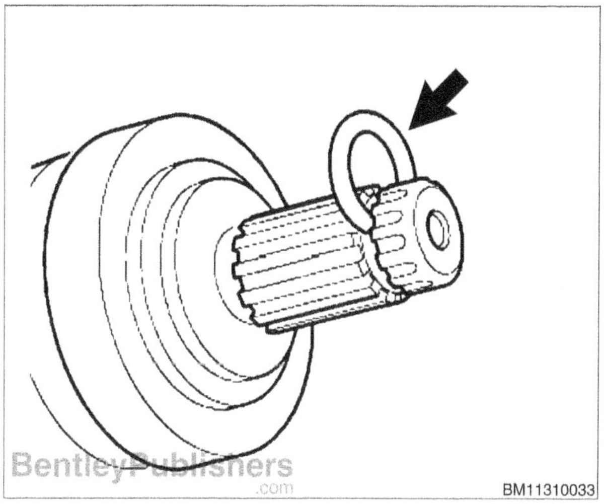

When installing, replace spring circlip (**arrow**) on inner joint shaft.

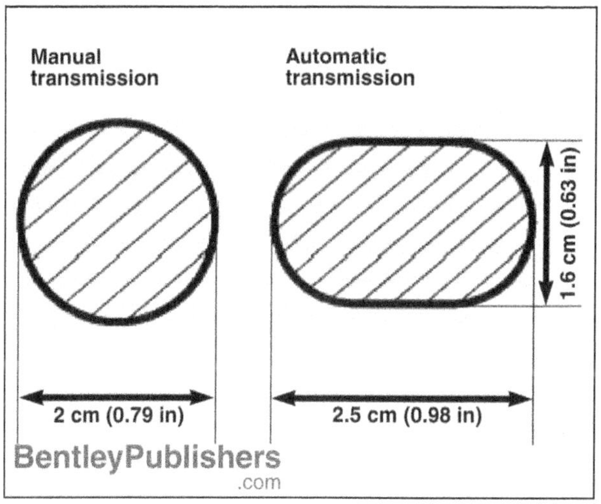

Note that drive axle shafts for manual and automatic transmission are fitted with different cross-section spring circlip.

– Before installing drive axle, replace drive axle flange seal in differential. See **230 Manual Transmission** or **240 Automatic Transmission**.

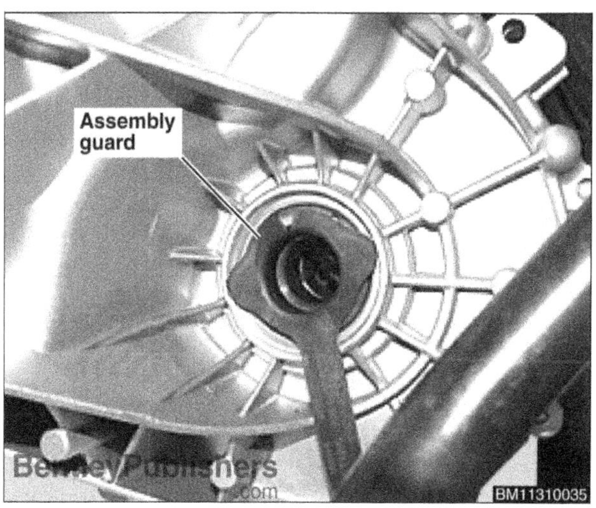

Insert assembly guard into drive axle shaft seal. Assembly guard, supplied with new shaft seal, fits over seal lip and protects sealing ring from damage during drive axle installation.

– Push drive axle shaft into differential case past resistance of snap ring until an audible snap is heard. Remove assembly guard.

– Remainder of installation is reverse of removal. Remember to:
  - Replace self-locking fasteners.
  - Fill transmission with fluid and check for leaks.
  - When installing self-locking flanged axle nut:
    -Use new nut.
    -Lightly oil contact surface of nut and wheel bearing hub. Do not allow oil to get on threads.
    -Final tightening with vehicle on ground and assistant with foot on brake pedal.
    -Stake nut to lock in place.

Tightening torques	
Ball joint, outer, to control arm (replace M14 x 1.5 self-locking nuts with new)	175 Nm (129 ft-lb)
Ball joint, outer, to steering knuckle (replace M12 x 1.5 ZNS3 self-locking nut with new)	70 Nm (ft-lb) + additional 90°
Drive axle outer CV joint to bearing hub (replace M22 x 1.5 flange nut with new)	182 Nm (134 ft-lb)
Intermediate shaft bearing housing to engine oil pan (M10 x 45 mm ZNS3 bolts)	38 Nm (28 ft-lb)
Road wheel to hub	140 ± 10 Nm (103 ± 7 ft-lb)
Tie rod end to steering knuckle (replace M12 x 1.5 self-locking nut)	65 Nm (48 ft-lb)

CV joint boot, removing and installing

> ***CAUTION—***
> - *Do not attempt to remove outer CV joint from drive axle. It is pressed on and not serviceable.*
> - *The inner (transmission side) CV joint may be flexed a maximum of 18°. The outer (wheel side) joint may be flexed a maximum of 45°.*
> - *Do not reuse old CV joint clamps.*
> - *Do not overfill CV joint boots with grease.*

Replace drive axle CV joint boots in case of grease leak or damage.

– To replace CV joint boots, first remove drive axle. See **Drive axle, removing and installing** in this repair group. Clamp axle in workshop vise with plastic protective jaws.

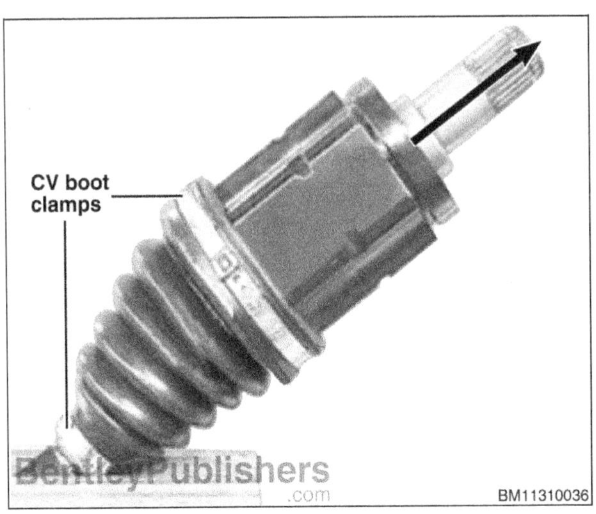

◄ Cut off inner CV boot retaining clamps and peel boot away from CV joint housing. Slide housing off (**arrow**).

– Use dry and lint-free shop cloth to clean old grease out of housing and tripod joint.

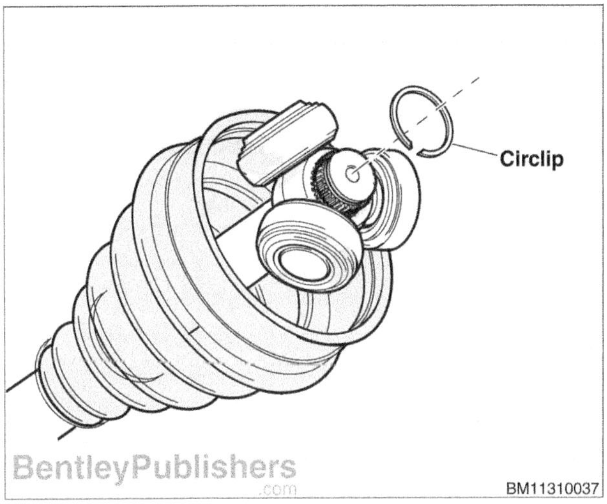

◄ Remove inner tripod locking circlip.

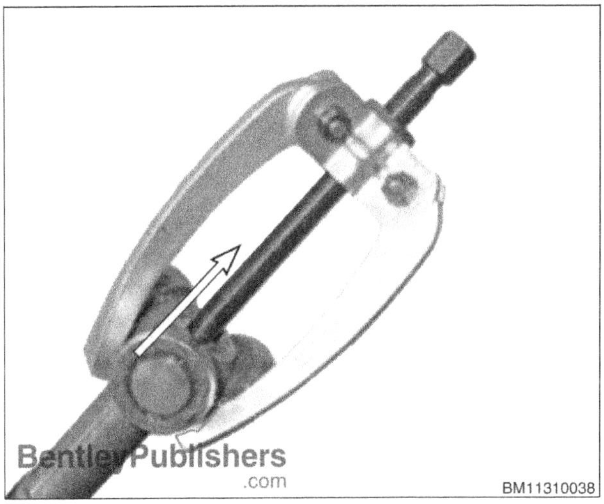

◄ Use puller to remove tripod from axle shaft.

– Inspect joint housing and inner parts carefully and discard in case of water contamination or dirt.

– Remove inner boot and discard. Clean shaft carefully.

– Manual transmission: Press drive axle vibration damper, if fitted, off shaft and discard. Do not reuse.

310

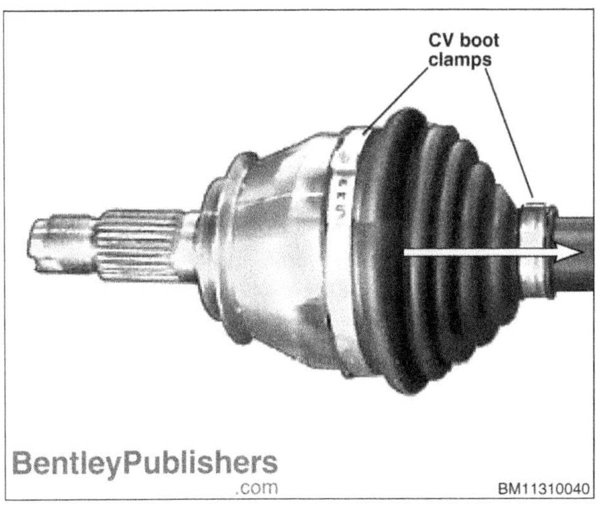

◄ Cut off outer CV boot retaining clamps and slide boot off CV joint housing (**arrow**).

– Use dry and lint-free shop cloth to clean old grease out of accessible parts of CV joint.

– Inspect CV joint for wear and contamination. Discard complete axle shaft if CV joint is unusable.

– Slide new outer boot over axle shaft and fill with 80 g of CV joint grease. Do not overfill. Only use amount of grease provided with boot kit.

– Slide boot over lip of outer CV joint, making sure sealing surfaces are clean, dry and grease-free.

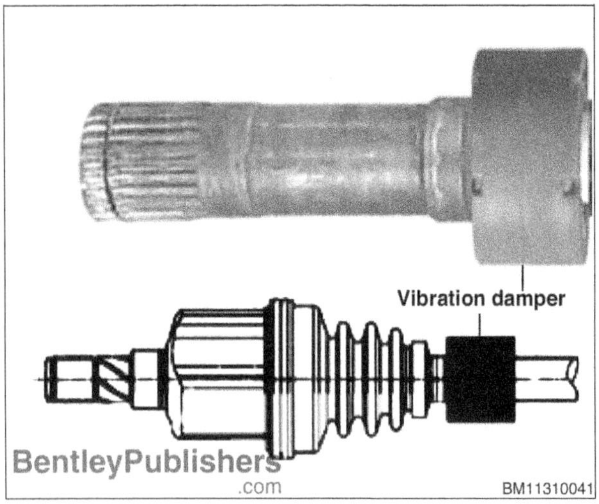

Manual transmission: Fit new drive axle vibration damper, even if vibration damper was not previously supplied.

- Coat vibration damper and axle shaft with tire mounting paste.
- Slide vibration damper by hand up to shaft shoulder for inner joint.
- Allow to air-dry for at least 4 hours before driving vehicle.

– Slide inner CV joint boot up to axle shaft shoulder.

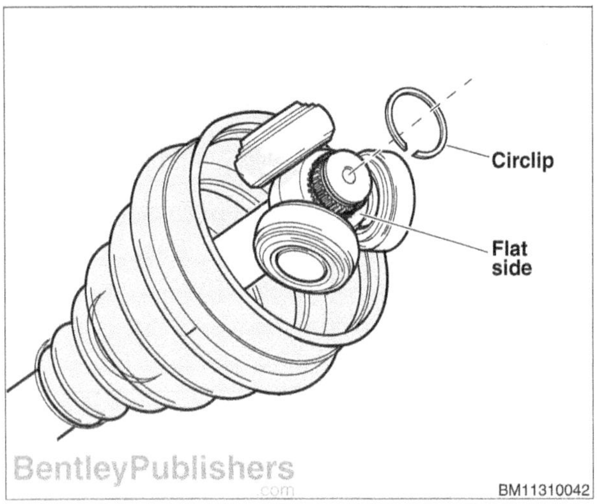

Align inner CV joint tripod so that flat side points to differential. Press tripod on axle shaft until circlip groove is exposed. Install new copper-colored circlip.

– Fill tripod joint outer housing with 130 g of CV joint grease. Do not overfill. Only use amount of grease provided with kit.

– Place outer housing over tripod and slide boot over lip of housing, making sure sealing surfaces are clean, dry and grease-free. Also, make sure boot fits housing contours correctly.

– Use new clamps to secure big ends of inner and outer CV joint boots to joint housings.

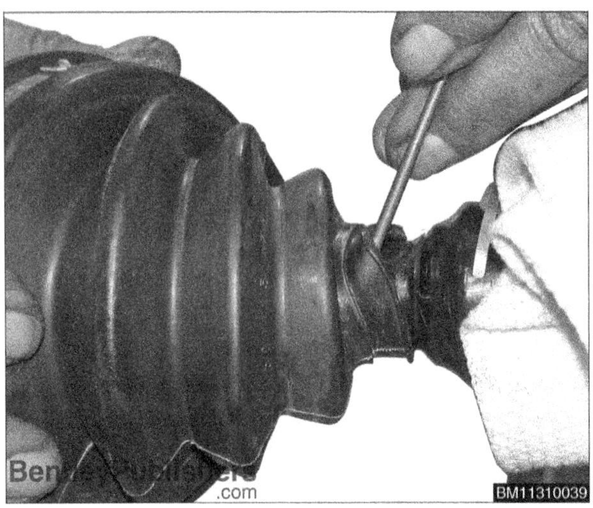

Before securing small boot clamps, be sure to burp each boot to relieve excess air pressure:

- Tilt CV joint to one side.
- Slide thin rod such as small screw driver or awl between boot small end and axle shaft.

– Use new clamps to secure small ends of CV joint boots.

– Reinstall drive axle. See **Drive axle, removing and installing** in this repair group. Note the following:

- Use correct style replacement circlip on inner joint shaft.
- Replace differential output flange seal(s) and use special assembly guard when inserting drive axle shaft in seal.
- Replace self-locking fasteners.
- Fill transmission with fluid and check for leaks.

320 Steering and Wheel Alignment

TABLES

General

This repair group covers steering wheel and column removal and steering system service, including wheel alignment information.

See also:

- **612 Switches** for steering column and ignition switch.
- **721 Airbag System (SRS)** for driver airbag removal.

Steering system

The steering system consists of electromechanical power-assisted rack-and-pinion steering (EPS) and connecting linkage to the road wheels. The steering wheel connects to an adjustable upper steering column. A shaft with two universal joints connects the upper column to the steering rack.

The steering rack requires no maintenance other than alignment and periodic inspection for worn parts. Inspect rubber rack boots and tie rod end boots for tear or damage and replace if necessary.

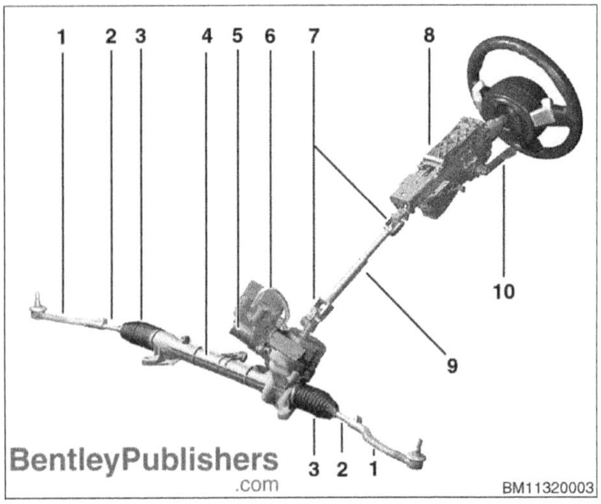

Steering system components

1. Tie rod end
2. Tie rod
3. Rack boot
4. Steering rack
5. Electric motor
6. EPS control module
7. Steering spindle universal joints
8. Steering column
9. Steering spindle
10. Steering column adjuster lever

EPS overview

Electromechanical power steering (EPS) dispenses entirely with hydraulic power assist to the steering. Vehicle steering properties are programmed into the EPS control module.

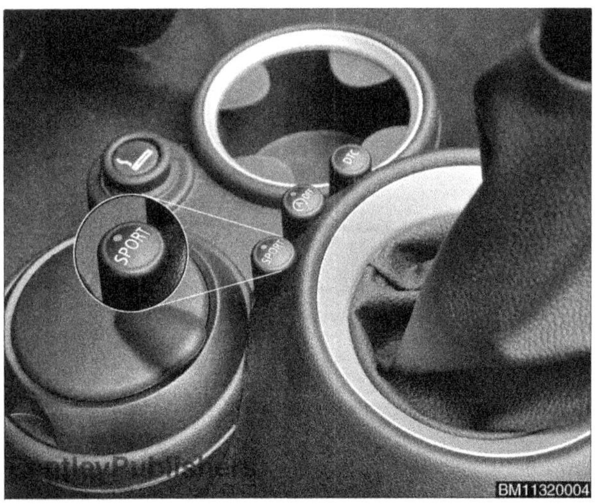

The optional driving dynamics control (SPORT) button adapts EPS module programming to allow for different driver needs:

- **Standard mode** results in low parking effort but less precise steering at high speeds.
- **Sports mode** programs for a sport-oriented driving style and provides greater steering precision and improved handling characteristics at high speeds. This is achieved by reducing steering torque assistance so that the driver exerts greater steering effort.

Omission of steering hydraulic components provides cost-saving potentials and simplifies vehicle assembly procedure. This system also helps avoid environmental risks in the form of hydraulic oil leaks. Since the electric steering motor is activated only when required (when steering but not when driving straight ahead), fuel consumption is reduced and available engine power output increased.

EPS features

Among the features and advantages of EPS are the following:

- Improved driving dynamics:
  -Precise steering provided by direct power steering.
  -Active return (centering).
  -Selectable steering characteristics (sports mode).
  -Use of lightweight sports steering wheel.
- Greater driving comfort:
  -Steering shaft isolated from axle vibration while still transmitting road surface conditions information.
  -Reduced steering judder.
  -Controlled, speed-dependent steering torque assistance (for example, when parking).

- Greater driving safety:
  -Assistance in steering the correct line by providing a lower level of power steering assistance.
  -Steering wheel movement reduced by active speed-dependent damping. This function also reduces vehicle tendency to roll in response to abrupt steering wheel movements.
- Environmental benefits:
  -Fuel saving of approx. 0.23 liter / 100 km (approx 1.1 mpg) at highway speeds.
  -No leaks as it is a dry system.
- Economic benefits:
  -Reduced assembly and inspection complexity at the production plant.
  -Reduced range of components.
  -Easier programming of power steering assistance.
  -Potential for steer-by-wire and vehicle system networking.

EPS schematic

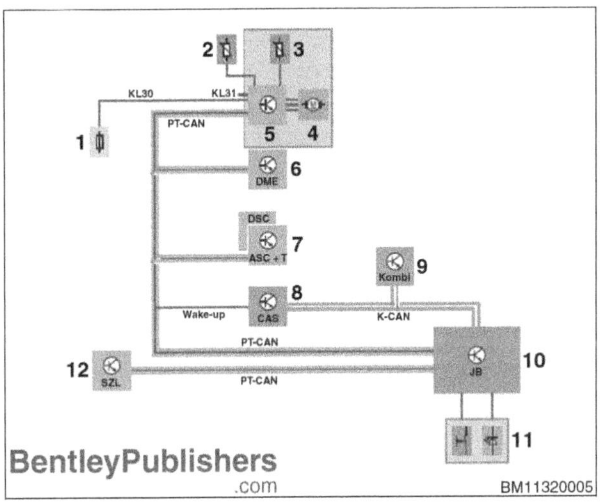

EPS, engine control module (ECM), ASC+T / DSC, steering column switch cluster, CAS and instrument cluster are networked via bus system.

1. **FL4** (80A) EPS power supply fuse in front power distribution box.
   - See **ECL Electrical Component Locations** for fuse locations and identification.
2. **Steering torque sensor** in EPS assembly, top of steering rack.
3. **Motor position sensor** in EPS assembly, top of steering rack.
4. **EPS motor**, top of steering rack.
5. **EPS control module** in EPS assembly, top of steering rack.
6. **Engine control module (ECM)**, left rear engine compartment.
7. **ASC + T / DSC control module**, left rear engine compartment.
8. **Car access system (CAS)**, underneath left side dashboard.
9. **Instrument cluster** control module.
10. **Junction box (JB)**, right kick panel behind trim.
11. **SPORT button** behind selector lever.
12. **Steering column switch cluster (SZL)**.

Warnings and Cautions

WARNING—

- *The airbag mounted in the steering wheel is an explosive device. Treat it with extreme caution. Follow the airbag removal procedure in* **721 Airbag System (SRS)**.
- *Serious injury may result if airbag system service is attempted by persons unfamiliar with the MINI airbag system and its approved service procedures. MINI specifies that all inspection and service be performed by an authorized MINI dealer.*
- *Wait 5 seconds after the battery cable is disconnected to work on airbag electrical connections. This allows the system capacitor to discharge.*
- *Prior to disconnecting the battery, read the battery disconnection cautions given in* **001 Warnings and Cautions**.
- *Store a removed airbag with soft pad facing up. If stored facing down, accidental deployment of airbag could propel it violently into the air, causing injury.*
- *Do not reuse self-locking fasteners. They are designed to be used only once and may fail if reused. Replace with new.*
- *Do not install bolts and nuts coated with undercoating wax, as correct tightening torque cannot be assured. Clean the threads with solvent before installation, or install new parts.*
- *Do not attempt to weld or straighten steering components. Replace damaged parts.*

CAUTION—

- *Be sure to have wheel alignment carried out if fasteners for any of the following components are loosened and retorqued:*
  -Steering rack.
  -Ball joint.
  -Tie rod end to tie rod.
- *Have wheel alignment performed if any of the following components are replaced:*
  -Front subframe.
  -Steering rack.
  -Steering rack boot.
  -Steering knuckle.
  -Ball joint.
  -Control arm.
  -Tie rod end.
  -Tie rod.
- *To avoid marring interior trim, use a plastic prying tool or a screwdriver with the tip wrapped with masking tape.*
- *Static discharge may permanently damage an electronic component such as control module or sensor. Handle electronic components using accepted static prevention equipment and techniques. See* **600 Electrical System–General**.

TROUBLESHOOTING

Table a lists symptoms of steering system problems, probable causes, and suggested corrective actions. Unless noted otherwise, relevant repairs are described in this repair group.

Table a. Steering system troubleshooting

Symptom	Probable cause	Repairs
Steering wheel play or vibration	Lower steering spindle universal joint pinch bolt loose or missing.	Replace pinch bolt. See **Steering column, removing and installing** in this repair group.
	Steering rack inner shaft deformed.	Replace steering rack and tie rods.
	Steering rack damaged.	Replace steering rack.
	Tie rod end worn.	Check and replace tie rod ends as needed.
	Tie rod worn.	Check and replace tie rods as needed.
Steering wheel spokes not centered	Tie rod end worn.	Check and replace tie rod ends as needed.
	Front alignment incorrect.	Check and adjust wheel alignment.
Steering power assist reduced or failed	EPS control module or motor electrical contacts corroded.	Check EPS electrical harnesses, connectors and plugs. Clean or replace corroded contacts.
	EPS electrical cable damaged.	Check EPS electrical harness insulation and repair or replace as necessary.
Interference noise when steering under load	Steering rack worm gears or bearings damaged.	Replace steering rack.

Corrosion damage

Inspect tie rod ends and steering rack boots periodically for damage or tears. Replace as necessary. See **Tie rod end, removing and installing** and **Steering rack boot, replacing** in this repair group.

Note that rodent bite damage to rack boot is difficult to detect. Bite marks are often small and only on one side. Leakage at one rack boot can result in corrosion of rack shaft metal at both ends.

If steering rack inner shaft metal shows sign of corrosion, replace complete rack. See **Steering rack, removing and installing** in this repair group.

STEERING WHEEL

Steering wheel components

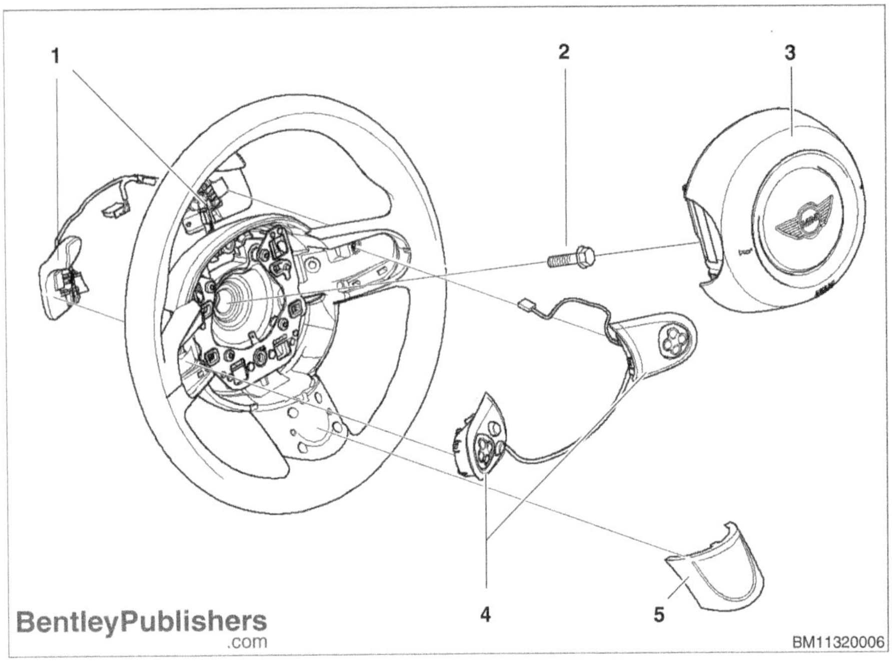

1. Paddle shifters
2. M14 x 1.5 bolt
   - Tighten to 63 Nm (46 ft-lb)
3. Horn button and driver airbag
4. Multifunction switches
5. Third spoke cover

Several functions can be performed without the driver having to remove hands from the steering wheel.

- Horn button is at the center of the wheel.
- Cruise control buttons are in the left multifunction switch cluster.
- Audio system and telephone control are in the right spoke cluster.

Some models are equipped with shifter paddles above the steering wheel horizontal spokes.

The driver airbag is in the center of the steering wheel. Before working on or around components or systems that involve any part of the airbag system, be sure to read **Warnings and Cautions** in this repair group and also in **721 Airbag System (SRS)**.

Steering wheel, removing and installing

– Center steering wheel. Make sure front wheels are pointed straight ahead. During remainder of this procedure, do not alter position of steering column or front wheels.

– Extend steering wheel as far back and as low as possible.

– Disconnect negative (-) cable from battery.

> ***CAUTION—***
> - *Prior to disconnecting the battery, read the battery disconnection cautions given in* **001 Warnings and Cautions**.

– Remove driver airbag in steering wheel. See **721 Airbag System (SRS)**.

> ***WARNING—***
> - *The steering-wheel-mounted airbag is an explosive device. Treat with extreme caution. Follow the airbag removal procedure in* **721 Airbag System (SRS)**.
> - *Airbag removal may set SRS fault codes. The SRS indicator light remains ON until problems are corrected and fault memory cleared.*

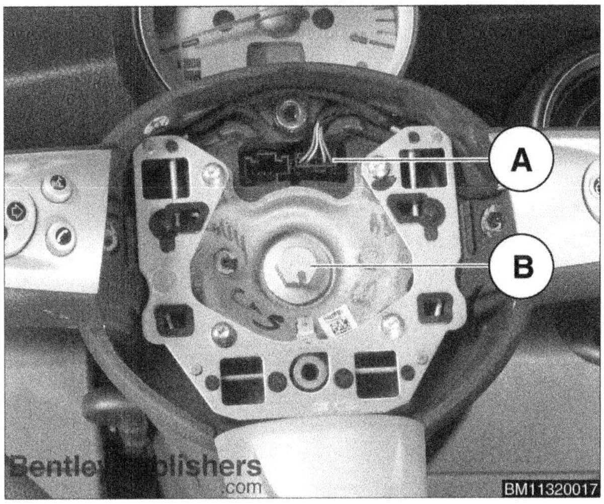

Detach electrical connector (**A**), remove steering wheel mounting bolt (**B**) and lift wheel off steering shaft.

– When installing:
- Make sure locating notch on steering wheel hub matches notch on column shaft.
- Connect harness connectors, making sure harnesses are laid down as before.

Tightening torque	
Steering wheel to steering column shaft (M14 x 1.5 bolt)	63 Nm (46 ft-lb)

– Install airbag, being careful to not pinch electrical wire harnesses and connectors.

320

Steering angle sensor

Where equipped, the steering angle sensor is integral with the steering column switch cluster (SZL). For more information, including removal and installation of the SZL, see **612 Switches**.

To reset steering angle sensor, connect vehicle to MINI scan tool and carry out steering angle sensor adjustment system function.

STEERING COLUMN

Steering column components

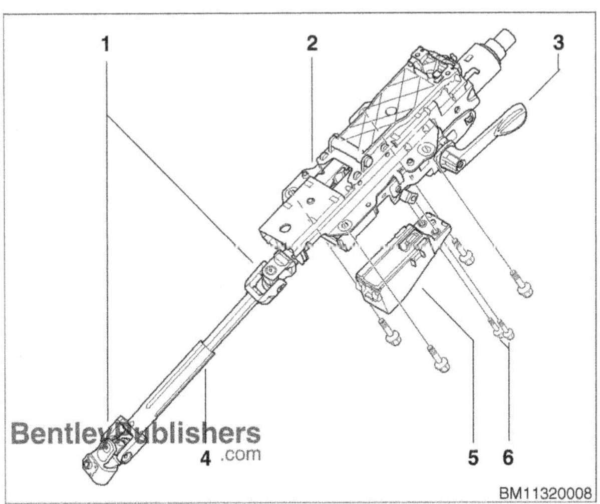

1. Upper and lower steering spindle universal joints
2. Adjustable steering column assembly
3. Adjuster lever
4. Telescoping steering shaft
5. Electronic steering lock
6. Shear bolts
   - M6 x 18 mm

Steering column, removing and installing

– Center steering wheel. Make sure front wheels are pointed straight ahead.

– Extend steering wheel as far back and as low as possible.

– Disconnect negative (-) cable from battery.

> ***CAUTION—***
> • *Prior to disconnecting the battery, read the battery disconnection cautions given in* **001 Warnings and Cautions**.

– Remove steering wheel. See **Steering wheel, removing and installing** in this repair group.

> ***WARNING—***
> • *The steering wheel mounted airbag is an explosive device. Treat with extreme caution. Follow the airbag removal procedure in* **721 Airbag System (SRS)**.
> • *Airbag removal may set SRS fault codes. The SRS indicator light remains ON until problems are corrected and fault memory cleared.*

– Remove upper and lower steering column trim covers. See **513 Interior Trim**.

– Remove instrument cluster (tachometer) above steering column. See **620 Instruments**.

◄ Working underneath left side dashboard, remove steering universal joint pinch bolt (**arrow**) and detach upper from lower steering shaft.

> ***CAUTION—***
> • *While steering spindle is detached from steering rack, make sure steering wheel is not rotated. Rotating the wheel may damage the airbag and horn contact spring.*

◄ Working underneath steering column, detach steering column switch cluster (SZL) electrical connectors (**arrows**).

◄ Working underneath steering column:

- If applicable, detach electrical steering lock connector.
- Remove column mounting bolts (**arrows**) and tilt column backward to remove.

– When installing:

- Replace column mounting bolts with new.
- Replace steering universal pinch bolt with new.

Tightening torques	
Steering column to dashboard cross member (replace self-threading bolts with new)	35 Nm (26 ft-lb)
Universal joint pinch bolt to steering shaft (replace M8 x 1.25 x 41 mm bolt with new)	28 Nm (21 ft-lb)

Steering lock, removing and installing

– Remove lower steering column trim cover. See **513 Interior Trim**.

◄ Working underneath steering column:

- If applicable, detach electrical steering lock connector.
- Drill out steering lock shear bolts (**arrows**). Alternately, use center punch and hammer to spin bolts out.
- Lower steering lock.

– Installation is reverse of removal. Tighten mounting bolts until heads shear off.

Steering System Repairs

Steering system components

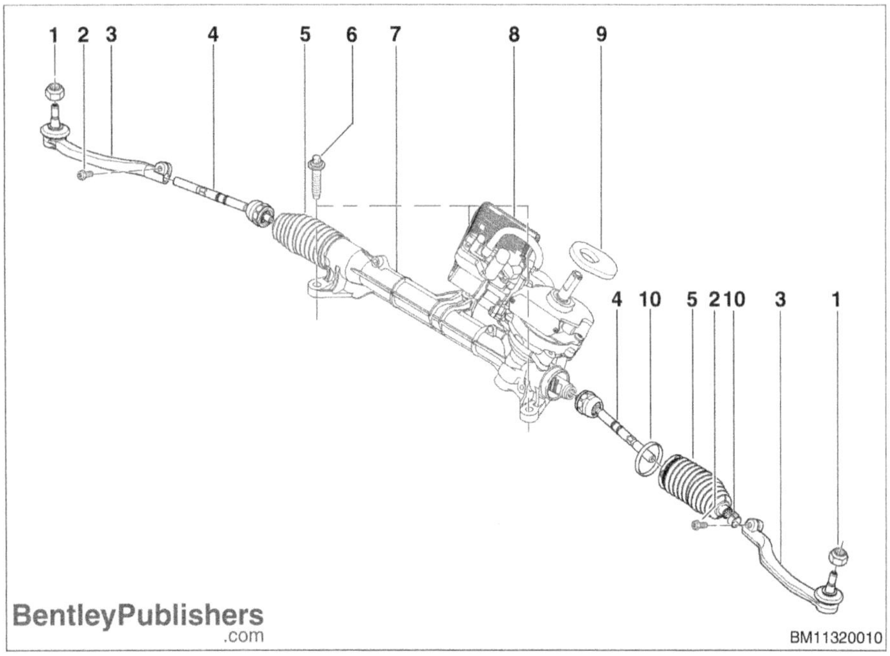

1. Tie rod end nut
   - Replace M12 x 1.5 self-locking nut with new
   - Tighten to 65 Nm (48 ft-lb)
2. Tie rod end to tie rod pinch bolt
   - Tighten to 28 Nm (21 ft-lb)
3. Tie rod end
4. Tie rod
   - Tighten to 77 Nm (57 ft-lb)
5. Rack boot
6. Steering rack mounting bolt
   - M12 x 1.5 x 60 mm 10.9 grade ZNS3
   - Tighten to 100 Nm (74 ft-lb)
7. Steering rack
8. EPS module
9. Bulkhead gasket (foam rubber seal)
10. Rack boot clamps

Steering rack, removing and installing

The steering rack, electomechanical power steering (EPS) module and EPS motor are bolted as a unit to the front subframe. Lower subframe for access to these components.

– Disconnect negative (-) battery cable and cover battery terminal to keep cable from accidentally contacting terminal.

> *CAUTION—*
> - *Prior to disconnecting the battery, read the battery disconnection cautions given in* **001 Warnings and Cautions**.

– Raise front of vehicle and support safely. Remove front wheels.

> *WARNING—*
> - *Make sure the vehicle is stable and well supported at all times. Use a professional automotive lift or jack stands designed for the purpose. A floor jack is not adequate support.*

– Remove tie rod ends. See **Tie rod end, removing and installing** in this repair group.

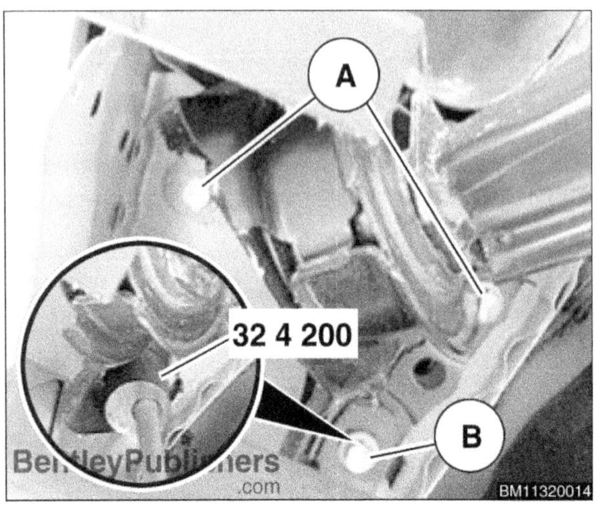

◄ Lower front subframe approx. 120 mm (5 in). See **310 Front Suspension, Drive Axles**.

> *CAUTION—*
> - *When lowering subframe, make sure steering rack stub shaft (***arrow***) does not collide with bulkhead.*

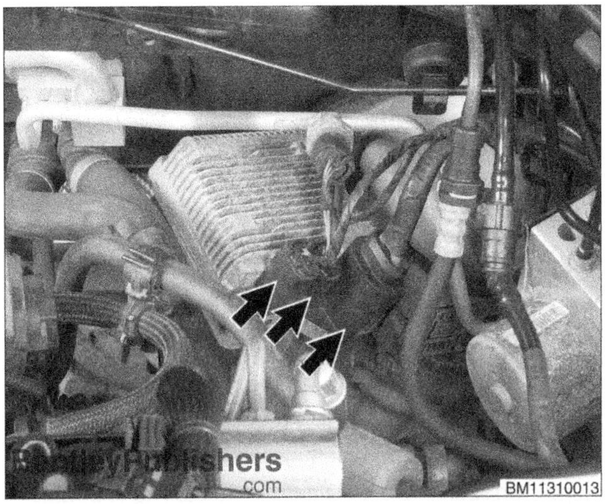

Detach EPS module electrical connectors (**arrows**).

> ***CAUTION—***
> - *Check EPS electrical connectors and harness plugs for corrosion and clean if necessary.*
> - *Replace steering rack if advanced corrosion is found on plugs.*
> - *Replace wiring harness if corrosion has advanced into the harness insulation.*

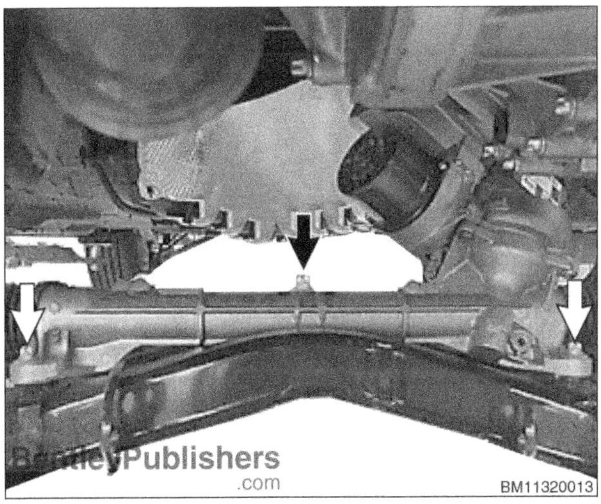

Remove steering rack mounting bolts (**arrows**).

- Carefully swing left end of rack forward, then remove entire rack by sliding forward. Do not allow rack stub shaft to collide with other components.
- If replacing steering rack with new, transfer components from old rack before reassembly.

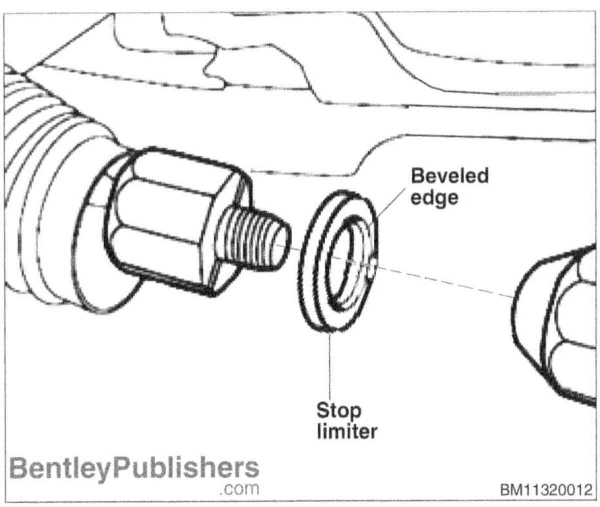

Vehicle with 18" tires: Reinstall tie rods using stop limiters. Make sure beveled edge of limiter butts up to steering rack inner shaft.

Tightening torque	
Tie rod to steering rack	77 Nm (57 ft-lb)

320

Be sure to replace steering rack foam rubber seal (**arrow**) before reassembly.

– Carefully fit rack to subframe and install bolts.

Tightening torque	
Steering rack to subframe (M12 x 1.5 x 60 mm 10.9 grade ZNS3 bolts)	100 Nm (74 ft-lb)

– Check subframe mounting bolt hole threads. Repair with Helicoil if necessary.

– When reinstalling:
- Raise subframe into place, watching carefully to make sure no electrical lines, hoses or components become snagged.
- Attach engine stabilizer link loosely before reinstalling subframe mounting bolts.
- Use correct length subframe bolts in different locations:
  -Front bolts = M12 x 85 mm
  -Control arm bushing carrier bolts = M10 x 60 mm
  -Rear bolts = M12 x 62 mm
- Replace self-locking fasteners.

Tightening torques	
Ball joint, inner, to subframe (replace M14 x 1.5 ZNS3 self-locking nut with new)	70 Nm (ft-lb) + additional 90°
Ball joint, outer, to steering knuckle (replace M12 x 1.5 ZNS3 self-locking nut with new)	70 Nm (ft-lb) + additional 90°
Bumper bracket to front subframe (M8 x 18 mm bolts)	19 Nm (14 ft-lb)
Control arm bracket and stabilizer anchor to subframe (M14 x 1.5 x 105 mm self-tapping bolts)	165 Nm (122 ft-lb)
Engine stabilizer link to bracket (M12 x 90 mm bolts)	108 Nm (80 ft-lb)
Exhaust pipe to catalytic converter (V-band clamp)	25 ± 3 Nm (18 ± 2 ft-lb)
Front subframe to chassis: • Front bolts M12 x 85 mm • Rear bolts M12 x 62 mm • Control arm bushing carrier bolts (replace M10 x 60 mm 10.9 grade bolts with new)	 100 Nm (74 ft-lb) 100 Nm (74 ft-lb) 59 Nm (44 ft-lb) + additional 90°
Road wheel to hub	140 ± 10 Nm (103 ± 7 ft-lb)
Stabilizer link to stabilizer bar (replace M10 self-locking nuts)	56 Nm (41 ft-lb)
Tie rod end to steering knuckle (replace M12 x 1.5 self-locking nut)	65 Nm (48 ft-lb)
Universal joint pinch bolt to steering shaft (replace M8 x 1.25 x 41 mm bolt with new)	28 Nm (21 ft-lb)

– Have vehicle professionally aligned.

– Vehicle with DSC: Use MINI scan tool to adjust steering angle sensor.

– If new rack with ESP control module is installed: Use MINI scan tool for programming and coding of new module.

Tie rod end, removing and installing

– Raise front of vehicle and support safely. Remove front wheel.

> ***WARNING—***
> - *Make sure the vehicle is stable and well supported at all times. Use a professional automotive lift or jack stands designed for the purpose. A floor jack is not adequate support.*

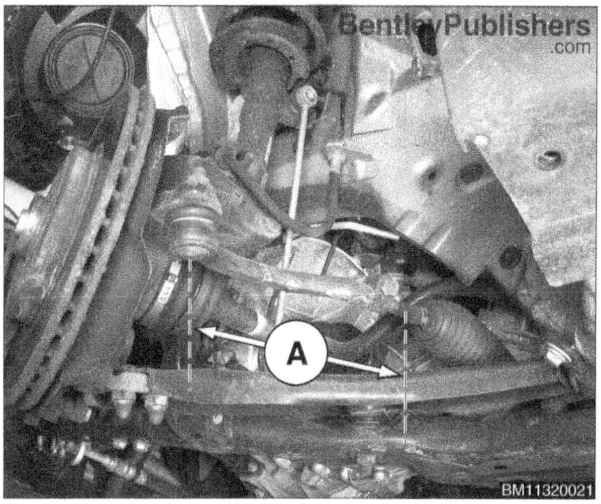

< Make reference measurement (**A**) of outer tie rod end to tie rod. Record measurement.

- Accurate measurement of tie rod end with reference to tie rod helps approximate correct wheel alignment when new parts are installed.

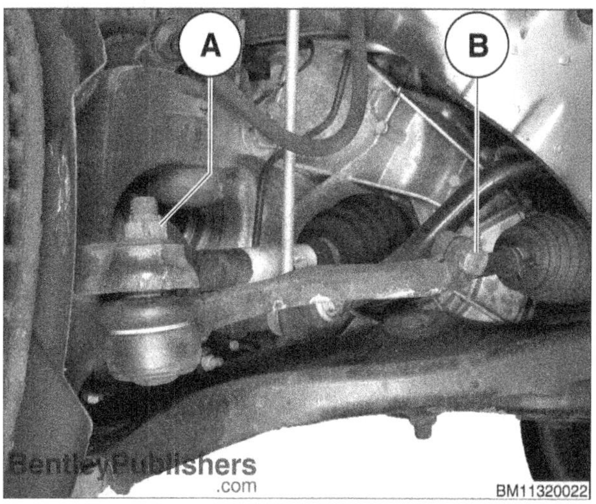

< Working at outer end of tie rod:

- Remove tie rod end nut (**A**) at steering knuckle. If necessary, counterhold tie rod ball joint shaft with Torx tool.
- Loosen tie rod end pinch bolt (**B**).

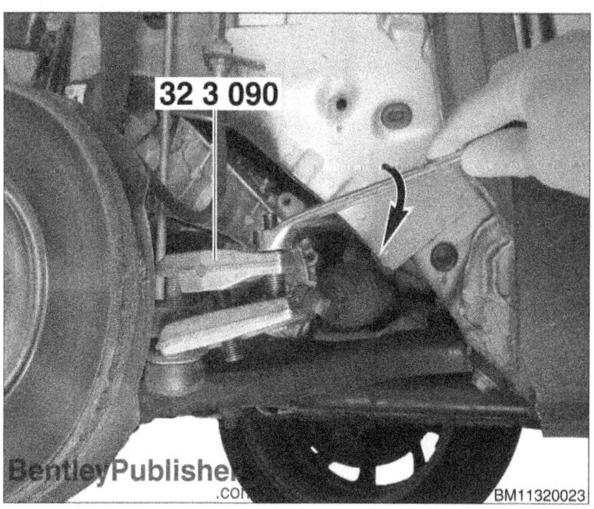

< Use MINI special tool 32 3 090 or equivalent to press tie rod end out of steering knuckle.

– Unscrew tie rod end from tie rod.

– Check steering rack boot carefully for damage. Replace if necessary. See **Steering rack boot, replacing** in this repair group.

- Installation is reverse of removal.
  - Make sure threaded parts are clean.
  - Use anti-seize paste on inner tie rod threads.
  - Use new self-locking nuts where applicable.
  - Use previously made reference marks to set toe temporarily.

Tightening torques	
Tie rod end to tie rod (M8 pinch bolt)	28 Nm (21 ft-lb)
Tie rod end to steering knuckle (replace M12 x 1.5 self-locking nut with new)	65 Nm (48 ft-lb)

- Have vehicle professionally aligned.

- Vehicle with DSC: Use MINI scan tool to adjust steering angle sensor.

Tie rod, removing and installing

- Raise front of vehicle and support safely. Remove front wheel.

> ***WARNING—***
> - *Make sure the vehicle is stable and well supported at all times. Use a professional automotive lift or jack stands designed for the purpose. A floor jack is not adequate support.*

- Make length reference mark on tie rod, then remove outer tie rod end. See **Tie rod end, removing and installing** in this repair group.

- Cut rack boot clamps and slide boot off. Check boot carefully for damage. Replace if necessary. See **Steering rack boot, replacing** in this repair group.

BM11320011

◄ Use MINI special tool **32 3 210** or equivalent deep 38 mm socket to loosen and remove inner tie rod shaft from steering rack.

> ***CAUTION—***
> - *To avoid damage to steering rack while removing tie rod, turn steering until end of rack is as far as possible inside rack housing.*

- Inspect end of rack inner shaft. If polished surface is corroded or otherwise damaged, replace complete steering rack. See **Corrosion damage** in this repair group.

- Lightly grease toothed section of rack inner shaft.

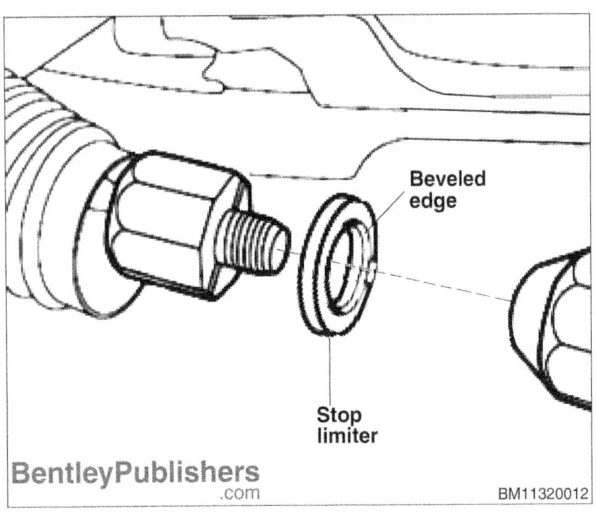

BM11320012

◄ Vehicle with 18" tires: Reinstall tie rod using stop limiter. Make sure beveled edge of limiter butts up to steering rack inner shaft.

Tightening torque	
Tie rod to steering rack	77 Nm (57 ft-lb)

- Grease tie rod taper so that small end of rack boot slides on tie rod when tie rod is tightened, preventing rack boot from twisting.

– Slide new boot into position and replace band clamps. Reinstall tie rod end using previously made reference marks.

Tightening torques	
Tie rod end to tie rod (M8 pinch bolt)	28 Nm (21 ft-lb)
Tie rod end to steering knuckle (replace M12 x 1.5 self-locking nut with new)	65 Nm (48 ft-lb)

– Have vehicle professionally aligned.

– Vehicle with DSC: Use MINI scan tool to adjust steering angle sensor.

Steering rack boot, replacing

Replace rack boot if ripped or damaged.

– Raise front of vehicle and support safely. Remove front wheel.

> ***WARNING —***
> *• Make sure the vehicle is stable and well supported at all times. Use a professional automotive lift or jack stands designed for the purpose. A floor jack is not adequate support.*

– Remove tie rod end. See **Tie rod end, removing and installing** in this repair group.

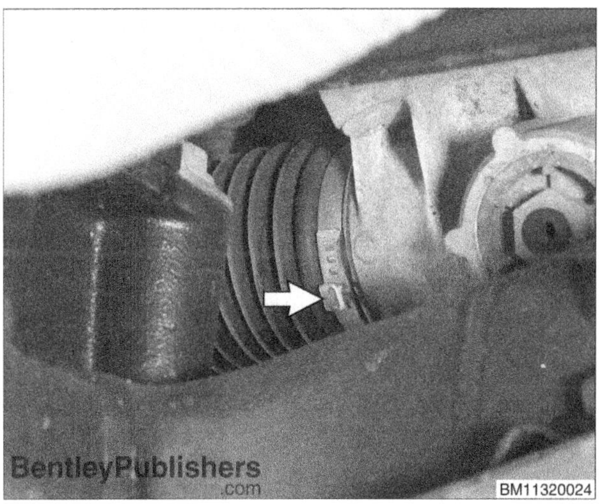

◄ Cut off inner rack boot clamp (**arrow**).

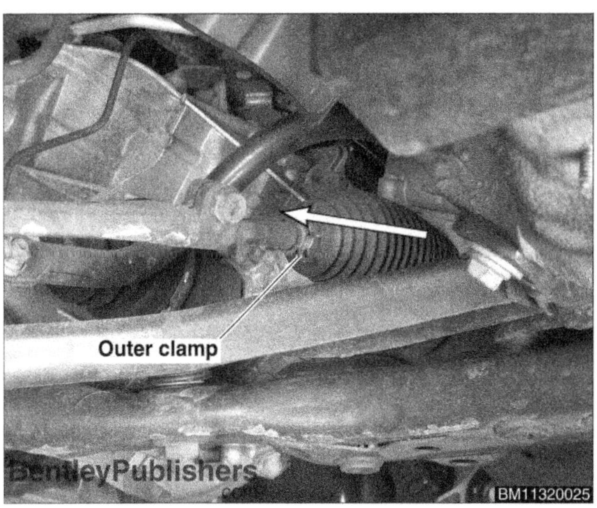

◄ Loosen or cut outer rack boot clamp and slide boot off rack (**arrow**) and tie rod.

– Once boot is removed, inspect end of rack inner shaft. If polished surface is corroded or otherwise damaged, replace complete steering rack. See **Corrosion damage** in this repair group.

– Lightly grease toothed section of rack inner shaft.

– Grease tie rod taper so that small end of rack boot slides on tie rod when tie rod is tightened, preventing rack boot from twisting.

– Slide new boot into position and replace band clamps. Reinstall tie rod end using previously made reference marks.

Tightening torques	
Tie rod end to tie rod (M8 pinch bolt)	28 Nm (21 ft-lb)
Tie rod end to steering knuckle (replace M12 x 1.5 self-locking nut with new)	65 Nm (48 ft-lb)

– Have vehicle professionally aligned.

– Vehicle with DSC: Use MINI scan tool to adjust steering angle sensor.

Wheel Alignment

> *CAUTION—*
> - *If vehicle suspension geometry is affected by an accident, repair the damage by repairing or replacing components. Do not attempt to compensate for accident damage using wheel alignment.*

Proper handling, stability and driving ease depend upon the correct alignment of all four wheels. The front axle is aligned in relation to the rear axle, then the front wheels are aligned in relation to one another. This is known as a four-wheel or thrust-axis alignment.

MINI vehicles use sophisticated suspension geometry at the front and rear of the car. Proper alignment requires computerized alignment equipment.

For front and rear alignment specifications, see **Alignment specifications** in this repair group.

Preparing for alignment

The following conditions are necessary before wheel alignment:

- Correct wheels and identical tires in good condition and inflated correctly. Tire tread depth for each axle may differ left to right by max. 1 - 2 mm (0.04 - 0.08 in).
- Steering and suspension parts and bushings undamaged and showing no signs of abnormal wear.
- Wheel bearings in good condition.
- Ride height in accordance with specifications. See **Ride height, measuring** in this repair group.
- Vehicle in normal loaded position. See **Normal loaded position** in this repair group.

Normal loaded position

MINI defines vehicle normal loaded position as follows.

Normal loaded position	
Each front seat, seats in central position	68 Kg (150 lb)
Center of rear seat	68 Kg (150 lb)
Cargo compartment	14 Kg (37.5 lb)
Fuel tank	Full

Ride height, measuring

– Preconditions for measuring ride height:
  - Correct wheel bearing play. See **310 Front Suspension, Drive Axles** and **330 Rear Suspension.**
  - Car in normal loaded position on the ground. See **Normal loaded position** in this repair group.

< Measure front ride height (**A**) from lower edge of front wheel arch to bottom center of lower wheel rim flange.

< Measure rear ride height (**A**) from lower edge of front wheel arch to bottom center of lower wheel rim flange.

– Determine a mean value for each wheel after rocking vehicle body up and down and derive a mean value for each axle.

– If ride height is outside specification listed in **Table b**, install new springs. Suspension spring removal and installation is covered in **310 Front Suspension, Drive Axles** and **330 Rear Suspension**.

320

Table b. Ride height

	Front ride height in mm (in)		**Rear ride height in mm (in)**				
Tire size	**All models except low-slung suspension**	**Low-slung suspnesion**	**Normally aspirated model**	**Turbocharged model**	**Turbocharged Sport except low-slung suspension**	**JCW except low-slung suspension**	**JCW Sport with low-slung suspension**
15 in	560 (22.28)	-	544 (21.42)	542 (21.34)	542 (21.34)	-	-
16 in	573 (22.56)	-	556 (21.89)	554 (21.81)	554 (21.81)	-	-
17 in	588 (23.15)	578 (22.76)	572 (22.52)	570 (22.44)	570 (22.44)	572 (22.52)	561 (22.09)
18 in	601 (23.66)	591 (23.27 in)	585 (23.03)	582 (22.91)	585 (23.03)	584 (22.99)	574 (22.60)
Maximum variation between sides: 10 mm (0.04 in)							
Maximum deviation from specifications: 2 mm (0.008 in)							

Front toe and camber, adjusting

Camber and toe influence each other. Adjust front toe first in order to simplify the adjustment procedure.

– Raise vehicle and support safely.

> ***WARNING—***
> - *Make sure the vehicle is stable and well supported at all times. Use a professional automotive lift or jack stands designed for the purpose. A floor jack is not adequate support.*

– Move steering wheel into straight-ahead position and secure with steering wheel stop.

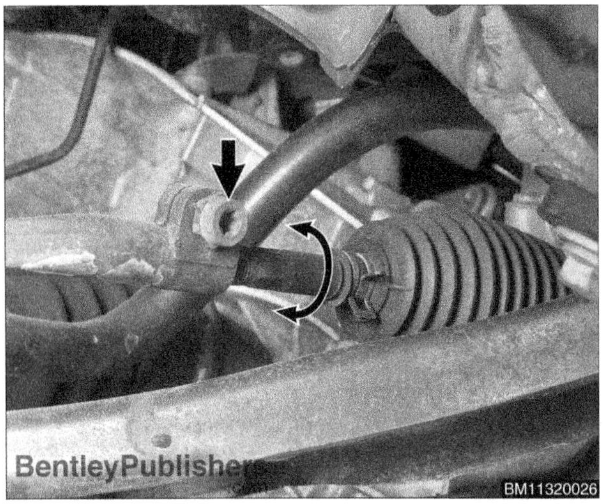

◄ Adjusting front toe:

- Loosen tie-rod end pinch bolt (**arrow**).
- Rotate (**curved arrow**) tie rod using open-end wrench on hexagon to shorten or lengthen toe. Adjust both tie rods equally to keep steering wheel centered.
- Make sure rack boot does not twist with tie rod and become damaged.
- Tighten tie rod pinch bolt.

Tightening torque	
Tie rod end to tie rod (M8 pinch bolt)	28 Nm (21 ft-lb)

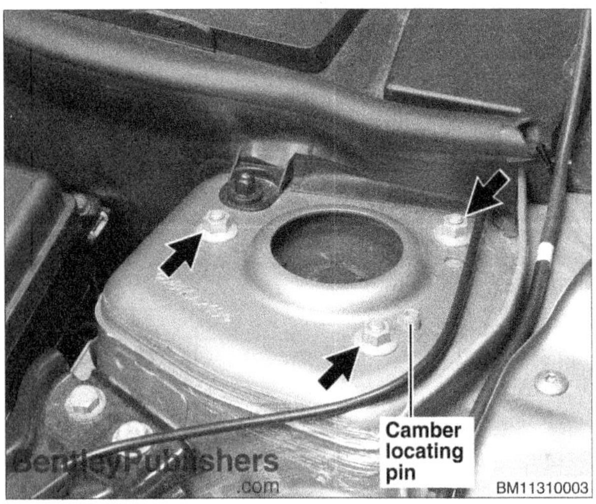

◄ Adjusting front camber:

- If camber is outside specified tolerances after toe is adjusted, use chisel to drive out camber locating pin.
- Loosen strut upper bearing mounting nuts (**arrows**).
- Adjust camber by moving upper strut bearing in inboard or outboard direction.
- Replace mounting nuts with new before retorquing.

Tightening torque	
Strut upper bearing to strut tower (replace M8 collar nuts with new)	34 Nm (25 ft-lb)

– Recheck toe.

Rear camber and toe, adjusting

Camber and toe influence each other. Adjust rear camber first in order to simplify the adjustment procedure.

– Raise vehicle and support safely.

> ***WARNING—***
> - *Make sure the vehicle is stable and well supported at all times. Use a professional automotive lift or jack stands designed for the purpose. A floor jack is not adequate support.*

Adjusting rear camber:

- Loosen camber adjuster lock bolt (**arrow**).
- Rotate eccentric disc (**curved arrow**) to adjust camber to maximum positive.
- Tighten bolt (**arrow**) finger tight.
- Lower vehicle to ground and bounce suspension a few times.
- Loosen bolt (arrow) and adjust camber by turning eccentric disk (2) to setpoint value.
- Torque lock bolt (**arrow**).

Tightening torque	
Rear camber adjuster (tighten M12 bolt in normal loaded position with vehicle on the ground)	100 Nm (74 ft-lb)

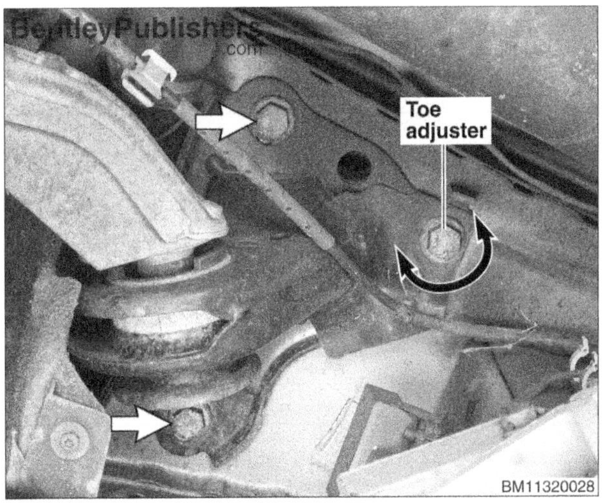

Adjusting rear toe:

- Loosen trailing arm bracket bolts (**arrows**) by approx. 1.5 turns.
- Place special tool 32 4 200 (eccentric socket wrench) on toe adjusting bolt and rotate (**curved arrow**) to correct rear toe.
- Torque trailing arm bracket bolts.

Tightening torque	
Trailing arm bracket to body (M12 bolts)	100 Nm (74 ft-lb)

– Recheck camber.

Alignment specifications

Table c. Alignment specifications, R55, R56		
Parameter	**Specifications**	
	Standard & Sport suspension	**Sport low-slung suspension**
Front axle		
Total toe	0° 12' ± 12'	0° 12' ± 12'
Adjustment total toe*	0° 12' ± 4'	0° 12' ± 4'
Toe difference between right and left**	max 12'	max 12'
Camber (difference between left / right, maximum 30')	-30' ± 30'	-35' ± 30'
Adjustment camber*	-30' ± 25'	-35' ± 25'
Toe angle difference with 20° lock on inside wheel (difference between left / right, maximum 30')	1° 49' ± 30'	1° 49' ± 30'
Front caster (difference between left / right maximum 30')		
Front wheel displacement	0° 15'	0° 15'
Maximum wheel lock		
Inside wheel (approx.)	36° 23'	36° 02'
Outside wheel (approx.)	30° 32'	30° 31'
Rear axle		
Total toe	0° 24' ± 12'	0° 24' ± 12'
Adjustment total toe*	0° 24' ± 4'	0° 24' ± 4'
Camber (difference between left / right maximum 30')	-1° 45' ± 25'	-2° 00' ± 25'
Adjustment camber*	-1° 45' ± 5'	-2° 00' ± 5'
Geometrical axis deviation	0° ± 12'	0° ± 12'
* To minimize inaccuracy, use a narrower tolerance ** Front axle toe difference = criterion for steering wheel inclination		

330 Rear Suspension

General

This repair group covers repair and replacement of rear suspension components.

See also:

- **300 Suspension, Steering and Brakes–General** for general suspension information.
- **320 Steering and Wheel Alignment** for wheel alignment specifications.

Rear suspension overview

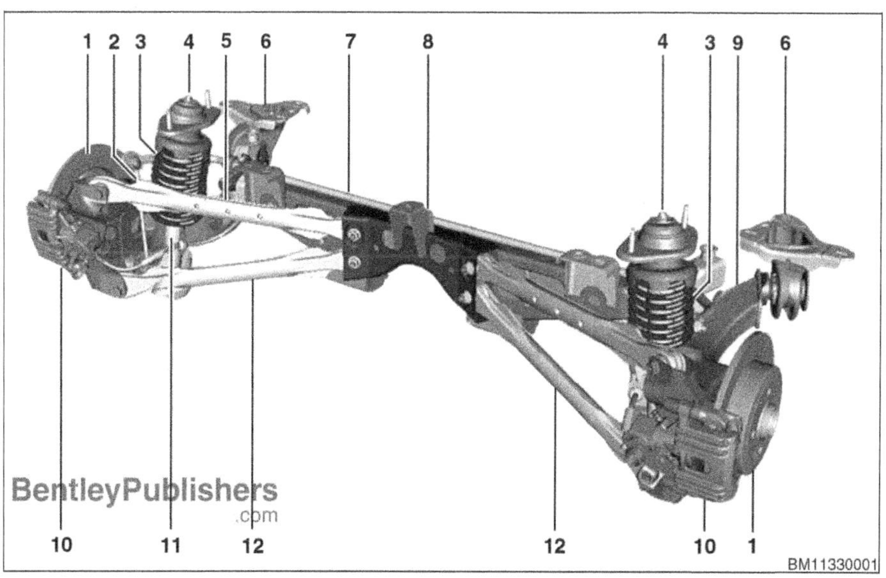

1. Brake disk
2. Stabilizer bar link
3. Coil spring
4. Strut assembly upper bearing
5. Upper control arm
6. Trailing arm mounting bracket
7. stabilizer bar
8. Rear subframe
9. Trailing arm
10. Brake caliper
11. Suspension strut (shock absorber)
12. Lower control arm

Rear suspension design

Weight-optimized trailing arms and control arms and a rigid subframe bolted to the body are the primary rear suspension components. The control arms are made of cast steel and the trailing arms are aluminum. Control arms, wheel bearing carriers, brake calipers and strut assemblies are mounted to the trailing arms.

The rear subframe is bolted to the car body and provides the inner mounting point for the four control arms and the stabilizer bar.

Coil springs and shock absorbers are combined into MacPherson strut assemblies.

A stabilizer bar is mounted to the subframe and attached at each end to the trailing arms.

Wheel bearing are incorporated in bolt-on wheel bearing carriers.

Rear axle geometry, which has a decisive influence on vehicle handling characteristics, is formed by the layout of the control arms and their mounting points on the subframe and trailing arms as well as the layout of the trailing arms and their mounting points on the car body.

Rear toe may be adjusted at the trailing arm mount on the car body. Rear camber is also adjustable using an eccentric control arm mounting bolt.

Warnings and Cautions

WARNING—
- *Do not reuse self-locking fasteners. They are designed to be used only once and may fail if reused. Replace with new.*
- *Do not attempt to disassemble the strut assembly without a spring compressor designed specifically for the job.*
- *Subframe removal may upset the balance of the vehicle on a lift.*

CAUTION—
- *Be sure to have wheel alignment carried out if fasteners for any of the following components are loosened and retorqued:*
  -Trailing arm bracket to body.
  -Upper control arm to subframe.
  -Lower control arm to subframe or trailing arm.
- *Have a wheel alignment performed if any of the following components are replaced:*
  -Rear subframe.
  -Upper control arm,
  -Lower control arm
  -Trailing arm or trailing arm bracket.

Shock Absorbers and Springs

The rear suspension shock absorbers are MacPherson struts. The strut is a major component of the suspension and supports the coil spring. Most strut assembly components are available as replacement parts.

Usually, both struts on one axle are replaced as a pair. However, if the shock absorbers have less than 50,000 km (30,000 miles), the manufacturer recommends replacing only the shock absorber that is worn or defective.

If a coil spring is broken due to corrosion, replace both springs as well as the spring pads at top and bottom. Be sure to select correct coil springs, which are allocated based on vehicle VIN and optional equipment.

Rear strut, upper strut mount or coil spring replacement is a two-step procedure:

- Removal of strut assembly from vehicle.
- Disassembly and replacement of components on work bench.

Rear strut assembly, removing and installing

– Raise rear of car and support in a safe manner. Remove rear wheel.

> ***WARNING—***
> - *Make sure the car is stable and well supported at all times. Use a professional automotive lift or jack stands designed for the purpose. A floor jack alone is not adequate support.*

◄ Pry wheel speed sensor lead and brake hose protective grommets off strut bracket (**arrows**).

◄ Support trailing arm from below using an adjustable jack stand. Remove strut lower mounting bolt (**arrow**).

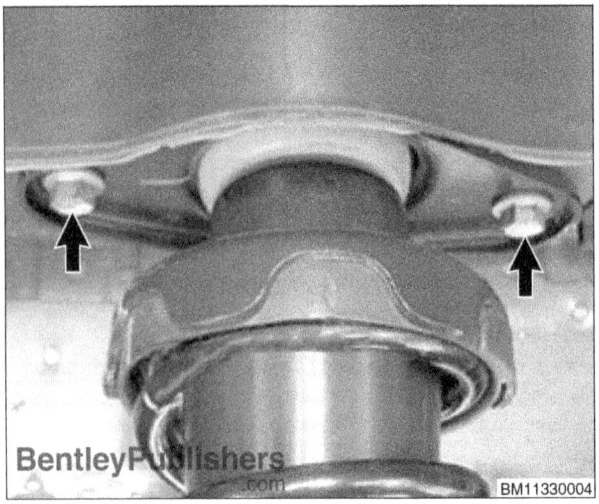

- While supporting strut assembly, remove strut upper mounting bolts (**arrows**).

- Lower strut assembly from car and place on work bench to disassemble. See **Rear strut, disassembling and assembling** in this repair group.

- Installation is reverse of removal. Keep in mind the following:
  - Remove metal chips from threads of trailing arm caused when fitting self-tapping lower shock absorber bolt for the first time.
  - After upper strut mount is attached, rotate strut to align lower bolt.
  - Make sure rubber grommets are correctly seated at strut bracket.
  - Replace self-locking bolts at strut upper mount.

Tightening torques	
Rear strut lower mount to trailing arm (M14 x 105 mm x 1.5)	165 Nm (122 ft-lb)
Rear strut upper mount to body (replace self-locking M10 bolts with new)	56 Nm (41 ft-lb)

Rear strut assembly components

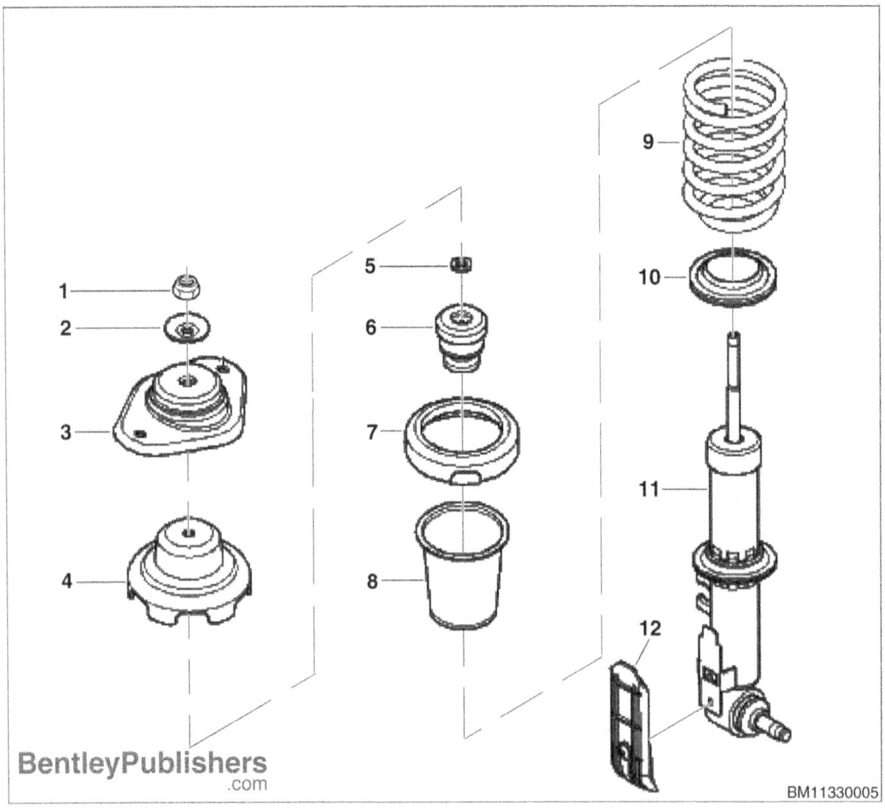

1. Center strut retaining nut
   - M10 x 1 self-locking
   - Tighten to 30 Nm (22 ft-lb)
2. Cupped washer
3. Upper mount
4. Upper spring plate
5. Spacer ring
6. Rubber bump stop
7. Upper spring pad
8. Dust boot
9. Coil spring
10. Lower spring pad
11. Strut
12. Chip guard

Rear strut, disassembling and assembling

Replacing the strut, upper strut mount or coil spring requires that the strut assembly first be removed from the car. See **Rear strut assembly, removing and installing** in this repair group.

Disassembly and component replacement is covered in this procedure. For a guide to parts used during component replacement, see **Rear strut assembly components** in this repair group.

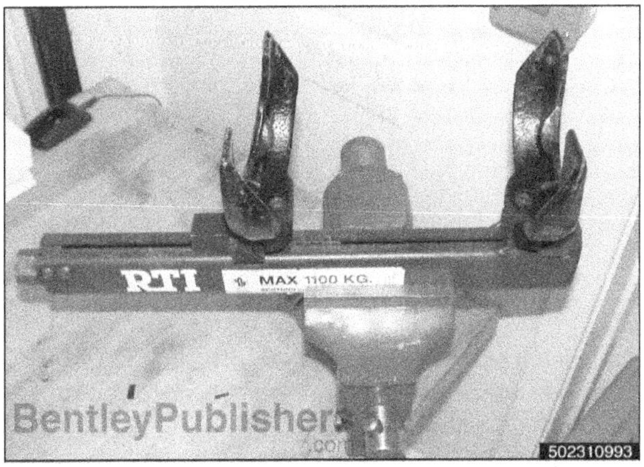

502310993

< Clamp spring compressor (MINI special tool 31 3 340 and 31 3 350 or equivalent) in shop vise.

> ***WARNING—***
> - *Do not attempt to disassemble the strut assembly without a spring compressor designed specifically for this job.*
> - *Prior to each use, check spring compressor for functionality. Do not use a damaged tool. Do not make any modifications to tool.*
> - *Use the correct size of spring retainers when compressing coil spring.*
> - *When assembling MINI spring compressor, make sure spring retainer plates are felt and heard snapping into place. Check seating of spring retainers carefully.*

min330011

< Position coil spring between spring holders so that coils lie in spring holder grooves (**arrows**). Compress spring.

> ***WARNING—***
> - *When tensioned, make sure spring coils rest completely in the spring holder recesses.*
> - *Do not tighten or loosen spring compressor with an impact tool.*
> - *Tighten down the coil springs until stress on the thrust bearing is relieved.*
> - *Loosen strut center nut only if spring coils are completely inserted in the spring holder grooves. If necessary, loosen compressor, reposition and recompress.*

> ***CAUTION—***
> - *Do not damage strut dust boot.*

min330012

< Remove center strut retaining nut. Counterhold strut shaft using 6 mm Allen wrench.

– Remove upper strut mount and related components.

– If new spring is being installed, relieve tension on spring compressor and remove coil spring.

Replace springs in pairs. Also replace spring pads at top and bottom. Be sure to select correct coil springs, which are allocated based on vehicle VIN and optional equipment.

330

Rear strut, disassembling and assembling

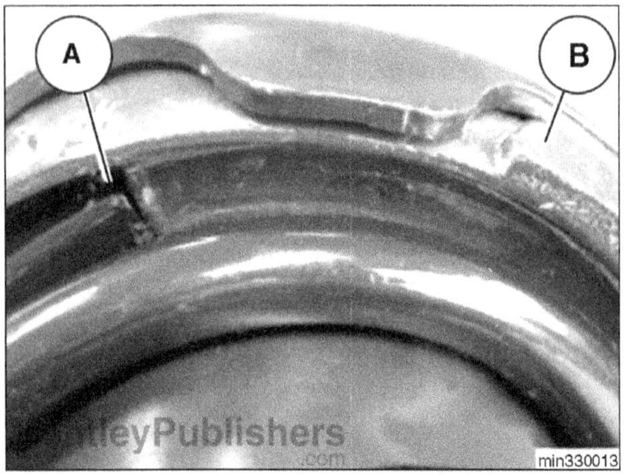

When placing spring in spring compressor:

- Smaller end of spring points to bottom of strut.
- Spring end touches stop on spring pad (**A**).
- Spring pad locator tab (**B**) rests in recess of spring seat.

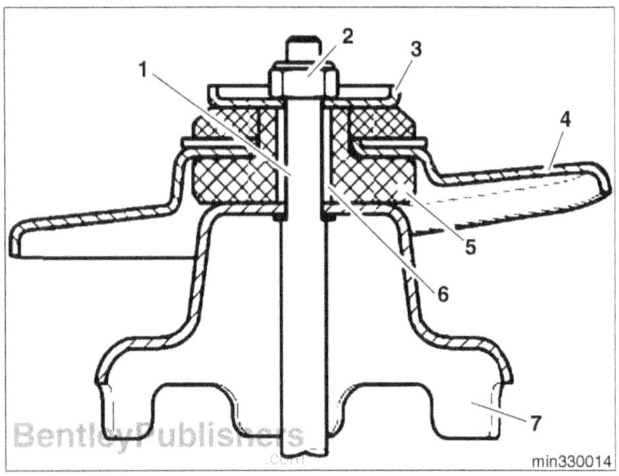

Reinstall upper strut mounting plate. Refer to illustration for order of components.

> ***CAUTION—***
> - *Make sure that the correct upper mounting plate is installed on each side of the vehicle. Match left and right markings on strut and upper mounting plate.*

1. Strut piston rod
2. Center strut nut
   - Replace self-locking M10 x 1 nut with new
   - Tighten to 30 Nm (22 ft-lb)
3. Cupped washer
4. Upper mount plate
5. Elastomer bushing
6. Inner sleeve (steel insert)
7. Upper spring seat

Tightening torque	
Upper strut mount to strut shaft (replace self-locking M10 x 1 nut with new)	30 Nm (22 ft-lb)

Rear Subframe

Subframe removal is a complex job. Read the procedure through before starting the job.

Rear subframe, removing and installing

- Raise vehicle and support in a safe manner. Remove rear wheels.

> *WARNING—*
> - *Make sure the car is stable and well supported at all times. Use a professional automotive lift or jack stands designed for the purpose. A floor jack alone is not adequate support.*
> - *Rear subframe removal may upset vehicle balance on the lift. Support the vehicle securely and place a 100 kg (200 lb) weight in the trunk to prevent the vehicle from toppling or slipping off the lift.*

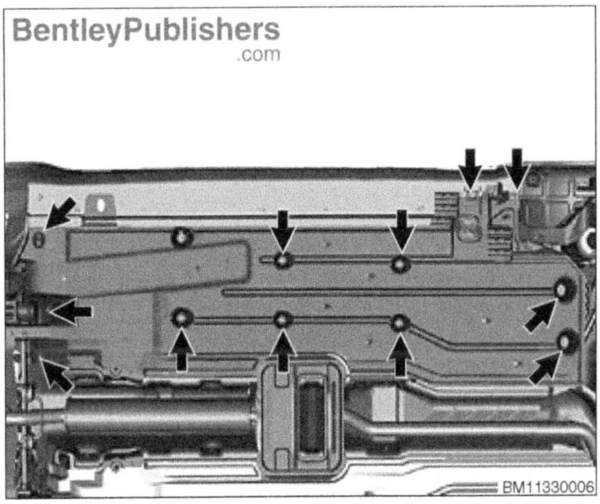

◄ Working underneath vehicle, remove left and right rear underbody panelling screws, plastic nuts and expanding rivets (**arrows**) and detach cover panels.

- Remove exhaust system and heat shields. See **180 Exhaust System**.
- Detach parking brake Bowden cables from brake calipers and from rear subframe. See **340 Brakes**.
- If applicable, remove rear ride height sensor and bracket.
- Remove upper and lower rear control arms. See **Rear Suspension Arms** in this repair group.

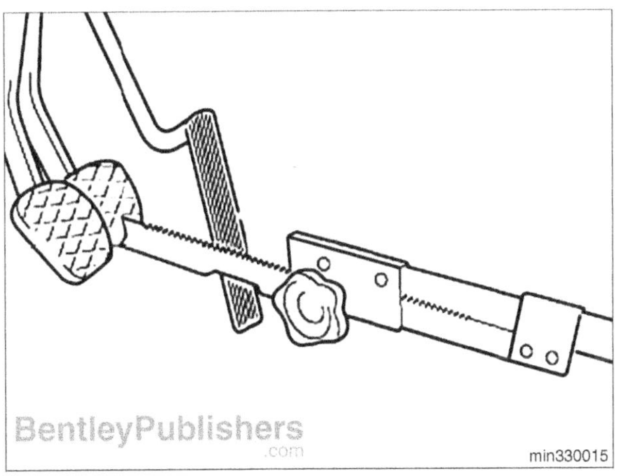

◄ Insert pedal prop and depress brake pedal slightly. This prevents brake fluid from escaping when brake lines are detached.

330

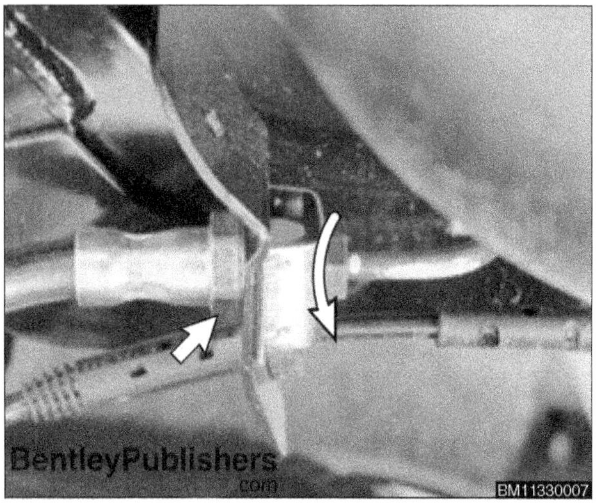

◄ Working at brake line junctions at left and right:

- Counterhold brake hose at hexagonal end (**arrow**).
- Loosen and detach brake line (**curved arrow**). Be prepared to catch dripping brake fluid.

– Detach wheel speed sensor and brake pad sensor harnesses from rear subframe.

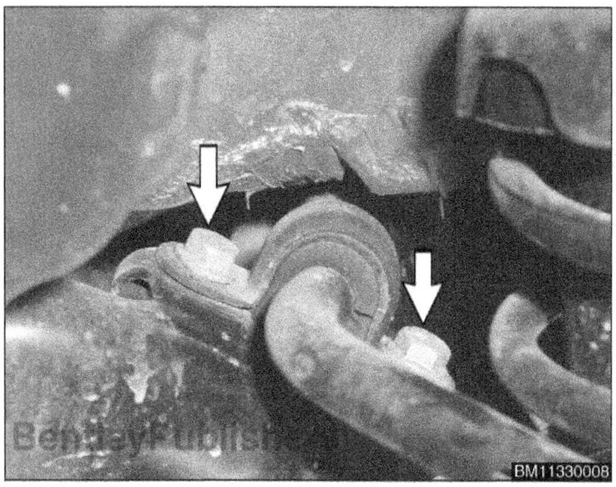

◄ Remove stabilizer bar mounting bolts (**arrows**).

– Support fuel tank from below.

– Support rear subframe with transmission jack or equivalent.

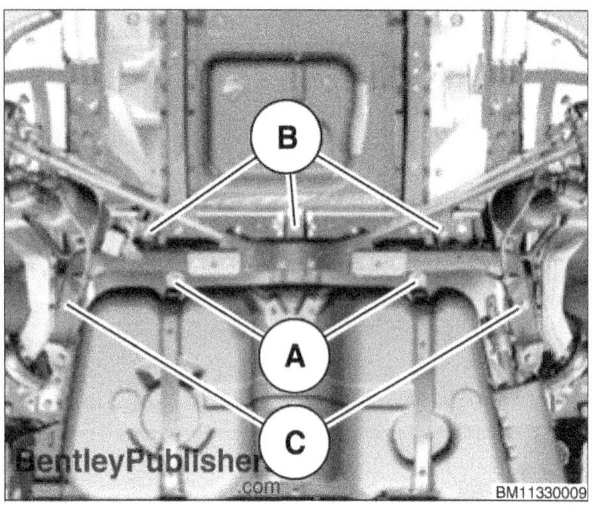

◄ Working underneath rear subframe:

- Remove fuel tank rear mounting bolts (**A**).
- Remove subframe mounting bolts (**B**, **C**).
- Note length difference of subframe bolts:
  -**B** = M12 x 62 mm
  -**C** = M12 x 132 mm

– Lower rear subframe while watching carefully to make sure no cables, electrical harnesses or other components become snagged.

- Disengage parking brake cables while lowering.

– Installation is reverse of removal. Bear in mind the following:
  - If installing new subframe, transfer parts from old part. Clip on new plastic brake line retainers.
  - Raise subframe into position carefully, checking for any wiring, hoses or mechanical parts that might get pinched.
  - Replace self-locking fasteners.
  - Replace defective underbody panelling retainers as necessary.
  - Tighten control arm mounting bolts only after vehicle is placed on ground and loaded normally.

Tightening torques	
Brake line to brake hose	14 Nm (10 ft-lb)
Control arm to rear subframe and to trailing arm (M12 x 1.5) (tighten in normal loaded position)	100 Nm (74 ft-lb)
Fuel tank to subframe (M8 bolts)	19 Nm (14 ft-lb)
Rear subframe to chassis (replace self-locking M12 bolts with new)	100 Nm (74 ft-lb)
Stabilizer bar to rear subframe (M8 bolts)	19 Nm (14 ft-lb)
Ride height sensor to rear subframe (M6 x 1)	8 Nm (71 in-lb)

– Adjust parking brake and bleed brakes. See **340 Brakes**.

– Have vehicle professionally aligned.

REAR SUSPENSION ARMS

This section covers removal, installation and repairs to rear suspension arms and bushings. Each side of the rear suspension is equipped with:

- A single trailing arm with integrated bushing. The bushing cannot be replaced separately.
- An upper and a lower control arm, each with integrated rubber bushings.

Rear suspension components

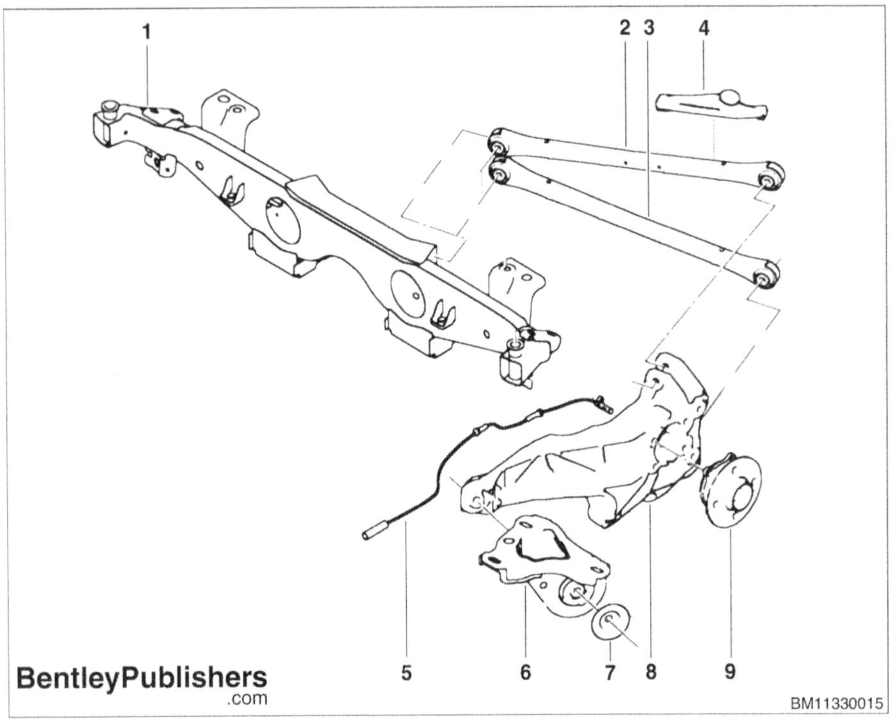

1. Rear subframe
2. Upper control arm
3. Lower control arm
4. Stop buffer
5. Wheel speed sensor and harness
6. Trailing arm bracket
7. Cup washer
8. Trailing arm
9. Wheel bearing hub

Trailing arm and bracket, removing and installing

Remove trailing arm and trailing arm mounting bracket together, then separate them if necessary. See **Trailing arm or mounting bracket, replacing** in this repair group

– Raise vehicle and support in a safe manner. Remove rear wheel.

> ***WARNING —***
> - *Make sure the car is stable and well supported at all times. Use a professional automotive lift or jack stands designed for the purpose. A floor jack alone is not adequate support.*

– Working underneath rear of car:
  - Remove brake caliper mounting bolts. Use stiff wire to hang caliper from body to relieve strain on brake hose. Do not detach brake fluid hose or parking brake cable. See **340 Brakes**.
  - Remove brake disc. See **340 Brakes**.
  - Remove wheel speed sensor mounting bolt from trailing arm. Detach sensor and set aside.
  - Detach stabilizer bar link from trailing arm. See **Stabilizer bar link, removing and installing** in this repair group.

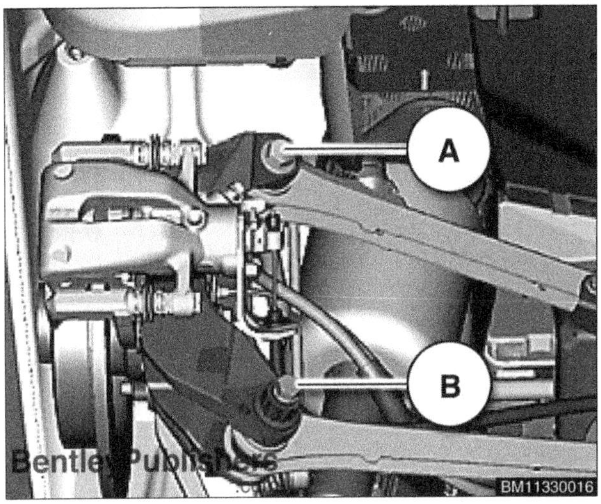

◄ While supporting trailing arm from below using an adjustable jack stand:
  - Remove upper control arm to trailing arm mounting fasteners (**A**).
  - Working at lower control arm fastener (**B**): Mark position of rear toe adjusting eccentric washer to trailing arm to simplify subsequent assembly. Then remove lower control arm mounting fasteners.

◄ Remove strut lower mounting bolt (**arrow**).

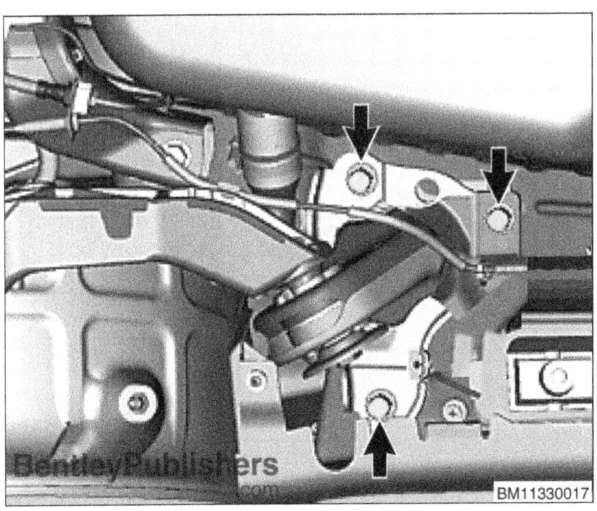

◄ Working underneath trailing arm:
  - Mark position of trailing arm bracket to body to simplify subsequent adjustment.
  - Remove bolts (**arrows**)
  - Lower trailing arm and bracket while watching carefully to make sure no cables, electrical harnesses or other components become snagged.

– Installation is reverse of removal. Keep in mind the following:
  - Replace self-locking fasteners.
  - Align trailing arm bushing bracket with previously made marks on body before tightening mounting bolts.
  - Use previously made mark to align lower control arm eccentric washer.
  - Tighten control arm mounting bolts only after vehicle is placed on ground and loaded normally.

Tightening torques	
Brake caliper to trailing arm (M10)	65 Nm (48 ft-lb)
Brake disc to wheel hub (Torx M10 x 13 mm)	27 Nm (20 ft-lb)
Control arm to trailing arm (M12 x 1.5) (tighten in normal loaded position)	100 Nm (74 ft-lb)
Rear strut lower mount to trailing arm (M14 x 105 mm x 1.5)	165 Nm (122 ft-lb)
Road wheel to hub	140 ± 10 Nm (103 ± 7 ft-lb)
Stabilizer bar link to trailing arm (replace M10 nut)	56 Nm (41 ft-lb)
Trailing arm bracket to body (M12 x 1.5)	100 Nm (74 ft-lb)
Wheel bearing hub to trailing arm (M10)	56 Nm (41 ft-lb)
Wheel speed sensor to trailing arm (M6)	8 Nm (71 in-lb)

– Have vehicle professionally aligned.

Trailing arm or mounting bracket, replacing

– Raise vehicle and support in a safe manner. Remove rear wheel.

> ***WARNING —***
> - *Make sure the car is stable and well supported at all times. Use a professional automotive lift or jack stands designed for the purpose. A floor jack alone is not adequate support.*

– Remove trailing arm and bracket from vehicle. See **Trailing arm and bracket, removing and installing** in this repair group.

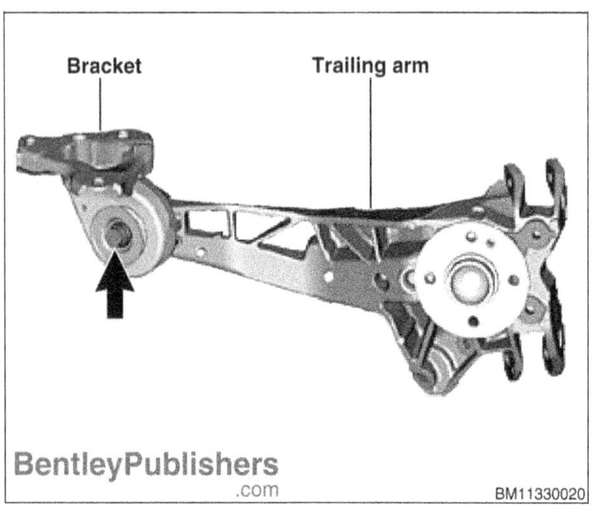

< Remove trailing arm to bracket mounting bolt (**arrow**) and separate components.

– Replace components as needed. Bracket bushing is integrated with bracket and is not available separately.

– Use new self-tapping bolt to mount bracket to trailing arm. If replacing trailing arm, tap new threads into arm using new bolt.

Tightening torque	
Trailing arm to bracket (replace M14 self-tapping bolt with new, tighten in normal loaded position)	165 Nm (122 ft-lb)

– Installation is reverse of removal.

Upper control arm, removing and installing

– Raise vehicle and support in a safe manner. Remove rear wheel.

> ***WARNING—***
> - *Make sure the car is stable and well supported at all times. Use a professional automotive lift or jack stands designed for the purpose. A floor jack alone is not adequate support.*

– If necessary, remove spare tire.

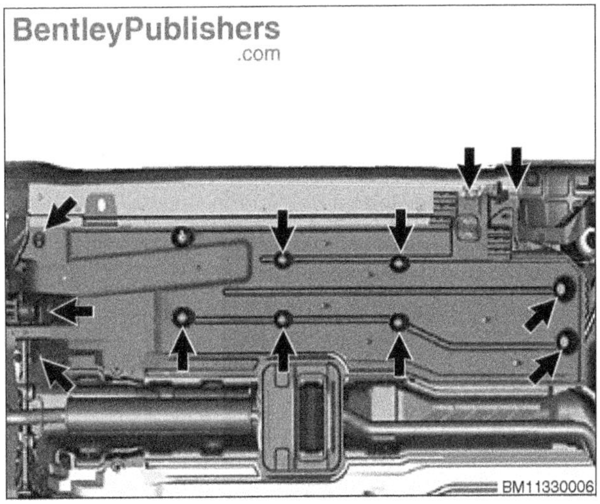

◄ Working underneath vehicle, remove left and right rear underbody panelling screws, plastic nuts and expanding rivets (**arrows**) and detach cover panels.

– If necessary, remove exhaust system heat shield.

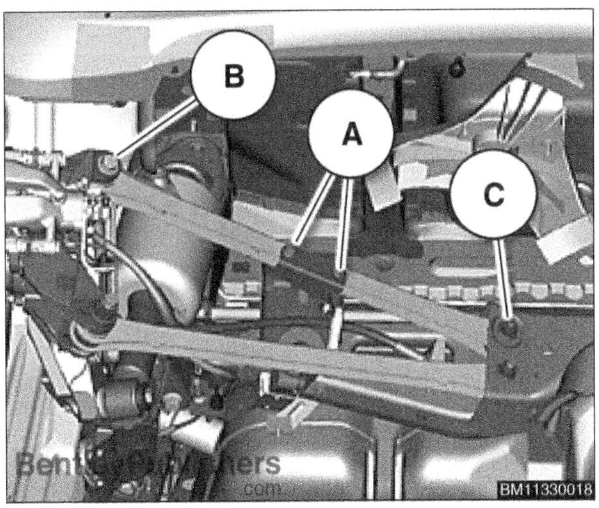

◄ Working underneath vehicle:

- Left side: If equipped, remove ride height sensor bracket screws (**A**).
- Remove control arm mounting fasteners at trailing arm (**B**).
- Remove fastener at subframe (**C**) and lower control arm.

– Installation is reverse of removal.

- If control arm is replaced, be sure to transfer stop buffer from old part and modify for new. Replace damaged or worn buffer.
- Replace self-locking fasteners.
- Tighten control arm mounting bolts only after vehicle is placed on ground and loaded normally.

Tightening torque	
Control arm to rear subframe and to trailing arm (replace M12 x 1.5 self-locking nuts with new)	100 Nm (74 ft-lb)
Road wheel to hub	140 ± 10 Nm (103 ± 7 ft-lb)
Vehicle level sensor to control arm (M8 screw)	6 Nm (53 in-lb)

– Have vehicle professionally aligned.

Lower control arm, removing and installing

– Raise vehicle and support in a safe manner. Remove rear wheel.

> ***WARNING —***
> • *Make sure the car is stable and well supported at all times. Use a professional automotive lift or jack stands designed for the purpose. A floor jack alone is not adequate support.*

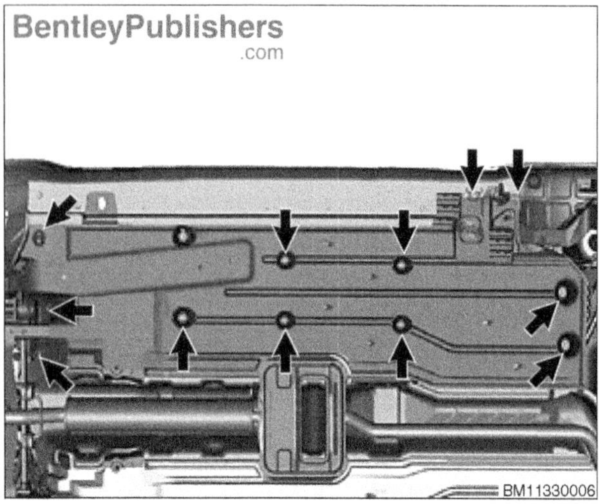

< Working underneath vehicle, remove left and right rear underbody panelling screws, plastic nuts and expanding rivets (**arrows**) and detach cover panels.

– Convertible: Remove cross-shaped reinforcement from underneath rear of body.

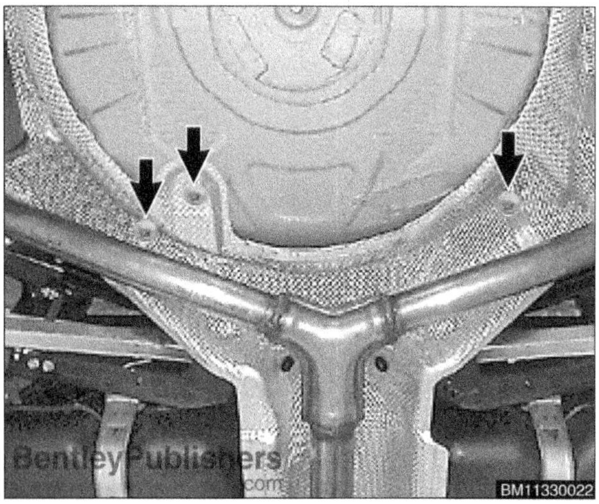

< Working underneath rear of vehicle:

- Support exhaust system. Remove rear exhaust hangers and allow exhaust pipes to sag down slightly.
- Remove heat shield fasteners (**arrows**) and remove rear heat shield.
- Illustration shows Cooper S Clubman. Other models similar.

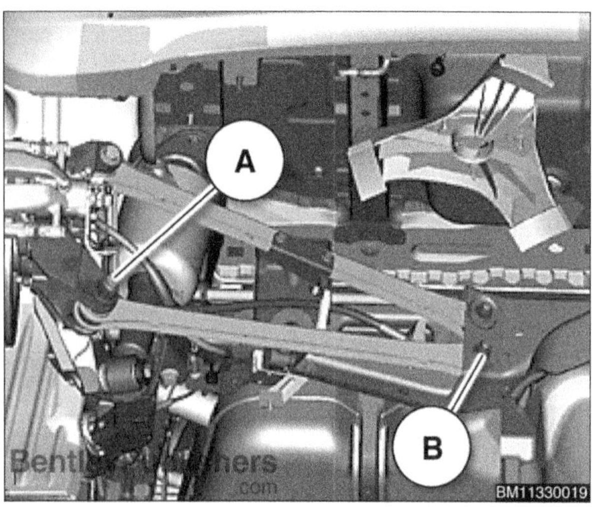

< Working underneath rear of vehicle:

- Mark position of rear toe adjusting eccentric washer (**A**) to trailing arm to simplify subsequent assembly. Remove control arm to trailing arm mounting fasteners.
- Remove control arm to subframe mounting fasteners (**B**).

– Installation is reverse of removal.

- Replace self-locking fasteners.
- Use previously made mark to align lower control arm eccentric washer.
- Tighten control arm mounting bolts only after vehicle is placed on ground and loaded normally.

Tightening torque	
Control arm to rear subframe and to trailing arm (replace M12 x 1.5 self-locking nuts with new)	100 Nm (74 ft-lb)
Road wheel to hub	140 ± 10 Nm (103 ± 7 ft-lb)

– Have vehicle professionally aligned.

STABILIZER BAR

Rear stabilizer bar, removing and installing

Rear stabilizer bar removal and installation requires that the rear subframe be loosened and lowered approx. 40 mm (1⅔ in). Read the procedure through before starting work.

– Raise vehicle and support in a safe manner. Remove rear wheels.

> ***WARNING—***
> - *Make sure the car is stable and well supported at all times. Use a professional automotive lift or jack stands designed for the purpose. A floor jack alone is not adequate support.*
> - *Rear subframe removal may upset vehicle balance on the lift. Support the vehicle securely and place a 100 kg (200 lb) weight in the trunk to prevent the vehicle from toppling or slipping off the lift.*

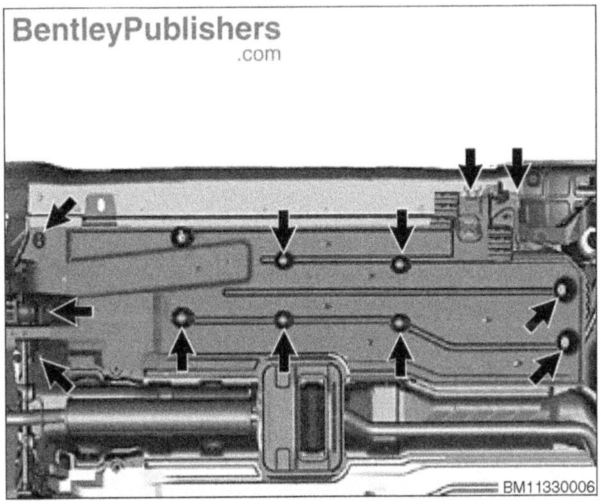

Working underneath vehicle, remove left and right rear underbody panelling screws, plastic nuts and expanding rivets (**arrows**) and detach cover panels.

– Convertible: Remove cross-shaped reinforcement from underneath rear of body.

– Convertible: Remove both rear strut assemblies. See **Rear strut assembly, removing and installing** in this repair group.

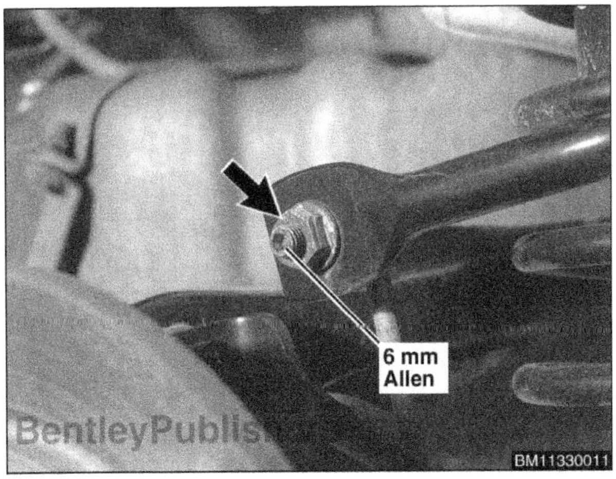

Working at stabilizer bar end, remove stabilizer bar link mounting nut (**arrow**). Counterhold link stud using 6 mm Allen key. Detach stabilizer bar link from stabilizer bar.

Rear stabilizer bar, removing and installing

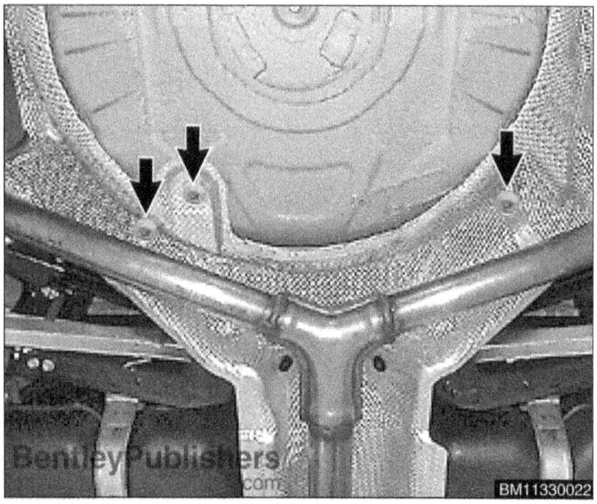

Working underneath rear of vehicle:

- Support exhaust system. Remove rear exhaust hangers and allow exhaust pipes to sag down slightly.
- Remove heat shield fasteners (**arrows**) and remove rear heat shield.
- Illustration shows Cooper S Clubman. Other models similar.

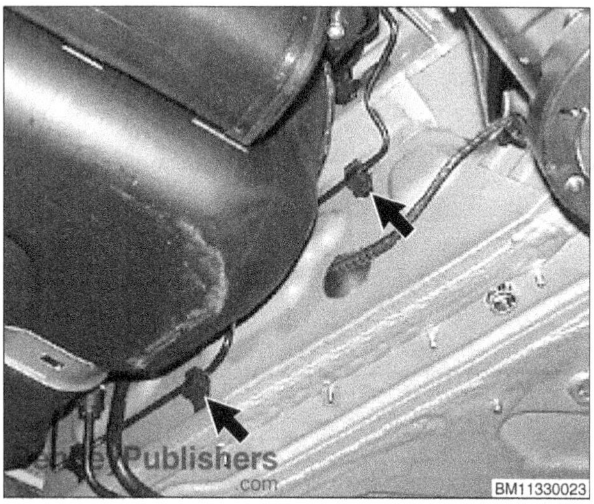

Working underneath vehicle at right side of floor plate in front of fuel tank, detach brake line from plastic clips (**arrows**).

– Support fuel tank from below.

– Support rear subframe with transmission jack or equivalent.

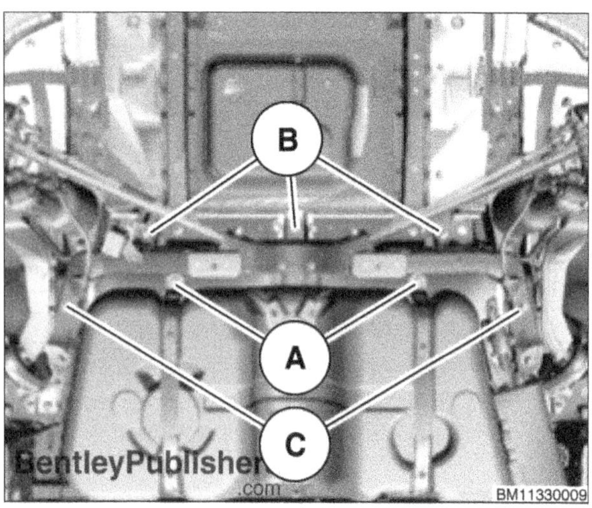

Working underneath rear subframe:

- Remove fuel tank rear mounting bolts (**A**).
- Remove subframe mounting bolts (**B**, **C**).
- Note length difference of subframe bolts:
  -**B** = M12 x 62 mm
  -**C** = M12 x 132 mm

– Tip front of rear subframe down by approx. 40 mm (1⅔ in).

◄ Remove stabilizer bar mounting bolts (**arrows**) and slide bar out one side.

– Installation is reverse of removal. Bear in mind the following:
  - Keep stabilizer and rubber mounts clean and grease-free.
  - Raise subframe into position carefully, checking for any wiring, hoses or mechanical parts that might get pinched.
  - Replace self-locking fasteners,
  - Replace defective underbody panelling retainers as necessary.

Tightening torques	
Fuel tank to subframe (M8 bolts)	19 Nm (14 ft-lb)
Rear subframe to chassis (replace self-locking M12 bolts with new)	100 Nm (74 ft-lb)
Road wheel to hub	140 ± 10 Nm (103 ± 7 ft-lb)
Stabilizer bar to rear subframe (M8 bolts)	19 Nm (14 ft-lb)
Stabilizer bar link to stabilizer bar (replace M10 nuts with new)	56 Nm (41 ft-lb)

Stabilizer bar link, removing and installing

– Raise vehicle and support in a safe manner. Remove rear wheel.

> ***WARNING —***
> - *Make sure the car is stable and well supported at all times. Use a professional automotive lift or jack stands designed for the purpose. A floor jack alone is not adequate support.*

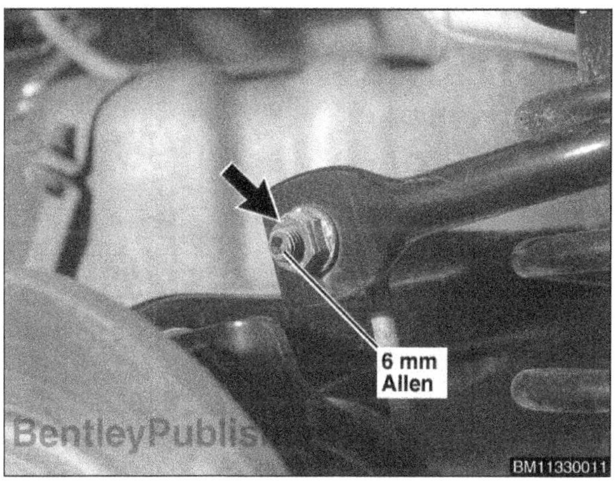

◄ Working underneath rear of car at stabilizer bar end, remove stabilizer bar link mounting nut (**arrow**). Counterhold link stud using 6 mm Allen key. Detach stabilizer bar link from stabilizer bar.

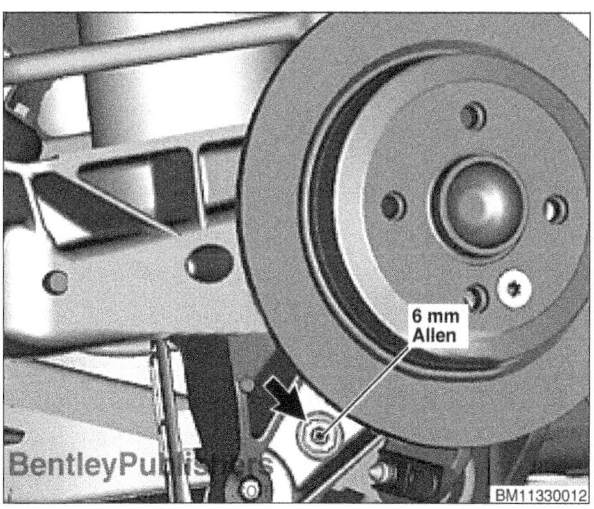

◄ Working at trailing arm, remove stabilizer bar link mounting nut (**arrow**). Counterhold link stud using 6 mm Allen key. Detach stabilizer bar link from trailing arm.

330

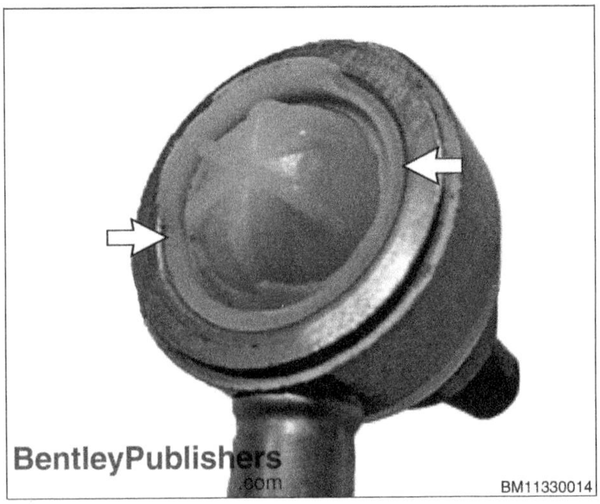

◀ If stabilizer link ball joint plastic cover (**arrows**) is broken or damaged, replace complete link.

– Installation is reverse of removal. Bear in mind the following:
  - Replace self-locking fasteners.
  - To avoid damaging link ball joint boot, counterhold link stud using 6 mm Allen key. Do not use impact wrench to tighten nut.

Tightening torques	
Road wheel to hub	140 ± 10 Nm (103 ± 7 ft-lb)
Stabilizer bar link to stabilizer bar and to trailing arm (replace M10 nuts with new)	56 Nm (41 ft-lb)

Stabilizer bar rubber mounts, replacing

– Raise vehicle and support in a safe manner. Remove rear wheels.

> ***WARNING —***
> - *Make sure the car is stable and well supported at all times. Use a professional automotive lift or jack stands designed for the purpose. A floor jack alone is not adequate support.*

◀ Remove stabilizer bar mounting bolts (**arrows**).

– Remove mount bracket. Remove split rubber mount.

– Installation is reverse of removal. Install rubber mounts dry.

Tightening torques	
Road wheel to hub	140 ± 10 Nm (103 ± 7 ft-lb)
Stabilizer bar to rear subframe (M8 bolts)	19 Nm (14 ft-lb)

Rear Wheel Bearings

The rear wheel bearings are permanently sealed and require no maintenance. The bearing is integral with the wheel hub. The wheel bearing and hub assembly is bolted to the trailing arm. Rear wheel bearing and wheel hub are not available separately.

Rear wheel bearing, removing and installing

- Raise vehicle and support in a safe manner. Remove rear wheel.

> ***WARNING —***
> - *Make sure the car is stable and well supported at all times. Use a professional automotive lift or jack stands designed for the purpose. A floor jack alone is not adequate support.*

- Remove brake caliper mounting bolts. Do not detach brake fluid hose or parking brake cable. Use stiff wire to hang caliper from body to relieve strain on brake hose. See **340 Brakes**.
- Remove brake disc. See **340 Brakes**.

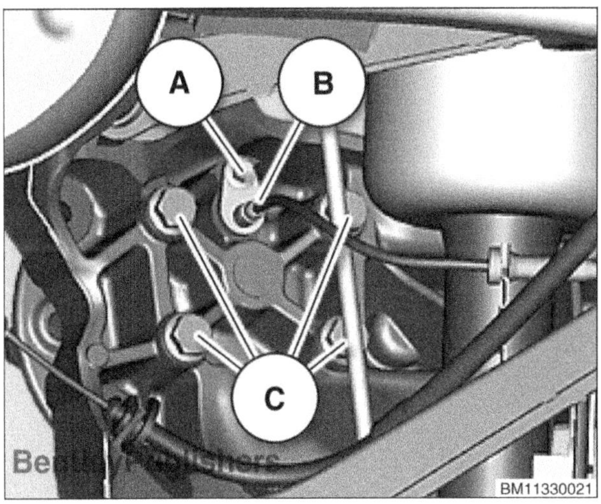

Working underneath vehicle at wheel hub:

- Remove wheel speed sensor mounting bolt (**A**) from trailing arm. Detach sensor (**B**) and set aside.
- Remove wheel hub mounting bolts (**C**). Remove wheel hub.

- Installation is reverse of removal.

Tightening torques	
Brake caliper to trailing arm (M10)	65 Nm (48 ft-lb)
Brake disc to wheel hub (Torx M10 x 13)	27 Nm (20 ft-lb)
Road wheel to hub	140 ± 10 Nm (103 ± 7 ft-lb)
Wheel hub to trailing arm (M10 x 75 mm)	56 Nm (41 ft-lb)
Wheel speed sensor to trailing arm (M6)	8 Nm (71 in-lb)

340 Brakes

TABLES

GENERAL

This repair group covers brake service:

- Brake pads, calipers and rotors (disks) repairs.
- Master cylinder, brake booster and parking brake repairs.
- ABS / ASC and ABS / DSC component replacement.

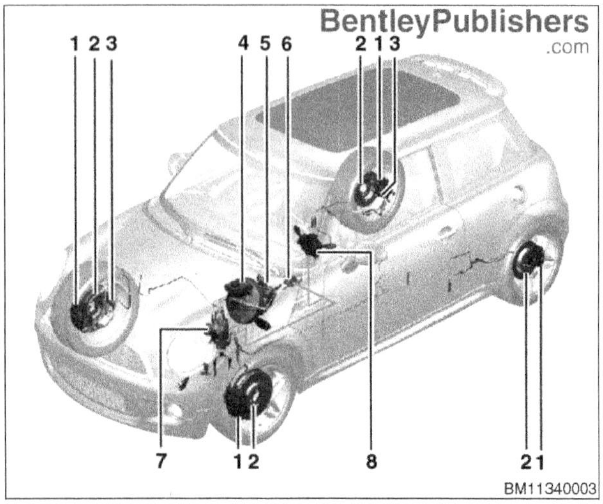

MINI brake system

1. Brake caliper
2. Brake rotor (disc)
3. ABS wheel speed sensor
4. Brake booster, master cylinder, brake fluid reservoir
5. Brake light switch
6. DSC acceleration sensor
7. ABS / ASC or ABS / DSC hydraulic unit, control module
8. Steering column switch center (SZL) with steering angle sensor

MINI models are equipped with vacuum-assisted four-wheel disc brakes. A 2-stage brake pad wear sensor for each axle activates the condition-based service (CBS) warning function and indicates when brake pads need replacement.

The parking brake system is integrated with the rear brake calipers and rotors.

Electronic braking, stability and traction control features on MINI models are as follows:

- Antilock brakes (ABS), standard in all models.
- Automatic stability control + traction (ASC + T), standard in all models.
- Dynamic stability control (DSC) or traction control, standard or optional, depending on model.

Antilock brakes (ABS)

ABS prevents road wheel lock-up during brake application, thus maintaining vehicle stability and steering control under emergency conditions. Fitted to all MINI models, ABS is always active when the vehicle is being driven and cannot be disabled by the driver.

Automatic stability control + traction (ASC + T)

ASC + T prevents the driven front wheels from spinning when engine torque is applied and maintains an equal torque distribution to both wheels. This ensures that optimum traction and directional stability are maintained.

Dynamic stability control (DSC)

DSC supplement ASC +T by monitoring the following:

- Steering wheel movement (*via* steering angle sensor)
- Lateral (side to side) forces and vehicle rotational speed (yaw) *via* DSC acceleration sensor.

Based on sensor inputs, the DSC control module analyzes the desired motion of the vehicle and compares it with the actual motion. If instability is detected, DSC is programmed to take corrective action using braking, engine management and transmission components.

For marketing reasons, DSC is labelled dynamic traction control or DTC.

ABS indicator and warning lights

The traction control or ABS warning light flashes during control intervention. Warning lights illuminate continuously if the ABS control module detects a malfunction or if DSC is deactivated.

In case of such malfunction, connect MINI scan tool and interrogate fault memory.

Disabling traction control

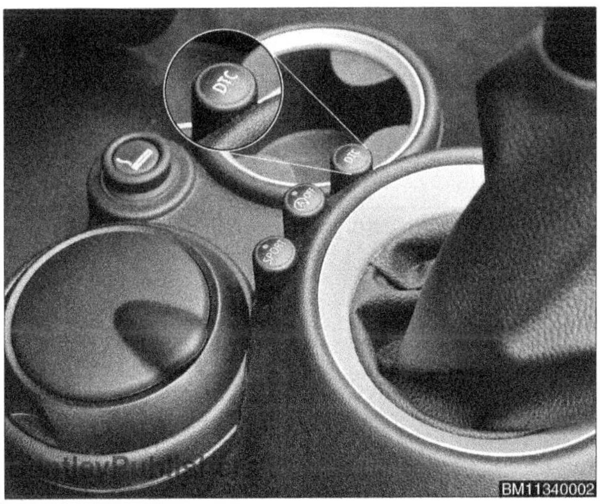

In situations where it is useful to deactivate DSC, press the DTC button for at least 3 seconds: the DSC indicator lights illuminates and DSC is deactivated.

This may be useful when driving on snow-covered inclines, in slush, on uncleared snowy roads or when rocking the vehicle free, driving out of deep snow or on loose surfaces when driving with snow chains.

To activate, press the button gain.

Troubleshooting

Brake performance is mainly affected by three things:

- Level and condition of brake fluid.
- Ability of brake system to create and maintain pressure.
- Condition of friction components.

Fluid contaminated by moisture or dirt can corrode the system. Inspect the brake fluid inside the reservoir. If it is dirty or murky, or is more then two years old, replace fluid. See **Bleeding Brakes** in this repair group.

To check the function of the master cylinder, hold the brake pedal down hard with the engine running. If the pedal slowly falls to the floor, either the master cylinder is leaking internally, or fluid is leaking externally.

Inspect the rotors for glazing, discoloration and scoring. Steering wheel vibration while braking at speed is often caused by warped rotors, but can also be caused by worn suspension components.

When troubleshooting, keep in mind that tire inflation, wear and temperature can all have an affect on braking. See

300 Suspension, Steering and Brakes–General for tire size and inflation information.

Table a lists symptoms of brake problems, probable causes, and suggested corrective actions. Unless noted otherwise, relevant repairs are described later in this repair group.

Table a. Brake system troubleshooting

Symptom	Probable cause	Repairs
Brake squeal	Brake pad carriers dirty or corroded or pads loose (poor fit) in pad carrier	Remove brake pads and clean calipers. Use original equipment pads for proper fit and use anti-squeal compound during installation.
	Brake pads heat-glazed or oil-soaked	Replace brake pads. Clean rotors. Replace leaking calipers as required.
	Wheel bearings worn (noise most pronounced when turning)	Replace worn bearings. See **310 Front Suspension, Drive Axles** or **330 Rear Suspension**.
	Incorrectly installed brake pads or other brake parts	Check component installation. Check and replace anti-rattle springs.
Pedal goes to floor when braking	Brake fluid loss due to system leaks	Check fluid level and inspect for signs of leakage.
	Master cylinder or traction control system faulty	Replace master cylinder. Diagnose traction control system using factory or compatible diagnostic tool.
Low pedal after system bleeding	Master cylinder faulty	Replace master cylinder.
Pedal spongy or brakes work only when pedal is pumped	Air in brake fluid	Bleed system using factory or compatible diagnostic tool.
	Master cylinder or traction control system faulty	Replace master cylinder. Diagnose traction control system using factory or compatible diagnostic tool.
Excessive braking effort	Brake pads wet	Use light pedal pressure to dry pads while driving.
	Brake pads heat-glazed or fluid-soaked	Replace brake pads and rotors. Replace leaking calipers.
	Vacuum booster or vacuum hose connections to booster faulty	Inspect vacuum lines. Test vacuum booster and replace as required. Test vacuum check valve for one-way flow.
Brakes pulsate, chatter or grab	Warped brake rotors	Resurface or replace rotors.
	Brake pads worn	Replace brake pads.
	Brake pads heat-glazed or oil-soaked	Replace brake pads. Clean rotors. Replace leaking calipers.
Uneven braking, car pulls to one side, rear brakes lock	Incorrect tire pressures or worn tires	Inspect tire condition. Check and correct tire pressures.
	Brake pads on one side of car heat-glazed or fluid-soaked	Replace brake pads. Clean rotors. Replace leaking calipers.
	Caliper or brake pads binding	Clean and recondition brakes.
	Worn suspension components	Inspect for worn or damaged suspension components. See **310 Front Suspension, Drive Axles** or **330 Rear Suspension**.
Brakes drag, bind or overheat	Brake caliper or brake pads binding	Clean or replace caliper.
	Master cylinder or traction control system faulty	Replace master cylinder. Diagnose traction control system using factory or compatible diagnostic tool.

Warnings and Cautions

WARNING—
- *MINI vehicles require special MINI service equipment to properly bleed the brake hydraulic system. See* **Bleeding Brakes** *in this repair group.*
- *Semi-metallic and metallic brake friction materials in brake pads produce dangerous dust. Wash or vacuum to remove dust. Do not blow away with compressed air.*
- *Brake fluid is poisonous, corrosive and dangerous to the environment. Wear safety glasses and rubber gloves when working with brake fluid. Do not siphon brake fluid with your mouth. Dispose of brake fluid properly.*
- *Do not reuse self-locking nuts or bolts. They are designed to be used only once and may fail if reused. Replace with new self-locking fasteners.*

CAUTION—
- *Immediately clean brake fluid spilled on painted surfaces and wash with water, as brake fluid will remove paint.*
- *Use new brake fluid from a fresh, unopened container. Brake fluid absorbs moisture from the air. This leads to corrosion problems in the braking system, and lowers the fluid boiling point.*
- *Do not allow petroleum products or brake cleaner fluid to enter the brake system. These substances cause brake failure.*
- *Route brake lines and hoses so that they do not chafe on the body or other components.*
- *To prevent damage to brake line couplings, loosen and tighten with special flare wrenches.*
- *Plug open lines and brake fluid ports to prevent contamination.*
- *Only tighten brake hoses on front wheels with wheels in straight ahead position.*
- *If carrying out electric welding work, be sure to disconnect electrical harness connector from electronic control modules.*
- *Do not expose electronic control modules to high sustained heat, such as in a paint drying booth. Maximum heat exposure:*
  -95°C (203°F) for short periods of time
  -85°C (185°F) for long periods of time (approx. 2 hours)

340

Bleeding Brakes

Brake bleeding is usually done for one of two reasons: either to replace old brake fluid as part of routine maintenance or to expel trapped air in the system that resulted from opening the brake hydraulic system during repairs.

Use MINI scan tool to bleed the brake system.

> ***WARNING—***
> - *ABS, ASC and DSC use electronic controls and a sophisticated hydraulic unit. When flushing brake fluid from the system, use extreme care to not let the brake fluid reservoir run dry. Once air enters the hydraulic unit, it is very difficult to remove using traditional methods. For this reason, bleed brakes using a pressure bleeder and a large supply of fluid.*

> ***CAUTION—***
> - *Use MINI scan tool to recode the ABS control module if it is removed or replaced:*

Use new brake fluid from an unopened container. Bleed the entire system when any part of the hydraulic system is opened. Be sure to have the special equipment needed before beginning the job.

When bleeding the brakes, start at the wheel farthest from the master cylinder and progress in the following order:

- right rear brake
- left rear brake
- right front brake
- left front brake

Brakes, bleeding

Brake fluid application	
All models	DOT 4

– Connect MINI scan tool to OBD II diagnostic plug (under left side of dashboard). Call up service function Bleeding ABS/ASC or service function Bleeding ABS/DSC, depending on installed system.

min340013

◄ Top off brake fluid in reservoir. Connect pressure bleeder to brake fluid reservoir. Turn on pressure.

> ***CAUTION—***
> - *Do not exceed a pressure of 2 bar (29 psi) when pressure bleeding the brake system. Higher pressure may damage the fluid reservoir.*

– Raise vehicle and support in a safe manner.

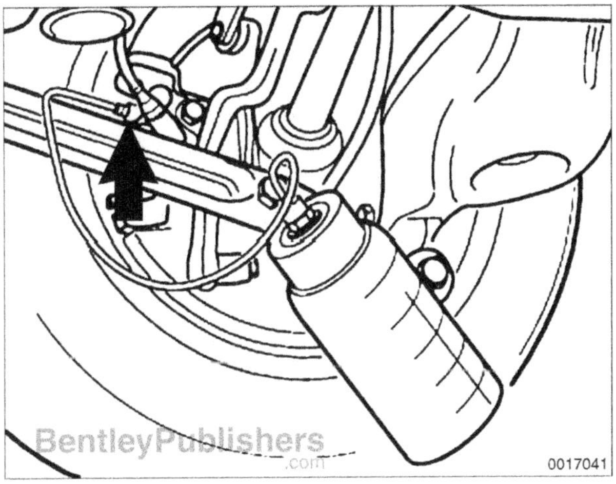

> **WARNING—**
> • *Make sure the car is stable and well supported at all times. Use a professional automotive lift or jack stands designed for the purpose. A floor jack is not adequate support.*

- Starting at right rear wheel, connect bleed hose and fluid receptacle to brake caliper bleeder screw (**arrow**).
- Open bleeder valve and flush until clear brake fluid emerges without bubbles. Close bleeder valve.

> **CAUTION—**
> • *Make sure bleeder hose remains submersed in clean brake fluid whenever the bleeder valve is open.*

- Follow same procedure to bleed left rear, right front and left front brakes.
- Return to right rear caliper and follow scan tool instructions to bleed. Repeat on left rear.
- Connect bleed hose and fluid receptacle to brake caliper bleeder screw at right front. Open bleeder valve.
- Fully depress brake pedal at least 5 times until clear brake fluid emerges without bubbles.
- Hold brake pedal in fully depressed position and close bleeder valve. Release brake pedal.
- Follow same procedure to bleed left front brakes.
- Remove brake bleeding apparatus. Top up brake fluid. Check to make sure brake fluid reservoir cap seal is intact.

Brake Pads, Calipers and Rotors

Front brake assembly components (ATE)

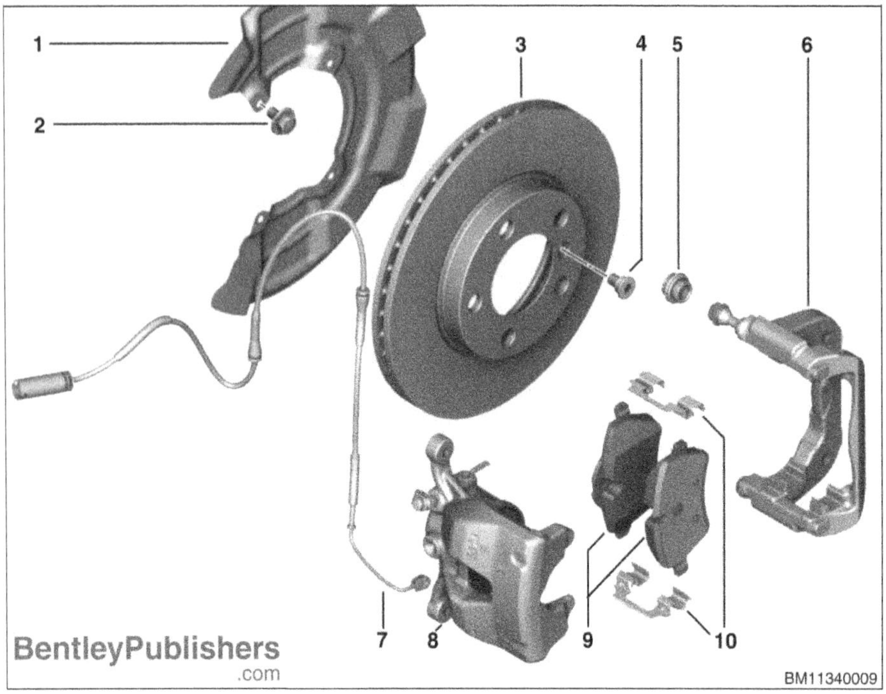

1. Backing plate
2. M6 x 10 mm bolt
   - Tighten to 8 Nm (6 ft-lb)
3. Brake rotor (disc)
   - Diameter 280 mm (11 in)
   - Thickness 22 mm (0.87 in)
4. M10 x 13 mm Torx bolt
   - Tighten to 27 Nm (20 ft-lb)
5. Sealing collar
6. Brake pad carrier
7. 2-stage brake pad wear sensor
   - Left front only
8. Brake caliper
9. Brake pads
10. Anti-rattle clips

Front brake assembly components (Brembo)

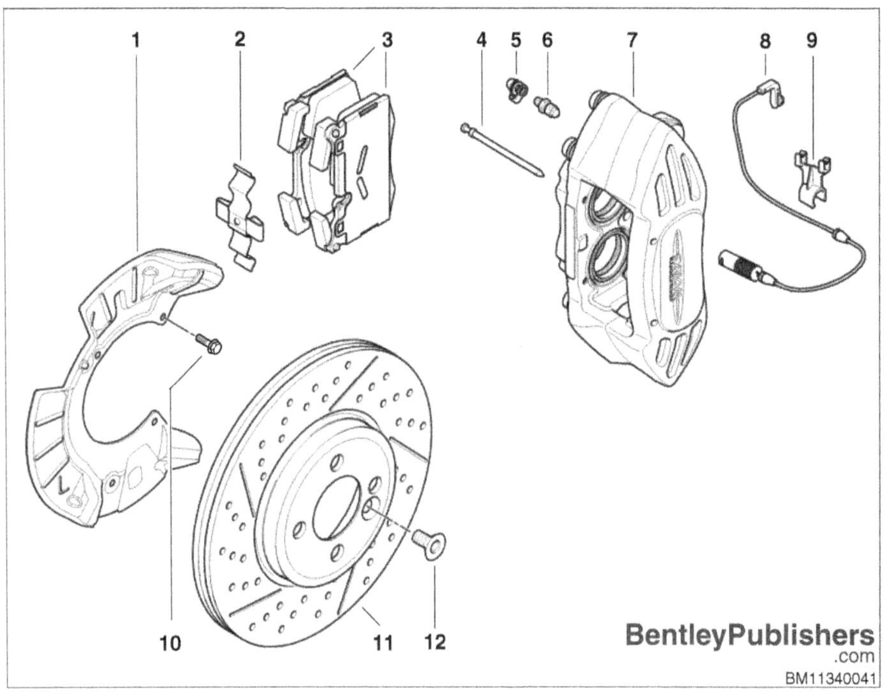

1. Backing plate
2. Spring clip
3. Brake pads
4. Retaining pin
5. Dust cap
6. Bleeder screw
7. Brake caliper
8. 2-stage brake pad wear sensor
9. Clip
10. M6 x 10 mm bolt
    - Tighten to 8 Nm (6 ft-lb)
11. Drilled and grooved brake rotor (disc)
    - Diameter 294 mm (11.6 in)
    - Diameter JCW Sport 316 mm (12.4 in)
    - Thickness 22 mm (0.87 in)
12. M10 x 13 mm Torx bolt
    - Tighten to 27 Nm (20 ft-lb)

Rear brake assembly components

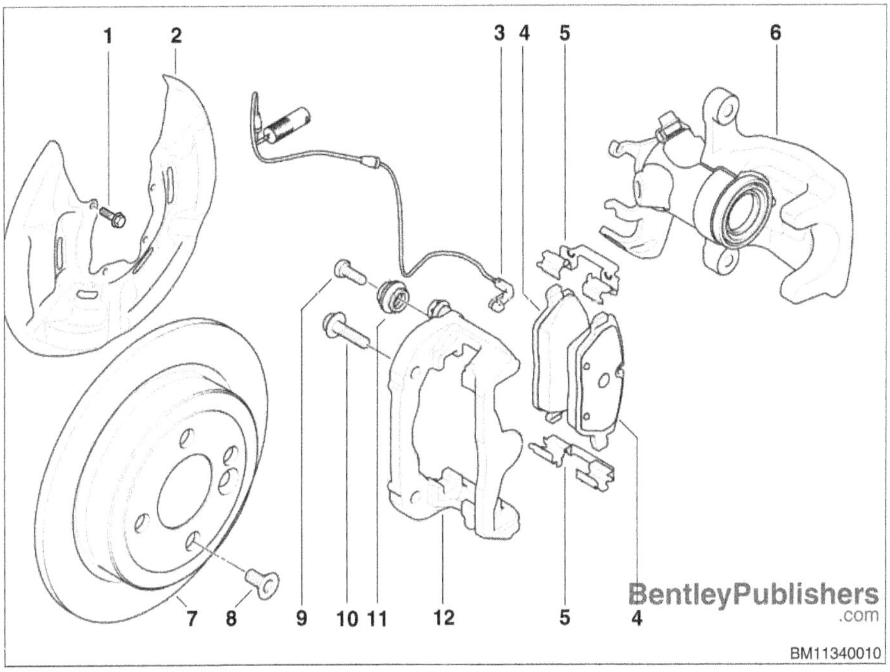

1. M6 x 10 mm bolt
   - Tighten to 8 Nm (6 ft-lb)
2. Backing plate
3. 2-stage brake pad wear sensor
   - Right rear only
4. Brake pads
5. Anti-rattle clips
6. Brake caliper
7. Brake rotor (disc)
   - Diameter 259 mm (10.2 in)
   - Diameter JCW Sport 280 mm (11 in)
   - Thickness 10 mm (0.4 in)
8. M10 x 13 mm Torx bolt
   - Tighten to 27 Nm (20 ft-lb)
9. M8 bolt
10. M10 x 32 mm collar bolt
    - Tighten to 65 Nm (48 ft-lb)
11. Sealing collar
12. Brake pad carrier

340

Brake pad lining thickness, checking

To determine thickness of outer brake pad without removing road wheel:

- Rotate wheel until inspection bore (**arrow**) in brake pad backing plate can be seen through rim.
- Insert MINI special tool 34 1 260 through rim into bore and read off pad thickness.

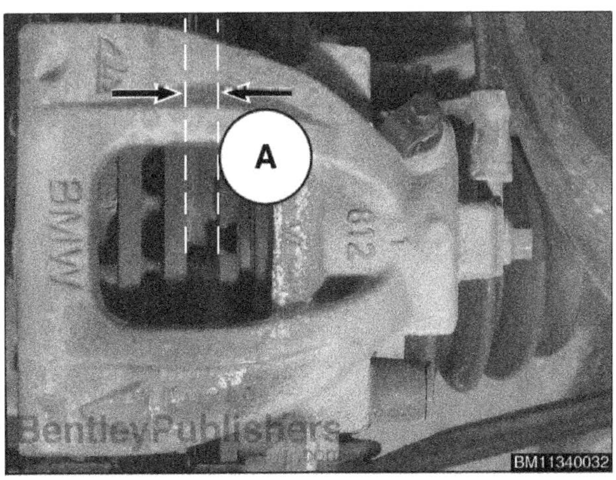

To determine brake pad thickness after removing wheel but without removing brake caliper:

- Measure pad thickness through caliper opening.

Brake pad thickness (A)	
Minimum thickness	3 mm (0.11 in)

Brake pad wear sensor, replacing

> ***CAUTION—***
> - *Replace the brake pad wear sensor once it is detached from the brake pad. The sensor spring clips lose their capacity to stick to the brake pad.*
> - *Replace the brake pad sensor whenever brake pads are replaced.*

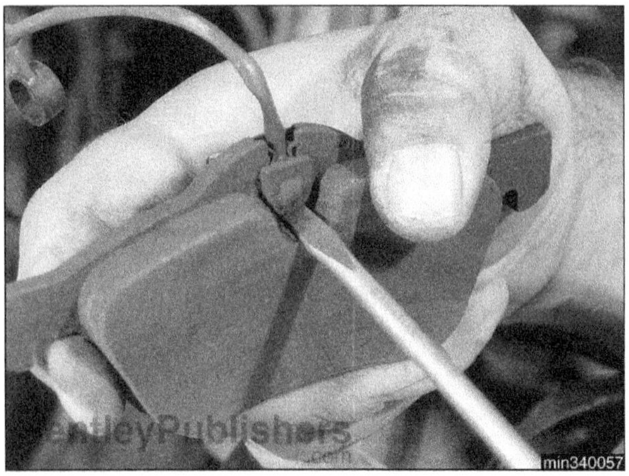

- Working at left front or right rear brake pad, carefully pry pad wear sensor from pad.

  Photo shows sensor removal with brake pad removed. Sensor can also be pried off installed brake pad easily.

- Front sensor: Separate sensor electrical coupling (**arrow**) and remove sensor.

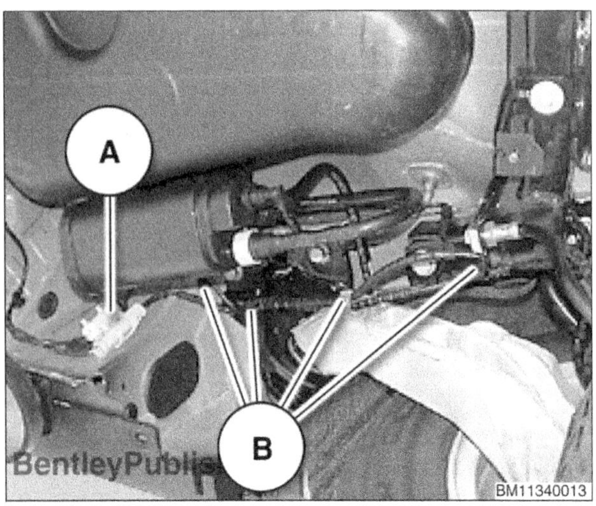

- Rear sensor:
  - Remove right rear underbody splash shield.
  - Unclip sensor harness from clips (**B**).
  - Separate sensor electrical coupling (**A**) and remove sensor.

– Installation is reverse of removal.
  - Insert sensor into cutout in new pad.
  - Route pad wear sensor wiring properly to avoid snagging.

Front brake pads, removing and installing (ATE)

Replace brake pads in axle sets.

– Raise vehicle and support in a safe manner. Remove front wheels.

> ***WARNING—***
> - *Make sure the car is stable and well supported at all times. Use a professional automotive lift or jack stands designed for the purpose. A floor jack is not adequate support.*

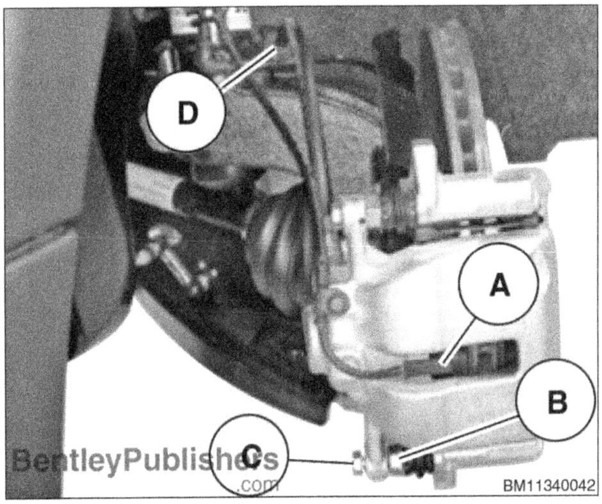

Working at front caliper:

- Left front wheel: Remove brake pad wear sensor (**A**) and discard. See **Brake pad wear sensor, replacing** in this repair group.
- Remove lower caliper mounting bolt (**C**). If necessary, counter hold at hexagon head (**B**).
- Detach brake hose from bracket (**D**).
- Tilt brake caliper upwards, swiveling it on upper mounting bolt.

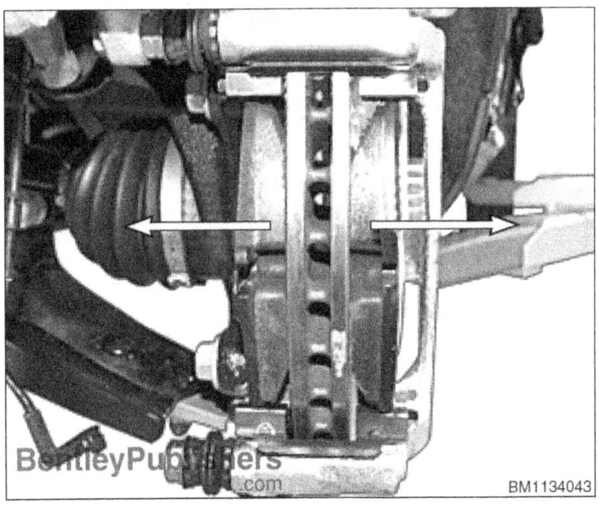

Remove brake pads from pad carrier (**arrows**).

– Measure brake rotor for minimum thickness. See **Brake rotor (disc), checking** in this repair group. Replace rotors if necessary. See **Front brake rotor (disc), removing and installing** in this repair group.

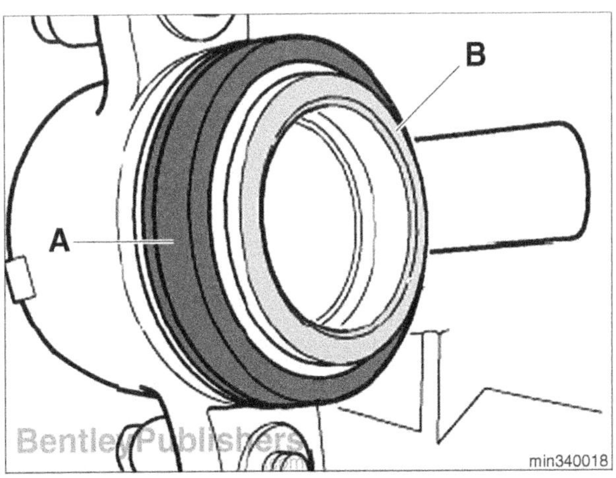

Inspect brake caliper for damage or signs of leakage.

- Check that caliper piston slides smoothly into caliper.
- Check dust seal (**A**) for damage. Replace if necessary.
- Replace caliper if leaks are found. See **Front brake caliper, removing and installing** in this repair group.
- Apply thin coat of Plastilube® or equivalent antisqueak compound to flat surface (**B**) of caliper piston.

> ***CAUTION—***
> - *Make sure antisqueak compound does not some into contact with rubber dust seal.*

Press in caliper piston fully with MINI special tool **34 1 050** or equivalent.

CAUTION—
- *Pressing caliper piston in may cause brake fluid reservoir to overflow and damage paint work. To prevent this, use a clean syringe to first remove some fluid from reservoir.*

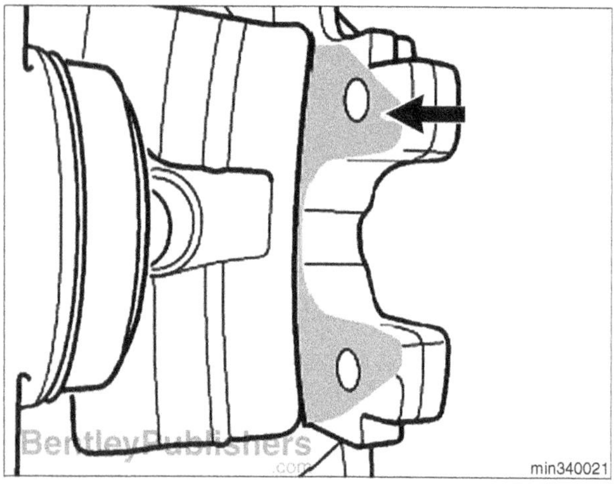

Clean contact face (**arrow**) of brake caliper. Apply thin coating of Plastilube® or equivalent antisqueak compound.

CAUTION—
- *If necessary, use brake cleaning solvent to clean caliper surfaces. To avoid corrosion and rust, do not scrape or use a metal brush on these surfaces.*

Working at brake pad carrier:
- Remove pad retaining anti-rattle clips (**arrows**).
- Use brake cleaning solvent on contact faces, then reinstall clean or new spring clips. Apply thin coating of Plastilube® or equivalent antisqueak compound.

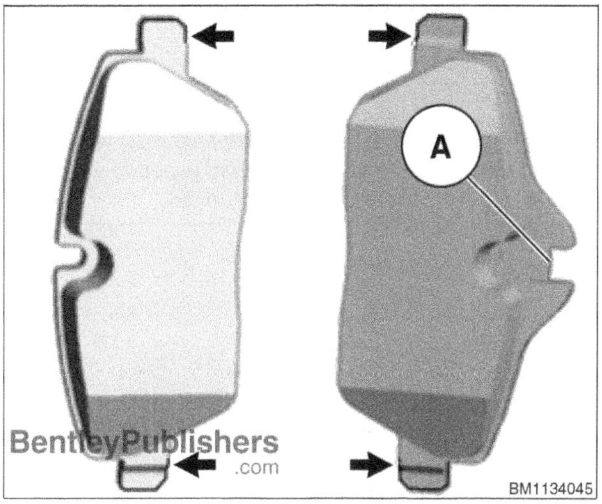

◀ Before installing new pads:

- Apply thin coating of Plastilube® or equivalent antisqueak compound to contact tabs (**arrows**).
- Do not grease brake pad backing plates.
- Left front wheel: Make sure brake pad wear sensor notch (**A**) is installed on piston side of caliper.

– Installation is reverse of removal.

- Replace caliper mounting bolts with new.
- Install new brake pad wear sensor.
- Top off brake fluid reservoir.

Tightening torques	
Brake caliper to brake pad carrier (replace self-locking M8 bolts with new)	35 ± 3 Nm (26 ± 2 ft-lb)
Road wheel to hub	140 ± 10 Nm (103 ± 7 ft-lb)

– After reassembly, pump brake pedal several times to make sure that brake pads are firmly pressed against brake rotors.

– If applicable: Reset CBS display in instrument cluster.

Front brake pads, removing and installing (Brembo)

Replace brake pads in axle sets.

– Raise vehicle and support in a safe manner. Remove front wheels.

> ***WARNING—***
> - *Make sure the car is stable and well supported at all times. Use a professional automotive lift or jack stands designed for the purpose. A floor jack is not adequate support.*

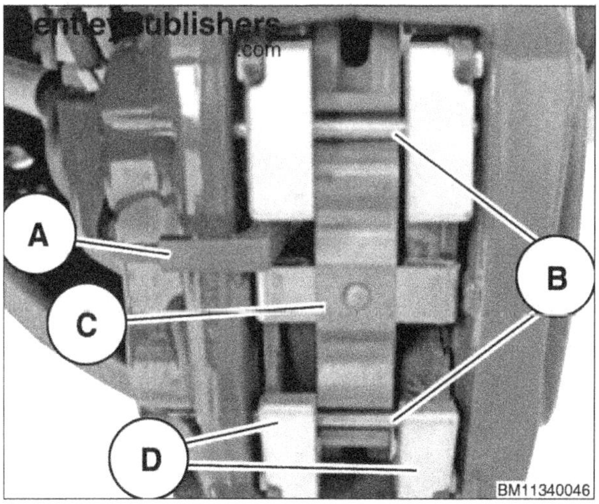

◀ Working at front caliper:

- Left front wheel: Remove brake pad wear sensor (**A**) and discard. See **Brake pad wear sensor, replacing** in this repair group.
- Use pin punch to drive out retaining pins (**B**).
- Remove spring clip (**C**).
- Remove brake pads (**D**).

– Measure brake rotor for minimum thickness. See **Brake rotor (disc), checking** in this repair group. Replace rotors if necessary. See **Front brake rotor (disc), removing and installing** in this repair group.

◀ Inspect brake caliper for damage or signs of leakage.

- Check dust seal for damage. Replace if necessary.
- Replace caliper if leaks are found. See **Front brake caliper, removing and installing** in this repair group.

Front brake pads, removing and installing (Brembo)

Press in caliper piston fully with MINI special tool **34 1 050** or equivalent.

> ***CAUTION—***
> - *Pressing caliper piston in may cause brake fluid reservoir to overflow and damage paint work. To prevent this, use a clean syringe to first remove some fluid from reservoir.*

Clean brake pad grooves (**arrows**) in brake caliper. Apply thin coating of Plastilube® or equivalent antisqueak compound.

> ***CAUTION—***
> - *If necessary, use brake cleaning solvent to clean caliper surfaces. To avoid corrosion and rust, do not scrape or use a metal brush on these surfaces.*

– Installation is reverse of removal.
  - Do not grease brake pad backing plates.
  - Left front wheel: Make sure brake pad wear sensor notch is installed on piston side of caliper.
  - Replace pad retaining spring clips and pins with new.
  - Install new brake pad wear sensor.
  - Top off brake fluid reservoir.

Tightening torque	
Road wheel to hub	140 ± 10 Nm (103 ± 7 ft-lb)

– After reassembly, pump brake pedal several times to make sure that brake pads are firmly pressed against brake rotors.

– If applicable: Reset CBS display in instrument cluster.

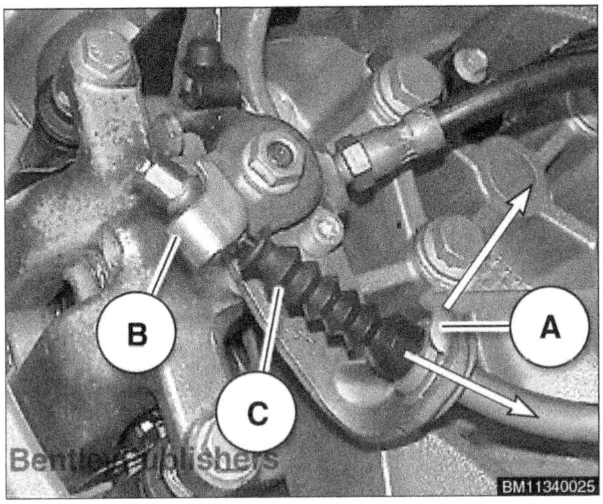

Rear brake pads, removing and installing

Replace brake pads in axle sets.

– Raise vehicle and support in a safe manner. Remove rear wheels.

> ***WARNING —***
> - *Make sure the car is stable and well supported at all times. Use a professional automotive lift or jack stands designed for the purpose. A floor jack is not adequate support.*

◄ Working at rear brake caliper:

- Pull off parking brake cable sheath locking clip (**A**).
- Detach cable end (**B**) from rear caliper actuating lever.
- Slide cable housing (**C**) out of caliper bracket.

– Right rear wheel: Remove brake pad wear sensor and discard. See **Brake pad wear sensor, replacing** in this repair group.

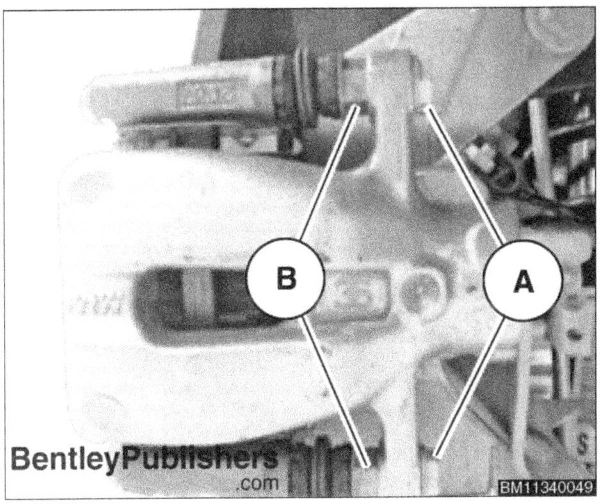

◄ Remove caliper mounting bolts (**A**). If necessary, counter hold at hexagon heads (**B**).

– Slide caliper toward rear and tie up.

> ***CAUTION—***
> - *To avoid stretching brake fluid hose, use stiff wire to hang caliper securely from suspension.*

340

◄ Remove brake pads from pad carrier (**arrows**).

– Measure brake rotor for minimum thickness. See **Brake rotor (disc), checking** in this repair group. Replace rotors if necessary. See **Rear brake rotor (disc), removing and installing** in this repair group.

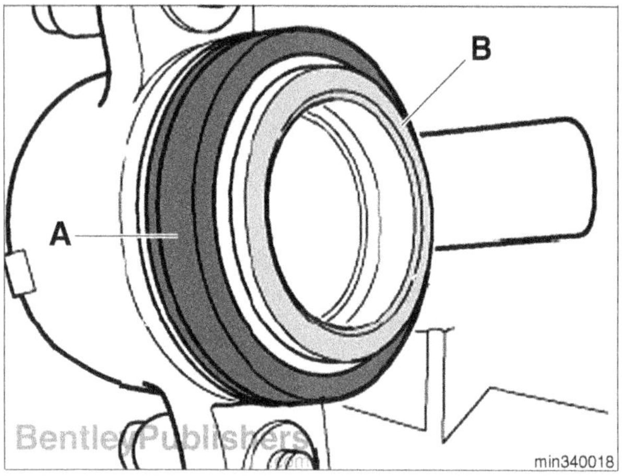

Inspect brake caliper for damage or signs of leakage.

- Check that caliper piston slides smoothly into caliper.
- Check dust seal (**A**) for damage. Replace if necessary.
- Replace caliper if leaks are found. See **Rear brake caliper, removing and installing** in this repair group.
- Apply thin coat of Plastilube® or equivalent antisqueak compound to flat surface (**B**) of caliper piston.

> *CAUTION—*
> - *Make sure antisqueak compound does not some into contact with rubber dust seal.*

Press in each rear caliper piston fully with MINI special tool set **34 6 300** or equivalent. The special tool turns the piston while pressing in.

> *CAUTION—*
> - *Make sure that the caliper piston dust seal does not twist and get damaged during this process.*
> - *Pressing caliper piston in may cause brake fluid reservoir to overflow and damage paint work. To prevent this, use a clean syringe to first remove some fluid from reservoir.*

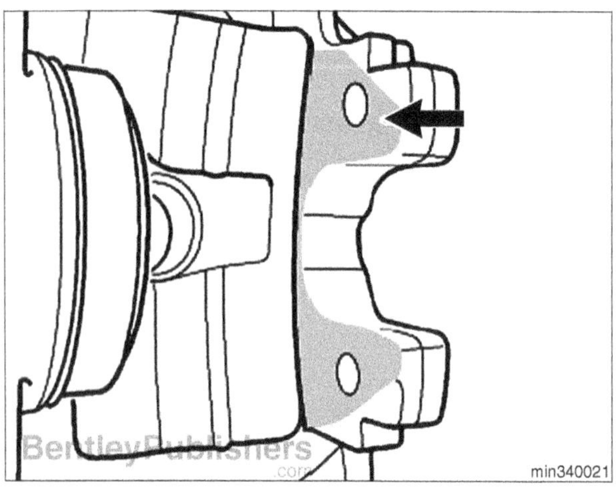

Clean contact face (**arrow**) of brake caliper. Apply thin coating of Plastilube® or equivalent antisqueak compound.

> *CAUTION—*
> - *If necessary, use brake cleaning solvent to clean caliper surfaces. To avoid corrosion and rust, do not scrape or use a metal brush on these surfaces.*

◄ Working at brake pad carrier:

- Remove pad retaining anti-rattle clips (**arrows**).
- Use brake cleaning solvent on contact faces, then reinstall clean or new spring clips.

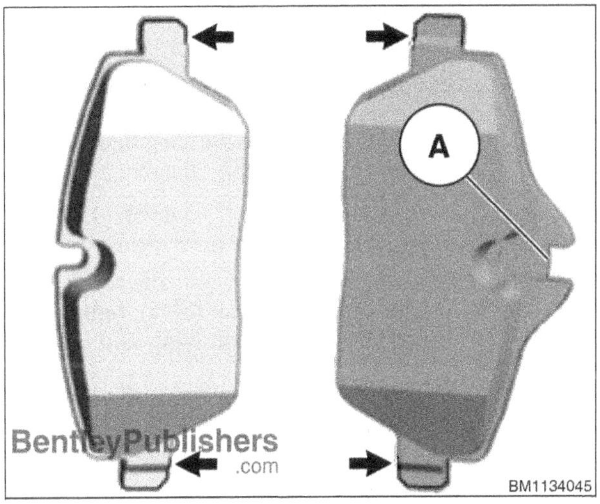

◄ Before installing new pads:

- Apply thin coating of Plastilube® or equivalent antisqueak compound to contact tabs (**arrows**).
- Do not grease brake pad backing plate.
- Right rear wheel: Make sure brake pad wear sensor notch (**A**) is installed on piston side of caliper.

340

– Installation is reverse of removal.

- Replace caliper mounting bolts with new.
- Install new brake pad wear sensor.
- Reattach parking brake cable.
- Top off brake fluid reservoir.

Tightening torques	
Brake caliper to brake pad carrier (replace self-locking M8 bolts with new)	35 ± 3 Nm (26 ± 2 ft-lb)
Road wheel to hub	140 ± 10 Nm (103 ± 7 ft-lb)

– After reassembly, pump brake pedal several times to make sure that brake pads are firmly pressed against brake rotors.

– If applicable: Reset CBS display in instrument cluster.

Brake rotor (disc), checking

– Raise vehicle and support in a safe manner. Remove wheels.

> ***WARNING—***
> • *Make sure the car is stable and well supported at all times. Use a professional automotive lift or jack stands designed for the purpose. A floor jack is not adequate support.*

– Inspect rotor for cracks, signs of overheating and scoring.

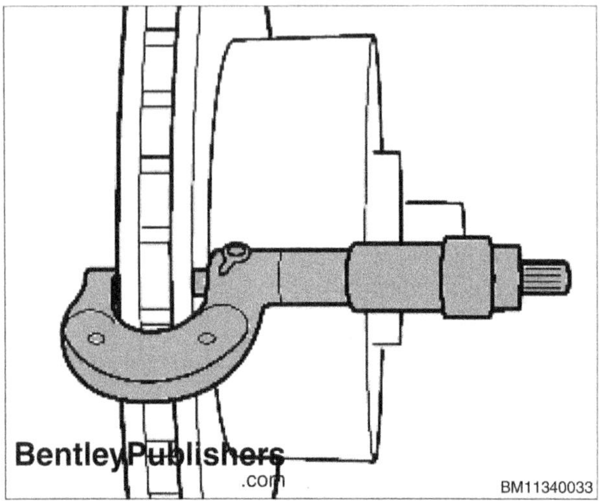

On original equipment rotor, minimum allowable thickness is stamped on rotor hub or edge. Measure friction surface thickness with a micrometer at eight to ten different points and use the smallest measurement recorded. See **Table b**.

Table b. Brake rotor thickness specifications	
Front brake rotor thickness: • New • Wear limit	 22.0 mm (0.866 in) 20.4 mm (0.803 in)
Rear brake rotor thickness: • New • Wear limit	 10.0 mm (0.394 in) 8.4 mm (0.331 in)
Max. machining limit per friction side	0.8 mm (0.031 in)

– If rotor does not pass minimum thickness requirements or is damaged, replace rotor. See **Front brake rotor (disc), removing and installing** or **Rear brake rotor (disc), removing and installing** in this repair group.

Front brake rotor (disc), removing and installing

Replace brake rotor without disassembling wheel hub or bearing. Replace both brake rotors on one axle. Fit new rotors together with new pads.

– Raise vehicle and support in a safe manner. Remove front wheel.

> ***WARNING—***
> • *Make sure the car is stable and well supported at all times. Use a professional automotive lift or jack stands designed for the purpose. A floor jack is not adequate support.*

– Left front wheel: Remove brake pad wear sensor and discard. See **Brake pad wear sensor, replacing** in this repair group.

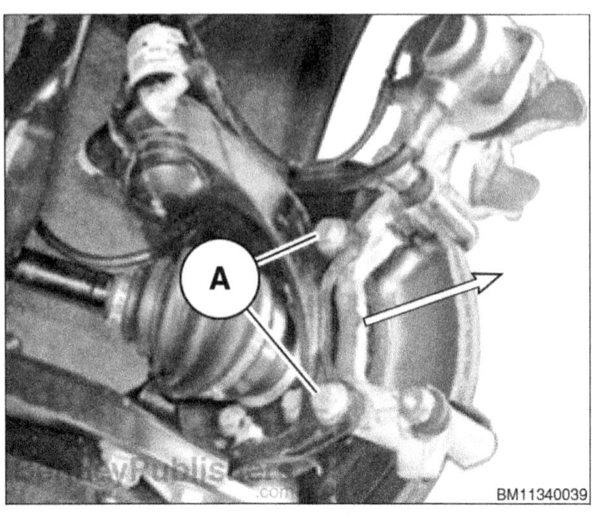

ATE system:

- Remove caliper lower mounting bolt from brake pad carrier and tilt caliper up to remove brake pads.
- Remove pad carrier mounting bolts (**A**). Slide pad carrier and brake caliper off steering knuckle (**arrow**) and tie up.

> ***CAUTION—***
> • *To avoid stretching brake fluid hose, use stiff wire to hang pad carrier and caliper securely from suspension.*

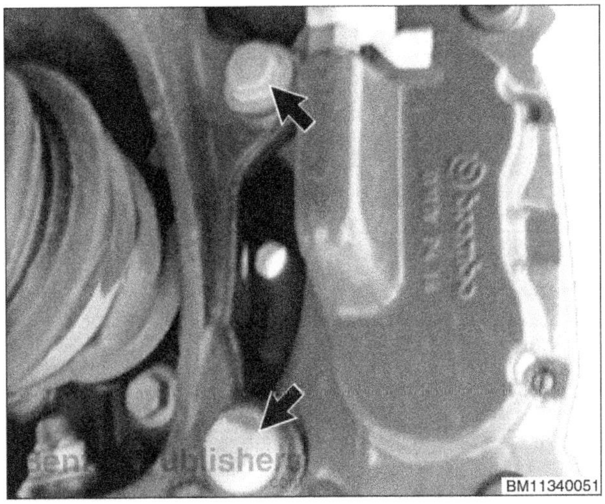

◄ Brembo system:

- Remove brake pads. See **Front brake pads, removing and installing (Brembo)** in this repair group.
- Remove caliper mounting bolts (**arrows**).
- Slide caliper off brake rotor and tie up.

> ***CAUTION—***
> - *To avoid stretching brake fluid hose, use stiff wire to hang caliper securely from suspension.*

◄ Remove brake rotor mounting screw (**arrow**). Remove rotor.

> ***CAUTION—***
> - *Do not strike the rotor friction surface with a hammer or other hard object. If rotor is stuck on wheel bearing hub, carefully tap with a rubber hammer.*

– When reinstalling:

- Make sure rotor contact surface with wheel bearing hub is clean and smooth. Remove rust traces, if necessary.
- Replace brake rotor retaining bolt.
- Replace brake pads. See **Front brake pads, removing and installing (ATE)** or **Front brake pads, removing and installing (Brembo)** in this repair group.

Tightening torques	
ATE brake caliper to brake pad carrier (replace self-locking M8 bolts with new)	35 ± 3 Nm (26 ± 2 ft-lb)
ATE brake pad carrier to steering knuckle (replace self-locking M12 bolts with new)	110 Nm (81 ft-lb)
Brembo brake caliper to steering knuckle (replace self-locking M12 bolts with new)	110 Nm (81 ft-lb)
Brake rotor to wheel hub (replace Torx M10 x 13 bolt with new)	27 Nm (20 ft-lb)
Road wheel to hub	140 ± 10 Nm (103 ± 7 ft-lb)

– After reassembly, pump brake pedal several times to make sure that brake pads are firmly pressed against brake rotors.

Rear brake rotor (disc), removing and installing

Replace brake rotor without disassembling wheel hub or bearing. Replace brake rotors in pairs. Fit new rotors together with new pads.

– Raise vehicle and support in a safe manner. Remove rear wheel.

> ***WARNING—***
> - *Make sure the car is stable and well supported at all times. Use a professional automotive lift or jack stands designed for the purpose. A floor jack is not adequate support.*

– Right rear wheel: Remove brake pad wear sensor and discard. See **Brake pad wear sensor, replacing** in this repair group.

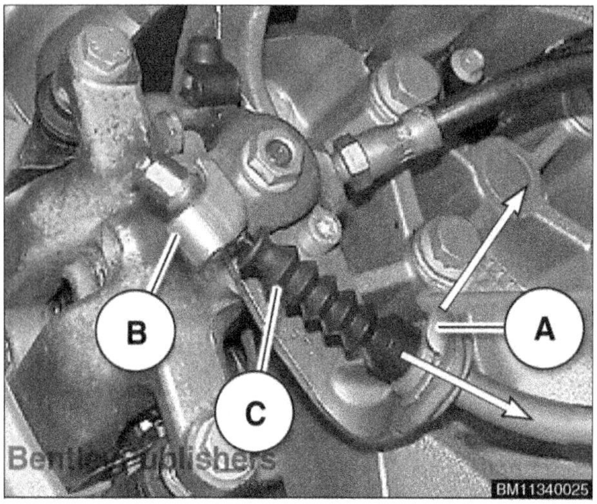

◄ Working at rear brake caliper:
- Pull off parking brake cable sheath locking clip (**A**).
- Detach cable end (**B**) from rear caliper actuating lever.
- Slide cable housing (**C**) out of caliper bracket.

– Remove brake caliper and pads. See **Rear brake pads, removing and installing** in this repair group.

> ***CAUTION—***
> - *To avoid stretching brake fluid hose, use stiff wire to hang caliper securely from suspension.*

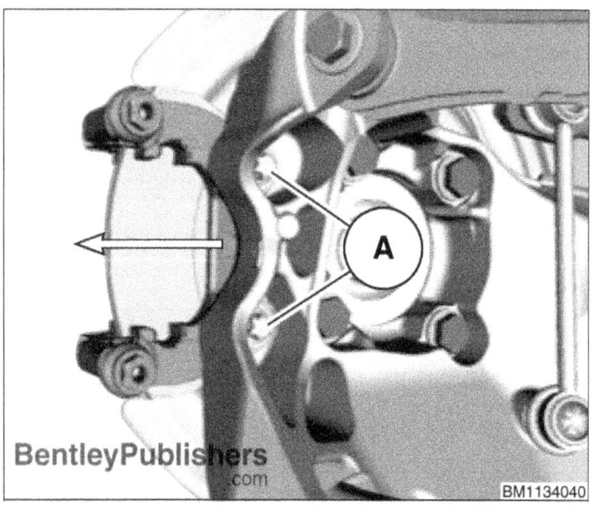

◄ Remove pad carrier mounting bolts (**A**). Slide pad carrier off wheel bearing hub (**arrow**).

◄ Remove brake rotor mounting screw (**arrow**). Remove rotor.

> ***CAUTION—***
> - *Do not strike the rotor friction surface with a hammer or other hard object. If rotor is stuck on wheel bearing hub, carefully tap with a rubber hammer.*

- When reinstalling:
  - Make sure rotor contact surface with wheel bearing hub is clean and smooth. Remove rust traces, if necessary.
  - Replace brake rotor retaining bolt.
  - Replace brake pads. See **Rear brake pads, removing and installing** in this repair group.

Tightening torques	
Brake caliper to brake pad carrier (replace self-locking M8 bolts with new)	35 ± 3 Nm (26 ± 2 ft-lb)
Brake pad carrier to steering knuckle (replace self-locking M10 bolts with new)	65 Nm (48 ft-lb)
Brake rotor to wheel hub (replace Torx M10 x 13 bolt with new)	27 Nm (20 ft-lb)
Road wheel to hub	140 ± 10 Nm (103 ± 7 ft-lb)

- Reattach parking brake cable.
- After reassembly, pump brake pedal several times to make sure that brake pads are firmly pressed against brake rotors.

Front brake caliper, removing and installing (ATE)

- Raise vehicle and support in a safe manner. Remove front wheel.

> ***WARNING —***
> - *Make sure the car is stable and well supported at all times. Use a professional automotive lift or jack stands designed for the purpose. A floor jack is not adequate support.*

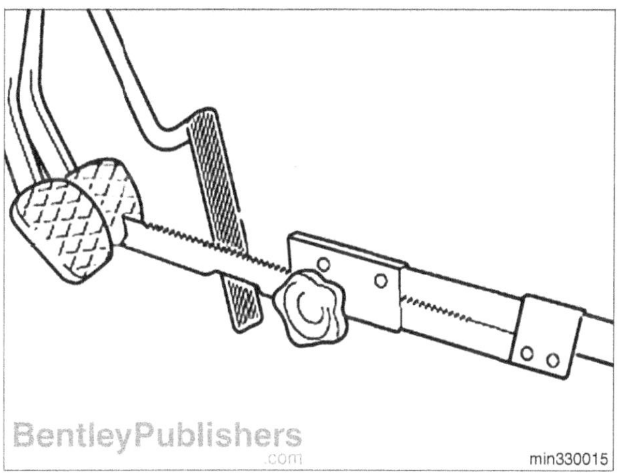

◄ Insert pedal prop and depress brake pedal slightly. This prevents brake fluid from escaping when brake lines are detached.

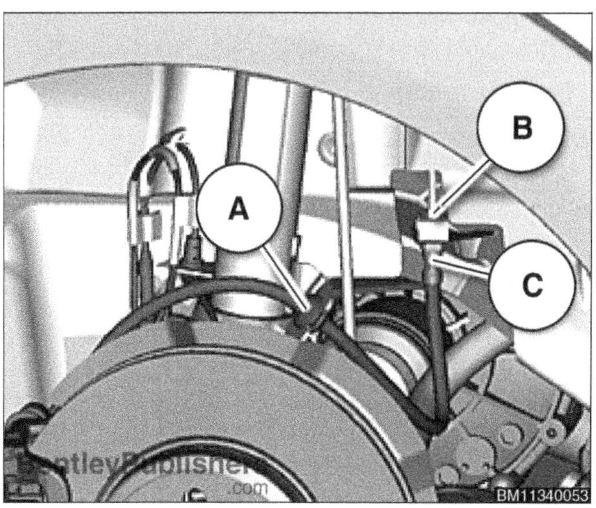

◄ Working inside front wheel housing:
- Pull brake hose out of bracket (**A**).
- Use flare-wrench to loosen and separate brake line (**B**). Counter hold brake hose hex (**C**) to prevent it front rotating in retaining bracket. Be prepared to catch dripping brake fluid.

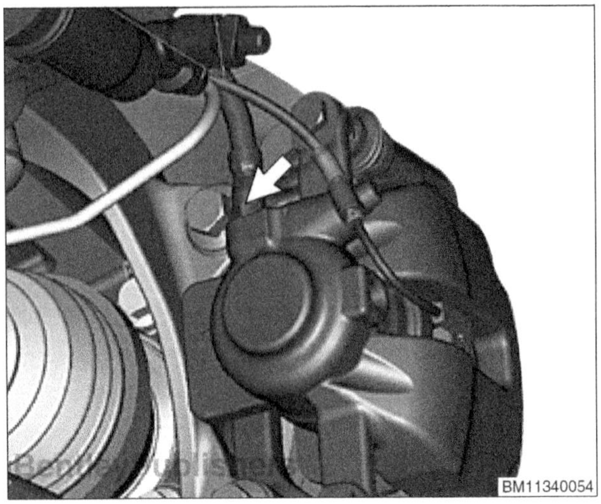

BM11340054

Detach brake hose (**arrow**) from brake calliper.

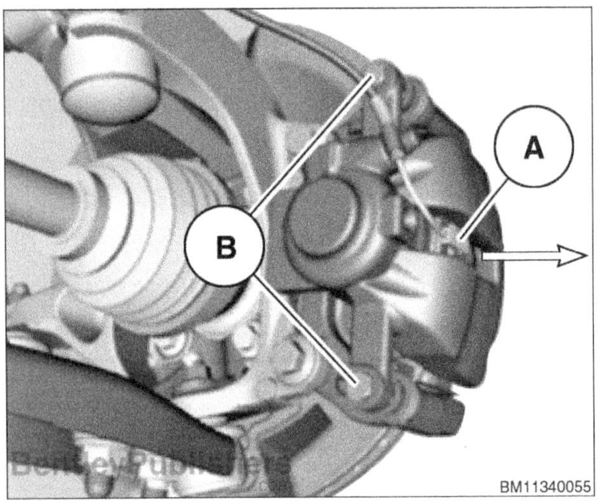

BM11340055

Working at brake caliper:

- Left front wheel: Remove brake pad wear sensor (**A**). See **Brake pad wear sensor, replacing** in this repair group.
- Remove caliper mounting bolts (**B**). If necessary, counter hold hexagon heads while loosening caliper bolts.
- Slide caliper off brake pads and brake pad carrier (**arrow**).

– When installing:

- Place caliper over brake pad carrier, install new mounting bolts and tighten.
- Install and tighten brake hose on brake calliper.
- Move wheels into straight-ahead position. Insert brake hose in retaining bracket and attach brake line.

> ***CAUTION—***
> - *Do not twist brake fluid hose when removing or install ling.*
> - *Make sure brake hose does not contact body parts when steering is rotated lock to lock.*

Tightening torques	
Brake caliper to brake pad carrier (replace self-locking M8 bolts with new)	35 ± 3 Nm (26 ± 2 ft-lb)
Brake fluid line to brake hose	14 Nm (10 ft-lb)
Brake fluid hose to brake caliper	24 ± 2 Nm (18 ± 1 ft-lb)
Road wheel to hub	140 ± 10 Nm (103 ± 7 ft-lb)

– When finished, bleed brakes. See **Bleeding Brakes** in this repair group.

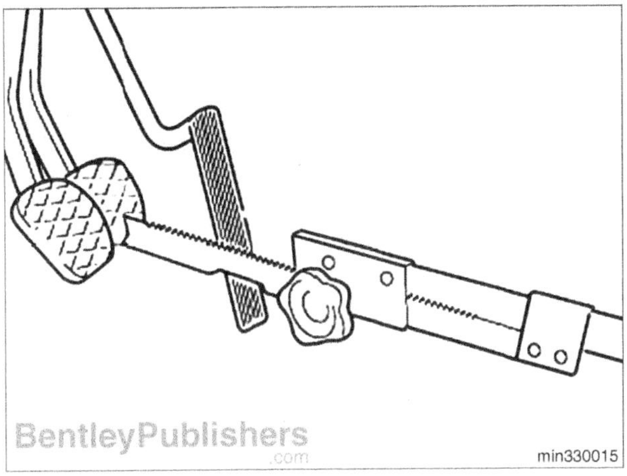

Rear brake caliper, removing and installing

- Raise vehicle and support in a safe manner. Remove rear wheel.

> ***WARNING —***
> - *Make sure the car is stable and well supported at all times. Use a professional automotive lift or jack stands designed for the purpose. A floor jack is not adequate support.*

< Insert pedal prop and depress brake pedal slightly. This prevents brake fluid from escaping when brake lines are detached.

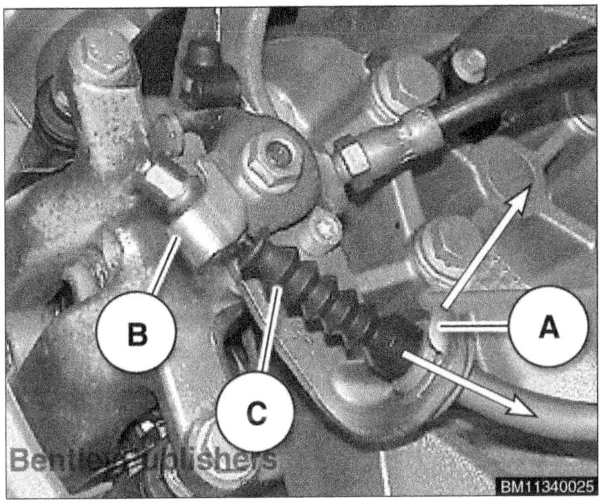

< Working at rear brake caliper:
- Pull off cable sheath locking clip (**A**).
- Detach cable end (**B**) from rear caliper actuating lever.
- Slide cable housing (**C**) out of caliper bracket.

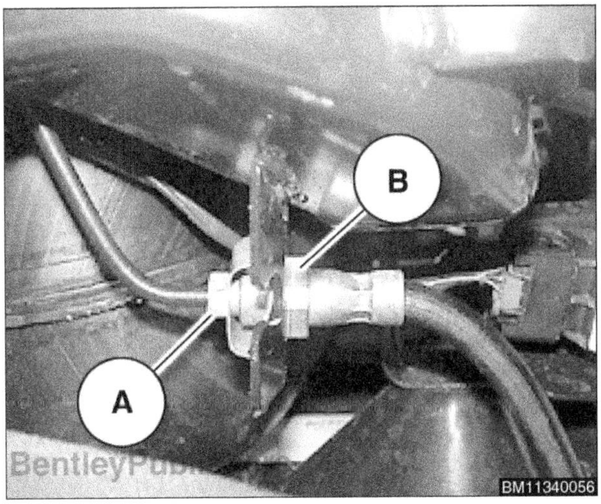

< Working inside rear wheel housing, use flare wrench to loosen and separate brake line (**A**). Counter hold brake hose hex (**B**) to prevent it front rotating in retaining bracket. Be prepared to catch dripping brake fluid.

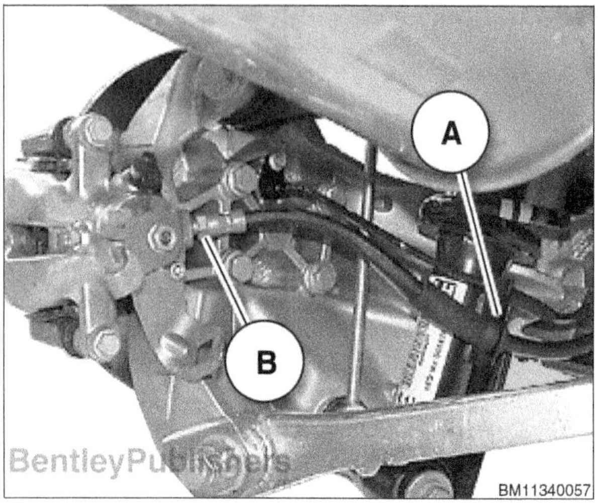

Working behind rear wheel hub assembly:

- Pull brake hose out of bracket (**A**).
- Detach brake hose (**B**) from brake caliper.

– Right rear wheel: Remove brake pad wear sensor (**A**). See **Brake pad wear sensor, replacing** in this repair group.

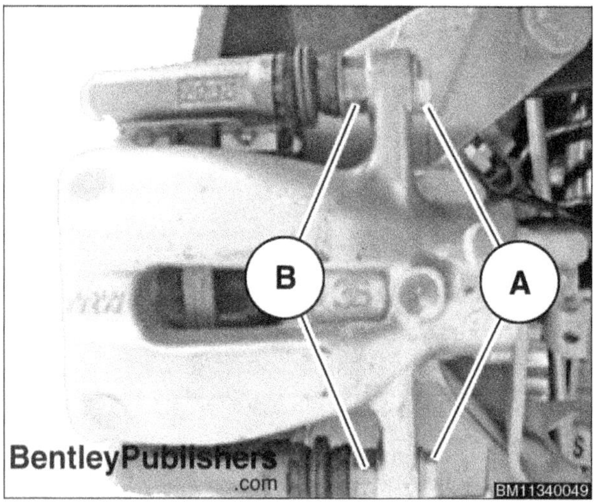

Remove caliper mounting bolts (**A**). If necessary, counter hold at hexagon heads (**B**). Slide caliper off brake pads and pad carrier.

– When installing:

- Place caliper over brake pad carrier, install new mounting bolts and tighten.
- Install and tighten brake hose on brake calliper.
- Insert brake hose in retaining bracket and attach brake line.
- Reattach parking brake cable.

> ***CAUTION—***
> - *Do not twist brake fluid hose when removing or installing.*
> - *Make sure brake hose does not contact body parts.*

Tightening torques	
Brake fluid hose to brake caliper	24 ± 2 Nm (18 ± 1 ft-lb)
Brake fluid line to brake hose	14 Nm (10 ft-lb)
Rear brake caliper to brake pad carrier (replace self-locking M8 bolts with new)	35 ± 3 Nm (26 ± 2 ft-lb)
Road wheel to hub	140 ± 10 Nm (103 ± 7 ft-lb)

When finished, bleed brakes. See **Bleeding Brakes** in this repair group.

Brake Master Cylinder

The brake master cylinder is mounted to the front of the vacuum booster on the left engine compartment bulkhead.

> ***WARNING—***
> - *If the brake master cylinder is replaced, make sure that the brake system is bled by an authorized MINI dealer or other qualified repair facility using the MINI scan tool. See* **Bleeding Brakes** *in this repair group.*

Brake master cylinder, removing and installing

– Remove right and left cowl covers. See **513 Interior Trim**.

– Using a clean syringe, empty brake fluid reservoir.

> ***CAUTION—***
> - *Brake fluid is highly corrosive and dangerous to the environment. Dispose of it properly.*
> - *Discard old brake fluid. Do not reuse.*

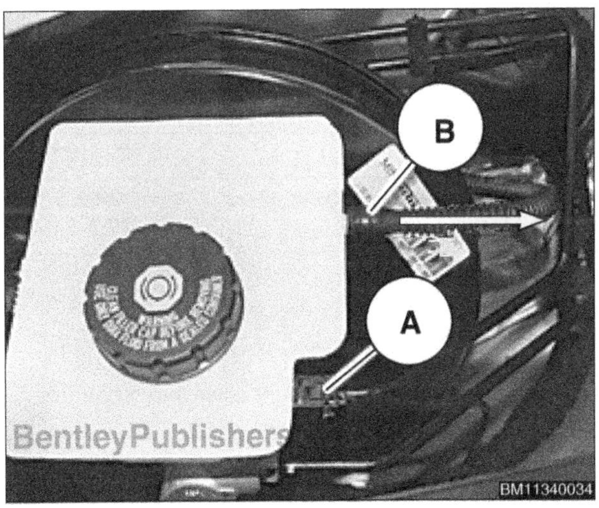

Working at left side of brake fluid reservoir:
- Detach brake fluid level switch electrical connector (**A**).
- If applicable: Pull off clutch master cylinder supply hose (**B**).

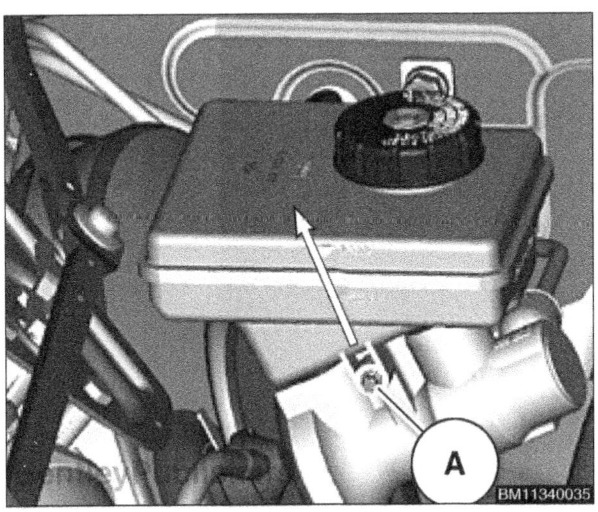

Remove reservoir retaining screw (**A**) and lift off reservoir (**arrow**).

Brake master cylinder, removing and installing

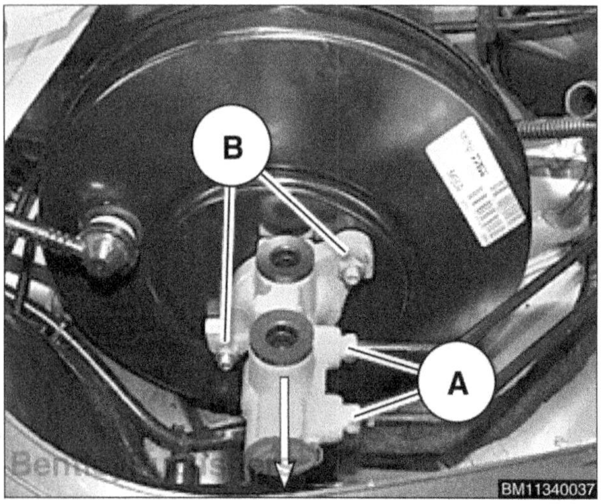

Working at brake master cylinder:

- Loosen and detach brake lines (**A**).
- Remove master cylinder mounting nuts (**B**).
- Lift out master cylinder.

> ***CAUTION—***
> - *Catch dripping brake fluid in shop towels. Do not allow to drip on vehicle painted surface.*
> - *Do not bend brake lines.*
> - *Plug open brake lines and ports immediately.*

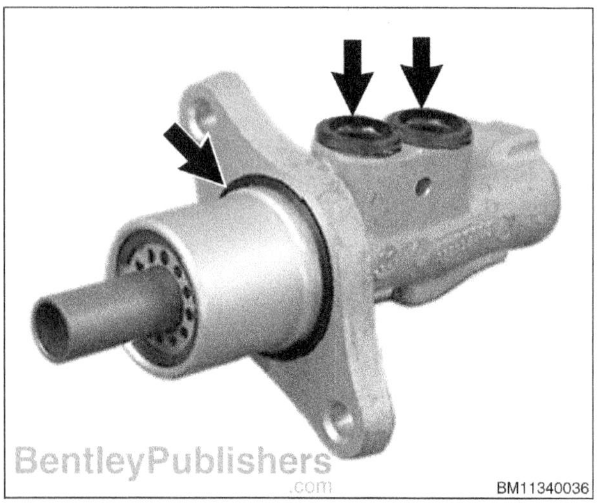

Prior to reinstallation, install new O-ring seals (**arrows**) on master cylinder.

– When installing:

- Make sure master cylinder pressure rod lines up with brake booster pressure rod.
- Replace master cylinder self-locking mounting nuts with new.

Tightening torques	
Brake lines to master cylinder	14 Nm (10 ft-lb)
Brake master cylinder to brake booster (M8)(replace self-locking nuts)	23 ± 3 Nm (17 ± 2 ft-lb)

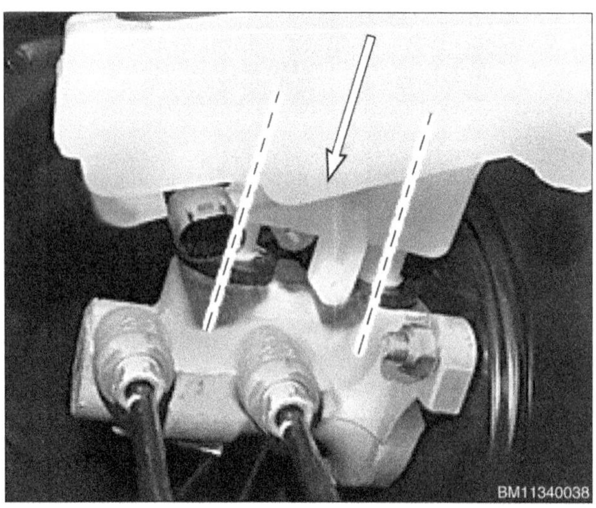

Reinstall fluid reservoir by pressing straight down into grommets.

Tightening torque	
Fluid reservoir to master cylinder	8 ± 2 Nm (6 ± 1 ft-lb)

– Top off fluid reservoir and bleed brakes. See **Bleeding Brakes** in this repair group.

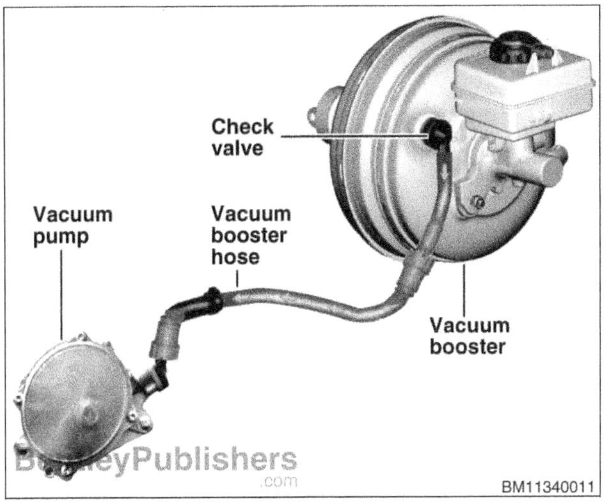

Brake Booster

The brake booster is mounted to the bulkhead on the left (driver) side of the engine compartment, directly behind the brake master cylinder.

Vacuum generated by an engine camshaft powered vacuum pump acts on a large diaphragm in the brake booster to reduce brake pedal effort.

A check valve in the brake booster vacuum line prevents contamination from engine backfires and other sources from entering the brake booster.

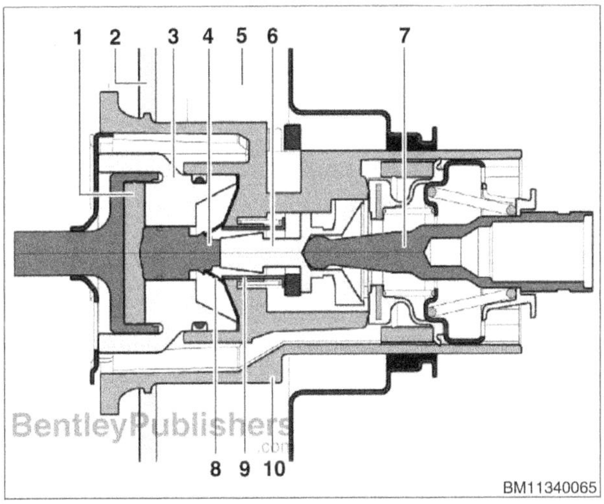

Mechanical braking assistant (MBA)

A brake booster with mechanical braking assistant (MBA) is used in models equipped with ASC + T (not DSC). This purely mechanical device improves braking in emergency situations by providing maximum braking power assistance.

1. Reaction plate
2. Brake booster diaphragm
3. Reaction housing
4. Sensing plate
5. Working chamber
6. Sensing piston
7. Brake master cylinder pushrod
8. Actuating spring
9. Deactivation sleeve
10. Valve housing

340

Brake booster, removing and installing

– Remove brake master cylinder. See **Brake master cylinder, removing and installing** in this repair group.

– Remove left footwell trim (pedal cluster trim).

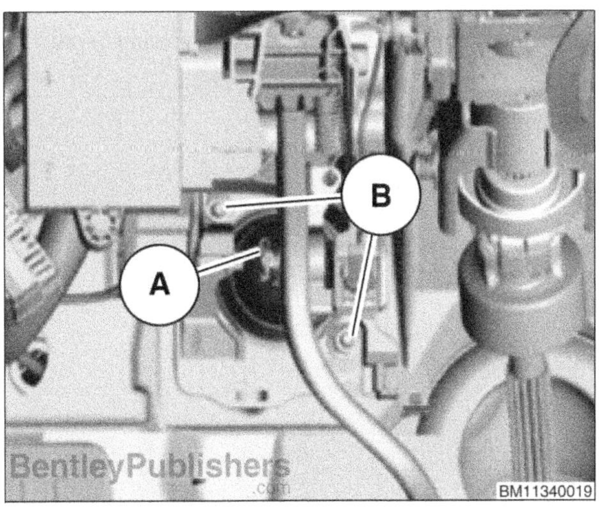

Working at brake booster flange underneath dashboard:

- Detach locking clip (**A**) from brake pedal pin, then pull out pin.
- Unscrew brake booster mounting nuts (**B**).

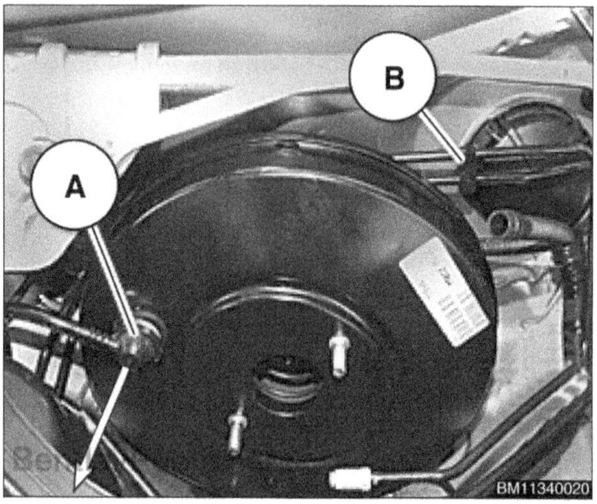

Working at left rear engine compartment bulkhead:

- Pull off one-way check-valve (**A**).
- Unclip brake lines from bracket (**B**).
- Tilt brake booster and remove, carefully checking to make sure no lines, hoses or other components become snagged.

> ***CAUTION—***
> - *To avoid damage to brake booster, do not use excessive force when removing.*
> - *Do not bend brake lines.*

– Installation is reverse of removal. Replace self-locking mounting nuts.

Tightening torque	
Brake booster to bulkhead (replace self-locking nuts)	21 Nm (15 ft--lb)
Brake lines to master cylinder	14 Nm (10 ft-lb)
Brake master cylinder to brake booster (M8)(replace self-locking nuts)	23 ± 3 Nm (17 ± 2 ft-lb)

Brake booster vacuum pump, removing and installing

– With engine fully cooled off, detach vacuum line from vacuum booster pump at left end of cylinder head.

Remove pump mounting screws (**arrows**) and pull pump out of cylinder head.

Installation is reverse of removal. Replace pump sealing O-ring (**arrow**).

Brake booster vacuum hose, removing and installing (N12 engine)

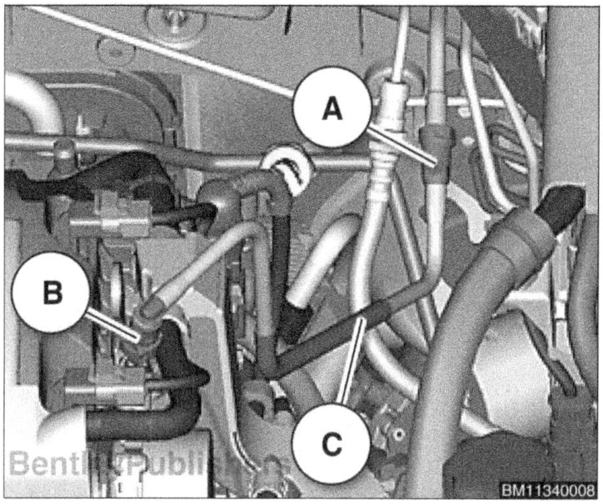

- Pump brake pedal several times to reduce vacuum in brake booster.
- Remove air filter housing from engine compartment. See **130 Fuel Injection**.
- ◁ Separate quick-connect couplings (**A**, **B**) and remove vacuum line (**C**).
- Installation is reverse of removal.

Brake booster vacuum hose, removing and installing (N16 engine)

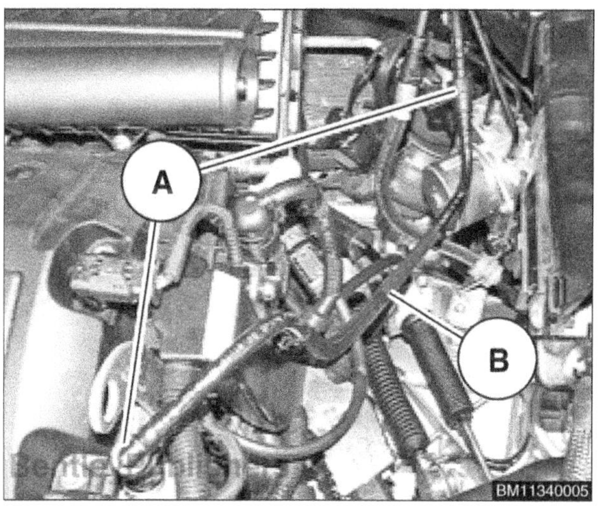

- Pump brake pedal several times to reduce vacuum in brake booster.
- Remove air filter housing from engine compartment. See **130 Fuel Injection**.
- ◁ Separate quick-connect couplings (**A**) and remove vacuum line (**B**).
- Installation is reverse of removal.

Brake booster vacuum hose, removing and installing (N14 engine)

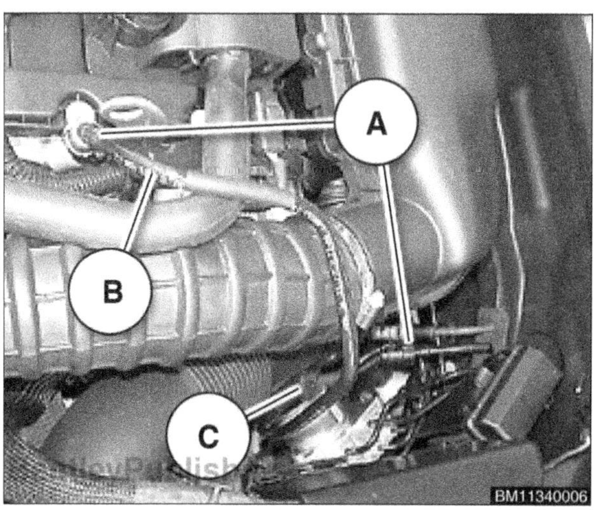

- Pump brake pedal several times to reduce vacuum in brake booster.
- Remove right and left cowl covers. See **513 Interior Trim**.
- ◁ Detach vacuum line at clip (**C**) and separate quick-connect couplings (**A**, **B**).
- Installation is reverse of removal.

340

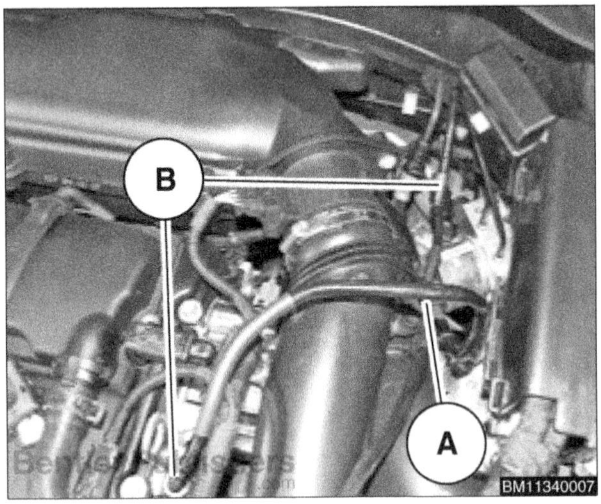

Brake booster vacuum hose, removing and installing (N18 engine)

– Pump brake pedal several times to reduce vacuum in brake booster.

– Remove right and left cowl covers. See **513 Interior Trim**.

Detach vacuum line at clip (**A**) and separate quick-connect couplings (**B**).

– Installation is reverse of removal.

Brake booster check valve, removing and installing

The check valve is integrated with a section of molded vacuum line. Replace both together.

– Pump brake pedal several times to reduce vacuum in brake booster.

– Remove air filter housing from engine compartment. See **130 Fuel Injection**.

– Remove right and left cowl covers. See **513 Interior Trim**.

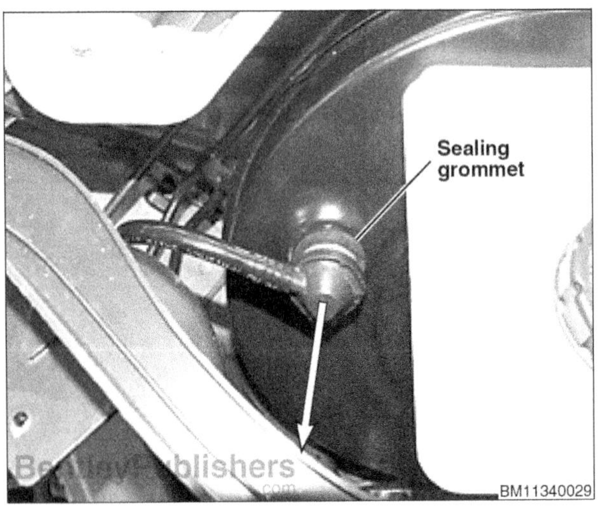

Carefully pry out check-valve from brake booster.

- Inspect check-valve sealing grommet in brake booster. Replace if necessary.

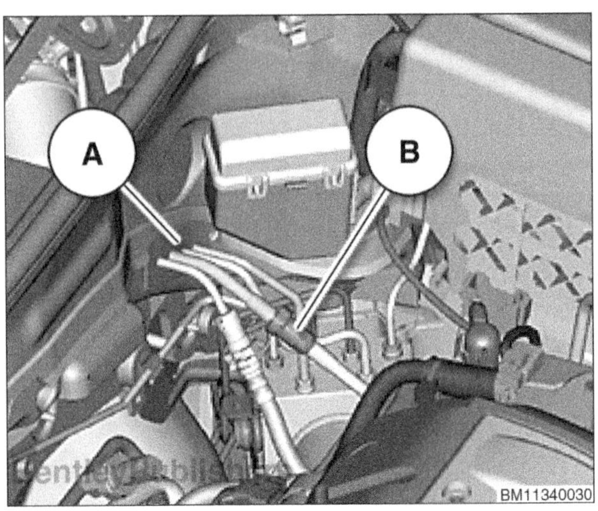

Working at bulkhead:

- Remove brake line bracket (**A**).
- Separate quick-release connector (**B**) and slide out vacuum line with check valve.

– Installation is reverse of removal.

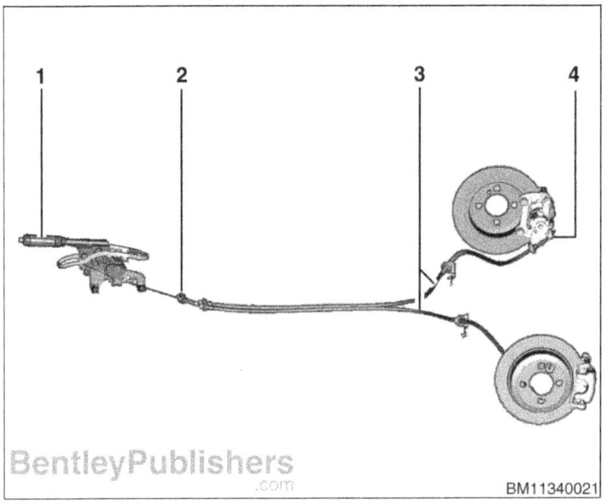

BentleyPublishers.com
BM11340021

PARKING BRAKE

Parking brake overview

The parking brake is operated from the parking brake lever (**1**) by two Bowden cables (**3**), with a left-right compensator (**2**) fitted behind the brake lever and an adjustment nut under the lever in the center console.

The parking brake stops the car via the rear caliper pistons (**4**) acting directly on the rear brake pads which, in turn, press on the brake rotors (discs).

The parking brake is self-adjusting, with the adjustment being made at the caliper piston. A special tool is required to wind back the piston when fitting new pads.

Parking brake, checking and adjusting

- Check parking brake as follows:
  - Release parking brake lever. Check to see that there is no drag at rear wheels.
  - Pull up brake lever 1 notch. Again check to see that there is no drag. Parking brake warning light, however, may be ON.
  - Pull up lever 2 notches. Check to see that there is a slight drag at each wheel and that parking brake warning light is ON.
  - Drive vehicle and pull up on parking brake handle. Check that rear wheels lock up.

> ***WARNING—***
> - *Perform this test where there is no other traffic.*

- Adjust parking brake under the following circumstances:
  - Brake lever pulls up more than 6 notches.
  - Rear brake pads or rotors are replaced.
  - Parking brake cables or handle are replaced.

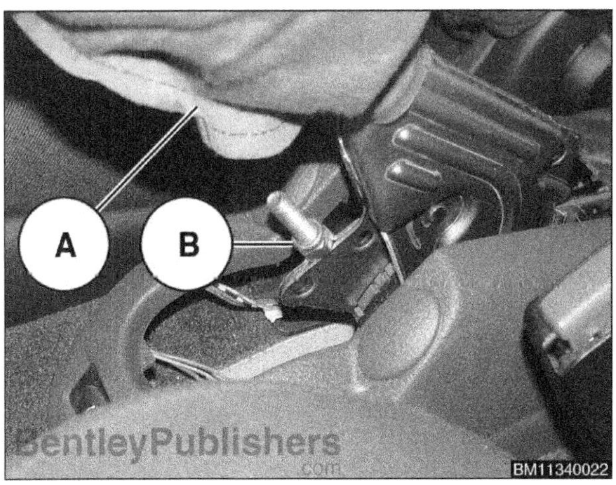

BentleyPublishers.com
BM11340022

Working inside vehicle between front seats:
- Unclip parking brake handle boot (**A**) and lift off to access end of parking brake cable.
  Back off self-locking nut (**B**) until cable is completely slack.

- Raise rear of vehicle and support in a safe manner. Remove rear wheels.

> ***WARNING—***
> - *Make sure the car is stable and well supported at all times. Use a professional automotive lift or jack stands designed for the purpose. A floor jack is not adequate support.*

- Remove rear calipers from brake pad carriers without disconnecting brake fluid hoses or parking brake cables. See **Brake caliper, removing and installing** in this repair group.

340

- Press in each rear caliper piston fully with MINI special tool set **34 6 300** or equivalent. The special tool turns the piston while pressing in.

> *CAUTION—*
> - *Make sure that the caliper piston dust seal does not twist and get damaged during this process.*
> - *Pressing caliper piston in may cause brake fluid reservoir to overflow and damage paint work. To prevent this, use a clean syringe to first remove some fluid from reservoir.*

- Reinstall brake calipers. Replace self-locking caliper mounting bolts with new.

Tightening torque	
Rear caliper to brake pad carrier (replace self-locking M8 bolts with new)	35 ± 3 Nm (26 ± 2 ft-lb)

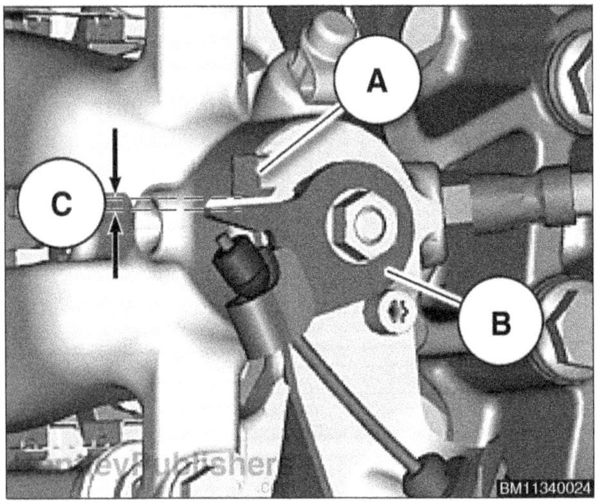

- Working inside vehicle, tighten parking brake cable adjusting nut until gap (**C**) at rear caliper is set between cable-actuated lever (**B**) and stop (**A**).

Parking actuator lever	
Gap **C** between actuator lever and stop	0.5 - 1.5 mm (0.02 - 0.06 in)

- Apply parking brake lever three times.
- Press brake pedal to floor at least three times.
- Check gap at actuator lever. Reset if necessary.
- Check function of parking brake.
- Check function of parking brake warning light.
- Top off brake fluid.

Parking brake handle, removing and installing

- Remove center console. See **513 Interior Trim**.

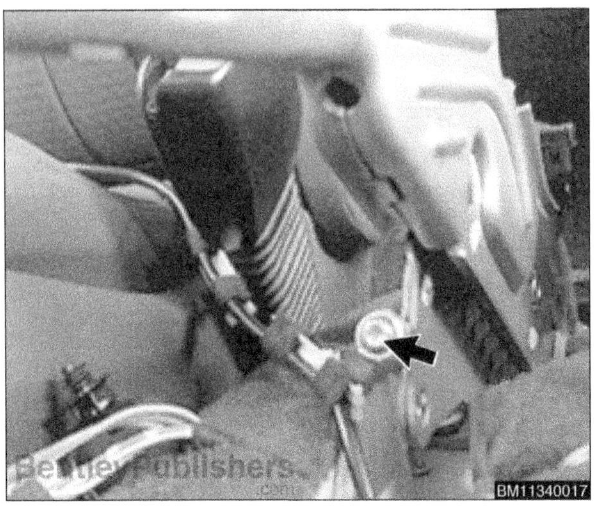

- Release parking brake handle adjustment unit (**arrow**).

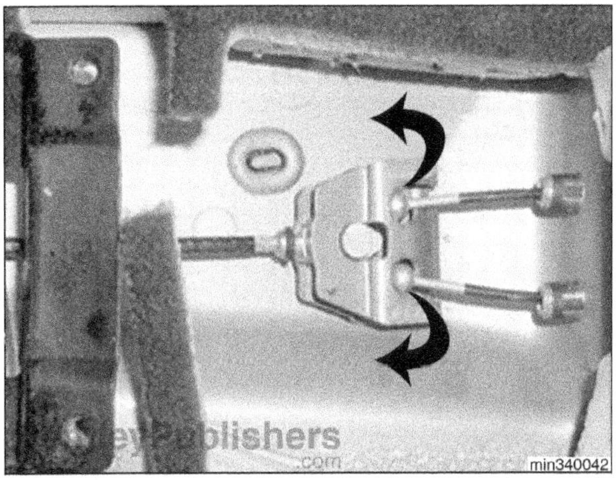

- Working behind parking brake handle, detach both brake cables from compensator arm.
- Disconnect parking brake warning switch connector.

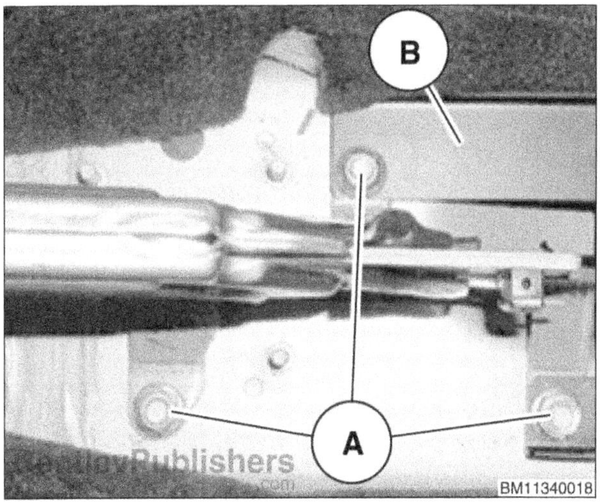

- Remove mounting fasteners (**A**) and lift off bracket (**B**). Remove parking brake handle.
- Installation is reverse of removal.

Tightening torque	
Parking brake handle bracket to floor (M8)	22 ± 3 Nm (16 ± 2 ft-lb)

Parking brake cables, replacing

- Remove center console. See **513 Interior Trim**.

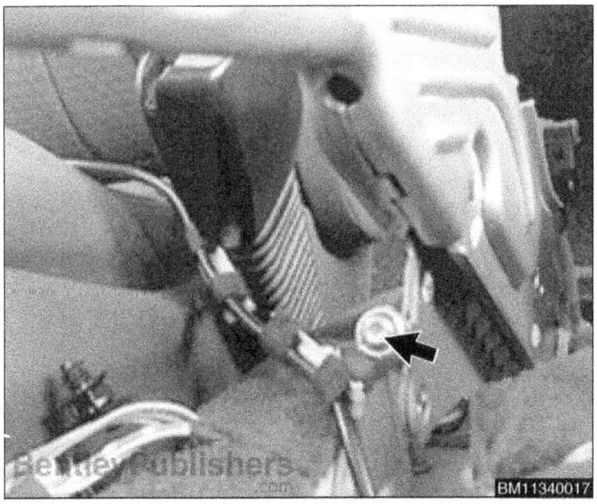

- Release parking brake handle adjustment unit (**arrow**).

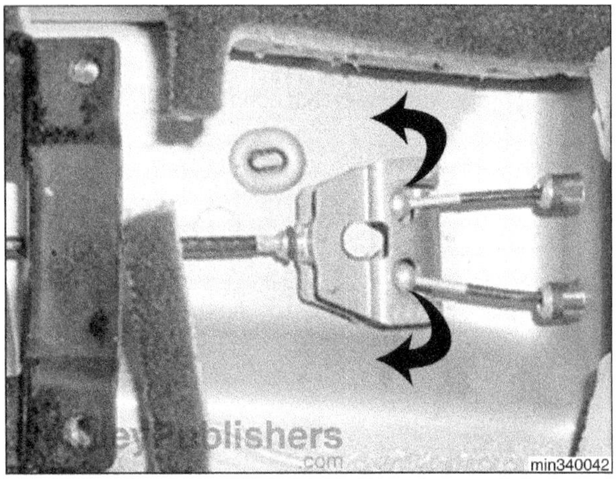

Working behind parking brake handle, detach both brake cables from compensator arm.

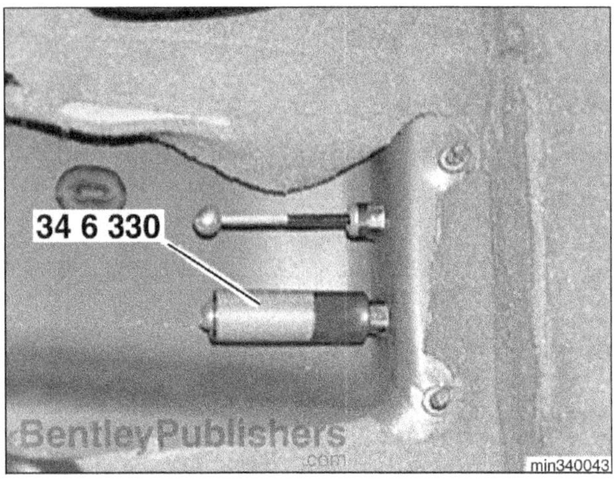

Use MINI special tool **34 6 330** to force parking brake cable barbs free of body. Push cables through.

– Raise rear of vehicle and support in a safe manner.

> ***WARNING —***
> - *Make sure the car is stable and well supported at all times. Use a professional automotive lift or jack stands designed for the purpose. A floor jack is not adequate support.*

– Lower exhaust system slightly and remove heat shield underneath rear subframe.

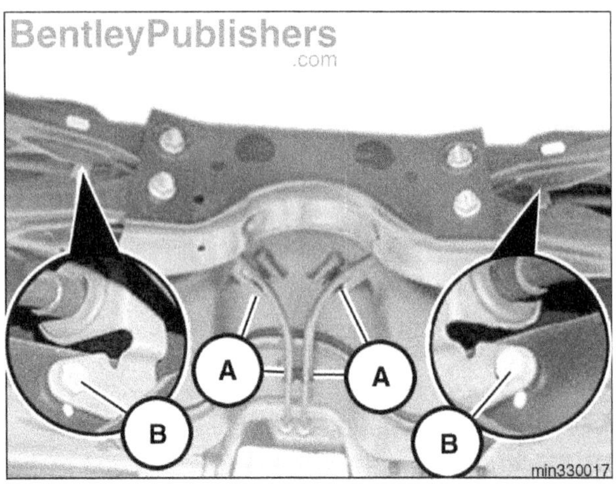

Detach parking brake cables:

- Pry cables off plastic retainers (**A**) on fuel tank.
- Remove cable bracket mounting bolts (**B**) at rear subframe.

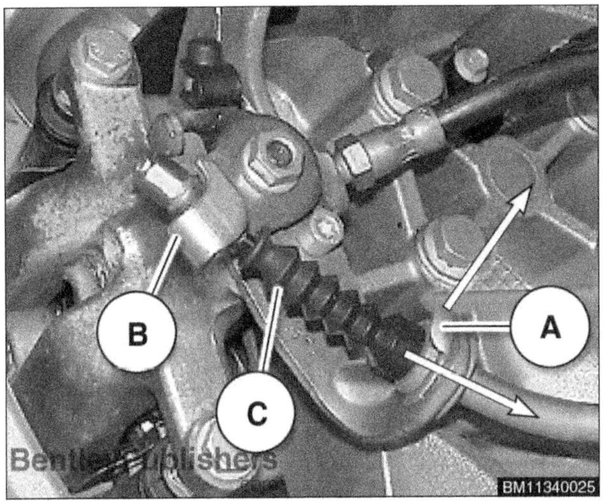

Working at rear brake caliper:

- Pull off cable sheath locking clip (**A**).
- Detach cable end (**B**) from rear caliper actuating lever.
- Slide cable housing (**C**) out of caliper bracket.

– Installation is reverse of removal. Adjust parking brake. See **Parking brake, checking and adjusting** in this repair group.

ABS Component Replacement

> ***WARNING—***
> - *The laws of physics cannot be repealed, even with DSC. Practice a responsible driving style. Avoid using the additional margin of safety afforded by traction control as an excuse for taking risks.*
> - *Do not make any modifications to the DSC system. Allow only authorized technicians to service DSC.*

> ***CAUTION—***
> - *If tires are different makes, the ABS / traction control system may over-react. Only fit tires of the same make and tread pattern.*
> - *In adverse conditions, such as rocking the car out of deep snow or other soft surface, or when snow chains are fitted, switch traction control OFF and allow the car's drive line to operate conventionally.*

MINI vehicles are equipped with antilock brakes (ABS) combined with automatic stability control (ABS / ASC) or dynamic stability control (ABS / DSC). This manual refers to these systems as ABS. ASC or DSC is specified when necessary.

Front ABS wheel speed sensor, removing and installing

– Raise vehicle and support in a safe manner. Remove front wheel.

> ***WARNING—***
> - *Make sure the car is stable and well supported at all times. Use a professional automotive lift or jack stands designed for the purpose. A floor jack is not adequate support.*

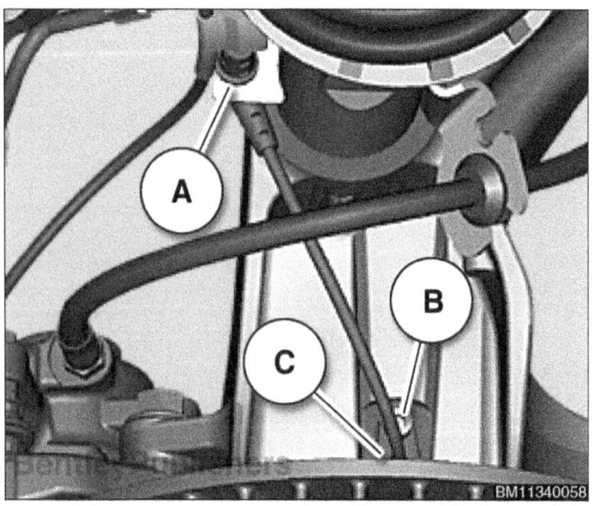

Working above wheel bearing housing:

- Detach wheel speed sensor harness from bracket (**A**) on strut housing.
- Remove speed sensor mounting bolt (**B**) and remove sensor (**C**).

Separate sensor electrical coupling (**arrow**) and remove sensor.

Installation is reverse of removal. Bear in mind the following:

- Clean wheel speed sensor mounting hole (**arrow**) and grease with Staburags® NBU 12/K or equivalent anti-seize compound.
- Make sure plug connections and plastic harness retainers are undamaged.
- Make sure sensor harness is routed as before.

Tightening torque	
Road wheel to hub	140 ± 10 Nm (103 ± 7 ft-lb)
Wheel speed sensor to wheel bearing housing	8 Nm (6 ft-lb)

Rear ABS wheel speed sensor, removing and installing

– Remove rear underbody splash shield.

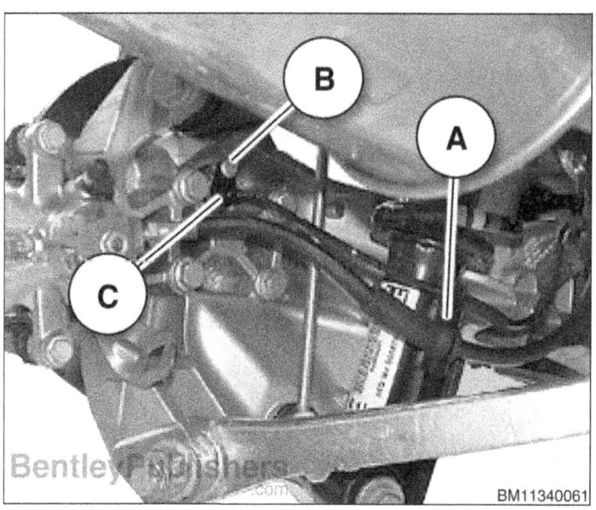

Working underneath vehicle, behind wheel bearing housing:

- Pull wheel speed sensor harness out of bracket (**A**).
- Remove speed sensor mounting bolt (**B**) and remove sensor (**C**).

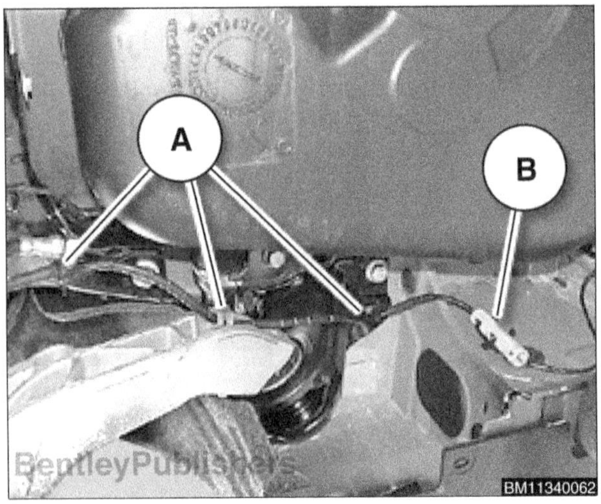

◀ Working underneath vehicle:

- Detach sensor harness from brackets (**A**).
- Separate sensor electrical coupling (**B**) and remove sensor.

— Installation is reverse of removal. Bear in mind the following:

- Clean wheel speed sensor mounting hole and grease with Staburags® NBU 12/K or equivalent anti-seize compound.
- Make sure plug connections, plastic harness retainers and rubber grommets are undamaged.
- Make sure sensor harness is routed as before.
- Use new wire ties as necessary.

Tightening torque	
Wheel speed sensor to wheel bearing housing	8 Nm (6 ft-lb)

DSC acceleration sensor, removing and installing

The DSC acceleration sensor combines the functions of the transverse acceleration sensor and the yaw (rotation rate) sensor.

— Remove passenger seat and lift up floor covering. See **520 Seats**.

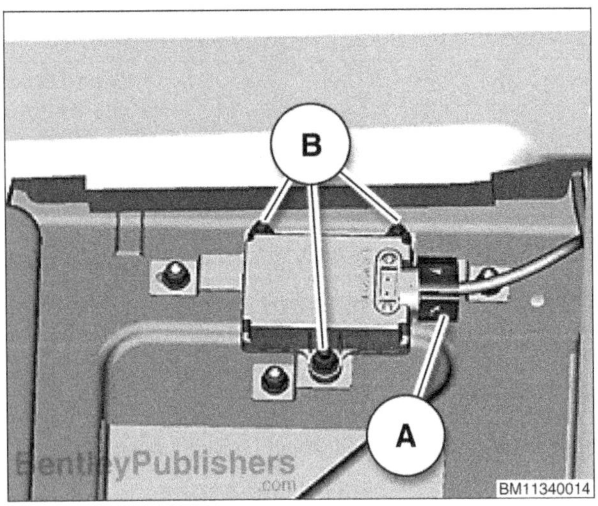

◀ Working at acceleration sensor unit:

- Detach electrical connector (**A**).
- Remove mounting nuts (**B**) and lift off sensor.

— Installation is reverse of removal.

> *CAUTION—*
> - *The sensor is vibration sensitive. Make sure mounting screws are torqued correctly and the harness connector is attached firmly. Otherwise there may be DSC malfunctions.*

Tightening torque	
DSC sensor to floor bracket (M6)	9 Nm (7 ft-lb)

Steering angle sensor

Where equipped, the steering angle sensor is integral with the steering column switch cluster (SZL). For more information, including removal and installation of the SZL, see **612 Switches**.

To reset steering angle sensor, connect vehicle to MINI scan tool and carry out steering angle sensor adjustment system function.

ABS control module, removing and installing

> *CAUTION—*
> - *Handle the electronic control module using accepted static prevention equipment and techniques. See* **600 Electrical System–General**.

The ABS electronic control module is attached to the ABS hydraulic unit. Remove combined unit and place on workshop bench to

BM11340015

separate. Be sure to plug open fluid lines and ports. See **ABS hydraulic unit, removing and installing** in this repair group.

Before starting work, use MINI scan tool to read out ABS control module fault memory.

< Remove module mounting screws (**arrows**) and pull module off hydraulic unit without tilting side to side.

CAUTION—
• *Do not allow hydraulic unit or control module contact surfaces to contact dirt or moisture.*

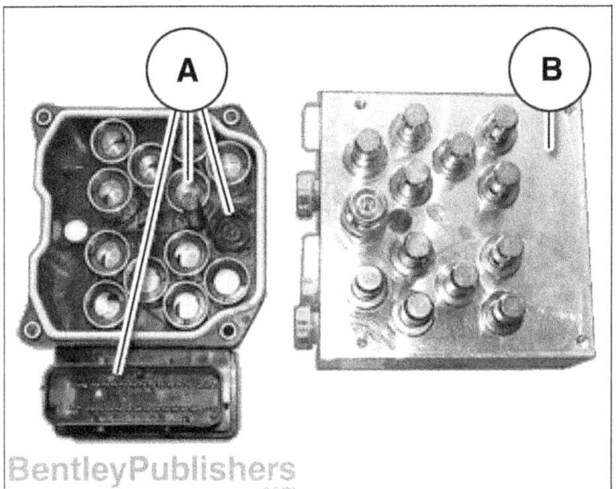

BM11340016

< When installing:

- Do not touch plug contacts (**A**) or hydraulic unit sealing surface (**B**) with bare fingers.
- Do not remove and reinstall control module. Module seal distorts during installation and cannot be reused.
- Clean hydraulic unit sealing surface (**B**) with lint-free cloth. Do not use hard or sharp objects.
- Replace mounting screws. Tighten in crisscross pattern in 2 stages.

Tightening torques	
ABS control module to hydraulic unit • Stage 1 • Stage 2	 1 Nm (9 in-lb) 3 Nm (27 in-lb)

– Code and program new module using MINI scan tool.

ABS hydraulic unit, removing and installing

The ABS electronic control module is attached to the hydraulic unit. Remove combined unit and place on workshop bench to separate. Be sure to plug open fluid lines and ports.

Before starting work, use MINI scan tool to read out ABS control module fault memory.

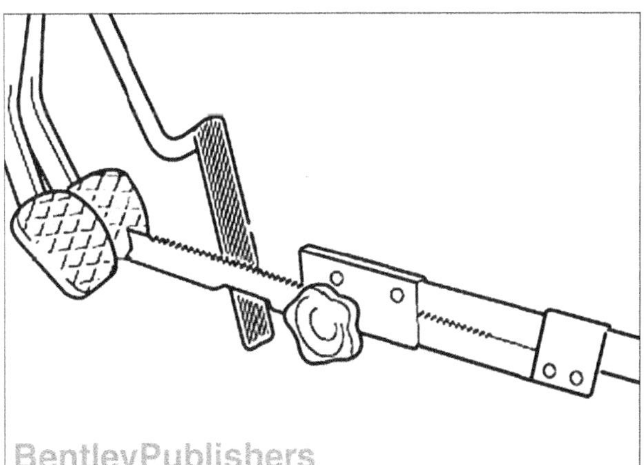

min330015

< Insert pedal prop and depress brake pedal slightly. This prevents brake fluid from escaping when brake lines are detached.

– If necessary, remove air filter housing.

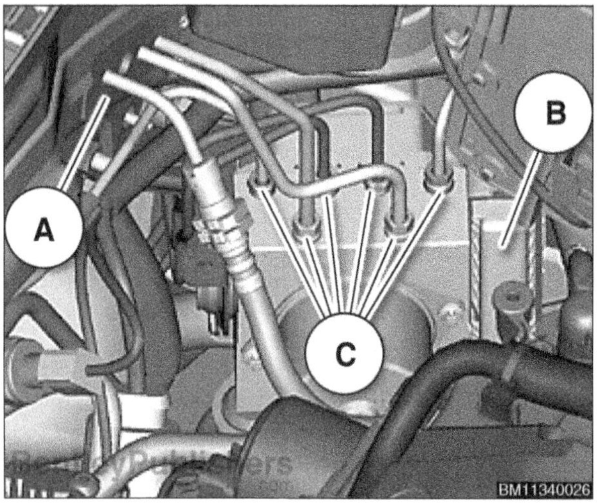

Working in rear engine compartment bulkhead:

- Remove brake line bracket (**A**).
- Detach control module connector (**B**).
- Loosen and detach brake lines (**C**).

> ***CAUTION—***
> - *Catch dripping brake fluid in shop towels. Do not allow fluid to drip on vehicle painted surface.*
> - *Mark brake lines before removing to avoid mixing them up.*
> - *Do not bend brake lines.*
> - *Plug open brake lines and ports immediately.*

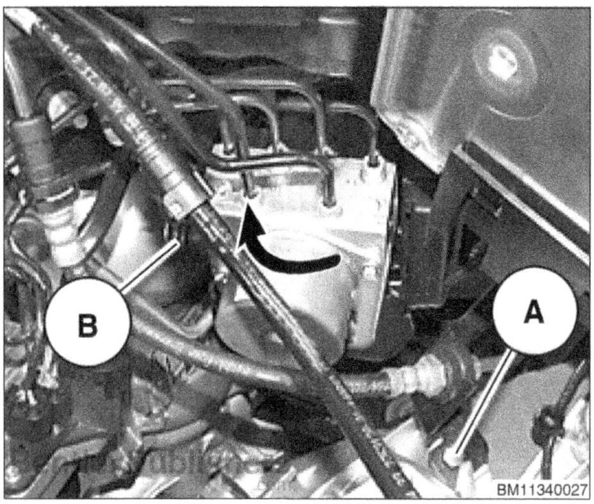

Working in rear engine compartment bulkhead:

- If applicable: Remove clutch cable bracket mounting screw (**A**).
- Remove hydraulic unit bracket mounting bolt (**B**).
- Push clutch cable aside. Raise hydraulic unit slightly, then swing out.

– Separate hydraulic unit from ABS control module. See **ABS control module, removing and installing**.

Installation is reverse of removal. Keep in mind the following:

- Make sure rubber mounts (**arrows**) are fitted correctly in bores.
- Install brake lines carefully to avoid mixing them up.
- Bleed brakes. See **Bleeding Brakes** in this repair group.
- Code and program ABS control module using MINI scan tool.

Tightening torques	
ABS hydraulic unit to bracket	8 Nm (6 ft-lb)
Brake line to hydraulic module (M10 or M12)	14 Nm (10 ft-lb)

400 Body–General

GENERAL

This section covers system descriptions and general information for the repair groups found in **4 Body** and **5 Body Equipment**.

BM11400001

Body Assembly

Body dimensions

Body dimensions vary slightly among models. Dimensions are given in inches (mm).

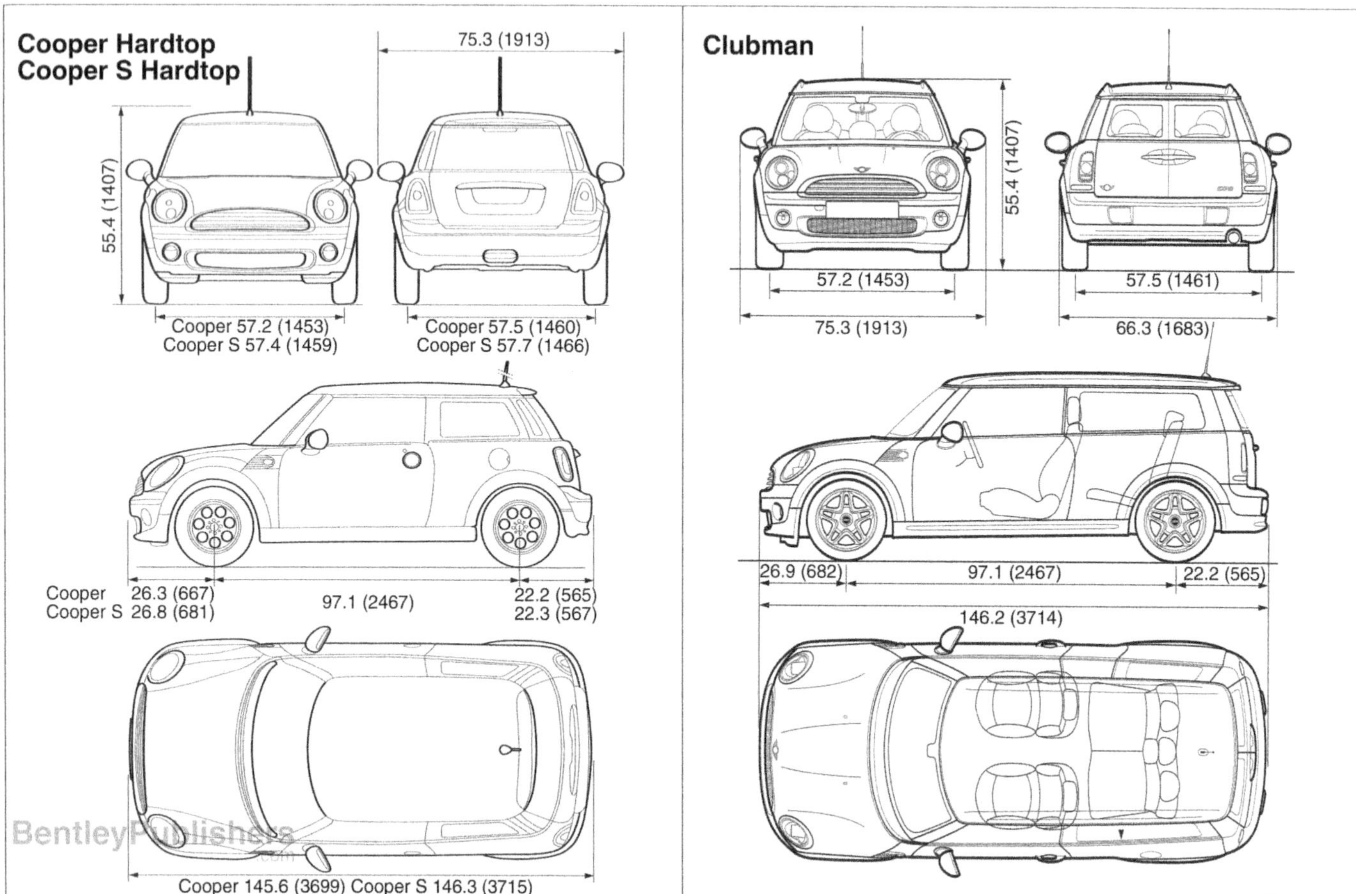

BM11400002

Body shell

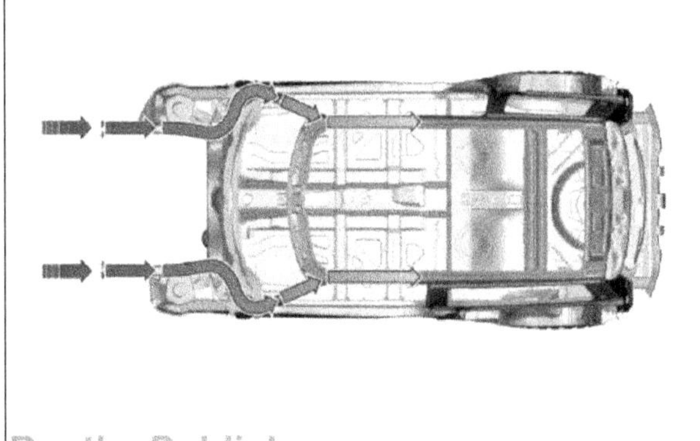
BM11400004

Engineered load paths transmit crash impact forces through the rigid body shell and safely around the passenger compartment.

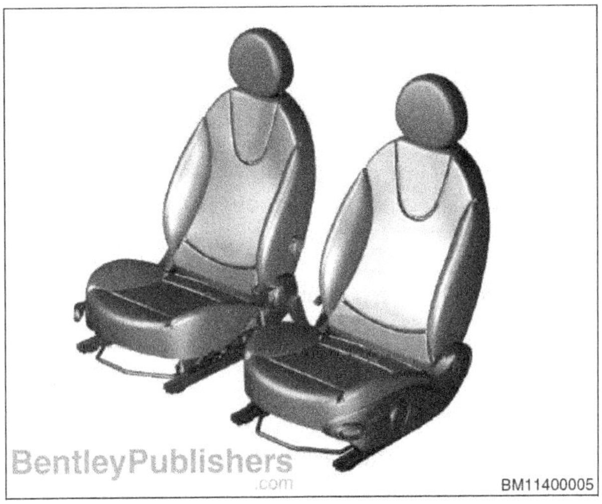

BM11400005

INTERIOR FEATURES

Seats

Each front seat is equipped with an airbag in the outboard seatback cushion. The cushions have rupture points from which the airbags are deployed.

Two-level seat heating is available as an option. Pressing the seat heating switch once or twice adjusts the level setting, indicated by LEDs in the switches. The switches are mounted on the center dashboard.

See **520 Seats** for seat removal information.

BM11400007

Panorama sunroof

The sliding and tilting panorama sunroof operates with two glass panels and two roof panel headliners. The rear glass panel lifts to ventilate the interior.

The sunroof control module is connected to other vehicle control modules with K-CAN.

See **540 Sunroof** for additional information.

400

Instruments and controls

Tilt steering column is operated manually. See **320 Steering and Wheel Alignment**.

Multi-function steering wheel contains two key pads containing controls for sound system and cruise control. See **612 Switches**.

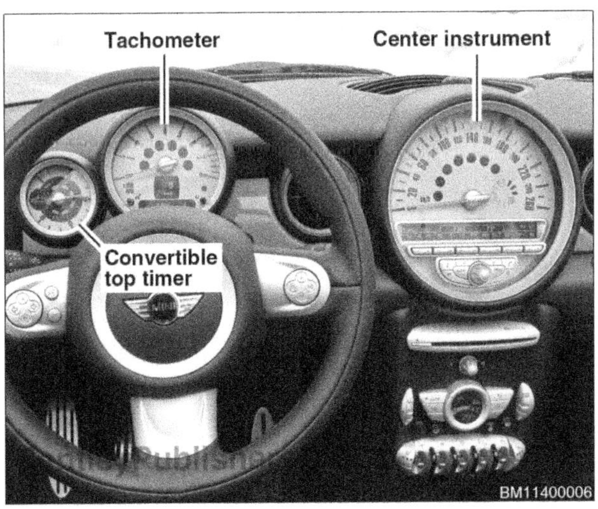

BM11400006

Instruments are large easy-to-read analog gauges. Each is removable as a unit without removing the dashboard. See **620 Instruments**.

On-board computer LCD displays are integrated into the tachometer. See **620 Instruments**.

Integrated on-board navigation system, based on global positioning system (GPS) technology, is optional. When navigation is installed, the car communication computer (CCC) in center dashboard displays radio, communications and other vehicle information and is controlled by a joystick. See **650 Radio**.

Park distance control (PDC) is an optional system which uses ultrasonic sensors in the bumper trim to warn the driver of approaching too close to obstacles when parking.

Tire pressure monitoring uses 2 different systems:

- Flat tire monitor or FTM (2007 models) uses ABS wheel speed sensors to monitor run-flat tire inflation and warn of under-inflation.
- Tire pressure monitor or TPMS (2008 and later models) monitors tire pressure directly using a sensor in each wheel.

See **300 Suspension, Steering and Brakes–General** for more information on tire pressure montoring.

Heating and air-conditioning

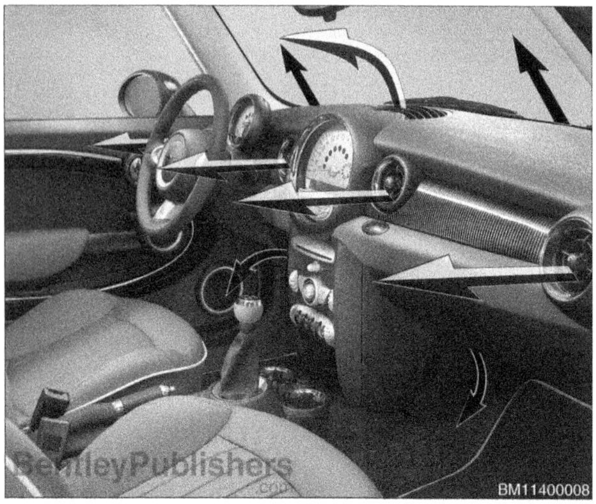

MINI models are equipped with one of two systems:

- **IHKS** is the manually operated heating and air-conditioning system.
- **IHKA** is the integrated automatic climate control system.

Repair information for the heating and air-conditioning system is in **640 Heating and Air-conditioning**.

410 Fenders, Engine Hood

GENERAL

This repair group covers removal and installation of the front fenders, wheel housing liners and engine hood. Also included is the procedure for placing the radiator support (front panel) in service position.

Cautions

CAUTION—
- *The body is painted at the factory after assembly. Realignment of body panels may expose unpainted metal. Paint all exposed metal once work is complete.*
- *To avoid marring trim pieces, use a plastic prying tool or a screwdriver with the tip wrapped with masking tape.*
- *Trim panel plastic clips are easily damaged. Have extra clips on hand before beginning repair procedures.*

FRONT FENDERS

Front fender, removing and installing

– Raise front of vehicle and remove wheel.

WARNING—
- *Make sure car is stable and well supported at all times. Use a professional automotive lift or jack stands designed for the purpose. A floor jack is not adequate support.*

– Remove side turn signal light by prying on front edge with a plastic prying tool and sliding bezel forward. Detach harness connector for light. See **630 Lights**.

– Remove wheel arch trim and wheel housing liner. See **Wheel housing liner, removing and installing** in this repair group.

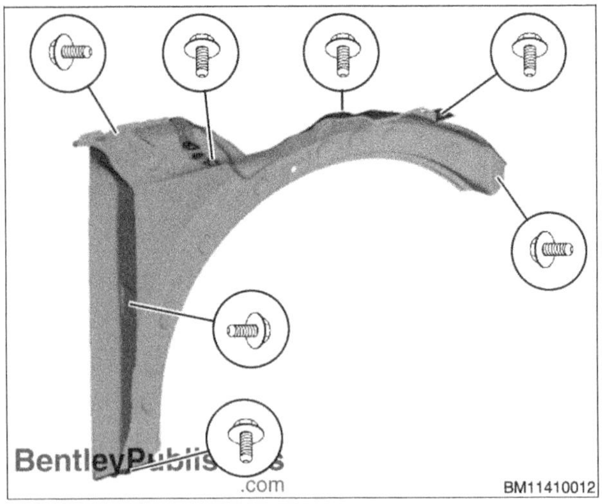

- Remove mounting bolts (**insets**) and lift off fender.
- Installation is reverse of removal.

Tightening torque	
Front fender to body (M6 bolt)	13 Nm (10 ft-lb)

Wheel housing liner, removing and installing

- Raise vehicle and support safely.

> *WARNING—*
> - *Make sure car is stable and well supported at all times. Use a professional automotive lift or jack stands designed for the purpose. A floor jack is not adequate support.*

Wheel housing liner, front

- Remove front wheel.
- Working underneath front wheel housing, remove nuts, screws and plastic rivets (**arrows**) and peel off wheel housing liner.
- Installation is reverse of removal. Replace broken body clips.

Wheel arch trim, front

- Detach side marker light electrical connector. See **630 Lights**.
- Detach front of wheel housing liner and peel back.
- Remove plastic rivet (**arrow**) between wheel arch trim and lower front bumper trim.
- Remove plastic rivet behind side marker light housing.

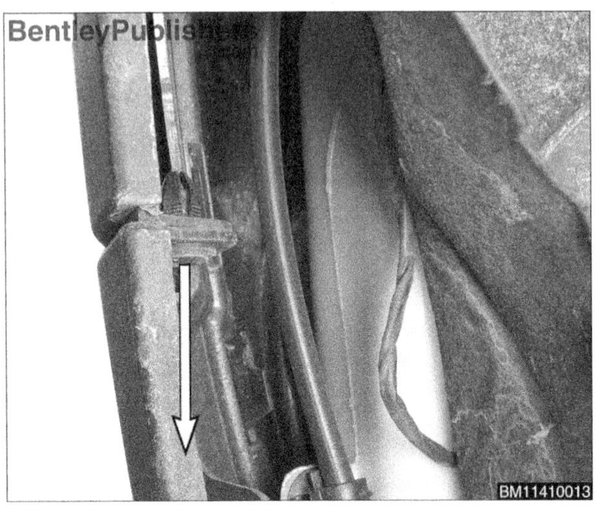

Tilt front of wheel arch trim (**curved arrow**) to disengage from seam between front fender and bumper. Release arch cover clips (**circles**) to disengage from fender.

– Installation is reverse of removal. Replace broken body clips.

Wheel housing liner, rear

– Remove rear wheel.

Working underneath rear wheel housing:

- Remove wheel housing liner fasteners (**arrows**).
- Peel back wheel housing liner to remove.

– Installation is reverse of removal. Replace broken body clips.

Engine Hood

Hood, removing and installing

> ***CAUTION—***
> *• The hood is heavy. Before removing hood mounting bolts, have an assistant help support the hood.*

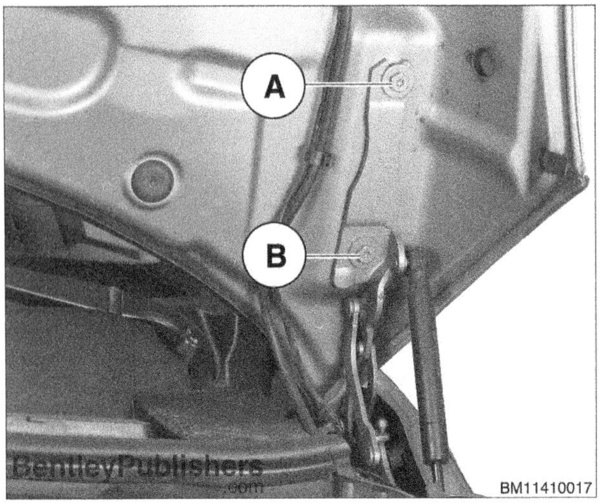

– Detach electrical connectors and washer nozzle fluid hoses.

< Working at hood hinge on left and right:
- Loosen upper hinge bolts (**A**).
- Remove lower hinge bolts (**B**).

– With help of assistant, lift hood off hinges.

– Installation is reverse of removal. Remember to:
- Tighten hinges bolts in previously installed positions to maintain hood alignment.
- Reconnect wires and hoses and replace wire ties.

Tightening torque	
Hood to hood hinge (M8 10.9 grade bolt)	22 Nm (16 ft-lb)

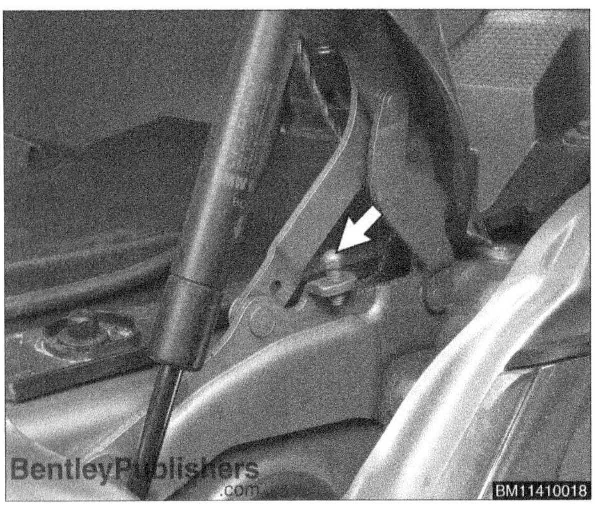

Hood, aligning

< Check hood hinge stop screw (**arrow**) to make sure that it is not damaged or missing. Replace hood hinge if necessary.

Note that stop screw position is preset. Do not use to adjust hood position.

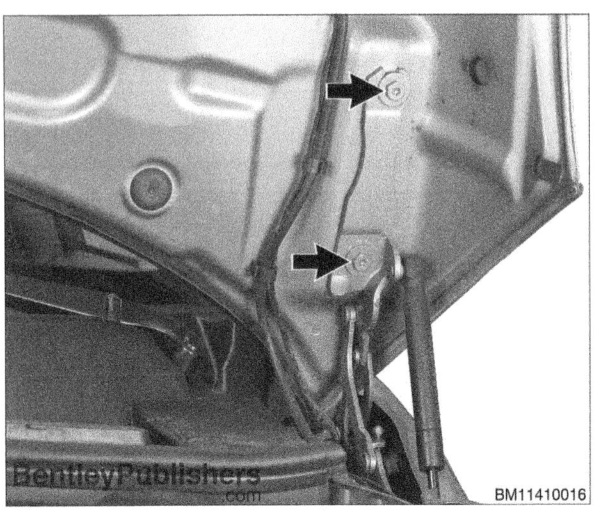

< Loosen hood hinge bolts on left and right (**arrows**) just enough to allow slight movement and adjust hood position.

– Once correct hood position is achieved with even gaps to other body panels on all sides, tighten bolts.

Tightening torque	
Hood to hood hinge (M8 10.9 grade bolt)	22 Nm (16 ft-lb)

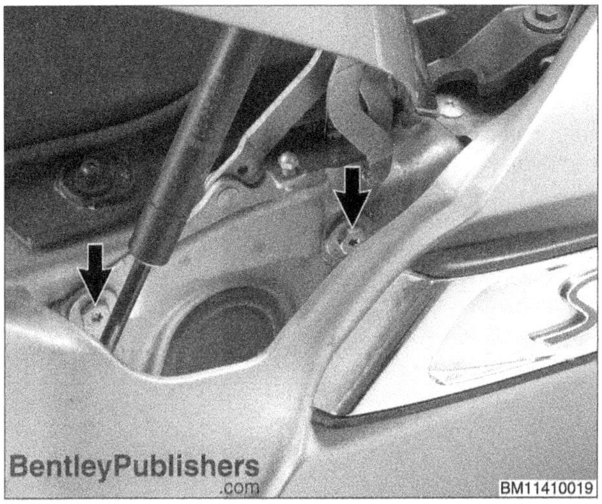

◀ If adjustment range is not sufficient, loosen left and right hinge bolts on body (**arrows**) and move hinge to adjust hood position. Retighten hinge bolts.

Tightening torque	
Hood hinge to body (M8 10.9 grade bolt)	22 Nm (16 ft-lb)

◀ Adjust hood height relative to front lock carrier:

- Remove left and right hood striker cover screws (**arrow**) and remove covers.
- Adjust height by rotating striker (**curved arrow**).

– If necessary, rotate bump stops to achieve correct hood height.

– When finished, make sure hood latches firmly.

Hood lock carrier, removing and installing

– Remove right headlight assembly. See **630 Lights**.

– Remove radiator grille. See **510 Exterior Trim, Bumpers**.

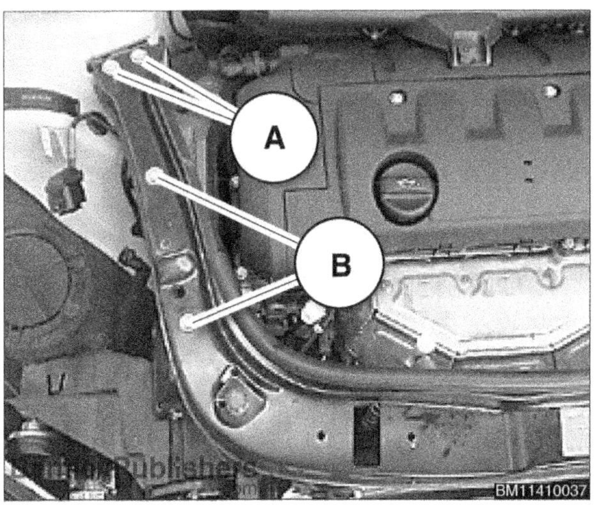

◀ Working at right side of lock carrier, remove bolts (**A**, **B**).

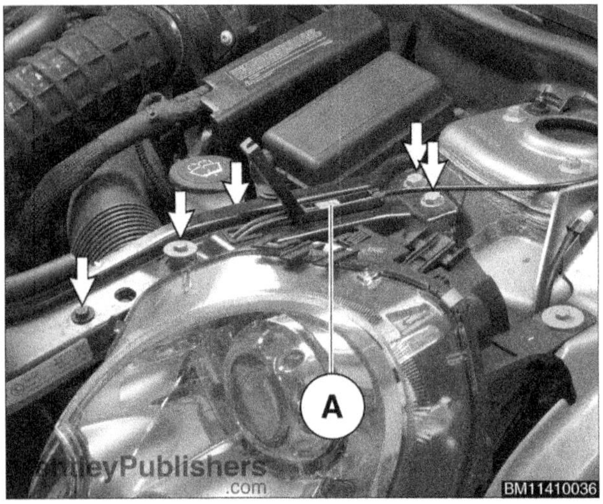

Working at left side of lock carrier:

- Open coupling and disengage hood release cable (**A**).
- Remove bolts (**arrows**).

– Tilt lock carrier, complete with hood lock assembles, and slide forward to remove.

– Installation is reverse of removal. Make sure headlight mounts seat accurately to avoid breakage of plastic tabs.

Tightening torque	
Lock carrier to body (M8 bolts)	22 Nm (16 ft-lb)

Radiator Support

Radiator support service position

A number of repair procedures, including engine removal, require that the radiator support (vehicle front panel) be placed in service position. In the following procedure, the radiator support is loosened and slid forward for clearance purposes. Coolant hoses, refrigerant lines and electrical harnesses are long enough to allow the components attached to the radiator support to be slid forward approx. 10 cm (2.5 in)

Additional notes at the end of this procedure are useful if the radiator support is to be removed from the vehicle altogether. In that case, drain coolant and discharge refrigerant first.

– Raise front of vehicle and support safely.

> ***WARNING—***
> - *Make sure the vehicle is stable and well supported at all times. Use a professional automotive lift or jack stands designed for the purpose. A floor jack is not adequate support.*

– Detach front section of front wheel housing liners and peel back. See **Wheel housing liner, removing and installing** in this repair group.

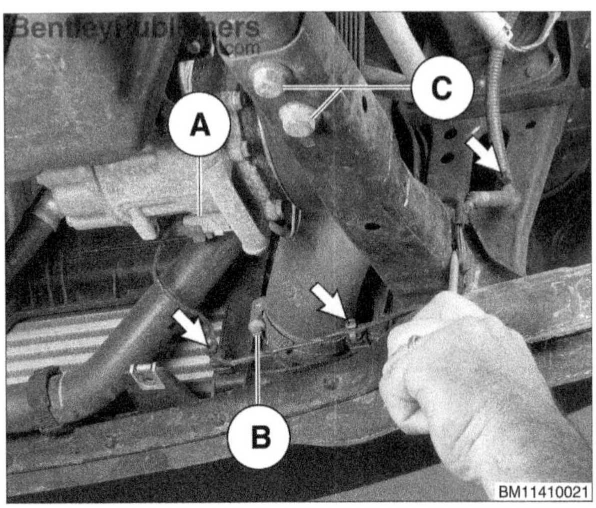

Working underneath right front of vehicle:

- Remove underbody splash shield, if equipped.
- Cut wire ties (**arrows**) and pry out harness holder to detach electrical harness from radiator support.
- Detach A/C compressor electrical connector (**A**).
- Turbo model: Loosen hose clamp (**B**) and detach intercooler hose.
- Remove subframe extension mounting bolts (**C**).

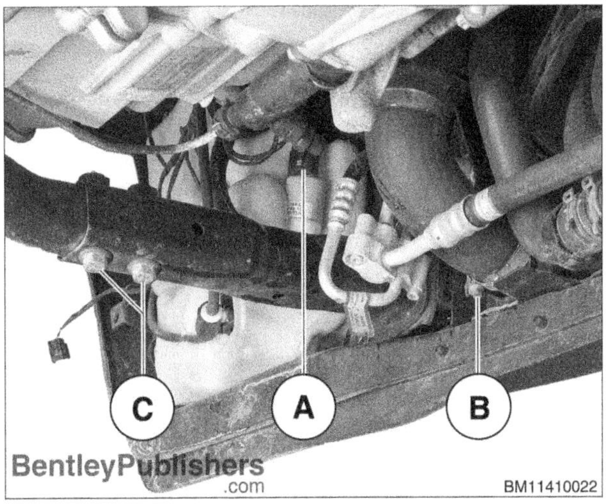

◄ Working underneath left front of vehicle:
- If applicable: Detach headlight washer pump connector (**A**).
- Turbo model: Loosen hose clamp (**B**) and detach intercooler hose.
- Remove subframe extension mounting bolts (**C**).

◄ Working at side of headlight, remove fender bolt (**arrow**).

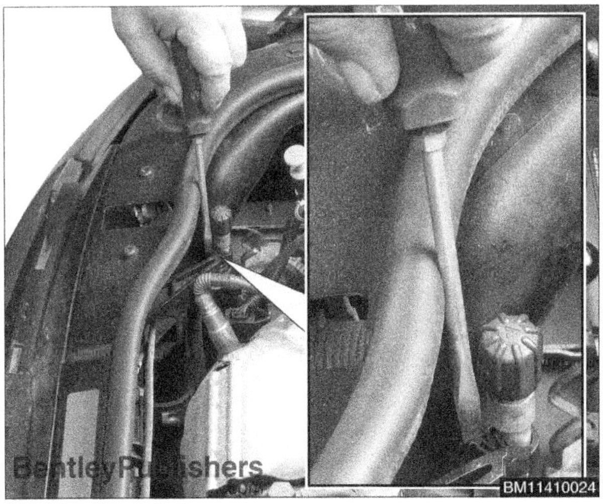

◄ Detach refrigerant line from bracket (**inset**).

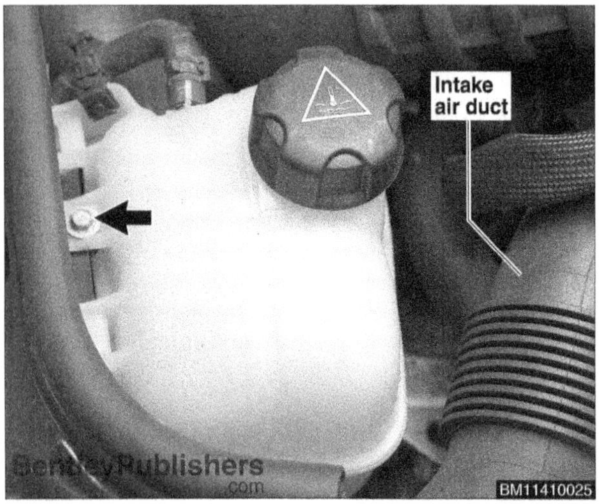

Working at upper left of engine compartment:

- Remove coolant expansion tank retaining screw (**arrow**) and place tank aside.
- Detach intake air duct from radiator support.

Working near left front strut tower:

- Open coupling and disengage hood release cable (**A**).
- Remove lock carrier mounting bolts (**B**).
- Remove headlight bracket mounting bolt (**C**).

Working near right front strut tower:

- Turbo model: Loosen hose clamp (**A**) and detach air duct.
- Turbo model: Detach mass air flow sensor connector (**B**).
- Detach headlight harness connector (**C**).
- Remove lock carrier mounting bolts (**D**).
- Remove headlight bracket mounting bolt (**E**).

Radiator support in service position

With help of assistant, carefully slide front end module including front bumper, radiator and radiator support and headlights forward (**arrow**) approx. 10 cm (2.5 in).

- Make sure subframe rail extensions remain on subframe rails. Weight of front module is supported by subframe rails.
- Insert one bolt in each side subframe extension to keep radiator support secure against tipping off vehicle.

Radiator support removal

– With radiator support in service position, drain cooling system. See **170 Radiator and Cooling System**.

– Connect an approved refrigerant recovery unit to A/C system and follow manufacturer's instructions to discharge system.

> ***WARNING —***
> - *Do not attempt to discharge or charge the A/C system without proper equipment and training. Damage to the vehicle and personal injury may result.*

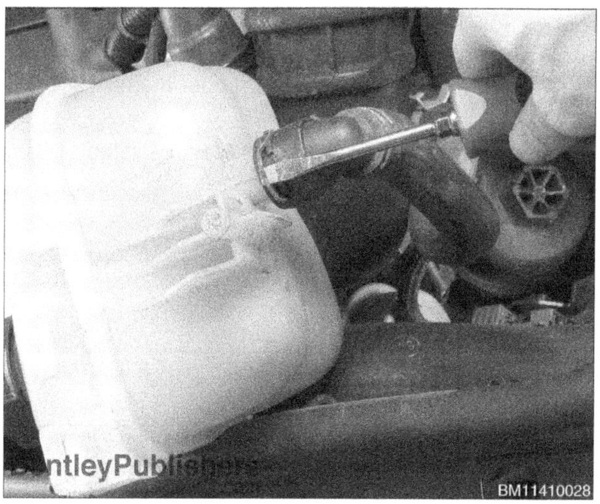

Remove coolant expansion tank mounting screw, lift tank up and pry off hose clip to detach coolant hose. Set tank aside.

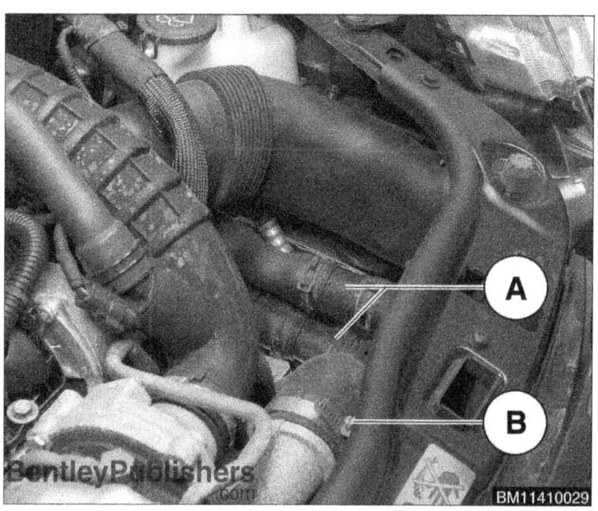

Working at left front of engine compartment:

- Use hose clamp pliers to release coolant hose clamps (**A**) and detach radiator hoses.
- Turbo model: Loosen hose clamp (**B**) and detach turbocharger duct.

Working underneath left fender:

- Detach washer pump and washer fluid sensor connectors and washer fluid hoses (**arrows**) from pumps. Be prepared to catch dripping fluid. See **611 Wipers and Washers**.

– Detach horn, foglight and left headlight harness connectors.

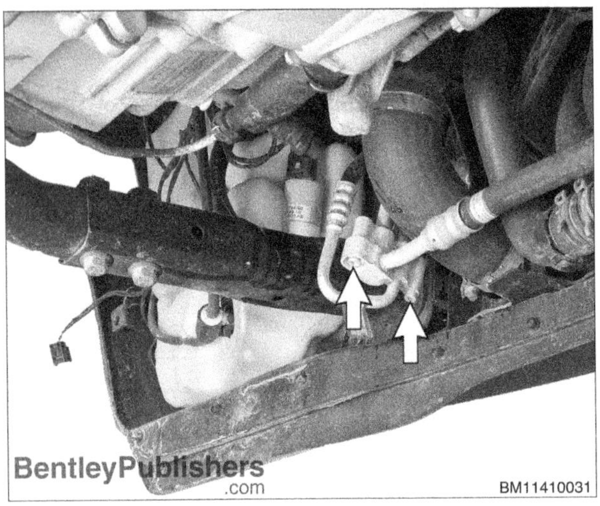

Remove refrigerant line mounting bolts (**arrows**) and detach lines from A/C condenser.

– With help of assistant, slide radiator support with lock carrier, headlight assemblies, radiator, and A/C condenser off subframe rails. Watch carefully to make sure no lines, hoses or harnesses become snagged.

– Installation is reverse of removal.

- With help of assistant, slide radiator support evenly on left and right. Make sure headlight mounts seat on fender arches accurately to avoid breakage of plastic tabs.
- Reattach refrigerant lines using new O-ring seals. Moisten seals with refrigerant oil. Recharge refrigerant.
- Fill and bleed cooling system.

Tightening torques	
Lock carrier to body	22 Nm (16 ft-lb)
Refrigerant lines to A/C condenser (use new O-ring seals, moisten with refrigerant oil)	8 Nm (6 ft-lb)
Subframe extension to subframe rail (M12)	100 Nm (74 ft-lb)

411 Doors

GENERAL

This repair group covers removal and installation of side doors and door trim panels.

See also:

- **412 Rear Hatch, Trunk Lid, Split Rear Doors** for Clubman split rear door repairs.
- **512 Door Windows** for power door windows and door glass replacement.
- **515 Central Locking and Anti-Theft** for door locks and cylinders.

Cautions

CAUTION—

- *The body is painted at the factory after assembly. Realignment of body panels may expose unpainted metal. Paint all exposed metal once work is complete.*
- *To avoid marring trim pieces, use a plastic prying tool or a screwdriver with the tip wrapped with masking tape.*

DOORS

Front door, removing and installing

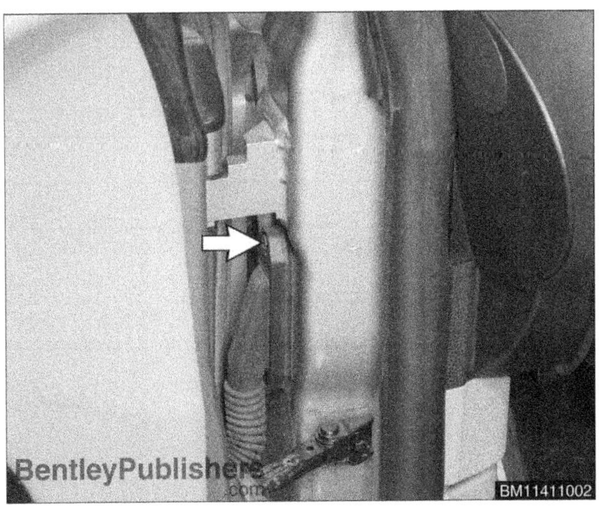

With door fully open, remove door harness connector mounting bolt (**arrow**) on A-pillar.

◀ Working at lower A-pillar, remove door check mounting bolt (**arrow**).

> ***CAUTION—***
> - *With door check disconnected, door can hyper-extend, damaging door and body panels.*

◀ With door fully open, unscrew pin bolts (**arrows**) from top and bottom hinges.

◀ Pull connector lock bar up (**arrow**) to separate door electrical connector.

– Lift door to remove from hinge halves and place on suitable surface.

– Installation is reverse of removal:
  - Check door alignment to see that panel gaps are even on all sides. If necessary, adjust door alignment. See **Door, adjusting** in this repair group.
  - Repair paint damage and paint exposed metal.

Tightening torques	
Door check to A-pillar	23 Nm (17 ft-lb)
Door harness connector to A-pillar (M5 x 16 mm bolt)	3 Nm (27 in-lb)
Door hinge bolt to door hinge (M6)	10 Nm (7 ft-lb)

Club door, removing and installing

– Disconnect negative (-) battery cable and cover battery terminal to keep cable from accidentally contacting terminal.

> ***CAUTION—***
> • *Prior to disconnecting the battery, read the battery disconnection cautions given in* **001 Warnings and Cautions**.

– Unbolt seat belt from club door. See **720 Seat Belts**.

– Remove club door inner trim panel. See **Club door trim panel, removing and installing** in this repair group.

– Remove speaker in club door.

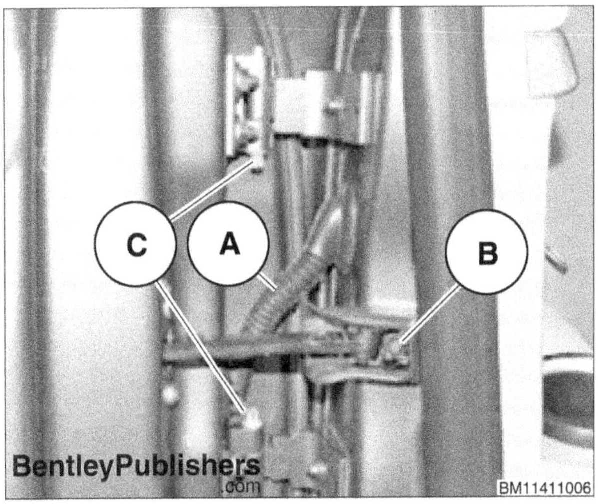

◄ With club door open:

- Carefully pull speaker wiring harness (**A**) out of door.
- Working at door post, remove door check mounting bolt (**B**).
- Unscrew pin bolts (**C**) from top and bottom hinges.

> ***CAUTION—***
> • *With door check disconnected, door can hyper-extend, damaging door and body panels.*

– Lift door to remove from hinge halves and place on suitable surface.

– Installation is reverse of removal:

- Check door alignment to see that panel gaps are even on all sides. If necessary, adjust door alignment. See **Door, adjusting** in this repair group.
- Inspect wiring harness grommet. Replace if damaged. To prevent water leak in door cavity, make sure grommet seats correctly in door.
- Repair paint damage and paint exposed metal.

Tightening torques	
Door check to door pillar	23 Nm (17 ft-lb)
Door hinge bolt to door hinge (M10 x 1.25)	23 Nm (17 ft-lb)

Door, adjusting

The body is painted at the factory after assembly. Realignment of body panels may expose unpainted metal. Paint all exposed metal once work is complete.

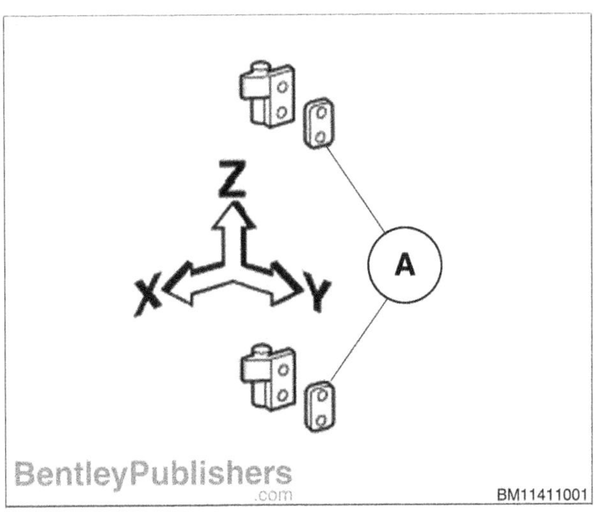

◄ Adjust door along X- and Z-axis are by repositioning door hinges. Adjust along Y-axis by installing shims (**A**) behind hinges.

Door, adjusting

Loosen door hinge attachment bolts (**arrows**) and slide door to adjust door position. Check that door lock striker does not scrape against door or interfere with door operation.

Check that, when closed, door panel lines up with adjacent panel.

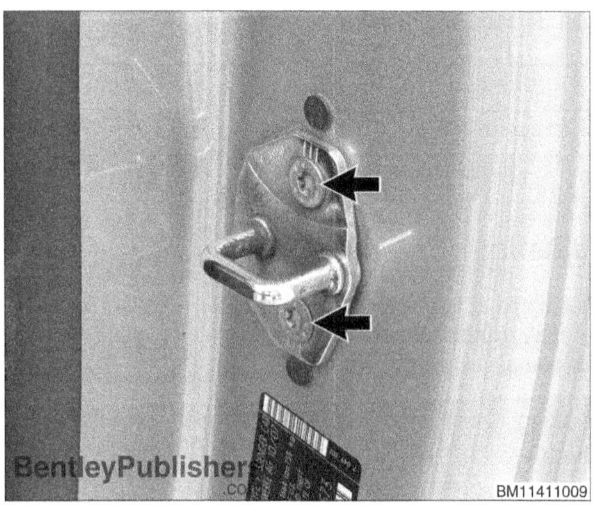

If necessary, adjust door lock striker position by loosening bolts (**arrows**). Note that club door has an upper and a lower striker.

- When satisfied with door and striker alignment, tighten bolts securely and touch up exposed metal surfaces.

Tightening torques	
Club door to door hinge (M8)	18.5 Nm (14 ft-lb)
Club door upper striker to body	24 Nm (18 ft-lb)
Front door lock striker to B-pillar	19 Nm (14 ft-lb)
Front door to door hinge (M8)	18.5 Nm (14 ft-lb)
Club door lower striker to front door	24 Nm (18 ft-lb)

Door Trim Panels

Trim panel plastic clips are easily damaged. Have extra clips on hand before beginning the removal procedure.

Front door trim panel, removing and installing

– Remove footwell light. See **630 Lights**.

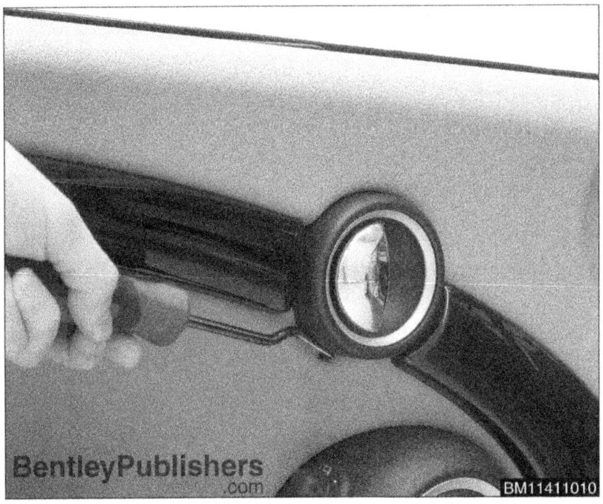

Working at inside door handle, pull out retaining pin. Do not unclip door handle trim cover. Trim cover is detached while removing door trim panel.

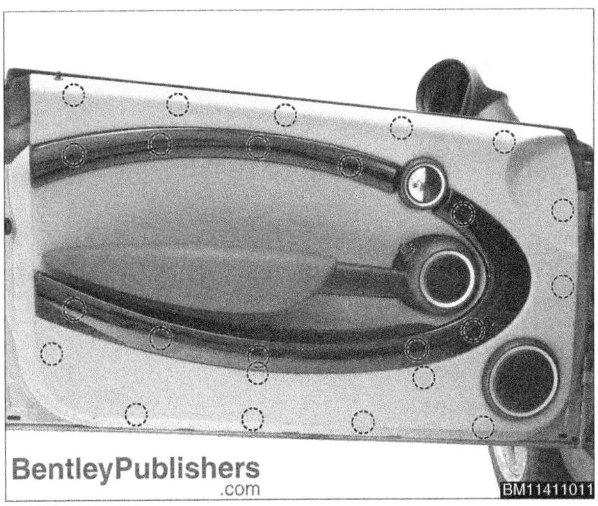

Use plastic prying tool to lever door panel clips out of door. Clip positions are indicated by **broken circles**.

Note that inside door handle trim cover is unclipped when detaching door trim panel. Make sure catches on trim cover (**arrows**) are not damaged.

411

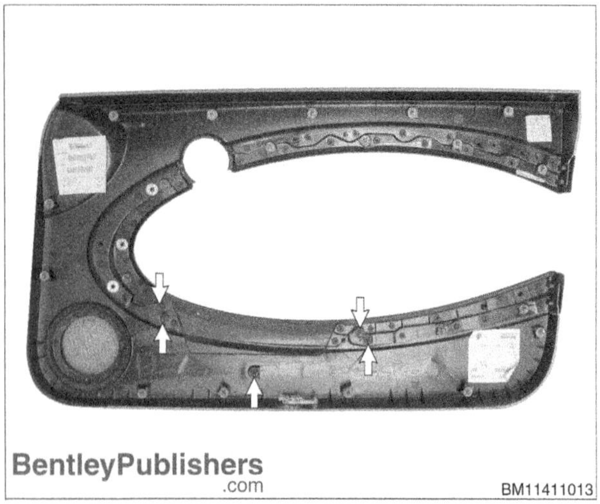

When installing:

- Replace damaged or missing plastic trim clips. Note that certain clips (**arrows**) are red or white and others are grey.
- Inside door handle trim can only be installed in one position.

Club door trim panel, removing and installing

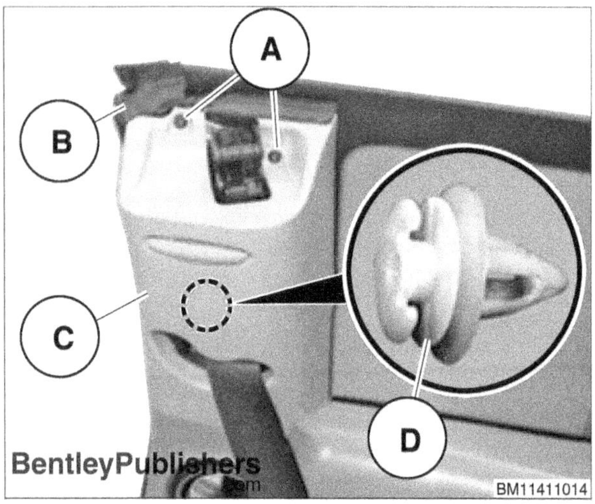

Working at upper club door trim panel:

- Remove belt guide mounting screws (**A**).
- Detach door seal (**B**) in working area.
- Pry belt guide (**C**) off with clip (**D**).

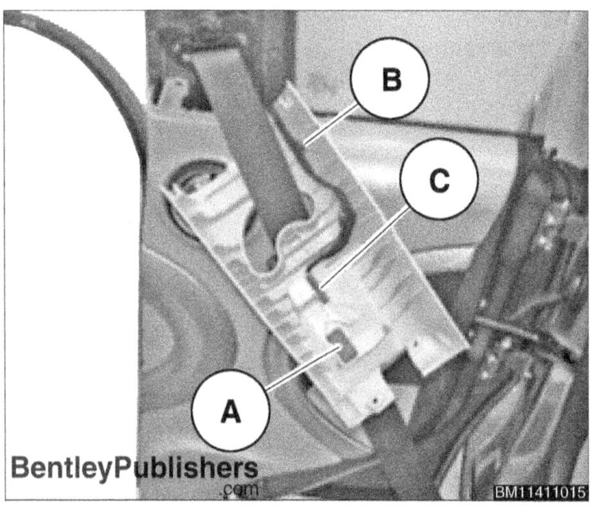

If equipped with LED (**A**): detach electrical harness (**B**) and connector (**C**).

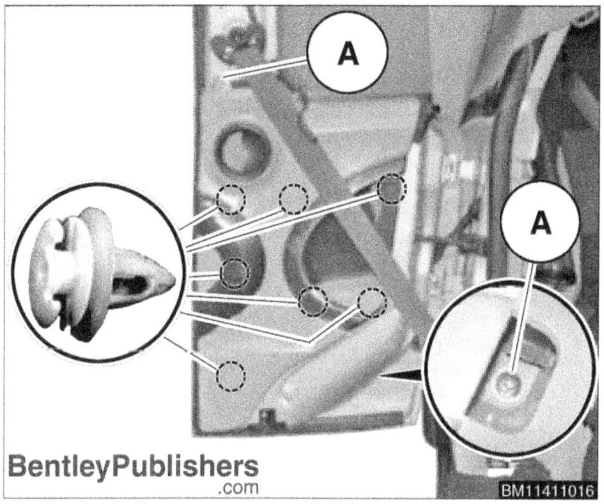

Working at lower club door panel:

- Remove panel mounting screws (**A**).
- Pry trim panel off with clips (**inset**).

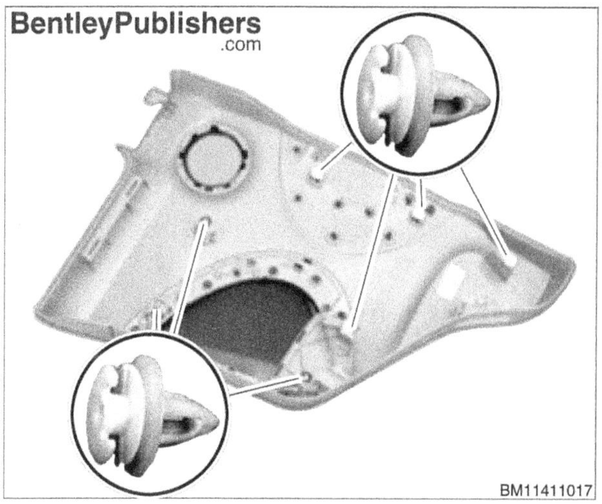

When installing:

- Replace damaged or missing plastic trim clips (**insets**).
- Make sure seat belt strap can move freely. If necessary, replace faulty plastic clip in upper cover.
- Reseat door seal firmly.
- Make sure LED is seated correctly.

Tightening torque	
Upper trim panel to door	2 Nm (18 in-lb)

412 Rear Hatch, Tailgate, Split Rear Doors

GENERAL

This repair group covers removal and installation of the Hardtop rear hatch, Clubman split rear doors, Convertible tailgate and associated trim panels. Also included is rear spoiler service

See also:

- **515 Central Locking and Anti-theft** for locks and cylinders.
- **611 Wiper and Washers** for rear window wiper-washer system.

Warnings and Cautions

> ***WARNING—***
> - *MINI cars are equipped with multiple airbags around the rear hatch. When servicing the hatch, disconnect the negative (-) battery terminal. See* **721 Airbag System (SRS)**.

> ***CAUTION—***
> - *The body is painted at the factory after assembly. Realignment of body panels may expose unpainted metal. Paint all exposed metal once work is complete.*
> - *To avoid marring trim pieces, use a plastic prying tool or a screwdriver with the tip wrapped with masking tape.*

Rear Hatch (Hardtop)

Rear hatch strut, removing and installing

> ***WARNING —***
> *• Support hatch with suitable prop rod before removing struts.*

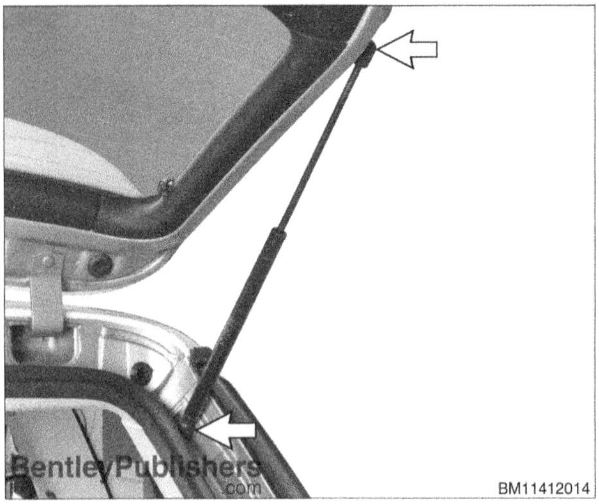

- Release top of gas strut from hatch by prying spring clip (**upper arrow**). Remove strut end from ball joint.
- Remove bottom (**lower arrow**) of gas strut from body in a similar manner.
- Install struts to hatch and body by firmly pressing strut sockets onto ball joints and locking spring clips into position.

Rear hatch, removing and installing

- Remove rear spoiler. See **Rear spoiler, removing and installing** in this repair group.
- Remove upper rear window frame trim. See **Rear hatch upper trim, removing** in this repair group.
- Support hatch with a suitable prop and remove hatch struts. See **Rear hatch strut, removing and installing** in this repair group.
- Detach rear hatch electrical connectors and washer hose.
- Close rear hatch but do not allow hatch lock to snap closed.

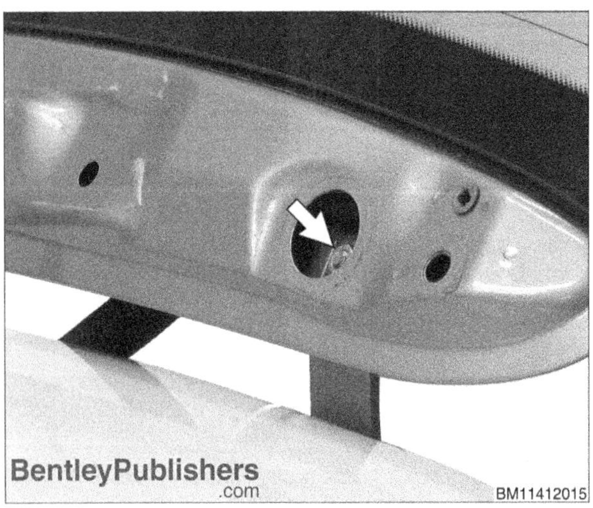

- With rear hatch open and supported:
  - Remove trim covers on upper hatch.
  - Remove rear hatch hinge mounting nuts (**arrow**) on left and right.
- With help of assistant, remove rear hatch. Make sure no wires or hoses interfere with removal.
- Installation is reverse of removal. Remember to:
  - Tighten hinges in previously installed positions to maintain hatch alignment. If necessary, see **Rear hatch, aligning** in this repair group.
  - Reroute wires and hoses and replace wire ties and clamps.
  - Repair paint damage and paint exposed metal.

Tightening torque	
Rear hatch to hinge (M8 nut)	18.5 Nm (14 ft-lb)

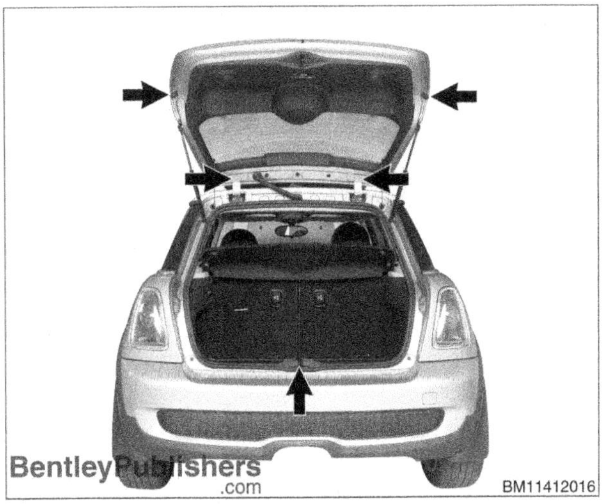

Rear hatch, aligning

◄ Hatch adjustment points (**arrows**) are at each hinge, buffer stops and lock striker.

– Remove rear spoiler. See **Rear spoiler, removing and installing** in this repair group.

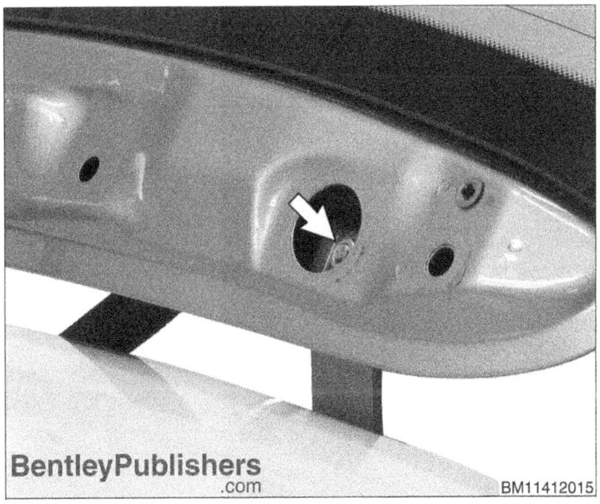

◄ Loosen hatch hinge bolts (**arrow**) enough to allow slight movement of rear hatch. Adjust hatch position, aiming for even gaps on all sides.

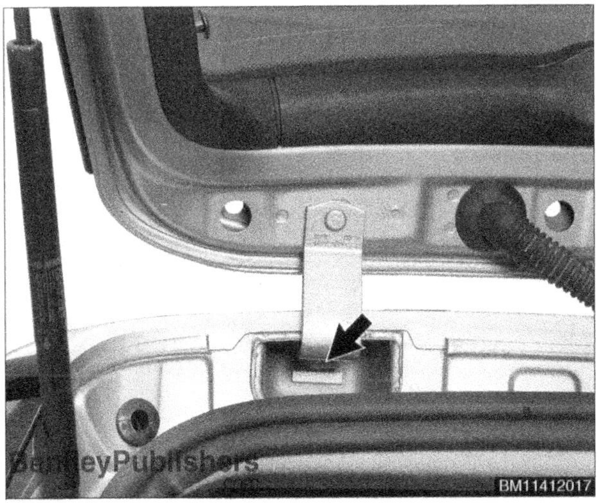

◄ If adjustment range is not enough, loosen hinge fasteners (**arrow**) on body.

– Tighten fasteners and seal cavities with sealant.

Tightening torques	
Rear hatch hinge to body (M8 nut)	20 Nm (15 ft-lb)
Rear hatch to hinge (M8 nut)	18.5 Nm (14 ft-lb)

– Reinstall rear spoiler, repair paint damage and paint exposed metal.

– Adjust rear latch if necessary. See **515 Central Locking and Anti-theft**.

Rear hatch lower trim, removing

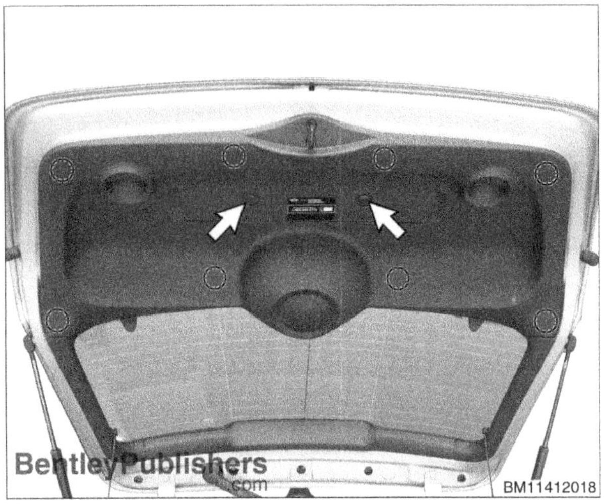

- With hatch open:
  - Remove tool kit attached to lower hatch trim.
  - Remove tool kit mounting screws (**arrows**).
  - Use plastic prying tool to lever trim panel clips out of rear hatch. Clip positions are indicated by **broken circles**.

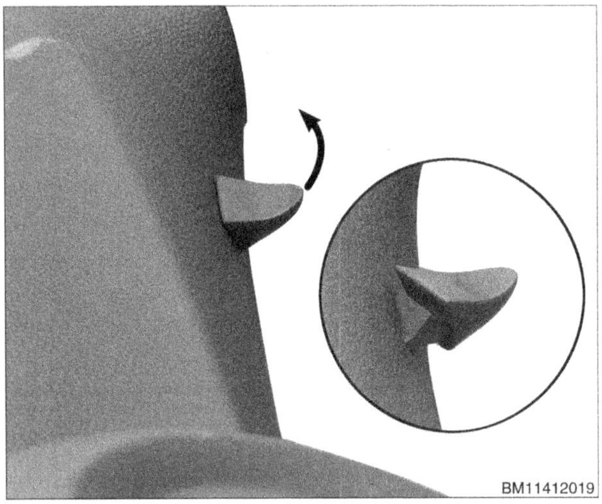

- Installation is reverse of removal. Remember to:
  - Replace damaged or missing trim clips.
  - If replacing rear panel, remove rear parcel shelf stops (**arrow**) and transfer to new panel.

Rear window side trim, removing

– Remove rear hatch lower trim. See **Rear hatch lower trim, removing** in this repair group.

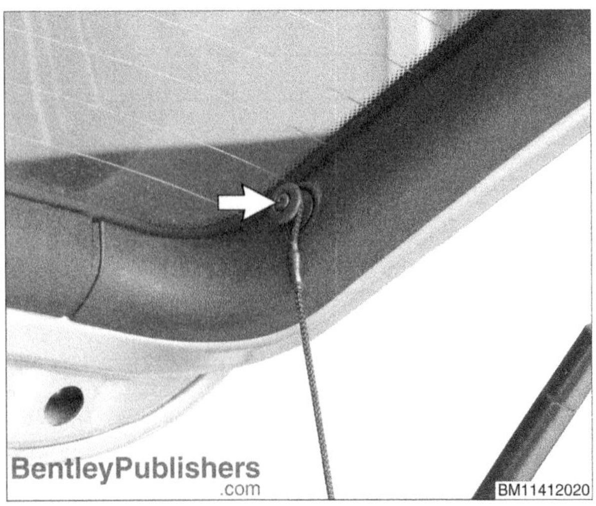

- Detach parcel shelf retaining strap and remove strap button mounting screw (**arrow**).

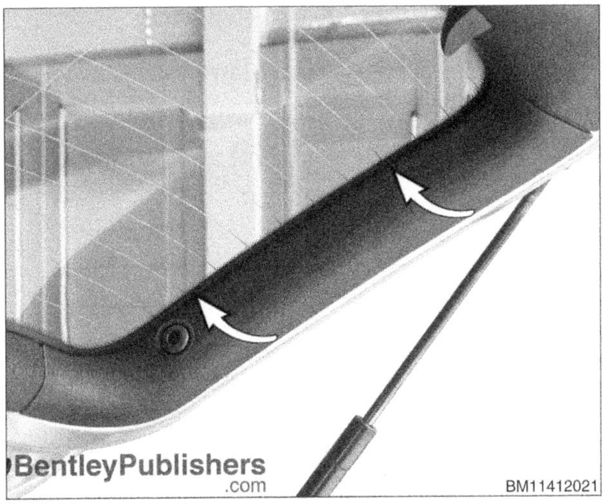

< Use plastic prying tool to lever off window side trim (**arrows**).

- Installation is reverse of removal. Replace damaged or broken trim clips.

Rear hatch upper trim, removing

- Remove rear window side trim. See **Rear window side trim, removing** in this repair group.

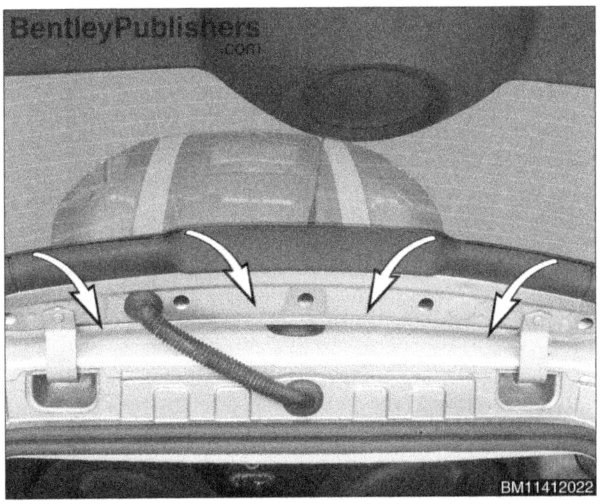

< Use plastic prying tool to lever off window upper trim (**arrows**).

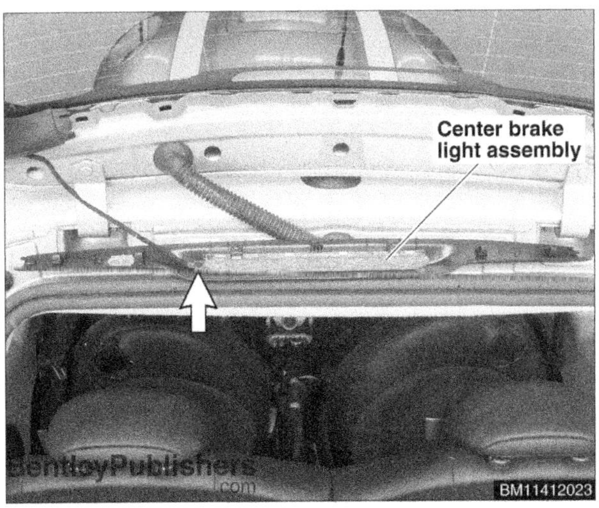

< Detach electrical connector (**arrow**) and remove center brake light assembly with trim.

- Installation is reverse of removal. Replace damaged or missing trim clips.

Rear spoiler, removing and installing

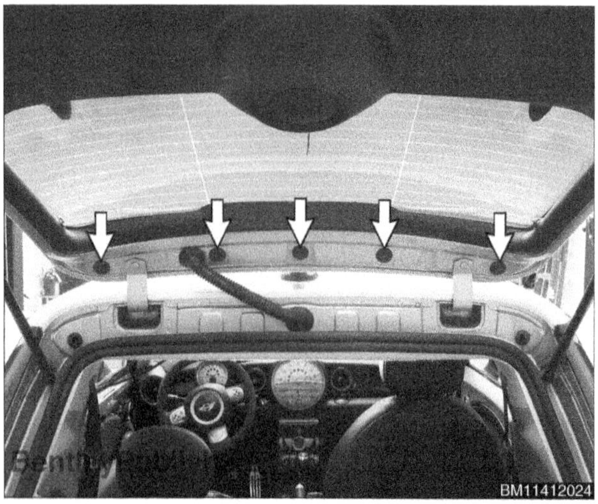

◄ With rear hatch open, pry out sealing plugs (**arrows**) above rear window and remove nuts underneath.

◄ Close rear hatch and pry up rear spoiler (**arrows**), detaching from clips (**broken circles**).

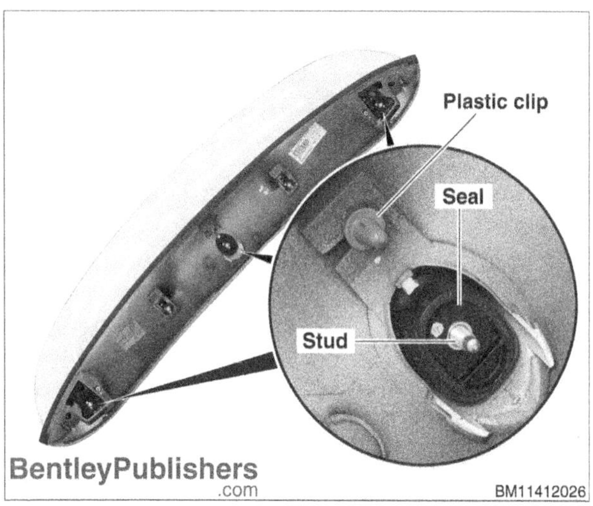

◄ When installing:

- Replace broken or damaged plastic clips.
- Make sure mounting studs and seals are not damaged.

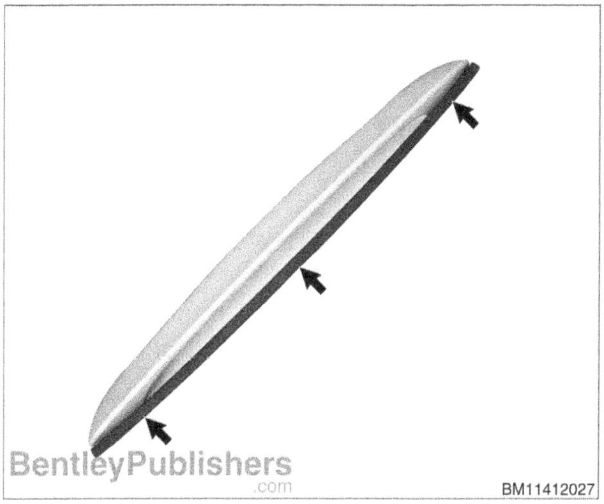

Clean rear spoiler adhesive area (**arrows**) with alcohol or turpentine. Allow air drying time of at least 1 minute.

- Glue new adhesive strip to adhesive area and press down using hand roller.
- Tighten nuts securing spoiler to hatch and replace sealing plugs.

Tightening torque	
Spoiler to rear hatch (M6)	4 Nm (35 in-lb)

TAILGATE (CONVERTIBLE)

Tailgate retractor cable, removing and installing

– Remove tailgate trim panel. See **Tailgate trim panel, removing and installing** in this repair group.

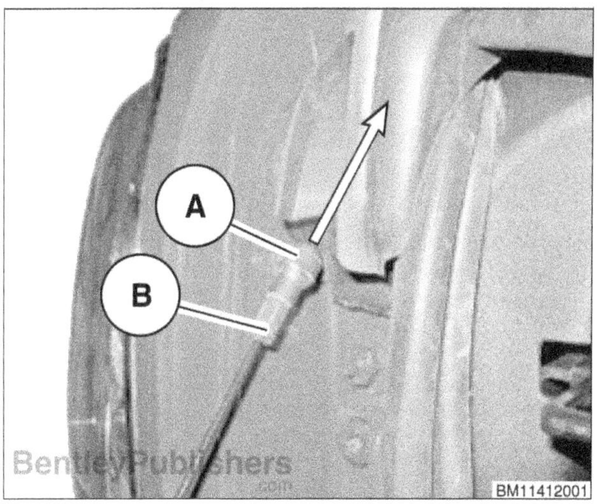

With trunk open, lift tailgate slightly and support in that position. Working at retractor cable end on body:

- Unlock retainer (**A**) and slide up (**arrow**).
- Slide retractor cable (**B**) off.

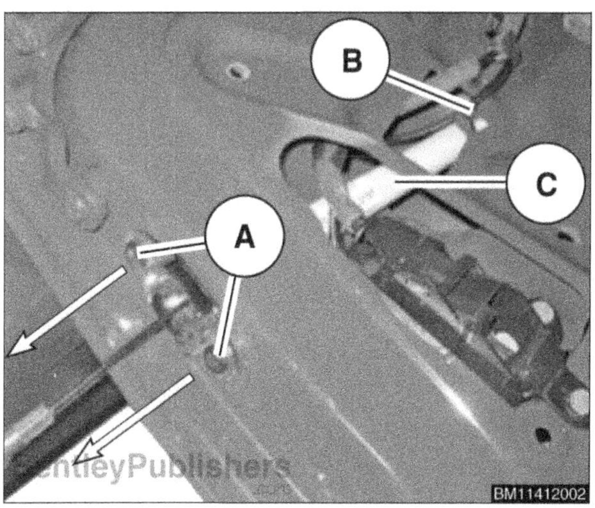

Working at tailgate:

- Remove retractor cable assembly mounting screws (**A**).
- Cut wire tie (**B**) and slide retractor cable (**C**) out of tailgate (**arrows**).

– Installation is reverse of removal.

- Make sure cable is seated correctly inside door.
- Replace wire tie.

Tightening torque	
Retractor cable to tailgate	8 Nm (6 ft-lb)

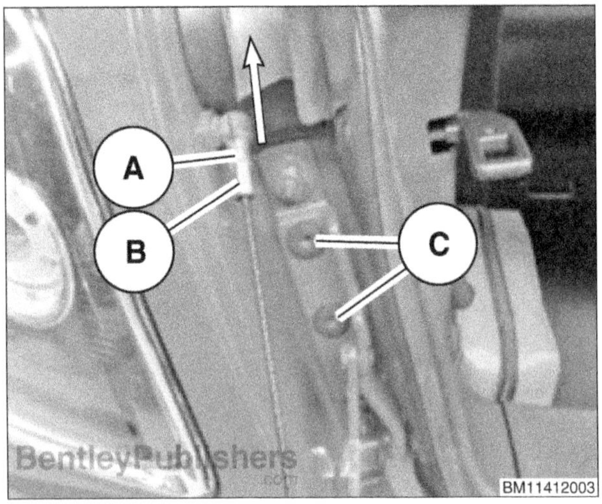

Tailgate, removing and installing

– Remove tailgate trim panel. See **Tailgate trim panel, removing and installing** in this repair group.

– Detach electrical connectors in tailgate and pull wiring harness out of lid.

◄ With trunk open, lift tailgate slightly and support in that position. Working at retractor cable end on body:

- Unlock retainer (**A**) and slide up (**arrow**).
- Slide retractor cable (**B**) off.
- Remove trunk hinge mounting bolts (**C**).

– Remove tailgate with help of assistant. Make sure no wires or hoses interfere with removal.

– Installation is reverse of removal. Remember to:

- Tighten hinges in previously installed positions to maintain tailgate alignment.
- Reroute wires and replace wire ties and clamps.
- Repair paint damage and paint exposed metal.

Tightening torque	
Tailgate hinge to hinge base (M8 x 12 mm bolt)	20 Nm (15 ft-lb)

Tailgate, aligning

– Loosen or remove tailgate lock striker.

◄ Loosen tailgate hinge to tailgate bolts (**arrows**) enough to allow slight tailgate movement. Adjust tailgate in or out to assure even, flat surface relative to adjacent body panels.

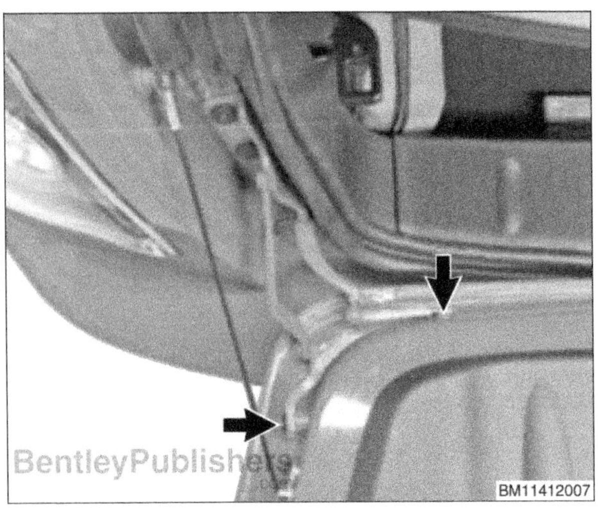

Tightening torque	
Tailgate hinge to tailgate (M8 x 30 mm bolt)	19 Nm (14 ft-lb)

◄ Loosen tailgate hinge to body bolts (**A**) enough to allow slight tailgate movement. Adjust tailgate sideways and vertically to achieve even gaps with adjacent panels.

Tightening torque	
Tailgate hinge to body (M8 nut)	19 Nm (14 ft-lb)

– Repair paint damage and paint exposed metal.

– Adjust tailgate latch if necessary. See **515 Central Locking and Anti-theft**.

Tailgate trim, removing

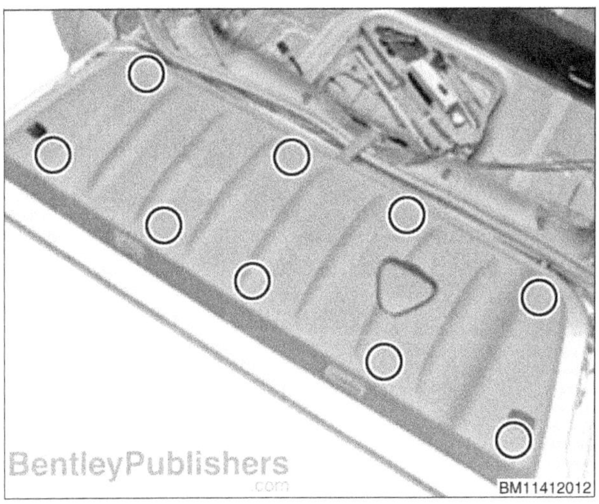

◄ With tailgate open, detach trim panel from tailgate. Trim clips are in **circled spots**.

– When installing, replace broken or damaged clips.

Split Rear Doors (Clubman)

Split rear door strut, removing and installing

– Mark left and right struts as they are not of equal length.

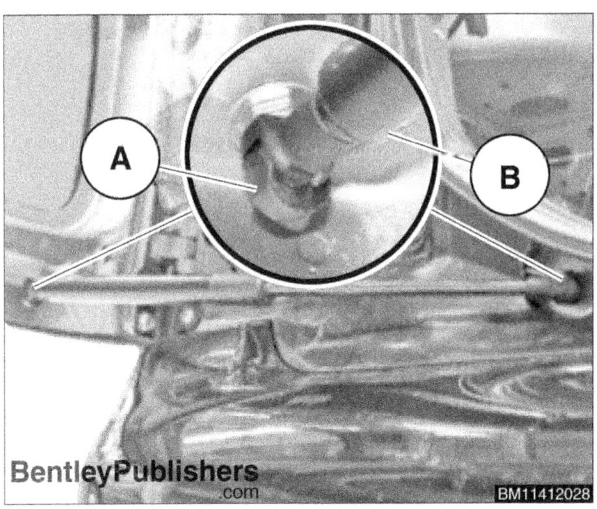

◄ Release ends of gas strut from door and from body by prying spring clips (**A**). Remove strut ends (**B**) from ball joints.

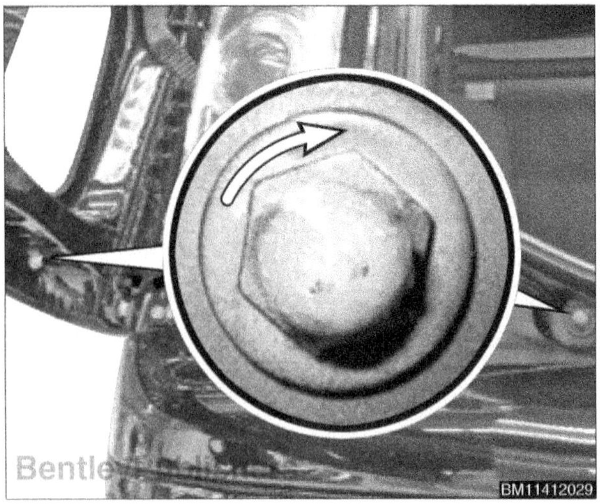

Check ball stud (**inset**) on body and split door for tightness. Tighten if necessary (**arrow**).

Tightening torque	
Door strut ball stud to split door or body	18.5 Nm (14 ft-lb)

- Install strut to door and body by firmly pressing strut sockets on ball joints and locking spring clips into position.

Split rear door, removing and installing

The following procedure is for left split door removal. Left and right split rear doors are similar.

- Remove taillight. See **630 Lights**.
- Remove split door strut. See **Split rear door strut, removing and installing** in this repair group.

Disengage electrical connector bracket and pull out through taillight opening (**arrow**).

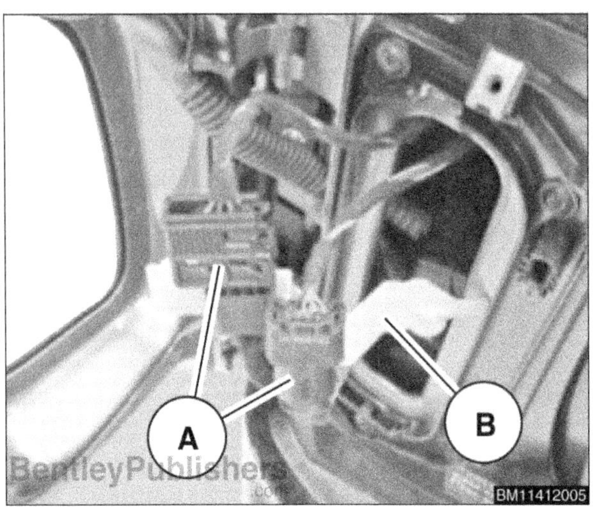

Separate harness connectors (**A**) and detach from bracket (**B**).

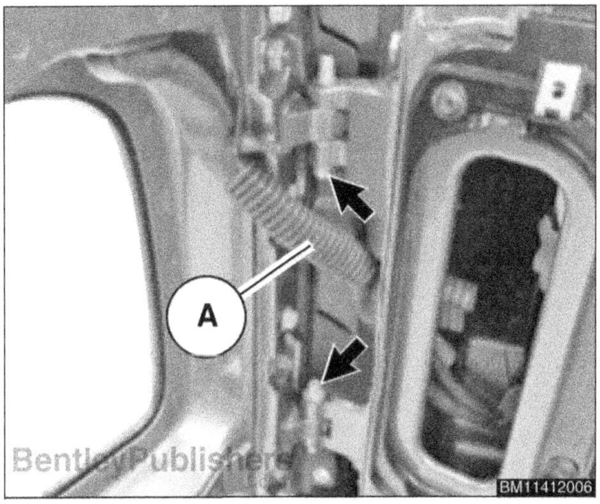

With split door open:

- Carefully pull door wiring harness (**A**) out of body.
- Remove split door hinge mounting bolts (**arrows**). Be sure to support split door while doing this.

– Lower split door with help of assistant. Make sure no wires or hoses interfere with removal.

– Installation is reverse of removal. Remember to:

- Reroute wires and hoses and replace wire ties and clamps.
- Inspect wiring harness grommet. Replace if damaged. To prevent water leak in body cavity, make sure grommet seats correctly in side of taillight housing.
- Check door alignment to see that panel gaps are even on all sides. If necessary, adjust door alignment. See **Split door, adjusting** in this repair group.
- Repair paint damage and paint exposed metal.

Tightening torque	
Split door hinge bolt to door hinge (M10 x 1.25)	23 Nm (17 ft-lb)

Split rear door, aligning

The following procedure is for left split door alignment. Left and right split rear doors are similar.

– Loosen or remove split door lock striker.

– Remove taillight. See **630 Lights**.

– Remove split door trim panel. See **Split rear door trim, removing and installing** in this repair group.

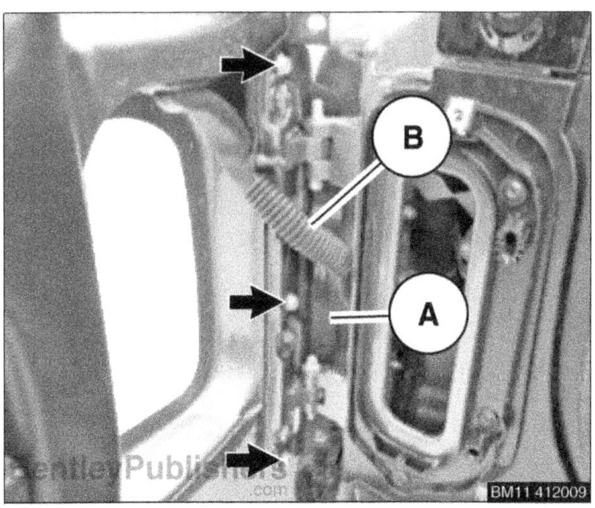

Working at split door hinge:

- Remove bolts (**arrows**) and detach holder (**A**).
- Detach wiring loom (**B**) from body.

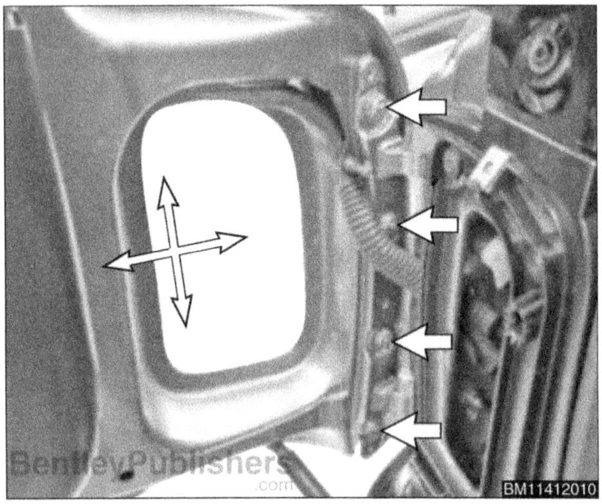

Loosen split door hinge bolts (**arrows**) enough to allow slight movement. Adjust split door sideways and vertically to achieve even gaps with adjacent panels.

Tightening torque	
Split door hinge to split door (M8 nut)	18.5 Nm (14 ft-lb)

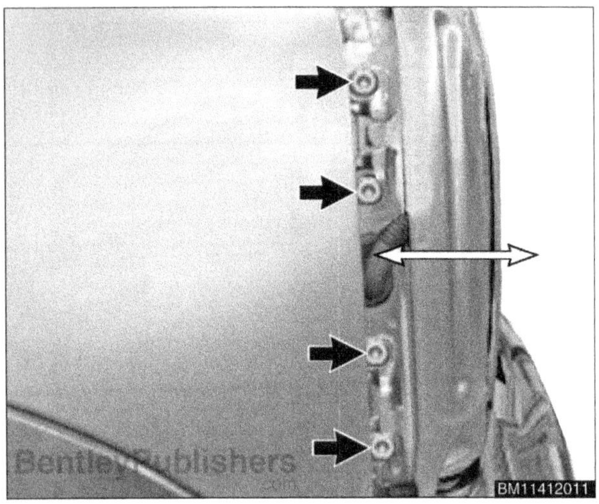

Close split door and loosen door hinge to body bolts (**arrows**) enough to allow slight movement. Adjust split door forward or back to assure even, flat surface relative to adjacent body panels.

Tightening torque	
Split door hinge to body (M10 bolt)	38 Nm (28 ft-lb)

- Reinstall taillight.
- Repair paint damage and paint exposed metal.
- Adjust split door latch and striker if necessary. See **515 Central Locking and Anti-theft**.

Split rear door trim, removing

The following procedure is for left split door trim panel removal. Left and right split rear doors are similar.

- Left split door: Remove upper lock striker. See **515 Central Locking and Anti-theft**.
- Remove door courtesy light. See **630 Lights**.

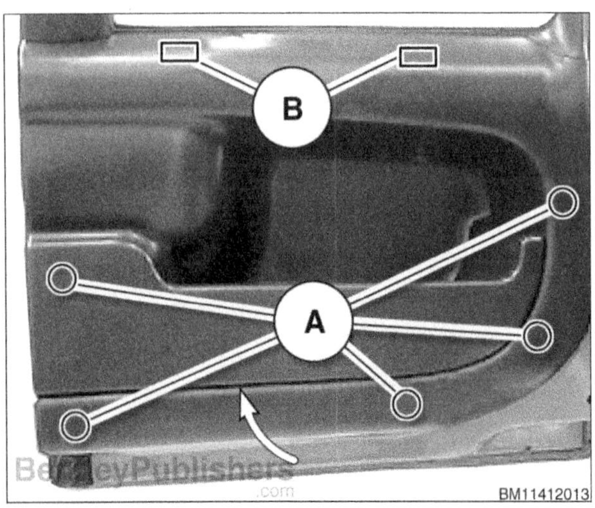

Detach split door panel:

- Use plastic prying tool to detach lower clips (**A**).
- Tilt panel up and detach upper clips (**B**).

- When installing, replace broken or damaged clips.

510 Exterior Trim, Bumpers

GENERAL

This repair group includes repair information for the outside rear view mirrors, front and rear bumpers and exterior trim parts.

See also:

- **410 Fenders, Engine Hood** for wheel housing liners and radiator support
- **412 Rear Hatch** for rear spoiler
- **515 Central Locking and Anti-theft** for door and luggage compartment locks

Exterior trim repairs

Exterior trim and finish panels are clipped or screwed into place. Many of the trim retaining clips are designed to be used only once. When removing trim that is held in place with clips, be sure to have spares on hand before beginning the job.

Cautions

CAUTION—

- *The body is painted at the factory after assembly. Realignment of body panels may expose unpainted metal. Paint all exposed metal once work is complete.*
- *To avoid marring trim pieces, use a plastic prying tool or a screwdriver with the tip wrapped with masking tape.*

Outside Rear View Mirrors

The mirror glass can be replaced separately. Replacements parts are available from an authorized MINI dealer.

> ***CAUTION—***
> - *Make sure glass and plastic components are at room temperature during repairs. This helps avoid breakage.*

Outside mirror glass, replacing

> ***CAUTION—***
> - *If the mirror is not already broken, this procedure may shatter it. Protect eyes and hands from broken glass.*
> - *Secure mirror glass against falling out.*
> - *Mirror heating harness wires (if equipped) are very short. Do not pull on mirror before disconnecting*

Working at door mirror:
- Turn door mirror glass to maximum inward position.
- Tape off glass with adhesive tape.
- Use plastic prying tool to unclip mirror glass (**arrow**).

– If equipped: Detach mirror heating harness wires.

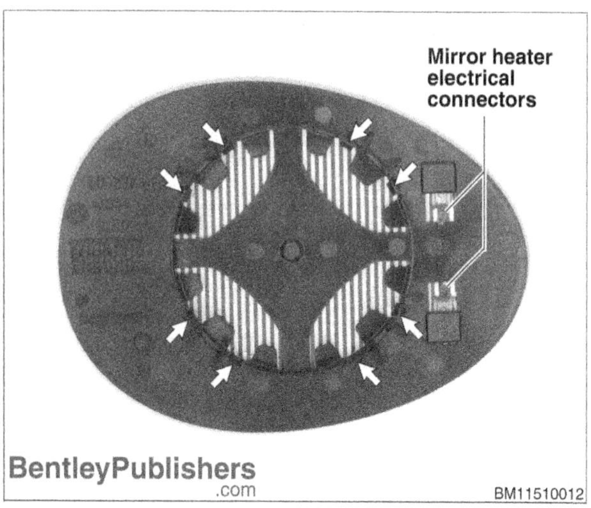

When installing:
- Make sure plastic retaining lugs (**arrows**) and mirror heater electrical connectors (if equipped) are not damaged.
- Place mirror glass retaining lugs flush on mirror motor and press to clip into place. Make sure mirror locks in place.
- Check mirror function.

Outside rear view mirror, removing and installing

◀ Rotate mirror housing and remove mounting bolts (**arrows**).

◀ Pull mirror away from door and disconnect electrical harness connector (**arrow**) at base of mirror. Remove mirror.

– When installing, make sure seal at mirror base is undamaged.

Tightening torque	
Mirror to door (M6 x 65 mm screws)	8 Nm (6 ft-lb)

Outside mirror motor, replacing

– Remove mirror glass. See **Outside mirror glass, replacing** in this repair group.

◀ Remove motor mounting screws (**arrows**) and pull out motor. Detach electrical connectors.

– Installation is reverse of removal.

510

Outside mirror housing, replacing

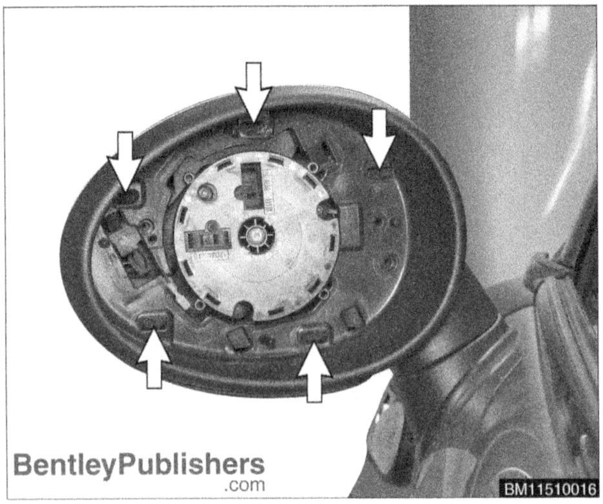

– Remove mirror glass. See **Outside mirror glass, replacing** in this repair group.

◄ Release door mirror cover clips (**arrows**) and remove cover.

– Installation is reverse of removal.

FRONT BUMPER

Front bumper assembly

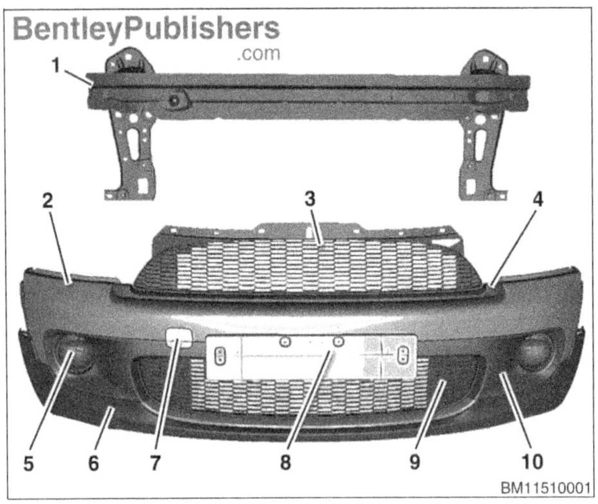

1. Bumper carrier. See **Front bumper carrier, removing and installing** in this repair group.
2. Bumper cover. See **Front bumper cover, removing and installing** in this repair group.
3. Radiator grille. See **Radiator grille, removing and installing** in this repair group.
4. Grille trim.
5. Front foglight. See **630 Lights**.
6. Front spoiler.
7. Towing eye cover.
8. License plate frame.
9. Lower grille.
10. Lower bumper trim.

Radiator grille, removing and installing

◄ Carefully pry lower chrome trim away from front grille. Start out by pressing out extended tab at one end of chrome (**arrow**).

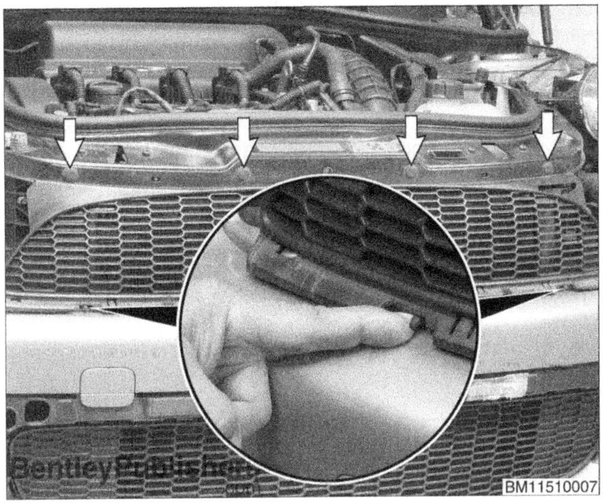

◄ Working in front of grille:
- Pry out plastic rivets (**arrows**).
- Press lower grille locking clips to side (**inset**) and slide grille forward.

– Installation is reverse of removal. Replace damaged plastic rivets.

Front bumper cover, removing and installing

– Remove radiator grille. See **Radiator grille, removing and installing** in this repair group.

– Raise front of vehicle and support in a safe manner.

> ***WARNING—***
> - *Make sure the car is stable and well supported at all times. Use a professional automotive lift or jack stands designed for the purpose. A floor jack is not adequate support.*

– Detach front wheel housing liners from front bumper cover. See **410 Fenders, Engine Hood**.

◄ Working underneath front of vehicle, remove bumper cover plastic rivets and sheet metal screws (**arrows**).

◄ Working in front of bumper, remove bumper cover screws (**arrows**) at left and right.

– Pull bumper cover slightly forward with aid of helper and detach electrical connectors from behind. Lower cover to ground.

– Installation is reverse of removal.
- Make sure foam insert is correctly positioned on bumper carrier
- Position bumper cover to achieve even gaps with adjacent panels.

Front bumper carrier, removing and installing

– Remove front bumper cover. See **Front bumper cover, removing and installing** in this repair group.

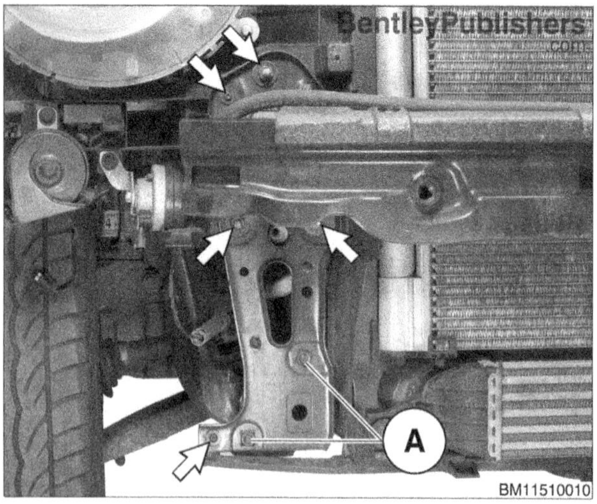

◄ Working at left and rights ends of bumper carrier:

- Remove lower bumper mounting bolts (**A**) from behind.
- Remove mounting bolts to subframe extenders (**arrows**) and screw.
- Remove bumper carrier from vehicle.

– Installation is reverse of removal. Make sure foam insert is positioned correctly.

Tightening torque	
Front bumper carrier to frame extensions (M8 nuts)	22 Nm (16 ft-lb)

Rear Bumper

Rear bumper assembly

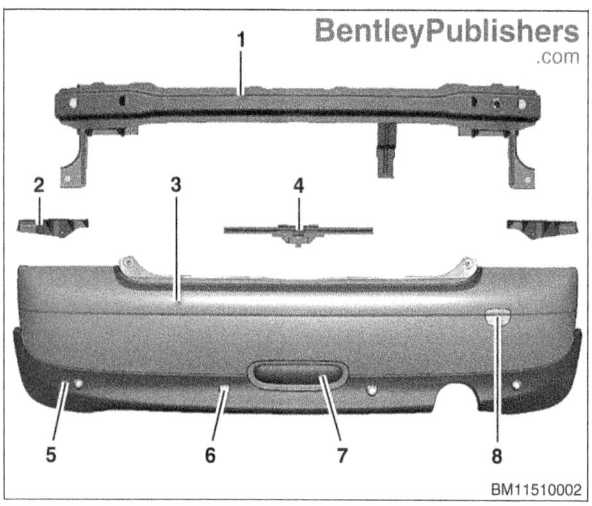

◄ Hardtop and Convertible rear bumper:

1. Bumper carrier. See **Rear bumper carrier, removing and installing** in this repair group.
2. Side bracket
3. Bumper cover. See **Rear bumper cover, removing and installing** in this repair group.
4. Center bracket.
5. Lower spoiler.
6. Park distance control (PDC) ultrasonic sensor.
7. Rear foglight.
8. Towing eye cover.

Cooper S rear bumper is similar.

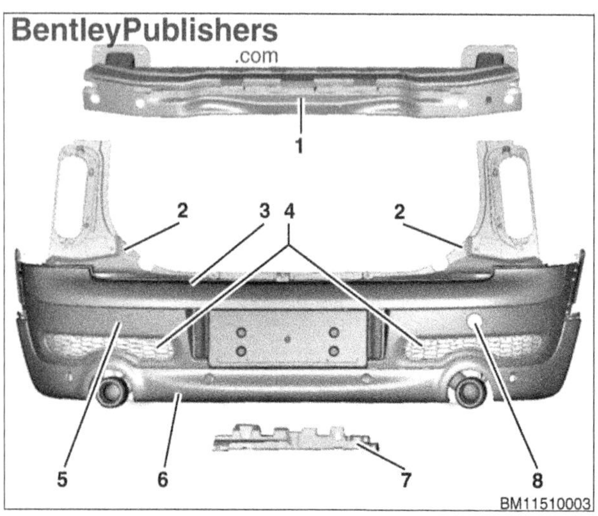

◄ Clubman rear bumper:

1. Bumper carrier. See **Rear bumper carrier, removing and installing** in this repair group.
2. Upper guide. See **Rear bumper upper guide, removing and installing (Clubman)** in this repair group.
3. Bumper cover. See **Rear bumper cover, removing and installing** in this repair group.
4. Trim.
5. Trim.
6. Lower spoiler.
7. Lower guide.
8. Towing eye cover

Rear bumper cover, removing and installing

This procedure covers rear bumper cover removal for Cooper. Cooper S, Convertible and Clubman rear bumper removal procedures are similar.

– Raise rear of vehicle and support in a safe manner.

> ***WARNING—***
> - *Make sure the car is stable and well supported at all times. Use a professional automotive lift or jack stands designed for the purpose. A floor jack is not adequate support.*

– Detach rear wheel housing liners from rear bumper cover. See **410 Fenders, Engine Hood**.

If applicable:

- Remove rear mudguard mounting screws (**A**) on left and right.
- Pull mudguard off, prying clips (**B**) to release.

Use plastic prying tool to pry off rear wheel arch trim (**arrow**) on left and right. Detach side marker light electrical harness.

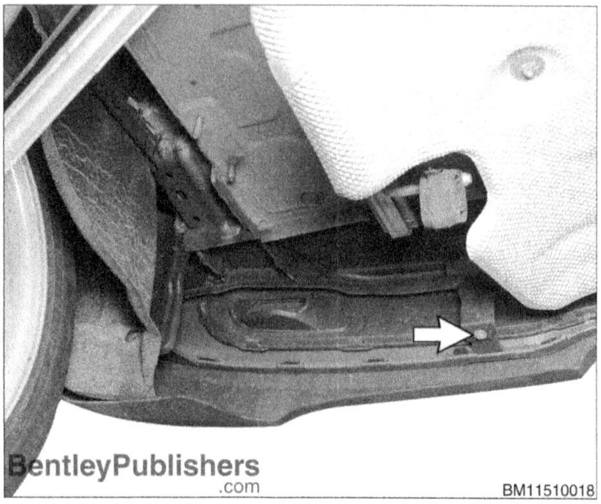

Working underneath rear of vehicle, remove bumper cover mounting screw (**arrow**). (Exhaust system removed for the purpose of illustration.)

Remove bumper cover mounting screws and plastic rivets (**arrows**) on left and right.

- Pull bumper cover slightly backward with aid of helper and detach electrical connectors (if equipped). Lower cover to ground.
- Installation is reverse of removal.
  - Make sure foam insert is correctly positioned on bumper carrier
  - Position bumper cover to achieve even gaps with adjacent panels.
  - Replace broken clips as necessary.

Rear bumper upper guide, removing and installing (Clubman)

- Remove rear bumper cover. See **Rear bumper cover, removing and installing** in this repair group.
- Remove taillight. See **630 Lights**.
- Remove split rear door strut. See **412 Rear Hatch, Tailgate, Split Rear Doors**.

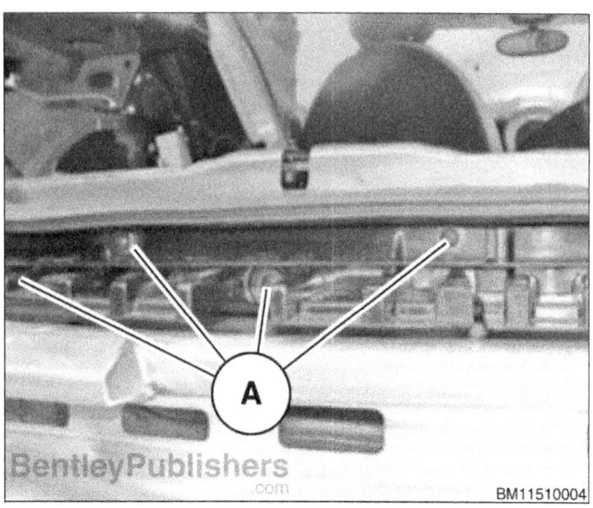

Remove upper guide mounting bolts (**arrows**).

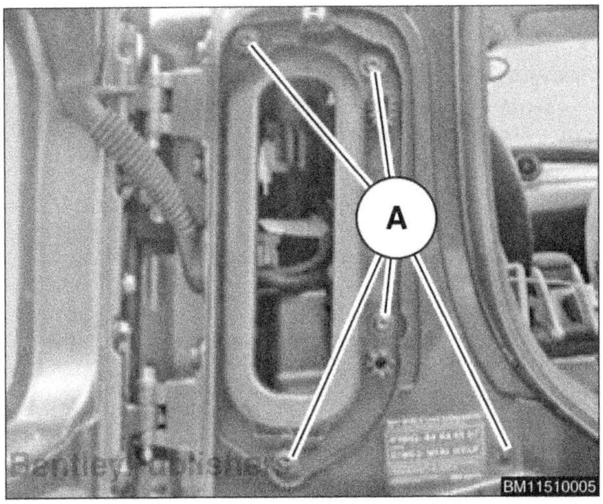

◀ Remove mounting screws (**A**) and remove bumper guide.

– Installation is reverse of removal.

Rear bumper carrier, removing and installing

– Remove rear bumper cover. See **Rear bumper cover, removing and installing** in this repair group.

– Clubman: Remove bumper upper guide. See **Rear bumper upper guide, removing and installing (Clubman)** in this repair group.

◀ Working at bumper carrier:

- If applicable, detach park distance control sensor harness.
- Remove bumper mounting bolts (**arrows**) and lower bumper.

– Installation is reverse of removal. Make sure foam insert is positioned correctly.

Tightening torque	
Rear bumper carrier to body (M8 bolts)	21 Nm (16 ft-lb)

510

EXTERIOR TRIM

Exterior trim is attached to the body with adhesive tape, plastic clips or fasteners. Be sure to have necessary 2-sided tape and fasteners on hand when reinstalling exterior trim pieces.

MINI emblem, removing and installing

The procedure below applies to both front and rear emblems.

The MINI emblem is attached with adhesive tape and located by two locating studs. Once removed, do not reuse emblem.

- Use heat gun to gently apply heat to emblem.
- Use plastic prying tool to lever off emblem.
- Clean off adhesive tape and residue from body panel.
- Install new emblem at room temperature.

512 Door Windows

General

This repair group covers door glass and power window motor repair information.

Service procedures for windshield, rear window glass and other fixed glass are not covered in this manual.

See also:

- **411 Doors** for door trim panel removal.
- **612 Switches** for window switch replacement.
- **ECL Electrical Component Locations** for fuse and ground information.

Window motor, initializing

> ***WARNING—***
> - *Window pinch protection (anti-trapping) is disabled during initialization.*

Each window motor needs to be taught the end points of window movement, as well as one-touch operation and pinch protection (anti-trapping). This is known as inilitializing the window motors. Window motor initialization may be performed using MINI scan tool or manually.

Initialize window motors under the following conditions:

- In case of malfunctions such as disabled one-touch function, no window opening or no comfort access function.
- After window motor replacement.
- After working on power window mechanism.
- After battery disconnection or other interruption of power to door window.
- After door window glass replacement or adjustment.
- After convertible top adjustment or repairs.
- After door seal or window seal replacement.

512

Each door window motor is initialized separately. The initialization process erases the previous initialization and then reinitializes (or reteaches) the window motor. Make sure the battery is fully charged. If necessary, recharge battery using a vehicle power supply.

Here is the procedure for manual initialization:

- Close doors and windows.
- Switch ignition ON.
- Move window approx. halfway down.
- Disconnect power to window motor:
  - Detach window motor electrical connector; or
  - Remove window motor fuse. **See ECL Electrical Component Locations** for fuse applications.
- Actuate power window switch several times to clear previous initialization. At this point pinch protection and one-touch control are disabled.
- Reconnect window motor power and test to make sure one-touch control is inoperative. If not, repeat previous step.
- With window motor powered, open window fully.
- After reaching lower position, move window without interruption to upper position and hold switch for at least 2 seconds in CLOSE position. This completes initialization.
- Check functions such as one-touch control, pinch protection and (if equipped) comfort access.

Door window assembly

Door window mechanism is attached to a separate integrated trim module (ITM) panel screwed to the door skin.

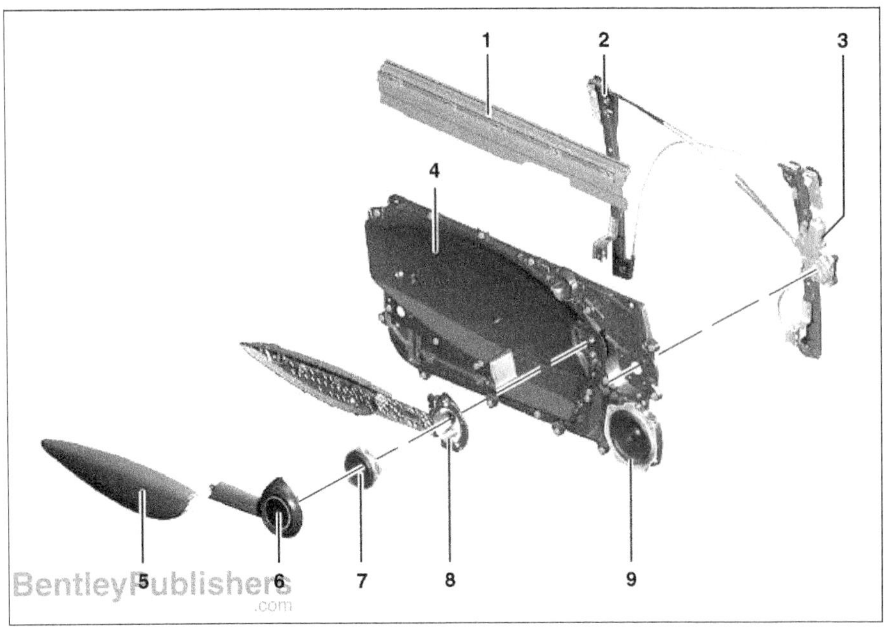

1. Aluminum brace
2. Window mechanism
3. Power window motor
4. Integrated trim module (ITM) panel
5. Armrest
6. Trim with outside rear view mirror switch
7. Mid-range speaker
8. Armrest bracket
9. Bass speaker

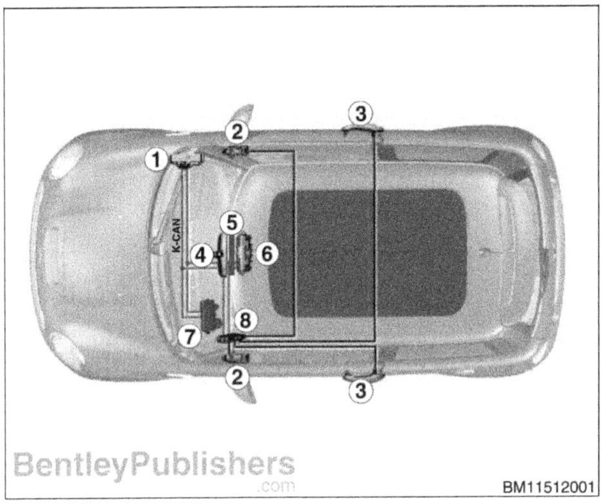

Power window components

1. Junction box (JB)
2. Power window motor
3. Door handle
4. Remote key antenna (in rear view mirror housing)
5. IHKA or IHKS control module
6. Power window switches
7. Car access system (CAS)
8. Footwell module (FRM)

Window drop down feature

When the door is opened, the window is automatically lowered slightly to clear the door seal. When the door is closed, the window is raised fully up into the seal. The amount of this movement in and out of the door seal is referred to as window retraction depth. Convertible top opening and closing activates this function via K-CAN.

Any time the window or window regulator is loosened or removed, adjust window retraction depth carefully to:

- Prevent glass breakage.
- Avoid high wind noise.
- Avoid water leaks.

See **Window glass position, adjusting** in this repair group.

Caution

CAUTION— • *To avoid marring trim pieces, use a plastic prying tool or a screwdriver with the tip wrapped with masking tape.*

Window Repairs

Window glass, removing and installing

◄ Starting at B-pillar end, use plastic prying tool to lever window edge strip off upper door edge (**arrows**).

– Remove door trim panel. See **411 Doors**.

Lower window glass to distance **A** above door frame.

Window glass removal height	
Rear edge of window above door (**A**)	approx. 75 mm (3 in)

- Disconnect negative (-) battery cable and cover battery terminal to keep cable from accidentally contacting terminal.

> ***CAUTION—***
> - *Prior to disconnecting the battery, read the battery disconnection cautions given in* **001 Warnings and Cautions**.

Working at inside door panel, pry out protective caps and reach through openings (**arrows**) to loosen window clamping screws far enough (approx. ¾ in) to release window glass.

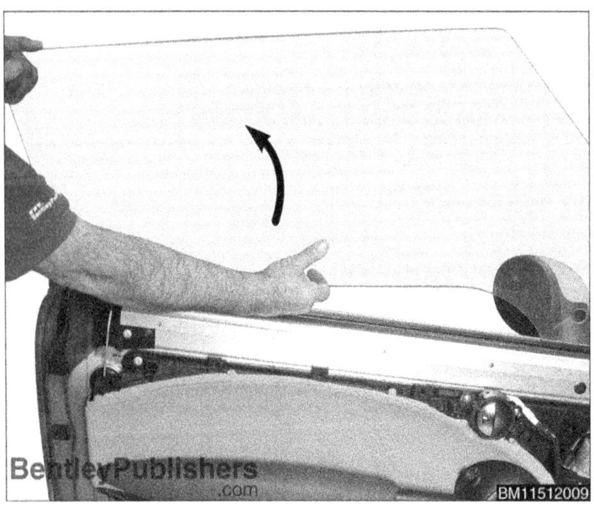

Lift glass out of door cavity.

- When installing, make sure base of window glass slides correctly in window clamping brackets. Tighten window clamping screws carefully, making sure glass is not damaged by protruding screw tips.
- Adjust window glass position. See **Window glass position, adjusting** in this repair group.
- Coat window cavity cover strip with water when installing.
- Reinitialize window motor. See **Window motor, initializing** in this repair group.

Window glass position, adjusting

Correct window adjustment reduces wind noise and prevents water leakage. Perform window position adjustments in this order:

- Adjust window inclination.
- Adjust window retraction depth (for correct window drop-down feature).
- Adjust window pretension.

Before adjusting glass position, check that door position is correct so that there are even gaps with adjacent panels. Make sure the vehicle is standing on its wheels.

Window inclination

Window inclination (inward rake of window) is measured with the door closed one notch.

With window closed completely, latch door to first catch on lock (**1**). Check to see that, with light finger pressure, top of window glass (**2**) just touches door seal (**3**).

If necessary:

- Pry out caps on lower door edge.
- Reach through bores to loosen window track screws (**inset**).
- Adjust window inclination by sliding track bracket in or out (**arrow**).
- Tighten screws.

– Recheck window inclination. If necessary, repeat procedure.

512

Window retraction depth

Up-down movement of window glass in door seal is referred to as window retraction depth. See **Window drop down feature** in this repair group.

Measure window retraction depth:

- Open door, engage special tool **51 3 080** on door window glass and close door.
- Measure retraction depth at 3 positions.

Window retraction depth	
Retraction depth measured using special tool **51 3 080**: • Position **1** • Position **2** • Position **3**	 5 ± 1 mm (0.197 ± 0.039 in) 5 ± 1 mm (0.197 ± 0.039 in) 4.5 ± 1 mm (0.177 ± 0.039 in)

If retraction depth adjustment is needed: Starting at B-pillar end, use plastic prying tool to lever window edge strip off upper door edge (**arrows**).

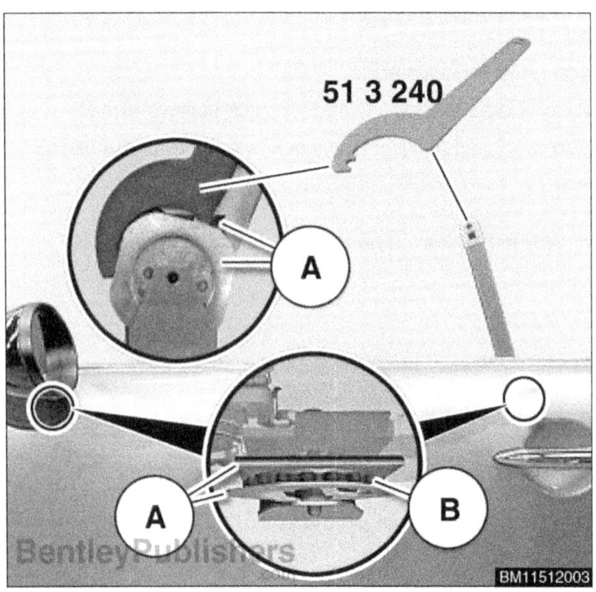

Use MINI special tool **51 3 240:**

- Reach through door edge opening to access space between glass supports (**A**).
- Loosen window adjusting nut (**B**) until window can be moved up and down.

> ***CAUTION—***
> - *Tape off door edge to protect against paint chipping caused by tool.*

– Slide window up or down, tighten adjusting nut and test retraction as before. Repeat if necessary.

Tightening torque	
Window retraction adjustment nut • With special tool • Without special tool	 10 Nm (7 ft-lb) 5 Nm (44 in-lb)

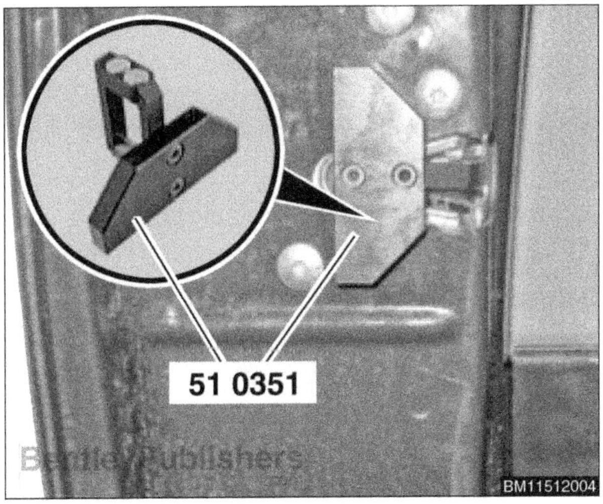

Window pretension

- Reinstall window cavity cover strip.

◄ Insert MINI special tool **51 0 351** in rotary latch of door lock and press in up to end stop.
  - Window glass then travels to uppermost position.
  - Carefully close door, leaving it slightly open until magnet makes contact with striker. Do not press on door or door glass during following measurement.

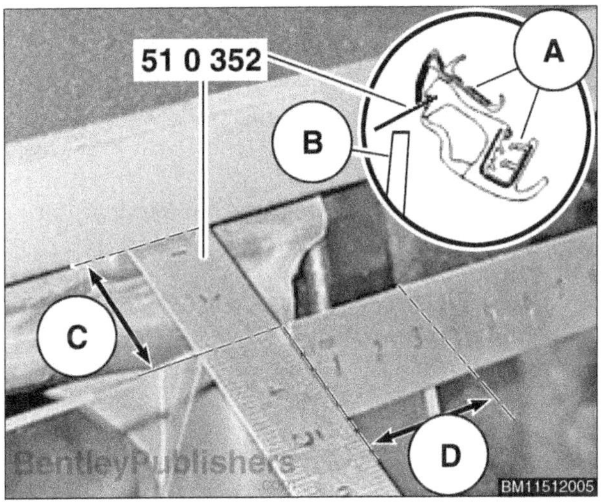

◄ Slide special tool **51 0 352** at distance (**D**) to door window glass side edge into seal (**A**) and measure dimension (**C**) to outside edge of glass (**B**).

Window pretension dimensions	
Dimension **C**	36 ± 2 mm (1.42 ± 0.08 in)
Dimension **D**	10 mm (0.39 in)

- Reinitialize window motor. See **Window motor, initializing** in this repair group.

Window electric motor, removing and installing

- Remove door trim panel. See **411 Doors**.

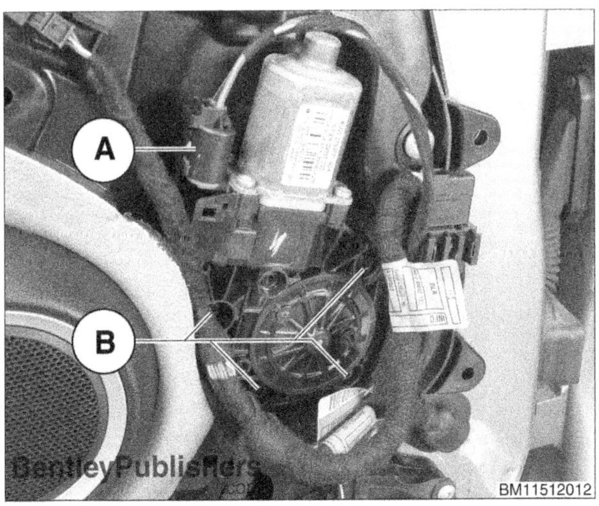

◄ Working near front edge of door:
  - Detach motor electrical connector (**A**).
  - Remove motor mounting screws (**B**).
  - Lift off window motor.

- Installation is reverse of removal. It is not necessary to reinitialize window motor.

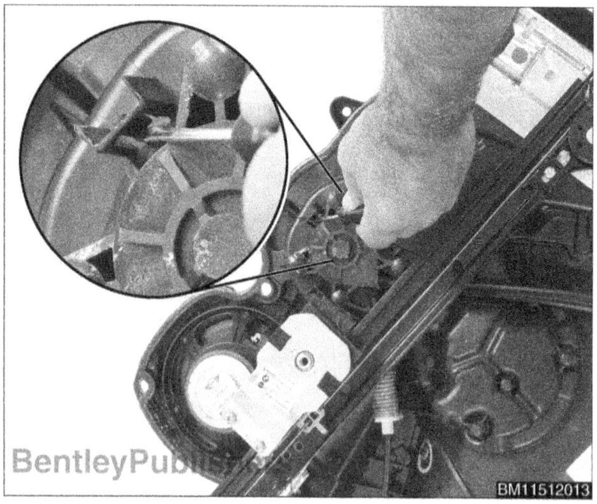

Power window drive, removing and installing

- Remove door trim panel. See **411 Doors**.
- Remove door window glass. See **Window glass, removing and installing** in this repair group.
- Remove window electric motor. See **Window electric motor, removing and installing** in this repair group.
- Remove ITM. See **Integrated trim module (ITM), removing and installing** in this repair group.

< Press catches (**inset**) and lift motor drive off ITM.

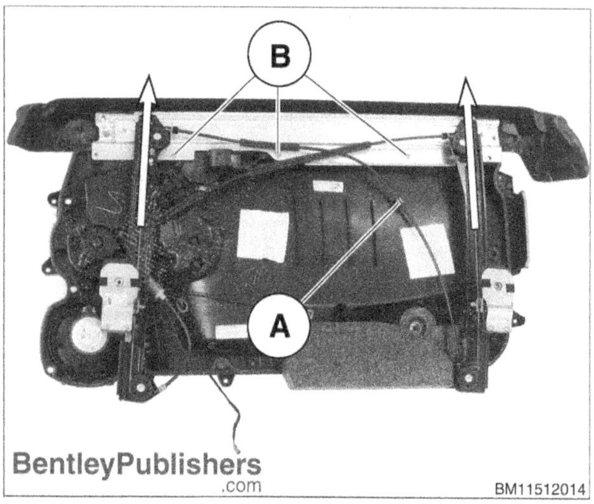

< Working at inside surface of ITM:

- Detach window drive cable sheath from plastic anchor (**A**).
- Remove window drive assembly mounting screws (**B**).
- Slide power window drive assembly off ITM (**arrows**).

- Installation is reverse of removal:
  - Make sure window drive assembly brackets fit correctly on ITM.
  - Coat window cavity cover strip with water when installing.
- Adjust window glass position after installing. See **Window glass position, adjusting** in this repair group.
- Reinitialize window motor. See **Window motor, initializing** in this repair group.

Integrated trim module (ITM), removing and installing

- Remove door trim panel. See **411 Doors**.
- Remove door window glass. See **Window glass, removing and installing** in this repair group.

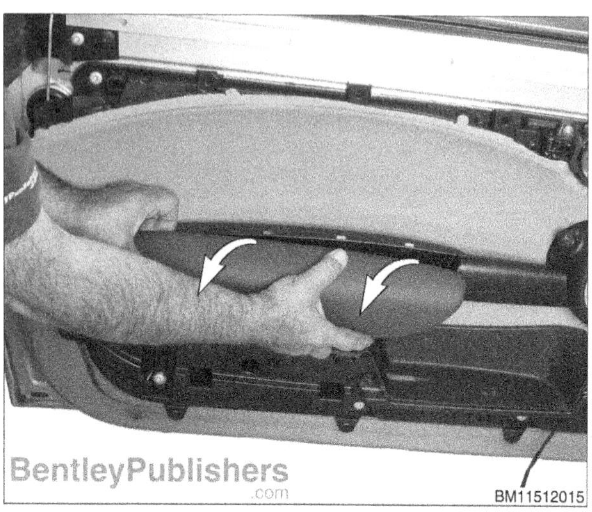

< Pull armrest off bracket (**arrows**).

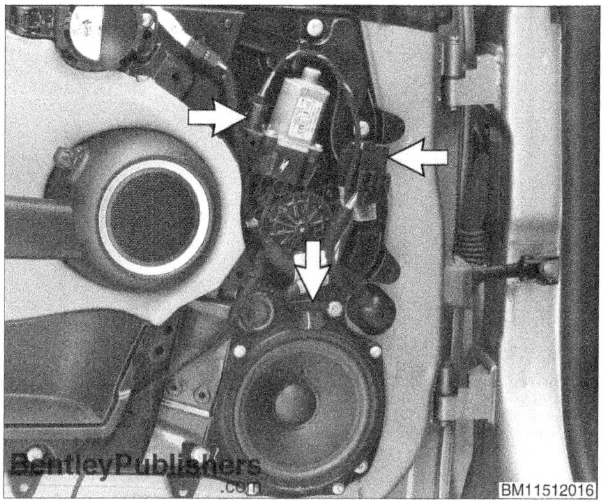

∢ Detach electrical connectors (**arrows**) at ITM.

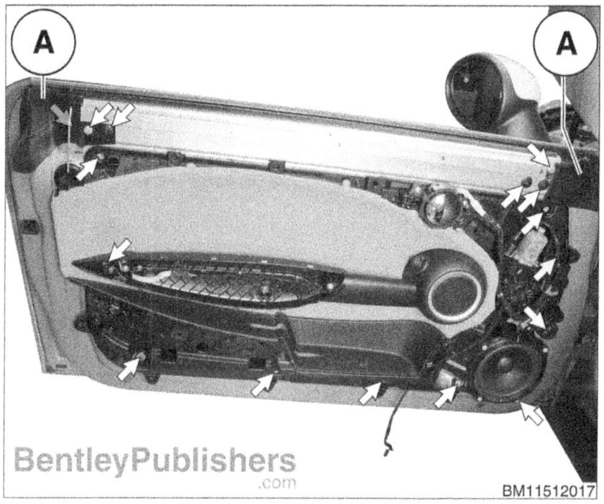

∢ Working at ITM mounting fasteners:

- First peel back upper edge front and rear rubber seals (**A**).
- Remove mounting bolts and nuts (**arrows**).
- Pull ITM upward slightly to disengage from door.

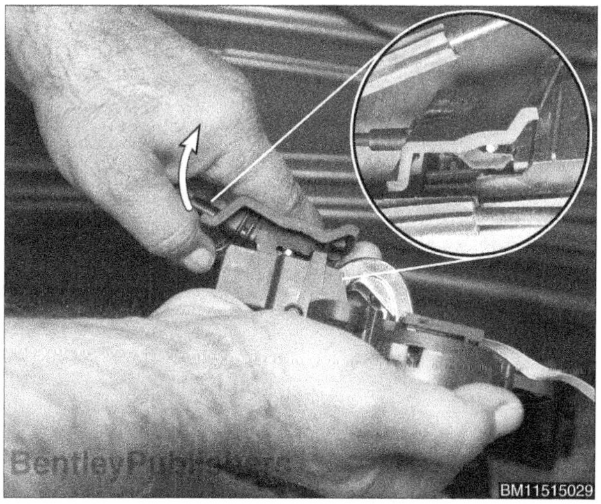

∢ Remove inner door handle and detach door lock Bowden cable from inner door lock release. Note routing of Bowden cable between window drive assembly and ITM panel.

– Lift ITM panel with window drive off door.

Integrated trim module (ITM), removing and installing

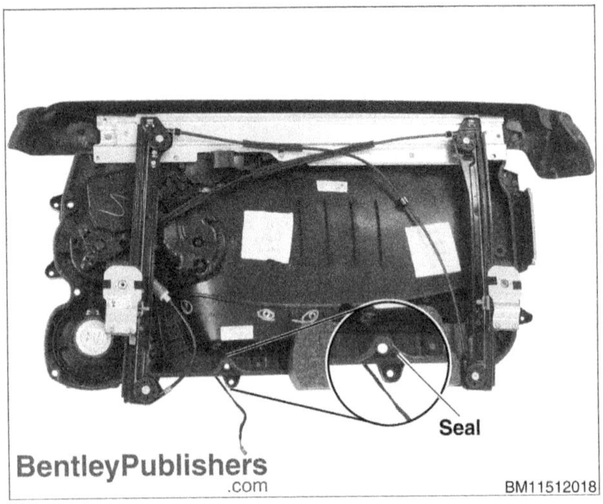

◄ Installation is reverse of removal.

- Check door seal at edges of ITM panel. Replace if necessary.
- Seat door lock Bowden cable on inner door handle and position correctly on ITM.

Tightening torques	
ITM panel to door skin • Screw • Nut	 8 Nm (6 ft-lb) 10 Nm (7 ft-lb)

– Adjust window glass position after installing. See **Window glass position, adjusting** in this repair group.

– Reinitialize window motor. See **Window motor, initializing** in this repair group.

513 Interior Trim

GENERAL

This repair group covers interior trim removal and installation procedures.

See also:

- **250 Gearshift Linkage** for shifter boot and bezel.
- **410 Fenders, Engine Hood** for wheel housing liners.
- **411 Doors** for door trim panels.
- **412 Rear Hatch, Split Rear Doors, Convertible Lid** for corresponding interior trim panels.
- **620 Instruments** for information displays and gauges
- **640 Heating and Air-conditioning** for climate control panel.

Interior trim repairs

Interior trim and finish panels are clipped or screwed into place. Many of the trim retaining clips are designed to be used only once. When removing trim that is held in place with clips, be sure to have spares on hand before beginning the job.

Warnings and Cautions

> ***WARNING—***
> - *MINI models are equipped with SRS airbags mounted in the following locations:*
>   *-Driver front airbag in steering wheel.*
>   *-Passenger front airbag in right side dashboard.*
>   *-Side protection (thorax) airbags in front seat backrest bolsters.*
>   *-Head protection airbags along roof edges from the A-pillars to the C-pillars.*
>   *Observe special precautions when servicing airbags or associated trim or components. Serious injury may result if system service is attempted by persons unfamiliar with SRS and its approved service procedures.*
> - *Before performing any work involving airbags, disconnect the negative (–) battery cable. See* **721 Airbag System (SRS)**.
> - *The SRS control module contains a back-up power supply. Allow a 5 second discharge period after the battery cable has been disconnected before working on airbag components.*

> ***CAUTION—***
> - *When working on electrical switches or lights, disconnect the negative (–) cable from the battery and insulate the cable end to prevent accidental reconnection.*
> - *Prior to disconnecting the battery, read the battery disconnection cautions in* **001 Warnings and Cautions**.
> - *To avoid marring trim pieces, use a plastic prying tool or a screwdriver with the tip wrapped with masking tape.*
> - *Plastic trim becomes brittle with age. When reinstalling, be prepared to replace broken or cracked plastic trim pieces.*

Cowl Covers

Cowl cover panels over the plenum chamber in the engine compartment isolate the battery, wiper assembly, brake fluid reservoir and other components from the main front part of the engine compartment.

Right cowl cover, removing

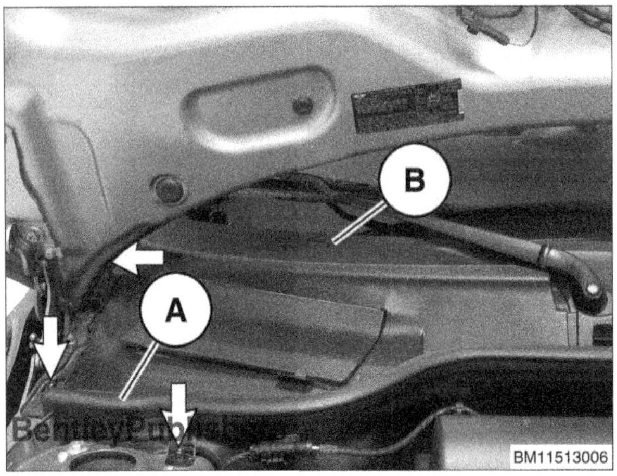

◄ With engine hood open:

- Remove cowl sealing gasket (**A**).
- Remove nuts and screw (**arrows**).
- Peel cover off retaining strip at bottom of windshield (**B**) and remove.

– When installing:

- Make sure cover seats correctly in retaining strip.
- Check sealing gasket. Replace if damaged.

Left cowl cover, removing

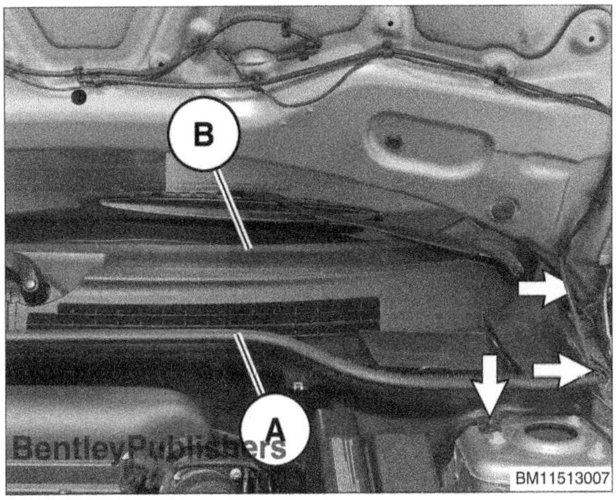

- Remove right cowl cover. See **Right cowl cover, removing** in this repair group.
- Remove wiper arms. See **611 Wiper and Washers**.

< Working at cowl cover:
- Remove cowl sealing gasket (**A**).
- Remove nuts and screw (**arrows**).
- Peel cover off retaining strip at bottom of windshield (**B**) and remove.

- When installing, make sure cover seats correctly in retaining strip.
- Check sealing gasket. Replace if damaged.

CENTER CONSOLES

Front console, removing and installing

- Remove gearshift knob by pulling knob straight up. See **250 Gearshift Linkage**.
- Slide parking brake handle grip off lever.

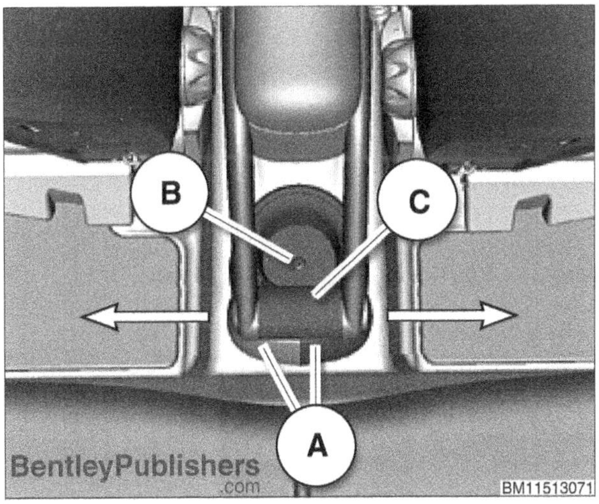

< Version with center armrest:
- Remove trim screws (**A**). Unclip trim and remove (**arrows**).
- Remove screw (**B**).
- Unclip and tilt cup holder (**C**) up to remove.

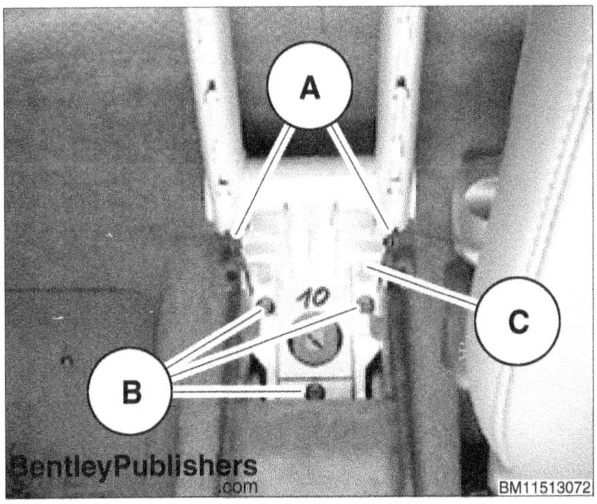

< Remove arm rest mounting screws (**A**, **B**) and lift out armrest (**C**).

513

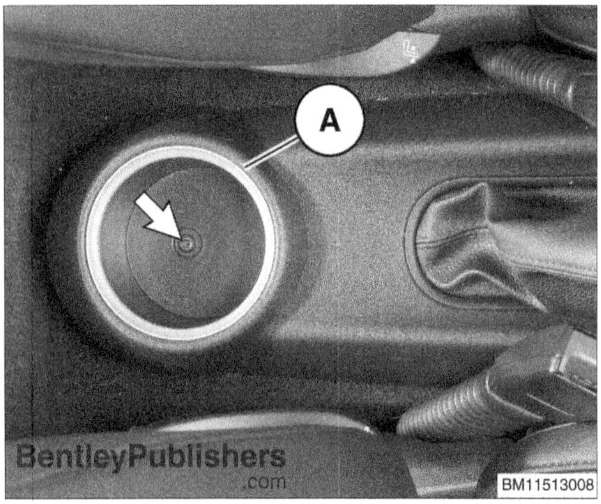

Version without center armrest: Remove screw (**arrow**) and remove cup holder (**A**).

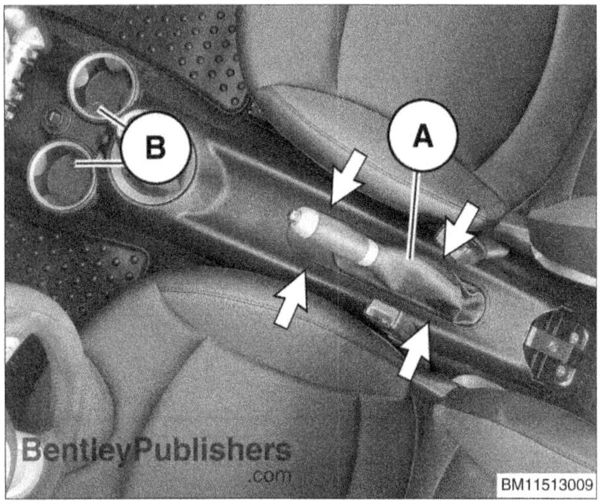

Working at storage compartment lid:

- Detach parking brake boot (**A**) by pulling up.
- Remove screws (**B**).
- Detach storage compartment at clips (**arrows**).

– Raise console slightly and detach electrical connectors from below. Lift console out of vehicle.

– Installation is reverse of removal.

Rear console, removing and installing

– Remove front console and (if applicable) arm rest.

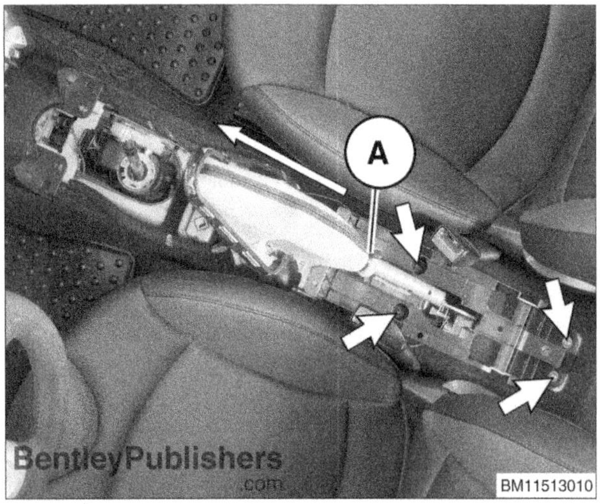

Working at rear console:

- Cut wire tie (**A**) and slide parking brake boot (**long arrow**) off handle.
- Remove console mounting nuts (**arrows**).
- If necessary, detach electrical harness bracket.

– Lift out console.

– Installation is reverse of removal.

Steering Column Covers

Upper steering column cover, removing and installing

– Remove driver airbag and steering wheel. See **320 Steering and Wheel Alignment** and **721 Airbag System (SRS)**.

> *WARNING—*
> - *The airbag is inflated by an explosive device. Handled improperly or without adequate safeguards, it can be very dangerous. Observe special precautions prior to any work at or near airbags.*
> - *Place removed airbag unit with the padded side facing upward. Do not leave airbag unattended.*

– Remove tachometer. See **620 Instruments**.

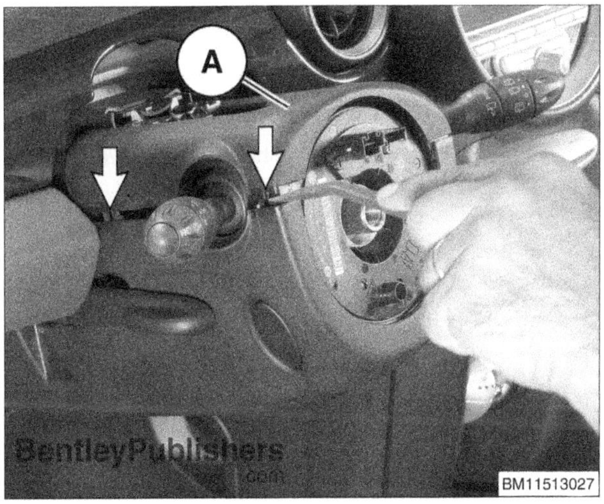

◄ Using plastic prying tool, unclip upper steering column cover (**A**) and pull up. To disengage plastic retainers, gently press in trim (**arrows**).

– Installation is reverse of removal.

513

Lower steering column cover, removing and installing

– Remove upper steering column cover. See **Upper steering column cover, removing and installing** in this repair group.

– Remove left lower dashboard trim (underneath steering column). See **Left lower dashboard trim, removing** in this repair group.

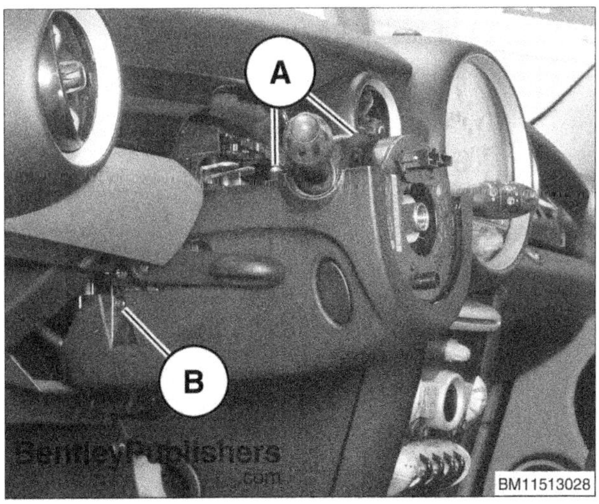

◄ Working above steering column, remove lower cover mounting screws (**A**).

– Working underneath steering column, remove lower cover mounting screws (**B**) at left and right.

– Installation is reverse of removal.

Tightening torque	
Lower cover to steering column (side screws)	4 Nm (35 in-lb)

Dashboard

Dashboard components

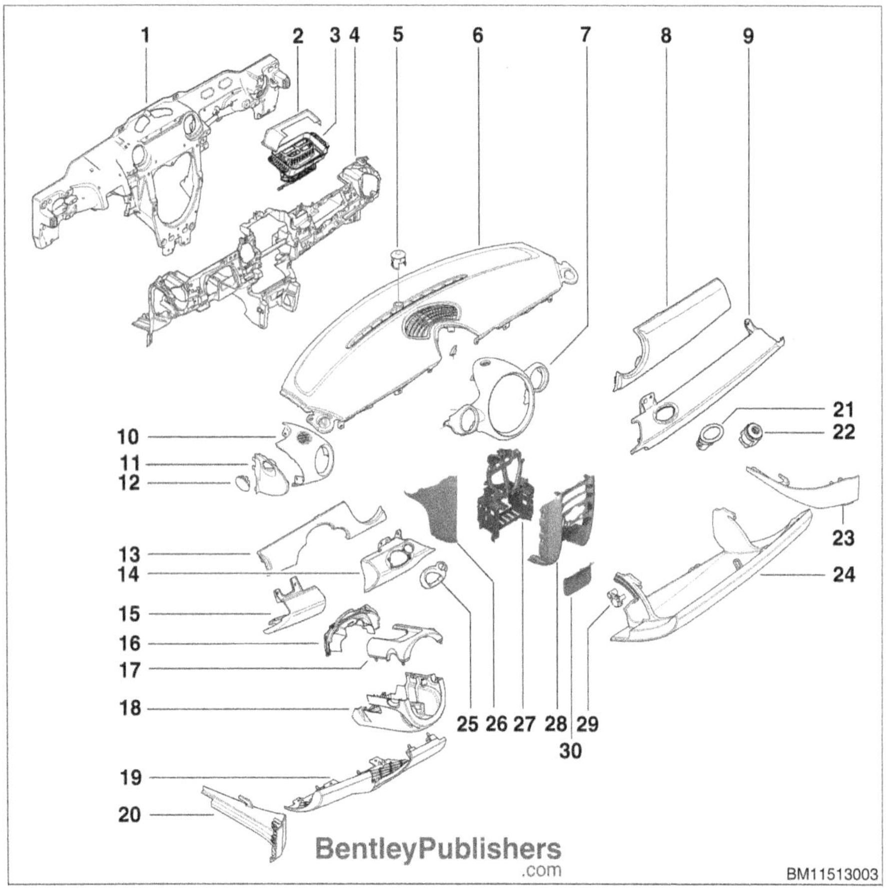

1. Dashboard carrier
2. Passenger airbag cover
3. Passenger airbag
4. Dashboard
5. Solar sensor
6. Upper dashboard cover
7. Center instrument bezel
8. Right side dashboard decorative strip
9. Passenger knee pad
10. Dashboard vent cover
11. Dashboard side cover
12. Trim cover
13. Left side dashboard decorative strip
14. Driver knee pad, right
15. Driver knee pad, left
16. Steering column rubber padding
17. Upper steering column trim cover
18. Lower steering column trim cover
19. Left lower dashboard trim
20. Left side dashboard cover
21. Glove compartment button bezel
22. Glove compartment button
23. Right side lower dashboard cover
24. Glove compartment lid
25. STOP - START (Ignition switch) bezel
26. Center dashboard side cover
27. Center dashboard module carrier
28. Center dashboard bezel
29. Glove compartment lid damper
30. Cover

Dashboard, removing and installing

Dashboard removal is a complicated job. Be sure to read the warning and cautions in **001 Warning and Cautions** and **721 Airbag System (SRS)**.

In the procedure that follows, the dashboard is separated and removed from the dashboard carrier in order to gain access to climate control components. Read the procedure through before starting work.

– Steer front wheels into straight ahead position and remove ignition key.

– Disconnect negative (-) battery cable and cover battery terminal to keep cable from accidentally contacting terminal.

> *CAUTION—*
> - *Prior to disconnecting the battery, read the battery disconnection cautions given in* **001 Warnings and Cautions**.

– Remove left lower dashboard trim (underneath steering column). See **Left lower dashboard trim, removing** in this repair group.

– Remove tachometer. See **620 Instruments**.

– Remove left and right dashboard decorative strips. See **Dashboard decorative strips, removing** in this repair group.

– Remove center instrument bezel. See **Center instrument bezel, removing** in this repair group.

– Remove center dashboard covers, switch center and climate control panel. See **Center dashboard covers, removing** in this repair group.

– Remove driver and passenger knee pads. See **Dashboard knee pads, removing** in this repair group.

– Remove center instrument. See **620 Instruments**.

– Remove dashboard vents and upper dashboard cover. See **Dashboard vent cover, removing** and **Upper dashboard cover, removing** in this repair group.

– Remove driver airbag and steering wheel. See **320 Steering and Wheel Alignment** and **721 Airbag System (SRS)**.

> *WARNING—*
> - *The airbag is inflated by an explosive device. Handled improperly or without adequate safeguards, it can be very dangerous. Observe special precautions prior to any work at or near airbags.*
> - *Place removed airbag unit with the padded side facing upward. Store airbag in secure location.*

– Remove steering column covers. See **Steering Column Covers** in this repair group.

– Remove steering column switch cluster. See **612 Switches**.

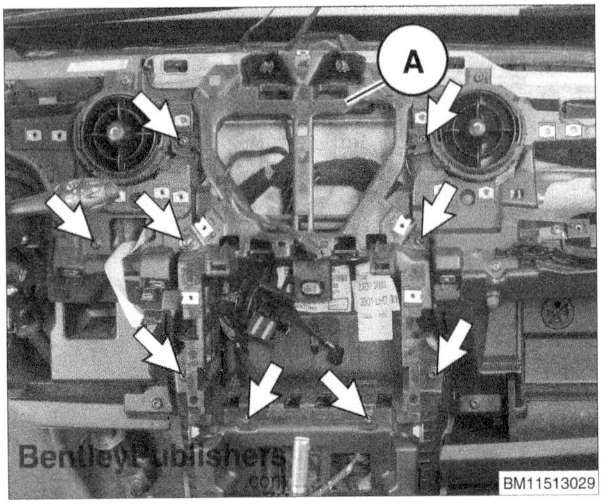

BM11513029

– Remove footwell and glove compartment lights. See **630 Lights**.

– If applicable: Remove dashboard mounted CD changer. See **650 Radio**.

◀ Remove center instrument mounting bracket fasteners (**arrows**) then remove mounting bracket (**A**).

BM11513030

◀ Working at left side of dashboard:

- Remove dashboard mounting screws (**arrows**).
- Detach OBD II plug electrical connector (**A**).

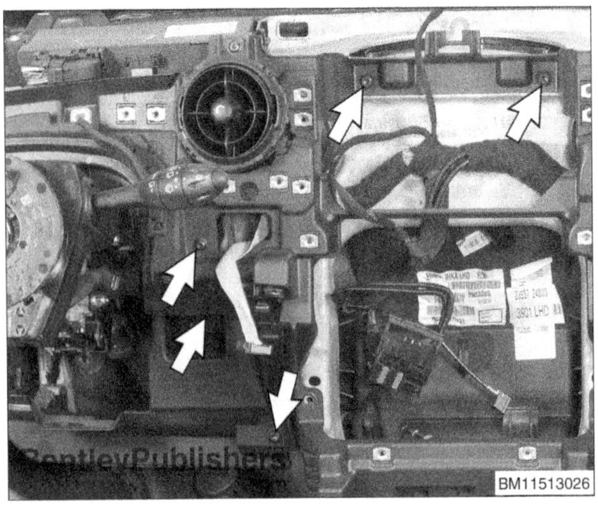
BM11513026

◀ Working at left side of center dashboard, remove mounting screws (**arrows**).

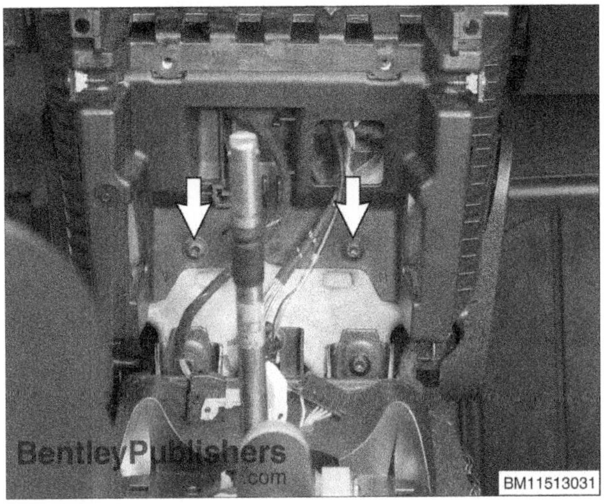

◄ Working at lower center dashboard, remove mounting screws (**arrows**).

◄ Working at right side of dashboard, remove mounting screws (**arrows**).

- Lift dashboard out of vehicle.
- When installing, make sure air ducts seat correctly.
- Vehicle with DSC: Reset steering angle sensor. Connect vehicle to MINI scan tool and carry out steering angle sensor adjustment system function.

Center instrument bezel, removing

- Remove tachometer. See **620 Instruments**.
- Remove dashboard decorative strips. See **Dashboard decorative strips, removing** in this repair group.

◄ Use plastic prying tool to unclip small covers (**arrows**) in center air outlet.

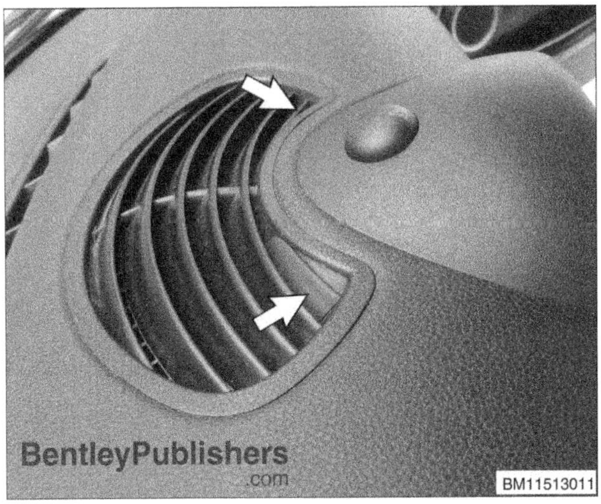

Center instrument bezel, removing

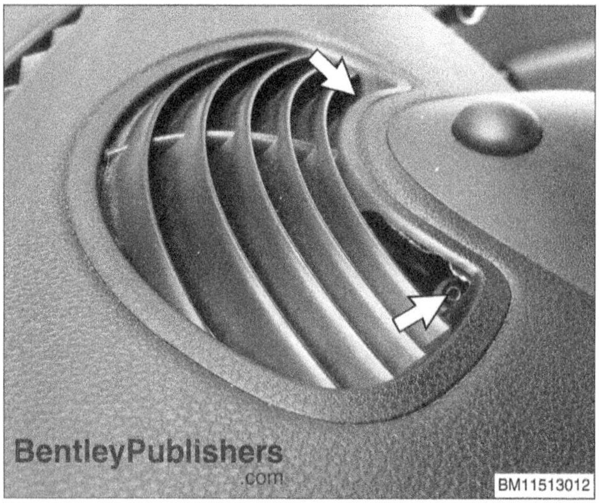

Remove instrument bezel mounting screws (**arrows**).

Working at center instrument:

- Remove bezel mounting screws (**arrows**).
- Pull bezel off instrument panel and tilt up.

Detach hazard warning lights switch electrical connector (**arrow**) and remove bezel.

– Installation is reverse of removal.

Dashboard knee pads, removing

Left side

- Remove left lower dashboard trim. See **Left lower dashboard trim, removing** in this repair group.
- Remove tachometer. See **620 Instruments**.
- Remove left side dashboard decorative strip. See **Dashboard decorative strips, removing** in this repair group.

Left of steering column

- Remove left side dashboard vent cover. See **Dashboard vent cover, removing** in this repair group.

◄ Remove knee pad mounting screws (**arrows**).

◄ Working underneath edge of dashboard, remove lower knee pad mounting screws (**arrows**) and remove knee pad (**A**).

Right of steering column

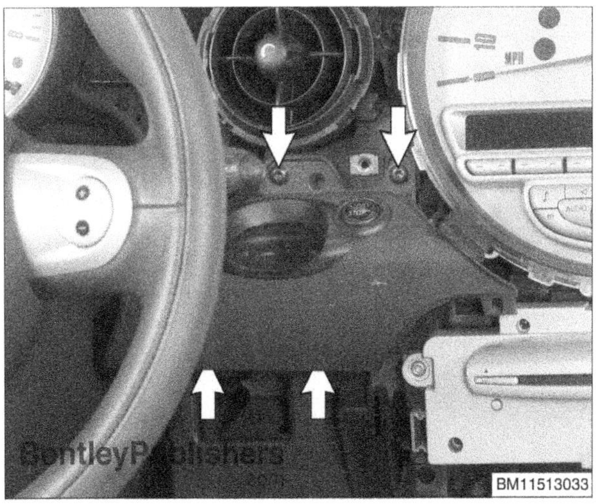

◄ Working at right side of steering wheel, remove knee pad fasteners (**arrows**).

Disconnect STOP - START module ribbon cable (**arrow**), then remove knee pad (**A**).

Right side

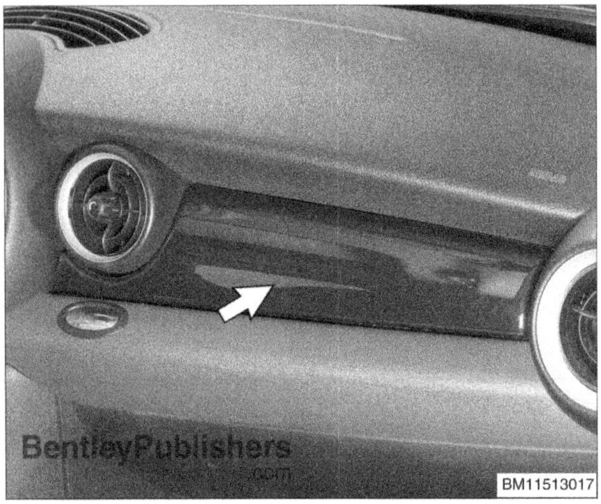

Open storage compartment (**arrow**).

- Remove right side dashboard decorative strip. See **Dashboard decorative strips, removing** in this repair group.
- Remove center instrument bezel. See **Center instrument bezel, removing** in this repair group.
- Remove center dashboard covers. See **Center dashboard covers, removing** in this repair group.
- Remove right A-pillar trim panel. See **A-pillar trim, removing** in this repair group.
- Remove right side dashboard vent cover. See **Dashboard vent cover, removing** in this repair group.

Open glove compartment lid and remove lid mounting screws (**arrows**).

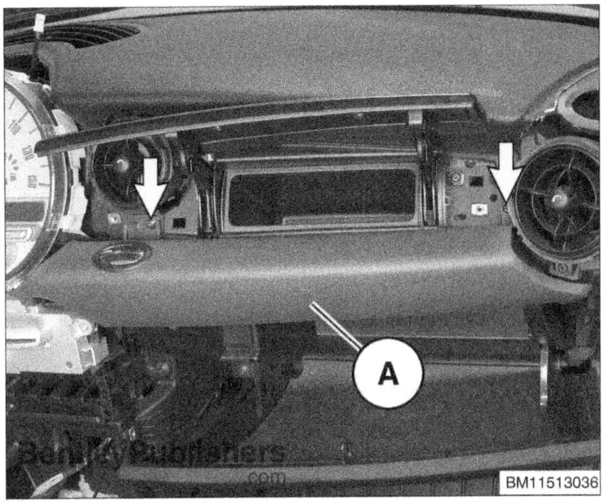

Remove knee pad mounting screws (**arrows**) and lift off knee pad (**A**).

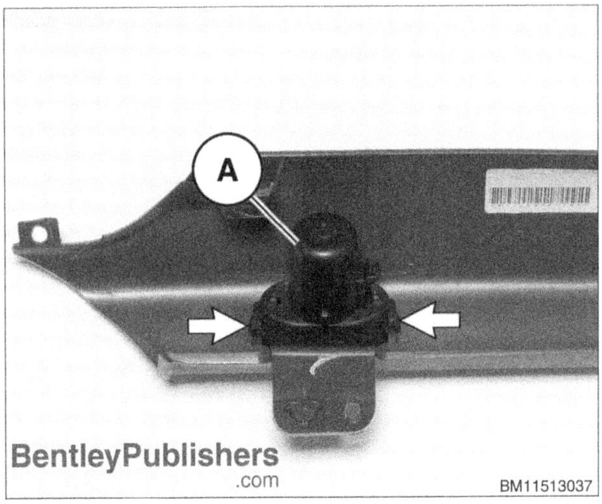

If necessary, detach glove compartment lid button (**A**). Squeeze spring clips (**arrows**) to disengage.

Dashboard decorative strips, removing

Left side

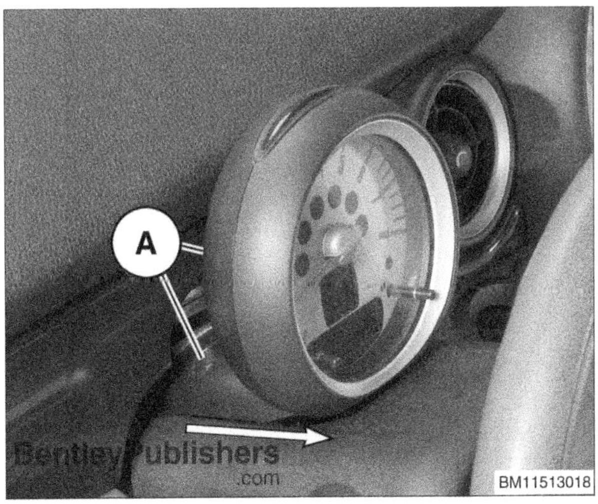

Remove tachometer fasteners (**A**) and slide tachometer forward (**arrow**) off mounting bracket.

◀ Use plastic prying tool (**A**) to lever left side dashboard decorative strip off dashboard clips (**arrows**).

Right side

◀ Open storage compartment (**arrow**).

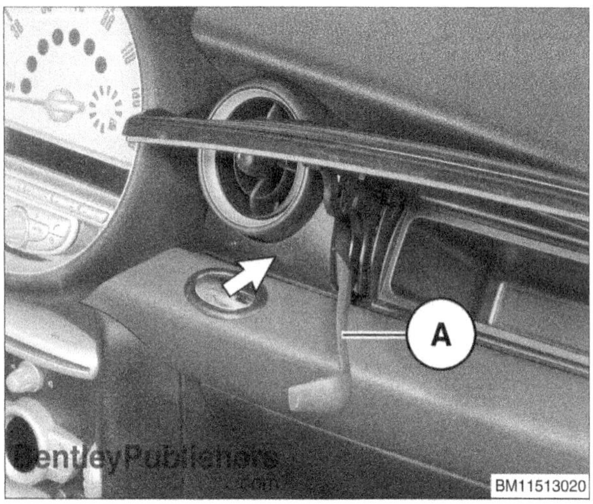

◀ Use plastic prying tool (**A**) to lever left trim off clips (**arrow**).

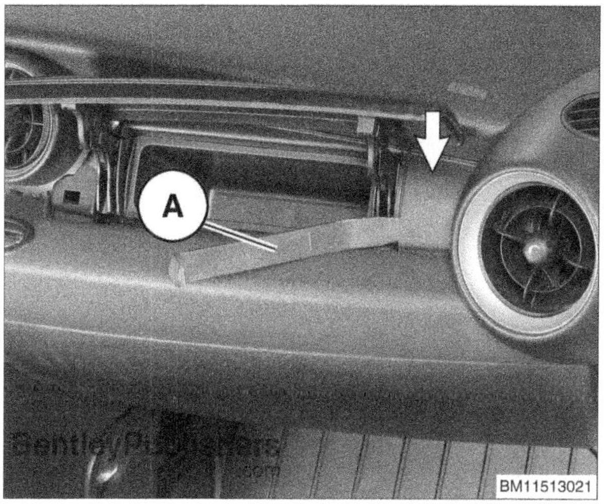

Use plastic prying tool (**A**) to lever right trim off clips (**arrow**).

Dashboard vent cover, removing

Left and right dashboard vent cover removal procedures are similar.

- Remove A-pillar trim. See **A-pillar trim removing and installing** in this repair group.
- Remove dashboard decorative strip. See **Dashboard decorative strips, removing** in this repair group.

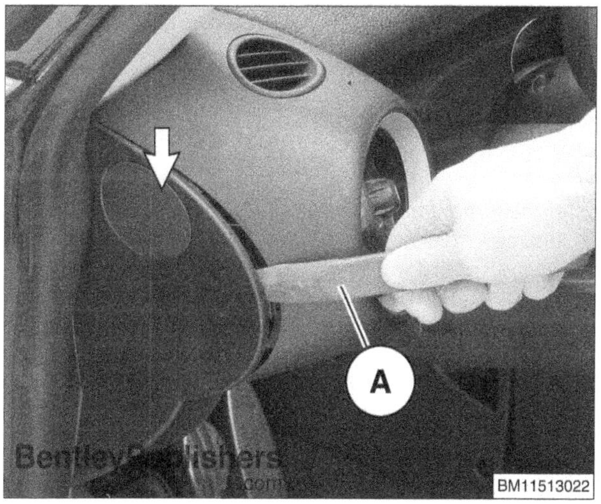

Use plastic prying tool (**A**) to lever off (**arrow**) dashboard end cover.

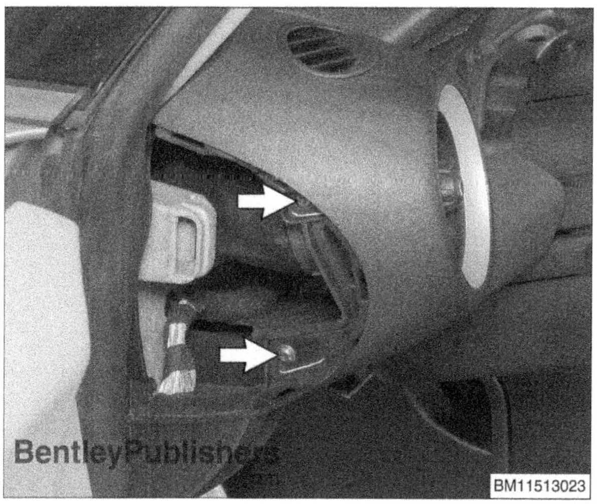

Remove air vent mounting screws (**arrows**) through dashboard end opening.

513

⮜ Remove remaining vent mounting screws (**arrows**) and slide vent off dashboard.

– Installation is reverse of removal. Make sure air duct seats correctly.

Upper dashboard cover, removing

– Disconnect negative (-) battery cable and cover battery terminal to keep cable from accidentally contacting terminal.

> ***CAUTION—***
> • *Prior to disconnecting the battery, read the battery disconnection cautions given in* **001 Warnings and Cautions**.

– Remove tachometer. See **620 Instruments**.

– Remove dashboard decorative strips. See **Dashboard decorative strips, removing** in this repair group.

– Remove center instrument bezel. See **Center instrument bezel, removing** in this repair group.

– Remove dashboard vent covers. See **Dashboard vent cover, removing** in this repair group.

⮜ Working above steering column, remove dashboard cover mounting screws (**arrows**).

< Working at center dashboard, remove dashboard cover mounting screws (**arrows**).

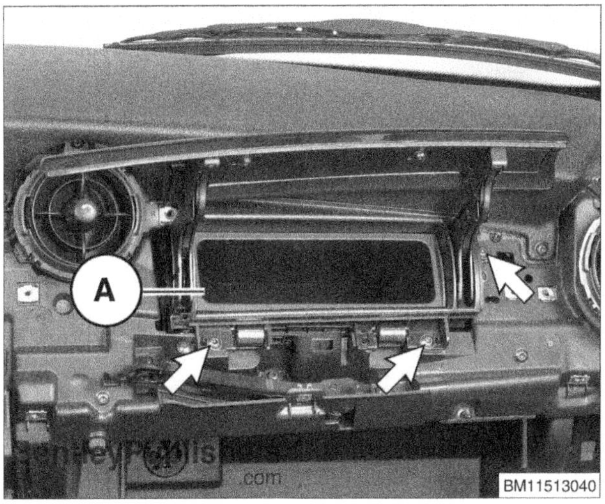

< Working at right side of dashboard:

- Remove fasteners (**arrows**) for storage compartment or CD changer.
- Remove storage compartment (**A**) or CD changer.

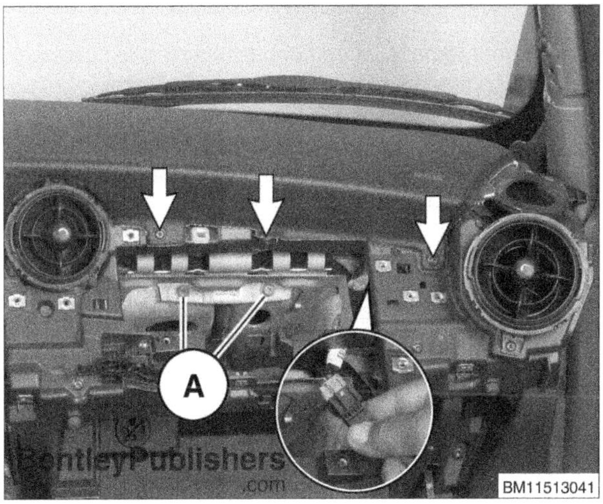

< Working at right side of dashboard:

- Remove dashboard cover mounting screws (**arrows**).
- Remove passenger airbag mounting bolts (**A**).
- Detach electrical connector (**inset**).

– Lift dashboard cover slightly and slide solar sensor harness out of opening.

Using a small screwdriver, remove solar sensor from dashboard, then disconnect electrical connector.

– Slide dashboard cover backward to remove.

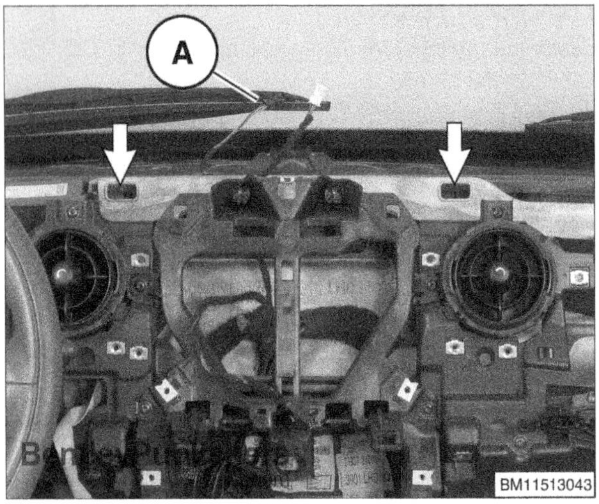

Installation is reverse of removal:

- Insert solar sensor harness through dashboard cover opening (**A**).
- Insert dashboard cover centering pins correctly into locators (**arrows**).

Tightening torque	
Passenger airbag to dashboard	6 Nm (53 in-lb)

Center dashboard covers, removing

This procedure covers removal of center dashboard side covers, switch cluster, A/C control panel and module carrier bezel.

– Remove left lower dashboard trim (panel underneath steering column). See **Left lower dashboard trim, removing** in this repair group.

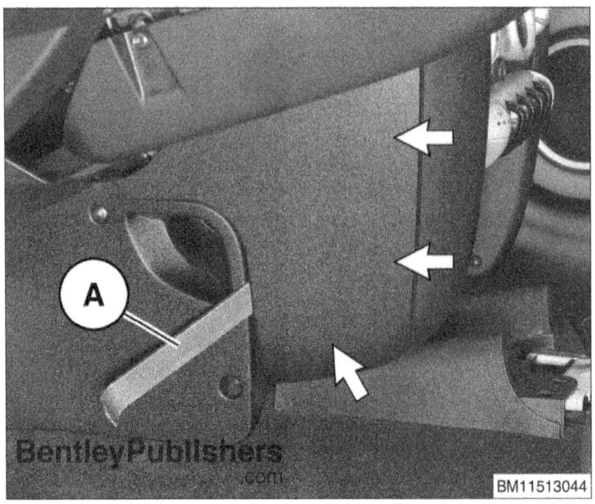

Use a plastic prying tool (**A**) to lever left side center dashboard cover from retaining clips (**arrows**).

– Remove center instrument bezel. See **Center instrument bezel, removing** in this repair group.

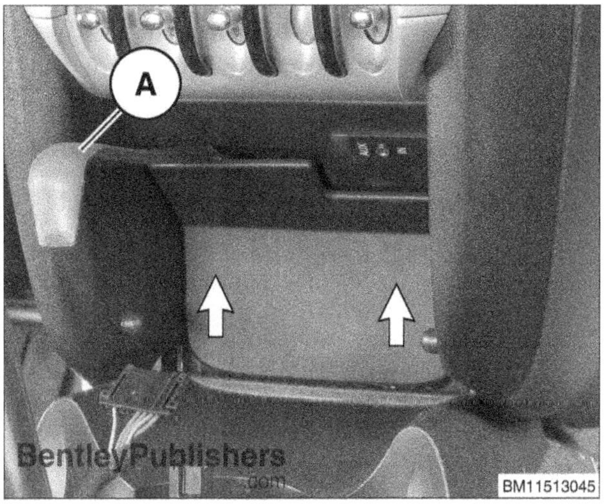

< Use plastic prying tool (**A**) to lever lower center dashboard cover from retaining clips (**arrows**).

< Remove bezel mounting screws (**arrows**) on both sides and at bottom of center dashboard.

– Remove front center console. See **Center Consoles** in this repair group.

< Using plastic prying tool (**A**), unclip center dashboard bezel (**arrows**) and pull back slightly.

< Detach electrical connectors (**arrows**) on back of switch cluster and climate control panel and lift out bezel.

– Installation is reverse of removal.

Left lower dashboard trim, removing

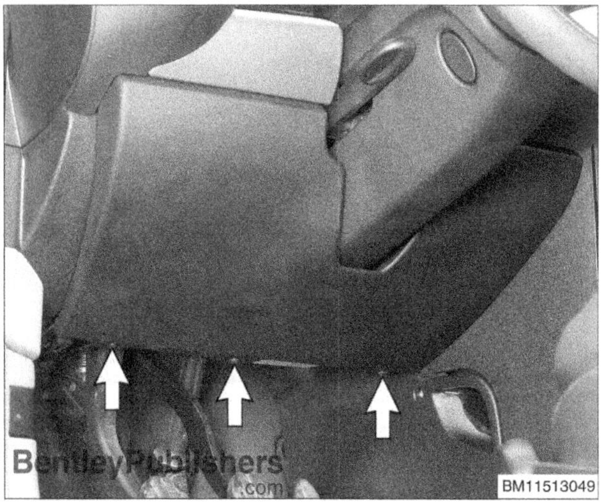

< Working underneath steering column, remove trim panel mounting screws (**arrows**).

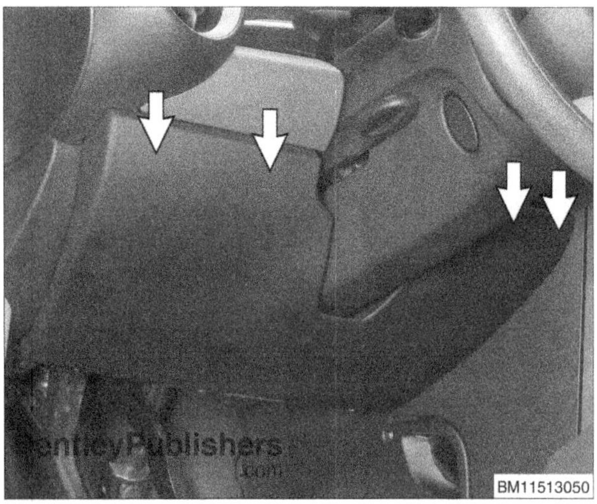

< Unclip trim panel from retaining clips (**arrows**) and remove.

– Installation is reverse of removal.

INTERIOR TRIM PANELS

Front door sill cover, removing

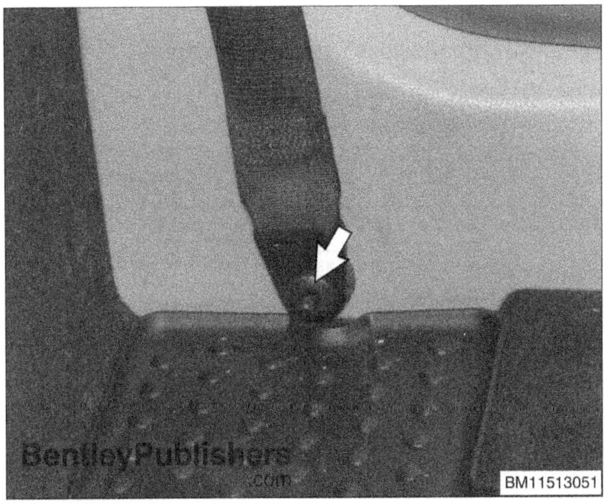

◀ Working below door opening, remove seat belt anchor bolt (**arrow**) and discard.

– Slide front seat forward.

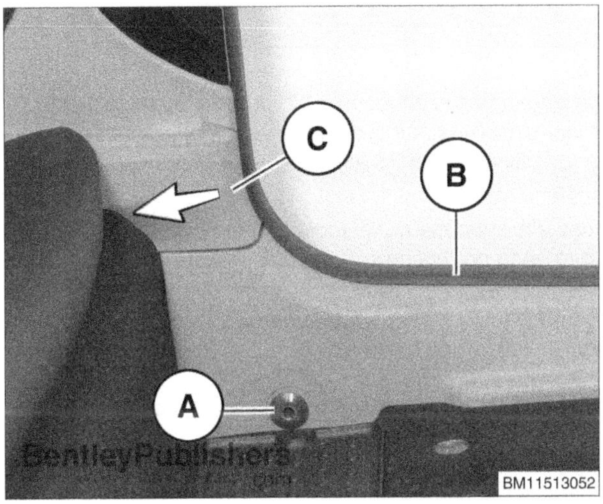

◀ Working at rear lower corner of door opening:

- Remove bushing (**A**).
- Disengage and remove rubber door seal (**B**)
- Detach (**arrow**) side trim panel (**C**). Inspect panel clip and discard if damaged.

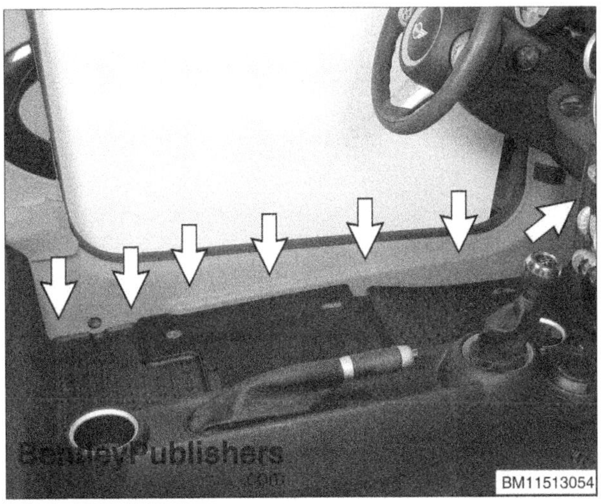

◀ Detach sill cover from clips (**arrows**). Bend cover slightly to remove. Seat removed for purposes of illustration.

> ***CAUTION—***
> *• To avoid scratching seat, cover with blanket.*

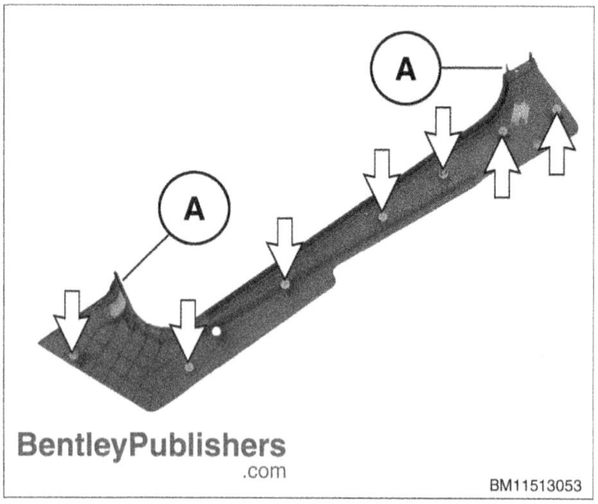

When installing:

- Check to see that corner guides (**A**) are undamaged
- Replace damaged clips (**arrows**).
- Use new mounting bolt for seat belt lower anchor.
- Use Locktite® when installing new seat belt bolt.

Tightening torque	
Seat belt anchor to body (replace M10 duo-taptite bolt with new, use Locktite®)	36 Nm (27 ft-lb)

A-pillar trim, removing

> ***WARNING —***
> - *Read the warnings and cautions in* **721 Airbag System (SRS)**. *Serious injury may result if system service is attempted by persons unfamiliar with SRS and its approved service procedures.*
> - *Do not use sharp-edged metal tools to remove A-pillar trim. Damage to head protection airbag could result.*

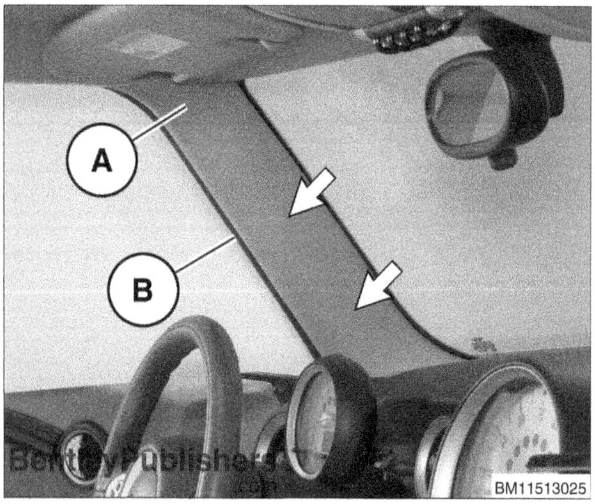

Use plastic prying tool to lever out trim plug (**A**). Remove screw underneath.

– Unclip trim (**arrows**) and remove. Detach tweeter electrical connector, if equipped.

– When installing, make sure door seal (**B**) seats correctly.

Tightening torque	
A-pillar trim to A-pillar	2 Nm (18 in-lb)

B-pillar trim, removing

Left and right B-pillar trim covers on Hardtop models are similar. On Clubman models, the right side B-pillar trim is different.

> ***WARNING—***
> - *Read the warnings and cautions in* **721 Airbag System (SRS)**. *Serious injury may result if system service is attempted by persons unfamiliar with SRS and its approved service procedures.*
> - *Do not use sharp-edged metal tools to remove B-pillar trim. Damage to head protection airbag could result.*

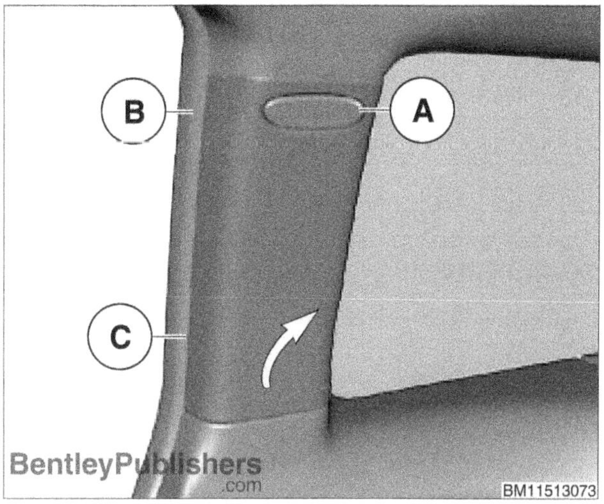

Clubman right side

- ◄ Use plastic prying tool to lever out trim plug (**A**). Remove screw underneath.
- Unclip trim (**B**) and remove (**arrow**).
- When installing, make sure rubber door seal (**C**) seats correctly.

Tightening torque	
B-pillar trim to B-pillar	2 Nm (18 in-lb)

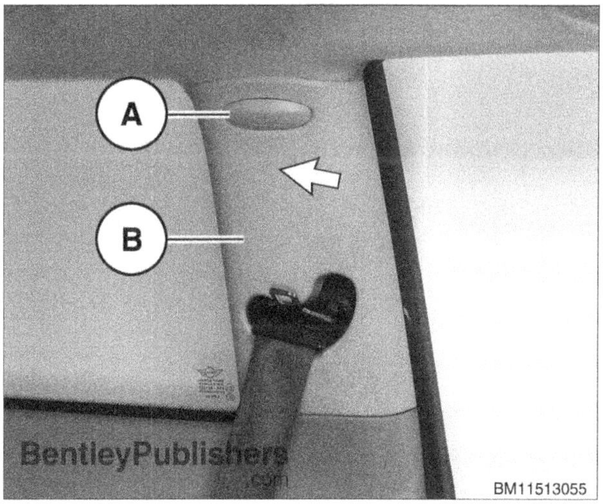

Hardtop, Clubman left side

- ◄ Use plastic prying tool to lever out trim plug (**A**). Remove screw underneath.
- Detach trim (**B**) from clip (**arrow**) and remove.

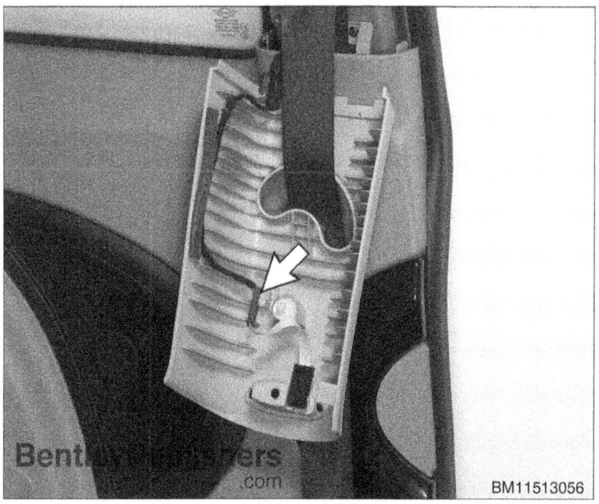

- ◄ Detach B-pillar light electrical connector (**arrow**) from trim.

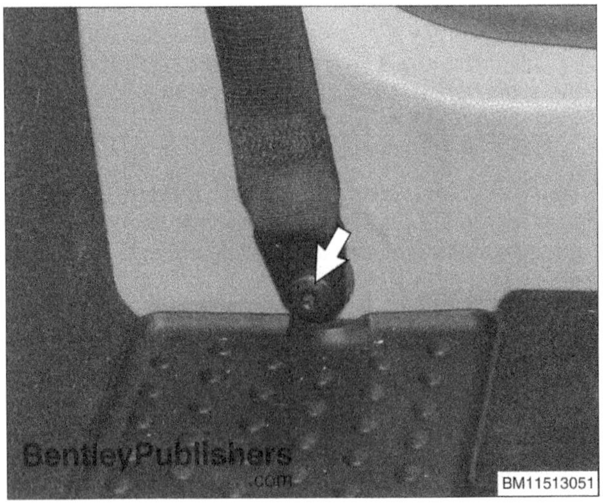

Working below door opening, remove seat belt anchor bolt (**arrow**) and discard. Slide B-pillar trim off belt to remove.

– When installing:
  - Use new mounting bolt for seat belt lower anchor.
  - Use Locktite® when installing new seat belt bolt.
  - Replace damaged trim clip.

Tightening torques	
Seat belt anchor to body (replace M10 duo-taptite bolt with new, use Locktite®)	36 Nm (27 ft-lb)
B-pillar trim to B-pillar	2 Nm (18 in-lb)

– Make sure door seal seats correctly.

C-pillar trim, removing

> ***WARNING —***
> - *Read the warnings and cautions in* ***721 Airbag System (SRS)****. Serious injury may result if system service is attempted by persons unfamiliar with SRS and its approved service procedures.*
> - *Do not use sharp-edged metal tools to remove C-pillar trim. Damage to head protection airbag could result.*

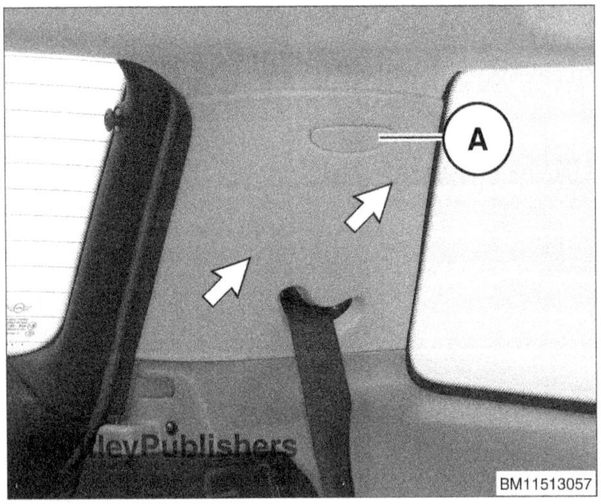

Use plastic prying tool to lever out trim plug (**A**). Remove screw underneath.

– Detach trim from clips (**arrows**) and remove.

– Remove rear seat. See **520 Seats**.

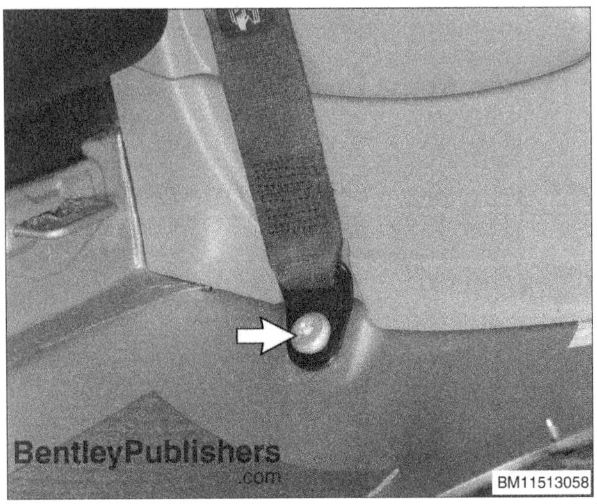

Remove rear seat belt anchor bolt (**arrow**) and discard. Slide B-pillar trim off belt to remove.

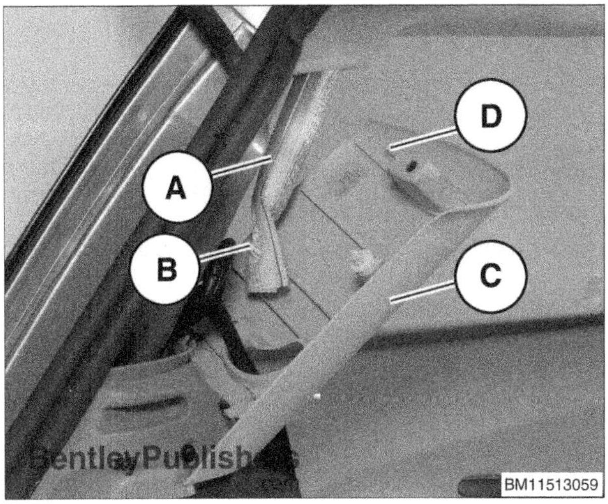

When installing:

- Replace damaged trim panel clips.
- Insert seat belt in trim panel slot.
- Secure rear end of head protection airbag (**A**) to panel clip (**B**).
- Place panel (**C**) with clips and guide (**D**) on C-pillar and clip into place.
- Use new mounting bolt for seat belt lower anchor.
- Use Locktite® when installing new seat belt bolt.

Tightening torques	
Seat belt anchor to body (replace M10 duo-taptite bolt with new, use Locktite®)	36 Nm (27 ft-lb)
C-pillar trim to C-pillar	2 Nm (18 in-lb)

– Make sure rear hatch or split door seal seats correctly.

Rear side trim panel, removing

– Remove rear seat cushion and backrest. See **520 Seats**.

– Remove B-pillar and C-pillar trim. See **B-pillar trim, removing** and **C-pillar trim, removing** in this repair group.

– Remove rear sill trim and cargo compartment side trim. See **Cargo Compartment Trim** in this repair group.

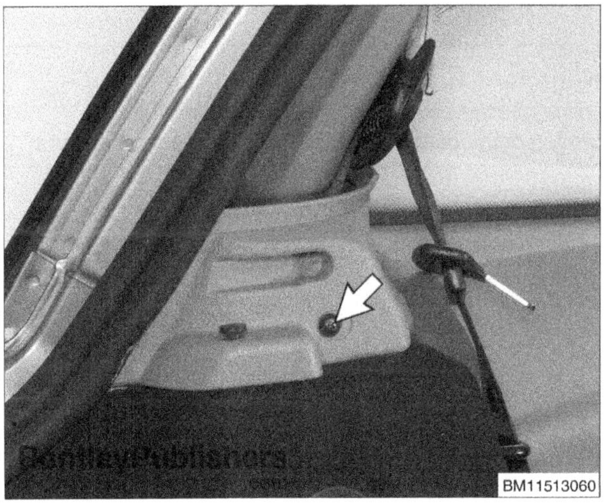

Working just inside rear hatch at C-pillar, remove trim screw (**arrow**).

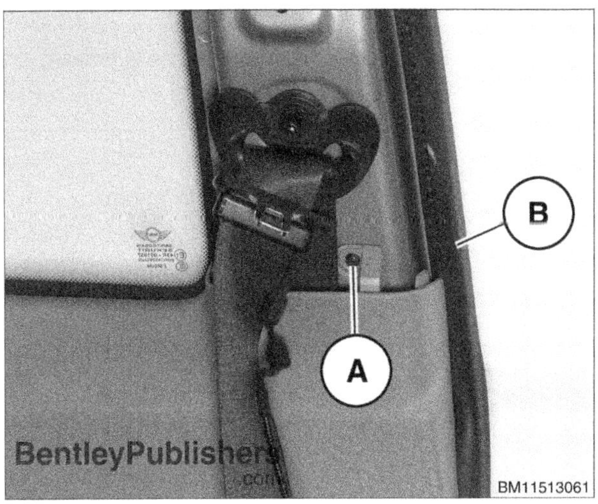

Working at B-pillar, remove trim screw (**A**) then detach door seal from trim panel (**B**).

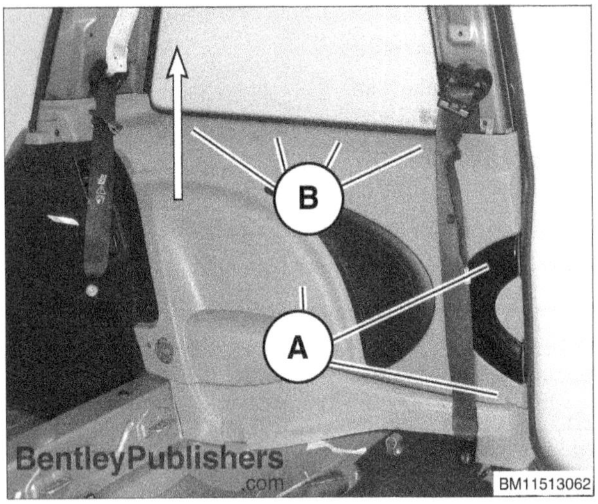

◄ Working at side trim panel:
- Detach panel from clips (**A**).
- Slide panel up (**arrow**) and off upper edge retainers (**B**).
- If necessary, disconnect tweeter electrical connector.

– When installing:
- Replace damaged retainers and clips.
- Engage panel in upper edge retainers first, then snap in lower clips.

– Make sure rubber door seal and rear hatch seal seat correctly.

Inside rear view mirror, removing and installing

The vehicle remote key antenna is contained in the inside mirror housing.

– Disconnect negative (-) battery cable and cover battery terminal to keep cable from accidentally contacting terminal.

> ***CAUTION—***
> - *Prior to disconnecting the battery, read the battery disconnection cautions given in* **001 Warnings and Cautions**.

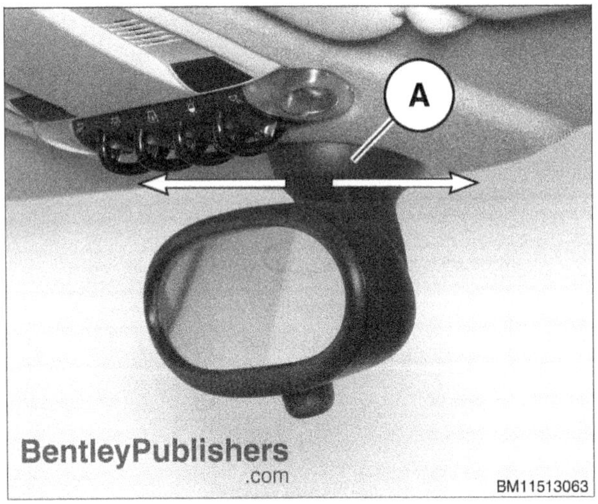

◄ Working at inside mirror:
- Separate halves of mirror base cover (**A**) in direction of **arrows**.
- Rotate mirror to remove covers (**A**).

◄ Snap mirror off mirror base by pressing steadily forward (toward windshield). Do not press abruptly.

> ***CAUTION—***
> - *To avoid windshield breakage, press mirror only toward windshield.*
> - *Do not twist mirror base when removing. Twisting the mirror base may damage the rear catch. This causes the mirror to be loose when installed.*

◀ Separate electrical harness (**arrow**).

◀ When installing:

- Fit mirror foot to headliner opening at 45°
- Rotate mirror until it engages.

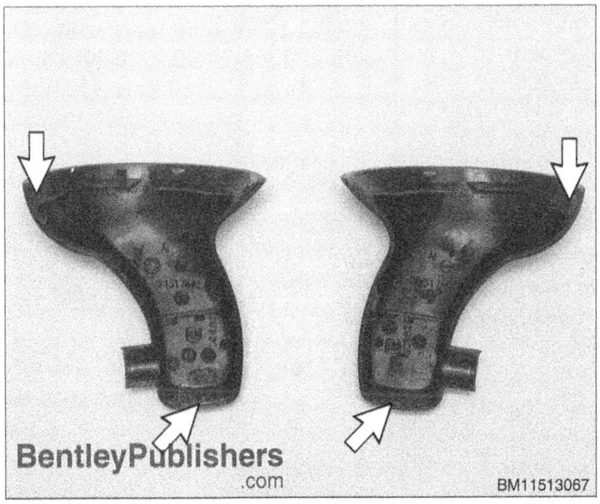

◀ Inspect mirror base covers. Replace if latch mechanisms (**arrows**) are damaged.

– Initialize remote keys. See **515 Central Locking and Anti-theft**.

HEADLINER

Headliner removal is a complex operation which requires the removal of multiple pieces of interior trim. Be sure to read the procedure through before starting work.

Headliner, removing (Hardtop)

> ***WARNING—***
> • *Read the warnings and cautions in* **721 Airbag System (SRS)**.

– Remove A-pillar, B-pillar and C-pillar trim panels. See **A-pillar trim, removing**, **B-pillar trim, removing** and **C-pillar trim, removing** in this repair group.

– Remove sun visors and sun visor supports.

– Remove roof-mounted switch center and front dome light fixture. See **540 Sunroof**.

– Remove inside rear view mirror. See **Inside rear view mirror, removing and installing** in this repair group.

– Remove rain and light sensor cover.

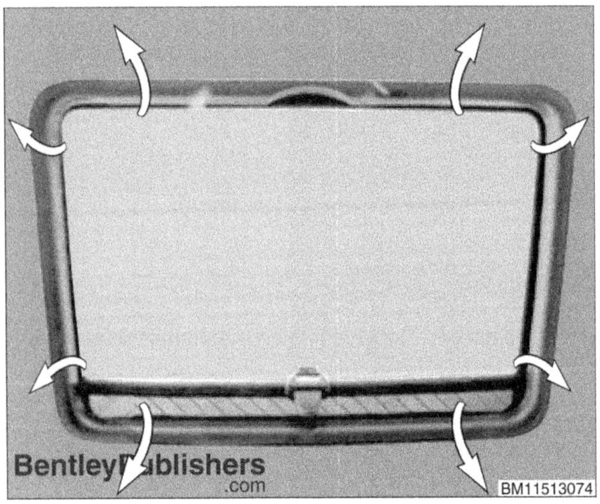

◄ Open sunroof inner panel. Unclip sunroof trim all around frame (**arrows**) and remove.

– Remove grab handles.

– Remove rear dome light fixture. See **630 Lights**.

– Remove left rear seat backrest. See **520 Seats**.

– Remove driver seat headrest.

– Remove rear sill trim and left side cargo compartment side trim. See **Rear sill trim, removing and installing (Hardtop)** and **Cargo compartment side trim, removing and installing (Hardtop)** in this repair group.

– Remove rear hatch seal.

◄ Slide driver seat fully forward and fold backrest back. Slide front passenger seat fully forward.

- To prevent headliner from marring inferior, tape off hatch opening at top right and bottom left (**arrows**).

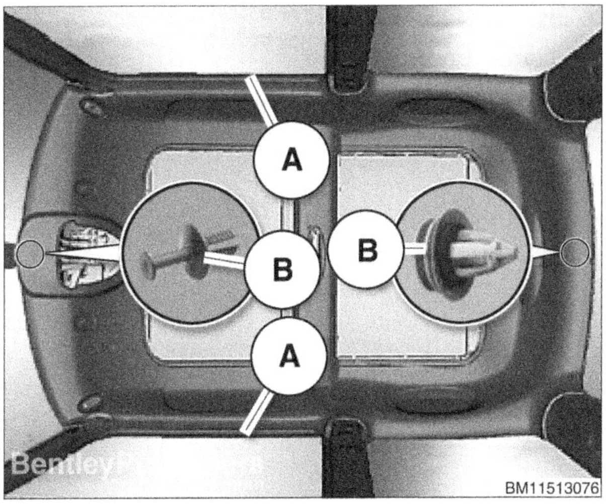

Detach headliner edge protection (**A**).

- Remove expanding rivet and plastic clip (**B**) to detach headliner.

Tilt headliner to left and, with help of assistant, slide out through rear hatch.

> ***CAUTION—***
> - *To prevent kinking of headliner, avoid bending it.*

– Installation is reverse of removal.

- Install mounting rivet and clip carefully to avoid tension creases in headliner.
- Make sure rear hatch and door seals seat correctly.

Headliner, removing (Clubman)

> ***WARNING—***
> - *Read the warnings and cautions in* **721 Airbag System (SRS)**.

– Remove A-pillar, B-pillar and C-pillar trim panels. See **A-pillar trim, removing**, **B-pillar trim, removing** and **C-pillar trim, removing** in this repair group.

– Remove sun visors and sun visor supports.

– Remove roof-mounted switch center and front dome light fixture. See **540 Sunroof**.

– Remove inside rear view mirror. See **Inside rear view mirror, removing and installing** in this repair group.

– Remove rain and light sensor cover.

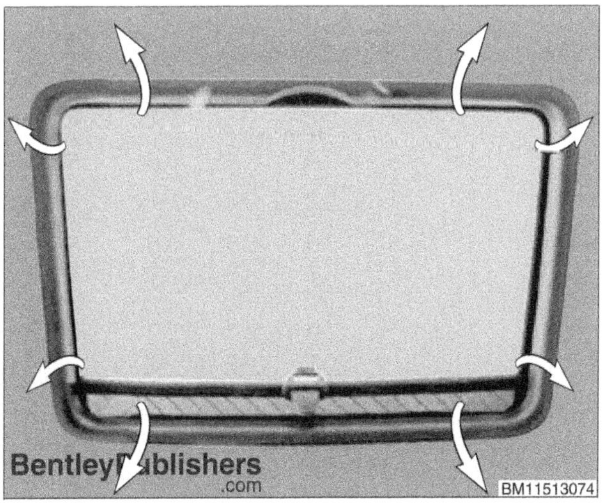

Open sunroof inner panel. Unclip sunroof trim all around frame (**arrows**) and remove.

– Remove grab handles.

– Rear rear dome light fixture. See **630 Lights**.

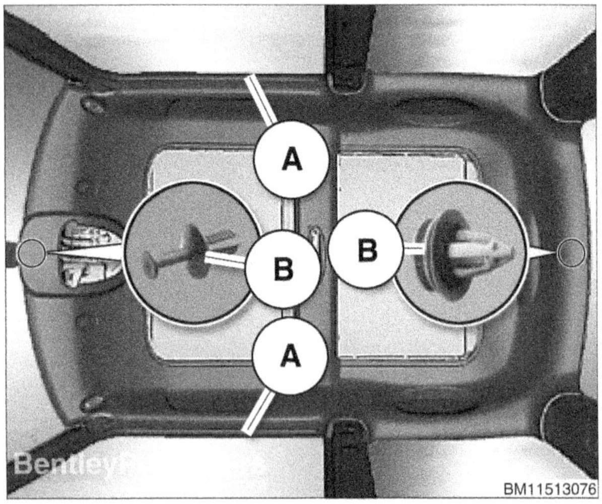

- Slide passenger seat fully forward and fold backrest back.
- Detach headliner edge protection (**A**).
  - Remove expanding rivet and plastic clip (**B**) to detach headliner.

- Tilt headliner and, with help of assistant, slide out right side door opening.

> ***CAUTION—***
> - *To prevent kinking of headliner, avoid bending it.*

- Installation is reverse of removal.
  - Install mounting rivet and clip carefully to avoid tension creases in headliner.
  - Make sure door seals seat correctly.

Cargo Compartment Trim

Rear sill trim, removing and installing (Hardtop)

- Open rear hatch and remove cargo compartment floor carpet.

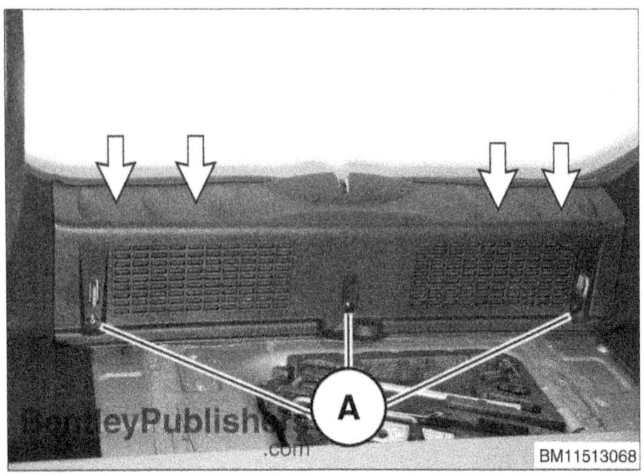

- Remove expansion rivets (**A**). Pull trim off clips (**arrows**) to remove.
- When installing, replace damaged rivets and clips.
- Make sure rubber rear hatch seal seats correctly.

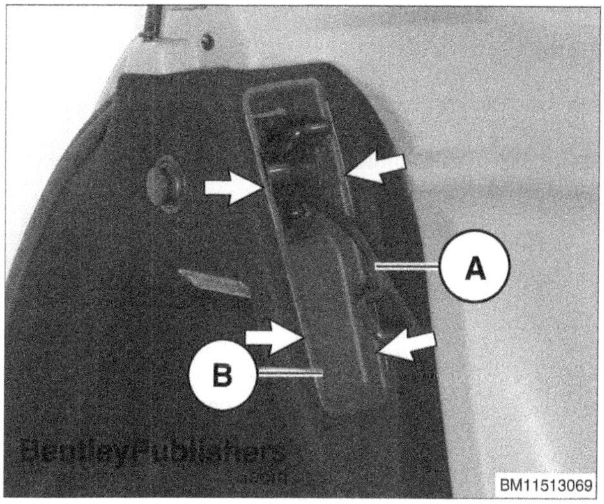

Cargo compartment side trim, removing and installing (Hardtop)

– Remove rear sill trim. See **Rear sill trim, removing and installing (Hardtop)** in this repair group.

– Remove cargo compartment light fixture. See **630 Lights**.

Version with storage hooks:

- Fold lock clip (**A**) half-way up.
- Fold open retaining tabs (**arrows**) and remove trim (**B**).

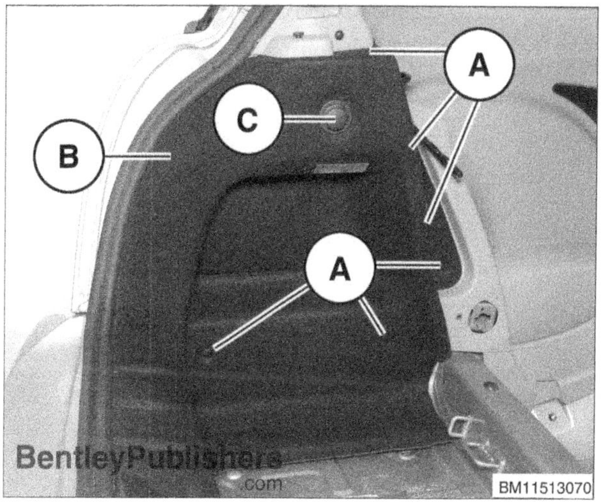

Working at compartment side trim:

- Remove expansion rivets (**A**).
- Pull back trim panel (**B**) and detach utility plug (**C**) electrical connector.

– When installing, replace damaged clips and tabs.

– Make sure rear hatch seal seats correctly.

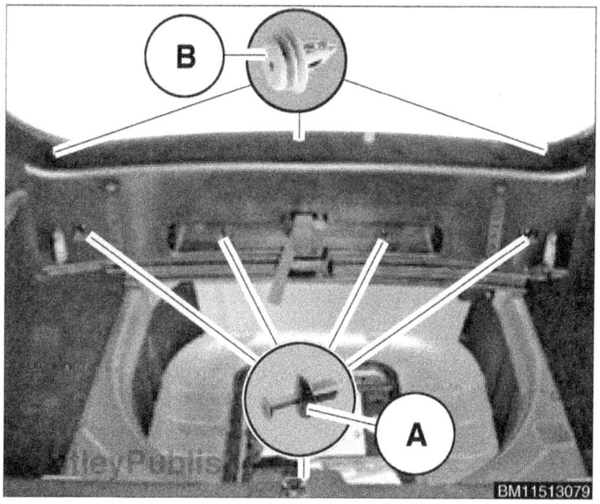

Rear sill trim, removing and installing (Clubman)

– Open split rear doors and remove cargo compartment floor covering and spare tire.

Remove expansion rivets (**A**). Pull trim off clips (**B**) to remove.

– When installing, replace damaged rivets and clips.

– Make sure split door seal seats correctly.

513

Cargo compartment side trim, removing and installing (Clubman)

– Remove cargo compartment roller cover.

– Remove C-pillar trim. See **C-pillar trim, removing** in this repair group.

– Remove rear sill trim. See **Rear sill trim, removing and installing (Clubman)** in this repair group.

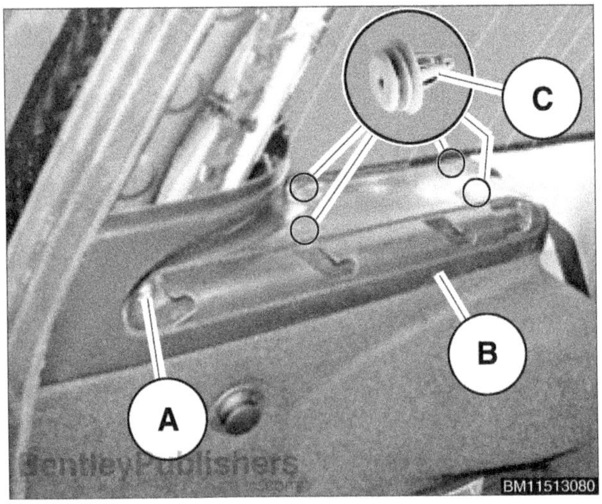

◄ Working at top of side trim:
- Remove plastic trim cap mounting screw (**A**).
- Use plastic prying tool to lever cap (**B**) off clips (**C**).

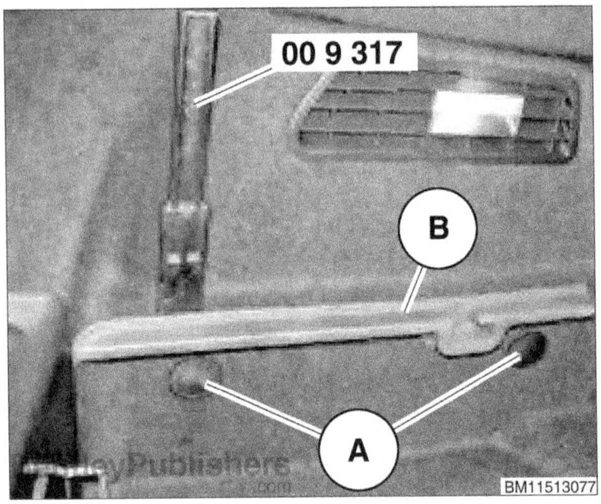

◄ Use plastic prying tool (MINI special tool **00 9 317** or equivalent) to lever out plastic clips (**A**) and remove trim (**B**).

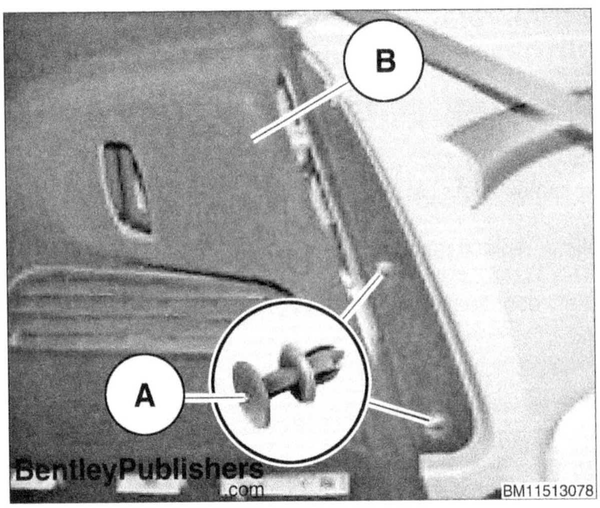

◄ Remove plastic rivets (**A**) and peel back trim panel (**B**). Detach electrical connectors from back of panel, if applicable.

– When installing, replace damaged clips and rivets.

– Make sure split rear door seal seats correctly.

515 Central Locking and Anti-theft

General

This repair group covers repair information for door, rear hatch and convertible lid locks. Also covered is the optional anti-theft alarm (DWA).

Electronic immobilizer (EWS) is integrated with the vehicle starting system and is not covered separately.

MINI cars are equipped with sophisticated and self-diagnostic electrical systems. When experiencing malfunctions relating to central locking or anti-theft systems, diagnose the system using MINI scan tool.

Additional information can be found in the following repair groups:

- **411 Doors** for door adjustment and trim panel removal
- **412 Rear Hatch, Tailgate, Split Rear Doors** for cargo compartment lid adjustment and trim panel removal
- **600 Electrical System–General** for an explanation of bus systems
- **ECL Electrical Component Locations**

BM11515002

Central locking description

Central locking, which controls side door locks, cargo compartment hatch, split doors or tailgate lock and fuel filler flap lock, can be operated via any of the following:

- Remote key(s).
- Mechanical key in driver door lock cylinder.
- Central lock button in center dashboard.
- Rear hatch, split rear door or tailgate exterior push button.
- Comfort access (optional): Vehicle access via microswitches in outside door handles.

In case of battery or electronic failure, the mechanical key can be used to lock or unlock the vehicle.

When unlocking the vehicle, windows and sunroof may be opened by pressing the UNLOCK button on the remote key for longer than 3 seconds. Similarly, turn the mechanical key for longer than 3 seconds in the unlock direction to open windows and sunroof.

When locking, turn the mechanical key in the lock direction for more than 3 seconds to close the windows and sunroof. However, for safety reasons, the remote key cannot be used for closing the windows and sunroof.

Vehicle key

Vehicle keys are electronic identification transmitters. The central locking system responds to the key radio signal to allow access and start authorization for the vehicle.

An antenna in the interior rear view mirror picks up the vehicle key radio signal and transmits it to the car access system (CAS) module. CAS evaluates this signal, and if the transmitted signal is identified as belonging to the vehicle, the vehicle is unlocked and may then be started.

The ID transmitter key battery is charged via the transponder in the START STOP unit (ignition switch). If a key is not used for an extended length of time, the internal battery may go dead.

The ID transmitter keys for a vehicle with comfort access are equipped with a battery with a service life of approximately 2 years. The reason for this is that the key does not need to be inserted into the START STOP unit and so it is not guaranteed that the battery would be kept adequately charged.

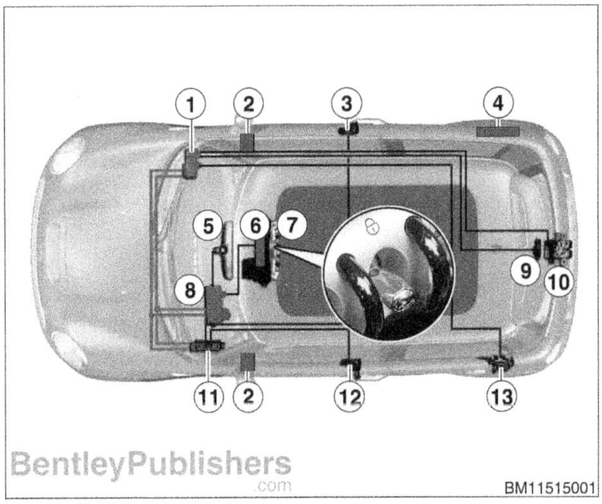

Central locking component overview

MINI Cooper Hardtop illustrated. Other models similar.

1. Junction box (A4010).
2. Window motor.
3. Right door handle, lock (and microswitch if equipped with comfort access).
4. Comfort access control module (optional).
5. Rear view mirror with remote key antenna.
6. Sunroof control module.
7. Central locking button.
8. Car access system (CAS).
9. Rear hatch push button.
10. Rear hatch lock release.
11. Footwell module (FRM) (A4011).
12. Left door handle, lock (and microswitch if equipped with comfort access).
13. Fuel filler flap lock.

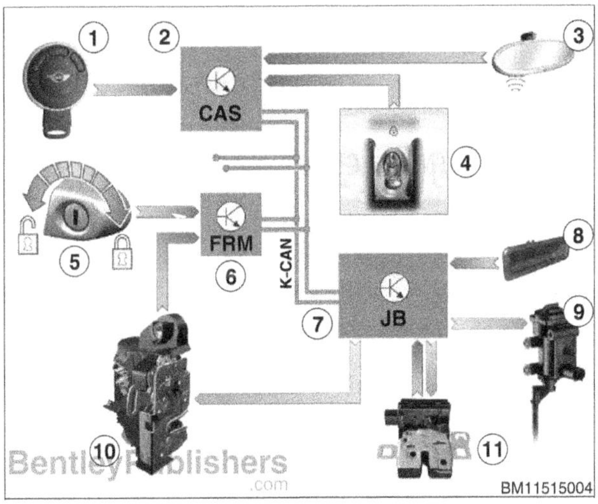

Central locking functions

1. Remote key.
   - Lock or unlock vehicle with push button.
   - Comfort access, if equipped: Lock, unlock, start vehicle if key is in proximity.
   - Mechanical key inside remote key can be used to lock, unlock, double lock, raise windows and close sunroof.
2. Car access system (CAS) module under left side dashboard.
   - Master control module for central locking.
   - When remote key signal is received, CAS authenticates signal as valid and belonging to vehicle.
   - Vehicle unlock request is then forwarded to junction box (JB).
3. Central locking antenna on inside rear view mirror.
   - Receives lock or unlock signal from remote key and forwards request to CAS.
4. Central locking switch, center dashboard switch cluster.
   - Lock or unlock vehicle from inside.
5. Driver door lock cylinder.
   - Locks and unlocks door mechanically.
   - Microswitches signal status to footwell module (FRM).
6. Footwell module (FRM) behind left footwell trim panel.
   - Monitors door lock and lock cylinder microswitch signals.
   - Forwards lock status information to junction box (JB) and car access system (CAS) via K-CAN
7. Junction box (JB) behind right footwell trim panel.
   - Slave module, executes lock and unlock requests for entire vehicle.
   - Lock and unlock relays for door locks, cargo compartment lock and fuel filler flap lock are located in this module.
8. Cargo compartment exterior unlock button.
9. Fuel filler flap lock.
10. Door lock.
11. Cargo compartment lock.

515

Personal profile

A variety of vehicle functions may be set individually and stored in the memory chip of the remote key currently in use. When the vehicle is later unlocked, the key is recognized and the memorized settings are implemented.

A maximum of three remote keys may be programmed in this way, each for a different driver with different preferences.

The programmable preferences are as follows:

- Response of the central locking system when the vehicle is unlocked, such as preferences on which parts of the vehicle to unlock and acoustic and optical signals when locking and unlocking.
- Automatic locking if the vehicle is left unattended or if driving above a certain speed.
- Triple turn signal activation: One flash or three flashes.
- Instrument display preferences such as 12 h / 24 h clock, miles vs kilometers in speedometer and odometer, measure of fuel consumption, date format and temperature in °C / °F
- Light settings such as pathway lighting and daytime running lights.
- Automatic climate control such as AUTO program, activating or deactivating cooling function and automatic recirculated air control, temperature settings, air flow rate and air distribution.
- Entertainment preferences such as audio volume, speed-dependent volume.

See owner's manual for details.

Comfort access

Comfort access (optional), is a passive access system which allows the vehicle to be unlocked by pressing a microswitch on the outer door handle, provided the remote key is located within 2 meters (6 ft) of the vehicle. The vehicle may be started with the START STOP button as long as the remote key is inside the vehicle.

A network of antennas within the vehicle exterior panels are used to detect the remote key.

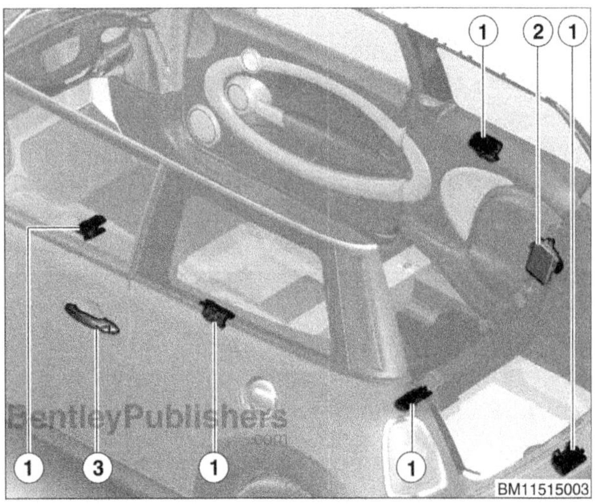

Comfort access system component are as follows:

1. Antenna.
2. Comfort access (CA) control module.
3. Driver door comfort access microswitch.

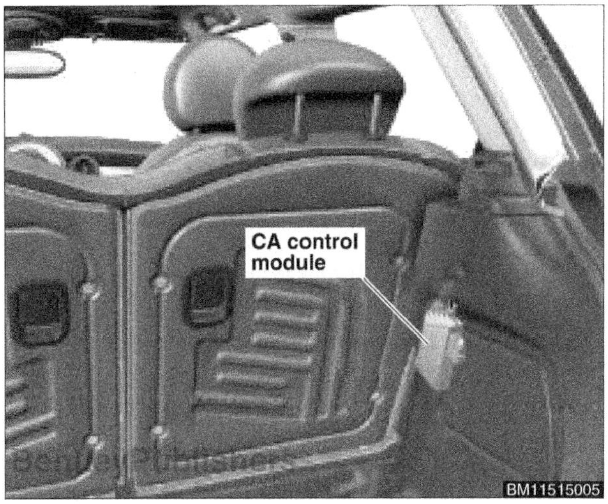

CA control module controls the transmit antennas and switches the remote control receiver to continuous reception. Microswitch signals from outer door handles are transferred to CAS.

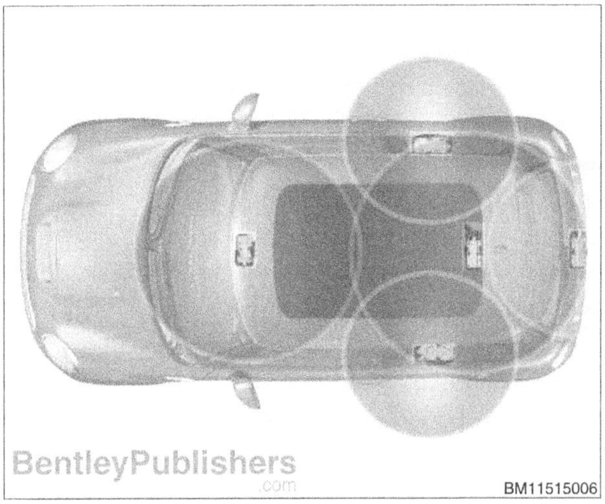

Antennas are designed to pick up remote key signals in overlapping spherical volumes.

The rear bumper antenna features a waterproof electrical connector.

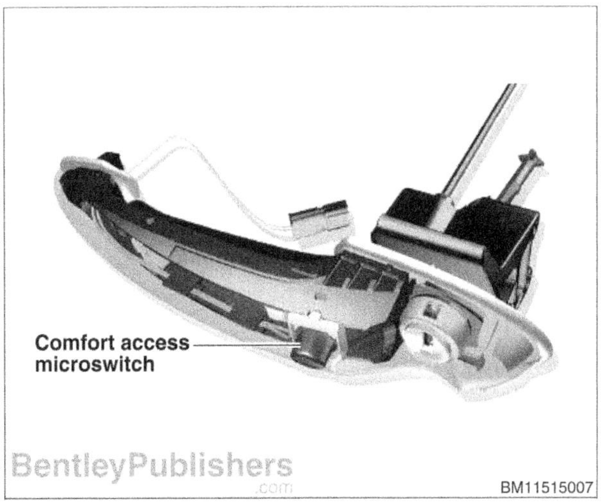

Microswitches are integrated in the outer door handles of the driver and passenger doors. The vehicle can be unlocked or locked by operating the microswitches provided a valid identification transmitter is detected outside the vehicle. Microswitch status is determined by CA control module.

When unlocking or locking the vehicle, auto remote opening and closing of windows and sunroof is initiated by pressing the door handle microswitch for longer than 3 seconds.

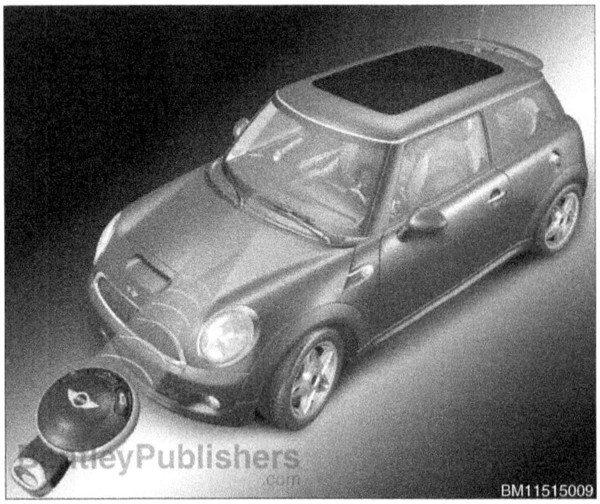

BM11515009

Anti-theft alarm (DWA)

The optional anti-theft alarm system (DWA) performs the following functions when armed:

- Detects break-in attempts and tampering with vehicle. An audible and visual alarm is triggered.
- Monitors doors, cargo compartment and engine hood for unauthorized opening.
- Signals an attempt to tamper with vehicle such as cutting the alarm siren power supply.
- Hardtop, Clubman: Monitors interior of vehicle with ultrasonic sensor in roof switch center.
- Convertible: Monitors interior of vehicle with ultrasonic sensors in doors.
- Detects attempted vehicle theft with a tilt sensor.

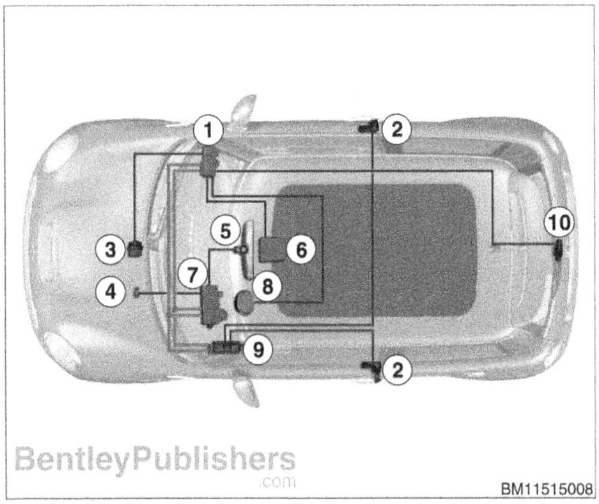

BM11515008

DWA components:

1. Junction box (JB) integrated with anti-theft alarm.
2. Door handle contact microswitch.
3. Alarm siren and tilt sensor.
4. Engine hood contact microswitch.
5. Remote key and alarm system antenna.
6. Ultrasonic interior motion sensor (USIS).
   - Convertible: Ultrasonic interior motion sensors in doors.
7. Car access system (CAS).
8. Steering column switch cluster with alarm system LEDs.
9. Footwell module (FRM).
10. Cargo compartment lock contact microswitch.

Cautions

CAUTION—
- *When working on electrical components, disconnect the negative (–) cable from the battery and insulate the cable end to prevent accidental reconnection.*
- *Prior to disconnecting the battery, read the battery disconnection cautions in* **001 Warnings and Cautions**.
- *Handle electrical modules and other electrical and electronic components using static prevention equipment and techniques. See* **600 Electrical System–General**.
- *To avoid marring trim pieces, use a plastic prying tool or a screwdriver with the tip wrapped with masking tape.*

Front Door Locks

Door trim panel removal is covered in **411 Doors**.

Front door lock components (left door)

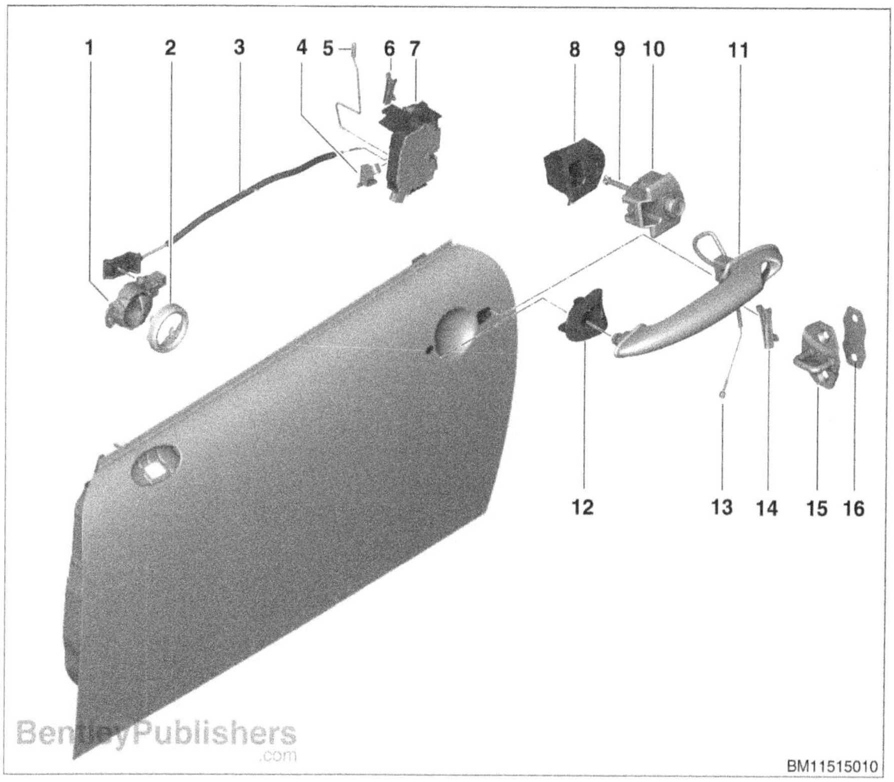

1. Inside door handle (right side shown)
2. Trim
3. Bowden cable, inner handle
4. Bowden cable bracket
5. Lock button
6. Bowden cable bracket
7. Lock
8. Gasket
9. Push button plunger paddle
10. Door handle push button
    - Left side: Includes lock cylinder
11. Door handle
12. Gasket
13. Bowden cable, outer handle
14. Bowden cable bracket
15. Striker
16. Gasket

Front door lock, removing and installing

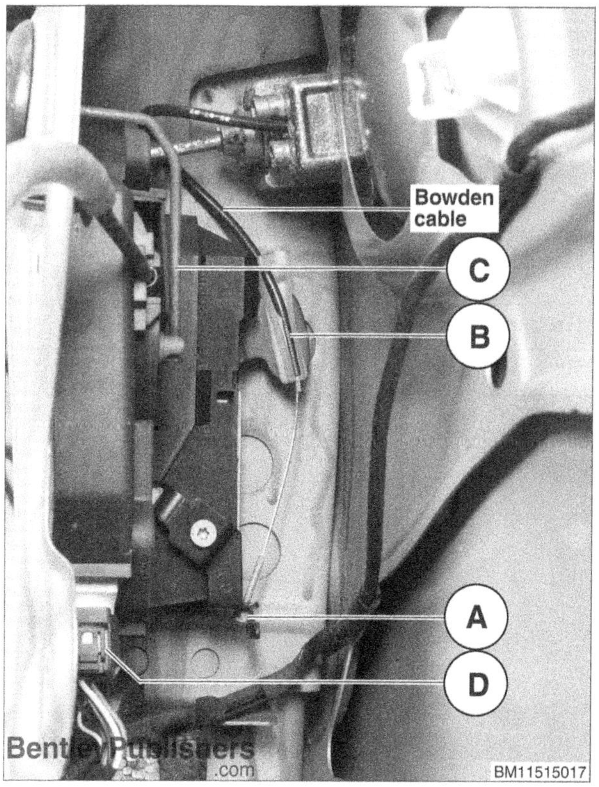

– Remove door trim panel. See **411 Doors**.

– Remove door window glass and integrated door trim module (ITM). See **512 Door Windows**.

◄ Working at door lock assembly inside door cavity:
- Detach outer handle Bowden cable from lock lever (**A**) and guide (**B**).
- Pry lock button rod (**C**) off lock assembly.
- Detach lock electrical connector (**D**).

Front door lock, removing and installing

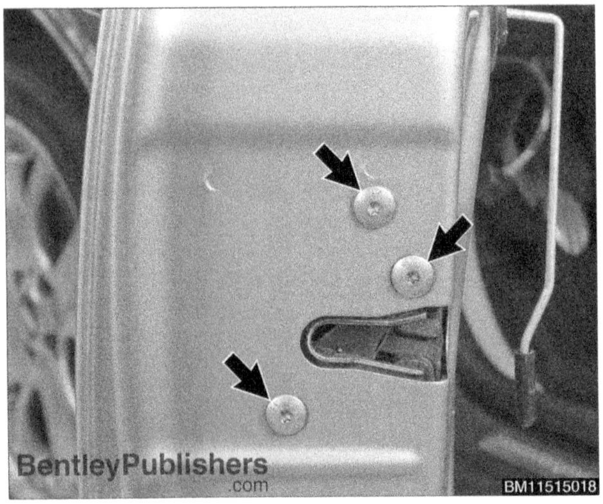

◄ Working at door edge, remove lock mounting screws (**arrows**).

– Disengage lock from door handle push button plunger in order to remove.

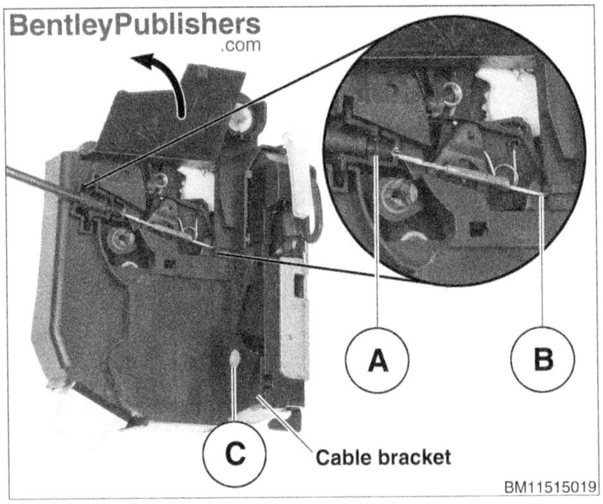

◄ With door lock assembly out of door cavity:

- Flip open (**arrow**) plastic cover.
- Detach inside handle Bowden cable from brackets (**A**, **B**)
- If installing new lock assembly: Remove screw (**C**) to detach outside handle Bowden cable bracket.

◄ When installing:

- Attach Bowden cables and brackets to lock assembly.
- Make sure door handle push button plunger paddle (**inset**) inserts correctly in door lock slot.
- Door lock assembly and door handle shown off the door for purposes of illustration.

Tightening torque	
Door lock lever to front door lock	3 Nm (27 in-lb)
Door lock to front door	6 Nm (53 in-lb)

Front door lock striker, replacing

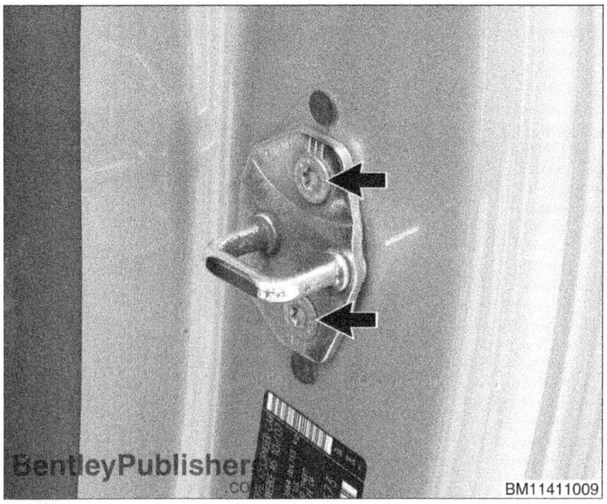

- Remove door striker mounting Torx bolts (**arrows**). Remove striker and retrieve gasket.

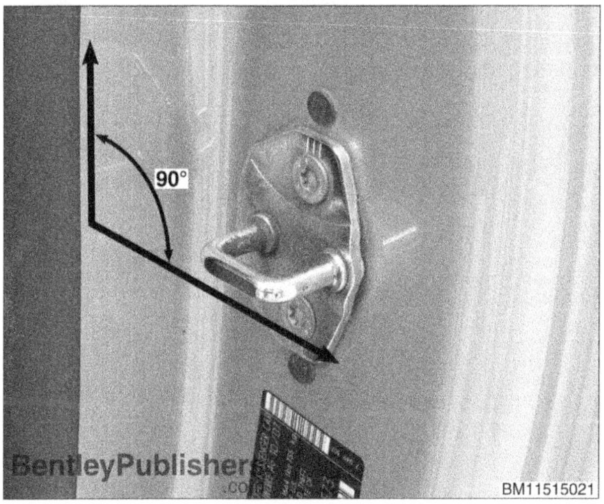

- When installing, adjust striker as follows:
  - Position striker approx. 90° to B-pillar inner edge.
  - Tighten bolts.
  - Close door and make sure striker does not touch or brush against door edge or lock. Look for scratches on paint.
  - Adjust striker position or door position if necessary. See **411 Doors**.

Tightening torque	
Door lock striker to B-pillar	19 Nm (14 ft-lb)

515

Outside door handle, removing and installing

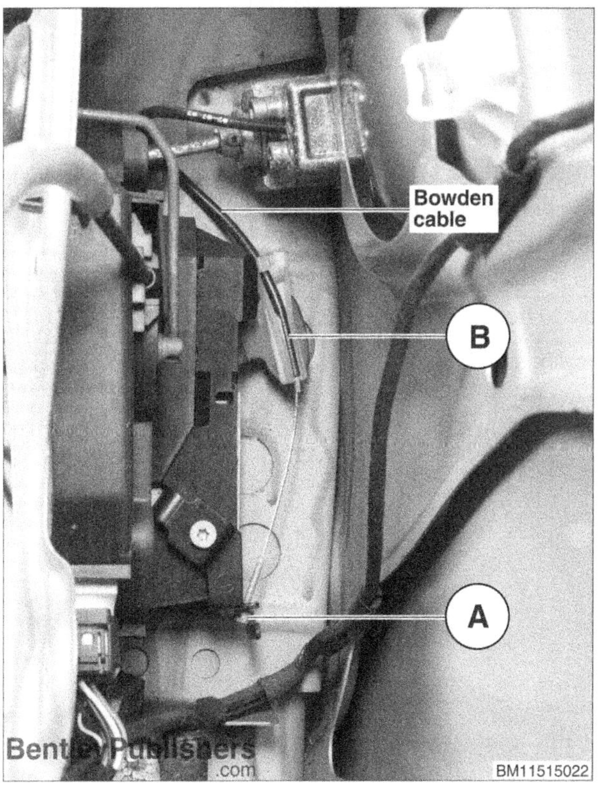

– Remove door trim panel. See **411 Doors**.

– Remove door window glass and integrated door trim module (ITM). See **512 Door Windows**.

- Working at door lock assembly, detach outer handle Bowden cable from lock lever (**A**) and guide (**B**).

Outside door handle, removing and installing

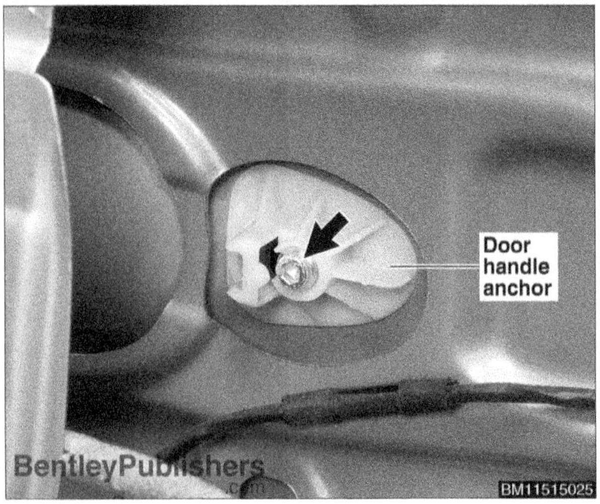

◀ Working inside door cavity, remove outer handle inboard mounting screw (**arrow**) and detach plastic door handle anchor.

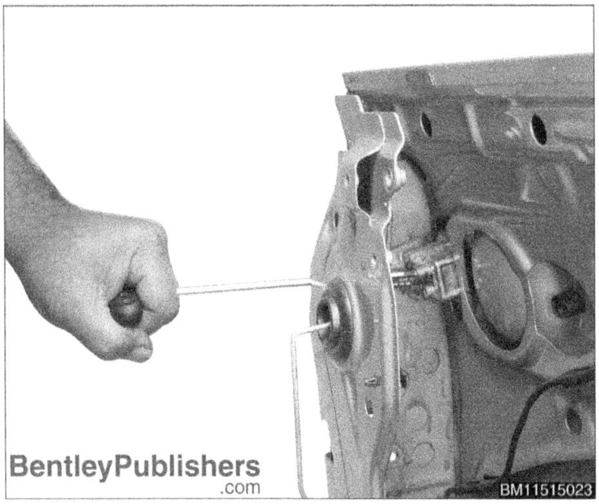

◀ Reach through hole in door edge to remove outer door handle outboard mounting screw. (Secures lock cylinder to door handle.)

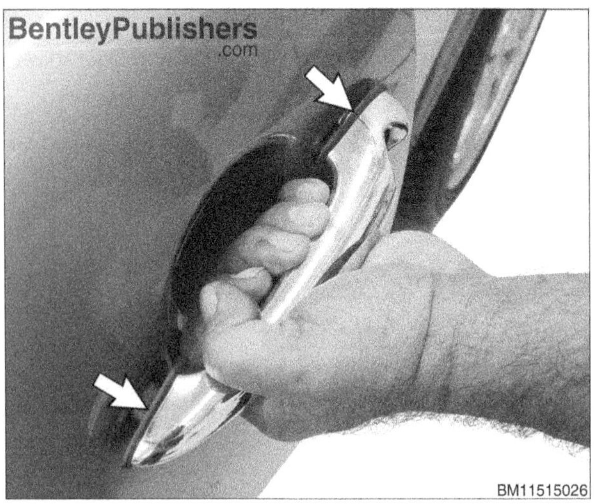

◀ Lift out door handle.

– When installing:
  - First insert self-tapping screws in door handle before installing.
  - Make sure door handle gaskets (**arrows**) seat correctly.

Tightening torque	
Door handle to front door	4 Nm (35 in-lb)
Door lock cylinder to front door handle	9 Nm (7 ft-lb)

Left door lock cylinder, removing and installing

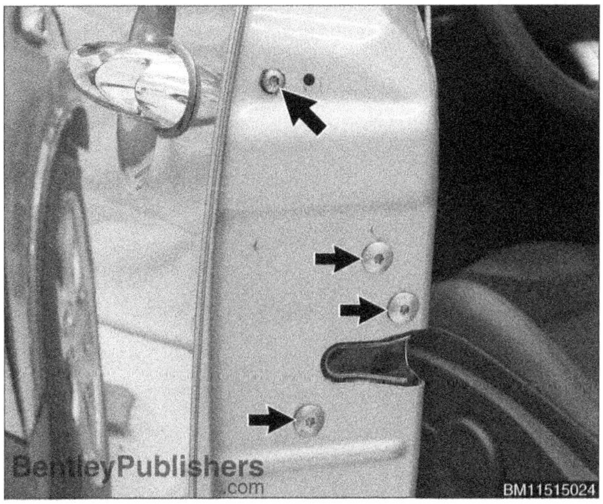

– Remove door trim panel. See **411 Doors**.

– Remove door window glass and integrated door trim module (ITM). See **512 Door Windows**.

– Remove outside door handle. See **Outside door handle, removing and installing** in this repair group.

< With door open, remove door handle mounting screw and lock assembly mounting screws (**arrows**).

– Slide door lock down slightly and disengage door handle push button plunger to remove lock cylinder.

< When installing:

- Make sure door handle push button plunger paddle (**inset**) inserts correctly in door lock slot.
- Door lock assembly and door handle shown off the door for purposes of illustration.

Tightening torque	
Door handle to front door	4 Nm (35 in-lb)
Door lock cylinder to front door	9 Nm (7 ft-lb)
Door lock cylinder to front door handle	9 Nm (7 ft-lb)
Door lock to front door	6 Nm (53 in-lb)

515

Inside door handle, removing and installing

– Remove door trim panel. See **411 Doors**.

< Working at inner door handle:

- Detach electrical connector (**A**).
- Remove mounting screws (**B**) and pull back handle.

– Remove door window glass. See **512 Door Windows**.

Partially remove and pull back top of integrated door trim module (ITM). See **512 Door Windows**. Disengage Bowden cable from back of inner door handle.

- Installation is reverse of removal.

Club Door Locks

Door trim panel removal is covered in **411 Doors**.

Club door lock components

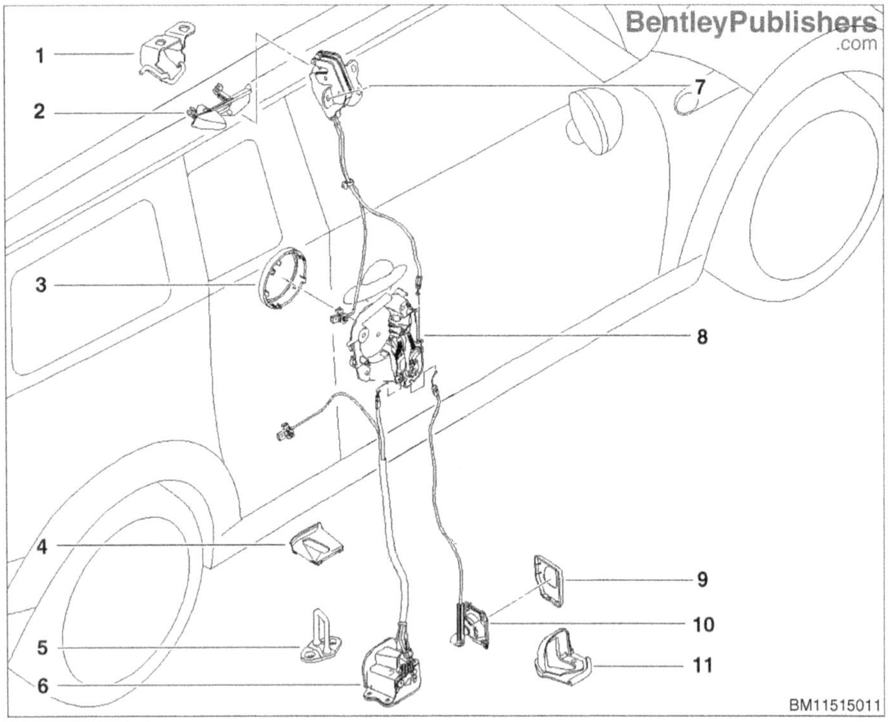

1. Upper striker
2. Upper striker cover
3. Trim
4. Lower striker cover
5. Lower striker
6. Lower lock
7. Upper lock
8. Inside door handle
9. Lower door opener cover
10. Cover
11. Gasket

Club door upper lock, removing and installing

- Remove club door inner panel. See **411 Doors**.
- Remove inside club door handle. See **Club door inside handle, removing and installing** in this repair group.

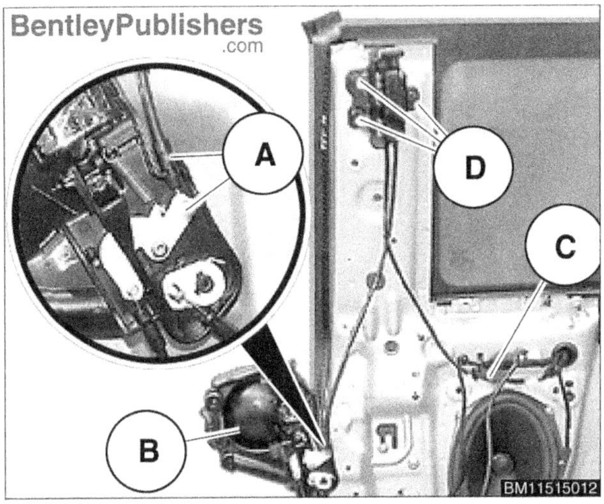

- With club door open:
  - Detach upper lock Bowden cable (**A**) from inner handle (**B**).
  - Detach electrical connector (**C**).
  - Remove upper lock mounting screws (**D**) and remove lock.
  - Seat belt reel shown removed for purpose of clarity.

- Installation is reverse of removal.
  - Make sure electrical harness and Bowden cable are routed as before.
  - If necessary, adjust door alignment. See **411 Doors**.

Tightening torque	
Door lock to club door	9 Nm (7 ft-lb)

Club door lower lock, removing and installing

- Remove club door inner panel. See **411 Doors**.

- Remove club door lower opener. See **Club door lower opener, removing and installing** in this repair group.

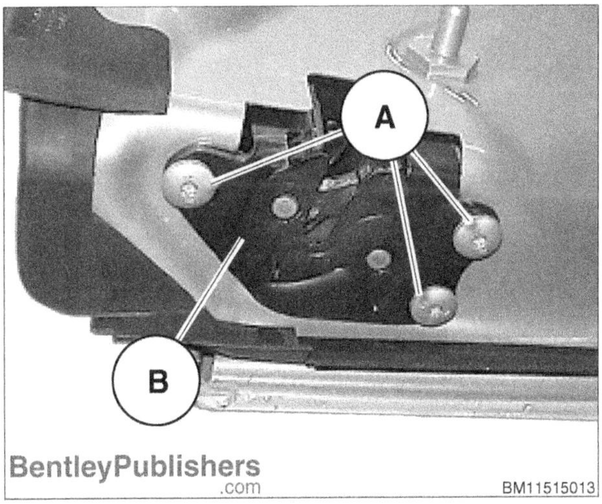

- Working at bottom of club door, remove lock mounting screws (**A**) and lower lock (**B**) out of door.

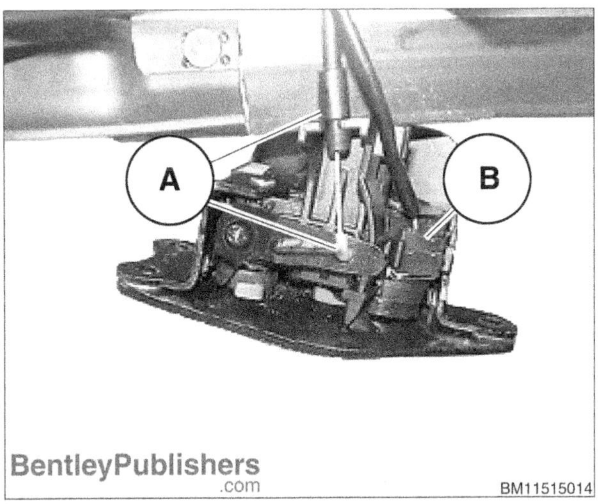

- Detach Bowden cable (**A**) and electrical connector (**B**) to remove lock.

- Installation is reverse of removal.

Tightening torque	
Door lock to club door	9 Nm (7 ft-lb)

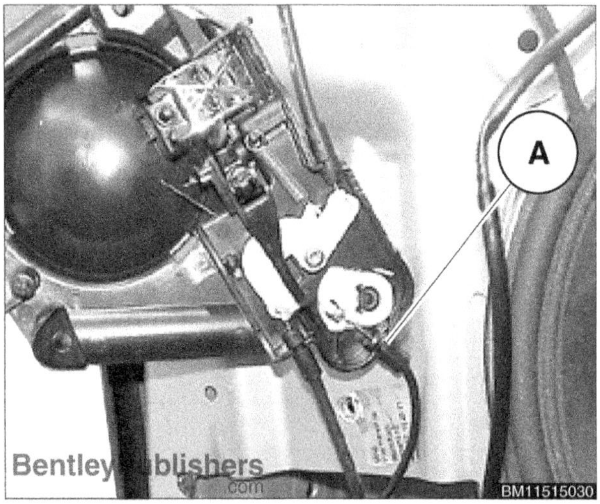

Club door lower opener, removing and installing

– Remove inside club door handle. See **Club door inside handle, removing and installing** in this repair group.

< Detach lower Bowden cable (**A**).

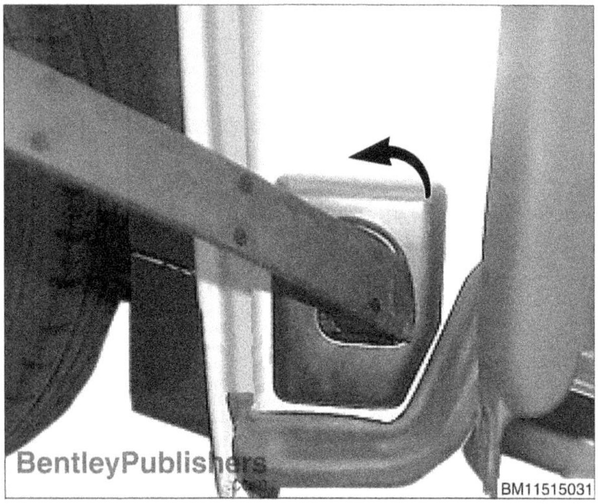

< Working at base of club door, use plastic prying tool to pry off door opener trim cover (**arrow**).

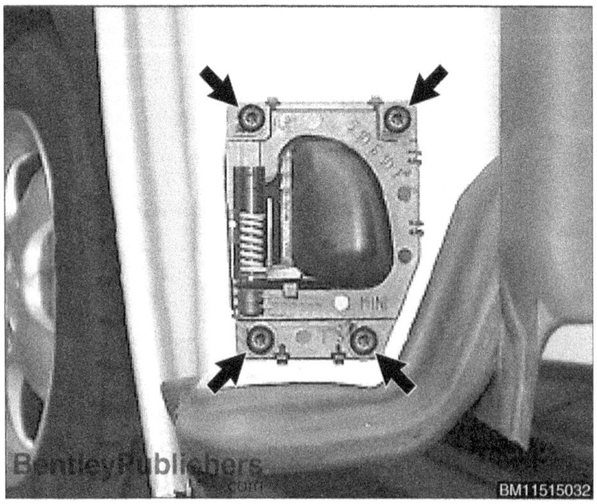

< Remove door opener mounting screws (**arrows**). Slide opener with Bowden cable out of bottom of door.

To detach Bowden cable, pry off trim cover on door opener, then disengage cable (**inset**).

Club door upper lock striker, removing and installing

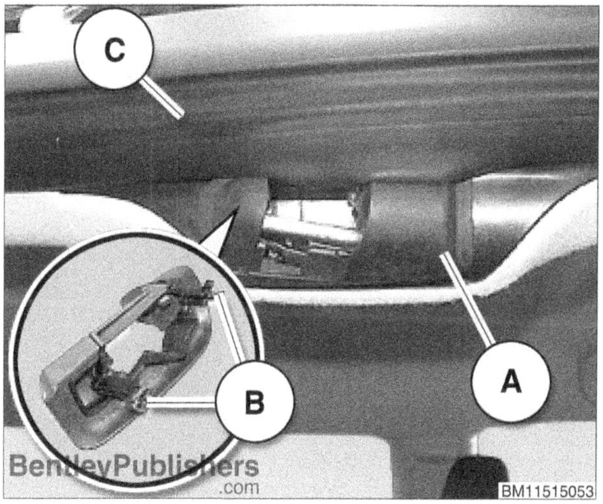

Open club door and pry off striker guard (**A**), making sure retaining lugs (**B**) are not damaged. If necessary, peel back rubber door seal (**C**).

Remove striker trim cover mounting screws (**arrow**). Remove cover.

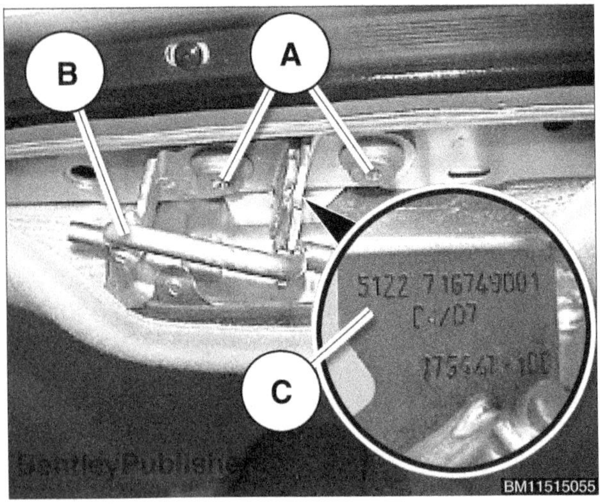

◀ Remove mounting bolts (**A**) and lift off striker (**B**).

– Installation is reverse of removal. Make sure striker label (**C** in **inset**) points forward.

Tightening torque	
Club door lock upper striker to body	24 Nm (18 ft-lb)

– Adjust striker position or door position if necessary. See **411 Doors**.

Club door lower lock striker, removing and installing

◀ Open club door and pry off (**arrow**) lower striker trim.

◀ Remove lock striker mounting bolts (**arrows**) and lift off striker.

– Installation is reverse of removal.

Tightening torque	
Club door lock lower striker to body	24 Nm (18 ft-lb)

– Adjust striker position or door position if necessary. See **411 Doors**.

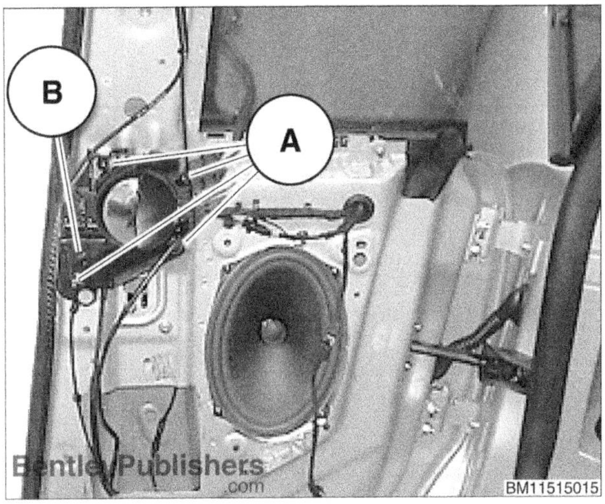

Club door inside handle, removing and installing

- Remove club door inner panel. See **411 Doors**.

< Remove door handle mounting screws (**A**) and remove door handle (**B**). Seat belt is shown removed for purpose of illustration.

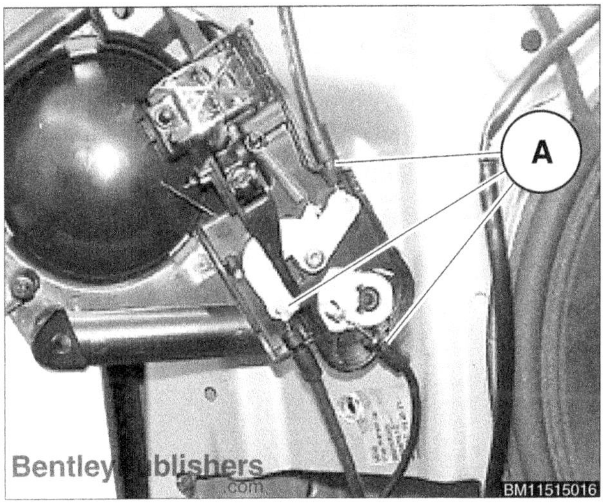

< Detach Bowden cables (**A**) and remove door handle.

- Installation is reverse of removal.

Tightening torque	
Door handle to club door	1.8 Nm (16 in-lb)

Rear Hatch Lock

Rear hatch lock, adjusting

Check and adjust rear hatch alignment before adjusting lock position. See **412 Rear Hatch, Tailgate, Split Rear Doors**.

- Screw in rear hatch bump stops completely.

- Remove cargo compartment sill trim panel. See **513 Interior Trim**.

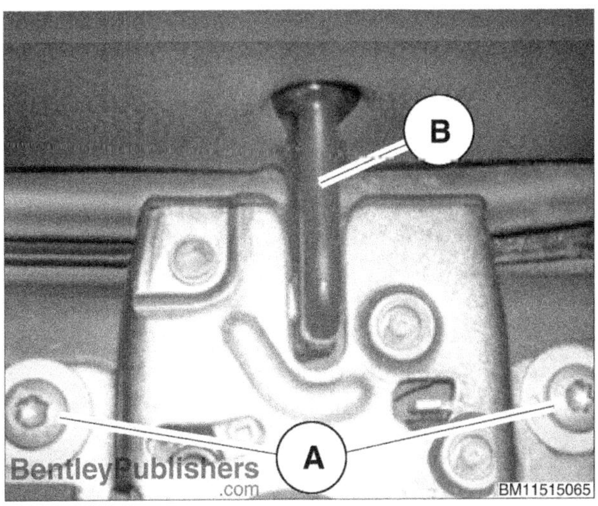

< Working at rear hatch lock assembly:

- Loosen lock mounting screws (**A**) until lock is just able to move.
- Close hatch and keep release button pressed until hatch lock snaps closed. This centers lock on striker (**B**).

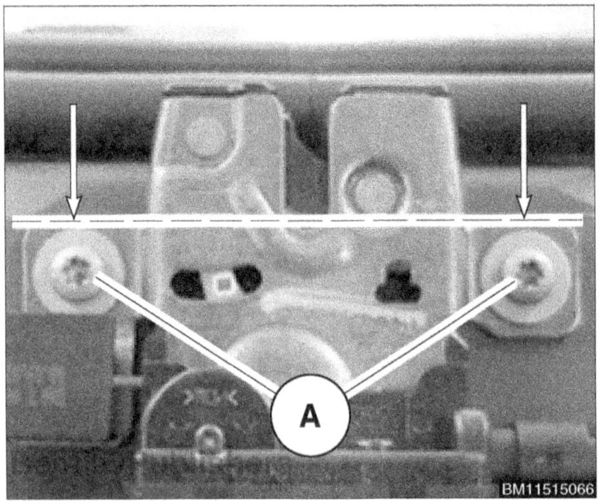

Open rear hatch and adjust position of lock so that lock rear edge (**arrows**) is flush with edge of flat washers. Tighten lock mounting screws (**A**).

Tightening torque	
Rear hatch lock to rear sill	9 Nm (7 ft-lb)

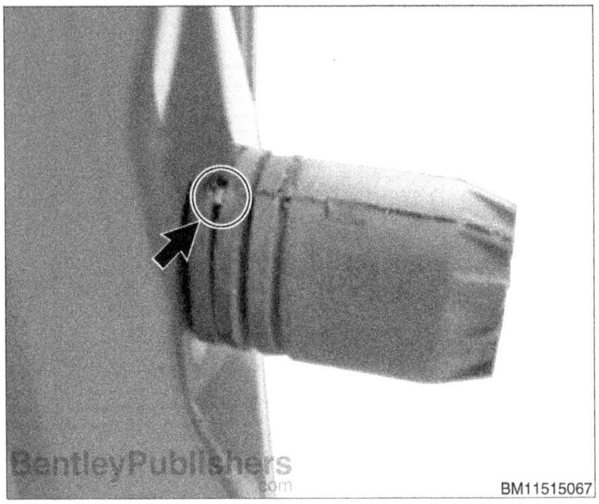

Rotate stop buffer until cast-on rib (**arrow**) rests on rear hatch.

– Check rear hatch and rear lock alignment and closing functions. Readjust stop buffers if necessary. Rotate buffer maximum of one full turn in either direction.

Rear hatch lock, removing and installing

– Remove cargo compartment sill trim panel. See **513 Interior Trim**.

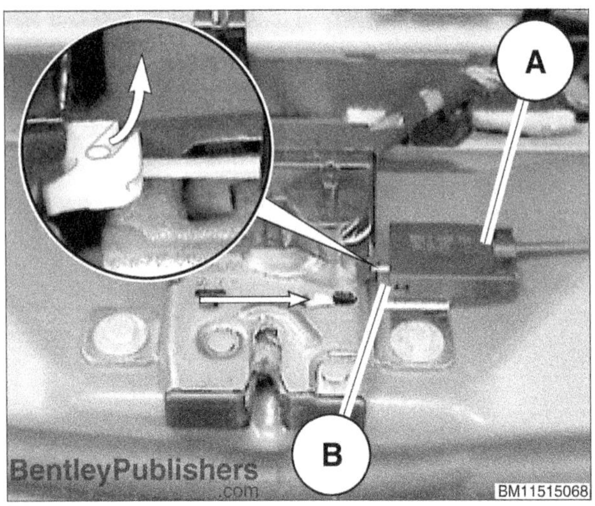

Working at rear hatch lock:

- Use thin screwdriver to press lock release lever to right (**straight arrow**).
- Detach Bowden cable end from lock release lever (**curved arrow**).
- This releases Bowden cable and end cover (**A**) from lock (**B**).

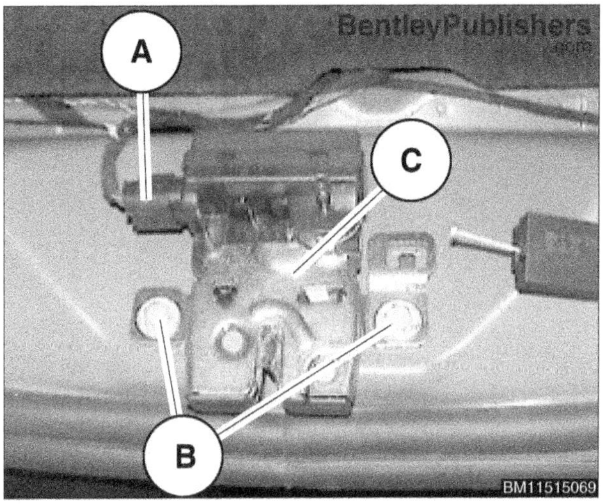

Working at rear hatch lock:

- Detach lock electrical connector (**A**).
- Remove lock mounting screws (**B**) and remove lock (**C**).

– Installation is reverse of removal. When finished, adjust rear hatch lock position. See **Rear hatch lock, adjusting** in this repair group.

Tightening torque	
Rear hatch lock to rear sill	9 Nm (7 ft-lb)

Rear hatch lock striker, replacing

– Remove rear hatch lower trim panel. See **412 Rear Hatch, Tailgate, Split Rear Doors**.

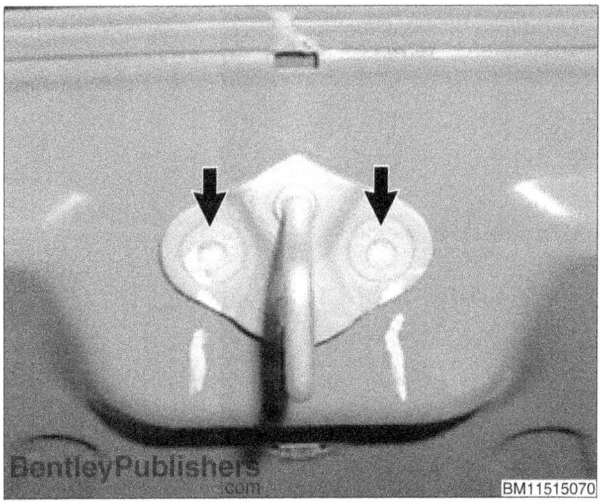

Remove striker mounting screws (**arrows**) and lift off striker.

– Installation is reverse of removal.

Tightening torque	
Rear hatch lock striker to rear hatch	9 Nm (7 ft-lb)

515

Rear hatch emergency release cable

The rear hatch emergency release cable handle is underneath the rear seat cushion.

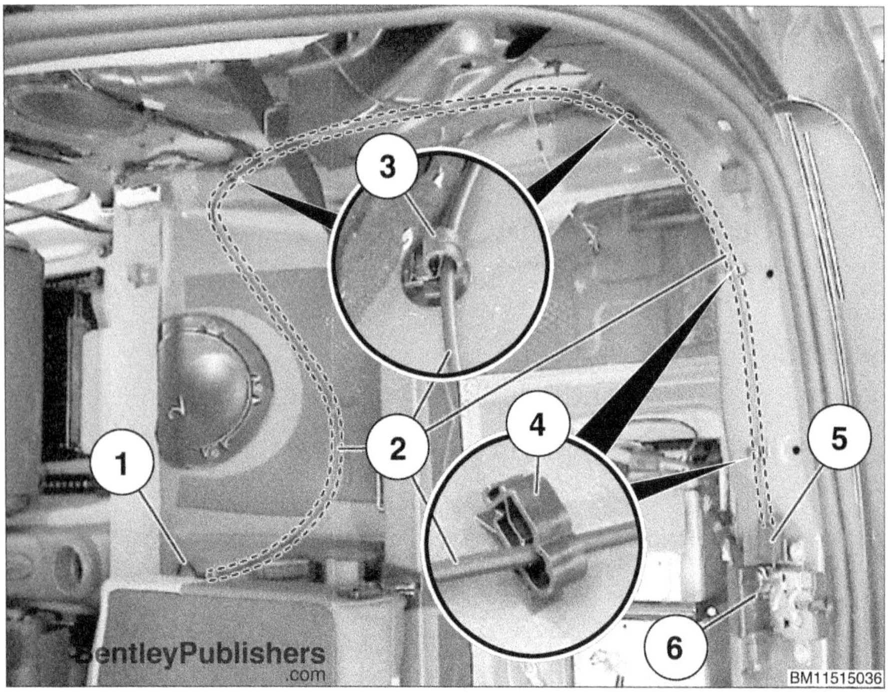

1. Release handle
2. Bowden cable
3. Bowden cable bracket
4. Bowden cable bracket
5. Bowden cable end cover
6. Rear hatch lock
   - Detaching Bowden cable from hatch lock is covered in **Rear hatch lock, removing and installing** in this repair group.

CONVERTIBLE TAILGATE LOCK

Tailgate lock components

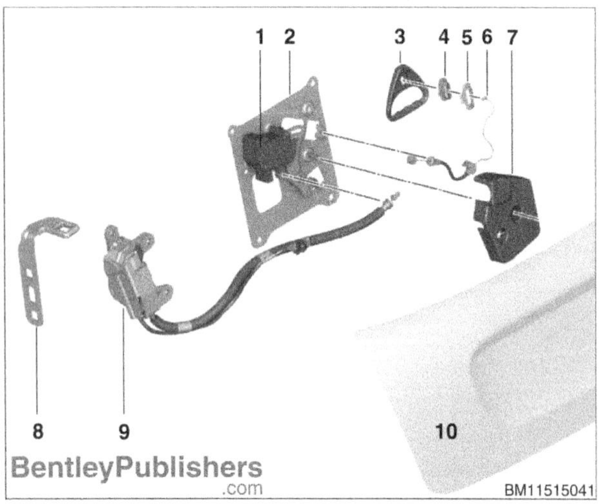

1. Electrical lock motor
2. Lock assembly carrier plate
3. Emergency release handle
4. Spacer
5. Clip
6. Emergency release Bowden cable
7. Security cover
8. Lock striker
9. Lock
10. Tailgate

Tailgate lock, adjusting

Check and adjust tailgate alignment before adjusting lock position. See **412 Rear Hatch, Tailgate, Split Rear Doors**.

With tailgate open, use plastic prying tool to lever out trim cover (**arrow**) over left side lock striker mounting bolts. Repeat for right side.

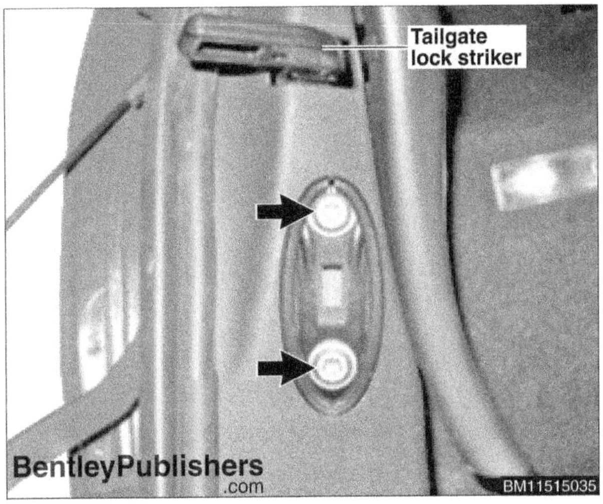

Loosen striker mounting screws (**arrows**) on left and right until strikers are just able to move.

- Close tailgate and keep release button pressed until tailgate lock snaps closed. This corrects position of lock strikers.
- Open tailgate and tighten striker mounting bolts.

Tightening torque	
Tailgate lock striker to body	20 Nm (15 ft-lb)

- Check tailgate and tailgate lock alignment and closing functions. Readjust stop buffers if necessary.

515

Tailgate lock, removing and installing

- Remove tailgate trim panel. See **412 Rear Hatch, Tailgate, Split Rear Doors**.

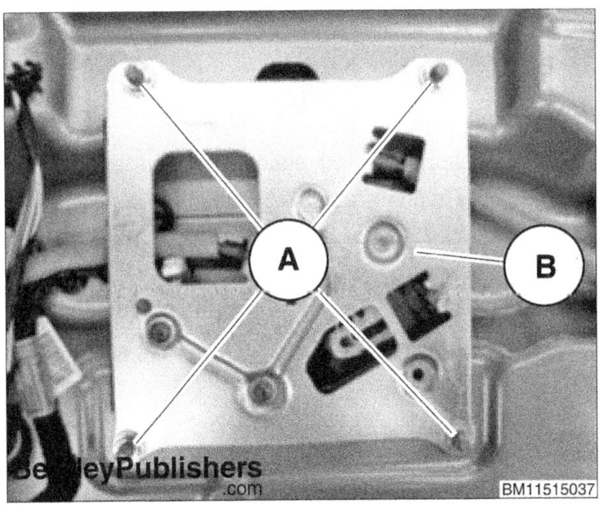

Working on left and right side locks: Remove lock assembly mounting bolts (**A**) and tilt lock carrier plate (**B**) over.

Tailgate lock, removing and installing

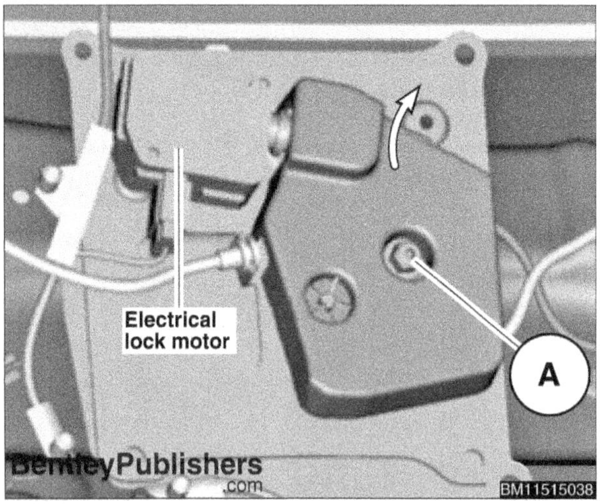

Working at rear of lock plate, remove nut (**A**) and detach security cover.

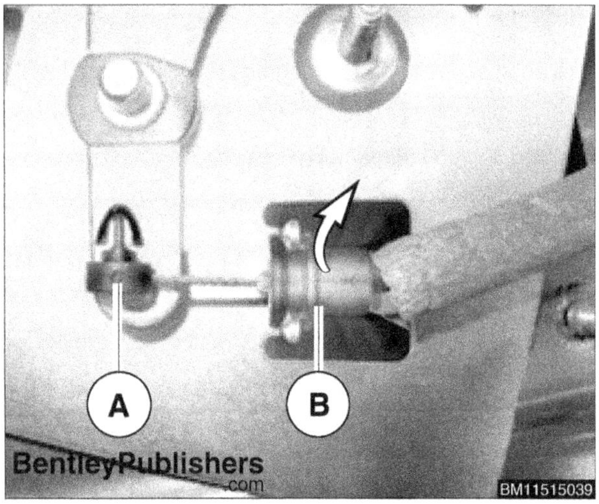

Working at Bowden cable between left and right locks:

- Detach cable end (**A**) from electrical lock actuator reversing lever.
- Lever out cable sheath (**B**) from counter support.

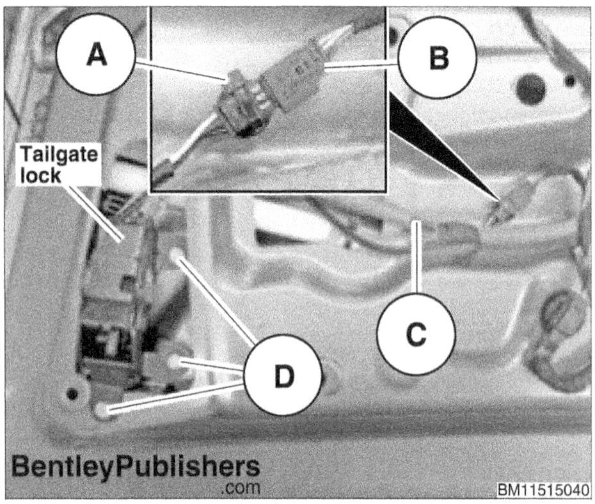

Working at tailgate lock assembly:

- Detach electrical connector (**A**) from left side lock electrical plug (**B**) or right side lock Bowden cable (**C**).
- Lift out lock assembly. If necessary, remove lock mounting screws (**D**) to separate lock from lock carrier plate. Remove lock with Bowden cable.

– When installing:

- Check that Bowden cable tabs and retainers are undamaged.
- Seat lock carrier plate securely and install mounting screws.

Tightening torques	
Tailgate lock plate to tailgate	8.5 Nm (75 in-lb)
Tailgate lock to lock plate	10 Nm (7 ft-lb)

– Check lock functions with tailgate open.

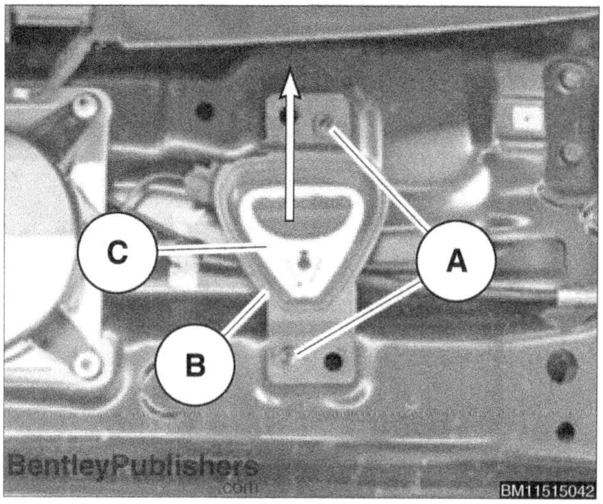

Tailgate emergency release, replacing

To reach tailgate emergency release (on tailgate trim panel) from inside vehicle, fold down rear seat backrest.

– Open tailgate and remove tailgate trim panel. See **412 Rear Hatch, Tailgate, Split Rear Doors**.

◄ Working at emergency release handle:

- Remove mounting screws (**A**) from release handle bracket (**B**).
- Pull handle (**C**) in direction of **arrow**.

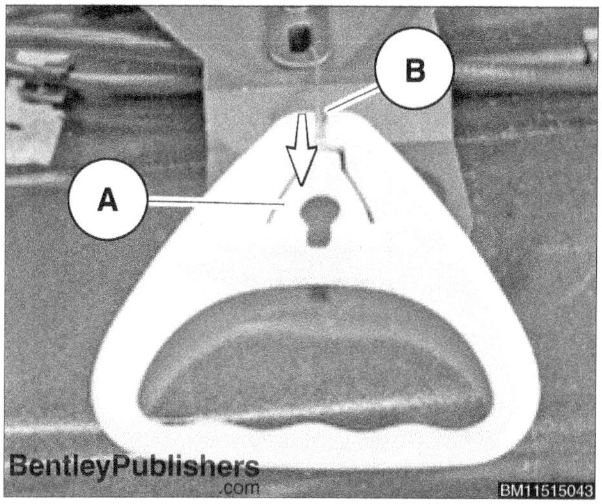

◄ Press on tab (**A**) and detach release handle (**B**) from Bowden cable.

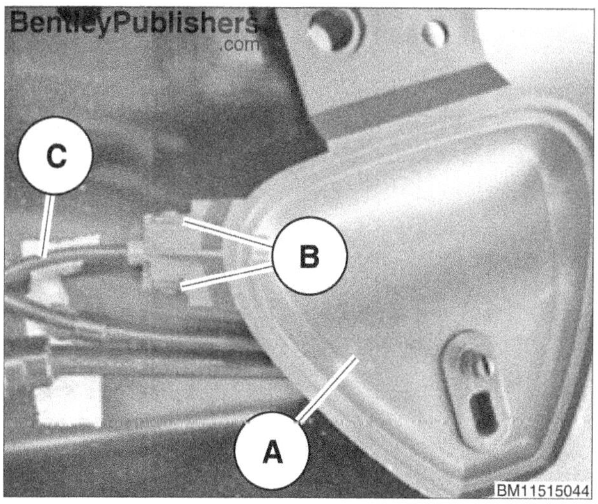

◄ Working at release handle bracket (**A**), squeeze catches (**B**) on Bowden cable (**C**) to separate cable from bracket.

– After installation, check lock functions with tailgate open.

Split Rear Door Locks

Split door lock components

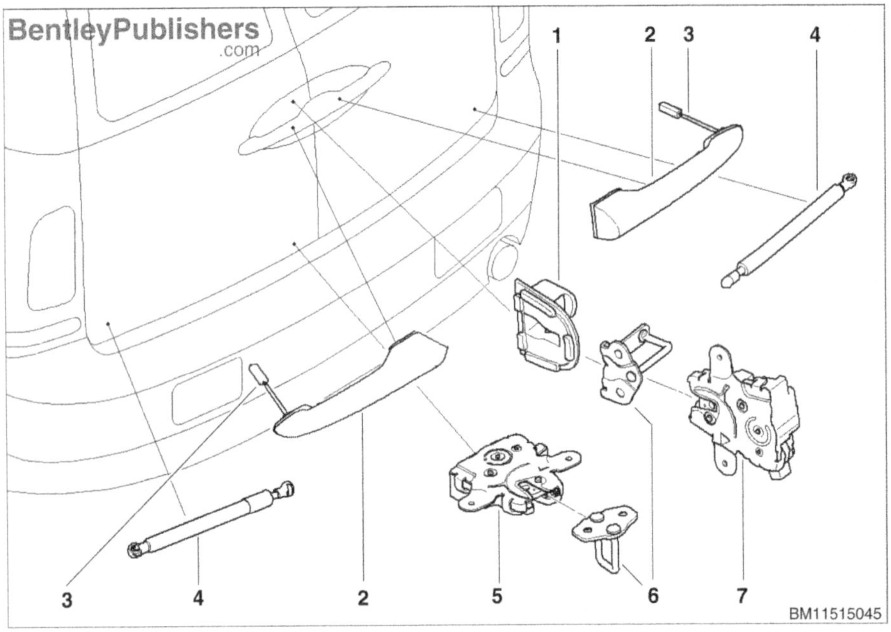

1. Striker guard
2. Outside handle
3. Outside handle electrical connector
4. Door strut
5. Left split door lock
6. Upper and lower lock strikers
7. Right split door lock

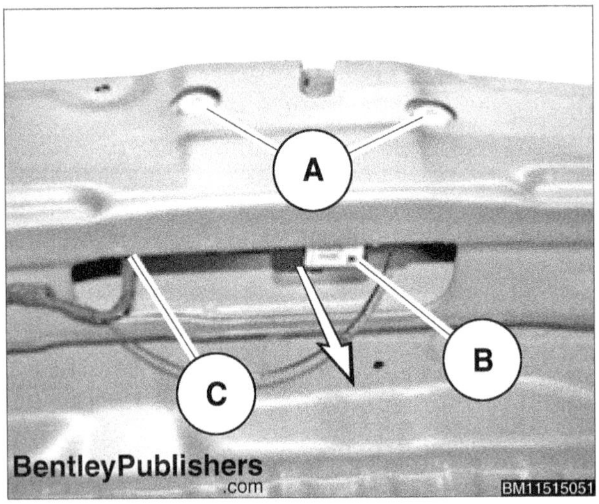

Left split door lock, removing and installing

– Remove rear sill trim. See **513 Interior Trim**.

Working at rear sill:
- Remove lock mounting bolts (**A**).
- Slide lock (**B**) forward out of sill (**arrow**).
- Detach electrical connector (**C**).

– Installation is reverse of removal.

Tightening torque	
Split door lock to rear sill	9 Nm (7 ft-lb)

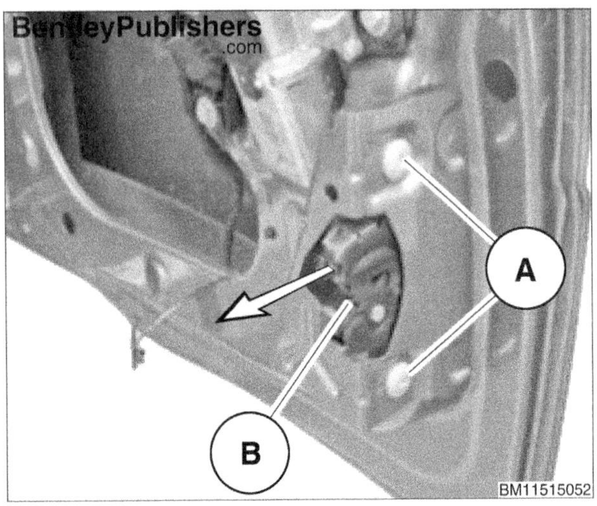

Right split door lock, removing and installing

– Remove right split door trim panel. See **412 Rear Hatch, Tailgate, Split Rear Doors**.

Working at lock on door edge:
- Remove lock mounting bolts (**A**).
- Slide lock (**B**) out of door (**arrow**).
- Detach electrical connector.

– Installation is reverse of removal.

Tightening torque	
Split door lock to rear sill	9 Nm (7 ft-lb)

– Adjust split door striker position(s). See **Split door lock strikers, removing and installing**.

Split door lock strikers, removing and installing

Upper striker

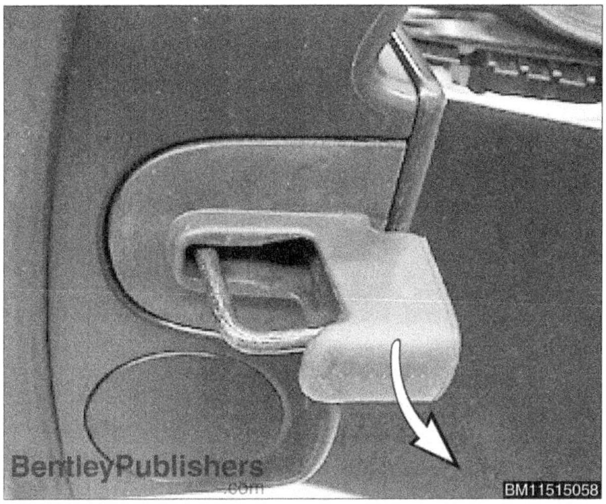

◄ Pry off striker guard (**arrow**).

◄ Remove mounting bolts (**arrows**) and lift off striker.

– Installation is reverse of removal.

Tightening torque	
Split door upper striker to door	19 Nm (14 ft-lb)

Lower striker

◄ Working underneath edge of left split door, remove mounting bolts (**arrows**) and lift off striker.

– Installation is reverse of removal. Installation:

Tightening torque	
Split door lower striker to door	19 Nm (14 ft-lb)

– Adjust split door alignment. See **412 Rear Hatch, Tailgate, Split Rear Doors**.

Left split door handle, removing and installing

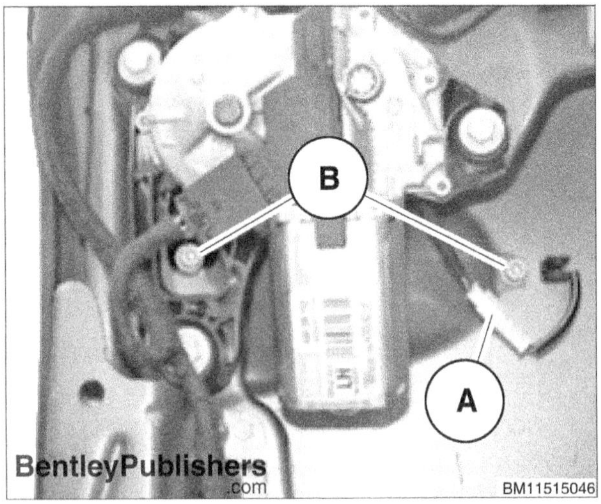

– Remove left split door trim panel. See **412 Rear Hatch, Tailgate, Split Rear Doors**.

◄ Working inside left split door cavity at door lock:
- Detach outside handle electrical connector (**A**).
- Remove handle mounting bolts (**B**).

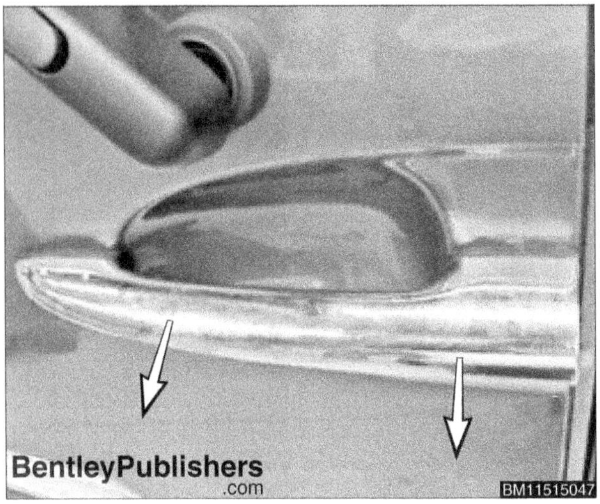

◄ Pull door handle off (**arrows**), carefully withdrawing electrical harness through insulating grommet in door.

– Installation is reverse of removal.

Tightening torque	
Outside handle to split door (M6 bolt)	5 Nm (44 in-lb)

Right split door handle, removing and installing

– Remove right split door trim panel. See **412 Rear Hatch, Tailgate, Split Rear Doors**.

◄ Working inside left split door cavity at door lock, remove handle mounting bolts (**A**).

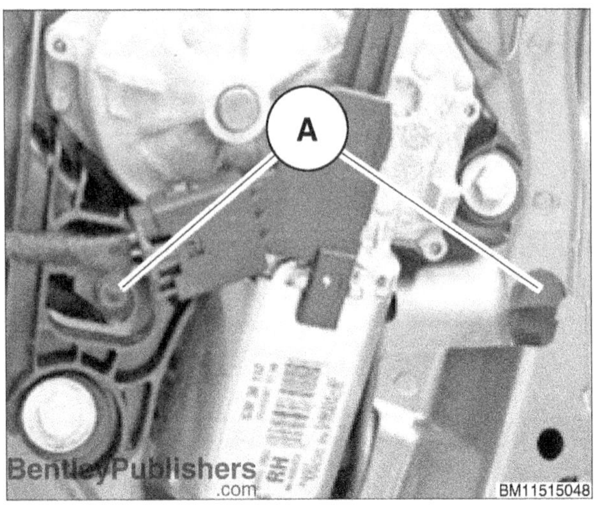

< Detach outside handle electrical connector (**arrow**).

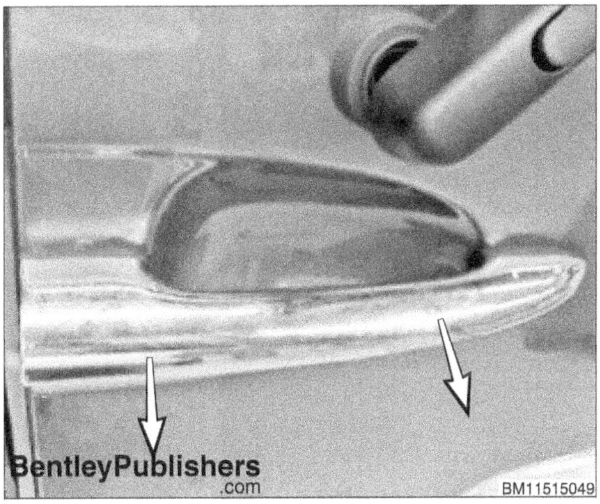

< Pull door handle off (**arrows**), carefully withdrawing electrical harness through insulating grommet in door.

– Installation is reverse of removal.

Tightening torque	
Outside handle to split door (M6 bolt)	5 Nm (44 in-lb)

ANTI-THEFT ALARM (DWA) COMPONENT REPLACEMENT

Alarm siren and tilt sensor, removing and installing

– Remove cowl covers. See **513 Interior Trim**.

– Remove wiper assembly. See **611 Wipers and Washer**.

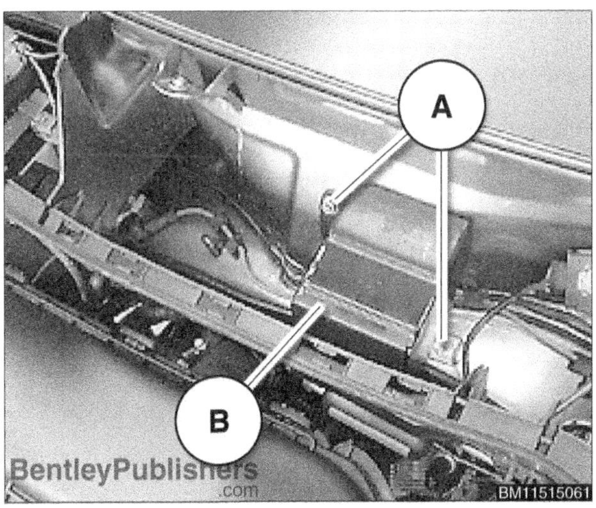

< Working inside cowl space:

- Remove mounting screws (**A**).
- Lift out siren and tilt sensor housing (**B**) and detach electrical connector.

◄ Remove nut (**arrow**) to separate tilt sensor from housing.

– Installation is reverse of removal.

Tightening torques	
Alarm siren and tilt sensor housing to body	3.5 Nm (31 in-lb)
Tilt sensor to housing	7 Nm (5 ft-lb)

– Use MINI scan tool to program and code new alarm sensor.

Interior motion sensor, removing and installing (Hardtop, Clubman)

– Remove roof switch center. See **612 Switches**.

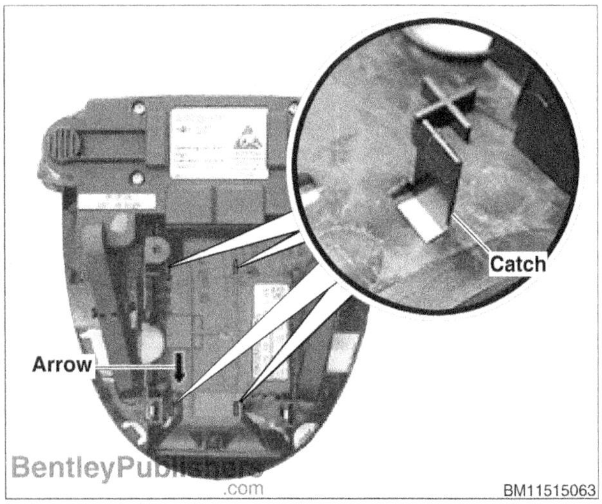

◄ Carefully unclip 4 catches (**inset)** and detach ultrasonic module from roof switch center.

– When installing;
- Do not damage any catches.
- Make sure **arrow** points to rear of vehicle.

Interior motion sensor, removing and installing (Convertible)

Convertible models are equipped with an interior motion sensor in each door behind the door trim panel.

– Remove interior door panel. See **411 Doors**.

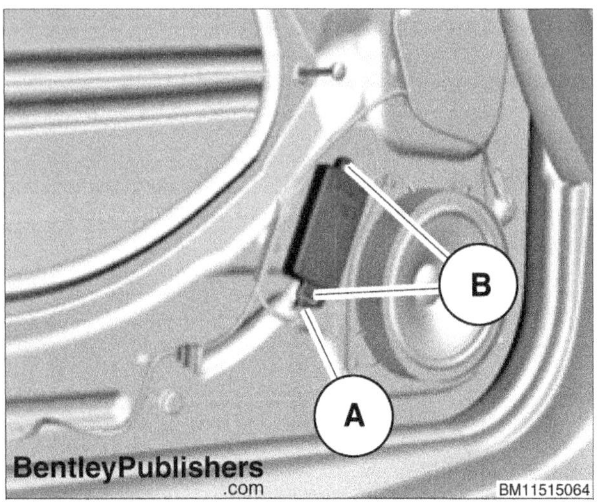

◄ Working at interior motion sensor:
- Detach electrical connector (**A**).
- Remove mounting screws (**B**) and lift off motion sensor.

– Installation is reverse of removal.

520 Seats

GENERAL

This repair group covers removal and installation procedures for front and rear seats. Seat repair procedures are beyond the scope of this manual.

See also:

- **720 Seal Belts** for pyrotechnic seat belt tensioner
- **721 Airbag System (SRS)** for passenger seat occupancy sensor

Seats overview

MINI models with normally aspirated engine are equipped with standard seats; sport seats are an option. Turbocharged models are equipped with sport seats as standard equipment.

On both the standard and sport seat, an adjustment lever makes it easier to swing the seat forward and move it in order to access the rear bench seat more easily. On the driver side, a mechanical memory device allows the seat to fall back into its former position without having to be readjusted afterwards.

Seat position, backrest position, seat height and head restraint height are manually adjustable.

Front seats are equipped with side ("thorax") airbags, controlled by the MRS control module. The seat backrests have designed rupture points from which the airbags are deployed. See **721 Airbag System (SRS)**.

The front seat belt buckle and pretensioner is mounted to the inboard edge of each front seat base. The pretensioner is a pyrotechnic (explosive charge) device controlled by the MRS control module. See **720 Seal Belts**.

Be sure to read the warnings and cautions in **721 Airbag System (SRS)** when working on front seats.

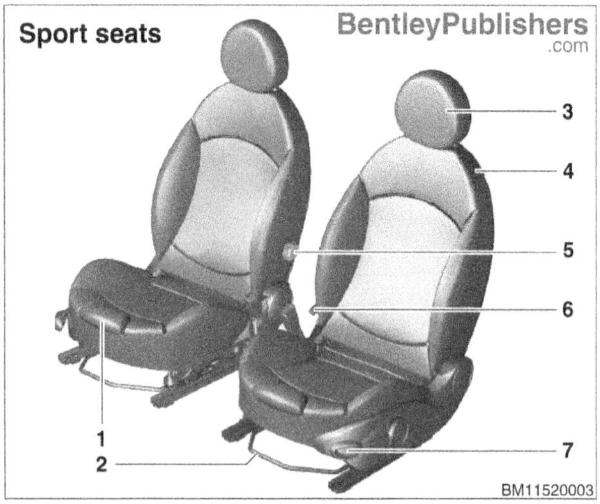

Sport front seats

1. Thigh support
2. Longitudinal position lever
3. Head restraint
4. Easy-entry lever
5. Lumbar support knob
6. Backrest angle lever
7. Height adjustment lever

Sport seats are available in fabric material, part leather, leather and artificial leather versions. They are bigger than the standard equipment seats and have strengthened side cushions and a thigh supports. Seat adjustment options are the same as on the standard equipment seat.

Lumbar support is available as an option on the sport seat. A rotary lumbar support adjuster is fitted to the backrest inboard edge.

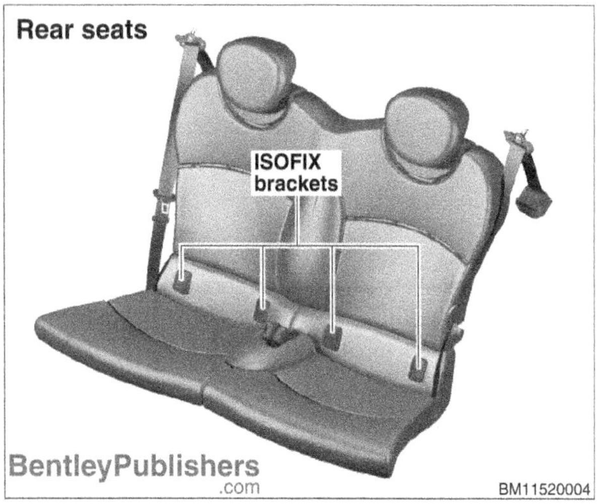

Rear seats

For additional luggage carrying capacity, the rear seat backrests can be folded down separately on right and left. A release lever on each seat allows the folding backrest to be locked or released.

The rear seats are equipped with ISOFIX brackets for installation of child seat(s).

Convertible rear seat backrests can be locked. This allows the top be left in the open position and still secure trunk contents.

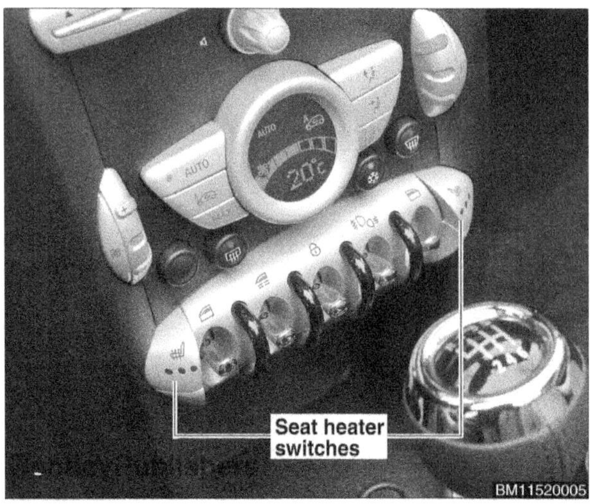

Seat heating

Front seat heaters are available as an option. Seat heater switches are in the center dashboard switch cluster.

Seat heating operates at three levels. Pressing the switch once, twice or three times changes the setting, indicated by LEDs in the switches.

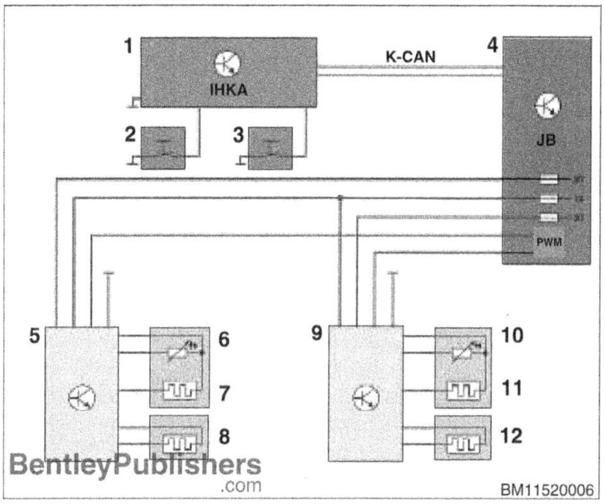

Seat heater system schematic

1. Automatic climate control (IHKA) control module
2. Driver seat heater switch
3. Passenger seat heater switch
4. Junction box (JB)
5. Driver seat heater module
6. NTC resistor
7. Heater mat, driver seat cushion
8. Heater mat, driver seat backrest
9. Passenger seat heater module
10. NTC resistor
11. Heater mat, passenger seat cushion
12. Heater mat, passenger seat backrest

Seat heater operation

Seat heating function is available when the ignition is switched ON (terminal 15 ON). Activation of the heating mats in the seats is controlled by two seat heating modules, on the bottom of each front seat. Seat heating modules are connected directly to the junction box (JB).

Seat heater switch status is sent via A/C control panel to JB via K-CAN. JB then sends pulse width-modulated (PWM) signals to the seat heating modules, with the pulse width depending on the heating level requested.

The seat heater module, regulated via an NTC resistor in the seat surface heating mat, adjusts heating to the temperature setting specified by JB.

Seat heating power reduction

Vehicle energy management system, preprogrammed in the engine control module (ECM), controls the power available for seat heating depending on vehicle energy requirements and battery condition. The junction box (JB) receives a seat heating power reduction request from ECM via PT-CAN and transmits this request directly to the seat heating modules.

Warnings and Cautions

WARNING—
- *Each front seat is equipped with both an airbag and a pyrotechnic seat belt pretensioner.*
- *The airbag is inflated by an explosive device. Handled improperly or without adequate safeguards, it is very dangerous. Observe special precautions prior to any work at or near the airbags. Read the warnings and cautions in* **721 Airbag System (SRS)**.
- *Before performing any work involving airbags, disconnect the negative (–) battery cable.*
- *Airbags contain a back-up power supply within the SRS control module. Allow a 5 second discharge period after the battery cable has been disconnected before working on airbag components.*

520

> *CAUTION—*
> - *When working on electrical components, disconnect the negative (–) cable from the battery and insulate the cable end to prevent accidental reconnection.*
> - *Prior to disconnecting the battery, read the battery disconnection cautions in* **001 Warnings and Cautions**.
> - *To avoid marring trim pieces, use a plastic prying tool or a screwdriver with the tip wrapped with masking tape.*

SEAT ELECTRICAL COMPONENTS

For seat power fuses, switches and controls, see **ECL Electrical Component Locations**.

Seat electrical components

Driver seat heating module

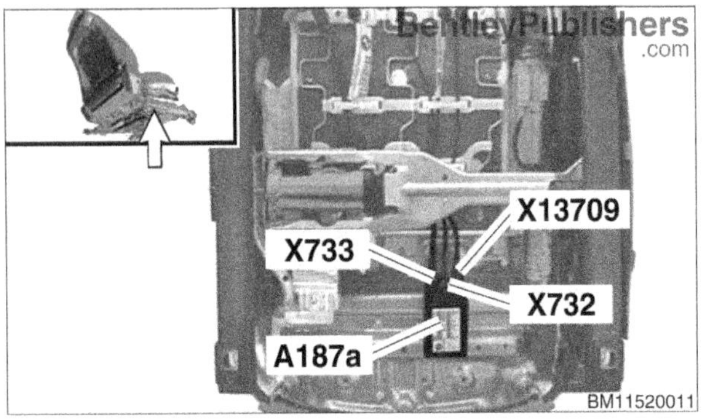

◄ Underneath passenger seat:
- **A187a** Driver seat heating module
- **X732** Connector
- **X733** Connector
- **X13709** Connector

Passenger seat heating module

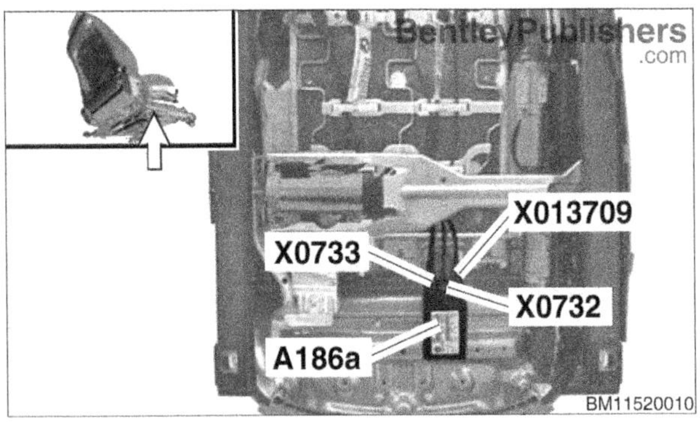

◄ Underneath passenger seat:
- **A186a** Passenger seat heating module
- **X0732** Connector
- **X0733** Connector
- **X013709** Connector

Seat position sensors

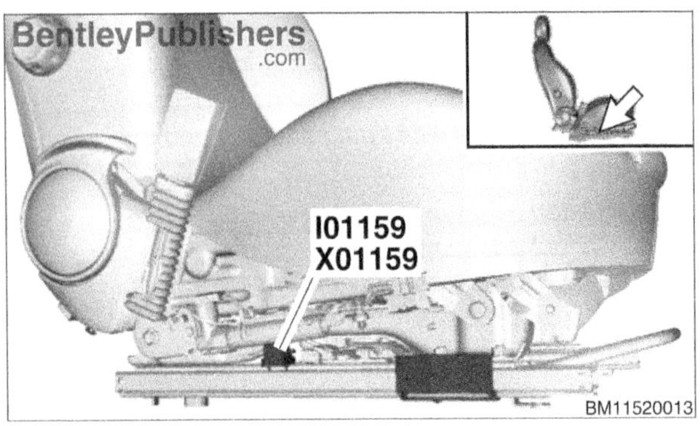

◄ On driver seat inboard seat rail:
- **I01159** Driver seat position sensor
- **X01159** Connector

On passenger seat rail (not illustrated):
- **I01184** Passenger seat position sensor
- **X01184** Connector

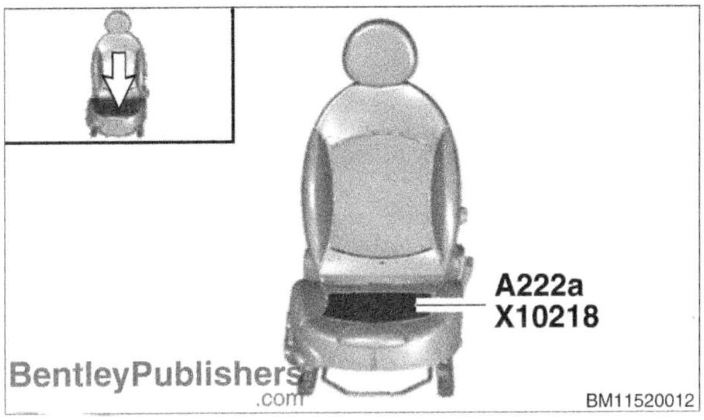

Passenger seat occupancy sensor

Underneath passenger seat cushion cover:

- **A222a** Passenger seat occupancy sensor
- **X10218** Connector

FRONT SEATS

Front seat, removing and installing

– Disconnect negative (-) battery cable and cover battery terminal to keep cable from accidentally contacting terminal.

> *CAUTION—*
> - *Prior to disconnecting the battery, read the battery disconnection cautions given in* **001 Warnings and Cautions**.

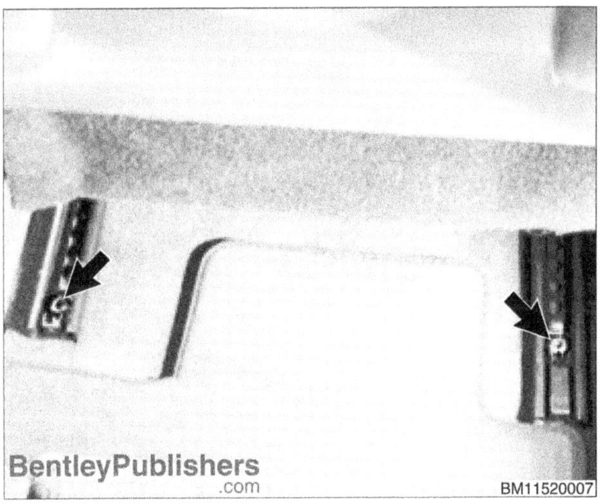

Slide front seat forward as far as it will go. Remove seat rail rear mounting bolts (**arrows**).

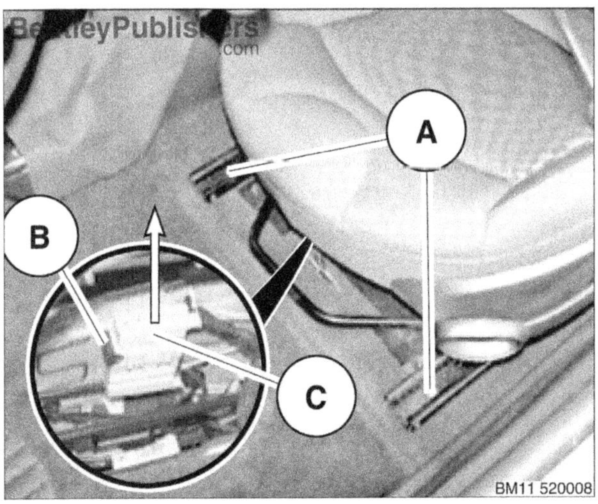

Slide seat as far back as it will go.

- Remove seat rail front mounting bolts (**A**).
- Tip seat back, separate electrical plug (**B**, **C**) and detach from seat (**arrow**).

– Lift seat out through door opening.

> *CAUTION—*
> - *Place blanket over door sill to avoid damage.*

– When installing, make sure carpet does not get pinched underneath seat rails. Install seat bolts using Locktite®

Tightening torque	
Seat to floor (use Locktite®)	35 Nm (26 ft-lb)

520

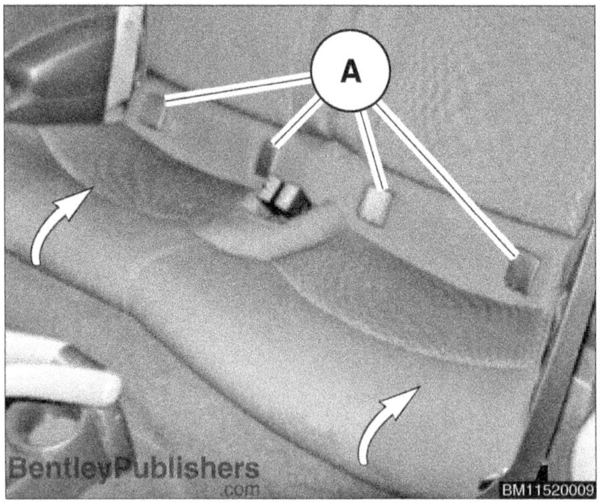

REAR SEAT

Rear seat cushion, removing and installing

◀ Working at rear seat:

- Pry out Isofix covers (**A**).
- Grasp front of seat cushion and pull up forcefully to remove (**arrows**).

– When installing, make sure seat belt buckles are positioned in center rear of seat cushion. Press seat down until it snaps audibly in place.

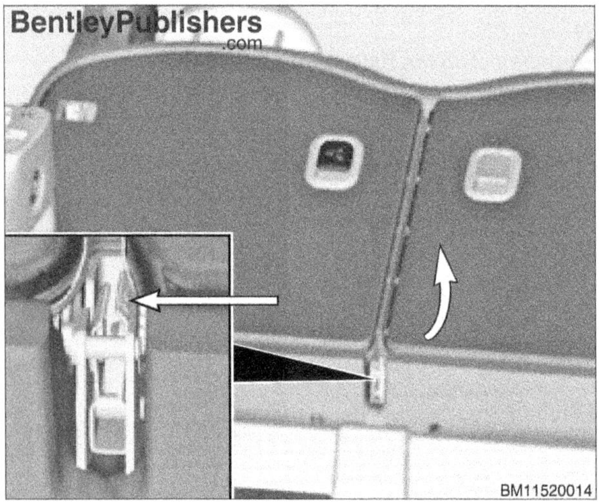

Rear seat backrest, removing and installing

– Fold down rear seat backrests.

◀ Right rear backrest: Press right retaining tab (**arrow**) between backrest halves and lift backrest out of center hinge.

– Procedure is similar for left rear backrest.

– When installing, press backrest down until it snaps audibly in place.

540 Sunroof

General

This repair group covers the slide-tilt panoramic sunroof assembly. Component locations, initialization, sunroof panel replacement, and sunroof height adjustment are covered.

Sunroof assembly removal and sunroof frame gasket replacement are beyond the scope of this manual.

See also:

- **612 Switches** for sunroof switch replacement.

Sunroof overview

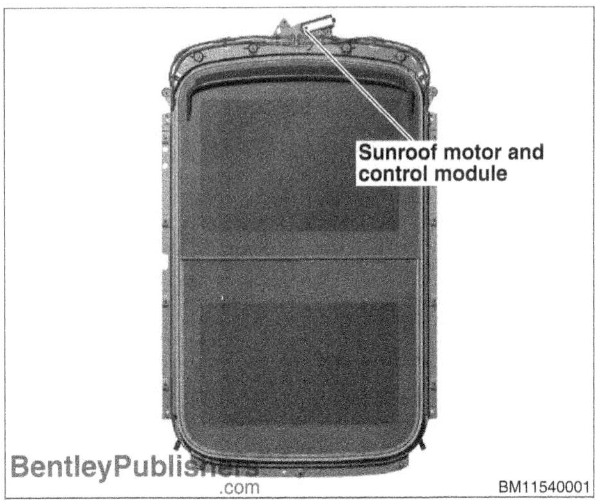

The panoramic sunroof is equipped with two glass panels and two movable headliner panels. A single motor drive is designed to slide open the front glass panel or to tilt both panels. When the roof is open, a net wind deflector reduces wind noise at road speeds.

The sunroof control module receives signals via K-CAN from other vehicle control modules for remote opening and closing. The following sunroof functions are controlled or monitored by the control module:

- Sunroof motor activation.
- Pinch protection (anti-trapping).
- Jam protection.
- Motor overheating protection.
- Panic mode closing.
- Interruption of power to sunroof during vehicle start.
- Initialization

Be sure to check drains in the corners of the sunroof assembly if water is entering the car through the headliner.

540

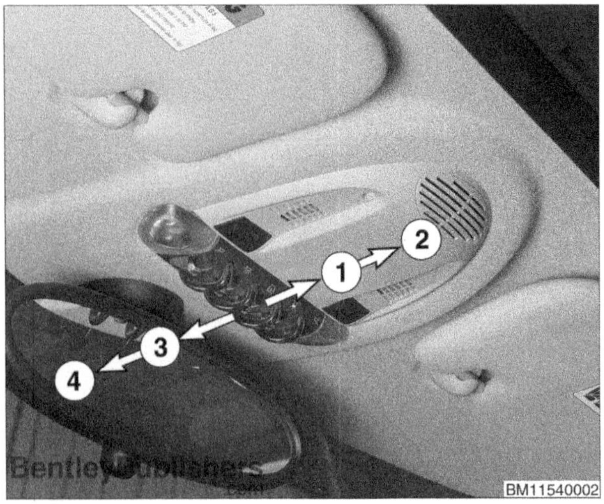

Sunroof operation

The sunroof switch in the roof switch center can be used in manual mode or automatic (one-touch) mode. If the switch is pressed forward or backward to the first detent, sunroof operation is manual. If the switch is pressed past the detent, sunroof operation is automatic.

1. Manual opening: Press backward to detent and hold.
   First press: Tilts both glass panels, then stops.
   Second press: Lowers both glass panels, slides front panel back.
2. Automatic opening: Press backward past detent and release.
   First press: Tilts both glass panels.
   Second press: Lowers both glass panels, slides front panel back.
3. Manual closing: Press forward to detent.
4. Automatic closing: Press forward past detent.

Sunroof motor initialization

Initialization activates the pinch protection (anti-trapping) and automatic (one-touch) features of the sunroof mechanism. Initialize sunroof after the following repairs:

- Sunroof control module and motor replacement.
- Sunroof emergency closing.
- Sunroof glass replacement.
- Sunroof drive unit disengagement.
- Failure of normal sunroof operation such as pinch protection, automatic opening or comfort opening.

To allow the sunroof to function normally, the following processes are implemented during initialization:

Normalization locates mechanical end positions for sunroof panels to regulate automatic opening and closing.

Memorizing the characteristic curve. Mechanical closing force for sunroof panel is stored for correct operation of pinch protection (anti-trapping).

A power interruption to the sunroof mechanism does not require initialization. However, if power supply is interrupted during initialization, initialization is not valid and must be repeated.

Initialization may fail if a fault is detected in one of the sunroof Hall sensors.

Carry out initialization manually by using the sunroof button or by using MINI scan tool.

Manual initialization

During this procedure, maintain uninterrupted pressure on sunroof switch. If the button is released, restart initialization.

> ***WARNING—***
> - *Sunroof pinch protection (anti-trapping) is disabled during initialization.*

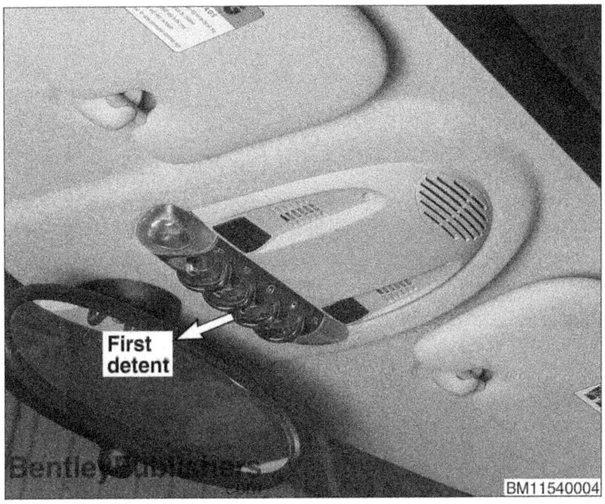

BM11540004

- Switch ignition ON (terminal 15 ON).
- ◀ Press sunroof switch forward to **first detent** (CLOSE direction).
- After approximately 30 seconds, glass sunroof panels close completely.
- Sunroof opens as far as end stop and then closes again.
- Once sunroof closes and electric motor switches off, release switch.
- Complete process lasts approx. 75 seconds. Once completed, use switch to test sunroof for normal operation. See **Sunroof operation** in this repair group.

Sunroof emergency closing

In case of electrical or mechanical failure, an open sunroof can be closed using the emergency Allen key from the vehicle tool kit.

- Remove roof switch center. See **612 Switches**.
- ◀ Insert a 4mm Allen key into drive in sunroof motor and turn to manually close sunroof.

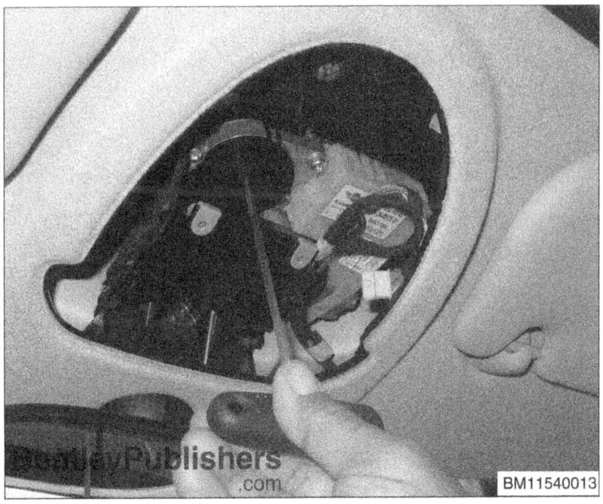

BM11540013

540

Cautions

CAUTION—
- *When working on electrical components, disconnect the negative (–) cable from the battery and insulate the cable end to prevent accidental reconnection.*
- *Prior to disconnecting the battery, read the battery disconnection cautions in* **001 Warnings and Cautions**.
- *Handle electrical modules and other electrical and electronic components using static prevention equipment and techniques. See* **600 Electrical System–General**.
- *To avoid marring trim pieces, use a plastic prying tool or a screwdriver with the tip wrapped with masking tape.*

Sunroof Repairs

Sunroof glass panel, removing and installing

Front and rear sunroof glass panels are held in place using self-locking, non-reusable Torx screws. Before starting work, be sure to have the correct M5 x 11 mm Torx screws on hand. Install with Loctite®.

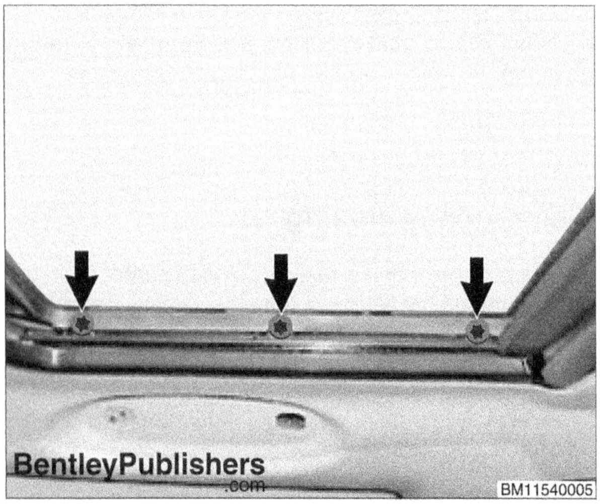

< Remove glass mounting screws (**arrows**) on left and right. Discard screws.

– Lift glass off vehicle roof.

– Reinstallation:
  - Clean threaded holes in sunroof frame.
  - Use new Torx screws with Loctite®. Tighten screws finger tight at first.
  - Adjust glass position. See **Sunroof glass panel, adjusting** in this repair group. Tighten screws to final torque within 20 minutes, before locking compound sets.

Tightening torque	
Glass to sunroof frame (replace M5 x 11 mm Torx screws with new, use Loctite®)	4 Nm (35 in-lb)

Sunroof glass panel, adjusting

Check sunroof glass panel position In case of wind noise, water leaks or glass panel replacement.

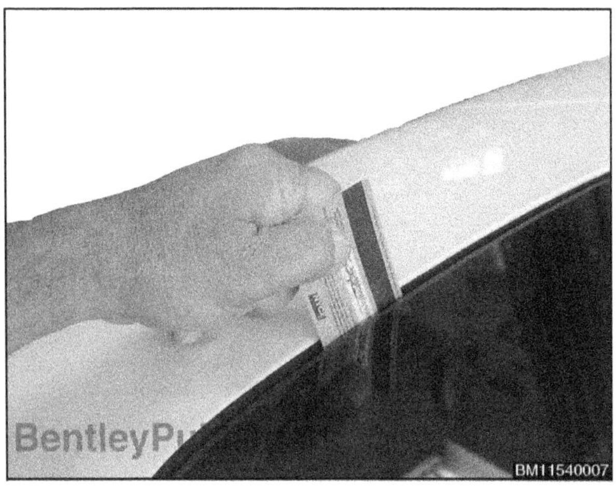

< With sunroof glass completely closed, use credit card to measure glass panel gaps. Make sure there is equal resistance to card as it is slid all around perimeter of both glass panels.

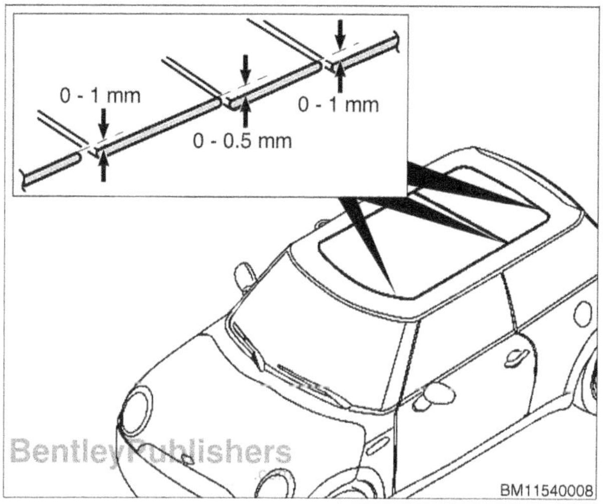

With sunroof fully closed, check height of front and rear glass panels against roof and stationary panel.

Sunroof height	
Rear of rear panel	flush to1 mm (0.04 in) higher than roof
Rear of front panel	flush to 0.5 mm (0.02 in) higher than front panel
Front of front panel	flush to1 mm (0.04 in) lower than roof

MINI special gap wedge tool 00 9 315 may be used to determine height of glass panel.

– If glass panel adjustment is needed:
  - Remove and discard Torx mounting screws. See **Sunroof glass panel, removing and installing** in this repair group.
  - Clean threaded screw holes in sunroof frame.
  - Install new Torx screws with Loctite®. Leave them finger tight.
  - Adjust position of glass panel(s) as necessary.
  - Tighten screws to final torque within 20 minutes, before locking compound sets.

Tightening torque	
Glass to sunroof frame (replace M5 x 11 mm Torx screws with new, use Loctite®)	4 Nm (35 in-lb)

Sunroof wind deflector, replacing

– Open sunroof.

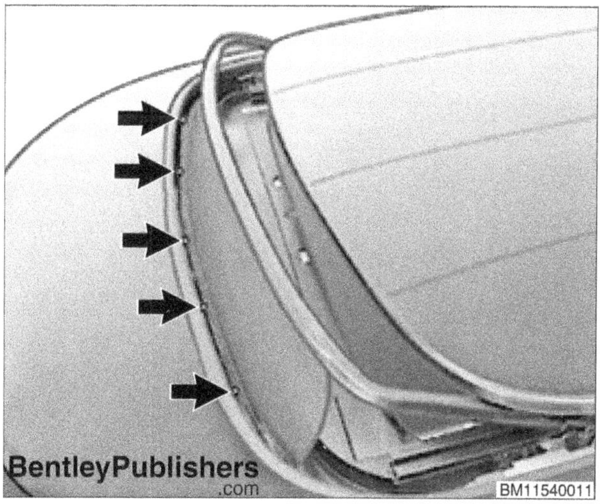

Remove wind deflector screws (**arrows**).

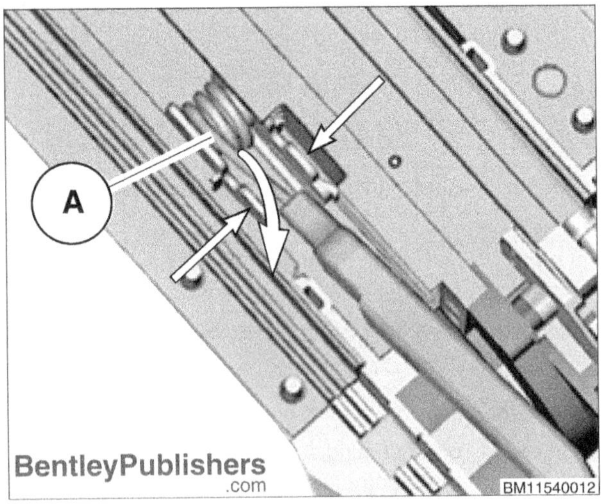

Working at wind deflector prop rod:

- Press in retaining lugs (**arrows**).
- Slide prop rod spring (**A**) forward and off sunroof frame (**curved arrow**).
- Repeat on other side.

– Installation is reverse of removal.

Sunroof motor, removing and installing

Sunroof control module and drive motor are replaced as one unit.

– Remove roof switch center. See **612 Switches**.

Working inside headliner cavity, remove sunroof module mounting screw (**arrow**).

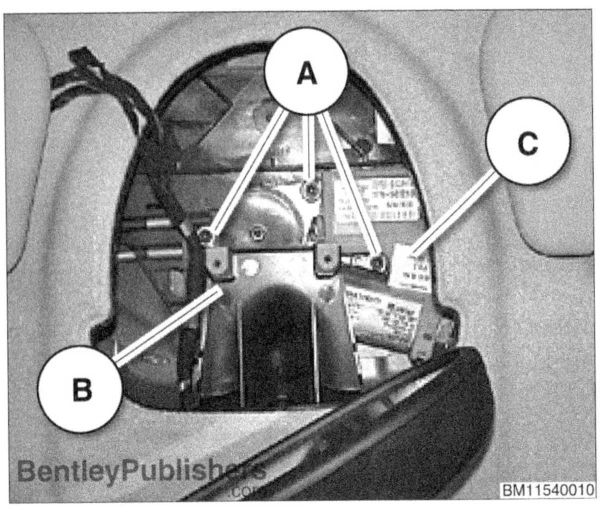

Working inside headliner cavity:

- Remove screws (**A**).
- Flex bracket **B** down slightly.
- Pull sunroof module and motor (**C**) out and detach electrical connector.

– Installation is reverse of removal. After installation:

- Use scan tool to program and code new control module.
- Initialize sunroof motor. See **Sunroof motor initialization** in this repair group.

541 Convertible Top

General

This repair group covers the automatic electrohydraulic convertible top. Information on the roll bar is also included.

Warnings and Cautions

WARNING—
- *Do not block the roll bar travel path with any objects.*
- *Faulty rollover protection system diagnosis or repairs may lead to failure of roll bar to deploy when needed.*

CAUTION—
- *If the roll bar is extended, do not operate the convertible top.*
- *If the roll bar deploys accidentally, diagnose using MINI scan tool.*
- *When working on convertible top hydraulic components, maintain strict cleanliness. Plug open lines and ports immediately to keep out dirt.*
- *When working on electrical components, disconnect the negative (–) cable from the battery and insulate the cable end to prevent accidental reconnection.*
- *Prior to disconnecting the battery, read the battery disconnection cautions in* **001 Warnings and Cautions**.
- *Handle electrical modules and other electrical and electronic components using static prevention equipment and techniques. See* **600 Electrical System–General**.
- *To avoid marring trim pieces, use a plastic prying tool or a screwdriver with the tip wrapped with masking tape.*

541

Convertible Top

Convertible top overview

The second generation MINI Convertible (factory code R57) was introduced to the North America market in the first quarter of 2009. The Convertible is equipped with a fully automatic soft top system and an active rollover safety system.

The Convertible is equipped with an open top timer ("always open" gauge) to left of the tachometer.

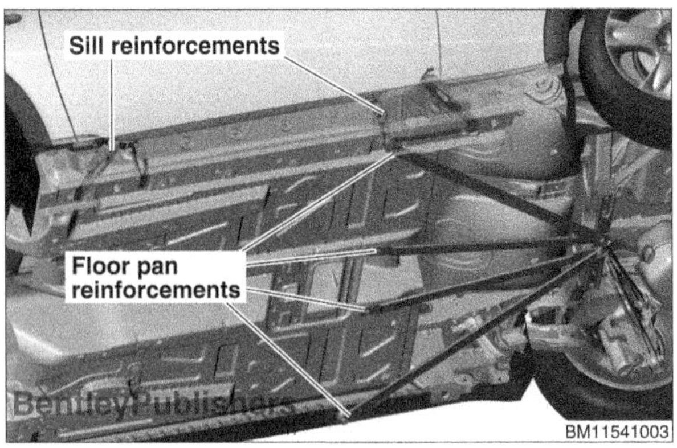

To increase torsional rigidity of the Convertible body and to brace the rear suspension, V-braces, tunnel braces and a rear-suspension bracing plate are bolted to the floor pan. The sills are reinforced with additional rigid plates.

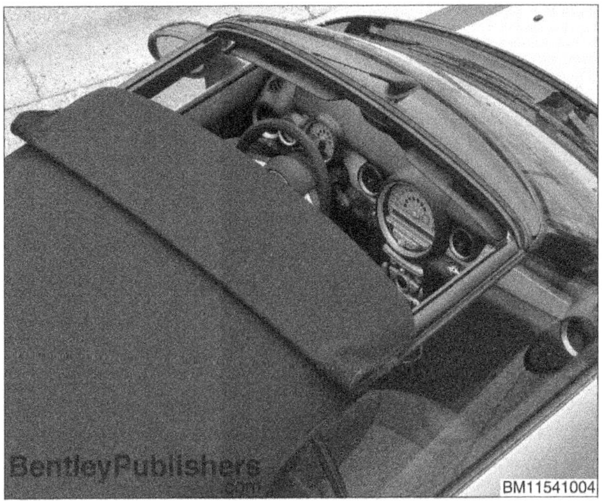

The convertible top has an integral folding sunroof. The drive for the sunroof is located in the center of the convertible top bow directly above the center armrest. Two Bowden cables operate the left and right sides of the sunroof.

Convertible top mechanical components

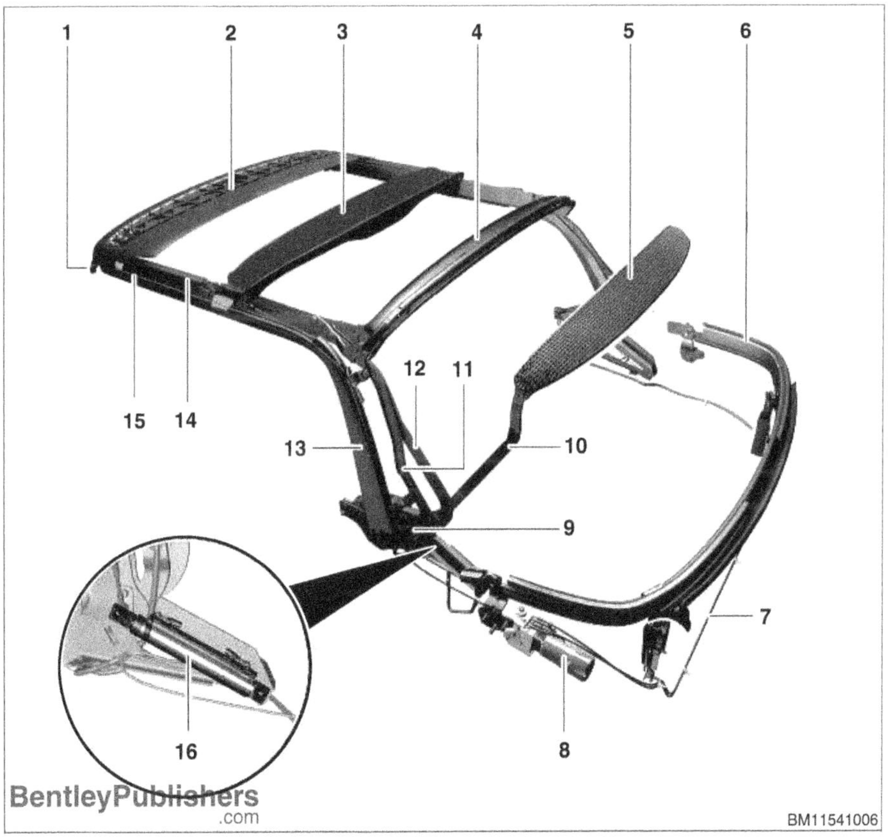

1. Front catch
2. Windshield header
3. Center bow with folding sunroof drive cassette
4. Main bow
5. Rear bow
6. Stretcher bar
7. Hydraulic lines
8. Hydraulic pump
9. Main mount
10. Left corner bow
11. Main pillar
12. Main control arm
13. Seal mount
14. Guide rail
15. Tensioning cable
16. Hydraulic cylinder

541

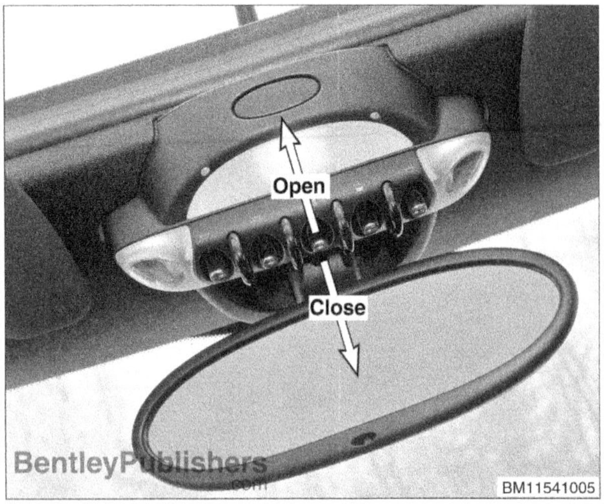

Convertible top function

The convertible top is operated by a switch in the roof switch center above the inside rear view mirror. Opening or closing the top takes about 15 seconds (with side windows, approximately 18 seconds).

For the convertible top to be folded down, the following conditions have to be met:

- Side windows opened by footwell module (FRM).
- Stretcher bar down and locked in position.
- Tailgate closed.

The convertible top is moved by hydraulic cylinders which are supplied with oil by a bidirectional hydraulic pump in the cargo compartment left side. The pump reverses direction of rotation to open or close the top.

To open the convertible top, pull the convertible top switch backward (OPEN) to open the sunroof first. This is possible up to a maximum speed of 74.5 mph (120 kph).

Operate the switch again after opening the sunroof. All the windows are lowered simultaneously. The front bow catches are then unlocked and the convertible top opened. The complete convertible top is stored and locked in the top storage compartment. If the switch is held in the OPEN direction, the windows are then closed.

Opening the convertible top is possible at a speed of up to 18.6 mph (30 kph). Above this speed, convertible top movement is stopped, a Check Control message is displayed and the green convertible top LED flashes.

Pressing the switch in the CLOSE direction first lowers the windows. After that, the top is raised and locked in position. If the button remains pressed, the sunroof and then the windows are closed.

Convenience opening and closing of the convertible top is possible using the mechanical key in the driver door lock. In addition, the top can be operated by the central locking switch in the center dashboard switch cluster.

Convertible top electrical components

Convertible top electrical components consist of the following:

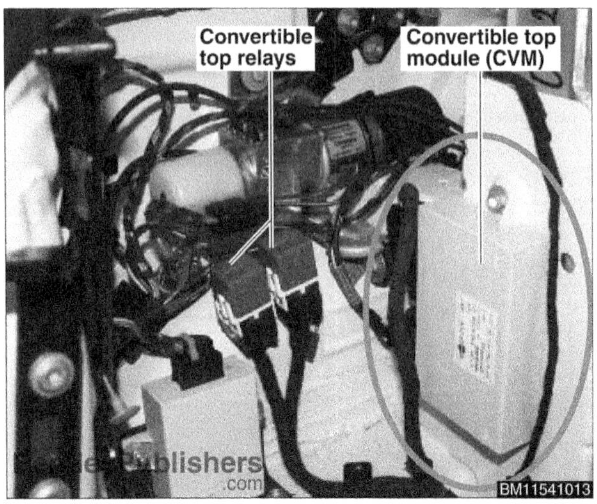

Convertible top module (CVM), located in left front of trunk behind trim, controls and monitors the mechanical and hydraulic portions of the convertible top. It is connected to other electronic components via K-CAN. It is self-diagnostic and stores fault codes relating to convertible top operation.

The convertible top hydraulic pump is activated via two relays by the CVM. A pump protection counter in CVM safeguards the hydraulic pump by inhibiting pump function after an operation time of 3.5 minutes. When pump protection activates, a fault is entered in CVM fault memory.

The sunroof motor, activated directly by CVM, is protected from overheating. Motor function is inhibited after operation time exceeds 2.2 minutes. When overheating protection activates, a fault is entered in CVM fault memory.

Footwell module (FRM) activates the front power windows. To open the convertible top, the side windows are lowered via bus signal from CVM.

Instrument cluster (KI) supplies vehicle road speed to CVM in order to determine whether the sunroof or convertible top can be operated. The cluster also displays CVM Check Control messages.

Junction box (JB) transmits the tailgate closed signal to CVM via bus. The convertible top does not open if trunk status is OPEN. The sunroof, however, functions regardless of tailgate position.

JB also supplies CVM with power (terminal 30g). In addition, CVM is supplied with ignition ON power (terminal 15) by car access system (CAS). If, for example, power supply from JB is interrupted, adequate communication with the diagnosis system is still possible.

Rear window heater is controlled by heating and A/C control panel and JB. The heater is only switched on if the top is up. JB receives convertible top position information from CVM via bus.

Hall sensors transmit convertible top position to CVM. This means that after loss of terminal 30 or emergency actuation the convertible top can still be operated without reinitialization.

Sunroof motor with incremental position sensors locks and unlocks the convertible top to the windshield header as well al opening and closing the sunroof. The incremental sensors (transducers) monitor the direction and speed of the motor.

The convertible top uses four Hall effect sensors to determine whether the top is latched, sunroof is open, top is stored or top main pillar is completely up.

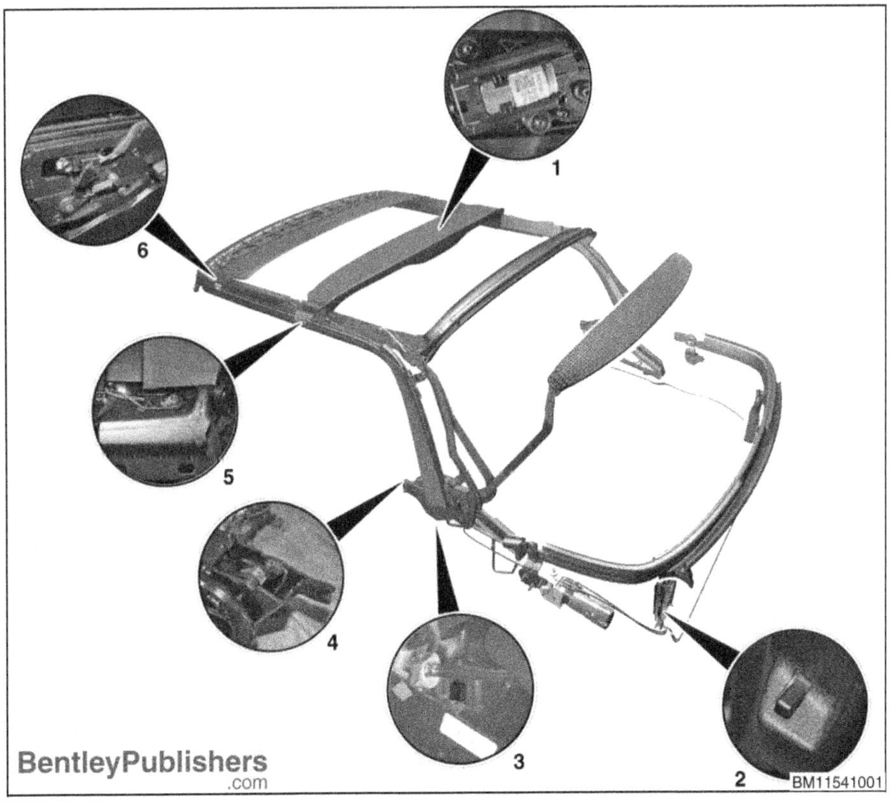

1. **M132a** Sunroof motor
   I01258 incremental sensor
   I01259 incremental sensor
2. **S239** Microswitch, rear parcel shelf
3. **I01230** Hall sensor, main pillar position (top up)
4. **I01229** Hall sensor, main pillar position (top stowed)
5. **I01234** Hall sensor, sunroof position (70% open)
6. **I01116** Hall sensor, sunroof position (100% open)

541

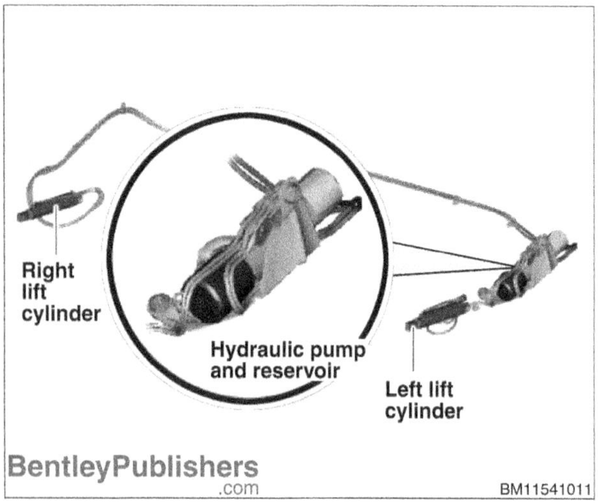

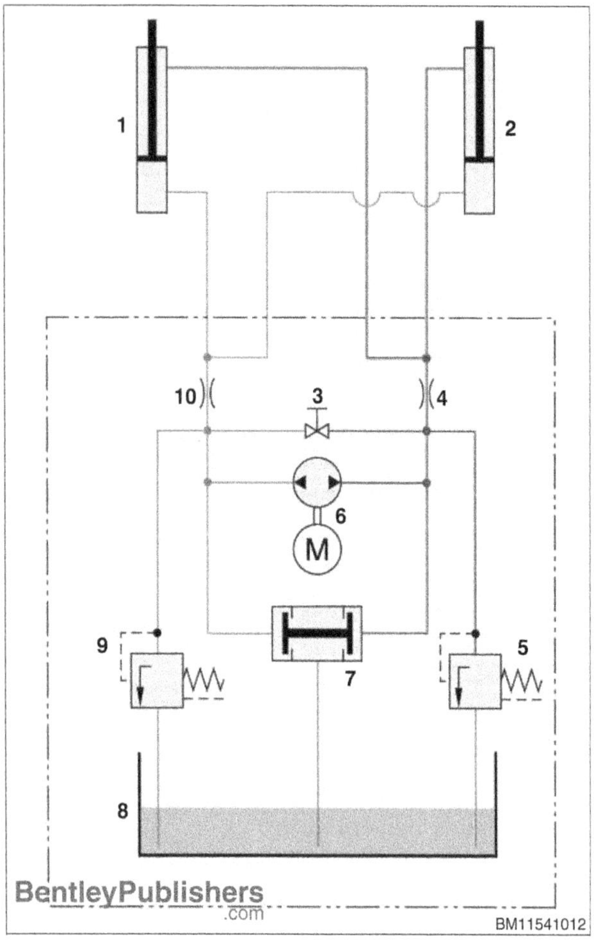

Convertible top hydraulic components

Convertible top hydraulic components consist of the hydraulic unit (pump and reservoir), right and left side lift cylinders and fluid lines.

The hydraulic system operates at pressures up to 140 bar (2031 psi) and is therefore robustly constructed. The hydraulic modulator pump is driven by an electric motor. Oil flow direction in the hydraulic system is reversed by changing the polarity of the electric motor. In this manner, the lift cylinders are made to move in or out. When the cylinders extend, the soft top folds down. When the cylinders retract, the soft top is raised.

The hydraulic system automatically self-bleeds by repeated top operation.

When topping up the fluid reservoir with hydraulic fluid, make sure that the transparent reservoir is as full as possible. Use approved hydraulic fluid or damage to the system may result. See **Hydraulic fluid level, checking or filling** in this repair group.

1. Left side lift cylinder
2. Right side lift cylinder
3. Emergency release valve
4. Right side flow restriction
5. Pressure limiting valve
6. Hydraulic pump
7. Changeover valve
8. Fluid reservoir
9. Pressure limiting valve
10. Left side flow restrictor

Convertible top repairs

The convertible top cannot be adjusted. The attachment points are fixed. If the convertible mechanism is defective, replace the complete unit.

The convertible top fabric can be replaced without the need to replace the mechanical frame.

The heated rear window cannot be separated. If repair is necessary, replace the entire convertible top.

Convertible top emergency closing

In case of electrical power failure, raise convertible top manually. The method shown below may also be used to lower the top manually, if needed.

- Open trunk tailgate and detach trunk light fixture.

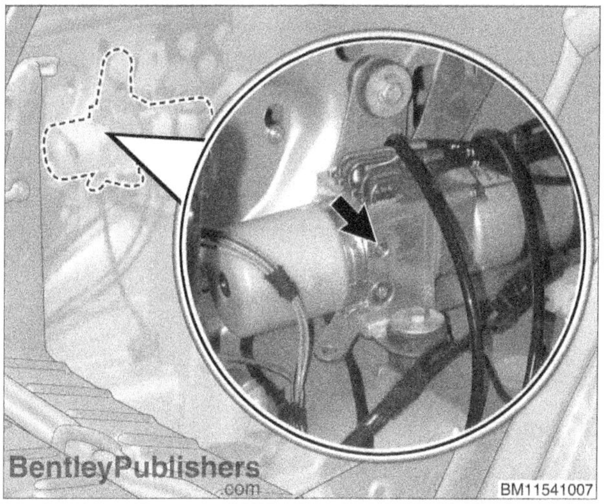

- Turn screw in convertible top motor (**arrow**) 1½ turns counterclockwise to release hydraulic pressure. Screw is accessible through light fixture opening.

- Working behind rear seat backrests, remove convertible top storage compartment cover from convertible top rails.

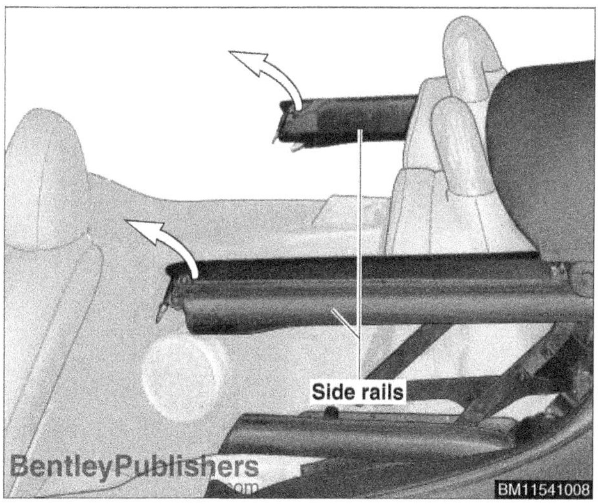

- Unscrew Allen head screws to release convertible top side rails. With help of assistant, lift both side rails forward and up (**arrows**) at the same time to windshield header.

- Remove stow bearing blocks.

541

Use plastic prying tool to lever cover off sunroof drive cassette on center convertible top bow. Insert Allen key from vehicle tool kit and press sunroof clutch up. Rotate clockwise to release convertible top locks.

- Remove holder catch.
- Release easy-load catch to reduce convertible top tension.
- Insert side rail pins in windshield header recesses.
- Press Allen key up and rotate counterclockwise to lock top.
- Lock easy-load catch (soft top is tensioned).

Hydraulic fluid level, checking or filling

The convertible top hydraulic system is filled with life time hydraulic fluid and does not require changing. No bleeding of the hydraulic system is necessary because system automatically self-bleeds by repeated top operation.

- Open convertible top.
- Close hydraulic system and remove hydraulic unit. Do not detach fluid lines. See **Hydraulic unit, removing and installing** in this repair group.

> ***CAUTION—***
> *• Protect trunk surfaces with shop towels to prevent oil staining.*

Topping up fluid

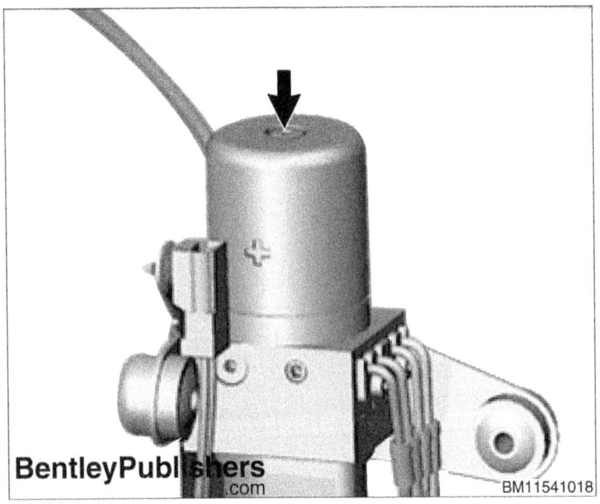

Place hydraulic unit vertically.

• Open oil filler plug (**arrow**).

• Pour in fluid, replace plug and check level. See below.

Convertible top hydraulic fluid	
Life time fluid	Aral Vitamol ZHM BMW part no. 54 34 0 394 395

Draining fluid

- Open oil filler plug.
- Tilt hydraulic unit and let fluid flow out. Alternatively, draw off fluid with suitable siphoning device.

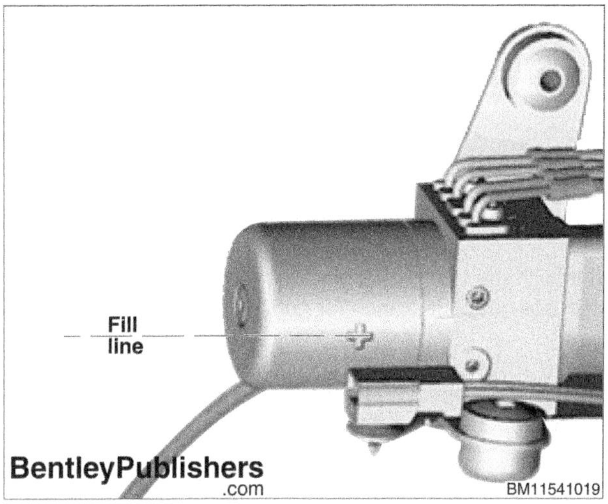

Checking fluid level

- With hydraulic unit horizontal, check that fluid level is at **fill line**. Top up as necessary.
- Reinstall hydraulic unit.
- Operate convertible top several times to self-bleed hydraulics. Note that some additional noise accompanies self-bleeding process while air bubbles circulate through system pump and seals.

Hydraulic unit, removing and installing

- Open convertible top.
- Open trunk and remove left side trunk liner.

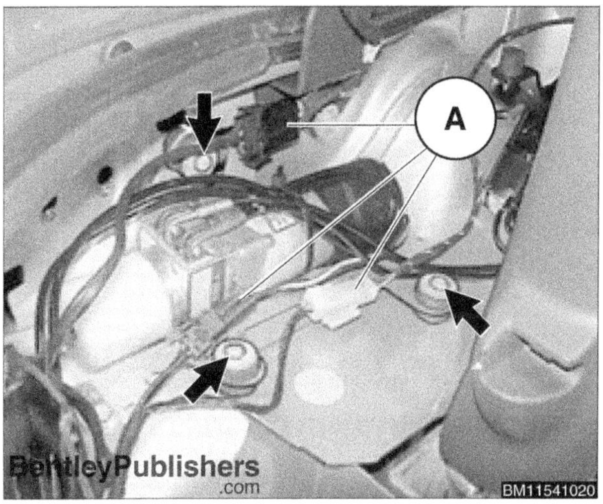

- Working at hydraulic unit:
  - Detach electrical connectors (**A**).
  - Remove mounting fasteners (**arrows**).
- Topping up fluid: Carefully rotate hydraulic unit to vertical position. Do not kink fluid lines. See **Hydraulic fluid level, checking or filling** in this repair group.

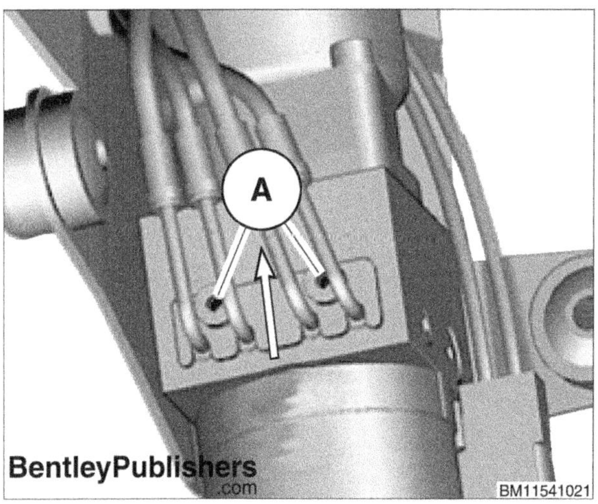

- Removing hydraulic unit:
  - Remove screws (**A**).
  - Slide lock late (**arrow**) and detach hydraulic lines. Plug open lines and ports immediately.
  - Lift out hydraulic unit.
- When installing:
  - Match numbered lines to ports.
  - Fill hydraulic reservoir. See **Hydraulic fluid level, checking or filling** in this repair group.
  - Do not kink fluid lines when installing.
- Fully open and close convertible top at least 10 times to self-bleed hydraulics. Note that some additional noise accompanies self-bleeding process while air bubbles circulate through system pump and seals.

Rollover Protection System

System overview

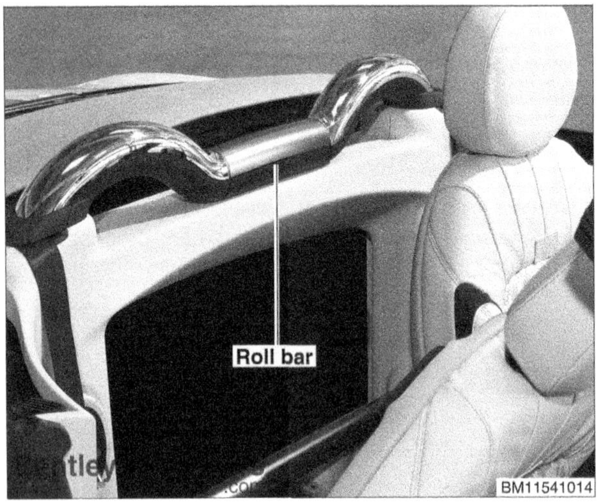

BM11541014

The rollover protection system is automatically activated during an accident or dangerous driving situation, during extreme vehicle tilt or during loss of ground contact. The roll bar, located behind the rear seat headrest, is projected within a fraction of a second.

In conjunction with the reinforced windshield frame, the roll bar ensures adequate head clearance for occupants in case of vehicle rollover.

Use MINI scan tool to inspect and test the rollover protection system every 30,000 miles as recommended by MINI.

Roll bar assembly

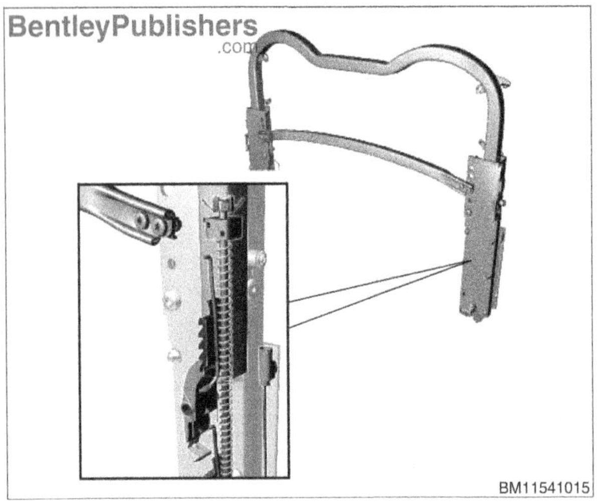

BM11541015

The roll bar assembly is spring loaded and triggered electromechanically. The total extension time under normal operating conditions is approximately 150 ms.

Rollover protection system control module

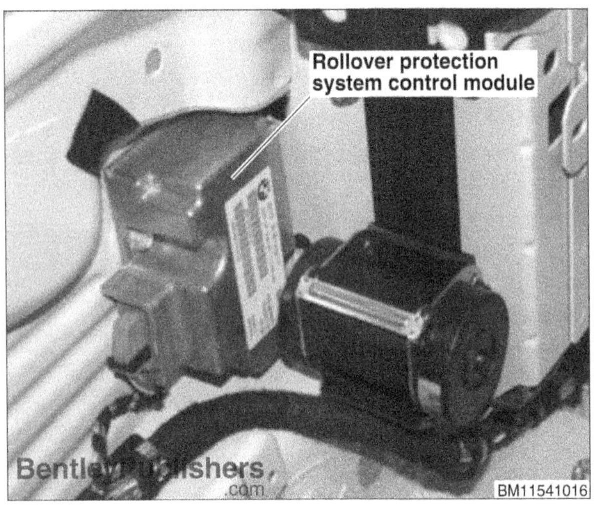

BM11541016

The rollover protection system control module is mounted behind the right rear seat back trim

Roll bar, resetting

When resetting a deployed roll bar:

- Remove plastic plug from top of roll bar tower on left and right of trunk. (Trim panel removed for the sake of photo.)
- Insert two screwdrivers, one on each side, into holes and press down (**arrow**) on ratcheting lever.
- Push down on roll bar until it locks in place.

600 Electrical System–General

TABLE

General

This section describes the principal parts of the electrical system. Also covered here are basic electrical system troubleshooting tips.

See also:

- **121 Battery, Starter, Alternator**
- **ECL Electrical Component Locations** for common MINI acronyms as well as component location information
- **ELE Electrical Wiring Diagrams**
- **OBD On-Board Diagnostics**

Electrical overview

The vehicle operates on a 12-volt direct current (DC) negative-ground system. Power is supplied by a lead-acid battery in right rear of the engine compartment, underneath the cowl cover.

The power supply fuse holder in the compartment next to the battery houses high-amperage fusible links.

A voltage regulator on the alternator controls system voltage at approximately 12 volts. All circuits are grounded by direct or indirect connection to the negative (–) terminal of the battery.

Ground connections throughout the vehicle body connect the wiring harness to chassis ground. These circuits are completed by the battery cable or ground strap between the body and the battery negative (–) terminal.

Fuses are color coded to indicate current capacities.

600

Electrical components connect using one of the following:

- Heavy cables with lug-type connectors (battery, starter, alternator).
- Electrical harnesses with keyed, push-on connectors that lock into place.
- Buses with modular connectors.

Relays and control modules are mounted in various places throughout the vehicle. See **ECL Electrical Component Locations**.

Bus basics

BM11600003

To handle electrical complexity, MINI vehicles are equipped extensively with bus systems. Signals are shared digitally among electrical components on a bus, eliminating the need for separate connections for each pair of components. The use of bus communication for controls and accessories reduces wiring complexity and improves system response time.

Data transfer over a bus is similar to a telephone conference. A component on the bus transmits a stream of data (data telegrams) which other components receive at the same time. Each component in the bus network is then able to use or ignore this telegram.

The benefits of the bus method of data transfer are as follows:

- As data and programs are modified and extended, only software modifications are necessary.
- Continuous verification of transmitted data leads to low error rate.
- Sensors and signal wires can be simplified or eliminated due to the transmittal of multiplexed digital data.
- Control modules transfer data at a high rate.
- Control module sizes and connector sizes are smaller.
- Bus architecture conforms to international standards. This facilitates data interchange between components of different manufacture.

Electrical test equipment

Many electrical tests described in this manual call for measuring voltage, current or resistance using a digital multimeter. A digital meter is preferred for precise measurements and for electronics work because it is generally more accurate than an analog meter (swing-needle). An analog meter may draw enough current to damage sensitive electronic components.

An LED test light is a safe, inexpensive tool that can be used to perform many simple electrical tests that would otherwise require a digital multimeter. The LED indicates when voltage is present between any two test-points in a circuit.

The integrated safety, comfort, security and handling systems in MINI vehicles are designed with self-diagnostic capabilities. The quickest way to diagnose many problems is to start with a scan tool read-out of diagnostic trouble codes (DTCs). See **Scan tool basics** in this repair group.

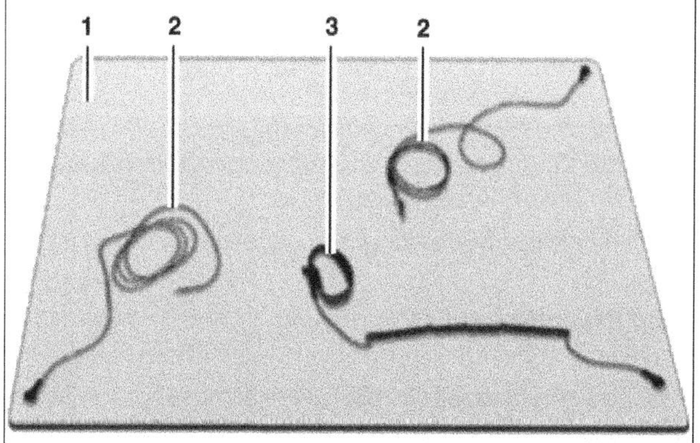

Static damage prevention

Electronic circuits can be easily damaged by static electricity held by the human body and vehicle electrical components. To avoid static electricity damage, take precautions to ground and discharge static electricity before touching any internal components of a circuit.

◀ Wearing an electrostatic discharge (ESD) wrist strap or working on an anti-static mat prevents static electricity damage.

1. ESD mat
2. Grounding lead
3. ESD wrist strap

Warnings and Cautions

WARNING—
- *Airbags and pyrotechnic seat belt tensioners utilize explosive devices. Handle with extreme care. See warnings and cautions in* **721 Airbag System (SRS)**.
- *The ignition system operates at lethal voltages. If you have a weak heart or wear a pacemaker, do not expose yourself to ignition system electric currents. Take extra precautions when working on the ignition system or when servicing the engine while it is running or the key is ON. See* **120 Ignition System** *for additional ignition system warnings and cautions.*
- *Keep hands, clothing and other objects clear of the electric engine cooling fan when working on a warm engine. The fan may start at any time, even when the ignition is switched OFF.*

CAUTION—
- *Do not disconnect the battery with the engine running.*
- *Prior to disconnecting the battery, read the battery disconnection cautions given in* **001 Warnings and Cautions**.
- *Switch the ignition OFF and remove the negative battery cable before removing any electrical components. Connect and disconnect electrical connectors and ignition test equipment leads only while the ignition is switched OFF.*
- *Relay and fuse positions are subject to change and may vary from vehicle to vehicle. If questions arise, an authorized MINI dealer is the best source for the most accurate and up-to-date information.*
- *Use a digital multimeter for electrical tests. Switch the multimeter to the appropriate function and range before making test connections.*
- *Many control modules are static sensitive. Static discharge damages them permanently. Handle the modules using static prevention equipment and techniques. See* **Static damage prevention** *in this repair group.*
- *To avoid damaging harness connectors or relay panel sockets, use jumper wires with flat-blade connectors that are the same size as the connector or relay terminals.*

> ***CAUTION—***
> - *To avoid damaging harness connectors or relay panel sockets, use jumper wires with flat-blade connectors that are the same size as the connector or relay terminals.*
> - *Do not try to start the engine of a vehicle which has been heated above 176°F (80°C) (for example, in a paint drying booth). Allow it to cool to normal temperature.*
> - *Disconnect the battery before doing any electric welding on the vehicle.*
> - *Do not wash the engine while it is running, or any time the ignition is ON.*
> - *Choose test equipment carefully. Use a digital multimeter with at least 10 MΩ input impedance or an LED test light. An analog meter (swing-needle) or a test light with a normal incandescent bulb may draw enough current to damage sensitive electronic components.*
> - *Do not use an ohmmeter to measure resistance on solid state components such as control modules.*
> - *Disconnect the battery before making resistance (ohm) measurements on a circuit.*

VEHICLE ELECTRICAL SYSTEM

Electrical system overview

Power supply components in engine compartment:

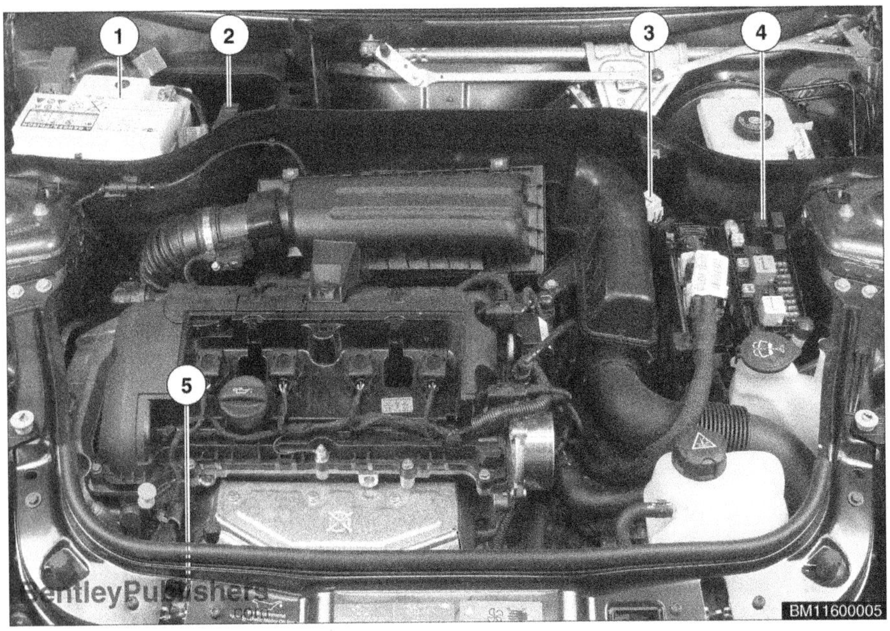

1. Battery
2. High-amperage power supply fusible links (A406)
3. Engine compartment relays
4. Power distribution fuses (engine compartment) (A400a)
5. Alternator

For detailed fuse and relay applications, see **ECL Electrical Component Locations**.

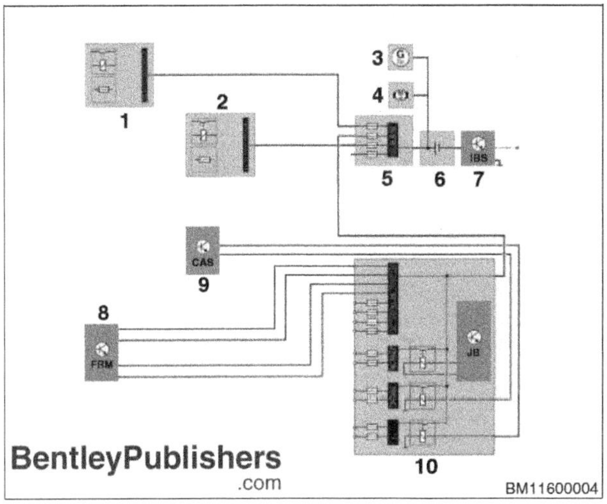

Power supply components, schematic:

1. Power distribution fuses (engine compartment) (A400a)
2. Relay carrier (engine compartment)
3. Alternator
4. Starter
5. High-amperage power supply fusible links (A406)
6. Vehicle battery
7. Intelligent battery sensor (IBS) (on vehicle with advanced power management or APM)
8. Footwell module (FRM)
9. Car access system (CAS)
10. Junction box fuses (right footwell)

Bus systems

Main buses

Main bus systems exchange data between control modules throughout the vehicle.

D-CAN (diagnosis CAN) is a high data-rate bus that connects the junction box module (JB) to the OBD II plug. A diagnosis request by scan tool is performed via the gateway in the junction box.

F-CAN (chassis CAN) is a high-speed bus used for rapid stability or traction control interventions. F-CAN connects the following:

- DSC acceleration sensor (DSC-SEN)
- DSC control module
- Steering angle sensor (in SZL)

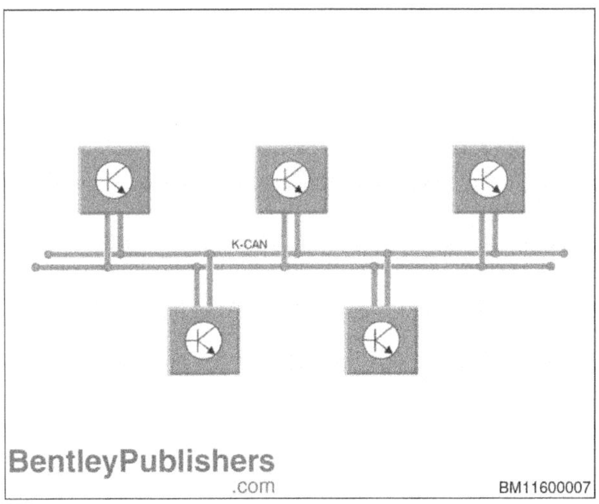

K-CAN (body CAN) is a multi-master bus. Each control module that is connected to the bus can send messages. The control modules communicate on an event-controlled basis. The control module wishing to send data sends a message when the bus is free. If the bus is not free, the message with the highest priority is sent.

Since there are no specified receive addresses, each control module receives every sent message and new receiver stations can be easily added to the system. The control modules on K-CAN are woken by the bus itself.

K-CAN connects the following:

- Central information display (CID)
- Central locking
- Climate control (IHKS or IHKA)
- Door contact switches
- Footwell module (interior and exterior lights)
- Instrument cluster (KOMBI)
- Joystick (MJOY)
- Multiple restraint system (MRS)

- Park distance control (PDC)
- Sound system amplifier (HIFI)
- Sunroof
- Tire pressure monitoring (TPM)

MOST (media oriented systems transport) is a fiber optic ring-shaped bus which connects satellite radio and digital radio with radio output or car communication computer. See also **650 Radio**.

PT-CAN (powertrain CAN) links powertrain, traction control and electromechanical power steering control modules. The change from CAN-bus (twisted pair of wires) is that PT-CAN has three wires, with the third wire used as the wake-up line for associated modules.

Table a summarizes main bus structure and data speeds.

Table a. Main bus system parameters

Bus system	Max. data rate	Structure
D-CAN	500 kBd	Linear, 2-wire
F-CAN	500 kBd	Linear, 2-wire
K-CAN	100 kBd	Linear, 2-wire
MOST	22.5 Mbd	Ring, fiber optic
PT-CAN	500 kBd	Linear, 3-wire

Sub-buses

Sub-bus systems exchange relatively smail quantities of data within one system.

BSD (bit-serial data interface) connects intelligent battery sensor (IBS) with alternator and engine control module (ECM).

CAS-bus connects the following:

- Car access system (CAS) and engine control module (ECM) for electronic immobilizer (EWS) function.
- Car access system (CAS) and electric steering lock (ELV).
- Car access system (CAS) and comfort access (CA) control modules.

K-bus connects the following:

- Multifunction steering wheel (MFL) switches to steering column switch cluster (SZL).
- Passenger seat occupant detection mat (OC3) to airbag module (MRS6 module).

LIN-bus (local interconnect network), a bidirectional single-wire bus system, is used to transmit the following:

- Rain and light sensor (RLS) data
- Tire pressure monitor (TPMS) data

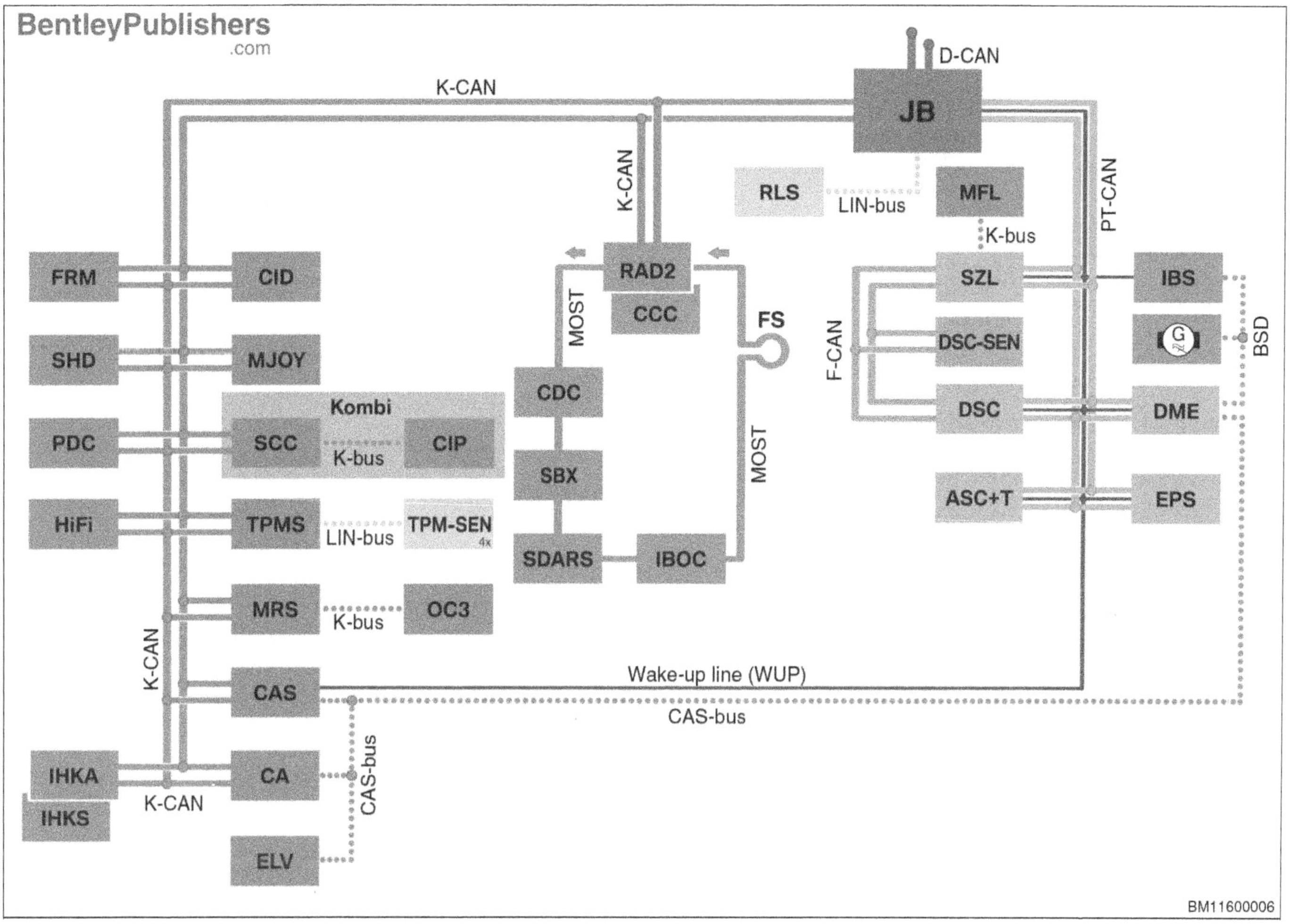

Bus systems and control modules overview

BSD Bit-serial data bus

CAS-bus Car access and comfort access bus

D-CAN Diagnosis CAN

F-CAN Chassis CAN

K-bus Body bus

K-CAN Body CAN

LIN-bus Local interconnect network

MOST Media oriented system transport (fiber optic bus)

PT-CAN Powertrain CAN

ASC + T Automatic stability control + traction

CA Comfort access

CAS Car access system

CCC Car communications computer

CDC CD changer

CID Central information display

CIP Central information panel

DME Engine control module (ECM)

DSC Dynamic stability control module

DSC-SEN DSC acceleration sensor

ELV Electric steering lock

EPS Electromechanical power steering

FRM Footwell module

FS MOST direct access

G Alternator

HIFI Sound system amplifier

IBOC Digital radio

IBS Intelligent battery sensor

IHKA Automatic climate control

IHKS Integrated climate control

JB Junction box control module

KOMBI Instrument cluster

MFL Multifunction steering wheel

MJOY Joystick controller

MRS Multiple restraint system

OC3 Passenger seat occupant detection mat

PDC Park distance control

RAD2 Radio

RLS Rain and light sensor

SBX Interface box

SCC Steering column cluster

SDARS Satellite radio

SHD Sunroof module

SZL Steering column switch cluster with steering angle sensor

TPMS Tire pressure monitor

TPM-SEN Tire pressure transmitter

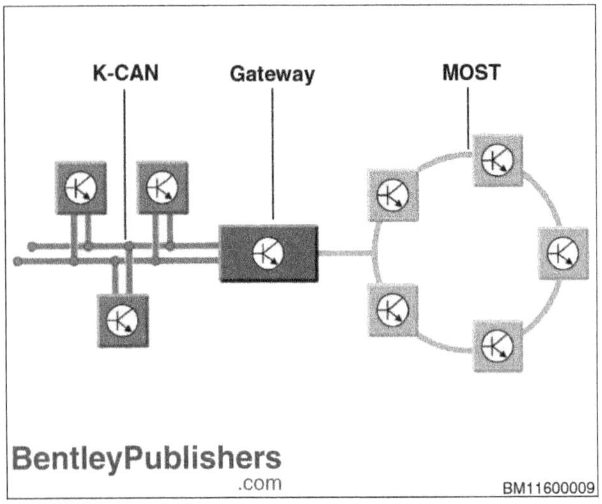

Bus gateways

The junction box module (JB) is the gateway that couples bus networks with different logic and physical properties. It ensures data exchange despite transmission rate differences among bus systems.

A second MOST bus gateway is in the radio or in the CCC.

POWER MANAGEMENT

Power management basics

Power management strategies in MINI vehicles are designed to monitor battery power, minimize power loss (dead battery) and diagnose battery state of charge (SoC) problems. Power management functions are integrated in engine control module (ECM) software.

Two types of power management are used in MINI vehicles, depending on vehicle equipment.

Basic power management (BPM)

The main functions of basic power management (BPM) are:

Idle speed adaptation. Idle speed is increased by up to 200 rpm as soon as current is drawn from the battery.

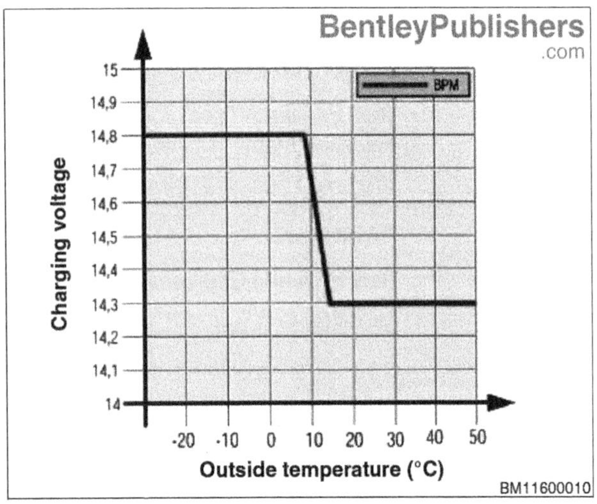

Charging voltage target value. Power management uses ambient temperature to calculate battery temperature and determines optimum battery charging voltage with the help of a preprogrammed map. This information is communicated to the alternator via bit-serial data interface (BSD).

Advanced power management (APM)

Requirements of advanced power management (APM) are intelligent battery sensor (IBS) and car communication computer (CCC). Similar to BPM, APM adapts idle speed and adjusts charging voltage. Instead of calculating battery temperature, IBS detects battery temperature directly.

APM additional functions are:

Electric load reduction based on battery SoC.

Electric load switch-off under the following conditions:
- Battery SoC in critical range
- Alternator fully utilized

Vehicle systems diagnosis. During vehicle operation, engine mechanical energy is converted by the alternator into electrical energy and made available to electric loads in the vehicle. Most electric loads receive power primarily via switched terminal 30g and terminal 30g_f (bistable) relays. In addition to idle speed and charging voltage regulation, APM controls the power usage of electric loads with relatively high power consumption; loads are switched OFF as required.

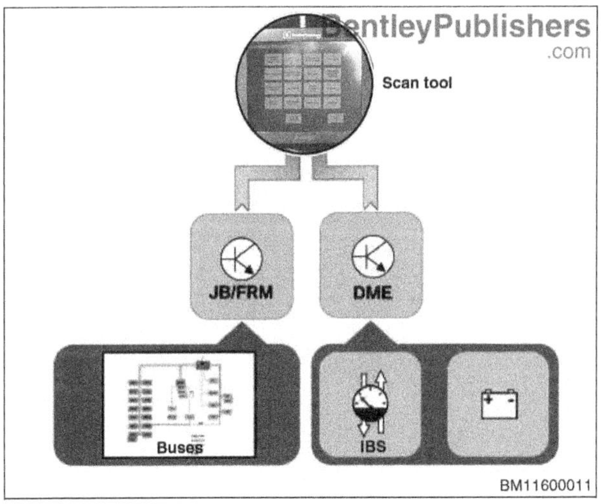

BM11600011

◄ The engine control module (ECM) and footwell module (FRM) provide a record of bus system and module functions during vehicle operation. This is stored in the junction box module (JB) and can be accessed by MINI scan tool.

Battery diagnosis. Vehicle battery SoC, is stored in the ECM.

Certain electric loads are supplied unswitched battery power directly by terminal 30 or terminal R. *Example:* Anti-theft alarm system is active when the vehicle is parked and engine is shut OFF. Therefore anti-theft power is from an unswitched terminal. This may cause no-start problems under certain conditions such as inadequate battery charging, weak battery or extreme cold.

APM monitors battery SoC and stores information about battery use with engine OFF to help with battery or charging system diagnosis.

Coding, Programming, Adaptation

Scan tool basics

BM11721025

◄ The scan tool, shown here connected to the OBD 2 plug (**arrow**), is an electronic device used for the following:
- Displaying vehicle data such as engine, engine management and transmission versions
- Viewing and clearing fault codes (diagnostic trouble codes or DTCs)
- Coding and programming new control modules and other electronic equipment
- Clearing and resetting adaptations

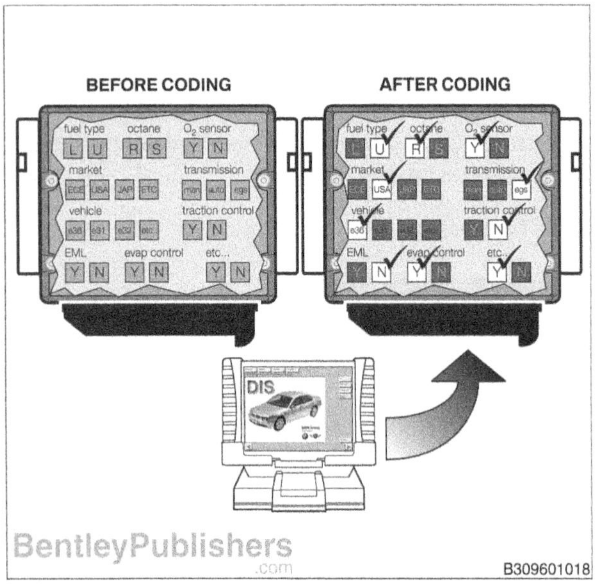

B309601018

Coding

The procedure of assigning one labeled group of data to the operating program of a control module or component is referred to as coding. A codable control module has a basic operating program already installed along with several variations of operating data. The coding process allows a specific set of operating data to be assigned to the basic operating program of that module or component to match a specific application.

Here are some types of operating requirements:

- Nominal values of device input signals (0.25 - 2.5 volts, 5 - 25 watts, etc.).
- Type of device input signal (PWM, square wave, analog).
- Operational parameters (device activation or deactivation time).
- Market specific operations (oxygen sensors, fuel type, emission standards).
- Country specific regulations (USA, Canada, Japan, UK, EU).
- Powertrain configurations (manual transmission, automatic transmission, normally aspirated, turbo-charged).

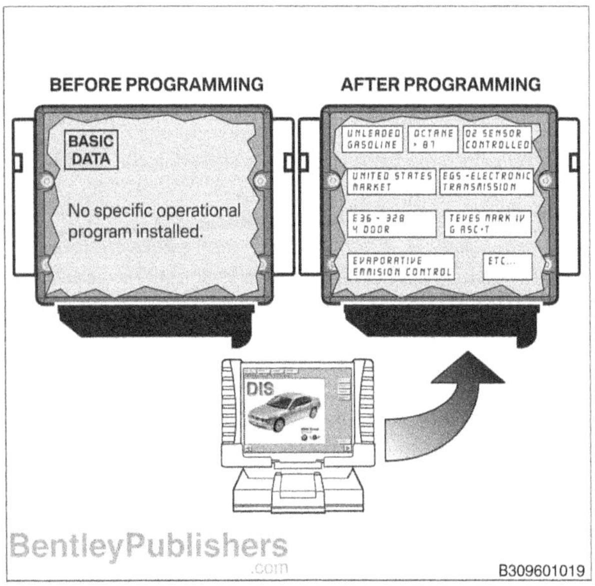

B309601019

Programming

Programming loads application and system-specific operating instructions (program) into a coded module. Programming is also a means of updating data and operating instructions previously installed in a control module.

Adaptation

The engine control model (ECM), automatic transmission control module (EGS), dynamic stability control (DSC) module and other modules and components in the vehicle are adaptive. The software in each module learns from use and becomes adapted to other accompanying modules.

When installing a new module or component, be sure to use the scan tool to clear old adaptations in order to allow the components to readapt.

Coding and programming procedure

Before starting the coding and programming procedure, use a vehicle power supply to maintain correct voltage throughout the electrical system. See **121 Battery, Starter, Alternator**.

For the most up-to-date data and program information, make sure the scan tool has the current version of CIP (coding, individualization and programming) installed.

Connect MINI scan tool to OBD II plug under dashboard. See **Scan tool basics** in this repair group.

ELECTRICAL TROUBLESHOOTING

Four things are required for current to flow in any electrical circuit:

- Voltage source.
- Wires or connections to transport voltage.
- Load or device that uses electricity.
- Connection to ground.

Most problems can be found using a digital multimeter (volt / ohm / ammeter) to check the following:

- Voltage supply.
- Breaks in the wiring (infinite resistance / no continuity).
- A path to ground that completes the circuit.

Electric current is logical in its flow, always moving from the voltage source toward ground. Electrical faults can usually be located through a process of elimination. When troubleshooting a complex circuit, separate the circuit into smaller parts. General tests outlined below may be helpful in finding electrical problems. The information is most helpful when used with wiring diagrams.

Be sure to analyze the problem. Use wiring diagrams to determine the most likely cause. Get an understanding of how the circuit works by following the circuit from ground back to the power source.

When making test connections at connectors and components, use care to avoid spreading or damaging the connectors or terminals. Some tests may require jumper wires to bypass components or connections in the wiring harness. When connecting jumper wires, use blade connectors at the wire ends that match the size of the terminal being tested. The small internal contacts are easily spread apart, and this can cause intermittent or faulty connections that can lead to more problems.

Voltage and voltage drop

Wires, connectors, and switches that carry current are designed with very low resistance so that current flows with a minimum loss of voltage. A voltage drop is caused by higher than normal resistance in a circuit. This additional resistance actually decreases or stops the flow of current. Excessive voltage drop can cause problems ranging from dim headlights to sluggish wipers. Some common sources of voltage drops are corroded or dirty switches, dirty or corroded connections or contacts and loose or corroded ground wires and ground connections.

A voltage drop test is a good test to perform if current is flowing through the circuit but the circuit is not operating correctly. A voltage drop test helps pinpoint a corroded ground strap or a faulty switch. Normally, there should be less than 1 volt drop across most wires or closed switches. Voltage drop across a connector or short cable varies but should not exceed 0.3 volt.

A voltage drop test is generally more accurate than a simple resistance check because the resistances involved are often too small to measure with most ohmmeters. For example, a resistance as small as 0.02 Ω results in a 3 volt drop in a typical 150 amp starter circuit. (150 amps x 0.02 Ω = 3 volts).

Keep in mind that voltage with ignition key ON and voltage with engine running are not the same. With ignition ON and engine OFF, fully charged battery voltage is approximately 12.6 volts. With engine running (charging voltage), normal voltage is approximately 14.0 volts. Measure voltage at battery with ignition ON and then with engine running to get exact measurements.

Voltage, measuring

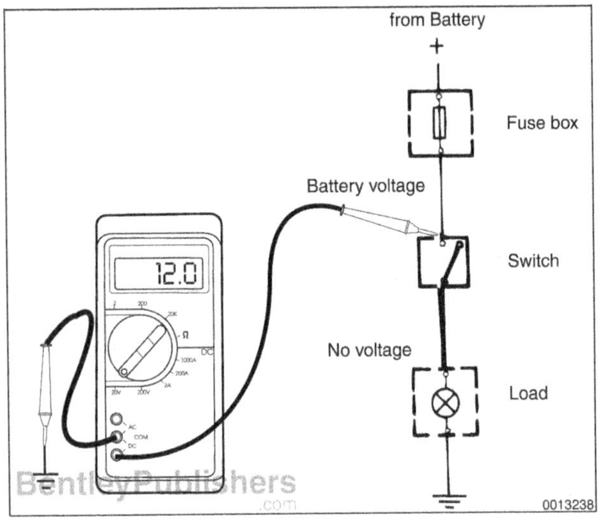

◄ Connect digital multimeter negative lead to a reliable ground point on vehicle. The negative (-) battery terminal is a good ground point.

– Connect digital multimeter positive lead to point in circuit you wish to measure.

– Check that voltage reading does not deviate more than 1 volt from voltage at battery. If voltage drop is more than this, check for a corroded connector or loose ground wire.

Voltage drop, testing

Check voltage drop only when there is a load on the circuit, such as when operating the starter motor or turning on the headlights. Use a digital multimeter to ensure accurate readings.

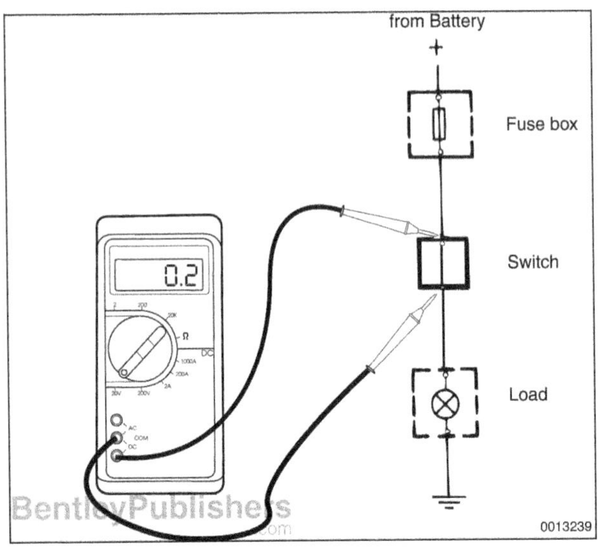

◄ Connect digital multimeter positive lead to positive (+) battery terminal or a positive power supply close to battery source.

– Connect digital multimeter negative lead to other end of cable or switch being tested.

– With power switched ON and circuit working, meter shows voltage drop (difference between two points). This value should not exceed 1 volt.

- Maximum voltage drop in an automotive circuit, as recommended by the Society of Automotive Engineers (SAE), is as follows:
  - 0 volt for small wire connections
  - 0.1 volt for high current connections
  - 0.2 volt for high current cables
  - 0.3 volt for switch or solenoid contacts
- On longer wires or cables, the drop may be slightly higher. In any case, a voltage drop of more than 1.0 volt usually indicates a problem.

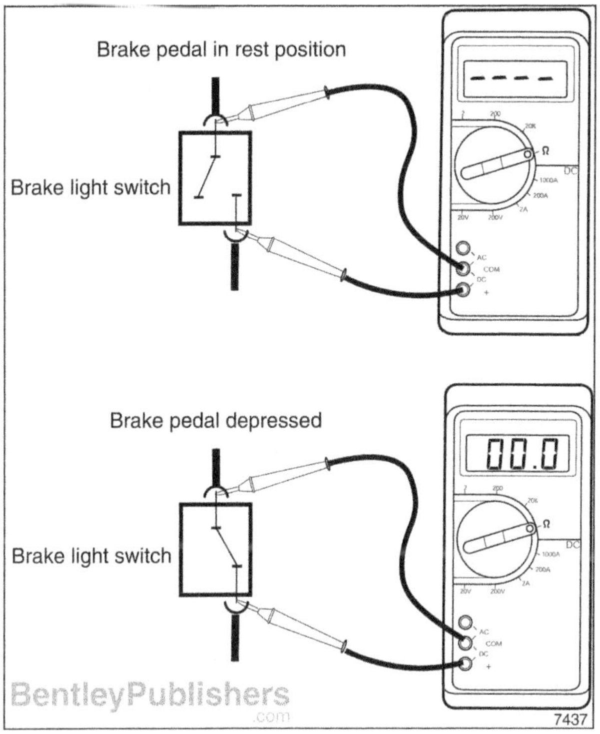

Continuity, checking

Use a continuity test to check a circuit or switch. Because most automotive circuits are designed to have little or no resistance, a circuit or part of a circuit can be easily checked for faults using an ohmmeter. An open circuit or a circuit with high resistance does not allow current to flow. A circuit with little or no resistance allows current to flow easily.

When checking continuity, switch ignition OFF. On circuits that are powered at all times, disconnect battery. Using the appropriate wiring diagram, test circuit for faulty connections, wires, switches, relays and engine sensors by checking for continuity.

Example: Test brake light switch for continuity:

- With brake pedal in rest position (switch open) there is no continuity (infinite Ω).
- With pedal depressed (switch closed) there is continuity (0 Ω).

Short circuits, testing

Short circuits are exactly what the name implies. Current in the circuit takes a shorter path than it was designed to take. The most common short that causes problems is a short to ground where the insulation on a positive (+) wire wears away and the metal wire is exposed. When the wire rubs against a metal part of the vehicle or other ground source, the circuit is shorted to ground. If the exposed wire is live (positive battery voltage), a fuse blows or the circuit may be damaged.

Short circuits vary in nature and are often difficult to locate. They can be found using a logical approach based on knowledge of the current path.

Use a digital multimeter to locate short circuits.

> ***CAUTION—***
> - *In circuits protected with high rating fuses (25 amp and greater), wires or circuit components may be damaged before the fuse blows. Check for wiring damage before replacing fuses of this rating. Also, check for correct fuse rating.*

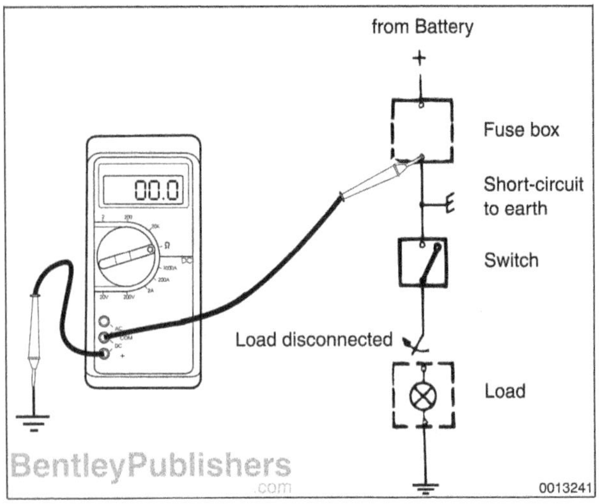

Testing with ohmmeter

- Remove blown fuse from circuit and disconnect cables from battery. Disconnect harness connector from circuit load or consumer.
- Using an ohmmeter, connect one test lead to load side of fuse terminal (terminal leading to circuit) and other test lead to ground.
- If there is continuity to ground, there is a short to ground.
- If there is no continuity, work from wire harness nearest to fuse and relay panel and move or wiggle wires while observing meter. Continue to move down harness until meter displays a reading. This is the location of the short to ground.
- Visually inspect wire harness at this point for any faults. If no faults are visible, carefully slice open harness cover or wire insulation for further inspection. Repair any faults found.

Testing with voltmeter

- Remove blown fuse from circuit. Disconnect harness connector from circuit load or consumer.

NOTE—

- *Most fuses power more than one consumer. Be sure all consumers are disconnected when checking for a short circuit.*

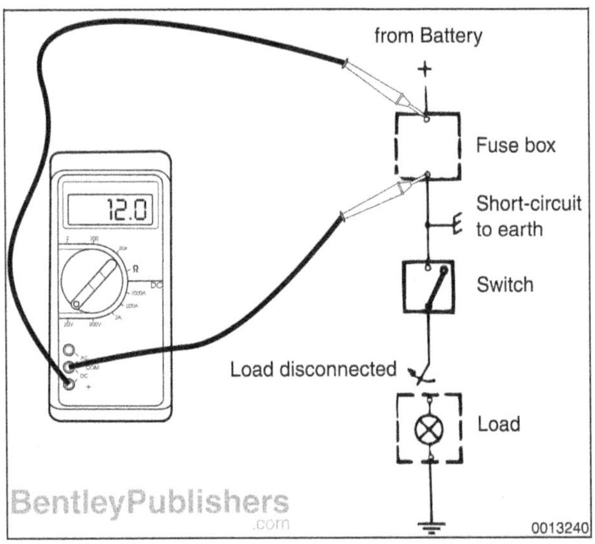

- Using a digital multimeter, connect test leads across fuse terminals. Make sure power is present in circuit. If necessary switch ignition ON.
- If voltage is present at voltmeter, there is a short to ground.
- If voltage is not present, work from wire harness nearest to fuse and relay panel and move or wiggle wires while observing meter. Continue to move down harness until meter displays a reading. This is the location of the short to ground.
- Visually inspect wire harness at this point for any faults. If no faults are visible, carefully slice open harness cover or wire insulation for further inspection. Repair any faults found.

611 Wipers and Washers

GENERAL

This repair group covers removal and installation of wiper blades, front and rear wiper assemblies and window and headlight washer components.

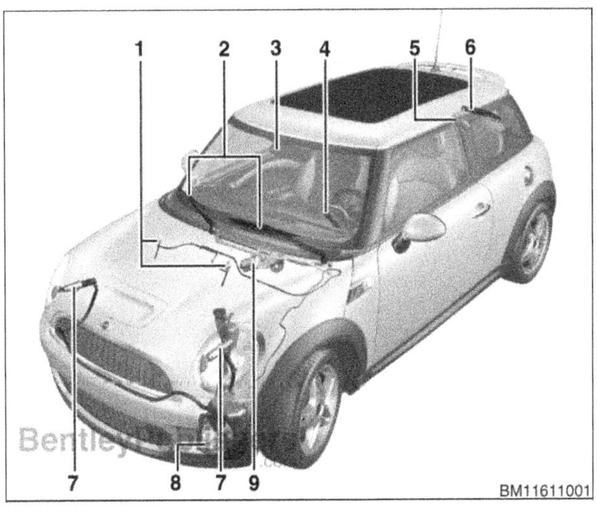

Wiper and washer system overview

1. Heated windshield washer nozzles
2. Windshield wiper arms and blades
3. Rain and lights sensor
4. Wiper and washer switch
5. Rear window wiper motor assembly
   • Hardtop model shown. Clubman model has two wiper assemblies on the rear windows (one on each split door)
6. Rear window wiper arm and blade
7. Headlight washer nozzle
8. Washer fluid reservoir
9. Windshield wiper motor assembly

Wiper switch

◄ Wiper and washer function requests begin at the wiper stalk switch on the right of the steering column. The stalk switch sends front and rear wipe, wash and speed requests to the junction box fuse and relay panel (A4010).

Wiper and washer system circuit schematic

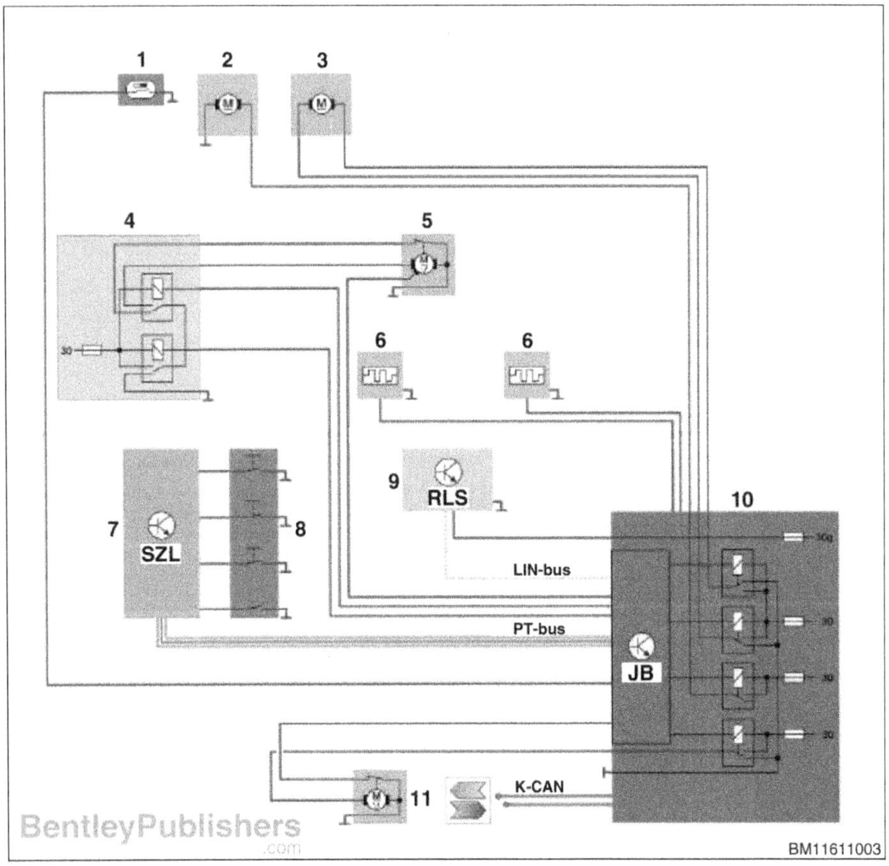

1. Washer fluid level sensor
2. Headlight washer pump
3. Windshield and rear window washer pump
4. Front power distribution panel, engine compartment
5. Windshield wiper motor
6. Heated windshield washer nozzles
7. Steering column switch cluster (SZL)
8. Wiper and washer switch
9. Rain and light sensor (RLS)
10. Junction box fuse and relay panel (JB)
11. Rear window wiper motor

Warnings and Cautions

CAUTION—

- *When working on electrical components, disconnect the negative (-) cable from the battery. Before disconnecting the battery, read battery disconnection cautions in* **001 Warnings and Cautions**.
- *When removing interior trim clips, avoid damaging trim pieces by using a plastic prying tool or a screwdriver with the tip wrapped with masking tape.*

WIPERS

Wiper blade cleaning problems

Common cleaning problems with wipers include streaking or sheeting, water drops after wiping and blade chatter.

Streaking is usually caused when wiper blades are coated with road film or car wash wax. Clean the blades using soapy water. If cleaning does not cure the problem, replace the blades. MINI recommends replacing the wiper blades twice a year, before and after the cold season.

Water drops that remain behind after wiping are usually caused by oil, road film or diesel exhaust residue on the glass. Use an alcohol or ammonia solution or other nonabrasive cleaner to clean the windshield.

Chatter may be caused by dirty or worn blades or by wiper arms that are out of alignment. Clean the blades and windshield as described above. Adjust the wiper arm so that there is even pressure along the blade, and so that the blade at rest is perpendicular to the windshield. If problems persist, replace the blades and wiper arms.

Windshield wiper blade, replacing (pre-LCI)

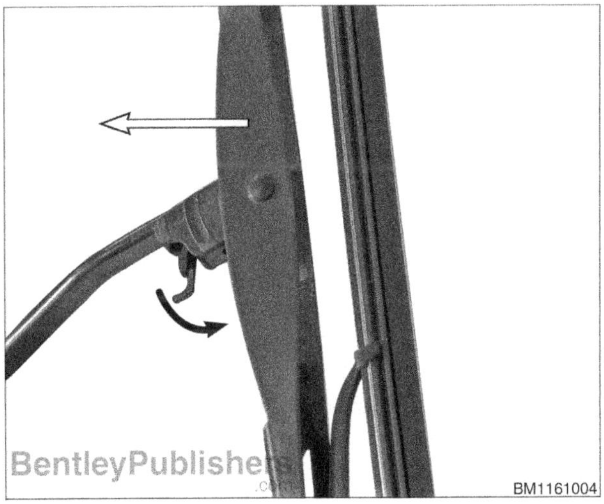
BM1161004

To replace wiper blade:

- Pivot wiper arm off windshield.
- Position wiper blade approximately perpendicular to wiper arm.
- Remove wiper blade from wiper arm by pushing on tab (**curved arrow**). Slide blade out of arm (**arrow**).
- Slide new blade on arm and snap into place.

611

Windshield wiper blade, replacing (LCI)

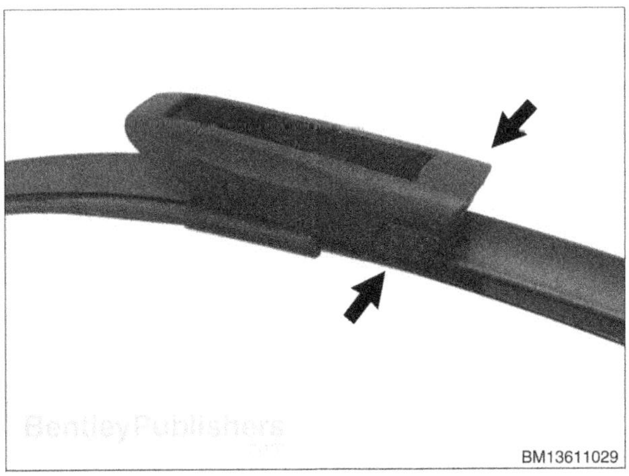
BM13611029

To replace wiper blade:

- Pivot wiper arm off windshield.
- Remove wiper blade from arm by pressing on tabs (**arrows**). Slide blade straight off of arm.
- Slide new blade on arm and snap into place.

Rear window wiper blade, replacing

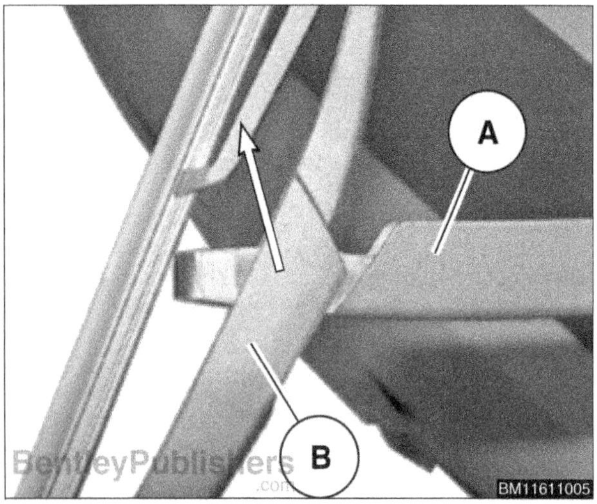

< Raise wiper arm (**A**) and push wiper blade (**B**) in direction of **arrow** to unclip.

– To install, snap blade firmly into place.

Windshield wiper arm, replacing

< Lift off protective cap from base of wiper arm (**arrow**).

– Make sure clips on underside of cap are not damaged. Replace if necessary.

< Remove nut (**arrow**) securing wiper arm to wiper pivot shaft.

– Remove wiper arm from pivot shaft using MINI special tool 61 6 130 or other suitable puller.

> ***CAUTION—***
> *• Do not bend wiper arm while removing.*

Installation is reverse of removal. Make sure wiper motor is in PARK position.

- Right wiper blade distance to windshield frame:
  A = 46 ± 2.5 mm (1.8 ± 0.1 in).
- Left wiper blade distance to windshield frame:
  B = 40 ± 2.5 mm (1.6 ± 0.1 in).
- Tighten nut and replace wiper arm nut protective cap.

Tightening torque	
Wiper arm to wiper shaft	20 Nm (15 ft-lb)

Rear window wiper arm, replacing

– Flip open protective cap at base of rear wiper arm.

Remove mounting nut (**arrow**). Remove wiper arm. If necessary, use special tool MINI 61 6 210 or another puller to remove arm from shaft.

– Check that protective cap is not damaged. Replace if necessary.

Installation is reverse of removal. Make sure wiper motor is in PARK position.

- Hardtop wiper blade distance to rear window frame:
  A = 38 ± 2.5 mm (1.5 ± 0.1 in).
- Clubman wiper blade distance to rear window frame:
  A = 18 ± 2.5 mm (0.7 ± 0.1 in).

Tightening torque	
Rear wiper arm to wiper motor	12 Nm (9 ft-lb)

611

WIPER MOTOR AND LINKAGE

Windshield wiper motor assembly, removing and installing

– Remove both windshield wiper arms and blades. See **Windshield wiper arm, replacing** in this repair group.

– Remove right and left cowl cover. See **513 Interior Trim**.

◄ Remove wiper motor lower mounting nut (**arrow**).

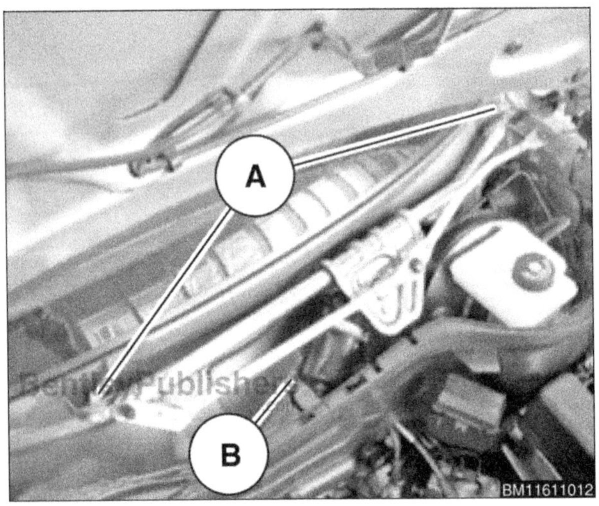

◄ Remove wiper assembly mounting bolts (**A**). Lift up wiper assembly and detach electrical connector (**B**).

– Installation is reverse of removal.

- Make sure wiper assembly isolating grommets are undamaged. Position assembly on grommets correctly.
- Make sure wiper motor is in PARK position before mounting wiper arms. See **Windshield wiper arm, replacing** in this repair group.

Tightening torques	
Wiper arm to wiper shaft	20 Nm (15 ft-lb)
Wiper assembly to bulkhead • M6 • M8	 10 Nm (7 ft-lb) 22 Nm (16 ft-lb)

Rear window wiper motor, removing and installing (Hardtop)

– Remove rear wiper arm. See **Rear window wiper arm, replacing** in this repair group.

– Remove lower hatch trim. See **412 Rear Hatch**.

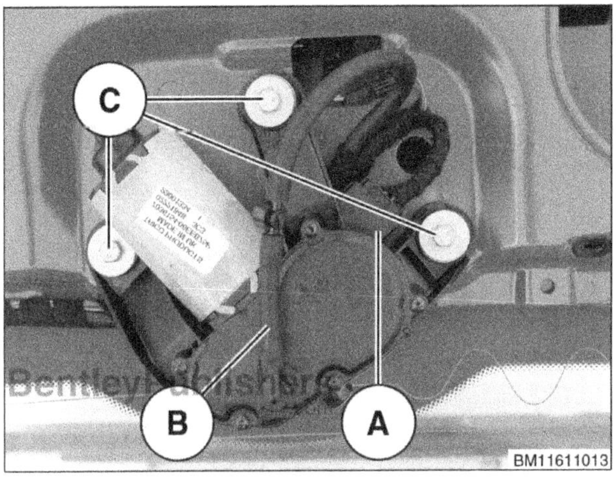

◄ Working at rear wiper motor assembly:

- Detach electrical connector (**A**).
- Detach washer fluid hose (**B**).
- Remove mounting bolts (**C**)
- Remove wiper motor.

– Installation is reverse of removal.

- Make sure wiper motor shaft and grommet are not damaged.
- Make sure wiper motor is in PARK before installing wiper arm. See **Rear window wiper arm, replacing** in this repair group.
- Adjust window washer nozzle. See **Rear window washer nozzle, adjusting**

Tightening torque	
Rear wiper motor to rear hatch	10 Nm (7 ft-lb)

Rear window wiper motor, removing and installing (Clubman)

Wiper motor removal procedures are similar for Clubman left and right split rear doors. The procedure below illustrates left split door wiper motor removal.

– Remove left rear wiper arm. See **Rear window wiper arm, replacing** in this repair group.

– Remove left split door inner trim panel. See **412 Rear Hatch, Split Rear Doors**.

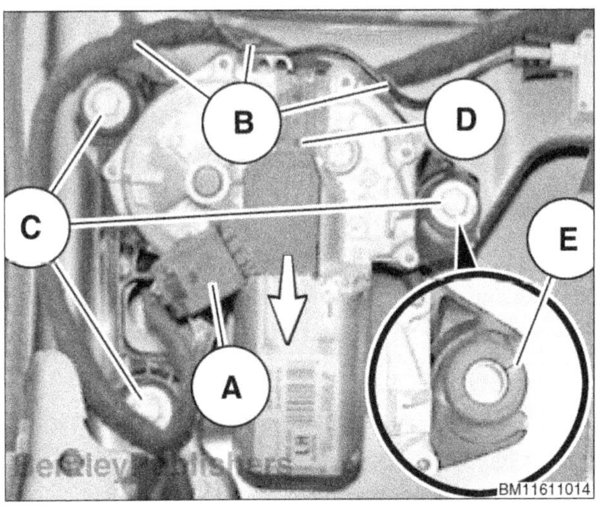

◄ Working at wiper motor assembly:

- Detach electrical connector (**A**) and release harness from bracket (**B**).
- Remove motor mounting bolts (**C**).
- Slide motor (**D**) out of door grommet (**arrow**).

– When installing:

- Make sure door grommet is undamaged. Replace if necessary.
- Make sure mounting bolt spacer sleeves (**E**) are seated correctly.
- Make sure wiper motor is in PARK before installing wiper arm. See **Rear window wiper arm, replacing** in this repair group.
- Adjust window washer nozzle. See **Rear window washer nozzle, adjusting**

Tightening torque	
Rear wiper motor to rear hatch	10 Nm (7 ft-lb)

611

Window and Headlight Washers

Windshield washer nozzle, replacing

– Open engine hood and remove sound insulation from under hood.

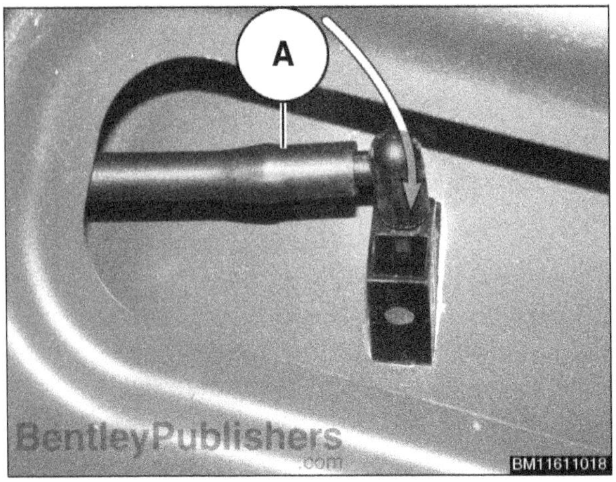

◄ Working at washer nozzle:

- Disconnect washer hose (**A**).
- Detach electrical connector for nozzle heater.
- Squeeze tab to release nozzle from hood (**arrow**).

– Installation is reverse of removal. To adjust nozzle, see **Windshield washer nozzle, adjusting** in this repair group.

Windshield washer nozzle, adjusting

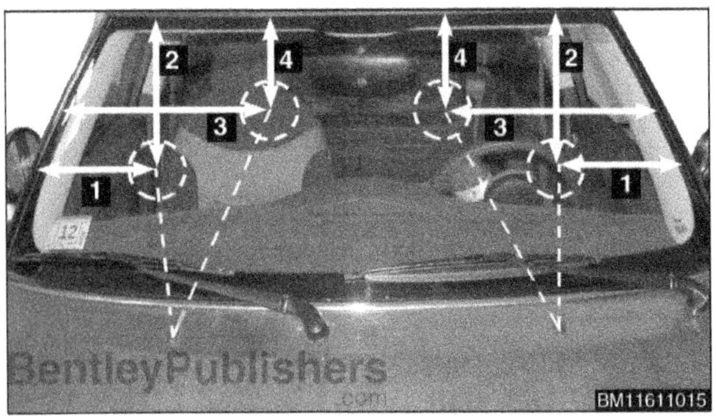

◄ Use MINI special tool 00 9 200 (or thin steel pin) to adjust washer nozzles so that spray contacts the windshield in accordance with the following values.

Measurement **1**	270 mm (10.6 in)
Measurement **2**	360 mm (14.2 in)
Measurement **3**	530 mm (20.9 in)
Measurement **4**	240 mm (9.4 in)

Rear window washer nozzle, replacing (Hardtop)

The Hardtop rear window washer nozzle is integrated with the wiper motor assembly.

Rear window washer nozzle, replacing (Clubman)

– Remove rear roof strip, detaching rear washer nozzle fluid hose in the process.

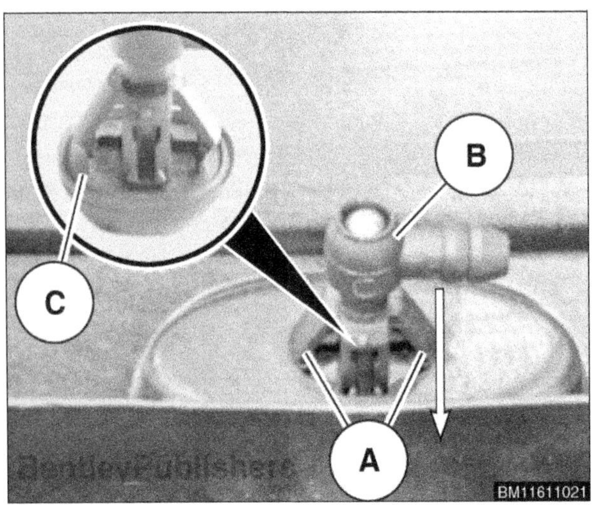

◄ With roof strip on working surface:

- Squeeze catches (**A**) and press nozzle (**B**) out of roof strip (**arrow**).
- When installing, make sure sealing O-ring is intact and installed correctly.

– Installation is reverse of removal. To adjust nozzle, see **Rear window washer nozzle, adjusting** in this repair group.

Rear window washer nozzle, adjusting

Hardtop:

- Make sure rear wiper is in PARK position.
- Adjust rear washer nozzle so that spray hits contact point.

Measurement **1**	490 mm (19.3 in)
Measurement **2**	55 mm (2.2 in)

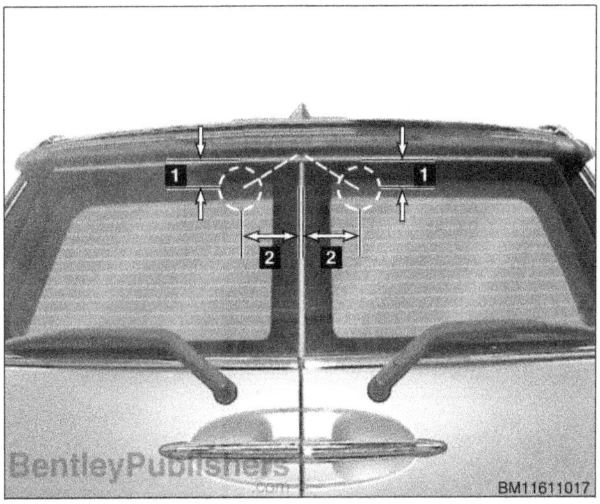

Clubman:

- Make sure rear wipers are in PARK position.
- Adjust rear washer nozzle so that spray hits contact points.

Measurement **1**	42 mm (1.7 in)
Measurement **2**	100 mm (3.9 in)

Adjust nozzle (**arrow**) using screwdriver in slot.

> ***CAUTION—***
> - *Do not insert pointed objects in washer nozzle.*

611

Headlight washer nozzle, replacing

- Remove headlight, detaching headlight washer fluid hose in the process. Be prepared to catch dripping washer fluid. See **630 Lights**.

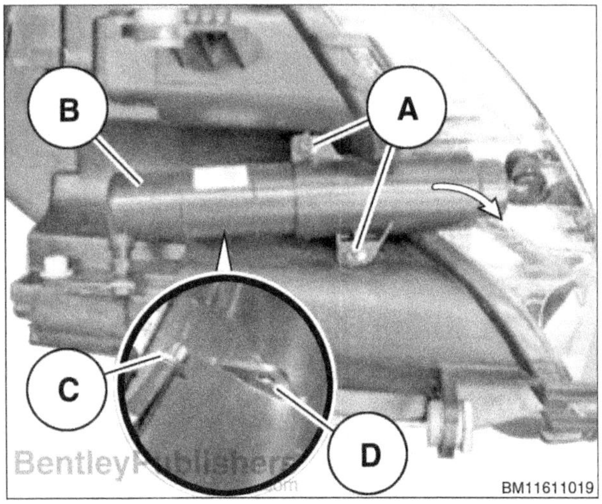

Working at side of headlight assembly:

- Remove washer nozzle mounting screws (**A**) and remove nozzle (**B**) in direction of **curved arrow**.
- Note that when correctly installed, guide (**C**) seats in lug (**D**).

Separate chrome trim from old washer nozzle by carefully prying catches off nozzle (**arrows**). Snap trim on new nozzle. Note that it is easy to break off plastic catches on trim or pins on nozzle.

- Installation is reverse of removal.

Tightening torque	
Washer nozzle to headlight	1.5 Nm (13 in-lb)

Washer fluid reservoir assembly

The washer fluid reservoir is underneath the left headlight. To gain access, remove left front wheel housing liner.

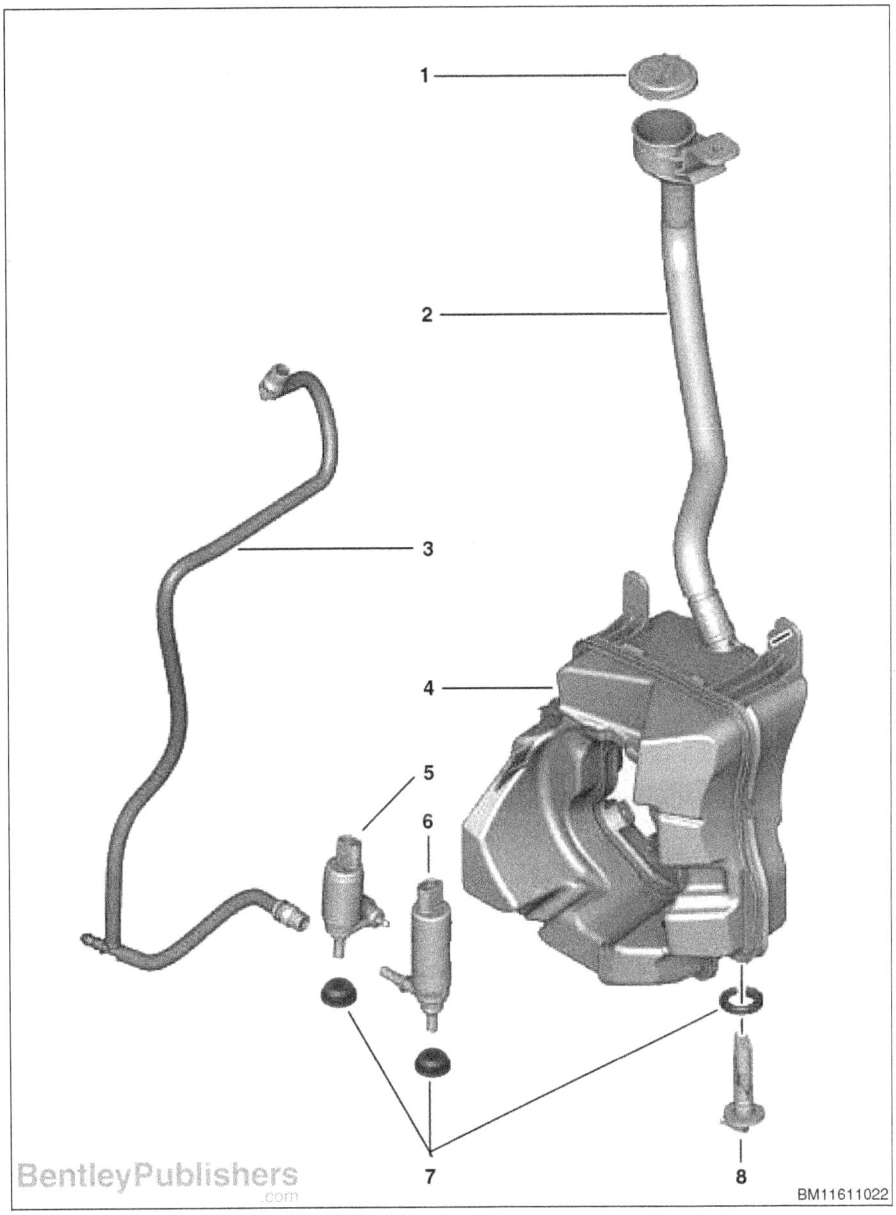

1. Washer fluid reservoir cap
2. Washer fluid reservoir filler neck
3. Fluid hose to headlight washers
4. Washer fluid reservoir
5. Dual washer pump (for windshield and rear window washers)
6. Headlight washer pump
7. Rubber grommets
8. Washer fluid level sensor

Washer pump, replacing

– Drain washer fluid reservoir and have a container ready to catch dripping fluid.

– Partially remove left front wheel housing liner. See **410 Fenders, Engine Hood**.

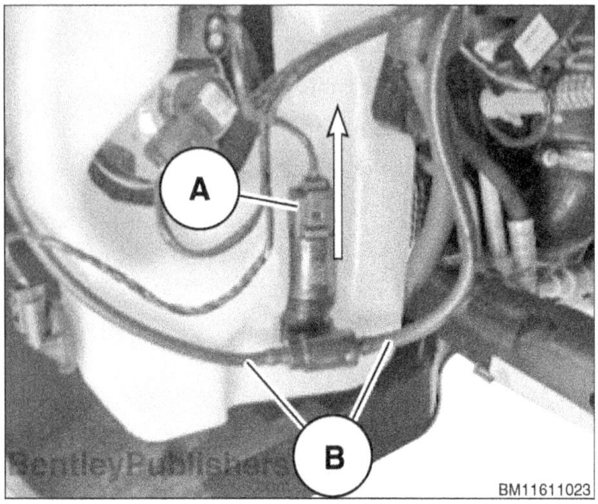

◀ Dual (windshield and rear window) washer pump:

- Detach electrical connector (**A**).
- Detach fluid hoses (**B**).
- Pull pump out of reservoir (**arrow**).

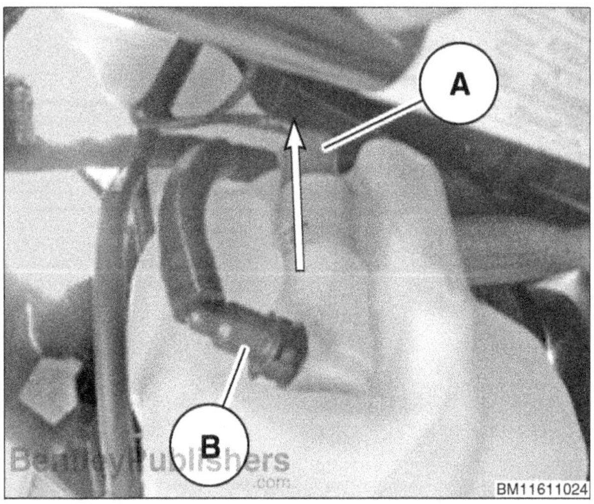

◀ Headlight washer pump:

- Detach electrical connector (**A**).
- Detach fluid hoses (**B**).
- Rotate pump clockwise and pull out of reservoir (**arrow**).

– Installation is reverse of removal. Remember to:

- Replace washer pump sealing grommet and strainer.
- Coat grommet with anti-seize paste.
- Make sure fluid hoses are not kinked.
- Refill washer fluid reservoir.

Washer fluid level sensor, replacing

- Drain washer fluid reservoir and have a container ready to catch dripping fluid.
- Partially remove left front wheel housing liner. See **410 Fenders, Engine Hood**.

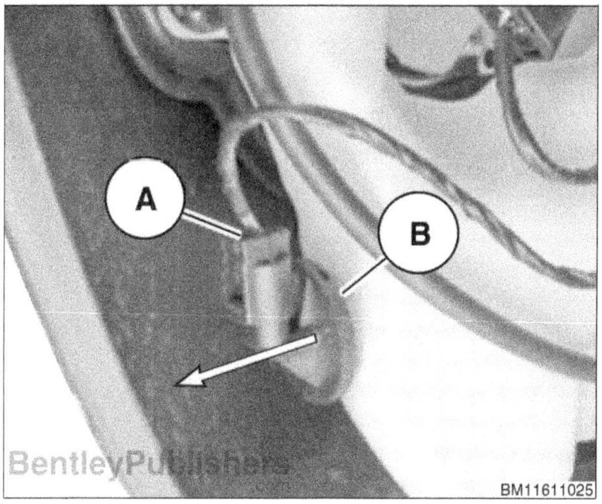

< Working at fluid level sensor:
- Detach electrical connector (**A**).
- Pull switch out of reservoir (**arrow**).
- Check sealing grommet (**B**). Replace of damaged.

- Installation is reverse of removal.
  - Install sealing grommet and coat with anti-seize compound before installing switch.
  - Refill washer fluid reservoir.

Washer fluid reservoir, replacing

- Drain washer fluid reservoir and have a container ready to catch dripping fluid.
- Partially remove left front wheel housing liner. See **410 Fenders, Engine Hood**.
- Remove left headlight. See **630 Lights**.

< Working in engine compartment, remove washer fluid reservoir retaining plastic rivet (**arrow**).

Washer fluid reservoir, replacing

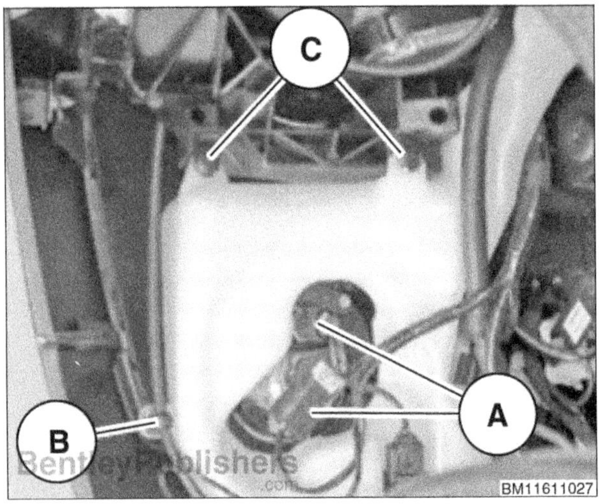

Working underneath left front fender at fluid reservoir:

- Detach electrical connectors (**A**).
- Remove reservoir retaining clip (**B**).
- Remove reservoir upper retaining screws (**C**).

– Remove windshield and headlight washer pumps. See **Washer pump, replacing** in this repair group.

– Remove washer fluid level sensor. See **Washer fluid level sensor, replacing** in this repair group.

– Use a suitable tool to carefully lever out washer fluid reservoir from base mount and remove from vehicle.

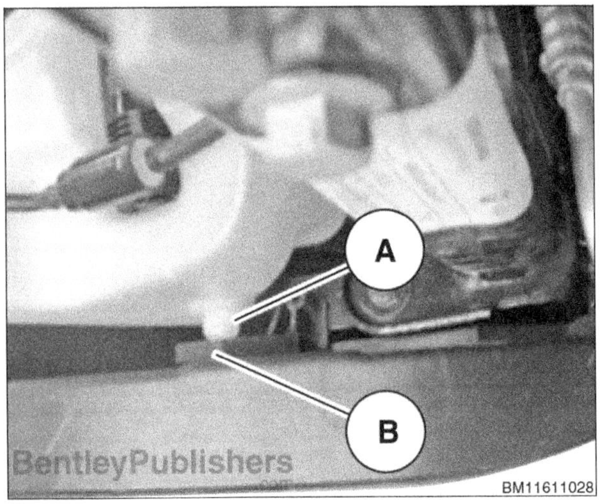

When installing:

- Make sure lower reservoir guide (**A**) is correctly seated in base mount (**B**).
- Replace washer pump sealing rings and strainers.
- Coat sealing rings with anti-seize paste.
- Make sure hoses are not kinked.
- Refill washer reservoir.

Tightening torque	
Washer reservoir to fender	4 Nm (3 ft-lb)

612 Switches

GENERAL

This repair group covers replacement of electrical switches at the steering wheel, steering column, dashboard, pedal cluster and front door.

See also:

- **230 Manual Transmission** for manual transmission back-up light switch.
- **240 Automatic Transmission** for automatic transmission back-up light switch function.
- **513 Interior Trim** for climate control panel and center dashboard switch cluster (SZL).
- **515 Central Locking and Anti-theft** for door and rear hatch lock switches.
- **630 Lights** for sunroof and convertible top switch.
- **721 Airbag System (SRS)** for horn button.
- **ECL Electrical Component Locations**

Switches overview

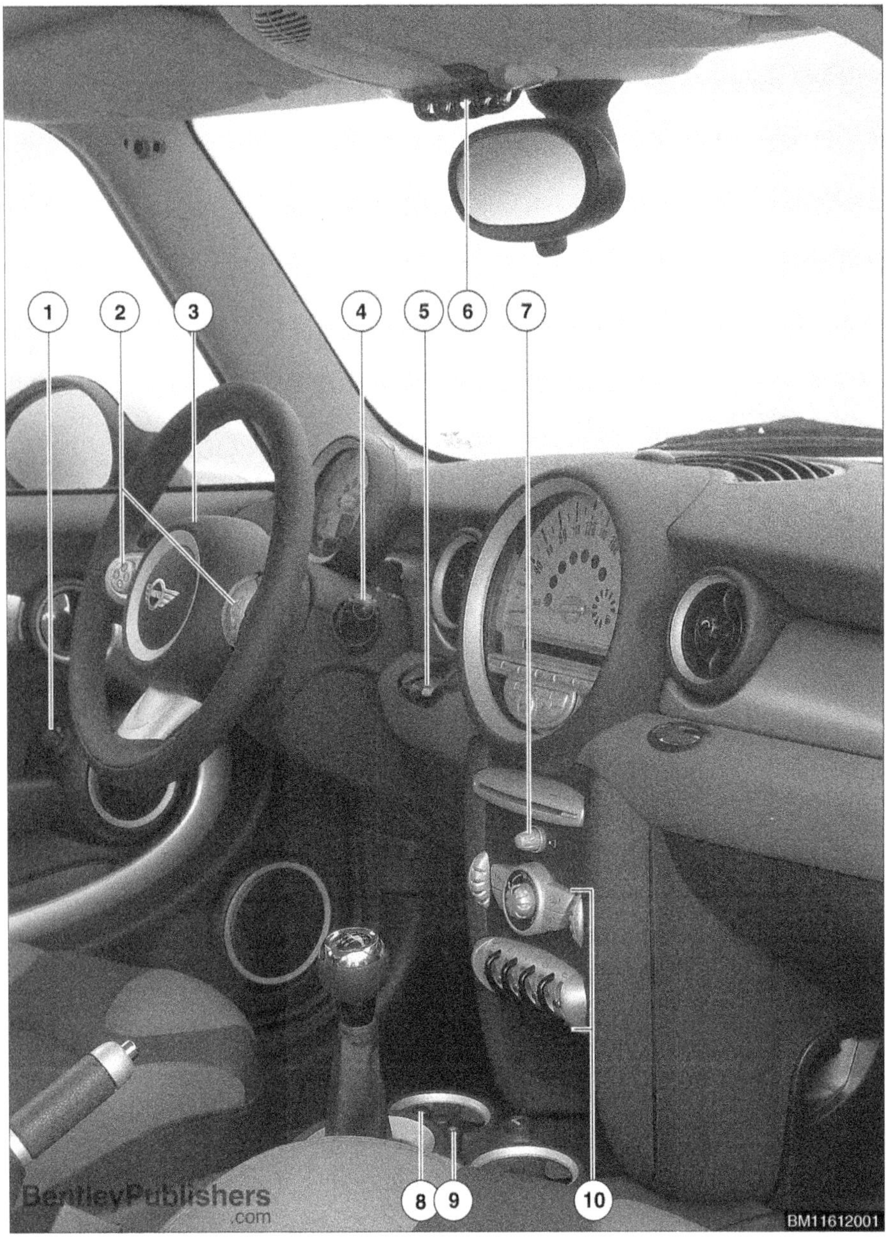

1. Outside mirror control switch
2. Steering wheel multifunction switches
3. Horn button, driver airbag
   - See **721 Airbag System (SRS)**
4. Wiper and washer switch
5. Ignition switch (remote key slot and START STOP button)
6. Roof switch center
   - Sunroof switch
   - Convertible top switch
   - Domelight switches
7. Volume control
8. Sport switch
9. DSC / DTC switch
10. Center dashboard switch cluster
    - Climate control panel
    - Seat heater switches
    - Power window switches
    - Central locking switch
    - Foglight switch
    - See **Center dashboard covers, removing** in **513 Interior Trim**

Cautions

CAUTION—

- *When working on electrical components, disconnect the negative (-) cable from the battery. Before disconnecting the battery, read battery disconnection cautions in* **001 Warnings and Cautions**.
- *Handle electrical modules and other electrical and electronic components using static prevention equipment and techniques. See* **600 Electrical System–General**.
- *When removing interior trim clips, avoid damaging trim pieces by using a plastic prying tool or a screwdriver with the tip wrapped with masking tape.*

Steering Wheel Switches

Multifunction steering wheel switches include:

- Horn button and airbag contact switch
- Cruise control switches
- Sound system controls

The steering wheel may also be equipped with automatic transmission shifter paddles.

Multifunction steering wheel switches, removing and installing

– Disconnect negative (-) cable from battery.

> *CAUTION—*
> - *Prior to disconnecting the battery, read the battery disconnection cautions given in* **001 Warnings and Cautions**.

– Remove driver airbag in steering wheel. See **721 Airbag System (SRS)**.

> *WARNING—*
> - *The steering-wheel-mounted airbag is an explosive device. Treat with extreme caution. Follow the airbag removal procedure in* **721 Airbag System (SRS)**.
> - *Airbag removal may set SRS fault codes. The SRS indicator light remains ON until problems are corrected and fault memory cleared.*

– Remove shift paddles from steering wheel. See **Steering wheel shift paddles, removing and installing** in this repair group.

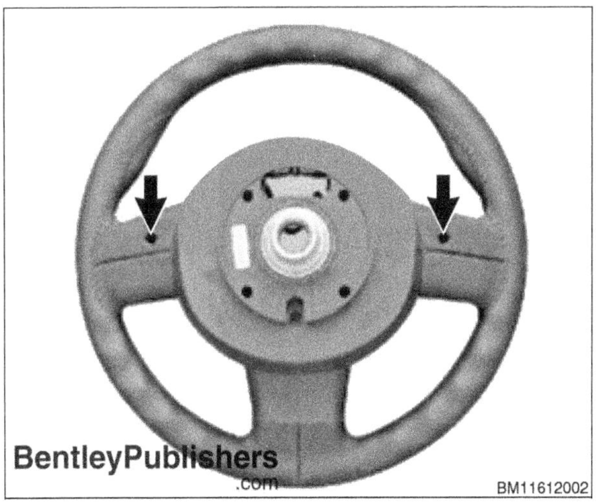

< Working behind left or right steering wheel spoke, remove multifunction switch mounting screw (**arrows**).

- Steering wheel shown removed for illustration.
- Left and right switches can be replaced independently.

Pull left or right switch off steering wheel (**arrows**).

Detach switch assembly wire harnesses (**arrows**) to remove switch.

– Installation is reverse of removal. Make sure switch harnesses are routed as before.

Steering wheel shift paddles, removing and installing

– Disconnect negative (-) cable from battery.

> ***CAUTION—***
> - *Prior to disconnecting the battery, read the battery disconnection cautions given in* **001 Warnings and Cautions**.

– Remove driver airbag in steering wheel. See **721 Airbag System (SRS)**.

> ***WARNING—***
> - *The steering-wheel-mounted airbag is an explosive device. Treat with extreme caution. Follow the airbag removal procedure in* **721 Airbag System (SRS)**.
> - *Airbag removal may set SRS fault codes. The SRS indicator light remains ON until problems are corrected and fault memory cleared.*

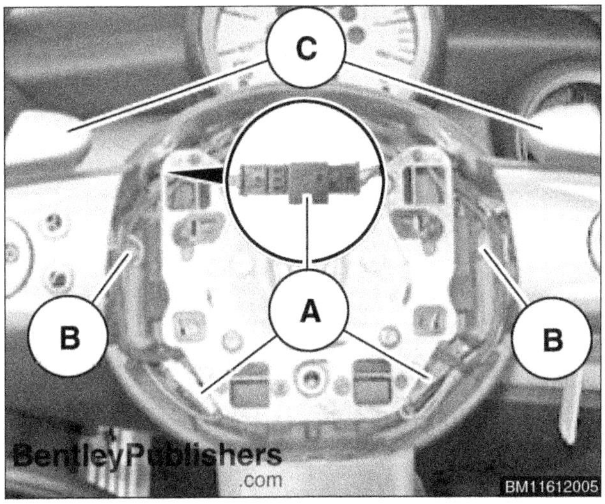

- Working at center of steering wheel:
  - Detach electrical connectors (**A**).
  - Remove shift paddle mounting screws (**B**).
  - Lift off shift paddles (**C**), carefully threading electrical harnesses out of steering wheel.

– Installation is reverse of removal. Make sure electrical harnesses are routed as before.

Steering Column Switches

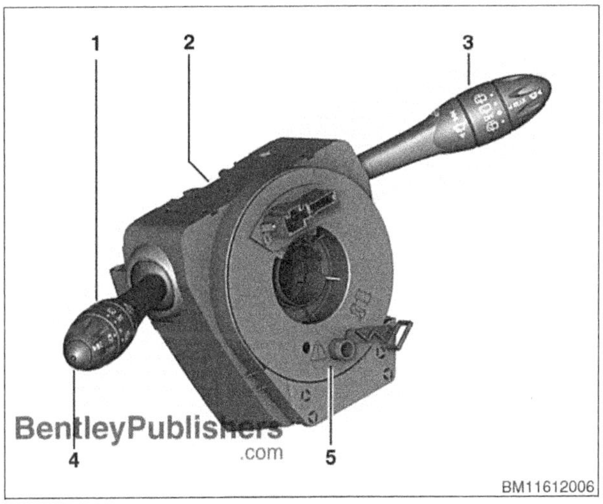

- The steering column switch cluster (SZL) contains the following integrated components:

1. Exterior lights control switch
   - Headlights, taillights
   - Headlight flasher
   - High beam switch
2. Steering angle sensor (if applicable)
3. Wiper and washer switch
   - Windshield wipers and washers
   - Rear window wiper(s) and washer(s)
   - Headlight washers
   - Intermittent wipe switch
4. On-board computer setting switch
5. Horn and driver airbag spring contact (volute spring)

These components are not available separately. In case of switch failure, replace SZL as a unit.

Steering column switch cluster (SZL), removing and installing

– Center steering wheel. Make sure front wheels are pointed straight ahead. During remainder of this procedure, do not alter position of steering column or front wheels.

– Extend steering wheel as far back and as low as possible.

– Disconnect negative (-) cable from battery.

> ***CAUTION—***
> - *Prior to disconnecting the battery, read the battery disconnection cautions given in* **001 Warnings and Cautions**.

612

– Remove driver airbag in steering wheel. See **721 Airbag System (SRS)**.

> ***WARNING —***
> - *The steering-wheel-mounted airbag is an explosive device. Treat with extreme caution. Follow the airbag removal procedure in* **721 Airbag System (SRS)**.
> - *Airbag removal may set SRS fault codes. The SRS indicator light remains ON until problems are corrected and fault memory cleared.*

– Remove steering wheel. See **320 Steering and Wheel Alignment**.

– Remove left lower dashboard trim (underneath steering column). See **513 Interior Trim**.

– Remove tachometer. See **620 Instruments**.

– Remove steering column upper and lower trim. See **513 Interior Trim**.

◄ Use adhesive tape (**arrow**) to prevent horn and airbag contact spring from rotating.

> ***CAUTION—***
> - *Do not allow airbag contact spring to rotate while steering wheel is off the steering column. Do so will damage the coil spring inside.*

◄ Working underneath steering column, detach SZL electrical connectors (**arrows**).

◀ Working at right side of SZL, loosen pinch collar screw (**arrow**).

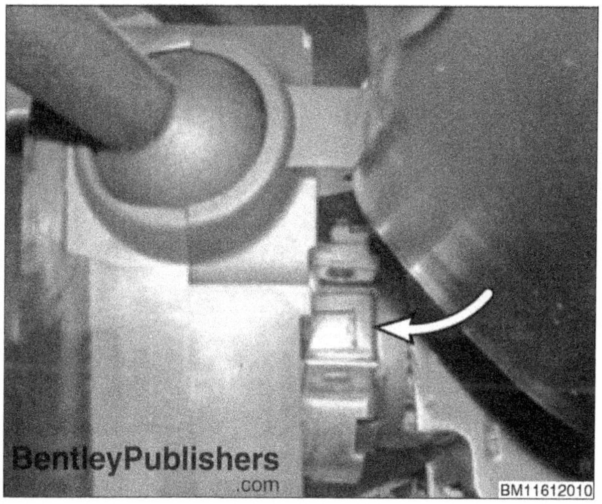

◀ Pry plastic tab carefully (**arrow**) on each side of steering column to disengage SZL. Slide SZL off column.

◀ When installing, check to make sure plastic tabs (**arrows**) are undamaged.

If horn and airbag contact spring is rotated for any reason, reset it to center position as follows:

- Rotate counterclockwise as far as it will go.
- Rotate clockwise as far as it will go. Count revolutions.
- Rotate back (counterclockwise) ½ number of revolutions, until centering pin (**straight arrow**) is at bottom center.
- Secure in this position with masking tape.

– Remainder of installation is reverse of removal.

Tightening torque	
Steering wheel to steering column shaft (M14 x 1.5 bolt)	63 Nm (46 ft-lb)

Vehicle with DSC: Reset steering angle sensor. Connect vehicle to MINI scan tool and carry out steering angle sensor adjustment system function.

DASHBOARD SWITCHES

The following dashboard switches are covered in this section:

- Ignition switch (remote key slot and START STOP button) at right of steering column.
- Hazard warning switch at top center of dashboard.
- Passenger airbag deactivation switch behind cover at right end of dashboard.

The switch cluster ganged in the center dashboard is covered in **640 Heating and Air-conditioning**. This cluster includes:

- Climate control panel.
- Seat heater switches.
- Power window switches.
- Foglight switch(es).
- Central locking switch.

START STOP unit, removing and installing

START STOP unit is the ignition switch.

– Disconnect negative (-) cable from battery.

> *CAUTION—*
> - *Prior to disconnecting the battery, read the battery disconnection cautions given in* **001 Warnings and Cautions**.

– Remove left lower dashboard trim (underneath steering column). See **513 Interior Trim**.

– Remove tachometer. See **620 Instruments**.

– Remove left dashboard decorative strip and center instrument bezel. See **513 Interior Trim**.

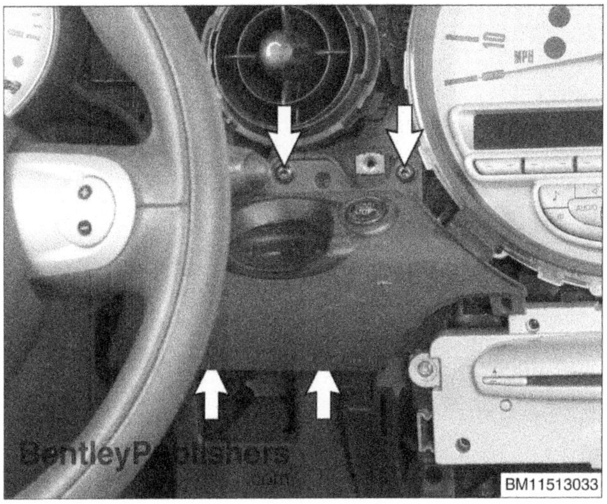

Working at right side of steering wheel, remove knee pad fasteners (**arrows**).

Disconnect STOP START unit ribbon cable (**arrow**), then remove knee pad (**A**).

Use thin plastic prying tool (such as credit card) to press in ignition switch cover trim locking tabs (**arrows**). Remove trim.

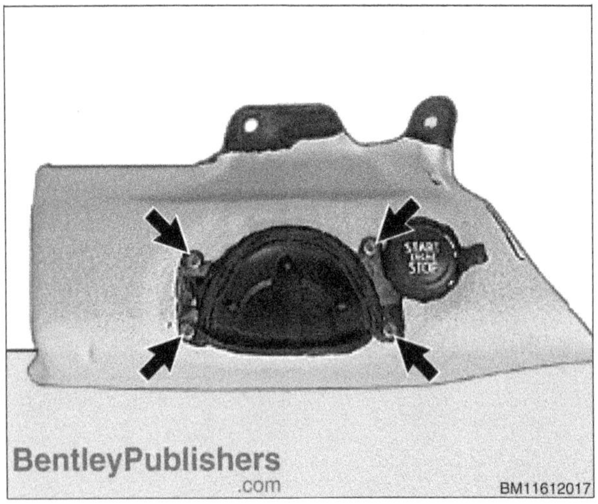

BM11612017

- Remove switch mounting screws (**arrows**) and slide switch off knee pad.

BM11612018

- When installing, make sure switch assembly locking tabs (**arrows**) are undamaged and click firmly in place.

Hazard warning switch, removing and installing

– Disconnect negative (-) cable from battery.

> *CAUTION—*
> - *Prior to disconnecting the battery, read the battery disconnection cautions given in* **001 Warnings and Cautions**.

– Remove left lower dashboard trim (underneath steering column). See **513 Interior Trim**.

– Remove tachometer. See **620 Instruments**.

– Remove left and right dashboard decorative strips and dashboard cover. See **513 Interior Trim**.

BM11612019

- Press switch out of dashboard cover (**arrow**).

– Installation is reverse of removal.

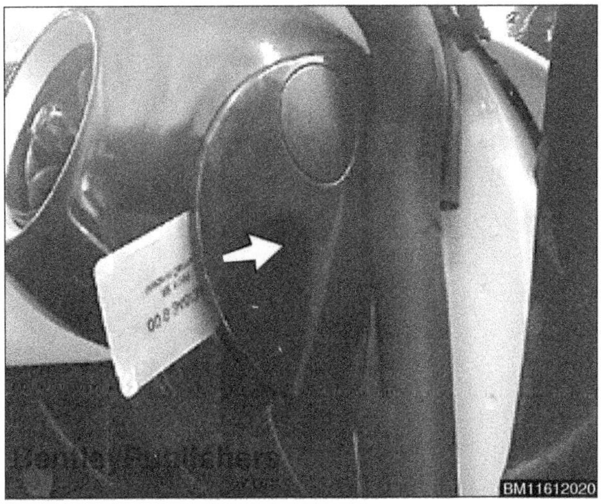

Passenger airbag deactivation switch, removing and installing

> ***WARNING—***
> - *The dashboard-mounted passenger airbag is an explosive device. Treat with extreme caution. See airbag warnings and cautions in* **721 Airbag System (SRS)**.

– Disconnect negative (-) cable from battery.

> ***CAUTION—***
> - *Prior to disconnecting the battery, read the battery disconnection cautions given in* **001 Warnings and Cautions**.

< Use thin plastic prying tool (such as credit card) to unclip (**arrow**) right side dashboard end cover and detach airbag deactivation switch electrical (yellow) connector.

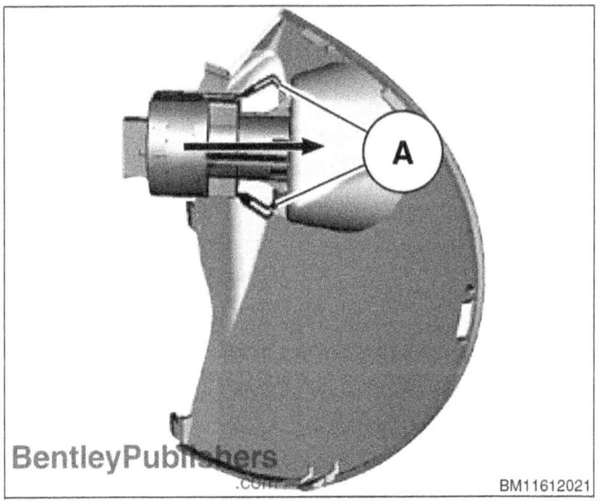

< Working at end cover:

- Squeeze switch catches (**A**).
- Slide switch (**arrow**) out of end cover.

– Installation is reverse of removal.

Roof Switch Center

Dome lights, front domelight switches and sunroof switch are in the roof switch center. Individual switches are not available separately.

Roof switch center, removing

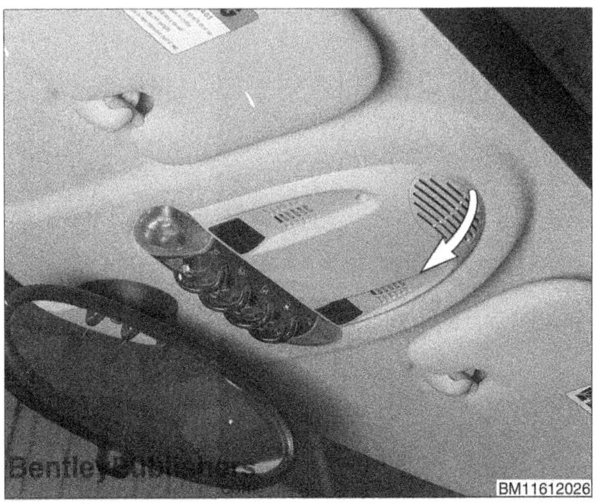

< Use plastic prying tool to lever switch center cover off headliner (**arrow**).

- If necessary, detach hand free microphone electrical connector.

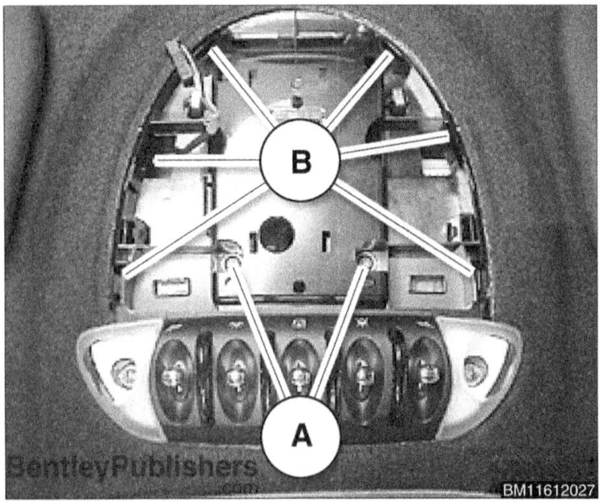

- Working at roof switch center:
  - Remove mounting screws (**A**).
  - Press catches (**B**) to release switch center.
  - Lower switch center and detach electrical connector.
- Installation is reverse of removal.

Pedal Cluster Switches

Brake light switch, removing and installing

The brake light switch is above the brake pedal.

- Remove left lower dashboard trim (underneath steering column). See **513 Interior Trim**.

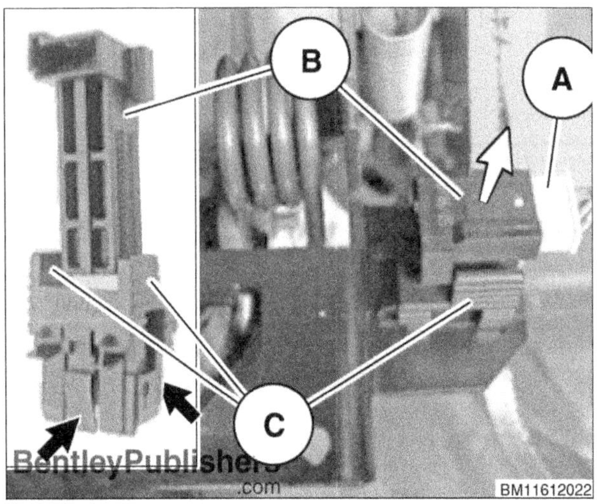

- Working above pedal cluster:
  - Disconnect plug connection (**A**).
  - Pull brake light switch (**B**) in direction of **flared arrow** out of brake light switch holder (**C**).
  - squeeze catches (**arrows**) and unclip brake light switch holder (**C**) from brake pedal.
- When installing:
  - Depress brake pedal.
  - Slide brake light switch (**B**) as far as it will go into brake light switch holder (**C**).
  - Grip brake light switch holder (**C**) and slowly return brake pedal to starting position, then pull back to stop.

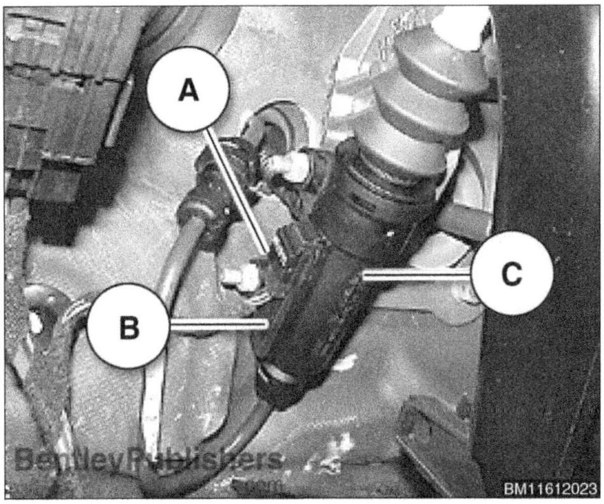

Clutch switch module, removing and installing

Where applicable, the clutch switch module is above the clutch pedal.

– Remove left lower dashboard trim (underneath steering column). See **513 Interior Trim**.

◄ Working above pedal cluster:

- Detach switch electrical connector (**A**).
- Use small screwdriver to lever out clutch switch module (**B**) from clutch master cylinder (**C**).

– When installing, note that switch module can only be installed one way. Make sure it snaps audibly into clutch master cylinder.

OUTSIDE MIRROR SWITCH

Outside mirror switch, removing and installing

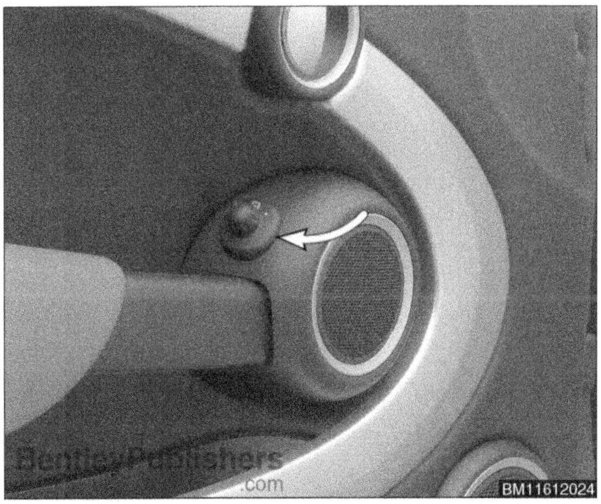

◄ Use plastic prying tool to pry outside mirror switch out of door trim panel (**arrow**).

◄ Pull switch back and detach electrical connectors (**arrows**).

– Installation is reverse of removal.

620 Instruments

GENERAL

This repair group covers removal and installation of the instrument cluster and tachometer.

See **020 Maintenance** for an explanation of condition based service (CBS) and Check Control, both of which use instrument cluster displays.

Instruments overview

BM11620005

MINI models are equipped with a tachometer with indicator and warning lights mounted on the steering column and a speedometer with indicator and warning lights in the center of the dashboard.

BM11620001

Convertible models are additionally equipped with the convertible top timer ("always open" gauge). Fitted left of the tachometer, the timer displays the time in hours and minutes that the soft top is down while the engine is running. This counter resets whenever the engine is shut off or the top is closed. It displays top-open time up to 7 hours.

620

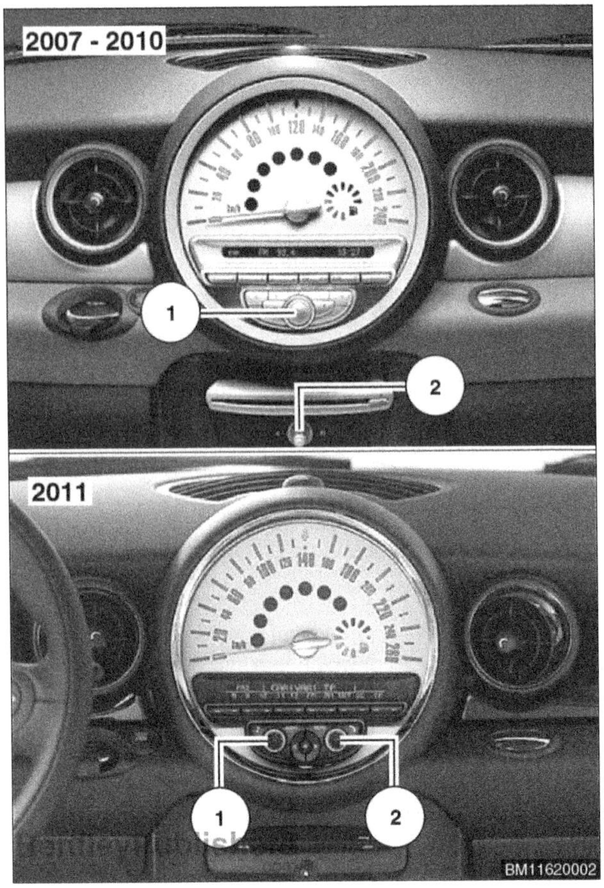

Beginning with 2011 LCI models, the center dashboard cluster and its control knobs are slightly reconfigured.

1. Menu selection
2. Volume control

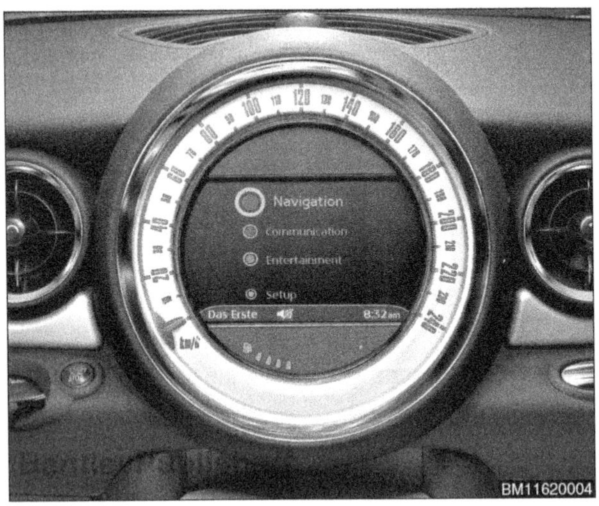

In MINI models equipped with the optional car communication computer and joystick, the center dashboard instrument (central information display or CID) shows vehicle speed, radio and sound system settings and navigation information.

Cautions

> ***CAUTION—***
> - *When working on electrical components, disconnect the negative (-) cable from the battery. Before disconnecting the battery, read battery disconnection cautions in* **001 Warnings and Cautions**.
> - *Static discharge may permanently damage an electronic component such as tachometer or instrument cluster. Handle electronic components using accepted static prevention equipment and techniques. See* **600 Electrical System–General**.
> - *When removing interior trim clips, avoid damaging trim pieces by using a plastic prying tool or a screwdriver with the tip wrapped with masking tape.*

TACHOMETER

The tachometer is an analog gauge mounted on top of the steering column. It also includes the following:

1. LCD display:
   - Vehicle speed
   - Indicator and warning lights
2. Instrument lighting control
3. LCD display:
   - Outside temperature
   - Automatic transmission gear range
   - On-board computer
   - Scheduled service reminder
   - Odometer, trip odometer
   - Tire pressure monitor initialization
   - Settings and information
   - Personal profile settings
4. Trip odometer reset

Tachometer, removing and installing

– Disconnect negative (-) cable from battery.

> ***CAUTION—***
> - *Prior to disconnecting the battery, read the battery disconnection cautions given in* **001 Warnings and Cautions**.

– Remove left lower dashboard trim (underneath steering column). See **513 Interior Trim**.

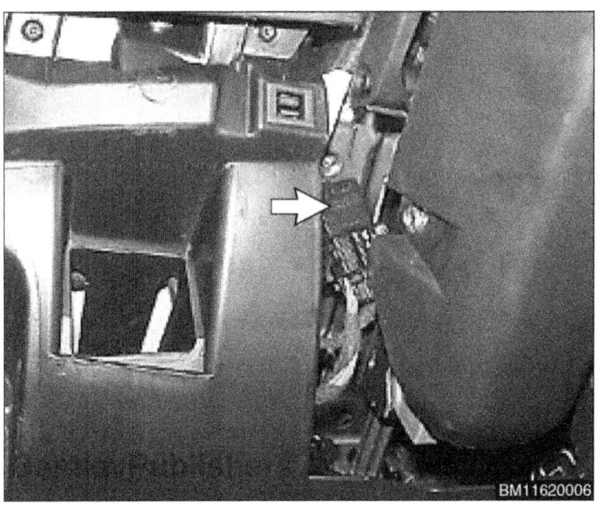

Separate tachometer electrical connector (**arrow**) and detach from bracket underneath steering column.

620

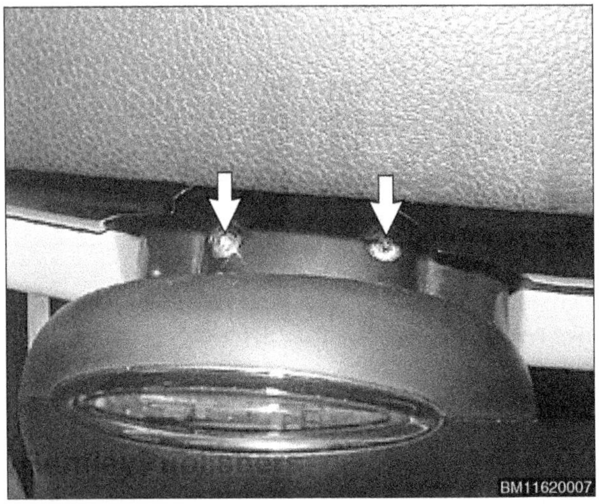

◄ Remove tachometer mounting screws and lift out tachometer, pulling connector harness carefully away from dashboard.

– Installation is reverse of removal. Make sure connector harness is routed as before.

CENTER INSTRUMENT

◄ The basic center dashboard instrument cluster is an analog speedometer which displays various warnings and indicators and includes the following:

1. Radio display
2. Radio controls
3. Fuel gauge

If equipped with car communication computer (CCC) or navigation, the center dashboard instrument is called central information display (CID).

Removal and installation of the two cluster types is similar.

Center instrument, removing and installing

– Disconnect negative (-) cable from battery.

> *CAUTION—*
> • *Prior to disconnecting the battery, read the battery disconnection cautions given in* **001 Warnings and Cautions**.

– Remove left lower dashboard trim (underneath steering column). See **513 Interior Trim**.

– Remove tachometer. See **Tachometer, removing and installing** in this repair group.

– Remove center instrument bezel and center dashboard covers. See **513 Interior Trim**.

– Partially remove passenger knee pad. See **513 Interior Trim**.

– Remove driver knee pad. See **513 Interior Trim**.

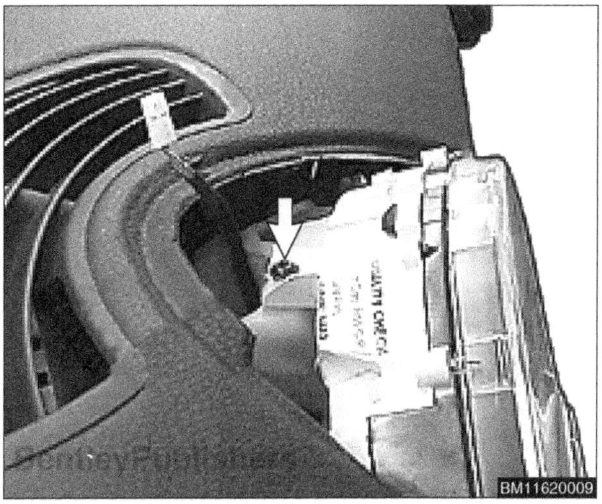

BM11620009

◀ Remove upper instrument cluster mounting screw (**arrow**).

BM11620010

◀ Remove lower cluster mounting screws (**arrows**).

- Pull back cluster, detach electrical connectors and remove cluster.

BM11620012

◀ To remove speedometer: Remove mounting screws (**arrows**) and separate speedometer from radio.

- Installation is reverse of removal.

Central information display (CID), removing and installing

– Disconnect negative (-) cable from battery.

> *CAUTION—*
> • *Prior to disconnecting the battery, read the battery disconnection cautions given in* **001 Warnings and Cautions**.

– Remove left lower dashboard trim (underneath steering column). See **513 Interior Trim**.

– Remove tachometer. See **Tachometer, removing and installing** in this repair group.

– Remove center instrument bezel and center dashboard covers. See **513 Interior Trim**.

– Partially remove passenger knee pad. See **513 Interior Trim**.

– Remove driver knee pad. See **513 Interior Trim**.

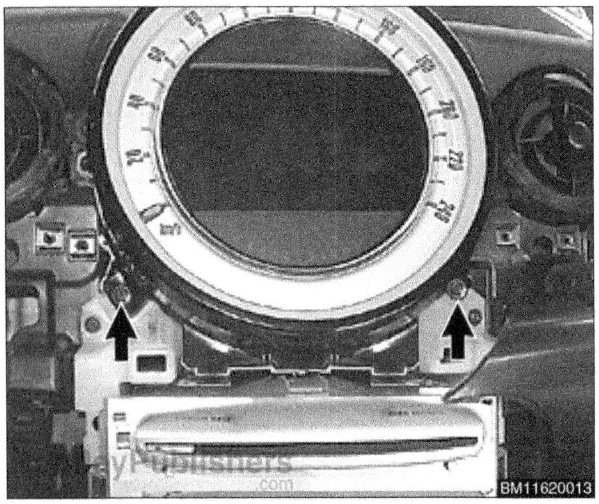

< Remove lower CID mounting screws (**arrows**).

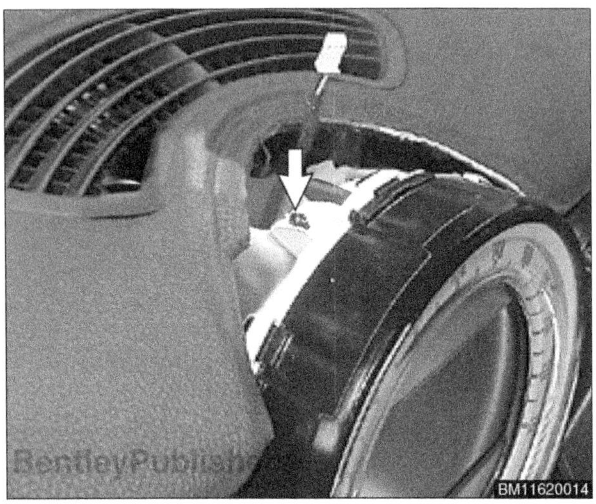

< Remove upper CID mounting screw (**arrow**).

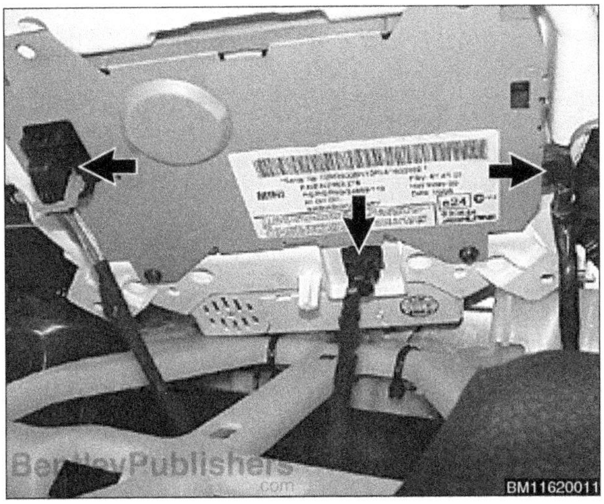

Pull back cluster, detach electrical connectors (**arrows**) and remove cluster.

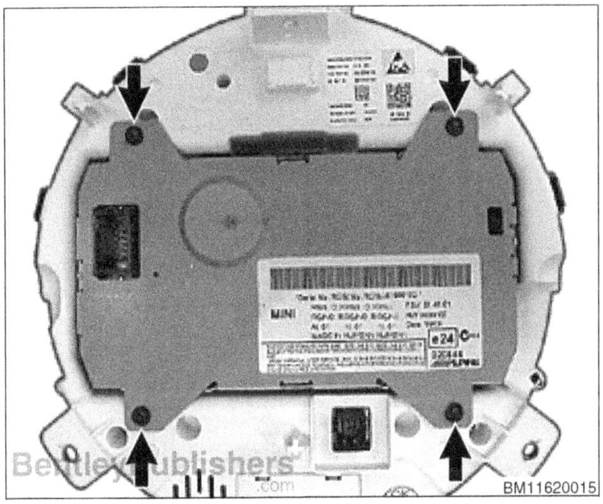

To remove CID: Remove mounting screws (**arrows**) and separate display from cluster.

– Installation is reverse of removal.

630 Lights

TABLE

General

This repair group covers exterior and interior lighting repairs and light bulb replacement.

The exterior light switch is a component of the steering column switch cluster (SZL). SZL replacement is covered in **612 Switches**.

Exterior lights overview

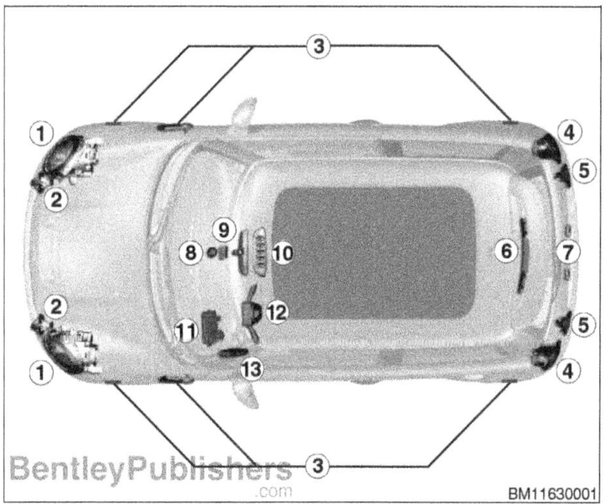

The footwell module (FRM) supplies power to and monitors exterior bulbs.

1. Headlight (standard halogen, optional bi-xenon)
2. Foglight
3. Side marker and turn signal lights
4. Taillight
5. Rear foglight
6. Center brake light
7. License plate light
8. Hazard warning switch
9. Rain and light sensor
10. Center dashboard switch cluster
11. Car access system (CAS)
12. Steering column switch cluster (SZL)
13. Footwell module (FRM)

630

Warnings and Cautions

WARNING—
- *There is high pressure inside a halogen bulb. Bulb temperature may exceed 700° Celsius. Wear safety glasses and gloves when removing and installing a halogen bulb.*
- *Do not look directly at an operating xenon bulb. The UV emissions are approximately 2.5 times that of a comparable halogen bulb.*
- *Some light bulbs, especially headlight bulbs, can become hot enough to cause injury. Allow bulbs to cool before working on the lighting system.*
- *Xenon bulbs contain mercury. Dispose of as hazardous waste.*

CAUTION—
- *Xenon bulbs and modules are electrically sensitive. Use an antistatic mat and work with caution. See* **600 Electrical System–General**.
- *Do not operate the xenon control module unless a bulb is connected.*
- *Before servicing the headlight system, switch electrical consumers OFF. Switch ignition OFF and remove ignition key.*
- *Do not handle bulb glass with bare fingers. Dirt and skin oils cause a bulb to fail prematurely. If necessary, wipe bulb using a clean cloth dampened with rubbing alcohol.*
- *When working on electrical systems, remove the fuse protecting the circuit under repair. See* **ECL Electrical Component Locations** *for fuse locations.*
- *Use only original equipment replacement bulbs. Non-original bulbs may cause false failure readings in Check Control.*
- *To avoid marring car paint or trim, work with plastic prying tools or wrap the tips of tools with tape.*
- *Only use a digital multimeter for electrical tests.*

HEADLIGHTS

Headlight components

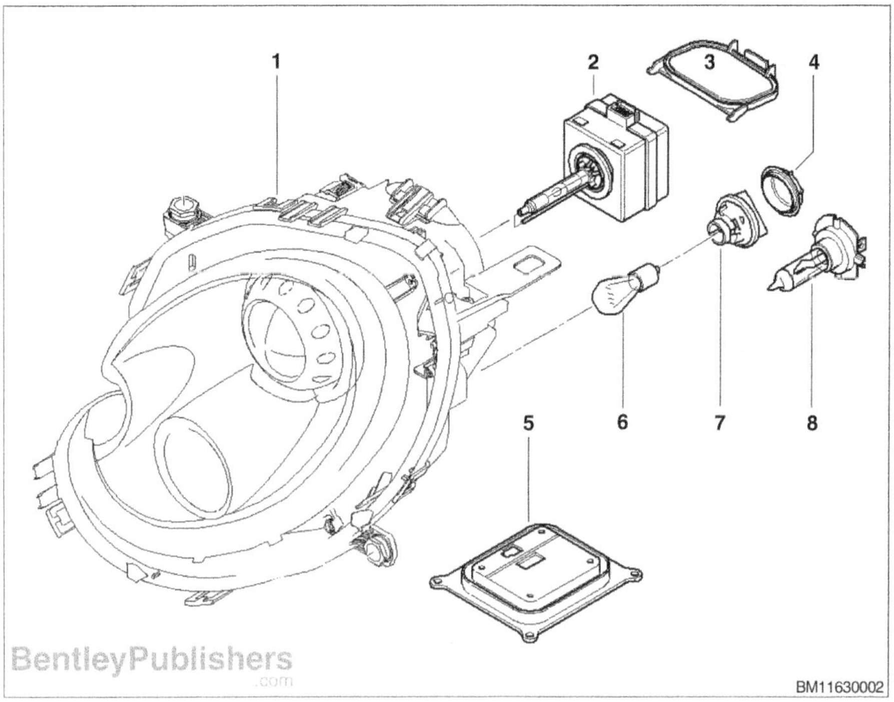

1. Headlight housing
2. D1S xenon bulb and igniter (optional)
3. Headlight bulb cover
4. Turn signal bulb cover
5. Xenon control module (optional)
6. 12V 21W turn signal bulb
7. Turn signal bulb socket
8. H13 halogen bulb

Headlight, adjusting aim

– Before adjusting headlight aim:
  - Fill fuel tank.
  - Correct tire pressures.
  - Replace faulty bulbs.
  - Park vehicle on level ground.
  - Load driver seat with approx. 75 kg (165 lb).
  - Xenon lights: Switch lights ON, wait approx. 80 seconds for self-leveling to stabilize.

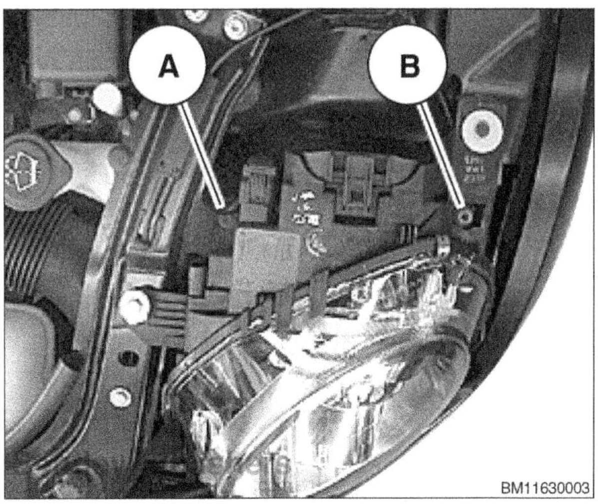

With engine hood open, rotate screws **A** and **B** to adjust headlight beam.
  - **A** = primarily vertical adjustment.
  - **B** = primarily horizontal adjustment.

Note that turning each adjuster affects both planes of adjustment.

630

Halogen headlight bulb, replacing

– Switch lights OFF.

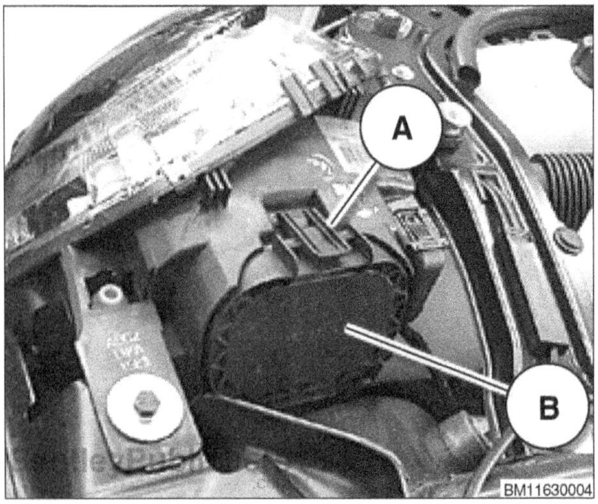

◄ With engine hood open, unlock headlight bulb cover retaining tab (**A**) and remove cover (**B**).

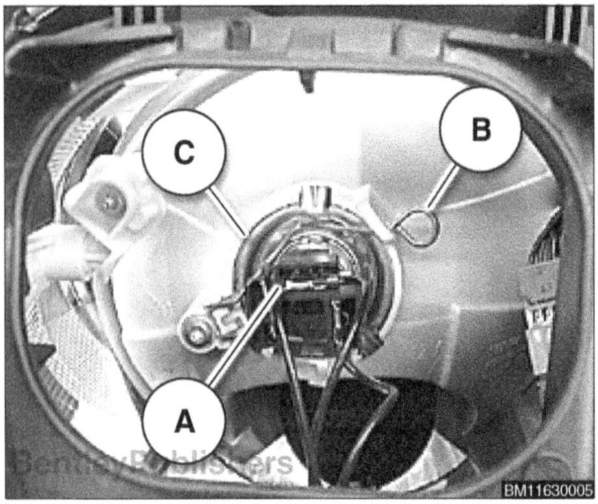

◄ Working inside rear of headlight housing:

- Pull off electrical connector (**A**).
- Unhook and remove bulb retaining spring (**B**) and pull halogen bulb (**C**) out of headlight.

– Installation is reverse of removal.

- Do not touch bulb surface with bare fingers. Oil from skin can cause bulb to fail prematurely.
- Make sure bulb is seated firmly and correctly in headlight housing.
- Make sure bulb cover is firmly seated and well-sealed on headlight. Replace cover if seal is damaged.
- Check headlight beam alignment. See **Headlight, adjusting aim** in this repair group.

Headlight bulb application	
Halogen bulb	H13

Xenon headlight bulb, replacing

The xenon bulb and xenon bulb igniter are an integrated unit. Do not attempt to separate.

– Switch lights OFF. Disconnect negative (-) battery cable and cover battery terminal to keep cable from accidentally contacting terminal.

> ***CAUTION—***
> - *Prior to disconnecting the battery, read the battery disconnection cautions given in* **001 Warnings and Cautions**.

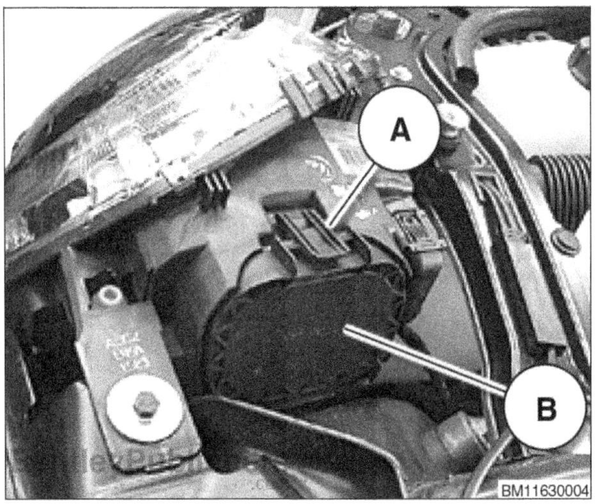

◄ With engine hood open, unlock headlight bulb cover retaining tab (**A**) and remove cover (**B**).

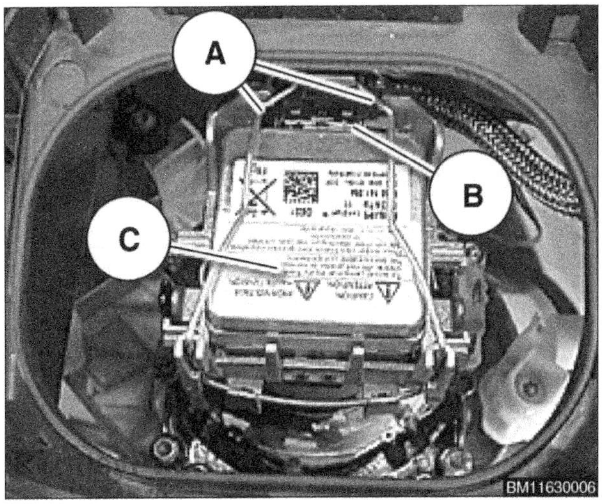

◀ Working inside rear of headlight housing:

- Unhook and remove bulb retaining spring (**A**).
- Pull off electrical connector (**B**).
- Pull igniter with halogen bulb (**C**) out of headlight.

– Installation is reverse of removal.

- Do not touch bulb surface with bare fingers. Oil from skin can cause bulb to fail prematurely.
- Make sure bulb and igniter are seated firmly and correctly in headlight housing.
- Make sure cover is firmly seated and well-sealed on headlight. Replace cover if seal is damaged.
- Check headlight beam alignment. See **Headlight, adjusting aim** in this repair group.

Headlight bulb application	
Xenon bulb	D1S

Xenon control module, removing and installing

– Switch lights OFF. Disconnect negative (-) battery cable and cover battery terminal to keep cable from accidentally contacting terminal.

> ***CAUTION–***
> - *Prior to disconnecting the battery, read the battery disconnection cautions given in* **001 Warnings and Cautions**.

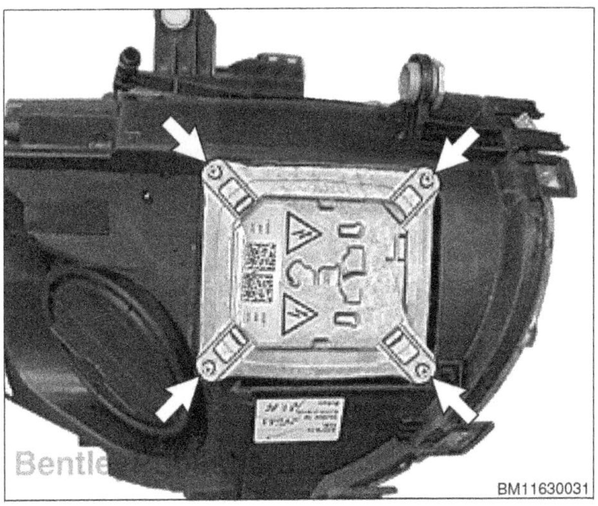

– Remove headlight assembly. See **Headlight, removing and installing** in this repair group.

◀ Remove xenon control module mounting screws (**arrows**) and tilt module out of headlight.

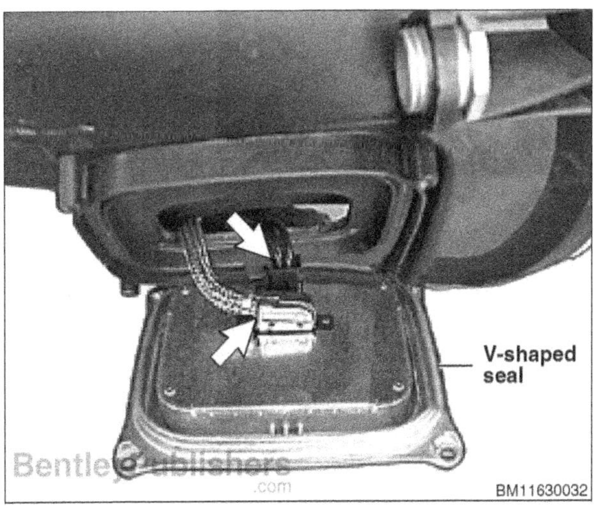

◀ Detach electrical connectors (**arrows**) to remove module.

– When installing, make sure V-shaped seal seats fully in headlight housing.

Tightening torque	
Xenon control module to headlight	1.0 ± 0.3 Nm (8.9 ± 2.7 in-lb)

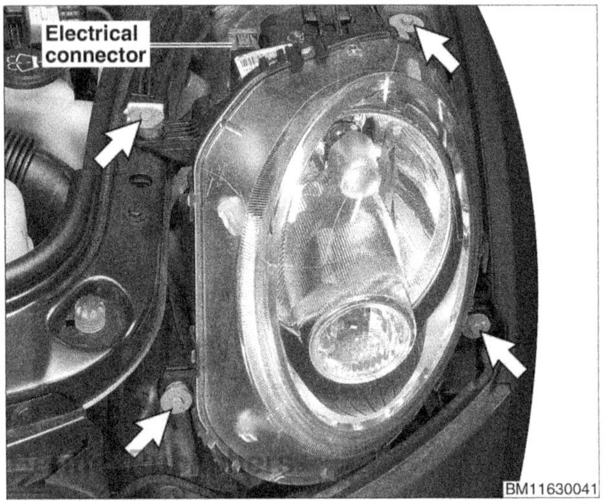

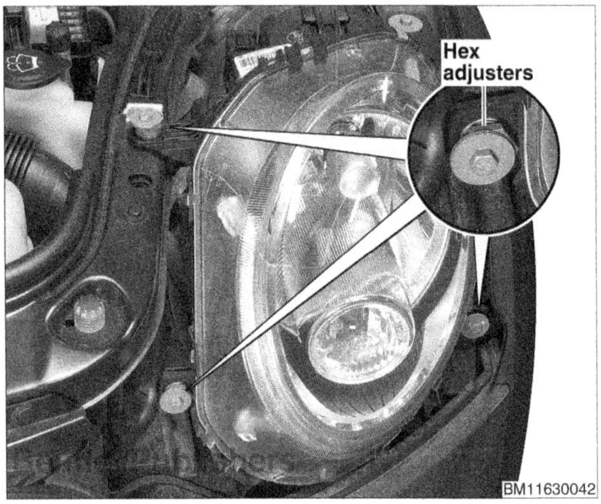

Headlight, removing and installing

Unless otherwise noted, procedures for removing halogen and xenon headlight are identical.

– Xenon: Switch lights OFF. Disconnect negative (-) battery cable and cover battery terminal to keep cable from accidentally contacting terminal.

> ***CAUTION—***
> • *Prior to disconnecting the battery, read the battery disconnection cautions given in* **001 Warnings and Cautions**.

◄ Open engine hood to work on headlight:

- Remove headlight mounting screws (**arrows**).
- Detach electrical connector.
- If equipped with headlight washers: Pull headlight forward slightly and detach high pressure washer hose from washer nozzle. Be prepared to catch dripping fluid.
- Lift out headlight.

◄ After reinstalling headlight, adjust placement and body gaps using hex adjusters (**inset**).

– Adjust headlight beam aim. See **Headlight, adjusting aim** in this repair group.

EXTERIOR LIGHTS

Front turn signal bulb, replacing

The front directional bulb is inside the headlight housing. Access to bulb is on the underside of the fender, through an access hole in the fender liner.

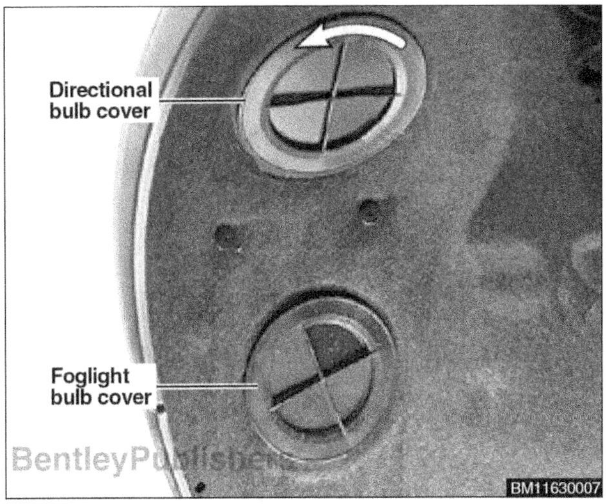

< Working underneath front fender, rotate directional bulb access cover counterclockwise (**arrow**) to remove.

< Rotate bulb cover on headlight housing counterclockwise (**arrow**) to remove.

< Rotate bulb holder (**arrow**) and pull bulb out of headlight housing.

- Unscrew bulb to replace.
- When reinstalling:
  - Do not touch new bulb surface with bare fingers. Oil from skin can cause bulb to fail prematurely.
  - Make sure headlight housing cover is firmly seated and well-sealed. Replace cover if seal is damaged.

Turn signal bulb application	
Front turn signal bulb (amber)	P(Y)21W

630

Front foglight, adjusting aim

– Before adjusting foglight aim:
  - Fill fuel tank.
  - Correct tire pressures.
  - Replace faulty bulbs.
  - Park vehicle on level ground.
  - Load driver seat with approx. 75 kg (165 lb).

< Use adjusting screw (**arrow**) to aim foglight.

Front foglight or parking light bulb, replacing

The front foglight and parking bulbs are inside the foglight housing. Access to bulbs is on the underside of the front fender, through an access hole in the wheel housing liner.

To replace foglight and parking light fixture, detach front wheel housing liner from front bumper cover. See **410 Fender, Engine Hood**.

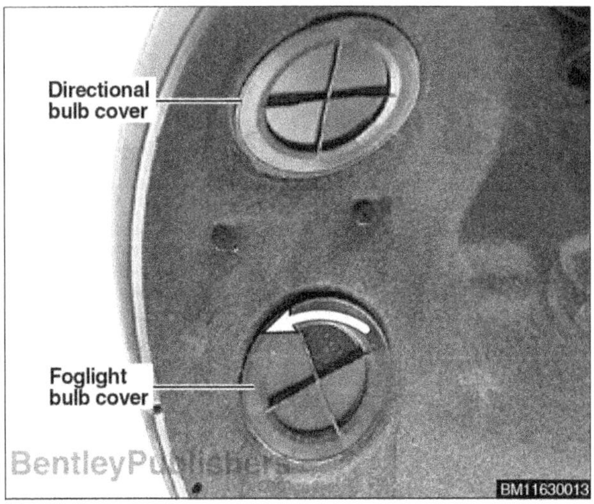

< Working underneath front fender, rotate foglight bulb access cover counterclockwise (**arrow**) to remove.

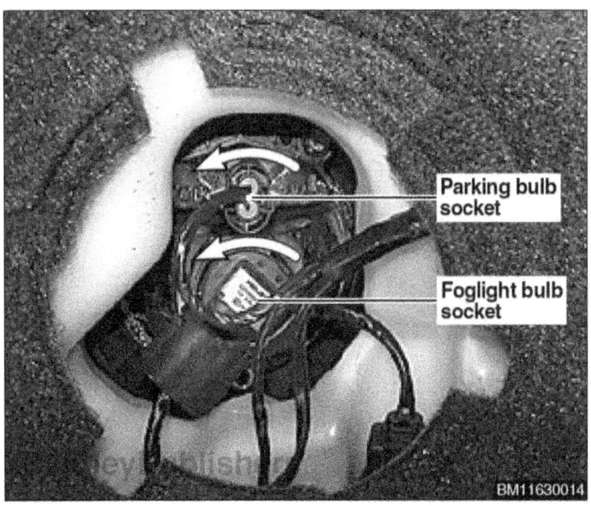

< Rotate bulb holder (**arrow**) and pull bulb socket out of foglight housing.
  - Foglight: Detach electrical connector.
  - Parking light: Pull on bulb to remove from socket.

– When reinstalling:
  - Do not touch new bulb surface with bare fingers. Oil from skin can cause bulb to fail prematurely.
  - Make sure bulb socket is firmly seated and well-sealed on housing.

Foglight and parking bulb applications	
Front foglight bulb	H8 35W
Front parking bulb	W5W

Side turn signal bulb, replacing

- Unlock engine hood.
- Slide trim containing side turn signal forward and remove (**arrow**).

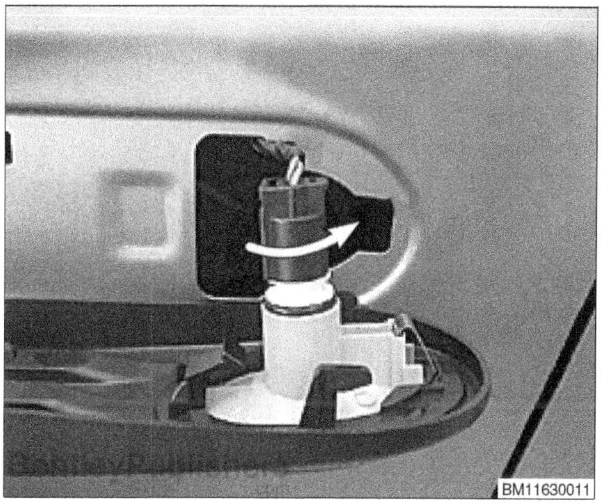

- Rotate bulb holder counterclockwise (**arrow**) to remove from turn signal fixture.
- Pull on bulb to remove from socket.
- When reinstalling:
  - Do not touch new bulb surface with bare fingers. Oil from skin can cause bulb to fail prematurely.
  - Make sure bulb holder is firmly seated and well-sealed on turn signal fixture.

Turn signal bulb application	
Side turn signal bulb	W(Y) 5W

Side marker light bulb, replacing

This procedure applies equally to side marker lights mounted on front or rear wheel housing liners

- Partially remove wheel housing liner. See **410 Fenders, Engine Hood**.

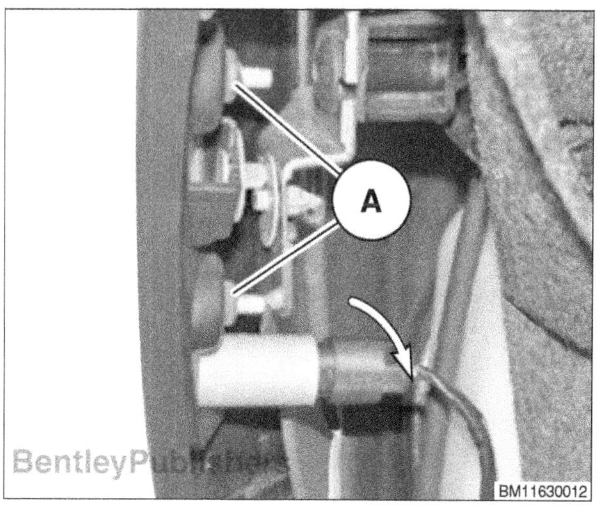

- Rotate bulb holder counterclockwise (**arrow**) to remove from fixture.
  - Front side marker fixture is attached to fender liner using 2 nuts (**A**). Rear fixture is attached using 1 nut.
- Pull on bulb to remove from socket.
- When reinstalling:
  - Do not touch new bulb surface with bare fingers. Oil from skin can cause bulb to fail prematurely.
  - Make sure bulb holder is firmly seated and well-sealed on fixture.

Side marker bulb application	
Side marker bulb	W3W

Taillight assembly (Hardtop, Convertible)

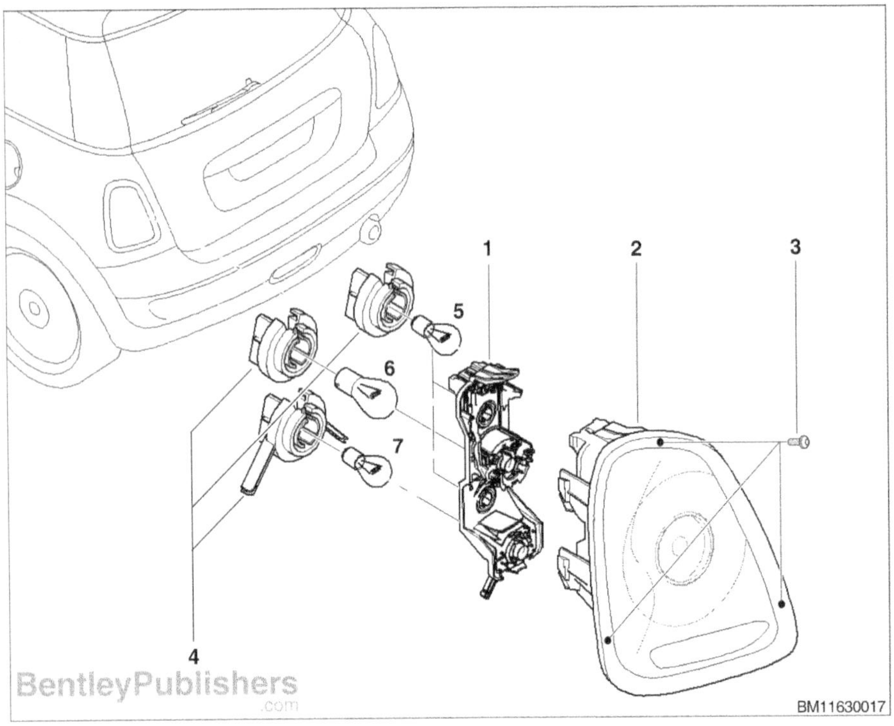

1. Bulb holder
2. Taillight lens
3. Collar screw
4. Bulb socket
5. W16W bulb
   • Brake light, parking light
6. P(Y)21W silver or amber bulb
   • Turn signal
7. W16W bulb
   • Back-up light

Taillight bulb, replacing (Hardtop, Convertible)

Taillight bulb access is through luggage compartment. Vehicle may be equipped with clear, silvered or amber bulbs. Remove burned out bulb and match for replacement.

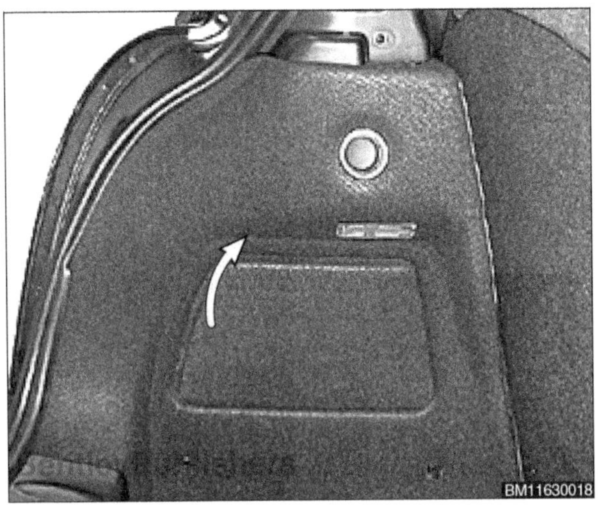

⮜ Open rear hatch and remove trim cover over taillight bulb sockets.

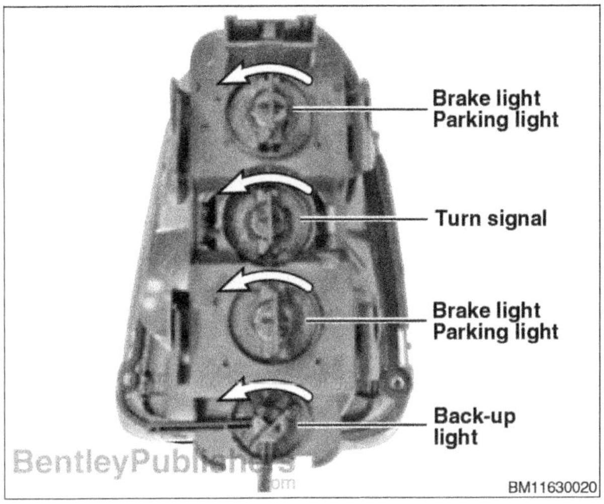

◀ Rotate bulb socket counterclockwise (**arrow**) to remove from taillight bulb holder.

– Rotate bulb and pull out of socket.

– When reinstalling:
  - Match bulb type carefully.
  - Do not touch new bulb surface with bare fingers. Oil from skin can cause bulb to fail prematurely.
  - Make sure bulb socket is firmly seated.

Taillight bulb applications (Hardtop, Convertible)	
Back-up light bulb	W16W
Brake light-parking light bulb	W16W
Turn signal bulb (amber)	P(Y)21W

Taillight assembly (Clubman)

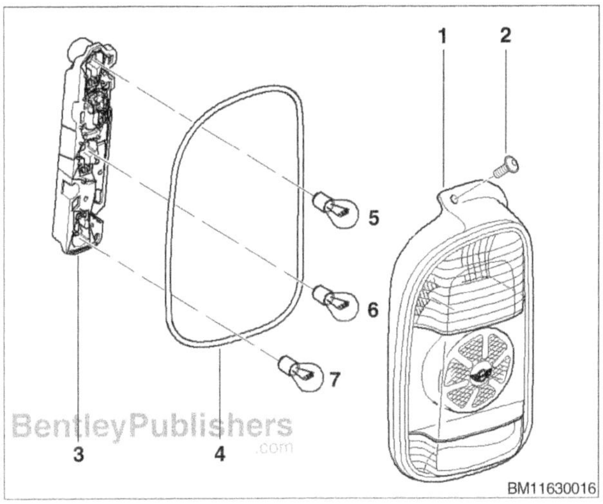

1. Taillight lens
2. Collar screw
3. Bulb holder and socket
4. Sealing gasket
5. P21W bulb
   - Brake light, parking light
6. P(Y)21W amber bulb
   - Turn signal
7. P21W bulb
   - Back-up light

Taillight bulb, replacing (Clubman)

To replace taillight bulb, remove taillight assembly from vehicle.

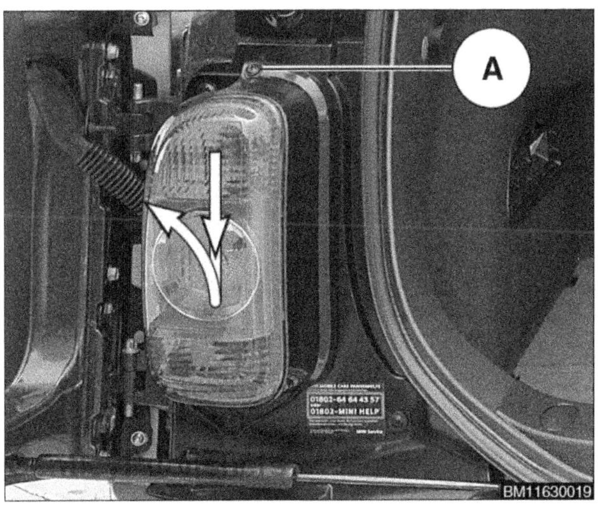

◀ Working at taillight assembly with rear split door open:
- Remove taillight lens mounting screw (**A**).
- Press taillight lens down, then tilt out and up to remove (**arrow**).

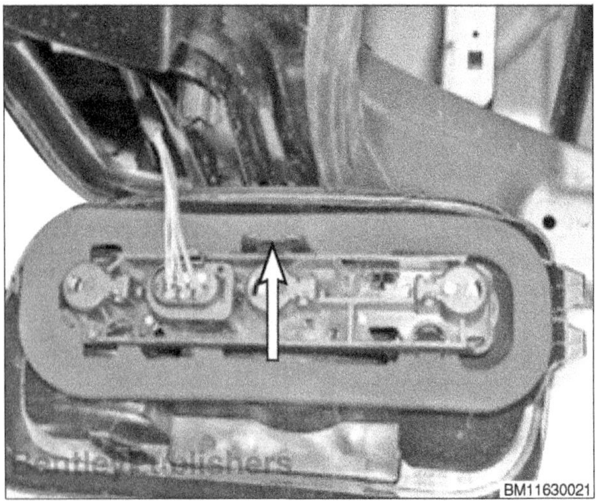

Press on plastic tab (**arrow**) to disengage bulb holder from taillight lens.

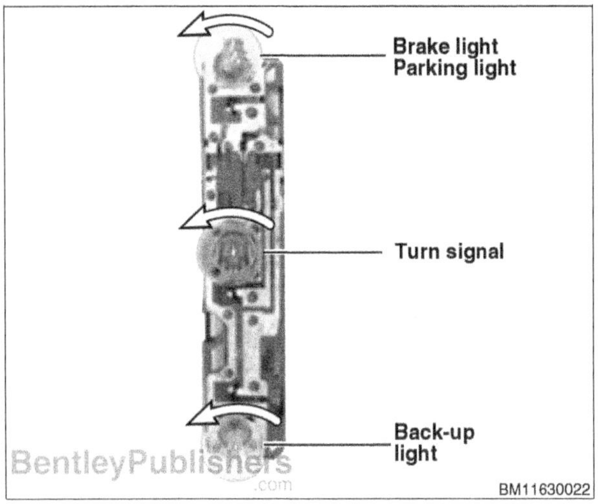

Rotate bulb (**arrow**) and pull out of socket.

- When reinstalling:
  - Match bulb type carefully.
  - Do not touch new bulb surface with bare fingers. Oil from skin can cause bulb to fail prematurely.

Taillight bulb applications (Clubman)	
Back-up light bulb	P21W
Brake light-parking light bulb	P21W
Turn signal bulb (amber)	P(Y)21W

- Reattach taillight lens, making sure taillight seal is positioned correctly on guides.

Tightening torque	
Taillight lens to body	1.5 Nm (13 in-lb)

Center brake light, replacing

Hardtop

- Open rear hatch and remove rear hatch upper trim. See **513 Interior Trim**.

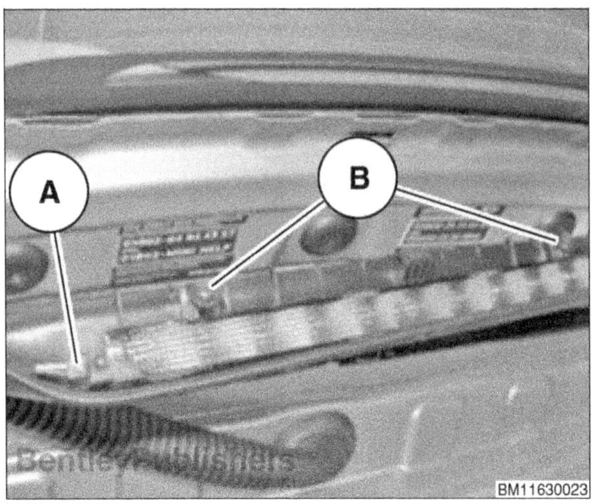

Detach center brake light electrical connector (**A**).

- Remove mounting screws (**B**) and lift off center brake light.

- Installation is reverse of removal.

Center brake light application	
Center brake light (replace as a unit)	strip of 12 LEDs

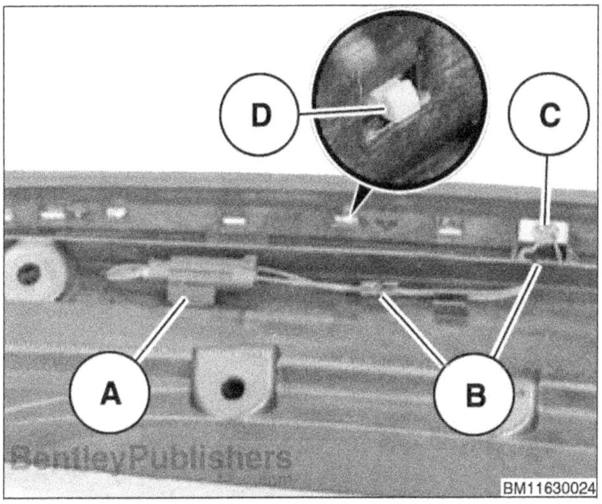

Clubman

– Open rear split doors and remove rear roof strip.

◄ Separate center brake light electrical connector (**A**) and remove from roof strip.

- Detach wiring harness from brackets (**B**).
- Carefully unlock retaining lugs (**D**) and remove center brake light (**C**).

– When installing, make sure retaining lugs (**D**) are not damaged.

Convertible

– Open trunk lid.

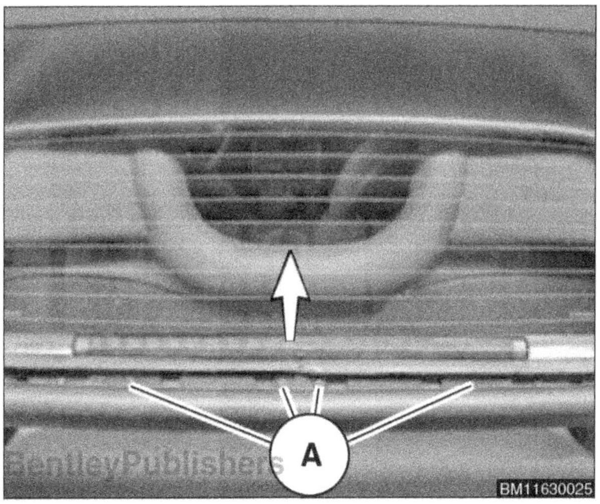

◄ Remove mounting screws (**A**) and pull out center brake light (**arrow**) slightly.

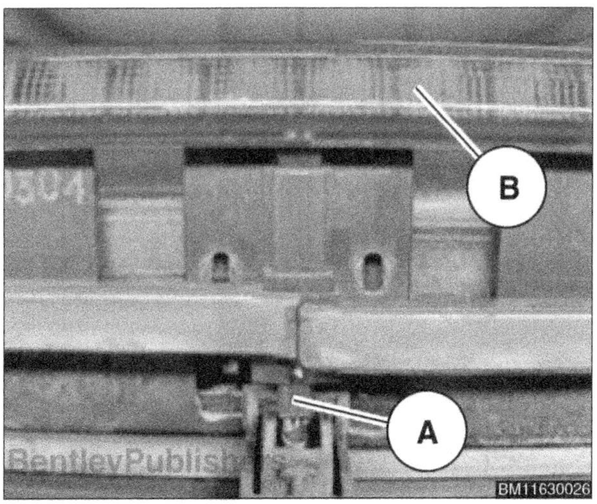

◄ Detach electrical connector (**A**) and lift off center brake light assembly (**B**).

When installing:

- Make sure electrical connector (**A**) seals correctly.
- Make sure sealing strips (**arrows**) mate and are correctly seated.

License plate light bulb, replacing

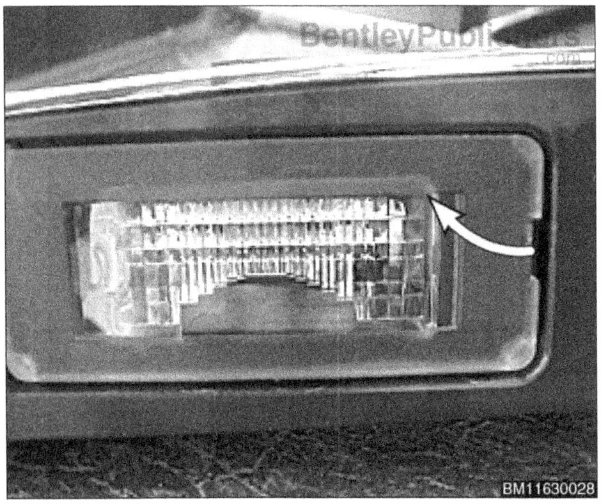

Use plastic prying tool to pry out (**arrow**) license plate light fixture.

Hardtop, Convertible: Spread bow contacts (**arrows**) and remove festoon bulb.

Clubman:

- Rotate bulb holder counterclockwise (**arrow**) to remove from license plate light fixture.
- Pull on bulb to remove from socket.

– When reinstalling:

- Do not touch new bulb surface with bare fingers. Oil from skin can cause bulb to fail prematurely.
- Clubman: Make sure bulb holder is firmly seated and well-sealed on license plate fixture.

License plate bulb applications	
Clubman	W5W
Hardtop, Convertible	Festoon C5W

Interior Lights

This section cover removal and installation of interior light fixtures for bulb replacement. Bulb applications are in **Table a**.

Table a. Interior bulb applications	
Location	**Type and wattage**
Cargo compartment or trunk light bulb	W5W
Dome light bulb (front or rear)	W6W
Footwell light bulb	W5W
Front storage compartment light bulb	1.2 W
Glove compartment light bulb	Festoon C5W
Vanity mirror light bulb	1.2W

Interior light bulbs, replacing

Front dome light bulb

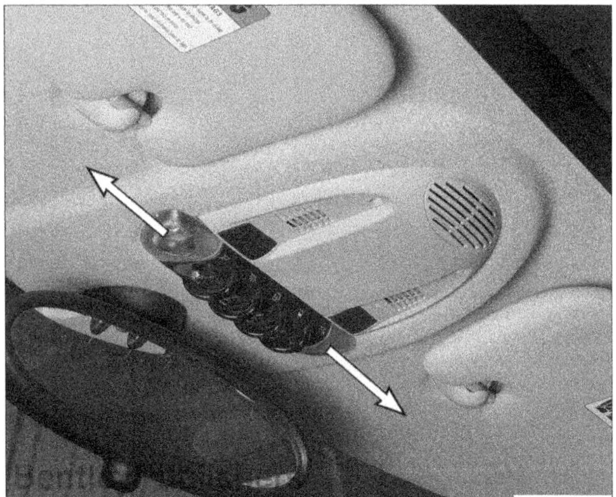
BM11630040

◄ Use plastic prying tool to lever (**arrows**) bulb cover trim off roof operating unit.

– Pull on bulb to remove from socket. See **Table a** for bulb applications.

– When reinstalling, do not touch new bulb surface with bare fingers. Oil from skin can cause bulb to fail prematurely.

Rear dome light bulb

– Use plastic prying tool to lever rear dome light fixture off head liner. Detach electrical connector.

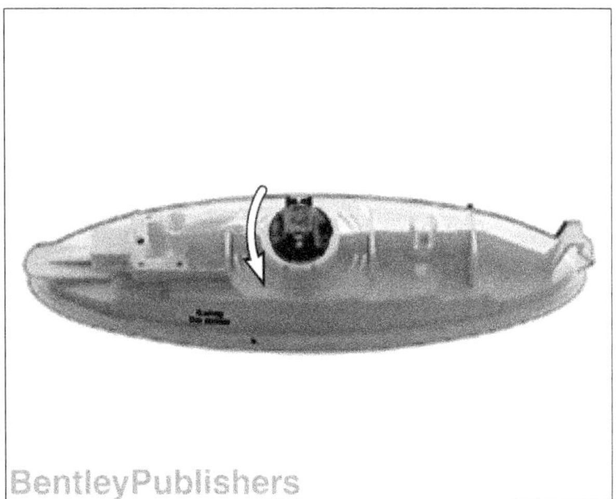
BM11630036

◄ Rotate bulb holder counterclockwise (**arrow**) to remove from dome light fixture.

– Pull on bulb to remove from socket. See **Table a** for bulb applications.

– When reinstalling, do not touch new bulb surface with bare fingers. Oil from skin can cause bulb to fail prematurely.

BM11630036

Cargo compartment, footwell and courtesy light bulbs

Bulb replacement procedure for a cargo compartment (trunk), footwell and door courtesy lights is similar.

< Use plastic prying tool to lever off light fixture (**arrow**).

– Detach electrical connector and remove fixture.

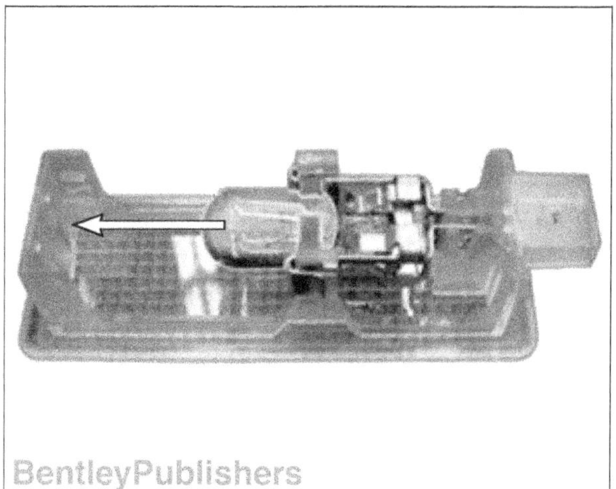

BM11630039

< Pull on bulb to remove from socket (**arrow**). See **Table a** for bulb applications.

– When reinstalling, do not touch new bulb surface with bare fingers. Oil from skin can cause bulb to fail prematurely.

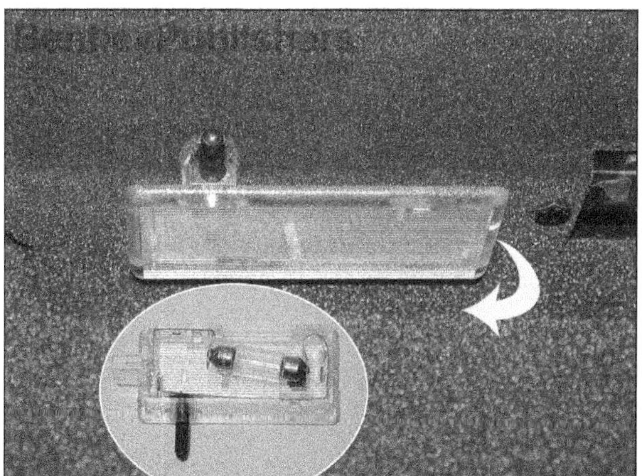

BM11630037

Glove compartment light bulb

< Use plastic prying tool to lever glove compartment light fixture from glove compartment (**arrow**).

- Detach electrical connector and remove light.
- Spread bow contacts and remove festoon bulb. See **Table a** for bulb applications.

– When reinstalling, do not touch new bulb surface with bare fingers. Oil from skin can cause bulb to fail prematurely.

630

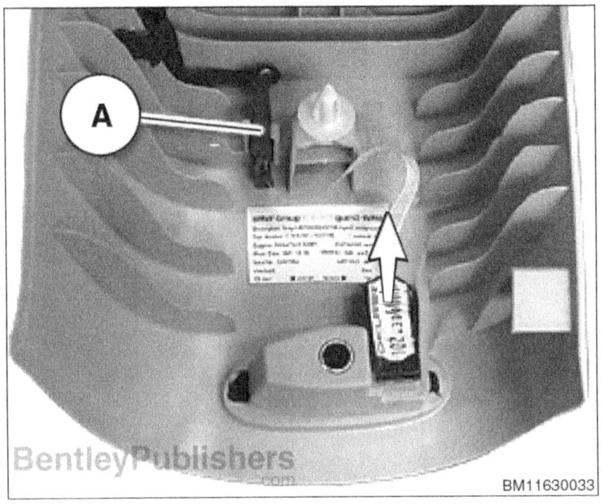

Upper B-pillar LED

– Remove upper B-pillar trim panel. See **513 Interior Trim**.

◄ Detach electrical connector (**A**) and slide out LED (**arrow**).

– Installation is reverse of removal.

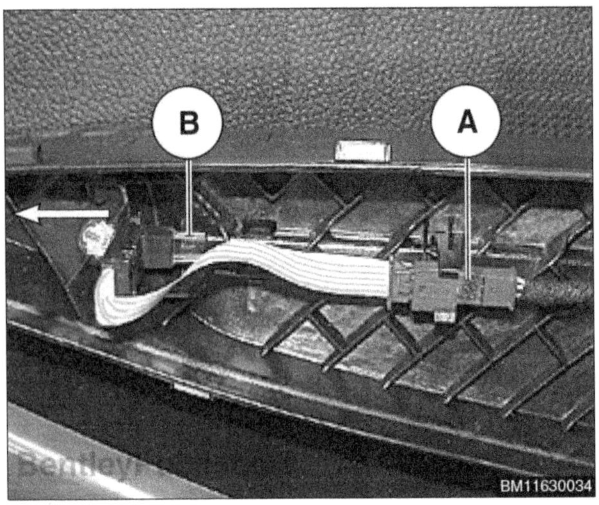

Door panel armrest LED

– Remove door panel armrest. See **411 Doors**.

◄ Working behind armrest:

- Separate LED electrical connector (**A**).
- Detach LED from fiber optic connector (**B**) by sliding forward (**arrow**).

– When installing, make sure LED snaps audibly to fiber optic connector.

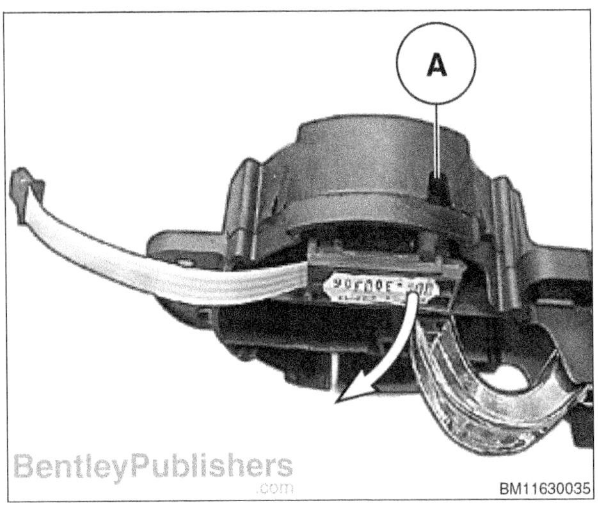

Inside door opener LED

– Remove inside door panel. See **411 Doors**.

– Detach LED electrical connector and remove inside door opener.

◄ Press on catch in opening (**A**) to detach LED from back of opener (**arrow**).

– When installing, make sure LED snaps audibly into door opener.

640 Heating and Air-conditioning

GENERAL

This repair group covers the heating and air-conditioning (A/C) system. Many of the procedures require that the A/C refrigerant charge be evacuated. A/C system recharging procedures are beyond the scope of this manual.

See also:

- **020 Maintenance** for belt replacement and interior microfilter service.
- **170 Radiator and Cooling System** for engine cooling system.

Heating and air-conditioning systems

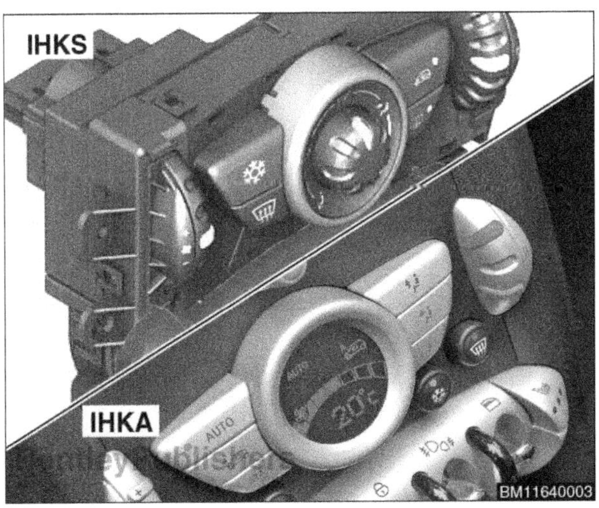

Two heating and A/C systems are available on MINI models:

- **IHKS**: Manually controlled system. Driver input is required to regulate temperature, air direction and air speed.
- **IHKA**: Optional automatic climate control. System controls air outlet temperature, air direction and air speed according to program set by driver. Automatic functions may be overridden by driver.

Both systems consist of a refrigerant system, a heater system and a control system. The refrigerant system is the same on both. Heater and A/C housing assembly and control systems are different.

Troubleshoot heating and air-conditioning system operation using MINI scan tool.

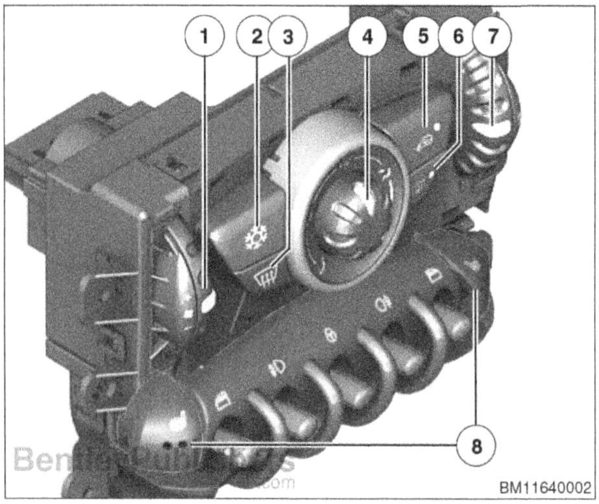

IHKS control panel

1. Blower control
2. A/C button
3. Windshield heater button
4. Air distribution control
5. Recirculation button
6. Rear window defogger button
7. Temperature control
8. Seat heater buttons

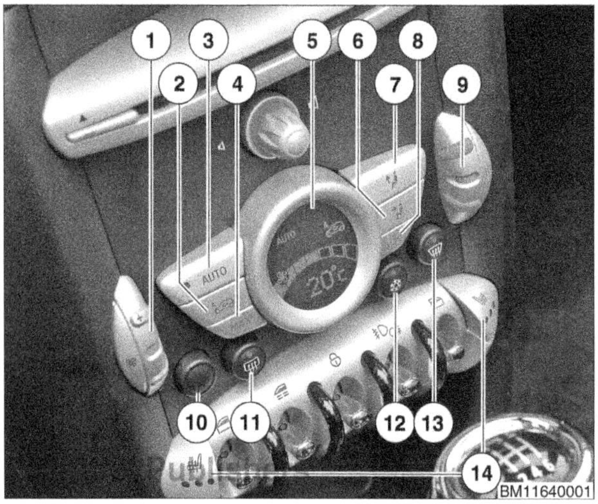

IHKA control panel

1. Blower control
2. Recirculation button
3. Automatic climate control button
4. Maximum cooling button
5. Control display
6. Face level button
7. Windshield defogger button
8. Foot level button
9. Temperature control
10. Windshield heater button
11. Rear window defogger button
12. A/C button
13. Window defogger button
14. Seat heater buttons

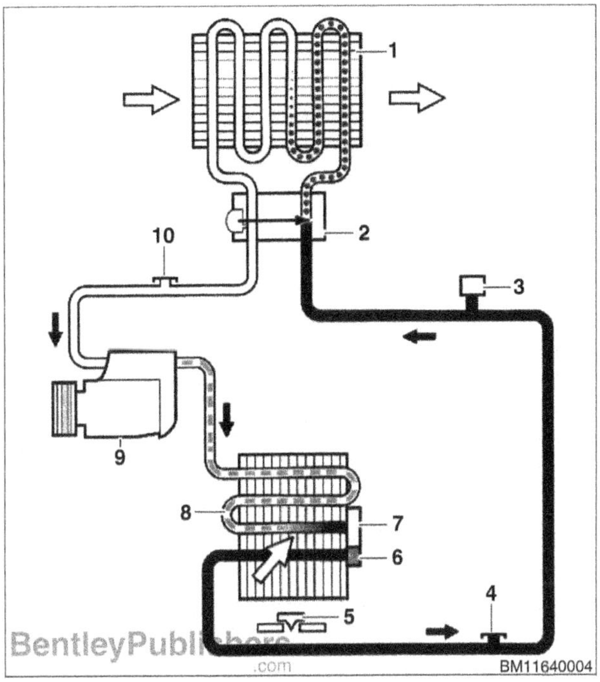

Refrigerant system (IHKS and IHKA)

The refrigerant system provides dehumidified cool air to the vehicle interior. The system is a sealed, closed loop, filled with R134a refrigerant as the heat transfer medium. Special oil is added to the refrigerant circuit to lubricate the internal components of the compressor.

1. Evaporator
2. Expansion valve
3. Refrigerant pressure sensor
4. High pressure service port
5. Blower
6. Filter
7. Receiver-dryer (desiccant)
8. Condenser
9. Compressor
10. Low pressure service port

Warnings and Cautions

WARNING—

- *Do not discharge or charge the A/C system without proper equipment and training. Damage to the vehicle and personal injury may result.*
- *Wear hand and eye protection (gloves and goggles) when working around the A/C system.*
- *If refrigerant comes in contact with your skin or eyes:*
  -Do not rub skin or eyes.
  -Immediately flush skin or eyes with cool water for 15 minutes.
  -Rush to a doctor or hospital.
  -Do not attempt to treat yourself.
- *Work in a well ventilated area. Switch on the exhaust ventilation system when working on the refrigerant system.*
- *Do not expose any component of the A/C system to high temperatures (above 80°C or 176°F) or open flames. Excessive heat causes a pressure increase which could burst the system.*
- *Keep refrigerant away from open flames. Poisonous gas is produced if it burns. Do not smoke near refrigerant gases for the same reason.*
- *The A/C system is filled with refrigerant gas which is under pressure. Pressurized refrigerant in the presence of oxygen may form a combustible mixture. Do not introduce compressed air into any refrigerant container (full or empty).*
- *Electric welding near refrigerant hoses causes R-134a to decompose. Discharge system before welding.*
- *At normal operating temperature the cooling system is pressurized. Allow the system to cool as long as possible before opening (a minimum of one hour), then release the cap slowly to allow safe release of pressure.*

640

CAUTION—

- *US law requires that any person who services a motor vehicle air-conditioner is properly trained and certified, and uses approved refrigerant recycling equipment. Technicians must complete an EPA-approved recycling course to be certified.*
- *Comply with state and local governments which may have additional requirements regarding air-conditioning servicing.*
- *A/C system service is best performed by an authorized MINI dealer or other qualified A/C service facility.*
- *Do not top off a partially charged refrigerant system. Discharge system, evacuate and then recharge system.*
- *The mixture of refrigerant oil (PAG oil) and refrigerant R-134a attacks some metals and alloys (for example, copper) and breaks down certain hose materials. Use only hoses and lines that are identified with a green mark (stripe) or marked with* R-134a.
- *Immediately plug open connections on A/C components and lines to prevent dirt and moisture contamination.*
- *Do not steam clean the A/C condenser or evaporator. Use only cold water or compressed air.*
- *To avoid damaging plastic interior trim, use a plastic prying tool or a screwdriver with the tip wrapped with masking tape.*

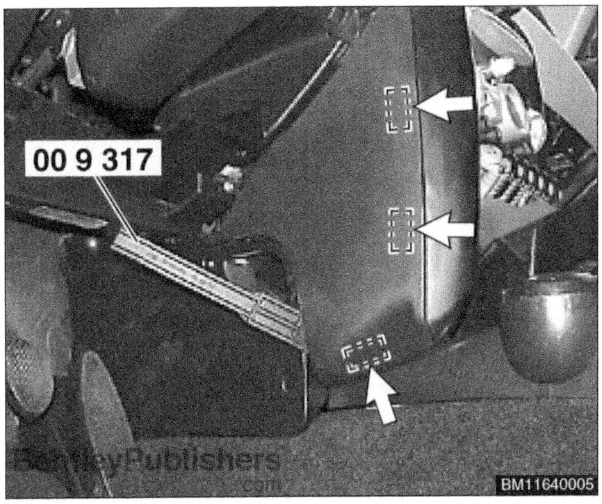

HEATING AND A/C CONTROLS

Heating and A/C control panel, removing and installing

- Remove tachometer. See **620 Instruments**.
- Remove left lower dashboard trim (panel underneath steering column). See **513 Interior Trim**.

◄ Use special tool **00 9 317** or equivalent plastic prying tool to lever left side center dashboard cover from retaining clips (**arrows**).

- Remove center instrument bezel. See **513 Interior Trim**.

◄ Use special tool **00 9 323** or equivalent plastic prying tool to lever lower center dashboard cover from retaining clips (**arrows**).

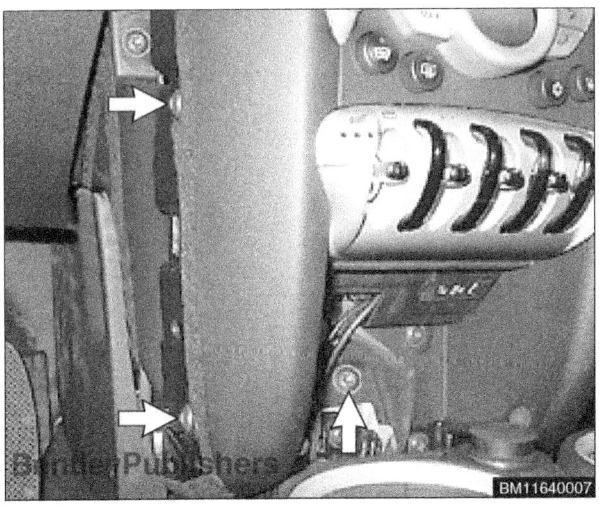

◄ Remove bezel mounting screws (**arrows**) on both sides and at bottom of center dashboard.

- Remove center console. See **513 Interior Trim**.

Heating and A/C control panel, removing and installing

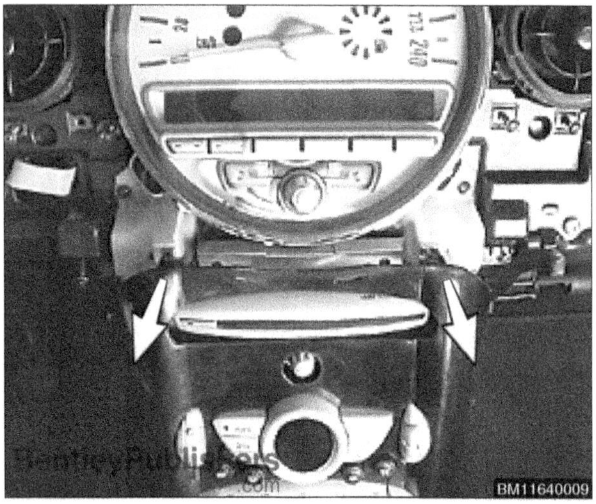

Unclip center dashboard bezel (**arrows**) and pull back slightly.

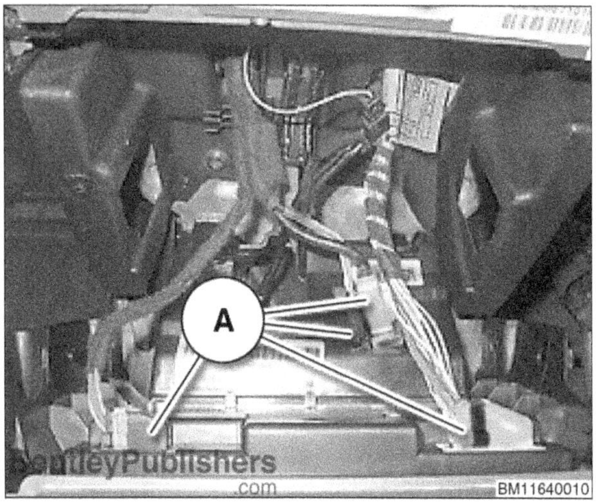

Detach electrical connectors (**A**) on back of climate control panel and lift out bezel.

Remove screws (**arrows**) to separate climat control control module from bezel.

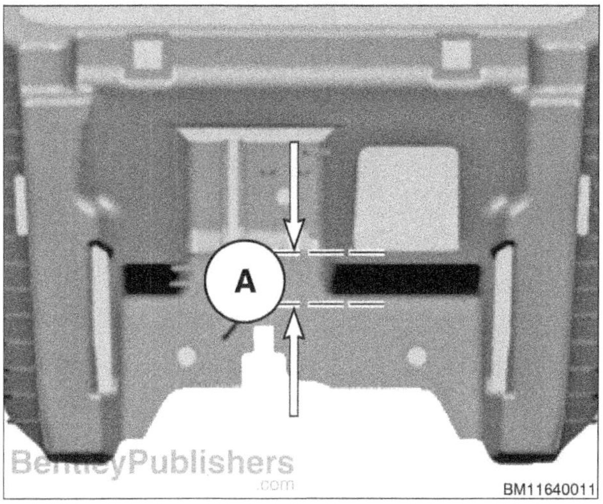

To 12 / 2007: If replacement climate control panel is installed, rework center dashboard opening to make opening **A** = 23 mm (0.91 in)

- Installation is reverse of removal.
- Use MINI scan tool to code and program new control module.

Evaporator temperature sensor, replacing

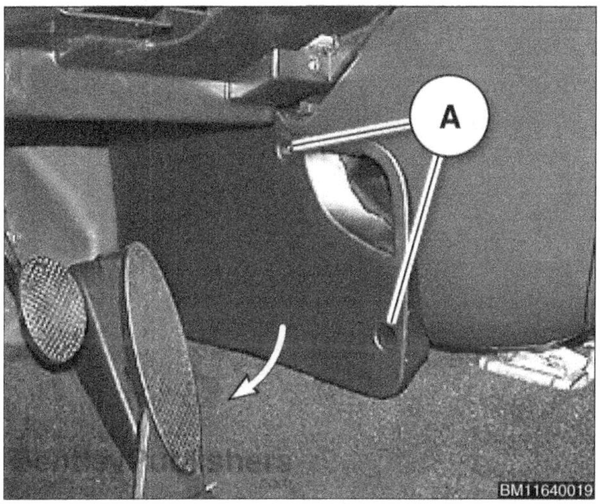

Working underneath left side dashboard, remove heater core cover screws (**A**). Slide cover down to remove (**arrow**).

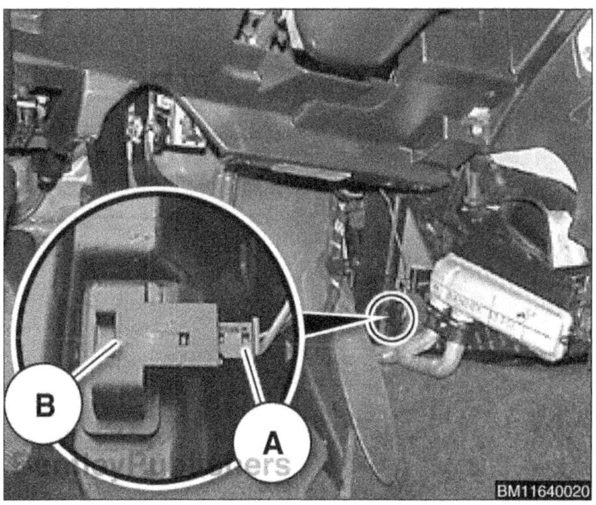

Detach electrical connector (**A**) and pull evaporator temperature sensor (**B**) out of housing.

- Installation is reverse of removal.

640

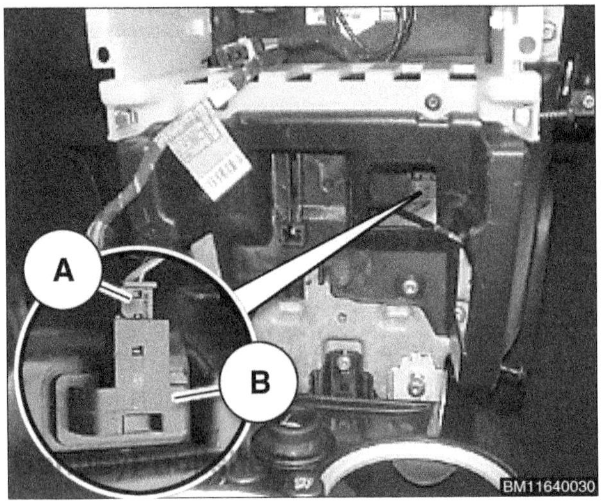

Heater core temperature sensor, replacing

- Remove center dashboard cover. See **513 Interior Trim**.

< Detach electrical connector (**A**) and pull heater core temperature sensor (**B**) out of housing.

- Installation is reverse of removal.

Refrigerant pressure switch, replacing

- Connect an approved refrigerant recovery unit to A/C system and follow manufacturer's instructions to discharge system.

> ***WARNING —***
> - *Do not attempt to discharge or charge the A/C system without proper equipment and training. Damage to the vehicle and personal injury may result.*

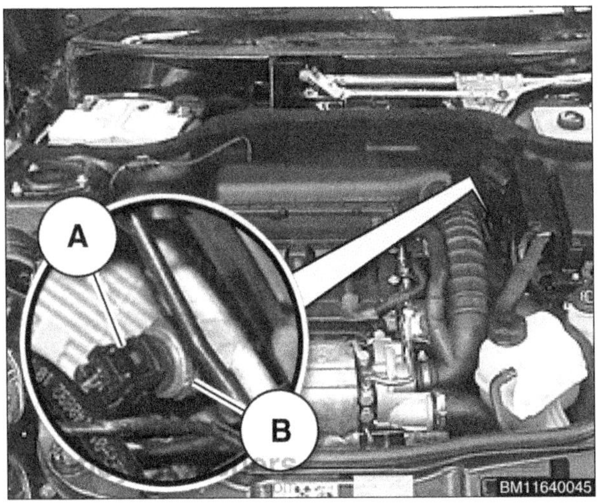

< Detach electrical connector (**A**) and unscrew pressure switch (**B**) from refrigerant line.

- Installation is reverse of removal.

Tightening torque	
Pressure switch to refrigerant line	10 Nm (7 ft-lb)

- Recharge A/C refrigerant and check for leaks.

Solar sensor, replacing

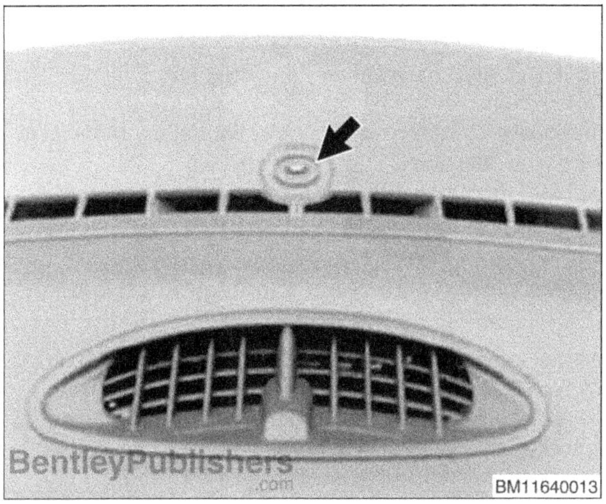

◄ Working at top center of dashboard, use plastic prying tool to pry out solar sensor cover and sensor (**arrow**).

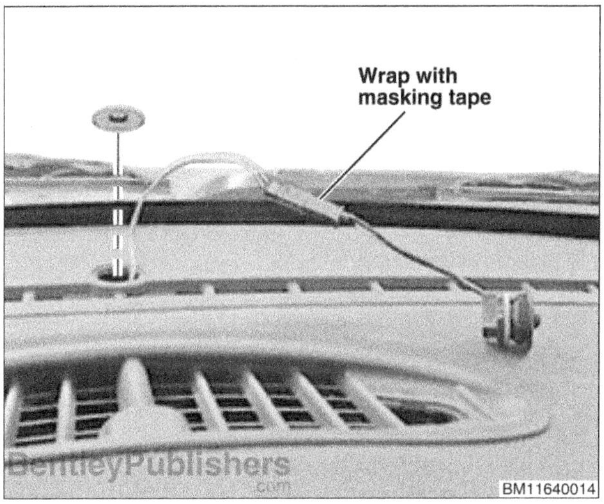

◄ Pull sensor up firmly to release adhesive tape inside dashboard. Disconnect electrical connector and remove solar sensor.

– When installing, wrap masking tape around connector to prevent rattling noises.

Air quality sensor, replacing

The climate control module uses input from the ambient (outside) air quality sensor to control automatic recirculation of passenger compartment air. The sensor is placed in the incoming air stream inside the plenum chamber, under the right side cowl cover.

– Remove right side cowl cover panel.

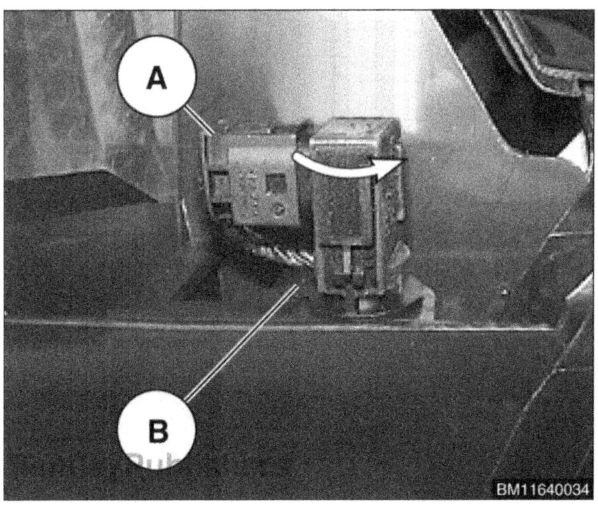

◄ Working at air quality sensor:

- Detach electrical connector (**A**).
- Press retaining clip (**B**) and rotate sensor counterclockwise (**arrow**) to remove.

– Installation is reverse of removal.

640

Air Flow Control Components

Blending flap servo motor, replacing

– Disconnect negative (-) battery cable and cover battery terminal to keep cable from accidentally contacting terminal.

> *CAUTION—*
> • *Prior to disconnecting the battery, read the battery disconnection cautions given in* **001 Warnings and Cautions**.

– Remove complete dashboard, including steering wheel and steering column switch cluster, center dashboard instrument cluster and cover and lower dashboard trim panels. See **320 Steering and Wheel Alignment** and **513 Interior Trim**.

> *WARNING—*
> • *Airbags mounted in the steering wheel and the dashboard are explosive devices. Treat with extreme caution. Follow the airbag removal procedure in* **721 Airbag System (SRS)**.

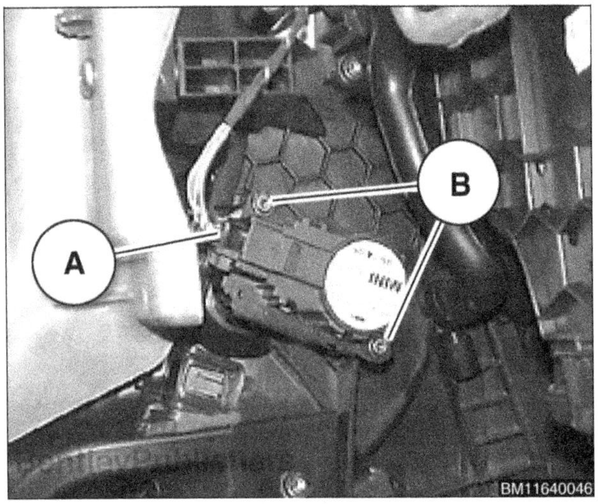

◄ Working at blending servo (stepper) motor:

• Detach electrical connector (**A**).

• Remove mounting screws (**B**) and remove servo motor.

– Installation is reverse of removal.

Center air distribution flap servo motor, replacing

– Disconnect negative (-) battery cable and cover battery terminal to keep cable from accidentally contacting terminal.

> *CAUTION—*
> • *Prior to disconnecting the battery, read the battery disconnection cautions given in* **001 Warnings and Cautions**.

– Remove center console, complete dashboard, including steering wheel and steering column switch cluster, center dashboard instrument cluster and cover and lower dashboard trim panels. See **320 Steering and Wheel Alignment** and **513 Interior Trim**.

> *WARNING—*
> • *Airbags mounted in the steering wheel and the dashboard are explosive devices. Treat with extreme caution. Follow the airbag removal procedure in* **721 Airbag System (SRS)**.

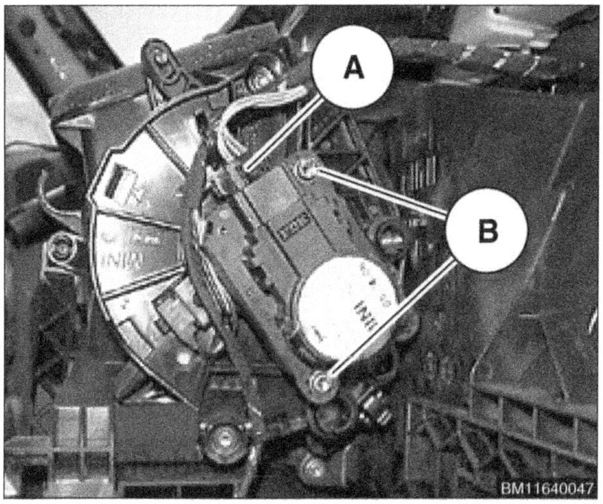

◀ Working at center air distribution servo (stepper) motor:

- Detach electrical connector (**A**).
- Remove mounting screws (**B**)and remove servo motor.

– Installation is reverse of removal.

Fresh-recirculated air flap servo motor, replacing

– Disconnect negative (-) battery cable and cover battery terminal to keep cable from accidentally contacting terminal.

> *CAUTION—*
>
> - *Prior to disconnecting the battery, read the battery disconnection cautions given in* **001 Warnings and Cautions**.

– Remove complete dashboard, including steering wheel and steering column switch cluster, center dashboard instrument cluster and cover and lower dashboard trim panels. See **320 Steering and Wheel Alignment** and **513 Interior Trim**.

> *WARNING—*
>
> - *Airbags mounted in the steering wheel and the dashboard are explosive devices. Treat with extreme caution. Follow the airbag removal procedure in* **721 Airbag System (SRS)**.

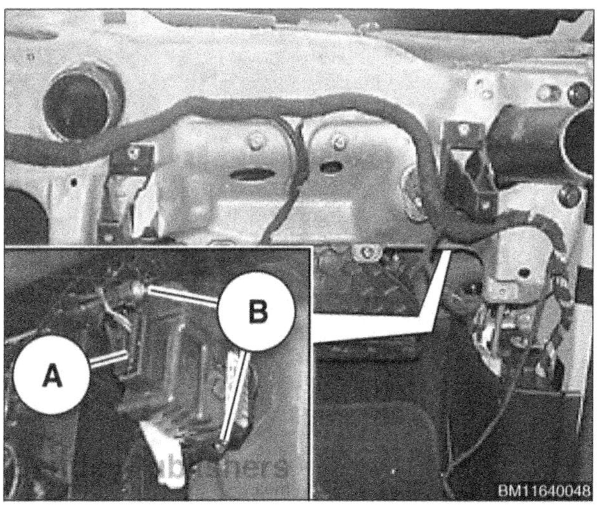

◀ Working at fresh-recirculated air distribution servo (stepper) motor:

- Detach electrical connector (**A**).
- Remove mounting screws (**B**) and remove servo motor.

– Installation is reverse of removal. Make sure deflection levers are seated correctly.

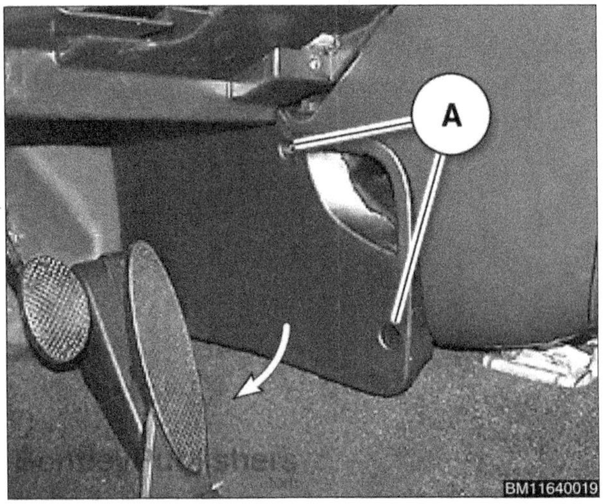

Blower Components

Blower, removing and installing

Working underneath left side dashboard, remove heater core cover screws (**A**). Slide cover down to remove (**arrow**).

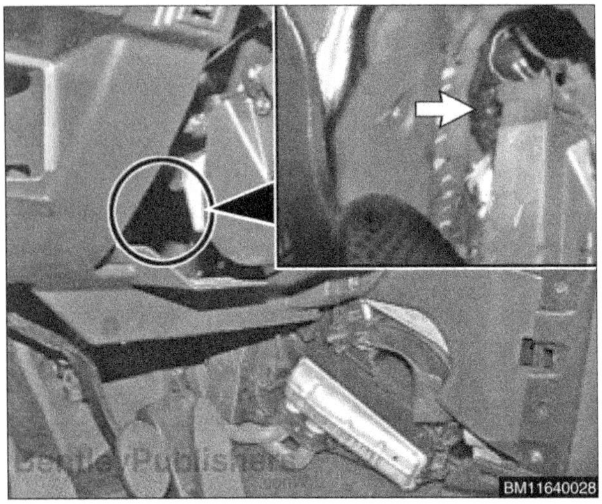

Blower (**arrow**) is above and forward of accelerator pedal at side of heater and A/C housing.

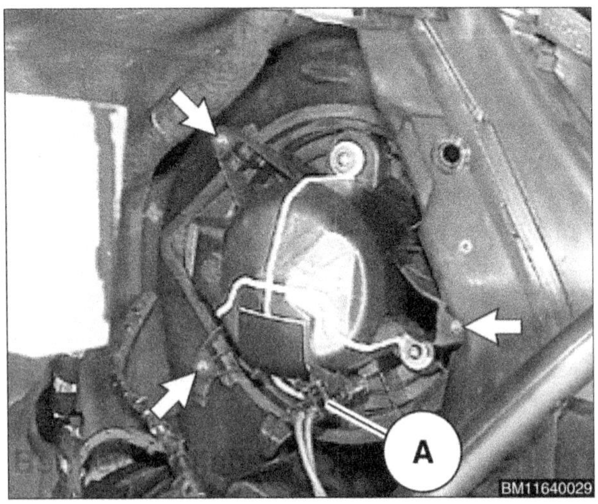

Working above accelerator pedal:

- Detach blower electrical connector (**A**).
- Remove blower mounting screws (**arrows**) and carefully extract blower from heater and A/C housing.
- Dashboard is shown removed for purposes of illustration.

– Installation is reverse of removal.

Blower resistor, replacing

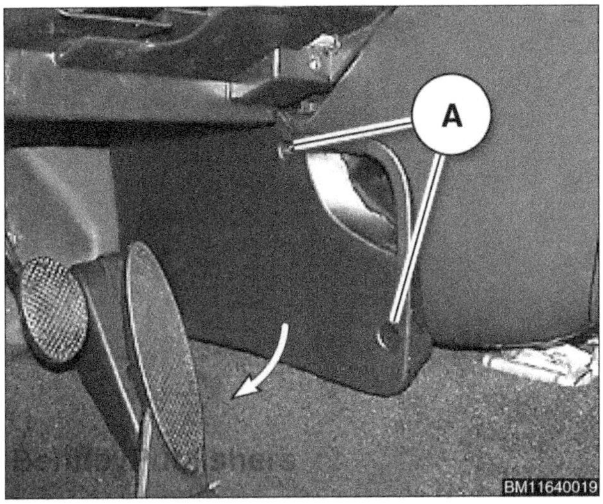

◄ Working underneath left side dashboard, remove heater core cover screws (**A**). Slide cover down to remove (**arrow**).

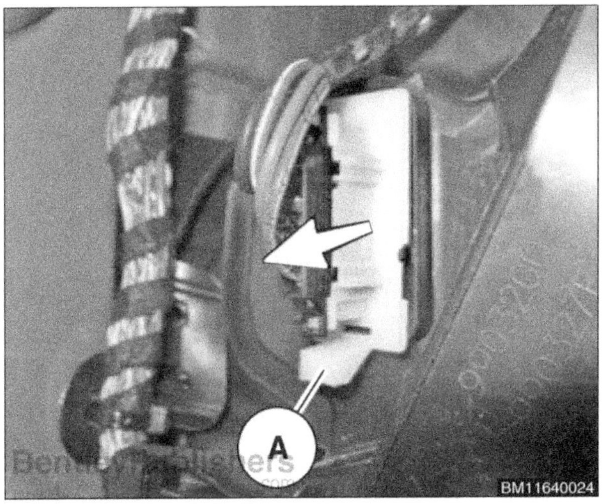

◄ Working above and forward of accelerator pedal at side of heater and A/C housing, unclip blower resistor retaining clip (**A**) and pull out of housing (**arrow**).

◄ Detach electrical connector (**arrow**) and remove sensor.

– Installation is reverse of removal.

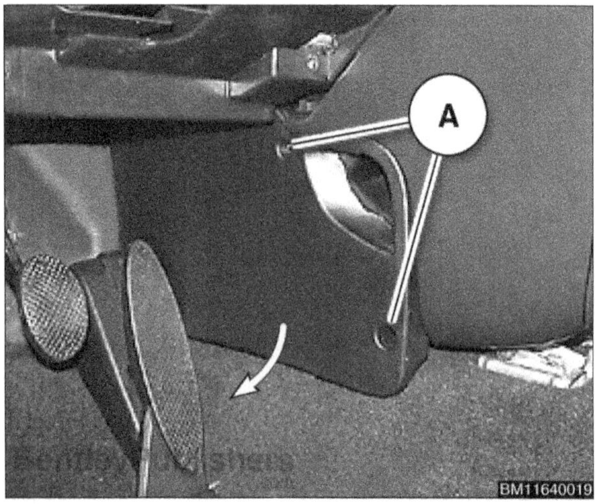

Blower final stage, replacing

Working underneath left side dashboard, remove heater core cover screws (**A**). Slide cover down to remove (**arrow**).

Working above and forward of accelerator pedal at side of heater and A/C housing:

- Detach electrical connector (**arrow**).
- Rotate final stage counterclockwise slightly to remove (**curved arrow**).

When installing, make sure markings on final stage and housing (**arrows**) line up.

HEATER COMPONENTS

Heating and A/C housing components

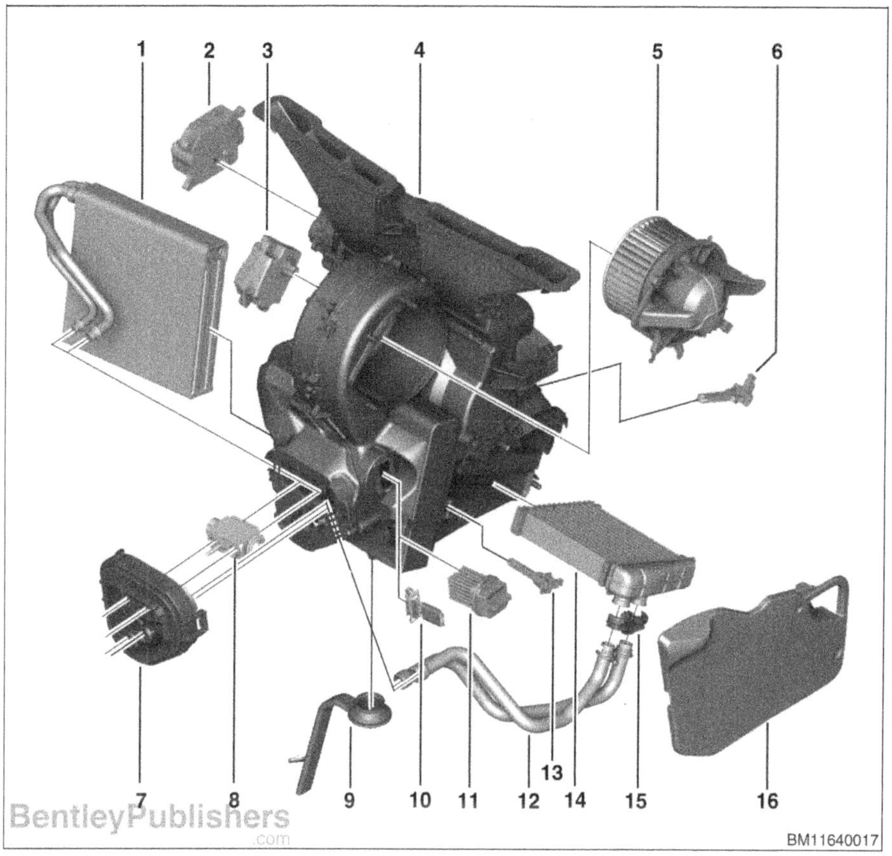

1. Evaporator
2. Recirculation servo (stepper) motor
3. Air distribution servo (stepper) motor
4. Heating and A/C housing
5. Blower
6. Heater core temperature sensor
7. Engine compartment end cover
8. Expansion valve
9. Condensate drain
10. Blower resistor
11. Blower final stage
12. Heater core coolant ducts
13. Evaporator temperature sensor
14. Heater core
15. Coolant duct clamp assembly
16. Heater core cover

Heating and A/C housing, removing and installing

Heater housing removal and installation is a complicated job. The procedure involves the following:

- Drain cooling system.
- Discharge air-conditioning refrigerant
- Remove airbag components.
- Remove complete dashboard.

Read through complete procedure before starting the job. Be sure to read airbag warnings and cautions in **721 Airbag System (SRS)**.

– Drain cooling system. See **170 Radiator and Cooling System**.

> ***WARNING—***
> - *To avoid personal injury, be sure the engine is cold before working on cooling system.*

– Connect an approved refrigerant recovery unit to A/C system and follow manufacturer's instructions to discharge system.

> ***WARNING—***
> - *Do not attempt to discharge or charge the A/C system without proper equipment and training. Damage to the vehicle and personal injury may result.*

– Remove heating and air-conditioning control panel. See **Heating and A/C control panel, removing and installing** in this repair group.

– Remove center console. See **513 Interior Trim**.

– Disconnect negative (-) battery cable and cover battery terminal to keep cable from accidentally contacting terminal.

> ***CAUTION—***
> - *Prior to disconnecting the battery, read the battery disconnection cautions given in* **001 Warnings and Cautions**.

– Remove complete dashboard, including steering wheel and steering column switch cluster, center dashboard instrument cluster and cover and lower dashboard trim panels. See **320 Steering and Wheel Alignment** and **513 Interior Trim**.

> ***WARNING—***
> - *Airbags mounted in the steering wheel and the dashboard are explosive devices. Treat with extreme caution. Follow the airbag removal procedure in* **721 Airbag System (SRS)**.

– Remove air filter housing in engine compartment. See **130 Fuel Injection**.

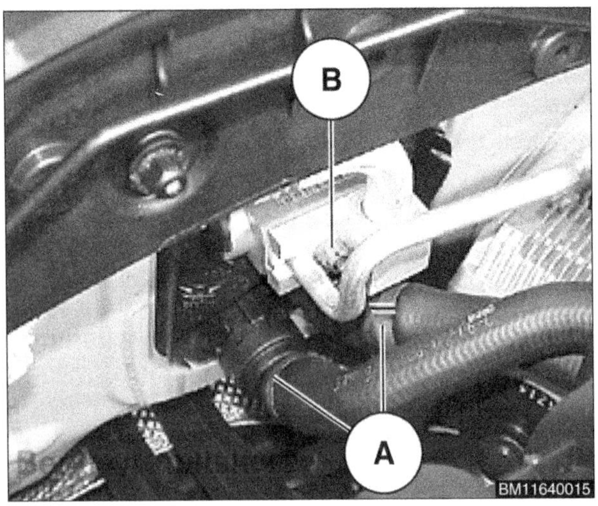

◄ Working at rear of engine compartment at bulkhead:

- Release hose clamps (**A**), detach heater hoses and blow compressed air through heater core to remove remaining coolant. Be prepared to catch dripping coolant.
- Remove nut (**B**) and detach refrigerant lines from expansion valve.

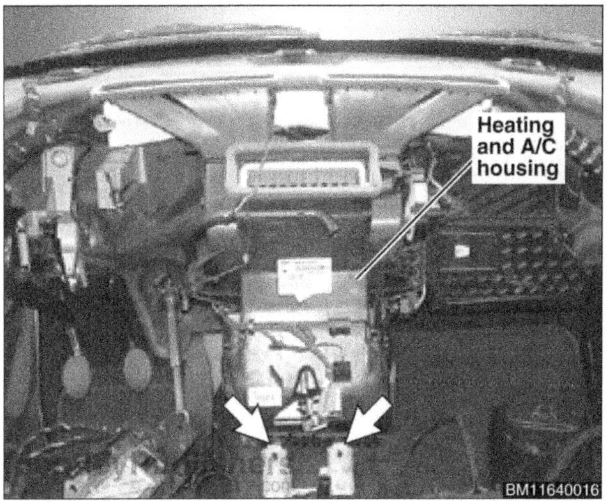

Working inside passenger compartment:

- Remove holders (**arrows**).
- With help of assistant, lift out heating and A/C housing, watching carefully to make sure no lines, hoses or electrical harnesses become snagged.

– With heating and A/C housing on work bench, repair or replace components as necessary.

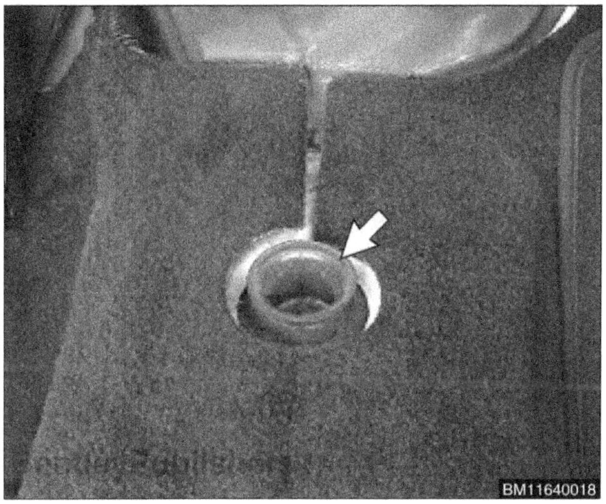

Installation is reverse of removal. Keep in mind:

- Seat heating and A/C housing correctly on condensate drain (**arrow**).
- Make sure heater is in correct position before installing dashboard support.
- Use new gaskets, sealing O-rings, hose clamps and wire ties as necessary. Fit seals over lines carefully to avoid damage to seals.

Tightening torque	
Refrigerant lines to expansion valve (replace sealing O-rings)	8 Nm (6 ft-lb)

– Fill and bleed cooling system. See **170 Radiator and Cooling System**. Check for leaks.

– If A/C system is open for more than 24 hours, replace receiver-dryer or desiccant insert. See **A/C receiver-dryer service** in this repair group.

– Recharge A/C refrigerant and check for leaks.

Heater core, removing and installing

– Drain cooling system. See **170 Radiator and Cooling System**.

> ***WARNING —***
> - *To avoid personal injury, be sure the engine is cold before working on cooling system.*

– Remove air filter housing in engine compartment. See **130 Fuel Injection**.

Working at rear of engine compartment at bulkhead, release hose clamps (**arrows**), detach heater hoses and blow compressed air through heater core to remove remaining coolant. Be prepared to catch dripping coolant.

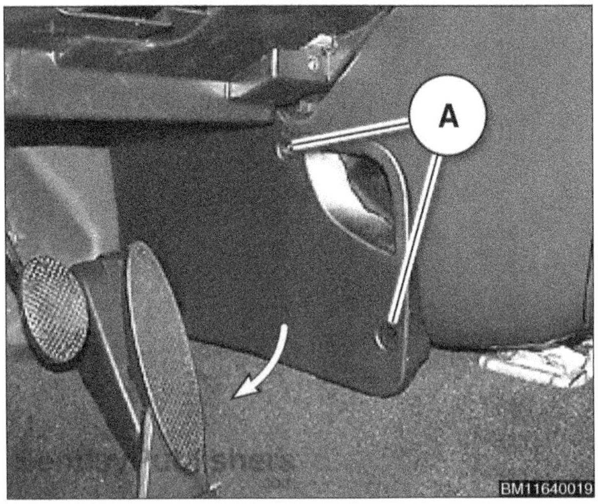

Working underneath left side dashboard, remove heater core cover screws (**A**). Slide cover down to remove (**arrow**).

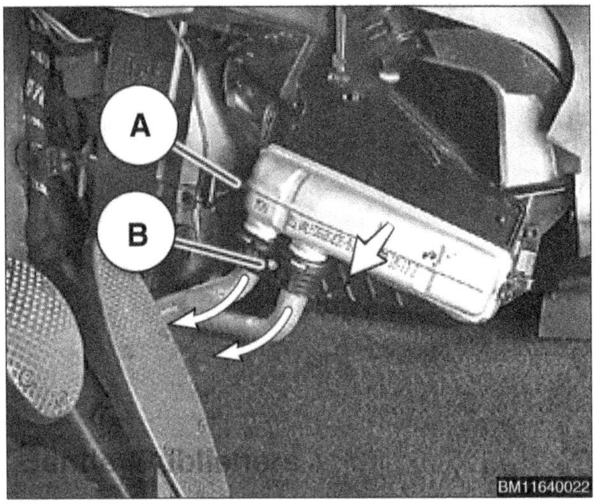

Working at heater core:

- Remove retaining screw (**A**) and pull heater core out approx 10 mm (3/8 in) (**flared arrow**).
- Remove screw (**B**) and remove coolant duct clamp assembly.
- Pull coolant ducts down (**curved arrows**) to disengage from heater core. Be prepared to catch spilling coolant.
- Pull heater core out of housing.

When installing:

- Fit heater core in housing in correct orientation, taking care to not damage heater core fins.
- Replace heater duct to heater core sealing O-rings (**arrows**). Fit seals over ducts carefully to avoid damage to seals. Make sure twin ducts seat correctly before installing duct clamp assembly.

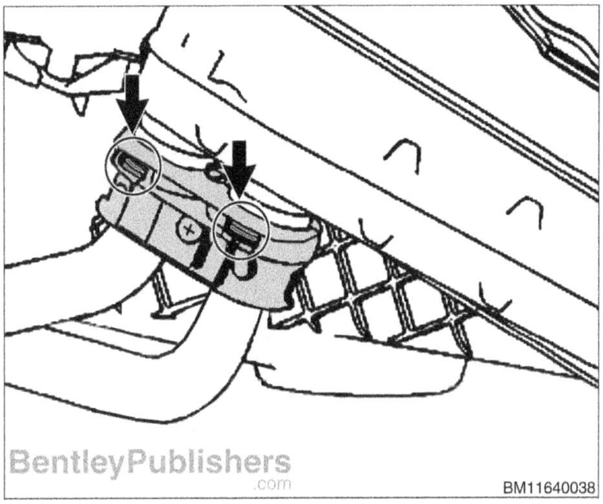

When clamp is correctly installed, lips of coolant ducts (**arrows**) are visible through apertures in outer clamp half.

AIR-CONDITIONING COMPONENTS

This section covers removal and installation of air-conditioning refrigerant components. A/C testing and diagnosis, refrigerant evacuation and recharge are not covered here. Be sure to read **A/C system warning and cautions** in this repair group.

The A/C condenser, compressor, desiccant insert, pressure sensor and expansion valve are accessible from the engine compartment.

The A/C evaporator is in the heater and A/C housing. Access to the evaporator is only possible by removing the dashboard assembly.

In the procedures that follow, use an approved refrigerant recovery unit to evacuate and refill A/C system refrigerant. Plug open refrigerant lines and ports immediately. If the system is open for more than 24 hours, replace the receiver-dryer or desiccant insert. See **A/C receiver-dryer service** in this repair group.

A/C compressor, removing and installing

> *WARNING—*
> - *To avoid personal injury, be sure the engine is cold before beginning work.*

– Connect an approved refrigerant recovery unit to A/C system and follow manufacturer's instructions to discharge system.

> *WARNING—*
> - *Do not attempt to discharge or charge the A/C system without proper equipment and training. Damage to the vehicle and personal injury may result.*

– With engine fully cooled off, remove upper exhaust manifold heat shield.

– Raise front of car and support in a safe manner.

> *WARNING—*
> - *Make sure the car is stable and well supported at all times. Use a professional automotive lift or jack stands designed for the purpose. A floor jack is not adequate support.*

– Remove splash shield from under engine. See **020 Maintenance**.

– Mark engine accessory belt with direction of rotation. Then remove belt. See **020 Maintenance**.

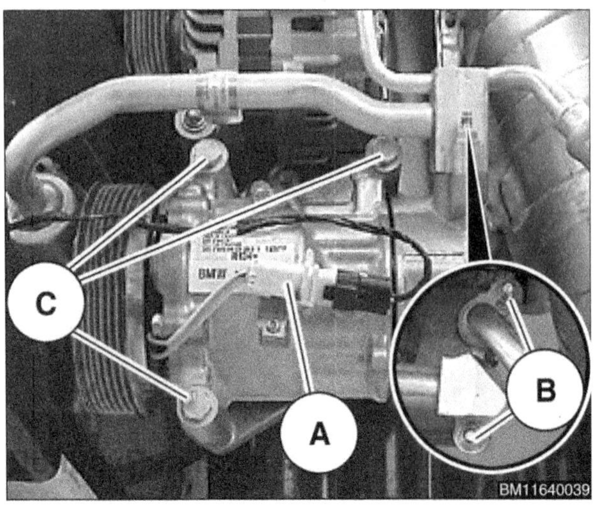

◄ Working at compressor at side of engine:
- Detach compressor electrical connector (**A**).
- Remove retaining nuts (**B**) and disconnect refrigerant lines. Plug open lines and ports immediately.
- Remove compressor mounting bolts (**C**) and lower compressor.

> *CAUTION—*
> - *Secure compressor against falling out.*

– Installation is reverse of removal. Keep in mind:
  - Use new sealing O-rings and wire ties as necessary. Fit seals over lines carefully to avoid damage to seals.
  - If installing new compressor or if A/C system is open for more than 24 hours, replace receiver-dryer or desiccant insert. See **A/C receiver-dryer service** in this repair group.
  - Install and tension engine drive belt, noting previously made direction mark. See **020 Maintenance**.

Tightening torques	
A/C compressor to engine (M8)	20 Nm (15 ft-lb)
Refrigerant lines to compressor (M8) (replace sealing O-rings)	20 Nm (15 ft-lb)

– Recharge A/C system following equipment manufacturer's instructions. Refer to rating plate on compressor to determine quantity of refrigerant oil required.

A/C system refrigerant	
R-134a capacity	490 ± 10 grams (15.8 ± 0.4 oz)

– When starting a new compressor for the first time, carry out the following break-in procedure:
  - Switch on A/C system.
  - Set all dashboard air vents to OPEN.
  - Start engine and allow to idle.
  - Set blower to at least 75% output.
  - Run A/C for at least 2 minutes at idle speed.

> ***CAUTION—***
> - *The new compressor may be damaged if run at higher than idle speed at start-up.*

A/C condenser assembly

The A/C condenser is in front of the radiator. A receiver-dryer canister with desiccant insert is integral with the condenser.

1. Receiver dryer with desiccant insert
2. Bracket
3. Condenser

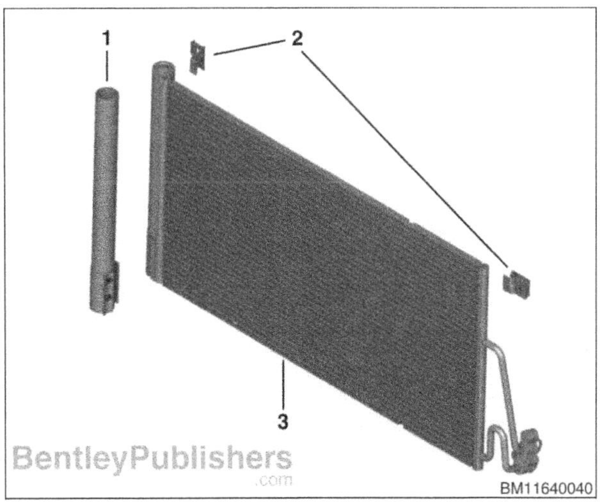

640

A/C condenser, removing and installing

– Connect an approved refrigerant recovery unit to A/C system and follow manufacturer's instructions to discharge system.

> ***WARNING —***
> *• Do not attempt to discharge or charge the A/C system without proper equipment and training. Damage to the vehicle and personal injury may result.*

– Remove front bumper cover. See **510 Exterior Trim, Bumpers**.

– Raise front of car and support in a safe manner.

> ***WARNING —***
> *• Make sure the car is stable and well supported at all times. Use a professional automotive lift or jack stands designed for the purpose. A floor jack is not adequate support.*

– Remove splash shield from underneath engine. See **020 Maintenance**.

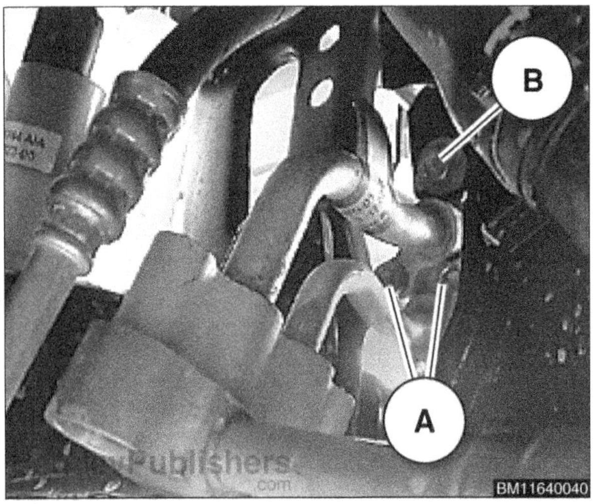

◄ Working underneath left front of engine:

• Remove retaining nuts (**A**) and disconnect refrigerant lines. Plug open lines and ports immediately.

• Remove mounting fastener (**B**).

◄ Remove mounting bolts (**arrows**) and lift out condenser.

– Service receiver-dryer as necessary. See **A/C receiver-dryer service** in this repair group.

– Installation is reverse of removal. Keep in mind:

• If A/C system is open for more than 24 hours, replace receiver-dryer or desiccant insert. See **A/C receiver-dryer service** in this repair group.

• Use new sealing O-rings and wire ties as necessary. Fit seals over lines carefully to avoid damage to seals.

Tightening torques	
A/C condenser to radiator (M6)	6 Nm (4 ft-lb)
Refrigerant lines to condenser (M6)	8 Nm (6 ft-lb)

– Recharge A/C refrigerant and check for leaks.

BM11640043

A/C receiver-dryer service

- Remove A/C condenser. See **A/C condenser, removing and installing** in this repair group.

◀ Remove desiccant insert cap (**arrow**).

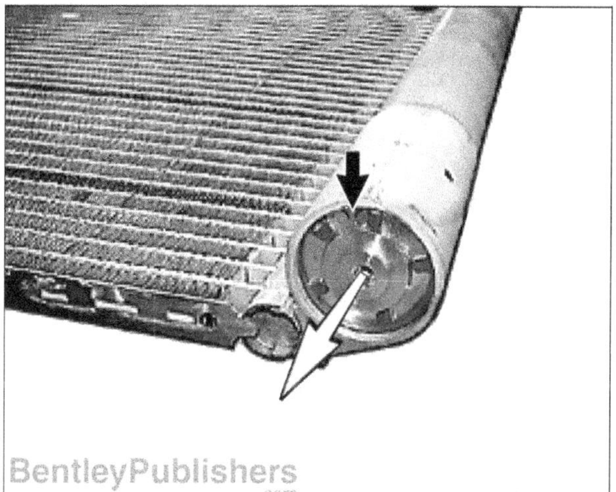

BM11640044

◀ Remove retaining circlip (**arrow**) and slide out desiccant insert (**flared arrow**).

- Installation is reverse of removal. Use new sealing O-rings. Fit seals carefully to avoid damage to seals.

- Recharge A/C refrigerant and check for leaks.

A/C evaporator, replacing

- Drain cooling system. See **170 Radiator and Cooling System**.

> ***WARNING—***
> - *To avoid personal injury, be sure the engine is cold before working on cooling system.*

- Connect an approved refrigerant recovery unit to A/C system and follow manufacturer's instructions to discharge system.

> ***WARNING—***
> - *Do not attempt to discharge or charge the A/C system without proper equipment and training. Damage to the vehicle and personal injury may result.*

- Remove heating and A/C housing from under dashboard. See **Heating and A/C housing, removing and installing** in this repair group.

- With heating and A/C housing placed on work bench. remove heater core from housing. See **Heater core, removing and installing** in this repair group.

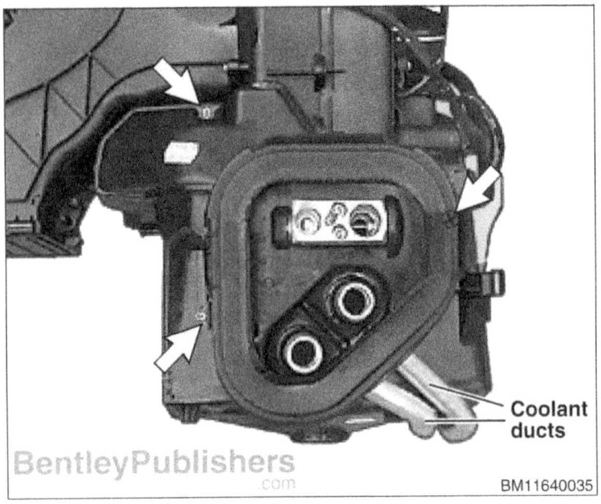

Remove front cover screws (**arrows**) and separate cover and dual coolant ducts from housing.

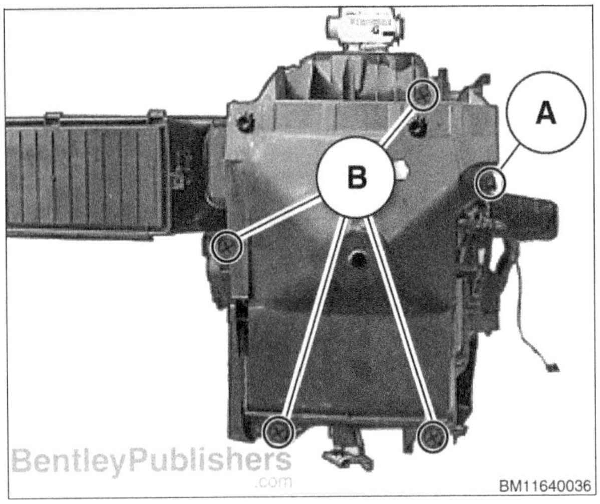

With heating and A/C housing on its side:

- Detach evaporator temperature sensor electrical connector (**A**).
- Remove screws (**B**) and separate evaporator cover.

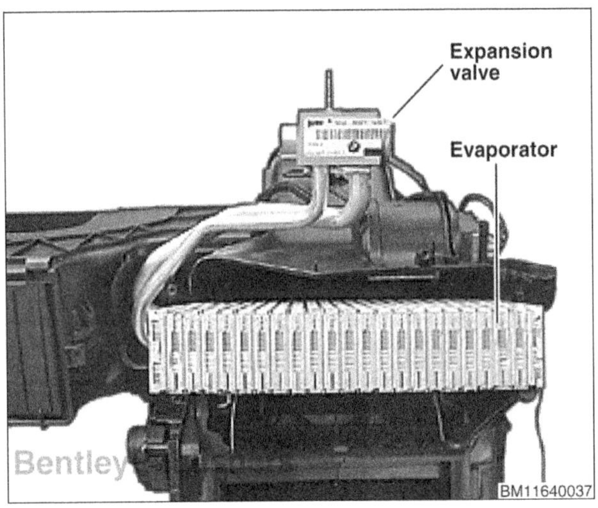

Pull evaporator, together with expansion valve, out of housing. If necessary, separate expansion valve. See **A/C expansion valve, replacing** in this repair group.

– Installation is reverse of removal. Use new gaskets, sealing O-rings, hose clamps and wire ties as necessary. Fit seals over lines carefully to avoid damage to seals.

Tightening torque	
Expansion valve to mounting plate (replace sealing O-rings)	5.5 Nm (4 ft-lb)
Refrigerant lines to expansion valve (replace sealing O-rings)	8 Nm (6 ft-lb)

– Fill and bleed cooling system. See **170 Radiator and Cooling System**. Check for leaks.

– If A/C system is open for more than 24 hours, replace receiver-dryer or desiccant insert. See **A/C receiver-dryer service** in this repair group.

– Recharge A/C refrigerant and check for leaks.

A/C expansion valve, replacing

- Connect an approved refrigerant recovery unit to A/C system and follow manufacturer's instructions to discharge system.

> ***WARNING—***
> - *Do not attempt to discharge or charge the A/C system without proper equipment and training. Damage to the vehicle and personal injury may result.*

- Remove air filter housing. See **130 Fuel Injection**.

◄ Remove nut (**arrow**) and detach refrigerant lines from expansion valve.

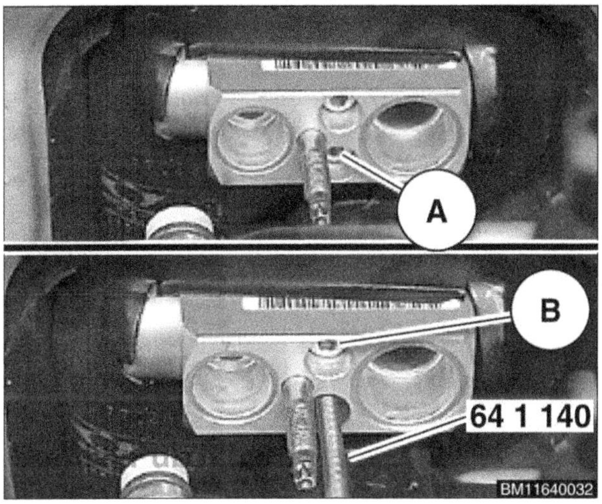

◄ Working at expansion valve:

- Remove lower mounting bolt (**A**) and screw MINI special tool **64 1 140** (long threaded M8 stud) into mounting plate behind expansion valve.
- Remove upper mounting bolt (B) and slide expansion valve forward off mounting plate.

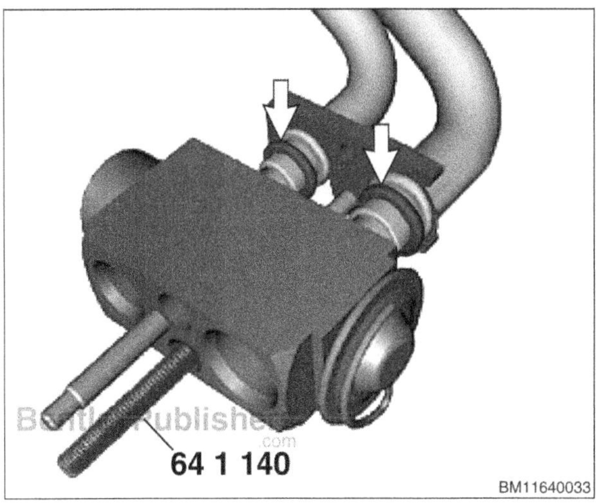

◄ When installing:

- Replace refrigerant line sealing O-rings (**arrows**). Fit seals over lines carefully to avoid damage to seals.
- Insert expansion valve using special tool **64 1 140** as guide, then secure valve with upper mounting bolt. Install lower mounting bolt after removing stud tool.
- Replace sealing O-rings on refrigerant supply lines.

Tightening torque	
Expansion valve to mounting plate (replace sealing O-rings)	5.5 Nm (4 ft-lb)
Refrigerant lines to expansion valve (replace sealing O-rings)	8 Nm (6 ft-lb)

- If A/C system is open for more than 24 hours, replace receiver-dryer or desiccant insert. See **A/C receiver-dryer service** in this repair group.

- Recharge A/C refrigerant and check for leaks.

650 Radio

General

This repair group covers MINI factory installed radio, CD changer, radio antenna and speaker service.

When installed, the navigation system uses the car communication computer (CCC) for GPS information processing. CCC service is also covered here.

Radio installations vary by year and model. For information on aftermarket sound equipment, contact the original installer.

See **620 Instruments** for central information display (CID) and navigation display removal.

Sound system overview

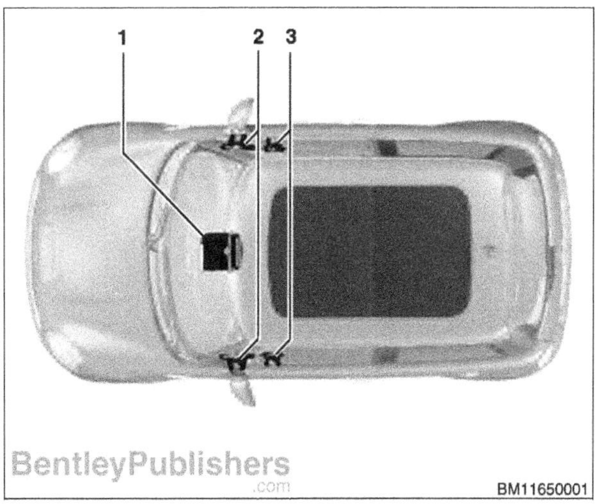

The 2-channel MINI Wave radio features a stereo receiver.

1. Radio
2. Bass speakers (150 mm)
3. Midrange speakers (100 mm)

- Bandwidth: 50 - 14,000 Hz
- Max. acoustic pressure: 98 dB
- Max. amplifier power: 2 x 15 W @4 Ω
- Connection: K-bus
- Antenna diversity: Not included

650

Sound system overview

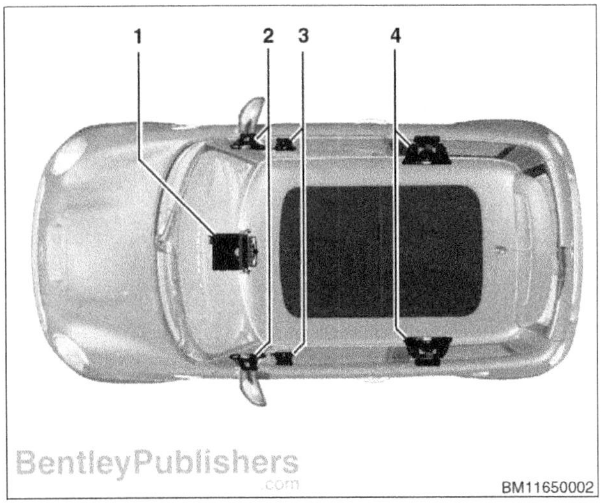

The 4-channel sound system, MINI Wave CD, is standard in models starting in 08 / 2007. It features 6 speakers.

1. Radio and CD player
2. Bass speakers (150 mm)
3. Midrange speakers (100 mm)
4. Woofers (6 x 8 in)

Speakers are driven directly by the radio output stages. Midrange and bass speakers in the front doors are driven on each side by a shared output stage. Woofers in the rear are driven on each side by a separate radio output stage.

- Bandwidth: 50 - 14,000 Hz
- Max. acoustic pressure: 98 dB
- Max. amplifier power: 4 x 15 W @4 Ω
- Connection: K-CAN

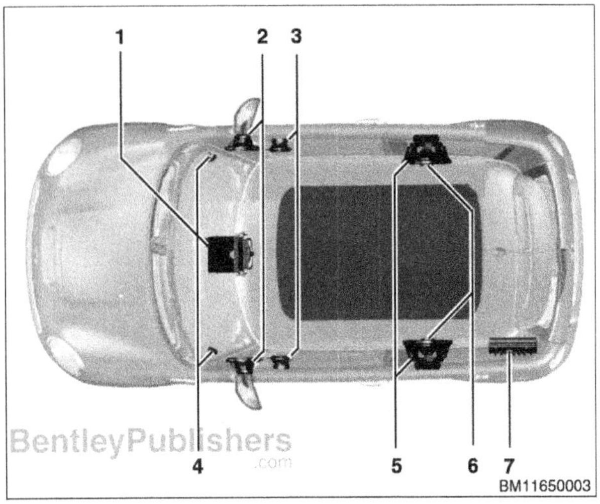

The optional 8-channel HiFi audio system (MINI Boost CD/Harman Kardon) is equipped with 10 speakers and a separate digital amplifier.

1. Radio and CD player
2. Bass speakers (150 mm)
3. Midrange speakers (100 mm)
4. Tweeters (26 mm)
5. Woofers (6 x 8 in)
6. Tweeters (26 mm)
7. Digital amplifier

- Bandwidth: 40 Hz to 20,000 Hz
- Max. acoustic pressure 104 dB
- Max. amplifier power:
  2 x 40 W @2 Ω bass
  6 x 25 W @2 Ω
- Connections: K-CAN, MOST-bus

The digital 8-channel HiFi amplifier, installed behind trim on the left side of the cargo compartment, is standard with the HiFi system but can be installed with any of the sound systems.

When equipped, the satellite radio receiver is under the passenger seat.

MINI models are equipped with a diversity antenna system that uses input from the rear window defroster element and the roof mounted mast aerial to improve radio reception.

Sound system components are interconnected via the vehicle bus system. Theft-proofing of the radio by code is not required as the radio does not function without the bus connection and a valid signal from the instrument cluster.

Car communication computer (CCC) and navigation system

When equipped, the car communication computer (CCC) is the central audio system controller and navigation computer. Radio and navigation information are both shown in the central information display (CID). A joystick on the center console is used to control CCC and navigation functions.

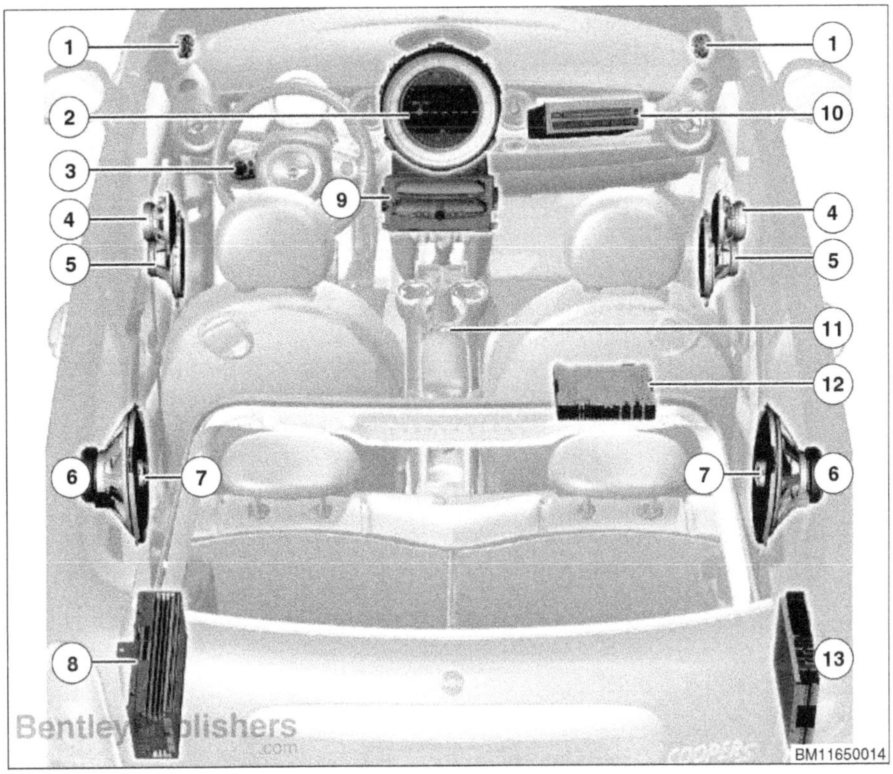

BM11650014

1. Front tweeter
2. Central information display (CID)
3. Multifunction steering wheel buttons
4. Front midrange speaker
5. Front bass speaker
6. Rear woofer
7. Rear tweeter
8. Audio amplifier
9. Car communication computer (CCC)
10. CD changer
11. Joystick
12. Satellite radio (SDARS)
13. High definition radio tuner (IBOC)

Navigation basics

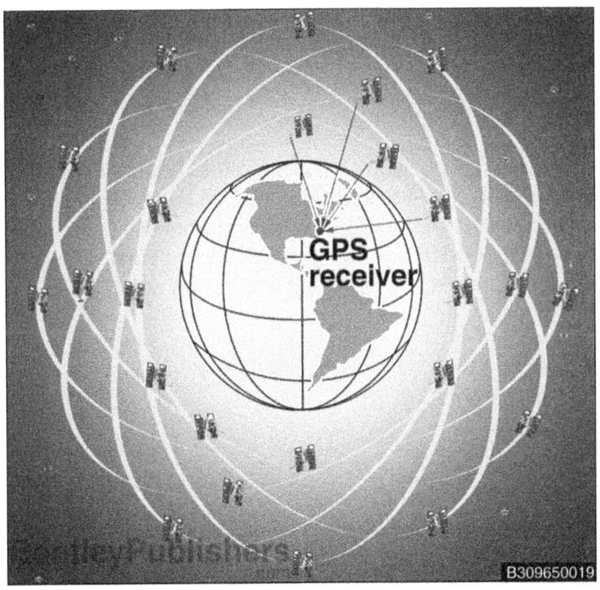

B309650019

Vehicle navigation systems utilize signals from global positioning system (GPS) satellites.

The MINI navigation system features the following range of functions:

- Navigation with map and arrow mode.
- Display on 6.5 in color central information display (CID).
- Input via joystick.
- Input optionally controlled by voice commands.

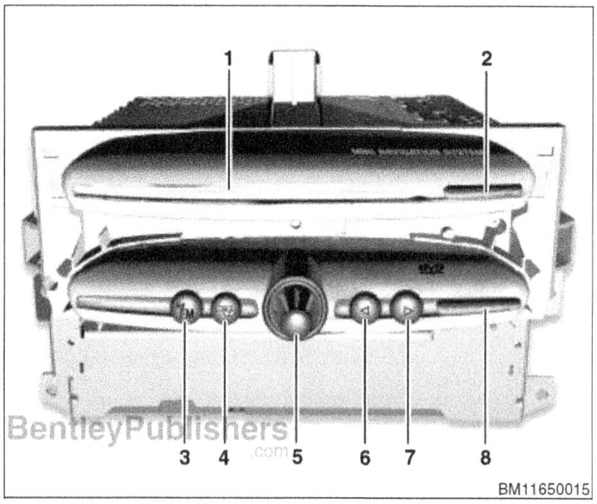

CCC overview

1. Audio CD player.
2. CD eject button.
3. FM button.
4. Mode button (next audio source).
5. Volume control.
6. Back search button.
7. Forward search button.
8. Detachable panel eject button.

The following functions are integrated in the CCC:
- Double radio tuner (AM / FM).
- Navigation system with maps on DVD.
- Voice activation system.
- Audio functions.
- MOST-CAN gateway functions.

A DVD player with navigation information is integrated in the CCC. Pressing the front panel eject button (**8**) exposes the opening for the DVD drive and the DVD eject button.

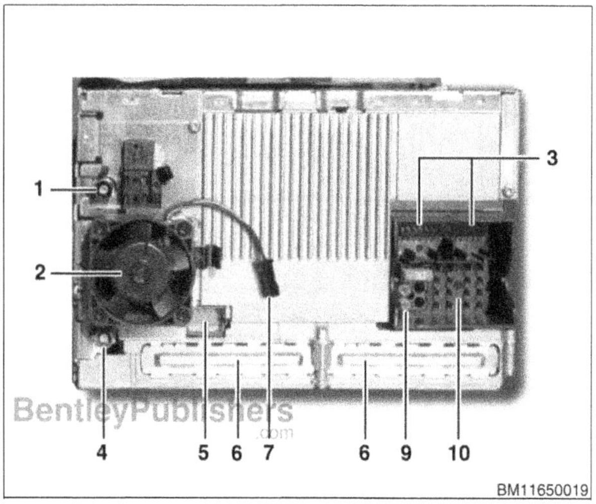

CCC rear panel connections

1. Radio antenna connector
2. Cooling fan
3. 12-pin connectors
4. GPS antenna connector
5. LVDS connector
6. Expansion card cover
7. Cooing fan connector
8. MOST-bus connector
9. 16-pin connector

HIP module

The host independence positioning (HIP) module is integrated with the CCC. This module converts and decodes signals received by the GPS antenna. In addition, telegrams containing wheel speed sensor signals are received via K-CAN from the DSC control unit. The HIP module uses these inputs to calculate the following data:
- Longitude
- Latitude
- Altitude above sea level
- Direction of travel
- Road speed

The navigation system continues to receive positioning information from satellites when the ignition is OFF. This ensures that vehicle position is known and the navigation system can function immediately after a start.

Yaw sensor

Because satellite signals are not received everywhere, a yaw sensor (rotation rate sensor) is integrated in the CCC. This allows the navigation system to have an independent means of verifying changes in vehicle direction.

MOST-bus (media oriented systems transport)

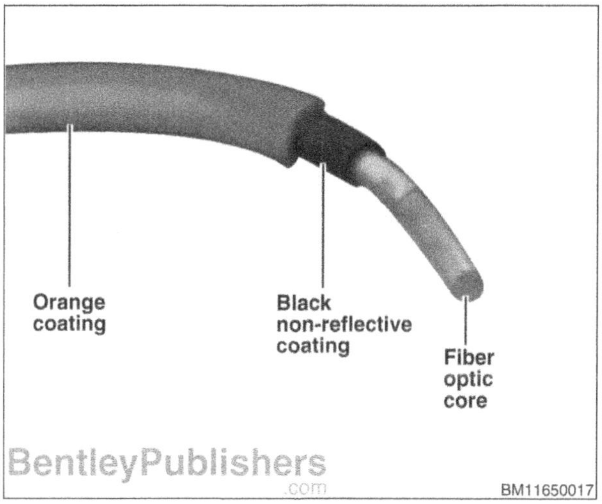

Sound system and navigation components are connected using MOST-bus, a fiber optic network.

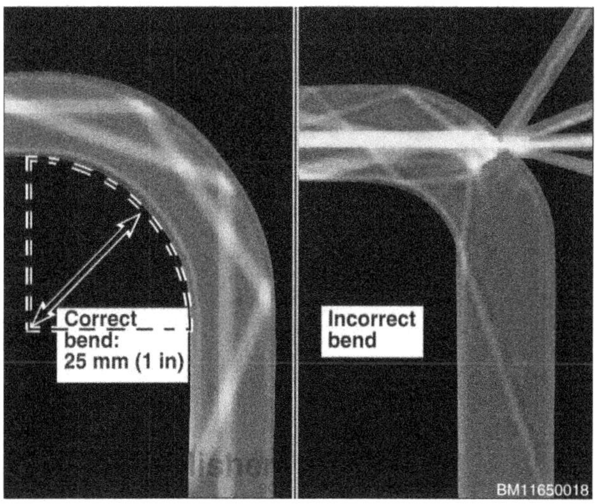

Make sure any bend in MOST-bus fiber optic harness has a radius of 25 mm (1 in) or greater.

> ***CAUTION—***
> - *Signal transmission through a fiber optic harness that is bent excessively is impaired.*

650

Cautions

> *CAUTION—*
> - *When working on electrical components, disconnect the negative (-) cable from the battery. Before disconnecting the battery, read battery disconnection cautions in* **001 Warnings and Cautions**.
> - *Static discharge may permanently damage an electronic component such as tachometer, instrument cluster or audio equipment. Handle using accepted static prevention equipment and techniques. See* **600 Electrical System–General**.
> - *To avoid marring trim pieces when removing interior trim clips, use a plastic prying tool or a screwdriver with the tip wrapped with masking tape.*

AUDIO SYSTEM

Radio, removing and installing

– Disconnect negative (-) cable from battery.

> *CAUTION—*
> - *Prior to disconnecting the battery, read the battery disconnection cautions given in* **001 Warnings and Cautions**.

– Remove left lower dashboard trim (underneath steering column). See **513 Interior Trim**.

– Remove tachometer. See **Tachometer, removing and installing**. See **620 Instruments**.

– Remove center dashboard trim. See **513 Interior Trim**.

– Remove left and right dashboard decorative strips and dashboard cover. See **513 Interior Trim**.

– Partially remove passenger knee pad. See **513 Interior Trim**.

– Remove driver knee pad. See **513 Interior Trim**.

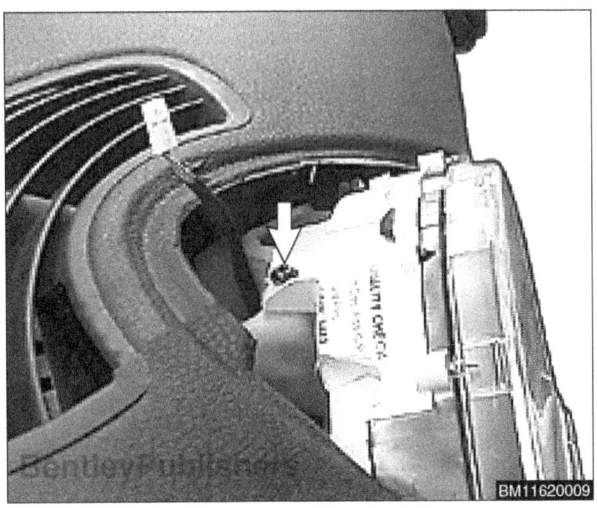

< Remove upper instrument cluster mounting screw (**arrow**).

Remove lower cluster mounting screws (**arrows**).

- Pull back cluster, detach electrical connectors and remove cluster.

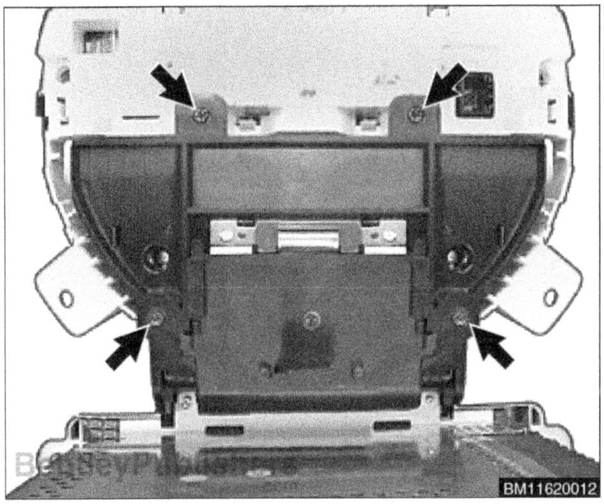

To remove radio: Remove mounting screws (**arrows**) and separate speedometer from radio.

- Installation is reverse of removal.
- Use MINI scan tool to code and program radio.

CD changer, removing and installing

This procedure applies to models with separate CD changer in the right side dashboard.

- Disconnect negative (-) battery cable and cover battery terminal to keep cable from accidentally contacting terminal.

> ***CAUTION—***
> - *Prior to disconnecting the battery, read the battery disconnection cautions given in* **001 Warnings and Cautions**.

- Remove passenger knee protection (panel underneath right side dashboard). See **513 Interior Trim**.

CD changer, removing and installing

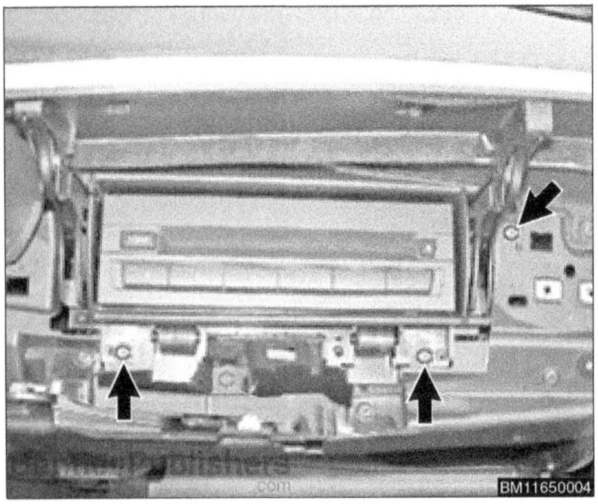

◀ Remove mounting screws (**arrows**) and pull out CD changer. Detach electrical connector.

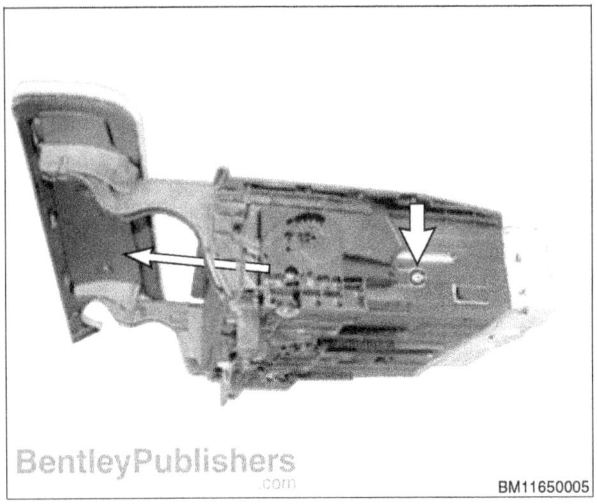

◀ Remove screws (**arrow**) and slide CD changer out of housing.

◀ Installation is reverse of removal. Make sure spring retainer (**arrow**) is seated correctly.

– Use MINI scan tool to code and program CD changer.

Audio amplifier, removing and installing

- Disconnect negative (-) battery cable and cover battery terminal to keep cable from accidentally contacting terminal.

> *CAUTION—*
> - *Prior to disconnecting the battery, read the battery disconnection cautions given in* **001 Warnings and Cautions**.

◄ Working in cargo compartment or trunk, remove left side trim cover (**arrow**) .

◄ Remove amplifier retaining screw (**arrow**) and slide amplifier up to disengage from bracket.

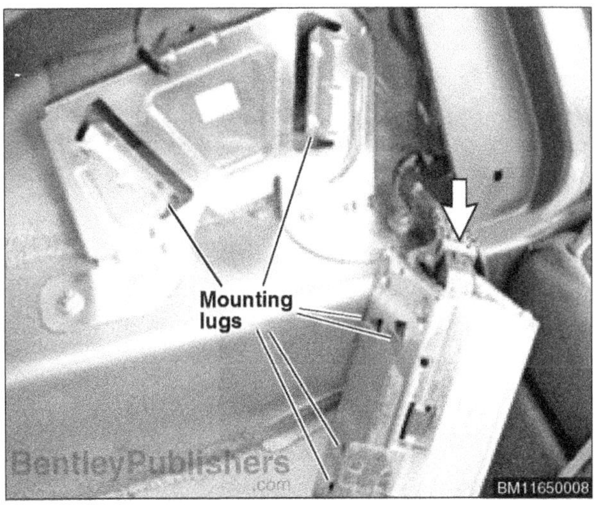

◄ Detach electrical connector (**arrow**) to remove amplifier.

- When installing, make sure amplifier and mounting bracket lugs engage securely.
- Use MINI scan tool to code and program audio amplifier.

Satellite radio, removing and installing

< Note that ID numbers (**arrows**) on radio tag are required to cancel service on old radio and to register new radio.

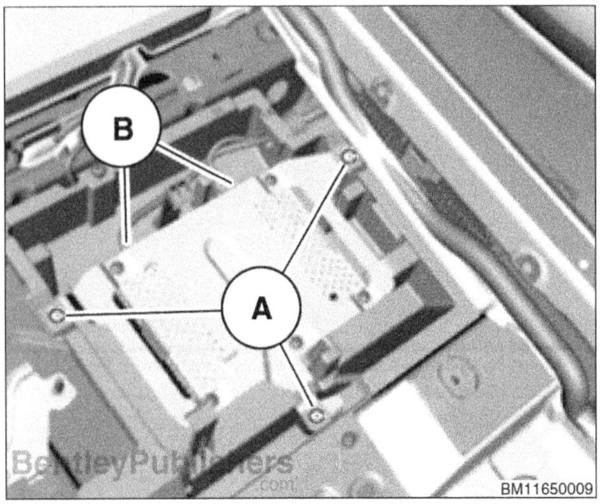

– Remove passenger seat. See **520 Seats**. Lift up carpet.

< Working in cavity underneath seat:

- Remove radio mounting screws (**A**).
- Lift up and detach electrical connectors (**B**).

– Installation is reverse of removal.

– Use MINI scan tool to code and program radio.

SPEAKERS

Front tweeter, removing and installing

– Remove A-pillar trim and separate speaker electrical connector. See **513 Interior Trim**.

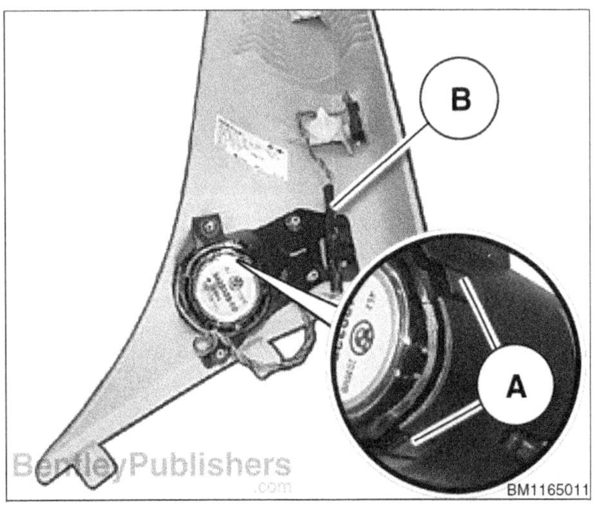

< Working at back side of trim panel:

- Unclip catches (**A**) carefully.
- Detach electrical harness (**B**).

– Installation is reverse of removal. Make sure sealing ring between speaker and panel is in place.

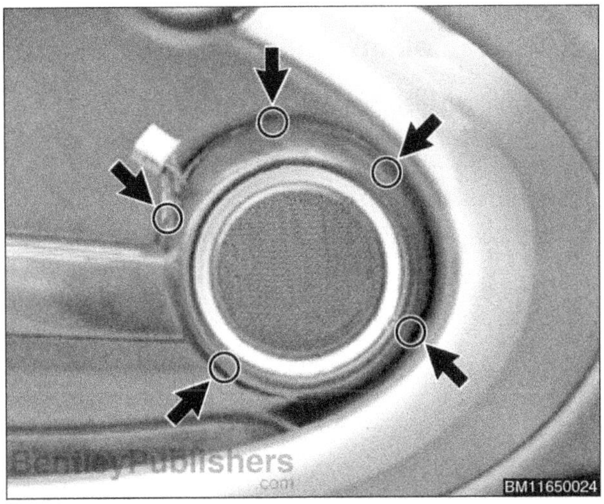

Front door midrange speaker, removing and installing

– Remove outside mirror switch. See **612 Switches**.

◄ Use plastic prying tool to lever speaker trim off plastic catches (**arrows**) and detach from door panel.

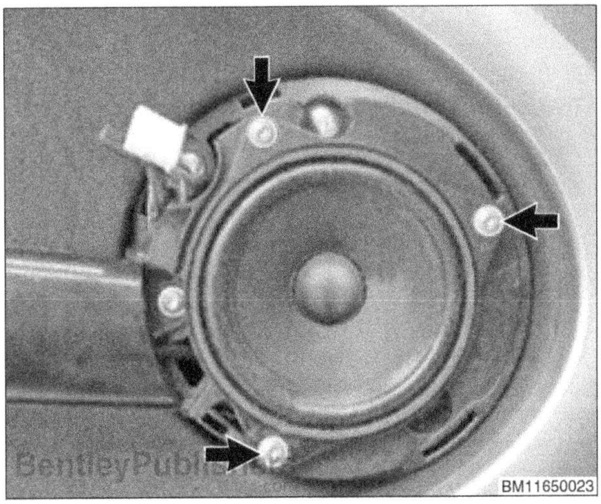

◄ Remove speaker mounting screws (**arrows**).

– Pull out speaker and detach electrical connector from behind.

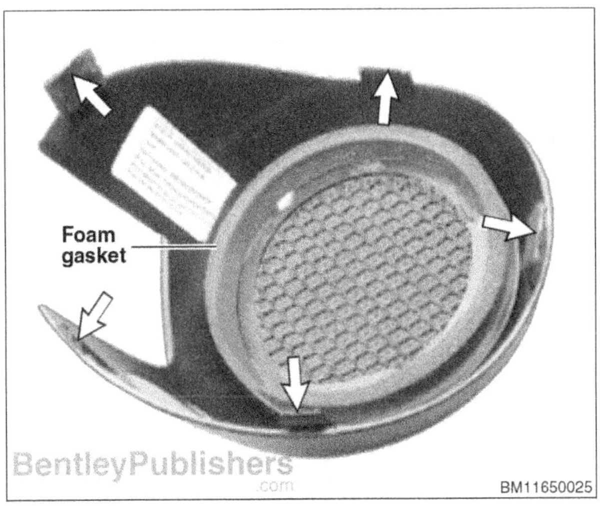

◄ When reinstalling, make sure plastic catches (**arrows**) and foam gasket on cover are undamaged.

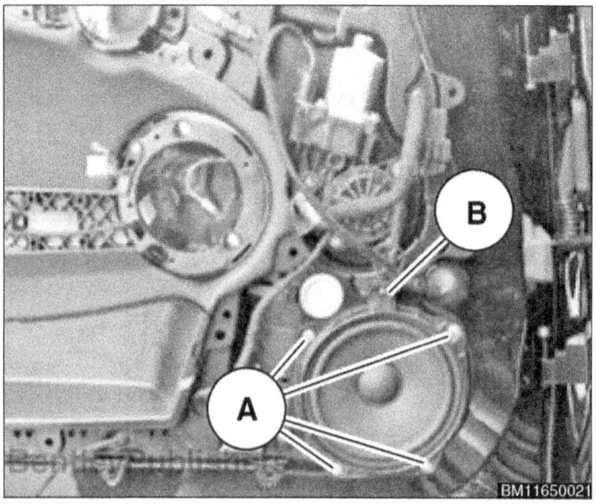

Front door bass speaker, removing and installing

- Remove front door trim panel. See **411 Doors**.

◄ Working at bass speaker:
  - Remove speaker mounting screws (**A**).
  - Detach electrical connector (**B**) and remove speaker.

- Installation is reverse of removal.

Rear tweeter, removing and installing

- Remove rear inside trim panel and separate speaker electrical connector. See **513 Interior Trim**.

◄ Working at back side of trim panel, unclip catches (**A**) carefully and lift off tweeter.

- Installation is reverse of removal. Make sure sealing ring between speaker and panel is in place.

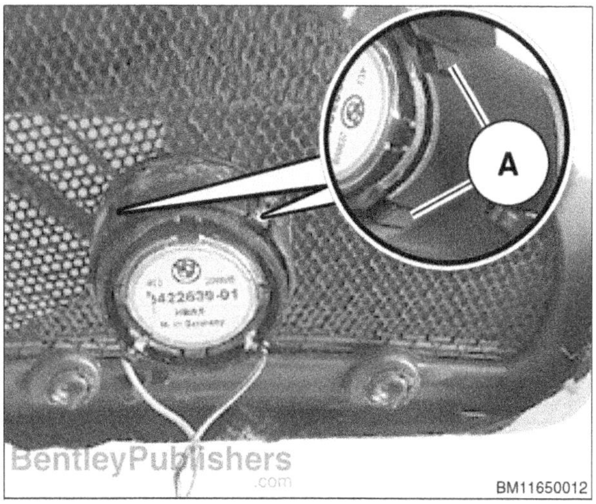

Rear woofer, removing and installing

- Remove rear inside trim panel. See **513 Interior Trim**.
- Clubman: Remove club door trim panel. See **513 Interior Trim**.

◄ Remove speaker mounting screws (**arrows**).

- Pull out speaker and detach electrical connector from behind.
- Installation is reverse of removal.

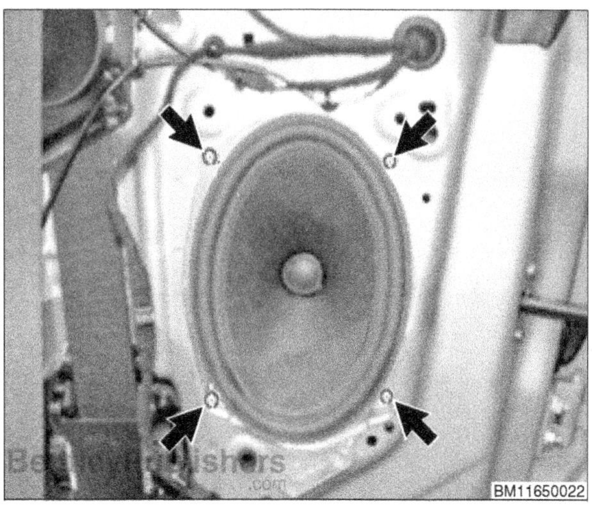

ANTENNA

Antenna components

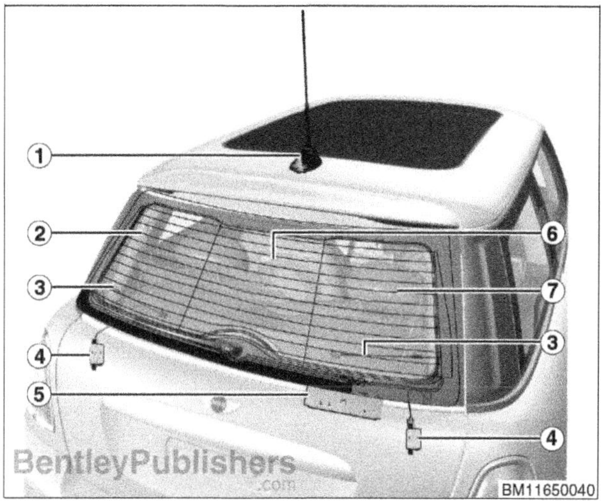

1. AM, FM1, telephone antenna mast
2. FM3 antenna
3. Connector
4. Wave trap
5. Antenna amplifier and diversity module
6. FM4 antenna
7. FM2 antenna

Hardtop model illustrated. Clubman is similar.

Antenna mast, removing and installing

◄ Rotate antenna mast counterclockwise (**arrow**) and remove.

– Installation is reverse of removal.

Roof antenna base, removing and installing

– Lower headliner just enough to gain access to antenna base. Note that full headliner removal requires removal of A-, B- and C-pillar trim panels, dome lights and grab handles.

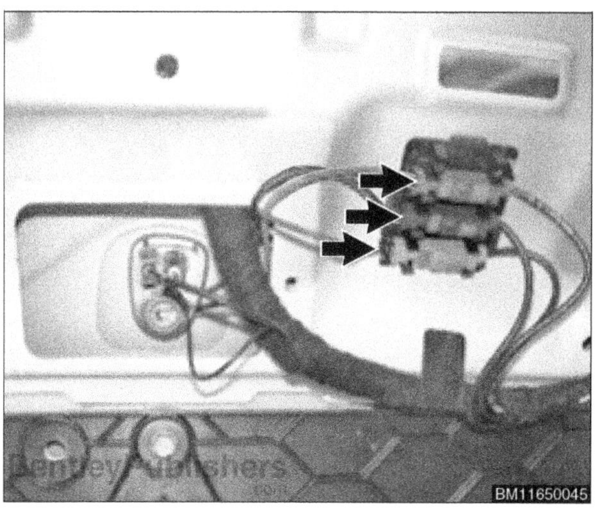

◄ Working above headliner, detach antenna electrical connectors (**arrows**) at multiple connector.

650

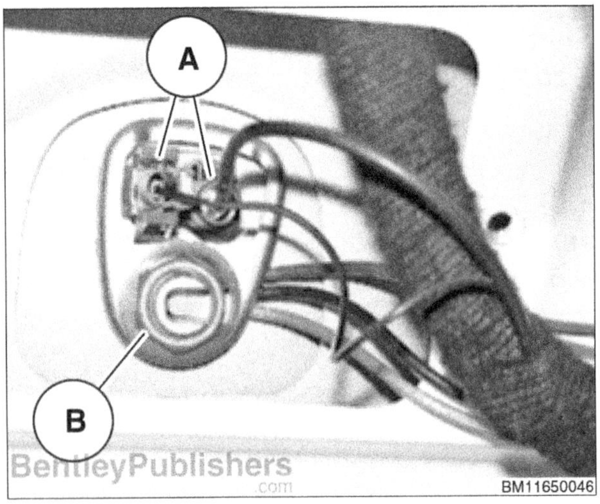

Working above headliner directly underneath antenna base:

- Detach electrical connectors (**A**).
- Remove retaining nuts (**B**) and lift off antenna base and wires from above.

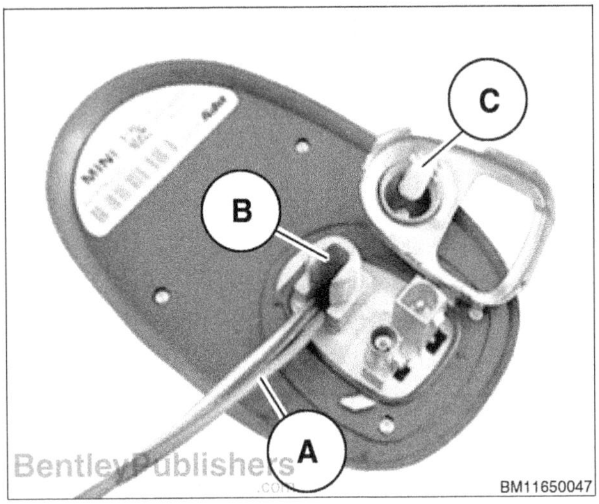

When installing, ensure correct cable routing:

- Antenna leads (**A**) are routed through groove (**B**).
- Retainer (**C**) secures leads in groove.

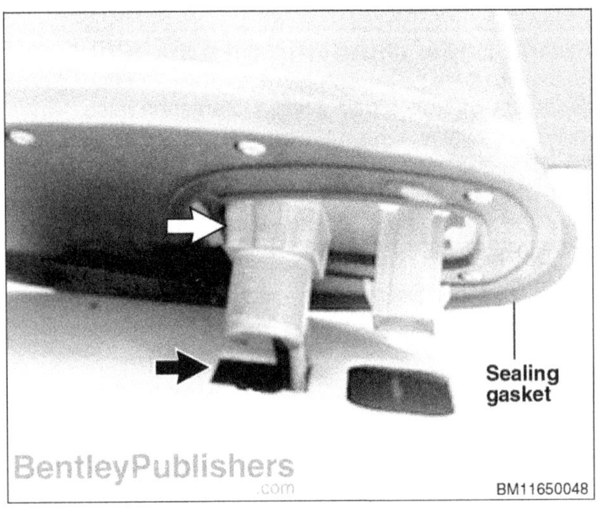

Align antenna base guide and roof opening (**arrows**). Make sure **sealing gasket** is undamaged.

Tightening torque	
Antenna base to roof (M13 x 1.0 nut)	5 Nm (4 ft-lb)

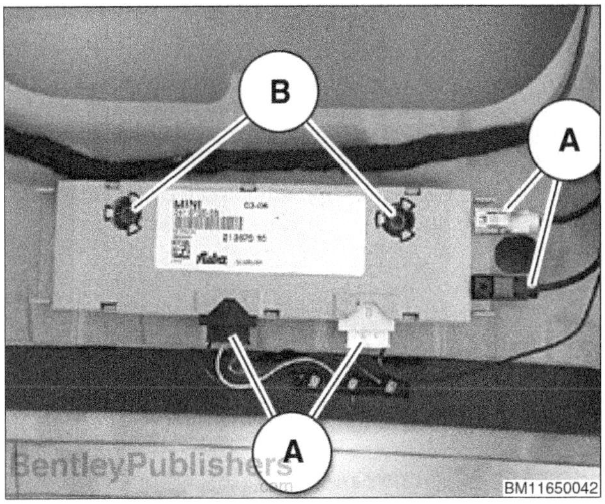

Antenna amplifier and diversity module, removing and installing

Hardtop

- Open rear hatch and remove hatch trim panel along lower edge of rear window. See **412 Rear Hatch, Split Rear Doors**.

- Working at module:
  - Detach electrical connectors (**A**).
  - Remove mounting screws (**B**) and lift off module.

- Installation is reverse of removal.

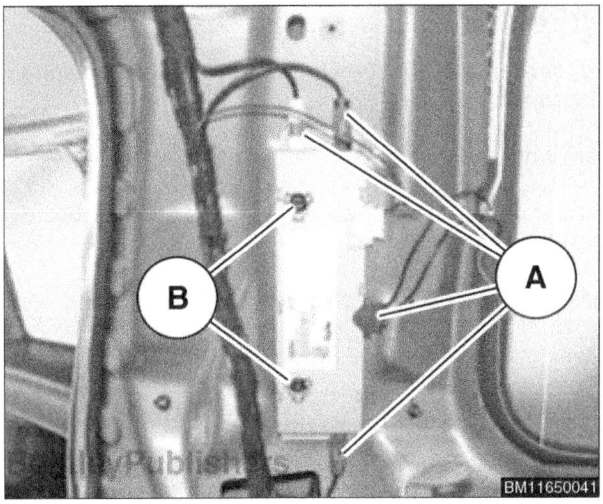

Clubman

- Open split rear doors and remove left side cargo compartment trim.

- Working at module:
  - Detach electrical connectors (**A**).
  - Remove mounting screws (**B**) and lift off module.

- Installation is reverse of removal.

Wave trap, removing and installing

- Hardtop: Open rear hatch and remove hatch trim panel along lower edge of rear window. See **412 Rear Hatch, Split Rear Doors**.

- Clubman: Open split rear doors. Remove right or left split door inner trim panel. See **412 Rear Hatch, Split Rear Doors**.

- Working at left or right wave trap:
  - Detach electrical connectors (**A**).
  - Remove mounting screw (**B**) and remove wave trap.

- Installation is reverse of removal.

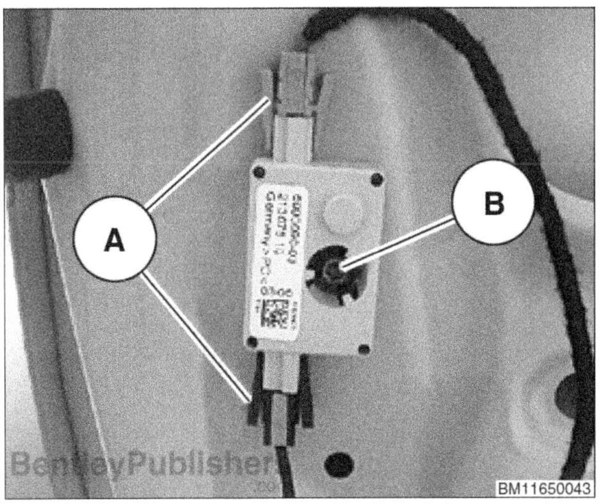

Car Communication Computer (CCC) Repairs

The car communications computer (CCC) controls joystick functions and houses a CD drive and the navigation DVD drive.

Be sure to handle CCC and associated equipment using accepted static prevention equipment and techniques. See **600 Electrical System–General**.

Car communication computer (CCC), removing and installing

– Disconnect negative (-) cable from battery.

> ***CAUTION—***
> - *Prior to disconnecting the battery, read the battery disconnection cautions given in* **001 Warnings and Cautions**.

– Remove left lower dashboard trim (underneath steering column). See **513 Interior Trim**.

– Remove tachometer. See **Tachometer, removing and installing**. See **620 Instruments**.

– Remove center dashboard trim. See **513 Interior Trim**.

– Remove left and right dashboard decorative strips and dashboard cover. See **513 Interior Trim**.

– Partially remove passenger knee pad. See **513 Interior Trim**.

– Remove driver knee pad. See **513 Interior Trim**.

– Remove central information display (CID). See **620 Instruments**.

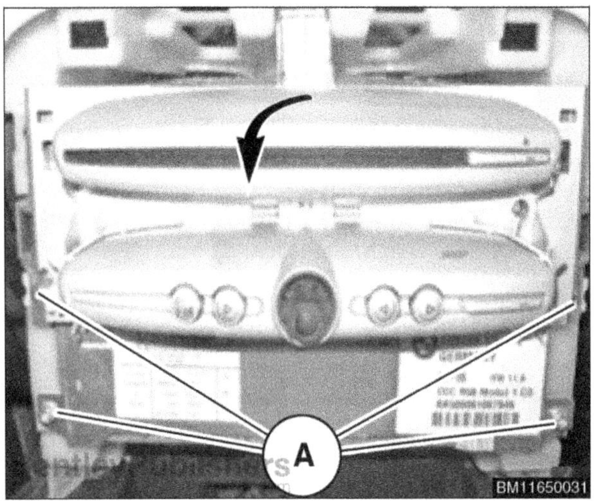

◄ Remove CCC mounting screws (**A**) and pull CC back slightly to detach electrical connectors. Then lift out CCC.

– Installation is reverse of removal. Make sure harnesses are routed as before.

– Use MINI scan tool to code and program CCC.

CCC CD drive, removing and installing

– Disconnect negative (-) cable from battery.

> ***CAUTION—***
> - *Prior to disconnecting the battery, read the battery disconnection cautions given in* **001 Warnings and Cautions**.

– Remove CCC from center dashboard. See **Car communication computer (CCC), removing and installing** in this repair group.

Use plastic prying tool (MINI special tool **64 1 020** or equivalent) to lever off CCC upper cover.

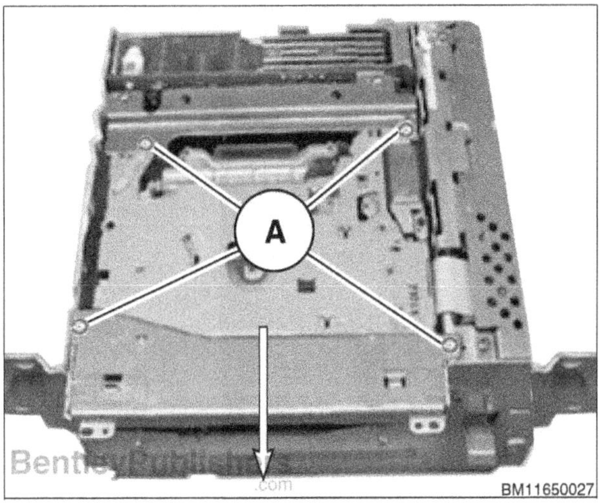

Remove CD drive mounting screws (**A**) and slide drive out of CCC (**arrow**).

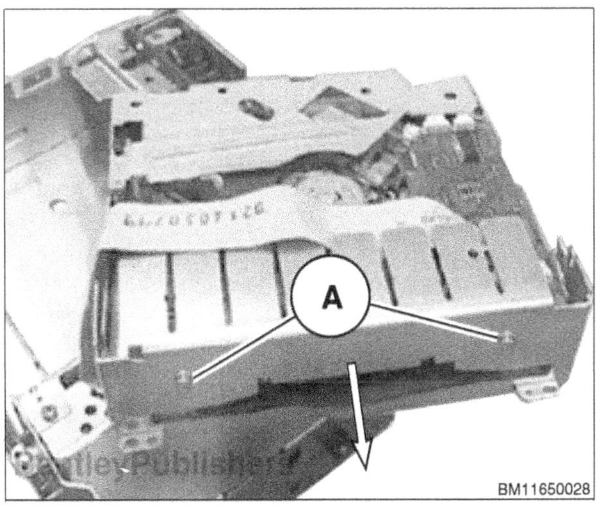

Remove CD drive trim mounting screws (**A**) and remove trim (**arrow**).

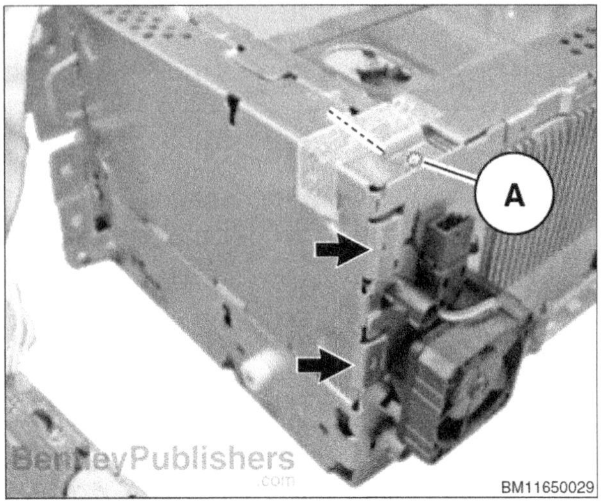

- Working at side of CCC:
  - Cut through warranty seal (**dashed line**).
  - Remove cover mounting screw (**A**).
  - Disengage side cover carefully from mountings (**arrows**) and place aside.

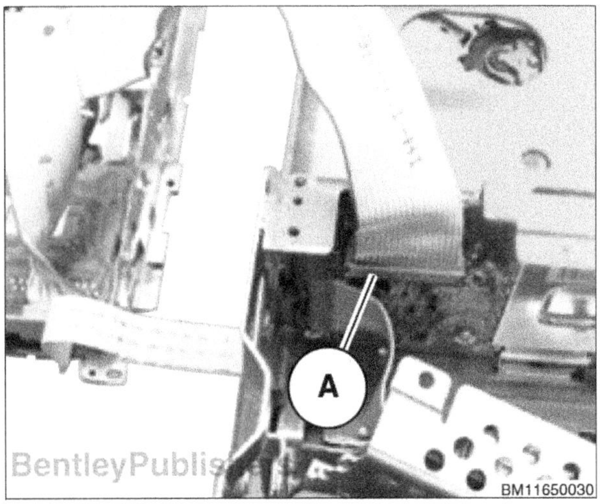

- Detach electrical ribbon connector (**A**) and remove CD drive.
- Installation is reverse of removal. Make sure harnesses are routed as before.
- Use MINI scan tool to code and program CCC.

Navigation DVD drive, removing and installing

- Disconnect negative (-) cable from battery.

> ***CAUTION—***
> - *Prior to disconnecting the battery, read the battery disconnection cautions given in* **001 Warnings and Cautions**.

- Remove CCC from center dashboard. See **Car communication computer (CCC), removing and installing** in this repair group.

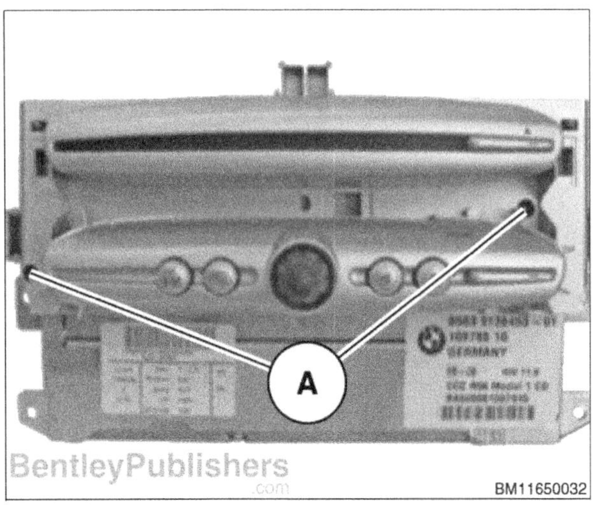

- Remove CCC front panel retaining screws (**A**) and tilt panel forward.

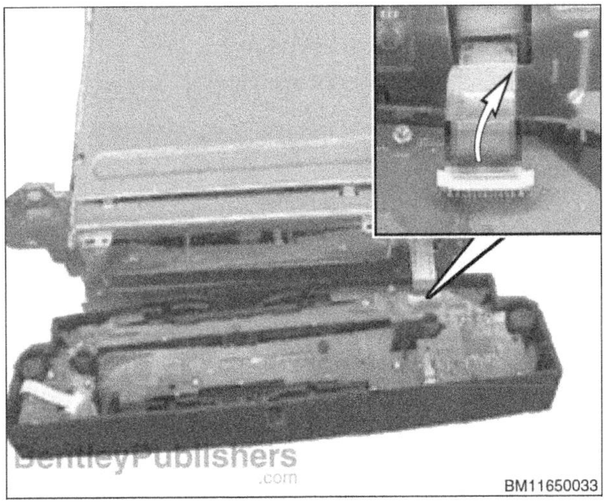

◄ Release electrical connector lock (**insert**) and pull off connector plug carefully. Remove CCC front panel.

– If necessary, cut through warranty seal at top of CCC unit.

◄ Use plastic prying tool (MINI special tool **64 1 020** or equivalent) to lever off CCC upper cover.

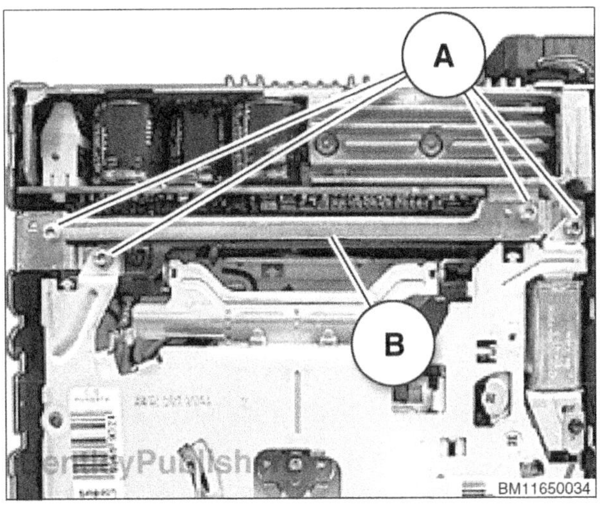

◄ Working above top of CCC, remove screws (**A**) and lift off bridge (**B**).

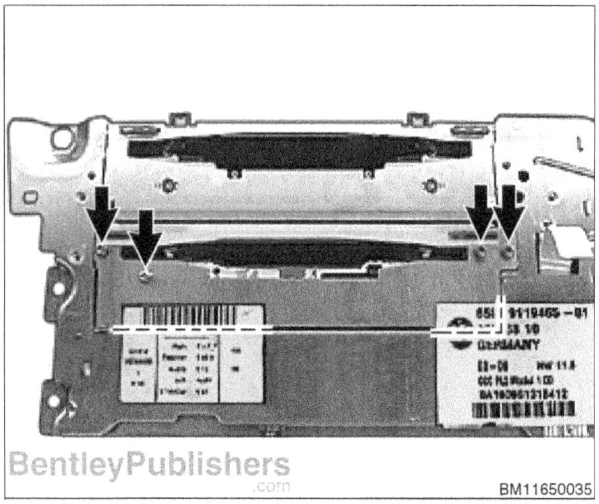

BM11650035

Working at DVD drive cover:

- Cut through product labels (**dashed lines**).
- Remove cover mounting screws (**arrows**) and lift off cover.

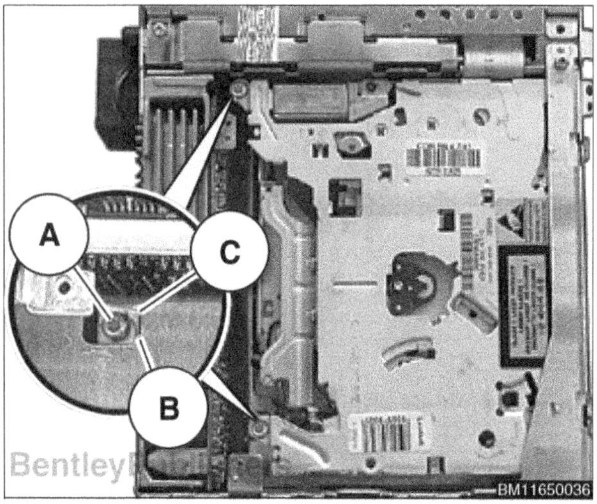

BM11650036

Working at bottom of CCC:

- Use magnetic screwdriver to remove mounting screws (**A**).
- Note that DVD drive tab (**B**) is above CCC tab (**C**).

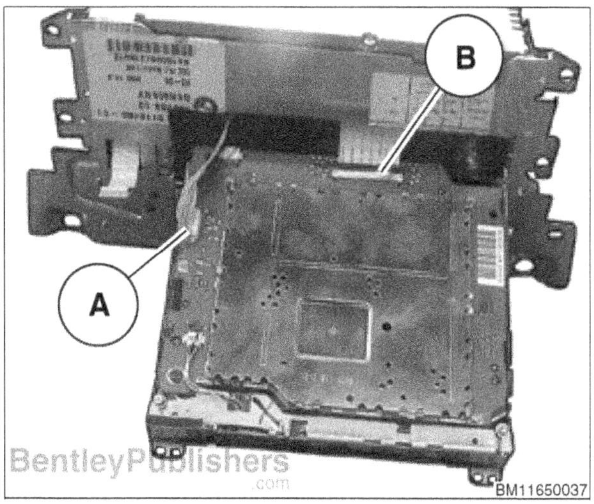

BM11650037

Tip DVD drive out of CCC, carefully detaching ribbon harness connectors (**A**, **B**). Note that ribbon cable **B** comes in two versions: Early version connects horizontally, current version connects vertically.

– Installation is reverse of removal. Make sure harnesses are routed as before.

– Use MINI scan tool to code and program CCC.

HIP module, removing and installing

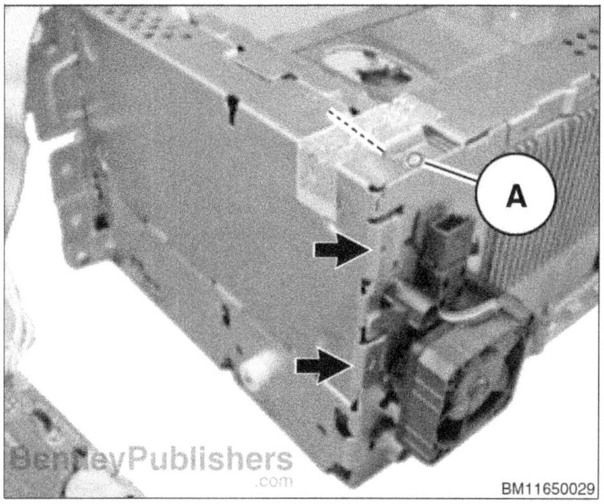

The host independence positioning (HIP) module, integrated with the CCC, is the primary interpreter of navigation data.

< Working at side of CCC:

- Cut through warranty seal (**dashed line**).
- Remove cover mounting screw (**A**).
- Disengage side cover carefully from mountings (**arrows**) and place aside.

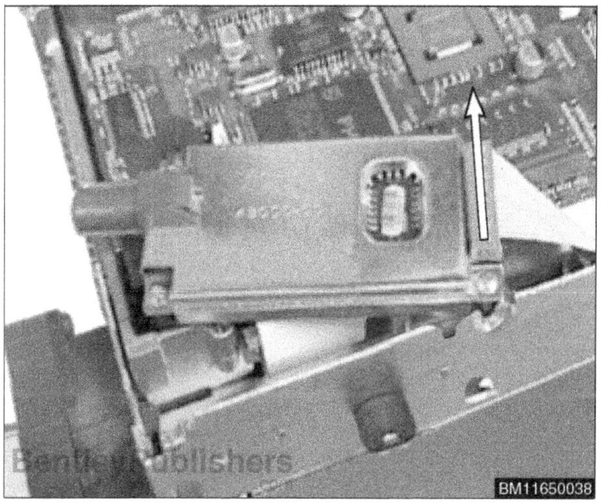

< Lift HIP module out of CCC housing and slide off ribbon connector (**arrow**).

– Installation is reverse of removal. Make sure harnesses are routed as before.

– Use MINI scan tool to code and program CCC.

– To initialize navigation: Drive vehicle for 15 - 20 minutes at 30 mph (50 kph),. Make sure GPS reception is good during this period. Watch navigation display until correct vehicle position is displayed. Note that during this initial drive, position of navigation pointer may change unpredictably until correct vehicle position is learned.

CCC cooling fan, removing and installing

– Disconnect negative (-) cable from battery.

> ***CAUTION—***
> - *Prior to disconnecting the battery, read the battery disconnection cautions given in* **001 Warnings and Cautions**.

– Remove CCC from center dashboard. See **Car communication computer (CCC), removing and installing** in this repair group.

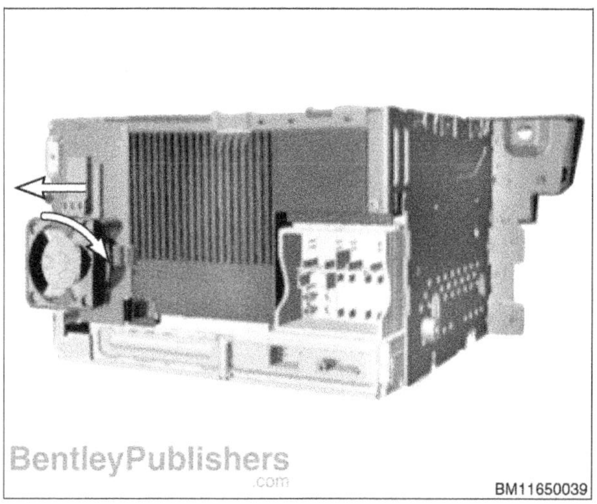

< working at cooling fan, raise retaining hook (**arrow**) and rotate fan clockwise (**curved arrow**) to remove.

– Installation is reverse of removal. Make sure harnesses are routed as before.

– Use MINI scan tool to code and program CCC.

720 Seat Belts

GENERAL

This repair group covers repairs to the seat belt system.

See also:

- **121 Battery Starter, Alternator**
- **600 Electrical System–General**
- **721 Airbag System (SRS)**

Front seat belt pyrotechnic tensioners

The front seat belt buckle is equipped with pyrotechnic (explosive charge) automatic tensioner. The tensioner is designed to automatically retract and tension the seat belt by 55 mm (approximately 2 inches) in case of accident.

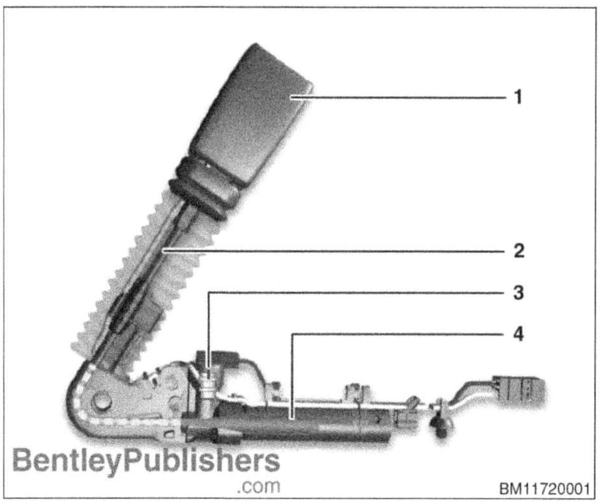

1. Seat belt buckle
2. Retractor cable
3. Pyrotechnic igniter
4. Locking element

The sixth generation multiple restraint system (MRS6) electronics coordinates the operation of pyrotechnic belt tensioners with the deployment of airbags. For more details about MRS6, see **721 Airbag System (SRS)**. For diagnostic work on MRS, use MINI scan tool.

720

Troubleshooting

Make sure that the seat belt reel locks when pulled quickly and sharply. The automatic reel cannot be serviced or repaired.

In case of trouble, see **Table a**

Table a. Seat belt troubleshooting

Problem	Probable cause	Repairs
Damage to seat belt.	Accident in which bumpers were permanently deformed.	Replace complete automatic seat belt. Also check and replace, if necessary, seat belt mounting on car body or seat runner.
Belt creased, unraveled, pinched, cut or melted. Belt buckle or belt lock plastic casing worn, damaged or missing.	Seat belts aged or worn.	Replace complete automatic seat belt.
Seat belt buckle is not ejected by spring pressure when red button on seat belt lock is pressed.	Seat belt buckle mechanism worn or damaged.	Replace seat belt buckle.
Seat belt automatic reel does not lock when pulled out suddenly.	Automatic reel defective.	Replace reel assembly.
Seat belt automatic reel jams when pulled out.	Automatic reel loose. Return spring broken inside reel.	Tighten reel mounting bolt. Replace reel assembly.
Seat belt does not retract automatically.	Automatic reel loose. Return spring broken inside reel.	Tighten reel mounting bolt. Replace reel assembly.
Automatic belt squeaks when fastened or unfastened.	Excessive friction in belt guides.	Replace reel assembly.
	Automatic reel loose.	Tighten reel mounting bolt.
	Return spring broken inside reel.	Replace reel assembly.
Seat belt pyrotechnic tensioner triggered.	Accident triggered pyrotechnic deployment.	Replace complete automatic seat belt. Also check retaining bracket of belt tensioner for twist.

Warnings and Cautions

WARNING—

- *Pyrotechnic seat belt tensioners are powerful devices. Handle with extreme care. Incorrect handling can trigger the tensioner and cause injury.*
- *MINI recommends that repair or replacement work on pyrotechnic devices be carried out by a qualified MINI technician.*
- *Disconnect the battery and wait 5 seconds before attempting to work on pyrotechnic devices.*
- *Pyrotechnic devices cannot be repaired. Replace if damaged.*
- *Do not treat pyrotechnic components with cleaning agents or grease.*
- *Do not expose pyrotechnic components to temperatures above 75°C (167°F).*
- *Do not fire a pyrotechnic gas generator prior to disposal. It must be fired by a special disposal company or shipped back to the manufacturer in the packaging of the new component.*
- *During body straightening and welding with an electric arc welder, disconnect the battery and the connection to the pyrotechnic gas generators.*
- *Replace seat belts and seat belt hardware if subjected to occupant loading in a collision.*
- *Do not modify or repair seat belts. Do not modify seat belt mounting points.*
- *Do not bleach or dye seat belt webbing. Webbing that is severely faded or redyed does not meet the strength requirements of a collision.*
- *Clean belts with a luke-warm soap solution only.*
- *Periodically inspect seat belts for webbing defects such as cuts or pulled threads.*
- *Immediately after replacing a damaged or worn seat belt, destroy the old belt to prevent it from being reused.*

CAUTION—

- *When working on electrical components, disconnect the negative (–) cable from the battery and insulate the cable end to prevent accidental reconnection.*
- *Prior to disconnecting the battery, read the battery disconnection cautions given in* **001 Warnings and Cautions**.
- *To avoid marring trim pieces when removing interior trim clips, use a plastic prying tool or a screwdriver with the tip wrapped with masking tape.*

Front Seat Belts

Front seat belt assembly

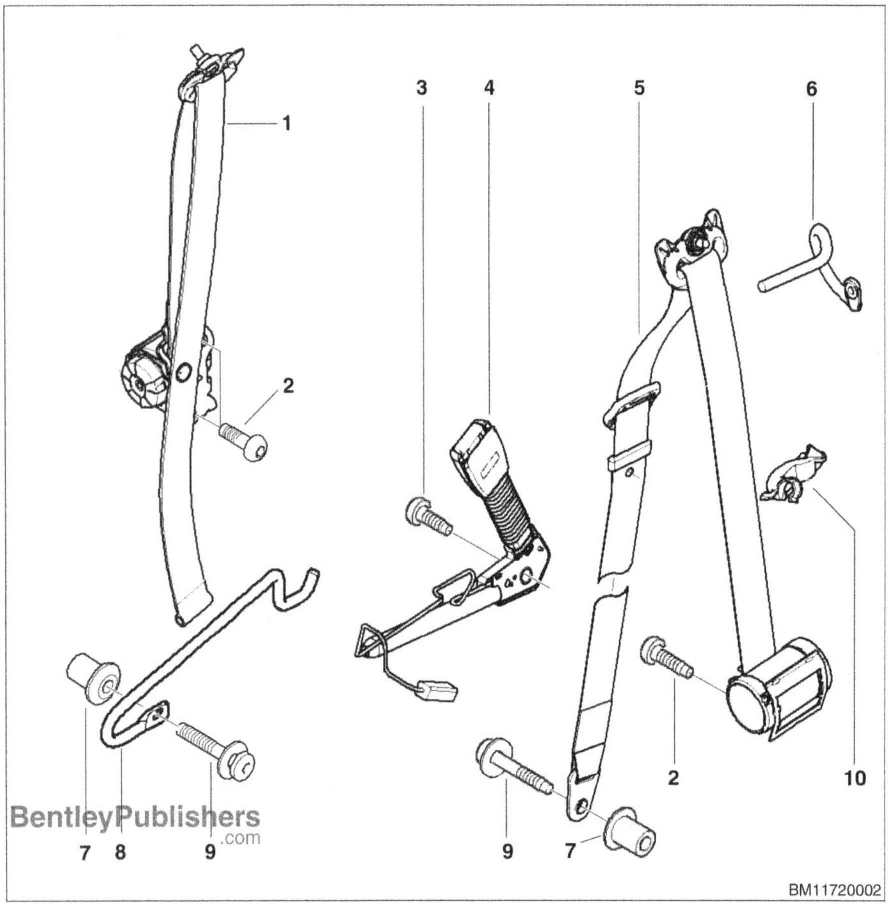

1. Clubman: Right front seat belt
2. Seat belt reel mounting bolt
   - Replace M10 x 28 mm bolt with new
   - Install with Locktite® and tighten to 36 Nm (27 ft-lb)
3. Seat belt buckle mounting bolt
   - M10 x 22 mm Torx
   - Install with Locktite® and tighten to 36 Nm (27 ft-lb)
4. Pyrotechnic tensioner and seat belt buckle assembly
5. Left front seat belt
   - Hardtop or Convertible: Right front seat belt is similar
6. Convertible: Belt deflection bar
7. Bushing
8. Clubman: Sliding belt bracket
9. Seat belt anchor or sliding bracket mounting bolt
   - Replace duo-taptite bolt with new
   - Install with Locktite® and tighten to 36 Nm (27 ft-lb)
10. Convertible: Belt guide

Left front seat belt, removing and installing (Hardtop, Clubman)

Left front seat belt service procedure on Hardtop and Clubman models is similar. Right front seat belt on Hardtop is similar to left front, whereas in Clubman models right front seat belt is different.

– Remove rear seat. See **520 Seats**.

– Remove B- and C-pillar trim panels and rear side trim panel. See **513 Interior Trim**.

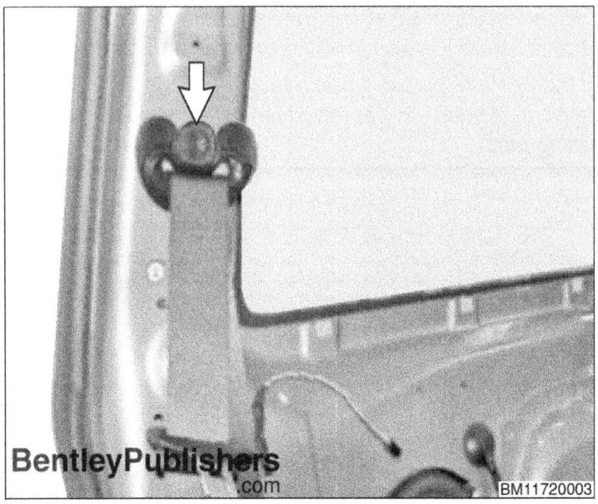

< Working at upper B-pillar, remove belt mounting bolt (**arrow**) and detach belt.

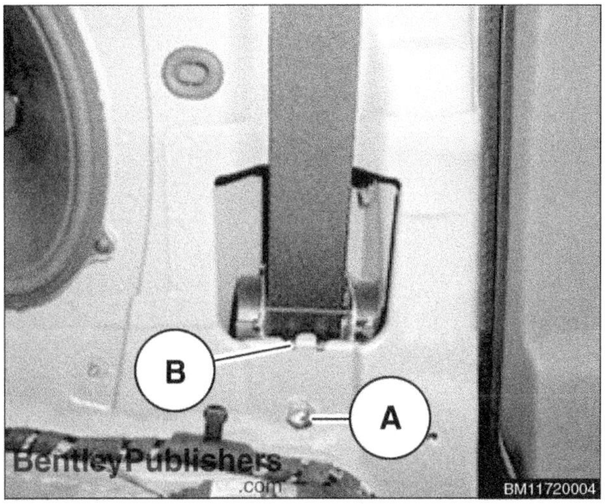

Working at base of B-pillar:

- Remove seat belt reel mounting bolt (**A**) and discard.
- Lift reel out of B-pillar opening.
- Note that locating tab on reel (**B**) prevents incorrect installation.

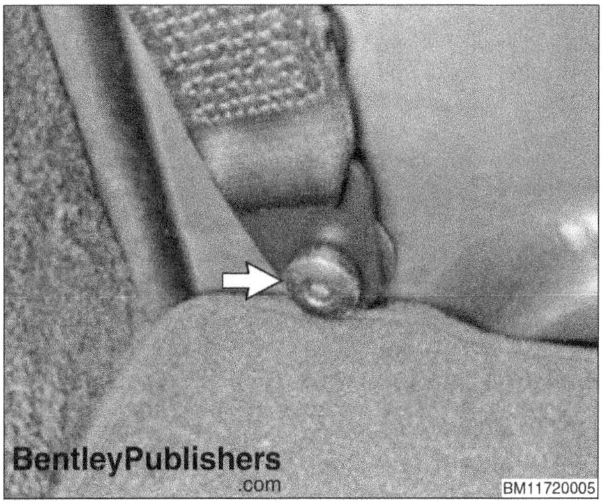

Working below door opening, remove belt anchor bolt (**arrow**) and discard. Remove belt from vehicle.

– Installation is reverse of removal.

- Use new mounting bolts for seat belt reel and lower anchor.
- Use Locktite® when installing new bolts.

Tightening torques	
Seat belt anchor to body (replace M10 duo-taptite bolt with new, use Locktite®)	36 Nm (27 ft-lb)
Seat belt reel to B-pillar (replace M10 bolt with new, use Locktite®)	36 Nm (27 ft-lb)
Seat belt to B-pillar	36 Nm (27 ft-lb)

Right front seat belt, removing and installing (Clubman)

The Clubman right front seat belt upper and anchor bolts are attached to the club door.

– Remove club door inner trim panel. See **411 Doors**.

Working at top of club door, remove belt mounting bolt (**arrow**) and detach belt.

Working at bottom of club door, remove seat belt reel mounting bolts (**arrows**) and detach seat belt. Discard bolts.

Working below door sill, remove sliding belt bracket bolt (**arrow**) and discard. Tilt bracket to remove from anchor hole (**circle**) under rear seat. Remove belt from vehicle.

- If seat belt is damaged due to accident, inspect bracket carefully for bending. Replace if necessary.
- Installation is reverse of removal.
  - Use new mounting bolts for seat belt reel and sliding bracket.
  - Use Locktite® when installing new bolts.

Tightening torques	
Seat belt reel to club door (replace M10 bolt with new, use Locktite®)	36 Nm (27 ft-lb)
Seat belt sliding bracket bolt (replace M10 duo-taptite bolt with new, use Locktite®)	36 Nm (27 ft-lb)
Seat belt to upper club door	36 Nm (27 ft-lb)

Front seat belt, removing and installing (Convertible)

- Remove rear seat. See **520 Seats**.
- Remove rear side trim panel. See **513 Interior Trim**.

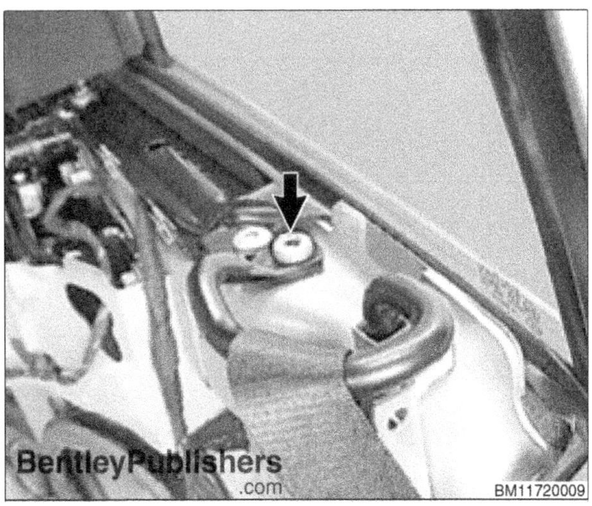

Working at top of B-pillar, remove belt deflection bar mounting bolt (**arrow**), lift up bar and detach belt.

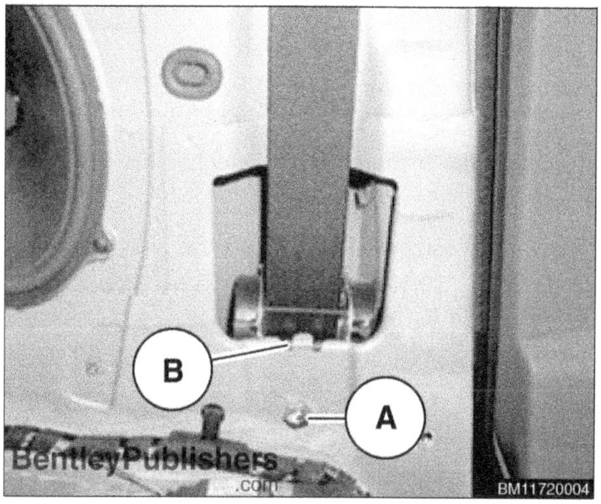

◄ Working at base of B-pillar:

- Remove seat belt reel mounting bolt (**A**) and discard.
- Lift reel out of B-pillar opening.
- Note that locating tab on reel (**B**) prevents incorrect installation.

◄ Working below door opening, remove belt anchor bolt (**arrow**) and discard. Remove belt from vehicle.

– Installation is reverse of removal.

- Use new mounting bolts for seat belt reel and lower anchor.
- Use Locktite® when installing new bolts.

Tightening torques	
Seat belt anchor to body (replace M10 duo-taptite bolt with new, use Locktite®)	36 Nm (27 ft-lb)
Seat belt reel to B-pillar (replace M10 bolt with new, use Locktite®)	36 Nm (27 ft-lb)

Front seat belt buckle, removing and installing

– Disconnect negative (–) cable from battery and cover negative terminal with insulating material.

> ***WARNING—***
> - *After disconnecting the battery, wait 5 seconds before beginning work on airbag components.*

> ***CAUTION—***
> - *Prior to disconnecting the battery, read the battery disconnection cautions given in* **001 Warnings and Cautions**.

– Remove front seat. See **520 Seats**.

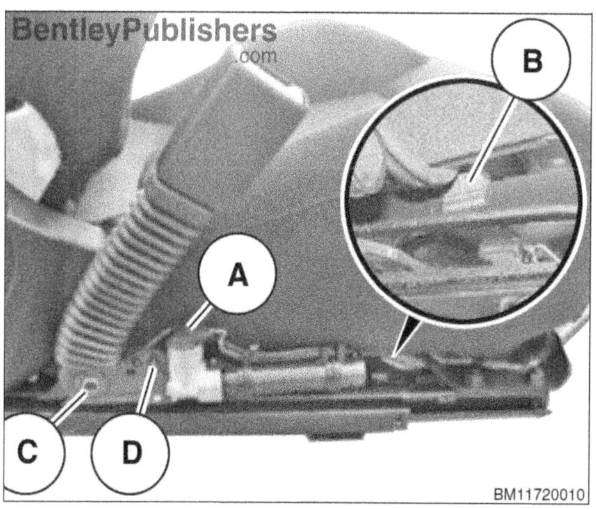

◄ Working at side of seat:

- Detach electrical connector (**A**).
- Pry retainer (**B**) out of seat mechanism.
- Remove belt buckle mounting bolt (**C**).
- Note that locating tab (**D**) prevents incorrect installation.

720

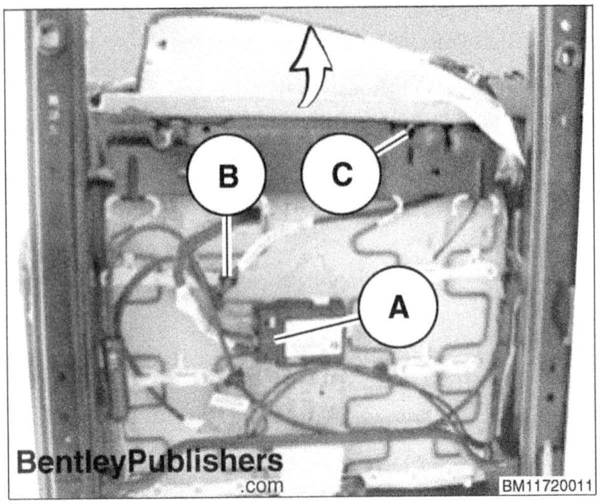

With seat placed upside down:

- Partially pull seat cover off seat rail (**curved arrow**).
- Detach electrical connector (**A**).
- Cut wire tie (**B**).
- Detach cable holder (**C**) and remove tensioner harness from seat mechanism.

– Installation is reverse of removal. Replace wire ties as necessary.

Tightening torque	
Seat belt buckle and tensioner to seat rail (replace M10 bolt with new, use Locktite®)	36 Nm (27 ft-lb)

Rear Seat Belts

Rear seat belt assembly

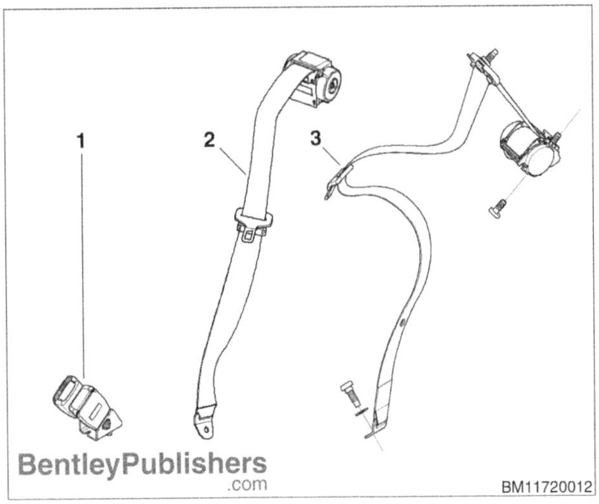

1. Seat belt buckle
2. Clubman rear seat belt
3. Hardtop or Convertible rear seat belt

Rear seat belt, removing and installing (Hardtop)

– Remove rear seat. See **520 Seats**.

– Remove C-pillar trim and cargo compartment side trim panel. See **513 Interior Trim**.

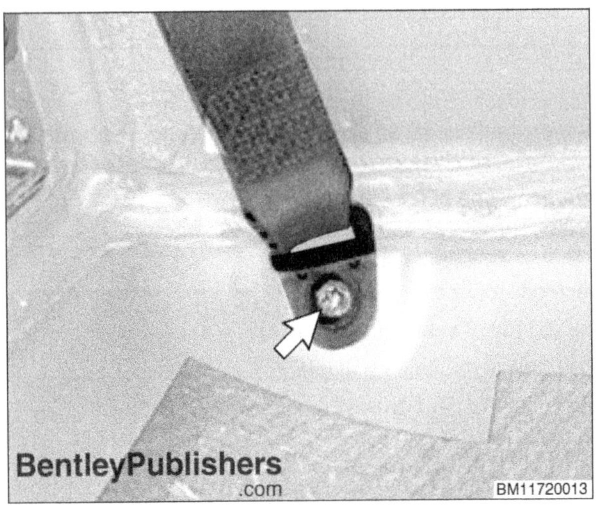

Remove seat belt anchor bolt (**arrow**) and discard.

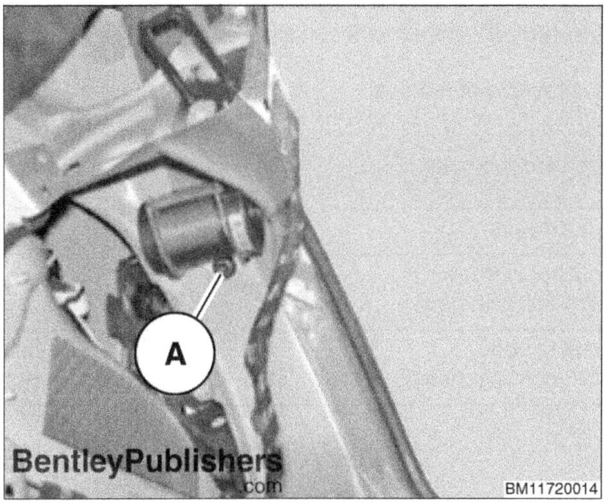

Working at lower C-pillar, remove seat belt reel mounting bolt (**A**).

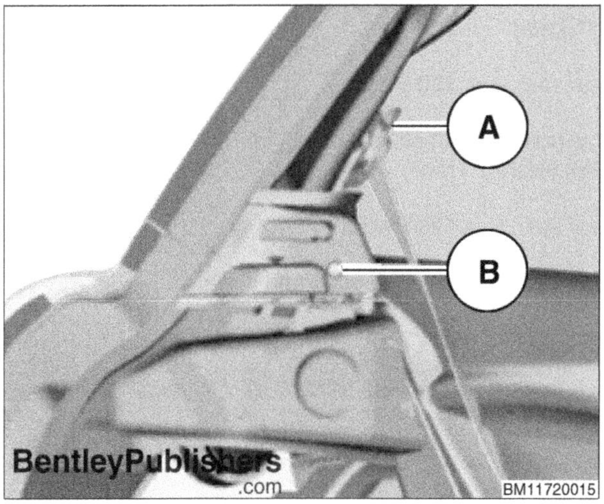

Working at upper C-pillar:

- Remove upper belt mounting bolt (**A**).
- Remove trim screw (**B**) and slide belt out from behind plastic trim.

– Installation is reverse of removal. Replace self-locking bolts.

Tightening torques	
Seat belt anchor to body (replace M10 bolt with new, use Locktite®)	36 Nm (27 ft-lb)
Seat belt reel to C-pillar (replace M10 bolt with new, use lock washer and Locktite®)	36 Nm (27 ft-lb)
Seat belt to upper C-pillar (use Locktite®)	36 Nm (27 ft-lb)

Rear seat belt, removing and installing (Clubman)

– Remove rear seat. See **520 Seats**.

– Remove C-pillar trim, cargo compartment side trim panel and plastic trim under rear window. See **513 Interior Trim**.

Remove seat belt anchor bolt (**arrow**) and discard.

720

◄ Working underneath rear side window, remove seat belt reel mounting bolt (**arrow**) and remove belt assembly from vehicle.

– Installation is reverse of removal.
  - Note that locating tab on reel prevents incorrect installation.
  - Replace self-locking bolts.

Tightening torques	
Seat belt anchor to body (replace M10 bolt with new)	36 Nm (27 ft-lb)
Seat belt reel to body (replace M10 bolt with new)	36 Nm (27 ft-lb)

Rear seat belt, removing and installing (Convertible)

– Remove rear seat. See **520 Seats**.

– Disassemble rear parcel shelf and interior trim surrounding seat belt reel. See **513 Interior Trim**.

– Remove rear seat. See **520 Seats**.

◄ Remove seat belt anchor bolt (**arrow**) and discard.

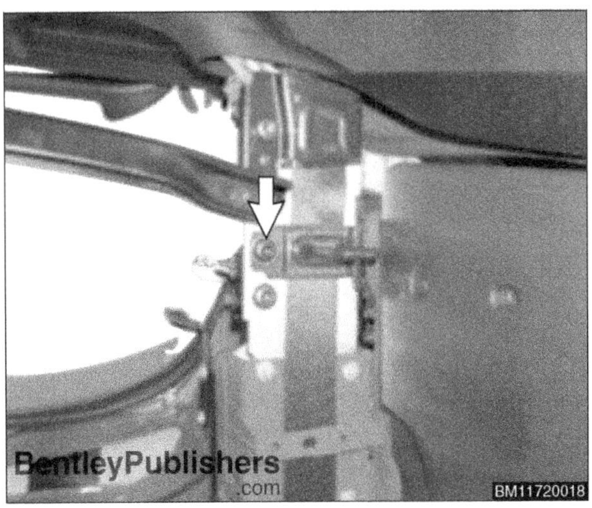

◄ Remove striker mounting bolt (**arrow**) and remove striker to free belt.

◄ Remove upper belt mounting bolts (**arrows**).

◄ Remove seat belt reel mounting bolt (**arrow**) and remove belt assembly from vehicle.

– Installation is reverse of removal. Replace self-locking bolts.

Tightening torques	
Seat belt anchor to body (replace M10 bolt with new, use Locktite®)	36 Nm (27 ft-lb)
Seat belt reel to body (replace M10 bolt with new, use Locktite®)	36 Nm (27 ft-lb)
Seat belt to upper C-pillar (use Locktite®)	36 Nm (27 ft-lb)

Rear seat belt buckle, removing and installing

– Remove rear seat. See **520 Seats**.

◄ Remove seat belt buckle mounting bolt cover (**arrow**).

720

Rear seat belt buckle, removing and installing

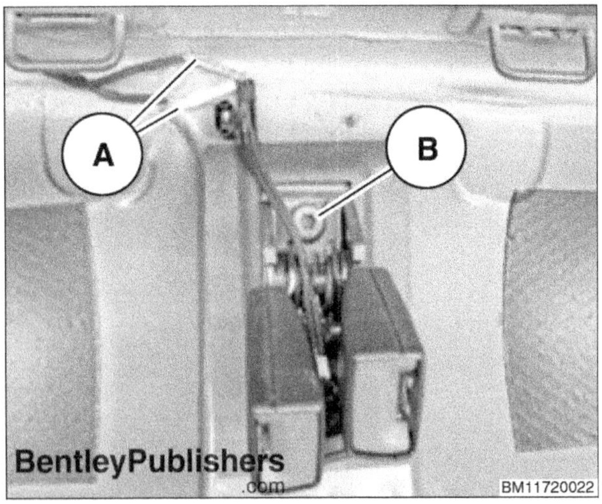

Detach electrical connectors (**A**) and remove mounting bolt (**B**). Remove seat belt buckle.

- Installation is reverse of removal. Replace mounting bolt with new.

Tightening torque	
Seat belt buckle to body (replace M10 duo-taptite bolt with new)	36 Nm (27 ft-lb)

721 Airbag System (SRS)

GENERAL

This repair group covers airbag components and replacement. Airbag system repair and fault diagnosis is not covered. Be sure that diagnostics, component testing and airbag system repair are carried out by a trained MINI service technician.

See also

- **600 Electrical System–General** for bus systems
- **720 Seat Belts** for front seat belt pyrotechnic tensioner
- **ELE Electrical Wiring Diagrams**

Multiple restraint system (MRS) overview

The airbag system is also called the supplemental restraint system (SRS). Airbags and pyrotechnic seal belt tensioners are part of the sixth-generation MINI multiple restraint system (MRS). The task of MRS is to detect accident situations and to activate the necessary restraint systems selectively corresponding to crash severity.

Using sensor inputs and preprogrammed software maps, the system is capable determining the direction and severity of a crash and triggering appropriate elements in the passive safety system.

MRS components

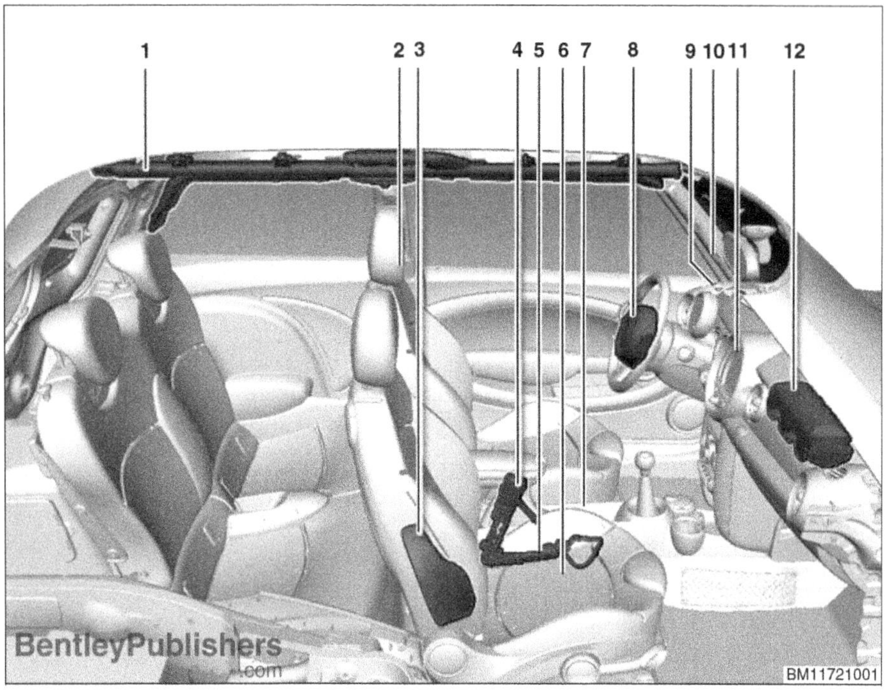

1. Head airbag
2. B-pillar satellite crash sensor
3. Side airbag
4. Seat belt buckle and pyrotechnic tensioner
5. Driver seat position sensor (US vehicles only)
6. Passenger seat occupancy sensor (OC3)
7. MRS control module
8. Driver airbag
9. Door cavity pressure sensor
10. Passenger airbag OFF warning light
11. Airbag warning light
12. Passenger airbag

MRS control module

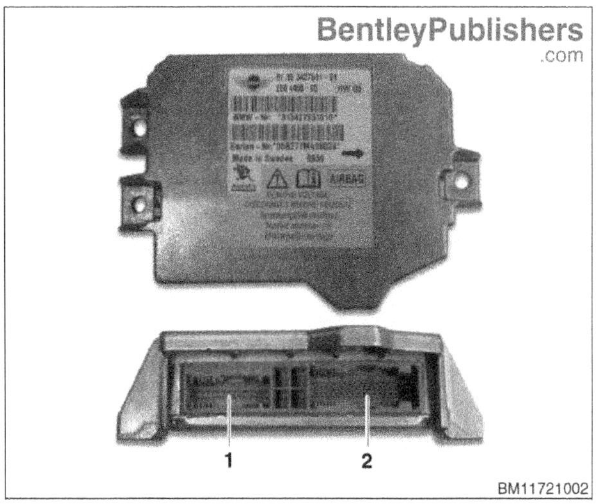

The MRS control module, underneath the center console behind the shifter, contains two acceleration sensors offset at an angle of 90°. These sensors measure vehicle longitudinal and transverse acceleration.

Control module electrical connectors:

1. 26-pin connector to cockpit wiring harness
2. 54-pin connector to vehicle wiring harness, including K-CAN

MRS functions

Impact detection. The control module determines whether there is a crash situation based on triggering algorithm and sensor threshold values. When two independent sensors signal an impact that exceeds their threshold values, a crash is detected.

For example, in a head-on crash, deceleration must be detected simultaneously by one of the B-pillar satellites and the longitudinal acceleration sensor in the MRS control module. Deceleration values determine crash severity and direction.

Based on this information, a software algorithm determines the triggering (firing) points and order of seat belt tensioner(s) vs. 2-stage airbag(s).

Crash details record. Airbag or tensioner triggering times are recorded in control module non-volatile memory. These memories cannot be erased and are used for analysis and device diagnosis by the vehicle development department. A maximum of three crash entries can be stored.The control module is capable of firing even after three crash entries.

System self-test. The control module performs a system self-test when the ignition is switched ON. The airbag warning light is illuminated 3 - 5 seconds during system self-test.

When self-test concludes and no fault is found, the airbag light goes OFF and the system is ready for operation. If faults are found, they are stored and the indication to the driver is that the warning light stays ON.

Cyclic monitoring. If system self-test is successfully concluded and the system is ready for operation, the control module performs cyclic monitoring for fault monitoring purposes. This continues as long as the ignition is ON.

Fault code storage. The MRS control module has a non-volatile fault code memory. The airbag warning light indicates any entry in the fault memory.

A distinction is made between internal and external faults when registering a fault code. Events such as triggering of the airbag or seat belt pretensioner are also stored in fault memory. The triggered pyrotechnic igniter is recognized as defective because its resistance is too high; it cannot be triggered again.

Diagnosis. Read fault code memory using MINI scan tool. See **OBD On-Board Diagnositics**. After rectifying faults or replacing triggered components, clear fault memory.

Crash telegram. In case of a crash involving triggering of at least one pyrotechnic restraint system, the MRS control module sends a crash telegram through K-CAN bus network. In response, control modules on K-CAN execute the following:

- **ECM** switches electric fuel pump and alternator OFF.
- **Junction box (JB)** releases central locking (all doors unlocked).
- **Footwell module (FRM)** switches hazard warning and interior lights ON.

Seat belt reminder. The seat belt reminder starts monitoring and warning when the vehicle is stationary. If the driver seat belt contact is not closed, an audible warning sounds and the seat belt warning light is activated for 6 seconds.

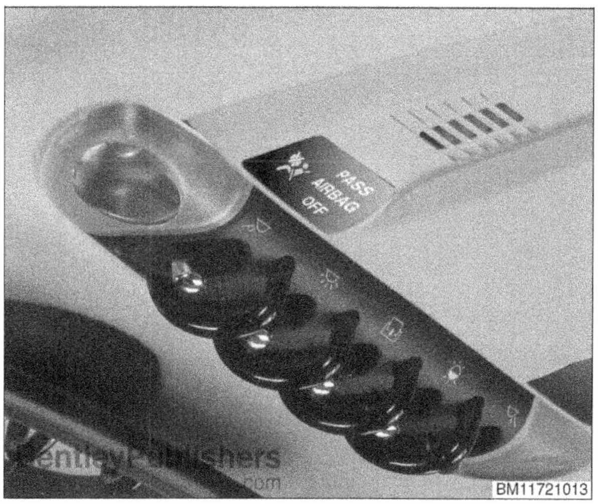

Passenger airbag OFF warning light. This light, next to the headliner switch module, illuminates when the passenger airbag is deactivated

Sensors and switches

In addition to the two accelerometers in the MRS control module, the system makes use of the following sensors and switches to evaluate crash severity and to determine triggering strategy:

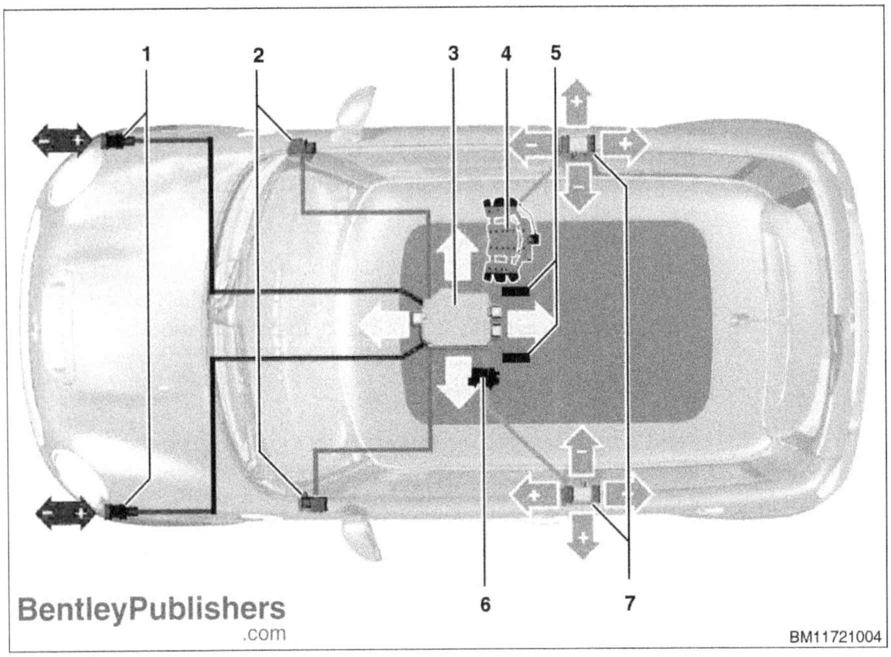

1. Front acceleration sensors (longitudinal) (US only)
2. Door pressure sensors
3. Acceleration sensors in MRS control module (longitiudinal, transverse)
4. Passenger seat occupancy sensor (OC3)
5. Front seat belt buckle microswitches
6. Driver seat position sensor (US only)
7. B-pillar satellite acceleration sensors (longitiudinal, transverse)

Longitudinal acceleration sensors underneath the front fenders detect front end and rear end crashes.

BM11721005

The B-pillar satellite sensor consists of a longitudinal acceleration sensor and a transverse acceleration sensor.

Transverse acceleration sensors detect side crashes.

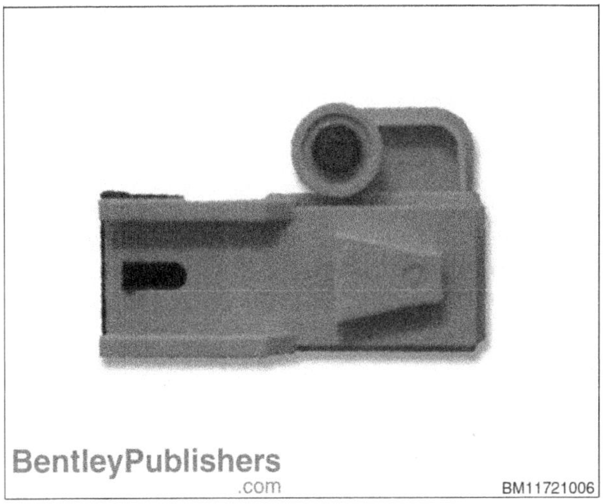

BM11721006

Door pressure sensor signals are used to verify the plausibility of acceleration sensor signal from B-pillar satellite and MRS module crash sensors during a side crash event.

Combining signals from 2 or more sensors allows the MRS control module to determine the direction of the impact and to deploy the correct, predetermined array of passive safety devices.

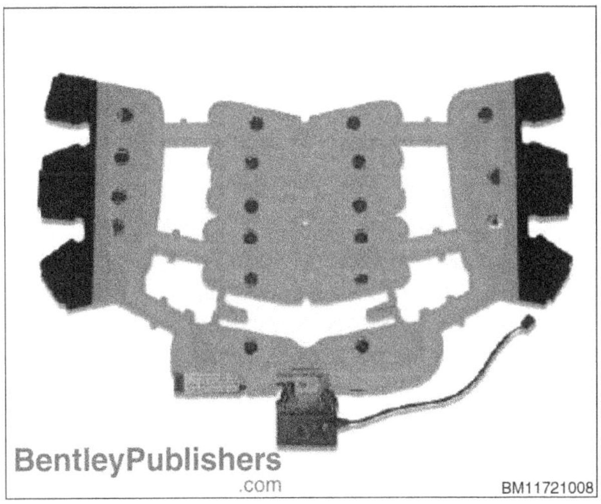

BM11721008

Passenger seat occupancy sensor (occupant classifier or OC3 mat) is integrated into the passenger seat cushion. The sensor consists of conductors with pressure-dependent resistor (force sensitive resistance or FSR) elements connected to an electronic analyzer module.

The purpose of the OC3 mat is to monitor the presence of a person or object placed on the front passenger seat and determine whether to deactivate passenger side airbags. The mat is capable of detecting a child seat that conforms to the NHTSA FMVSS 208 standard by the weight pattern on the seat.

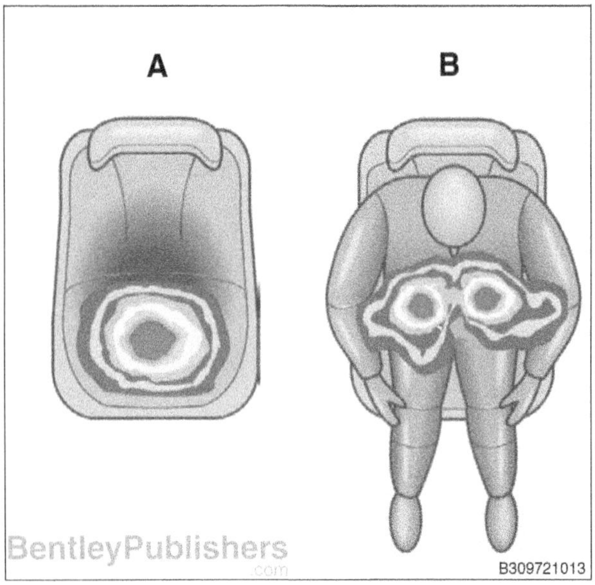

The FSR elements are wired in such a way that they can be sampled individually. When the mechanical load on a sensor element increases, its electrical resistance decreases and the current changes accordingly. By analyzing the signals from different sections of the mat, the analyzer maps the occupancy of the seat surface and can identify local concentrations of weight. The sensor mat is capable of distinguishing between a one-year-old child in a child's seat (**A**) and a light person (**B**). The distances between the areas where pressure is applied and the concentrations of pressure reveal who occupies the seat.

The OC3 sensor mat analyzer module transmits a message via K-bus to the MRS control module. If the sensor detects that the seat is unoccupied or that a child seat for a child up to one year old is fitted, the airbags on the passenger side are deactivated. If a child seat is detected, the passenger airbag deactivated light illuminates.

Seat belt buckle switch signals the MRS control module whether the seat belt is fastened or not. The signal is used for the seat belt reminder function and passive safety equipment deployment algorithm during a crash.

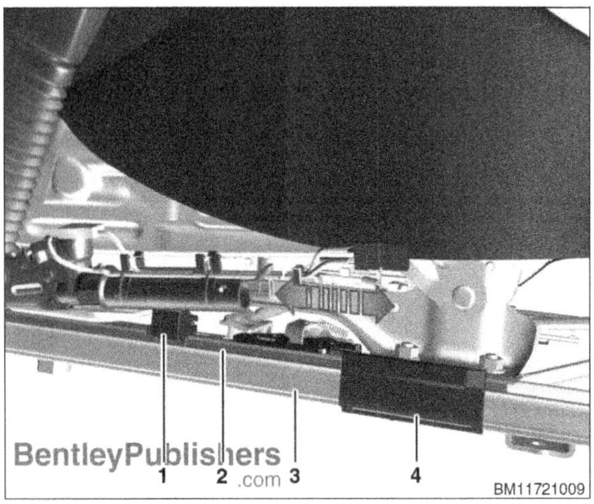

Driver seat position sensor (U.S. vehicles only) distinguishes between a relatively small person and an average-sized person within the fore-aft driver seat adjustment range. This leads to a refinement of deployment strategy for passive safety devices and an increase in driver safety.

1. Seat position sensor
2. Moving seat rail
3. Stationary seat rail
4. Magnet

Airbags

Driver front airbag, in conjunction with the seat belt, reduces the risk of serious injury to the driver's head or thorax during a head-on collision. This airbag, in the hub cushion of the steering wheel, is equipped with a 2-stage inflator assembly triggered by time delay depending on crash severity.

Passenger airbag, together with the seat belt, reduces the risk of serious injury to the front passenger during a head-on collision. Located under the dashboard, this airbag is equipped with a 2-stage inflator assembly.

Inflation of the front passenger airbag breaks the dashboard panel at defined points and opens a dashboard flap in the direction of the windshield. The airbag emerges upward and is supported against the windshield and dashboard.

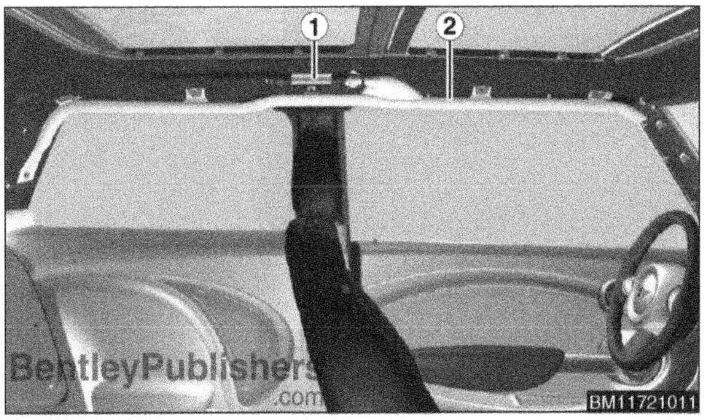

◄ **Head protection airbag** (or curtain airbag) extends from the A-pillar to the C-pillar on each side of the vehicle and covers the entire side section at head level. When deployed, the head airbag retains its structural strength and stability for several seconds, thus protecting the vehicle occupant(s) in case of side impact.

1. Airbag igniter and inflator
2. Head protection airbag (driver side)

◄ **Side airbag** is secured to the outboard edge of the front seat backrest frame and is covered by the seat cover. When deployed, the airbag tears a seam in the seat cover when it inflates.

In case of side impact, the side airbag reduces the risk of injury to the front seat occupant pelvis and rump.

Triggered airbags

The following is recommended after airbag deployment:

MRS control module. Fault memory can be cleared and reset twice. Replace module after 3rd incident.

Cables and plugs. Check for damage, replace if necessary.

Airbag modules. Replace if deployed.

Steering wheel. Replace if airbag deployed.

Steering column. Replace if damaged.

Passenger airbag cover. Replace if airbag deployed.

A-pillar, B-pillar, C-pillar trim, headliner. Replace if head protection airbag(s) deployed.

Warnings and Cautions

WARNING—

- *The airbag is inflated by an explosive device. Handled improperly or without adequate safeguards, it can be very dangerous. Observe special precautions prior to any work at or near airbags.*
- *Serious injury may result if system service is attempted by persons unfamiliar with the MINI airbag system and its approved service procedures. MINI advises that airbag system inspection and service be performed by an authorized MINI dealer.*
- *Special test equipment is required to retrieve airbag fault codes, diagnose system faults and switch the airbag warning light OFF. The warning light remains ON until problems are corrected and fault memory cleared.*
- *If the airbag warning light is ON, there is a risk that airbags will not trigger in case of an accident. Be sure to have the system inspected and repaired immediately.*
- *Disconnect the battery and cover the negative (–) battery terminal with an insulator before starting diagnostic, troubleshooting or service work associated with the airbags, and before doing any welding on the car.*
- *After disconnecting the battery, wait 1 minute before beginning work on airbag components.*
- *If an airbag deploys due to an accident, MINI specifies that airbag and airbag related components be replaced. For more information on post-collision airbag service, contact an authorized MINI dealer.*
- *Do not fire an airbag unit prior to disposal. It must be fired by a special disposal company or shipped back to MINI in the packaging of the new components.*
- *Wear gloves and avoid skin contact when removing a deployed airbag unit. In case of skin contact, wash with water.*
- *Do not allow airbag system components to come in contact with cleaning solutions or grease. Do not subject airbag components to temperatures above 75°C (167°F). When reconnecting the battery, make sure no person is inside the vehicle.*
- *Place a removed airbag unit with the padded side facing upward. Do not leave an airbag unit unattended.*
- *If the airbag unit or airbag control module is dropped from a height of ½ meter (1½ feet) or more, do not use it.*
- *Do not cover the front seat with an additional seat cover. Such a cover may interfere with the operation of the side airbag.*

CAUTION—

- *When working on electrical components, disconnect the negative (–) cable from the battery and insulate the cable end to prevent accidental reconnection.*
- *Prior to disconnecting the battery, read the battery disconnection cautions given in* **001 Warnings and Cautions**.
- *To avoid marring trim pieces when removing interior trim clips, use a plastic prying tool or a screwdriver with the tip wrapped with masking tape.*

AIRBAG ELECTRONICS

MRS control module, removing and installing

– Disconnect negative (-) battery cable and cover battery terminal to keep cable from accidentally contacting terminal.

> ***CAUTION—***
> *• Prior to disconnecting the battery, read the battery disconnection cautions given in* **001 Warnings and Cautions**.

– Remove center console. See **513 Interior Trim**.

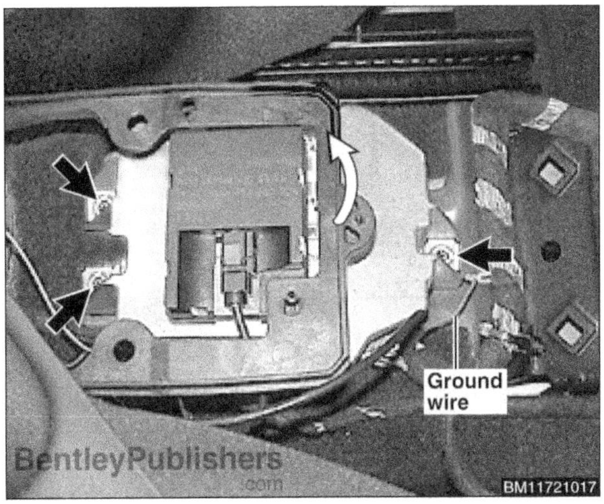

◄ Working above center tunnel:

- Remove control module mounting fasteners (**arrows**) and detach **ground wire**.
- Lift up center console bracket (**curved arrow**) and slide out control module.

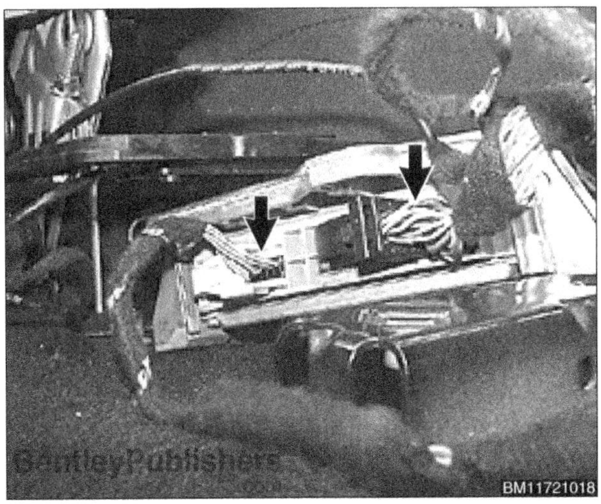

◄ Detach control module electrical harnesses (**arrows**) and lift out module.

– Installation is reverse of removal. Make sure harnesses are routed as before.

Tightening torque	
MRS control module bracket to center tunnel	8 Nm (6 ft-lb)

721

Front acceleration sensor, removing and installing

The front acceleration sensors are underneath the front fenders. Remove fender liner to access.

– Raise vehicle and support safely.

> ***WARNING —***
> • *Make sure the vehicle is stable and well supported at all times. Use a professional automotive lift or jack stands designed for the purpose. A floor jack is not adequate support.*

– Remove left or right front fender liner. See **410 Fenders, Engine Hood**.

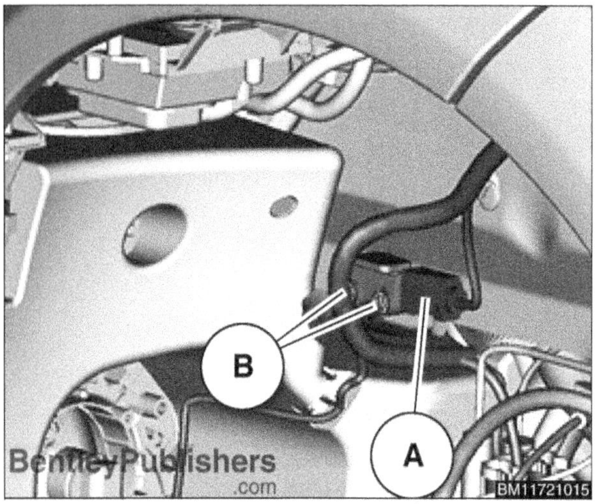

◄ Working at acceleration sensor:

• Detach sensor electrical connector (**A**).

• Remove mounting screws (**B**) and detach sensor from underbody.

– Installation is reverse of removal. Replace mounting screws with new.

Tightening torque	
Front acceleration sensor to underbody (replace screws with new)	8 Nm (6 ft-lb)

Door pressure sensor, removing and installing

– Disconnect negative (-) battery cable and cover battery terminal to keep cable from accidentally contacting terminal.

> ***CAUTION—***
> • *Prior to disconnecting the battery, read the battery disconnection cautions given in* **001 Warnings and Cautions**.

– Remove door trim panel. See **411 Doors**.

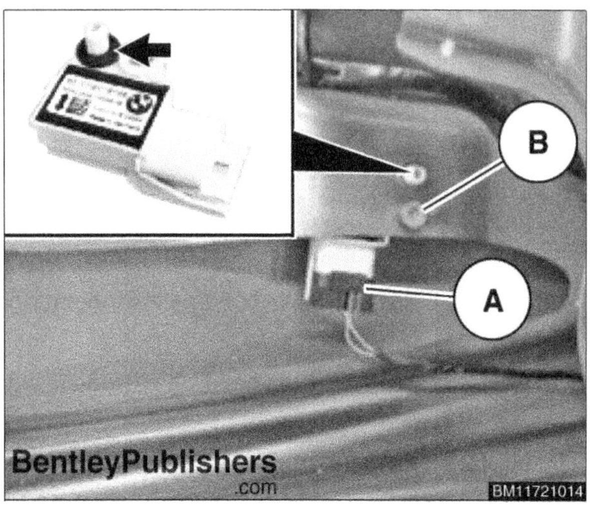

◄ Working at top front of door cavity:

• Detach pressure sensor electrical connector (**A**).

• Remove mounting bolt (**B**) and slide out of door cavity.

– When installing:

• Seat sensor in door securely.

• Make sure sealing O-ring (**arrow**) is undamaged.

Tightening torque	
Door pressure sensor to door	8 Nm (6 ft-lb)

B-pillar satellite acceleration sensor, removing and installing

– Disconnect negative (-) battery cable and cover battery terminal to keep cable from accidentally contacting terminal.

> *CAUTION—*
> • *Prior to disconnecting the battery, read the battery disconnection cautions given in* **001 Warnings and Cautions**.

– Remove rear side trim panel and B-pillar trim. See **513 Interior Trim**.

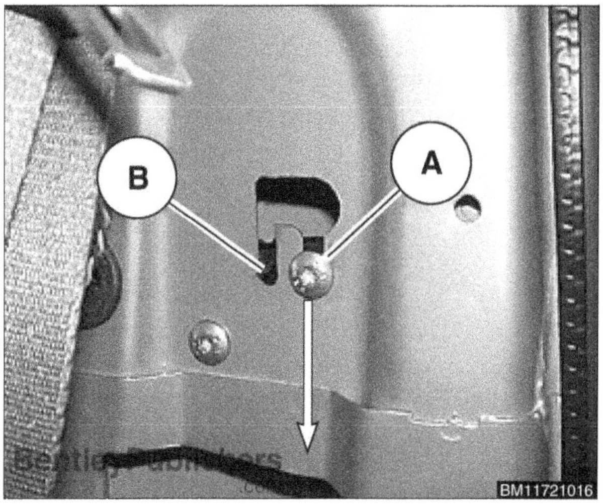

◄ Working at middle of B-pillar:

- Remove sensor mounting screw (**A**).
- Slide sensor down (**arrow**) to remove from B-pillar. Detach electrical connector.

– When installing, make sure locating lug (**B**) is undamaged.

Tightening torque	
Acceleration sensor to B-pillar	8 Nm (6 ft-lb)

721

Airbags

Replacement of the side airbag (in the front seat) is beyond the scope of this manual. Consult an authorized MINI dealer.

Driver airbag, removing and installing

– Disconnect negative (-) battery cable and cover battery terminal to keep cable from accidentally contacting terminal.

> *CAUTION—*
> - *Prior to disconnecting the battery, read the battery disconnection cautions given in* **001 Warnings and Cautions**.

Insert screwdriver into concealed opening on back side of steering wheel and press (**white arrow**) until spring resistance is felt.

- Increase pressure on spring (**black arrows**) until airbag unit is unlocked.

– Tilt back airbag.

Disconnect electrical connectors (**arrows**) and remove airbag unit.

> *WARNING —*
> - *Place removed airbag unit with the padded side facing upward. Store removed airbag safely.*
> - *Do not pinch airbag harness in center of steering wheel when removing or installing airbag.*

– Installation is reverse of removal:

- Make sure electrical harness is positioned correctly.
- Press airbag directly down on steering wheel until it snaps in place.

Passenger airbag, removing and installing

– Disconnect negative (-) battery cable and cover battery terminal to keep cable from accidentally contacting terminal.

> *CAUTION—*
> - *Prior to disconnecting the battery, read the battery disconnection cautions given in* **001 Warnings and Cautions**.

– Remove upper dashboard panel. See **513 Interior Trim**.

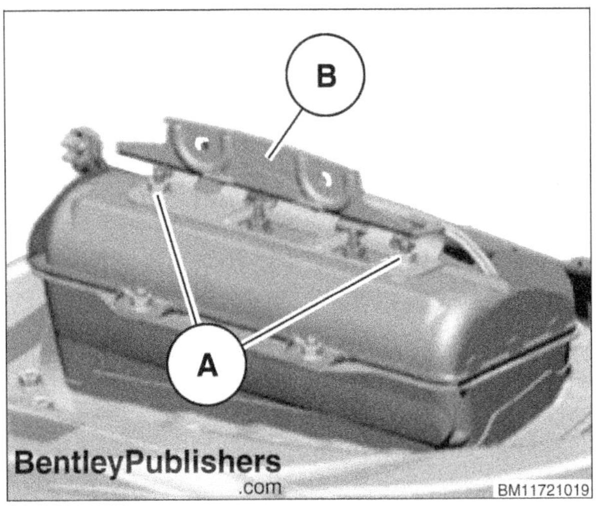

Working at airbag unit:

- Remove mounting nuts (**A**) and discard.
- Remove airbag bracket (**B**).

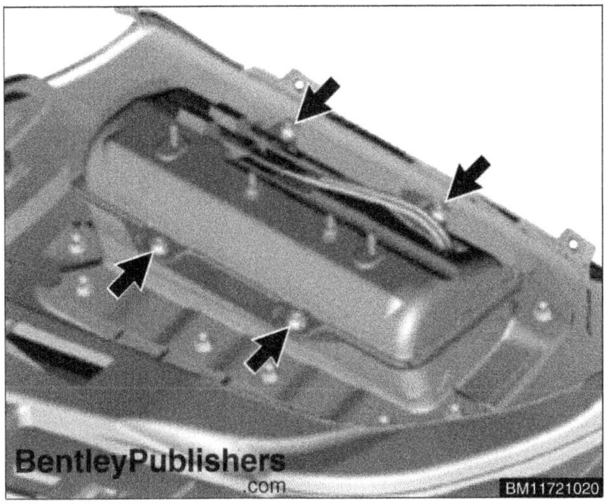

◄ Remove airbag mounting nuts (**arrows**) and discard. Lift airbag off dashboard and store in a safe place.

> ***WARNING—***
> *• Place removed airbag unit with the padded side facing upward. Do not leave airbag unattended.*

– Installation is reverse of removal.

Tightening torques	
Bracket to airbag (replace mounting nuts with new)	6 Nm (4 ft-lb)
Airbag to dashboard (replace mounting nuts with new)	9 Nm (7 ft-lb)

Head protection airbag (curtain airbag), removing and installing

Head protection airbag service is an extensive procedure and requires removal of A-, B- and C-pillar trim and the headliner. Be sure to read the procedure through before starting work.

– Disconnect negative (-) battery cable and cover battery terminal to keep cable from accidentally contacting terminal.

> ***CAUTION—***
> *• Prior to disconnecting the battery, read the battery disconnection cautions given in* **001 Warnings and Cautions**.

– Remove A-, B- and C-pillar trim panels. See **513 Interior Trim**.

– Remove dome light units and roof operating unit. See **630 Lights**.

– Remove interior rear-view mirror, sun visors, front and rear handles and headliner trim.

– Hardtop: Remove rear seat backrest. See **520 Seats**.

– Hardtop: Remove cargo compartment side trim panels.

– Remove headliner:
  - Hardtop: Headliner is removed through rear hatch.
  - Clubman: Headliner is removed through right side doors opening.

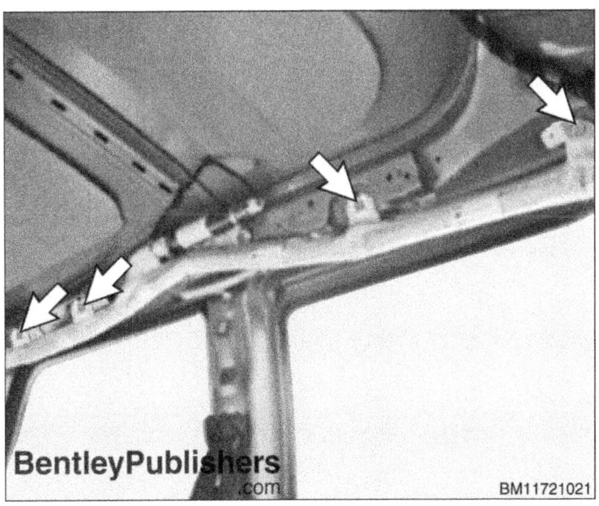

◄ Working inside passenger compartment, remove airbag mounting Torx screws (**arrows**) and discard.

Hardtop model illustrated. Clubman is similar.

Head protection airbag (curtain airbag), removing and installing

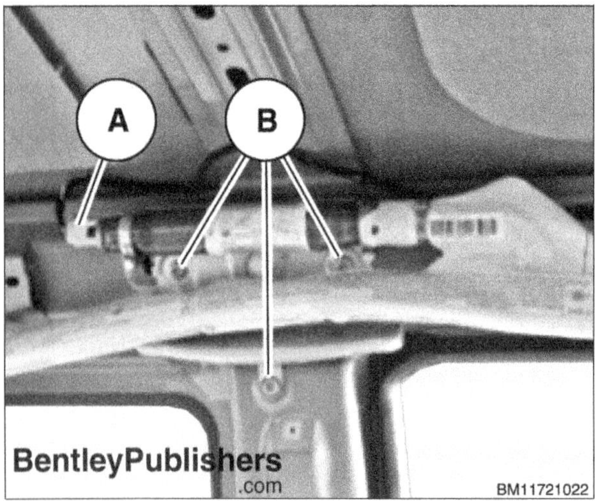

Working at airbag igniter unit:

- Detach electrical connector (**A**).
- Remove igniter Torx screws (**B**) and discard.
- Remove airbag.

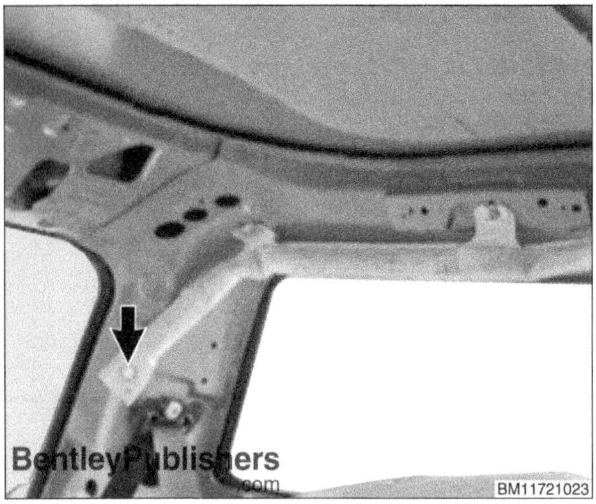

Installation is reverse of removal.

- Note that airbag locating tab prevents incorrect installation.
- Make sure tubular airbag is not twisted at folds. Hole in rear of tubular airbag (**arrow**) is secured by C-pillar trim clip.
- Replace mounting Torx screws with new and use Locktite®.

Tightening torque	
Head airbag to roof (replace Torx screws with new, use Locktite®)	8 Nm (6 ft-lb)

ECL Electrical Component Locations

TABLES

GENERAL

This repair group covers electrical component locations, fuse locations and ground points.

Electrical equipment and accessories included vary with model and model year. Confirm component identification whenever you begin work on the electrical system.

See also:

- **600 Electrical System–General** for bus systems
- **ELE Electrical Wiring Diagrams** for electrical schematics

Common acronyms

MINI uses many abbreviations and acronyms in the technical literature, service bulletins, wiring diagrams and parts bulletins it distributes. See **Table a** for common MINI, BMW and industry acronyms.

Common acronyms

Table a. Common acronyms	
Acronym	**Component or system**
A/C	air-conditioning
ABS	antilock brakes
AUC	automatic air recirculation control
CA	comfort access
CAN	controller area network (bus)
CAS	car access system (module)
CBS	condition-based service
CCC/M-ASK	car communication computer / audio system controller
CID	central information display
DME	digital motor electronics
DMTL	diagnosis module—tank leakage
DSC	dynamic stability control
DWA	anti-theft alarm
ECM	engine control module
ECT	engine coolant temperature (sensor)
EDK	electronic throttle
EEPROM	flash programmable read-only memory
EGS	electronic transmission control
ELV	electronic steering lock
EPROM	erasable-programmable read-only memory
EVAP	fuel tank evaporative control system
EWS	electronic immobilizer
FTM	flat tire monitoring
GPS	global positioning system
IAT	intake air temperature (sensor)
IBS	intelligent battery sensor
IHKA	automatic heating and air-conditioning
KL 15	battery positive, ignition switch on RUN
KL 30	battery positive (B+), power

Table a. Common acronyms	
Acronym	**Component or system**
KL 31	battery / chassis ground
KL 50	ignition start position, power
KL R	battery positive, ignition on accessories
LDP	fuel tank leak diagnosis pump
LED	light emitting diode
LEV	low emissions vehicle
LIN	local interconnected network (bus)
MAP	manifold absolute pressure
MIL	malfunction indicator light
MRS	multiple restraint system (airbags)
NAV	navigation system
NLEV	national low emissions vehicle
NTC	negative temperature coefficient (resistor)
OBD II	second generation on-board diagnostics
PDC	park distance control
PWG	accelerator pedal module
RAM	random access memory
RDC	tire pressure control
RXD	receive data line
SRS	supplemental restraint system (airbags)
SZL	steering column switch cluster
TCM	automatic transmission control module
TDC	top dead center
TI	injection signal
TLEV	transitional low emissions vehicle
TPM	tire pressure monitoring
TXD	transmit data line
ULEV	ultra low emissions vehicle
VANOS	variable camshaft timing

Warnings and Cautions

> ***WARNING—***
> - *Pyrotechnic seat belt tensioners and airbags utilize explosive devices. Handle with extreme care. See warnings and cautions in* **121 Battery, Alternator, Starter**, **720 Seat Belts**, *and* **721 Airbag System (SRS)**.

> ***CAUTION—***
> - *Prior to disconnecting the battery, read the battery disconnection precautions in* **001 Warnings and Cautions**.
> - *Relay and fuse positions vary from car to car. If questions arise, an authorized BMW dealer is the best source for the most accurate and up-to-date information.*
> - *To verify a relay position, compare the wiring colors at the relay socket to the colors indicated on the wiring diagrams in* **ELE Electrical Wiring Diagrams**.
> - *Switch the ignition OFF and remove the negative (-) battery cable before removing any electrical components. Connect and disconnect ignition system wires, multiple connectors and ignition test equipment leads while the ignition is switched OFF.*
> - *Use a digital multimeter for electrical tests.*

ELECTRICAL COMPONENTS

Electrical components table

Table b is a cross-referenced listing of electrical components.

- **Column 1**: Components alphabetized by names in current usage
- **Column 2**: BMW alphanumeric code (corresponds to code in electrical wiring diagrams)
- **Column 3**: Location in vehicle
- **Column 4**: Figure number in this repair group, if illustrated
- **Column 5**: Page number for illustration

Table b. Component locations

Component	Code	Location	Fig #	Pg #
A/C		*see also* IHKA entries		
A/C compressor clutch	Y8099	A/C compressor		
A/C compressor relay	K19	Junction box fuse and relay panel, right front behind kick panel trim cover	**Fig 36**	ECL-33
A/C refrigerant pressure sensor	B8	Rear of engine compartment at bulkhead	**Fig 17**	ECL-28
ABS		*see also* DSC entries		
ABS / DSC control module	A65a	Left rear engine compartment	**Fig 17**	ECL-28
ABS wheel speed sensor	B1, B2 B3, B4	Wheel hub, left front, right front, left rear, or right rear wheel	**Fig 29** **Fig 30**	ECL-31 ECL-32
Acceleration sensor		*see* Crash sensor		
Accelerator pedal sensor	B10	Integrated with accelerator pedal		
Air quality sensor	B414	Under center of windshield in cowl		

ECL

Table b. Component locations				
Component	**Code**	**Location**	**Fig #**	**Pg #**
Airbag crash sensor		*see* Crash sensor		
Airbag deactivation switch, passenger airbag	S559	Right end of dashboard		
Airbag gas generator, head protection, left, right	G17a G18a	Upper B-pillar, left, right		
Airbag sensor, right rear side airbag	A174a	*Clubman:* Behind Club door trim panel		
Airbag, driver	G5	In steering wheel under horn button		
Airbag, passenger	G6	Under right end of dashboard		
Airbag, side, driver, passenger	G14b G15b	Side bolster, front seats, left right		
Alarm		*see* Anti-theft		
Alternator (generator)	G6524	Right front engine compartment, bolted to engine block		
Amplifier	A18	Left rear quarter panel, behind trim	**Fig 35**	ECL-33
Antenna		*see also* Comfort access (CA) antennas *see also* Tire pressure control (RDC) antenna		
Antenna diversity module	A421	*Convertible:* Right side trunk behind trim		
Antenna diversity module	A421	*Clubman:* Left upper C-pillar		
Antenna diversity module	A421	*Coupe:* Right side rear hatch below rear window		
Antenna interference suppression filter	U410	Behind headliner, left rear		
Antenna noise suppression filters	Z1, Z2	*Coupe:* Rear hatch, below left and right corners of rear window		
Antenna noise suppression filters	Z3, Z4 Z14	*Clubman:* Right rear split door below rear window		
Anti-theft siren, tilt sensor	A14147	Engine bulkhead		
AUC sensor		*see* Air quality sensor		
Audio system controller / car communication computer (CCC/M-ASK)	N38a	Center dashboard		
Audio-visual socket	A18807	Lower center dashboard		
Automatic transmission control module (EGS)	A7000a	Right of steering column	**Fig 7**	ECL-25
Automatic transmission electronics: Gear position switch, shift solenoids, speeds sensors, temperature sensor	Y8532	Inside transmission housing		
Auxiliary sound system connector	A267a	Center dashboard below switch cluster		
Back-up light switch		*Automatic transmission:* Transmission range switch		
Back-up light switch	S8511	*Manual transmission:* Top of transmission, left side engine compartment		
Battery	G1	Right rear engine compartment	**Fig 1** **Fig 8** **Fig 25**	ECL-16 ECL-26 ECL-30
Battery sensor	B895a	Battery negative post, top of battery, right rear engine compartment	**Fig 25**	ECL-30

Table b. Component locations

Component	Code	Location	Fig #	Pg #
Bistable relay	I01139	Junction box fuse and relay panel, right front behind kick panel trim cover	**Fig 37**	ECL-34
Blower	M30	IHKA housing, under dashboard		
Blower output stage	N2	Under right side dashboard		
Blower resistor	R11	Under right side dashboard		
Bluetooth antenna	W18	Under center console	**Fig 7**	ECL-25
Brake fluid level switch	B18a	Brake fluid reservoir, left rear engine compartment	**Fig 8**	ECL-26
Brake light switch	S29	Above brake pedal		
Brake pad wear sensor, left front	B16a	Left front brake caliper		
Brake pad wear sensor, right rear	B17a	Right rear brake caliper		
Brake vacuum pressure sensor	B9529	Brake booster		
CA		*see* Comfort access (CA) entries		
Camshaft sensor, intake, exhaust	B6214a B6224a	Flywheel end of cylinder head	**Fig 8** **Fig 12** **Fig 17**	ECL-26 ECL-27 ECL-28
Car access system (CAS) module	A149a	Behind tachometer, underneath dashboard	**Fig 7**	ECL-25
Car communications computer		*see* Audio system controller / car communication computer (CCC/M-ASK)		
CAS		*see* Car access system (CAS) module		
CCC/M-ASK		*see* Audio system controller / car communication computer (CCC/M-ASK)		
CD changer	N22	Right side dashboard behind glove compartment		
Center brake light	H34a	*Clubman:* Above split rear doors		
Center brake light	H34a	*Convertible:* Top of trunk lid		
Center brake light	H34a	*Coupe:* Above rear window		
Center console switch cluster	S0241	Center console ahead of shifter		
Central information display	A165a	Center dashboard		
Clutch switch module	S805a	Above clutch pedal		
Comfort access (CA) control module	A215	Right rear quarter panel, behind trim	**Fig 7** **Fig 40**	ECL-25 ECL-34
Comfort access antenna, right side	W12a	*Clubman:* Club door, below window		
Comfort access antenna, right side	W12a	*Convertible:* Right rear behind trim		
Comfort access antennas		*Coupe:* Various locations inside vehicle	**Fig 7** **Fig 35** **Fig 40**	ECL-25 ECL-33 ECL-34
Compressor		*see* A/C compressor		
Convertible top module	A96b	Left side trunk		
Coolant		*see* Engine coolant		
Coolant pump friction wheel drive	M2327	Front of crankcase above vibration damper		
Crankcase breather heater	E65391	*Turbo engine:* Intake manifold	**Fig 20**	ECL-29

Electrical components table

Table b. Component locations

Component	Code	Location	Fig #	Pg #
Crankcase breather heating relay	K6539	Front power distribution and relay panel, left rear engine compartment	**Fig 10**	ECL-27
Crankshaft sensor	B6203a	Intake side lower crankcase, near engine bell housing		
Crash sensor, center		MRS control module, under center console		
Crash sensor, left front, right front	B10508a B10509a	Left front, right front wheel housing	**Fig 27**	ECL-31
Crash sensor, satellite		B-pillar		
Digital tuner	N48	Right rear quarter panel, behind trim	**Fig 40**	ECL-34
Diversity		*see* Antenna diversity module		
DME main relay	K6300	Front power distribution and relay panel, left rear engine compartment	**Fig 10**	ECL-27
Door handle button, comfort access	S263a S264a	Outer door handle, right, left door		
Door locks, Club door	S256b S256c	*Clubman:* Top and bottom of Club door		
Door locks, driver door, passenger door	S47, S49	Rear of door behind trim panel, left, right		
Door pressure sensor, front side airbag	B78, B79	Front of front door behind trim, left, right		
DSC		*see also* ABS entries		
DSC acceleration sensor	B9801	Underneath passenger seat, under carpet	**Fig 39**	ECL-34
DSC control module		*see* ABS / DSC control module		
DSC hydraulic unit		On DSC control module		
DSC steering angle sensor		On steering column switch center		
Dynamic stability control		*see* ABS / DSC *see also* DSC		
ECM		*see* Engine control module (ECM)		
Electric cooling fan	M135	Between radiator and engine		
Engine control module (ECM)	A6000	Left rear engine compartment	**Fig 7** **Fig 8** **Fig 17**	ECL-25 ECL-26 ECL-28
Engine control module relay		*see* DME main relay		
Engine coolant temperature (ECT) sensor	B6232	Flywheel end of cylinder head	**Fig 12** **Fig 23**	ECL-27 ECL-30
Engine cooling fan relays	K21, K22	Front power distribution and relay panel, left rear engine compartment	**Fig 10**	ECL-27
Engine hood contact switch	S19a	Left rear engine compartment		
Evaporative emissions control valve	Y6120	Intake side of engine	**Fig 14** **Fig 20**	ECL-28 ECL-29
Exhaust camshaft sensor		*see* Camshaft sensor		
Foglights, front	E9015 E9017	Front bumper ends		

Table b. Component locations				
Component	**Code**	**Location**	**Fig #**	**Pg #**
Foglights, rear	H26 H29 H16026	Rear bumper		
Footwell module	A4011	Left lower A-pillar behind kick panel	**Fig 7** **Fig 33**	ECL-25 ECL-33
Front cigar lighter	E28	Front of center console		
Front power distribution and relay panel	A400a	Left rear engine compartment	**Fig 1** **×** **Fig 10** **Fig 17**	ECL-16 ECL-16 ECL-27 ECL-28
Fuel filler flap lock motor	M16	Cargo compartment, left rear behind trim		
Fuel injector relay	K6327	Left rear engine compartment	**Fig 17**	ECL-28
Fuel injectors	Y6101 - Y6104	*Non-turbo engine:* Intake manifold *Turbo engine:* Cylinder head		
Fuel pump	M2	Underneath left rear seat cushion, top of fuel tank		
Fuel pump relay	K96	Junction box fuse and relay panel, right front behind kick panel trim cover	**Fig 37**	ECL-34
Fuel rail pressure sensor	B2261	*Turbo engine:* Left side of engine underneath intake manifold	**Fig 21**	ECL-29
Fuel tank flap emergency release cable		Trunk or cargo compartment, left side wall		
Fuel tank leakage diagnostic module (DMTL)	M119a	Underneath vehicle, ahead of right rear wheel housing	**Fig 30**	ECL-32
Fuel volume control valve	Y2381	*Turbo engine:* Top of cylinder head cover	**Fig 18**	ECL-29
Fuses		**Fuse Locations and Ratings** in this repair group		
Generator		*see* Alternator (generator)		
GPS		*see* Audio system controller / car communications computer (CCC/M-ASK) *see also* Navigation		
Grounds		*see* **Grounds** in this repair group		
Hall effect camshaft position sensor		*see* Camshaft sensor		
Hand-free microphone	B429a	Front dome light unit, left		
Hazard warning switch	S18a	Center dashboard, above instrument cluster		
Headlight washer pump		*see* Washer pumps		
Headlights	E126 E127	Front of engine compartment		
Heated oxygen sensor		*see* Oxygen sensor		
Heating		*see* **640 Heating and Air-conditioning**		
High beam headlight relay	K13562	Under driver seat beneath carpet		
Horn button	S4a	Steering wheel center		
Horn relay	K2a	Junction box fuse and relay panel, right front behind kick panel trim cover	**Fig 37**	ECL-34
Horns	H2b, H3b	Behind front bumper, right side		
Hot-film mass air flow sensor		*see* Mass air flow sensor		

Electrical components table

Table b. Component locations				
Component	**Code**	**Location**	**Fig #**	**Pg #**
IBS		*see* Battery sensor		
Ignition coils	T6151 - T6154	Top of cylinder head cover(s) above spark plugs		
Ignition coils suppression capacitor	I01046	Between ignition coil connectors 1and 2		
Ignition key slot	S2	Right side of steering column		
IHKA components		*see* **640 Heating and Air-conditioning**		
IHKA control panel	A11a	Center dashboard		
Instrument cluster	A2a	Center dashboard		
Intake air temperature-boost pressure sensor	B6123	*Turbo engine:* Intake duct, timing chain end of engine		
Intake camshaft sensor		*see* Camshaft sensor		
Intake pressure (MAP) sensor	B6239	*Turbo engine:* Top of intake manifold	**Fig 19**	ECL-29
Intake temperature-differential pressure sensor	B6286	Intake manifold	**Fig 11**	ECL-27
Intelligent battery sensor		*see* Battery sensor		
Interior movement sensor	A121b	Above center of windshield, behind trim	**Fig 7**	ECL-25
Joystick	A167b	Center console between front seats		
Junction box fuse and relay panel	A4010	Right lower A-pillar behind kick panel trim cover *see also* **Fuse Locations and Ratings** in this repair group.	**Fig 1** **Fig 6** **Fig 7** **Fig 36**	ECL-16 ECL-19 ECL-25 ECL-33
Knock sensor	B6241	Intake side of engine	**Fig 14** **Fig 21**	ECL-28 ECL-29
Light sensor		*see* Rain and light sensor		
MAP sensor		*see* Intake pressure (MAP) sensor entries		
Mass air flow sensor	B6207a	Air intake duct between air filter housing and throttle housing	**Fig 8** **Fig 16** **Fig 17**	ECL-26 ECL-28 ECL-28
Microphone	B400	Upper steering column trim		
Microwave sensors	B14143 - B14146	Right, left door, right, left quarter panel, behind trim panel		
Mirror, electrochromic (inside)	A22	Top center of windshield		
Mirror, outside, left, right	Y5 Y6	Driver door, passenger door		
MOST-bus connector	I14255	In front and right of glove compartment		
MRS control module	A12	Underneath center console	**Fig 7** **Fig 41**	ECL-25 ECL-35
Multifunction switch assemblies	I01032 I01033	Steering wheel spokes		
Neutral safety switch	Y19	Under selector lever bezel, center console		
Neutral sensor	B456	Flywheel end of engine block	**Fig 13**	ECL-27
O_2 sensor		*see* Oxygen sensors		

Table b. Component locations				
Component	**Code**	**Location**	**Fig #**	**Pg #**
Oil pressure switch	B6231	Flywheel end of cylinder head	**Fig 12** **Fig 23**	ECL-27 ECL-30
Outside mirror fold-in module	A208	Left rear quarter panel, behind trim	**Fig 35**	ECL-33
Outside mirror fold-in module	A208	*Convertible:* Left side trunk, behind rear seat back	**Fig 7**	ECL-25
Outside mirror switch	S17	Left door panel		
Outside temperature sensor	B21a	Behind front bumper		
Oxygen sensors	B62101 B62102	Top and bottom of catalytic converter, in engine exhaust system, front of engine compartment	**Fig 13**	ECL-27
Park / neutral position switch		*see* Neutral safety switch		
Park distance control (PDC) module	A81a	Right side cargo compartment behind panel	**Fig 7** **Fig 42**	ECL-25 ECL-35
Park distance control (PDC) sensors, rear	B34a - B37a	Rear bumper		
Park distance control (PDC) warning gong	H40	Right side cargo compartment behind panel	**Fig 42**	ECL-35
Parking brake warning switch	S31a	Base of parking brake handle, under center console	**Fig 41**	ECL-35
PDC		*see* Park distance control (PDC) entries		
Power steering		*see* Steering		
Power supply fuses	A406	Left of battery, right rear engine compartment	**Fig 1** **Fig 5**	ECL-16 ECL-18
Power-saving relay terminal 15	K6326	Front power distribution and relay panel, left rear engine compartment	**Fig 10**	ECL-27
Pressure sensor		*see* A/C refrigerant pressure sensor		
Radio	N9	Center dashboard		
Rain and light sensor	B57b	Above center of windshield	**Fig 7**	ECL-25
RDC		*see* Tire pressure control entries		
Rear hatch lock motor	M17b	*Coupe:* Cargo compartment floor		
Rear window defogger	E9, E9a, E9b	Rear window		
Rear window defogger relay	K13	Junction box fuse and relay panel, right front behind kick panel trim cover	**Fig 37**	ECL-34
Ride height sensor, front	B42b	Left front suspension		
Ride height sensor, rear	B64a	Right rear suspension	**Fig 29**	ECL-31
Rollover protection system module	N24	*Convertible:* Right side trunk		
Satellite radio	N47	Underneath passenger seat, under carpet	**Fig 39**	ECL-34
Seat belt buckle contact microswitches, front, rear	S58a S59a S107 S108	Front, rear seat belt buckles		
Seat belt tensioner, pyrotechnic, left, right, front, rear	G12 G13 G27 G28	Base of seat belt buckle, left, right, front or rear seats		

ECL

Table b. Component locations				
Component	**Code**	**Location**	**Fig #**	**Pg #**
Seat components		*see* **520 Seats**		
Seat heater coil, left front, right front	E56a - E59a	Seat cushion, backrest, left front, right front		
Seat heating module, front seat	A186a A187a	Underneath front seats	**Fig 7**	ECL-25
Seat occupancy detection module	A222	Passenger seat cushion		
Secondary air injection pump	M63	Lower right engine compartment		
Secondary air injection pump relay	K6304a	Left rear engine compartment	**Fig 17**	ECL-28
Selector lever position switch	S227	Under selector lever bezel, center console		
Shifter paddles	I01091 I01092	Steering wheel spokes		
Side marker lights, front	E21, E22	Front of front fenders		
Siren		*see* Anti-theft siren, tilt sensor		
Slide-in compartment		*see* Ignition key slot		
Solar sensor	B66	Front center of dashboard		
Sound system control panel cooling fan	M455	Sound system control panel, behind dashboard		
Speaker		*see* **650 Radio**		
Speed (rpm) sensor		*see* Crankshaft sensor		
Split rear door courtesy lights	I01221 I01222	Bottom of split rear door		
Split rear doors lock motors	M17c M17d	*Clubman:* Split rear doors		
SRS		*see* Airbag entries *see also* MRS entries		
START - STOP button	S2b	Dashboard, right of steering column		
Starter	M6510a	On crankcase bell housing flange under intake manifold		
Steering angle sensor		*see* Steering column switch center		
Steering column switch center	A72	Behind steering wheel on steering column		
Steering lock	M137a	Underneath steering column		
Steering torque sensor	I01096	Attached to steering control module, rear lower engine compartment		
Steering, electromechanical	A60233a	Rear lower engine compartment		
Stepper motor		*see* **640 Heating and Air-conditioning**		
Sunroof module	A33a	Above center of windshield, behind trim	**Fig 7**	ECL-25
Sunroof switch	S38a	Front dome light unit		
Tachometer	A9001	Left side dashboard		
Telephone		*see* **650 Radio**		
Telephone eject box	A117a	Center console, between front seats		
Telephone transceiver	U400a	Underneath driver seat, under carpet	**Fig 34**	ECL-33

Table b. Component locations				
Component	**Code**	**Location**	**Fig #**	**Pg #**
Temperature sensor, evaporator	B14a	IHKA housing, underneath top dashboard trim		
Temperature sensor, heater core	B11b	Top of IHKA housing, underneath top dashboard trim		
Terminal 15 relay	I01069	Junction box fuse and relay panel, right front behind kick panel trim cover	**Fig 37**	ECL-34
Terminal 30g relay	I01068	Junction box fuse and relay panel, right front behind kick panel trim cover	**Fig 36**	ECL-33
Thermostat, coolant	B6279	Coolant manifold, flywheel end of engine	**Fig 11** **Fig 18**	ECL-27 ECL-29
Throttle valve	Y63900	Intake side of engine	**Fig 14** **Fig 20**	ECL-28 ECL-29
Thrust air control valve	Y6040	Turbocharger boost duct	**Fig 24**	ECL-30
Tilt sensor		*see* Anti-theft siren, tilt sensor		
Tire pressure control central antenna	B46a	Under rear bumper	**Fig 31**	ECL-32
Tire pressure control module	A85a	Left side cargo compartment behind trim		
Tire pressure control sensor	B43b B44b B45b B47b	Inside each wheel		
Tire pressure control transmitter	B43a B44a B45a B47a	Left front, right front, left rear, right rear wheel housing	**Fig 27** **Fig 29**	ECL-31 ECL-31
Transmission		*see* Automatic transmission entries		
Turbocharger coolant pump	M220	Left rear engine, near bell housing flange	**Fig 24**	ECL-30
Turn signal / high beam switch		Steering column switch center, right side		
ULF module	U400e	Underneath driver seat, under carpet	**Fig 34**	ECL-33
USB hub	A873	Trunk, behind rear seat back		
Utility 12v socket	I01003	*Convertible:* Right front of trunk, underneath rear seat backrest		
Valvetronic actuator	M6351	Flywheel end of cylinder head	**Fig 11**	ECL-27
Valvetronic eccentric shaft sensor	B60213	Flywheel end of cylinder head	**Fig 12**	ECL-27
Valvetronic relay	K6319a	Left rear engine compartment	**Fig 8**	ECL-26
Vanity mirror lights	E35, E36	Above edges of windshield behind visors		
VANOS solenoid, exhaust	Y6276	Timing chain end of cylinder head	**Fig 8**	ECL-26
VANOS solenoid, intake	Y6275	Timing chain end of cylinder head	**Fig 15** **Fig 19**	ECL-28 ECL-29
Washer fluid level switch	S136	Washer fluid reservoir in front of left front wheel, behind wheel housing liner	**Fig 27**	ECL-31
Washer pump relay, headlight	K6	Junction box fuse and relay panel, right front behind kick panel trim cover	**Fig 36**	ECL-33
Washer pump relay, windshield	K5	Junction box fuse and relay panel, right front behind kick panel trim cover	**Fig 37**	ECL-34

ECL

Table b. Component locations

Component	Code	Location	Fig #	Pg #
Washer pumps	M4 M7	Washer fluid reservoir in front of left front wheel, behind wheel housing liner	**Fig 28**	ECL-31
Washer switch		Steering column switch center, right side		
Wastegate valve	Y6039	Turbocharger	**Fig 20**	ECL-29
Wheel speed sensor		*see* ABS wheel speed sensor *see also* **340 Brakes**		
Window regulator motor, left front, right front	M21 M23	Upper door, behind trim panel, left, right		
Window regulator motor, left rear, right rear	M20 M22	Rear quarter panels, behind trim panel, left, right,		
Window switch, driver		Door switch cluster, driver door		
Windshield heater relay	K416	Left lower A-pillar, behind kick panel		
Windshield washer nozzle heater, left, right	E51b E52b	Base of windshield on engine hood		
Wiper motor, rear window	M94a	*Coupe:* Base of rear window		
Wiper motor, rear window	M94a M94b	*Clubman:* Split rear door, below each window		
Wiper motor, windshield	M3a	Under center of windshield in cowl	**Fig 8**	ECL-26
Wiper relay, rear window	K91	Junction box fuse and relay panel, right front behind kick panel trim cover	**Fig 37**	ECL-34
Wiper relays, windshield	K36, K37	Front power distribution and relay panel, left rear engine compartment	**Fig 10**	ECL-27
Wiper switch		Steering column switch cluster, right side		
X connector		*see* **Harness connectors, wire splices (X connectors)** in this repair group.		
X ground		*see* **Grounds** in this repair group		

Harness connectors, wire splices (X connectors)

BMW electrical components connect to the electrical harness via X connectors. Most connectors bear the same numerical designation as the component they attach to. For example, the connector for Y6279 (coolant thermostat) is designated X6279.

Note that grounds are also designated with X numbers. See **Grounds** in this repair group.

Table c lists harness connectors and wire splices not attached directly to components and gives brief location information for each.

Table c. X connectors

Code	Description and application	Location	Fig #	Page #
X01159		On seat position sensor, driver side		
X01184		On seat position sensor, passenger side		
X01185		At steering wheel, between right paddle switch and steering column switch cluster		
X013816		Passenger side, base of A-pillar behind trim	**Fig 38**	ECL-34
X0465		Steering wheel, on right rocker switch		
X10237	Splice to grounds X45, X275: MRS module (A12)	Under center of dashboard		
X105		At driver side front windshield washer jet		
X106		At passenger side front windshield washer jet		
X1065	Splice to ground X2846: Car access system (CAS) Passenger seat occupancy sensor Sunroof module	Under top of dashboard		
X10665	4-pin connector, Connects to left wiper drive unit	Cargo compartment left side, behind trim panel		
X10666	4-pin connector, Connects to right wiper drive unit	Cargo compartment right side, behind trim panel		
X10669	8-pin connector, Connects to left wiper drive unit	Cargo compartment left side, behind trim panel		
X10670	8-pin connector, Connects to right wiper drive unit	Cargo compartment right side, behind trim panel		
X113	Right front wheel speed sensor	Right front wheel well behind headlight		
X11339	Splice to ground X2846: Automatic transmission control module Gear indicator lighting Selector lever position switch	Left side of center console, ahead of selector lever		
X114	Front left wheel speed sensor	Front left wheel well behind headlight	**Fig 27**	ECL-31
X123		Steering wheel, on left rocker switch		
X1265		On rear hatch, right side		
X1266		On rear hatch, left side		
X13566	18-pin connector (storage location C110) 10-pin connector (storage location D111)	Inside driver side door Next to power window motor		
X13567	10-pin connector (storage location D111), connected to midrange speaker	Inside passenger side door Next to power window motor		
X13696	CD changer Satellite receiver	Left side kick panel, beneath trim	**Fig 32**	ECL-32
X13895	Intelligent battery sensor	Engine compartment, attached to battery	**Fig 26**	ECL-30

Harness connectors, wire splices (X connectors)

Table c. X connectors				
Code	**Description and application**	**Location**	**Fig #**	**Page #**
X14118	AUX-IN connection USB connection	On dashboard, behind center instrument bezel		
X143	Rear right wheel speed sensor	Rear right wheel well towards front of vehicle		
X14430		Center console, under parking break		
X1487		Rear seat at left belt buckle		
X1488		Rear seat at right belt buckle		
X15	42-pin connector	Lower A-pillar behind trim	**Fig 33**	ECL-32
X15330	AUX-IN connection USB connection	On dashboard, behind center instrument bezel		
X16738		Engine compartment, attached to battery		
X18145		Front Left wheel well behind fog light	**Fig 27**	ECL-31
X18146		Rear right wheel well		
X18753	Splice to ground X13795: Split left rear door wiper, button (Clubman)	Left split rear door, above taillight socket		
X1883	Crash sensor, left	Inside driver door, adjacent to power window motor		
X1884	Crash sensor, right	Inside passenger door, adjacent to power window motor		
X18978	Oxygen sensor after catalytic converter	On engine, flywheel side	**Fig 13**	ECL-27
X19527	16-pin connector	Left side interior, underneath dashboard		
X2261	3-pin connector	On engine at rail pressure sensor	**Fig 21**	ECL-29
X2498	Starter Car access system	Right side engine compartment, in front of battery compartment		
X256	Door electronics connector, right door	Right door jamb		
X257	Door electronics connector, left door	Left door jamb		
X275	15-pin connector to ground X2042: Driver seat belt buckle contact Driver seat heating module Junction box (A4010) MRS module (A12)	Left front floor underneath carpet		
X279	15-pin connector to ground X2846: Passenger seat belt buckle contact Passenger seat heating module Passenger seat occupancy sensor Junction box (A4010) MRS module (A12)	Right front floor underneath carpet		
X3279	Splice to ground X2042: Car access system (CAS) Heating and air-conditioning Instrument cluster	Behind right side center dashboard		
X3456	Splice to ground X2042: CD changer Central information display OBD II socket	Behind right side center dashboard		
X376	Rear left wheel speed sensor	Rear Left wheel well towards front of vehicle	**Fig 29**	ECL-31
X380	Lockout circuit for rear window defogger, right (ground)	On right split door near lock mechanism, under trim		
X4545	18-pin connector	Center console		

Table c. X connectors				
Code	**Description and application**	**Location**	**Fig #**	**Page #**
X593	Splice to ground X2042: Center console switch cluster Footwell lights Hazard warning switch Steering column switch cluster	Behind right side center dashboard		
X6039	Waste gate valve	On engine, intake side	**Fig 20**	ECL-29
X6040	Thrust air control valve	On engine, at exhaust manifold	**Fig 24**	ECL-30
X6056	Heating and air-conditioning connector	Behind left side center dashboard		
X6203	Crankshaft sensor	Engine, intake side		
X62101	Oxygen sensors before catalytic converter, connector	Engine compartment, between transmission and radiator	**Fig 12** **Fig 13**	ECL-27 ECL-27
X6214	Intake camshaft sensor	Top of engine, flywheel end	**Fig 12**	ECL-27
X6224	Exhaust camshaft sensor	Top of engine, flywheel end		
X623	8-pin connector	At driver side mirror		
X624	8-pin connector	At passenger side mirror		
X6241	Knock sensor	On engine intake adjacent to starter	**Fig 14** **Fig 21**	ECL-28 ECL-29
X630	13-pin connector	In dashboard behind center instrument		
X6460	Splice to ground X175: Brake fluid level switch Turn signal, left front Washer fluid level switch	Left front engine compartment	**Fig 9**	ECL-26
X6461	Splice to ground X175: Front power distribution box Headlight, left	Left front engine compartment	**Fig 9**	ECL-26
X65399	Attached to engine breather heater 1	Left front engine compartment		
X6820 X6821	Splice connectors, engine management	Engine compartment, underneath air intake duct		
X782	Airbag inflator assemblies, passenger side	In dashboard, right side, next to storage compartment / CD changer		
X9230	6-pin connector Intelligent battery sensor	In battery compartment, behind negative terminal		
X9331	12-pin connector to ground X2042: Amplifier Heating and air-conditioning control panel Radio Speakers	Lower A-pillar behind trim	**Fig 33**	ECL-32
X9420	Midrange speaker, front right	Inside passenger door		
X9421	Midrange speaker, front right	Inside passenger door		

Fuse Locations and Ratings

Fuse panels

For a description of the power supply system, see **600 Electrical System–General**.

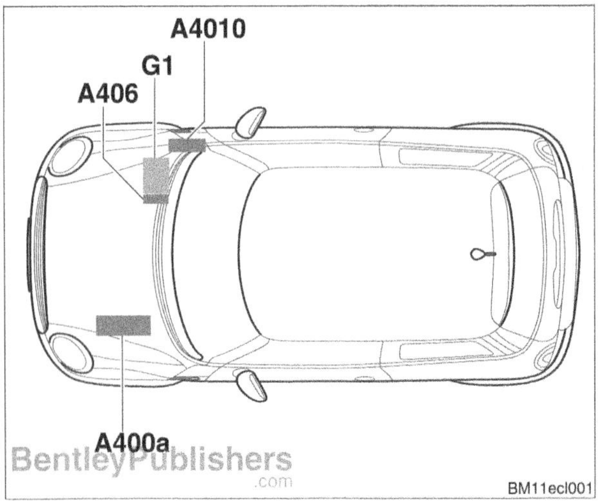

Fig 1 Power distribution

Fuse panels:

- **A400a** Front power distribution box, engine compartment:
  F01 - **F010** Engine management fuses
  FL1 - **FL12** High amperage fuses
  See **Front power distribution box (A400a)** in this repair group.
- **A406** Power supply fuses (fusible links), battery compartment:
  F60 - **F68**
  See **Power supply fuses** in this repair group.
- **A4010** Junction box, behind kick panel trim in right footwell:
  F1 - **F49** Fuses
  See **Junction box fuse panel** in this repair group.
- **G1** Battery, right rear engine compartment

Front power distribution box (A400a)

The front power distribution box in left side engine compartment contains fuses and relays. Fuses are in 3 rows:

- High amperage fuses **FL6** - **FL12** on right side of upper fuse box. See **×**.
- Fuses **F01** - **F010** on left side of upper fuse box.
- High amperage fuses **FL1** - **FL5** in lower fuse box.

Relays applications are covered in **Fig 10**.

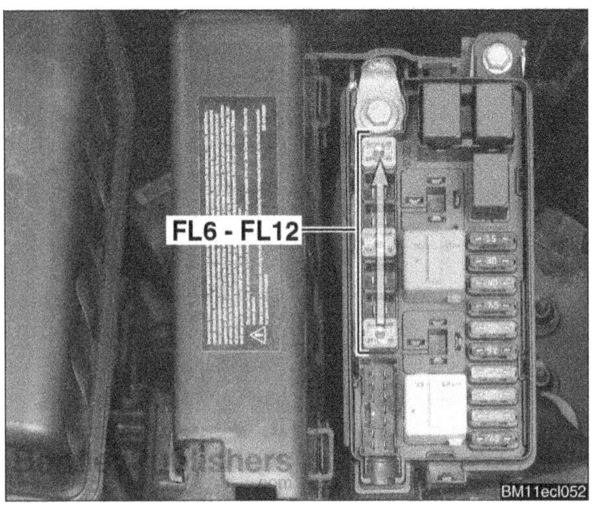

Fig 2 High amperage fuses FL6 - FL12 (A400a)

Fuse ratings for high amperage fuses **FL6** - **FL12** are in **Table d**.

Table d. Upper row high amperage fuses in front power distribution and fuse panel (A400a)

Fuse	Rating in amps	Protected circuits
FL6	40	Dynamic stability control (DSC)
FL7		not used
FL8	30	Dynamic stability control (DSC)
FL9	50	Engine cooling fan relay
FL10	50	Windshield heater relay
FL11		not used
FL12	30	DC / DC converter

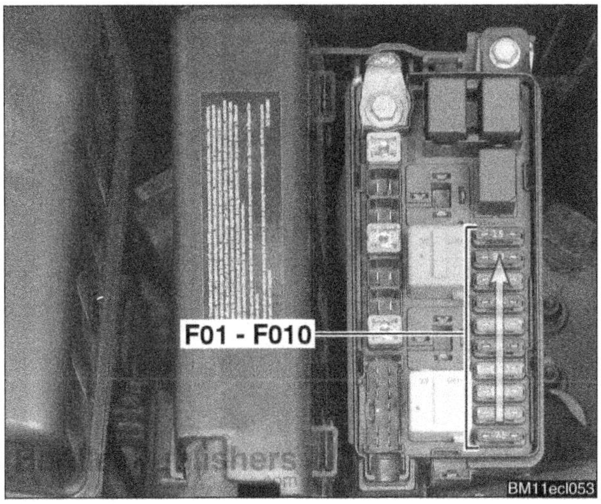

Fig 3 Fuses F01 - F010 (A400a)

Fuse ratings for engine management fuses **F01** - **F010** are in **Table e**.

Table e. Engine management fuses in front power distribution and fuse panel (A400a)		
Fuse	**Rating in amps**	**Protected circuits**
F01	5	Engine control module (ECM)
F02	25	Engine control module (ECM)
F03	20	Crankcase breather heating relay Engine control module (ECM)
F04	20	Engine control module (ECM)
F05	5 (to 02/2010) 7.5 (from 03/2010)	Cooper S: Fuel injector relay Fuel tank leakage diagnostic module Cooper: Valvetronic relay
F06	25	Dynamic stability control (DSC)
F07	7.5	Crankcase breather heating relay
F08	30 / 40	Engine cooling fan relay
F09	25	Wiper relays
F010	15	Power-saving relay terminal 15

To gain access to lower row of fuses (**FL1** - **FL5**):

- Remove power distribution box mounting screw and turn box on its side.
- Use small screwdriver to pry off lower fuse panel cover.

ECL

Fig 4 High amperage fuses FL1 - FL5 (A400a)

Fuse ratings for high amperage fuses **FL1** - **FL12** are in **Table f**.

Table f. Lower row high amperage fuses in front power distribution and fuse panel (A400a)		
Fuse	**Rating in amps**	**Protected circuits**
FL1		not used
FL2		not used
FL3		not used
FL4	80	Electromechanical power steering
FL5		not used

Power supply fuses

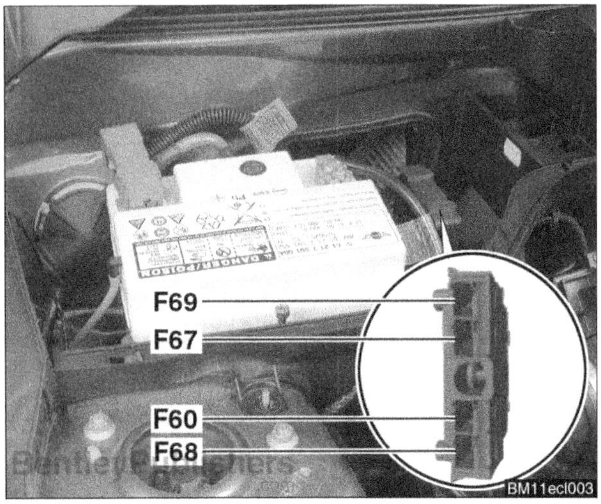

Fig 5 Power supply fuse panel (A406)

Fuse ratings for power supply fuses **F60** - **F69** are in **Table g**

Table g. Power supply fuses (fusible links) (A406)		
Fuse	**Rating in amps**	**Protected circuits**
F60	125	Junction box module (A4010) Fuses F1 - F49
F67	150	DME main relay (K6300) Fuses F01 - F010 Fuses FL1 - FL12
F68	40	Cooper: Valvetronic relay (K6319a) Cooper S: Fuel injector relay (K6327)
F69		not used

Fusible links are not available separately.

Junction box fuse panel

Remove junction box fuse panel (A4010) cover on kick panel in right footwell.

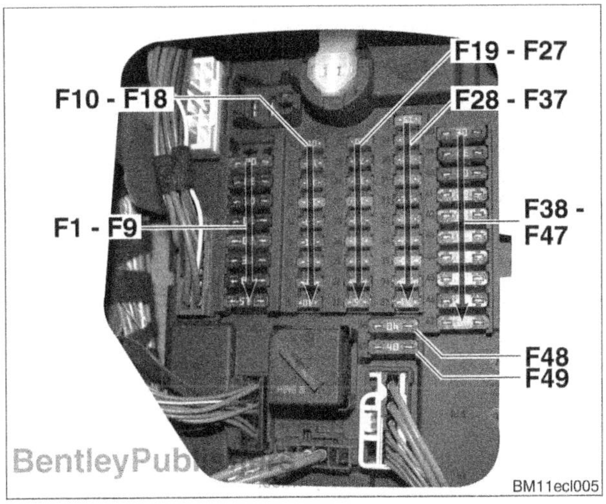

BM11ecl005

Fig 6 Junction box fuse panel(A4010)

Fuse ratings for junction box fuse panel fuses F1 - F49 are in **Table h**.

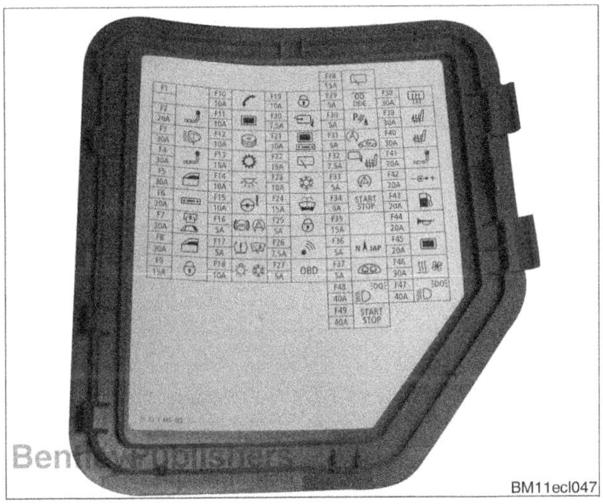

BM11ecl047

Note that fuse ratings are on inside of fuse panel cover.

Junction box fuse panel

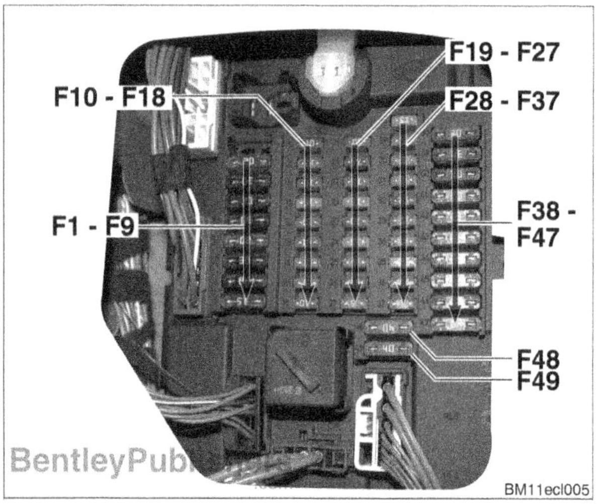

Table h. Junction box fuses (A4010) to 11 / 2008

Fuse	Rating in amps	Protected circuits
F1		not used
F2	20	Trailer socket
F3	30	Headlight washer relay
F4		not used
F5	30	Footwell module
F6	20	Amplifier
F7	30	Sunroof
F8	30	Footwell module
F9	15	Central locking
F10	10	Telephone ULF
F11	10	Digital tuner Satellite radio Video module
F12	10	CD changer
F13	15	Automatic transmission control (EGS)
F14	10	Cargo compartment lights Roof switch center Vanity mirror lights
F15	10	Steering column switch cluster
F16	5	Dynamic stability control (DSC)
F17	5	Automatic transmission control (EGS) Rain and light sensor Tire pressure control (RDC)
F18	10	Glove compartment light Heating and air-conditioning
F19	10	Comfort access control module
F20	7.5	Outside mirrors fold-in

Table h. Junction box fuses (A4010) to 11 / 2008

Fuse	Rating in amps	Protected circuits
F21	10	Central information display MINI joystick Radio
F22	15	Rear wiper relay
F23	10	A/C compressor relay
F24	15	Windshield washer pump
F25	5	Electrochromic rear view mirror Selector lever position switch
F26	7.5	Antitheft siren Tilt sensor
F27	5	OBD II socket
F28		not used
F29	5	Telephone connector
F30	5	Park distance control (PDC)
F31	5	Air quality (AUC) sensor DC / DC converter
F32	7.5	Electrochromic rear view mirror Outside mirror switch Seat heating
F33	5	DSC acceleration sensor
F34	5	Car access system (CAS)
F35		not used
F36	5	Navigation
F37	5	Tachometer
F38	30	Rear window defogger relay
F39	30	Passenger seat heating module
F40	30	Driver seat heating module
F41		not used
F42	20	Front cigar lighter Rear utility socket
F43	20	Fuel pump relay
F44	20	Horn relay Steering column switch cluster
F45	20	Car communication computer
F46	30	Blower Heating and air-conditioning
F47	40	Footwell module
F48	40	Footwell module
F49	40	Car access system (CAS)

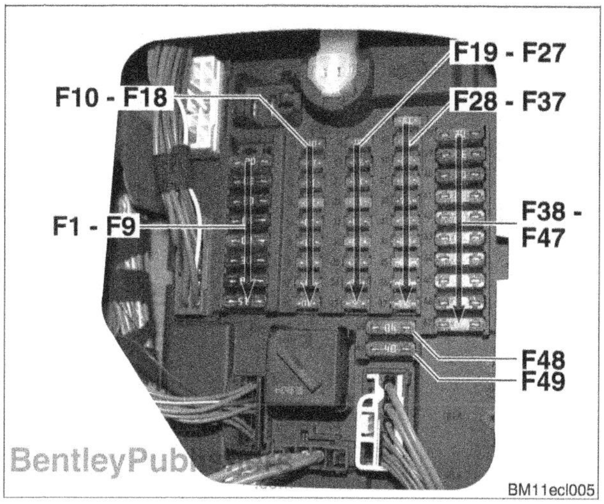

Table i. Junction box fuses (A4010) from 12 / 2008

Fuse	Rating in amps	Protected circuits
F1		not used
F2	30	Headlight washer relay
F3		not used
F4		not used
F5	20	Sound system amplifier
F6	30	Convertible top module Sunroof module
F7	20	Trailer socket
F8	30	Footwell module
F9		not used
F10		not used
F11	15	Central locking
F12	10	Telephone ULF
F13	10	Digital tuner Navigation Satellite radio Video module
F14	10	CD changer
F15	15	Automatic transmission control (EGS)
F16	10	Cargo compartment light Roof switch center Vanity mirror lights
F17	10	Steering column switch cluster
F18	5	Dynamic stability control (DSC)
F19	5	Automatic transmission control (EGS) Rain and light sensor Selector lever position switch Tire pressure control (RDC)
F20	10	Glove compartment light Heating and air-conditioning
F21	10	Comfort access control module
F22	7.5	Outside mirror fold-in
F23	10	Central information display MINI joystick Radio
F24	10	A/C compressor clutch
F25	15	Windshield washer pump
F26	5	Electrochromic rear view mirror Selector lever position switch
F27	7.5	Anti-theft siren Tilt sensor
F28	15	Central locking
F29	5	OBD II socket Tachometer
F30	15	Clubman: Rear window wiper relay 2
F31		not used
F32	5	Park distance control (PDC)
F33	5	Air quality (AUC) sensor DC / DC converter
F34	7.5	Electrochromic rear view mirror Outside mirror adjustment Seat heating
F35	5	DSC acceleration sensor
F36	5	Car access system (CAS)
F37		not used
F38	5	Navigation
F39		not used
F40	30	Rear window defogger
F41	30	Passenger seat heating
F42	30	Driver seat heating
F43	10	Rollover protection
F44	20	Front cigar lighter Rear utility socket
F45	20	Car communications computer
F46	20	Fuel pump
F47	20	Horns Steering column switch cluster
F48	30	Blower Heating and air-conditioning
F49	30	Sound system amplifier
F50	15	Rear window wiper
F51	40	Car access system (CAS)
F52	40	Footwell module
F53	40	Footwell module

Grounds

Ground distribution

Grounds are distributed throughout the vehicle body. Many are found under the interior carpets or behind trim panels. Several components grounds are often ganged. Ground positions vary among models. Lugs and connectors attached to ground are susceptible to damage and corrosion. Clean or renew as necessary.

Most ground positions are illustrated in photos in this repair group. See figure and page references in **Table j**. Ground applications are detailed in **Table k**.

Table j. Ground locations

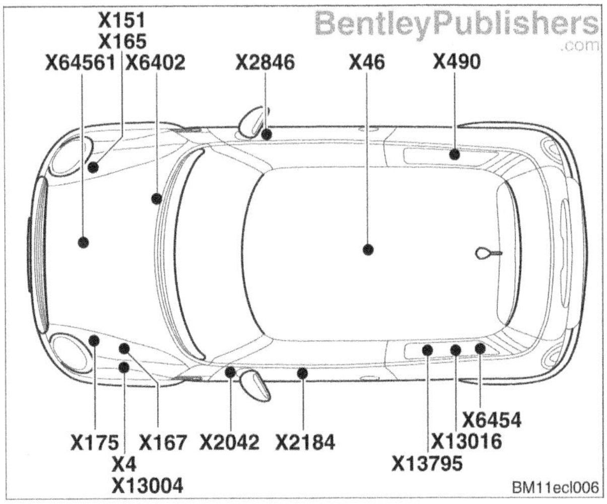

Code	Location	Fig #	Page #
X13004	Behind left wheel housing liner, electromechanical steering motor	**Fig 26**	ECL-31
X13016	Left rear quarter panel, behind trim	**Fig 35**	ECL-33
X13795	Left rear quarter panel, behind trim	**Fig 35**	ECL-33
X151	Underneath right front of vehicle, engine control module (ECM)		
X165	Underneath right front of vehicle, engine control module (ECM)		
X167	Underneath left front of vehicle, engine control module (ECM)		
X175	Left front engine compartment		
X2042	Left door sill	**Fig 33**	ECL-33
X2184	Left door sill	**Fig 34**	ECL-33
X2846	Right door sill	**Fig 38**	ECL-34
X4	Behind left wheel housing liner, DSC control module	**Fig 26**	ECL-31
X46	Under center console	**Fig 41**	ECL-35
X490	Right rear quarter panel, behind trim	**Fig 40**	ECL-34
X6402	Battery compartment, battery ground strap		
X6454	*Convertible:* left side trunk		
X64561	Center of cylinder head cover, ignition coil ground	**Fig 16**	ECL-28

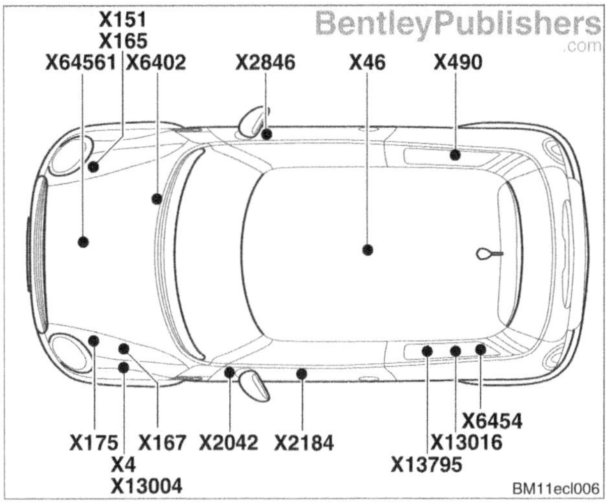

Ground applications

Table k. Ground applications

Circuit	Ground	Fig #	Pg #
A/C compressor clutch	X165		
Air quality (AUC) sensor	X175		
Amplifier	X13016	**Fig 35**	ECL-33
Anti-theft siren	X175		
AUC sensor	X165		
Automatic transmission control	X175		
Back-up light switch	X175		
Battery ground	X6402	**Fig 25**	ECL-30
Blower motor	X2042	**Fig 33**	ECL-33
Blower output stage	X2042	**Fig 33**	ECL-33
B-pillar light, left	X2184	**Fig 34**	ECL-33
B-pillar light, right	X2846	**Fig 38**	ECL-34
Brake fluid level switch	X175		
Brake fluid level switch	X165		
Brake light switch	X2042	**Fig 33**	ECL-33
Car access system (CAS)	X2042	**Fig 33**	ECL-33
CD changer	X2042	**Fig 33**	ECL-33
Center brake light	X13795	**Fig 35**	ECL-33
Center brake light	X490	**Fig 40**	ECL-34
Center console switch cluster	X2042	**Fig 33**	ECL-33
Central information display	X13016	**Fig 35**	ECL-33
Central information display	X2042	**Fig 33**	ECL-33
Cigar lighter, front	X2042	**Fig 33**	ECL-33
Clutch module	X2042	**Fig 33**	ECL-33
Comfort access control module	X490	**Fig 40**	ECL-34
Convertible top components	X6454		
Courtesy light, Club door	X490	**Fig 40**	ECL-34
Crankcase breather heater	X175		
Digital tuner	X13016	**Fig 35**	ECL-33
Dome light, rear	X2042	**Fig 33**	ECL-33
Door electronics, left	X2042	**Fig 33**	ECL-33
Door electronics, right	X2846	**Fig 38**	ECL-34
Door locks, Club door	X490	**Fig 40**	ECL-34
DSC acceleration sensor	X2846	**Fig 38**	ECL-34
DSC control module	X4	**Fig 26**	ECL-31
Electrochromic interior mirror	X2184	**Fig 34**	ECL-33
Electromechanical steering motor	X13004	**Fig 26**	ECL-31
Engine control module (ECM)	X151 X167		
Engine cooling fan	X175		
Foglight, left front	X175		
Foglights, rear	X490	**Fig 40**	ECL-34
Footwell lights	X2042	**Fig 33**	ECL-33
Footwell module	X2184	**Fig 34**	ECL-33
Footwell module	X2042	**Fig 33**	ECL-33
Front power distribution box	X175		
Fuel pump	X13795	**Fig 35**	ECL-33
Gear indicator lighting	X2846	**Fig 38**	ECL-34
Hazard warning switch	X2042	**Fig 33**	ECL-33
Headlight washer pump	X175		
Headlight, left	X175		
Headlight, right	X165		
Heating and air-conditioning system	X2042	**Fig 33**	ECL-33
Hood contact switch	X175		
Horns	X165		
Ignition coils	X64561	**Fig 16**	ECL-28
Instrument cluster	X2042	**Fig 33**	ECL-33
Interior movement detector	X2184	**Fig 34**	ECL-33
Joystick	X13016	**Fig 35**	ECL-33
Junction box	X2846	**Fig 38**	ECL-34
License plate lights	X490	**Fig 40**	ECL-34

Ground applications

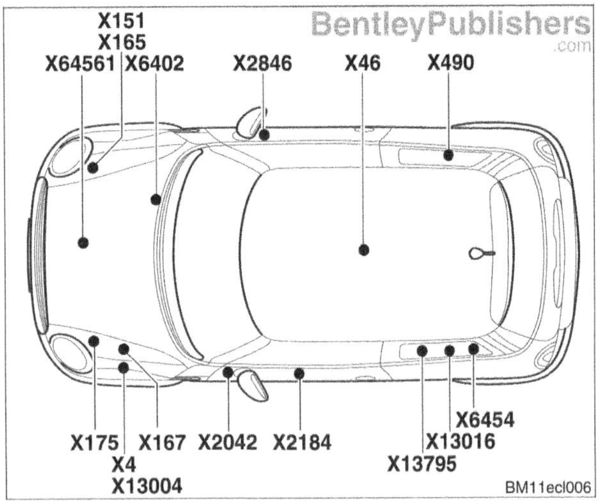

Table k. Ground applications

Circuit	Ground	Fig #	Pg #
MRS control module	X46	**Fig 41**	ECL-35
OBD II socket	X2042	**Fig 33**	ECL-33
Outside mirror fold-in module	X13795	**Fig 35**	ECL-33
Park distance control (PDC) module	X490	**Fig 40**	ECL-34
Passenger seat belt buckle contact	X2846	**Fig 38**	ECL-34
Passenger seat heating module	X2846	**Fig 38**	ECL-34
Passenger seat occupancy sensor	X2846	**Fig 38**	ECL-34
Radio	X13016	**Fig 35**	ECL-33
Radio	X2042	**Fig 33**	ECL-33
Rain and light sensor	X2184	**Fig 34**	ECL-33
Rear hatch lock (Hardtop)	X490	**Fig 40**	ECL-34
Rear window defogger	X490	**Fig 40**	ECL-34
Rear window defogger, left	X13795	**Fig 35**	ECL-33
Rear wiper drive (Hardtop)	X490	**Fig 40**	ECL-34
Satellite receiver	X13016	**Fig 35**	ECL-33
Seat belt buckles, rear seats	X2846	**Fig 38**	ECL-34
Selector lever position switch	X2846	**Fig 38**	ECL-34
Side marker light, left	X175		
Side marker lights, rear	X490	**Fig 40**	ECL-34
Side marker lights, right	X165		
Siren and tilt alarm sensor	X165		
Speakers	X2042	**Fig 33**	ECL-33
Split rear door button	X13795	**Fig 35**	ECL-33
Split rear door lock (Clubman)	X490	**Fig 40**	ECL-34
Steering column switch cluster	X2042	**Fig 33**	ECL-33
Sunroof	X2846	**Fig 38**	ECL-34
Sunroof module	X2184	**Fig 34**	ECL-33
Taillight, left	X13795	**Fig 35**	ECL-33
Taillight, right	X490	**Fig 40**	ECL-34
Telephone	X13016	**Fig 35**	ECL-33
Telephone eject box	X13016	**Fig 35**	ECL-33
Tilt alarm	X175		
Tire pressure control (RDC)	X13795	**Fig 35**	ECL-33
Trailer module	X490	**Fig 40**	ECL-34
Trailer socket	X490	**Fig 40**	ECL-34
Transmission control module	X2846	**Fig 38**	ECL-34
Turn signal repeater, left front	X175		
ULF interface	X13016	**Fig 35**	ECL-33
Utility socket, rear	X13795	**Fig 35**	ECL-33
Vanity light, left	X2042	**Fig 33**	ECL-33
Vanity light, right	X2846	**Fig 38**	ECL-34
Video module	X13016	**Fig 35**	ECL-33
Washer fluid level switch	X175		
Washer nozzle heaters	X175		
Windshield heater relay	X2042	**Fig 33**	ECL-33
Windshield heater, left	X2042	**Fig 33**	ECL-33
Windshield heater, right	X2846	**Fig 38**	ECL-34
Wiper drive, left rear (Clubman)	X13795	**Fig 35**	ECL-33
Wiper drive, right rear (Clubman)	X490	**Fig 40**	ECL-34
Wiper module	X165		

Component Location Illustrations

Fig 7 Control modules, antennas (Hardtop)

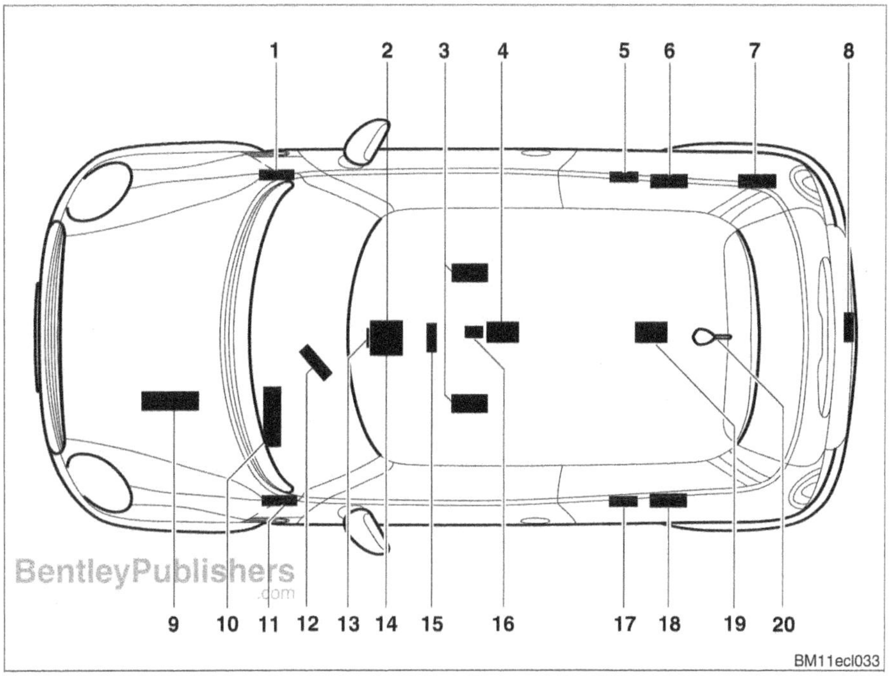

1. **A4010** Junction box fuse panel
2. **A33** Sunroof module
3. **A186a**, **A187a** Seat modules
4. **A12** MRS (airbag) control module
5. **W12a** Comfort access antenna
6. **A215** Comfort access module
7. **A81a** Park distance control (PDC) module
8. **W14** Comfort access antenna
9. **A6000** Engine control module (ECM)
10. **A149a** Car access system (CAS)
11. **A4011** Footwell module
12. **A7000** Transmission control module (EGS)
13. **B57b** Rain and light sensor
14. **A121b** Interior movement sensor
15. **W10b** Comfort access antenna
16. **W18** Bluetooth antenna
17. **W11a** Comfort access antenna
18. **A208** Folding mirror module
19. **W12** Comfort access antenna
20. **W16** Roof antenna

Engine compartment components

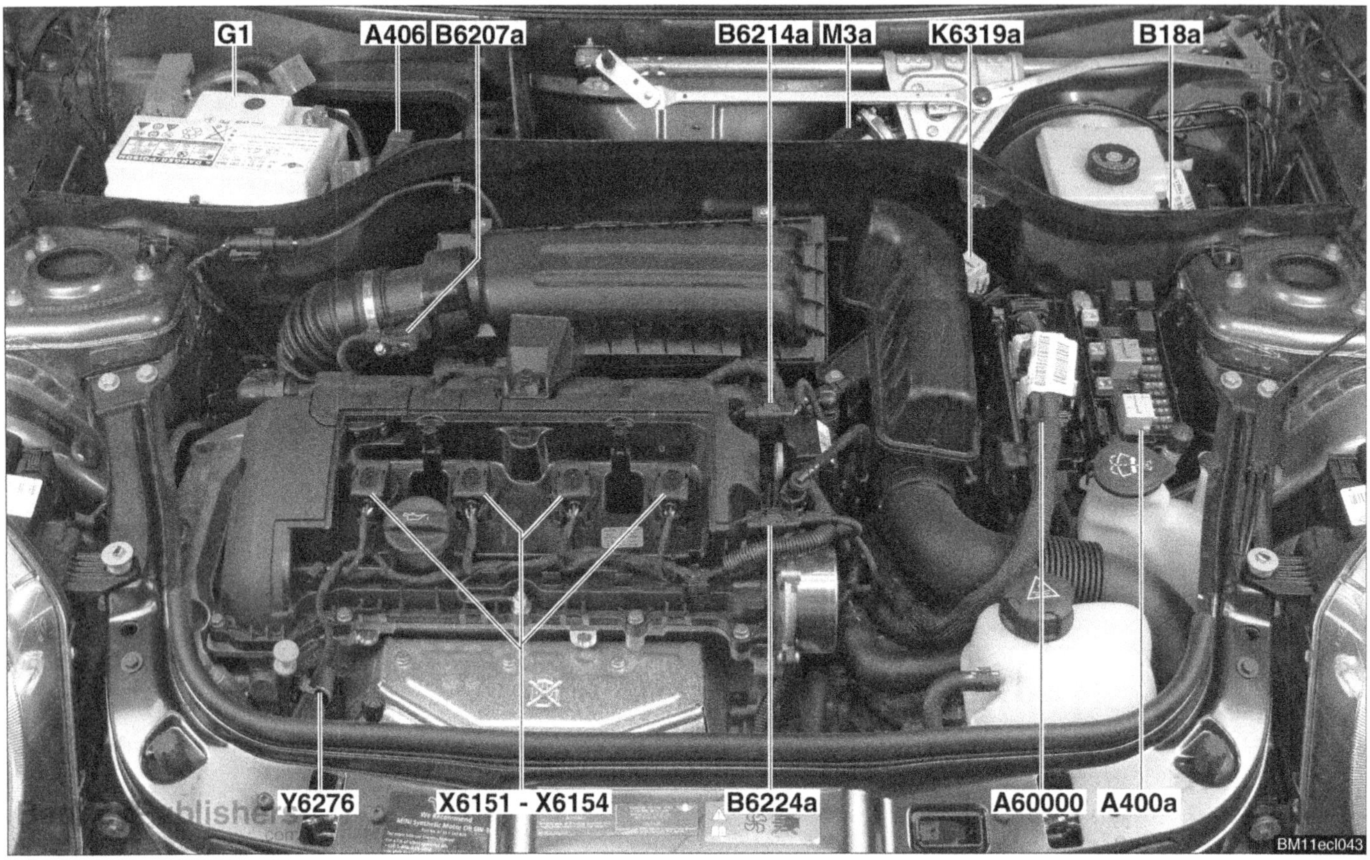

Fig 8 Engine compartment, covers removed

NOTE—

- *N12 engine shown, other engine configurations are similar*

1. **A400a** Front power distribution and fuse panel
2. **A406** Power supply fuse panel (fusible links)
3. **A6000** Engine control module (ECM)
4. **B18a** Brake fluid level switch
5. **B6207a** Mass air flow sensor
6. **B6214a** Intake camshaft sensor
7. **B6224a** Exhaust camshaft sensor
8. **G1** Battery
9. **K6319a** Valvetronic relay
10. **M3a** Windshield wiper motor
11. **Y6276** VANOS solenoid
12. **X6151** - **X6154** Ignition coil connectors

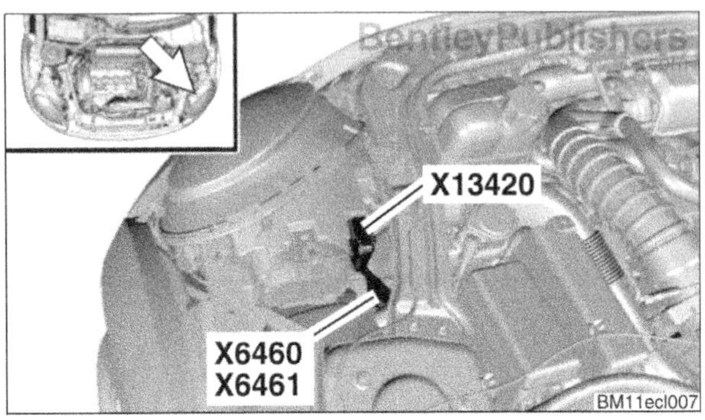

Fig 9 Left front engine compartment

Behind left headlight:

- **X6460** Ground splice
- **X6461** Ground splice
- **X13420** Connector

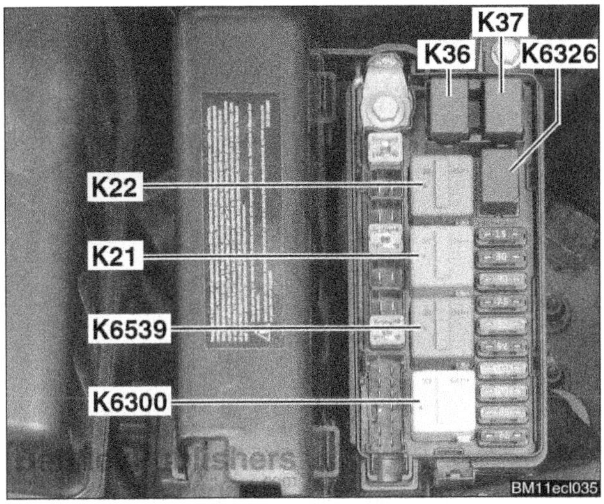

Fig 10 Left rear engine compartment

A400 Front power distribution box relays:

- **K21** Engine cooling fan relay 1
- **K22** Engine cooling fan relay 2
- **K36** Windshield wiper relay 1
- **K37** Windshield wiper relay 2
- **K6300** DME main relay
- **K6326** Power-saving relay terminal 15
- **K6539** Crankcase breather heating relay

For front power distribution box fuse applications, see **Front power distribution box (A400a)** in this repair group.

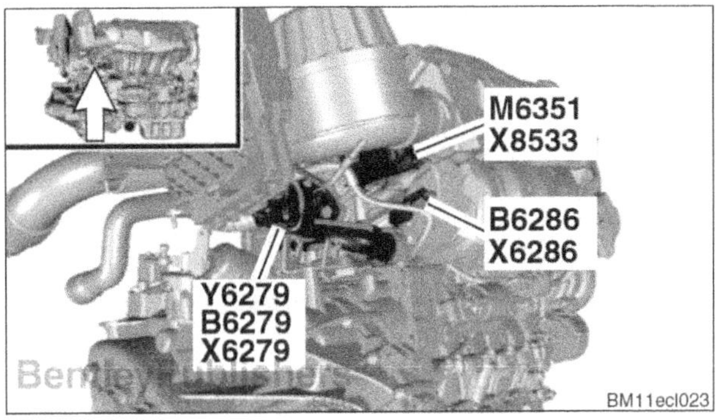

Fig 11 Engine flywheel end (non-turbo engine)

Left side and rear engine compartment:

- **B6279** Characteristic map coolant thermostat (N12 engine)
- **Y6279** Characteristic map coolant thermostat (N16 engine)
- **B6286** Intake temperature-differential pressure sensor
- **M6351** Valvetronic actuator
- **X6279** Connector
- **X6286** Connector
- **X8533** Connector

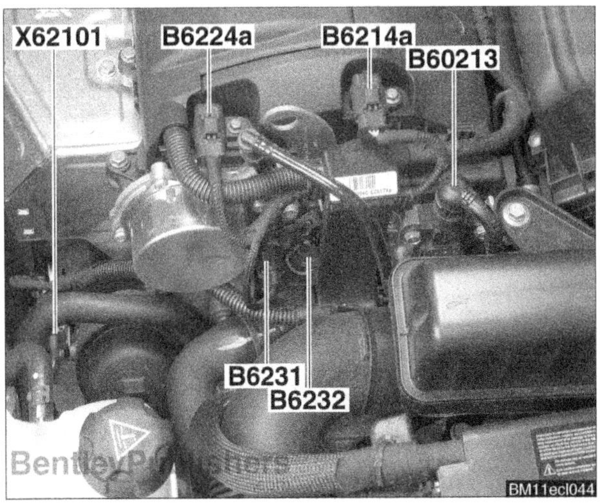

Fig 12 Engine flywheel end (N12 engine)

On end of cylinder head:

- **B6214a** Intake camshaft sensor
- **B6224a** Exhaust camshaft sensor
- **B6231** Oil pressure switch
- **B6232** Engine coolant temperature sensor
- **B60213** Valvetronic eccentric shaft sensor
- **X6214** Connector
- **X62101** Oxygen sensor connector

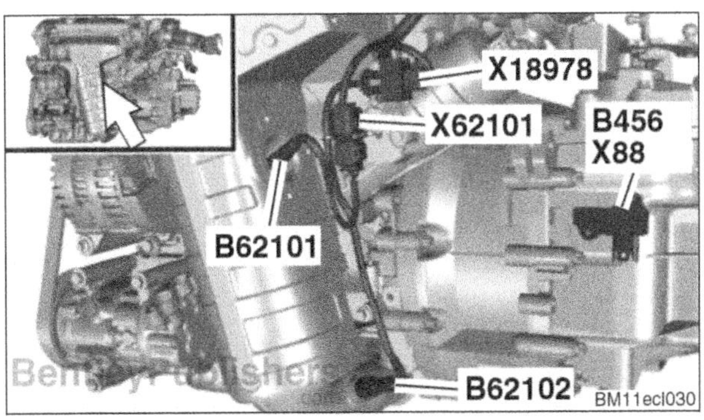

Fig 13 Engine exhaust side (non-turbo engine)

Front of engine compartment:

- **B456** Neutral sensor
- **B62101** Precatalyst oxygen sensor
- **B62102** Post-catalyst oxygen sensor'
- **X88** Connector
- **X18978** Connector
- **X62101** Oxygen sensor connector

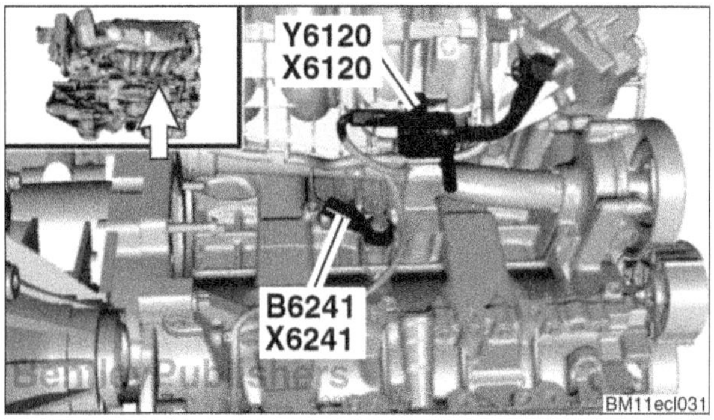

Fig 14 Engine intake side (non-turbo engine)

Rear engine compartment:

- **B6241** Knock sensor
- **Y6120** Evaporative control valve
- **X6120** Connector
- **X6241** Connector

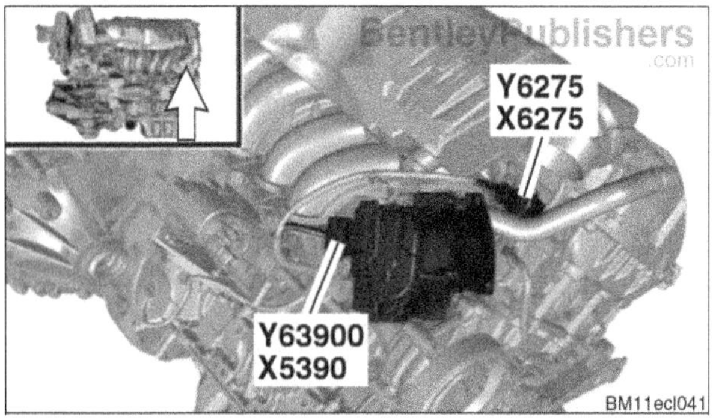

Fig 15 Engine intake side (non-turbo engine)

Rear engine compartment:

- **Y6275** Intake VANOS solenoid
- **Y63900** Electronic throttle valve
- **X5390** Connector
- **X6275** Connector

Fig 16 Engine timing chain end (non-turbo engine)

Front top of cylinder head:

- **B6207a** Mass air flow sensor
- **Y6275** Intake VANOS solenoid
- **Y6276** Exhaust VANOS solenoid
- **X6151**, **X6152**, **X6153** Ignition coil connectors
- **X64561** Ignition coils ground

Fig 17 Left rear engine compartment (turbo engine)

Components at flywheel end of engine and on engine bulkhead:

- **A65a** ABS / DSC control module and hydraulic unit
- **A400a** Front power distribution and relay panel
- **A6000** Engine control module (ECM)
- **B8** Refrigerant pressure sensor
- **B6207a** Mass flow sensor
- **B6214a** Intake camshaft sensor
- **K6304a** Secondary air injection relay
  K6327 Fuel injector relay
  K6319a Valvetronic relay (N18 engine)
  in bulkhead relay panel

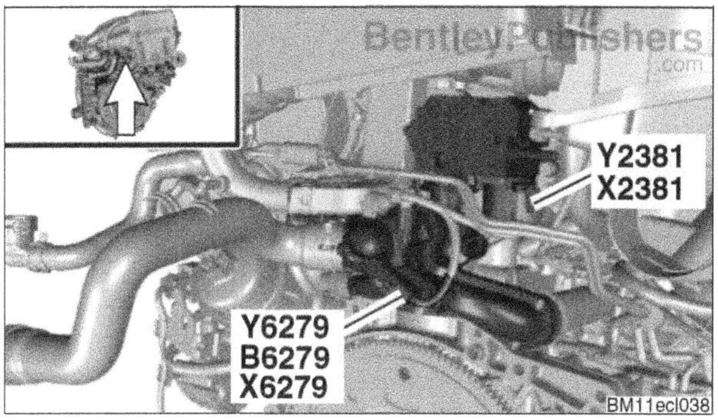

Fig 18 Flywheel end of engine (turbo engine)

Above bell housing flange:

- **B6279** Characteristic map coolant thermostat (N14 engine)
- **Y6279** Characteristic map coolant thermostat (N18 engine)
- **Y2381** Fuel volume control valve
- **X2381** Connector
- **X6279** Connector

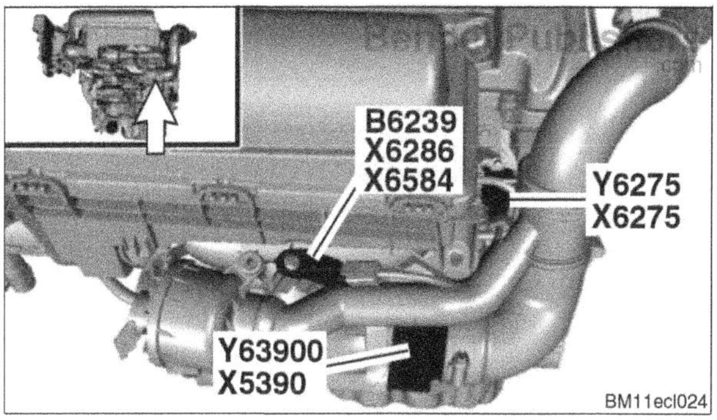

Fig 19 Engine timing chain end (turbo engine)

Right side and rear engine compartment:

- **B6239** Intake pressure (MAP) sensor
- **Y6275** Intake VANOS solenoid
- **Y63900** Electronic throttle valve
- **X5390** Connector
- **X6275** Connector
- **X6286** Connector (N14 engine)
- **X6584** Connector (N18 engine)

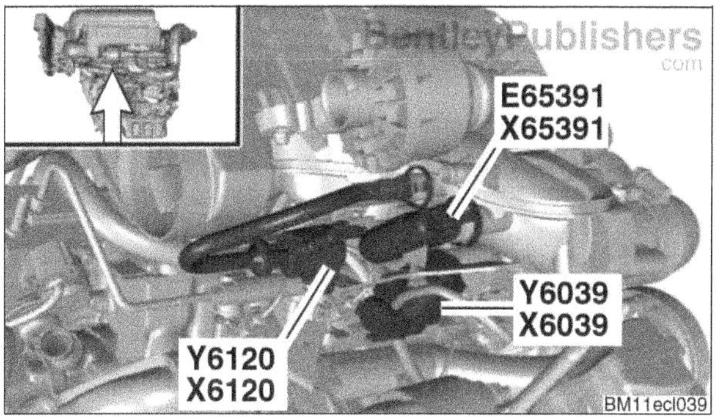

Fig 20 Intake manifold side (turbo engine)

Rear engine compartment:

- **E65391** Crankcase breather heater
- **Y6039** Waste gate valve
- **Y6120** Evaporative emissions control valve
- **X6039** Connector
- **X6120** Connector
- **X65391** Connector

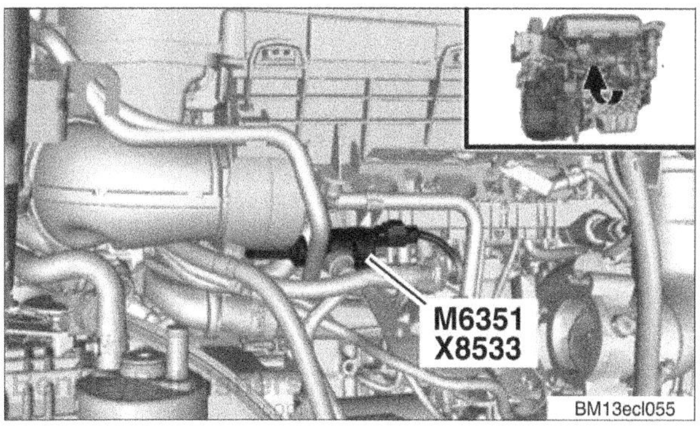

Fig 21 Underneath intake manifold (N18 engine)

Left side of engine compartment:

- **M6351** Valvetronic servo motor
- **X8533** Connector

ECL

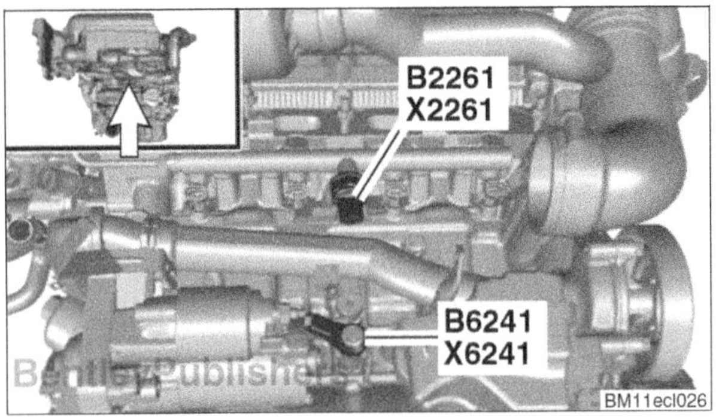

Fig 22 Underneath intake manifold (turbo engine)

Rear engine compartment:

- **B2261** Fuel rail pressure sensor
- **B6241** Knock sensor

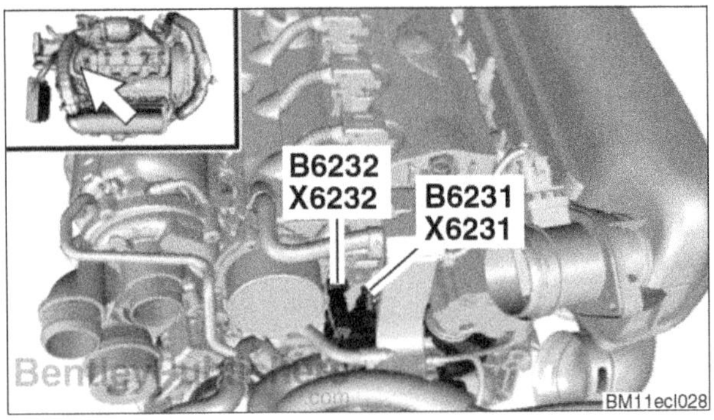

Fig 23 Engine flywheel end (N14 engine)

On cylinder head:

- **B6231** Oil pressure switch
- **B6232** Engine coolant temperature (ECT) sensor
- **X6231** Connector
- **X6232** Connector

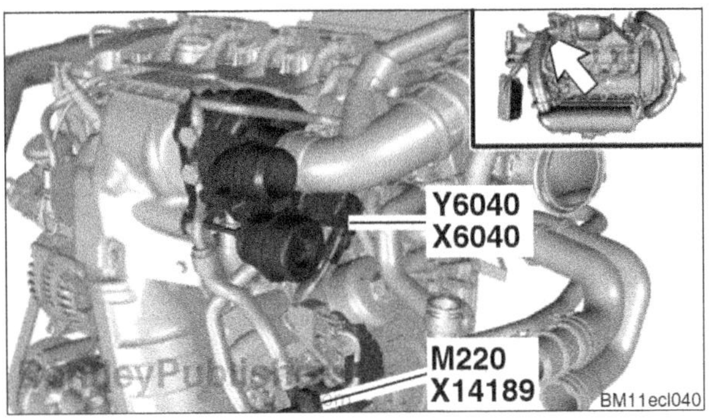

Fig 24 Engine flywheel end (turbo engine)

On turbo boost duct and near bell housing flange:

- **M220** Turbocharger coolant pump
- **Y6040** Thrust air control valve
- **X6040** Connector
- **X14189** Connector

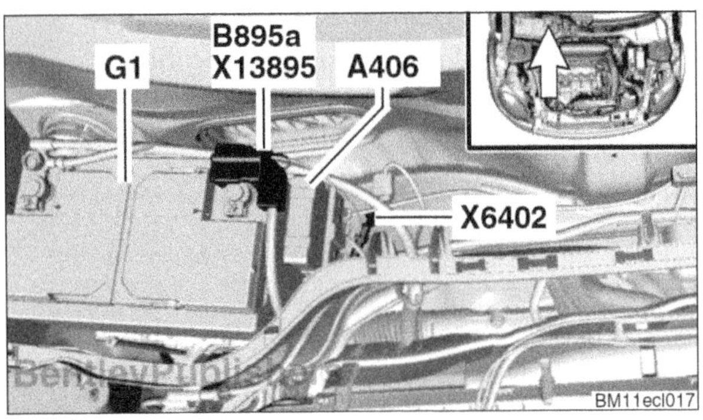

Fig 25 Rear engine compartment

Under cowl cover:

- **A406** Power supply fuses **F60** - **F68**
  See **Power supply fuses** in this repair group.
- **B895a** Intelligent battery sensor (IBS)
- **G1** Battery
- **X6402** Battery ground
- **X13895** Connector

Underneath vehicle

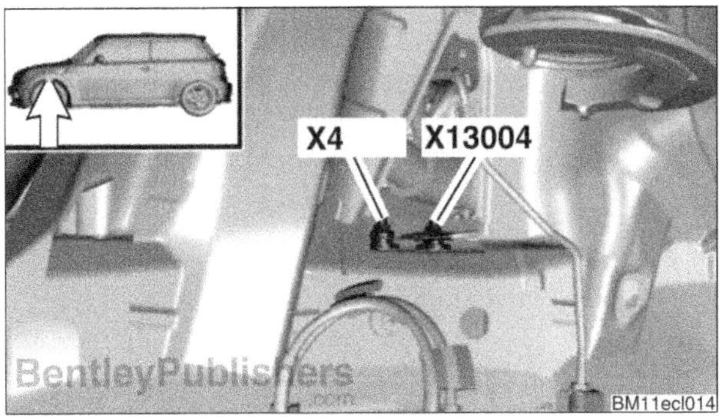

Fig 26 Left front wheel housing

Behind liner:

- **X4** Ground, DSC control module
- **X13004** Ground, electromechanical steering motor

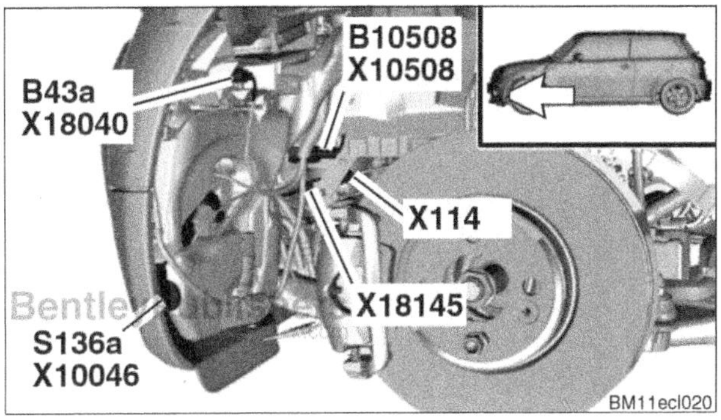

Fig 27 Left front wheel housing

Behind liner:

- **B43a** Tire pressure control (RDS) sensor
- **B10508** Crash sensor
- **S136a** Washer fluid level switch
- **X114** Connector
- **X10046** Connector
- **X10508** Connector
- **X18040** Connector
- **X18145** Connector

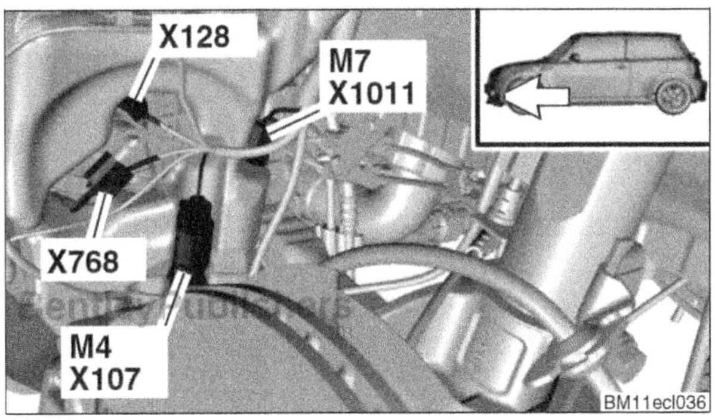

Fig 28 Left front wheel housing

Behind liner:

- **M4** Windshield washer pump
- **M7** Headlight washer pump
- **X107** Connector
- **X128** Connector
- **X768** Connector
- **X1011** Connector

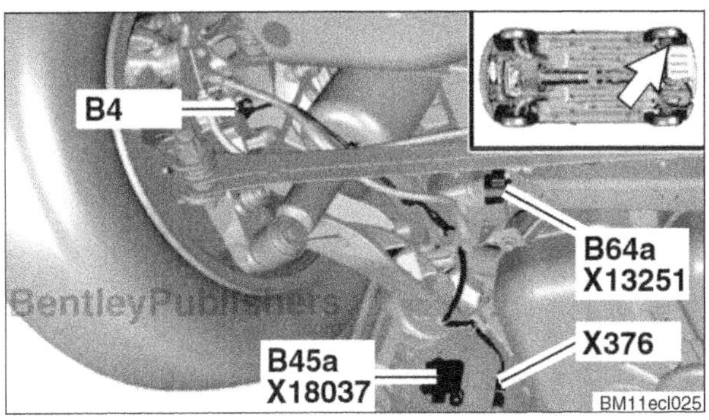

Fig 29 Left rear

Wheel housing, suspension:

- **B4** ABS wheel speed sensor
- **B45a** Tire pressure control (RDS) sensor
- **B64a** Rear ride height sensor
- **X376** Connector
- **X13251** Connector
- **X18037** Connector

Fig 30 Right rear

◄ Wheel housing

- **B3** ABS wheel speed sensor
- **M119a** Fuel tank leakage diagnosis module (DMTL)

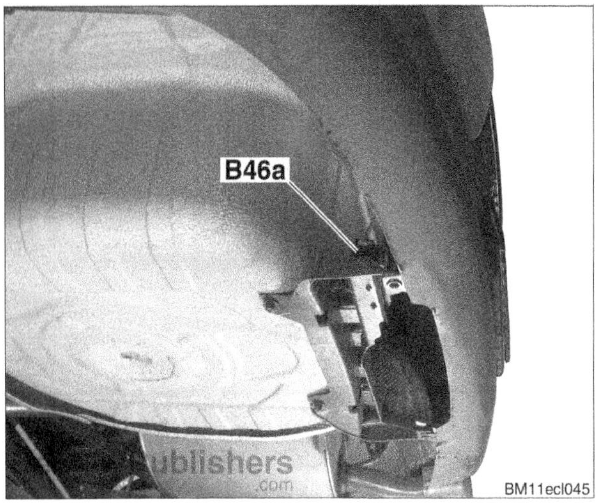

Fig 31 Underneath rear bumper

◄ 2010 Clubman:

- **B46a** Tire pressure control (RDC) antenna

Interior components

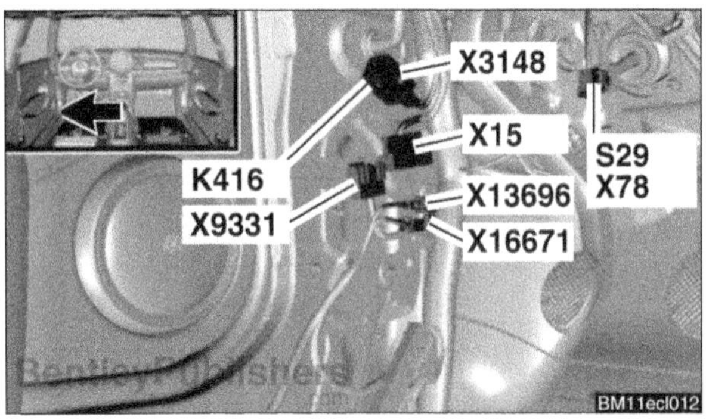

Fig 32 Left front

◄ Above pedals, behind kick panel trim:

- **K416** Windshield heater relay
- **S29** Brake light switch
- **X15** Connector
- **X78** Connector
- **X3138** Connector
- **X9331** Connector
- **X13696** Connector
- **X16671** Connector

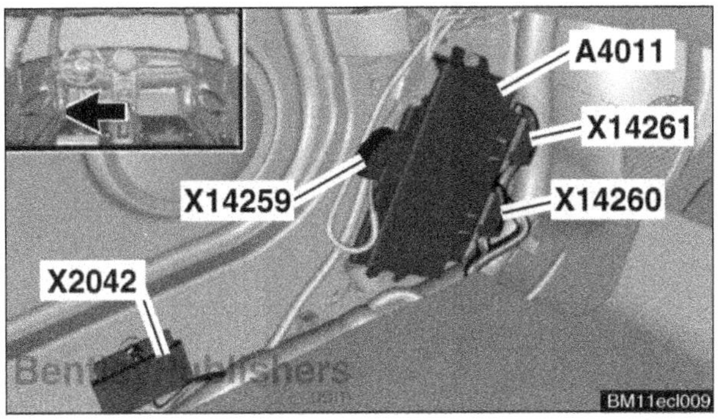

Fig 33 Left front

Behind left door sill, behind kick panel trim:

- **A4011** Footwell module
- **X2042** Ground
  See **Ground applications** for details.
- **X14259** Connector
- **X14260** Connector
- **X14261** Connector

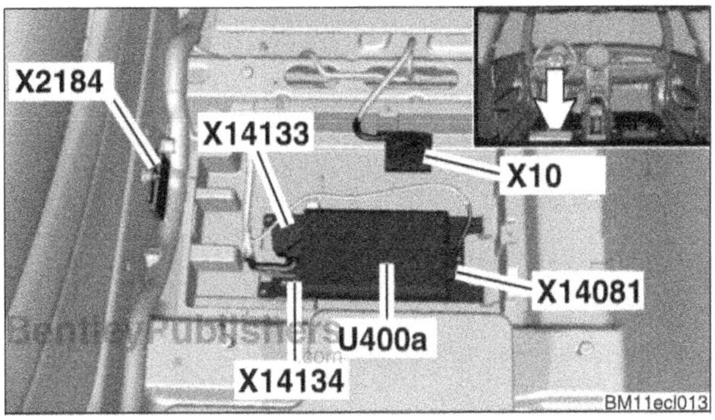

Fig 34 Left front

Underneath driver seat, behind sill cover:

- **U400a** Telephone transceiver
  U400e ULF module
- **X10** Connector
- **X2184** Ground
  See **Ground applications** for details.
- **X14081** Connector
- **X14133** Connector
- **X14134** Connector

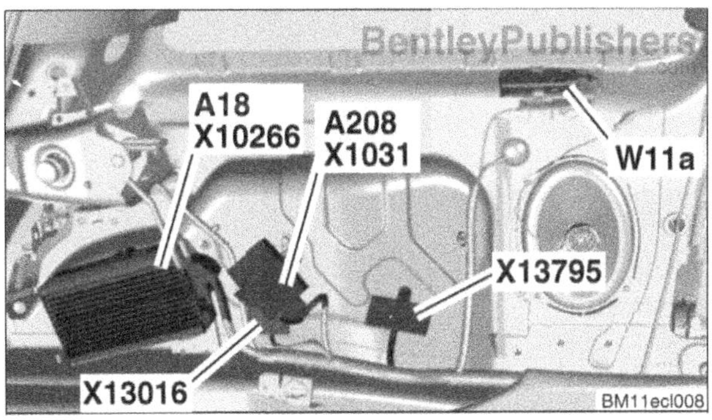

Fig 35 Left rear quarter panel (Hardtop, Clubman)

Behind trim:

- **A18** Amplifier
- **A208** Outside mirror fold-in module
- **W11a** Comfort access antenna, left rear
- **X1031** Connector
- **X10266** Connector
- **X13016** Ground
  See **Ground applications** for details.
- **X13795** Ground
  See **Ground applications** for details.

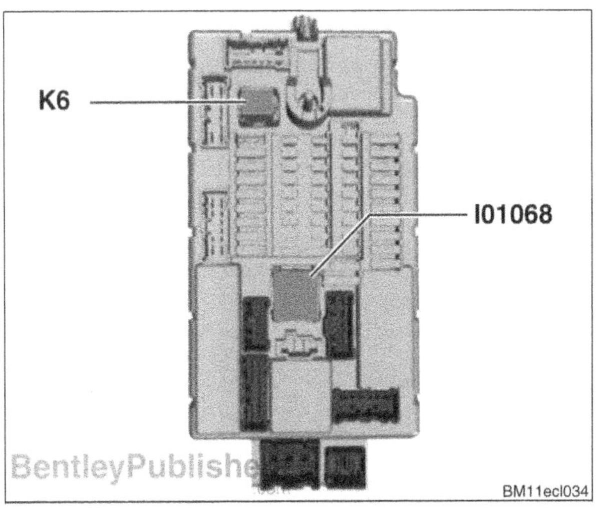

Fig 36 Right front

A4010 Junction box fuse and relay panel behind A-pillar kick panel cover:

- **I01068** Terminal 30G relay
- **K6** Headlight washer pump relay

For junction box fuse applications, see **Junction box fuse panel** in this repair group.

Interior components

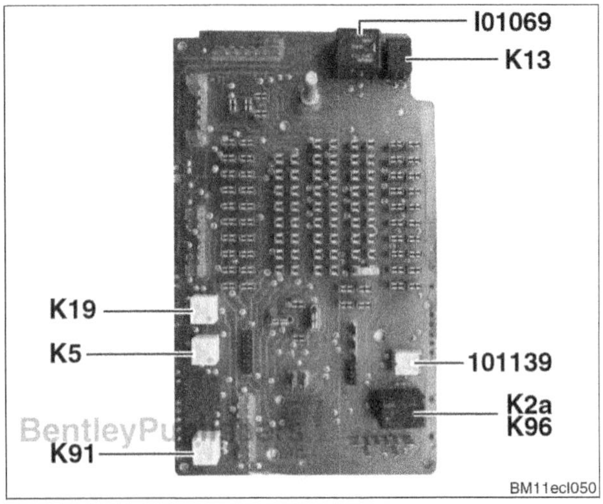

Fig 37 Right front

A4010 Junction box fuse and relay panel behind panel kick panel cover (soldered relays on back of circuit board):

- **I01069** Terminal 15 relay
- **I01139** Bistable relay
- **K2a** Horn relay
- **K5** Windshield washer pump relay
- **K13** Rear window defogger relay
- **K19** A/C compressor relay
- **K91** Rear window wiper relay
- **K96** Fuel pump relay

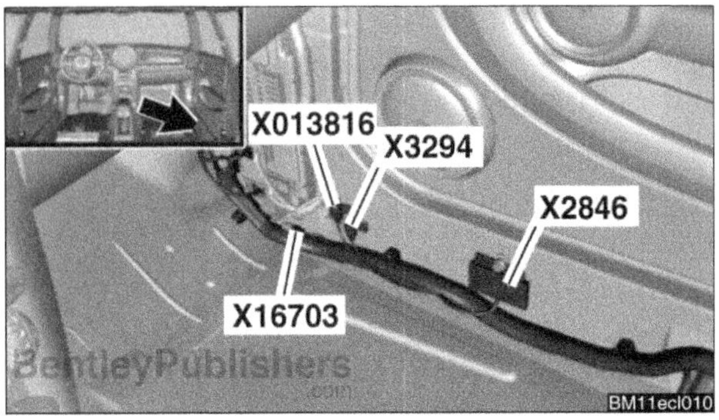

Fig 38 Right door sill

Behind trim:

- **X2846** Ground
  See **Ground applications** for details.
- **X3294** Connector
- **X16703** Connector
- **X013816** Connector

Fig 39 Right front

Underneath passenger seat:

- **B9801** DSC acceleration sensor
- **N47** Satellite radio

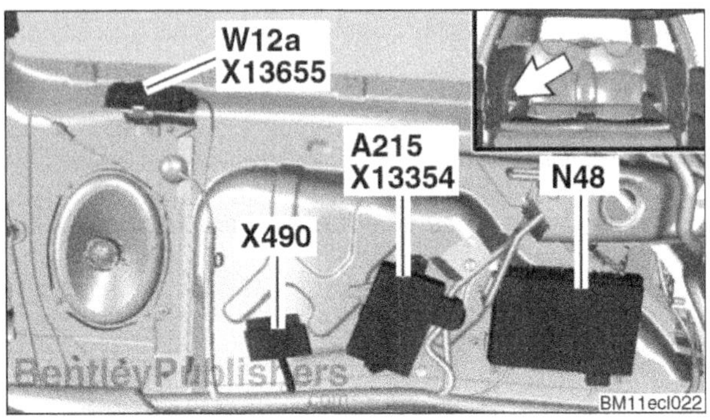

Fig 40 Right rear quarter panel (Hardtop)

Behind trim:

- **A219** Comfort access module
- **N48** Digital tuner
- **W12a** Comfort access antenna, right rear
- **X490** Ground
  See **Ground applications** for details.
- **X13354** Connector
- **X13655** Connector

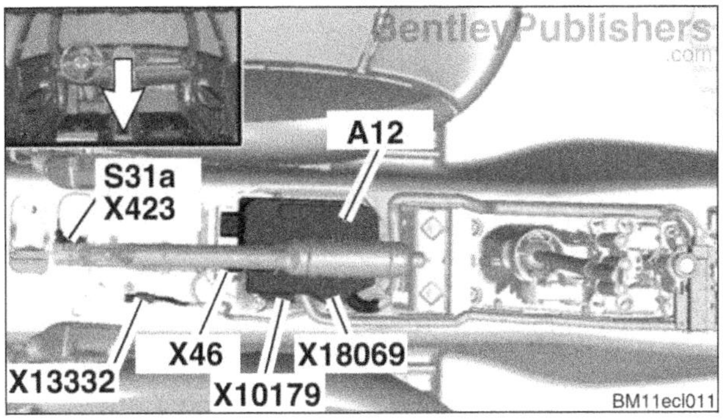

Fig 41 Underneath center console

Center console removed:

- **A12** MRS control module
- **S31a** Parking brake handle switch
- **X46** Ground
  See **Ground applications** for details.
- **X423** Connector
- **X10179** Connector
- **X13332** Connector
- **X18069** Connector

Cargo compartment components

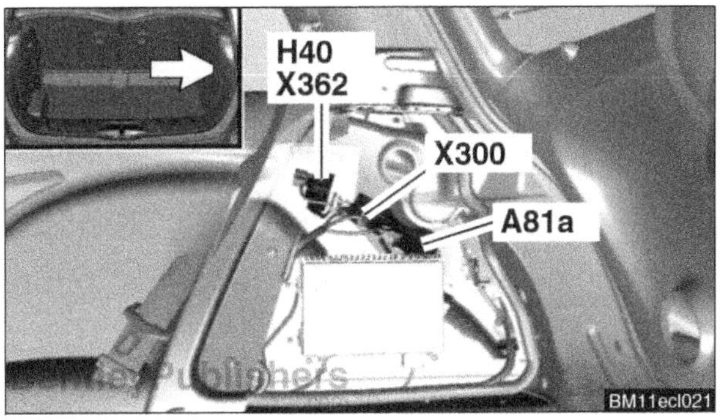

Fig 42 Right side

Behind trim:

- **A81a** Park distance control (PDC) module
- **H40** Park distance control warning speaker
- **X300** Connector
- **X362** Connector

ELE Electrical Wiring Diagrams

Electrical Component Locations (relay and fuse positions, ground locations) see **Repair Group ECL**

General

This section contains selected electrical wiring diagrams for E90 3 Series models. A detailed index of the diagrams and electrical component is listed in the opening pages.

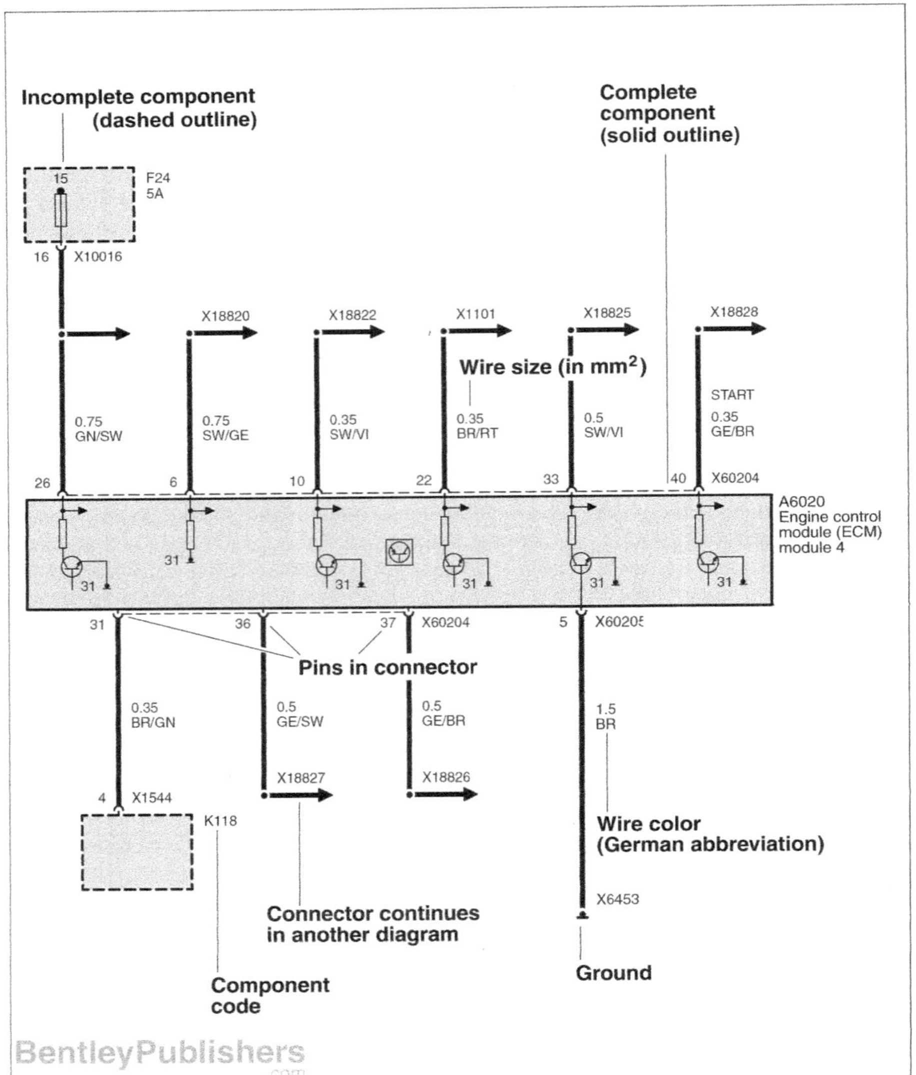

B701ele001

Schematic conventions

The schematics (wiring diagrams) divide the vehicle electrical system into individual circuits. Interacting electrical components are shown on one schematic.

Electrical components are represented in such a way that their general layout and function are self-explanatory. They are usually arranged in the diagrams so that the current path can be followed from positive at the top to negative at the bottom.

- If a schematic spans several pages, this fact is clearly indicated at the top of each page.
- A component (or connector) which is completely represented in the schematic is shown as a solid box.
- A component (or connector) which has other connectors in addition to the ones shown in the schematic is shown with a dashed line.
- Switches and relays are always shown in rest position (generally OFF).

Symbols

The schematics utilize simplified electrical symbols.

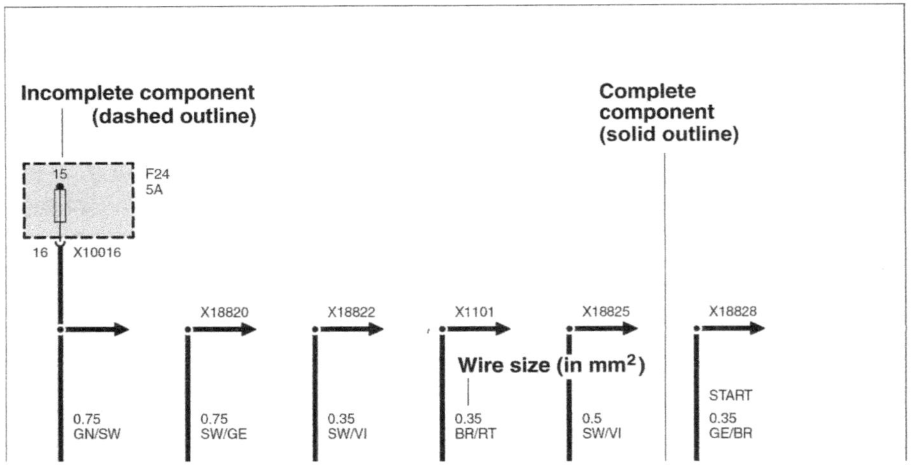

Wiring color code	
BL	= blue
BR	= brown
GE	= yellow
GN	= green
GR	= grey
RT	= red
SW	= black
WS	= white
VI	= violet

B701ele003

Wire insulation colors in this section are given with German color abbreviations.

Wire sizes follow the DIN (European) convention.

Example: 0.5 wire is ½ mm^2 in cross-section area. This corresponds to approx. SAE 16 gauge wire.

BMW identifies many electrical circuits with unique designations which follow the DIN standard. Also, BMW designates electrical components, junctions and grounds with a unique alphanumeric designation, most of which also follow the DIN standard.

Page

Engine management (fuel injection, ignition, turbocharger) (cont.)

Engine cooling system

Engine oil supply

Footwell module

Accelerator pedal module, brake light switch, clutch module

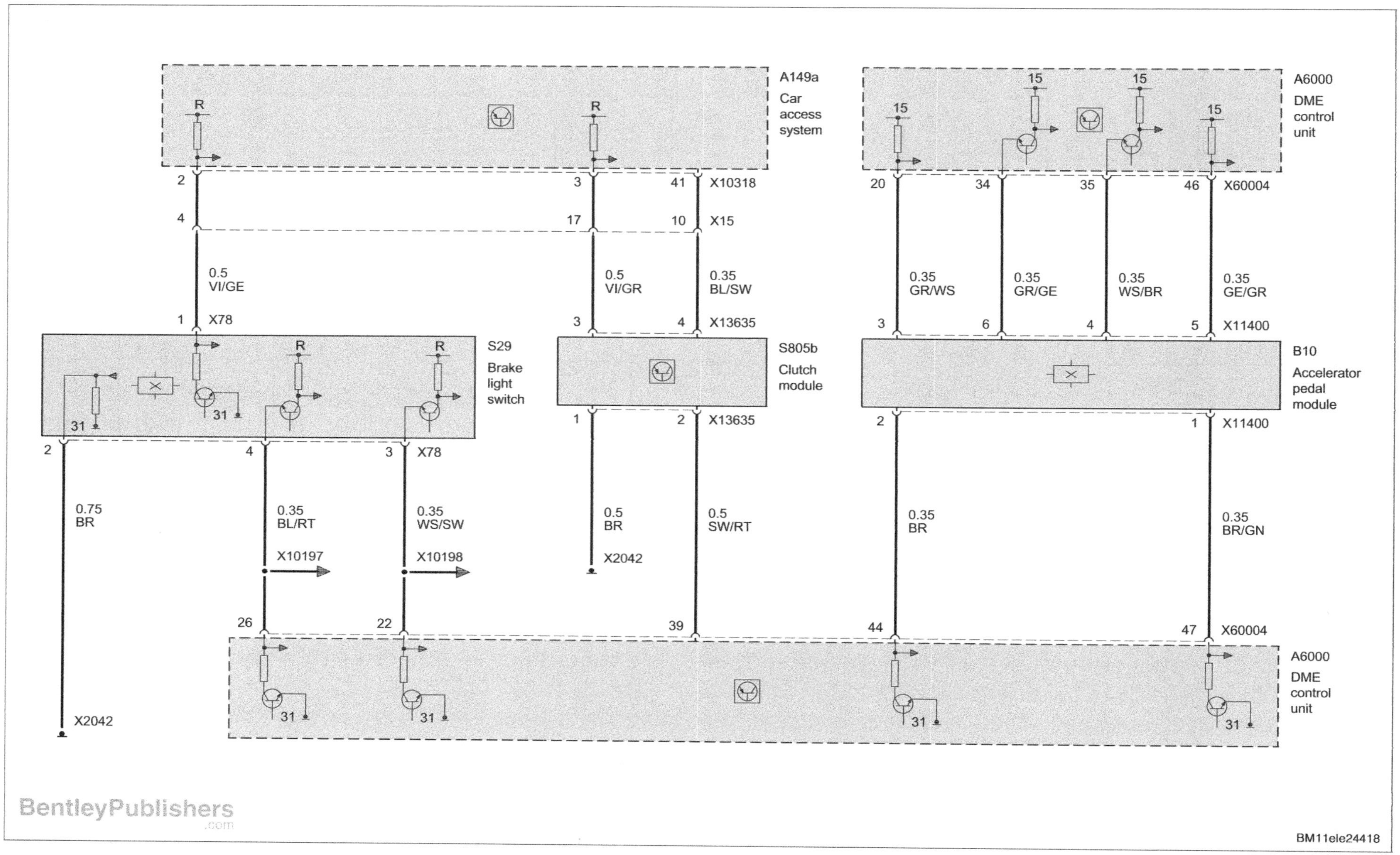

Air supply
N12 engine

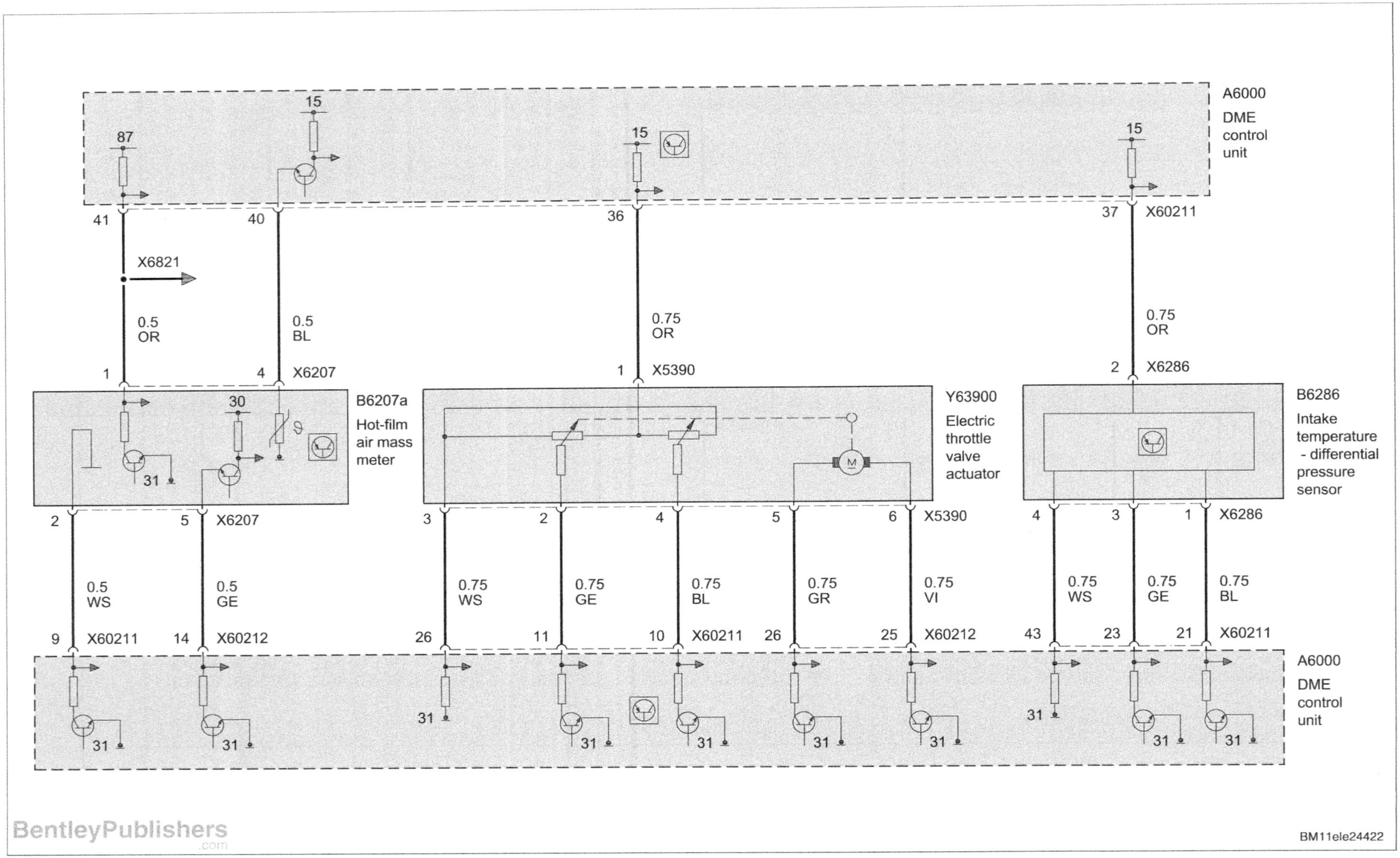

Camshaft position sensors, eccentric shaft sensor, valvetronic motor, valvetronic relay, VANOS valves N12 engine, page 1 of 2

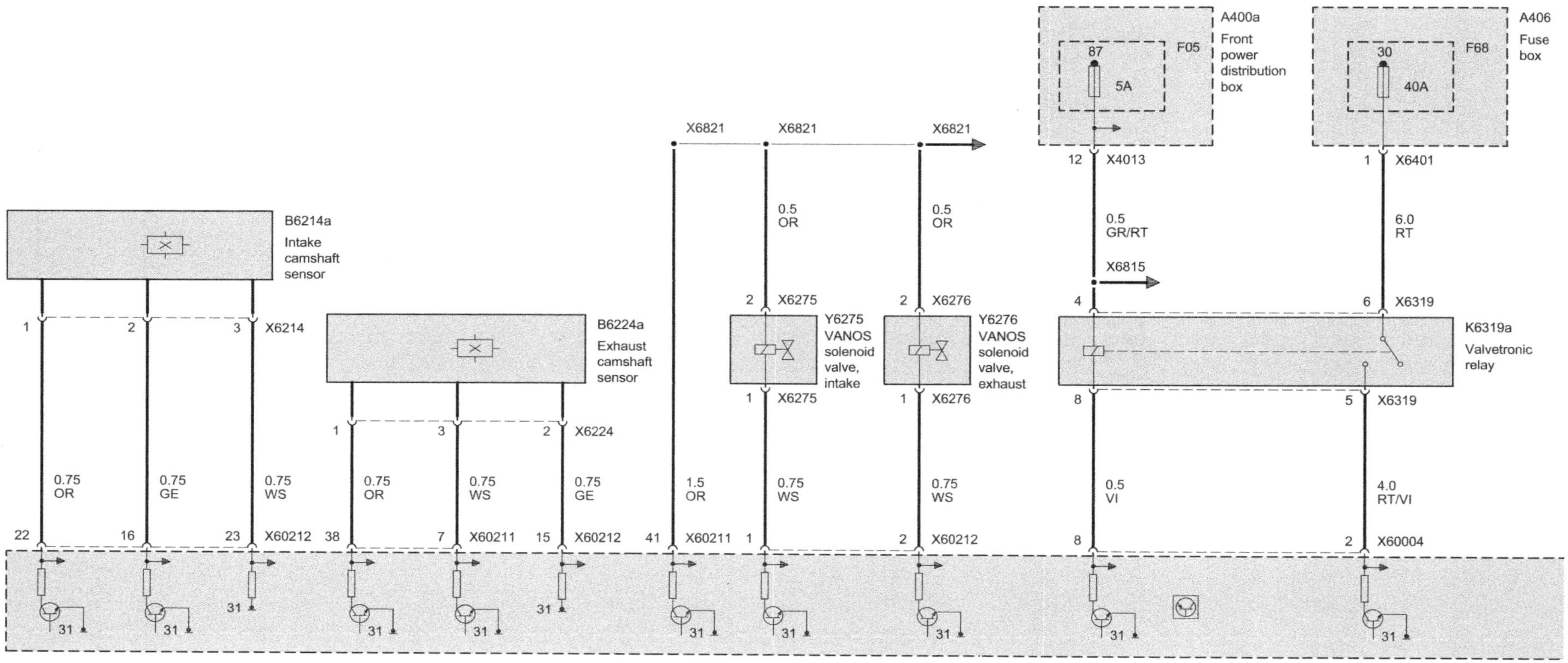

Camshaft position sensors, eccentric shaft sensor, valvetronic motor, valvetronic relay, VANOS valves N12 engine, page 2 of 2

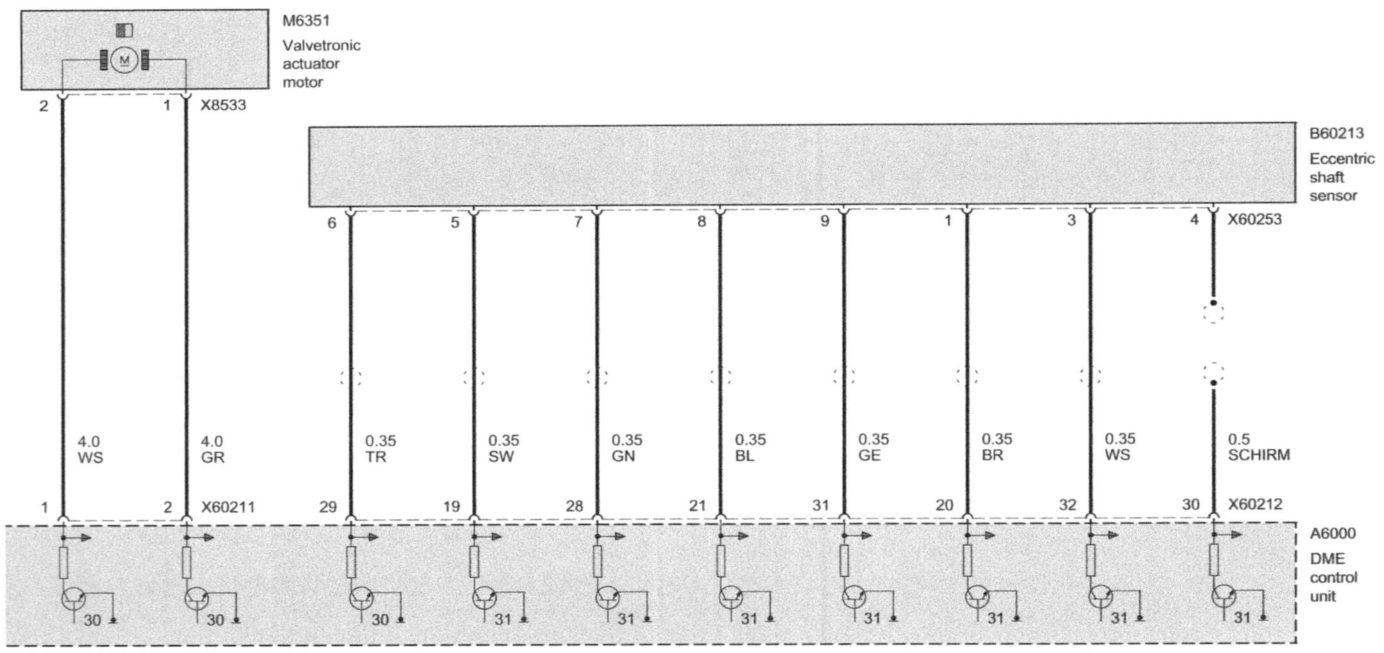

Crankshaft position sensor N12 engine, N16 engine

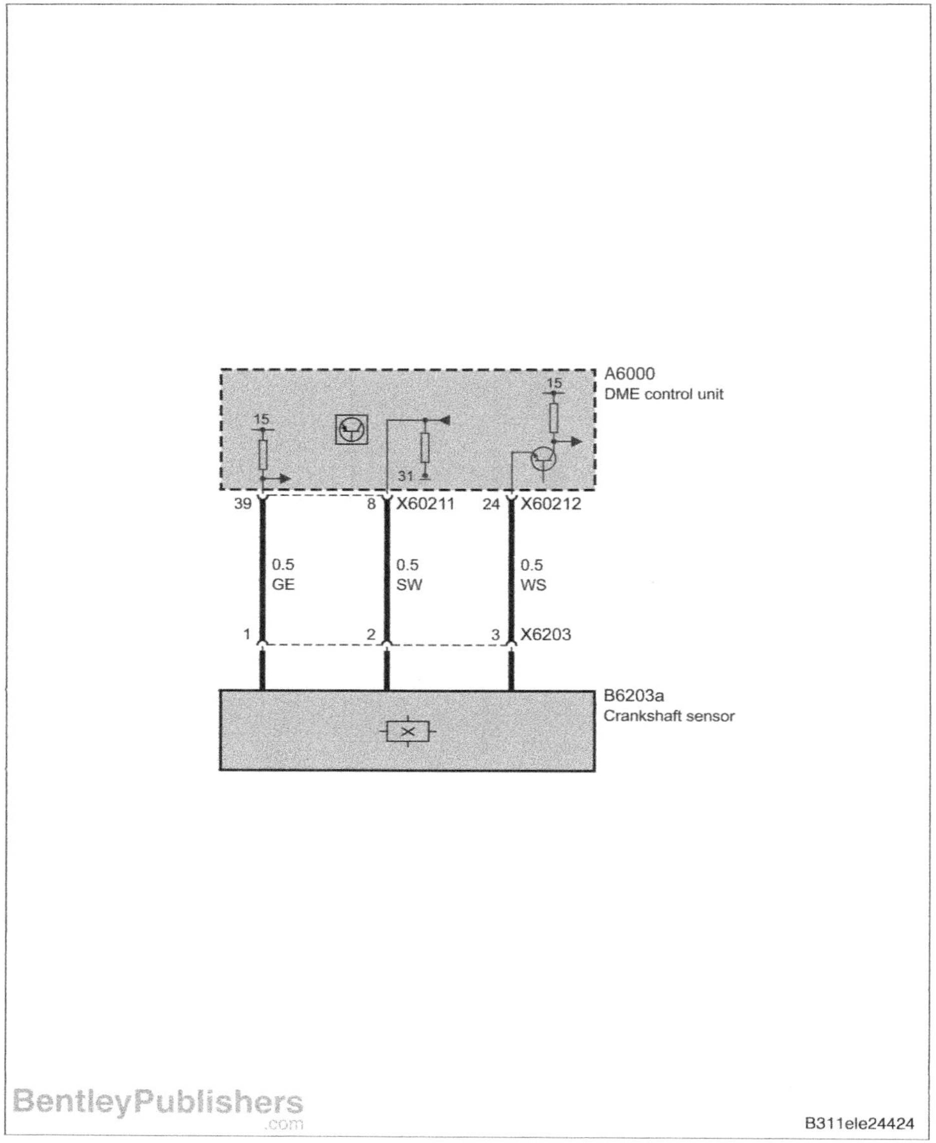

Diagnostic module for fuel tank leakage, fuel tank vent valve

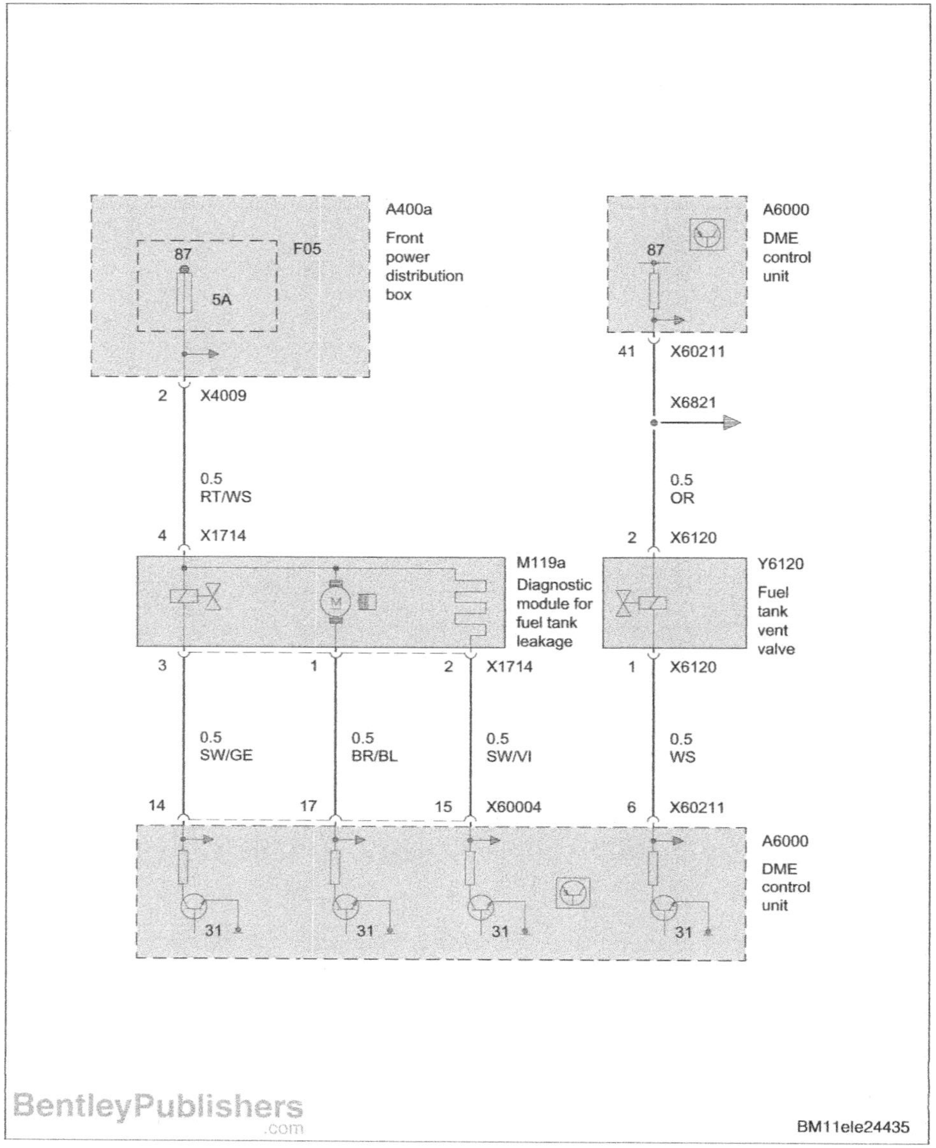

Power supply, DME control unit

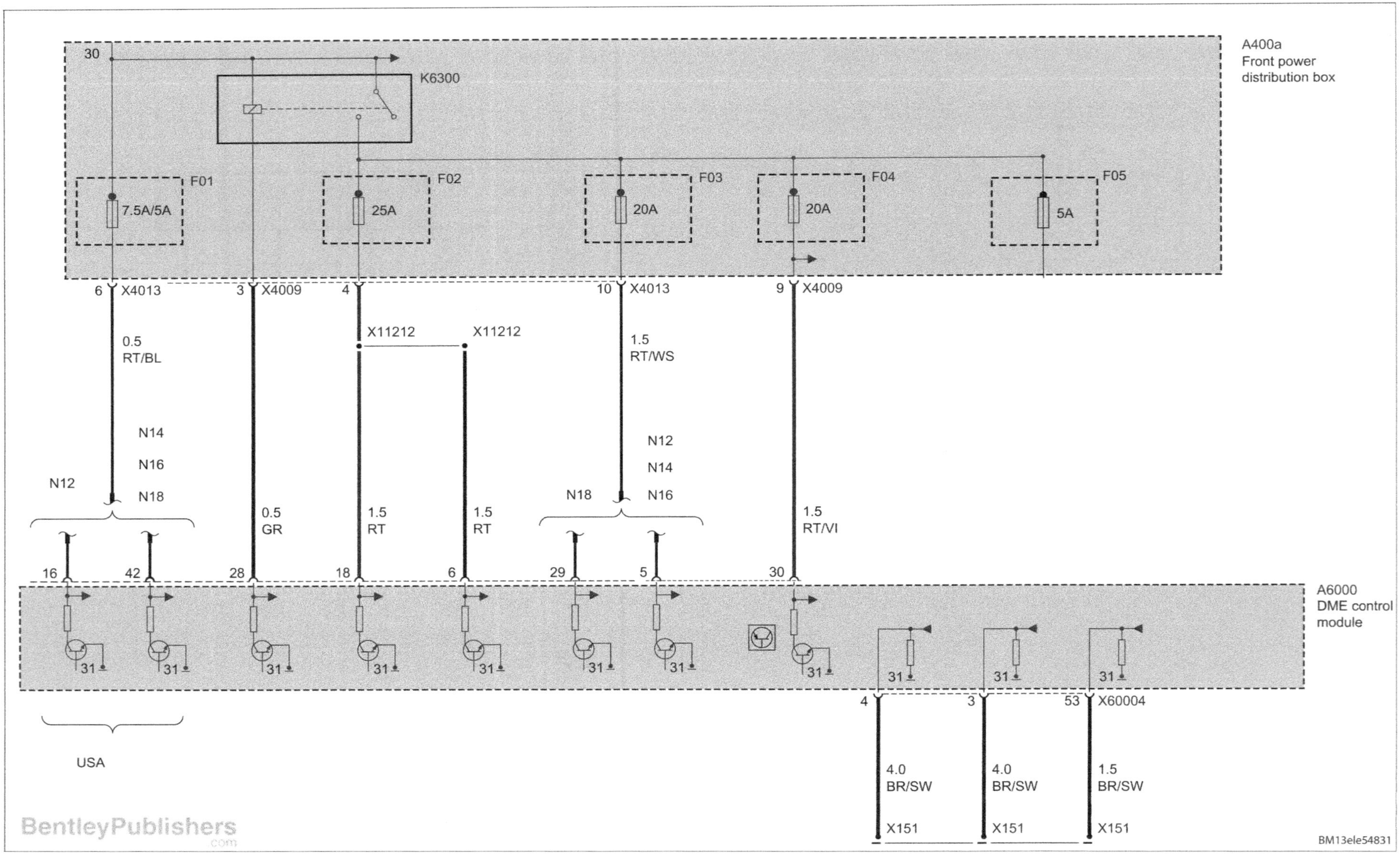

Fuel injectors
N12 engine, N16 engine

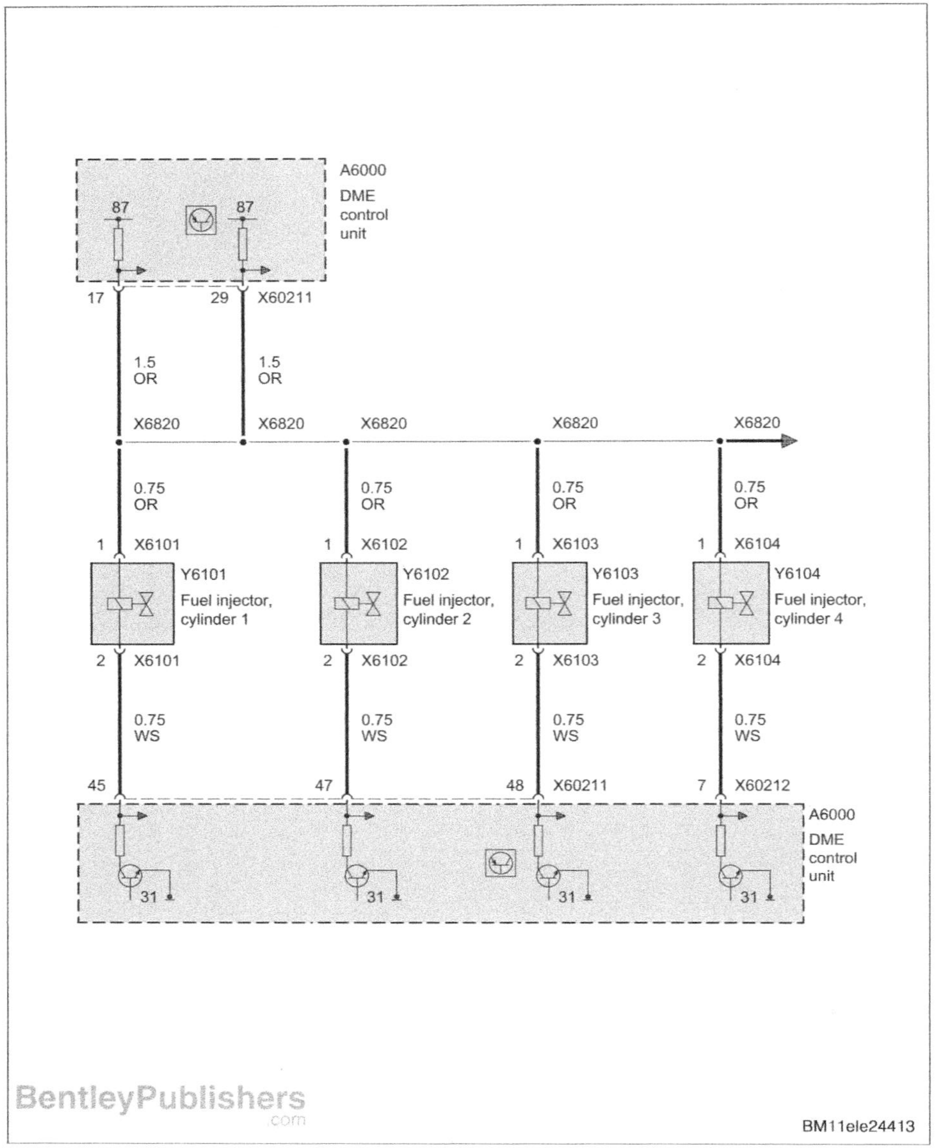

Brake lights, tail lights
N16 engine

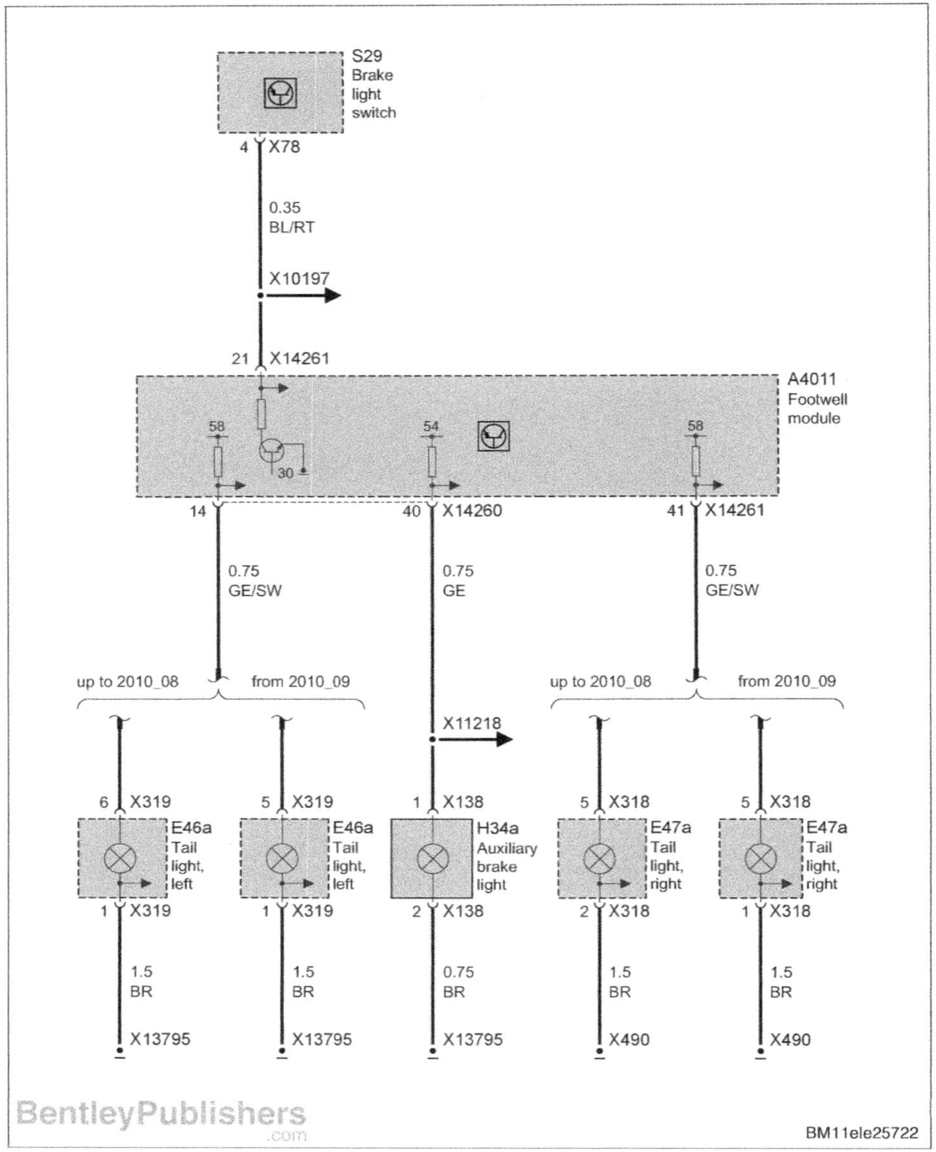

Fuel pump
N12 engine, N14 engine

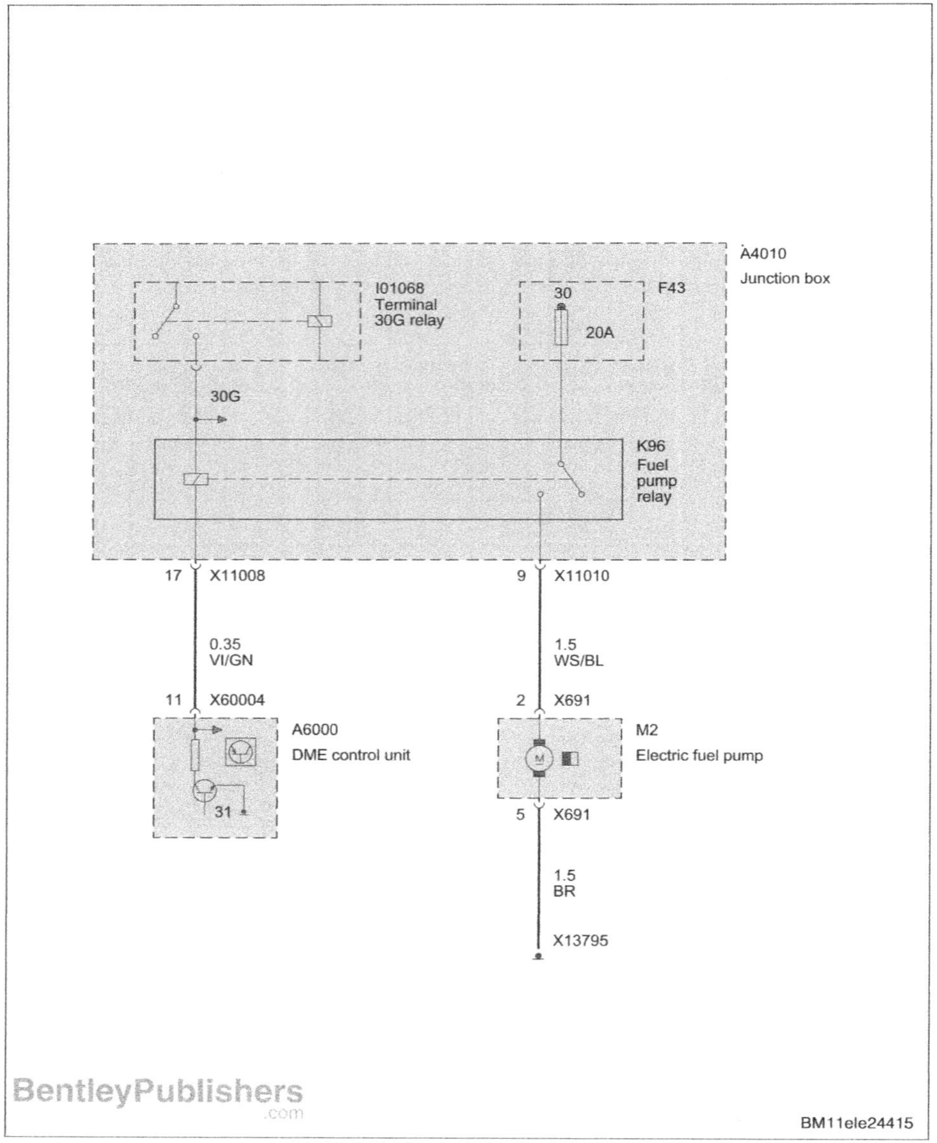

Knock sensor
N12, N16 engines

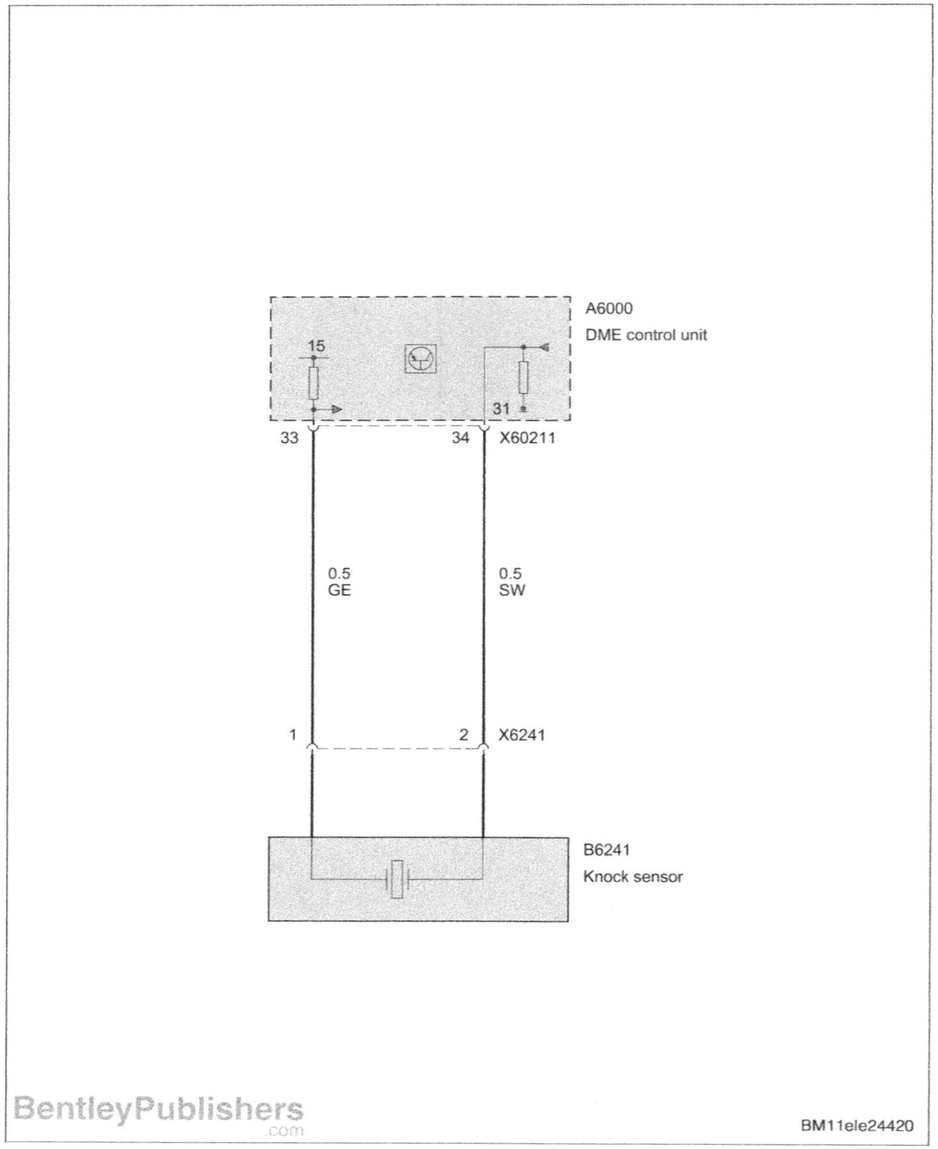

Ignition coils
N12 engine, N16 engine

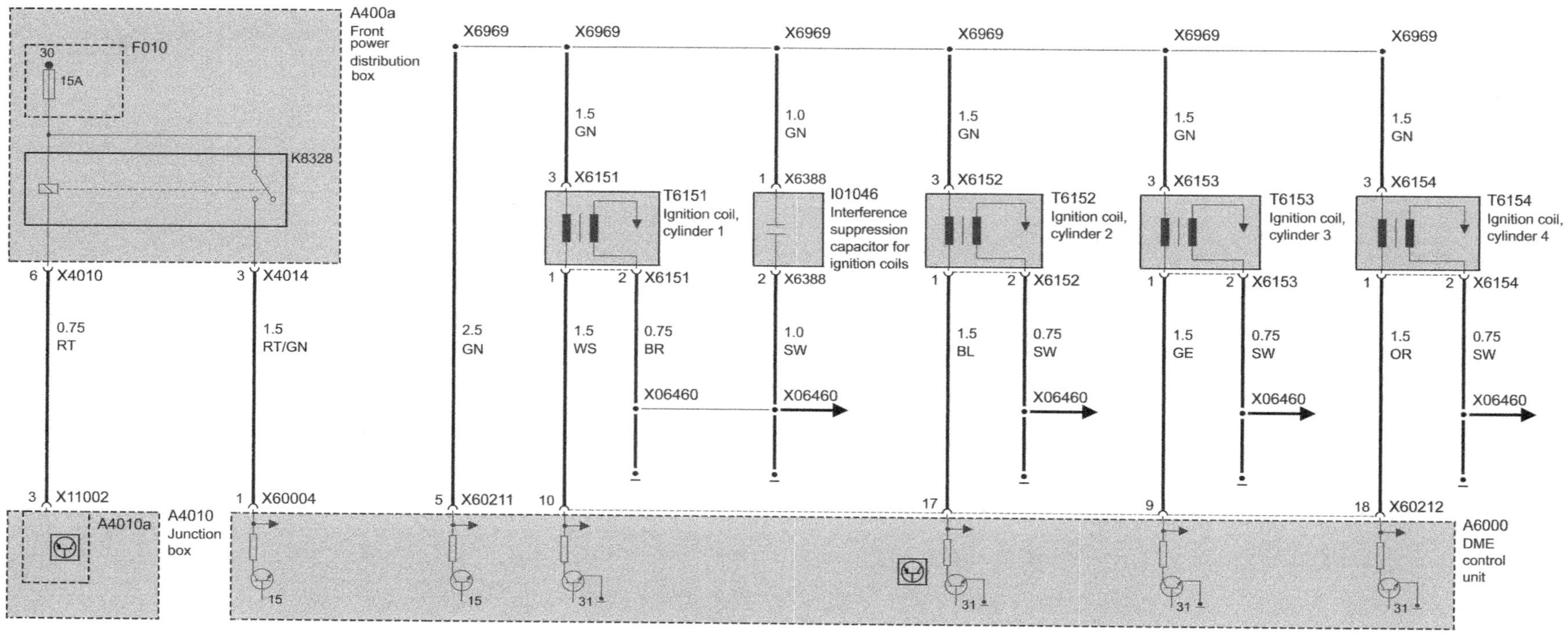

BM11ele26198

A/C refrigerant pressure sensor, IBS, CAS
N12 engine

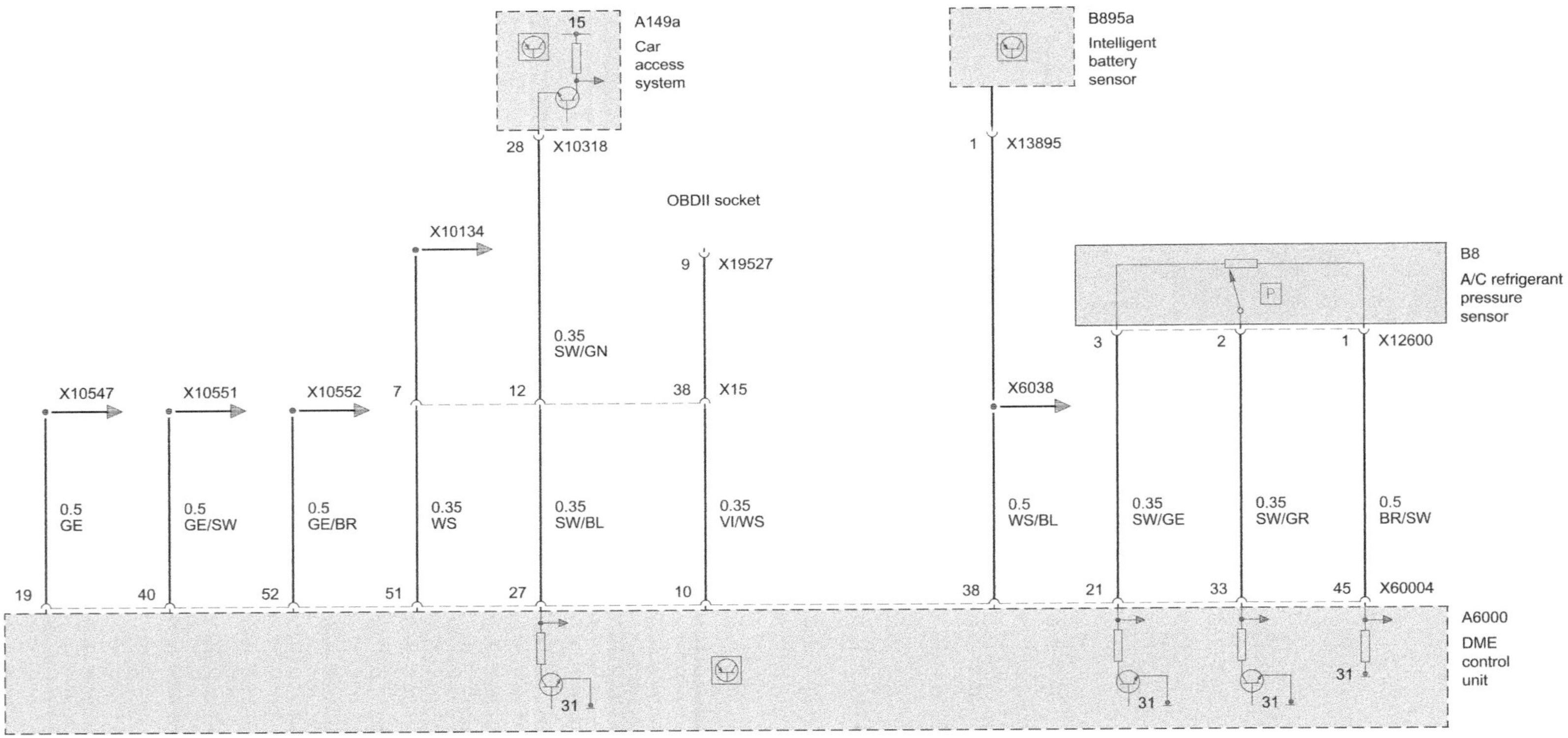

BM11ele24431

OBDII socket
N12, N14 engines

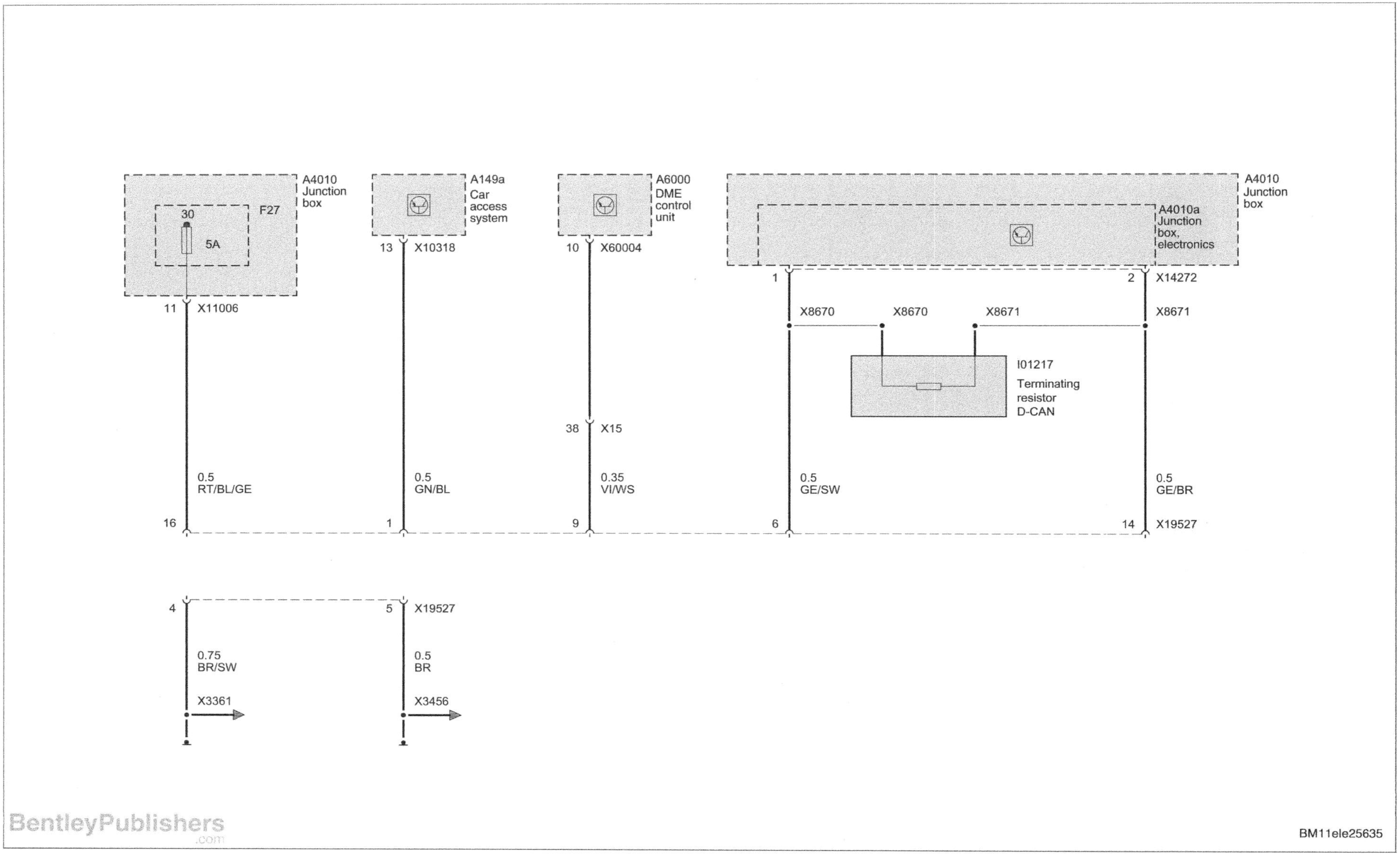

Oil supply, engine breather heater, oil pressure switch N12 engine, N14 engines

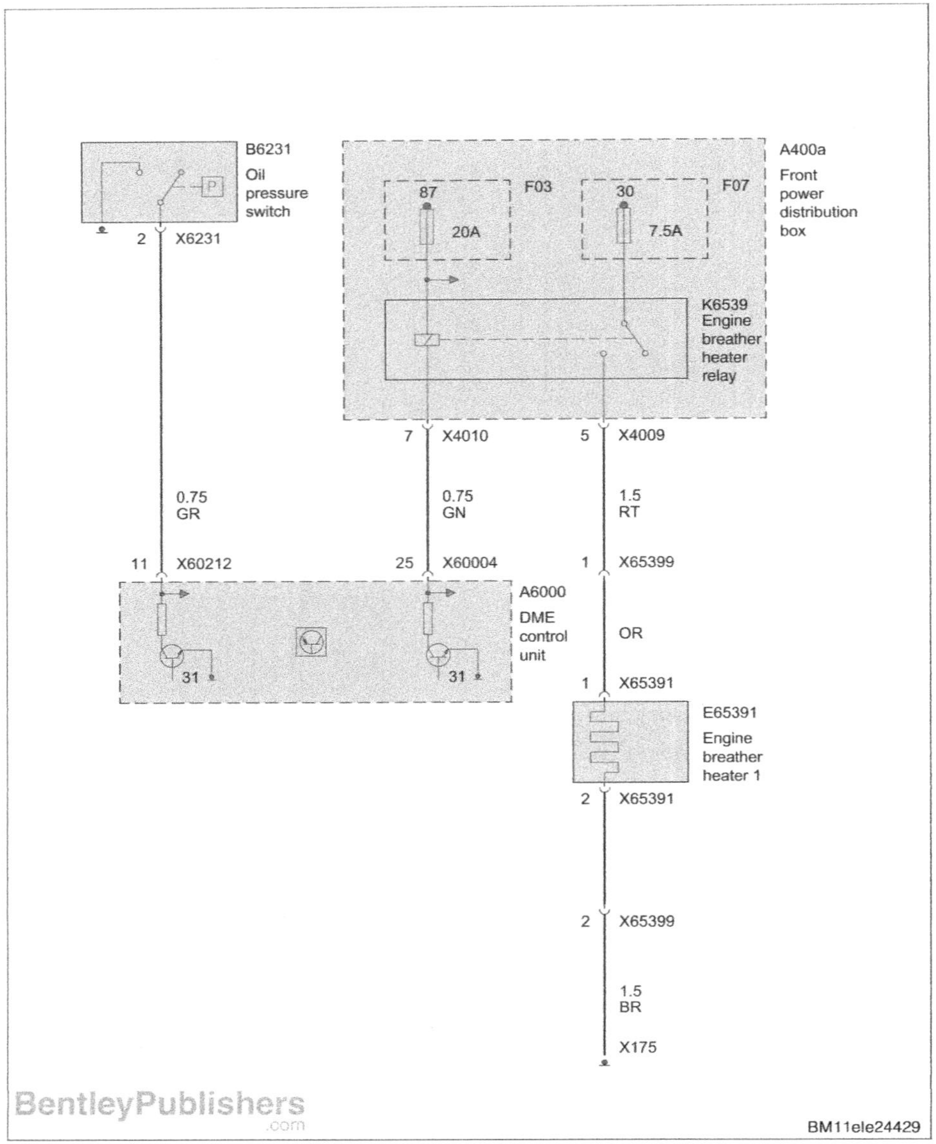

Fuel pump N16 engine, N18 engines

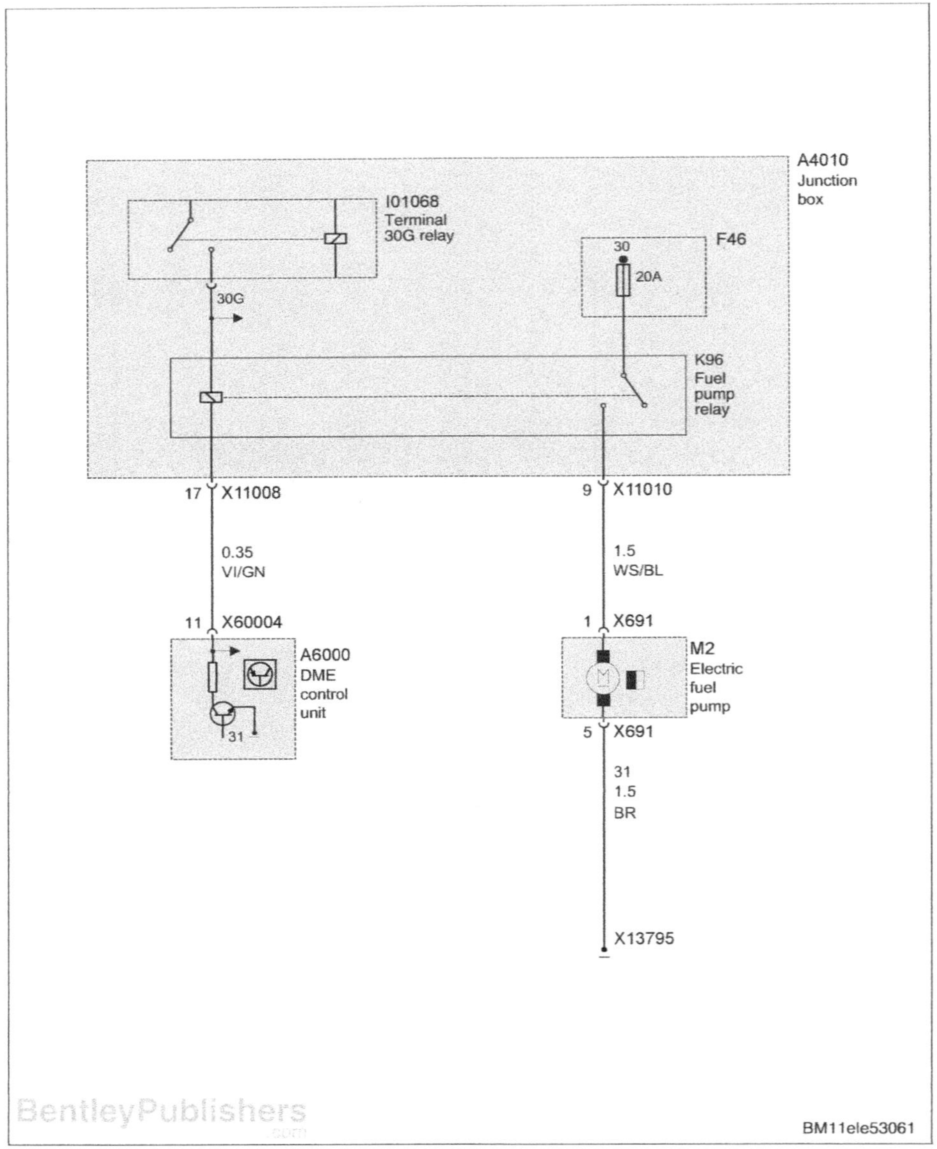

Brake light switch, tail lights, auxiliary brake light
N12 engine, N14 engines

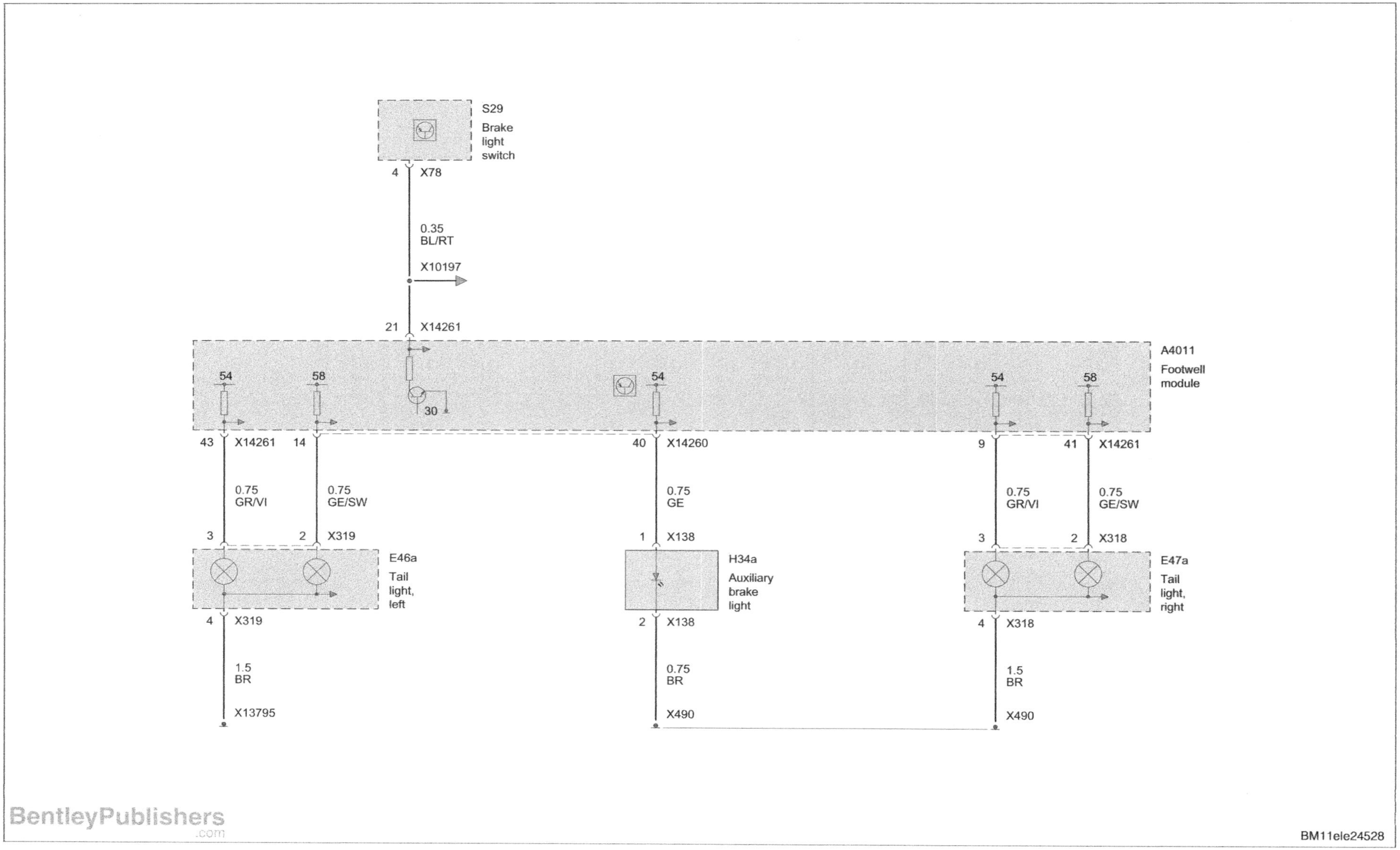

Thermostat, electric fan, electric fan relay (K21), engine temperature sensor, N12 engine

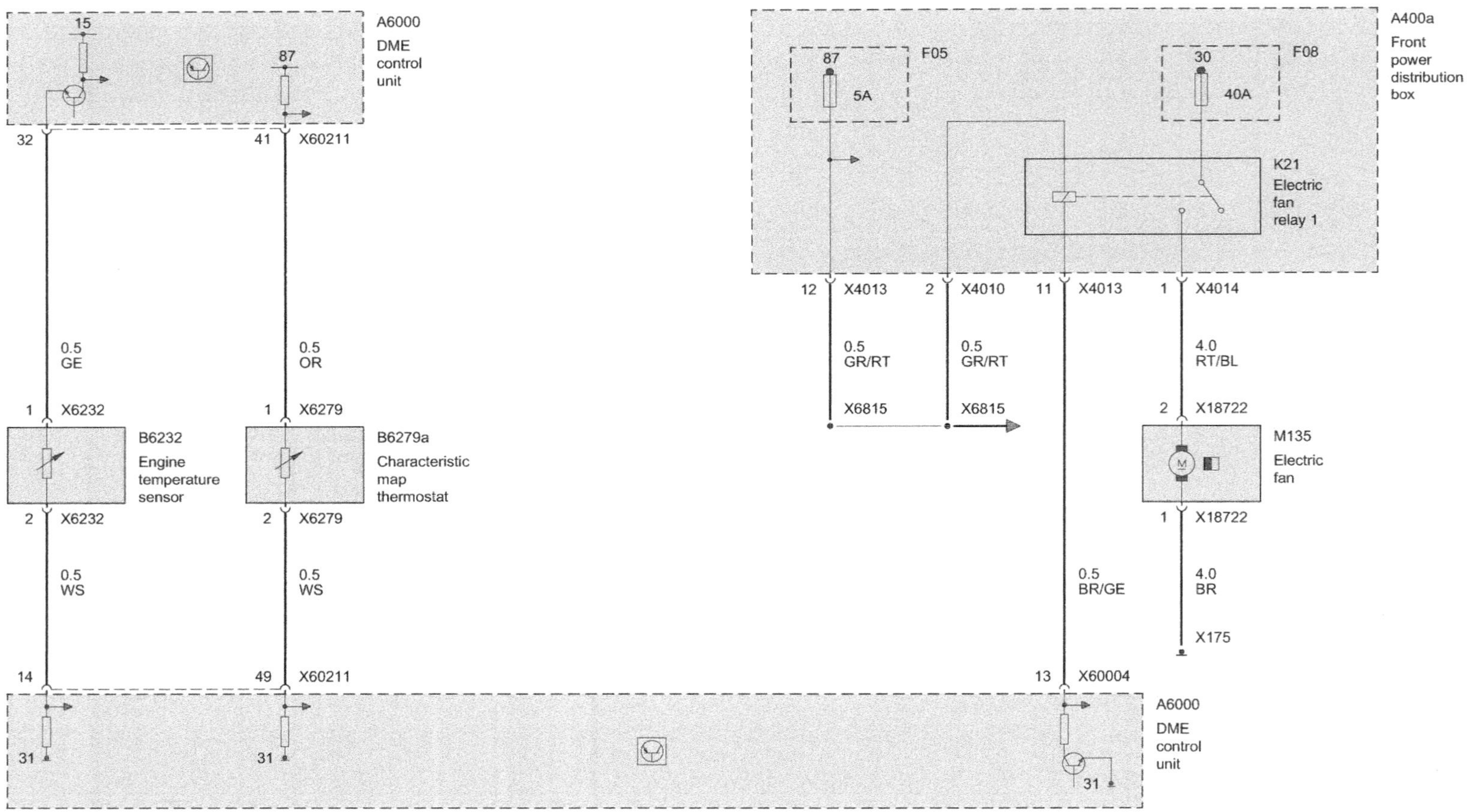

BM11ele26961

A6000 DME control unit, engine side
N12 engine
page 1 of 2

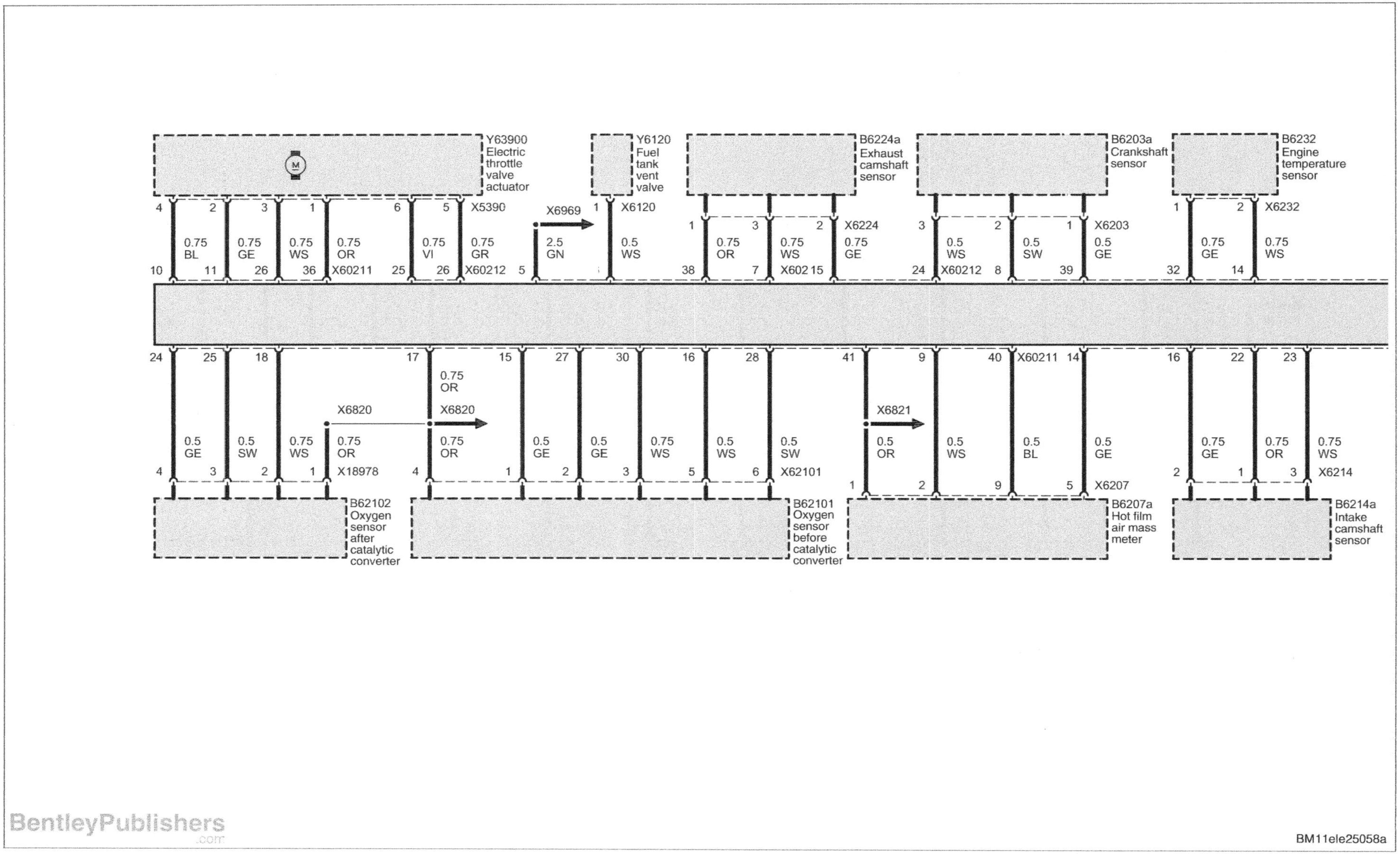

A6000 DME control unit, engine side
N12 engine
page 2 of 2

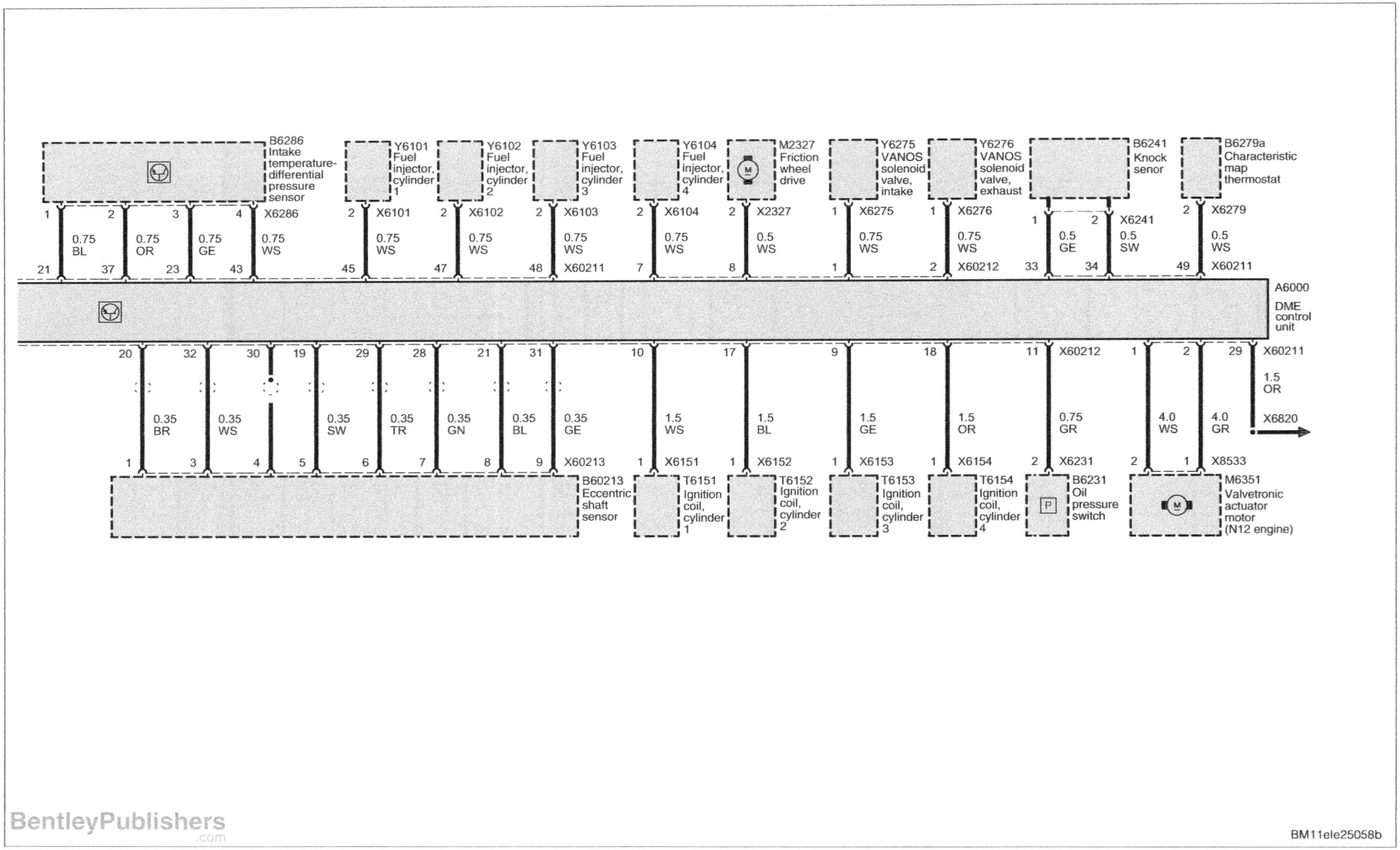

Fuel injectors, fuel injector relay
N14 engine

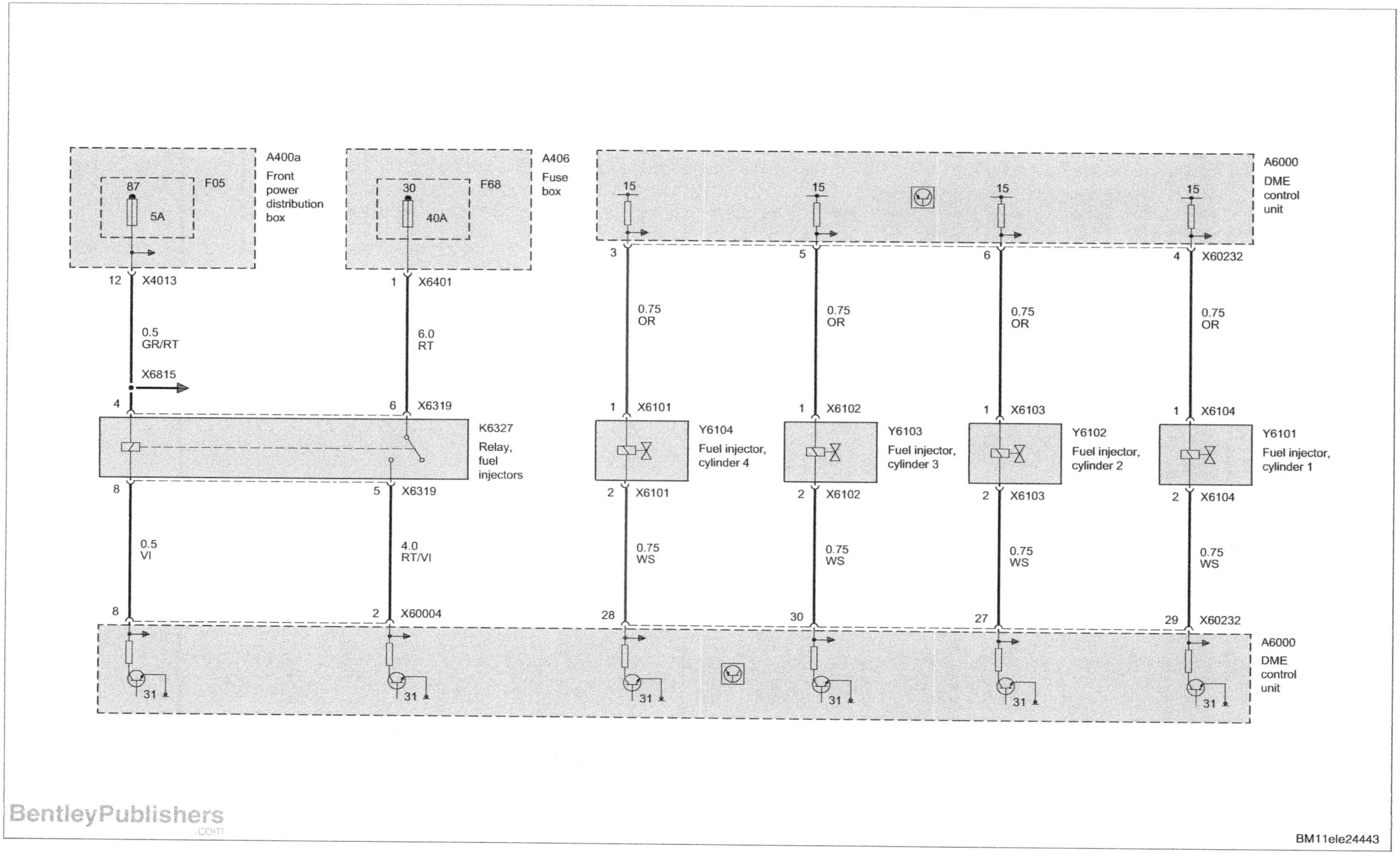

Knock sensor
N14 engine, N18 engines

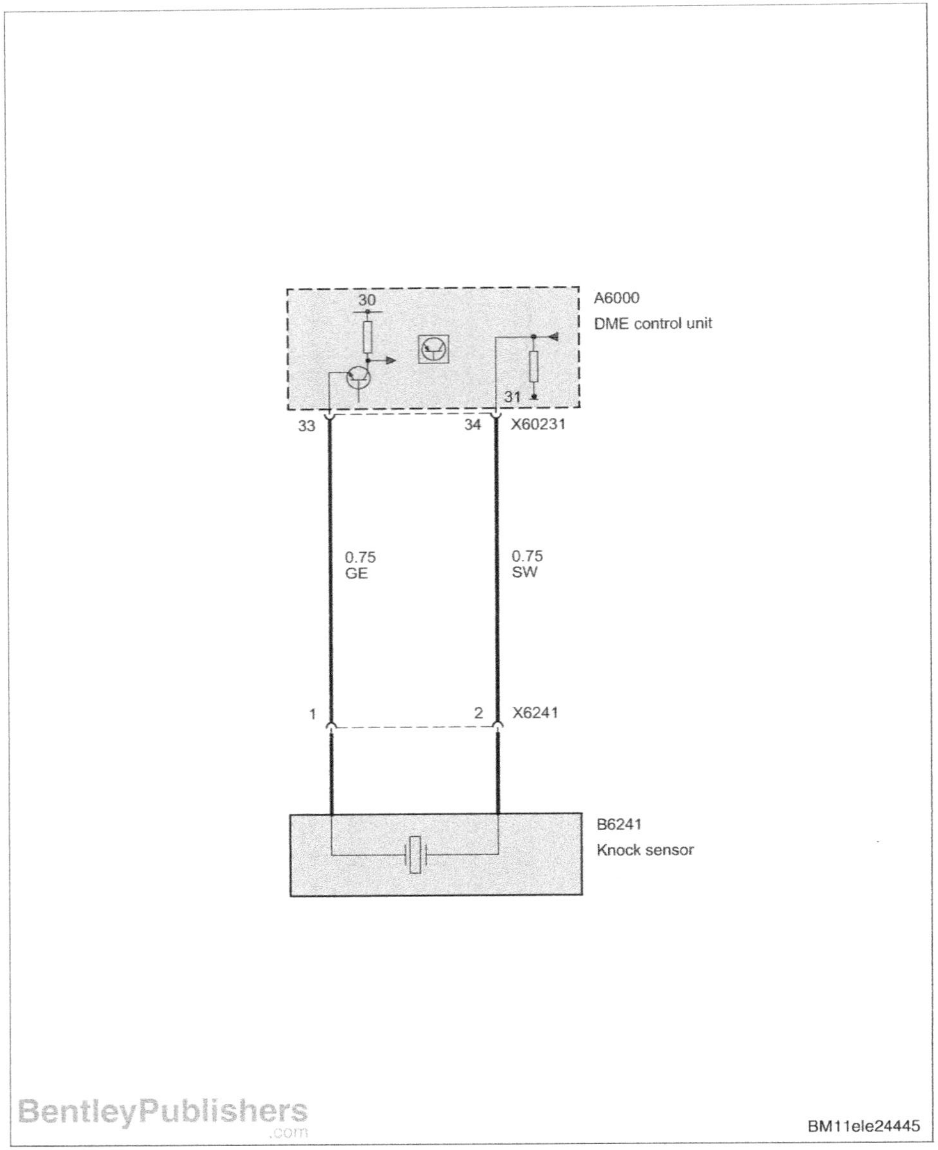

Crankshaft sensor
N14 engine

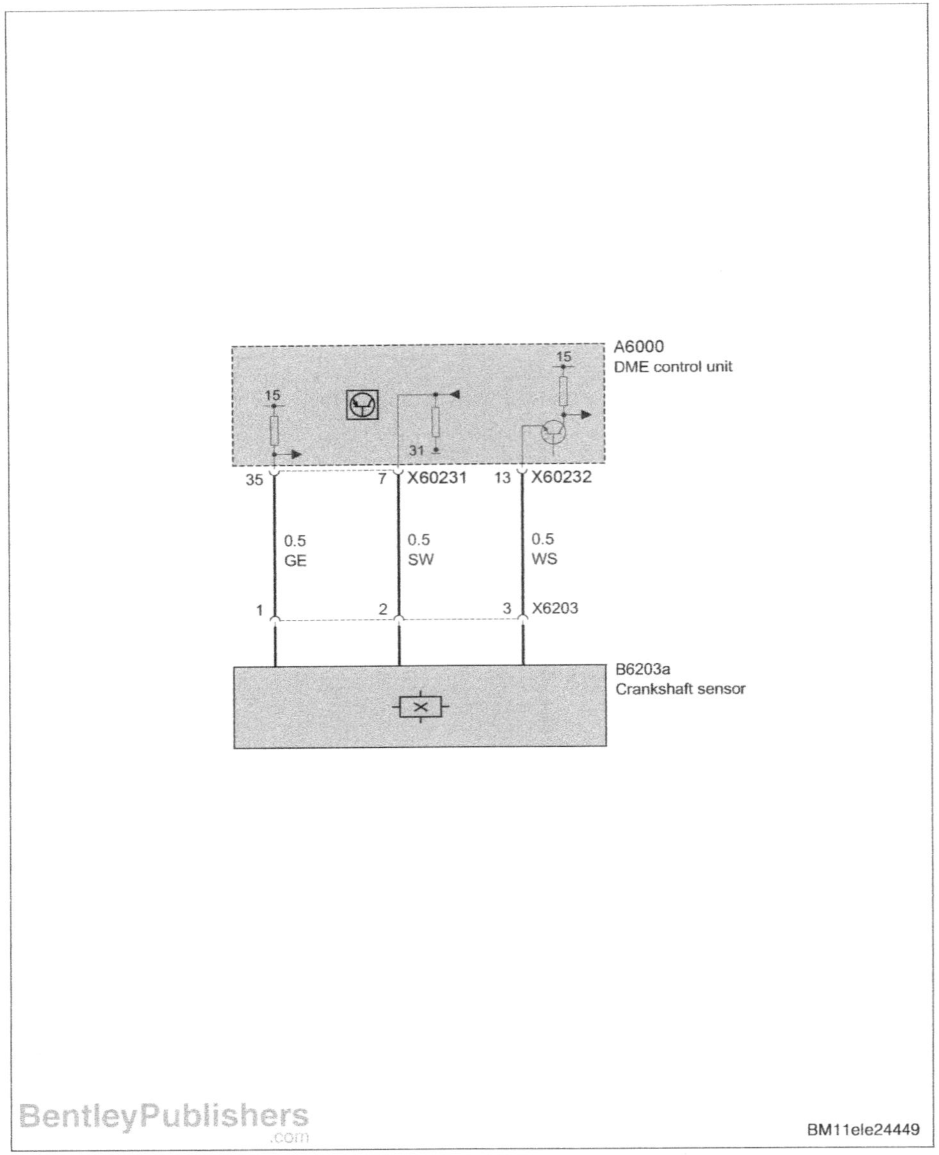

Fuel injectors
N18 engine

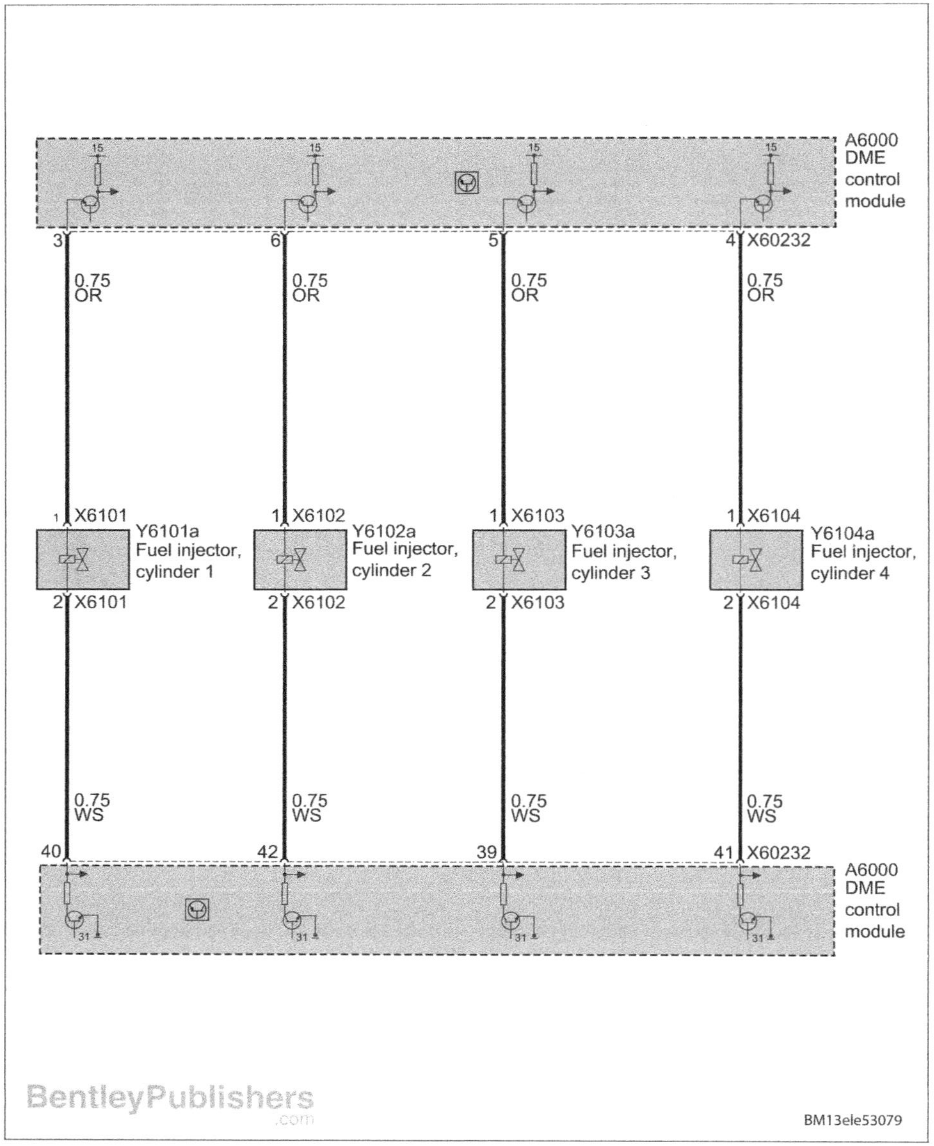

Engine breather heater, engine oil pressure sensor, oil pressure control valve
N16 engine

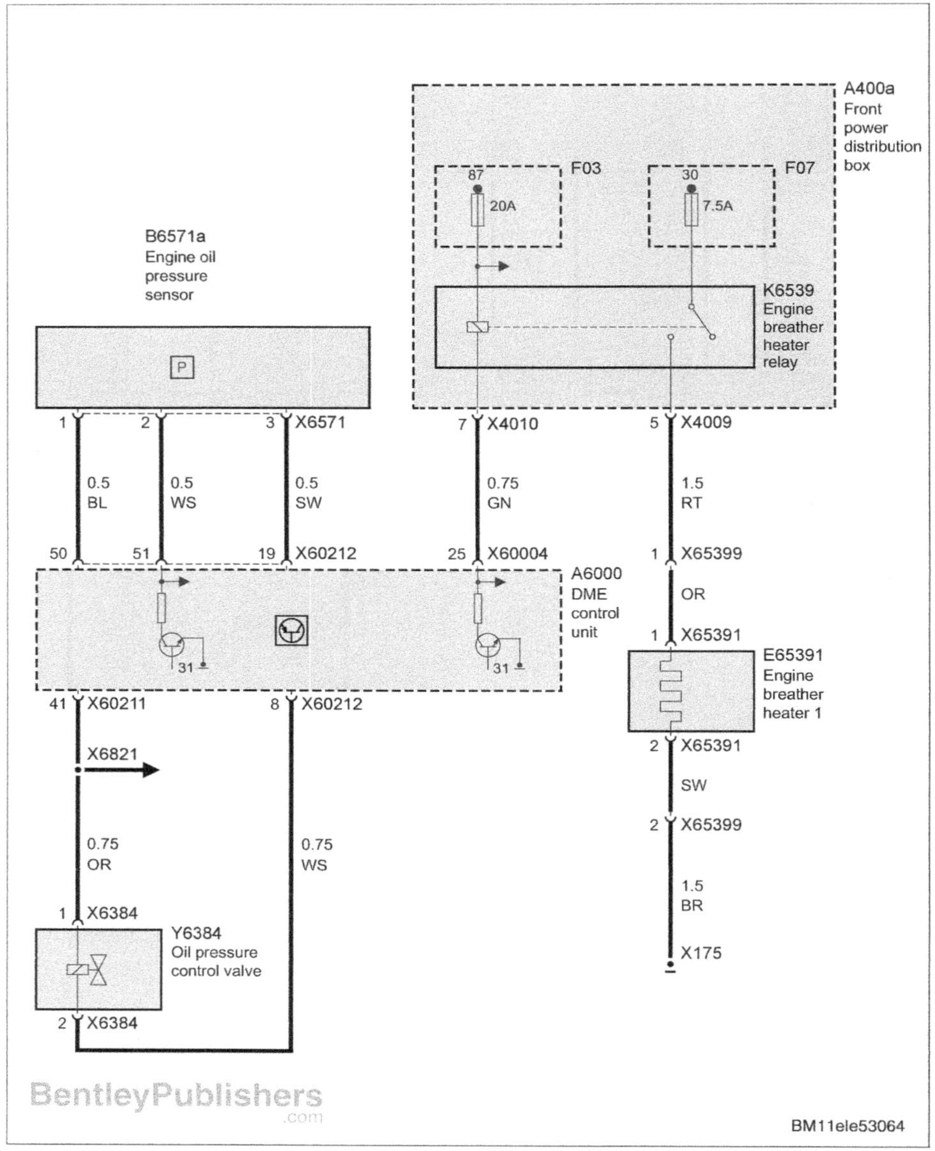

Air supply
N14 engine

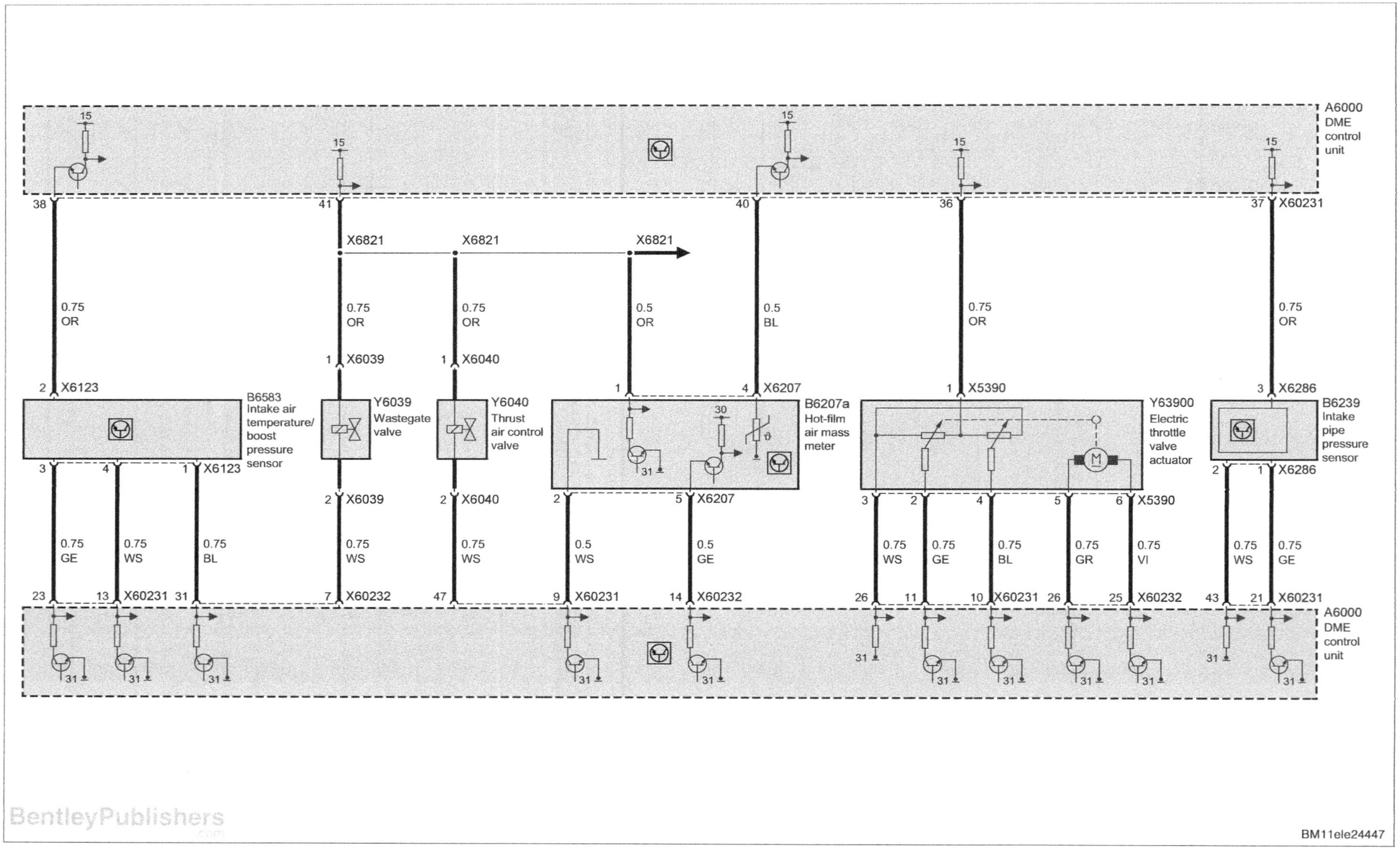

BM11ele24447

Intake camshaft sensor, intake VANOS solenoid valve N14 engine

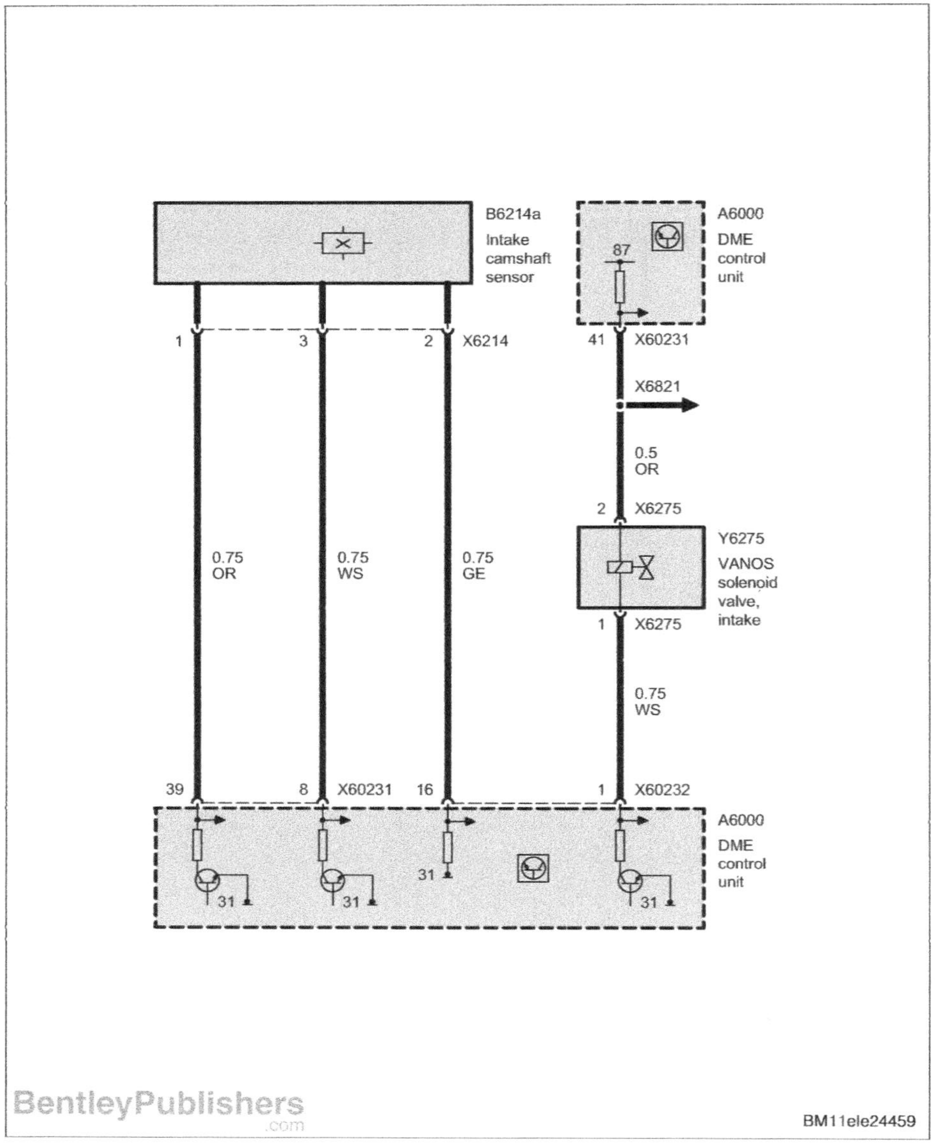

Fuel rail pressure sensor, volume control valve N14 engine

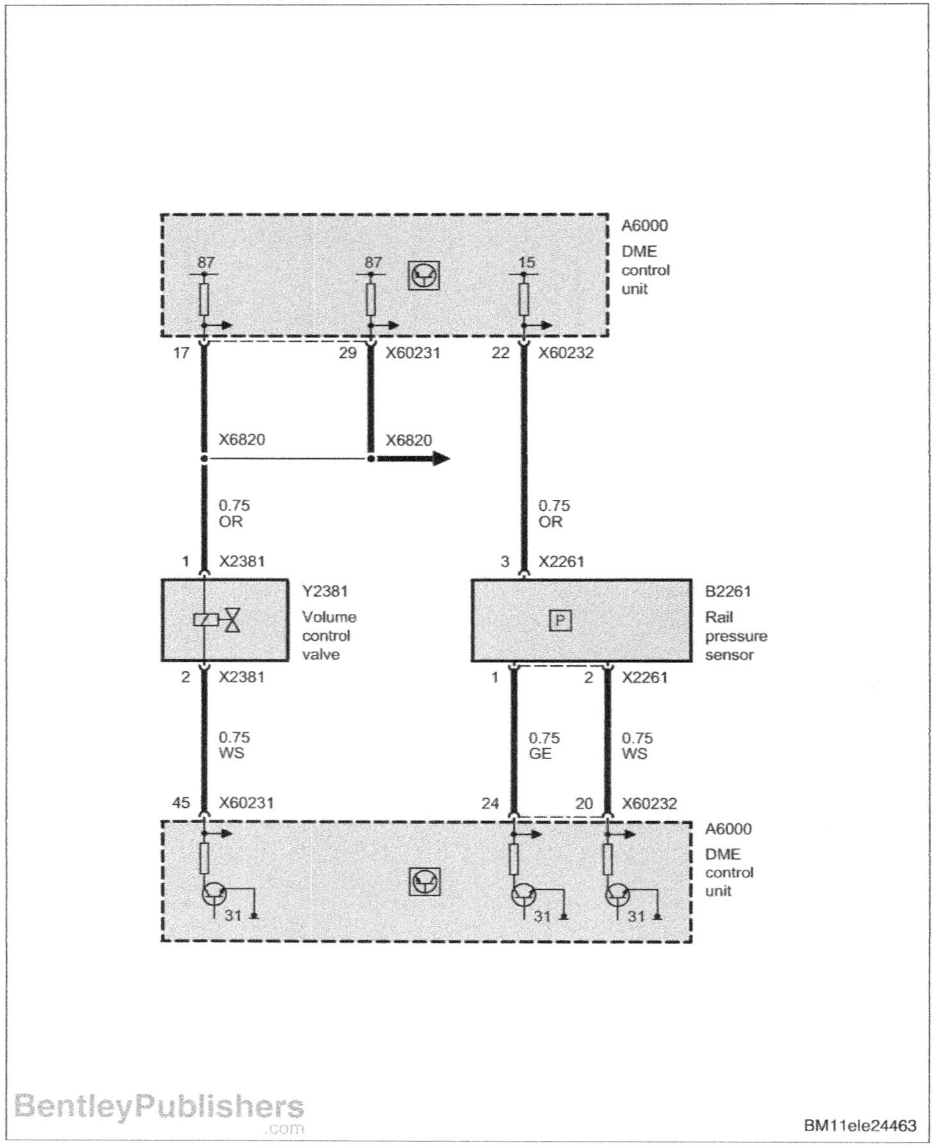

Car access system, A/C refrigerant pressure sensor
N14 engine, N16 engine, N18 engine

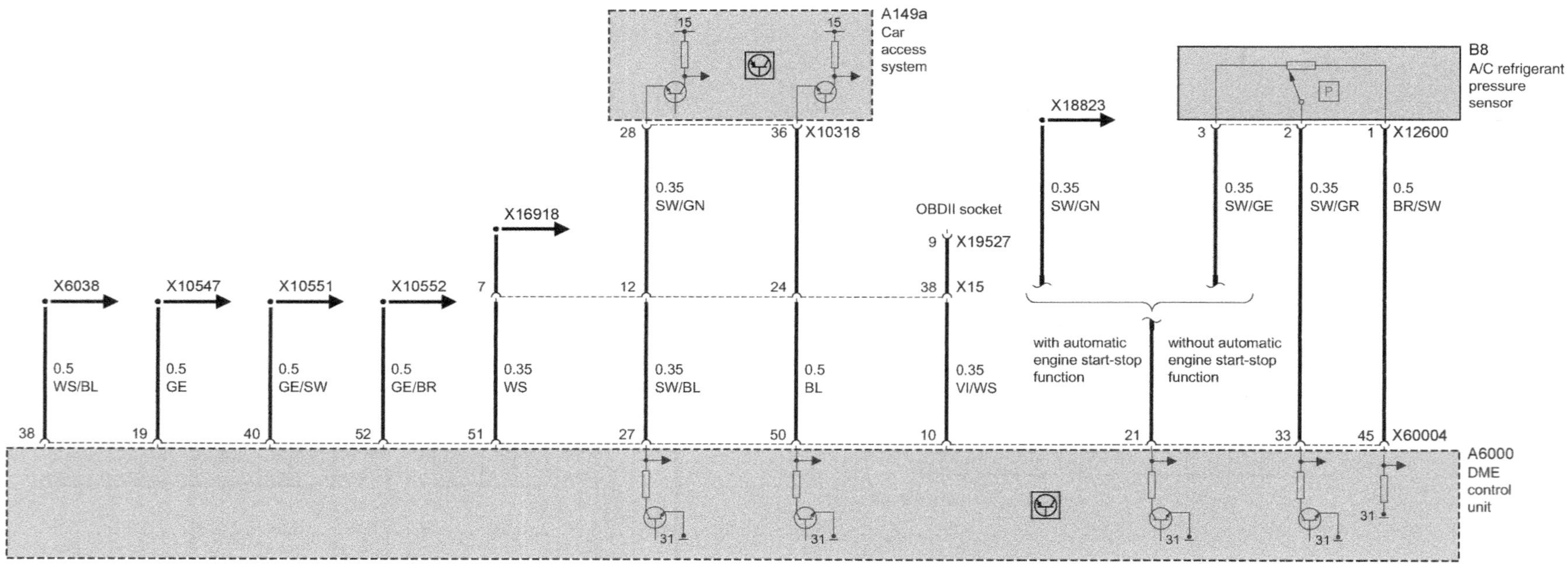

BM11ele26199

Ignition coils, N14 engine

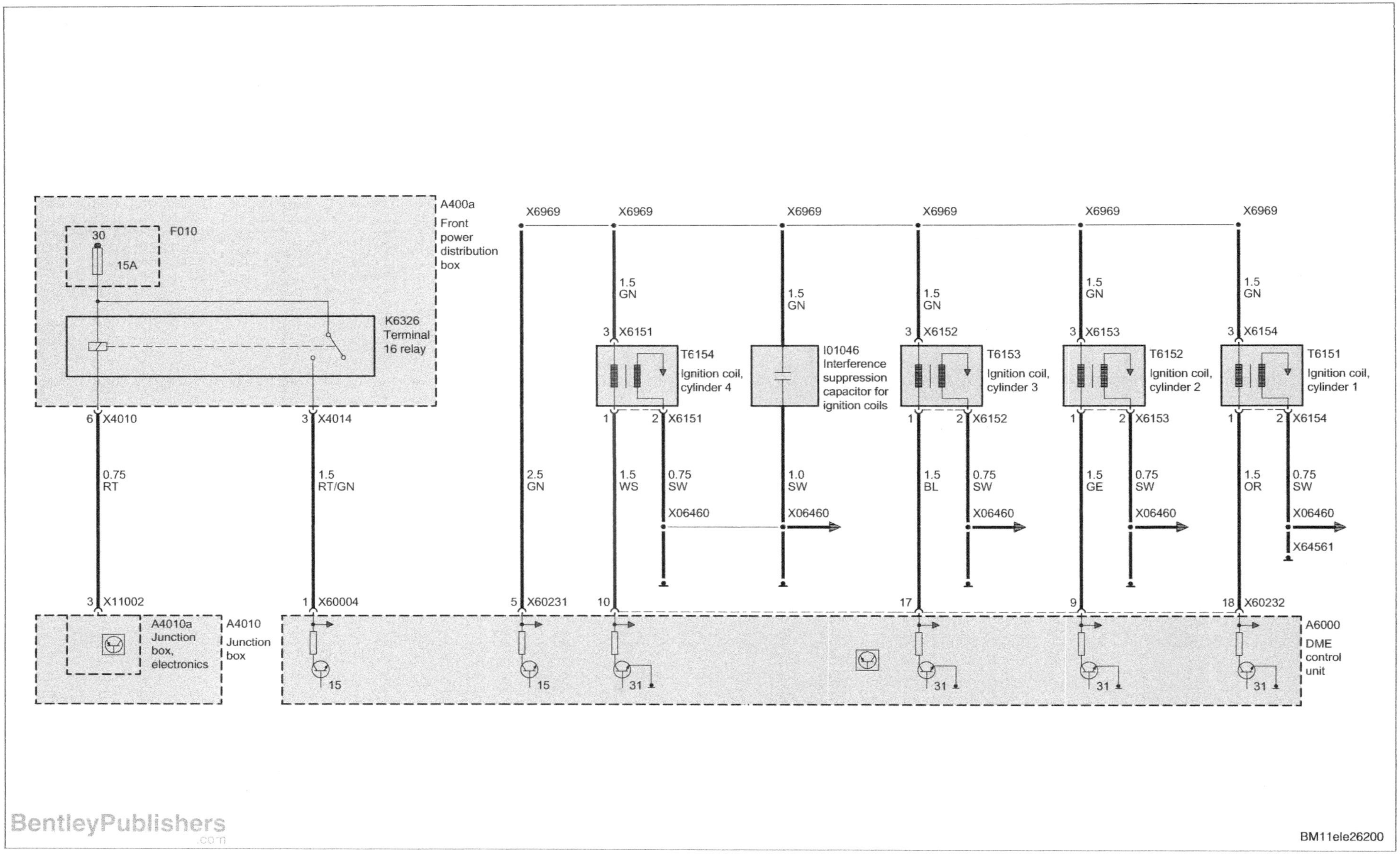

Thermostat, electric fan, engine temperature sensor, turbocharger coolant pump
N14 engine

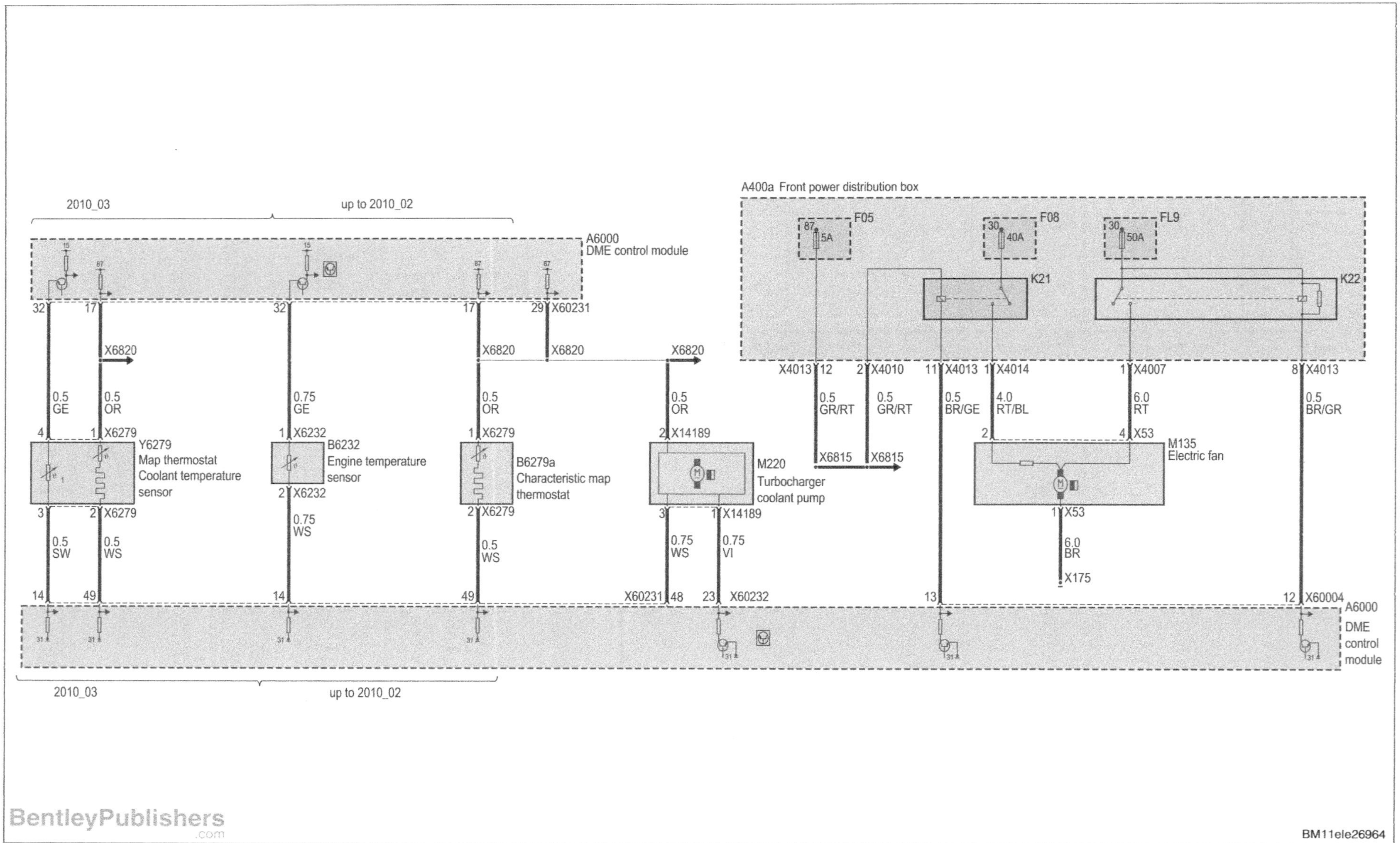

Convenient start

page 1 of 2

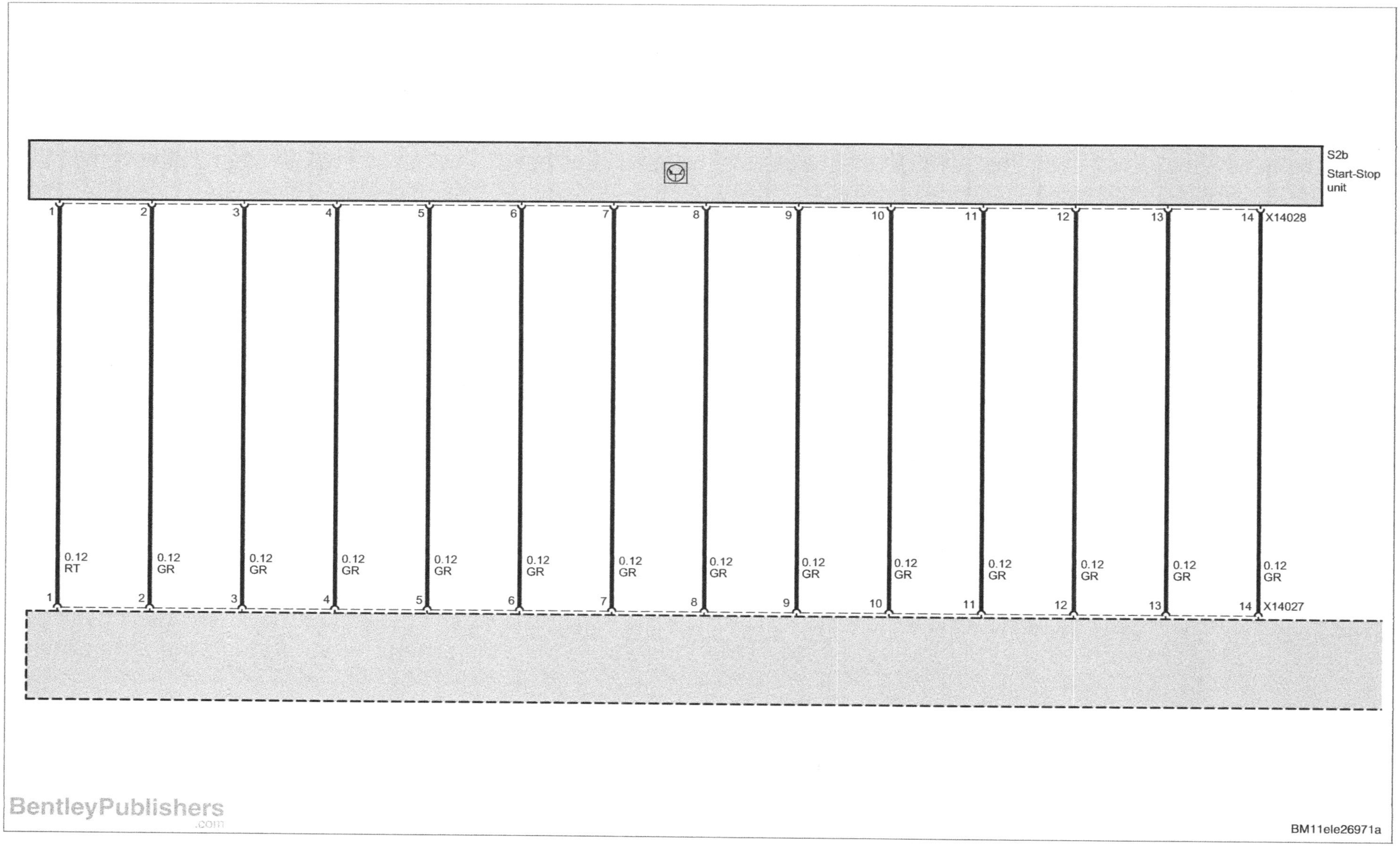

Convenient start

page 2 of 2

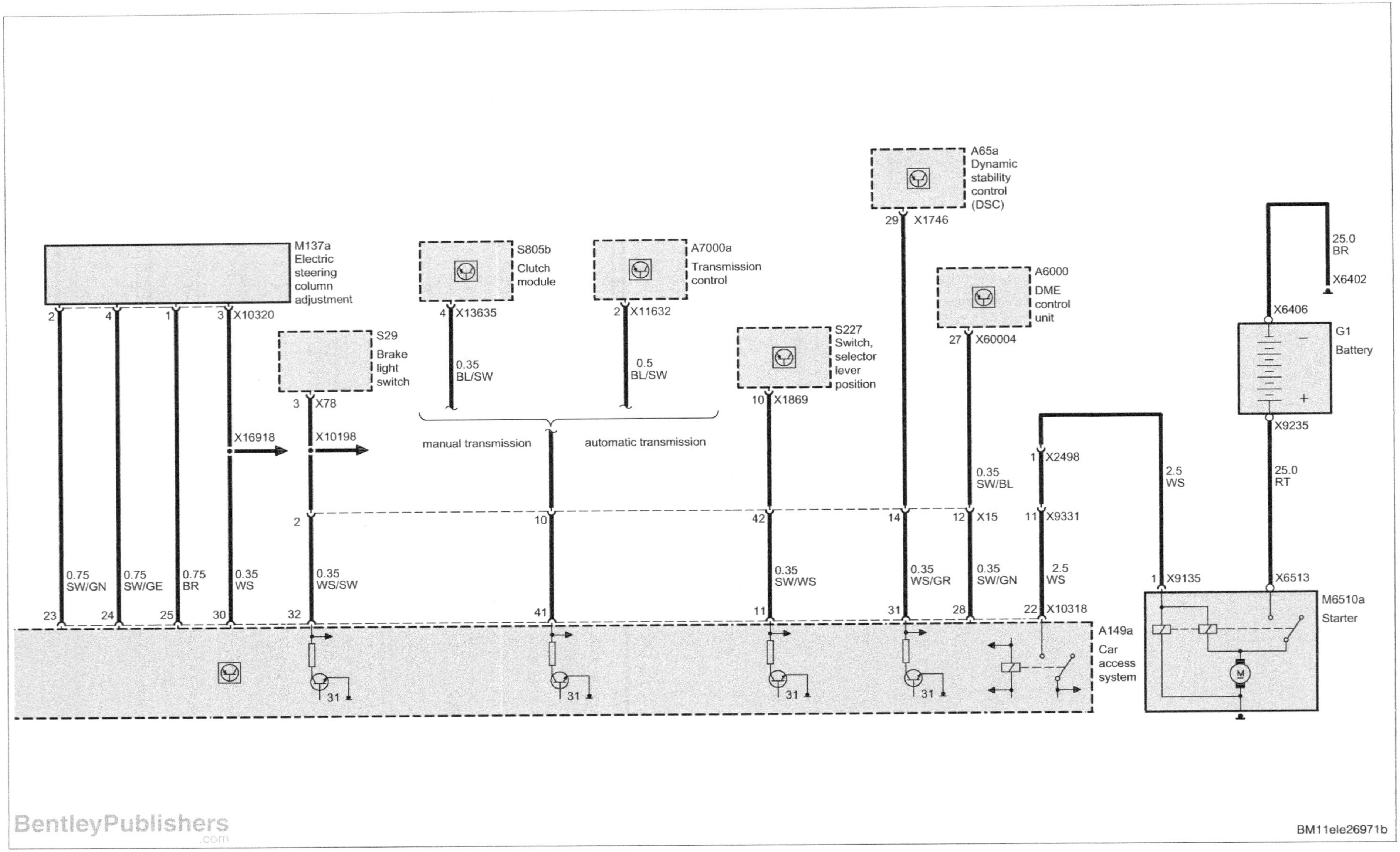

A6000 DME control unit, engine side
N14 engine
page 1 of 2

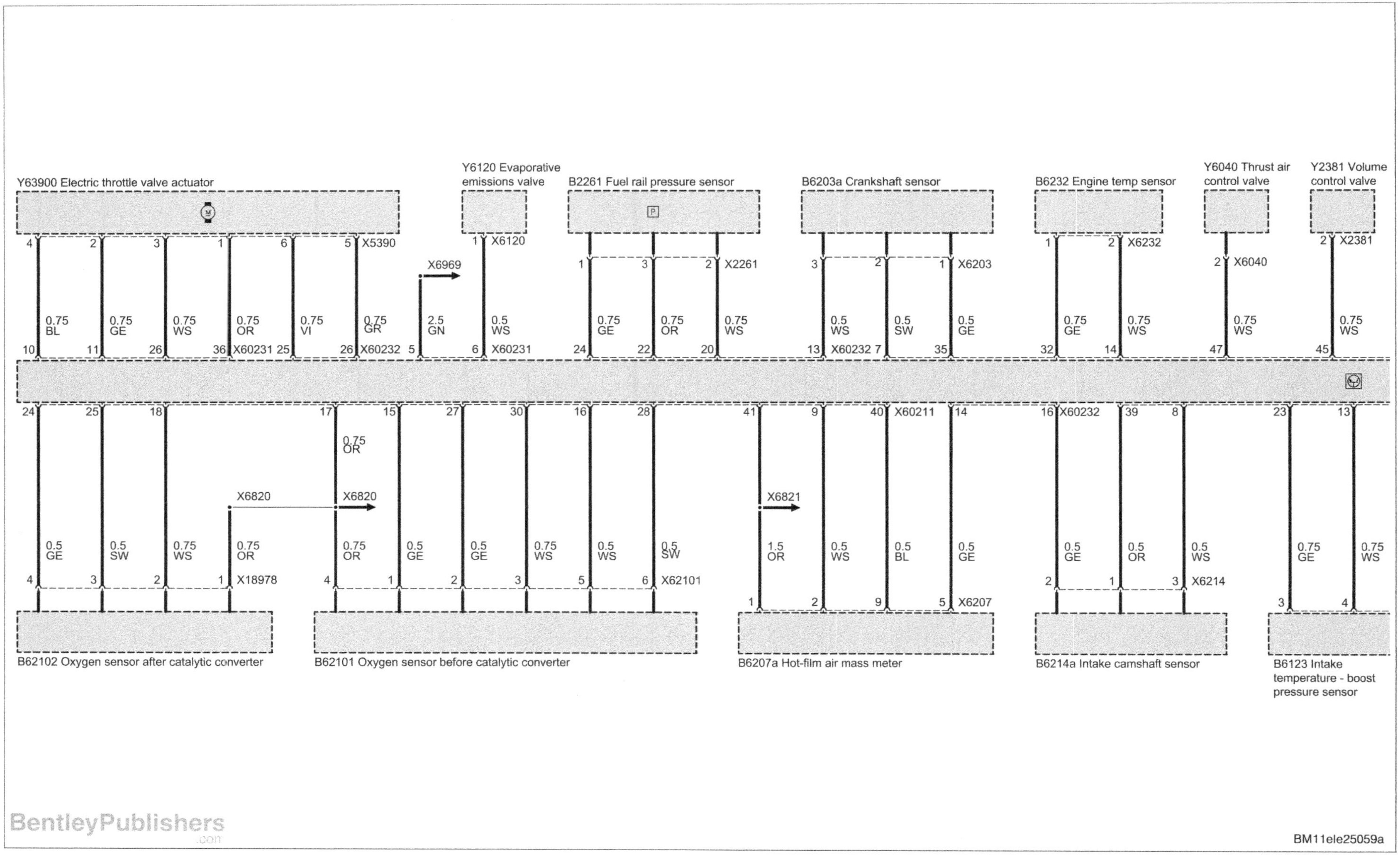

A6000 DME control unit, engine side
N14 engine
page 2 of 2

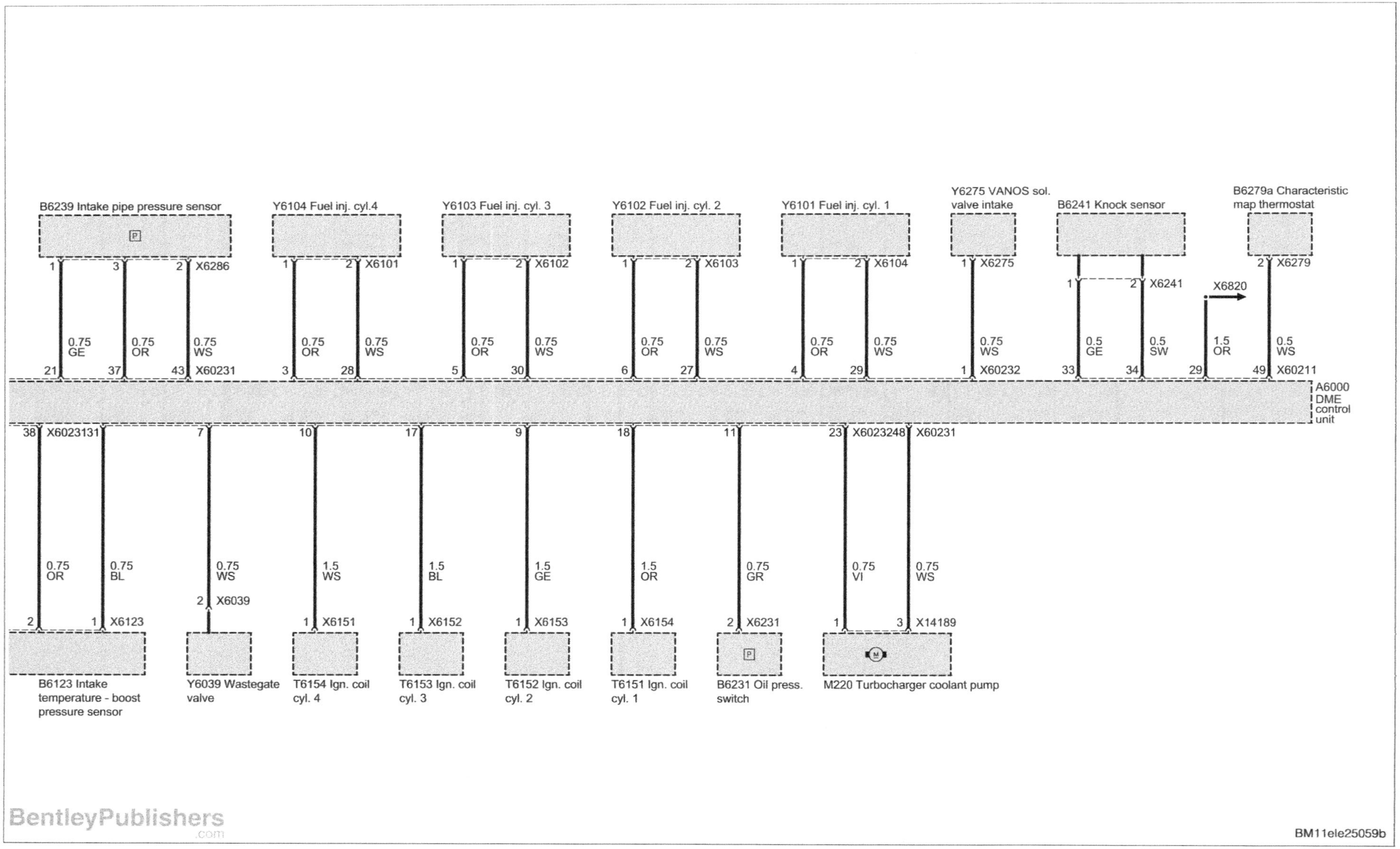

A6000 DME control unit, engine side
N18 engine
page 1 of 2

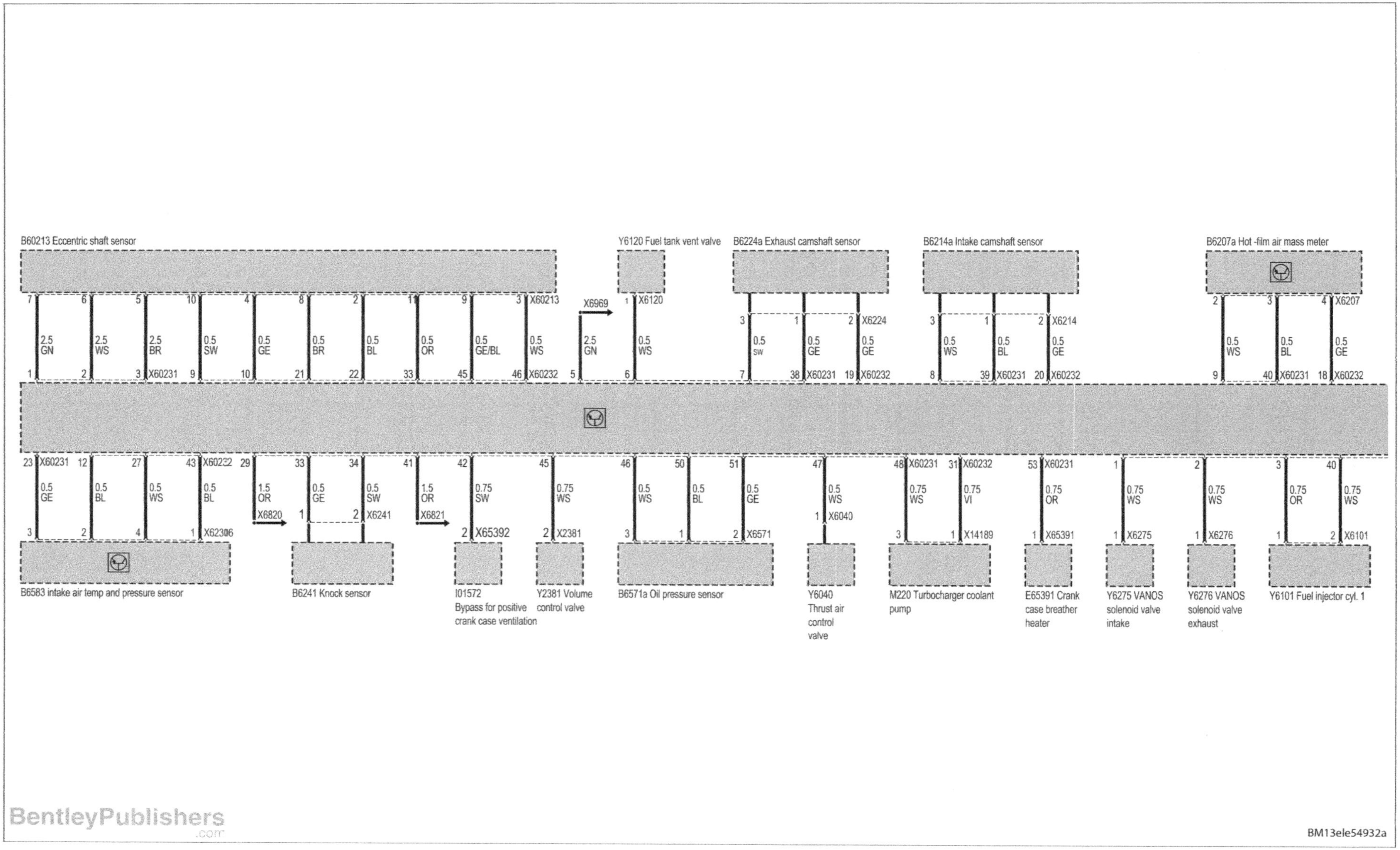

A6000 DME control unit, engine side
N18 engine
page 2 of 2

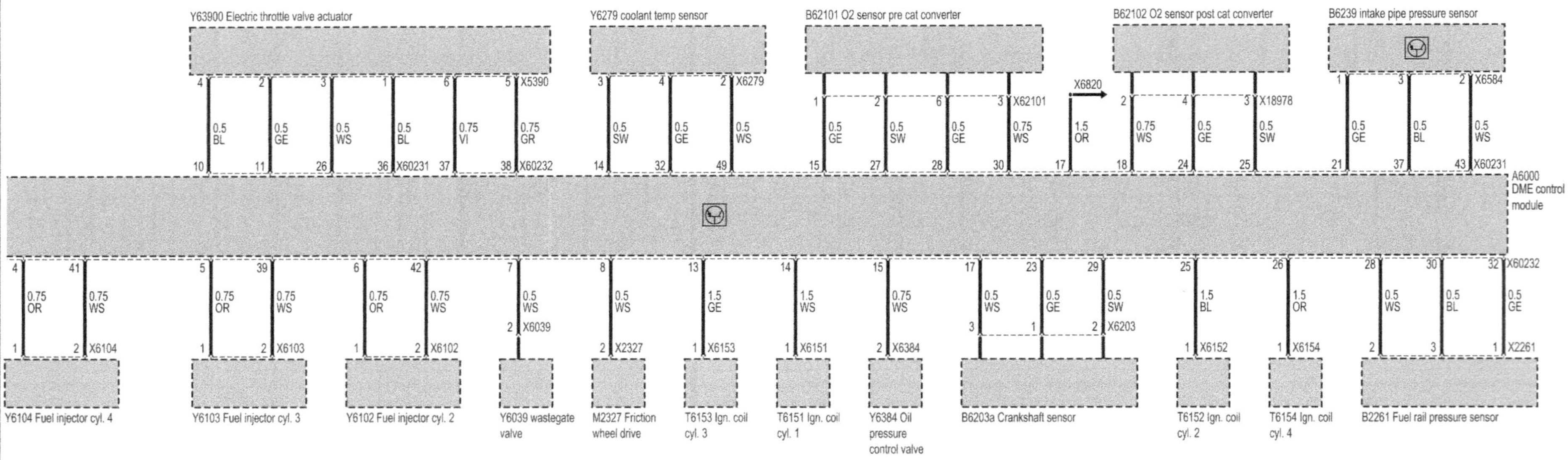

BM13ele54932b

A6000 DME control unit, engine side
N16 engine
page 1 of 2

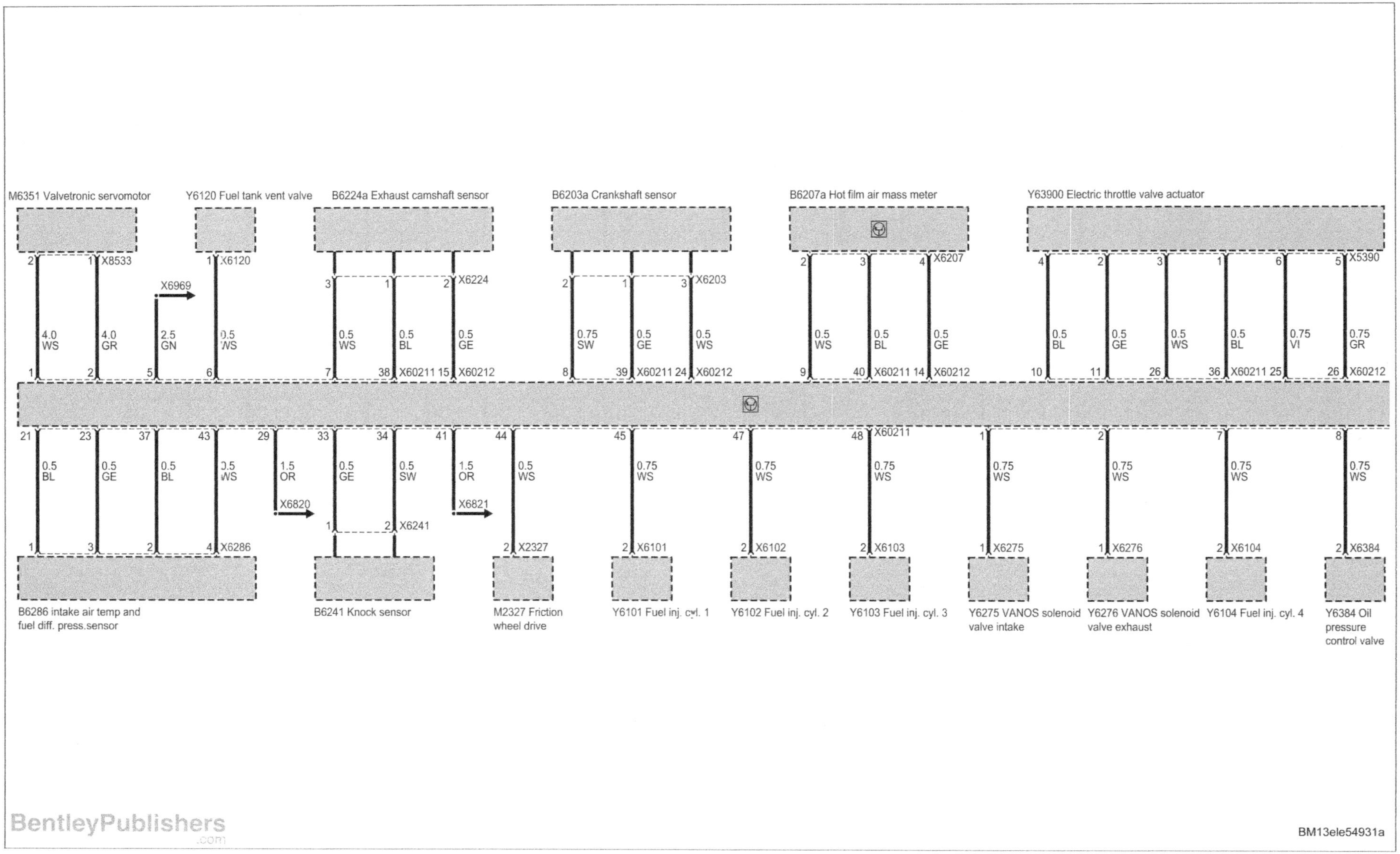

A6000 DME control unit, engine side
N16 engine
page 2 of 2

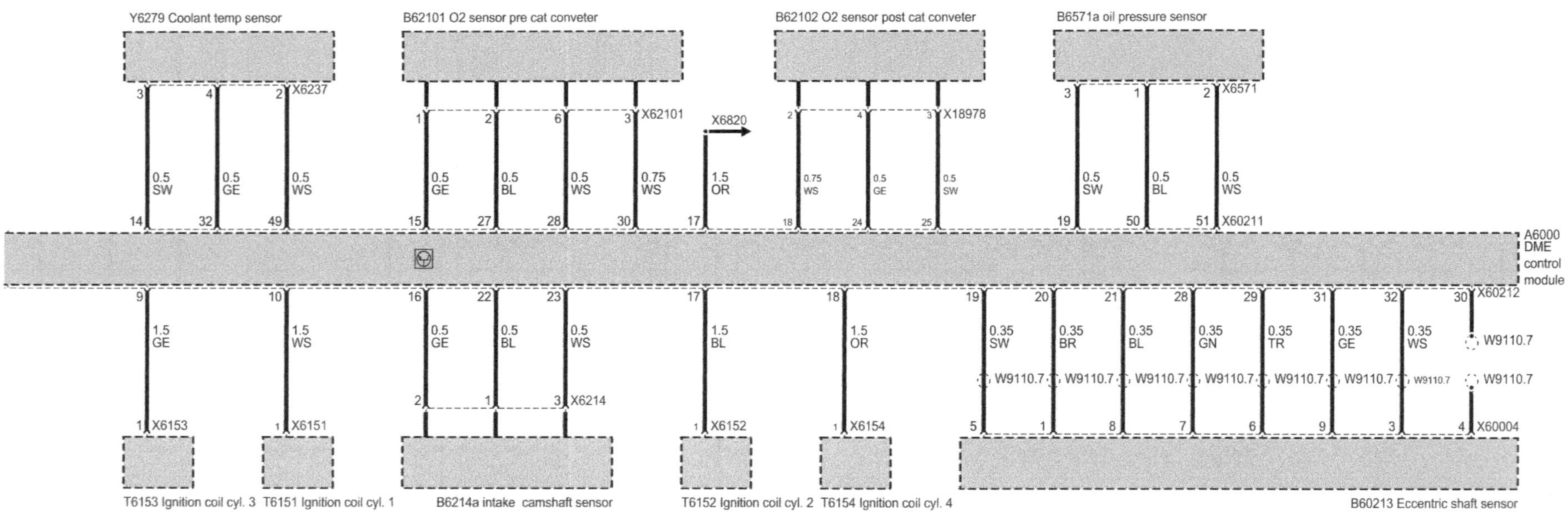

BM13ele54931b

A6000 DME control unit, body side
N12 engine, N14 engine
page 1 of 2

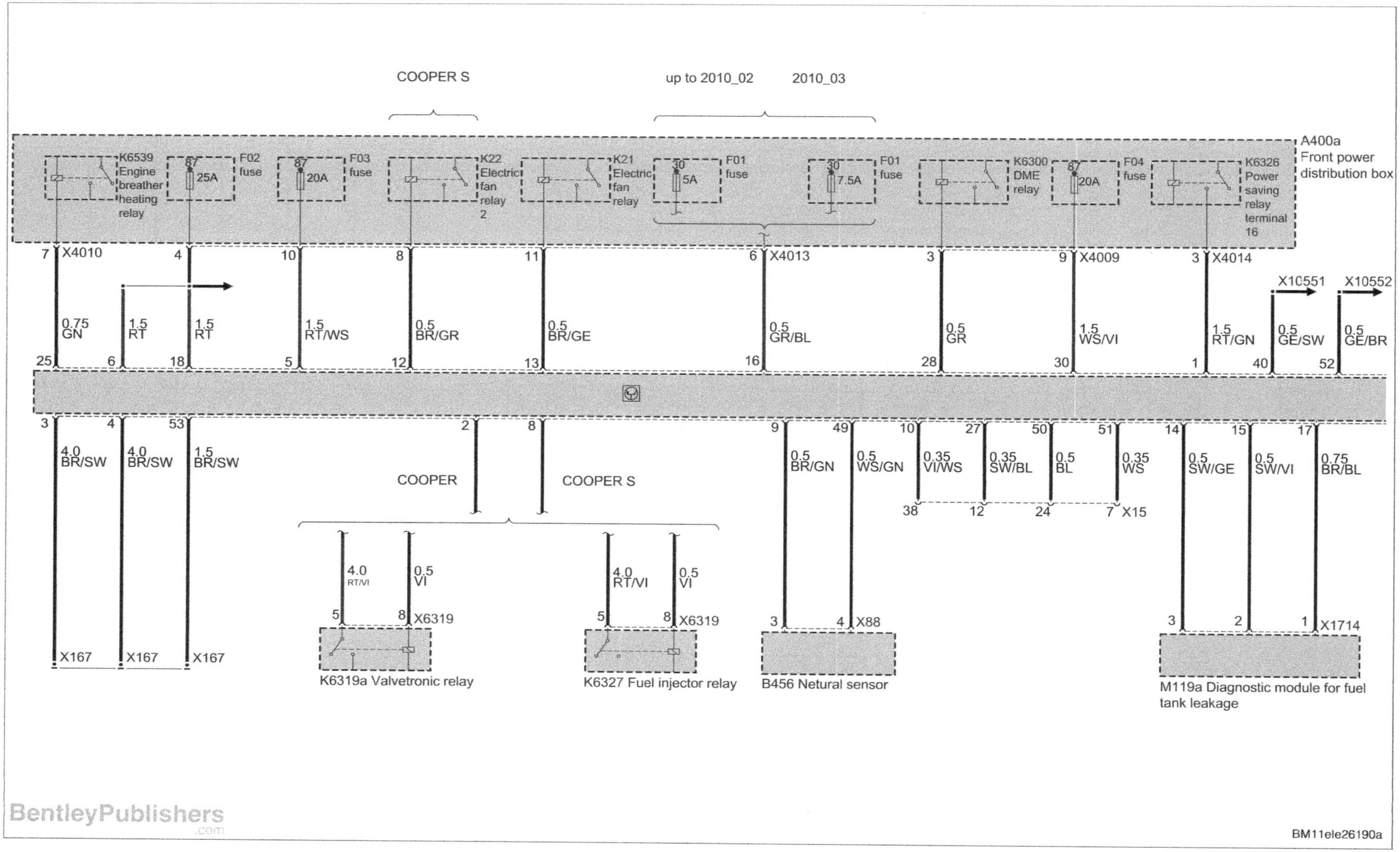

A6000 DME control unit, body side
N12 engine, N14 engine
page 2 of 2

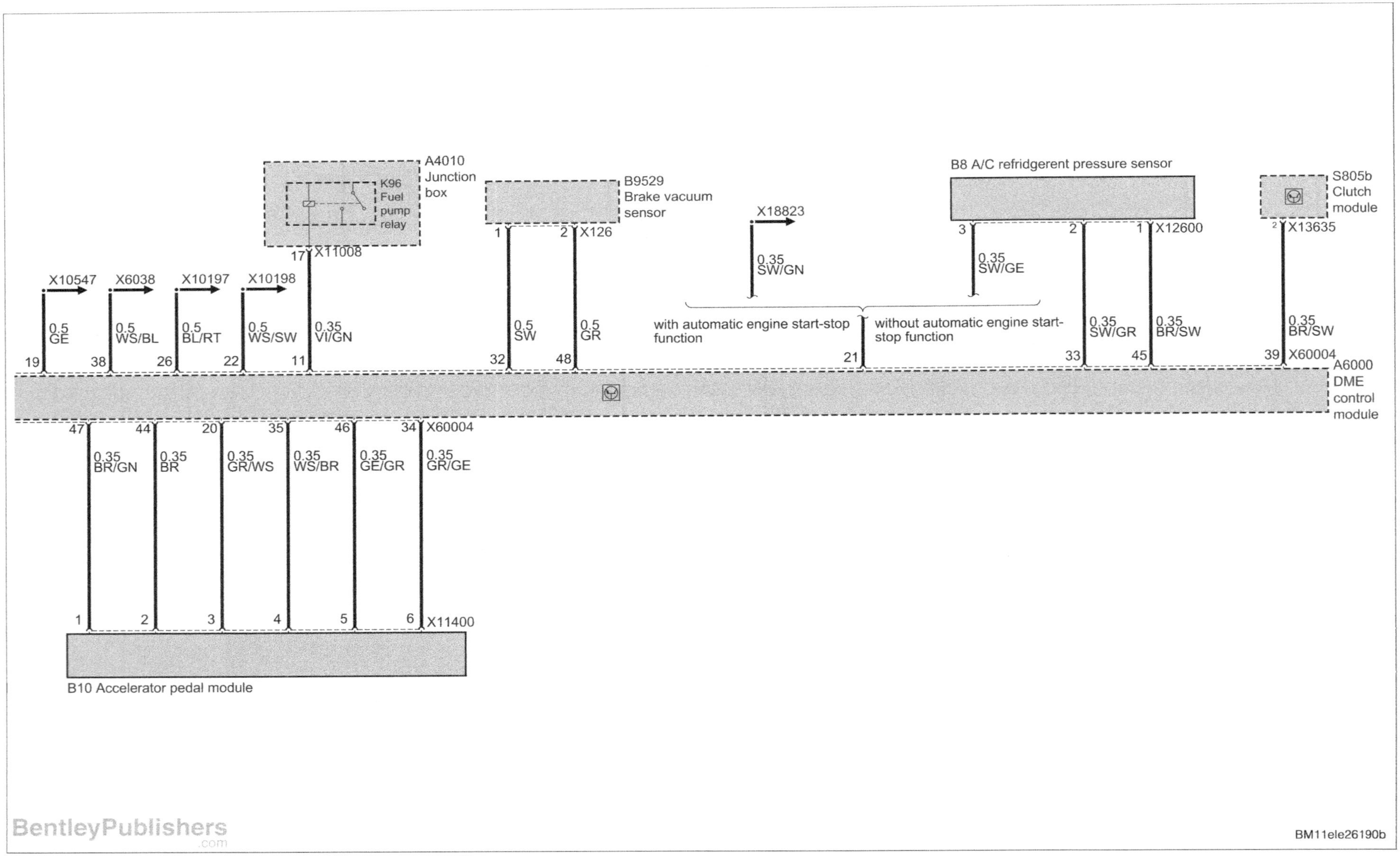

A6000 DME control unit, body side
N16 engine, N18 engine
page 1 of 2

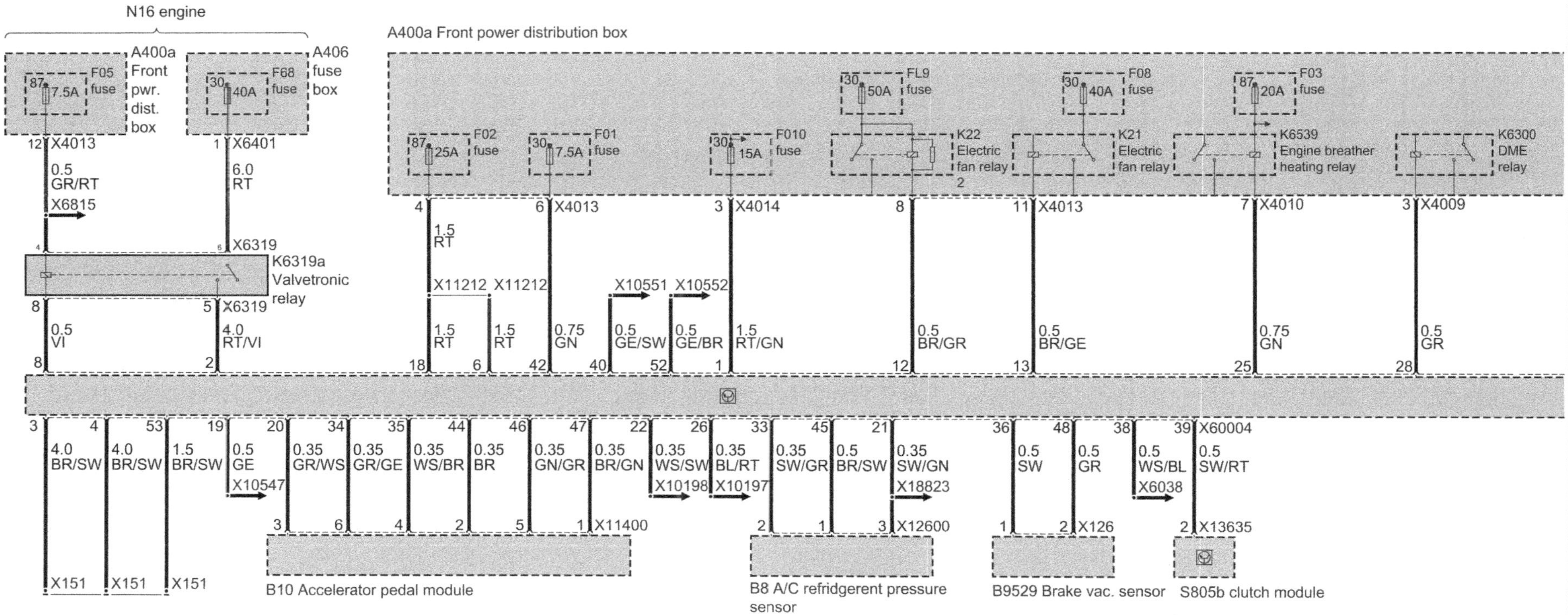

BM13ele54933a

A6000 DME control unit, body side
N16 engine, N18 engine
page 2 of 2

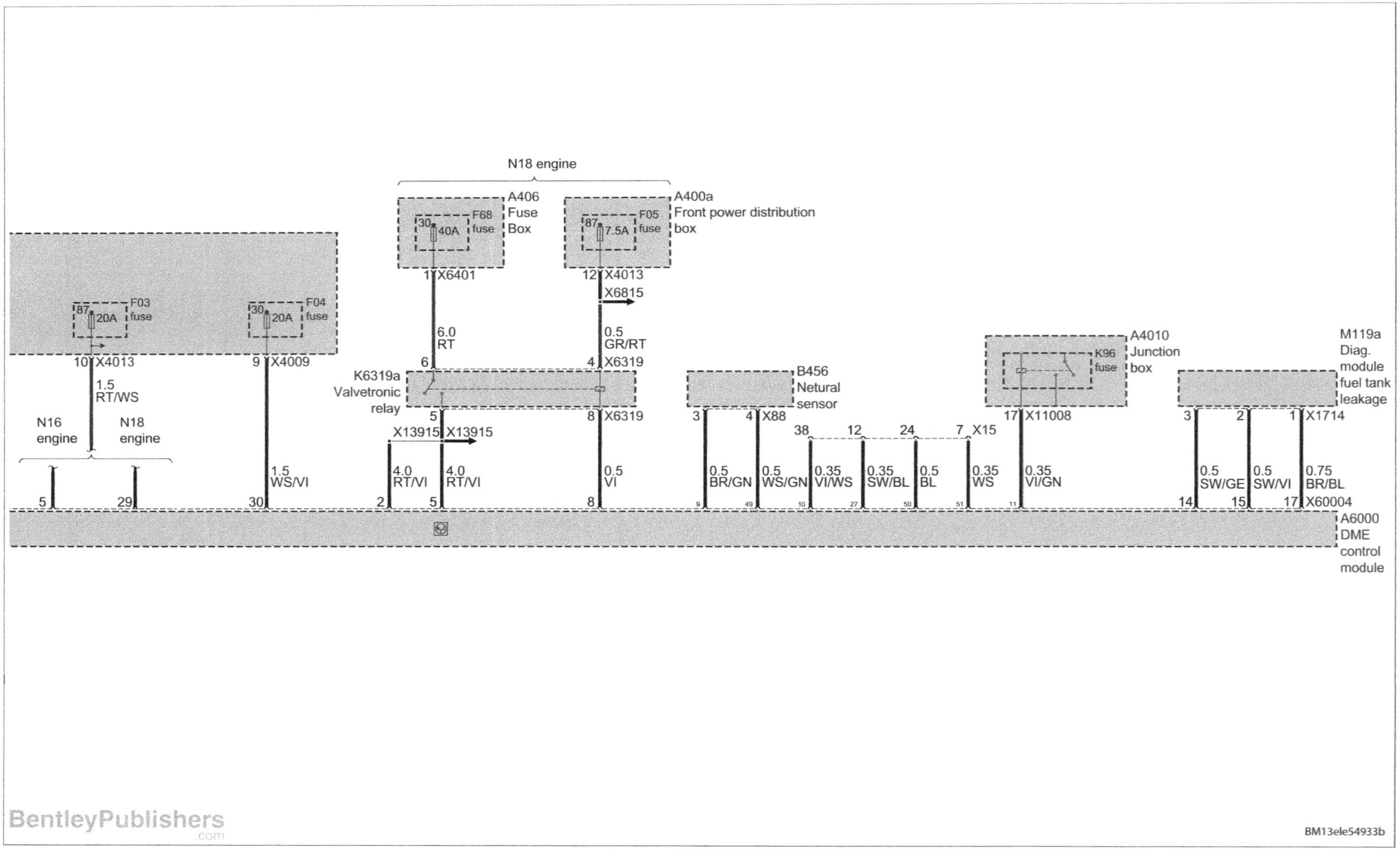

Oxygen sensors N12, N16, N18 engines

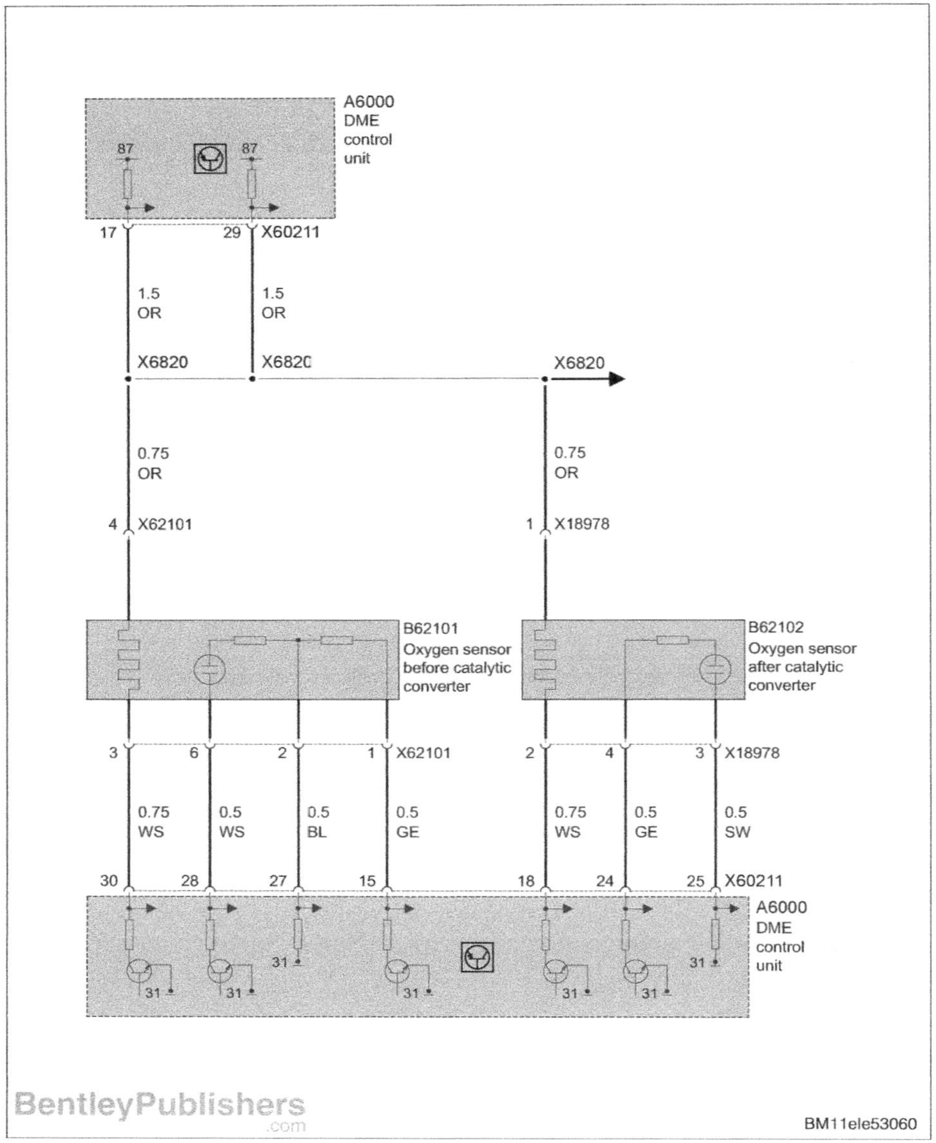

Oxygen sensors N14 engine

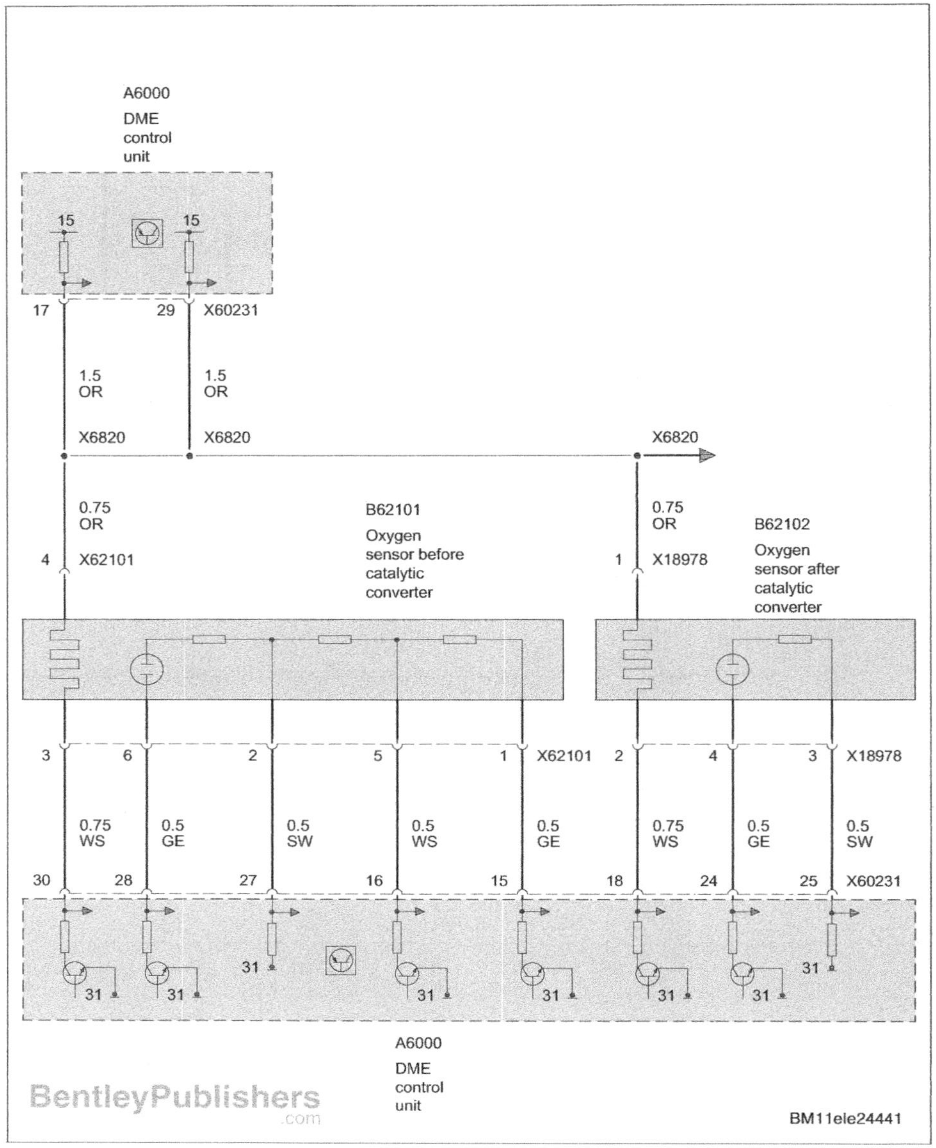

Air supply
N16 engine

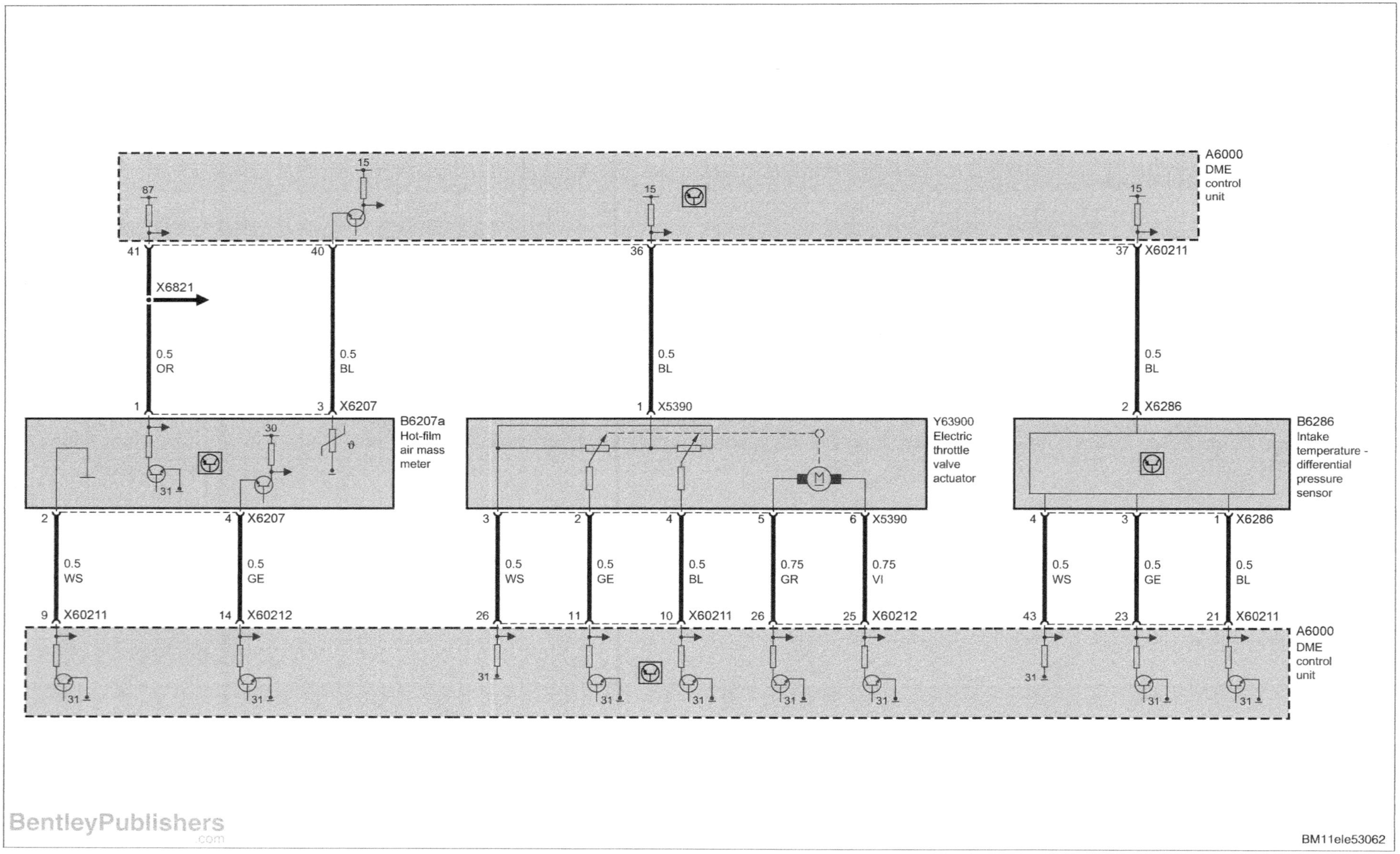

Thermostat, electric fan
N16 engine

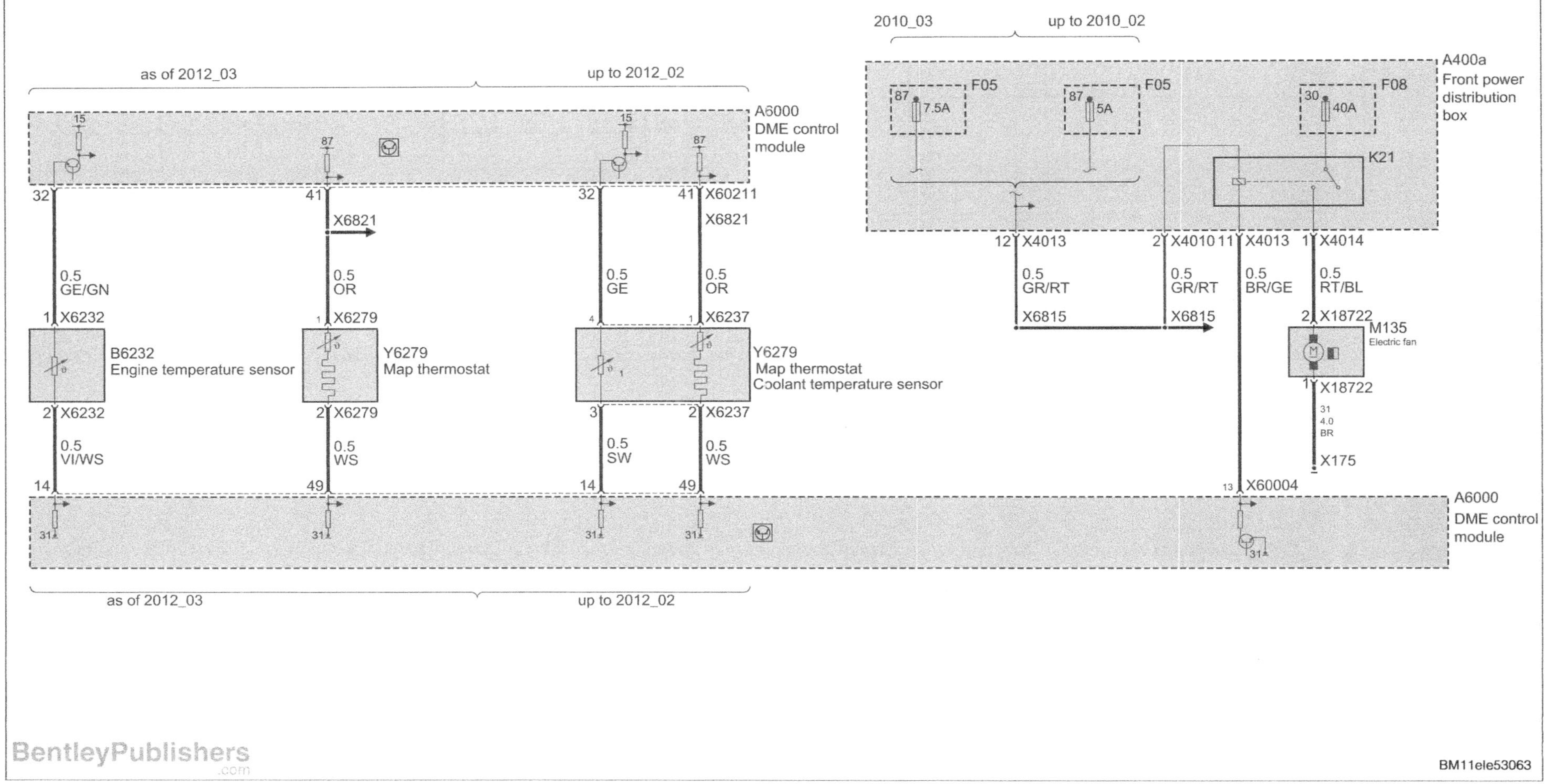

Camshaft sensors, eccentric shaft sensor, VANOS solenoid valves, Valvetronic relay, Valvetronic servomotor N16 engine, page 1 of 2

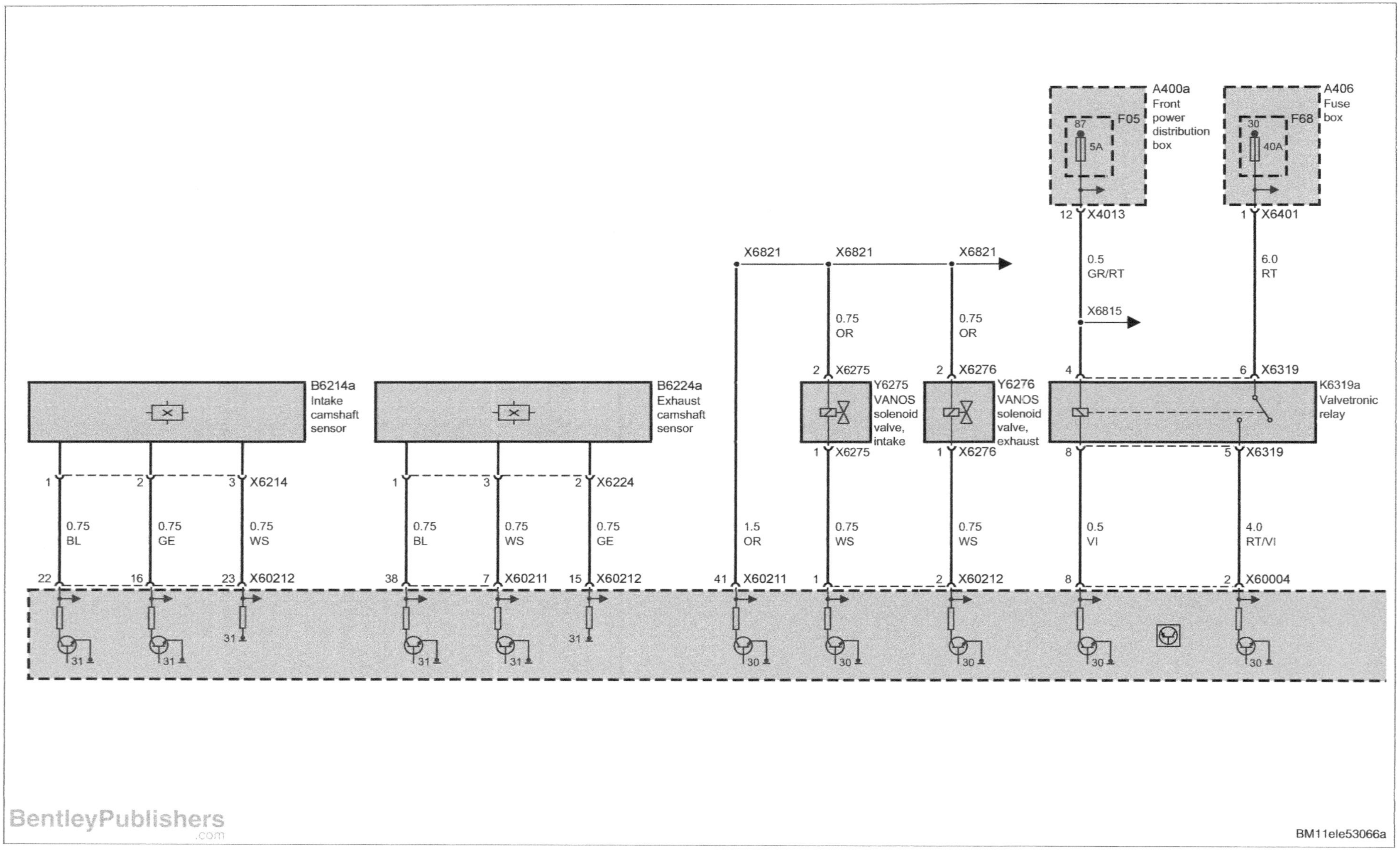

Camshaft sensors, eccentric shaft sensor, VANOS solenoid valves, Valvetronic relay, Valvetronic servomotor N16 engine, page 2 of 2

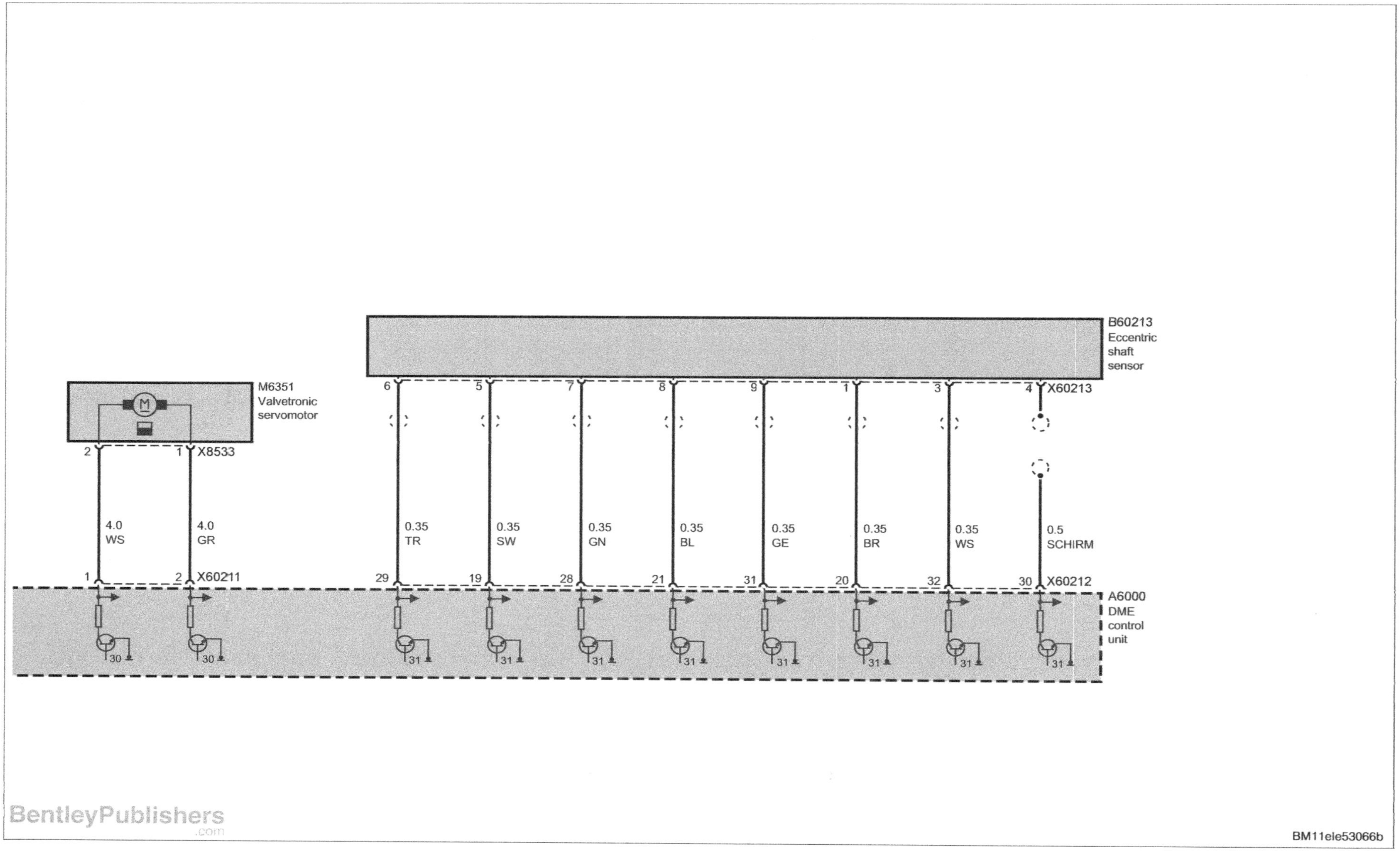

Thermostat, fuel tank vent valve, hot-film air mass meter, oil pressure control valve, VANOS solenoid valves N16 engine

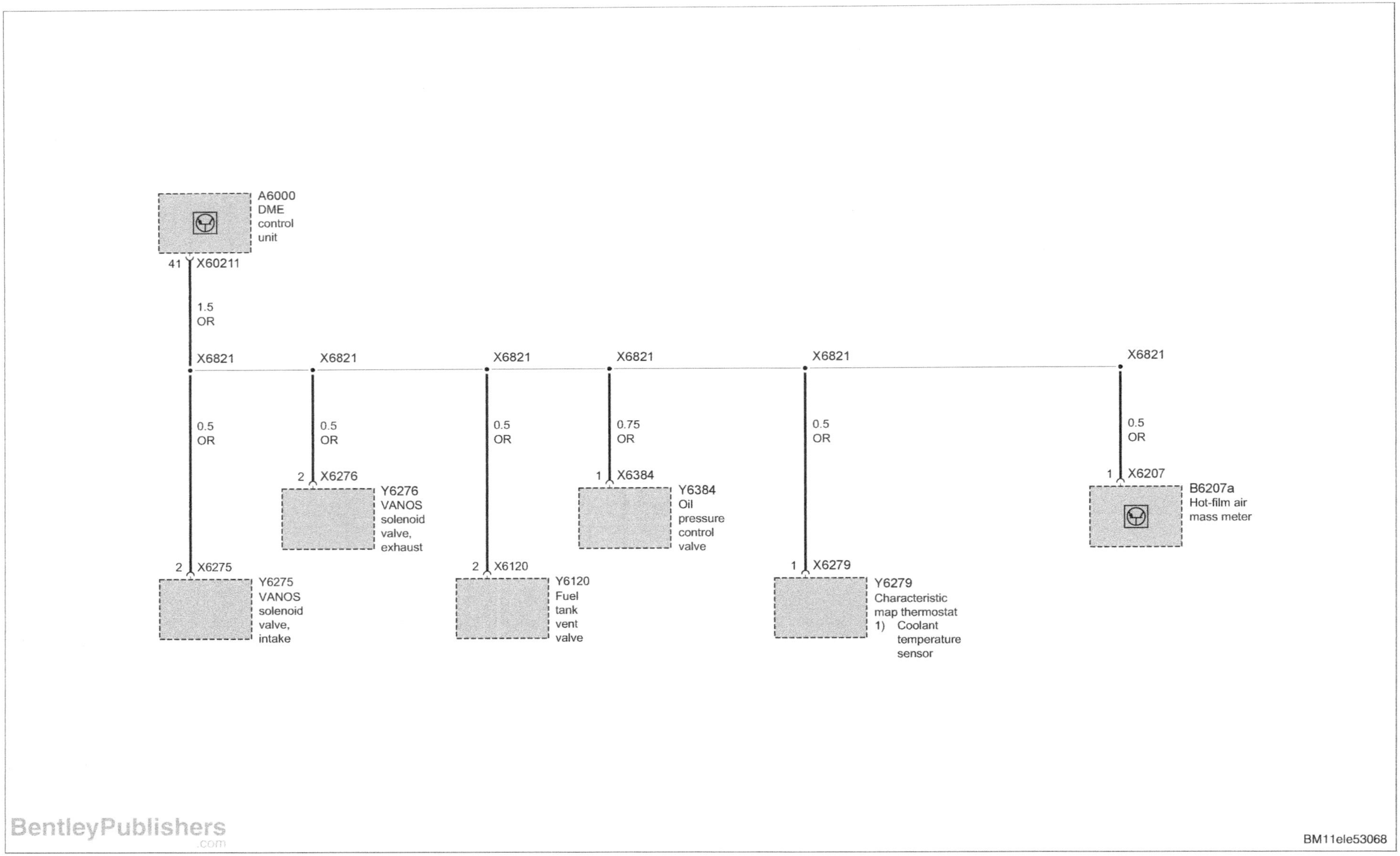

OBDII socket, terminating resistor D-CAN
N16, N18 engines

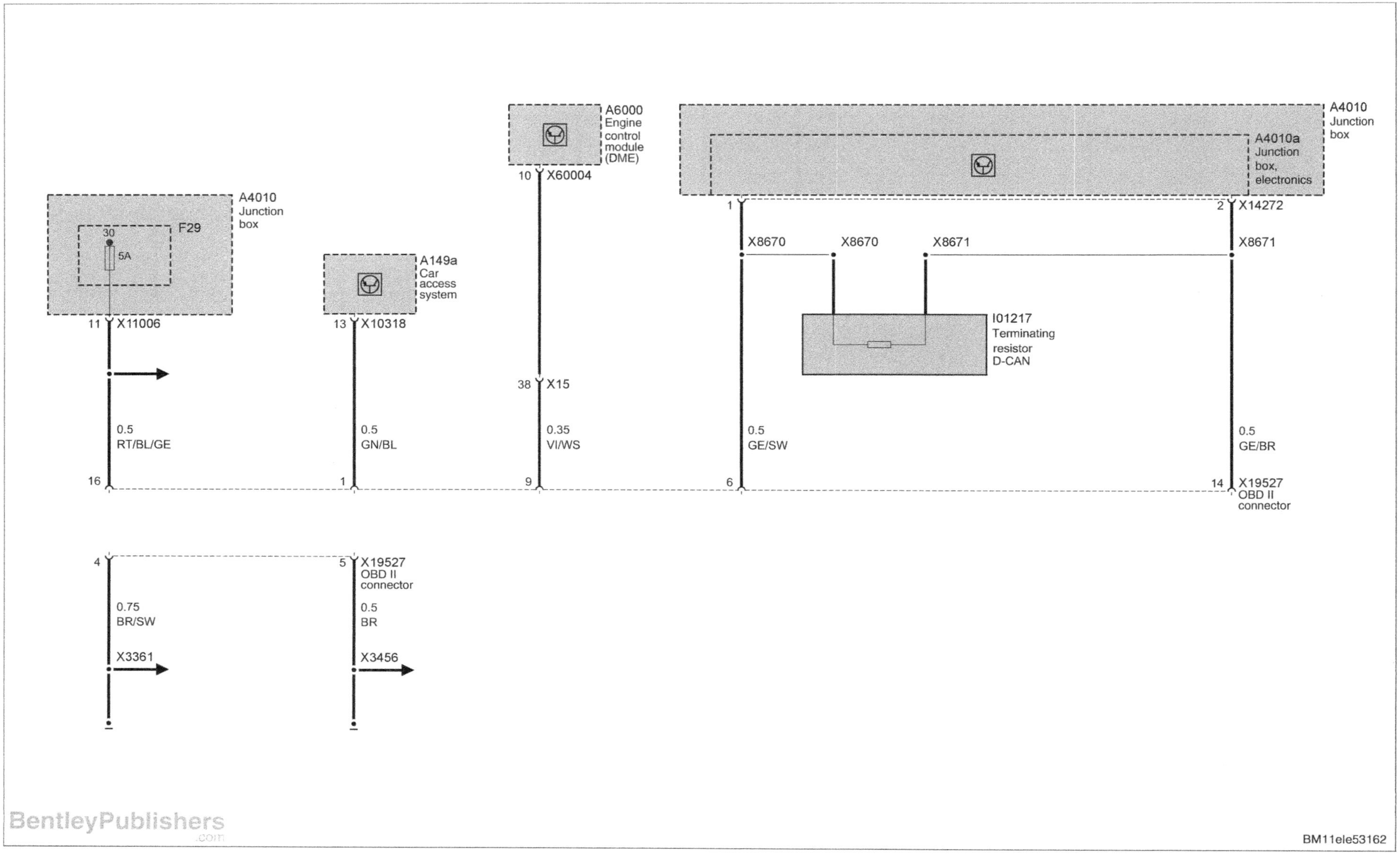

Battery, starter
N12 engine, N14 engine

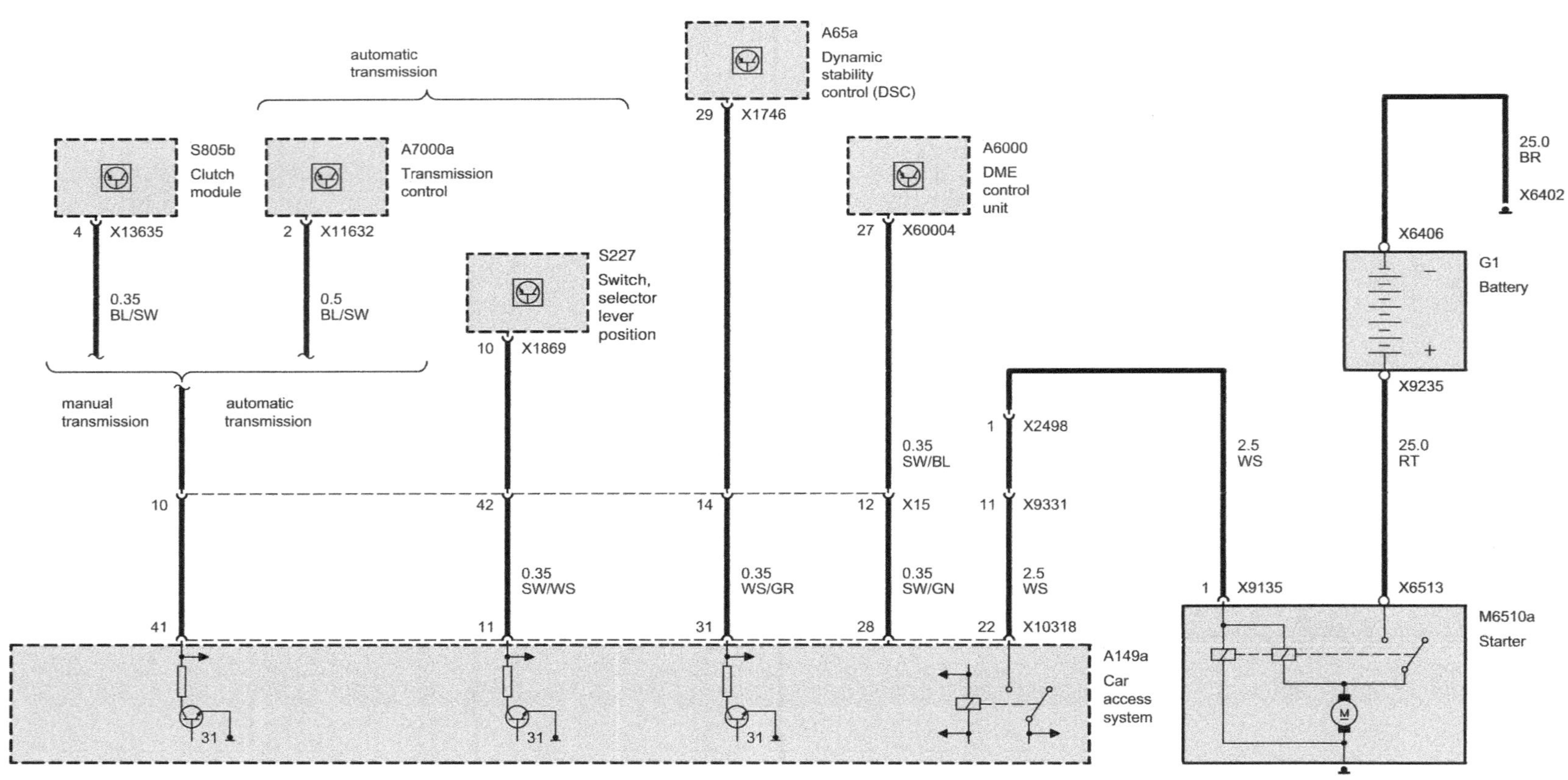

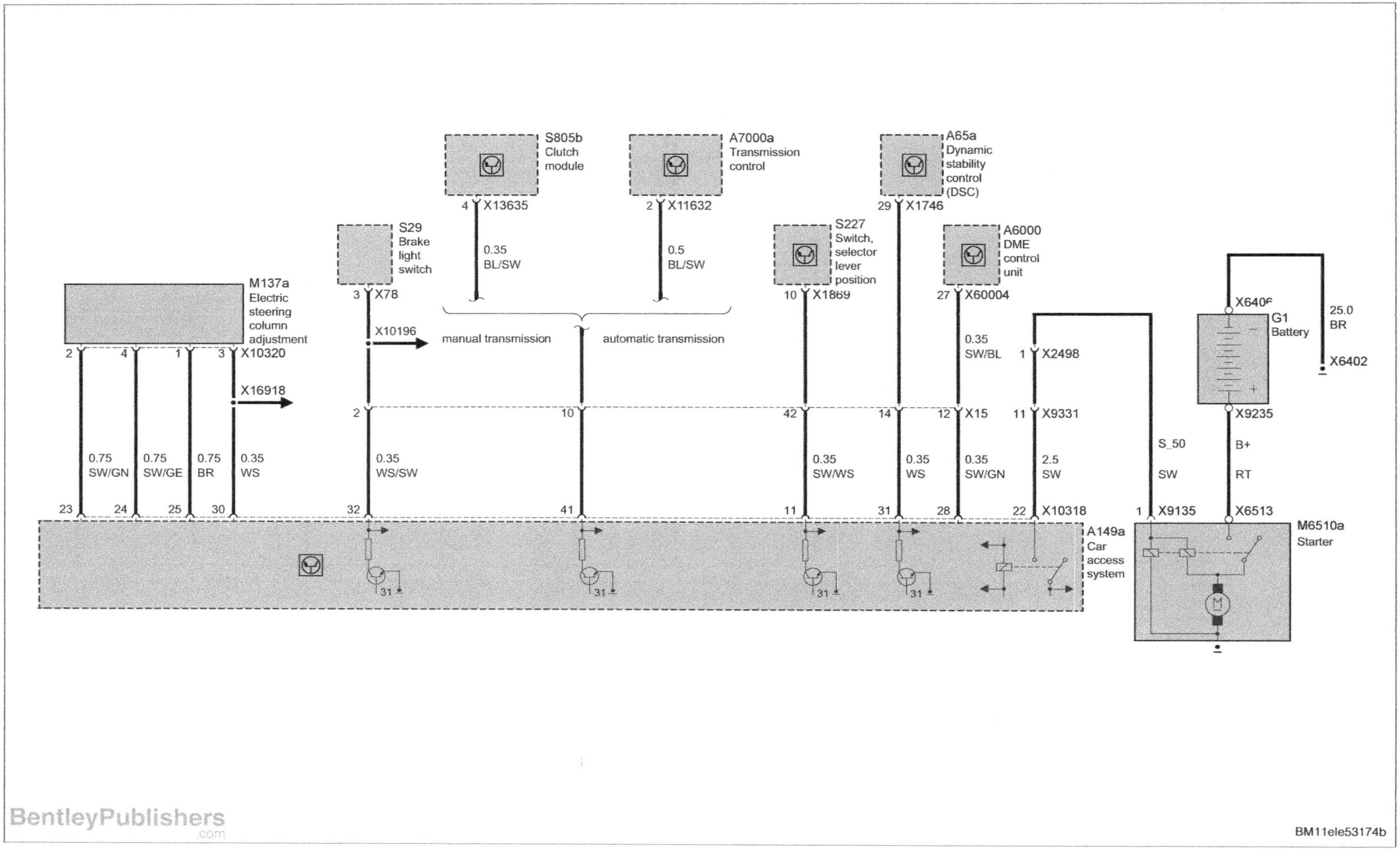
S805b
Clutch
module
4 X13635
0.35
BL/SW
A7000a
Transmission
control
2 X11632
0.5
BL/SW
manual transmission
automatic transmission
A65a
Dynamic
stability
control
(DSC)
29 X1746
S29
Brake
light
switch
3 X78
X10196
S227
Switch,
selector
lever
position
10 X1869
A6000
DME
control
unit
27 X60004
M137a
Electric
steering
column
adjustment
2 4 1 3 X10320
X16918
0.75
SW/GN
0.75
SW/GE
0.75
BR
0.35
WS
2
10
42
14
12 X15
11 X9331
1 X2498
0.35
WS/SW
0.35
SW/WS
0.35
WS
0.35
SW/BL
0.35
SW/GN
2.5
SW
23 24 25 30 32 41 11 31 28 22 X10318
31
A149a
Car
access
system
X640F
G1
Battery
25.0
BR
X6402
X9235
B+
RT
S_50
SW
1 X9135
X6513
M6510a
Starter
BM11ele53174b

CAN bus connector
page 1 of 2

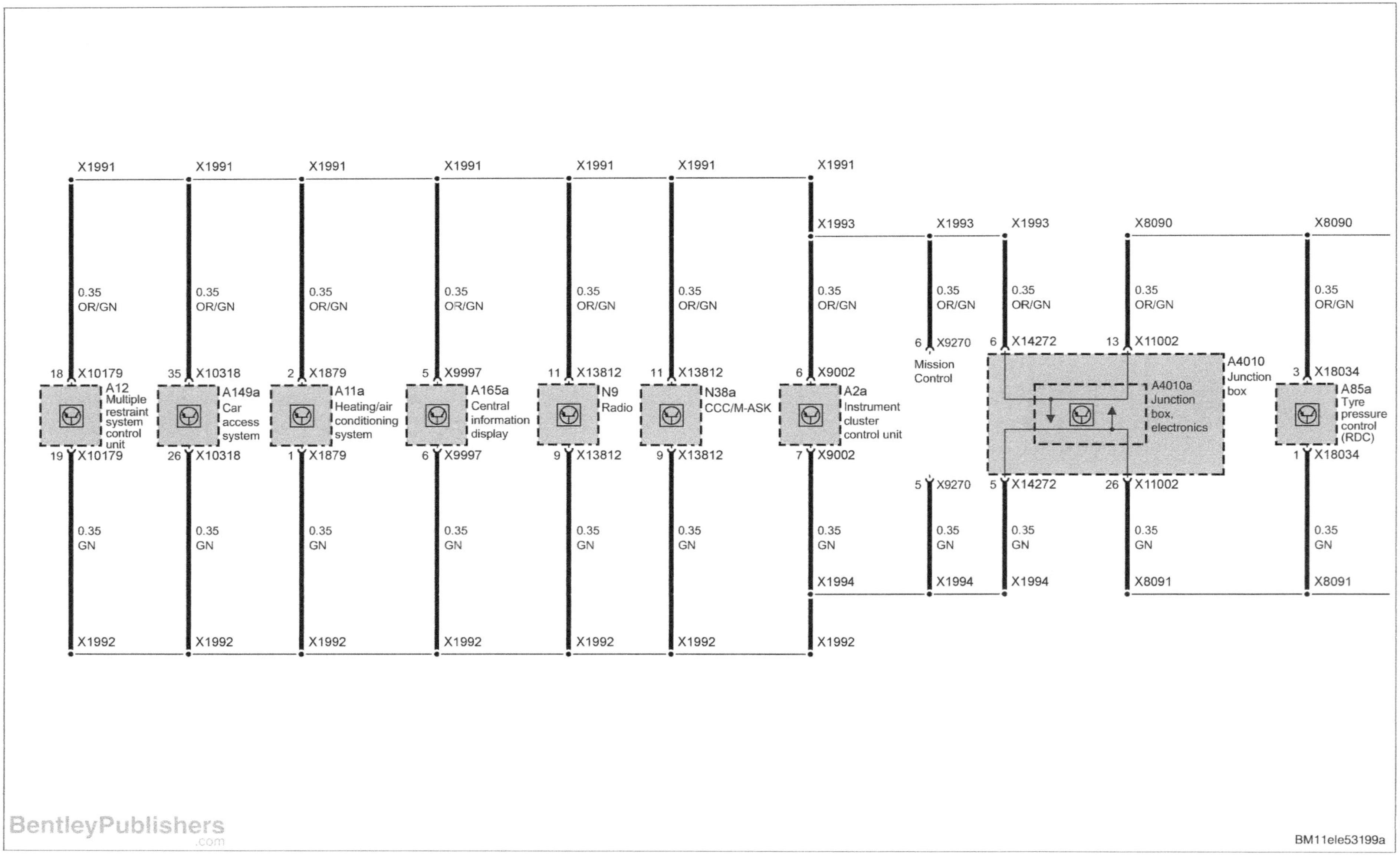

CAN bus connector
page 2 of 2

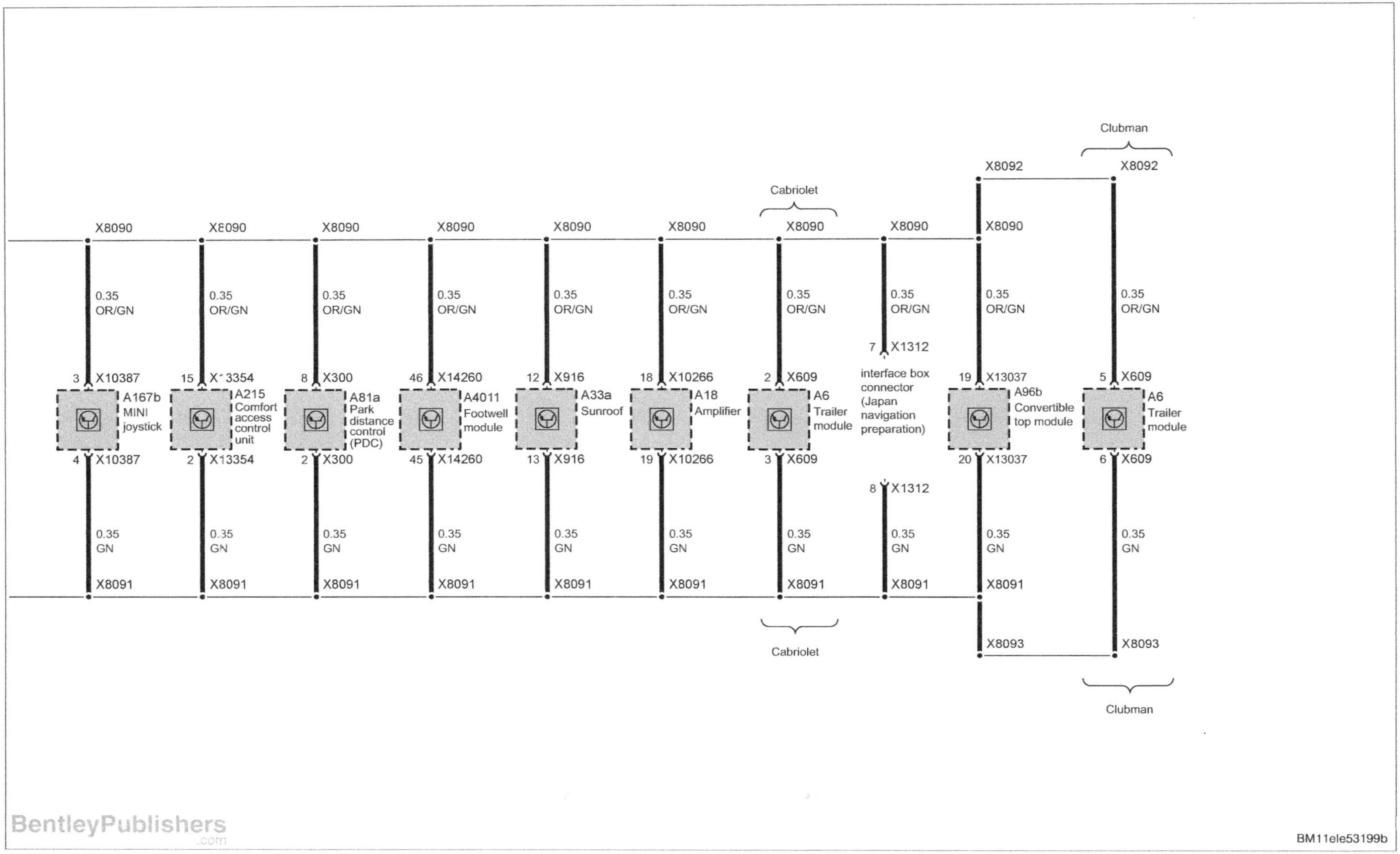

Heater, blower output stage N12 engine, N14 engine

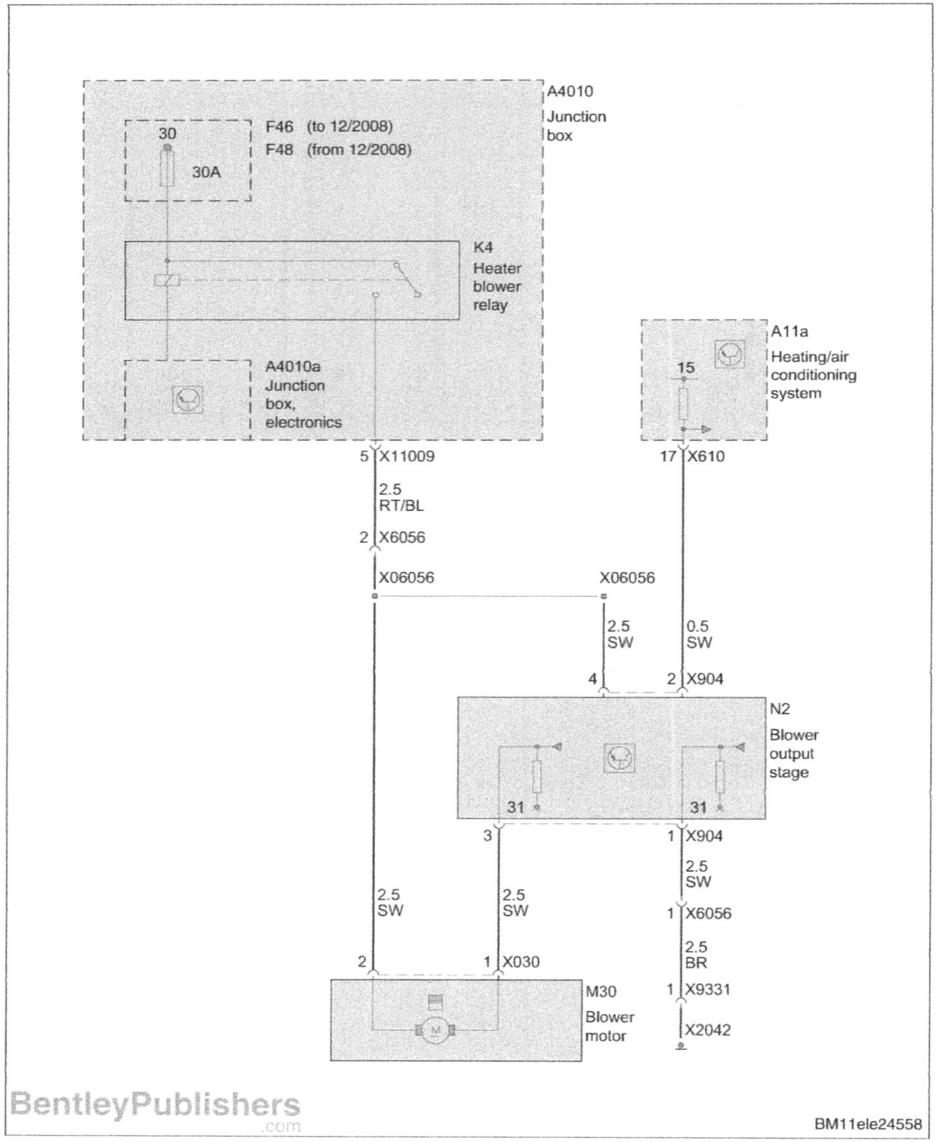

Heater, blower output stage N16 engine, N18 engine

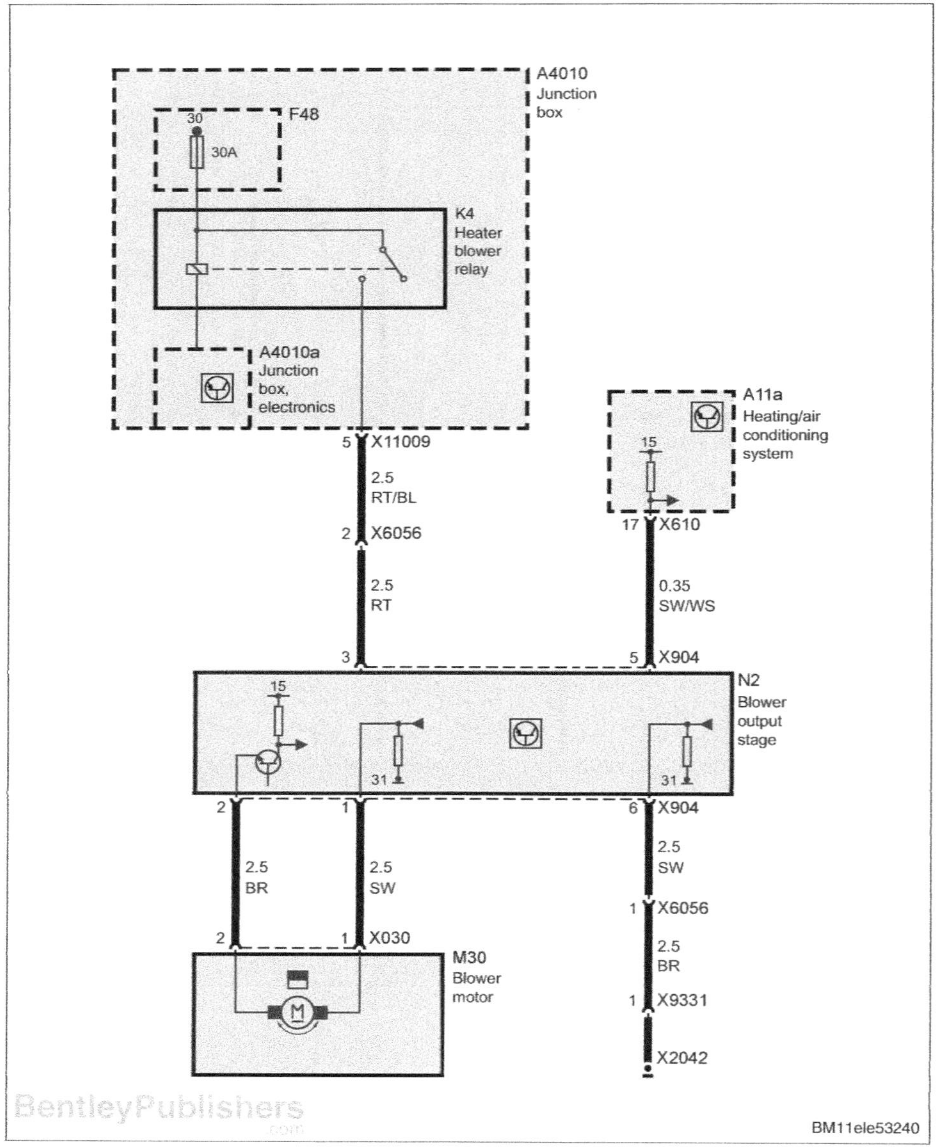

Alternator, battery, intelligent battery sensor
N12 engine, N14 engines

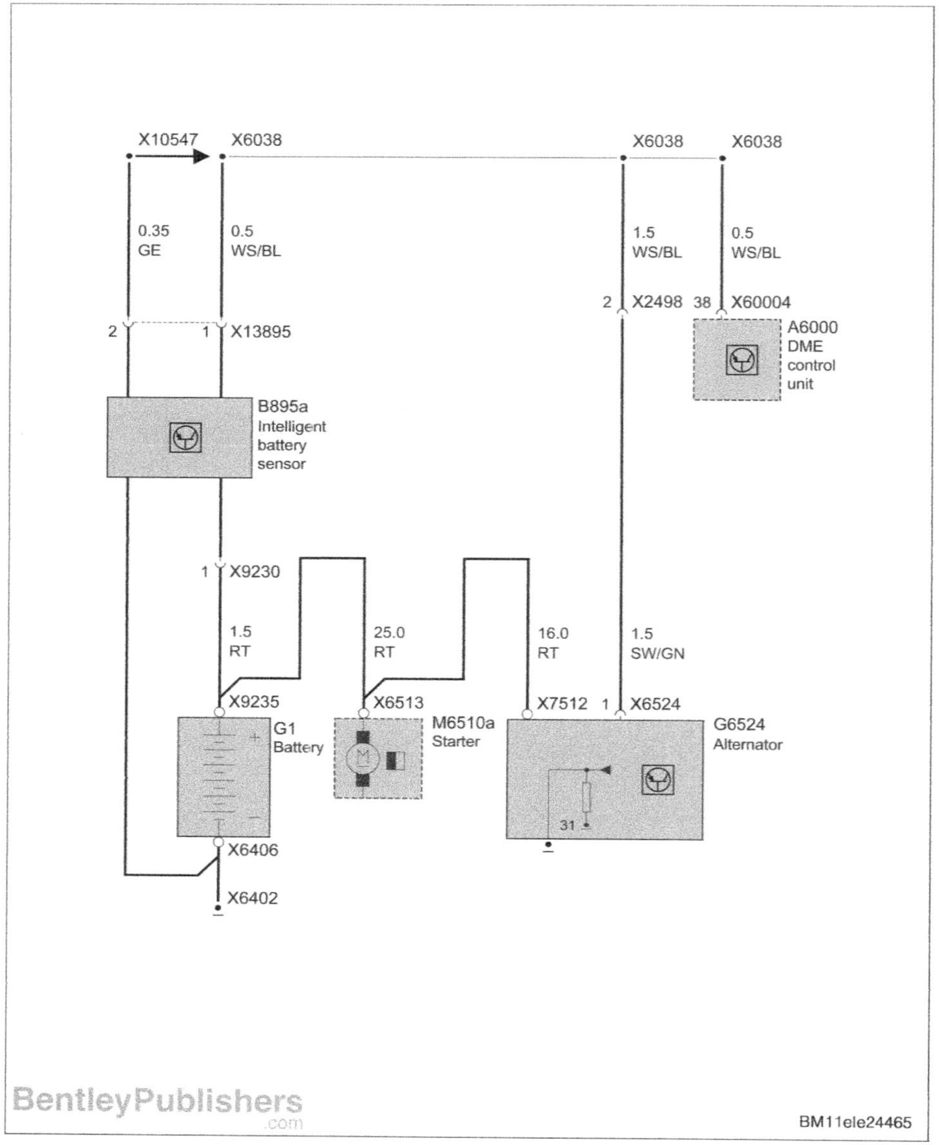

Alternator, battery, intelligent battery sensor, starter
N16 engine, N18 engines

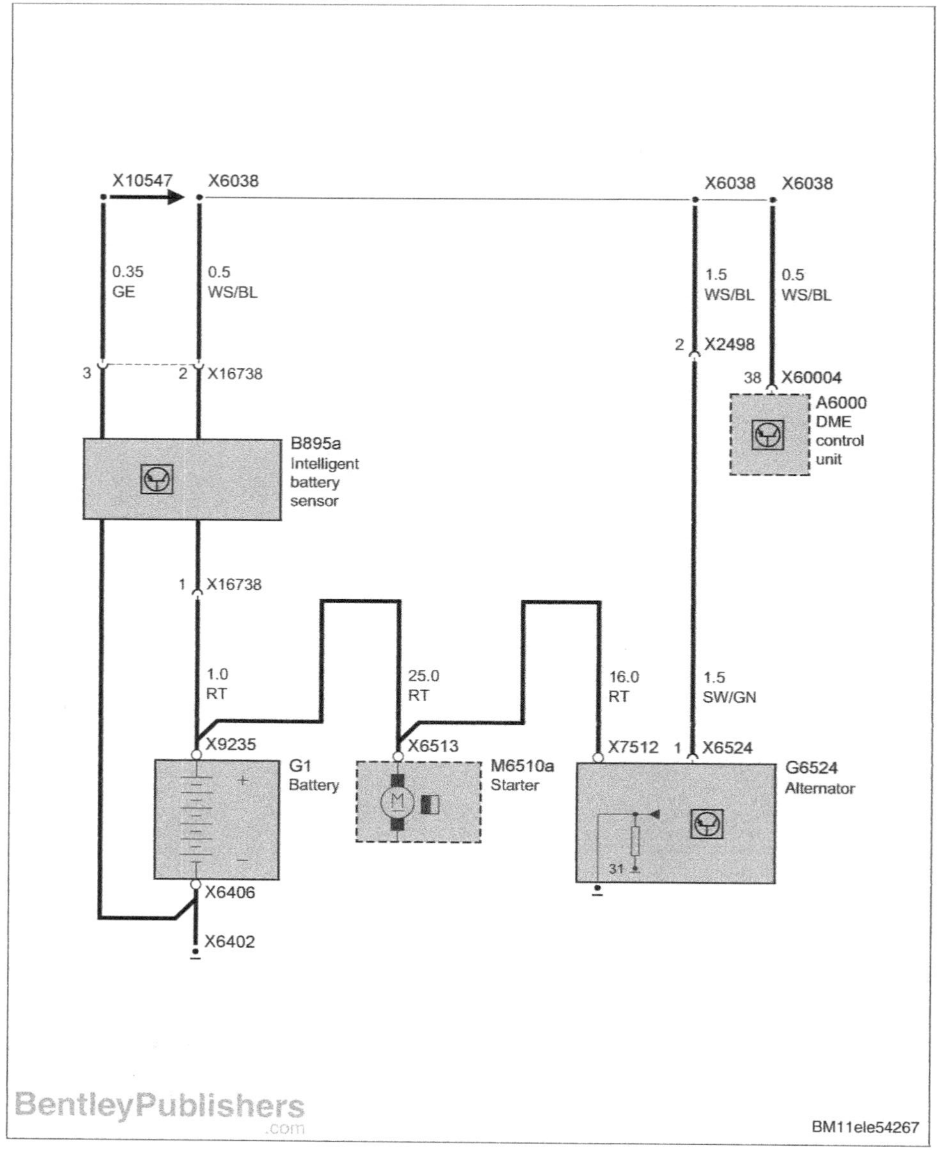

Auxiliary brake light, brake light switch, tail lights
N18 engine

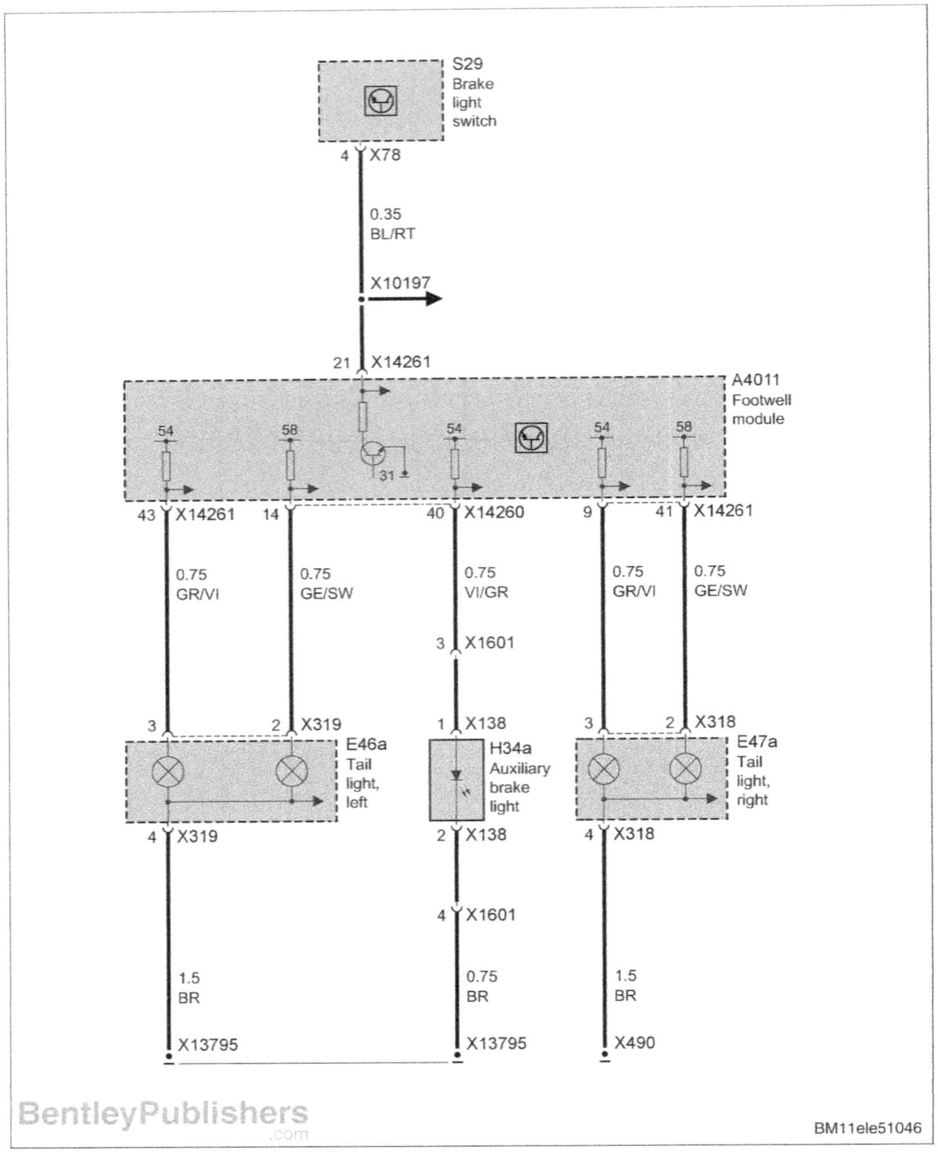

Fuel rail pressure sensor, volume control valve
N18 engine

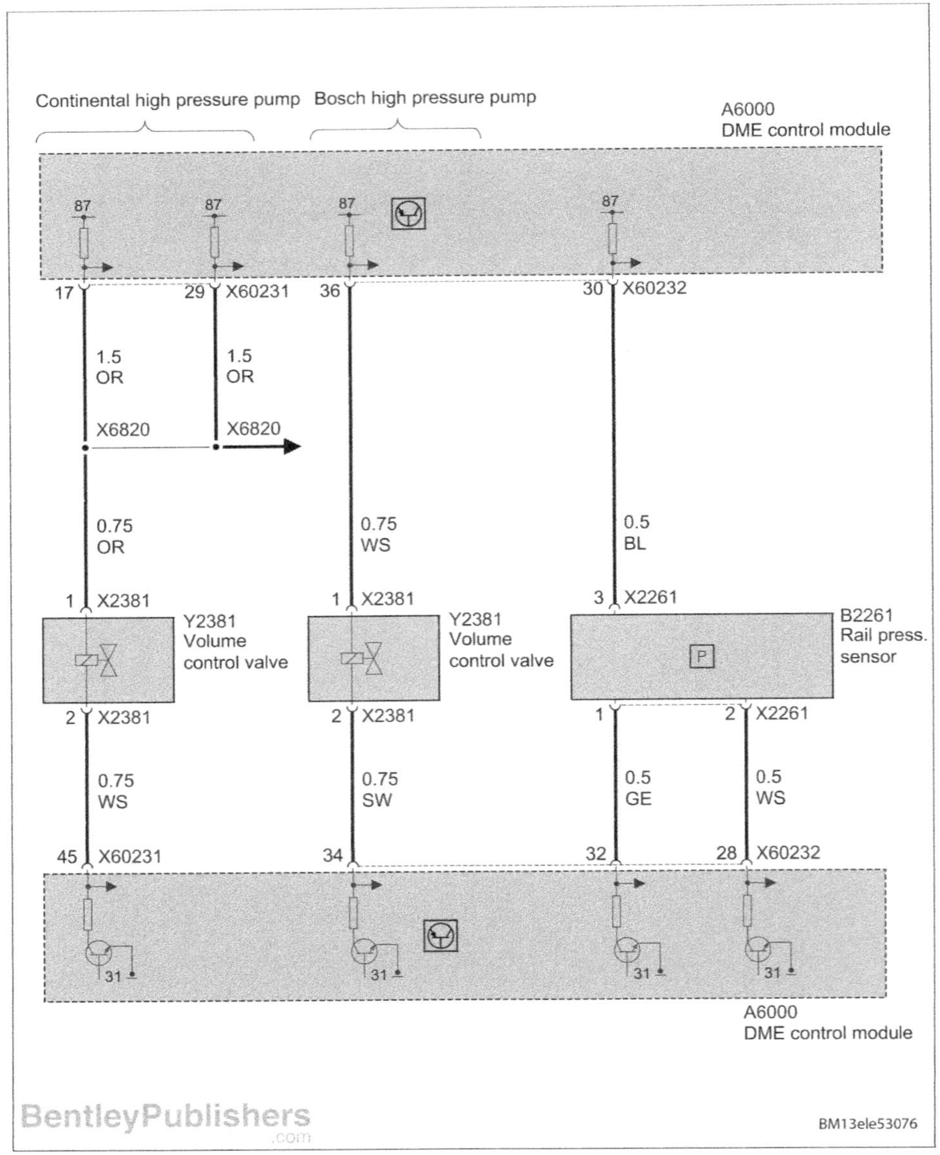

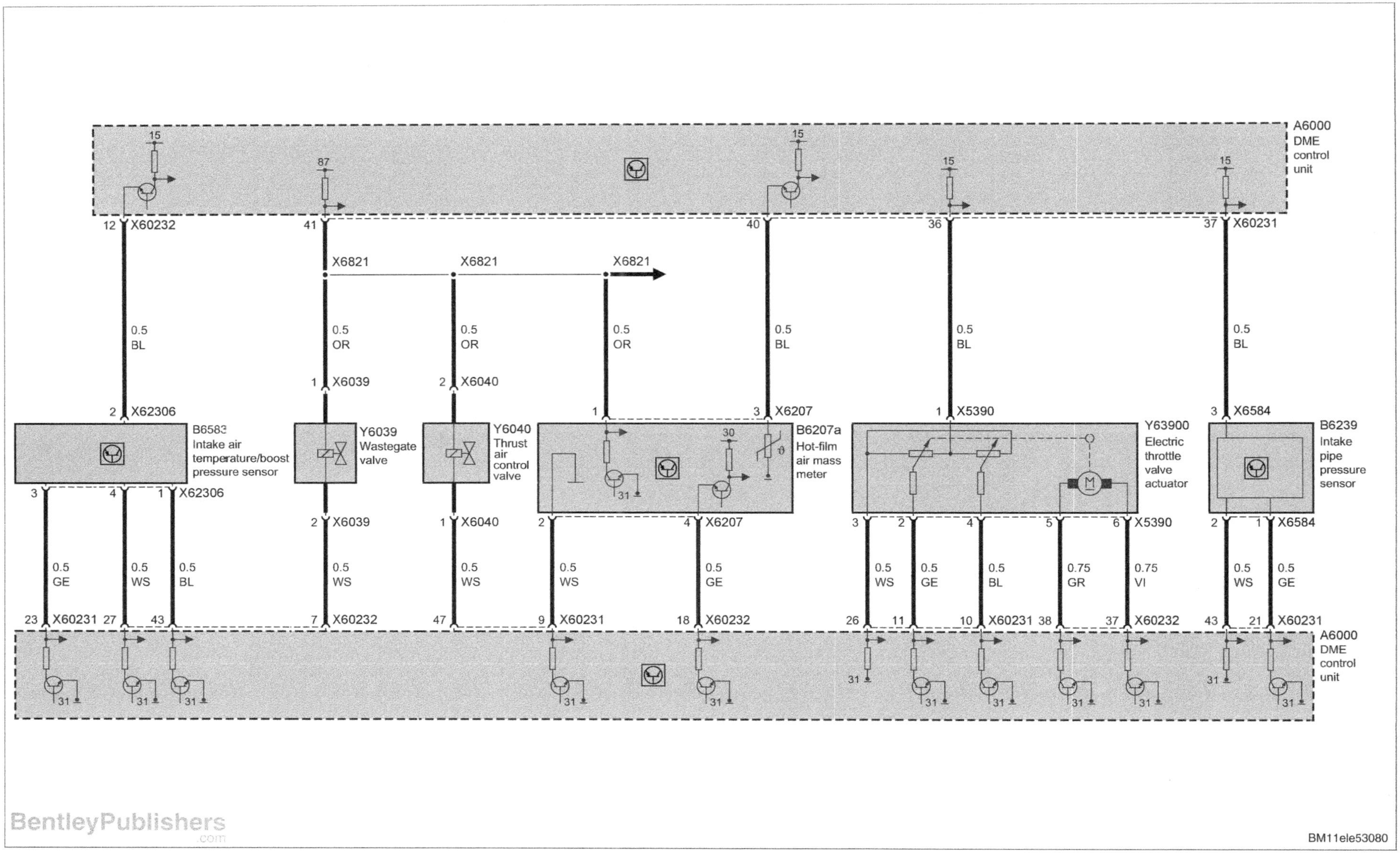
A6000
DME
control
unit
15
87
15
15
15
12 X60232
41
40
36
37 X60231
X6821
X6821
X6821
0.5
BL
0.5
OR
0.5
OR
0.5
OR
0.5
BL
0.5
BL
0.5
BL
1 X6039
2 X6040
2 X62306
1
3 X6207
1 X5390
3 X6584
B6583
Intake air
temperature/boost
pressure sensor
Y6039
Wastegate
valve
Y6040
Thrust
air
control
valve
30
B6207a
Hot-film
air mass
meter
Y63900
Electric
throttle
valve
actuator
B6239
Intake
pipe
pressure
sensor
3
4
1 X62306
2 X6039
1 X6040
2
4 X6207
3
2
4
5
6 X5390
2
1 X6584
0.5
GE
0.5
WS
0.5
BL
0.5
WS
0.5
WS
0.5
WS
0.5
GE
0.5
WS
0.5
GE
0.5
BL
0.75
GR
0.75
VI
0.5
WS
0.5
GE
23 X60231 27
43
7 X60232
47
9 X60231
18 X60232
26
11
10 X60231 38
37 X60232
43
21 X60231
A6000
DME
control
unit
31
BM11ele53080

Crankshaft sensor
N18 engine

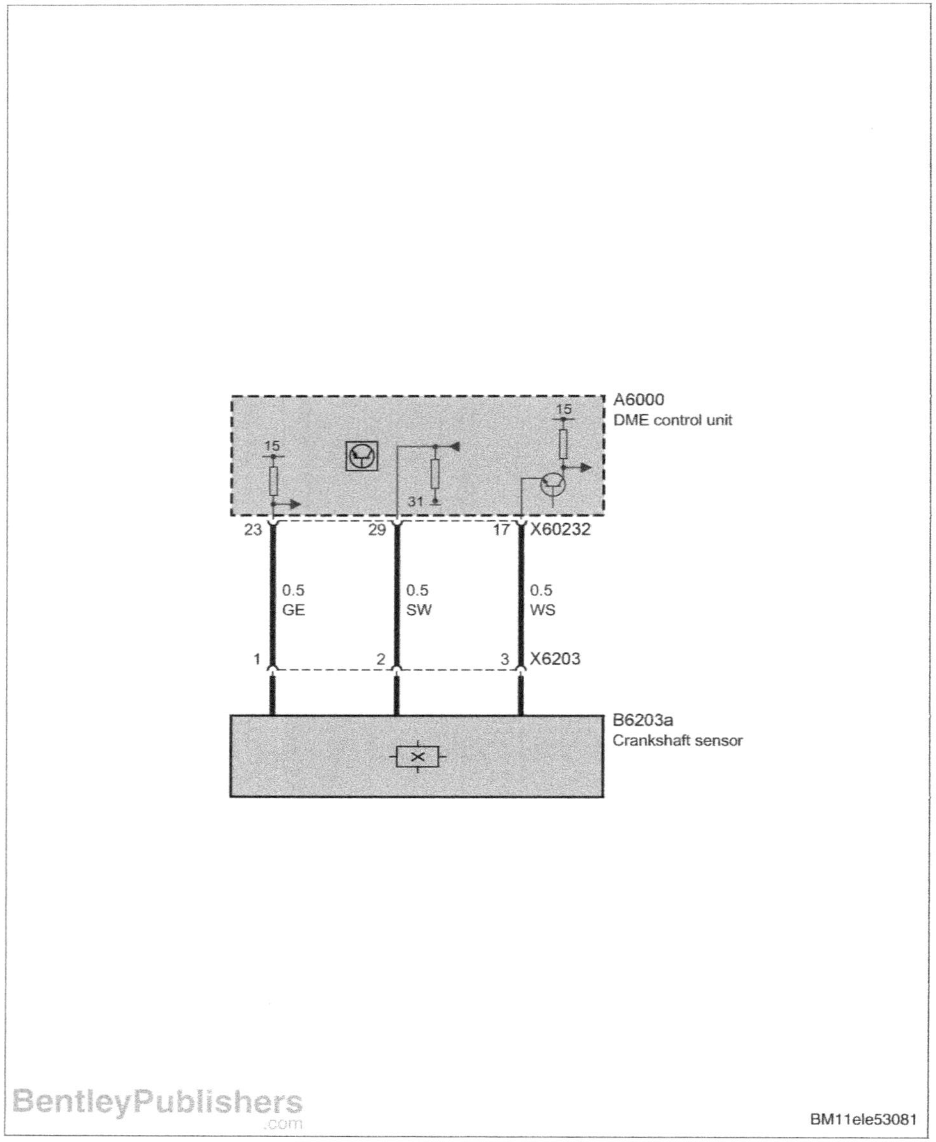

Engine breather heater, oil pressure control valve, oil pressure sensor
N18 engine

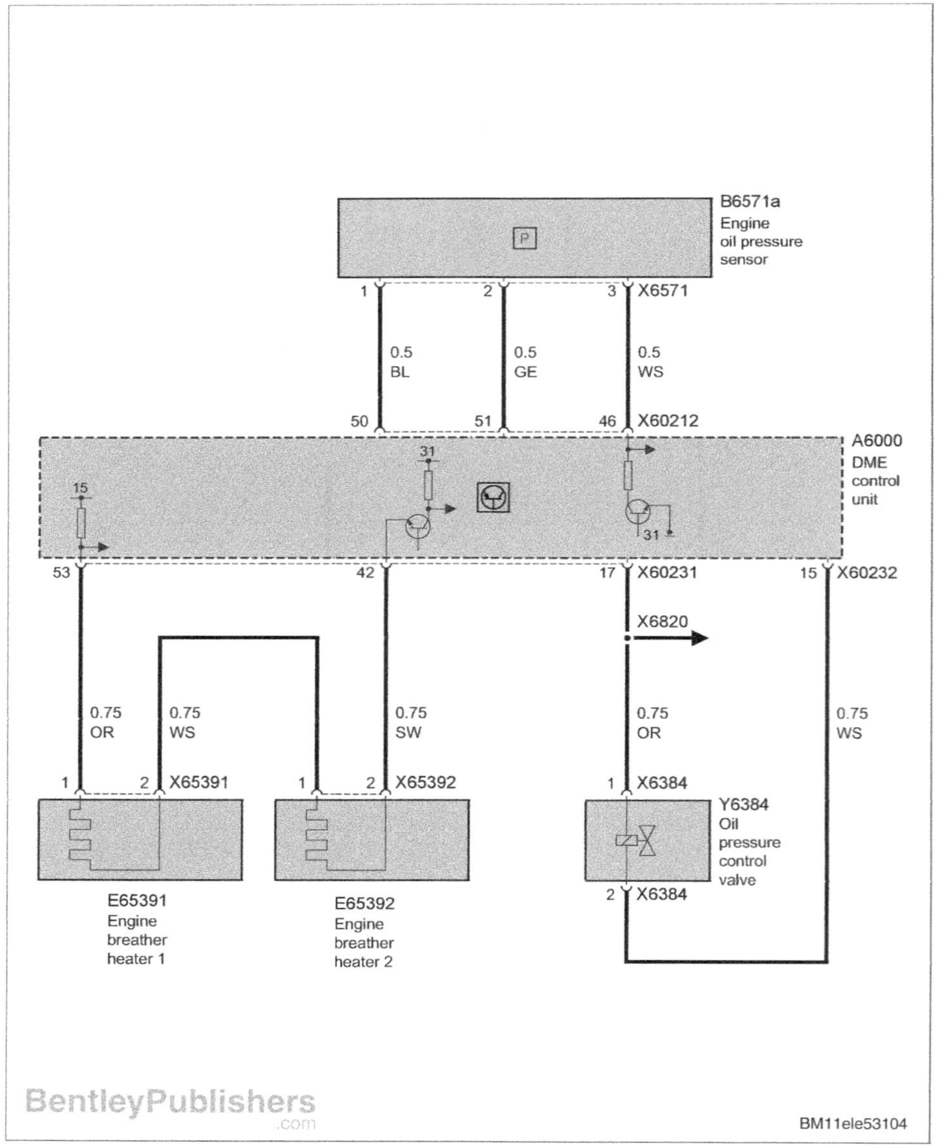

Thermostat, electric fan, turbocharger coolant pump
N18 engine

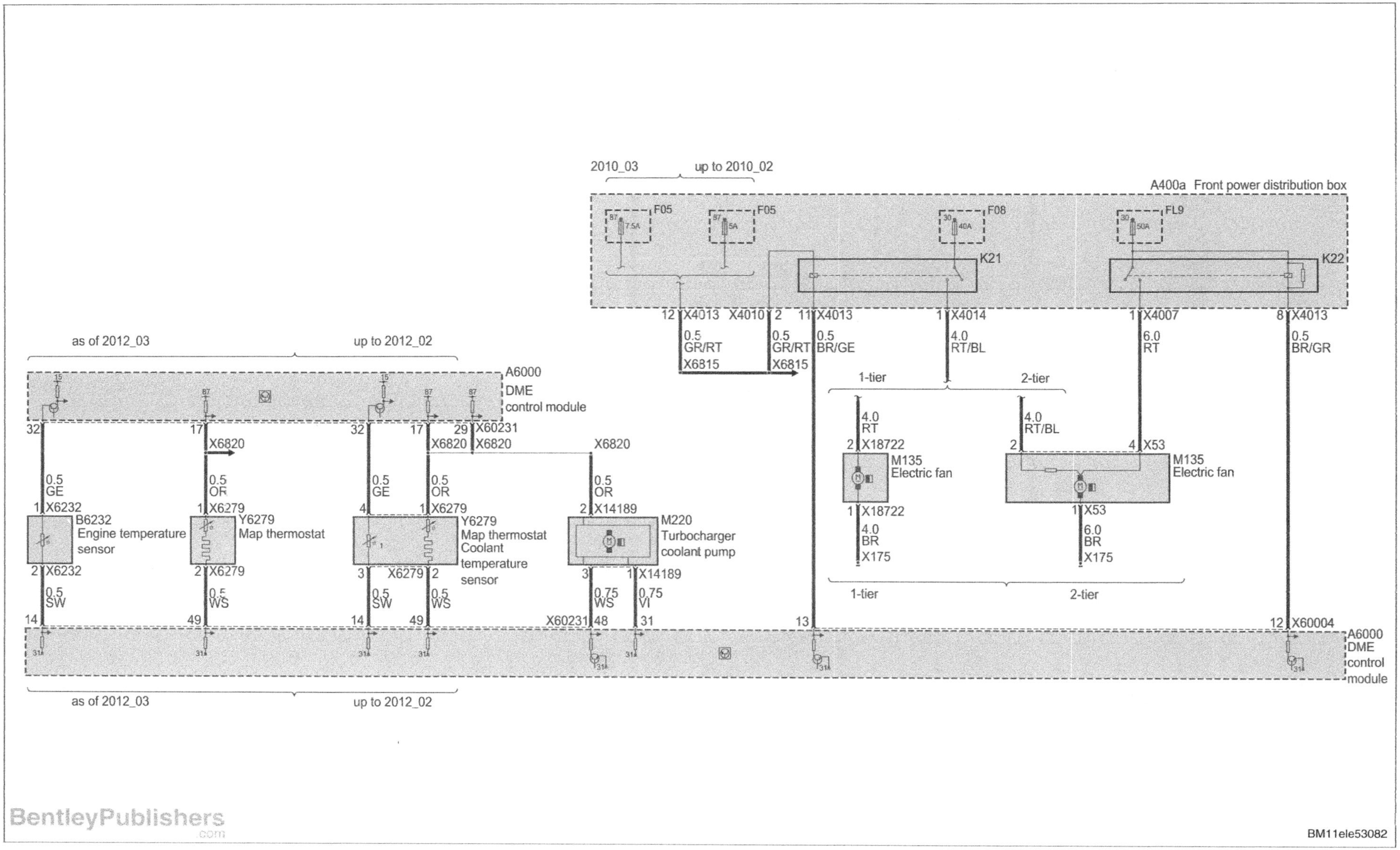

Camshaft sensors, VANOS solenoid valves, Valvetronic relay, Valvetronic servomotor, N18 engine
page 1 of 2

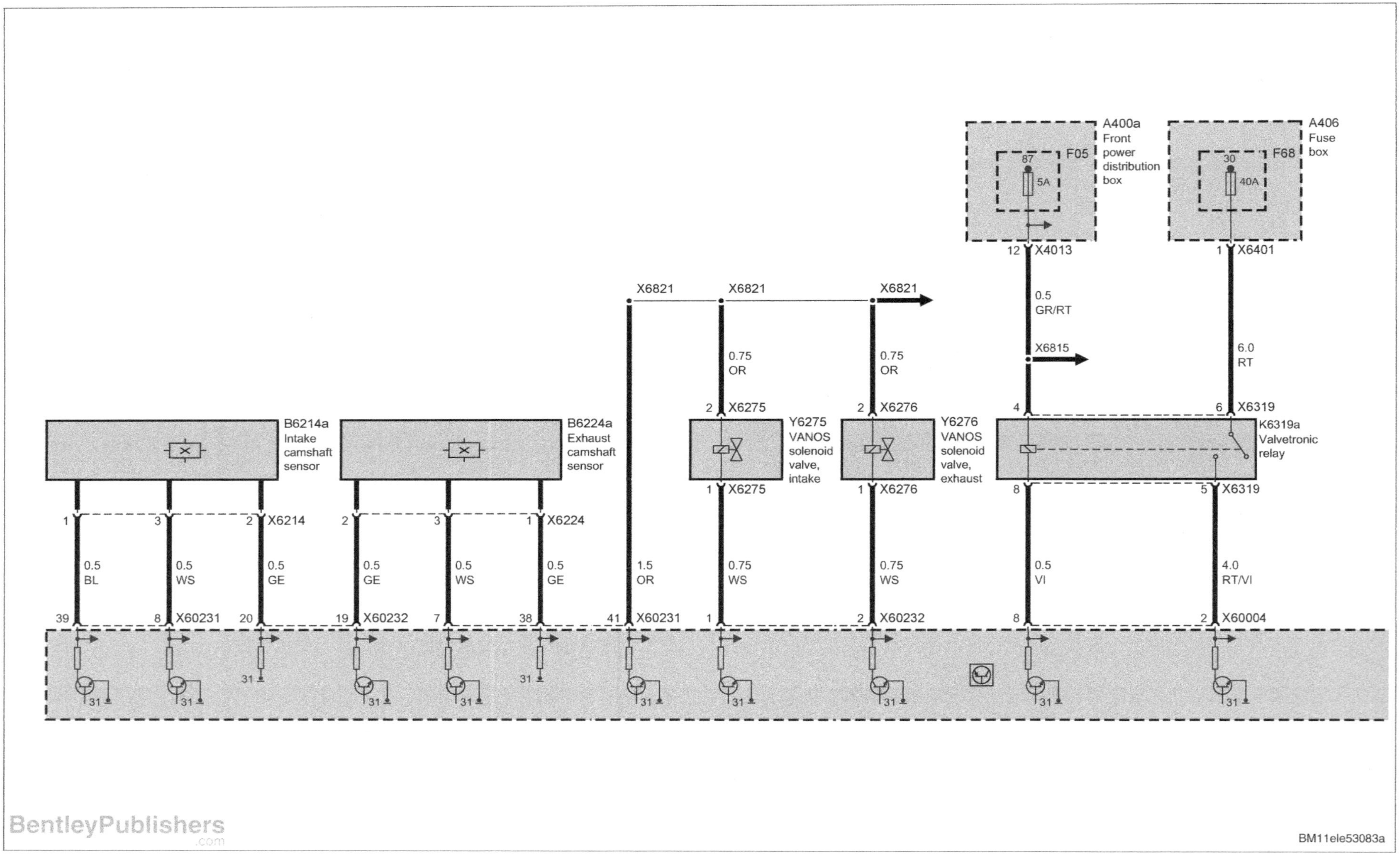

BM11ele53083a

Camshaft sensors, VANOS solenoid valves, Valvetronic relay, Valvetronic servomotor, N18 engine
page 2 of 2

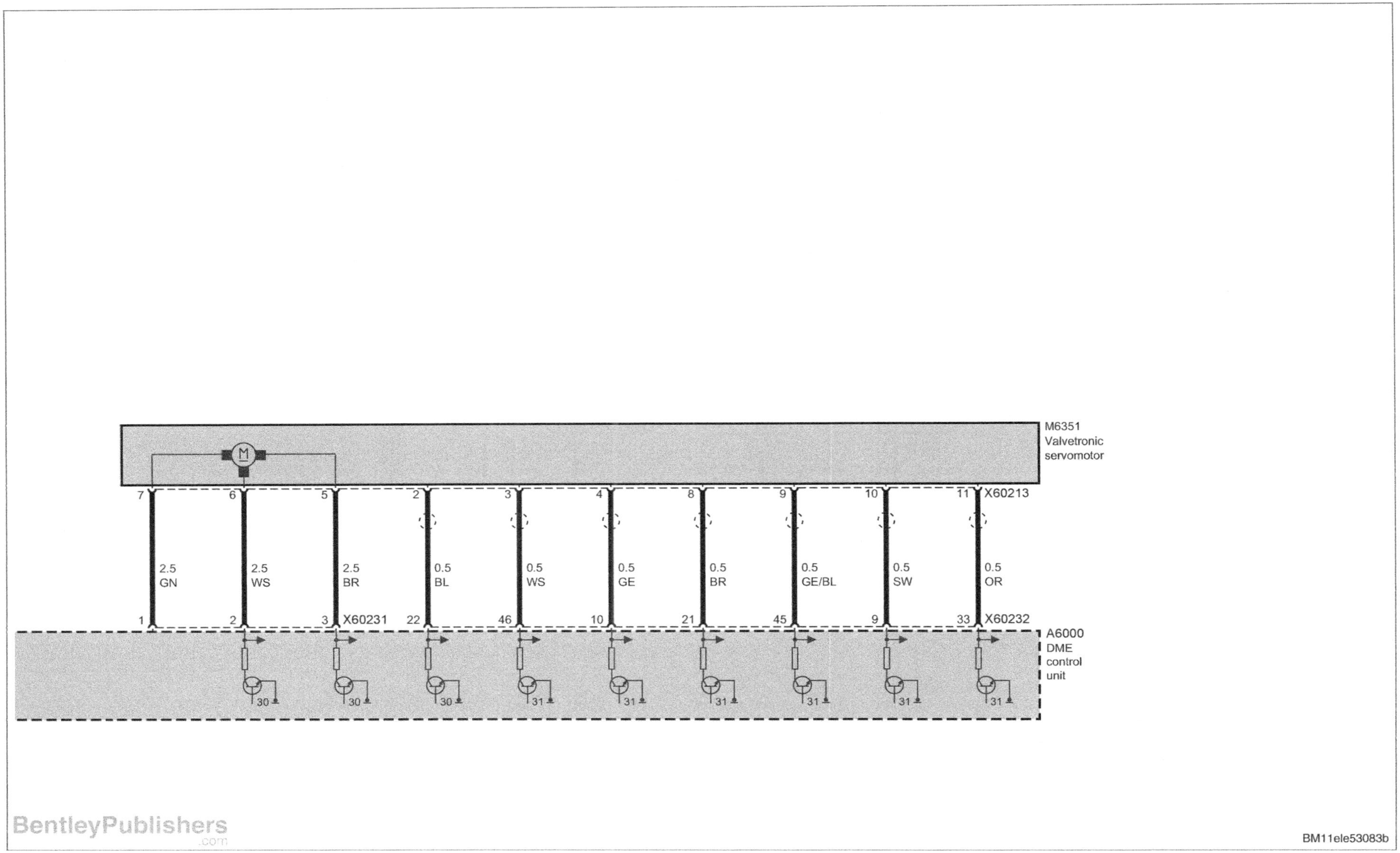

Ignition coils
N18 engine

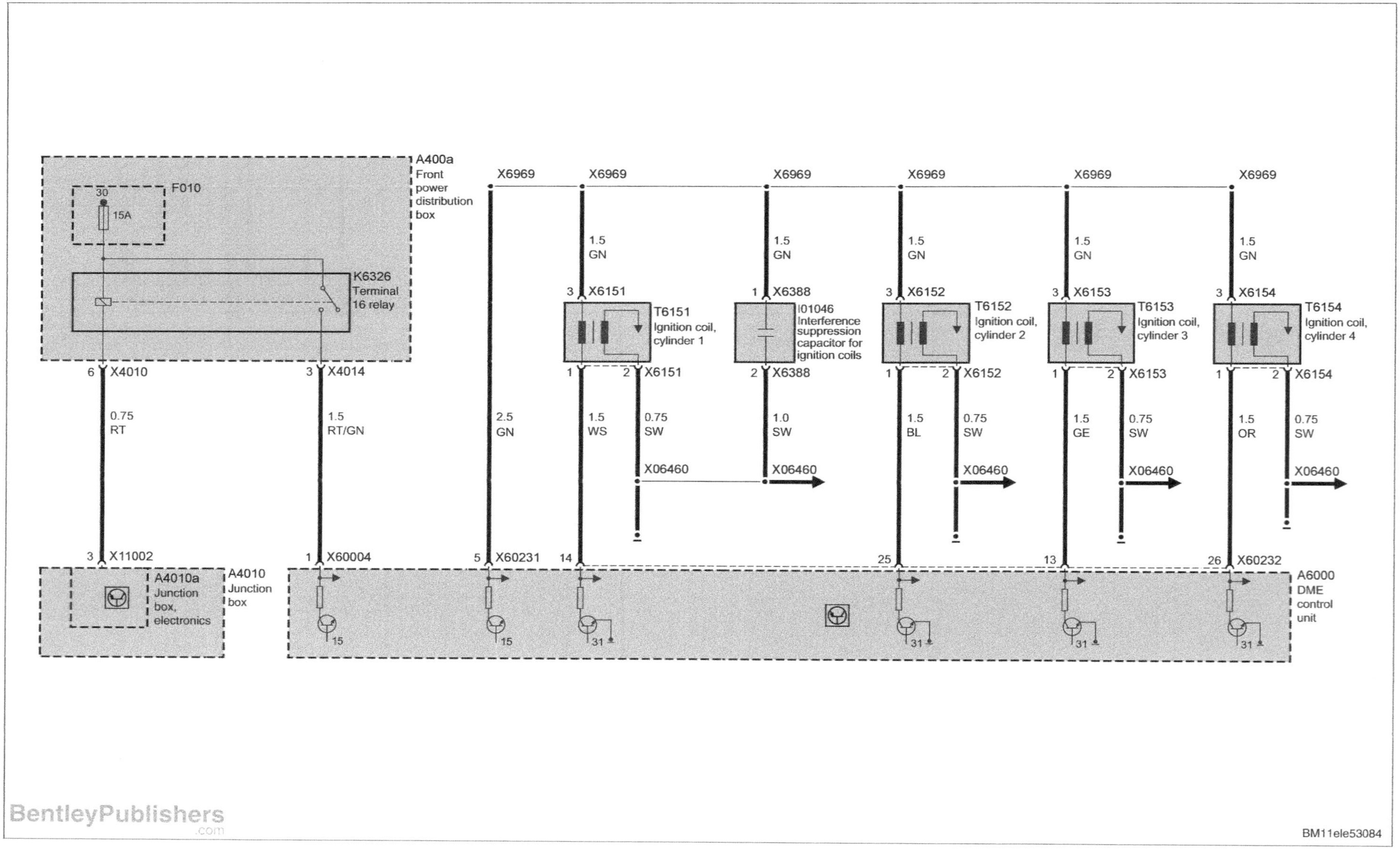

Fuse F01

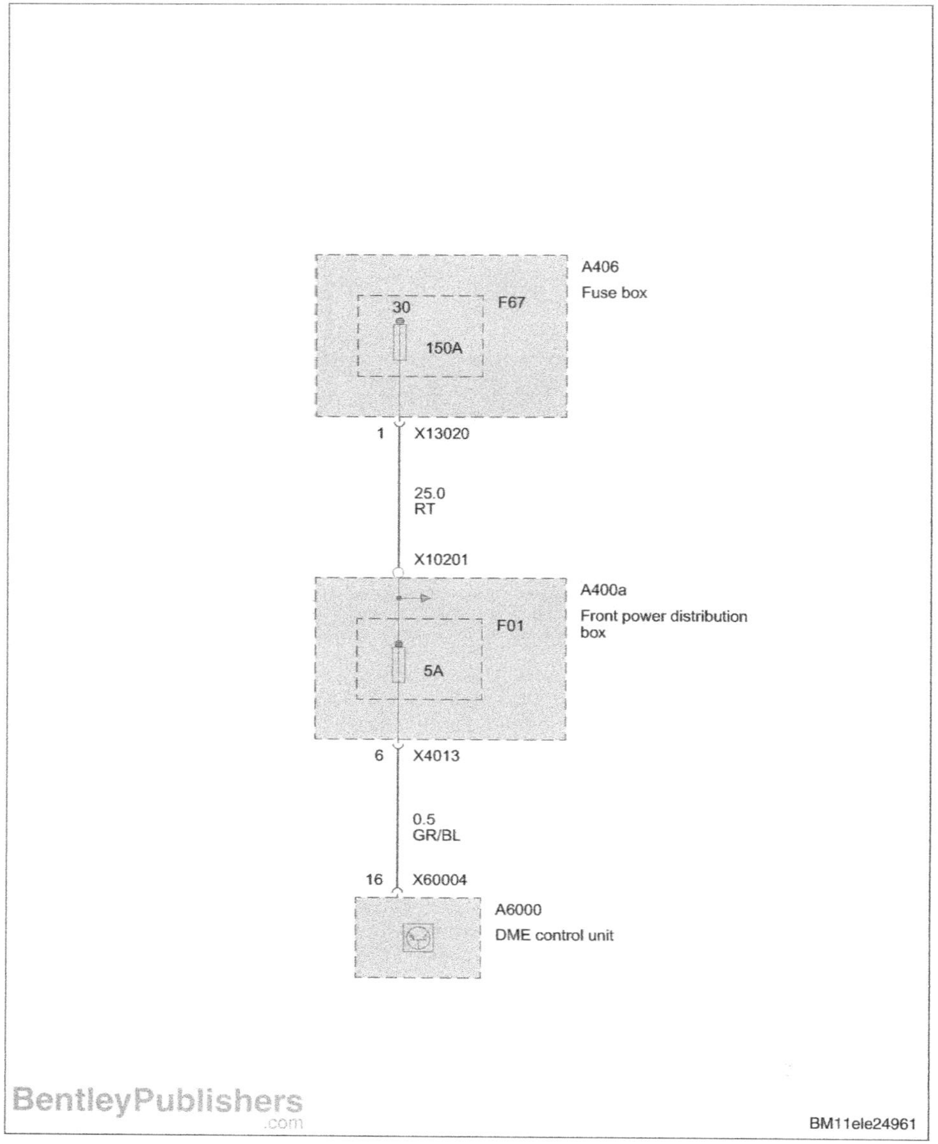

Fuse F02

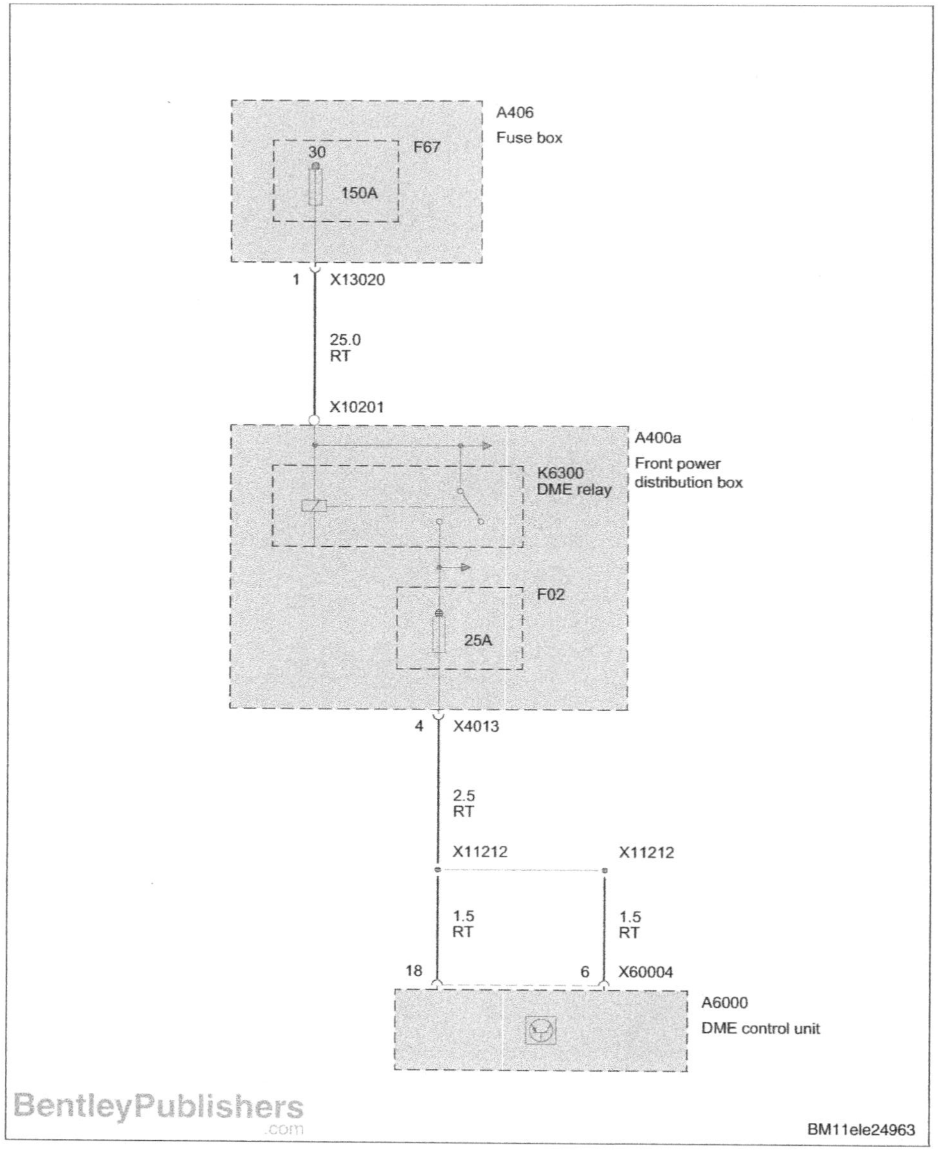

Fuse F03

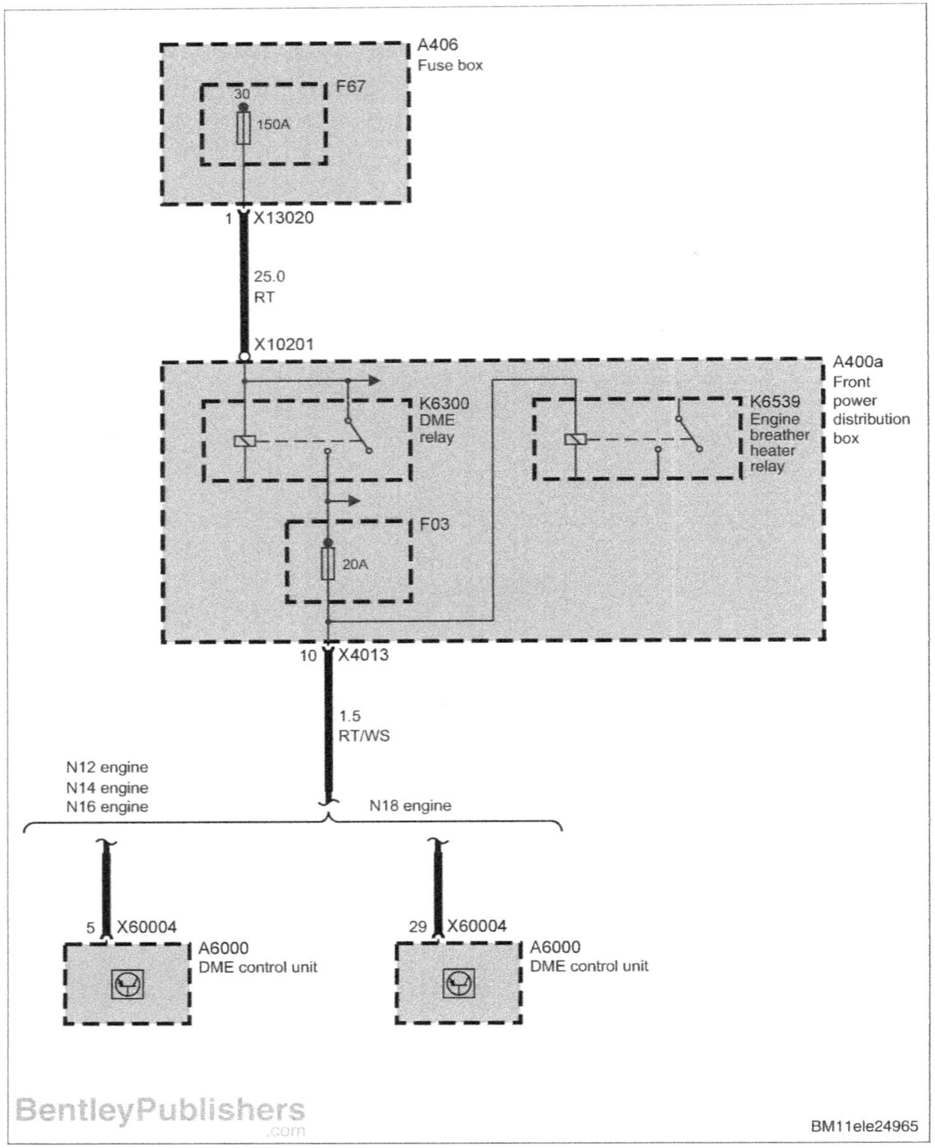

Fuse F04

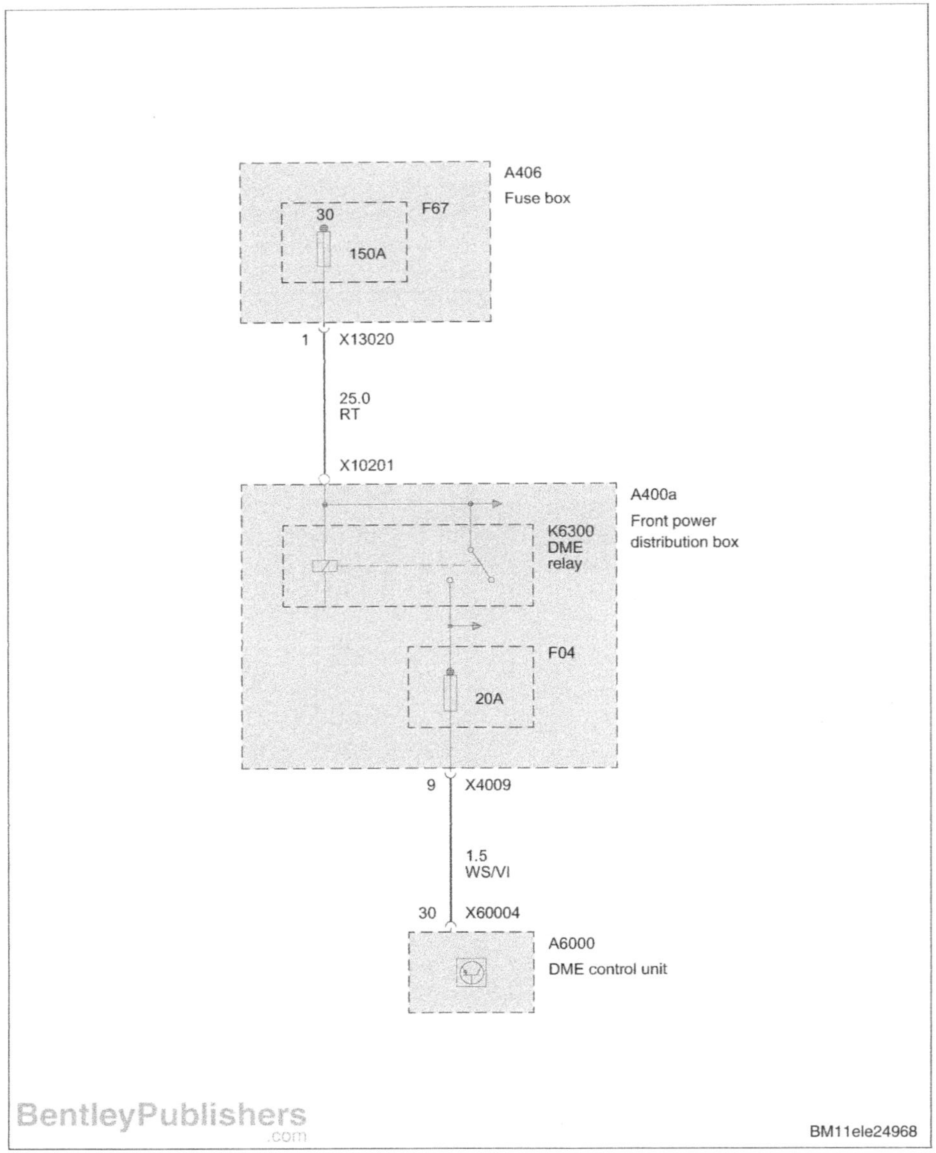

Fuse F05
except N18 engine

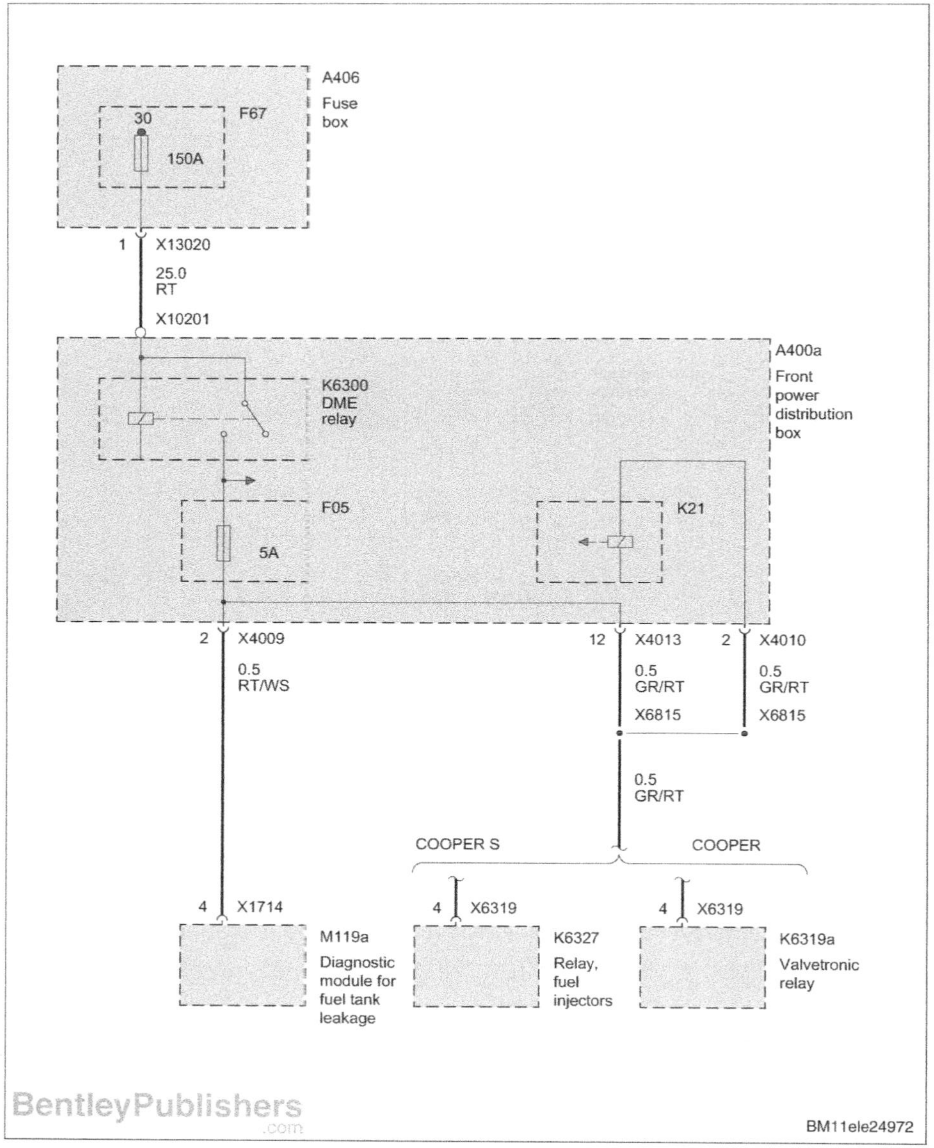

Fuse F05
N18 engine

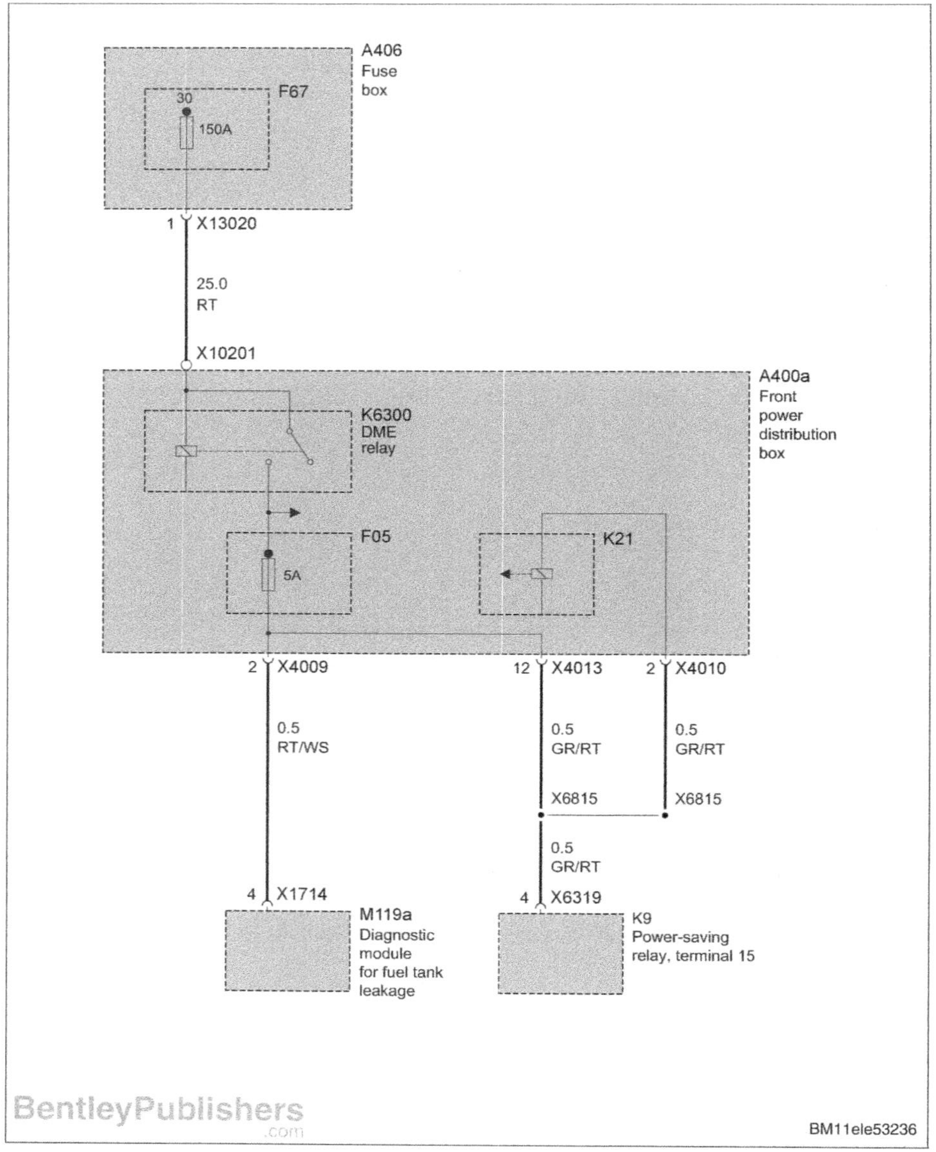

Fuse F06

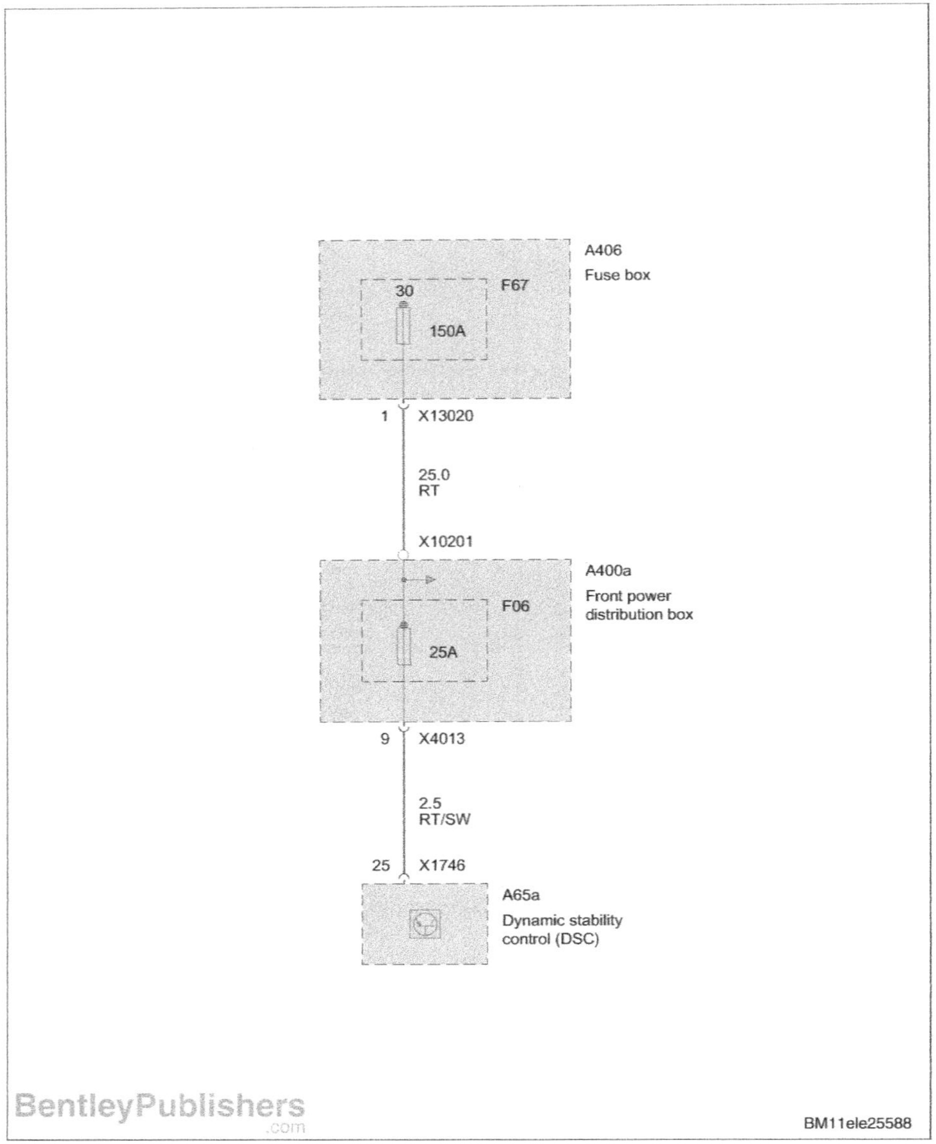

Fuse F07

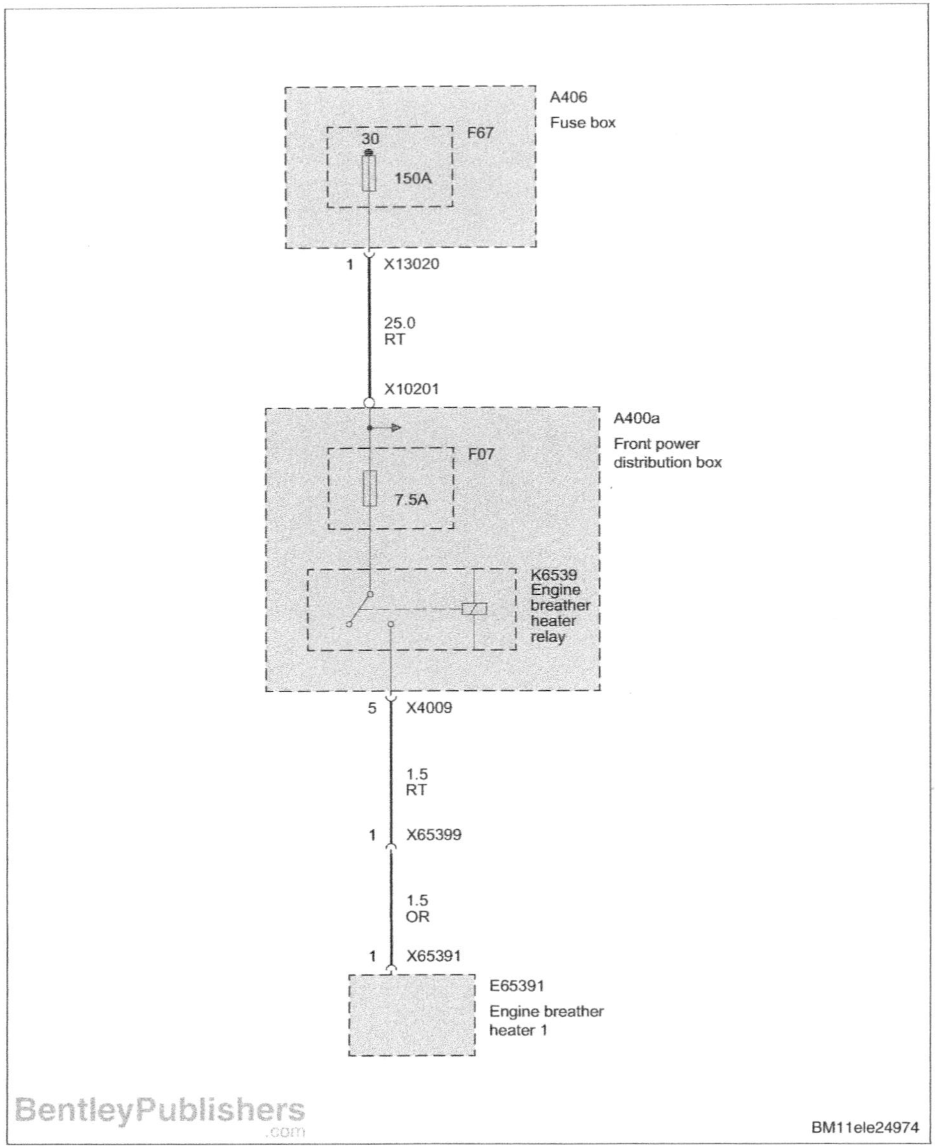

Fuse F08

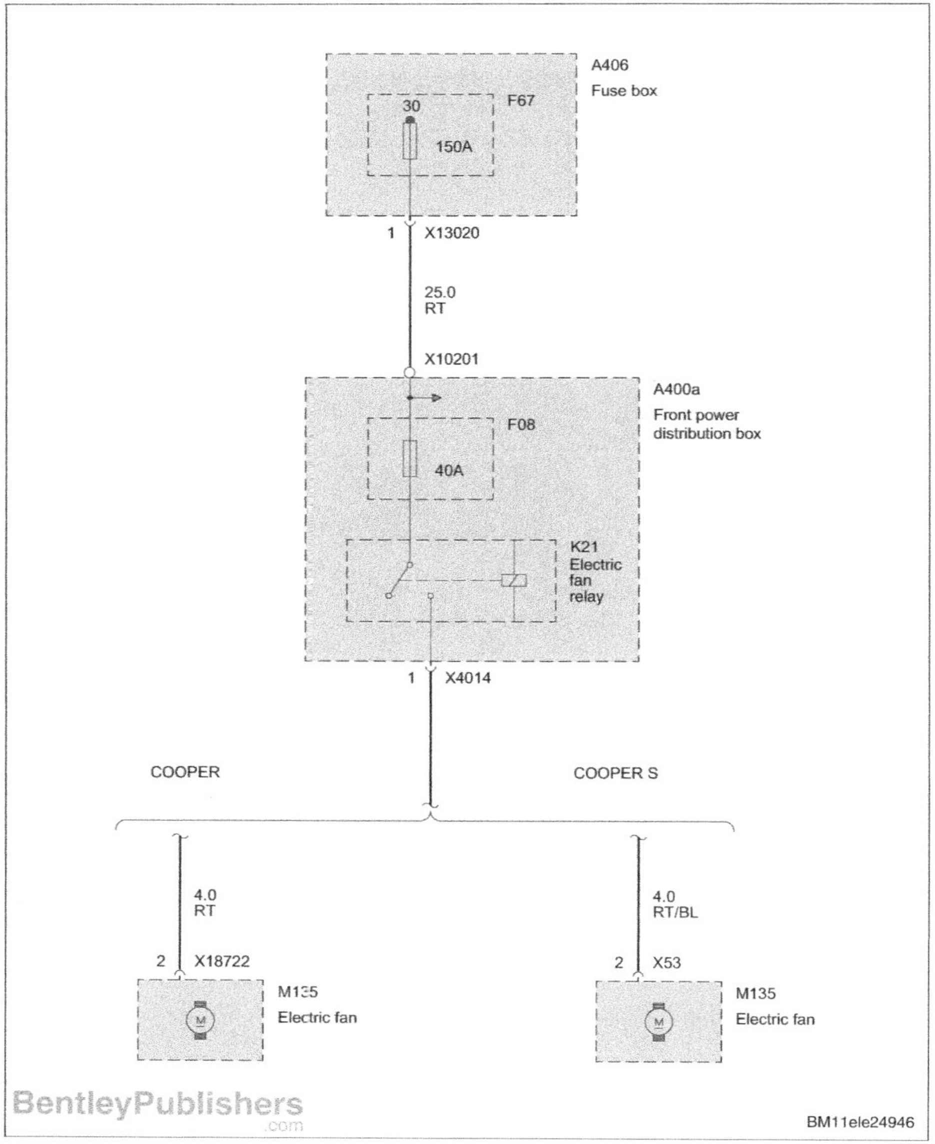

Fuse F09

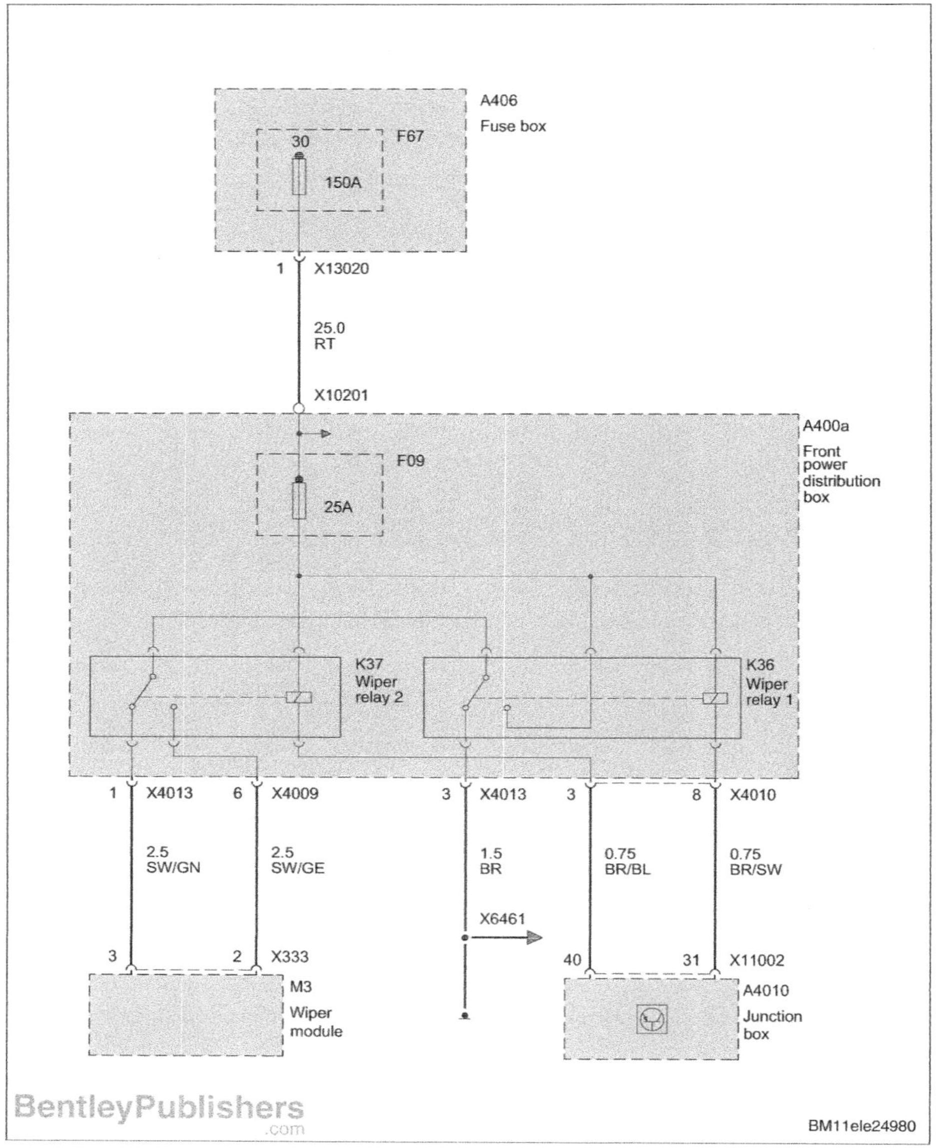

Fuse F010

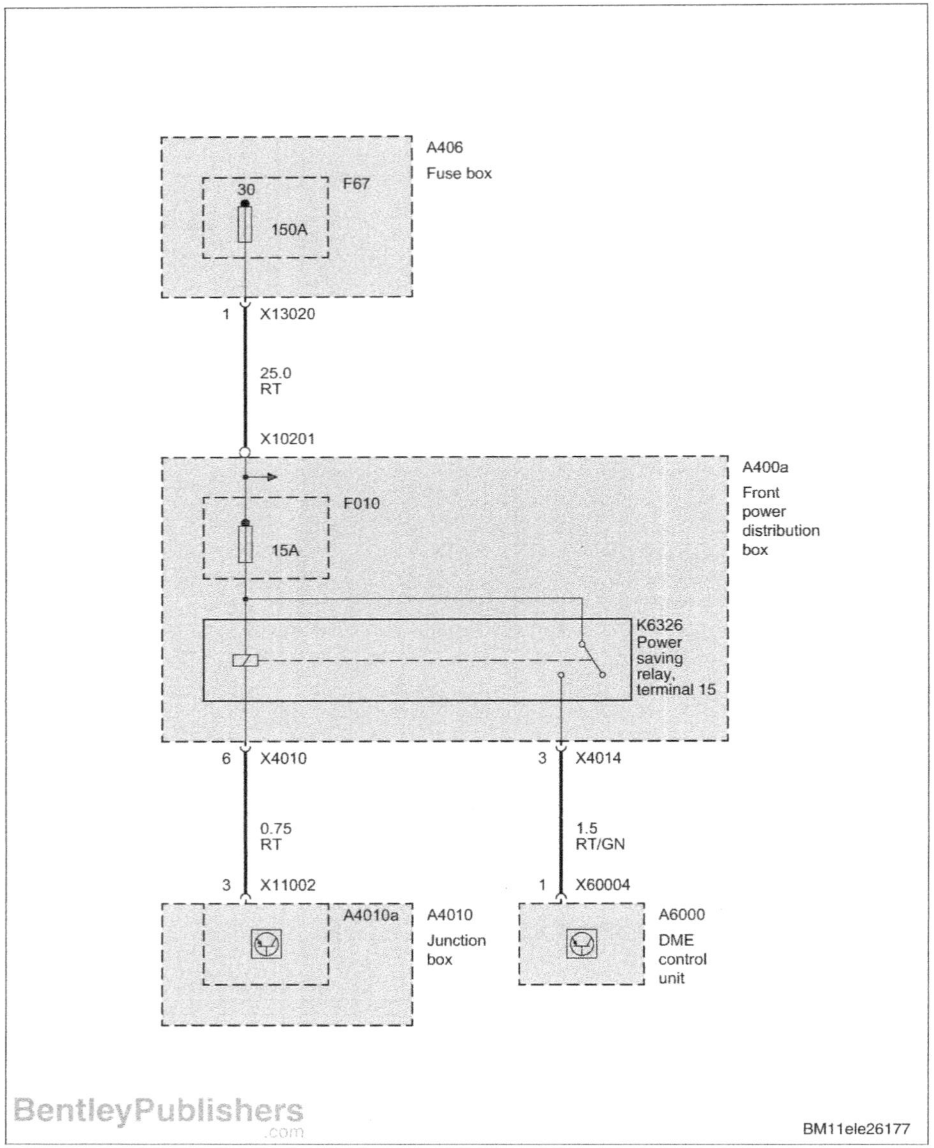

Fuse F2 to 11/2008

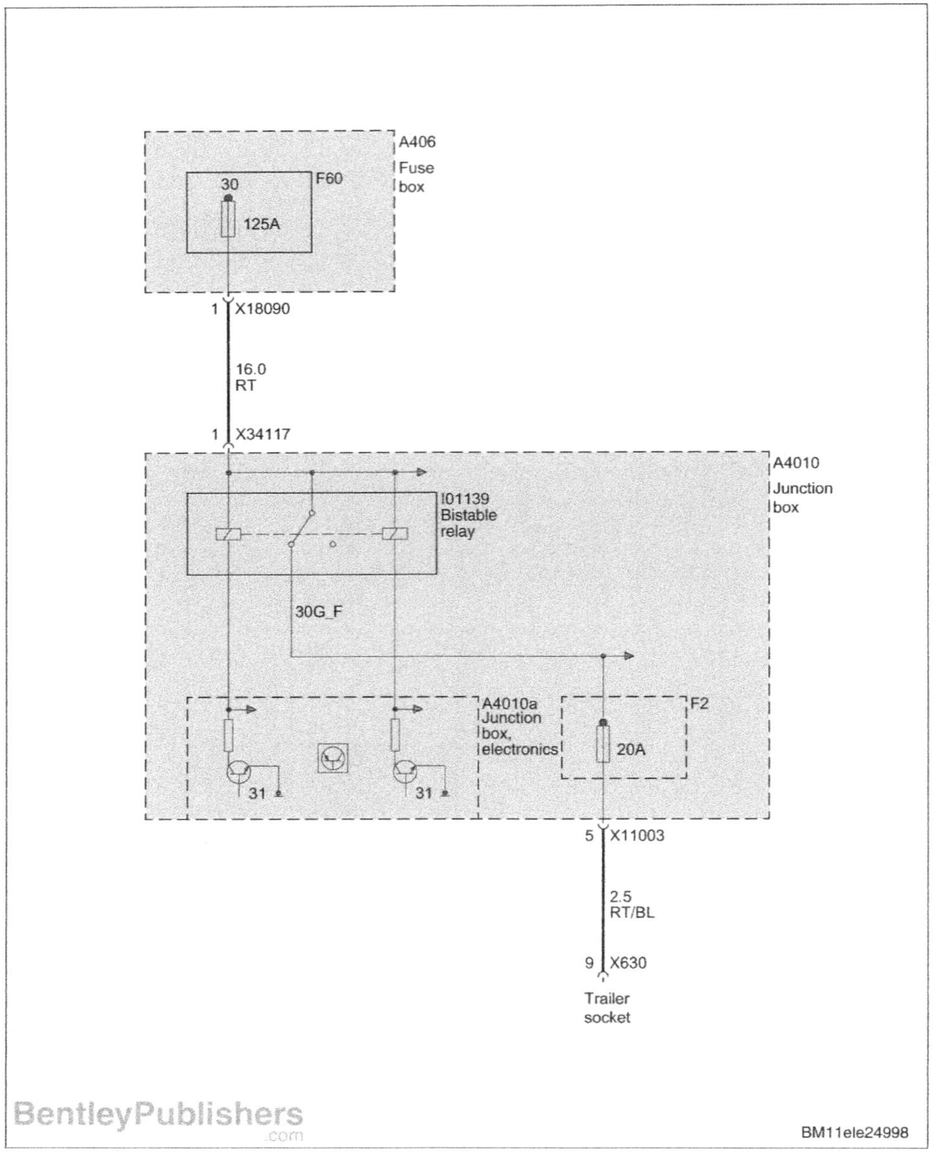

Fuse F2 from 12/2008

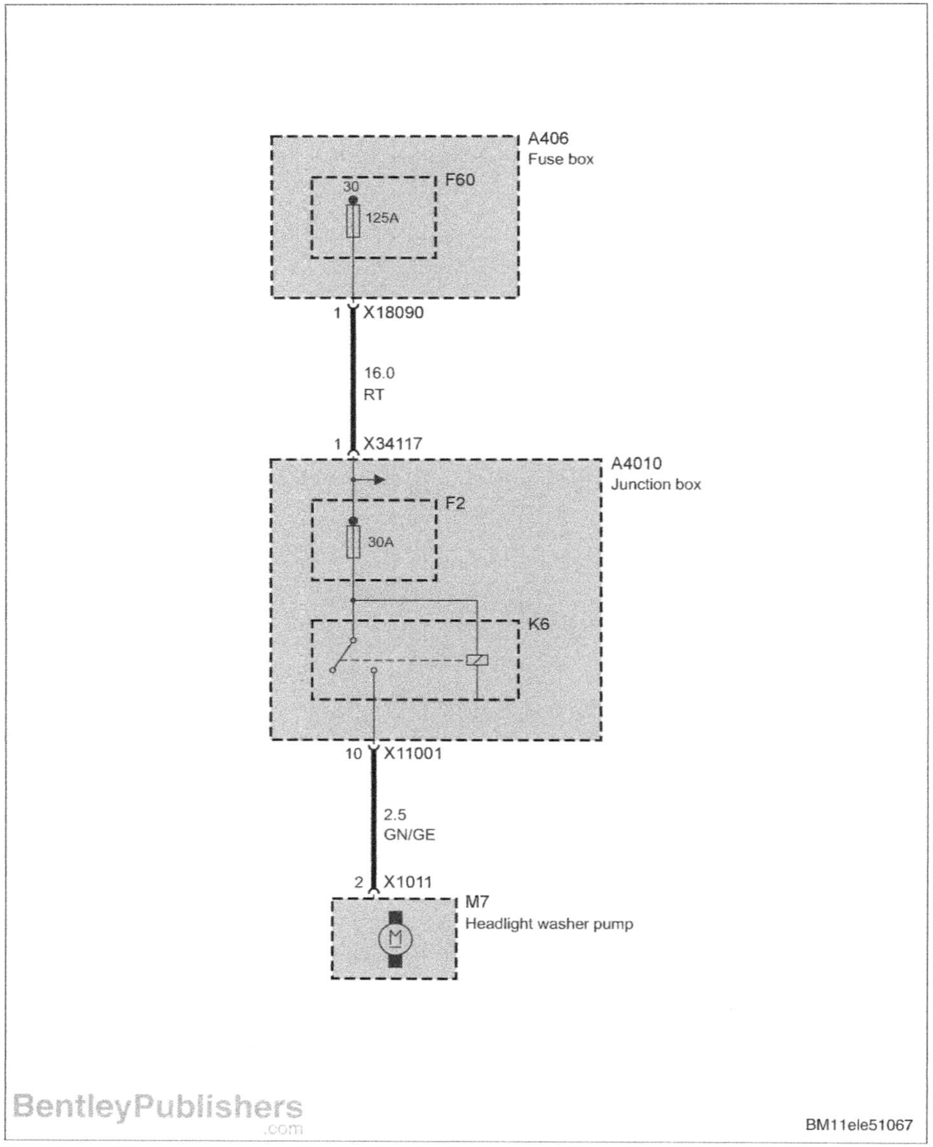

Fuse F3

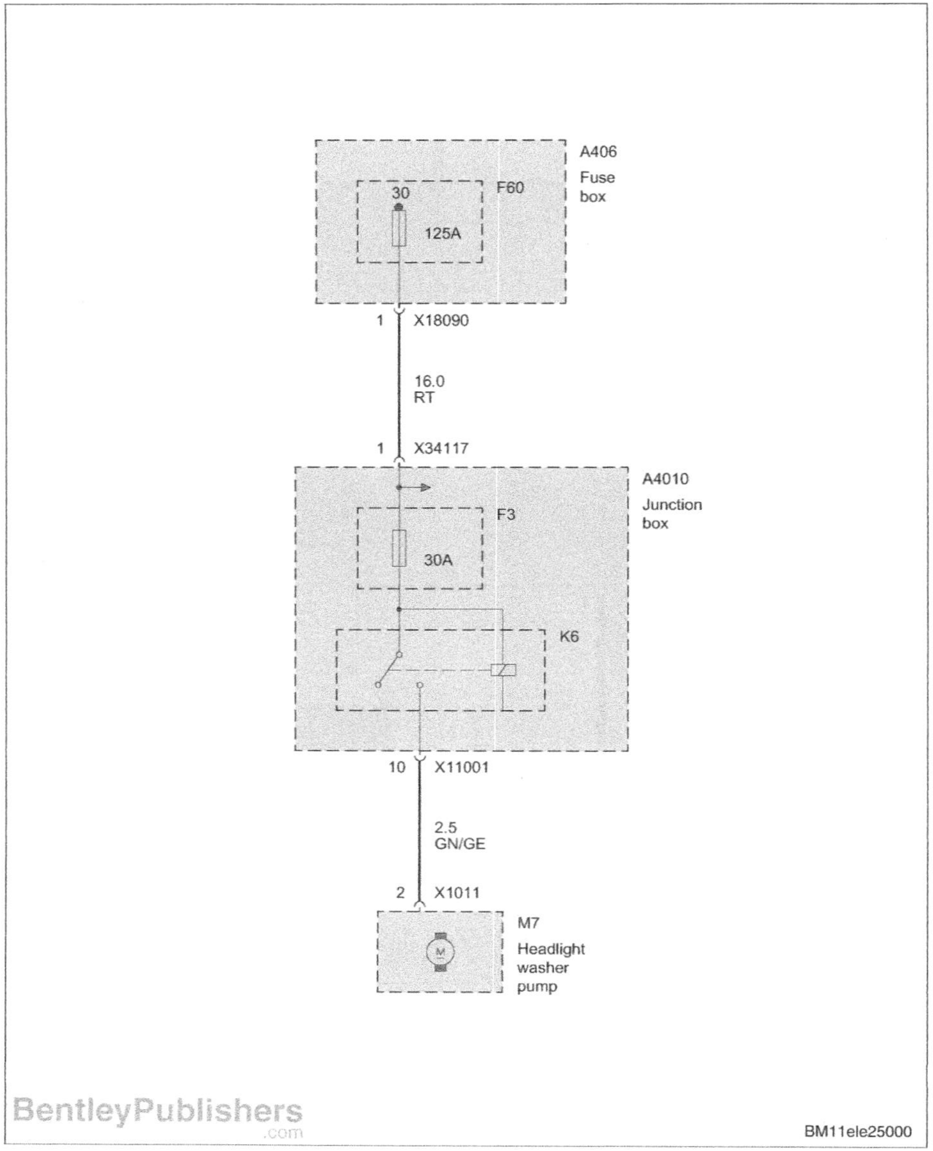

Fuse F5 to 11/2008

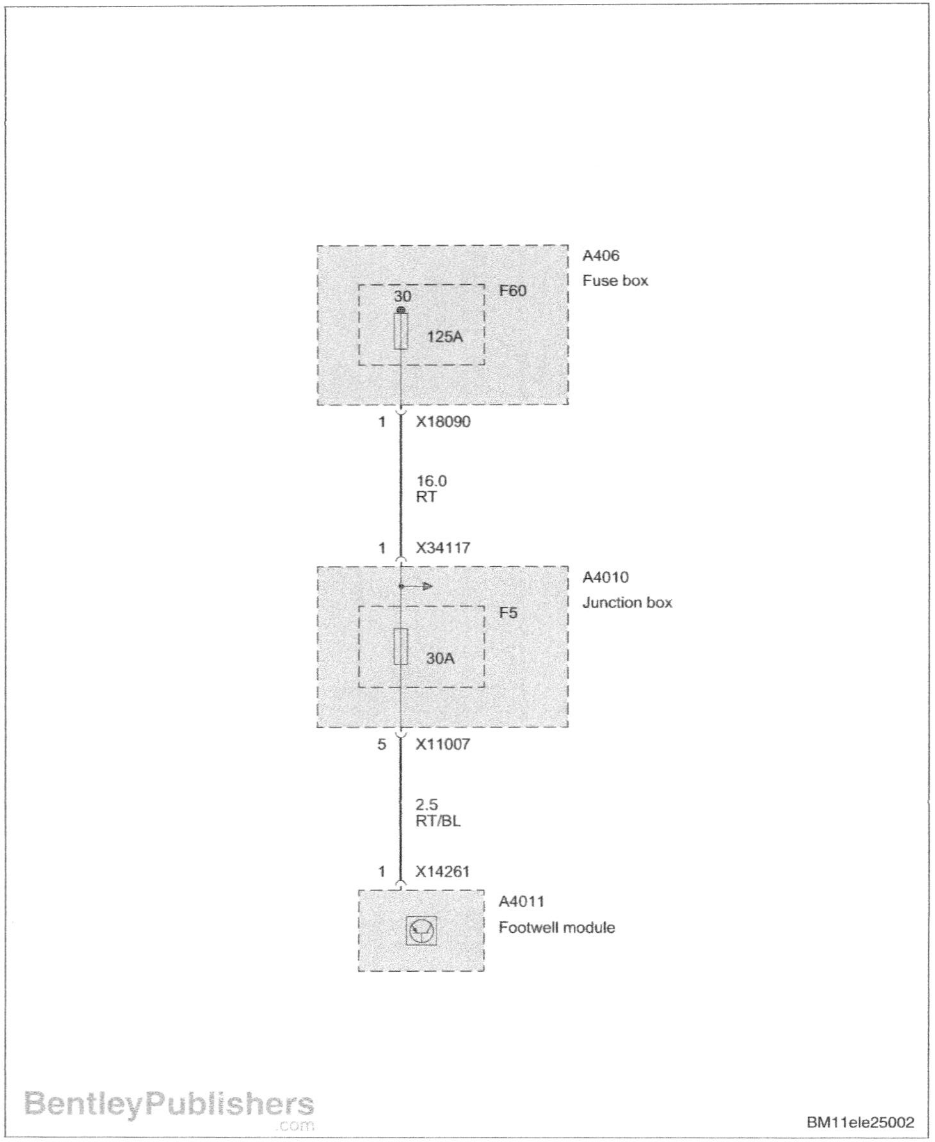

Fuse F5 from 12/2008

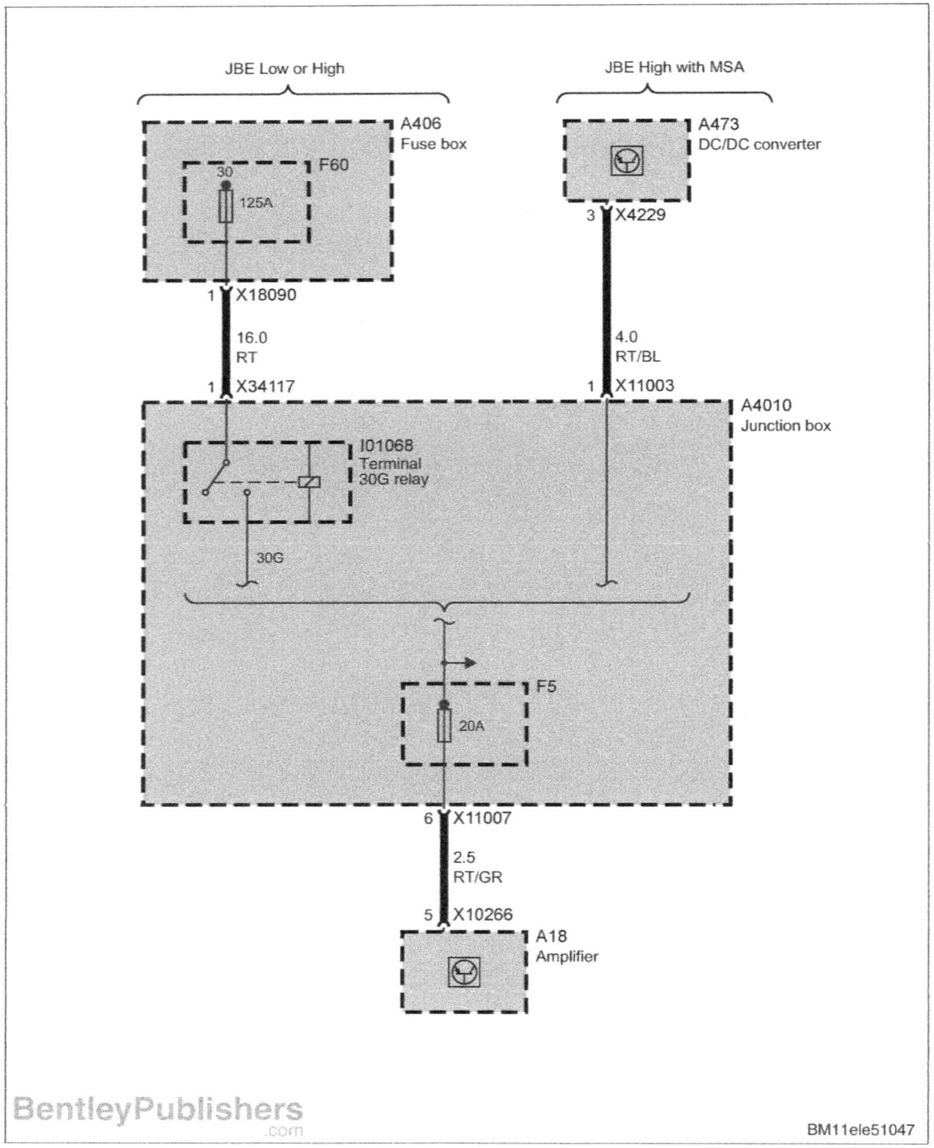

Fuse F6
to 11/2008

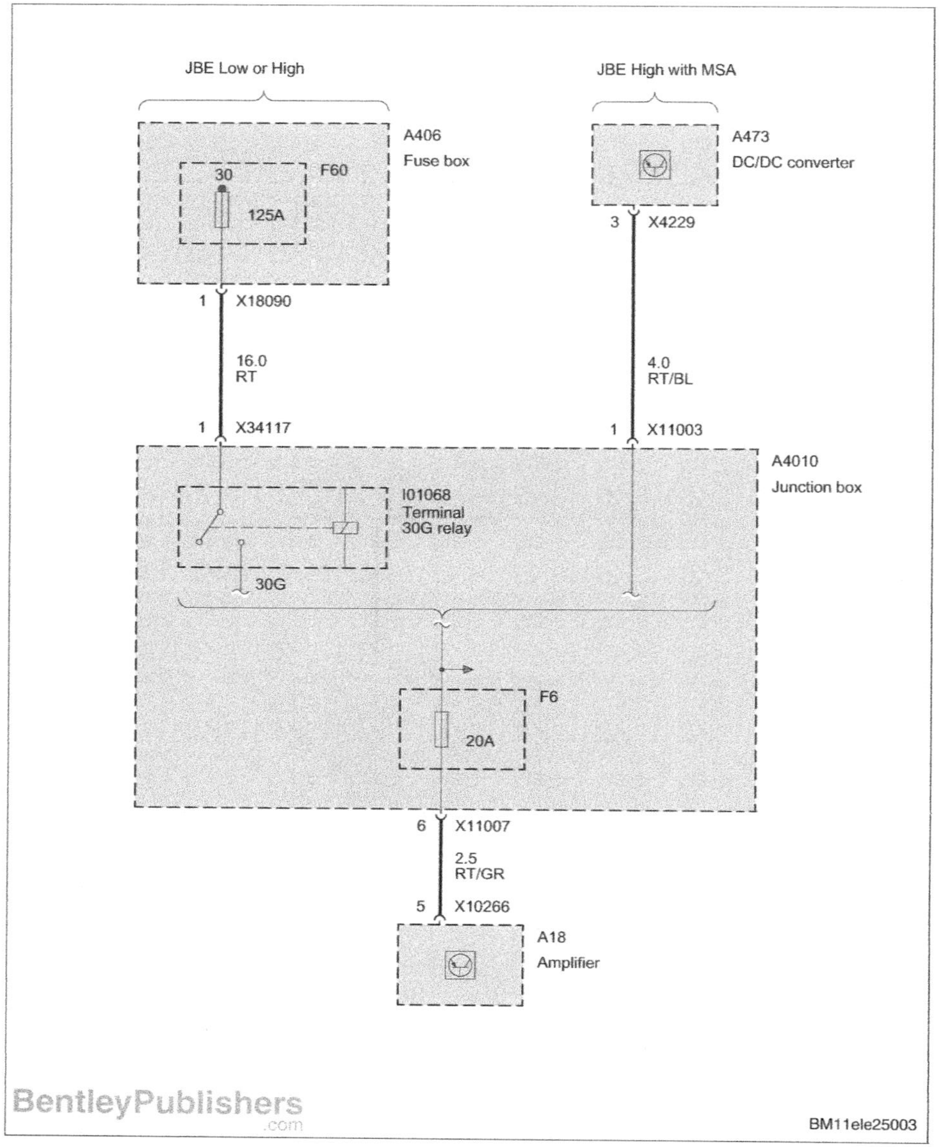

Fuse F6
from 12/2008, except Convertible

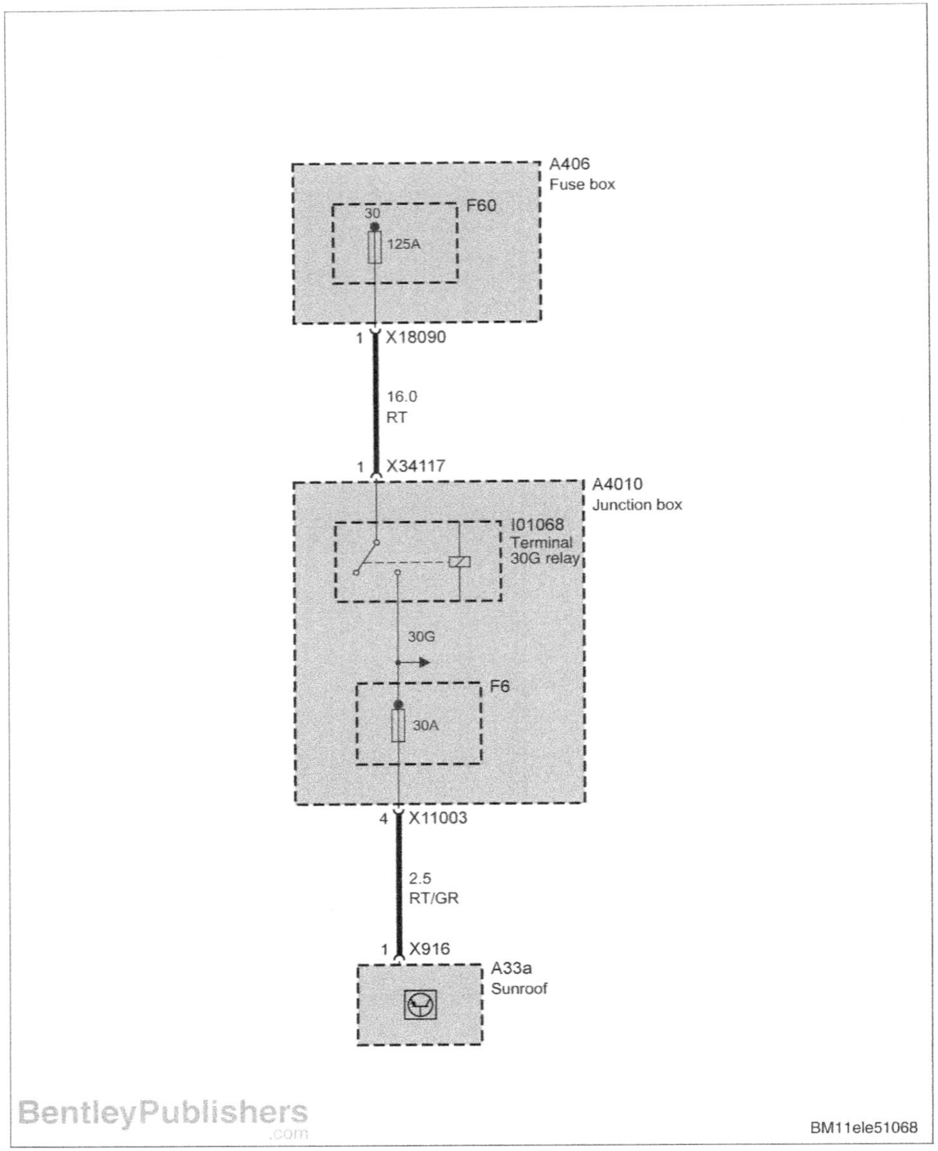

Fuse F6
from 12/2008, Convertible

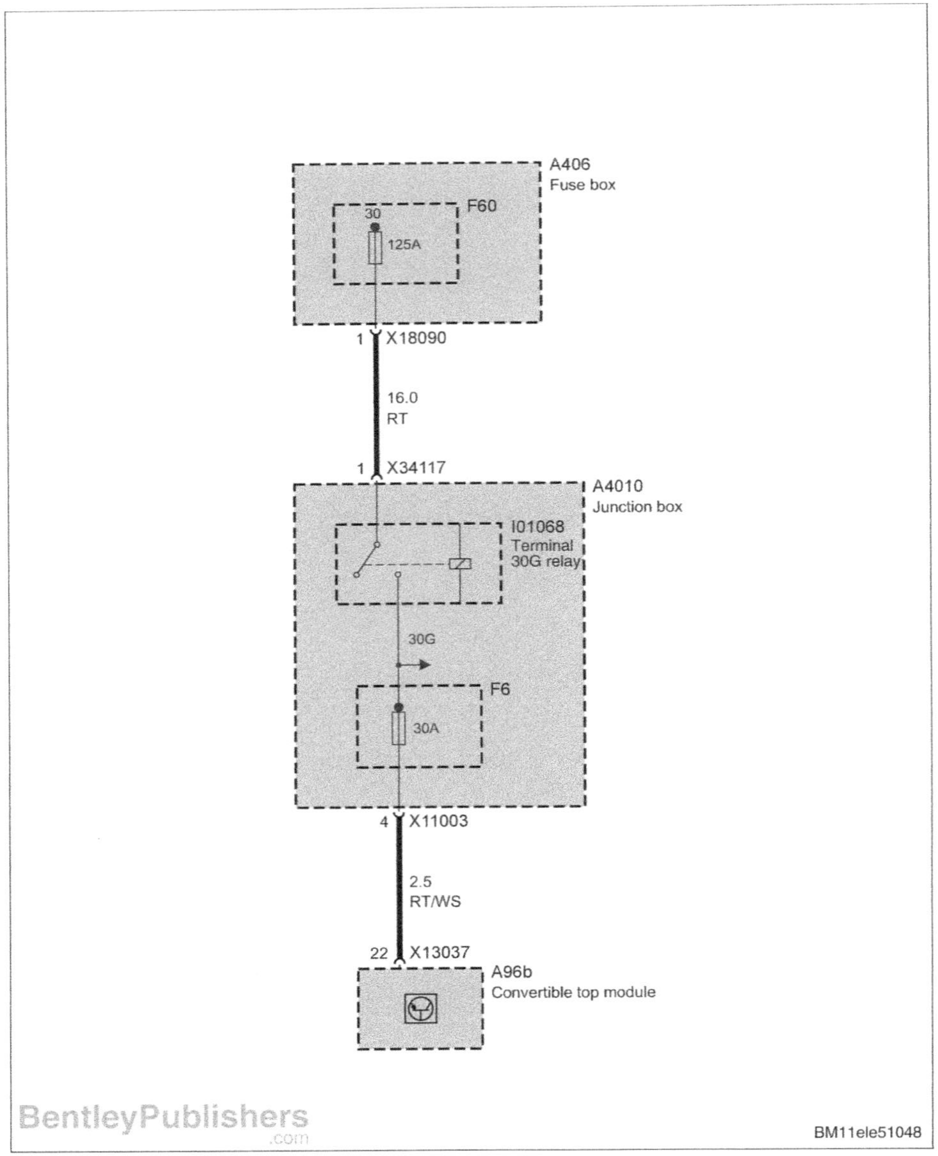

Fuse F7
to 11/2008

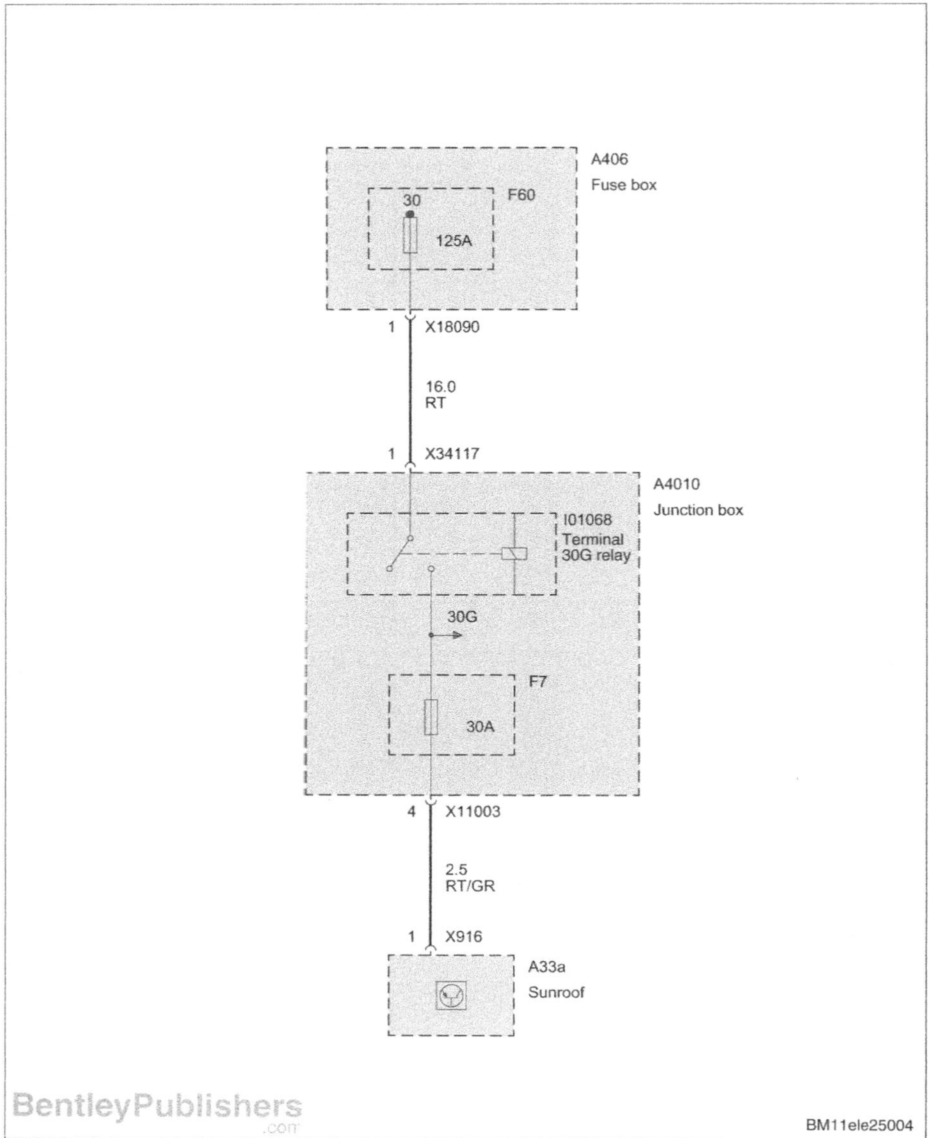

Fuse F7
from 12/2008

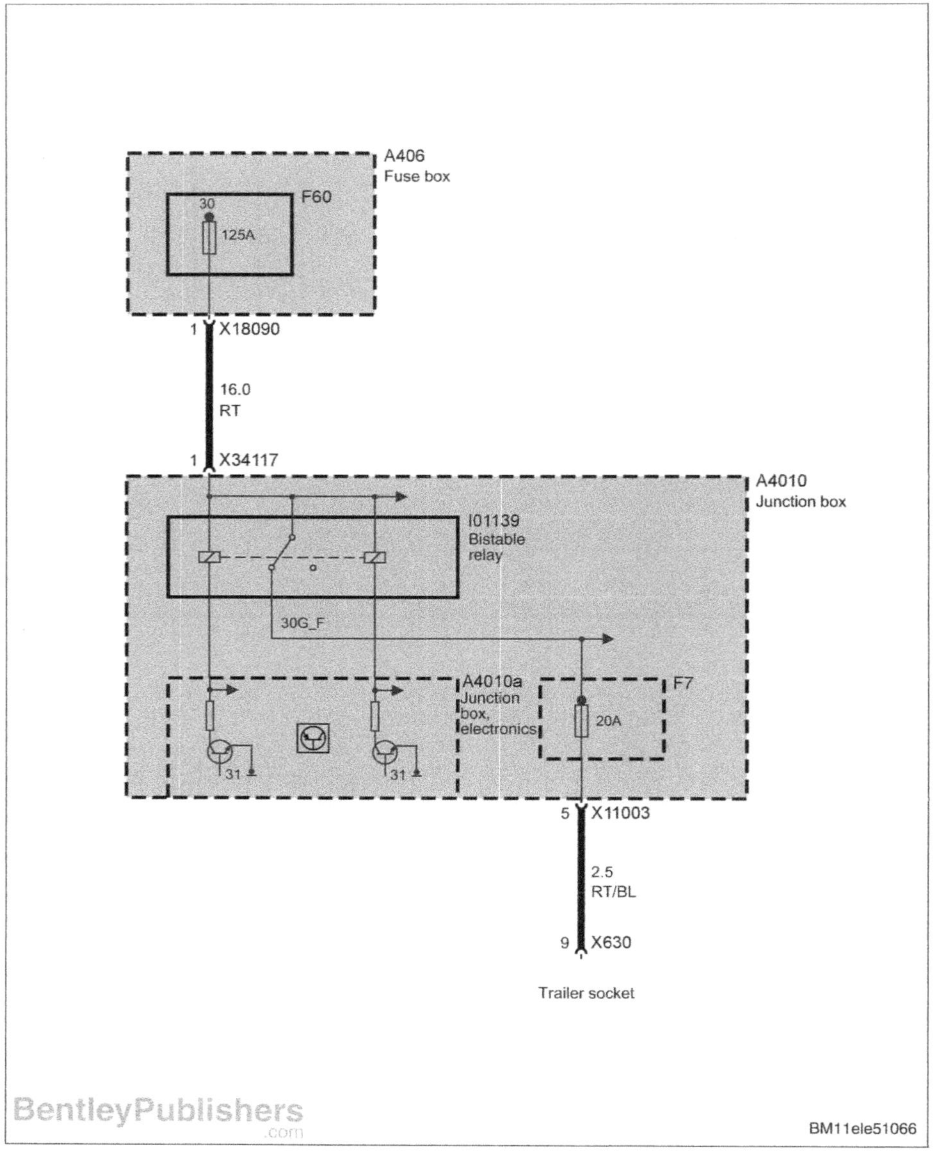

Fuse F8

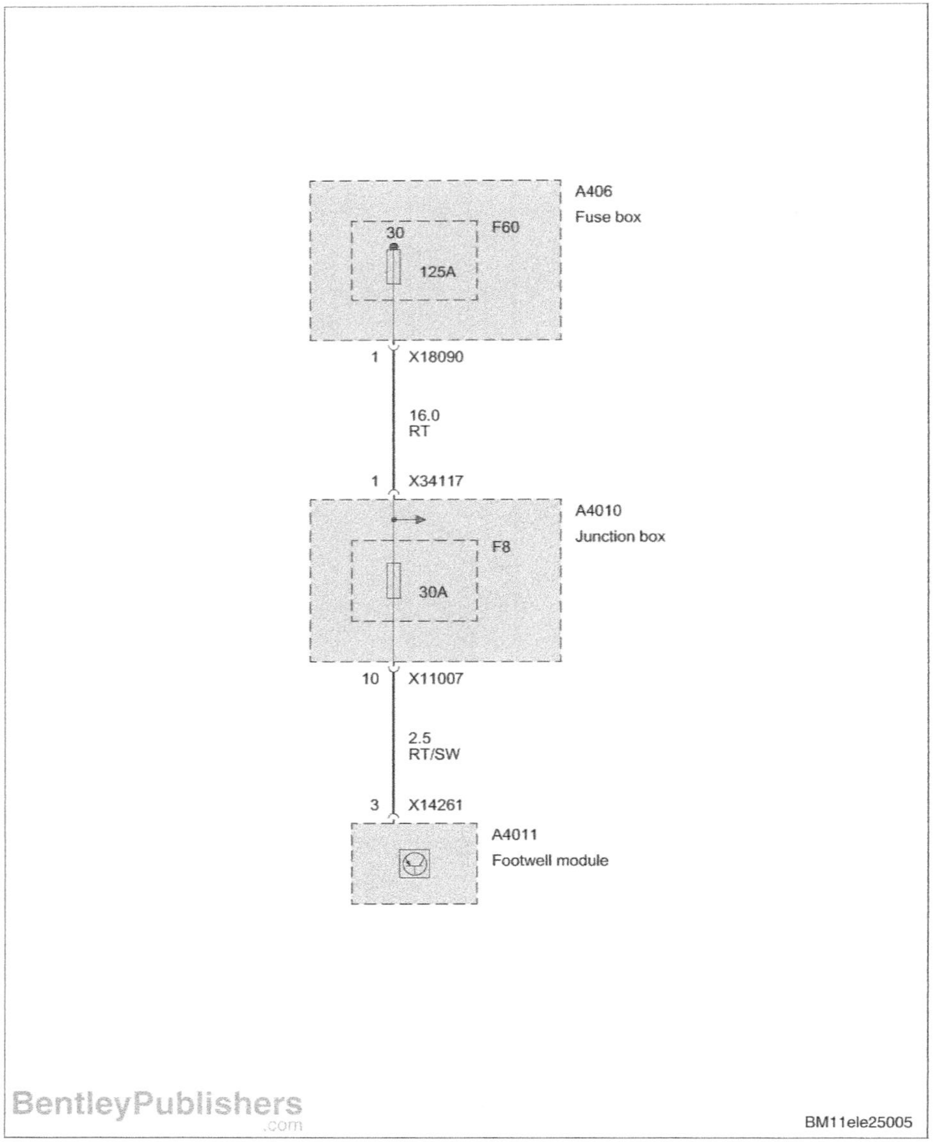

Fuse F9
to 11/2008

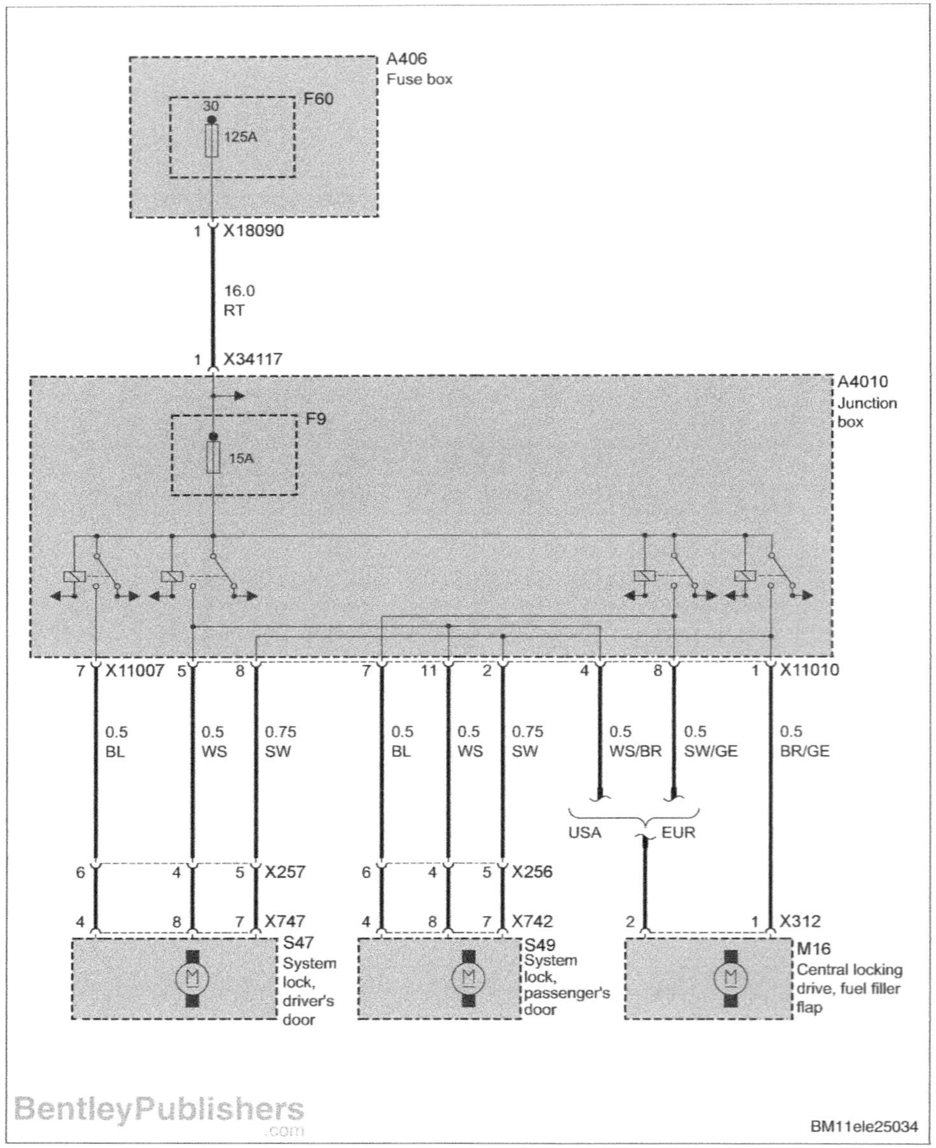

Fuse F10 to 11/2008

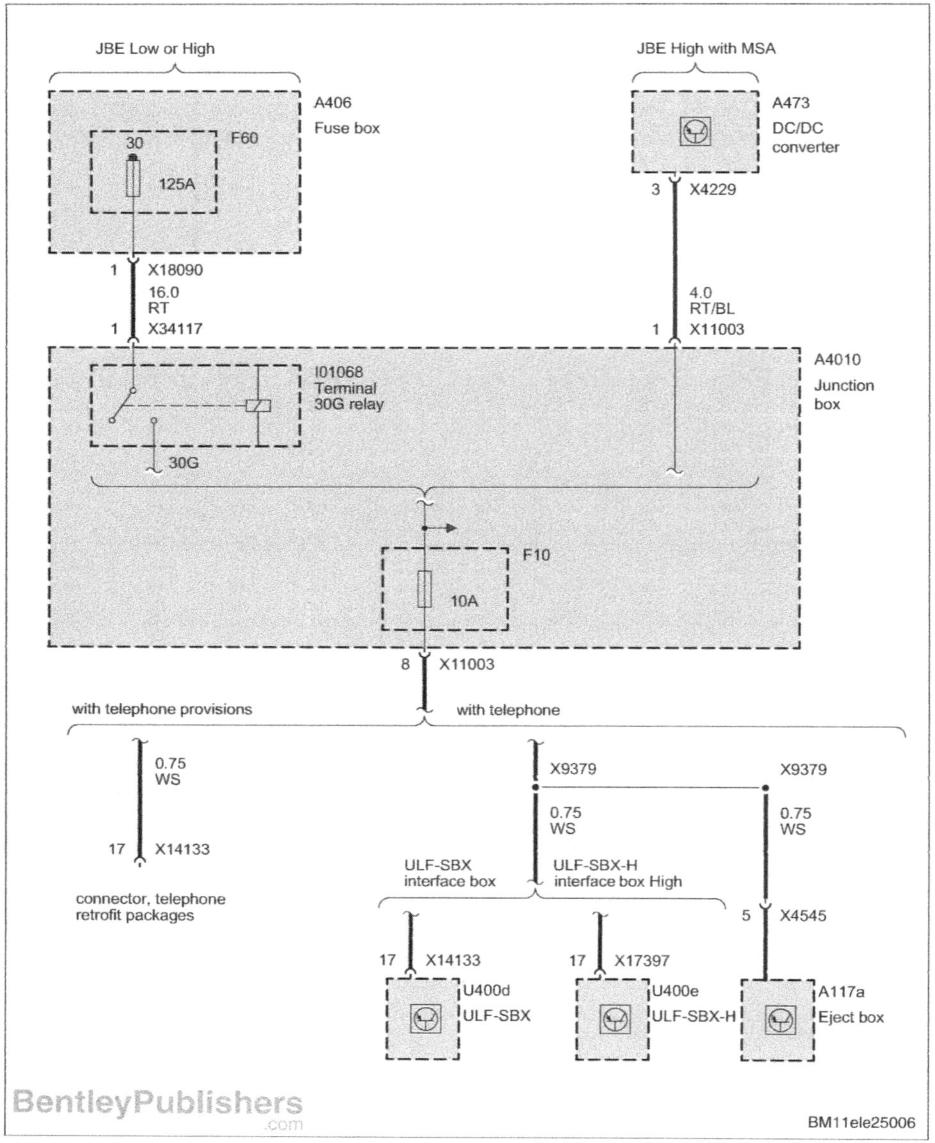

Fuse F11 to 11/2008

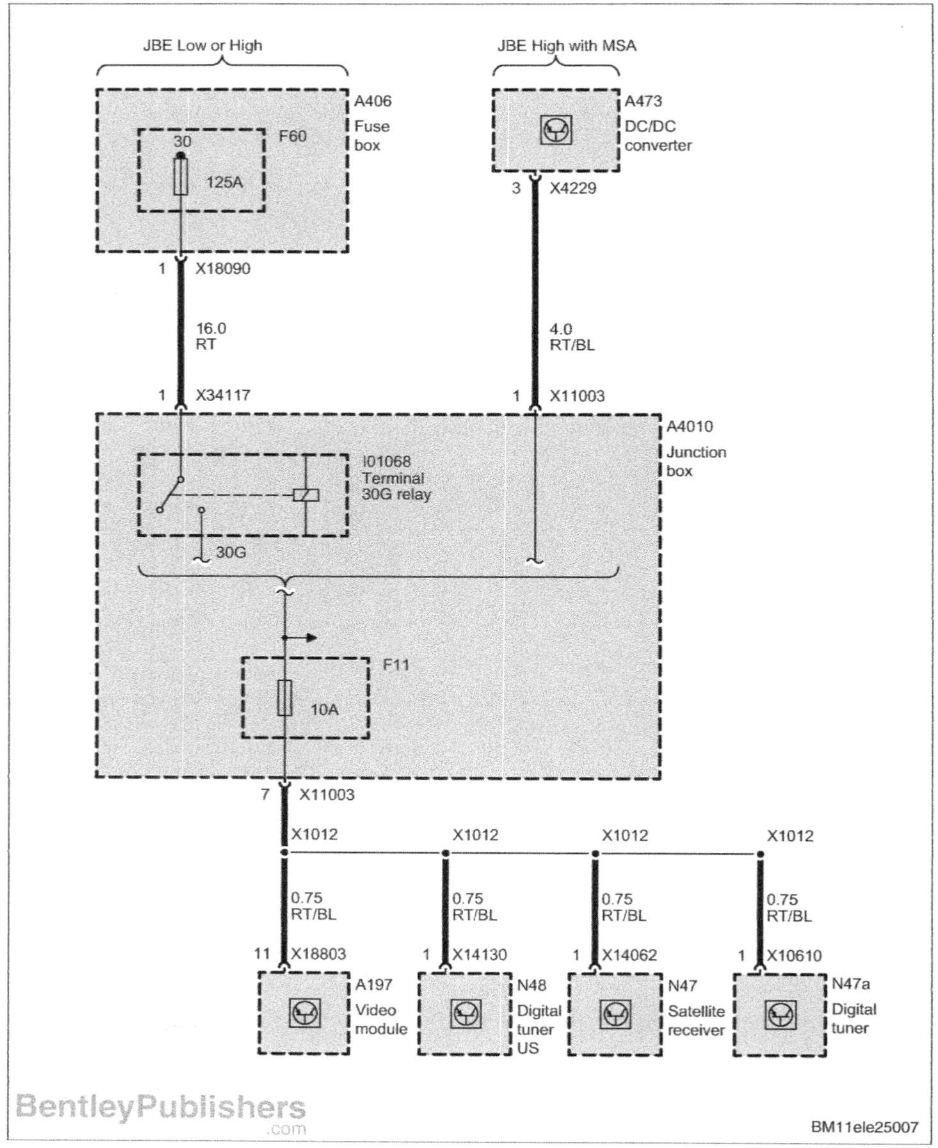

Fuse F11 from 12/2008

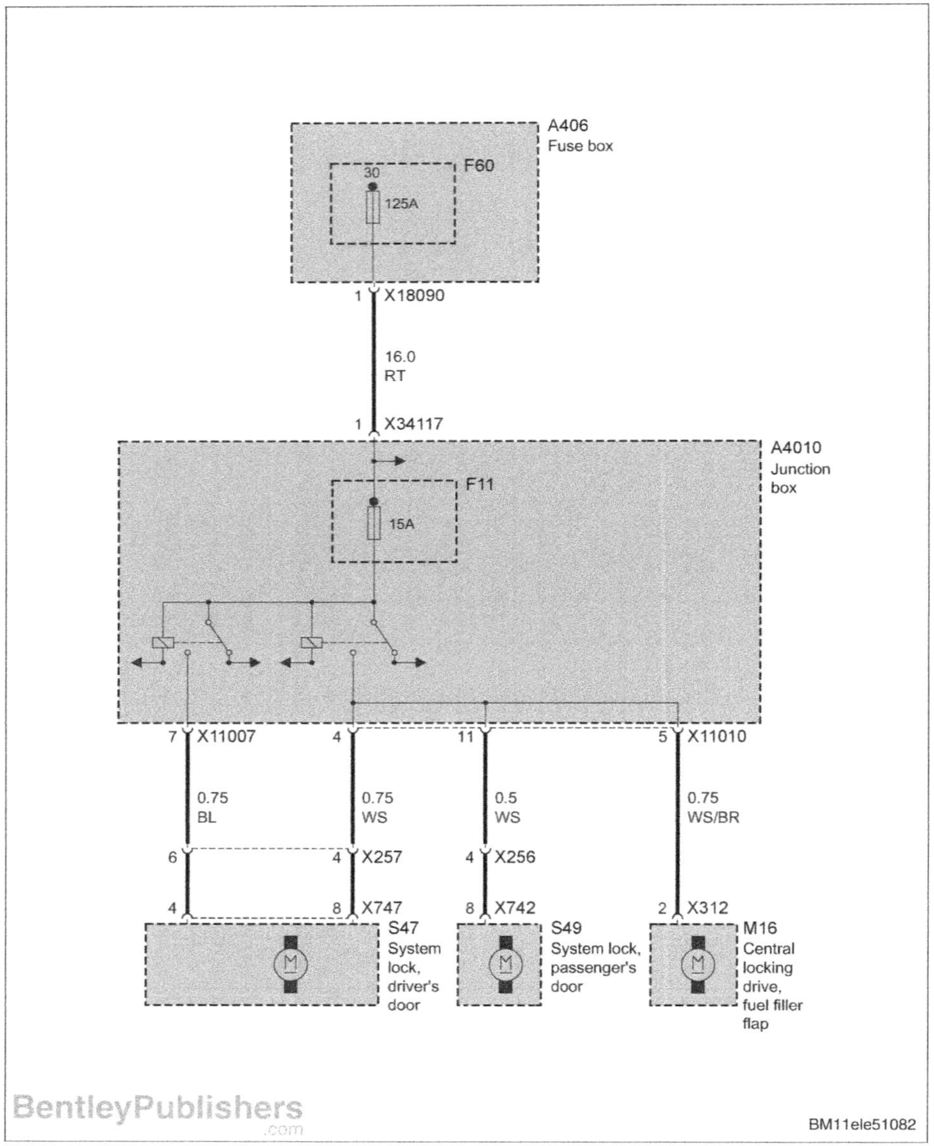

Fuse F12 to 11/2008

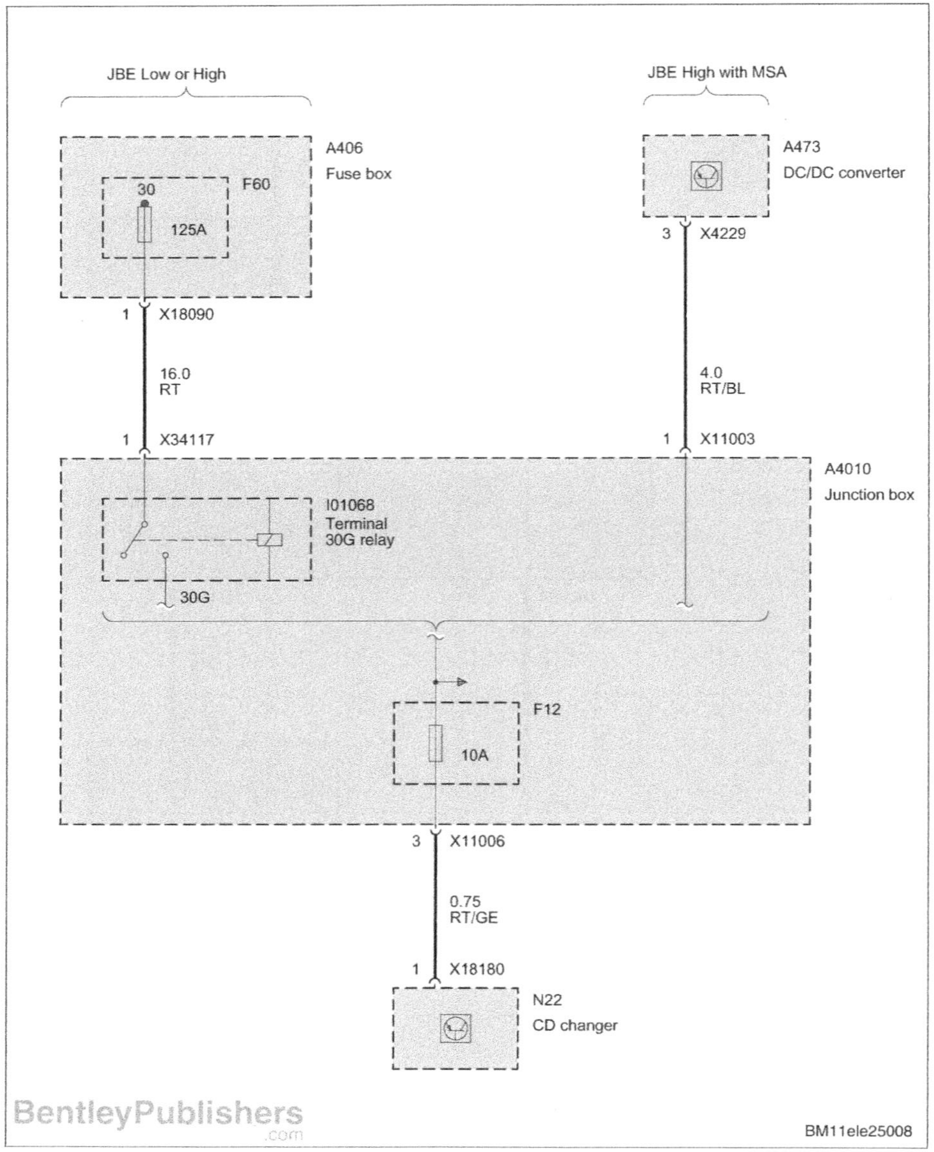

Fuse F12
from 12/2008

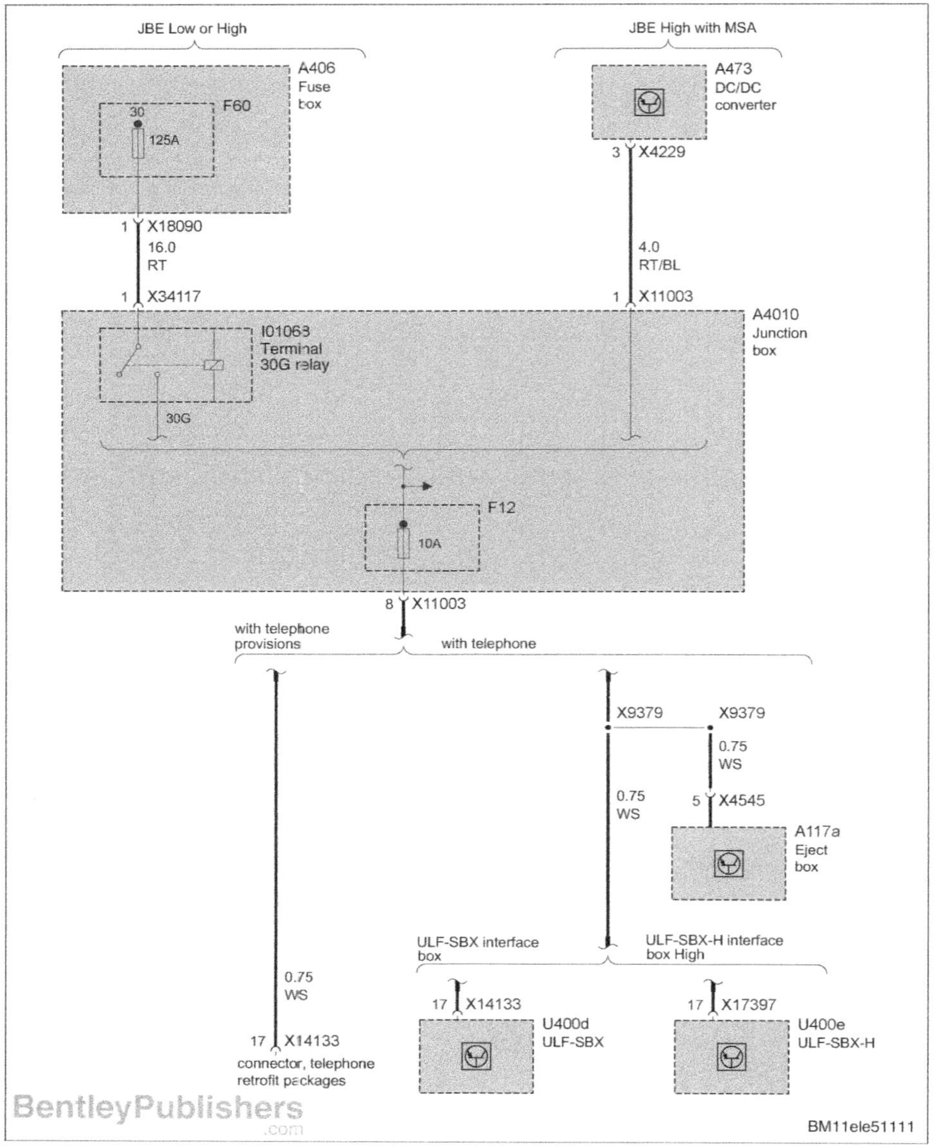

Fuse F13
to 12/2008

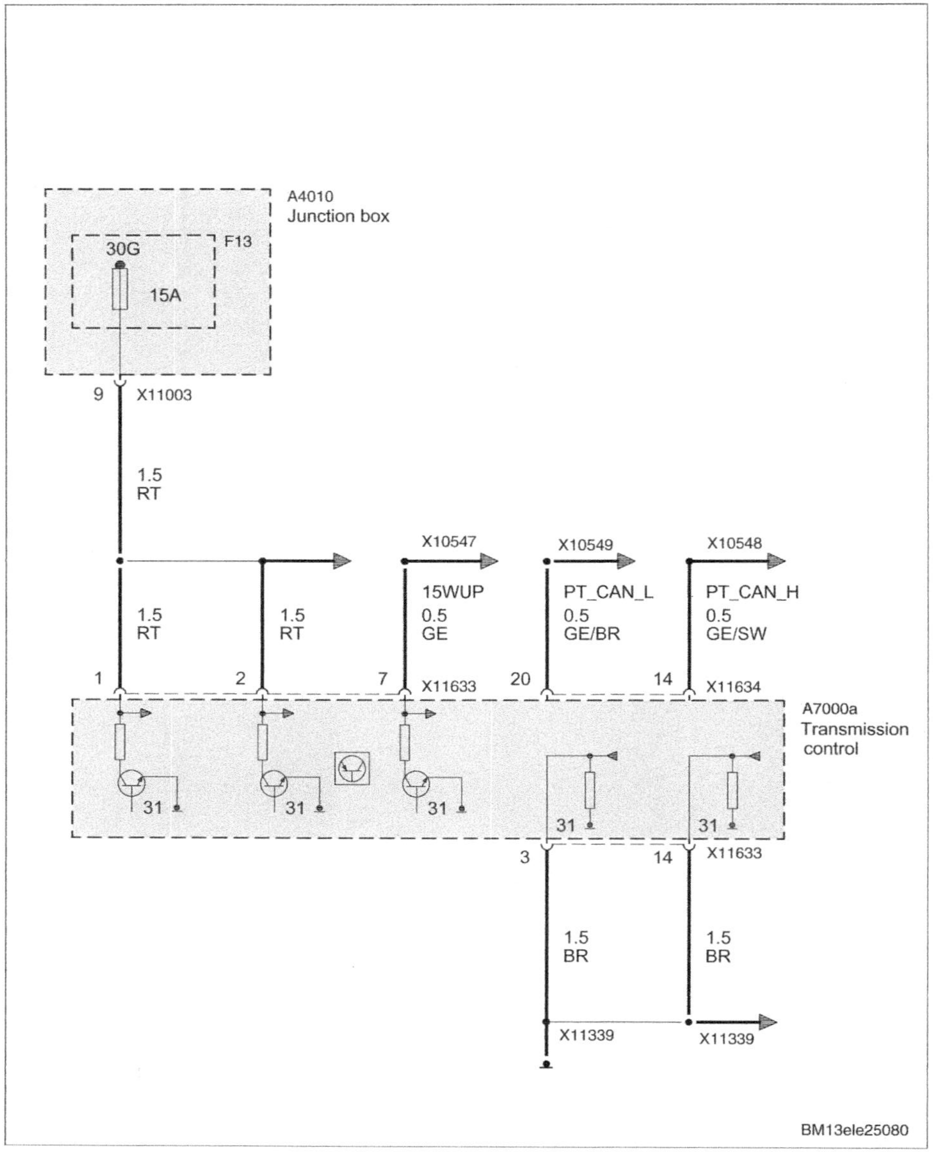

Fuse 13
from 12/2008

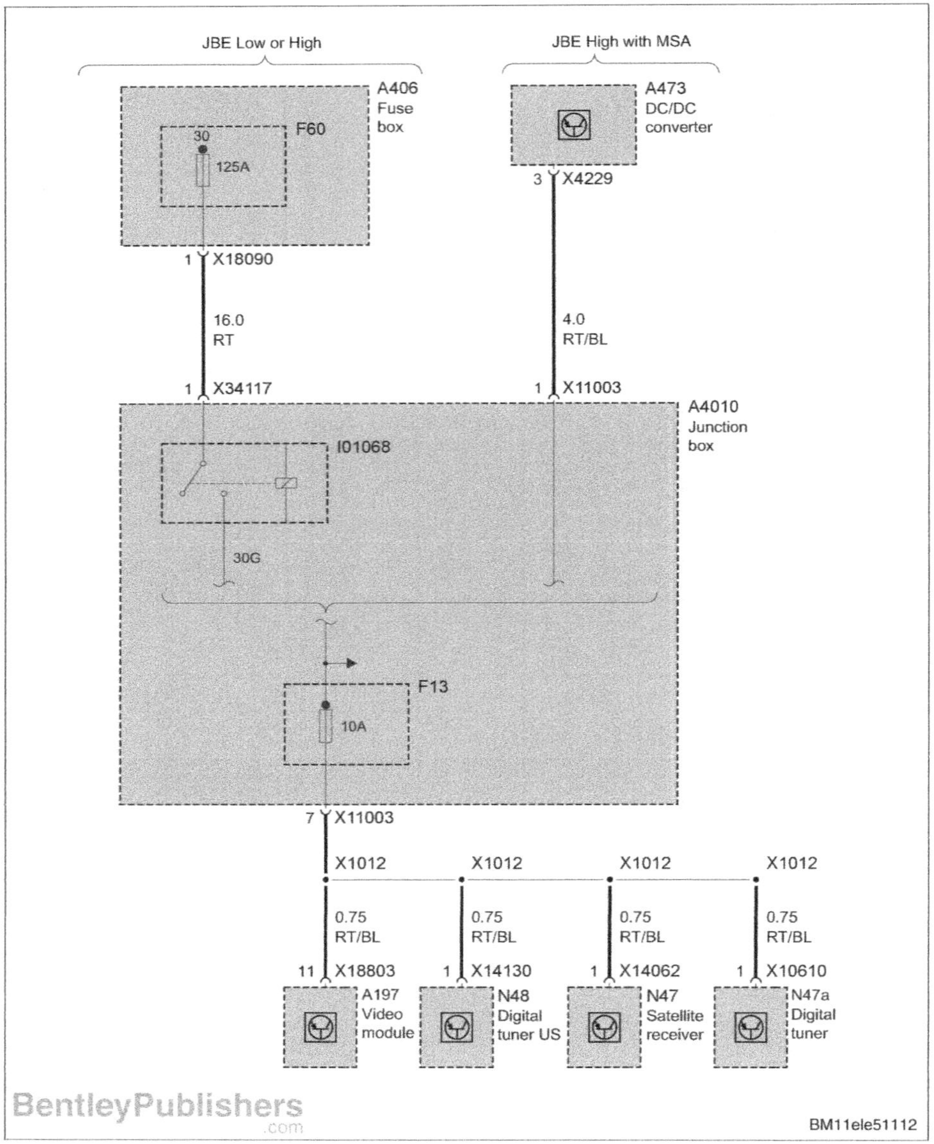

Fuse F14
to 11/2008

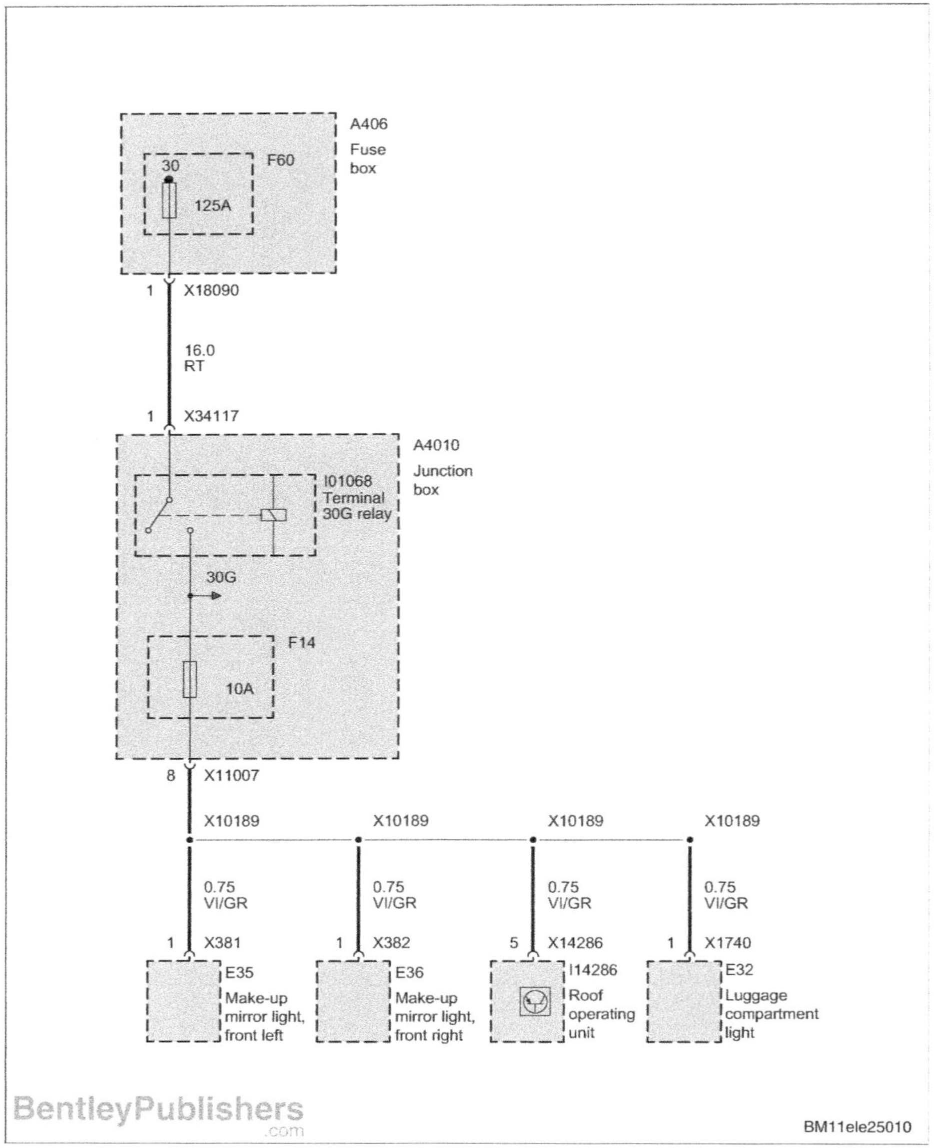

Fuse F14 from 12/2008

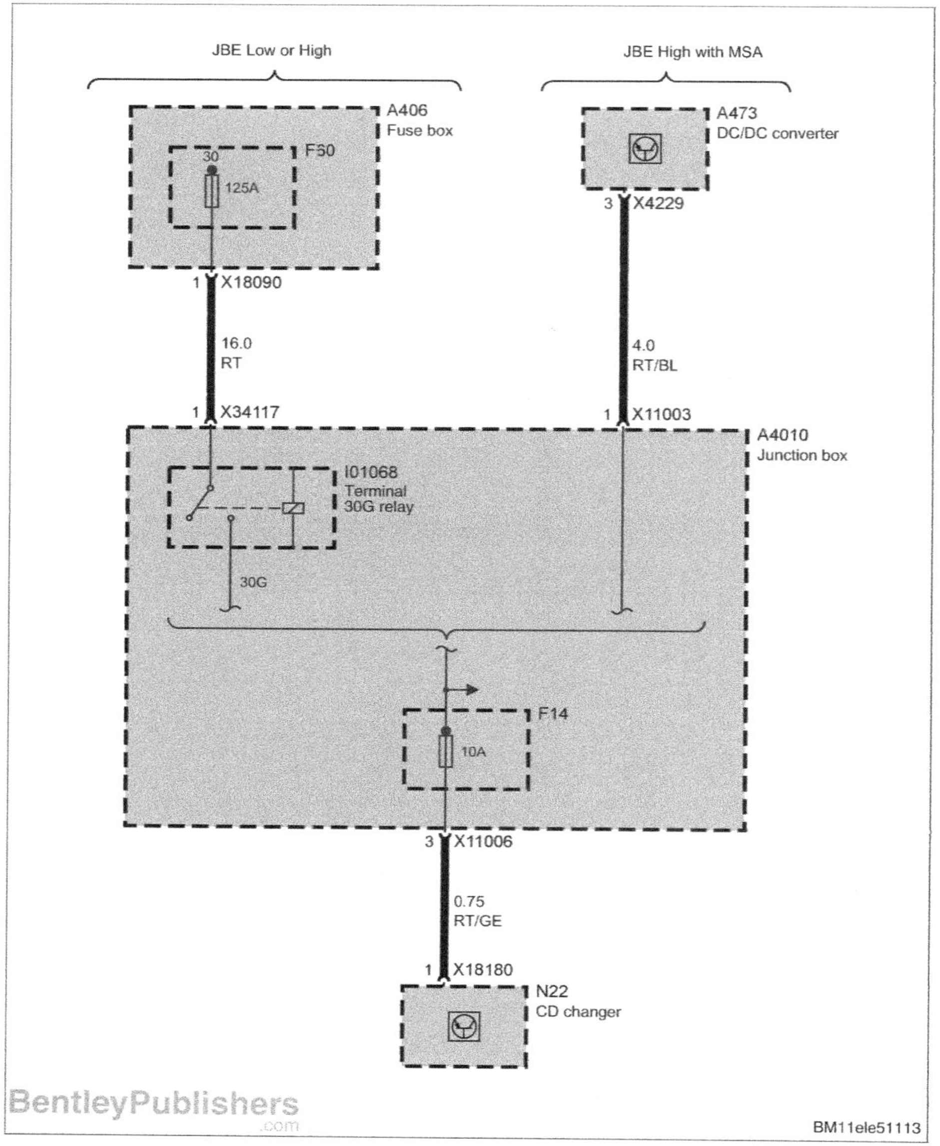

Fuse F15 to 11/2008

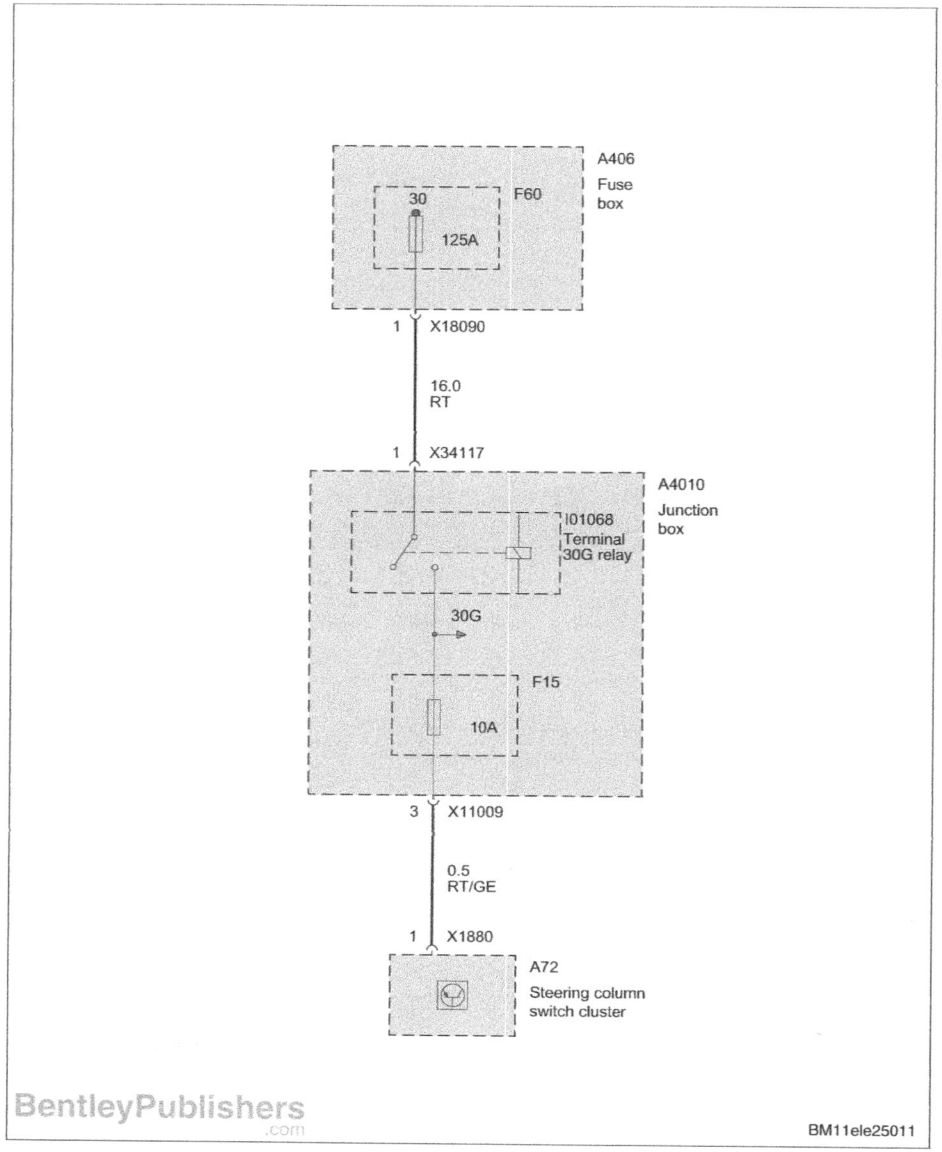

Fuse F15 from 12/2008

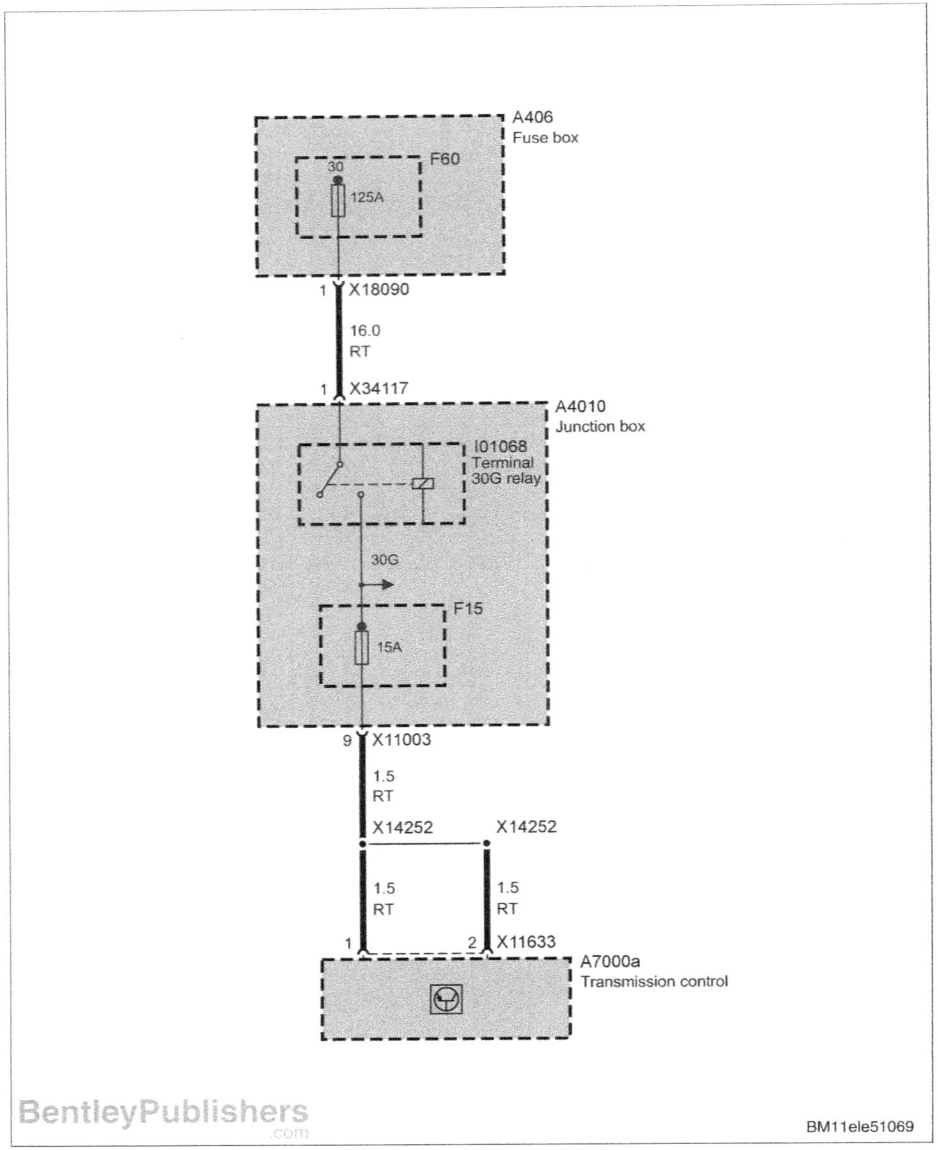

Fuse F16 to 11/2008

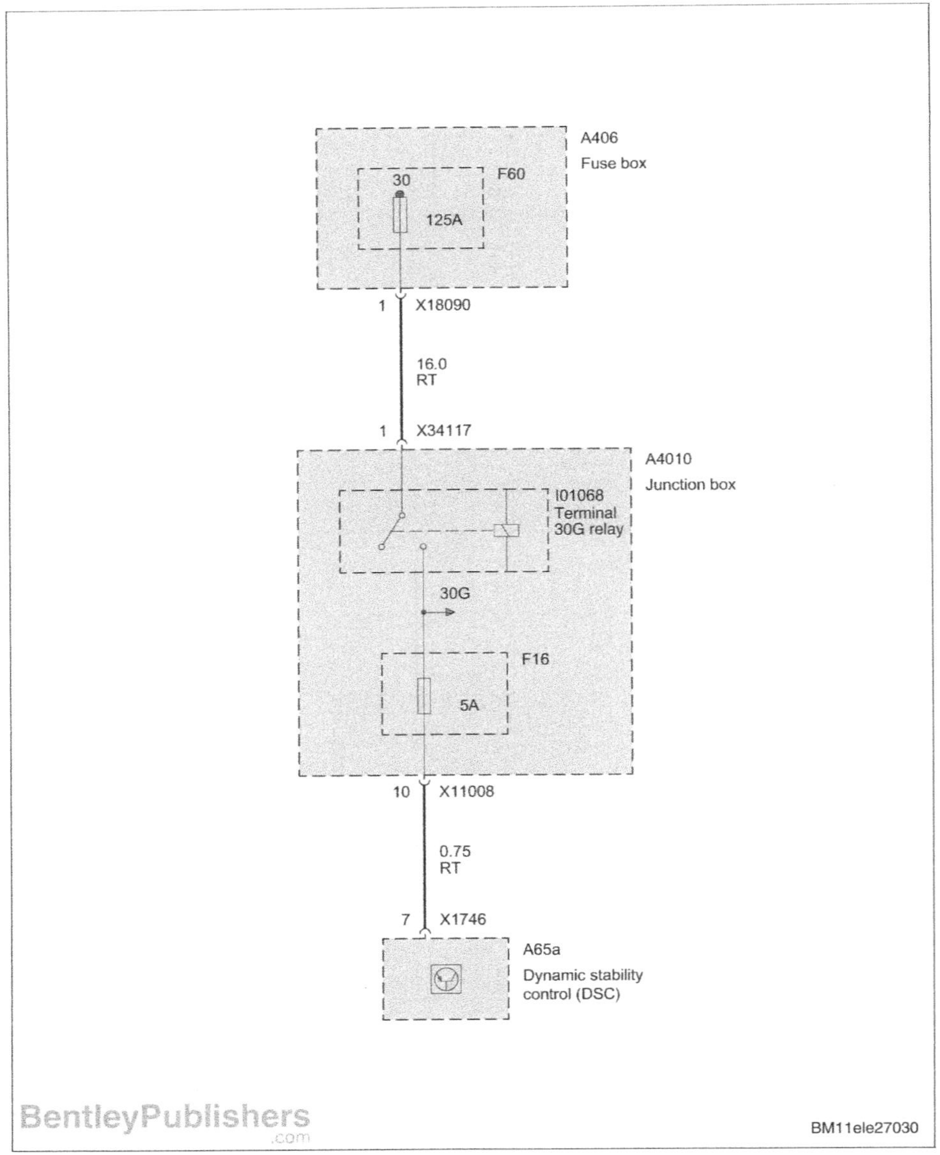

Fuse F16
from 12/ 2008, N16 engine

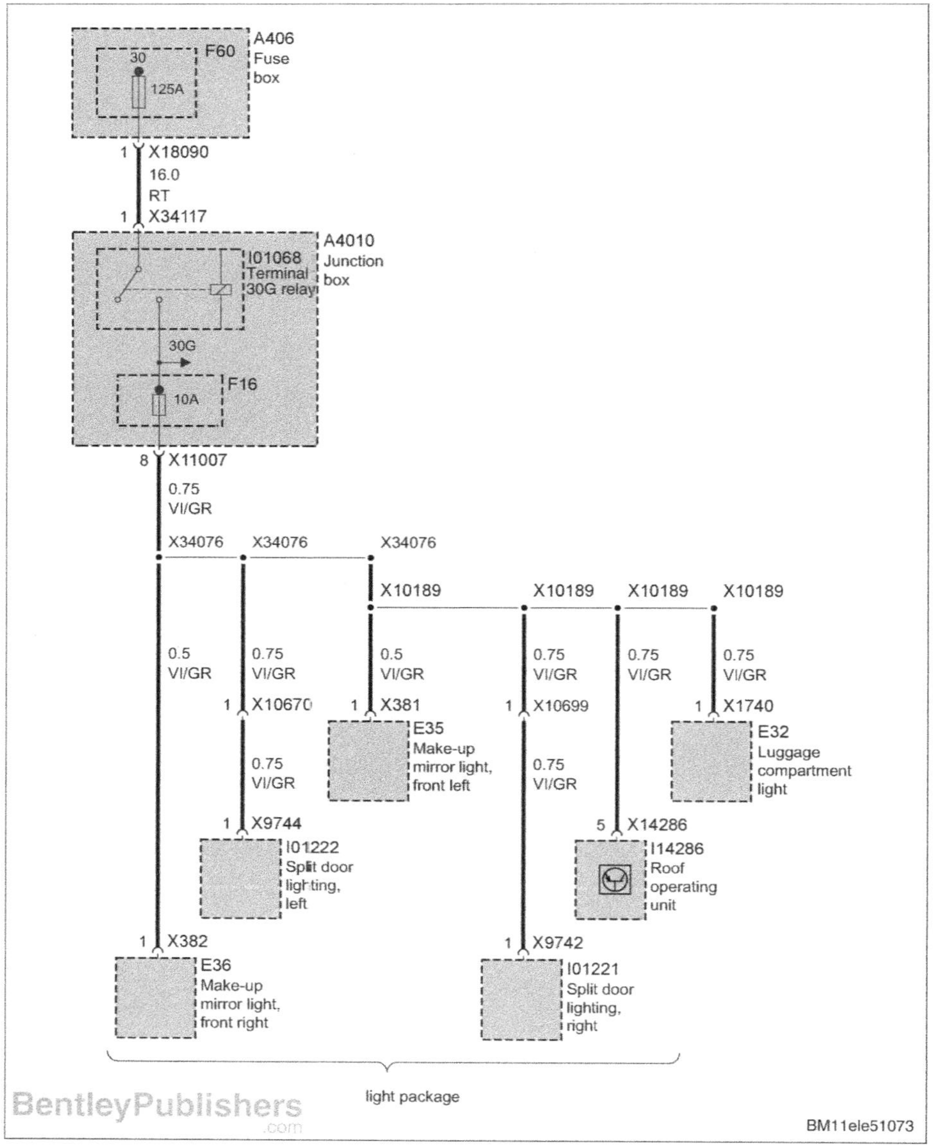

Fuse F17
to 11/2008

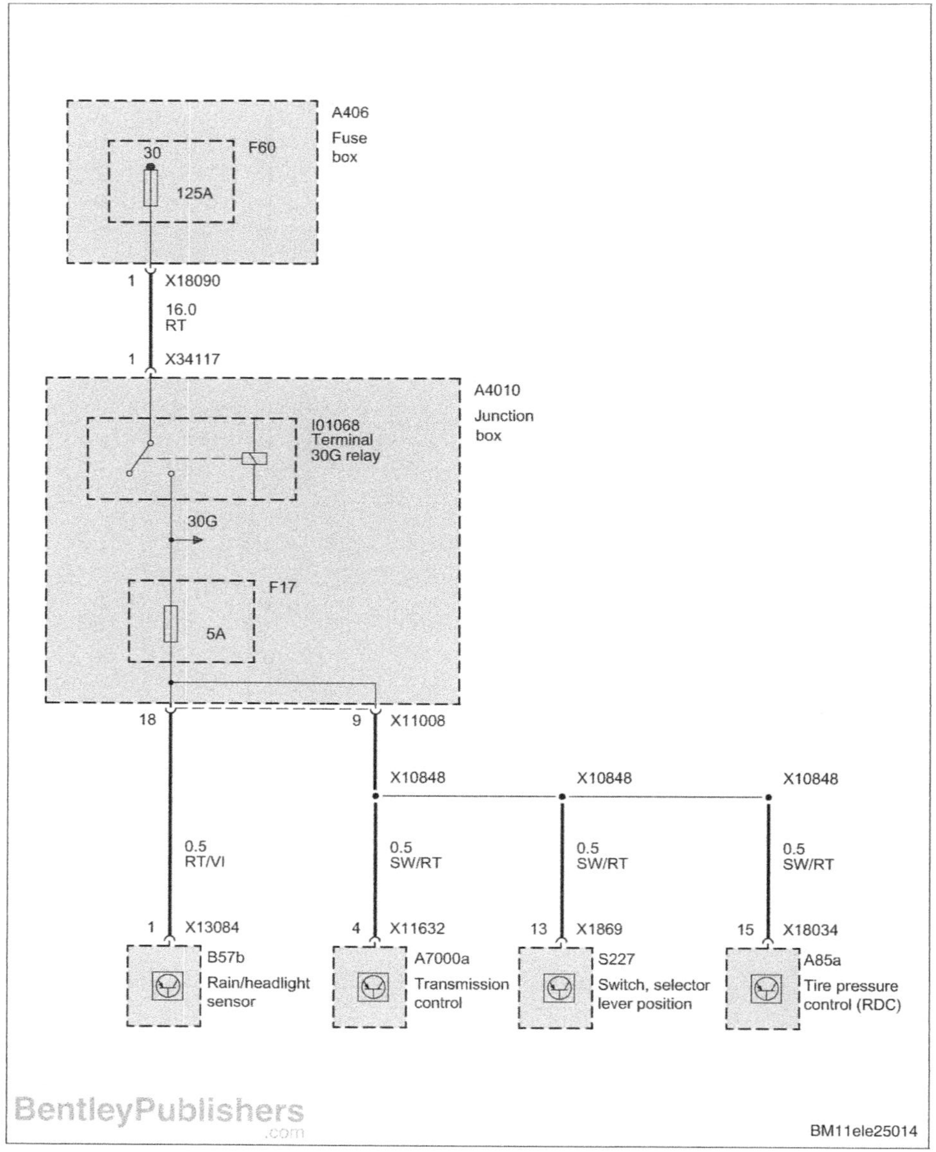

Fuse F17 from 12/2008

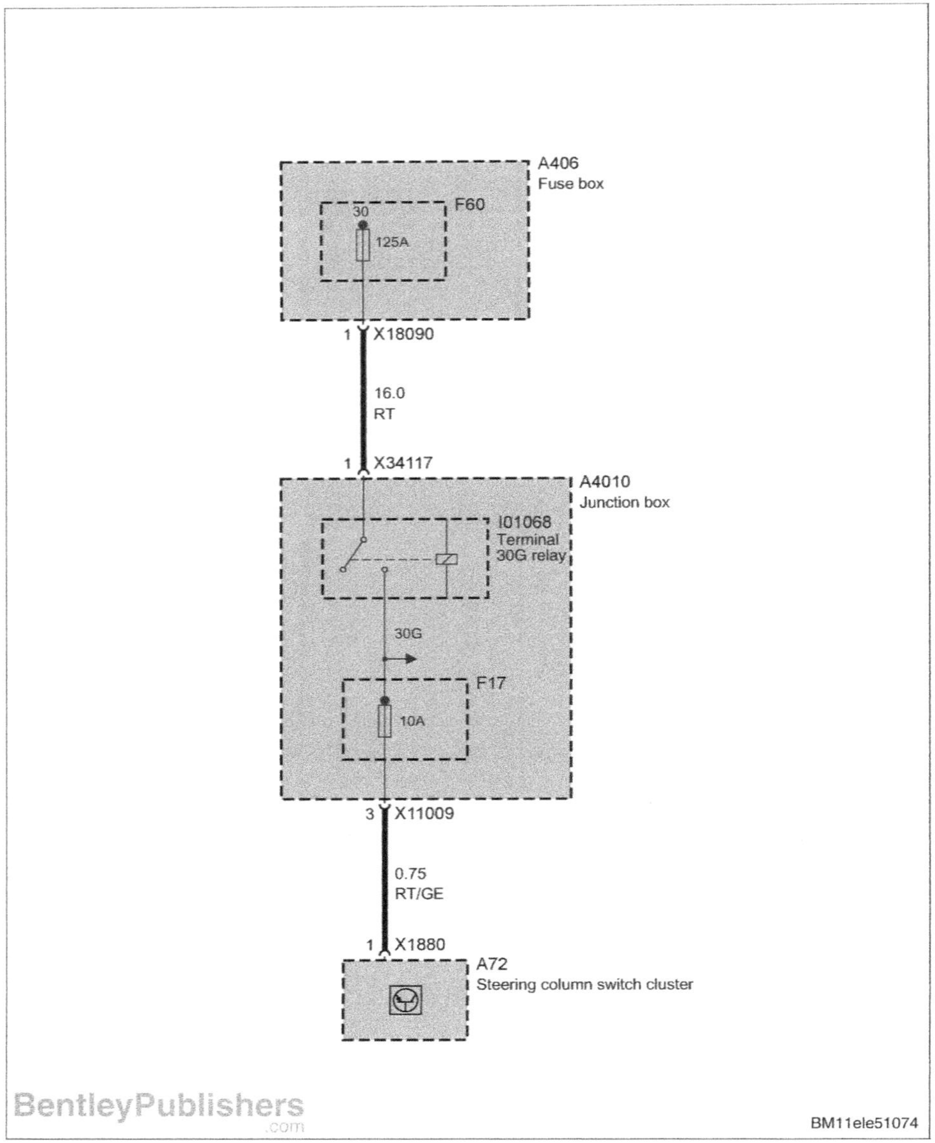

Fuse F18 to 11/2008

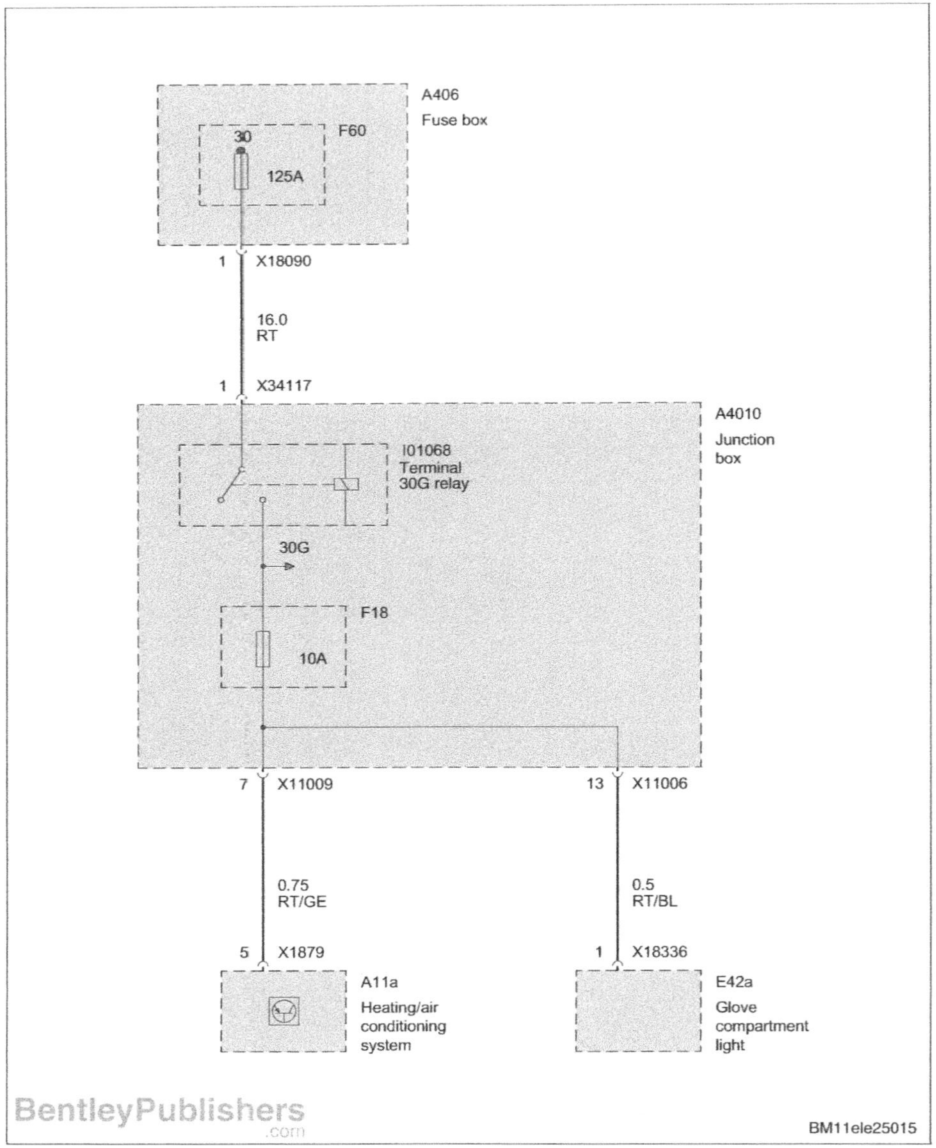

Fuse F18
from 12/2008

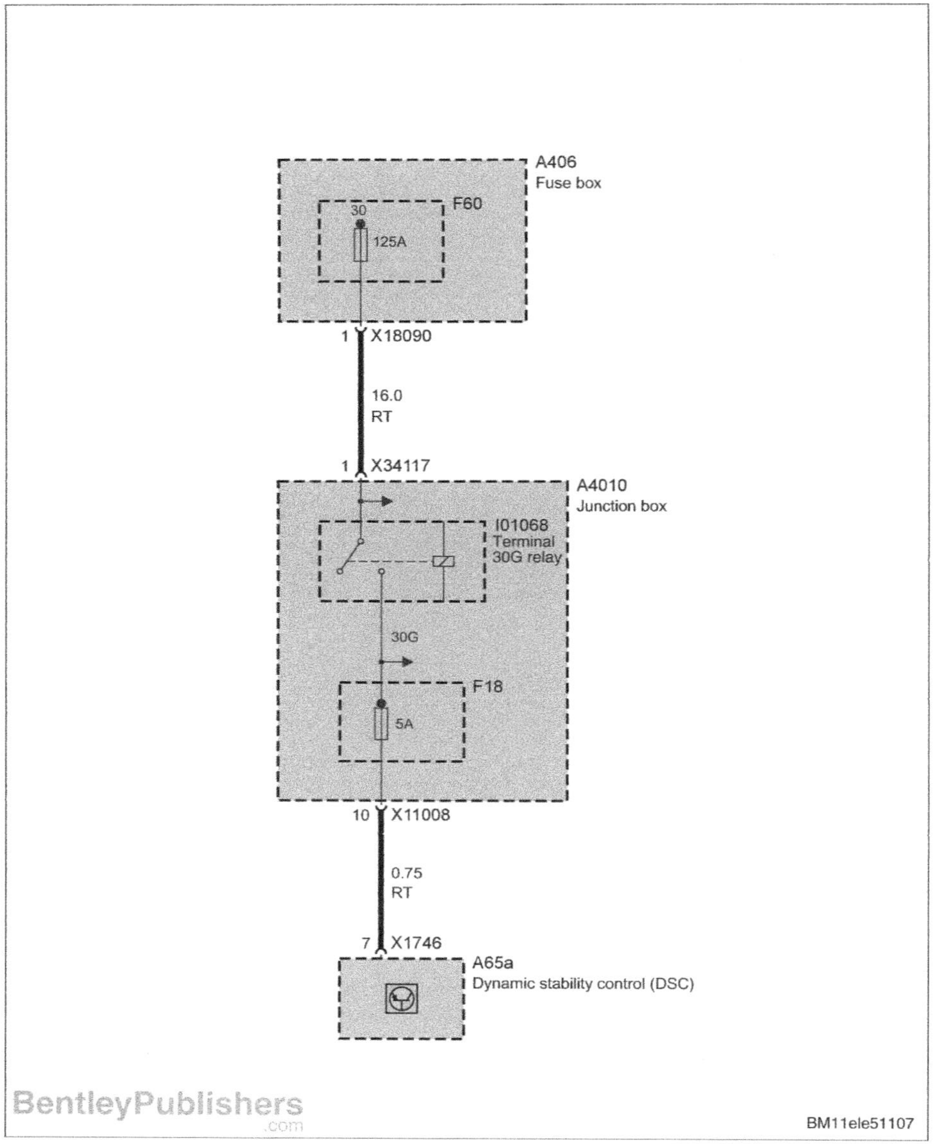

Fuse F19
to 11/2008

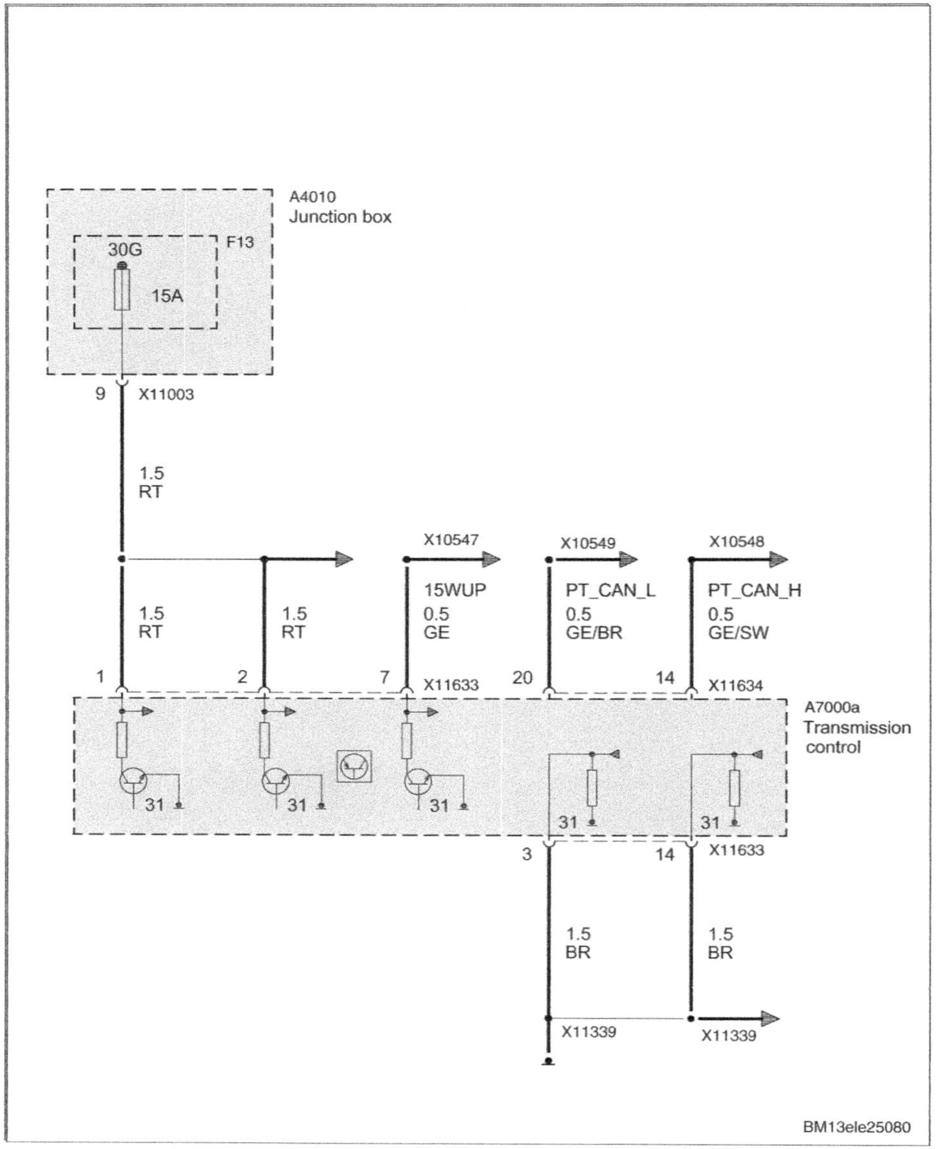

Fuse F19 from 12/2008

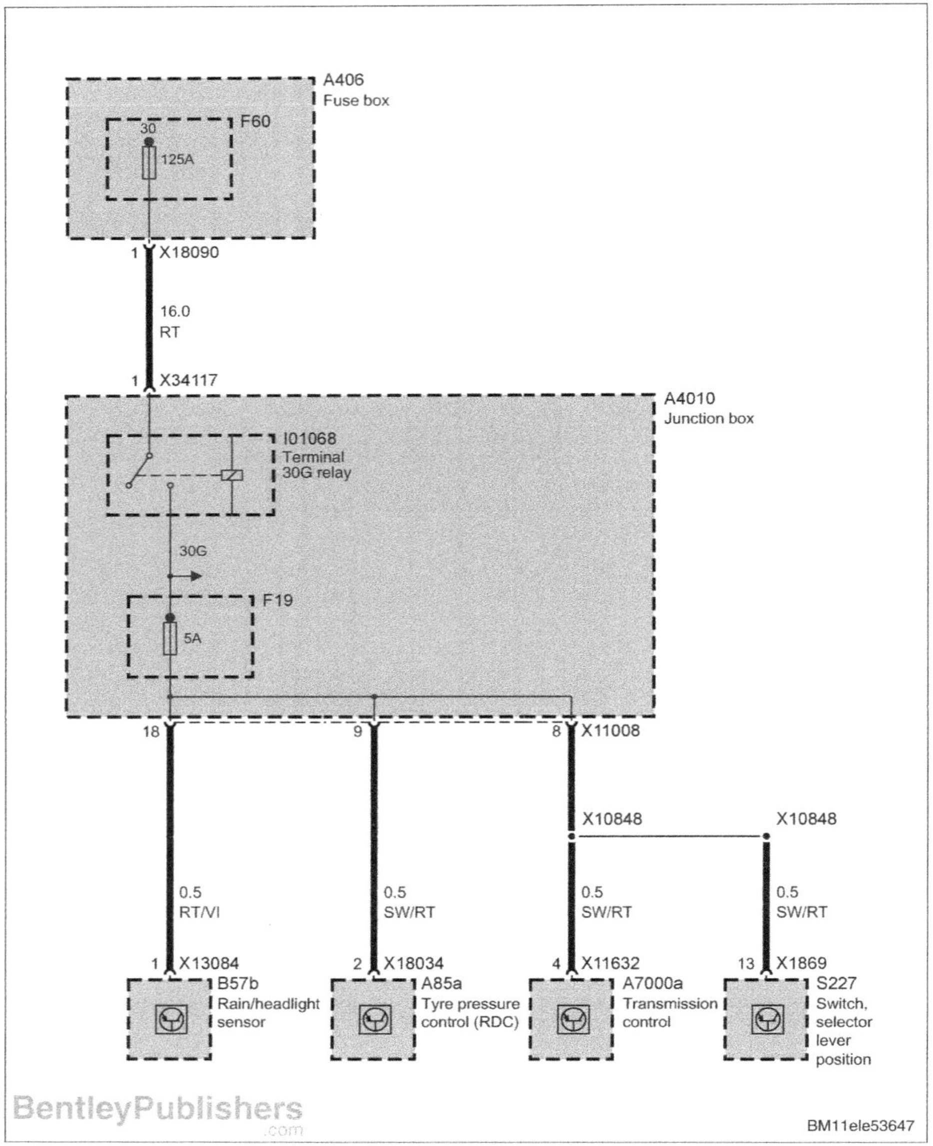

Fuse F20 to 11/2008

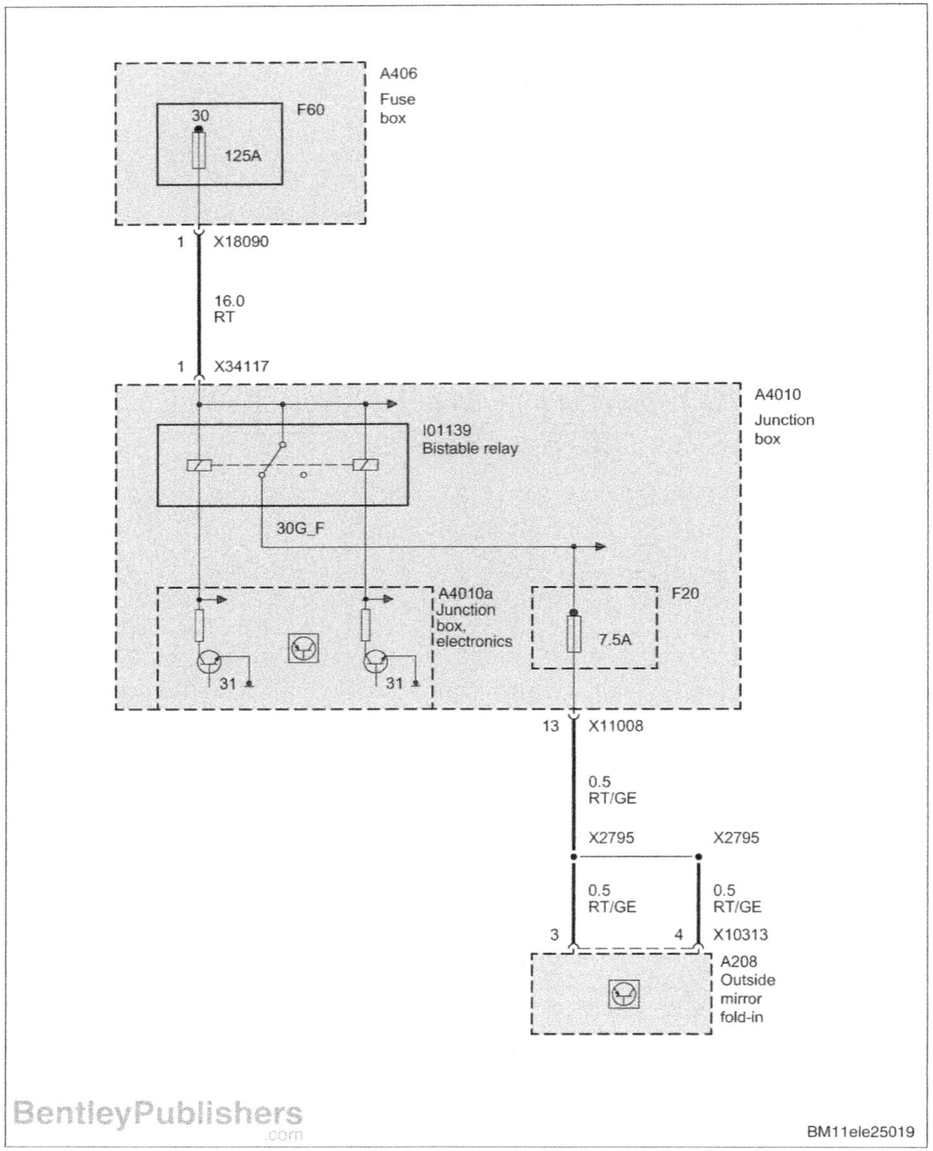

Fuse F20 from 12/2008

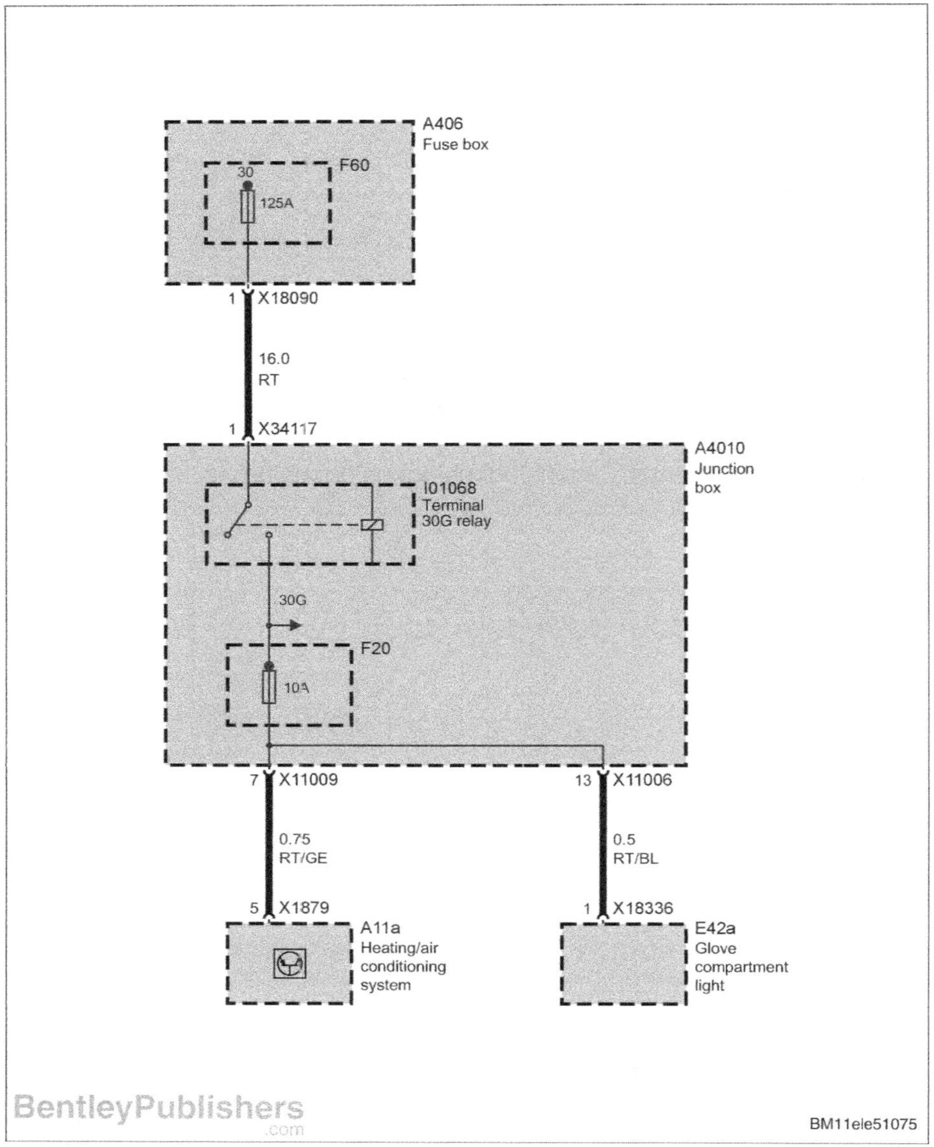

Fuse F21 to 11/2008

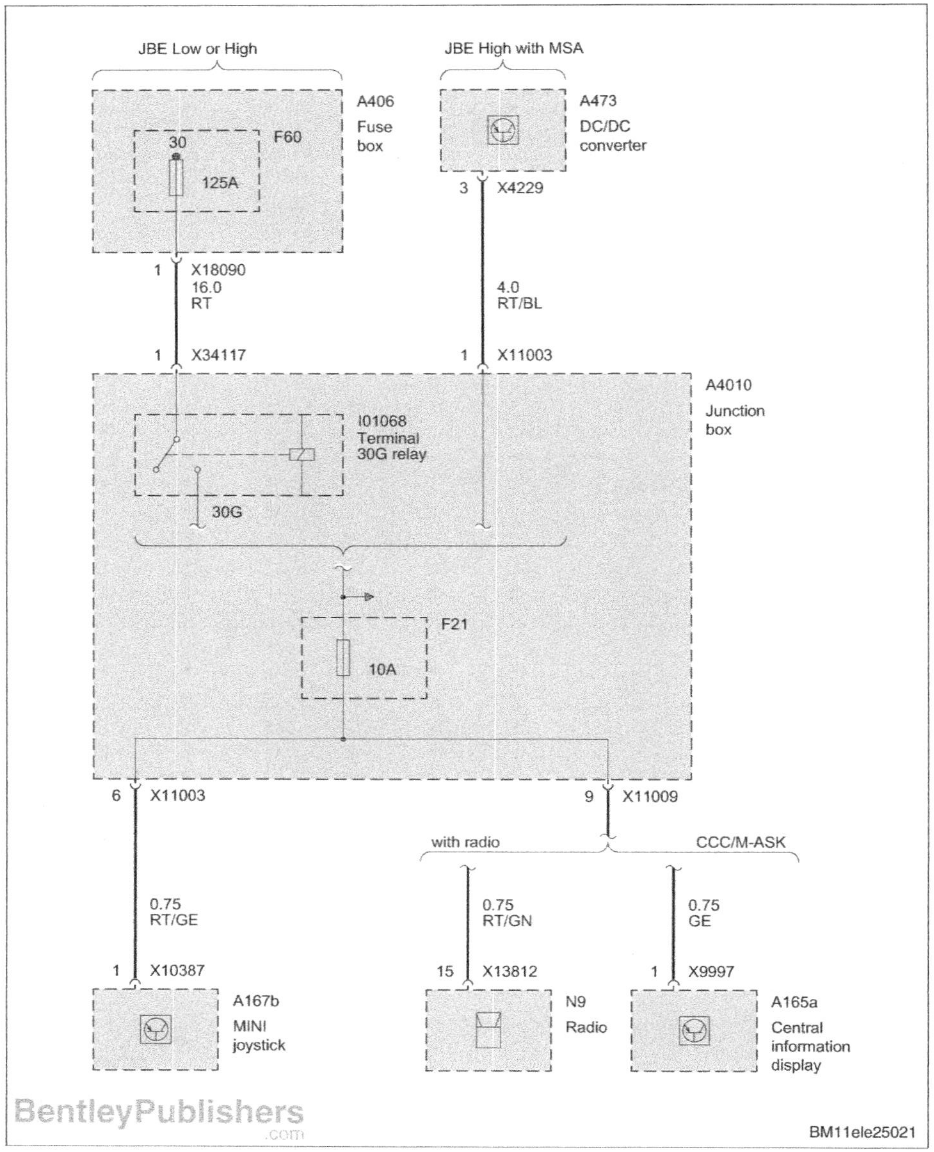

Fuse F21 from 12/2008

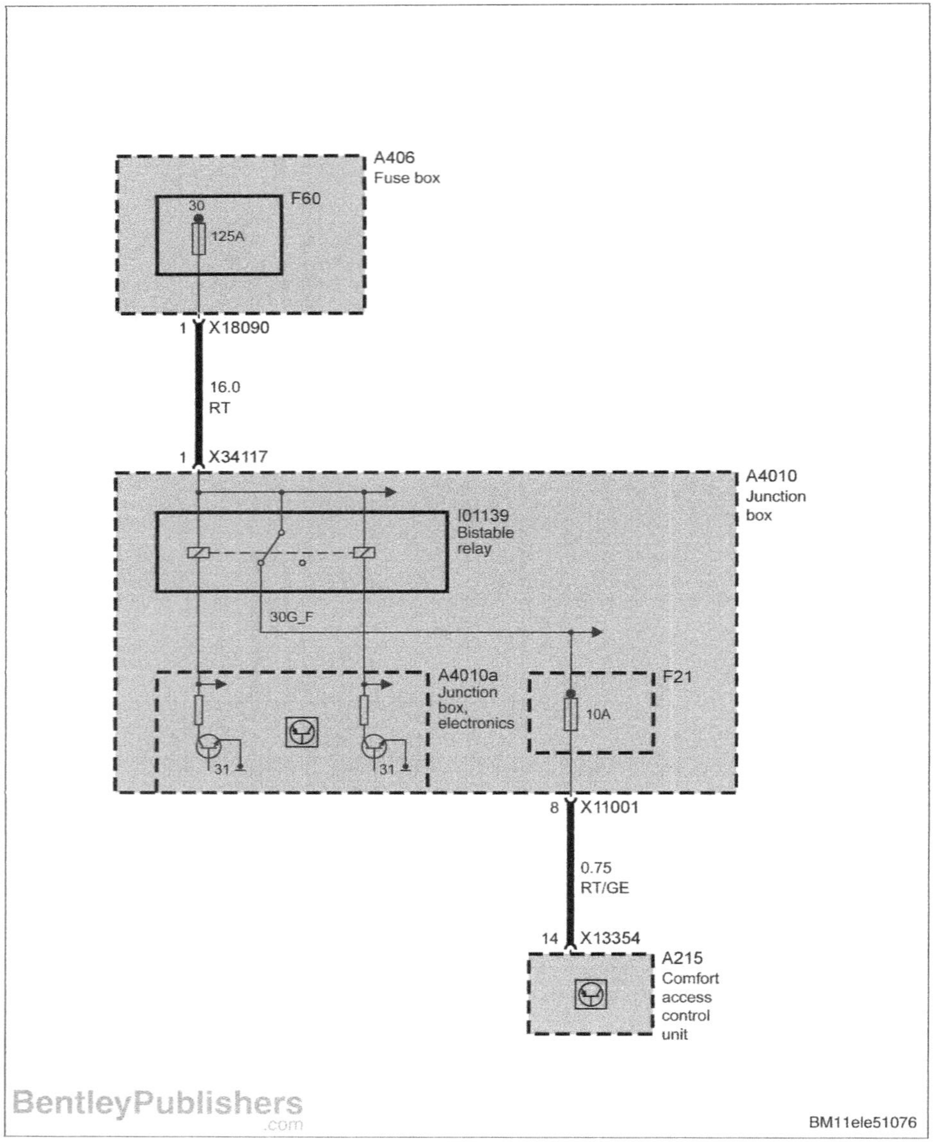

Fuse F22 to 11/2008

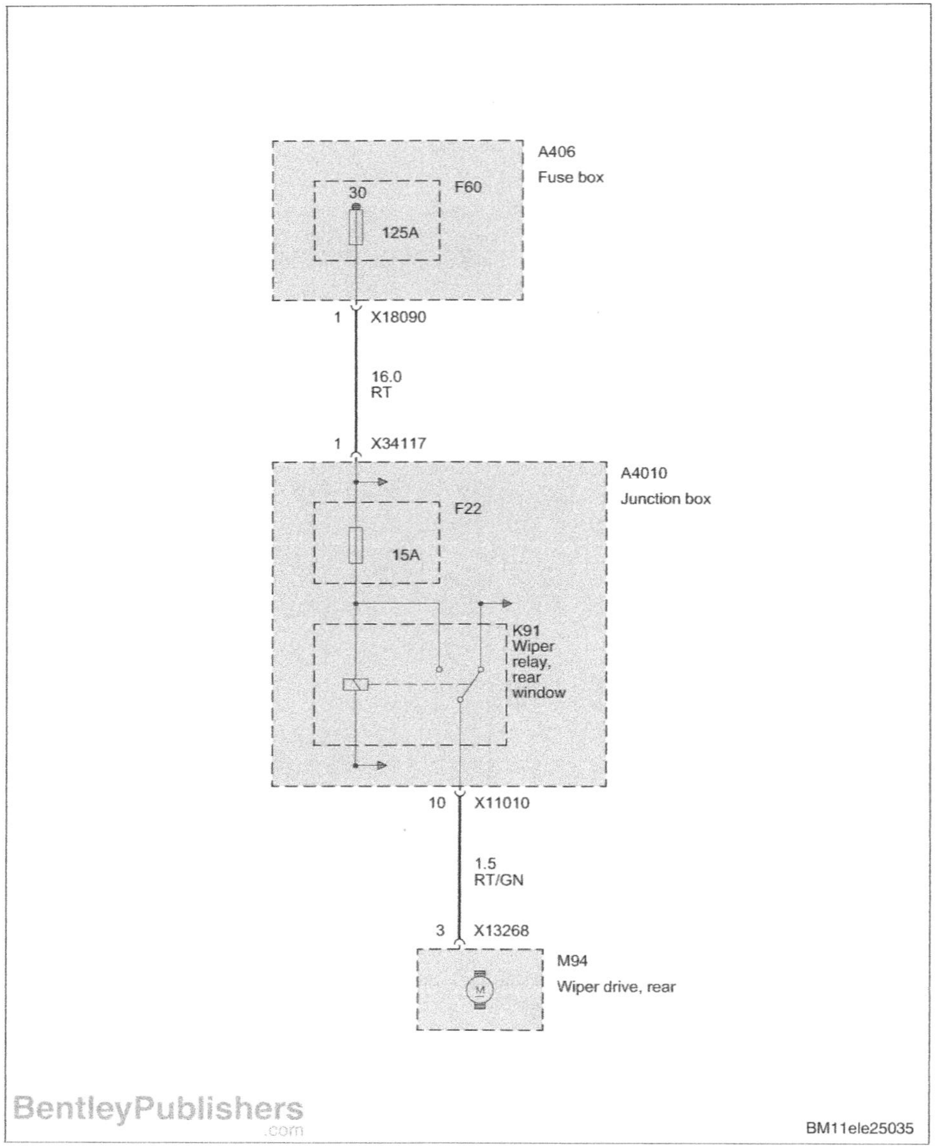

Fuse F22
from 12/2008

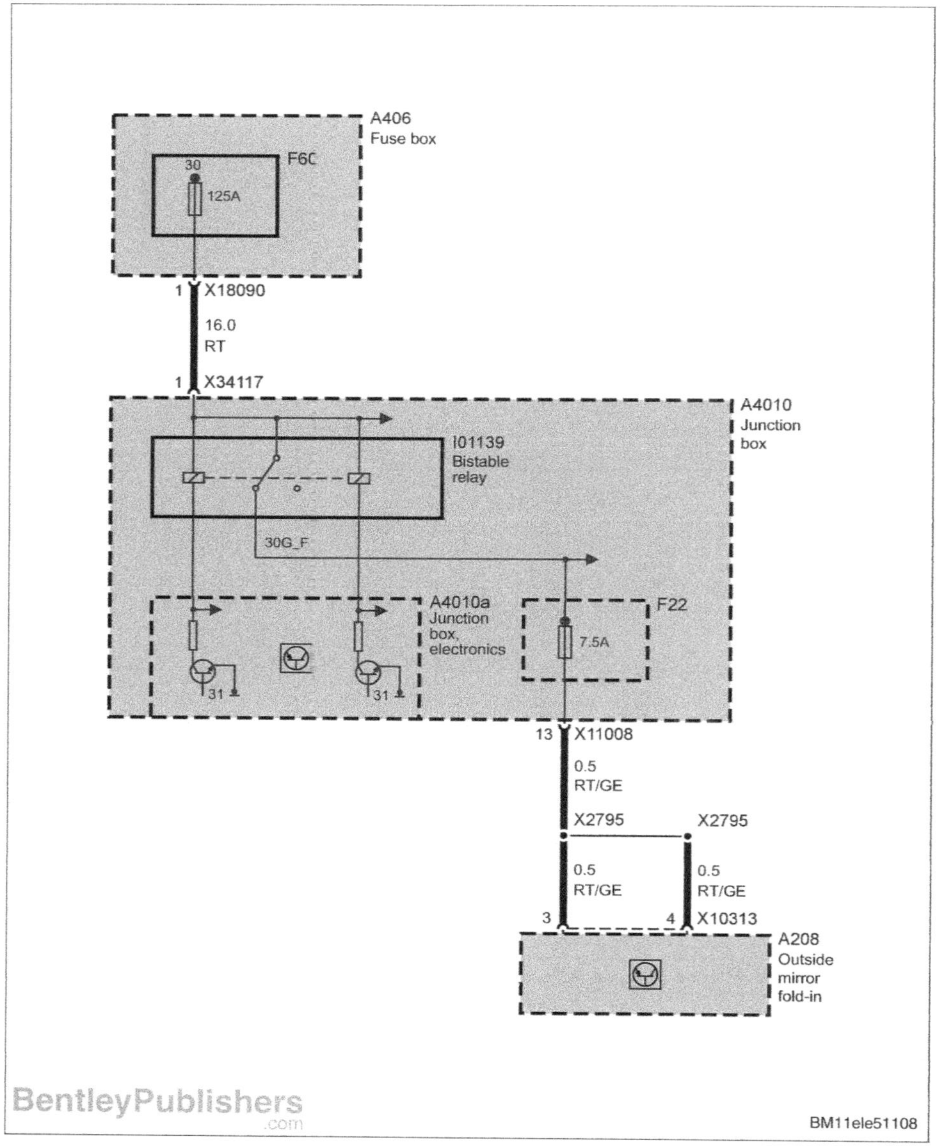

Fuse F23
to 11/2008

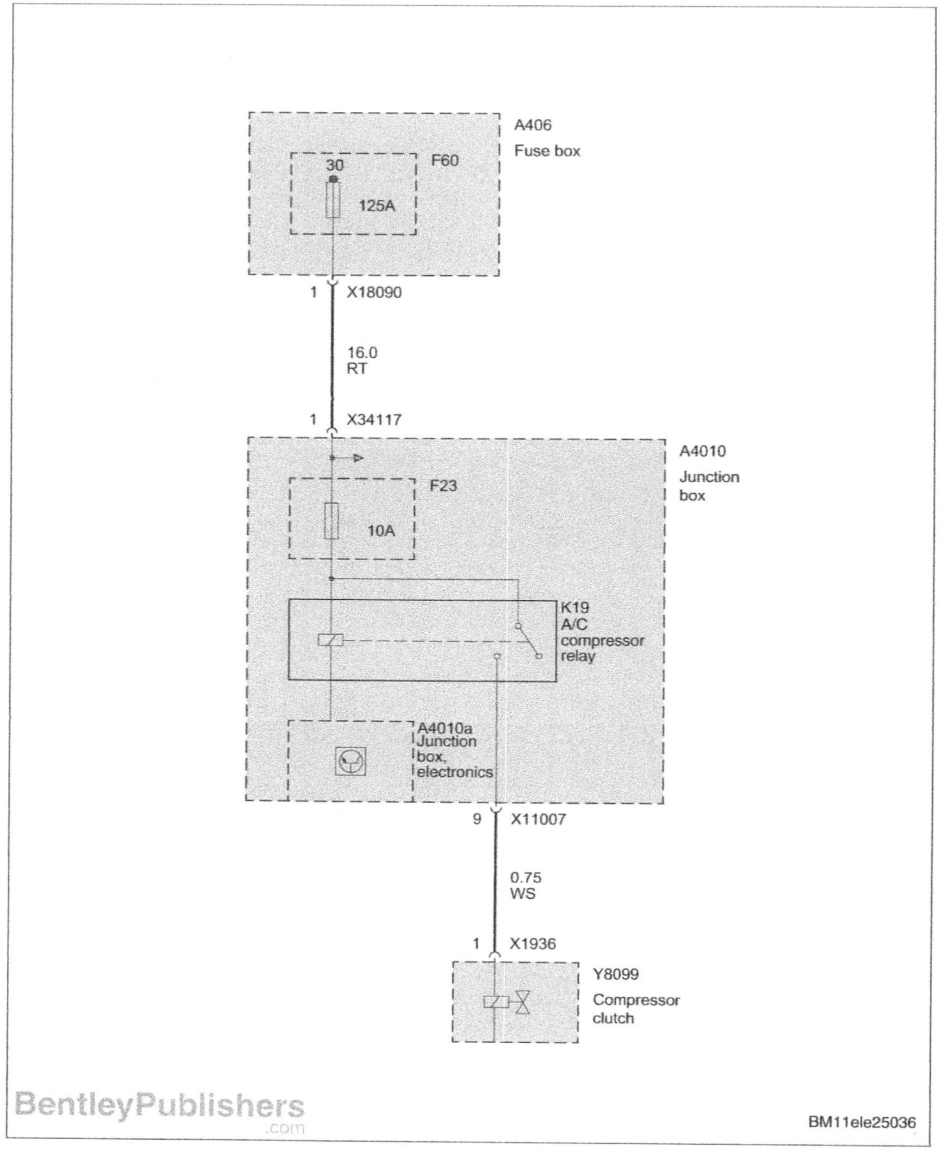

Fuse F23 from 12/2008

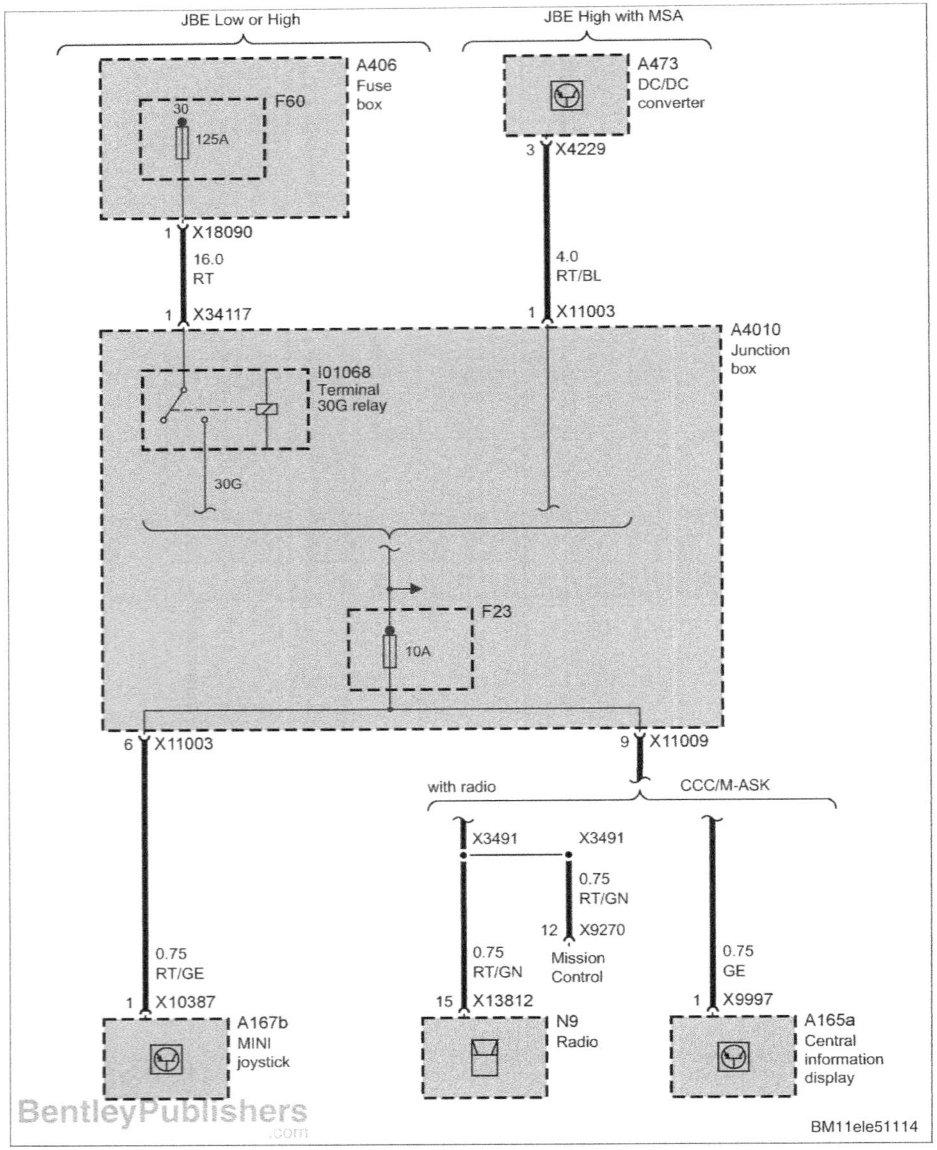

Fuse F24 to 11/2008

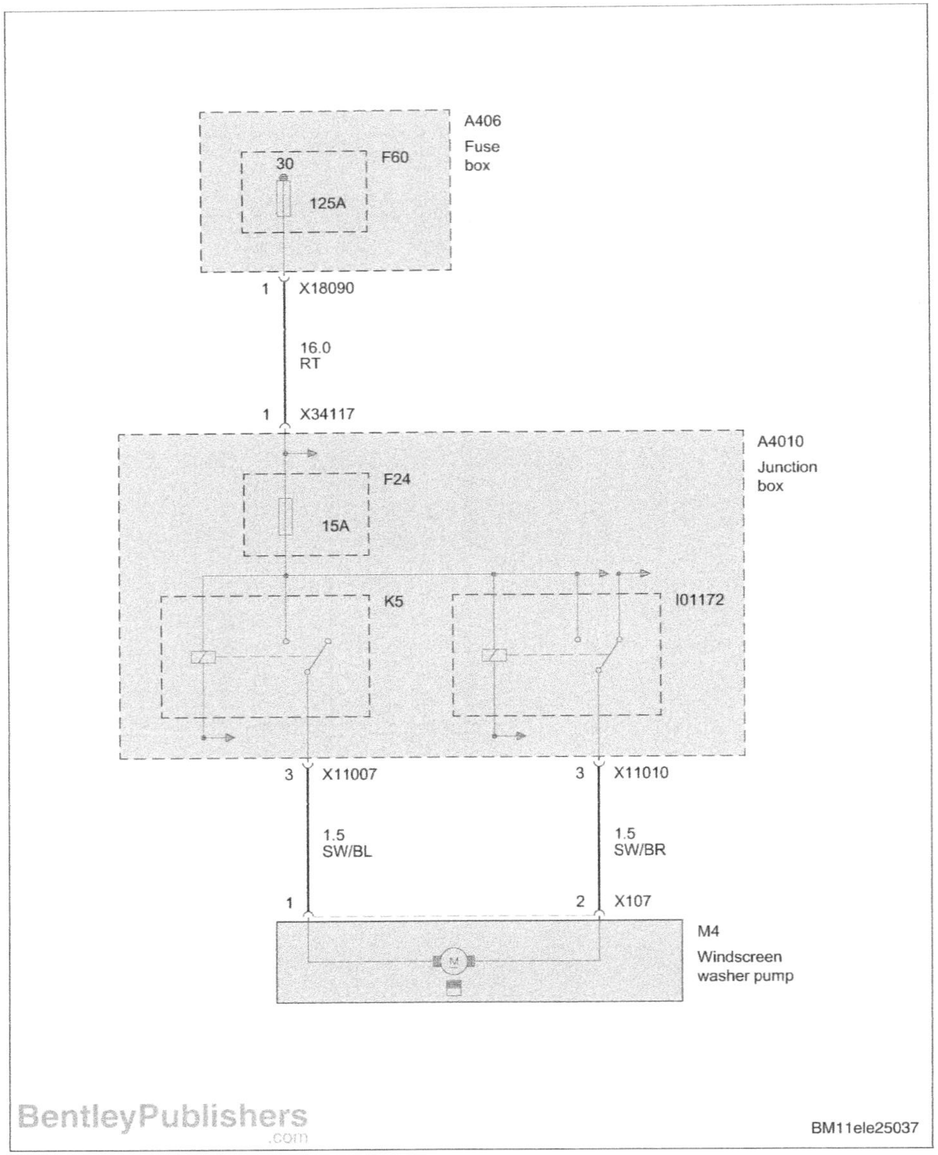

Fuse F24 from 12/2008

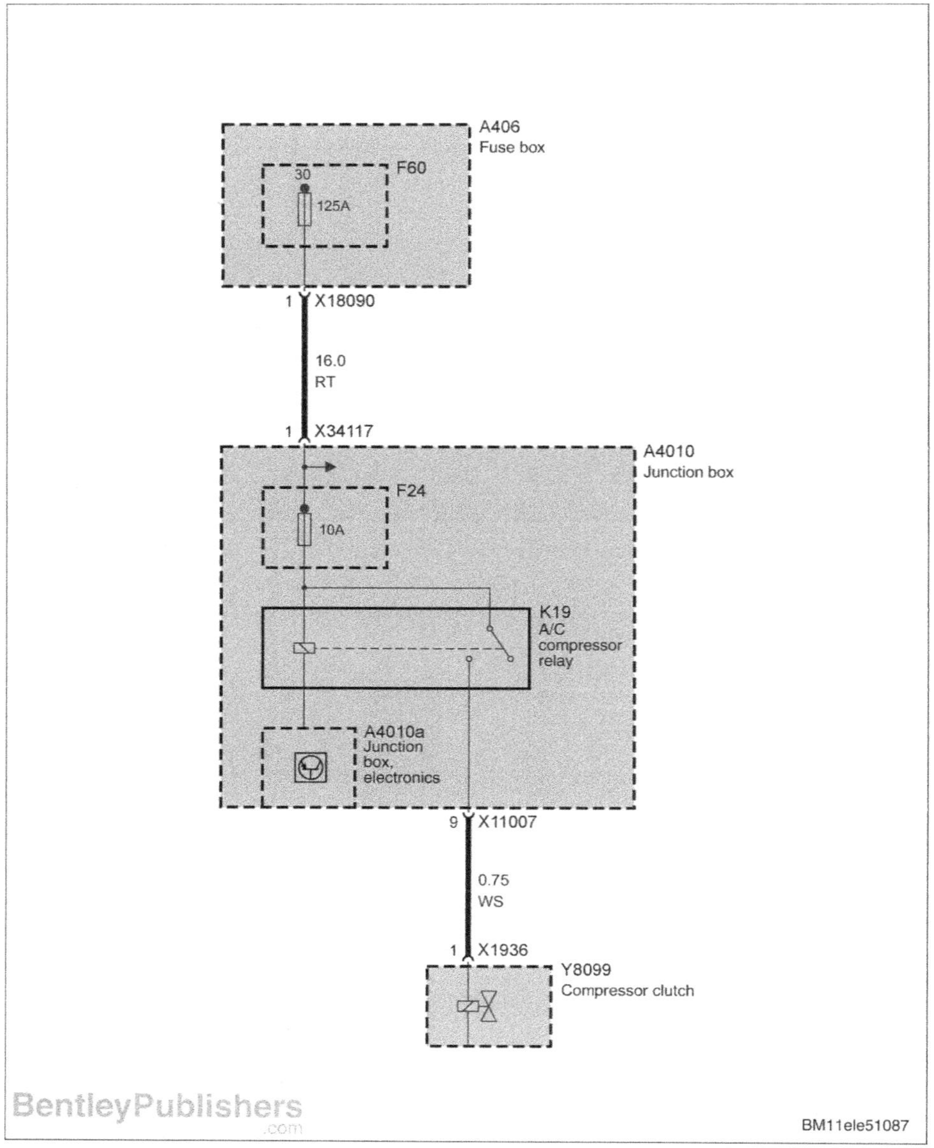

Fuse F25 to 11/2008

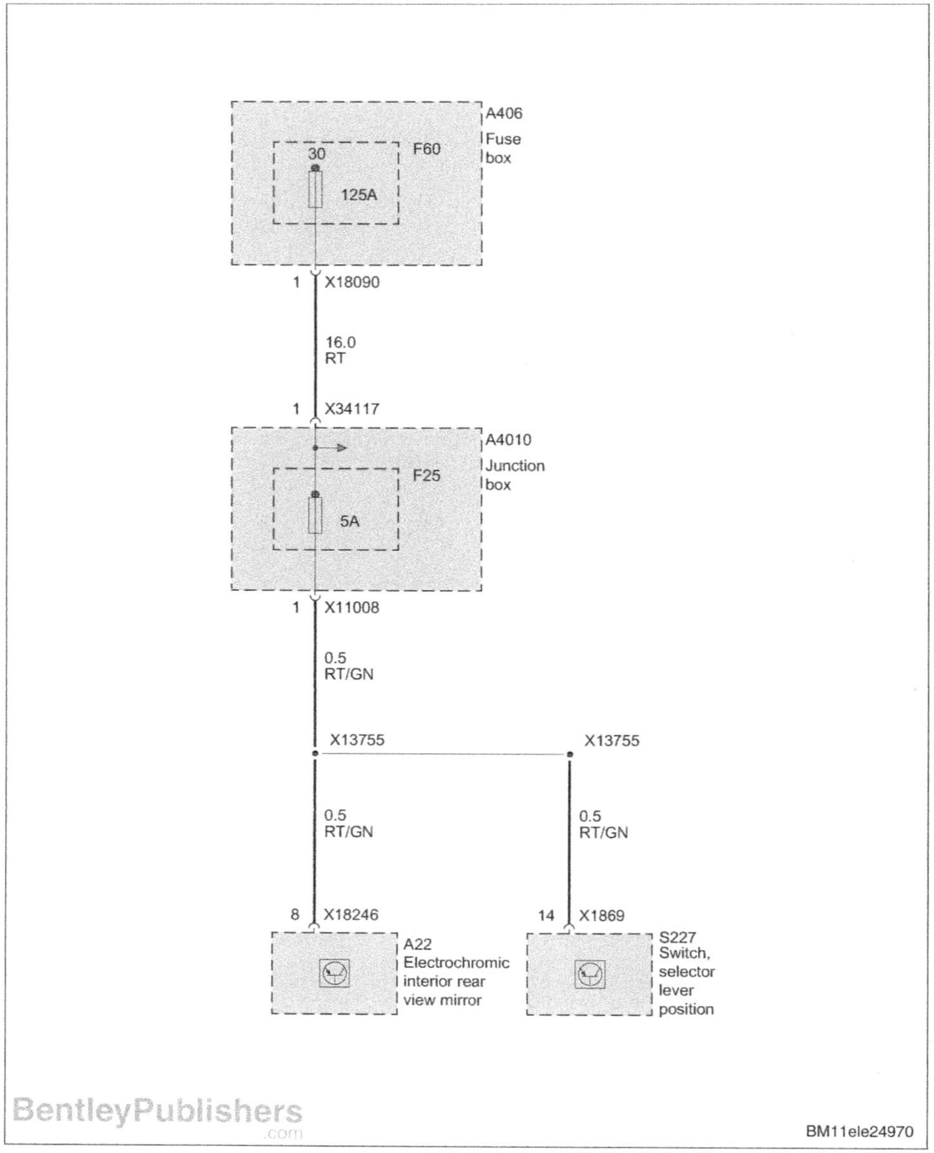

Fuse F25 from 12/2008

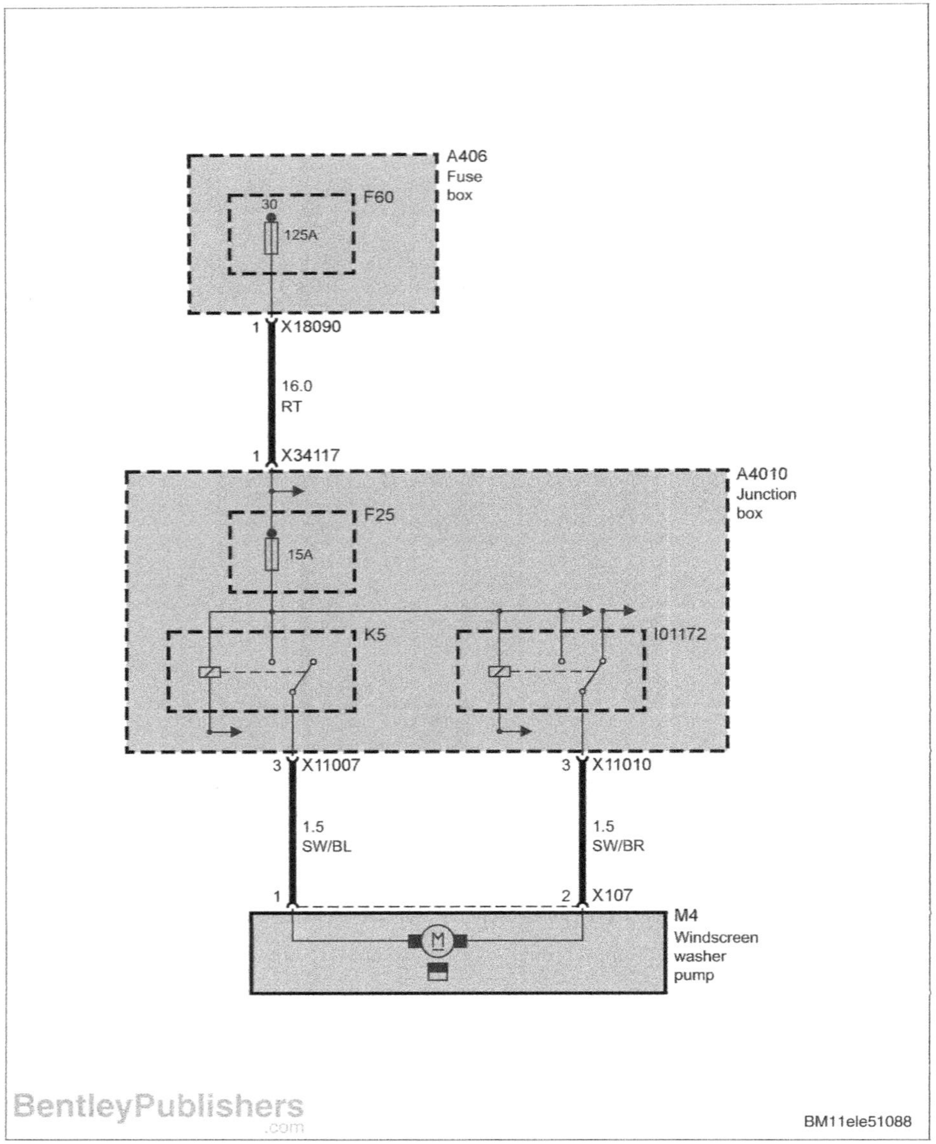

Fuse F26 to 11/2008

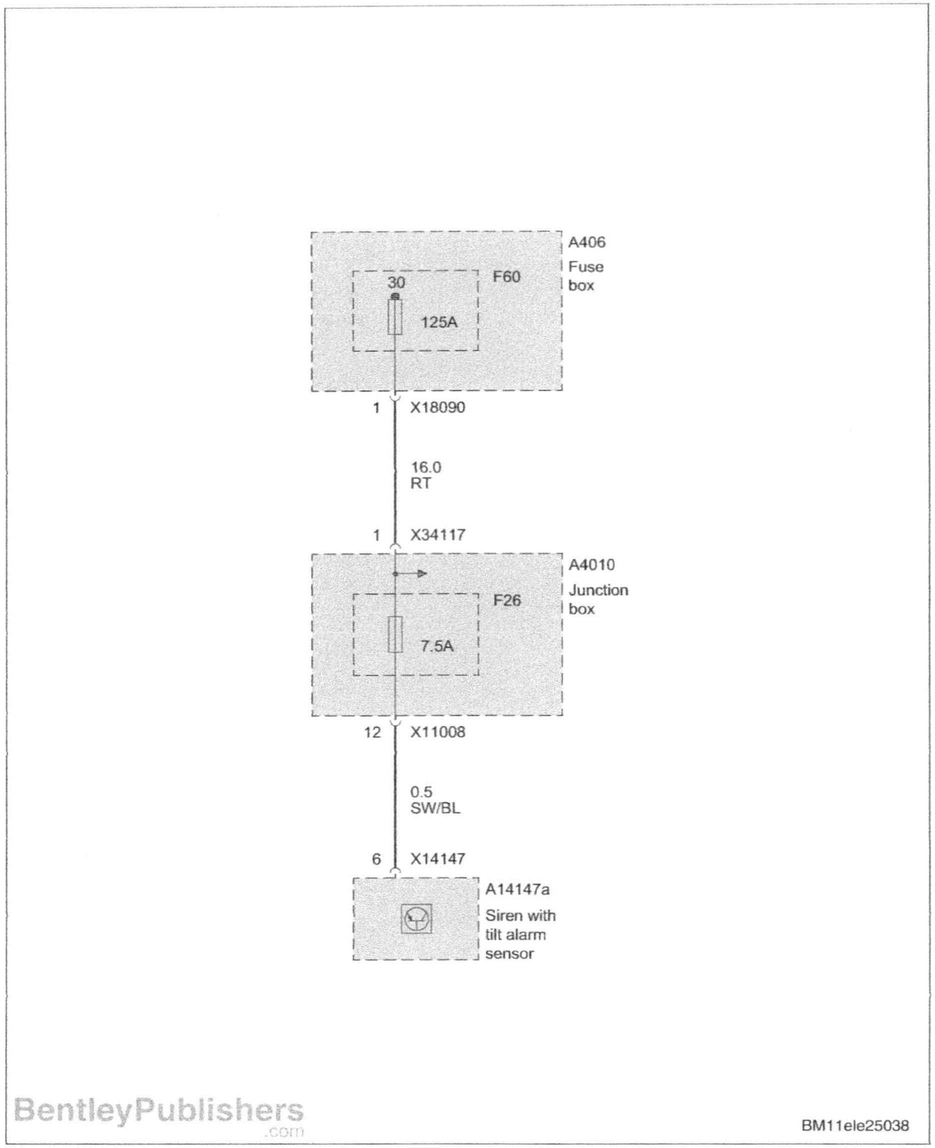

Fuse F26
from 12/2008

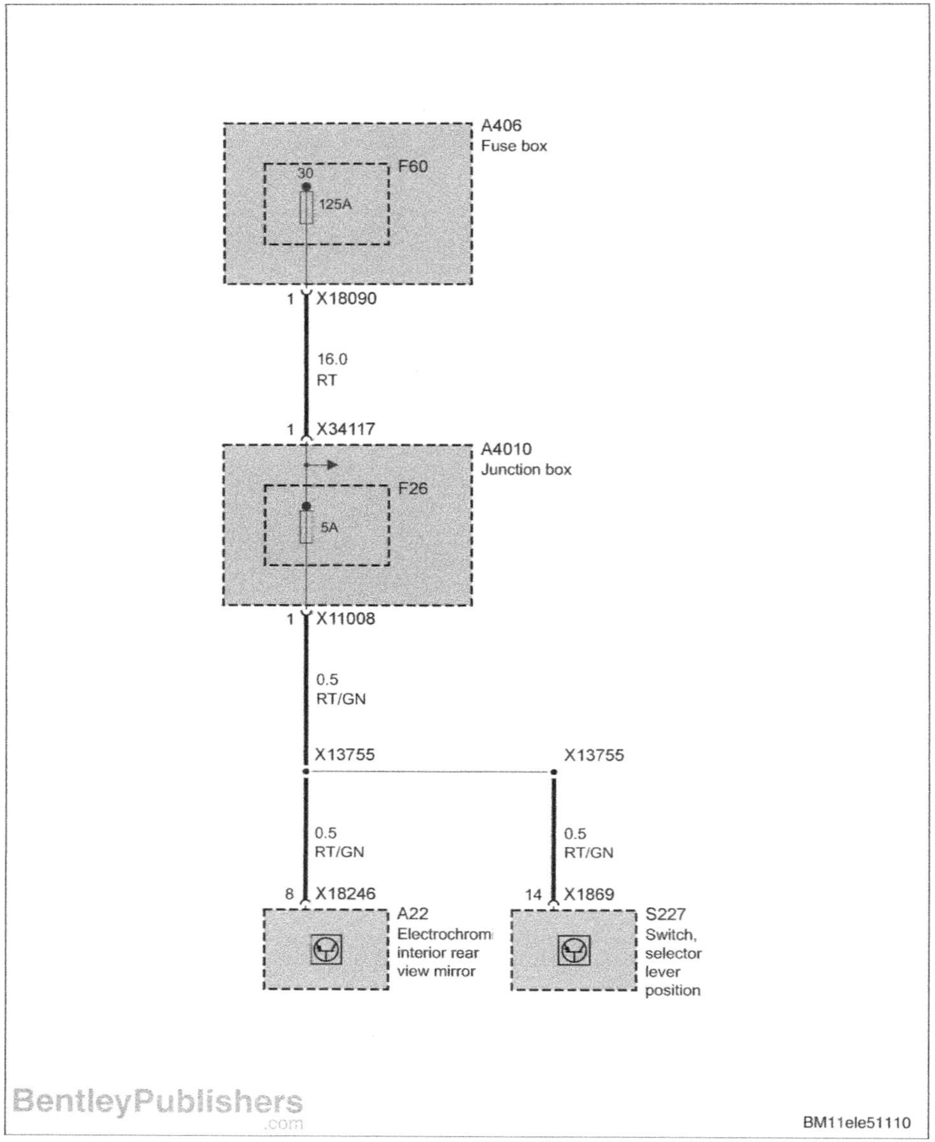

Fuse F27
to 11/2008

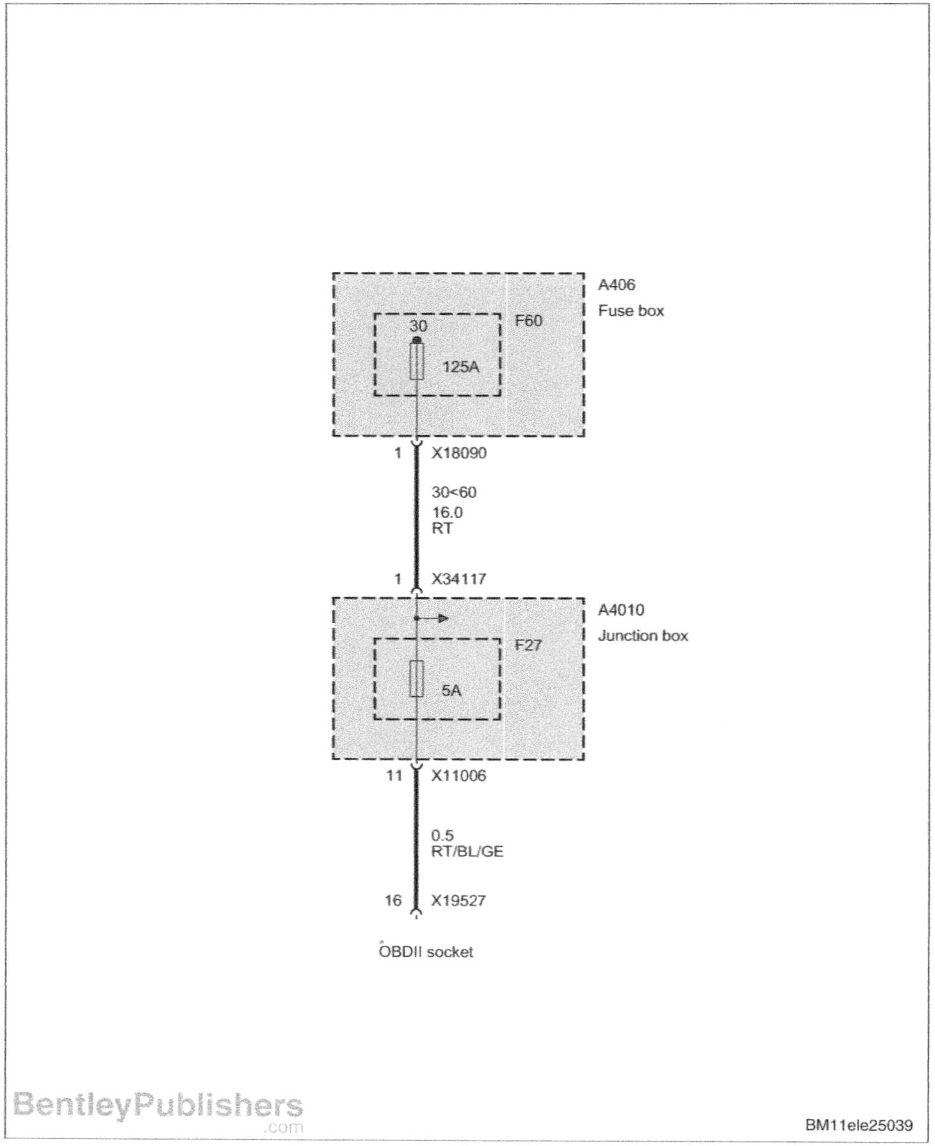

Fuse F27 from 12/2008

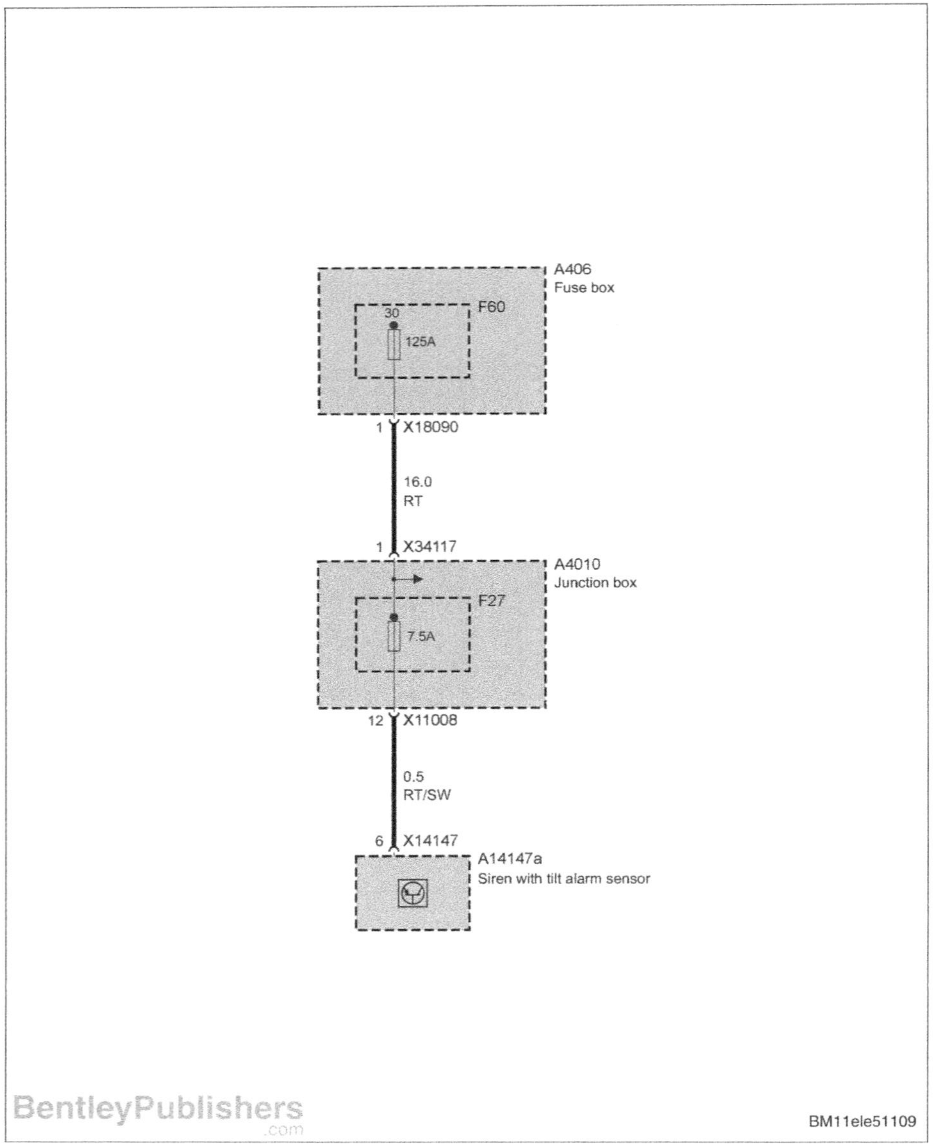

Fuse F28 from 12/2008

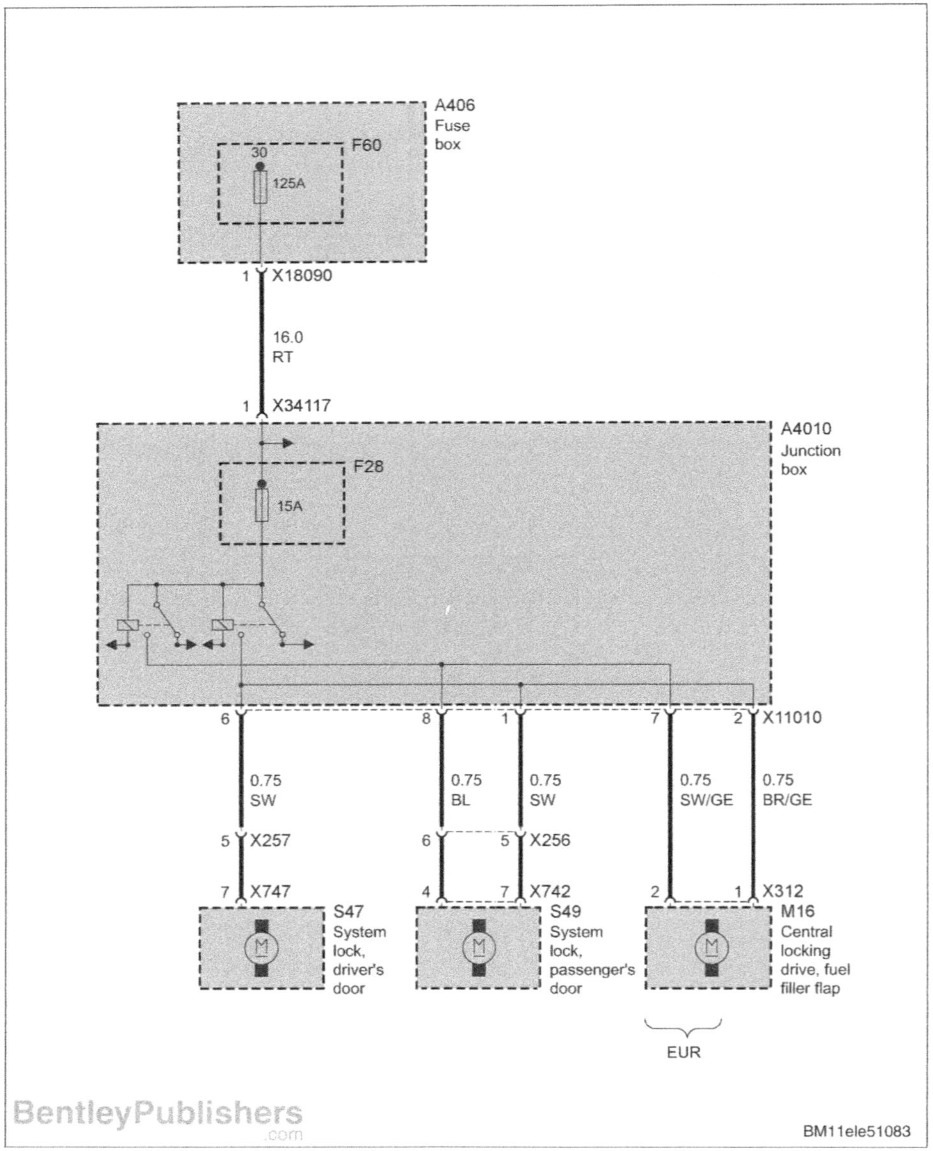

Fuse F29
to 11/2008

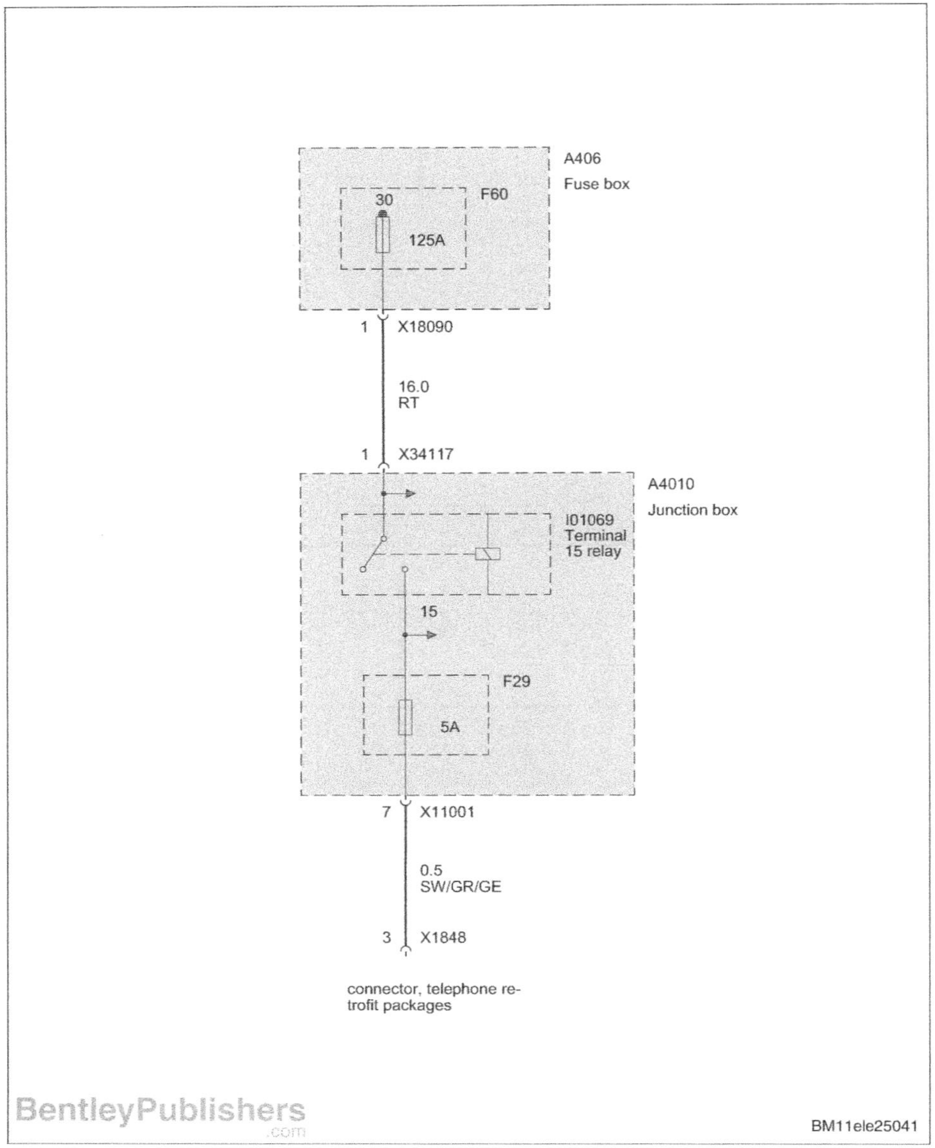

Fuse F29
from 12/2008

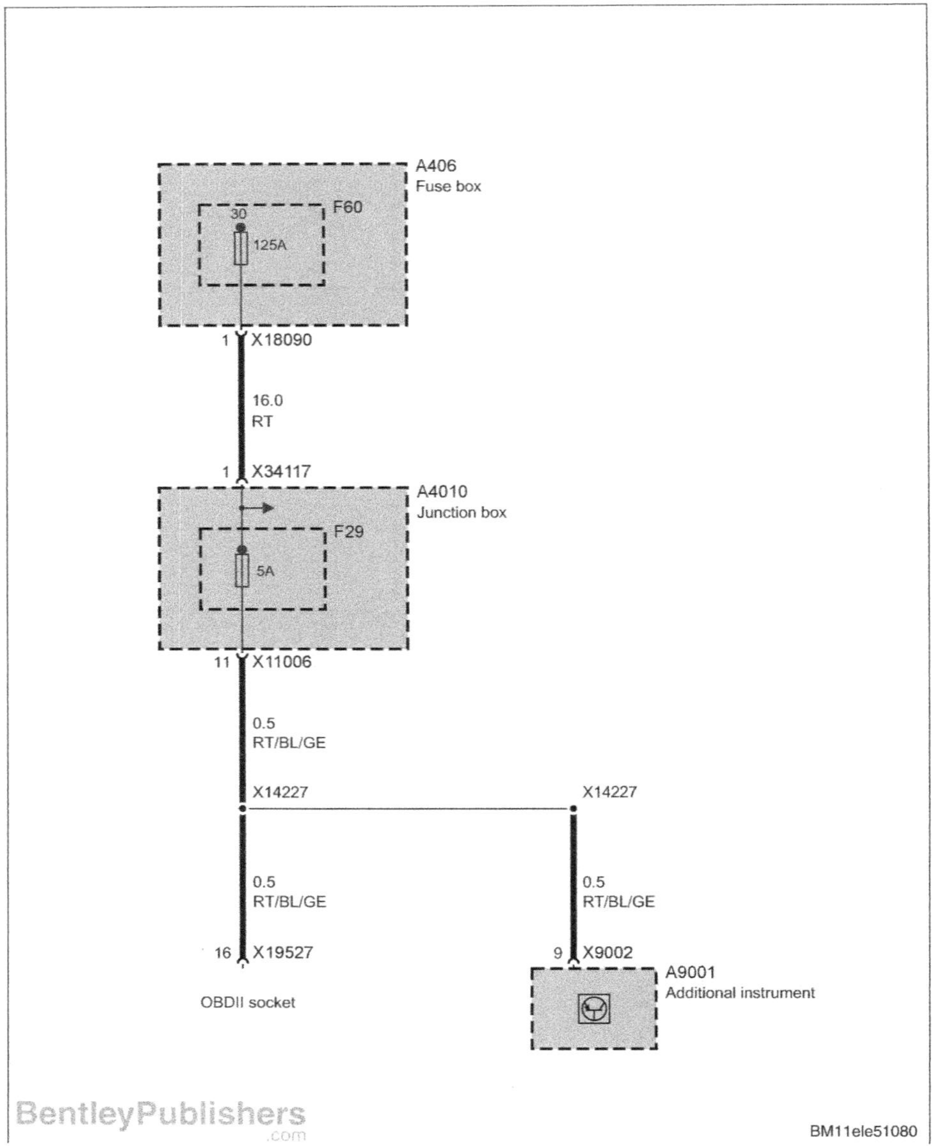

Fuse F30
to 11/2008

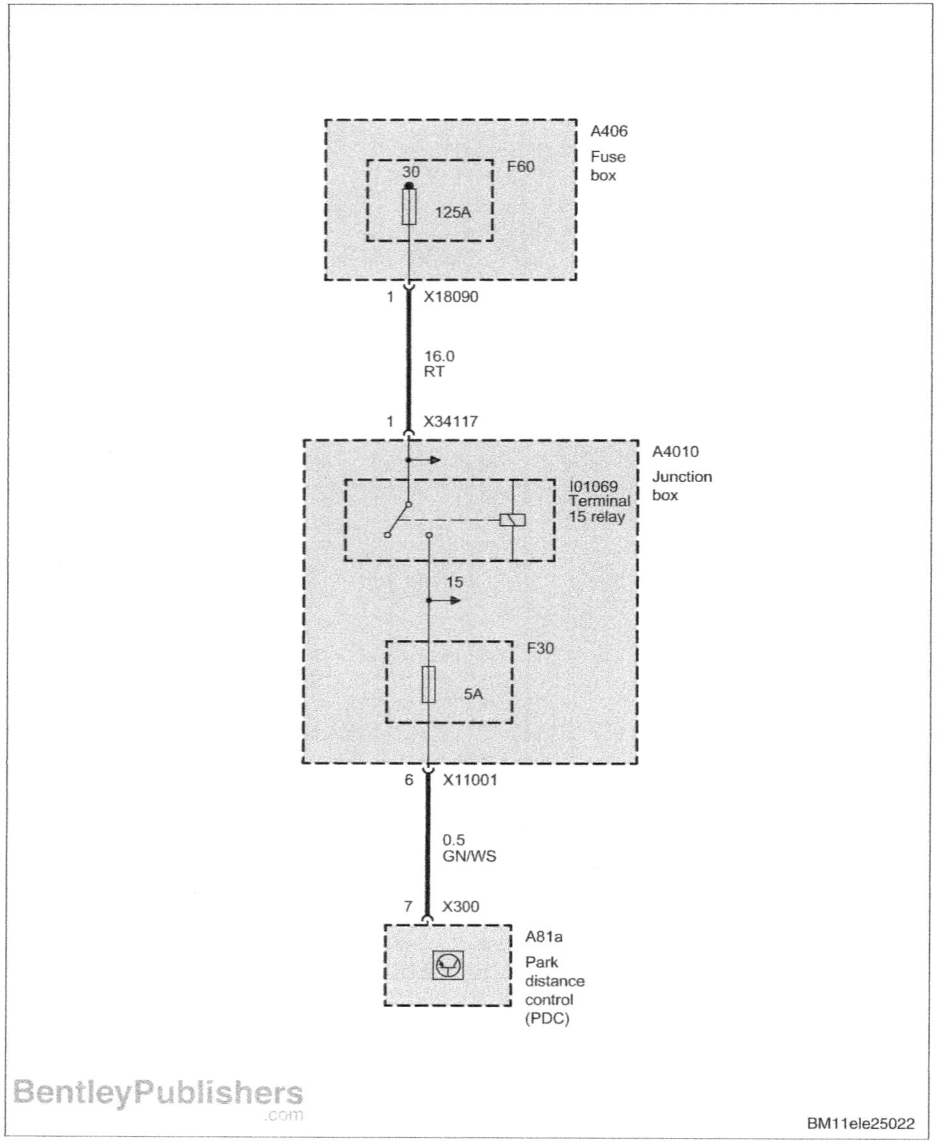

Fuse F30
Clubman from 12/2008

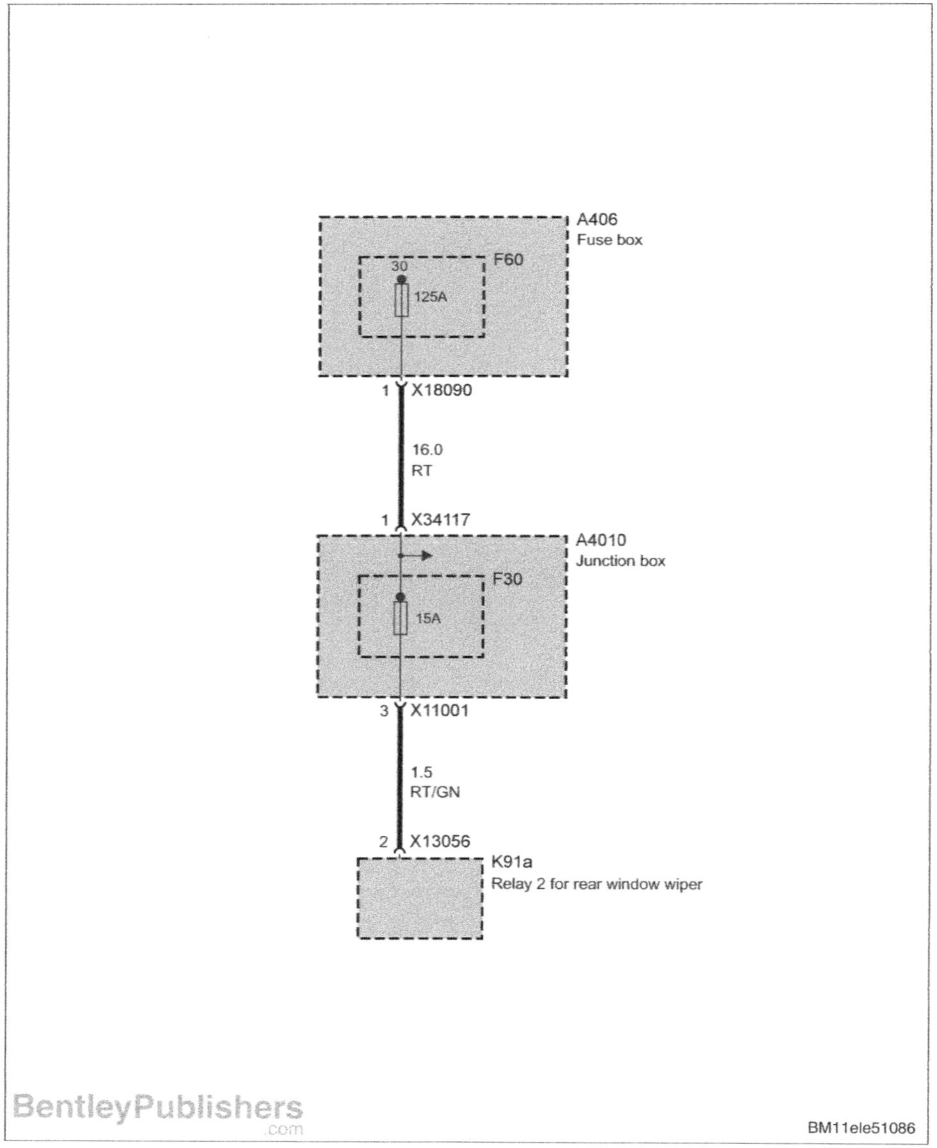

Fuse F31 to 11/2008

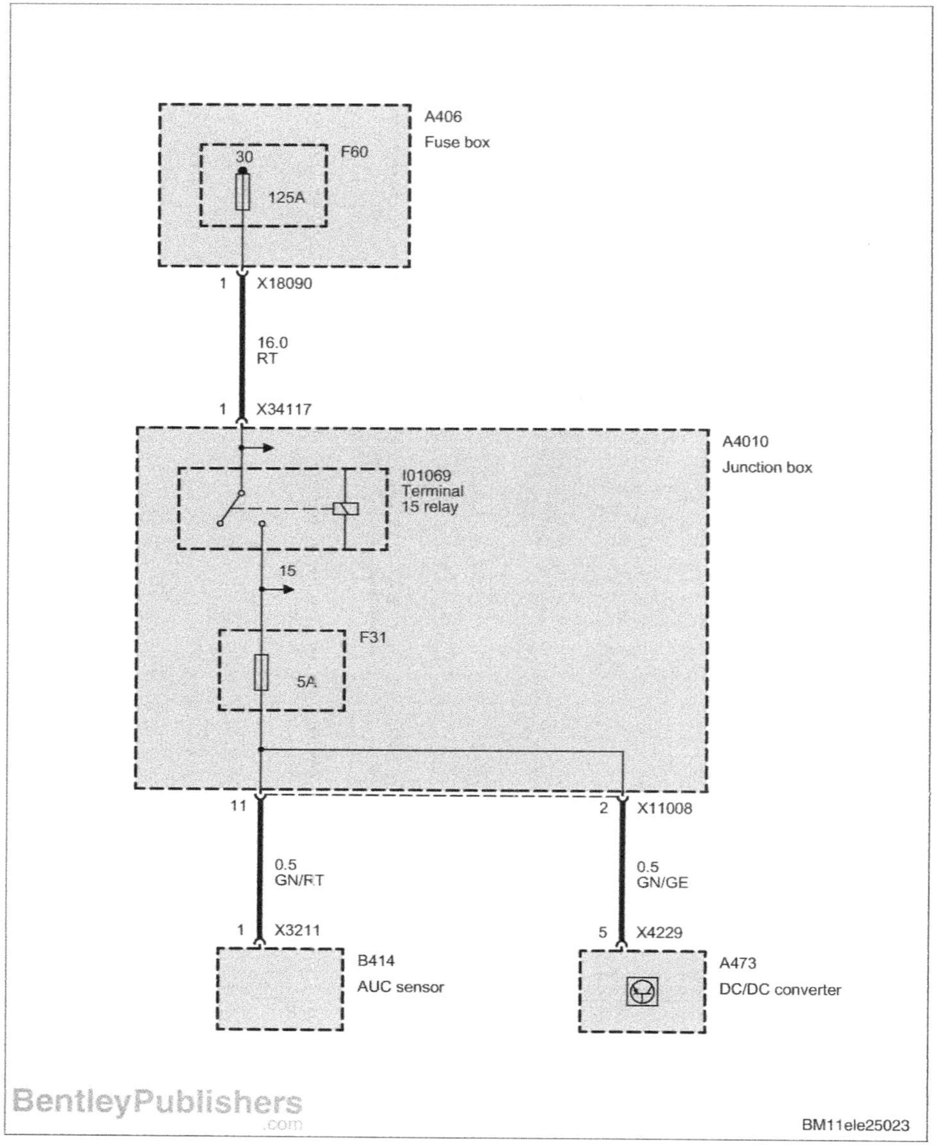

Fuse F32 to 11/2008

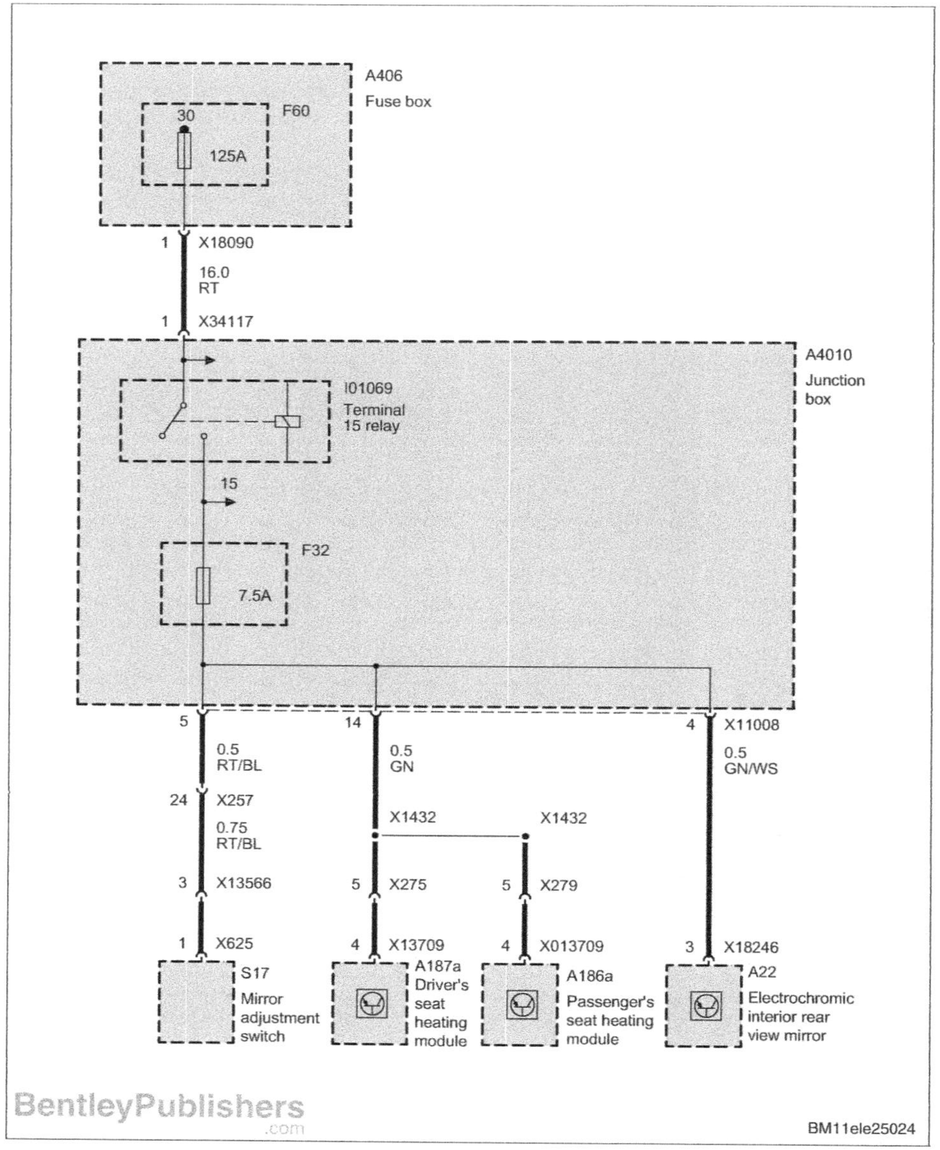

Fuse F32 from 12/2008

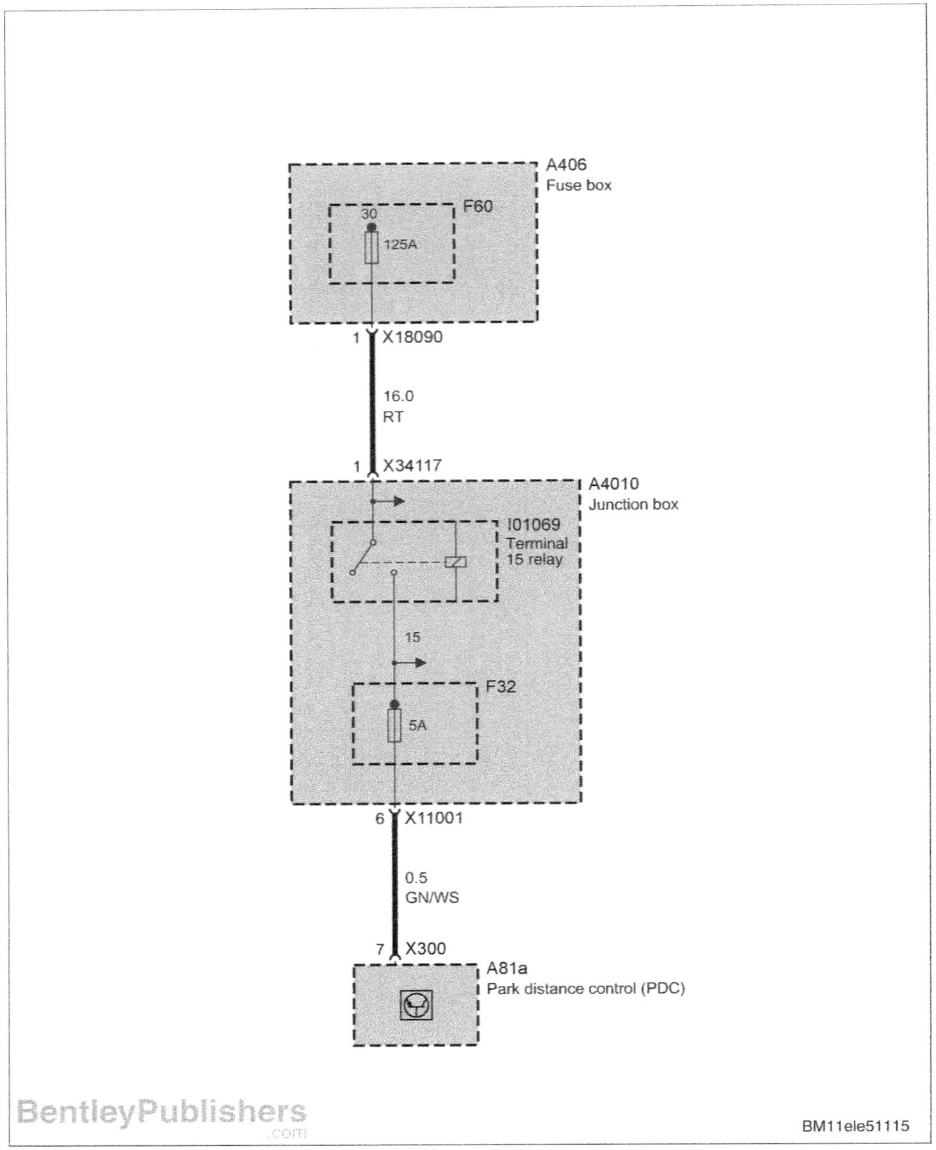

Fuse F33 to 11/2008

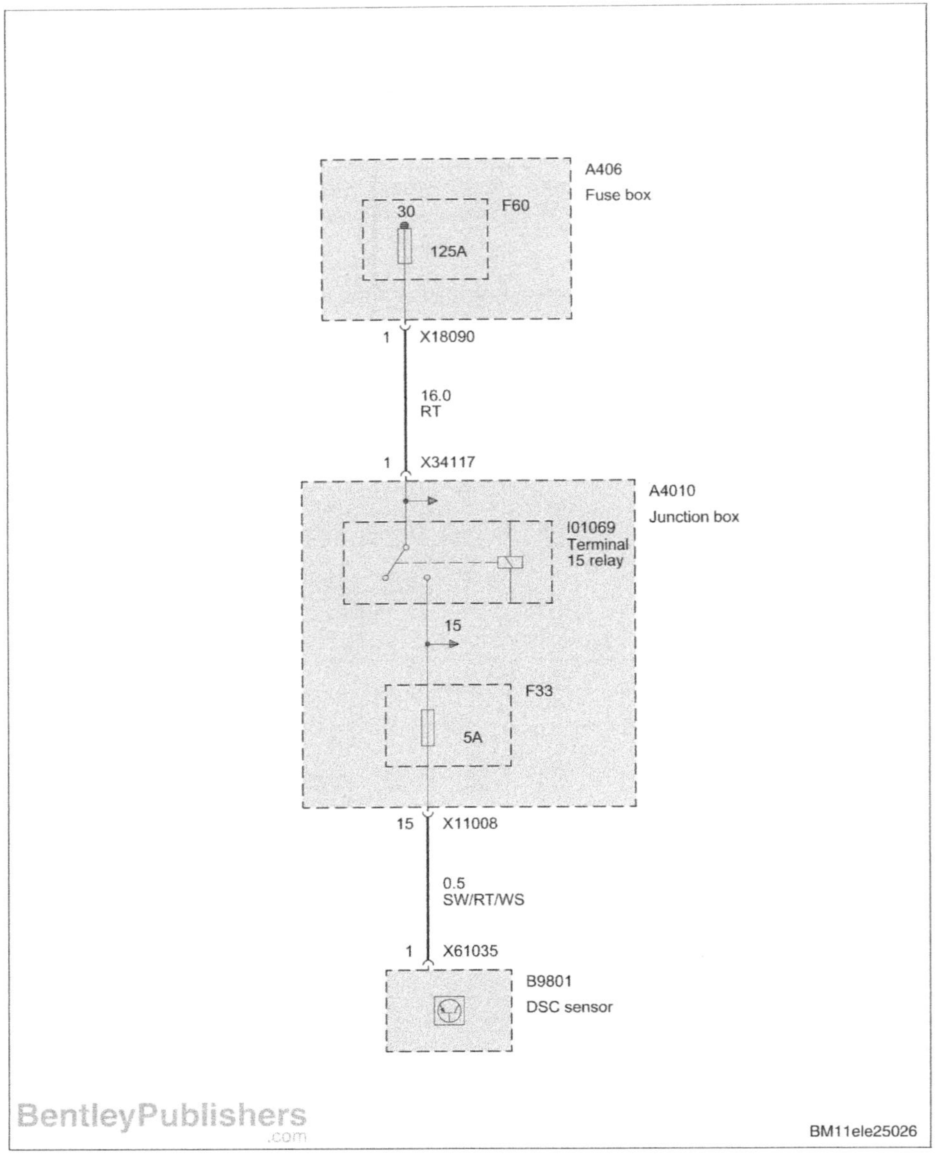

Fuse F33 from 12/2008

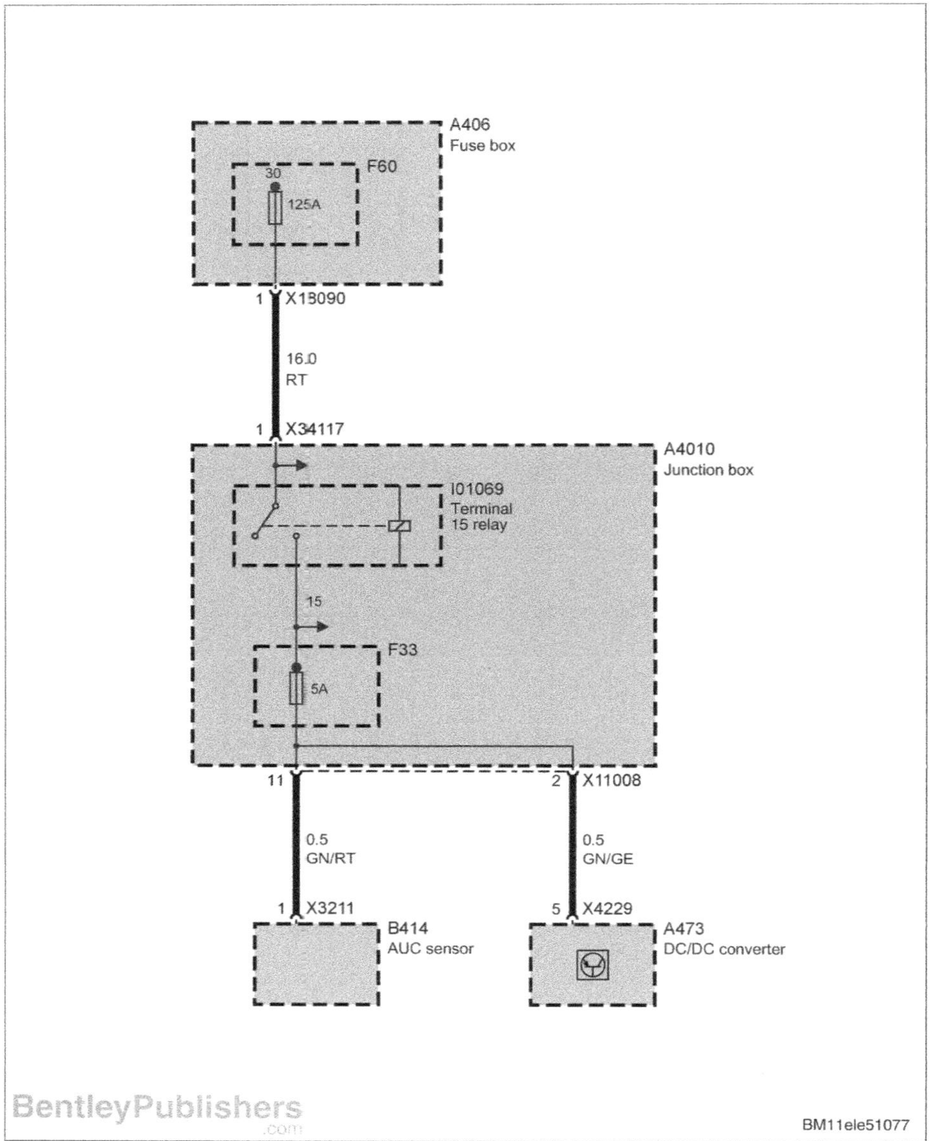

Fuse F34 to 11/2008

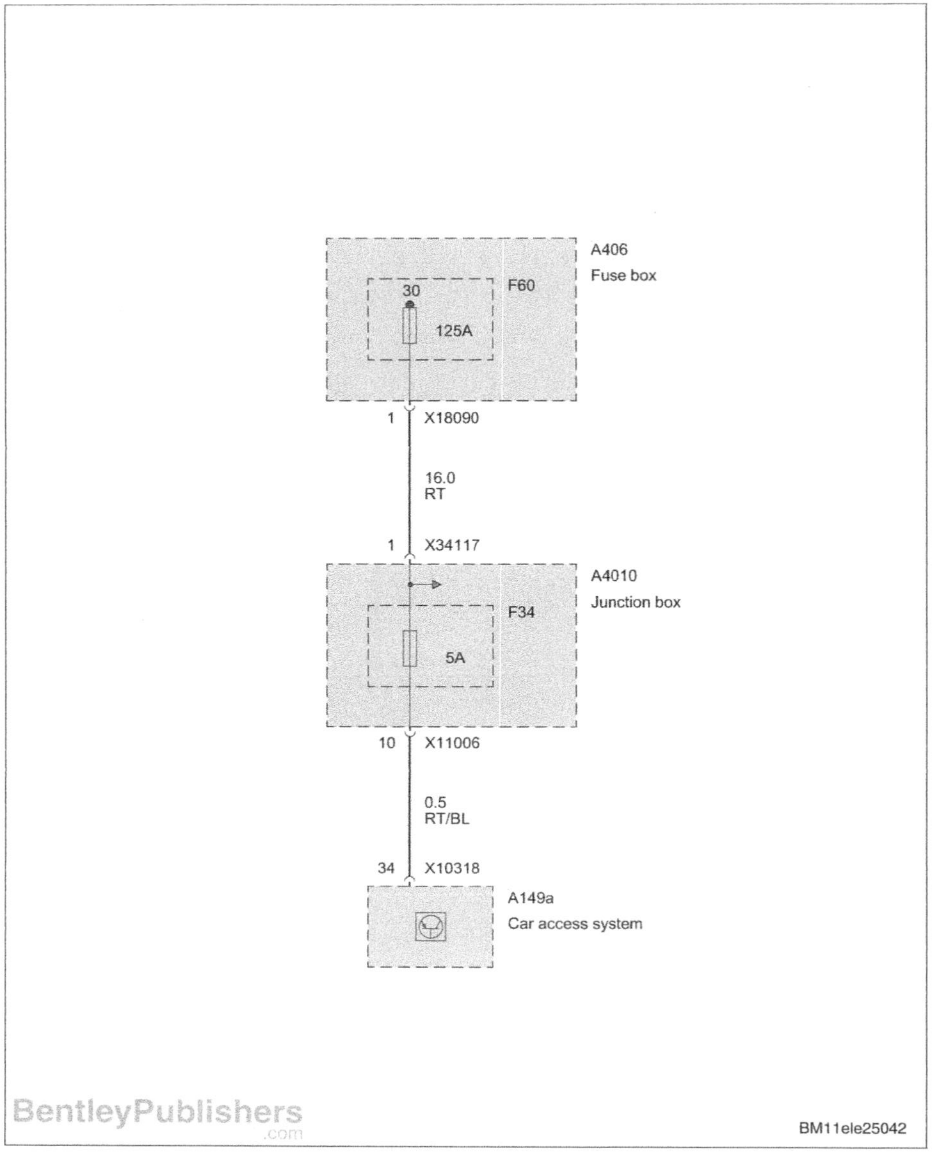

Fuse F34 from 12/2008

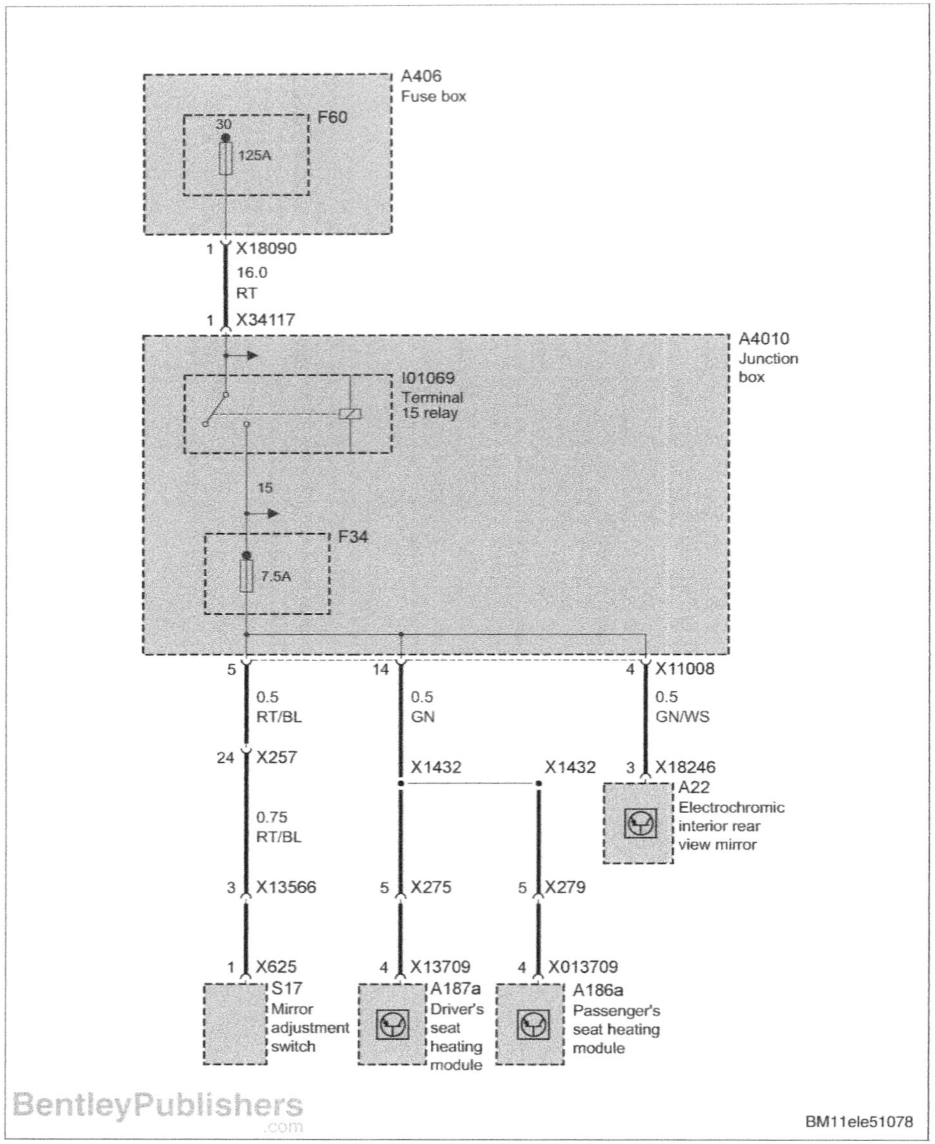

Fuse F35 from 12/2008

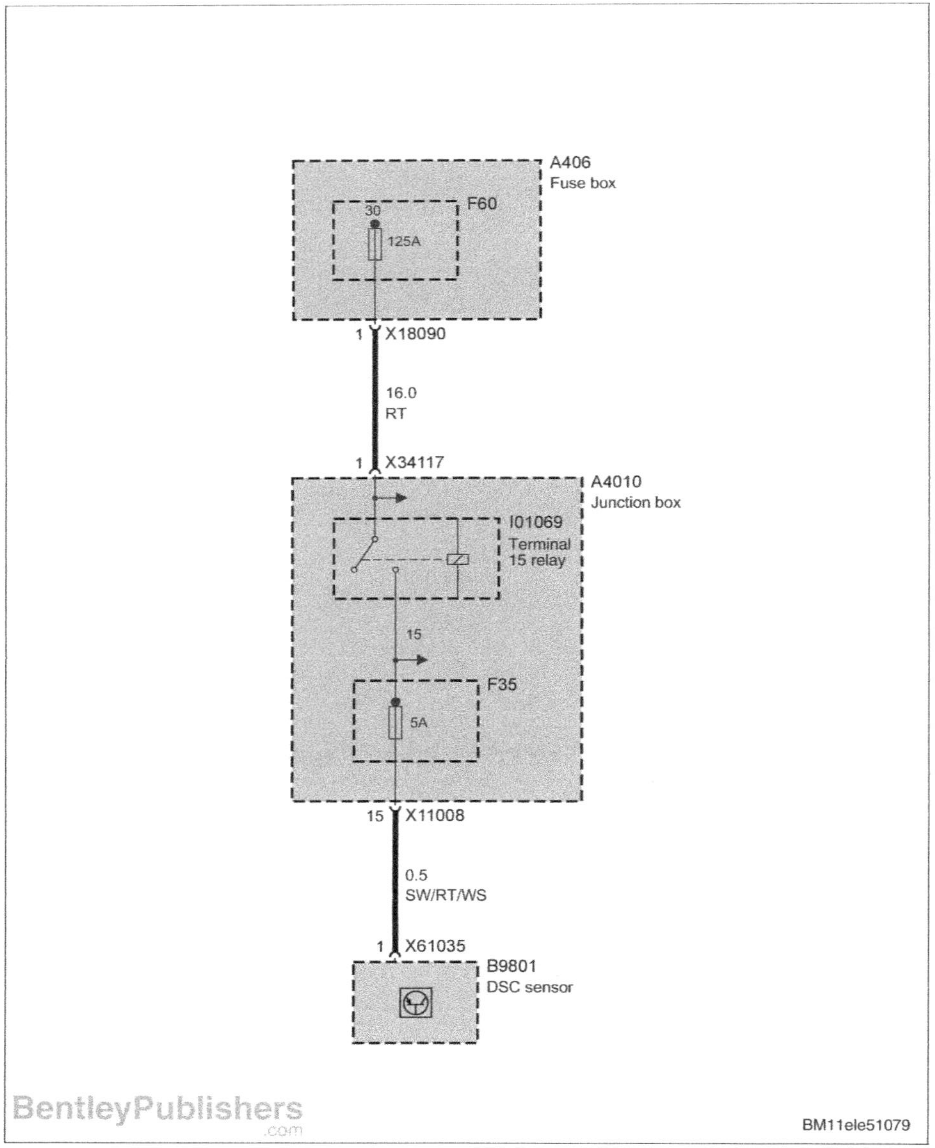

Fuse F36 to 11/2008

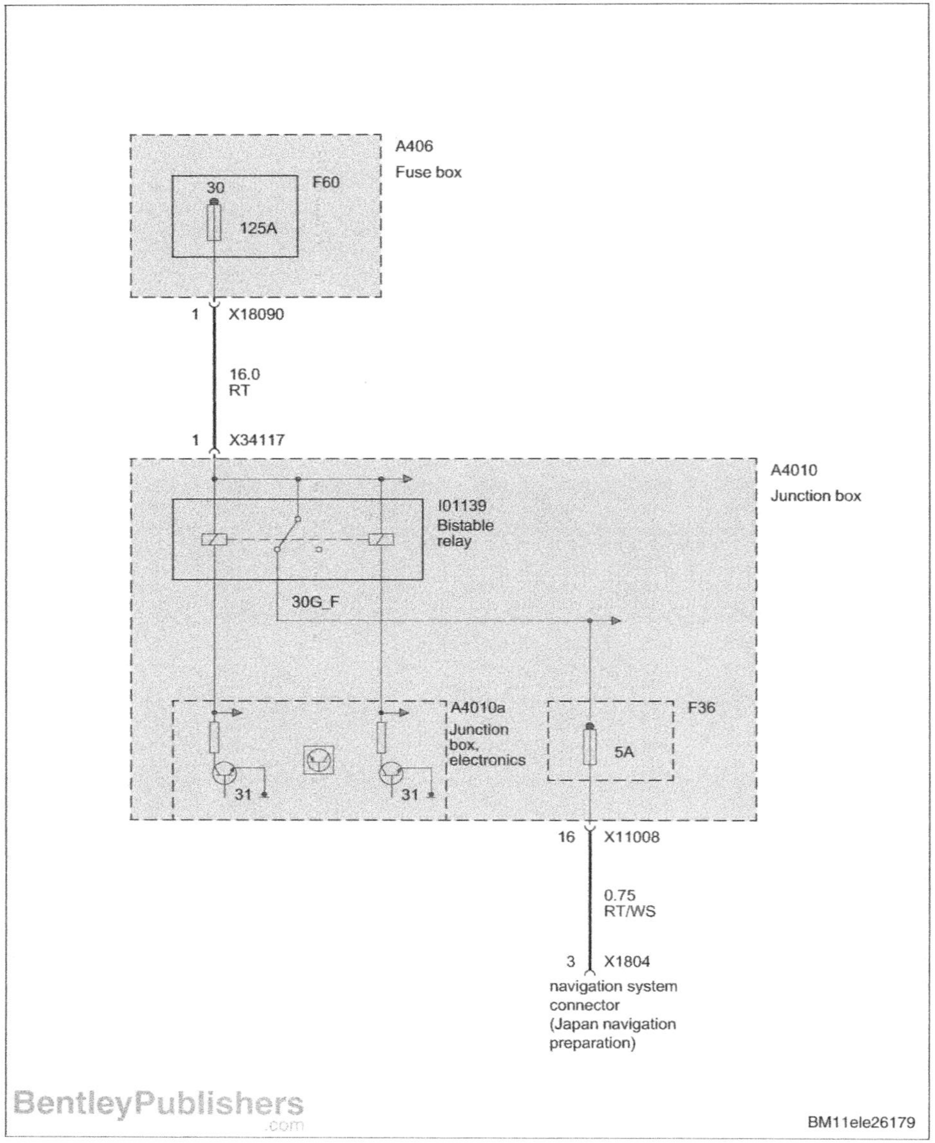

Fuse F36 from 12/2008

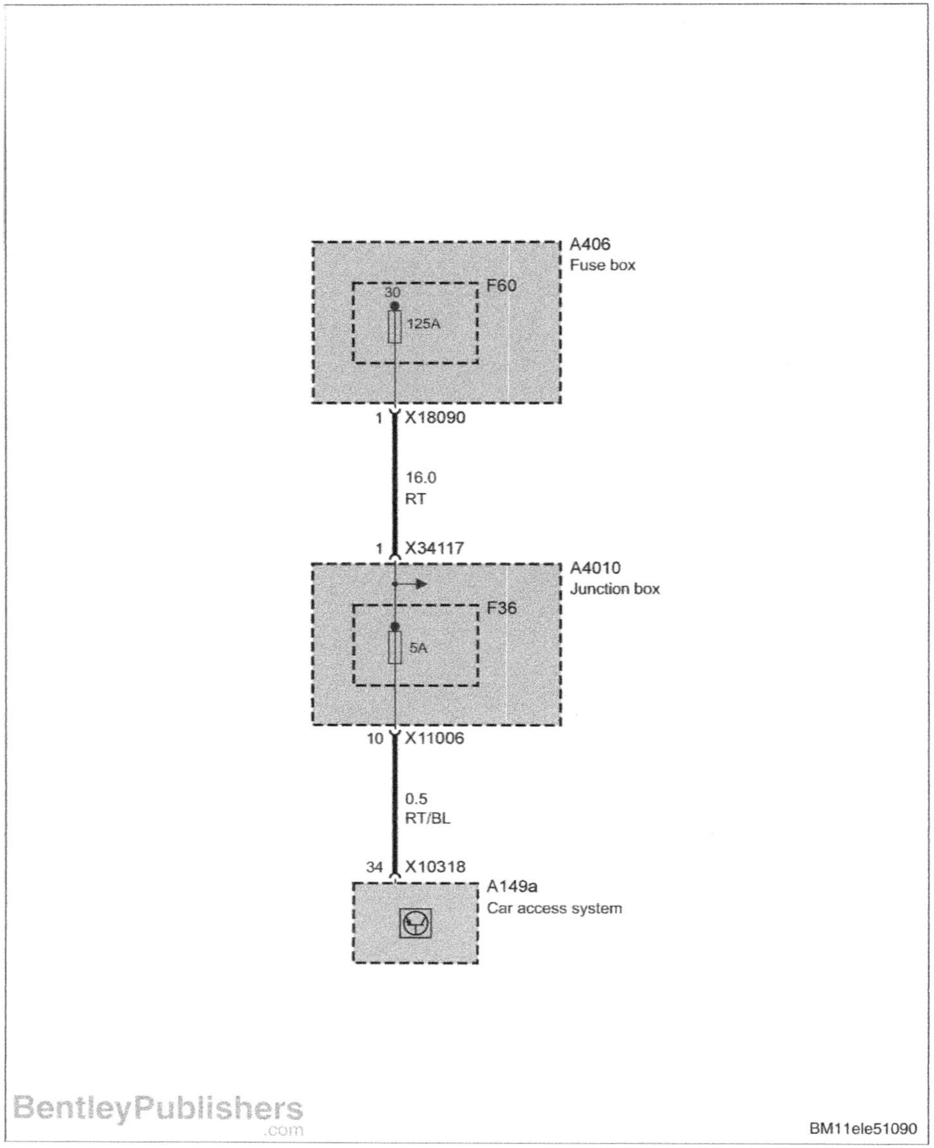

Fuse F37 to 11/2008

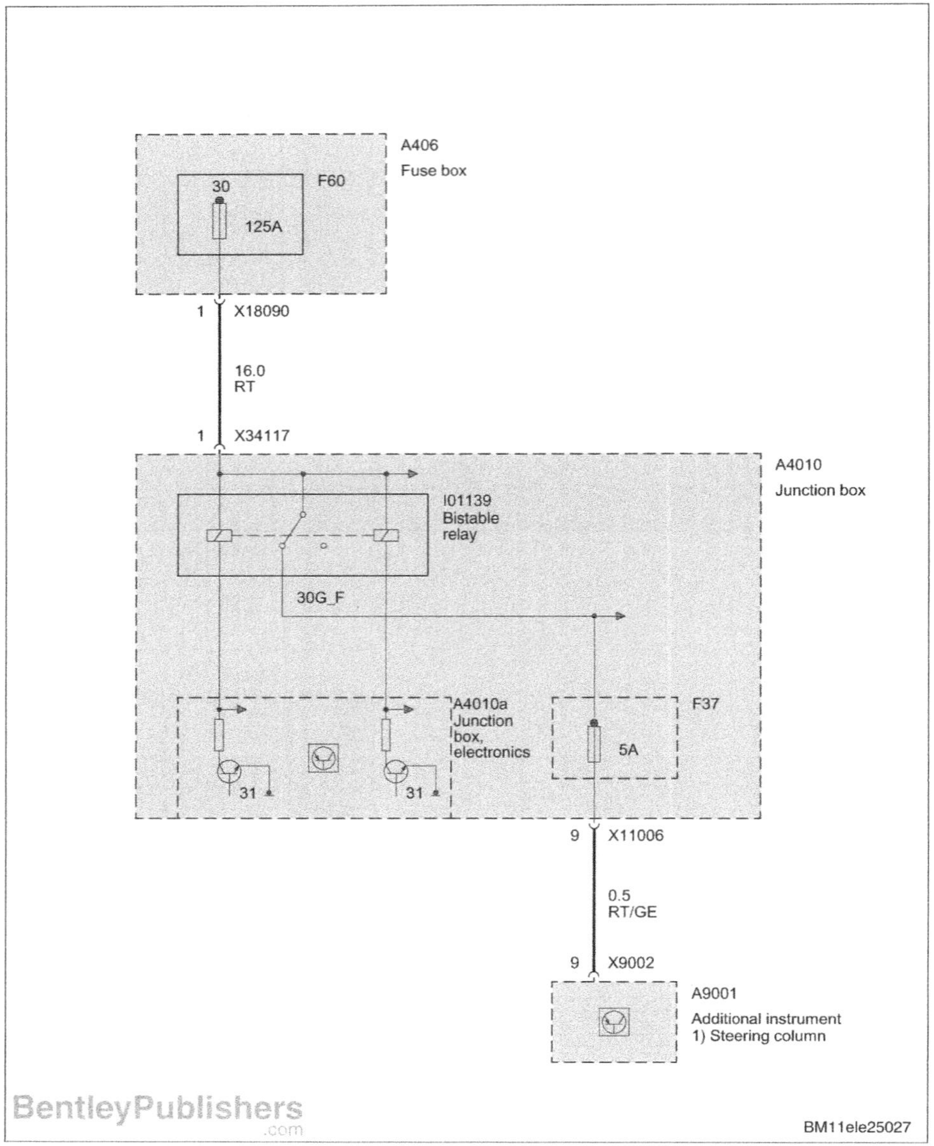

Fuse F38 to 11/2008

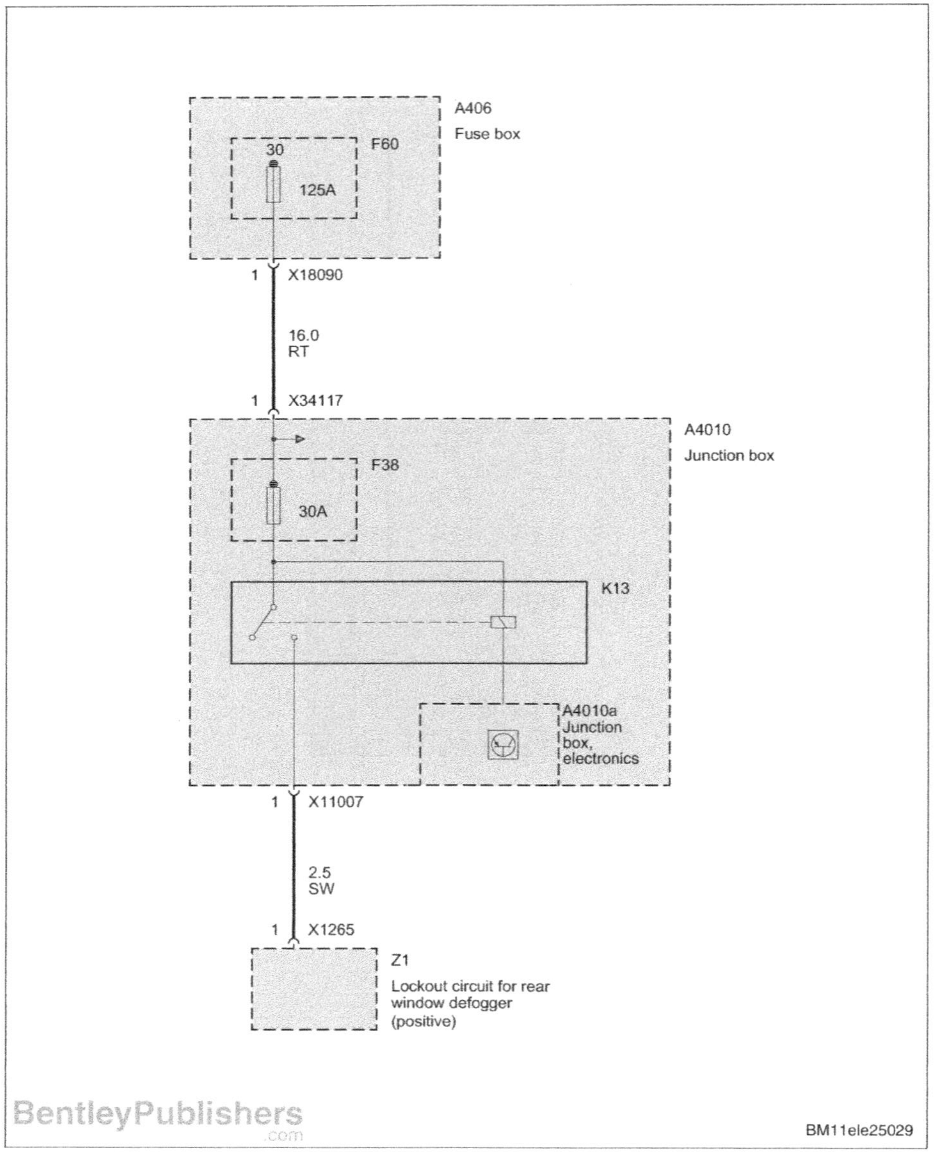

Fuse F38 from 12/2008

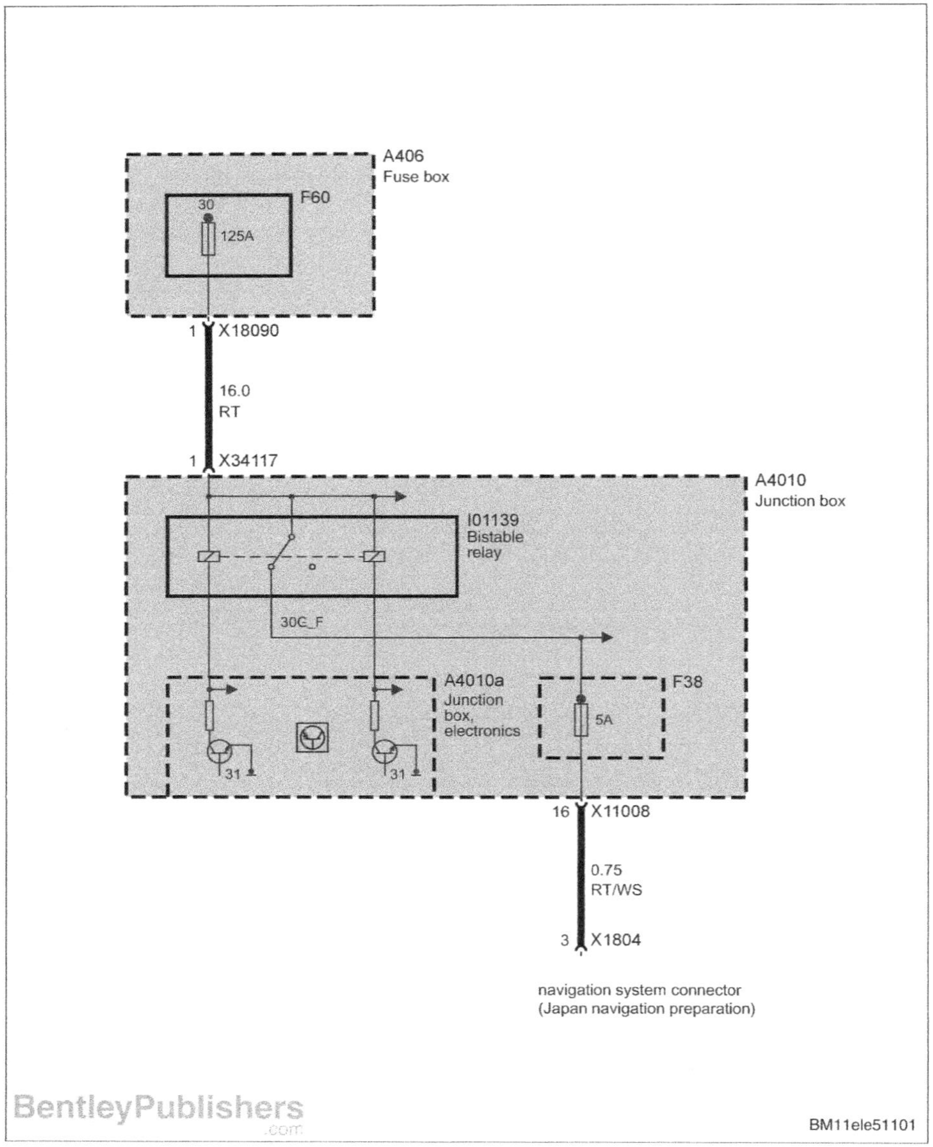

Fuse F39 to 11/2008

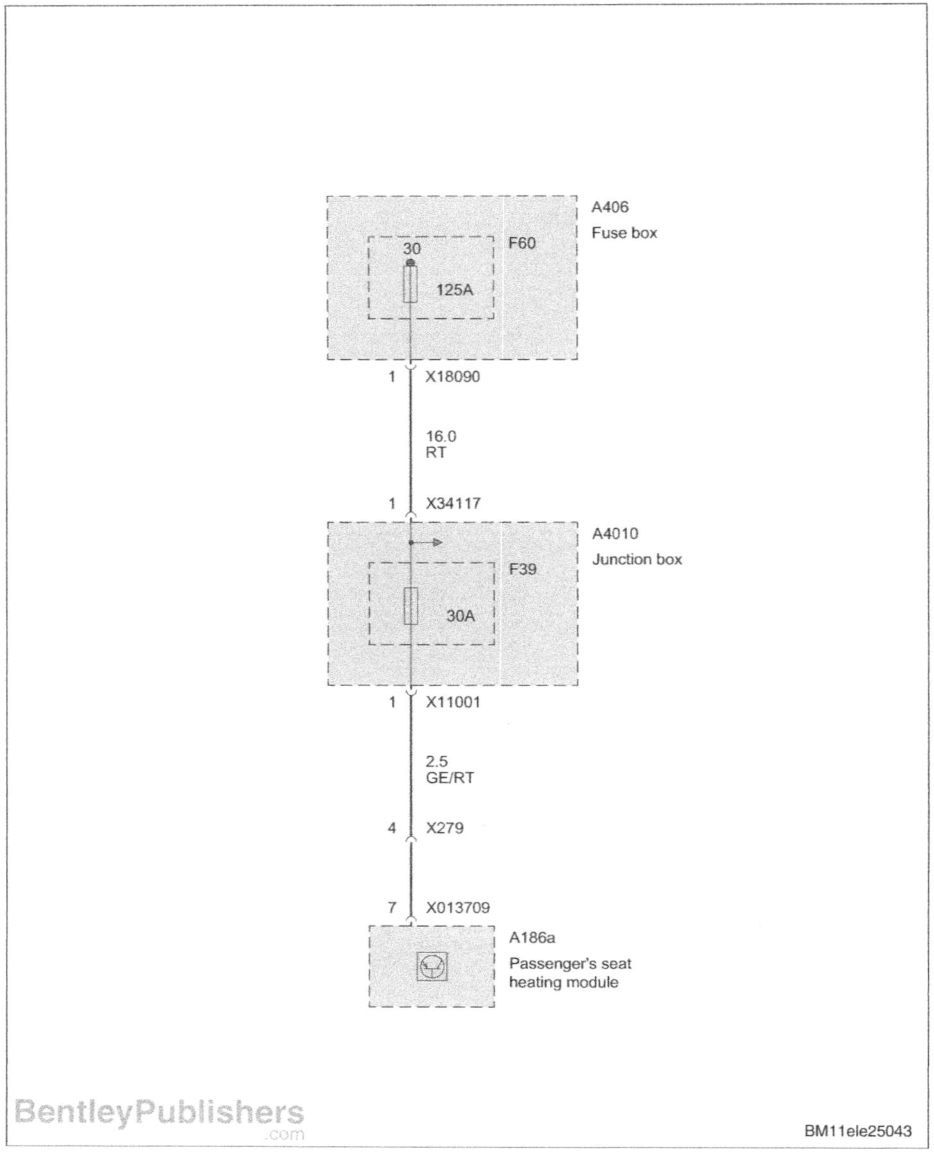

Fuse F40
to 11/2008

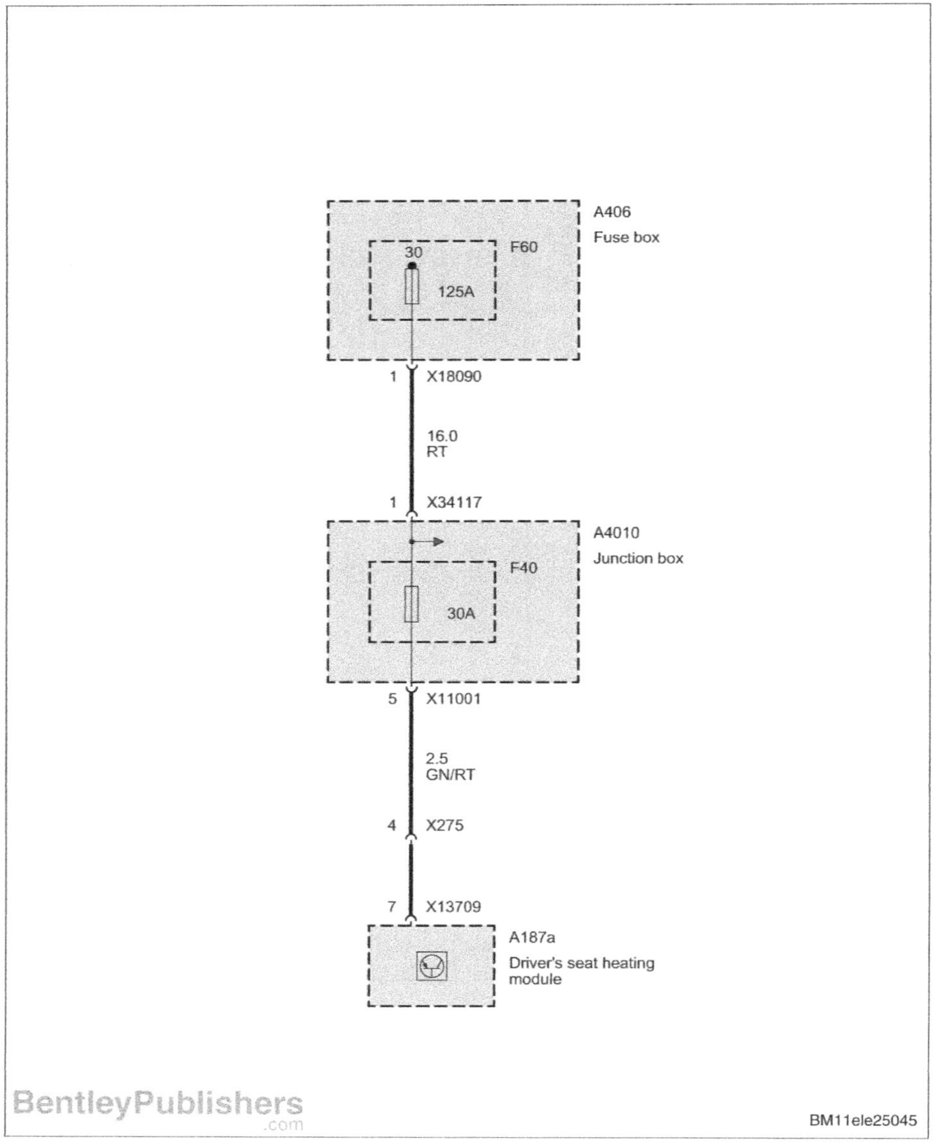

Fuse F40
Clubman from 12/2008

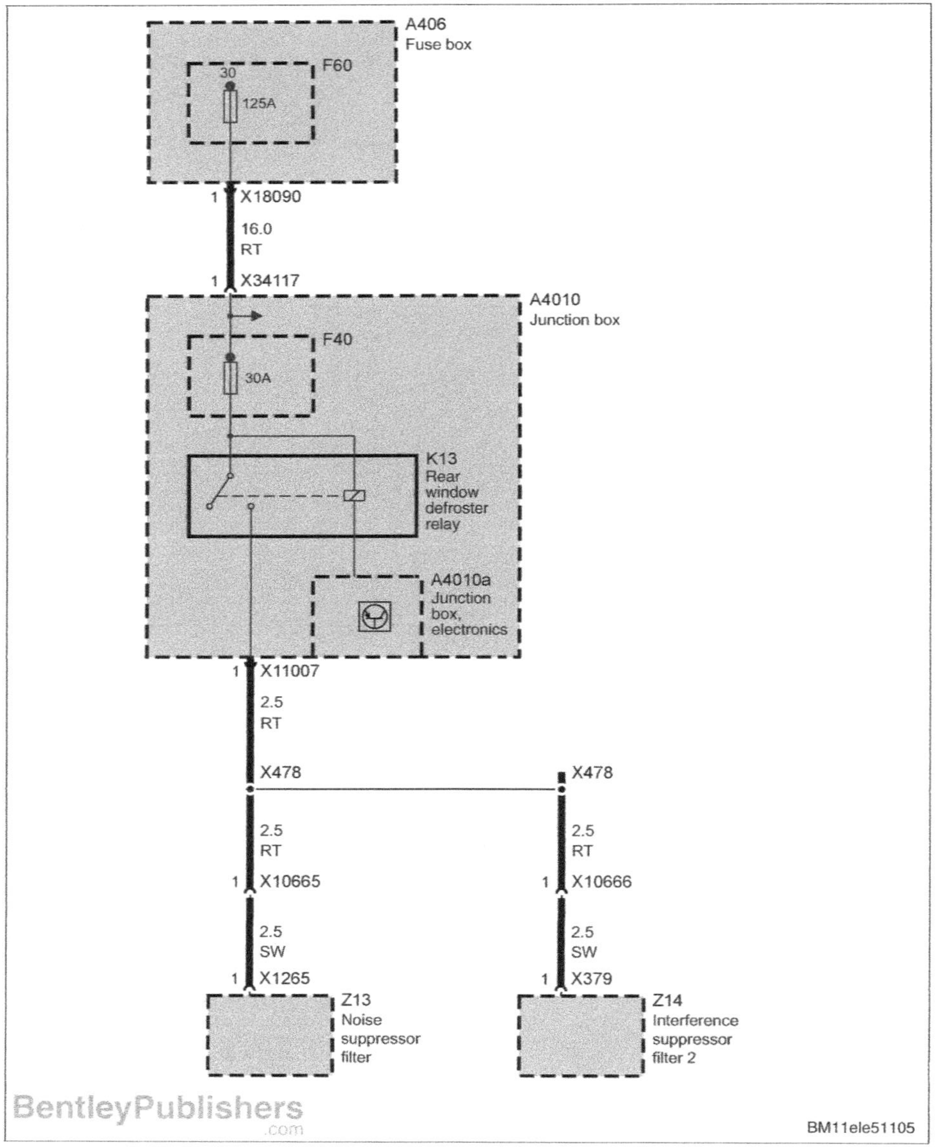

Fuse F40
from 12/2008, except Clubman

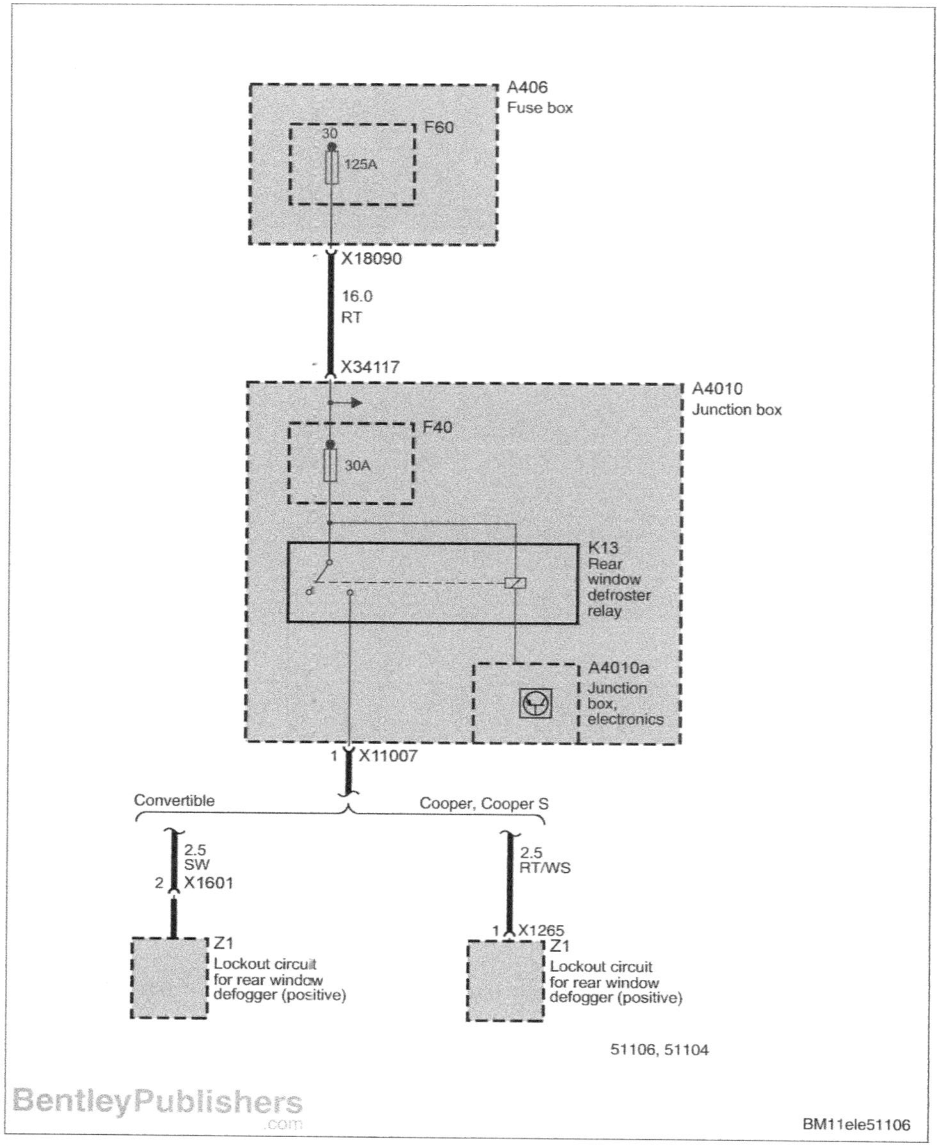

Fuse F41
from 12/2008

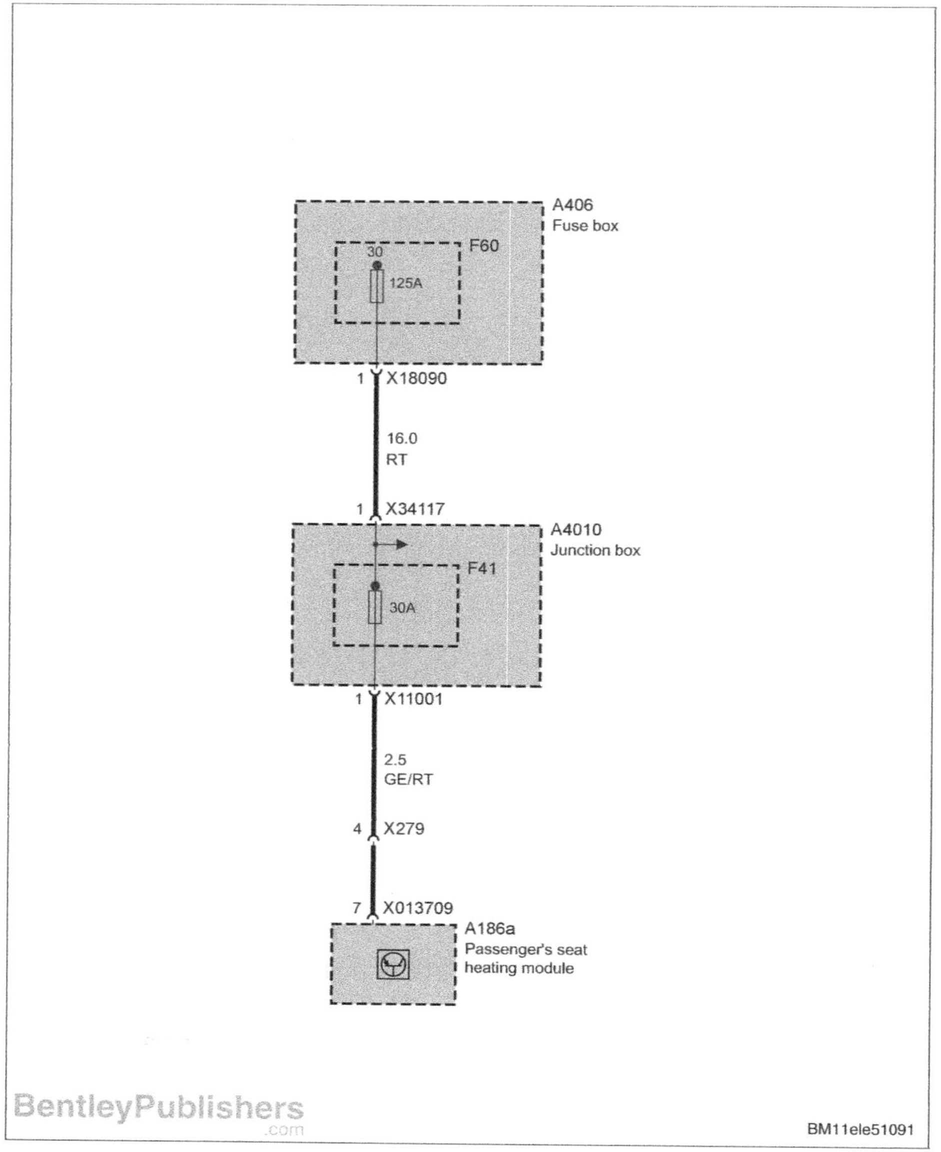

Fuse F42
to 11/2008

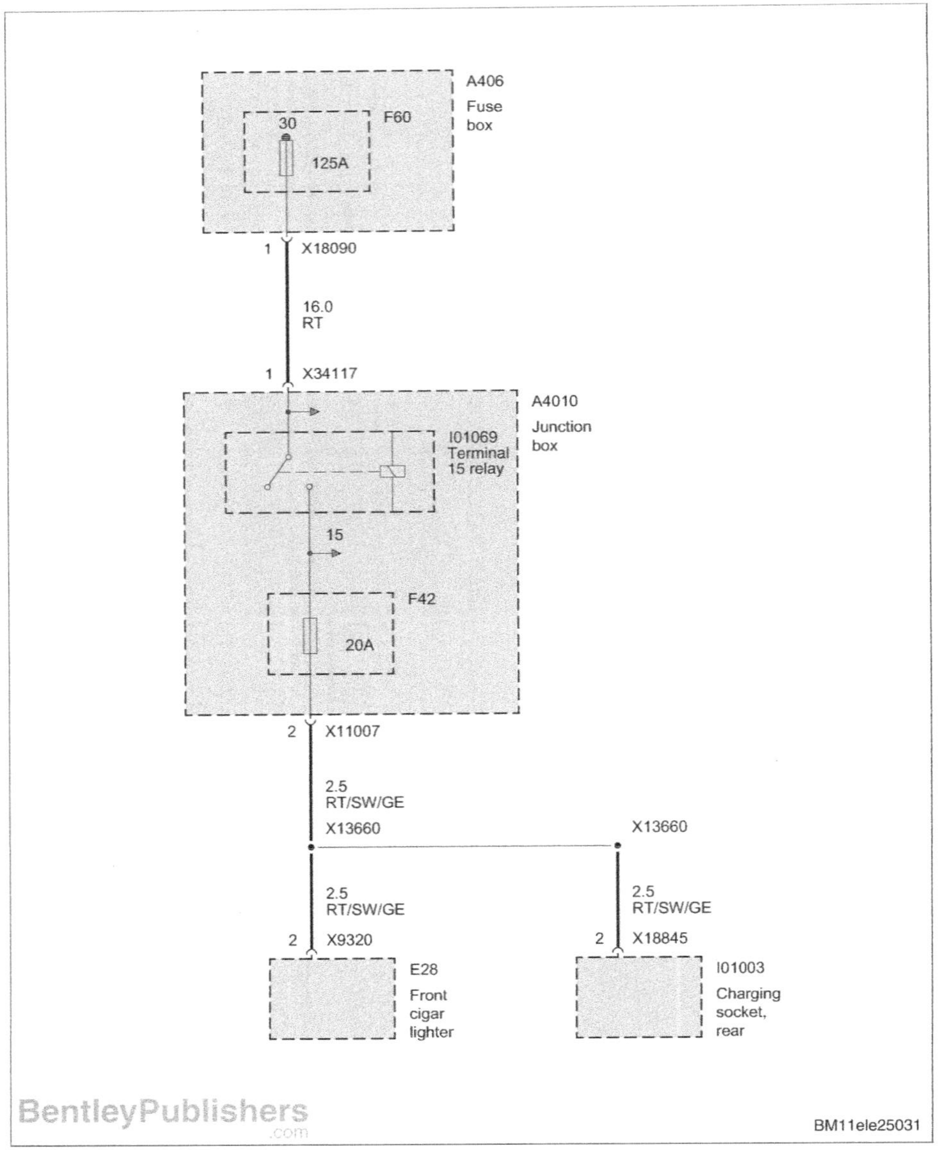

Fuse F42
from 12/2008

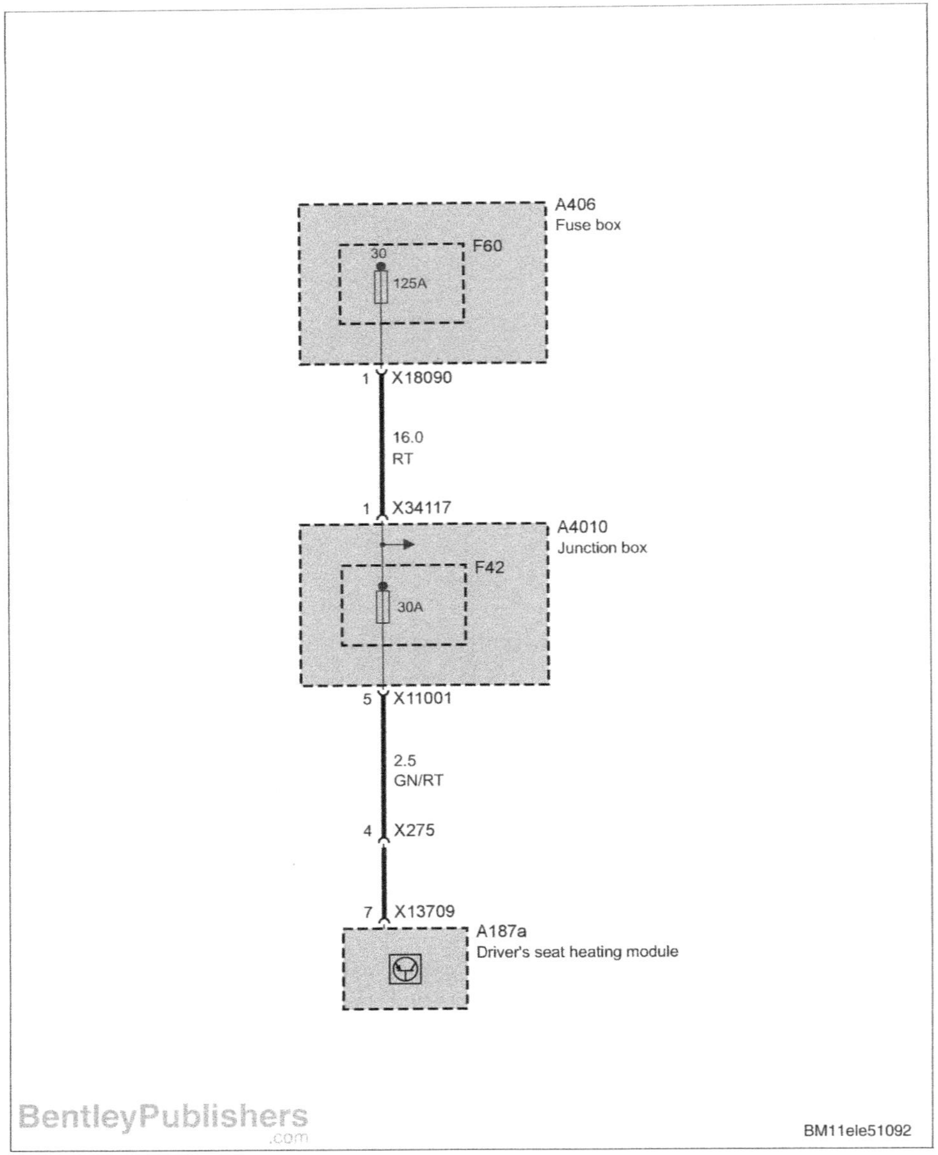

Fuse F43
to 11/2008

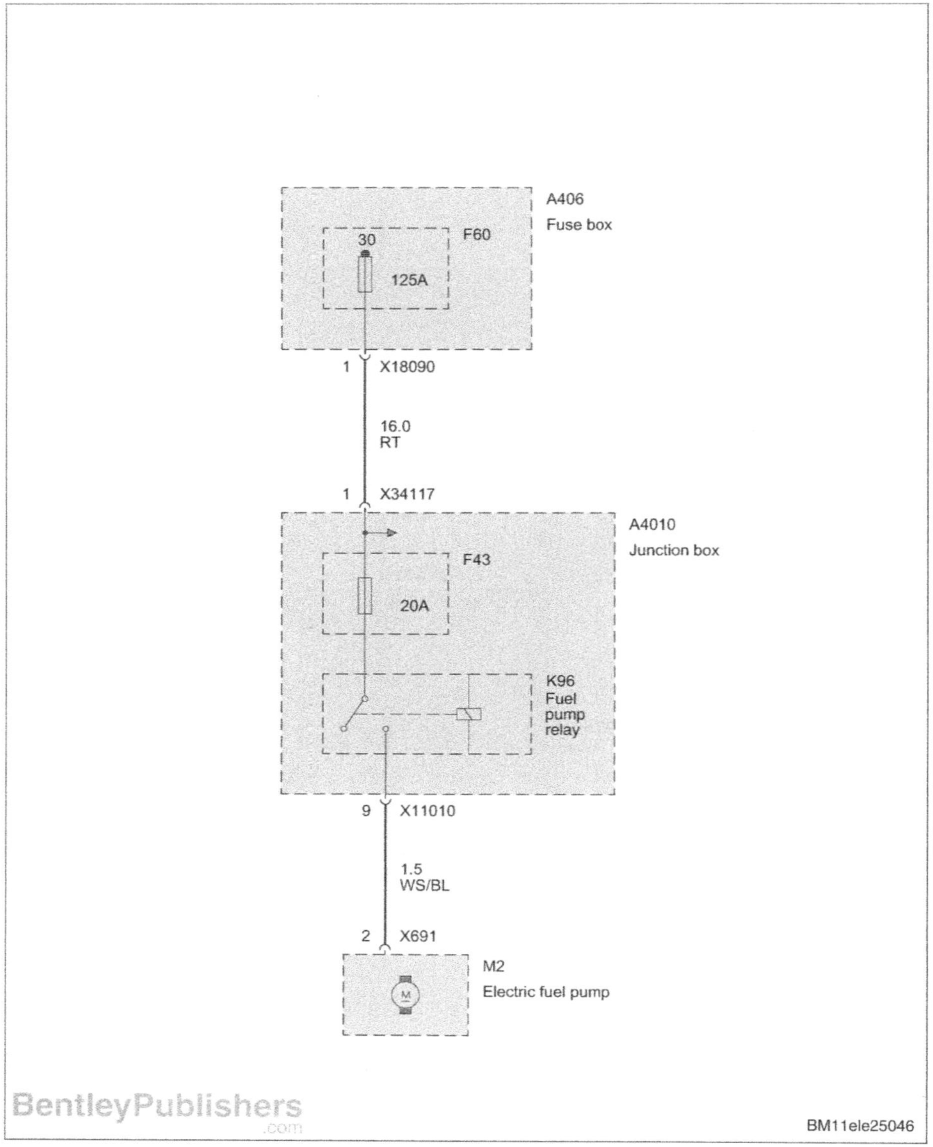

Fuse F43
Clubman from 12/2008

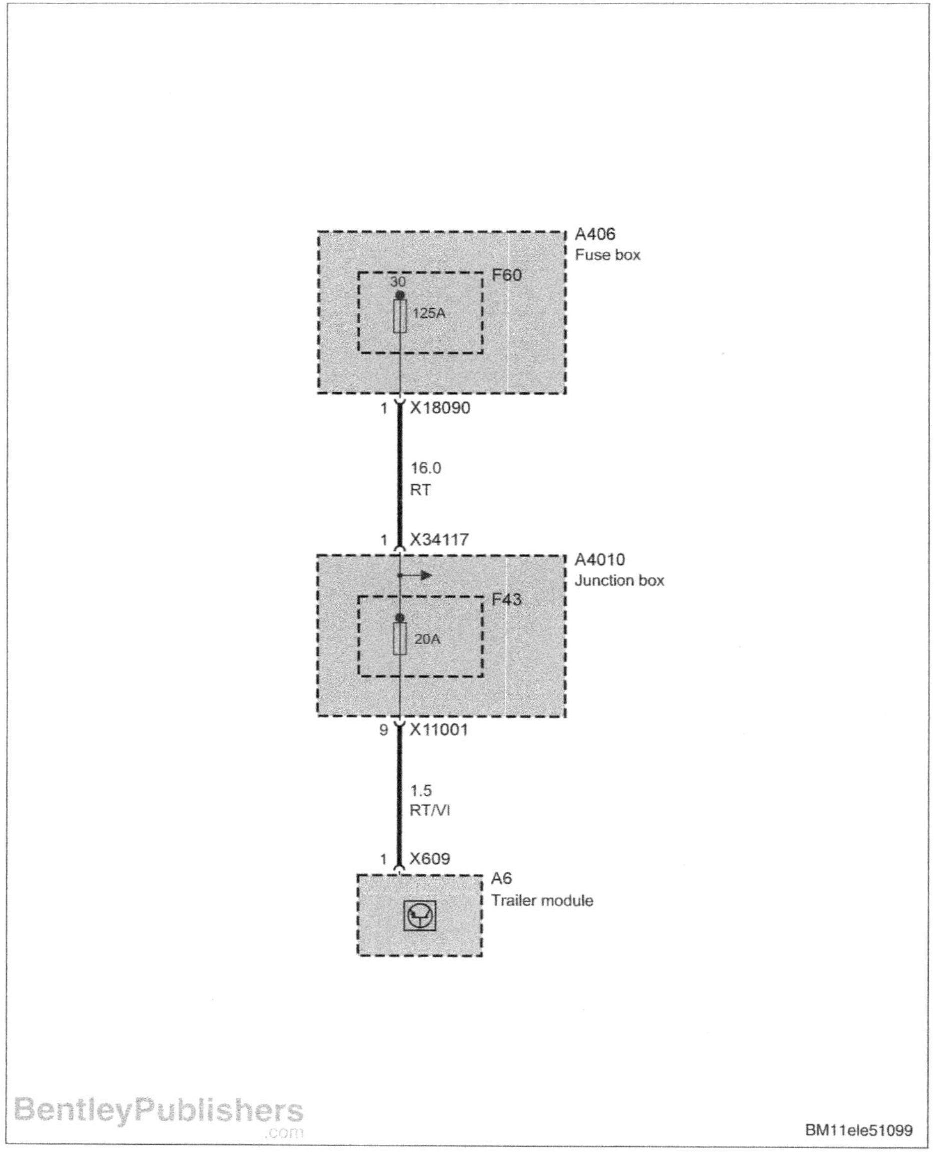

Fuse F43
Convertible from 12/2008

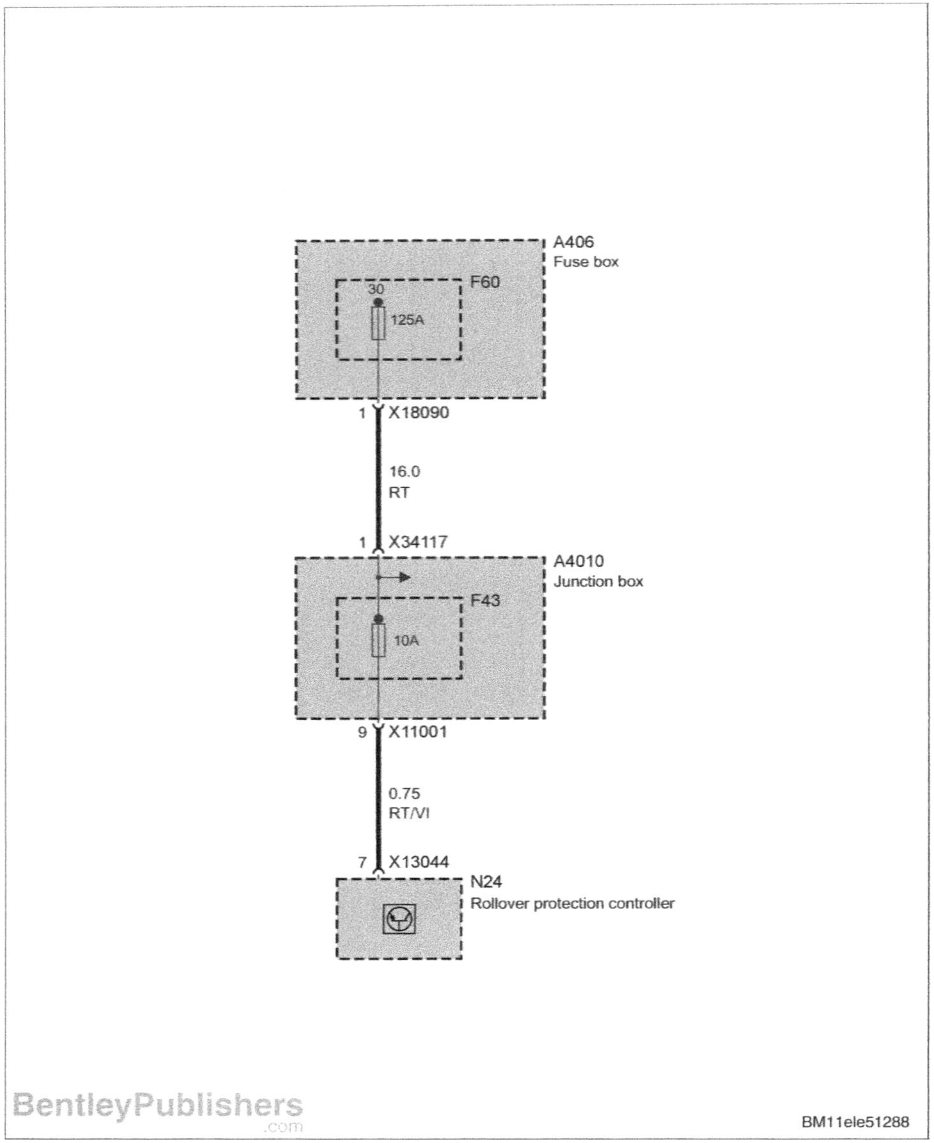

Fuse F44
to 11/2008

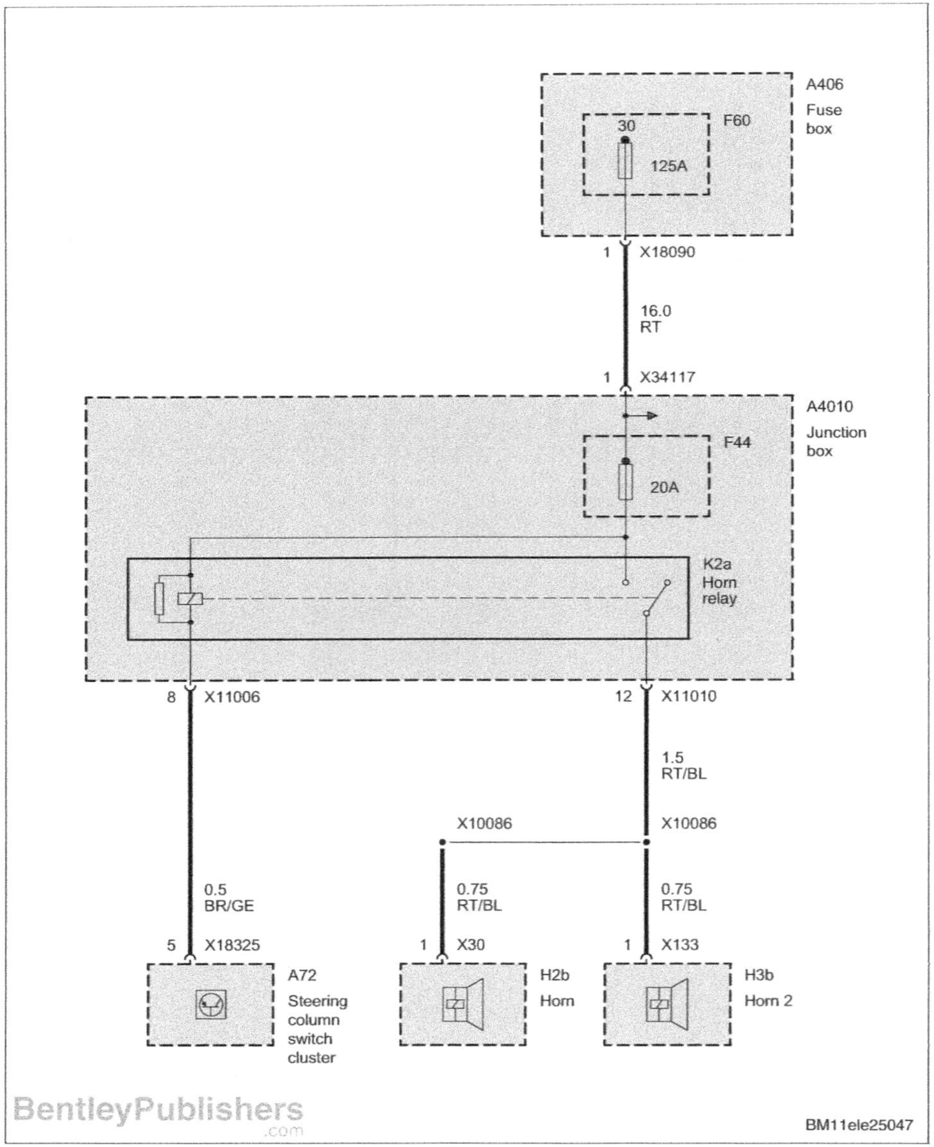

Fuse F44 from 12/2008

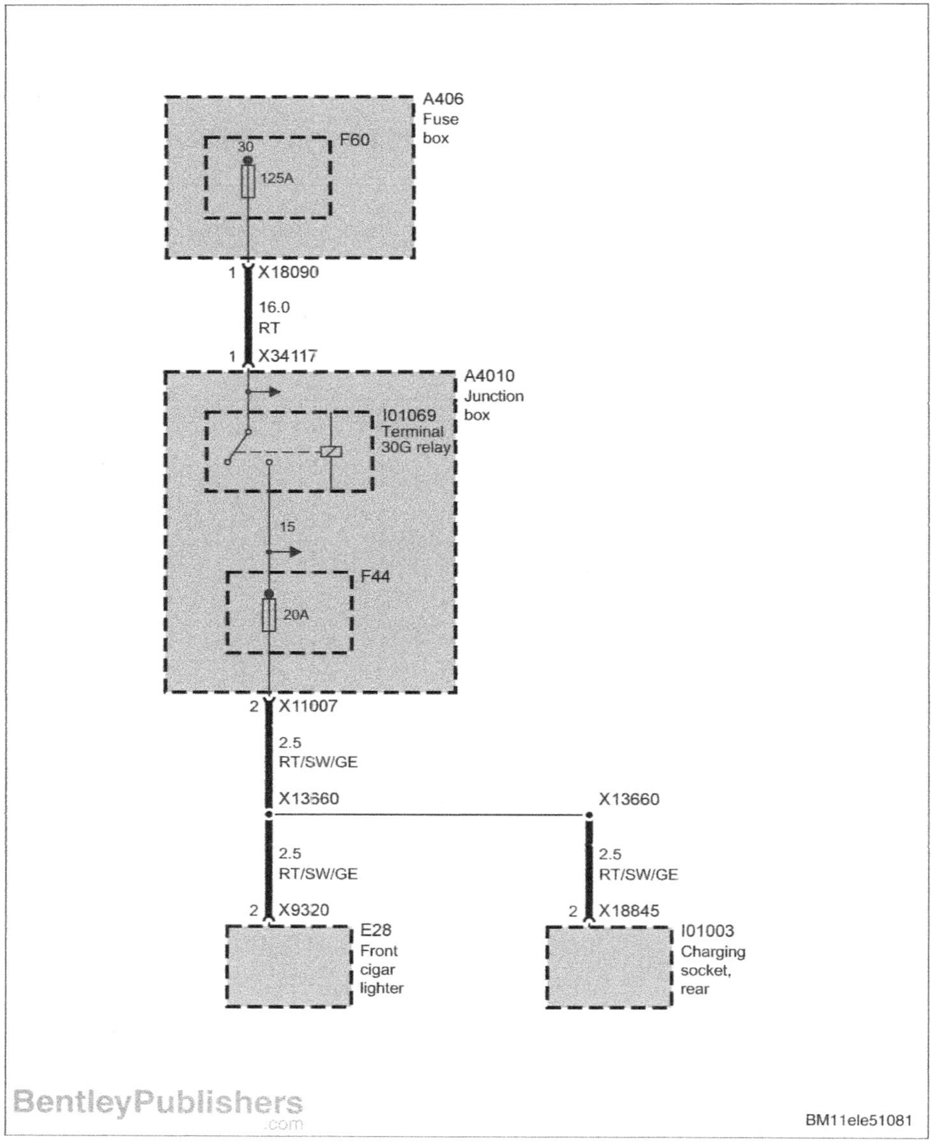

Fuse F45

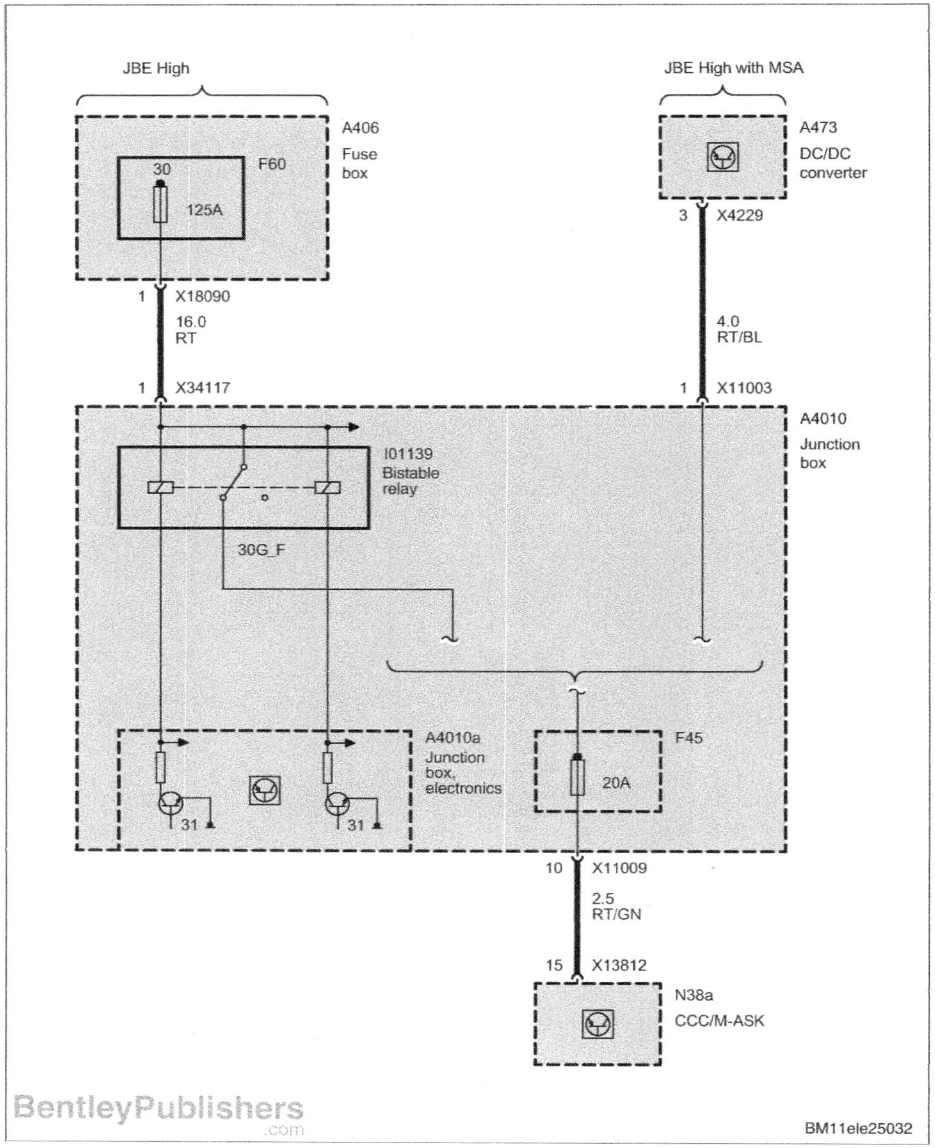

Fuse F46 to 11/2008

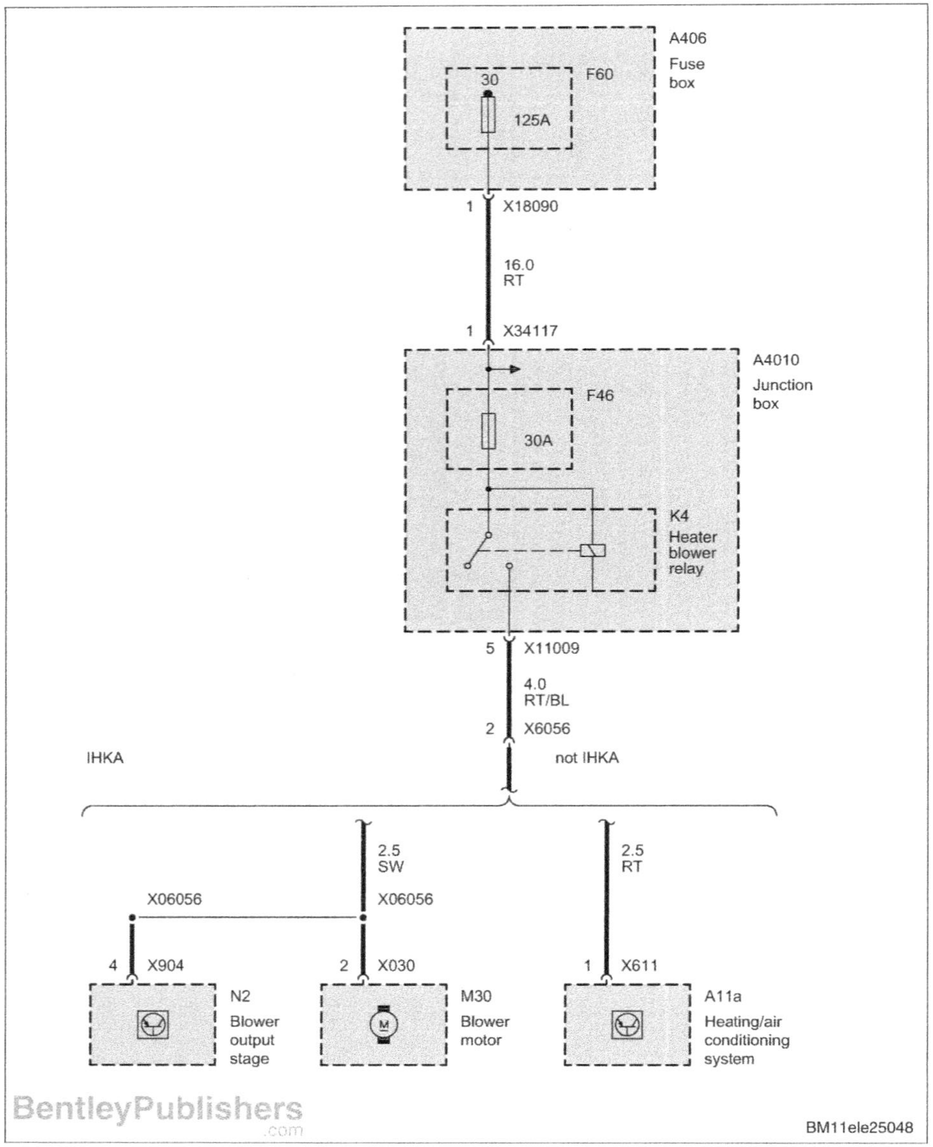

Fuse F46 from 12/2008

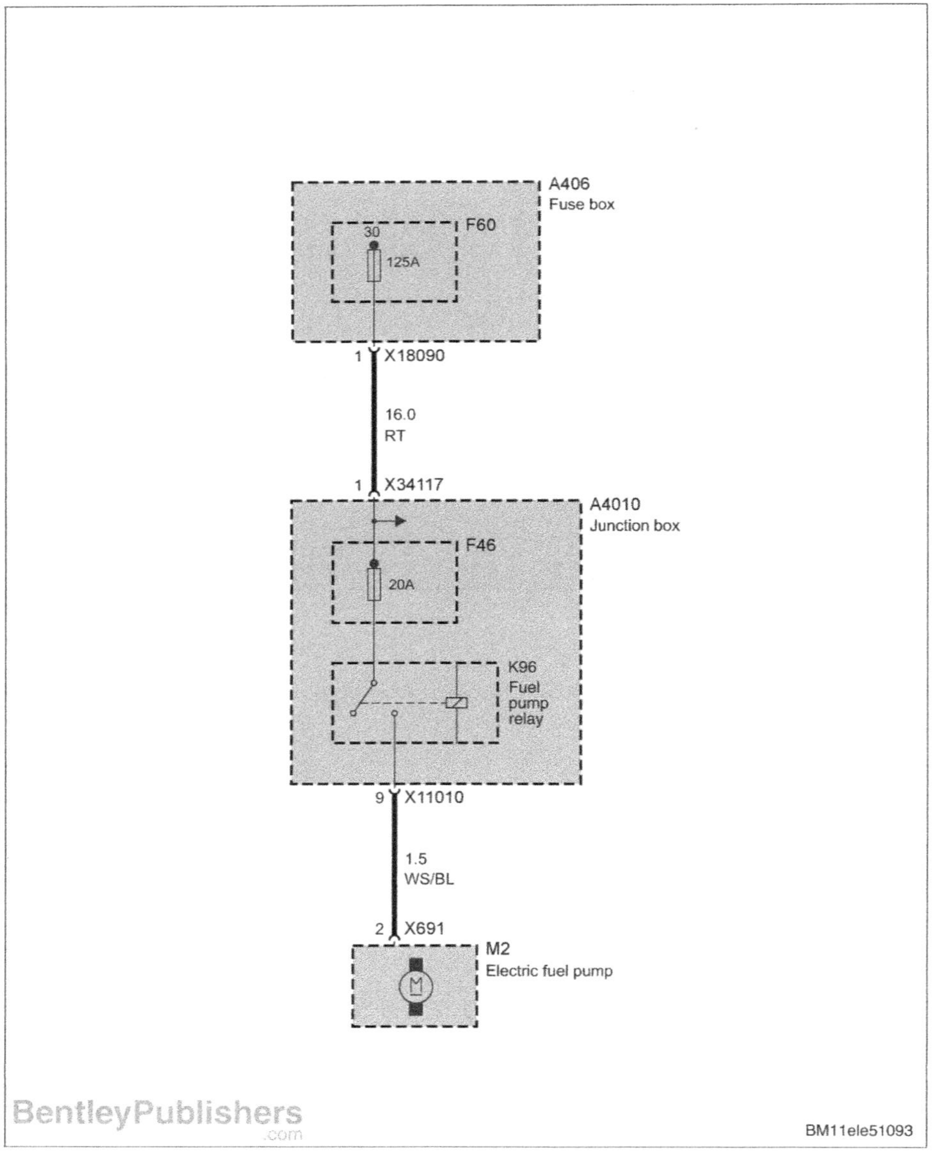

Fuse F47 to 11/2008

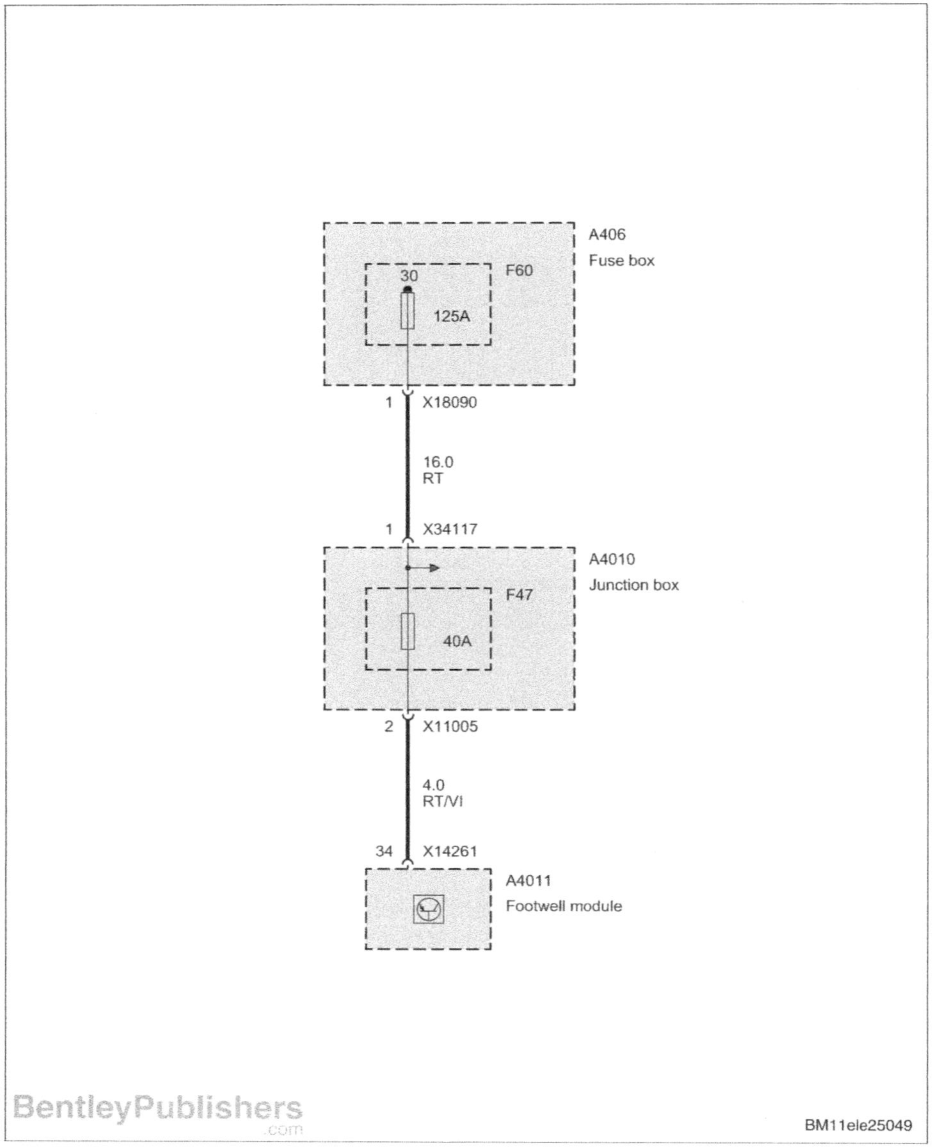

Fuse F47 from 12/2008

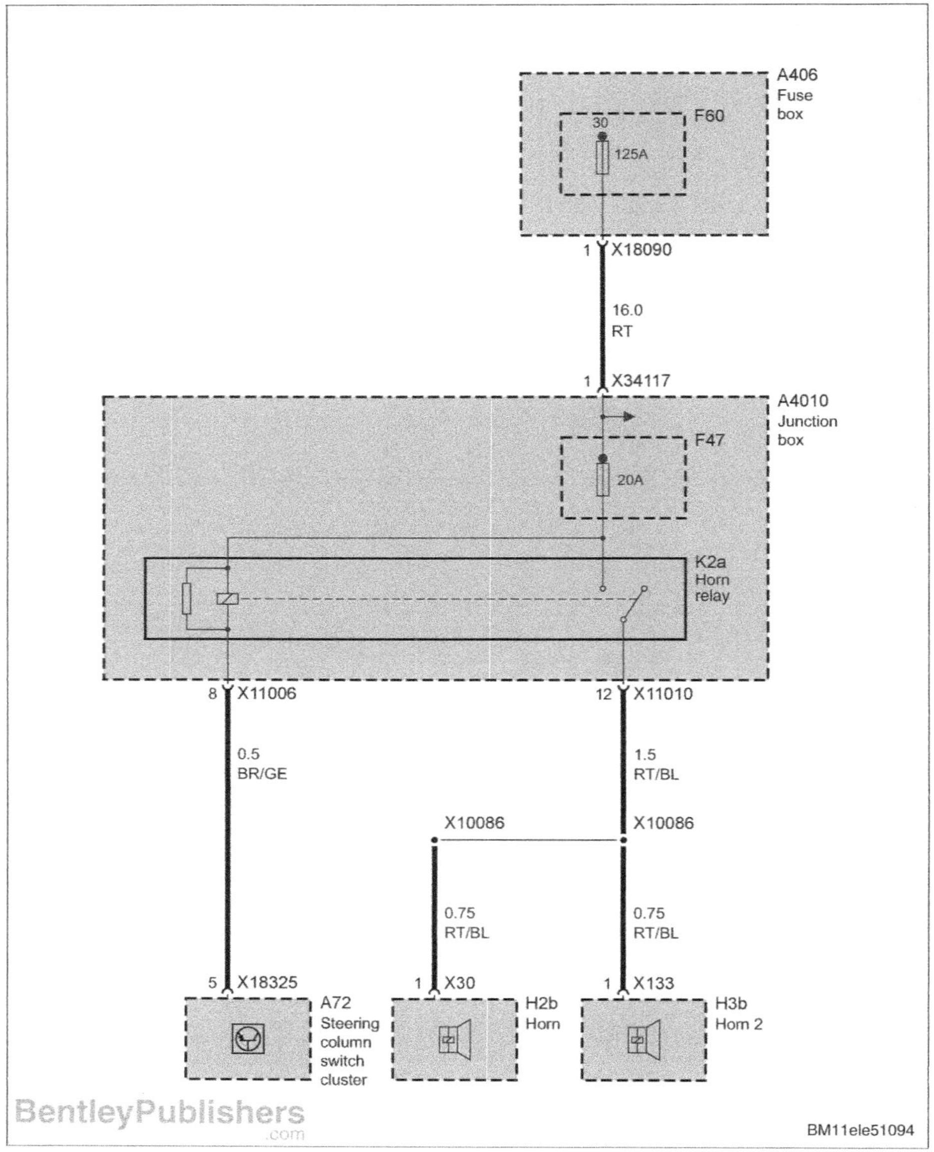

Fuse F48 to 11/2008

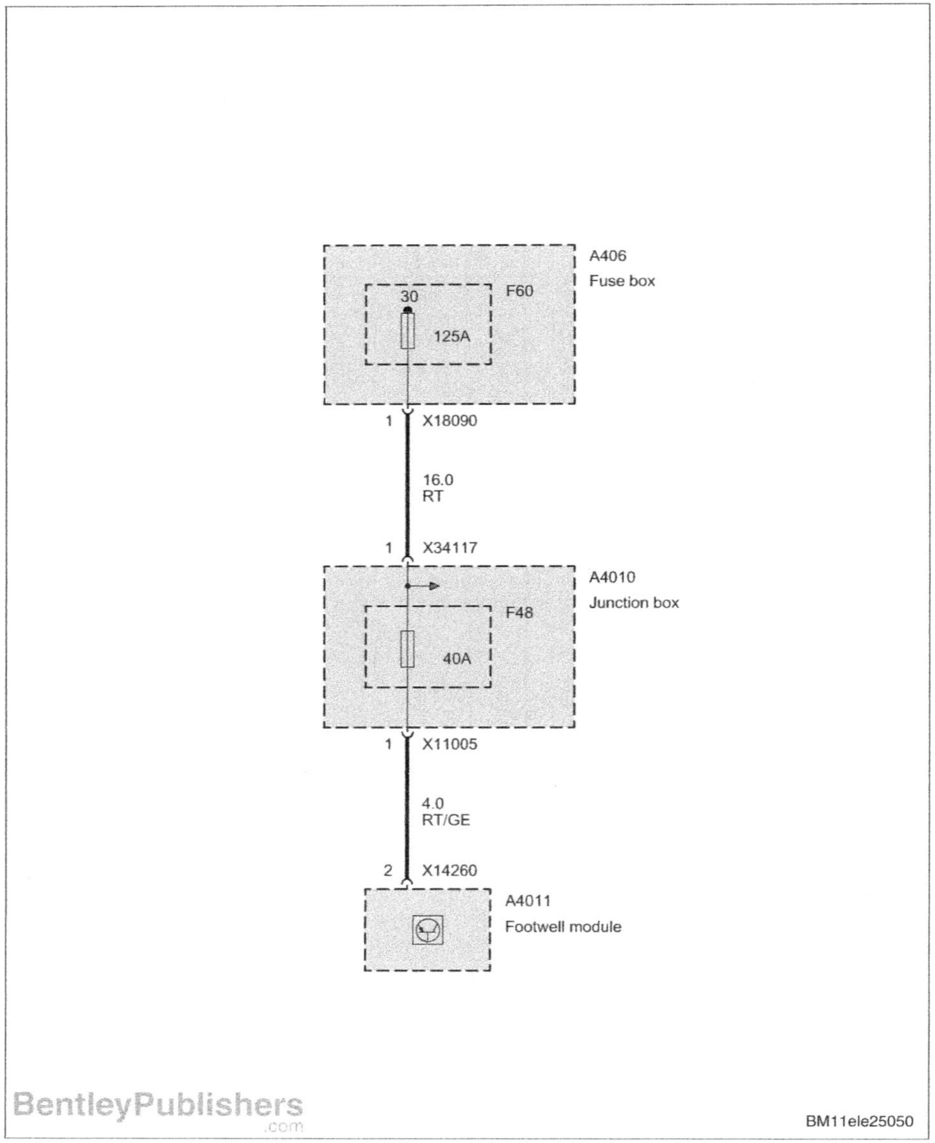

Fuse F48 from 12/2008

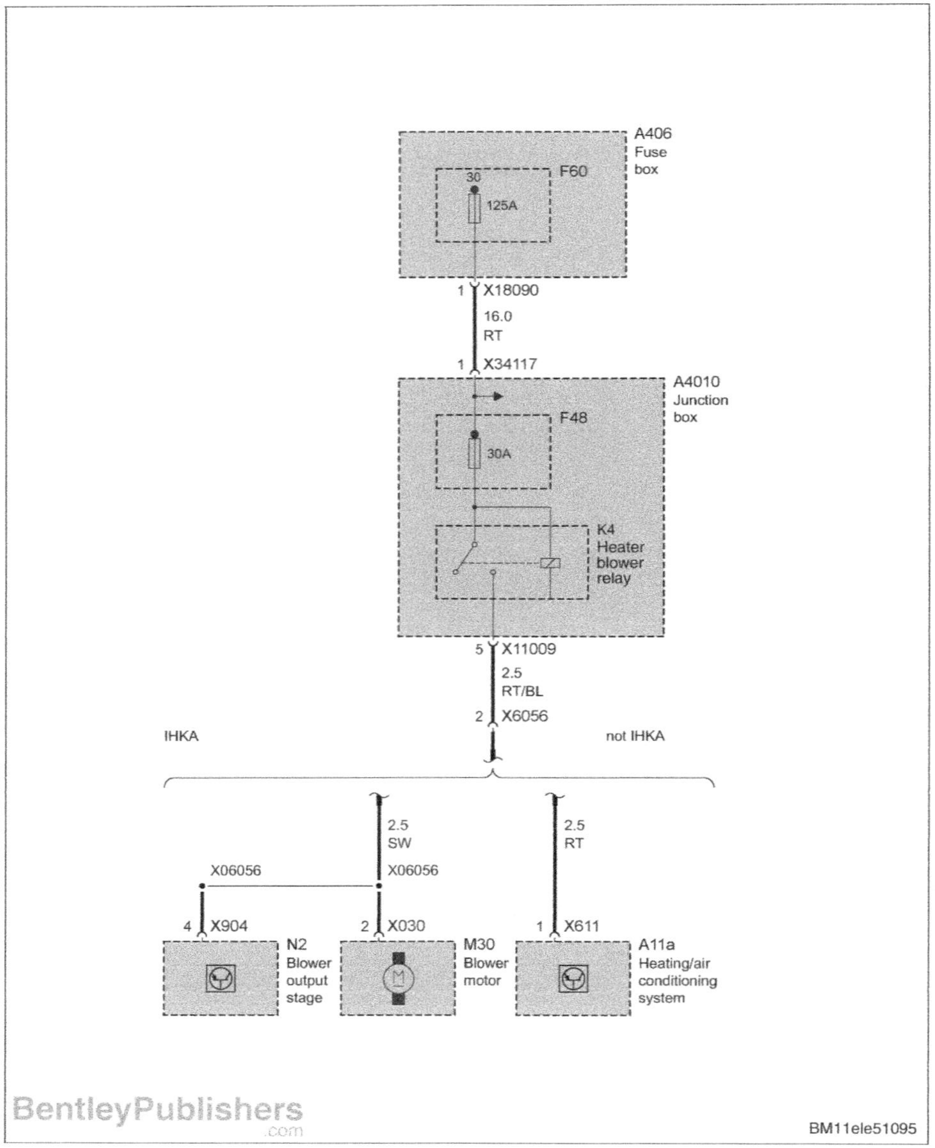

Fuse F49 to 11/2008

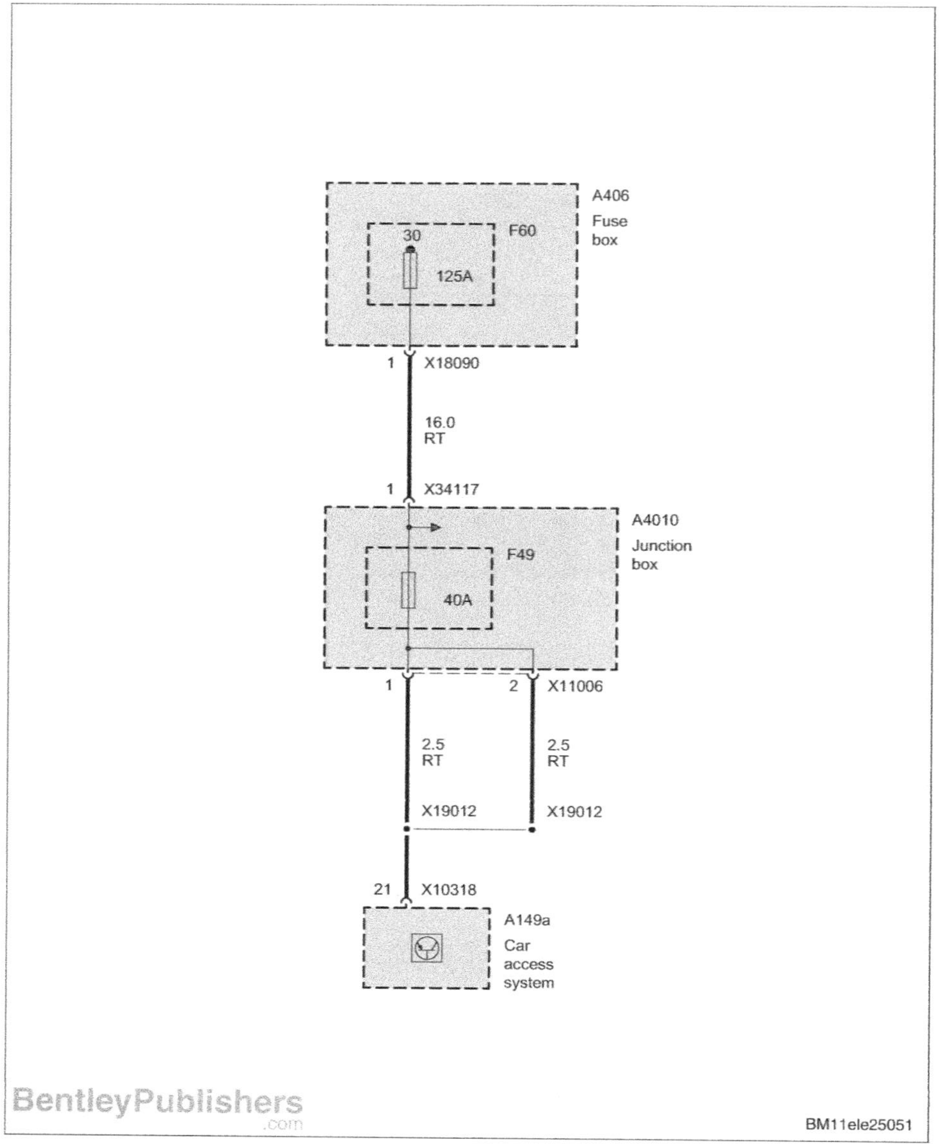

Fuse F49 from 12/2008

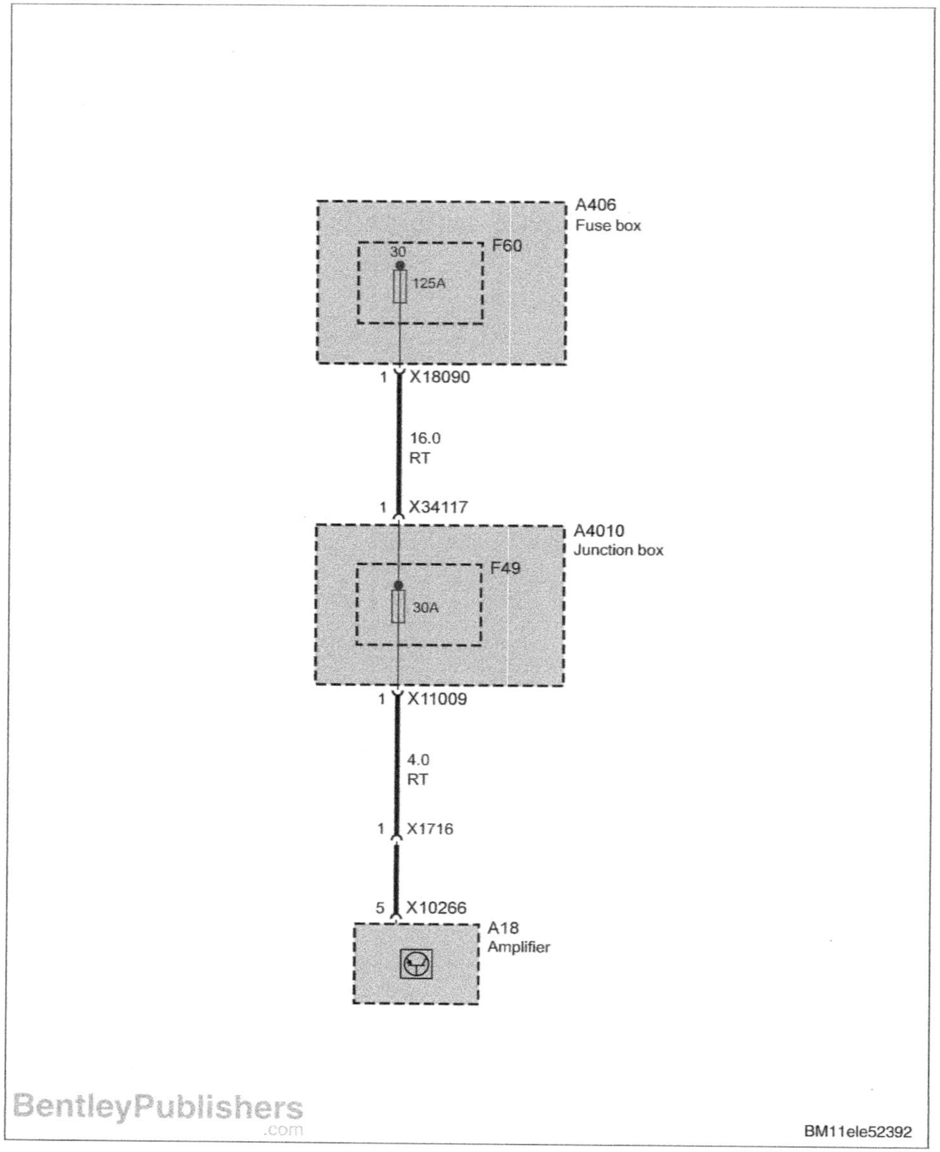

Fuse F50
from 12/2008 except Convertible

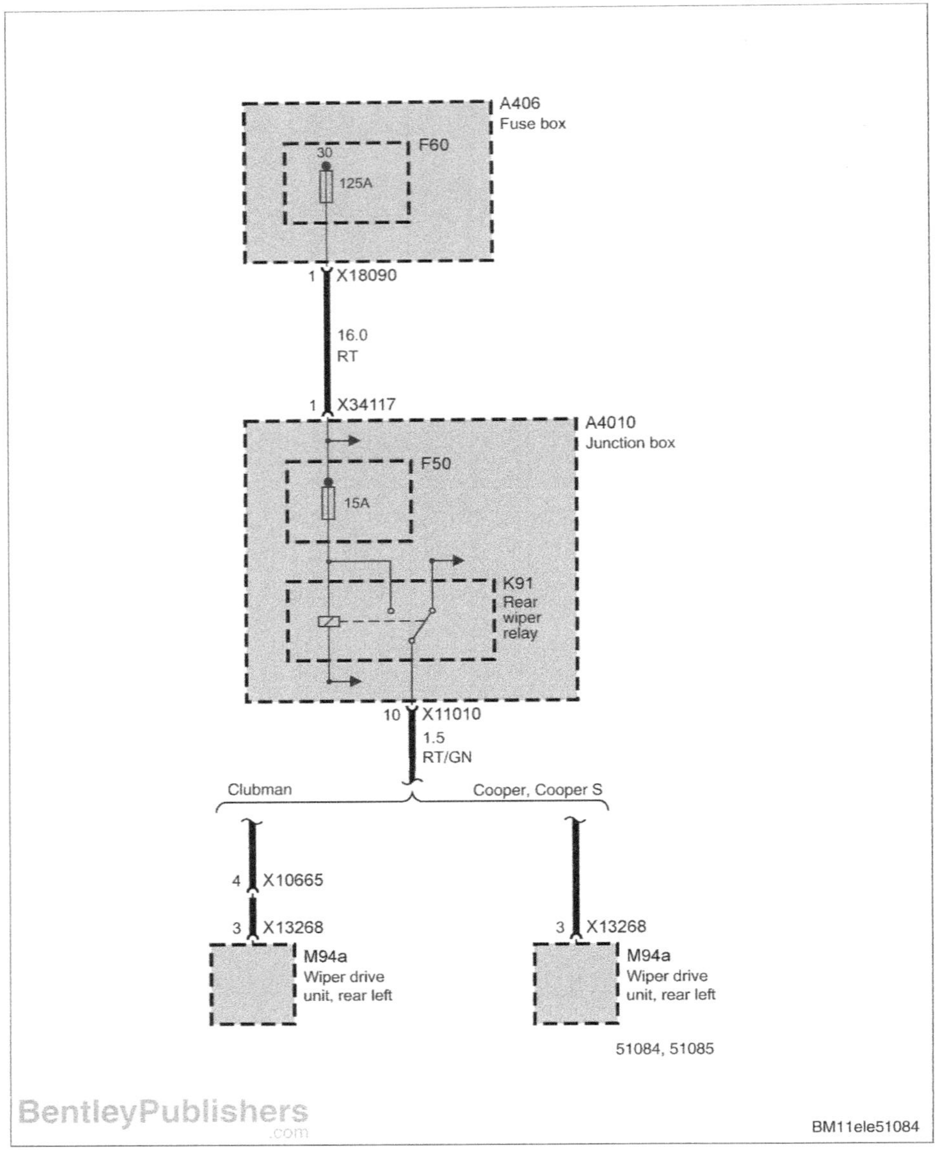

Fuse F51
from 12/2008

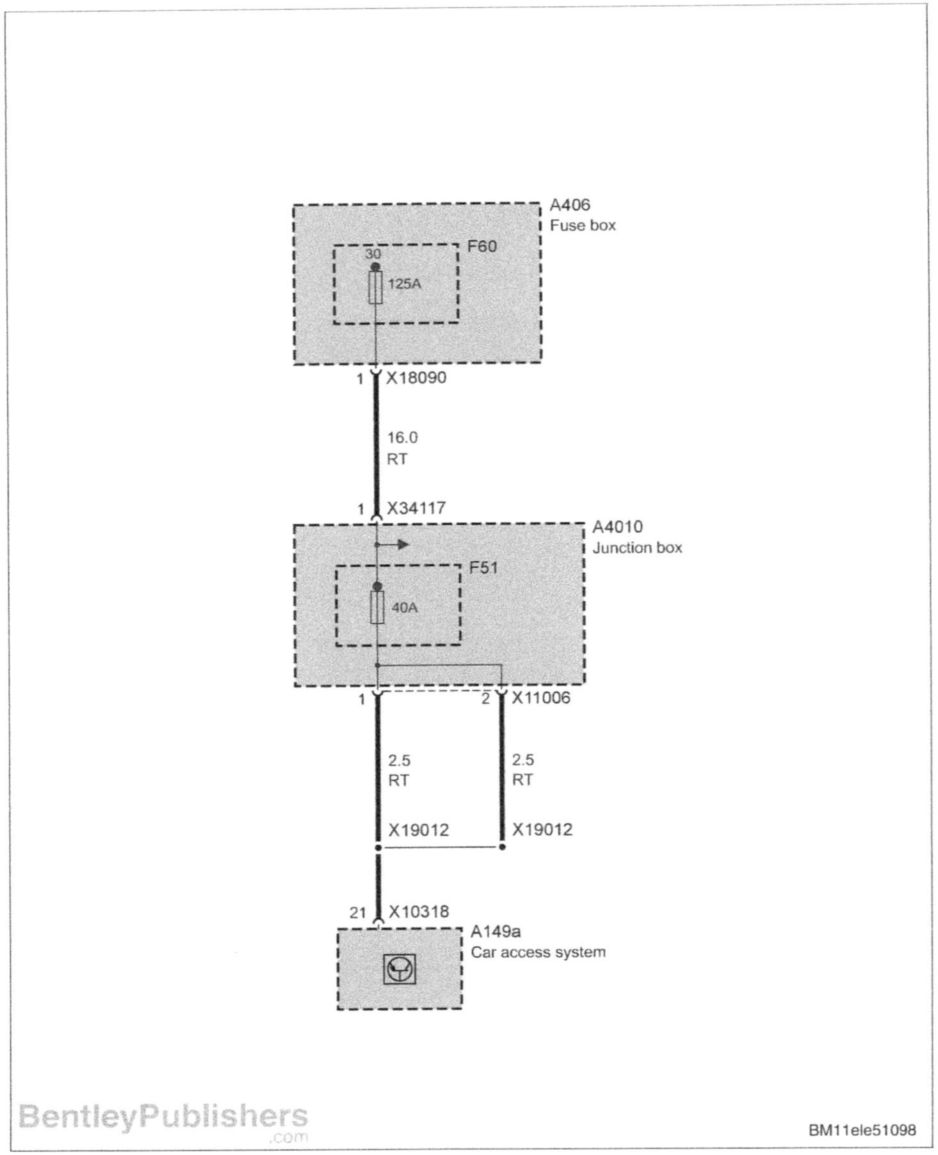

Fuse F52 from 12/2008

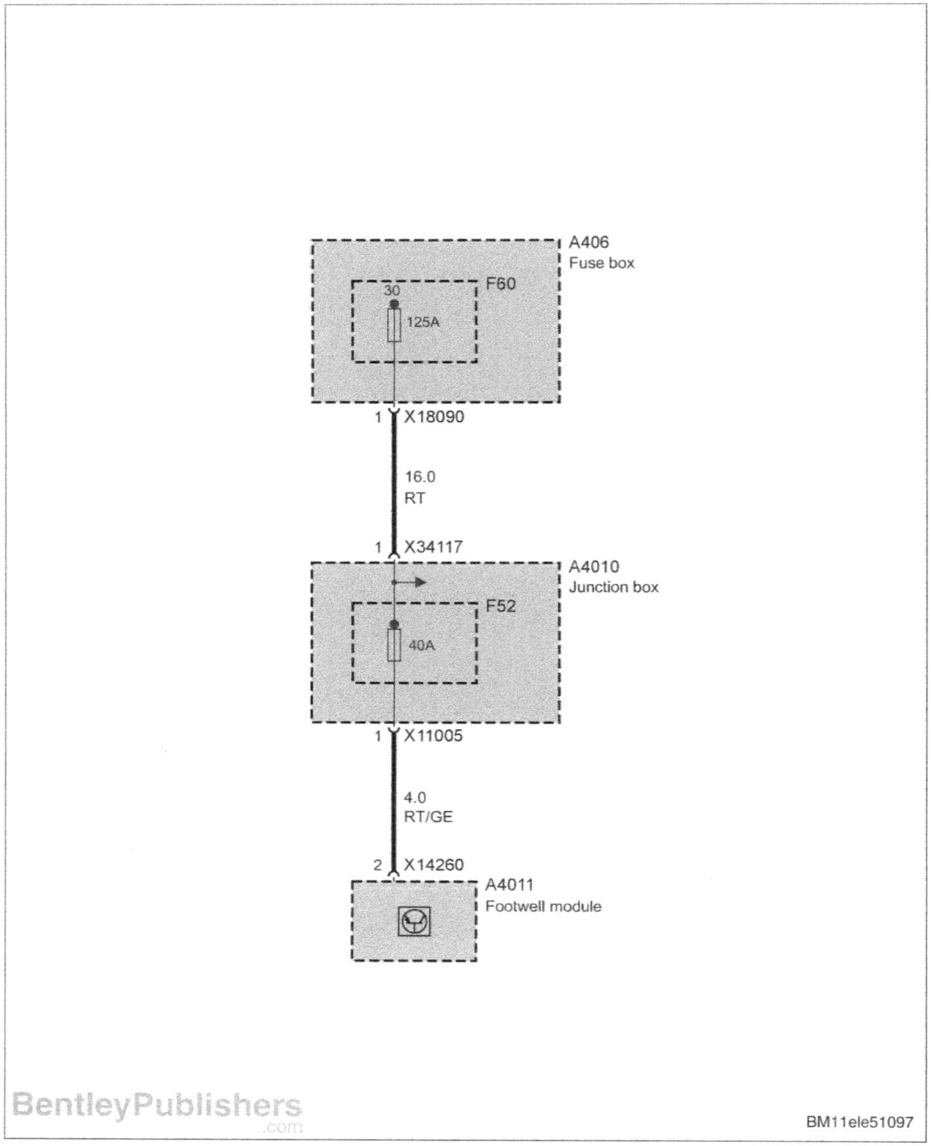

Fuse F53 from 12/2008

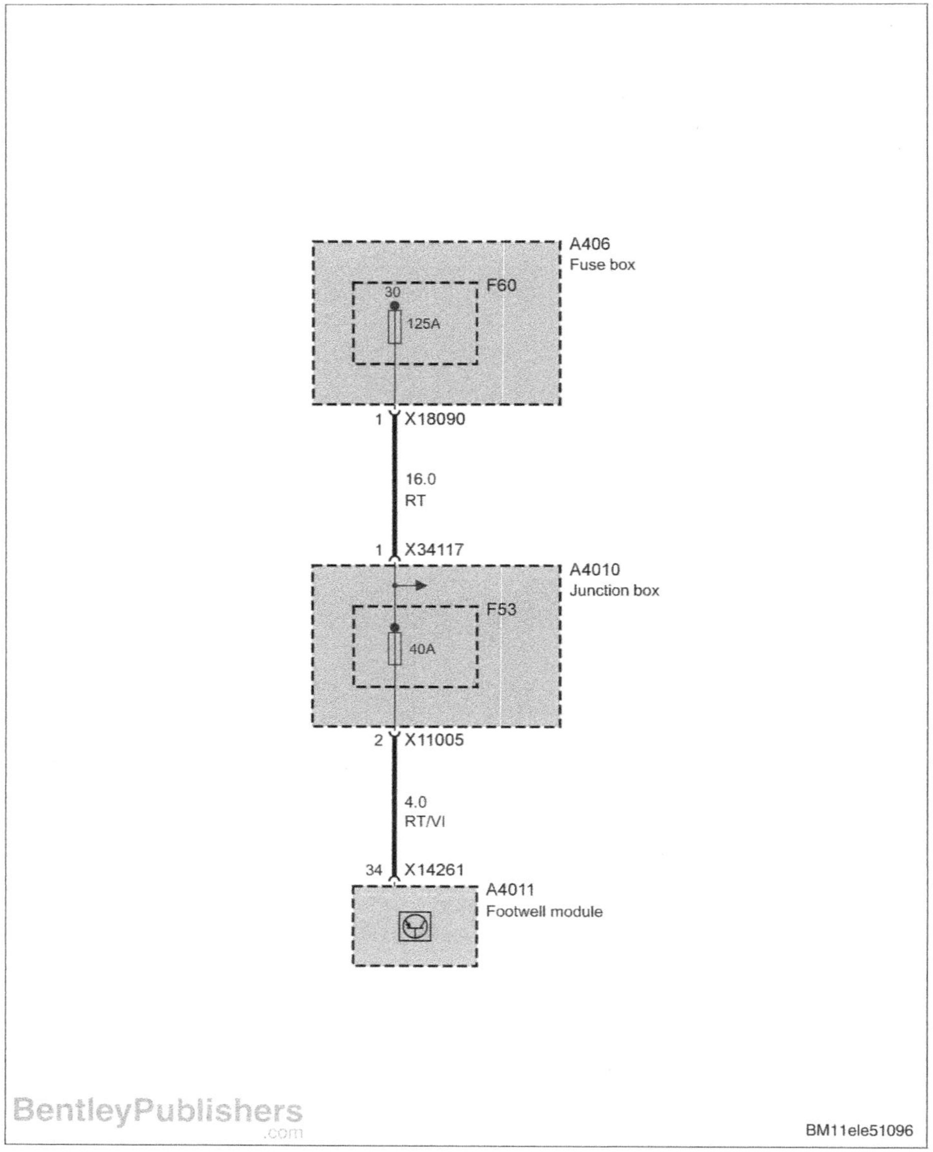

Fuse FL4

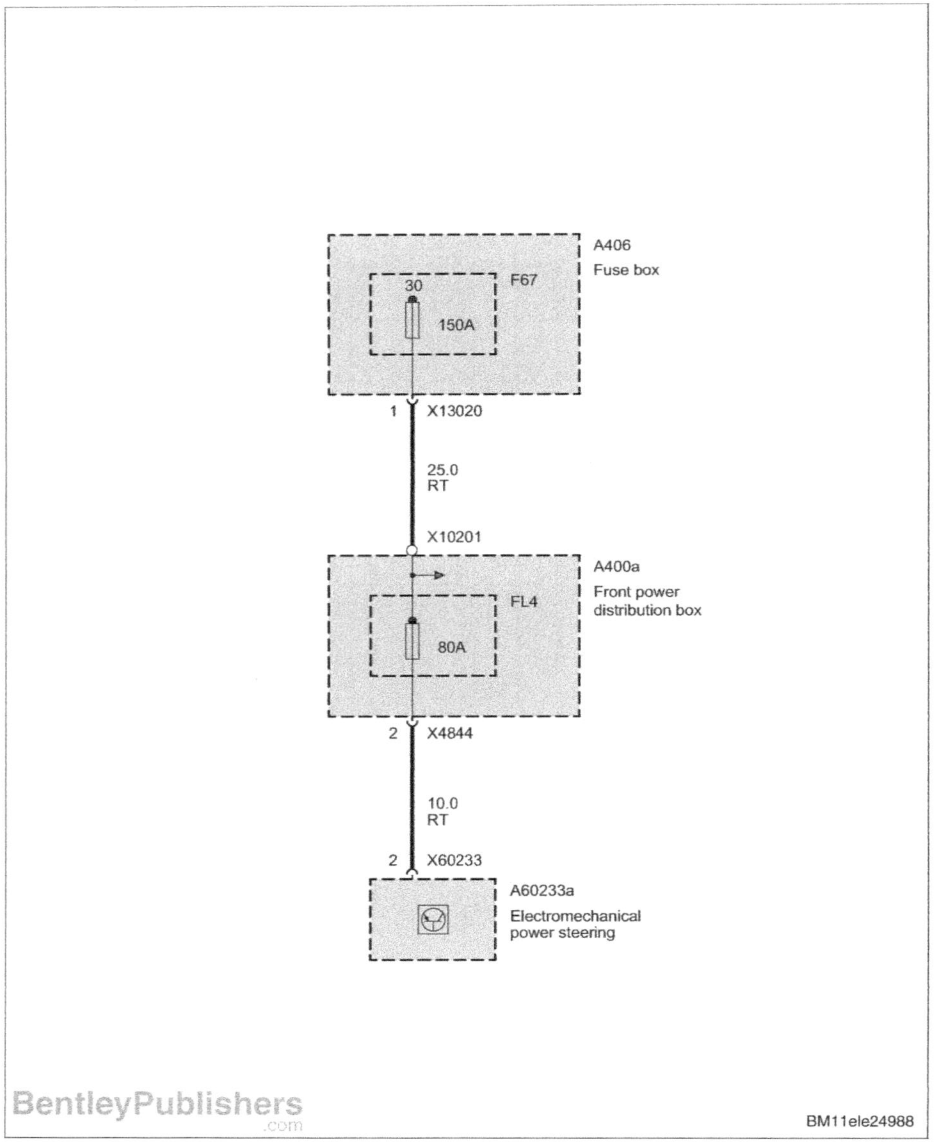

Fuse FL6

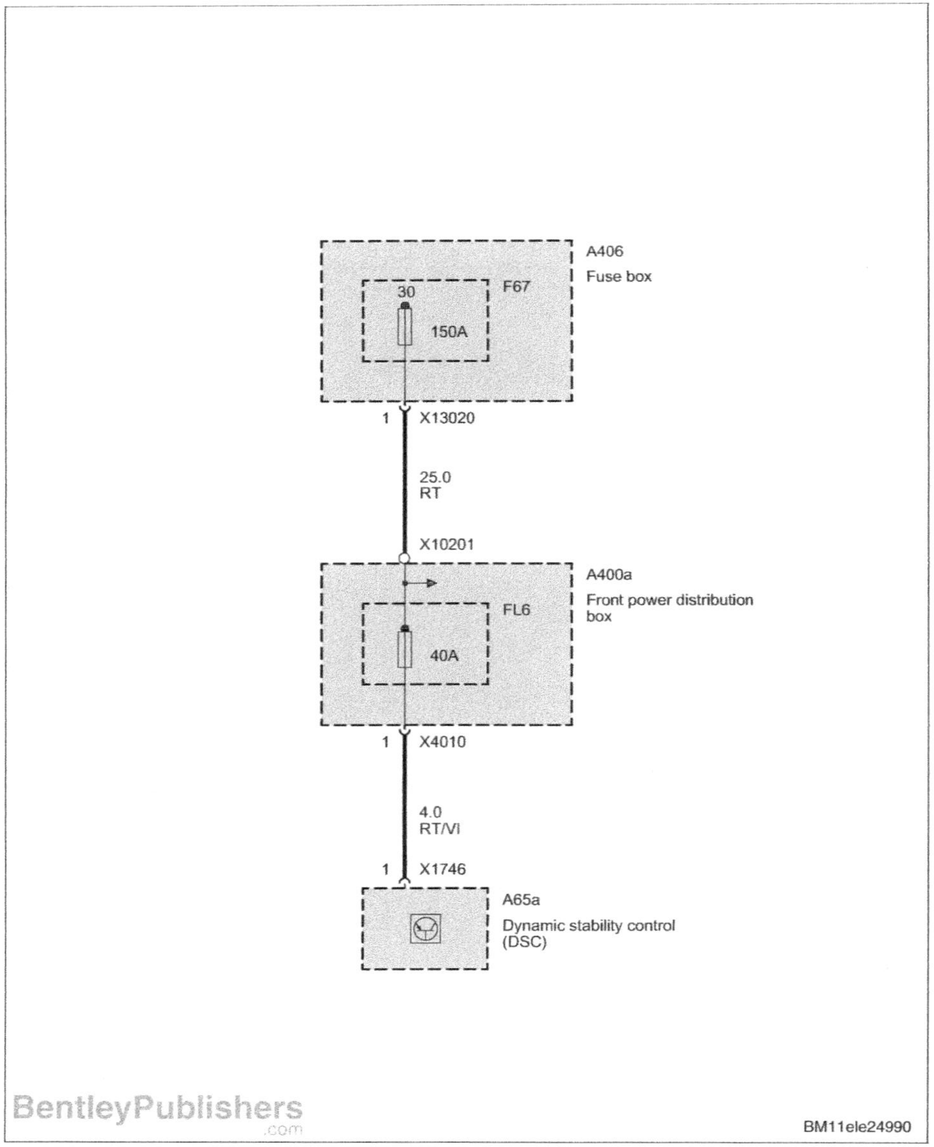

Fuse FL8

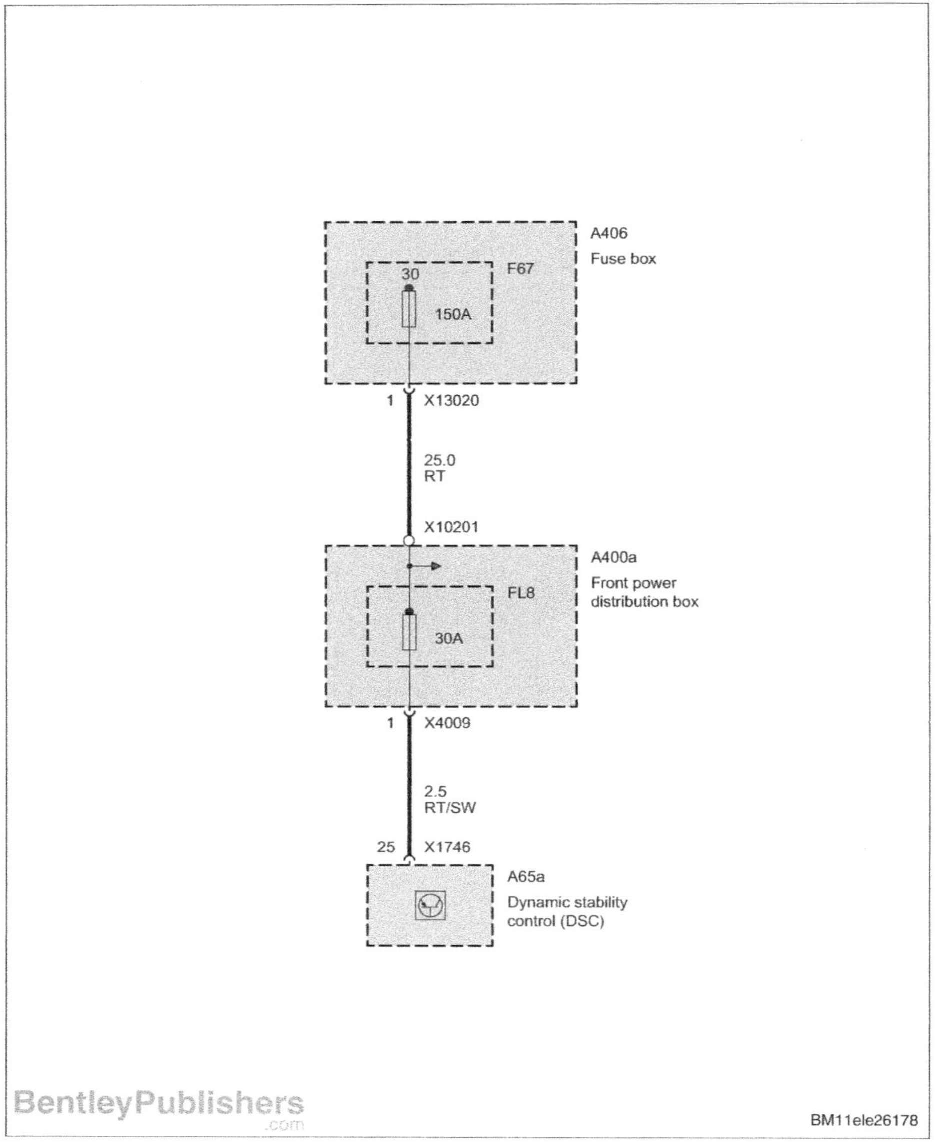

Fuse FL9

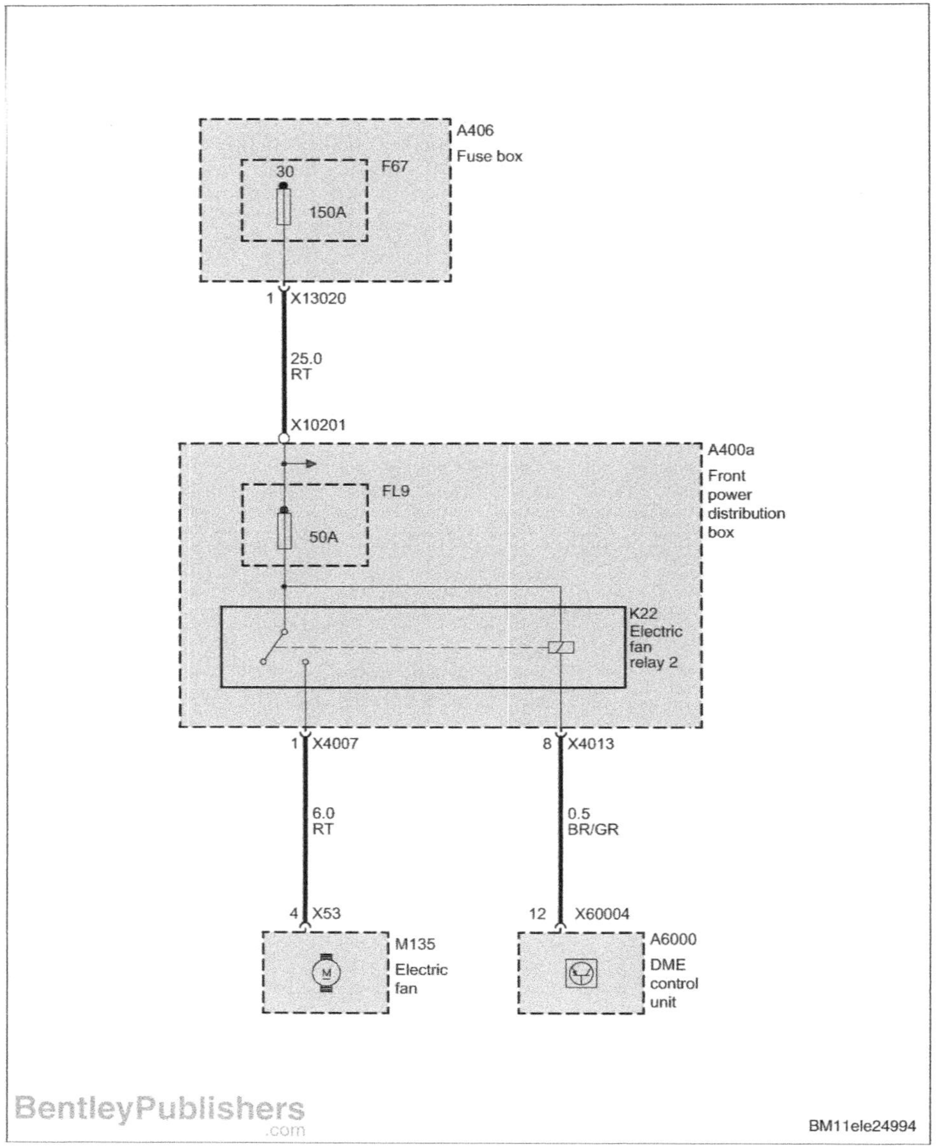

Fuse FL10 to 11/2008

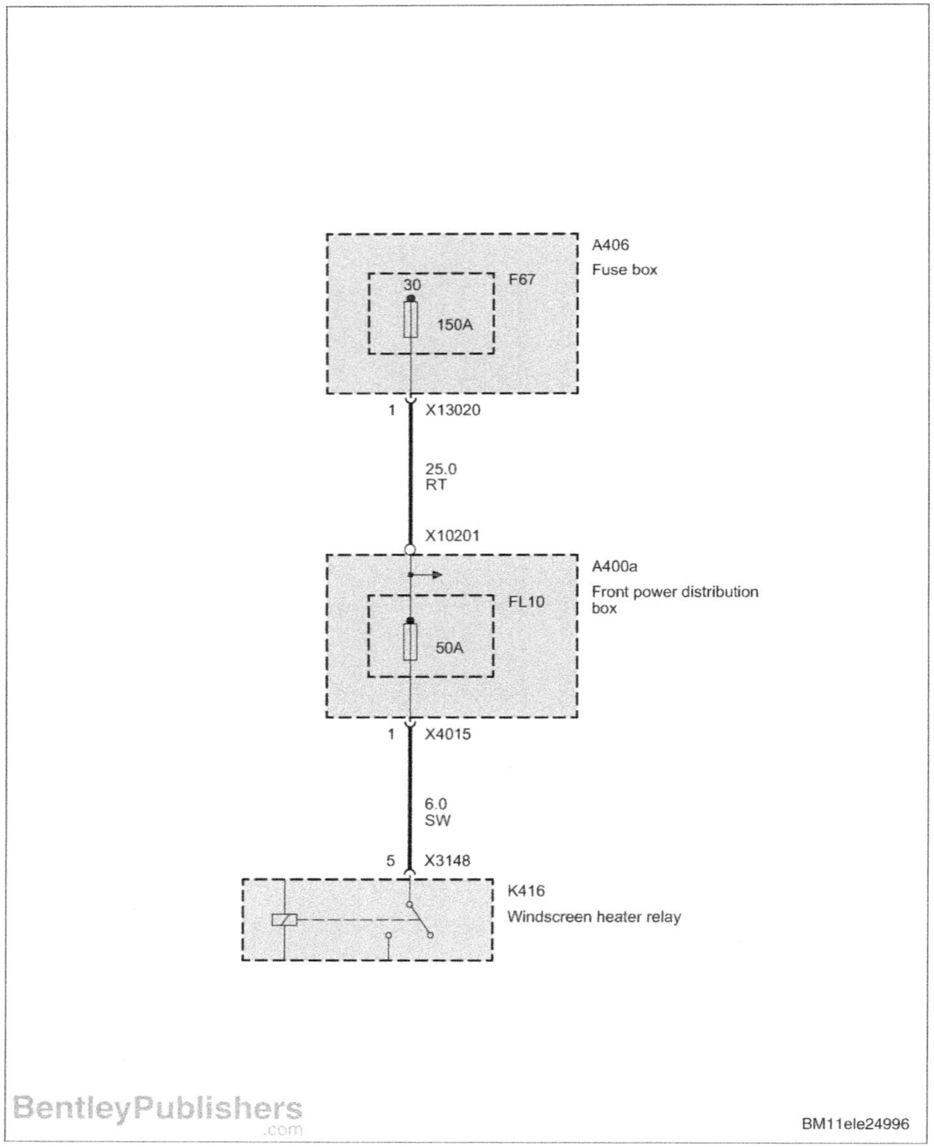

Fuse FL10 from 12/2008

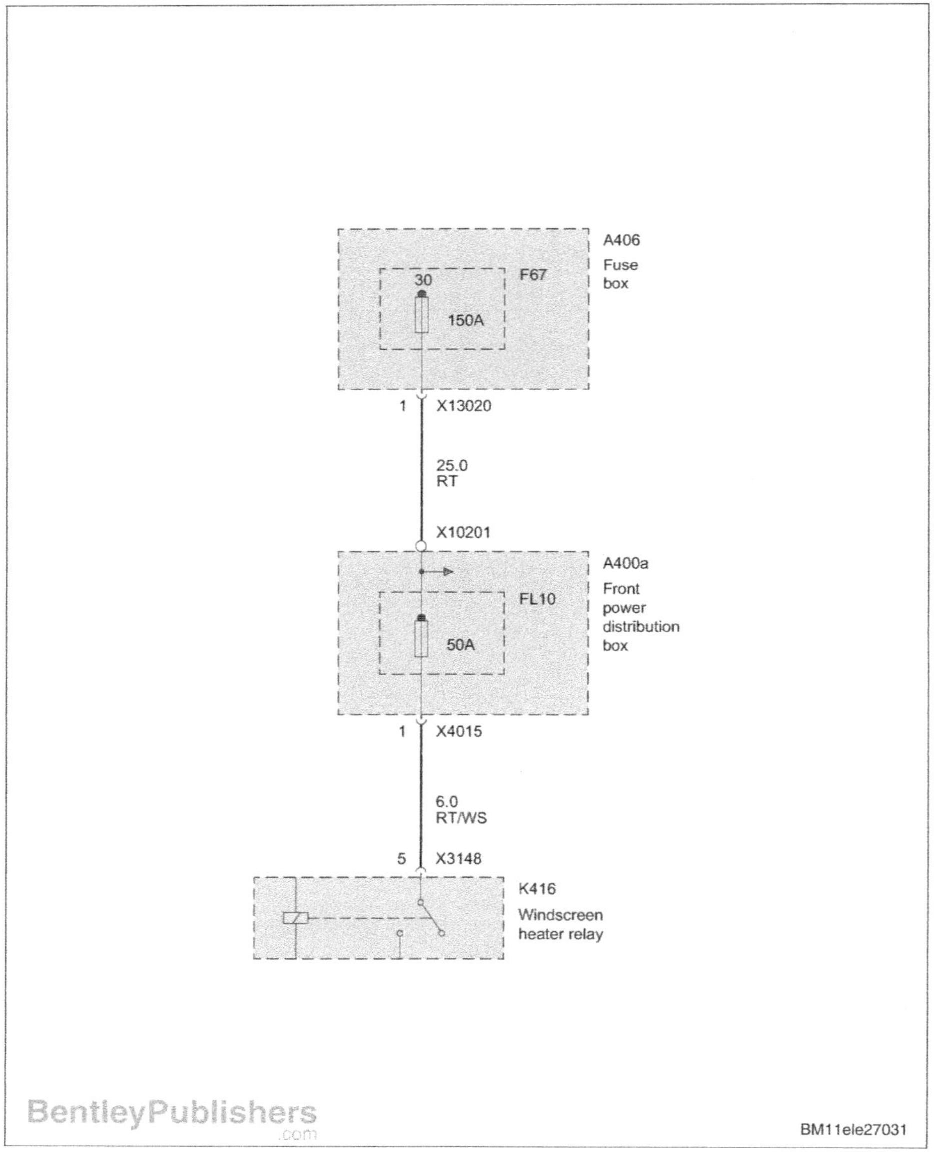

Fuse FL12
from 12/2008

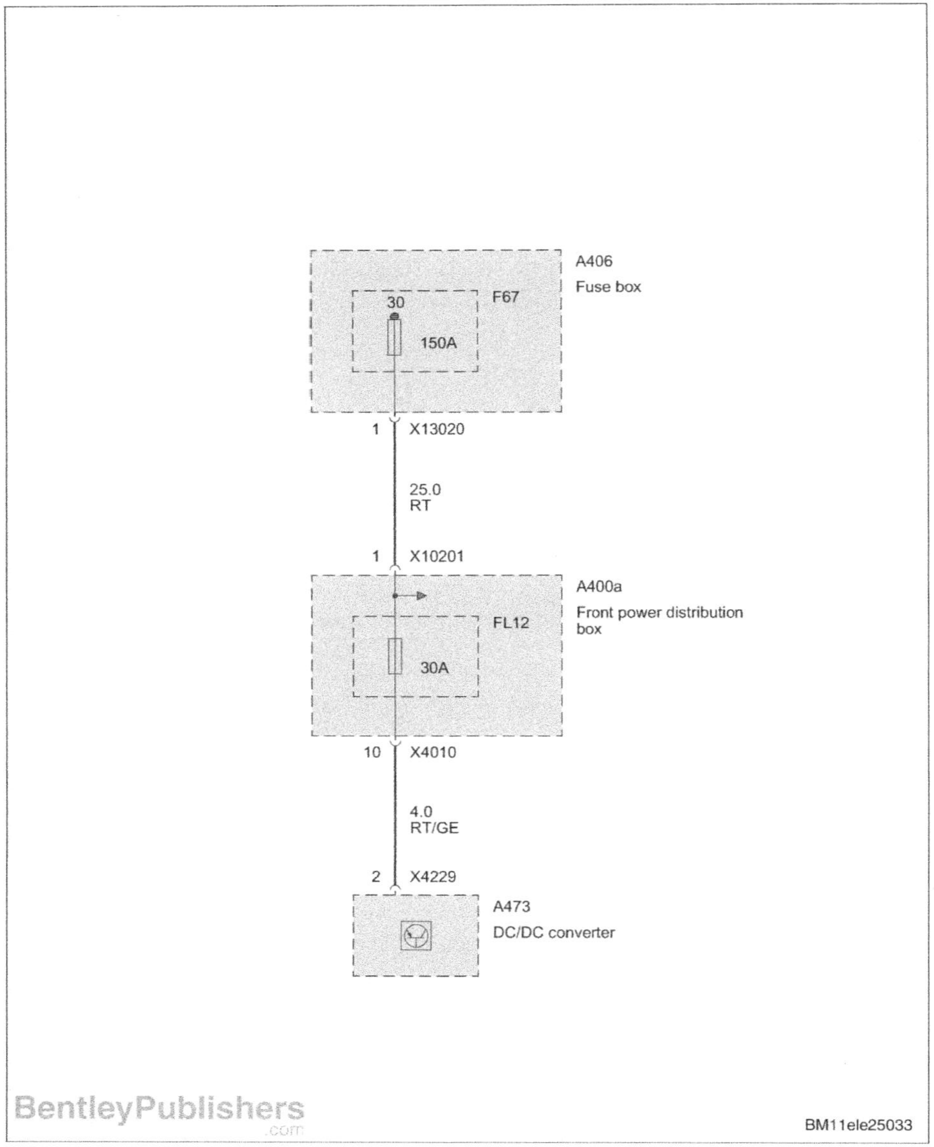

Car access system power supply
from 12/2008

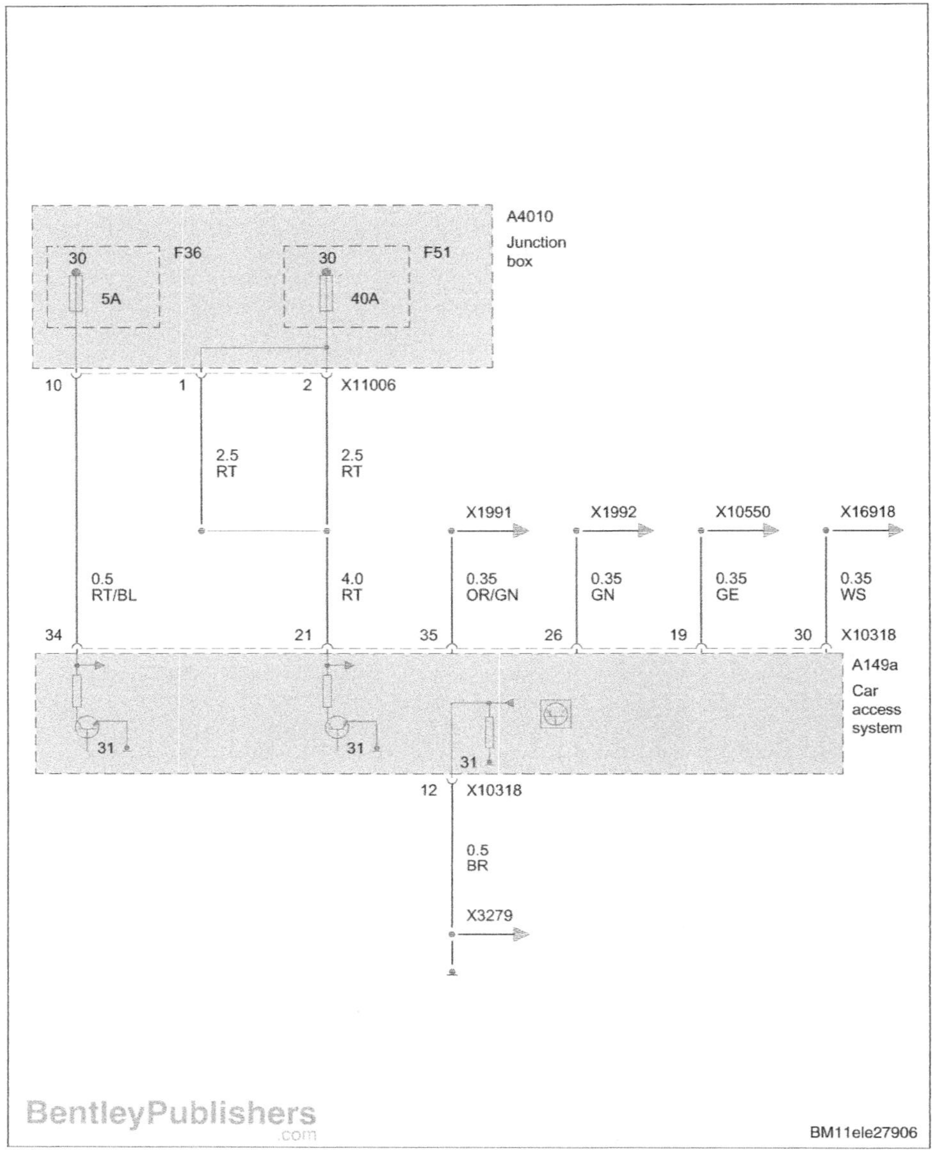

Car access system power supply to 12/2008

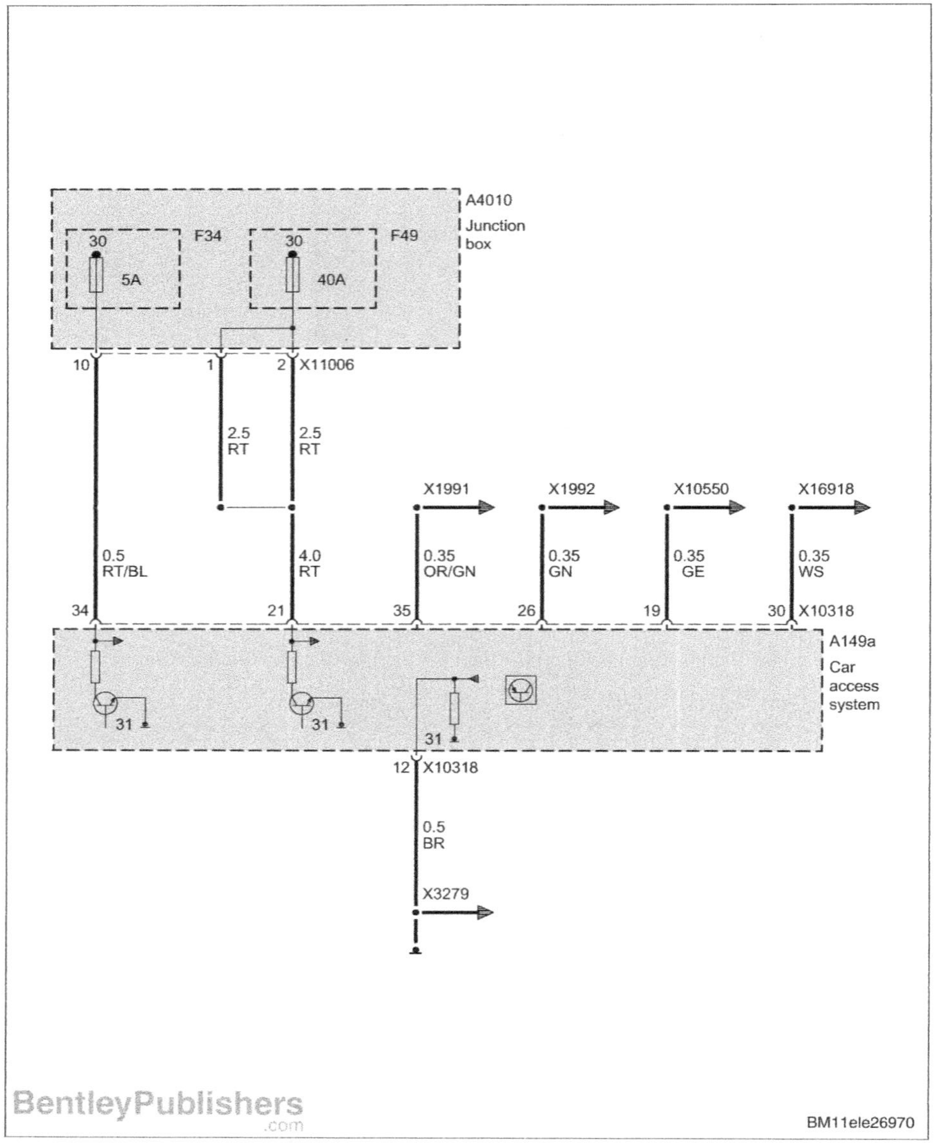

Transmission electronics to 12/2008 page 1 of 2

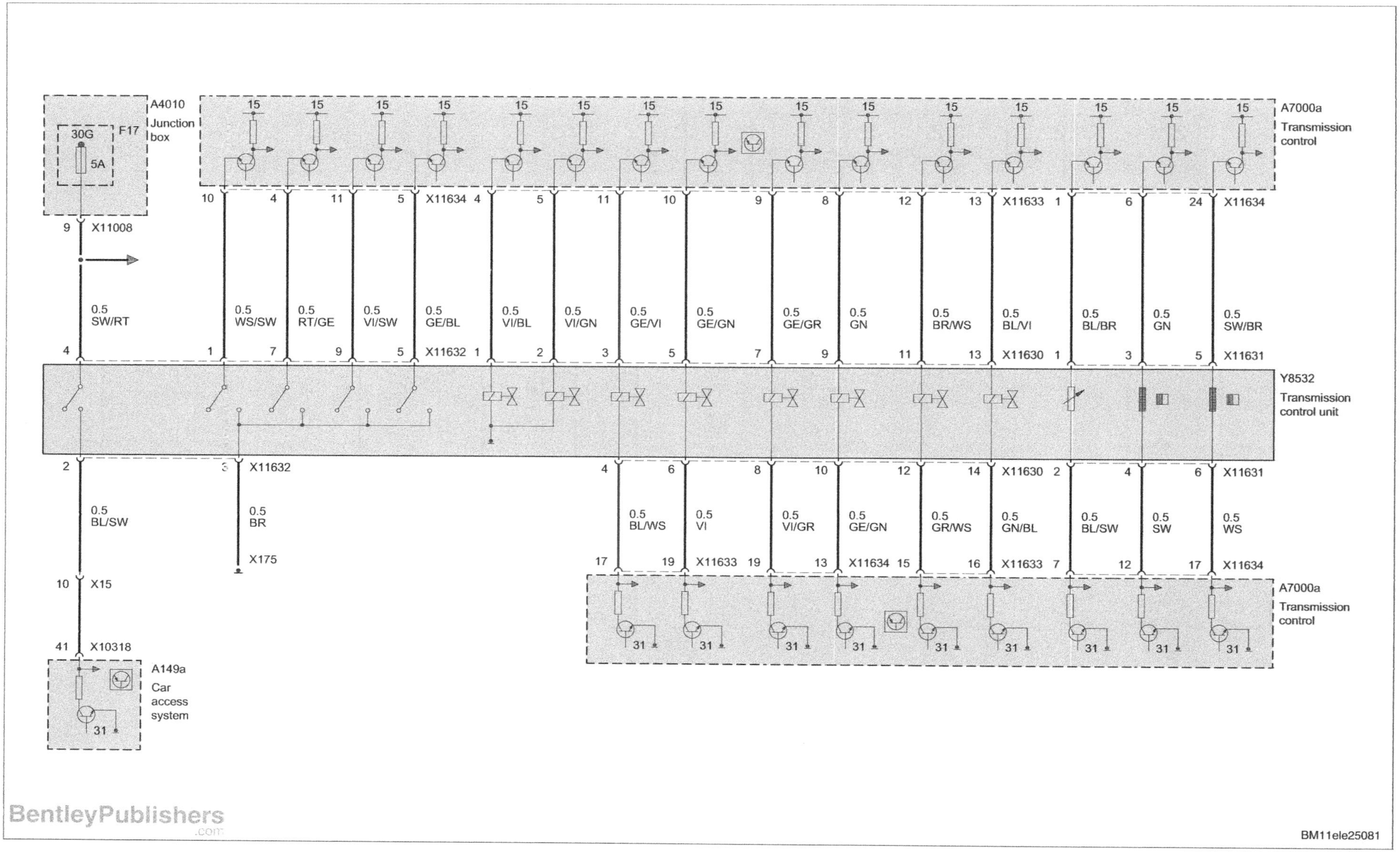

Transmission electronics to 12/2008 page 2 of 2

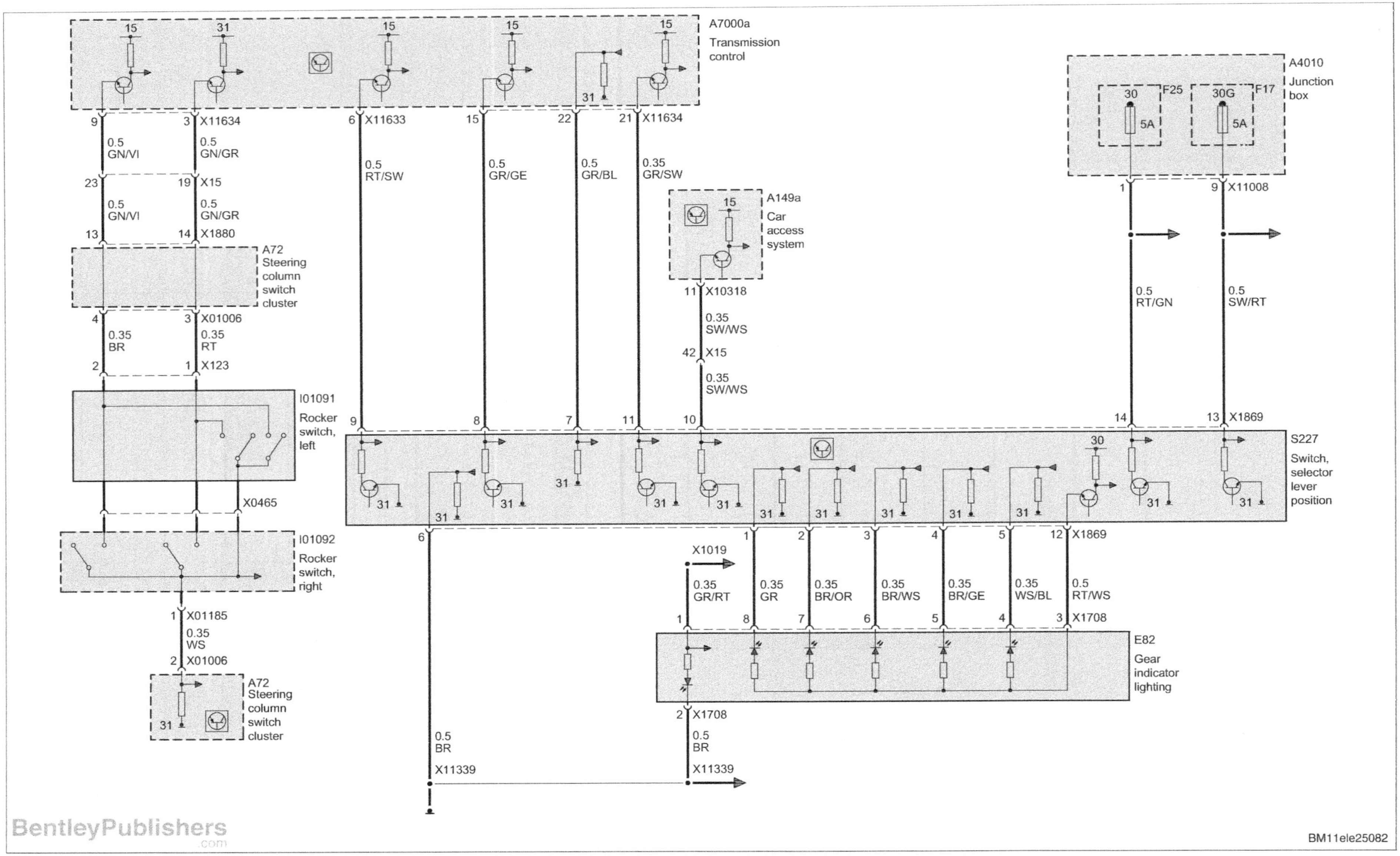

Transmission electronics from 12/2008 page 1 of 2

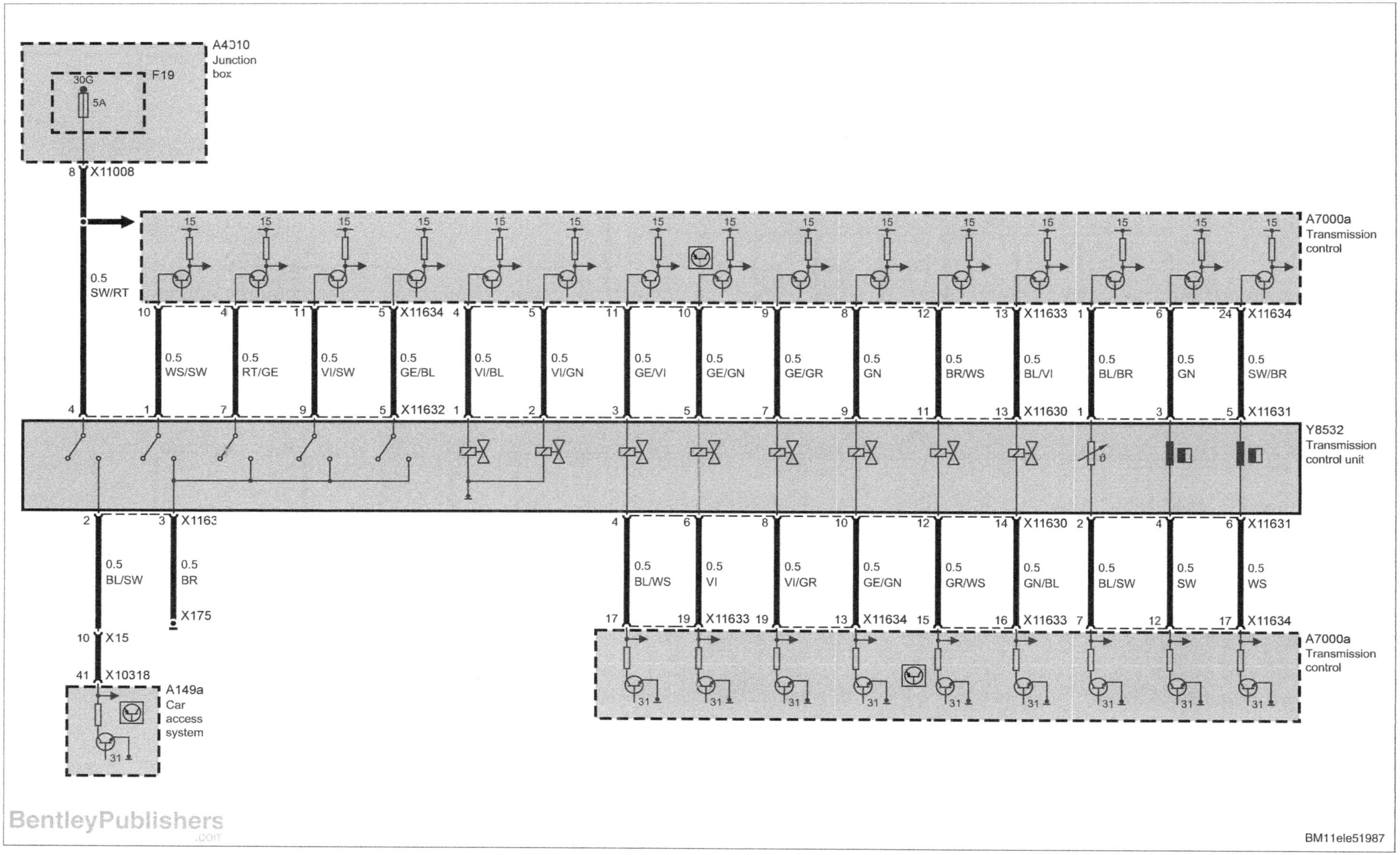

Transmission electronics from 12/2008 page 2 of 2

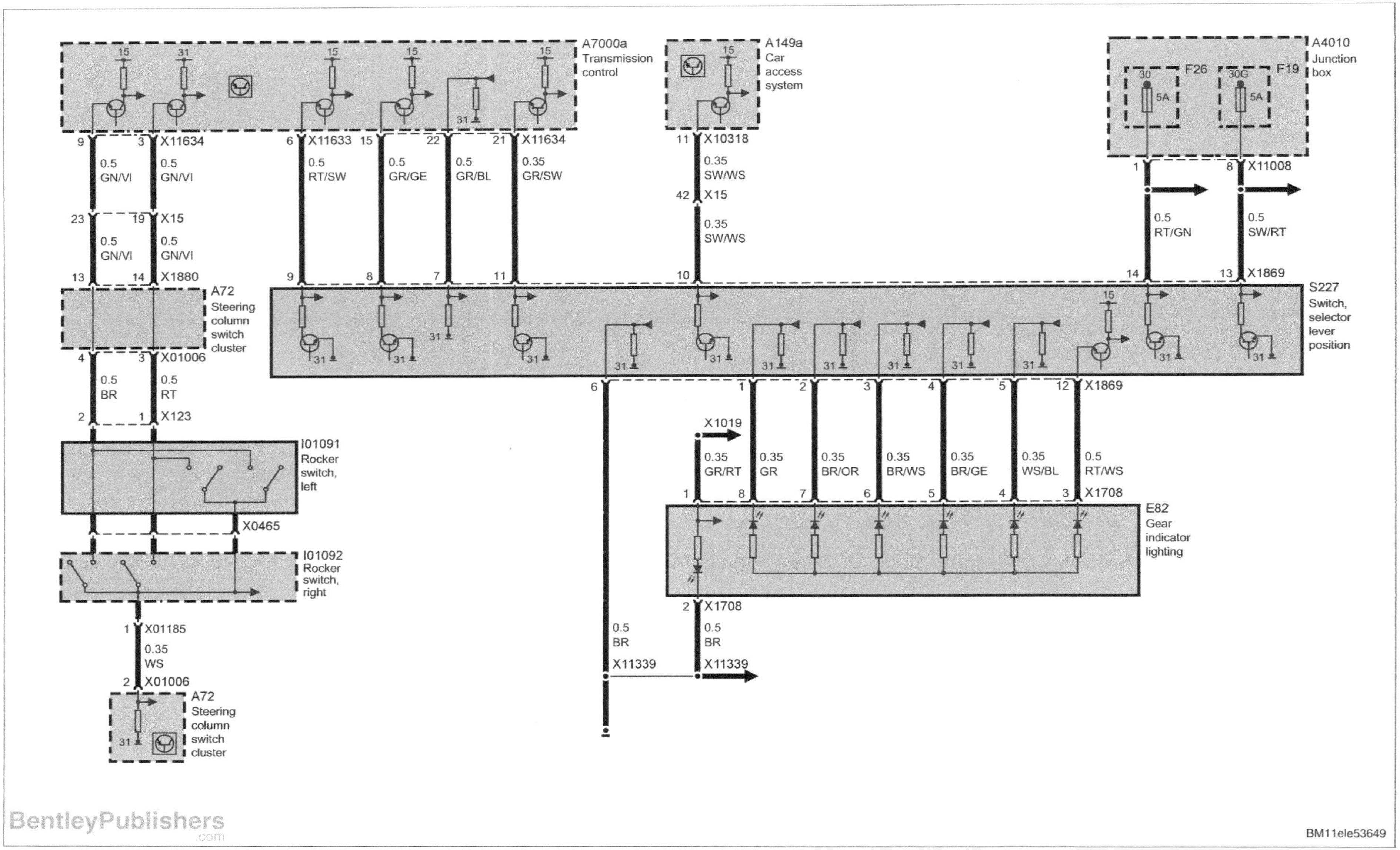

Transmission control module

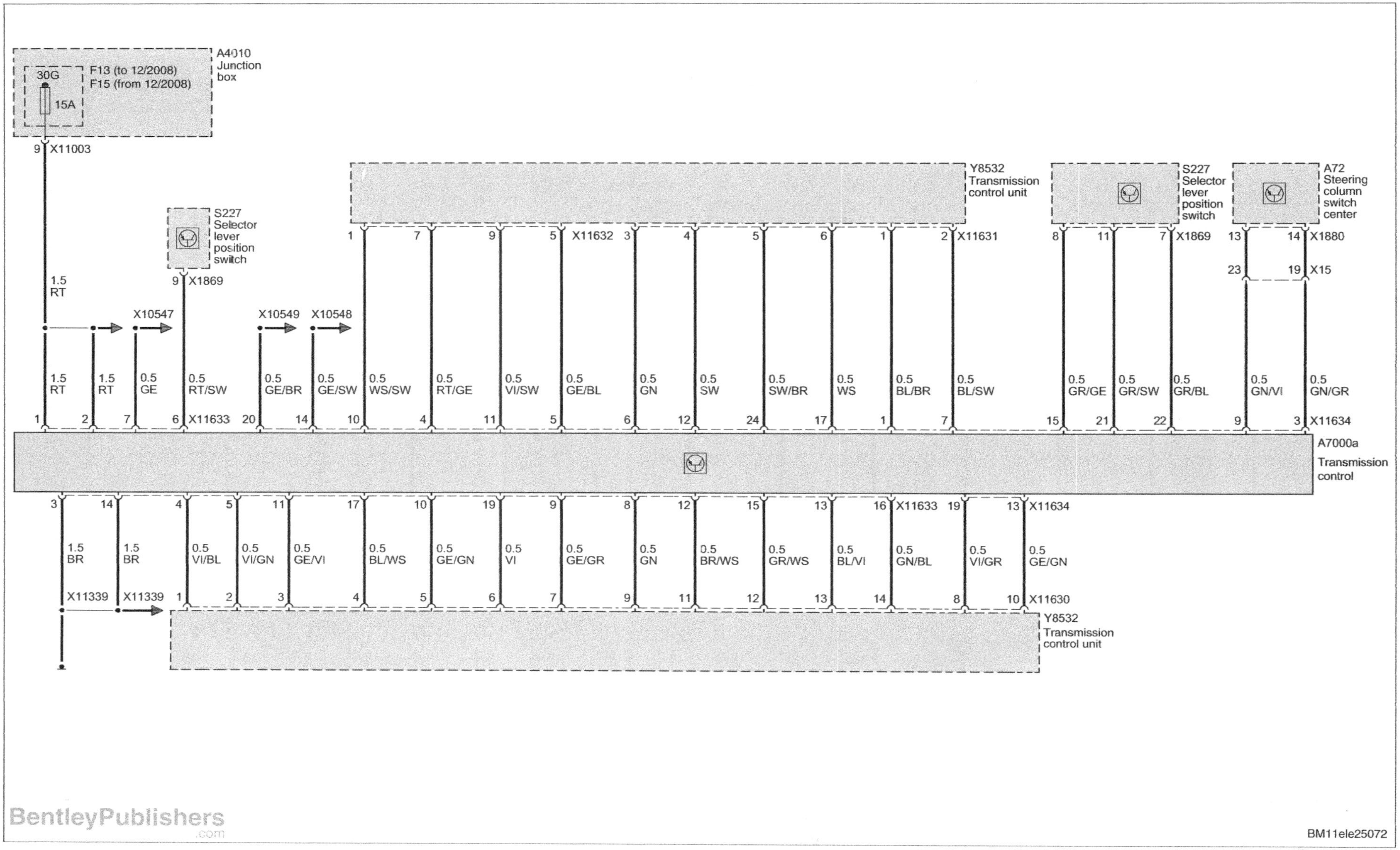

Car access system to 12/2008 page 1 of 2

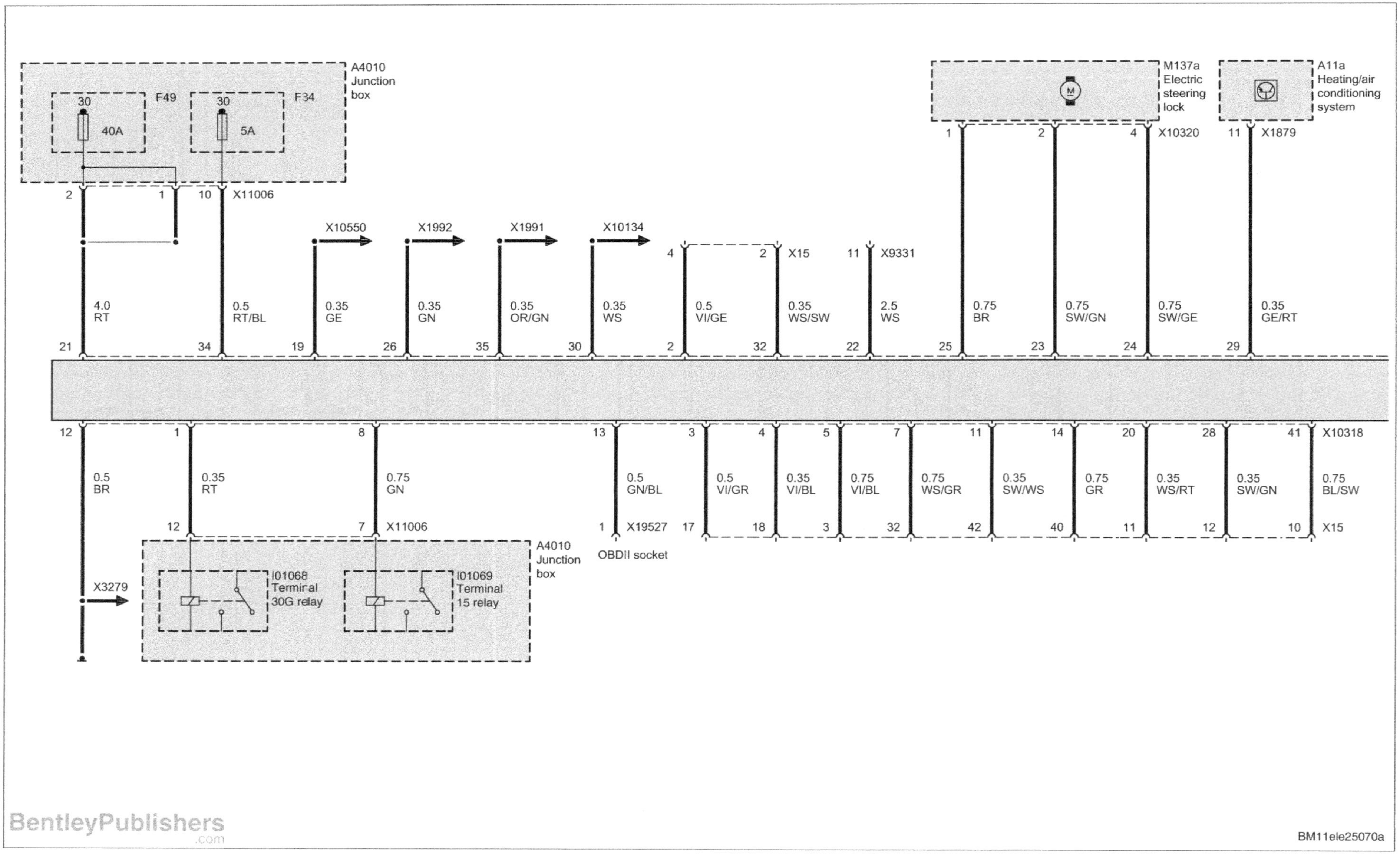

Car access system
to 12/2008
page 2 of 2

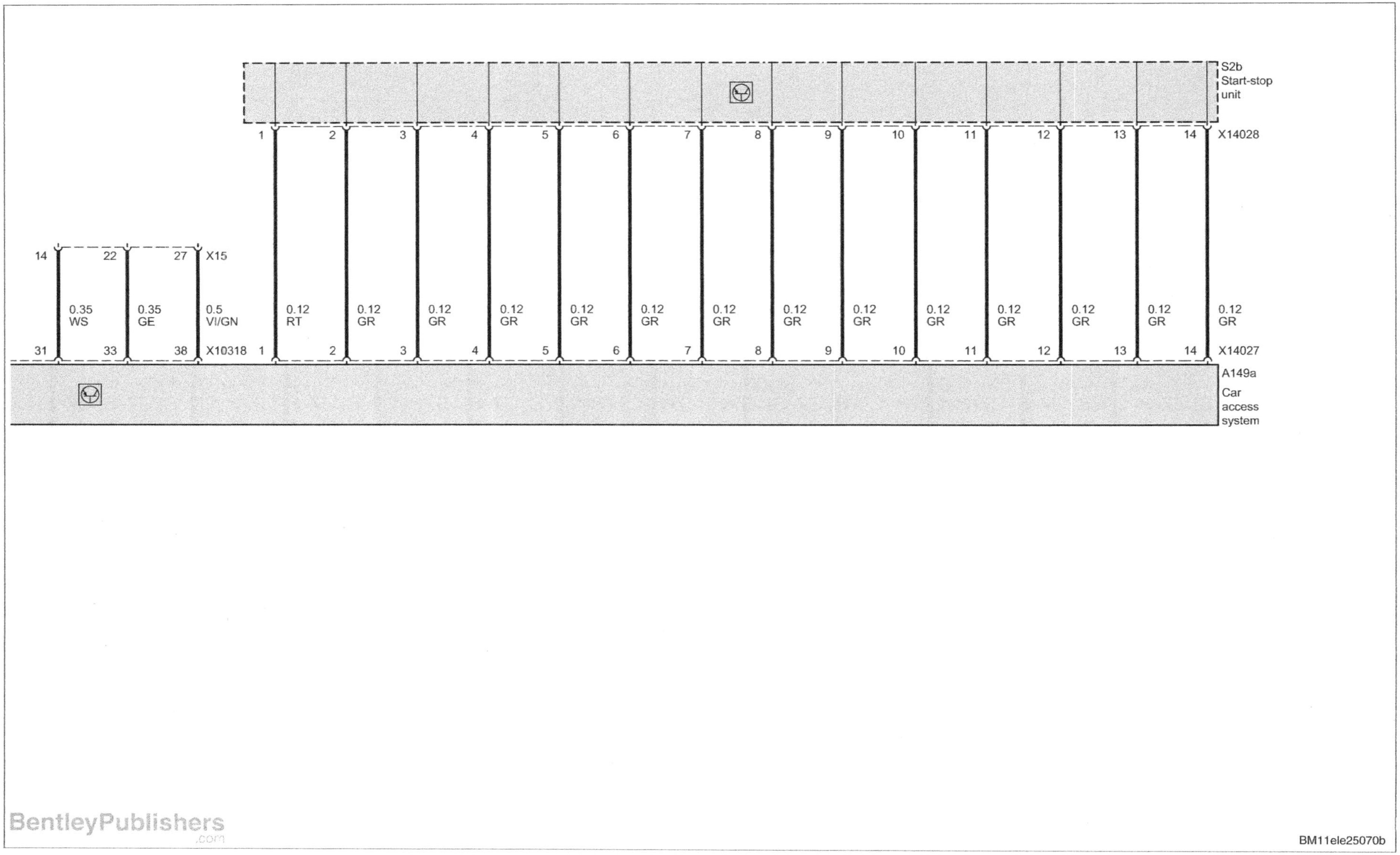

Car access system from 12/2008 page 1 of 2

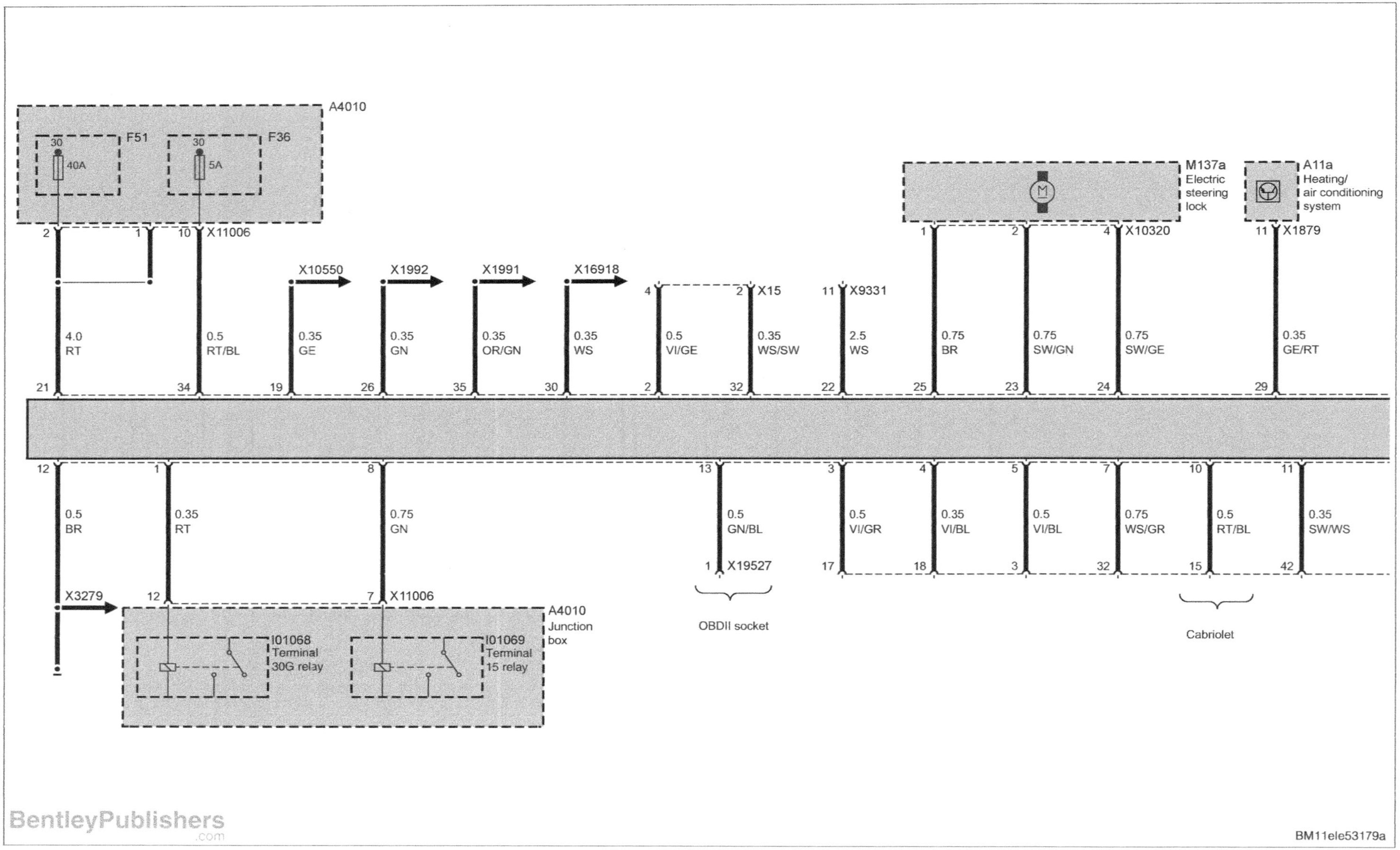

Car access system
from 12/2008
page 2 of 2

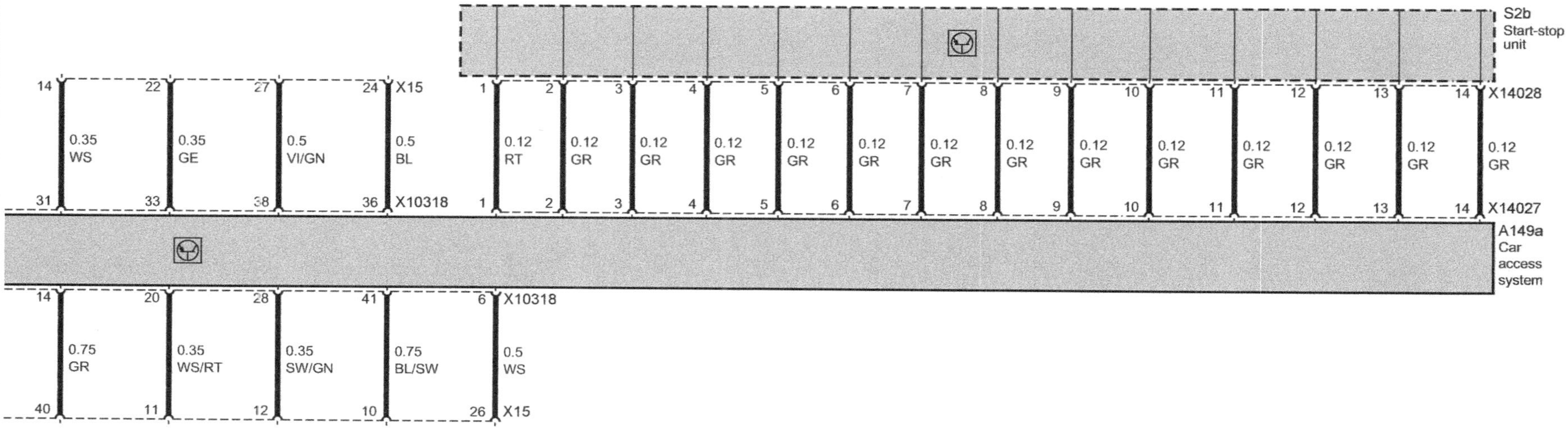

BM11ele53179b

Electric steering column adjustment to 08/2007

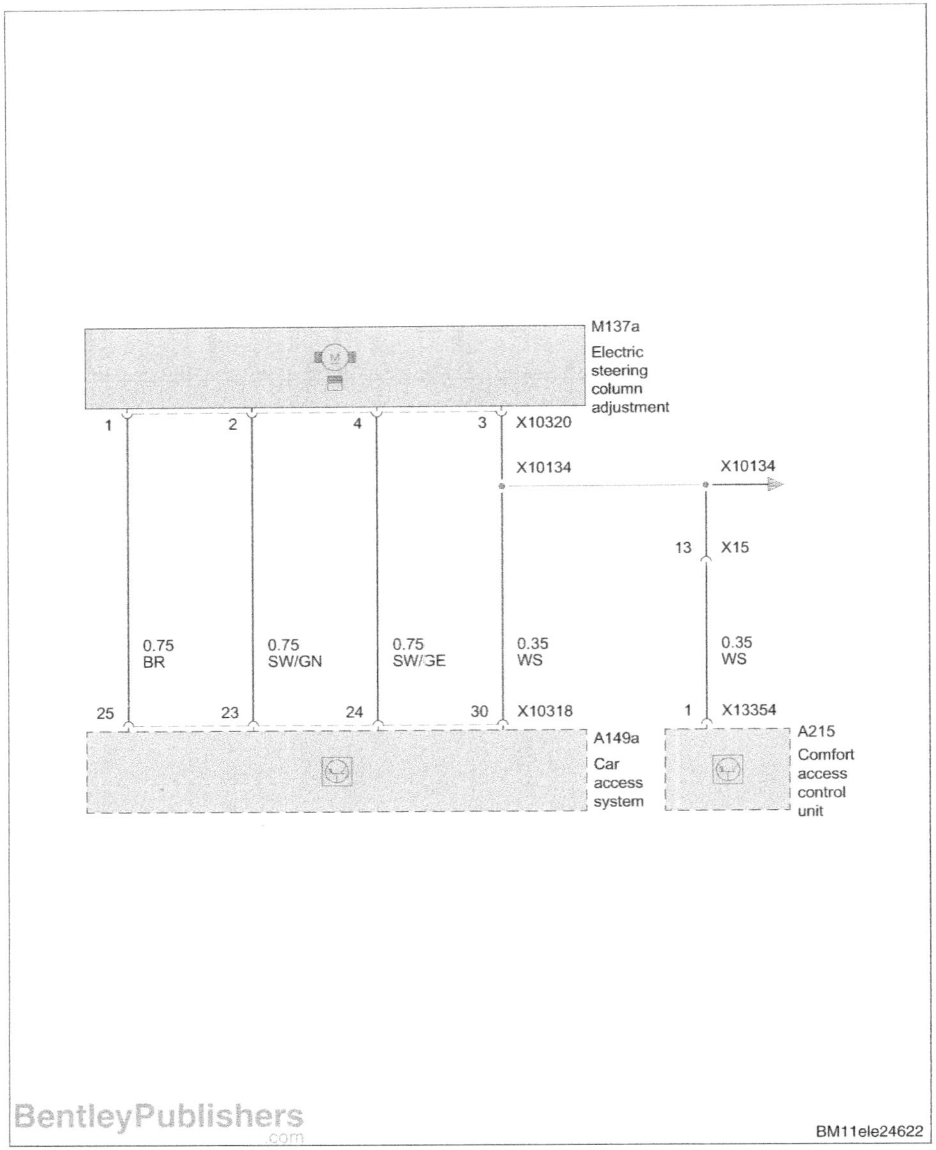

Electric steering column adjustment from 08/2007

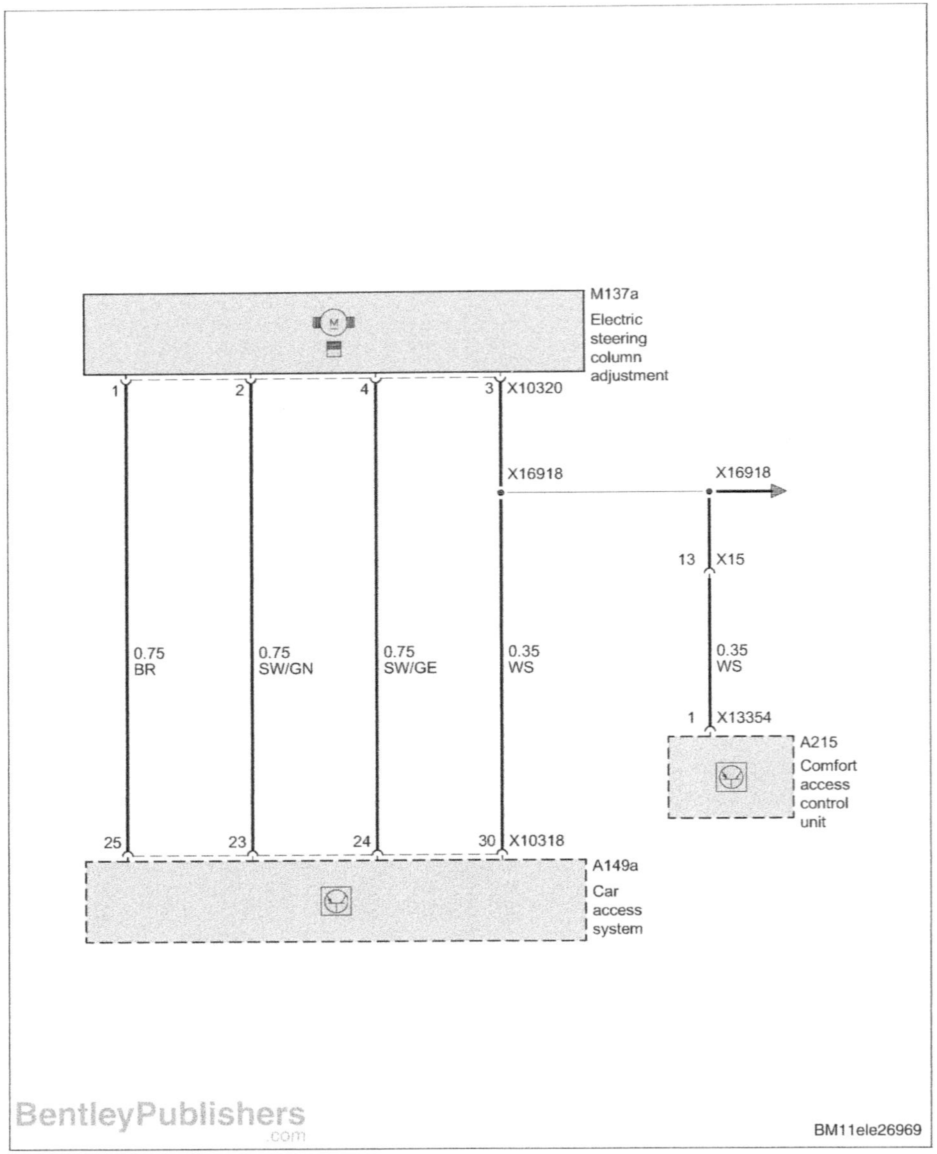

Rear fog lights

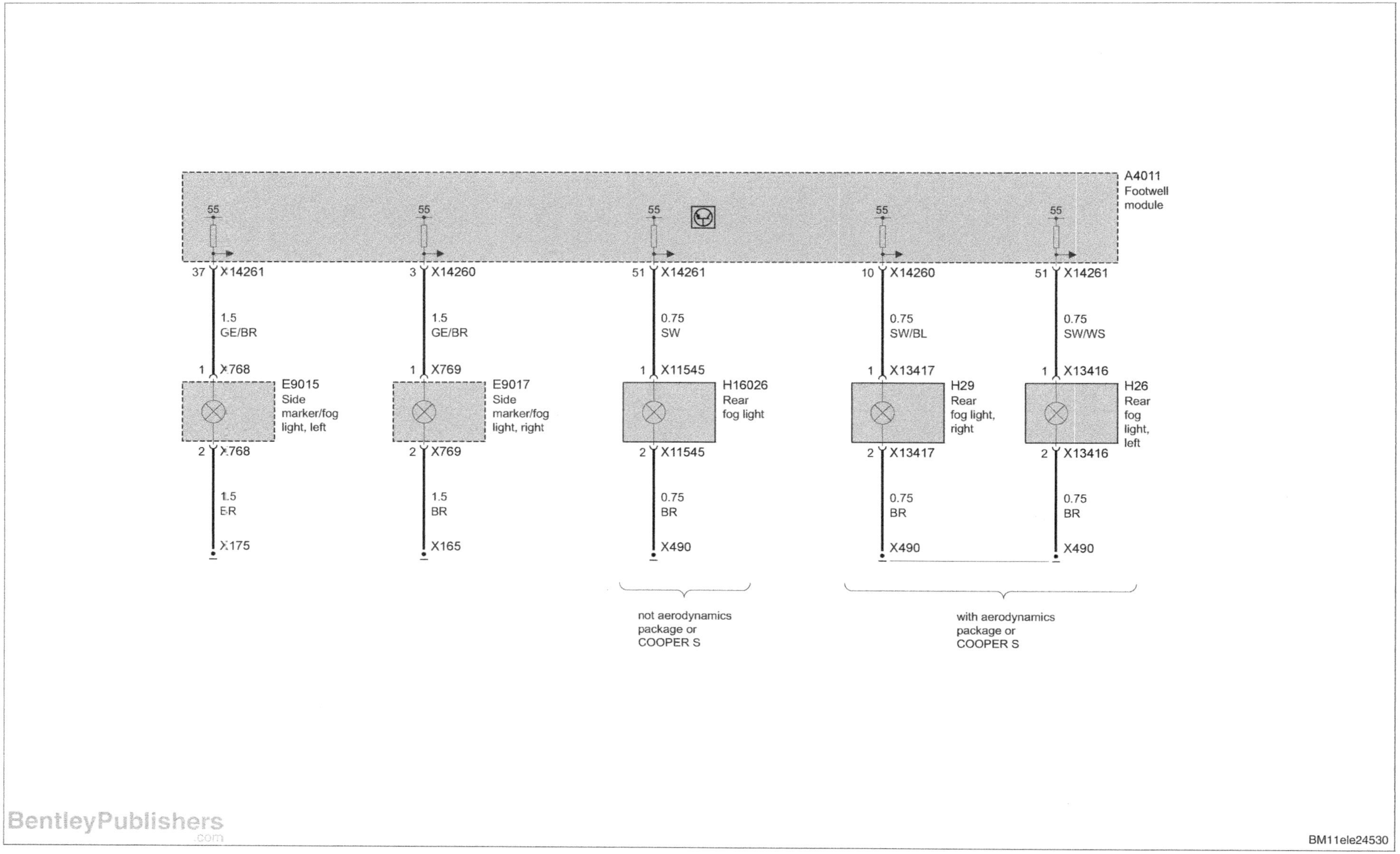

Footwell module power supply to 12/2008

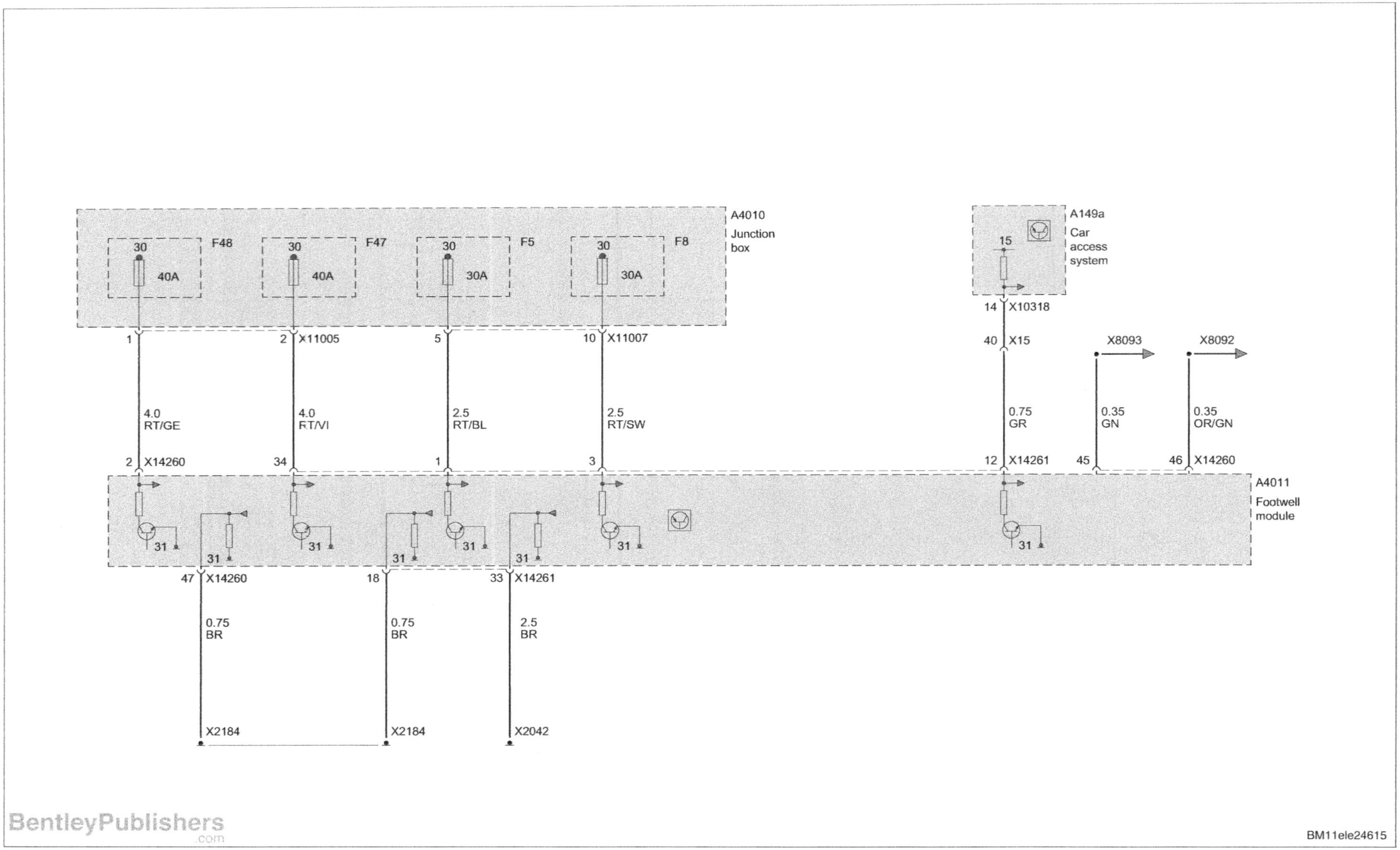

Footwell module
to 12/2008, except Clubman
page 1 of 3

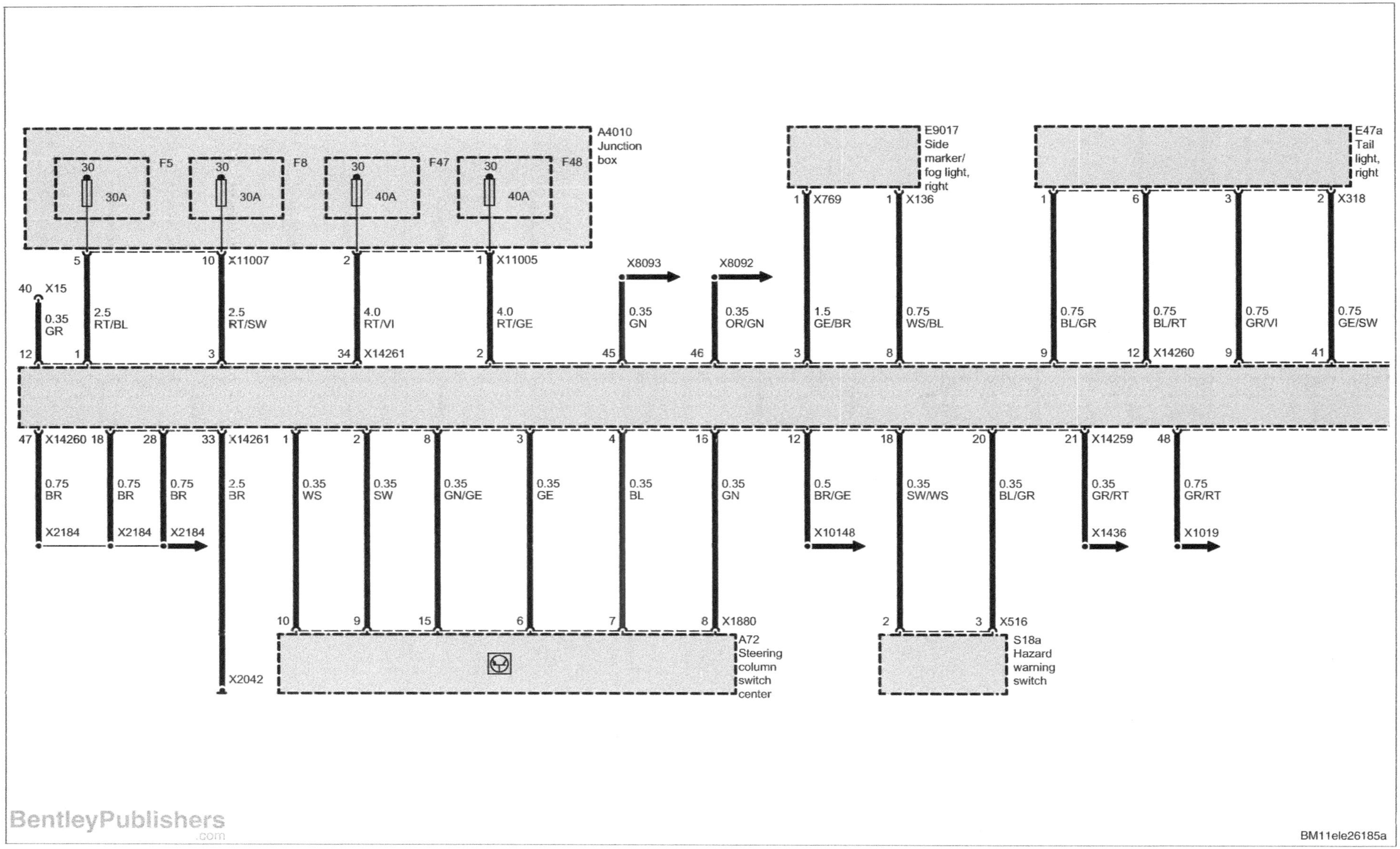

Footwell module
to 12/2008, except Clubman
page 2 of 3

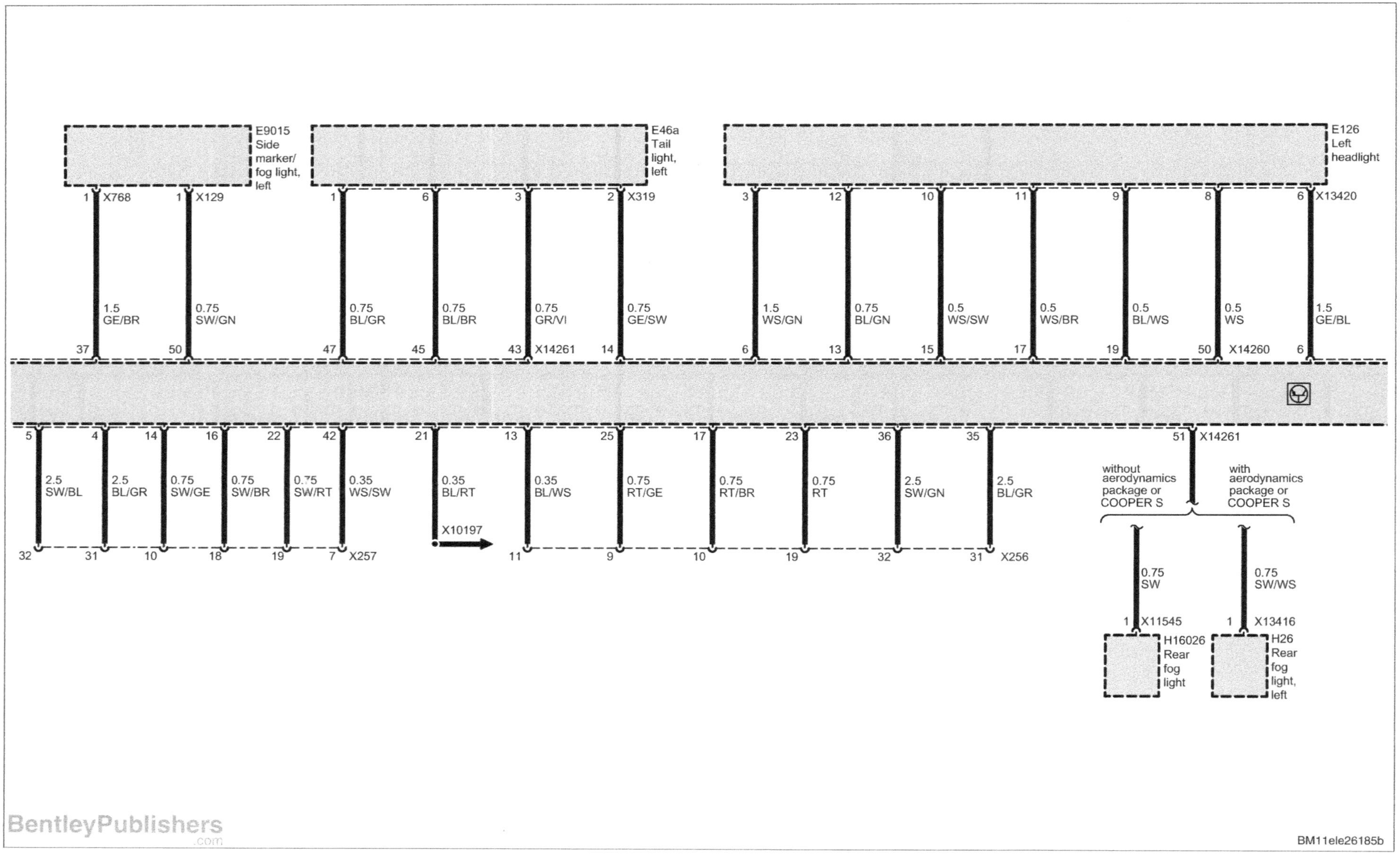

Footwell module
to 12/2008, except Clubman
page 3 of 3

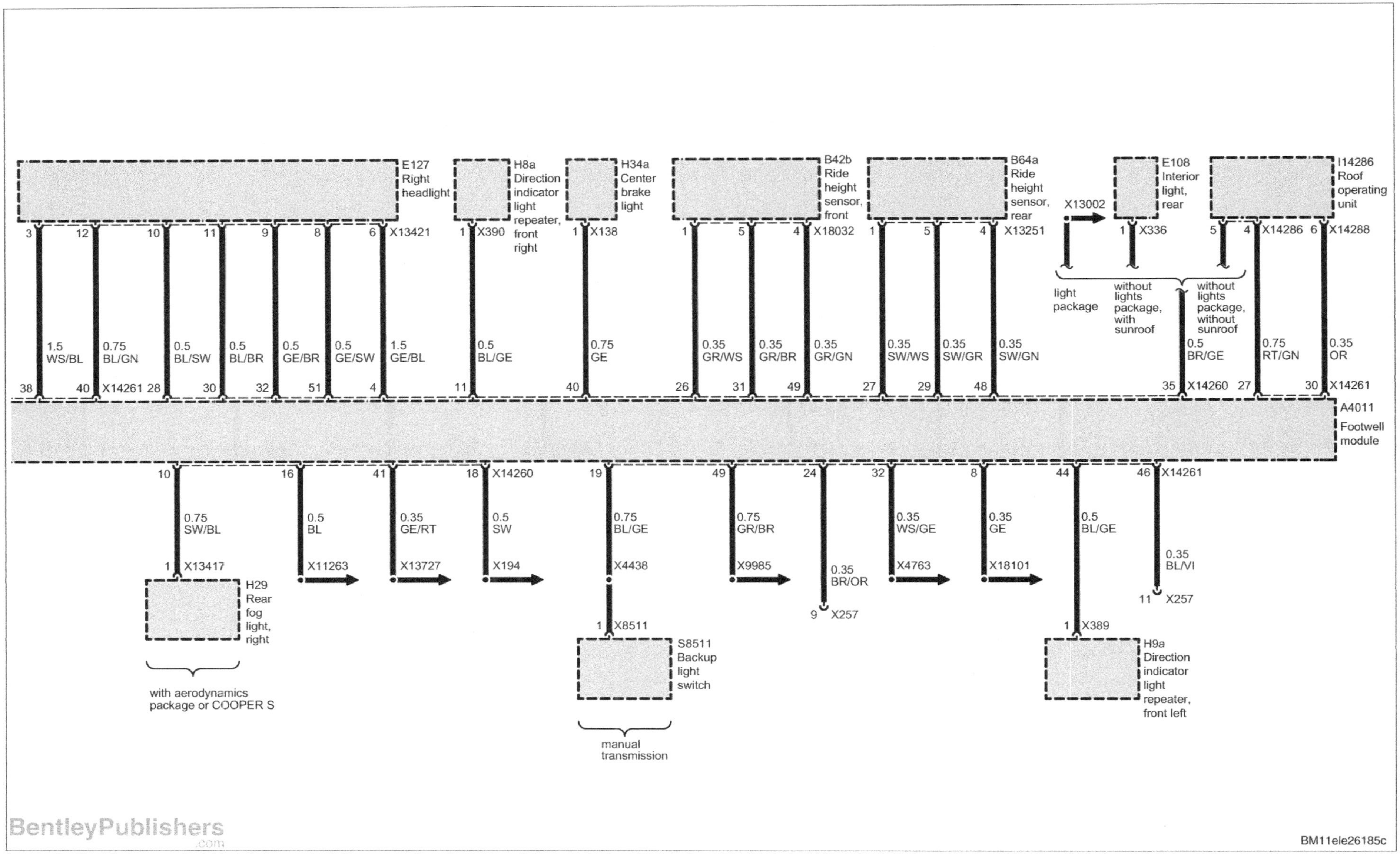

Footwell module power supply from 12/2008

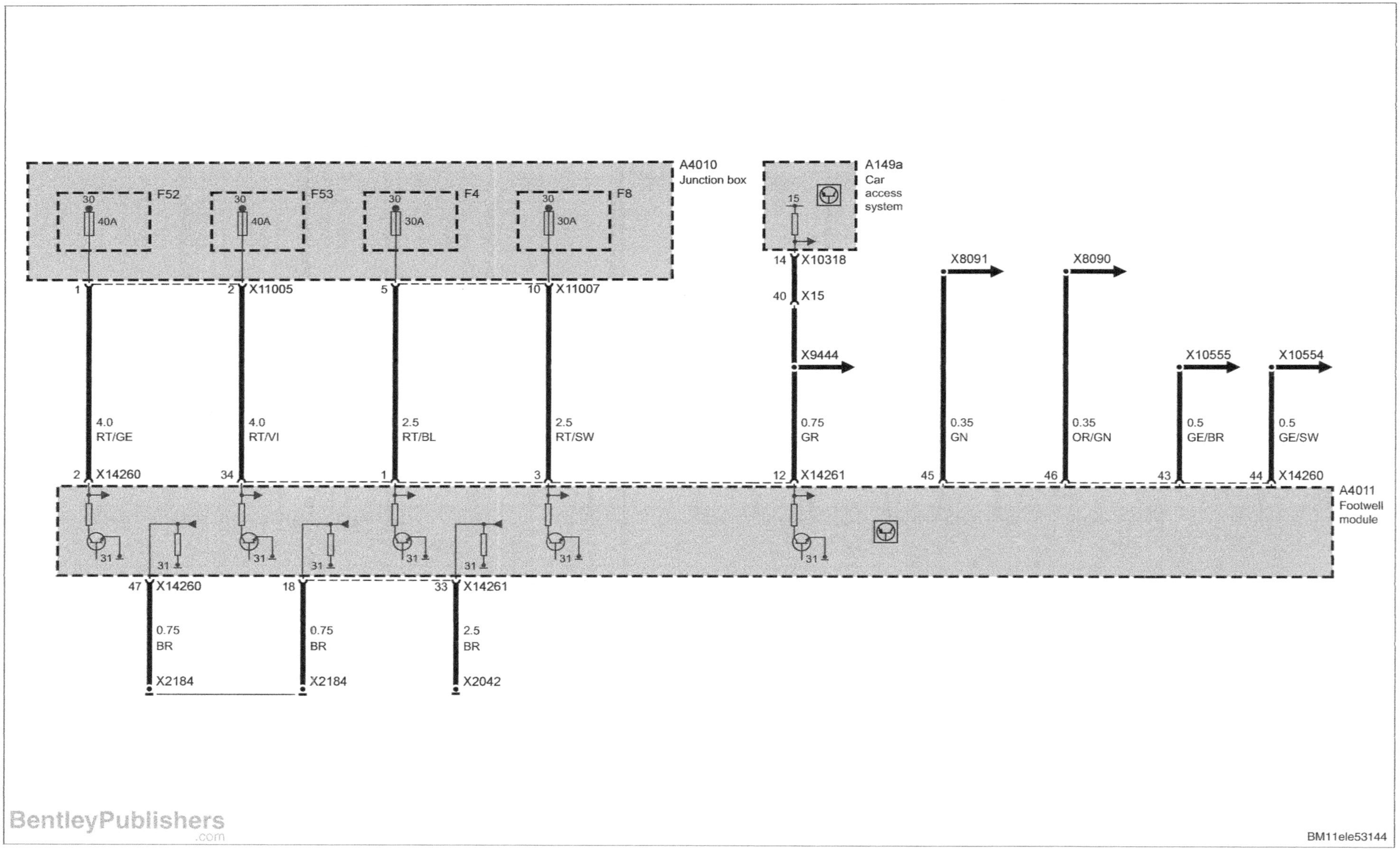

Footwell module
to 12/2008, Clubman
page 1 of 3

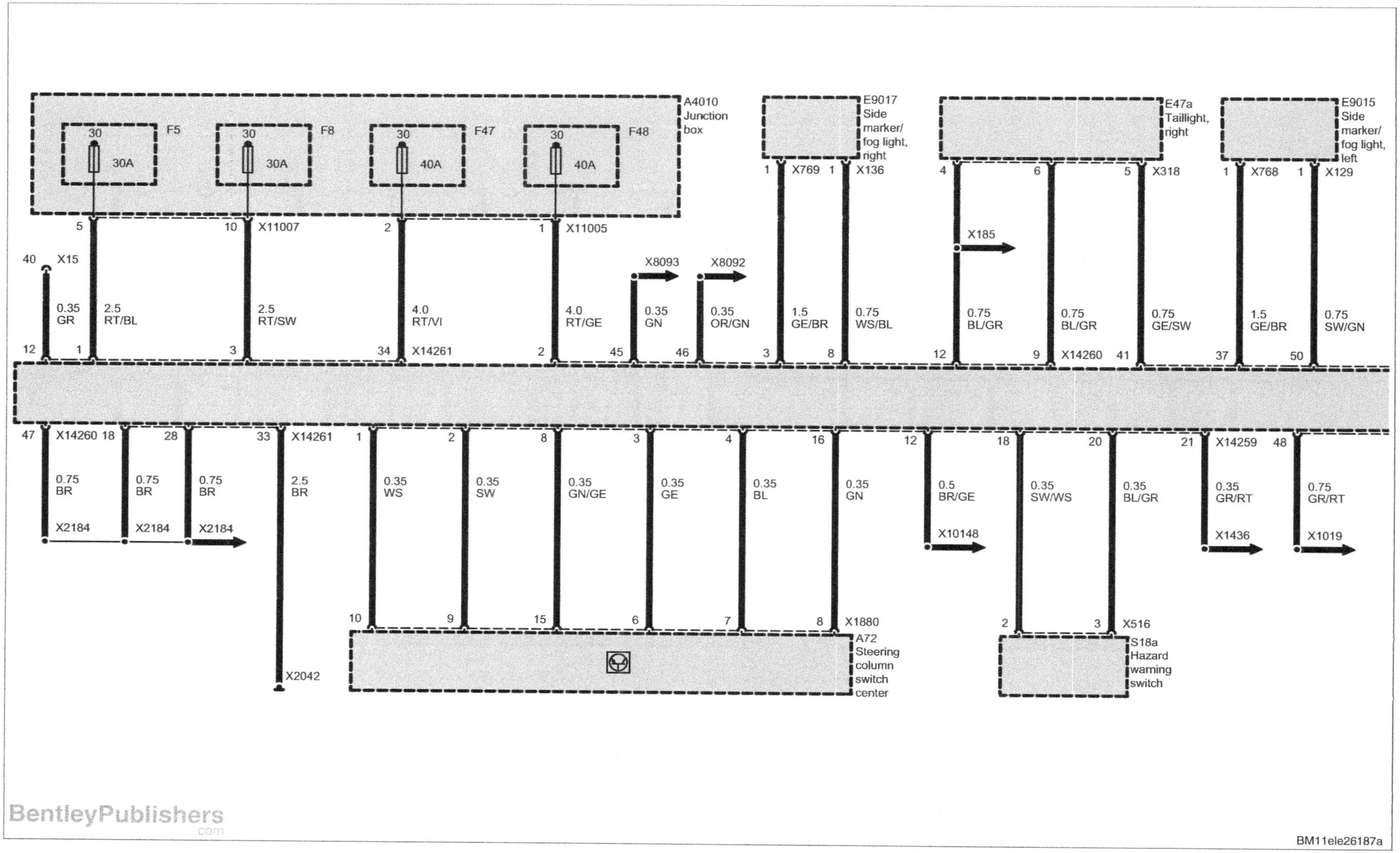

**Footwell module
to 12/2008, Clubman
page 2 of 3**

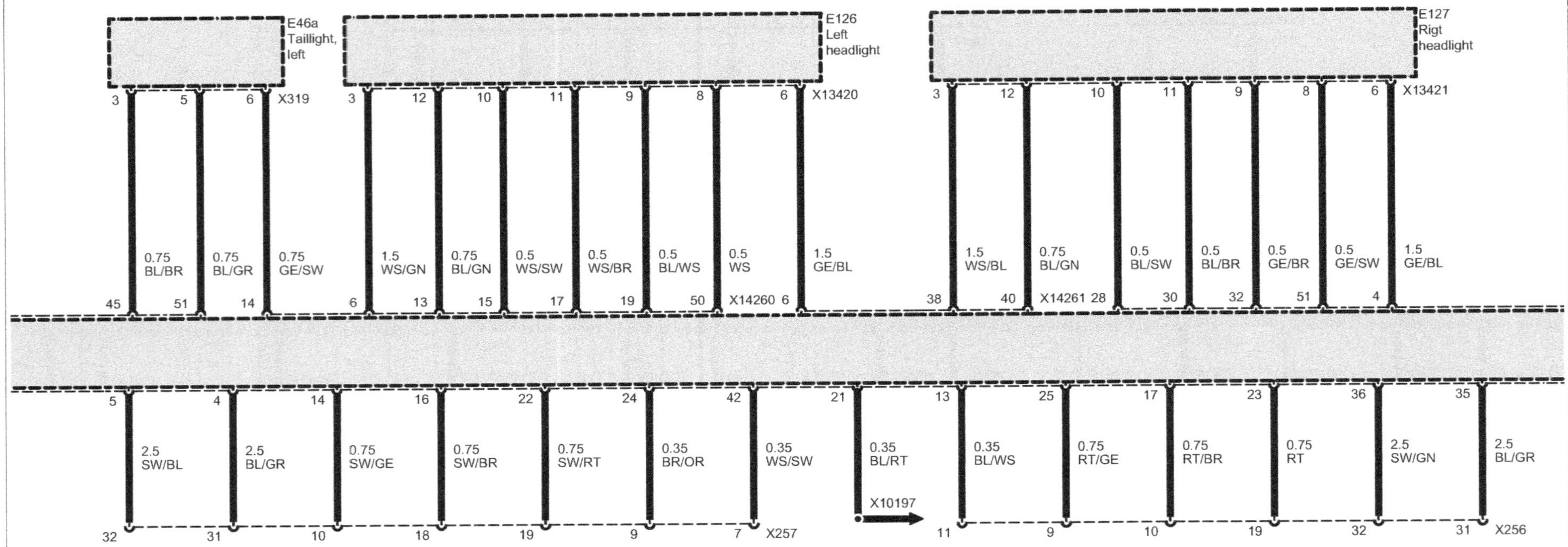

BM11ele26187b

Footwell module
to 12/2008, Clubman
page 3 of 3

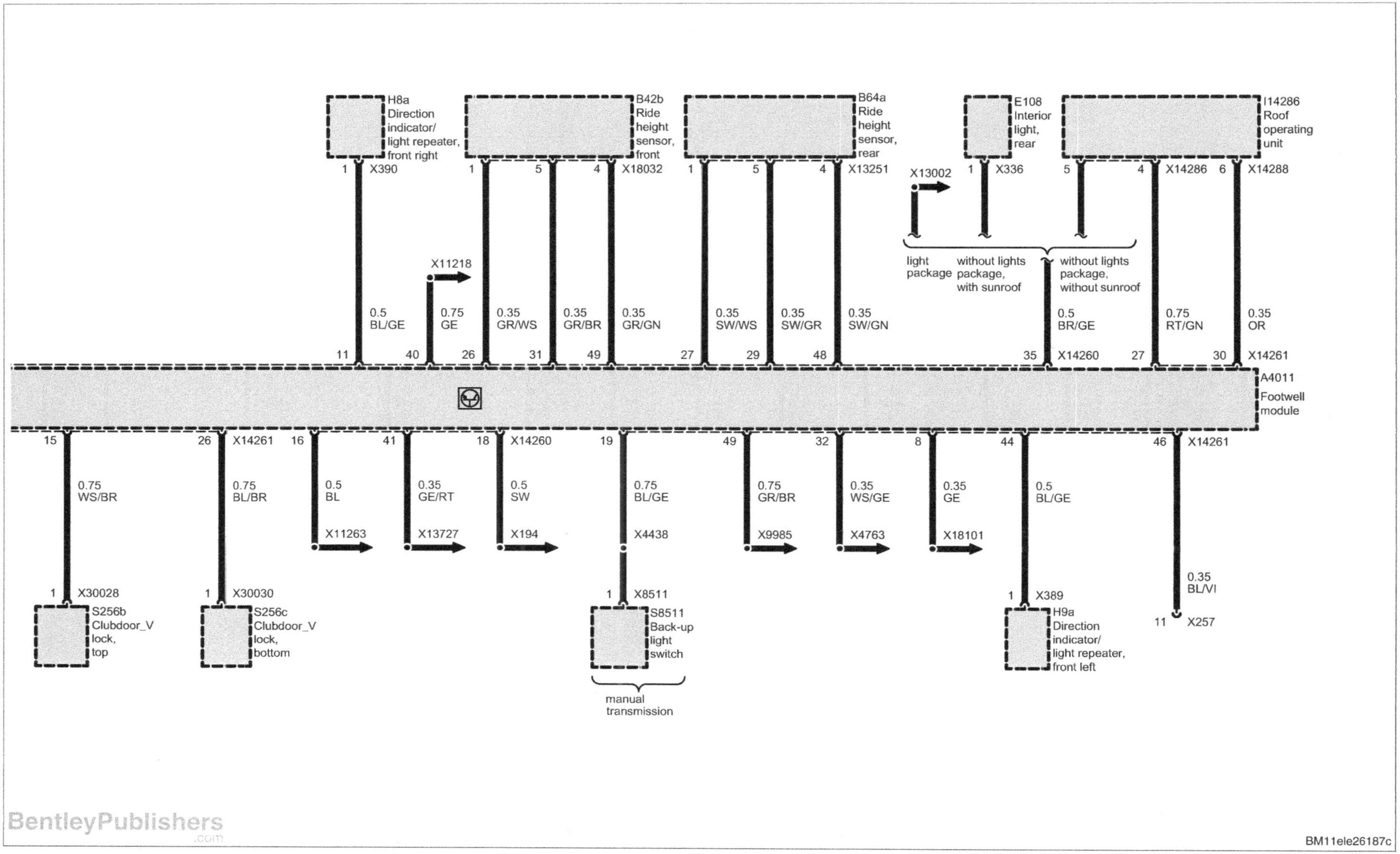

Footwell module from 12/2008, Clubman page 1 of 3

Footwell module
from 12/2008, Clubman
page 2 of 3

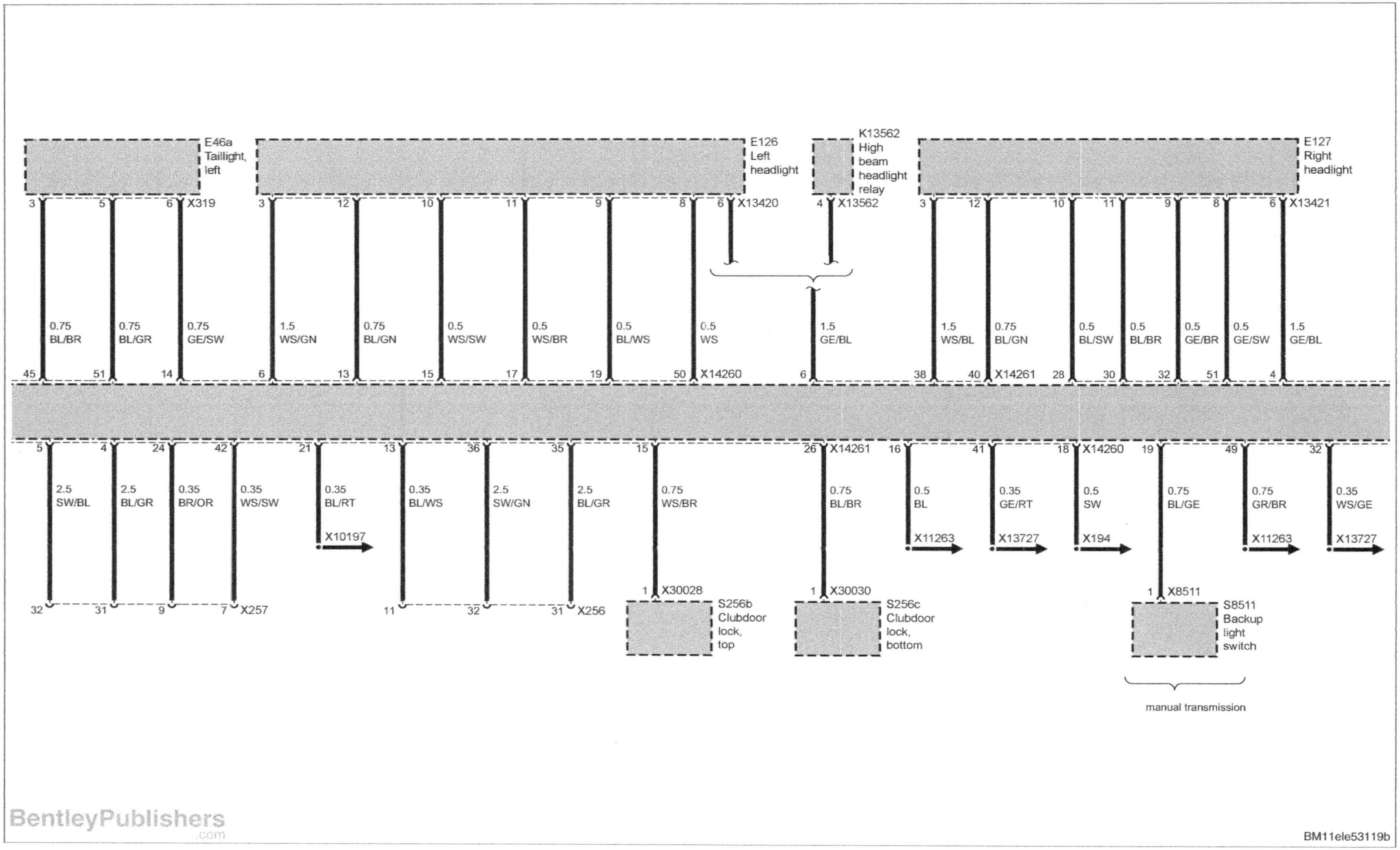

Footwell module from 12/2008, Clubman page 3 of 3

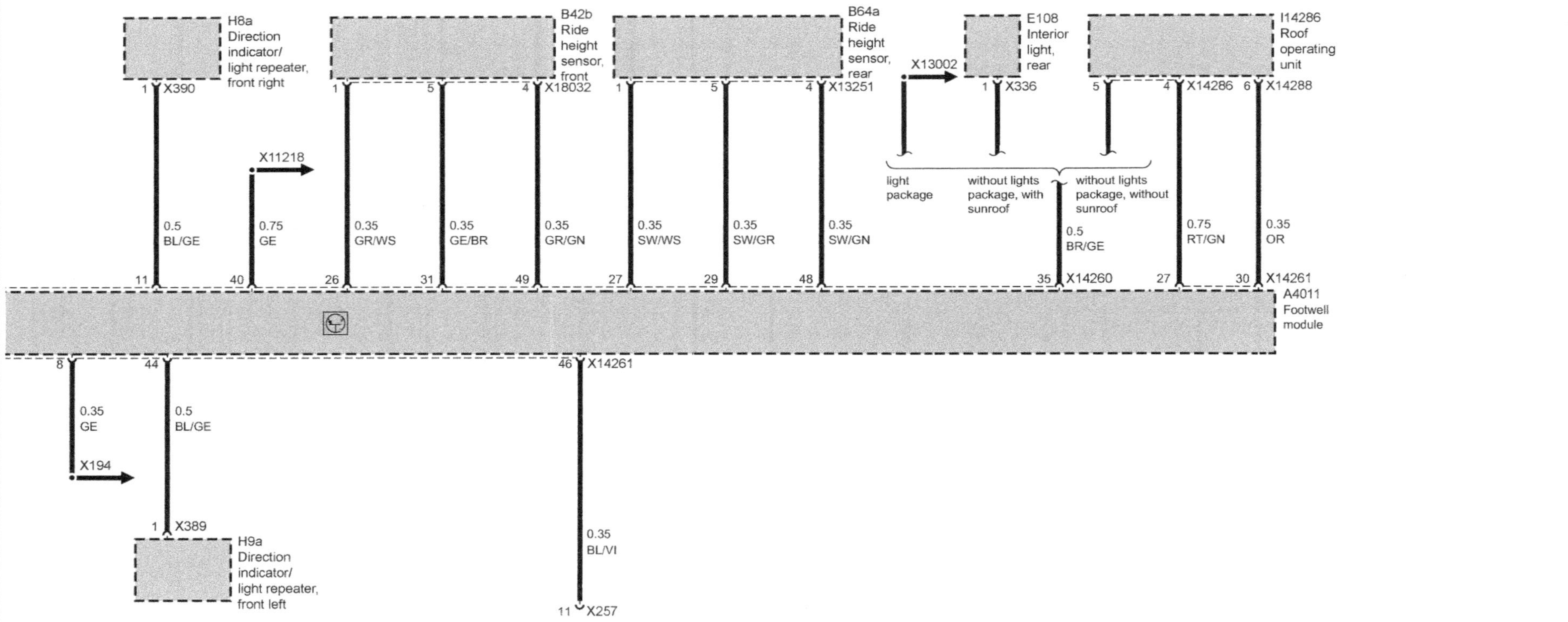

Footwell module, from 12/2008, except Clubman page 1 of 3

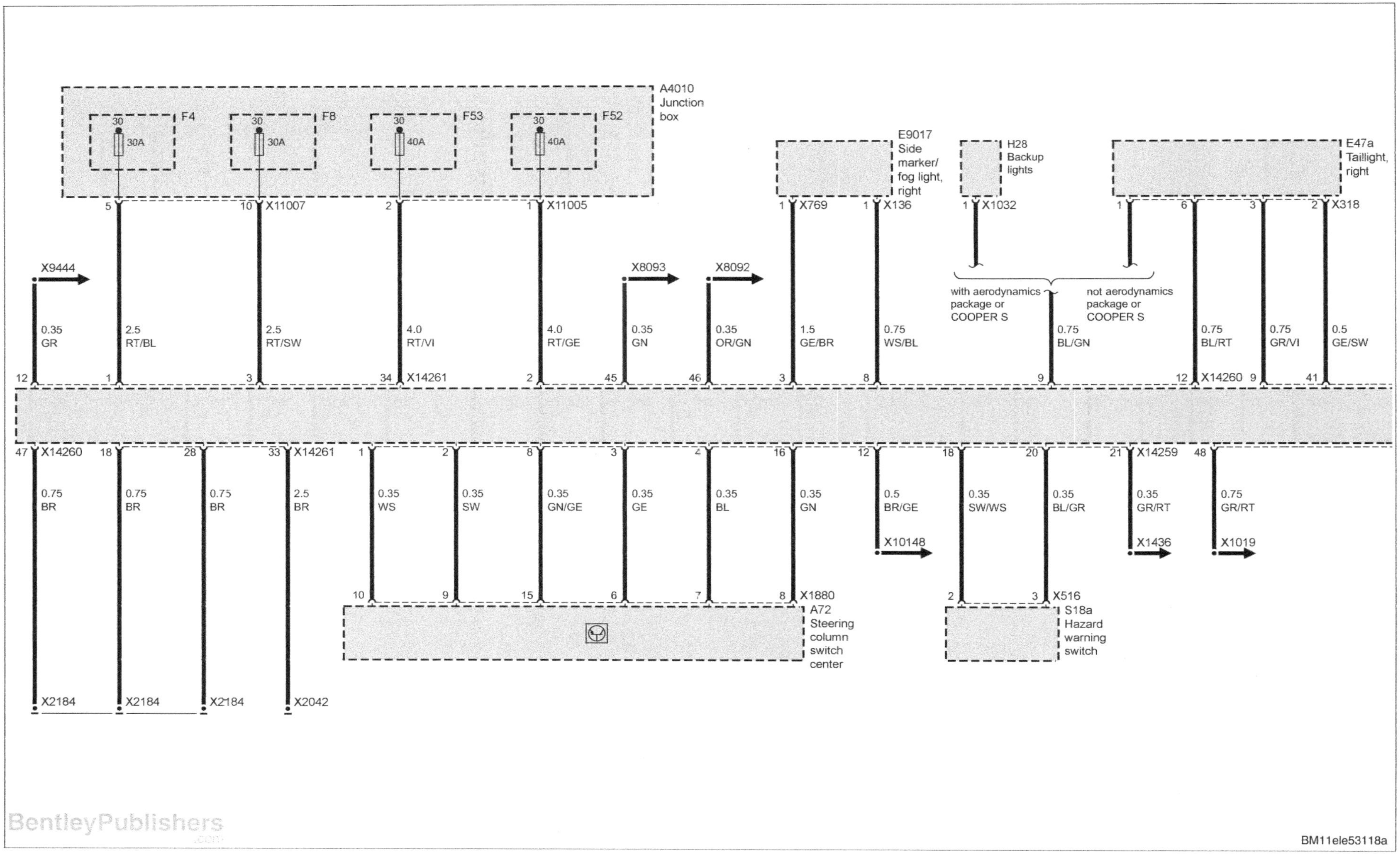

Footwell module, from 12/2008, except Clubman page 2 of 3

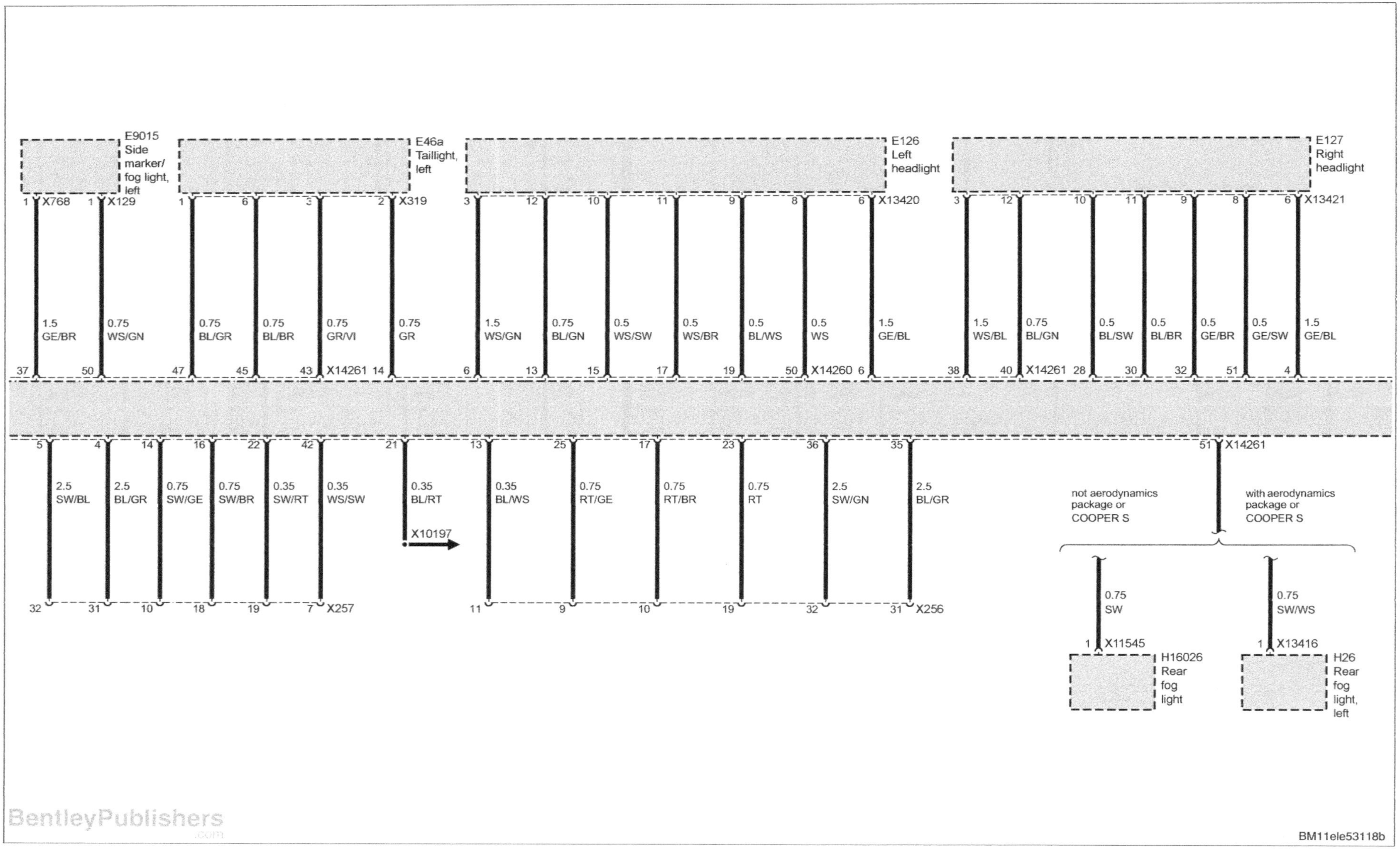

Footwell module, from 12/2008, except Clubman page 3 of 3

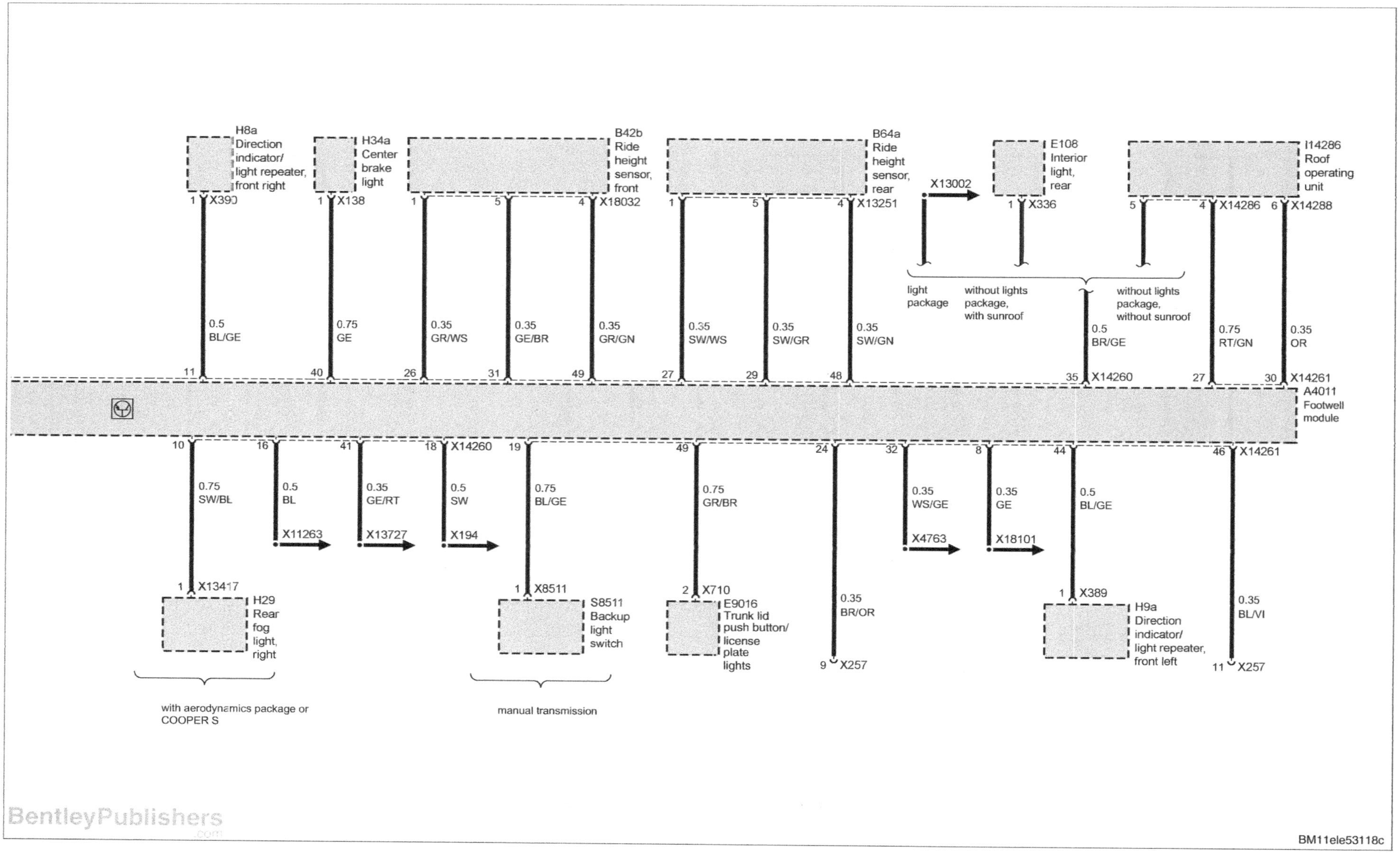

Connector X256
page 1 of 2

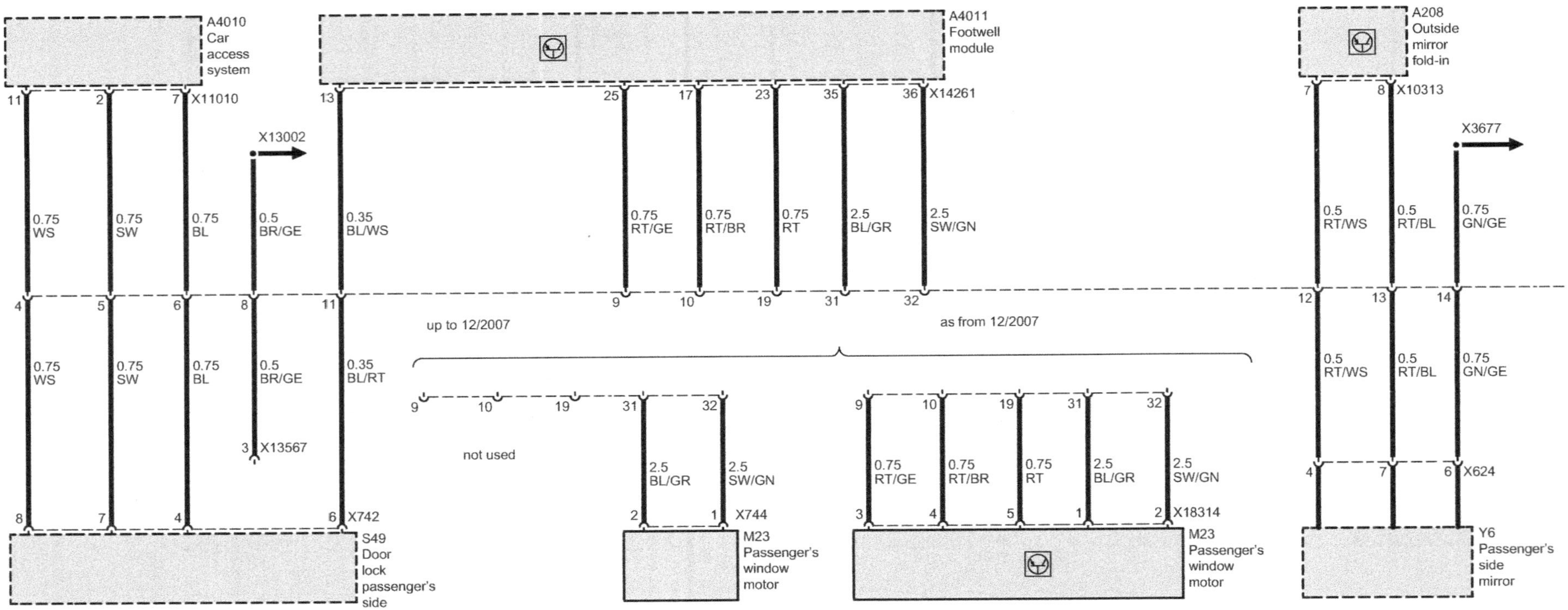

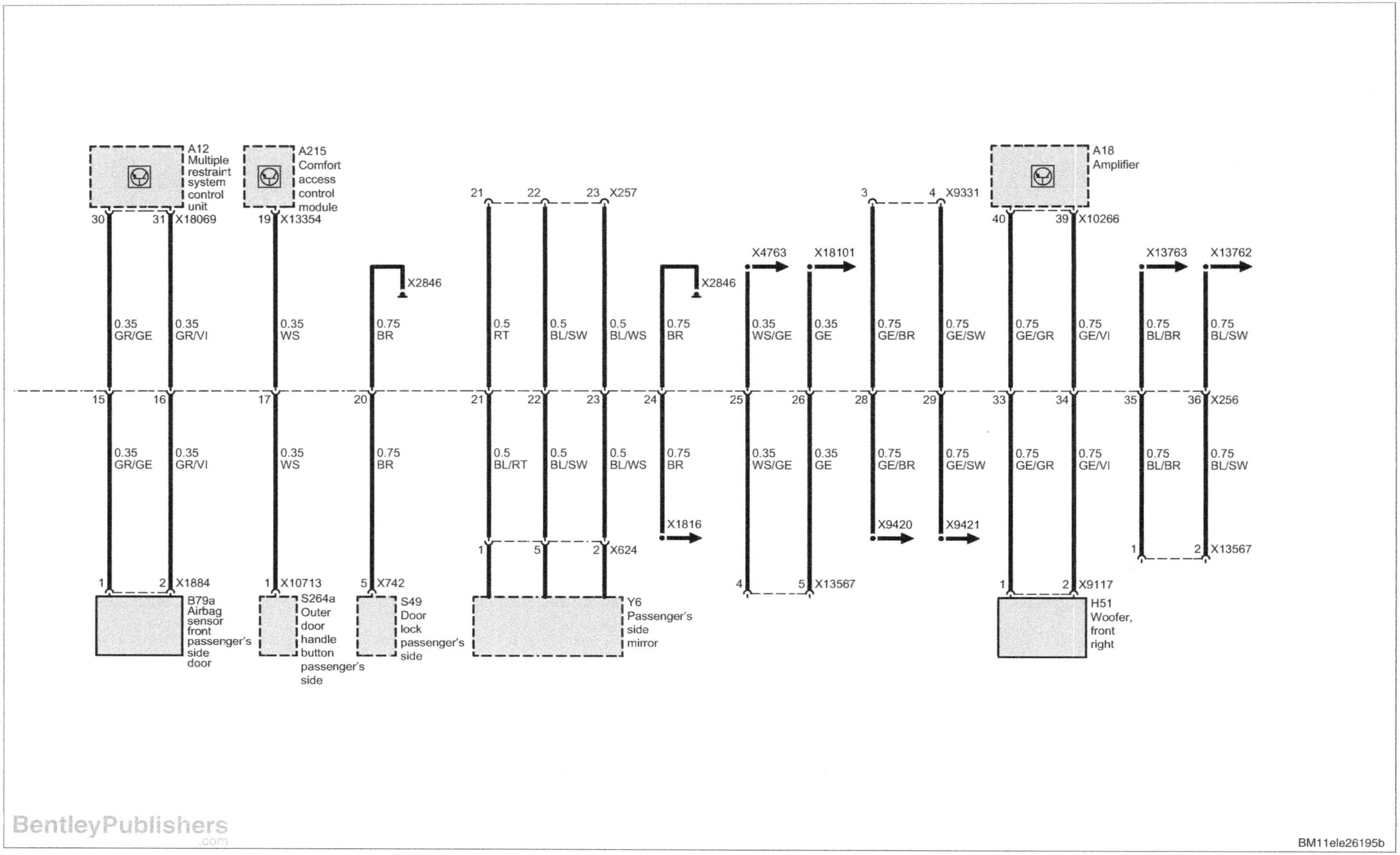
A12
Multiple restraint system control unit
30
31
X18069
A215
Comfort access control module
19
X13354
X2846
0.35
GR/GE
0.35
GR/VI
0.35
WS
0.75
BR
21
22
23
X257
0.5
RT
0.5
BL/SW
0.5
BL/WS
0.75
BR
X4763
X18101
0.35
WS/GE
0.35
GE
3
4
X9331
0.75
GE/BR
0.75
GE/SW
A18
Amplifier
40
39
X10266
0.75
GE/GR
0.75
GE/VI
X13763
X13762
0.75
BL/BR
0.75
BL/SW
15
16
17
20
21
22
23
24
25
26
28
29
33
34
35
36
X256
0.5
BL/RT
X1816
X9420
X9421
1
5
2
X624
1
2
X13567
1
2
X1884
B79a
Airbag sensor front passenger's side door
1
X10713
S264a
Outer door handle button passenger's side
5
X742
S49
Door lock passenger's side
Y6
Passenger's side mirror
4
5
X13567
1
2
X9117
H51
Woofer, front right
BentleyPublishers.com
BM11ele26195b

Connector X257
page 1 of 2

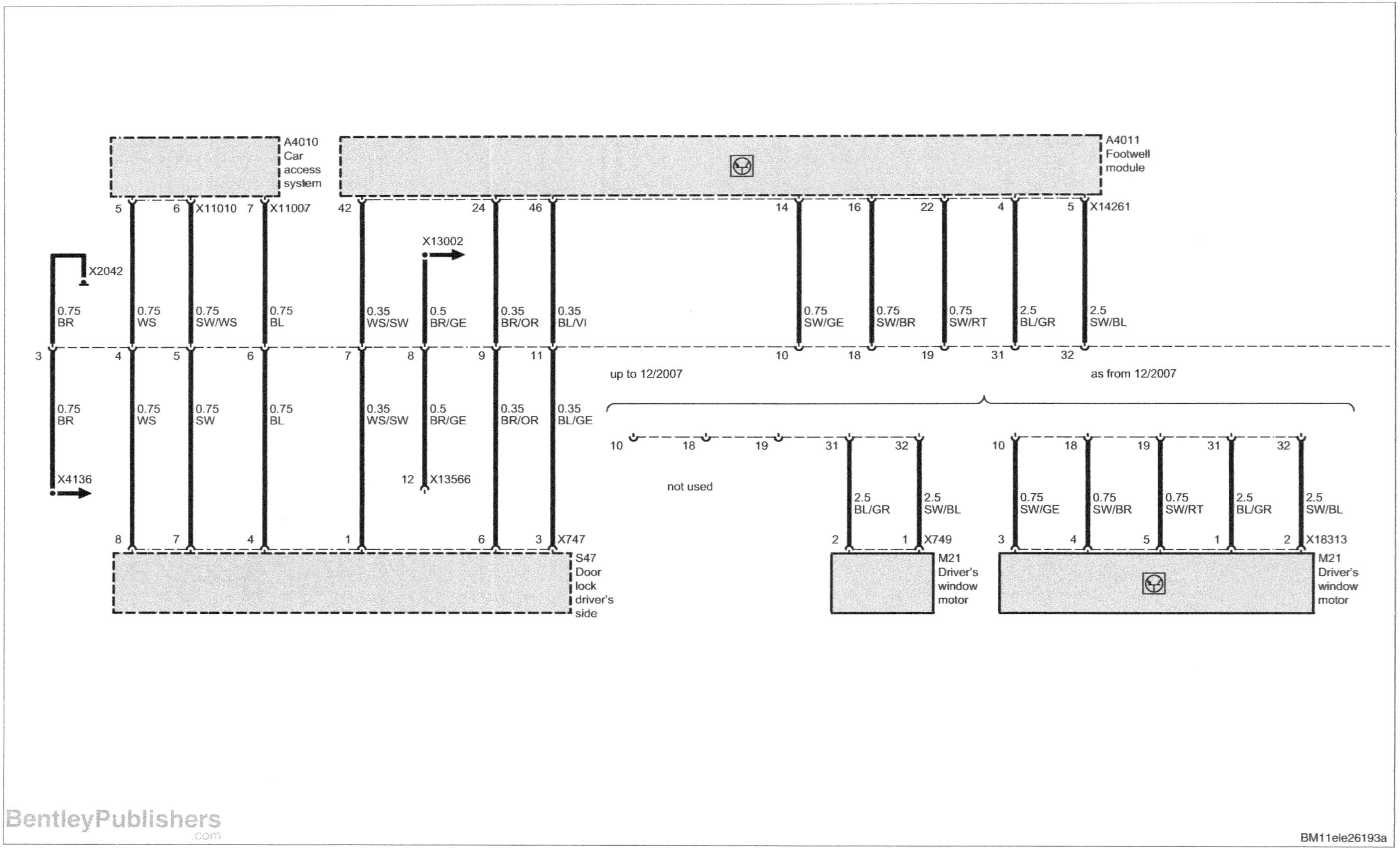

Connector X257
page 2 of 2

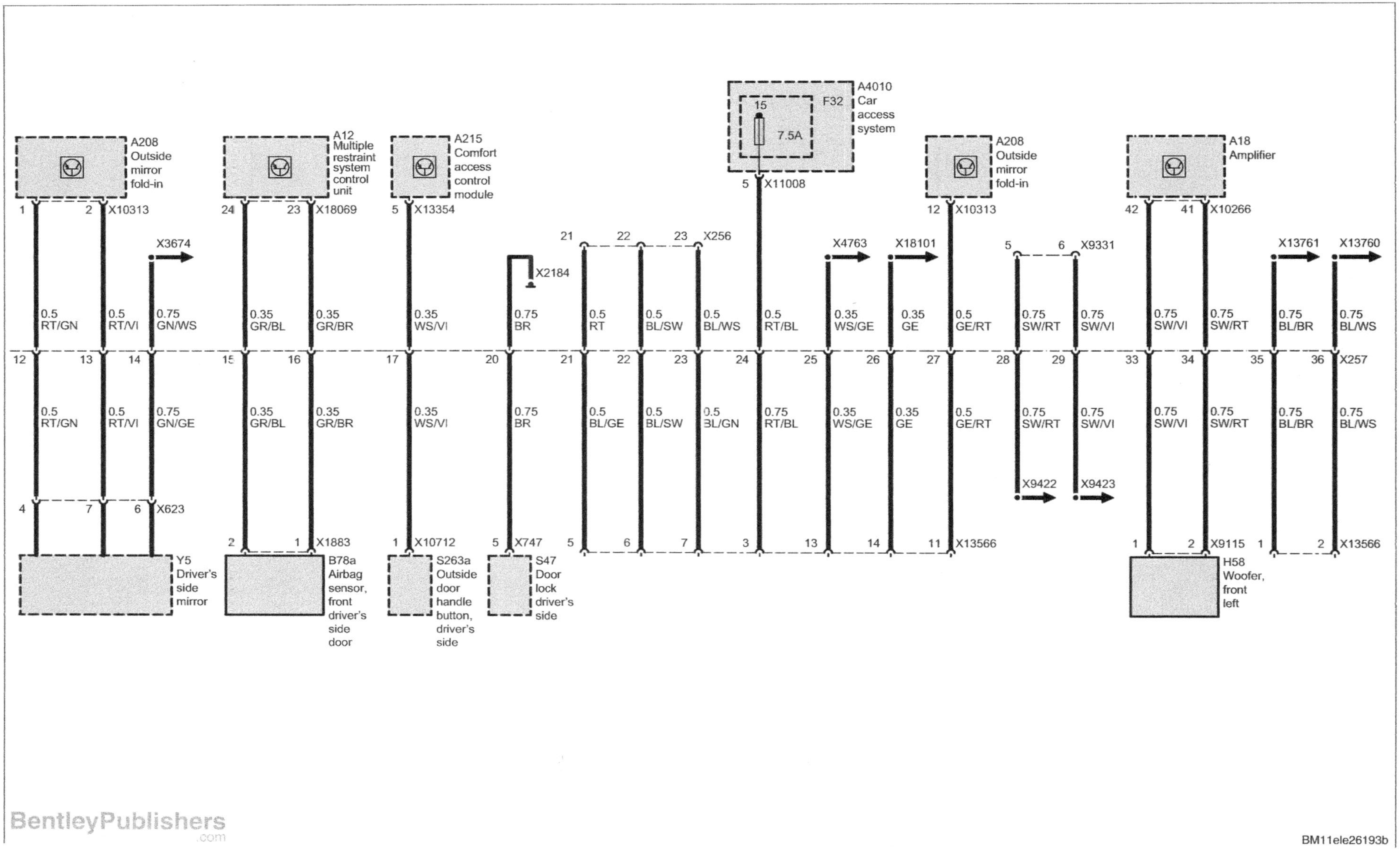

BM11ele26193b

Comfort access power supply to 12/2008

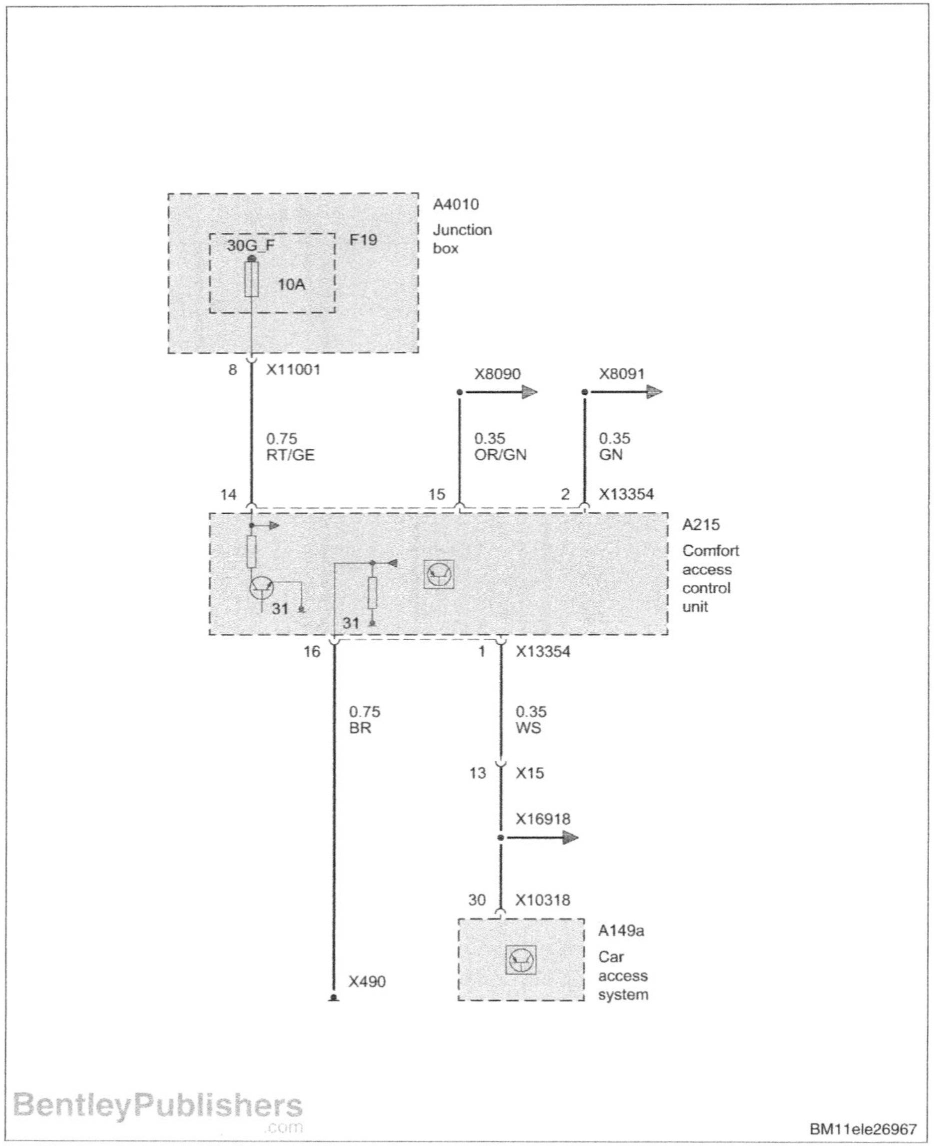

Comfort access power supply from 12/2008

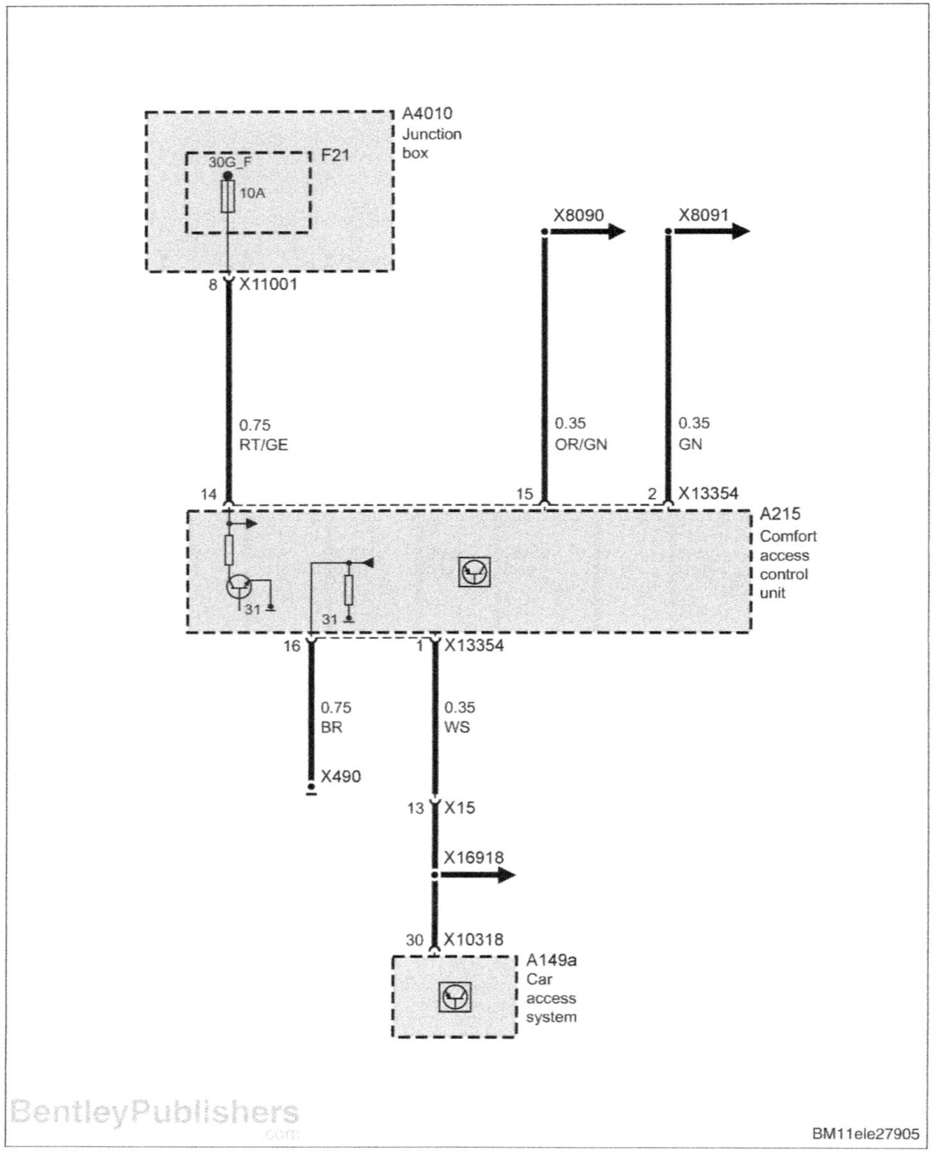

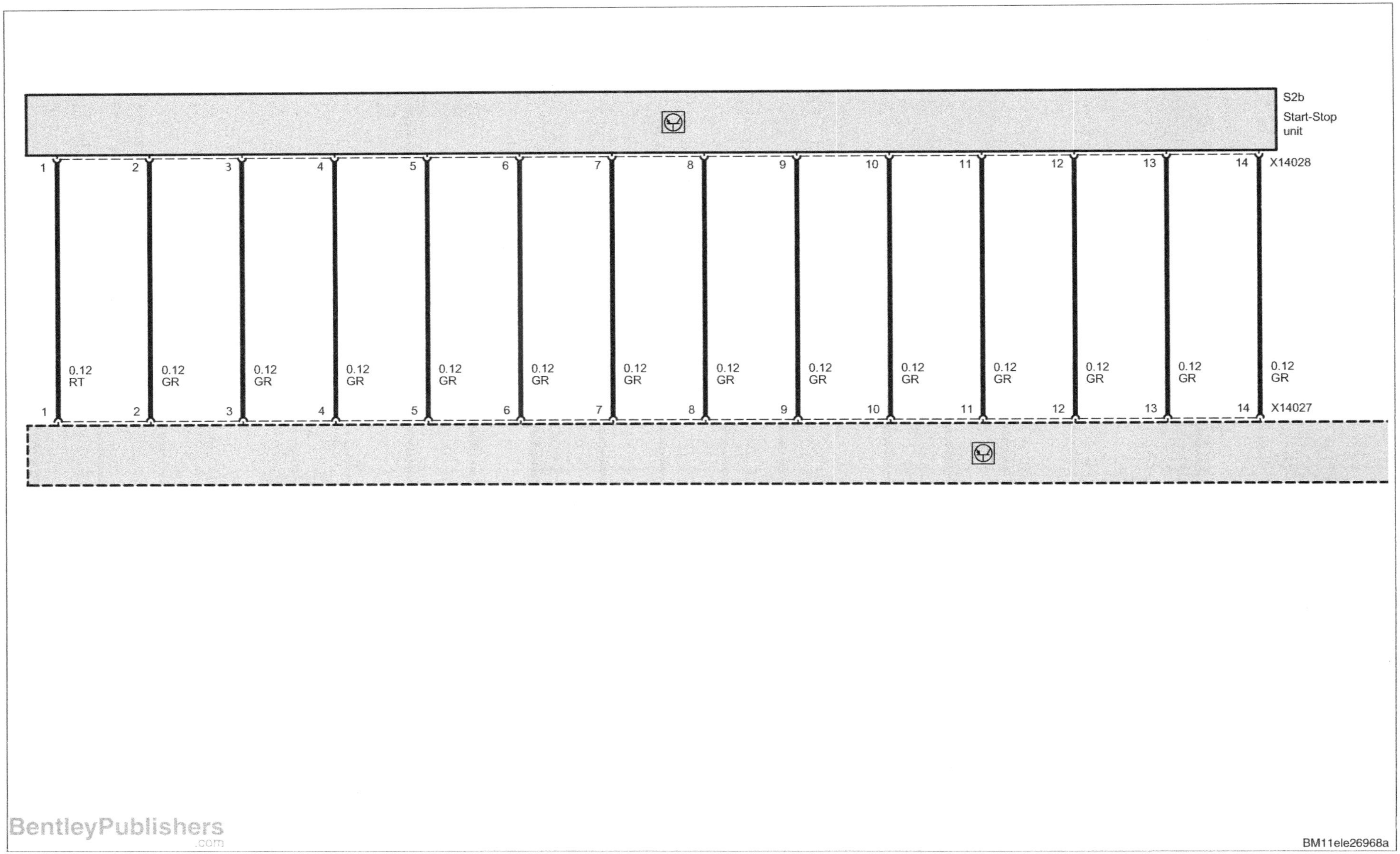
S2b
Start-Stop
unit
X14028
1 2 3 4 5 6 7 8 9 10 11 12 13 14
0.12
RT
0.12
GR
X14027
BentleyPublishers
.com
BM11ele26968a

Comfort access
page 2 of 2

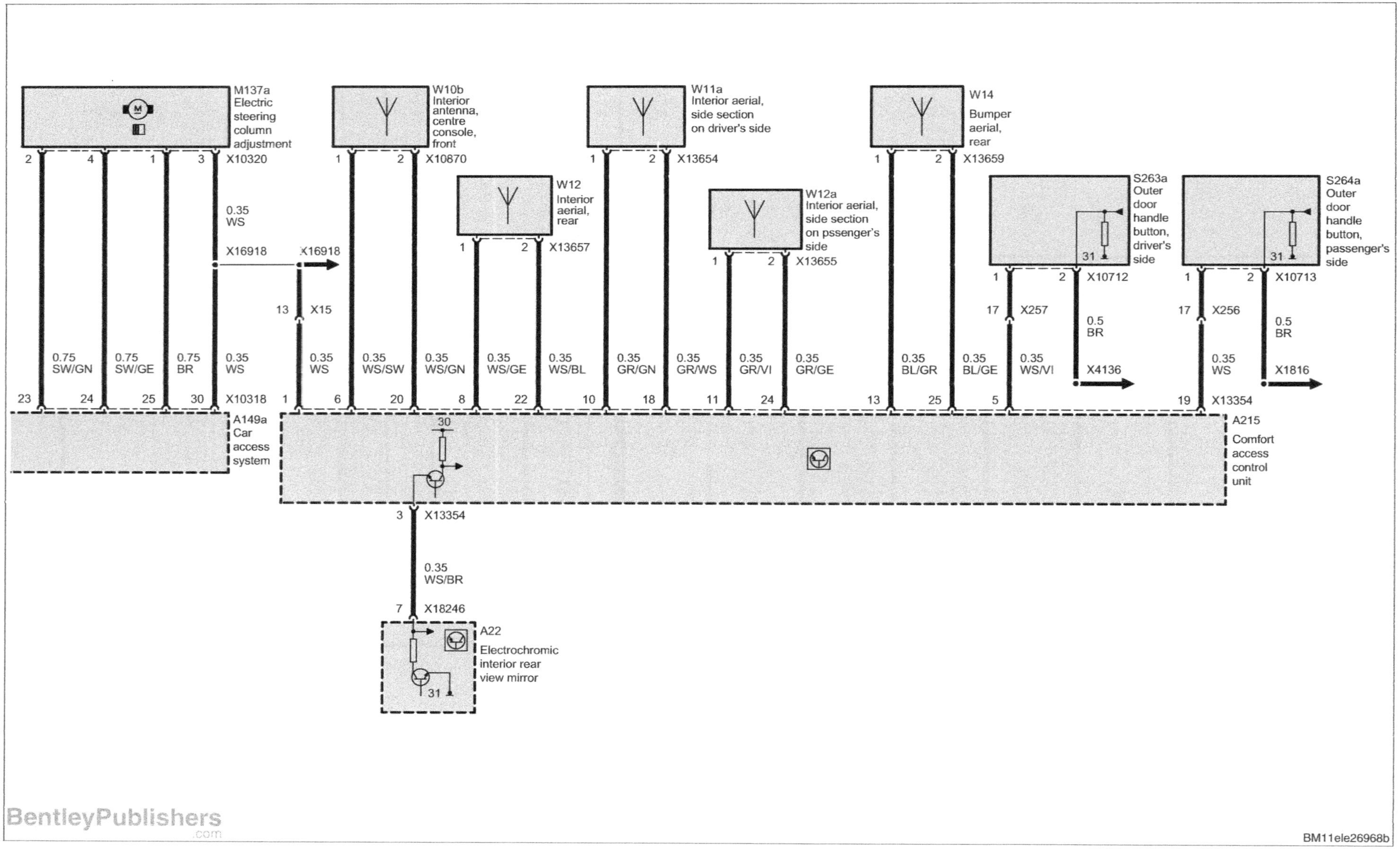

Adaptive headlights, level sensors

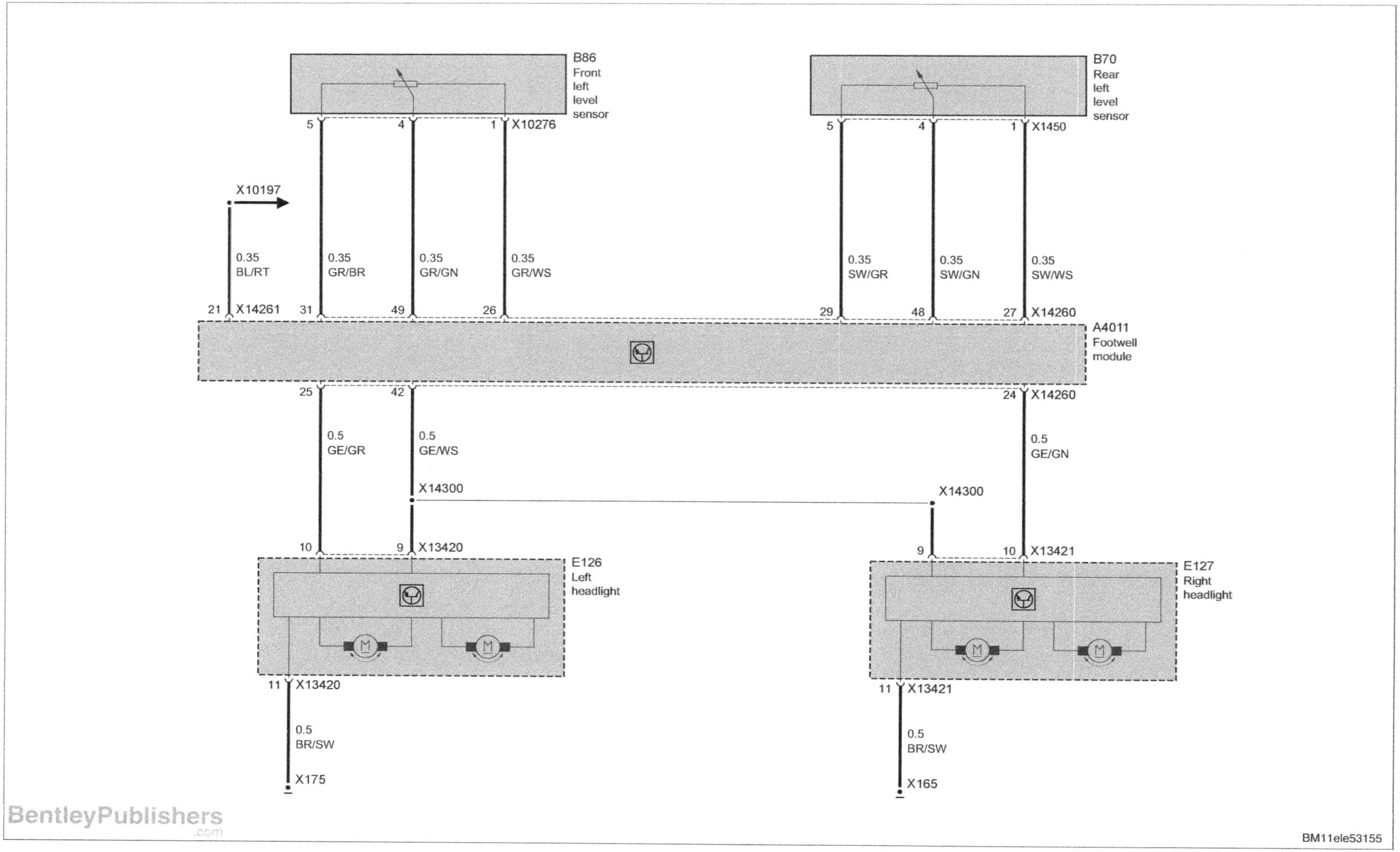

Automatic driving lights, headlights, rain/headlight sensor
non-xenon
to 12/2008

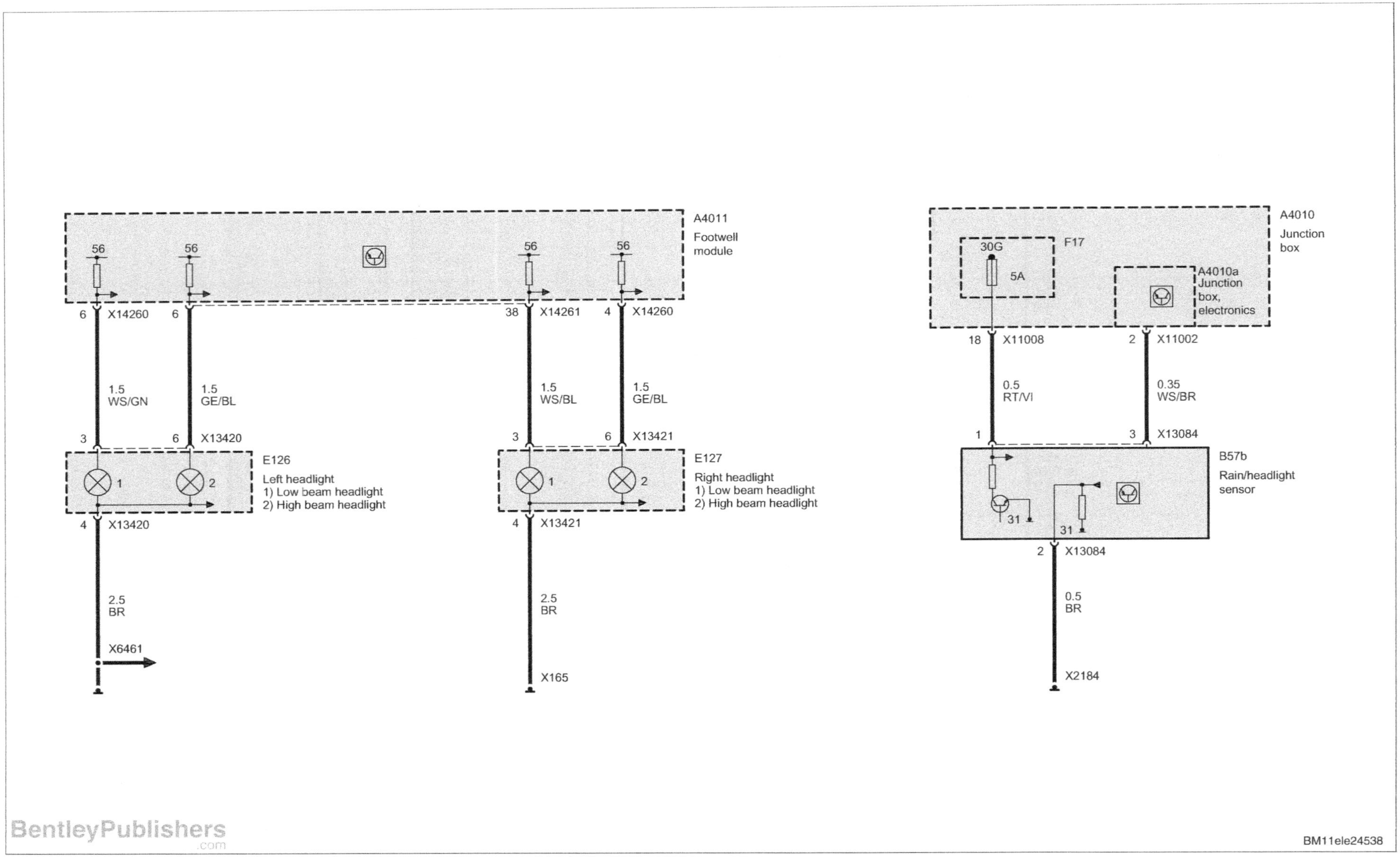

Automatic driving light control, headlights, rain/headlight sensor, xenon to 09/2010

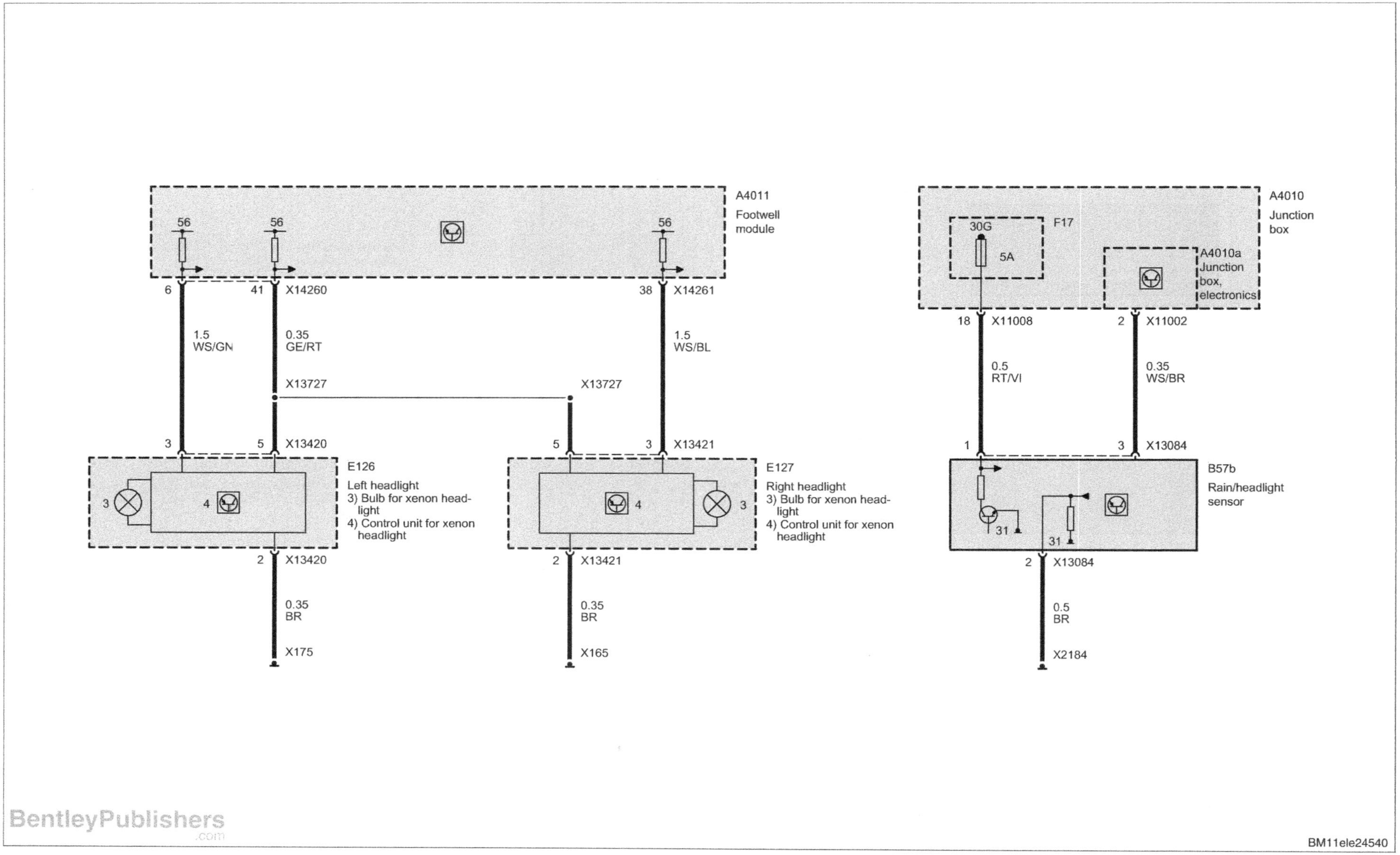

Automatic driving light control, headlights, rain/headlight sensor, non-xenon from 12/2008

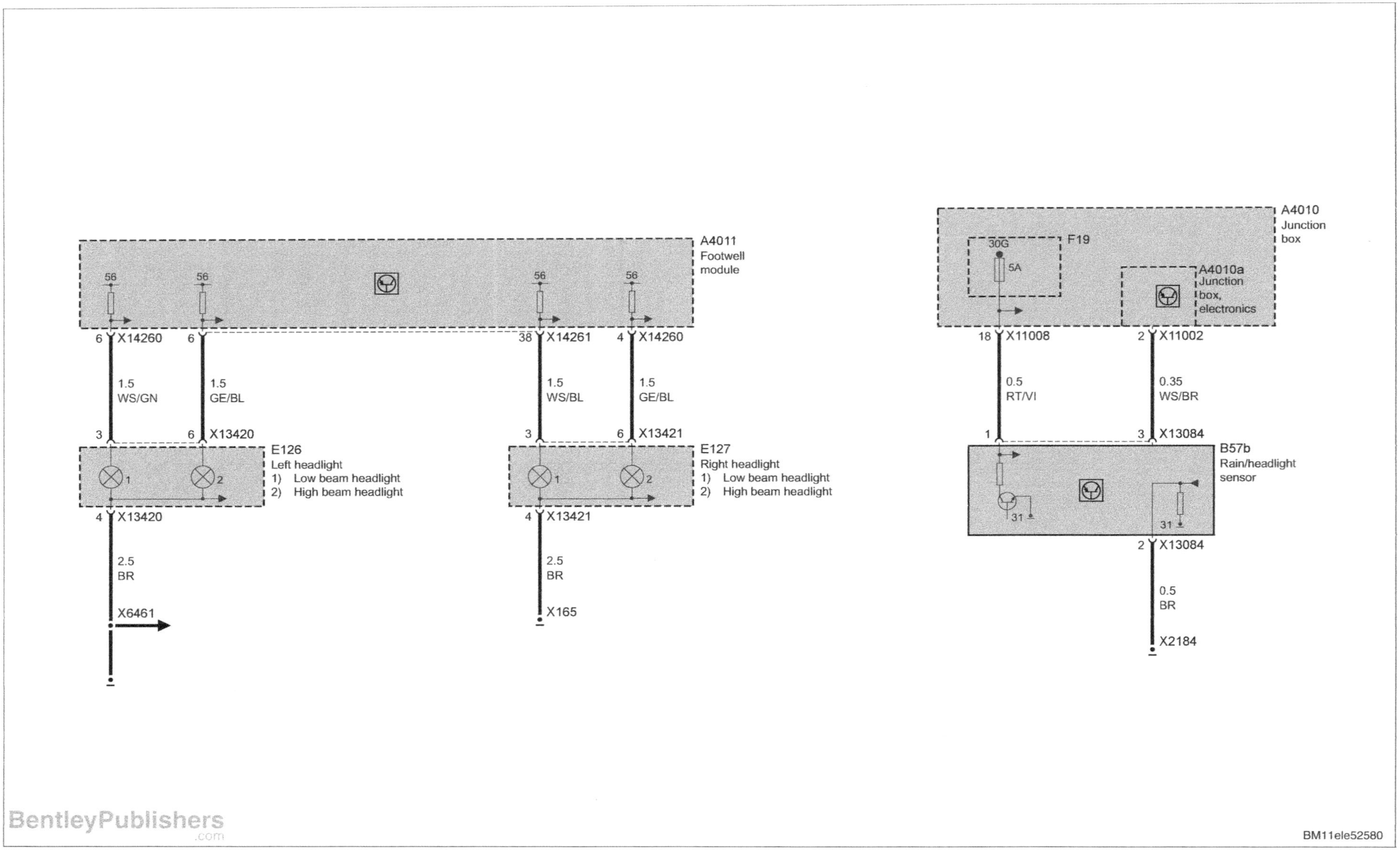

Automatic driving light control, headlights, rain/headlight sensor from 09/2010

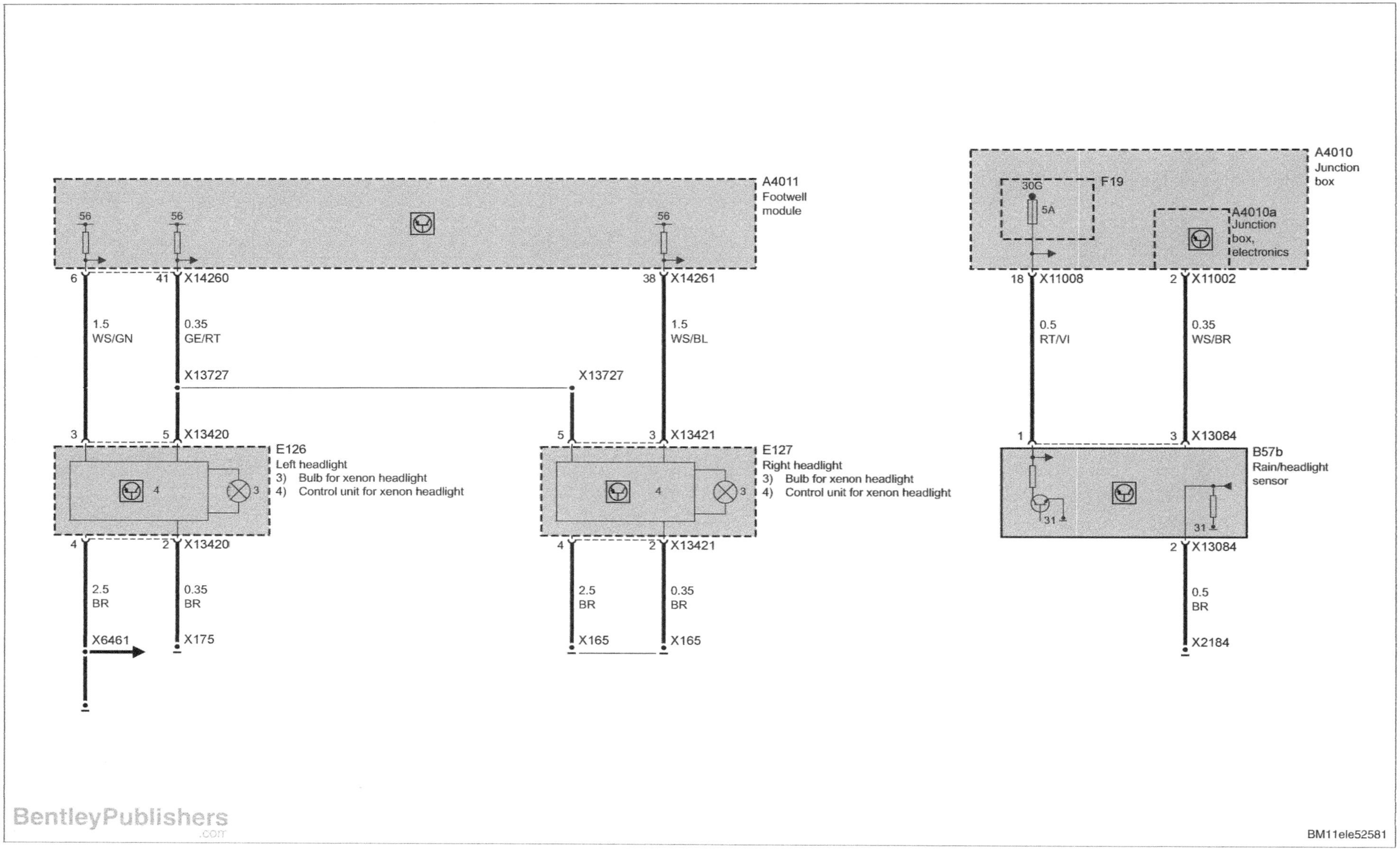

Automatic headlight vertical aim control, headlights, ride height sensors

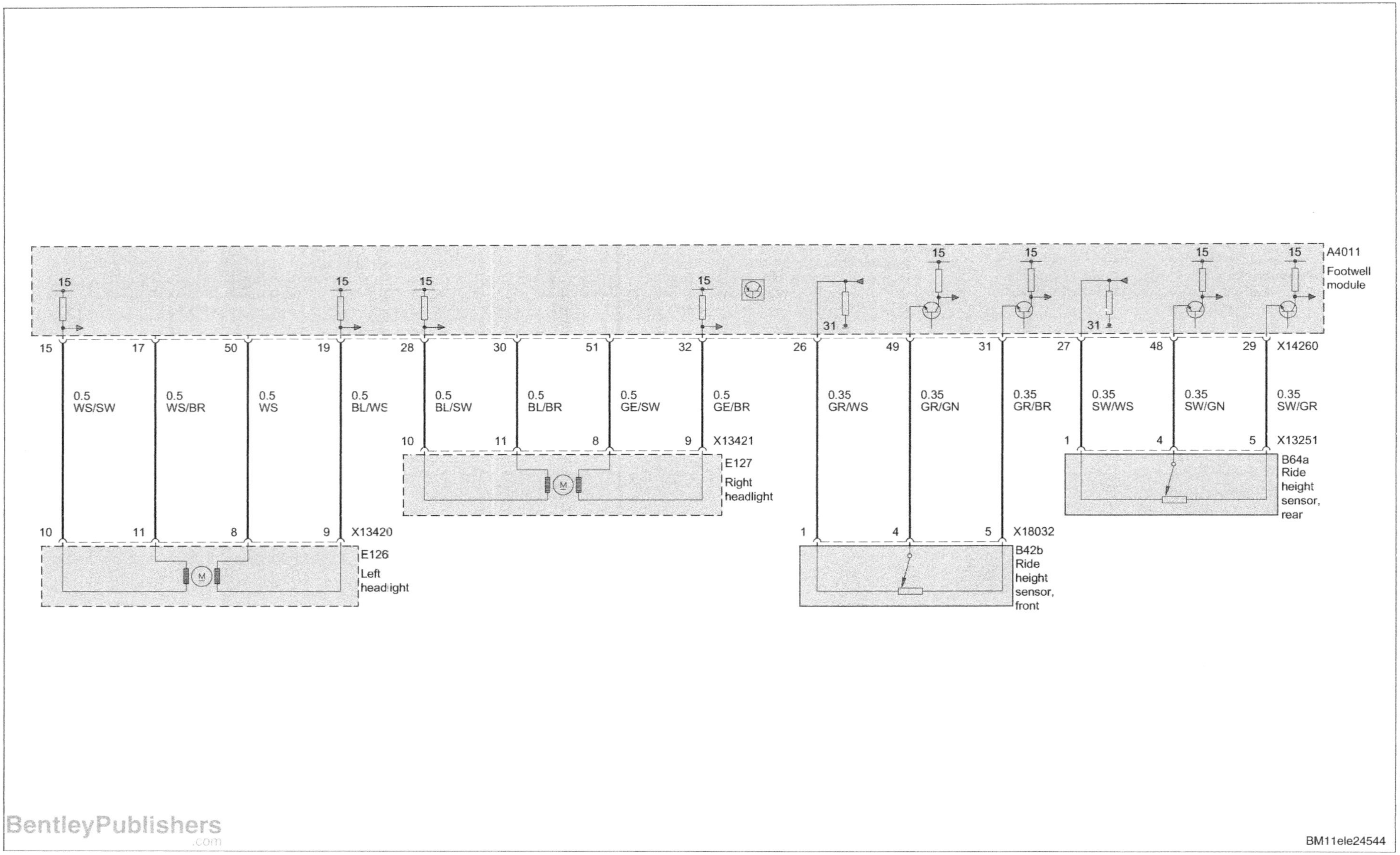

Low beam, high beam, xenon from 09/2010

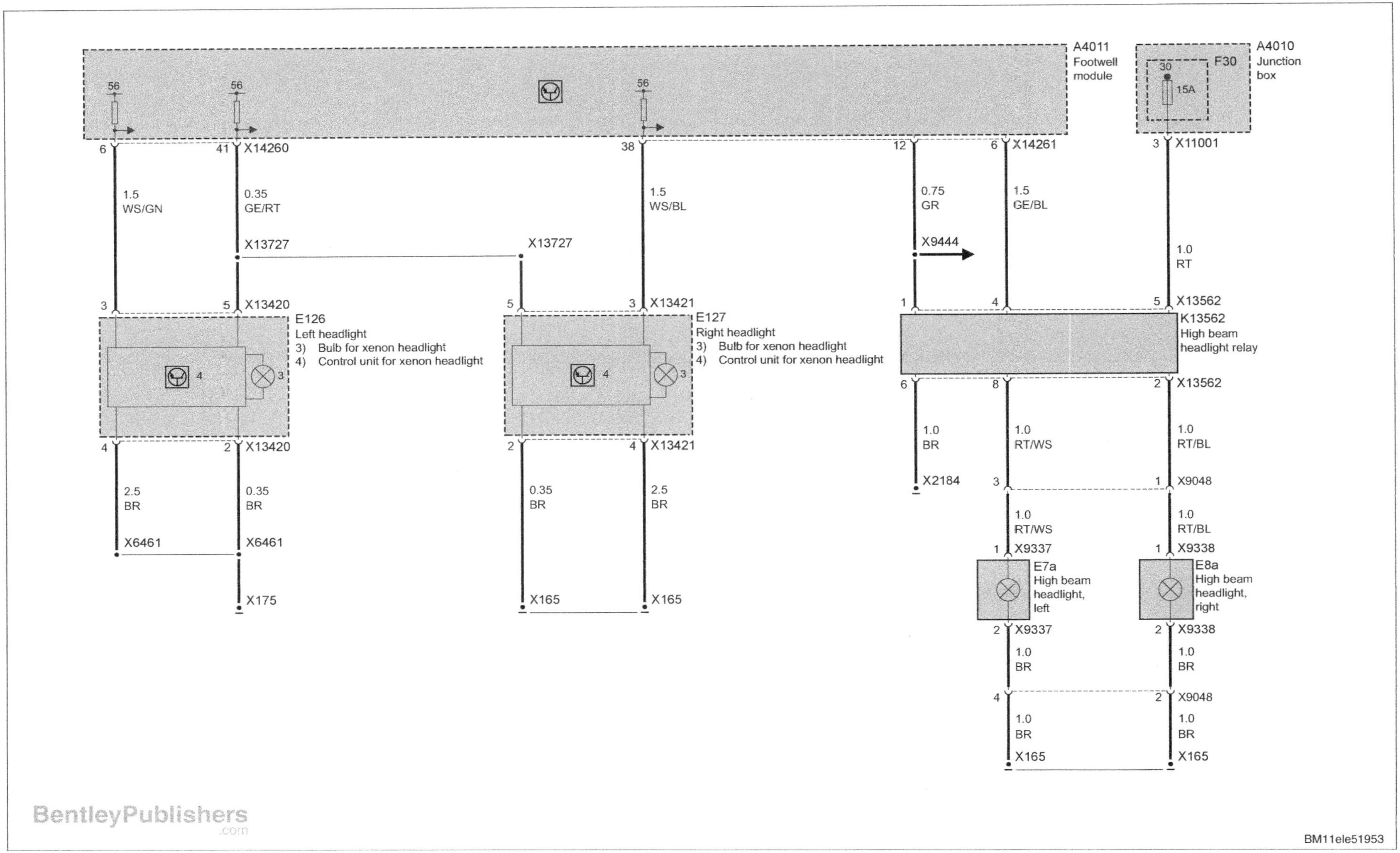

BM11ele51953

Low beam, high beam, non-xenon from 09/2009

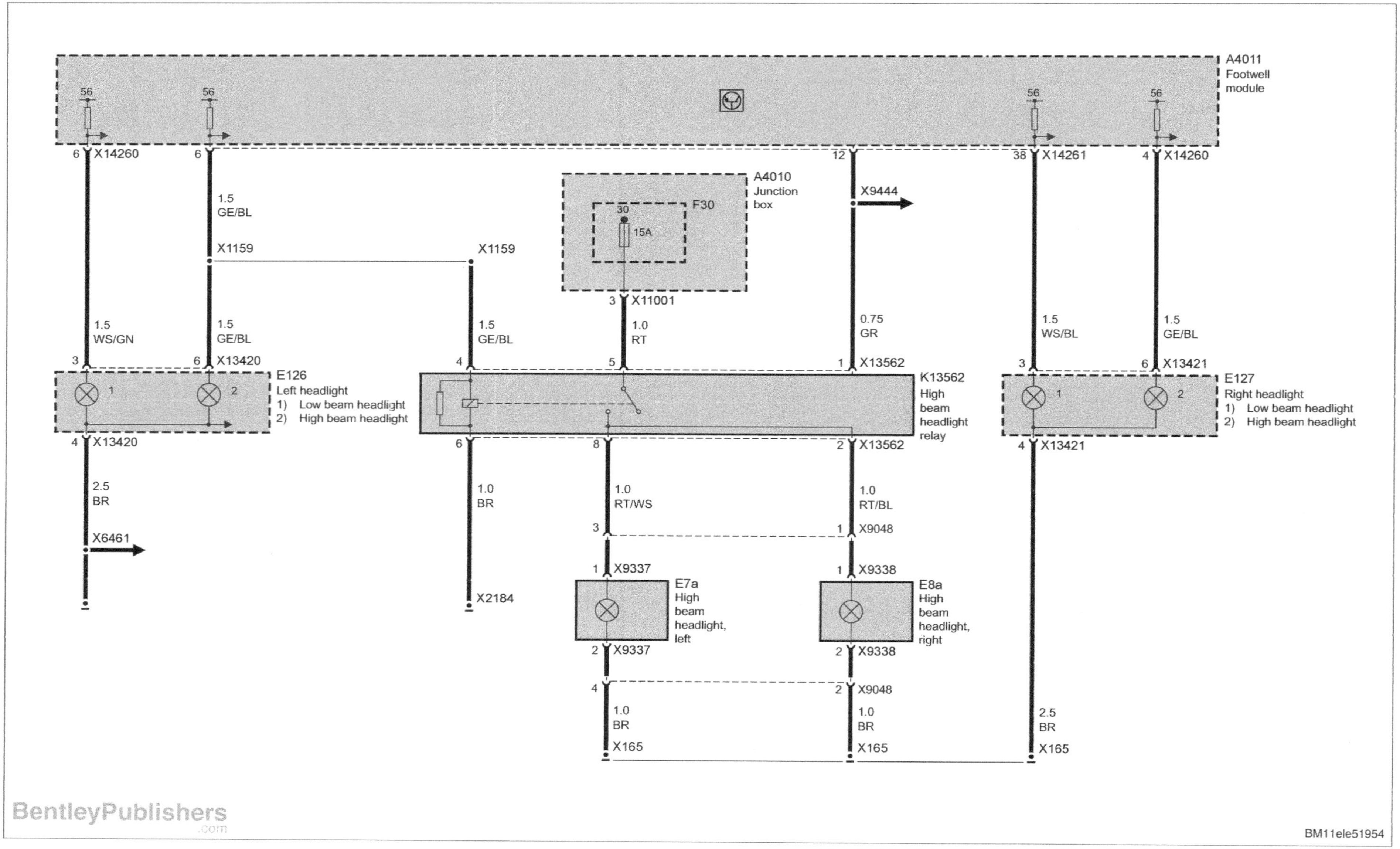

Backup lights to 09/2010

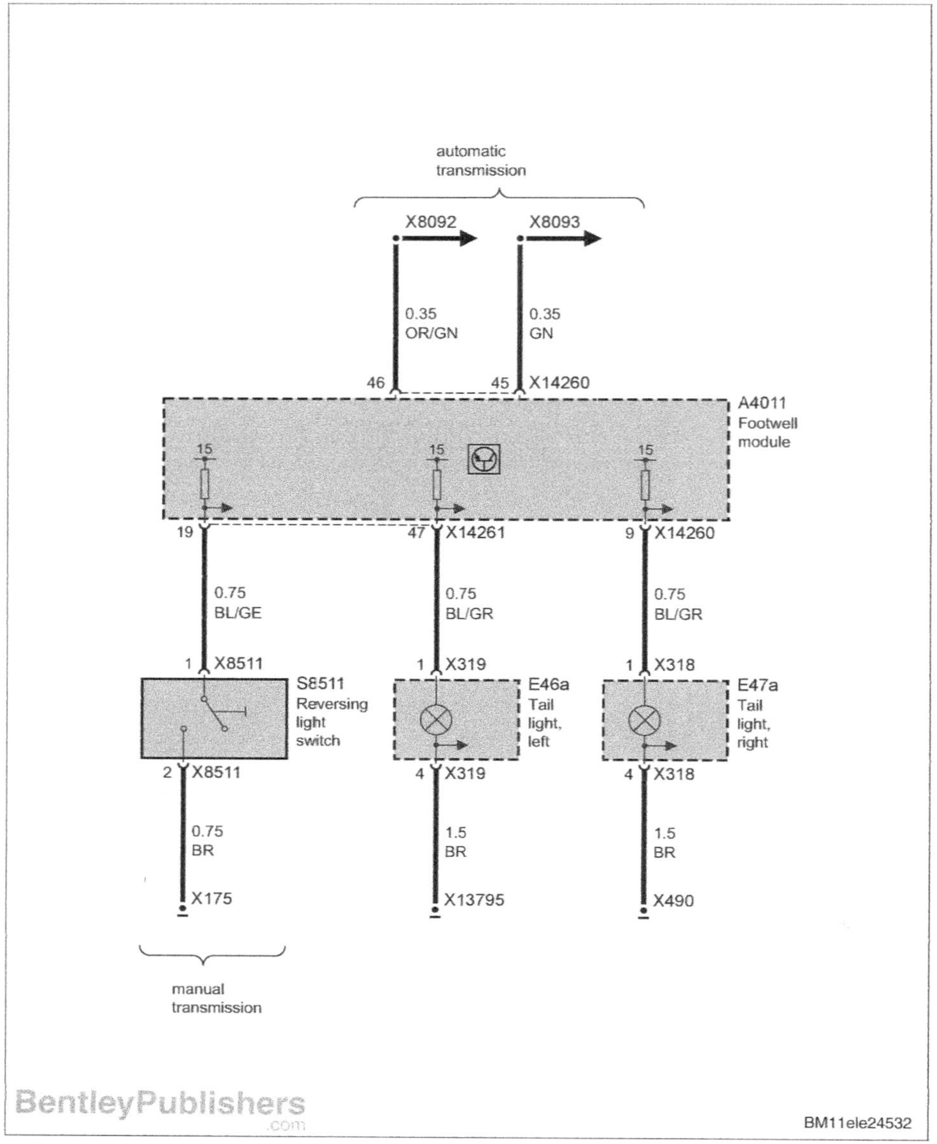

Backup lights from 09/2010

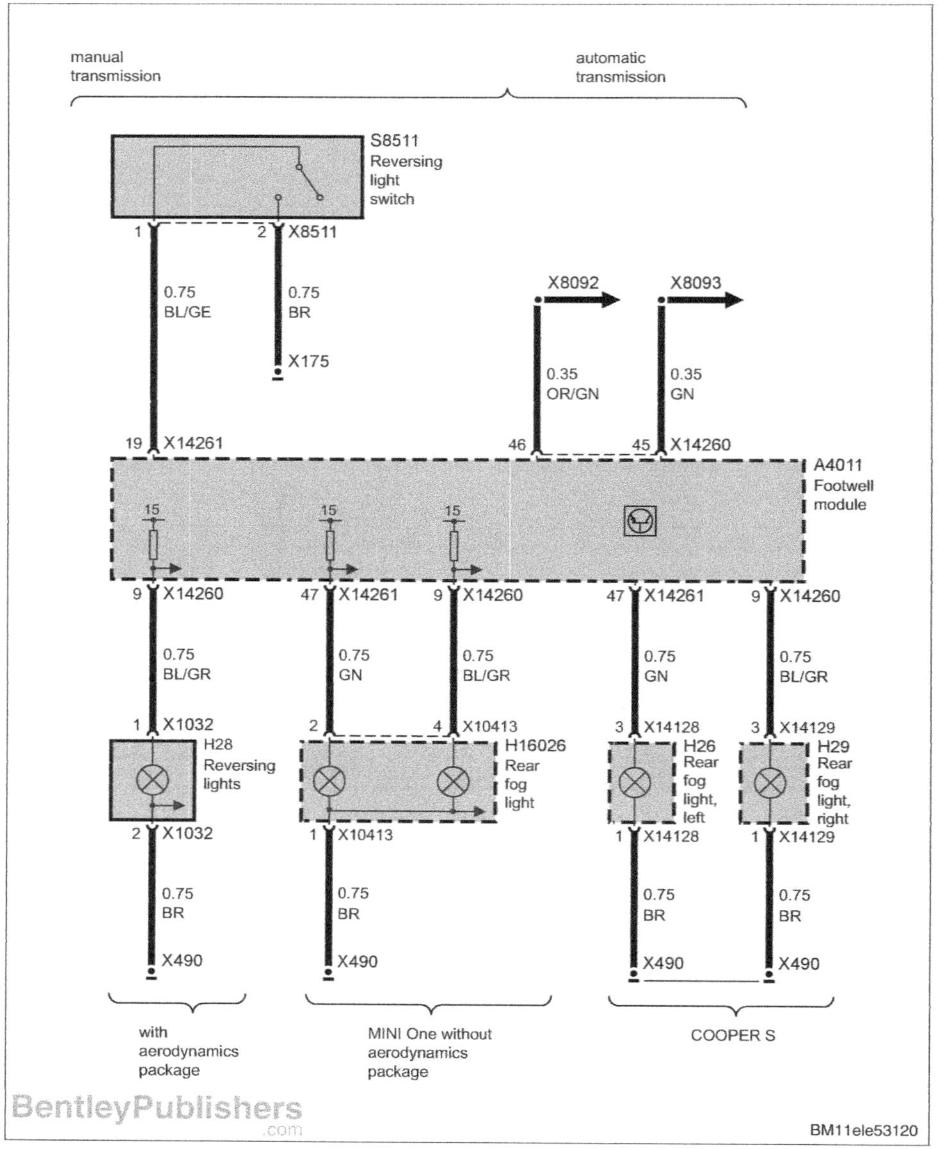

Side light, parking light, fog light, side marker, tail light to 09/2010 page 1 of 2

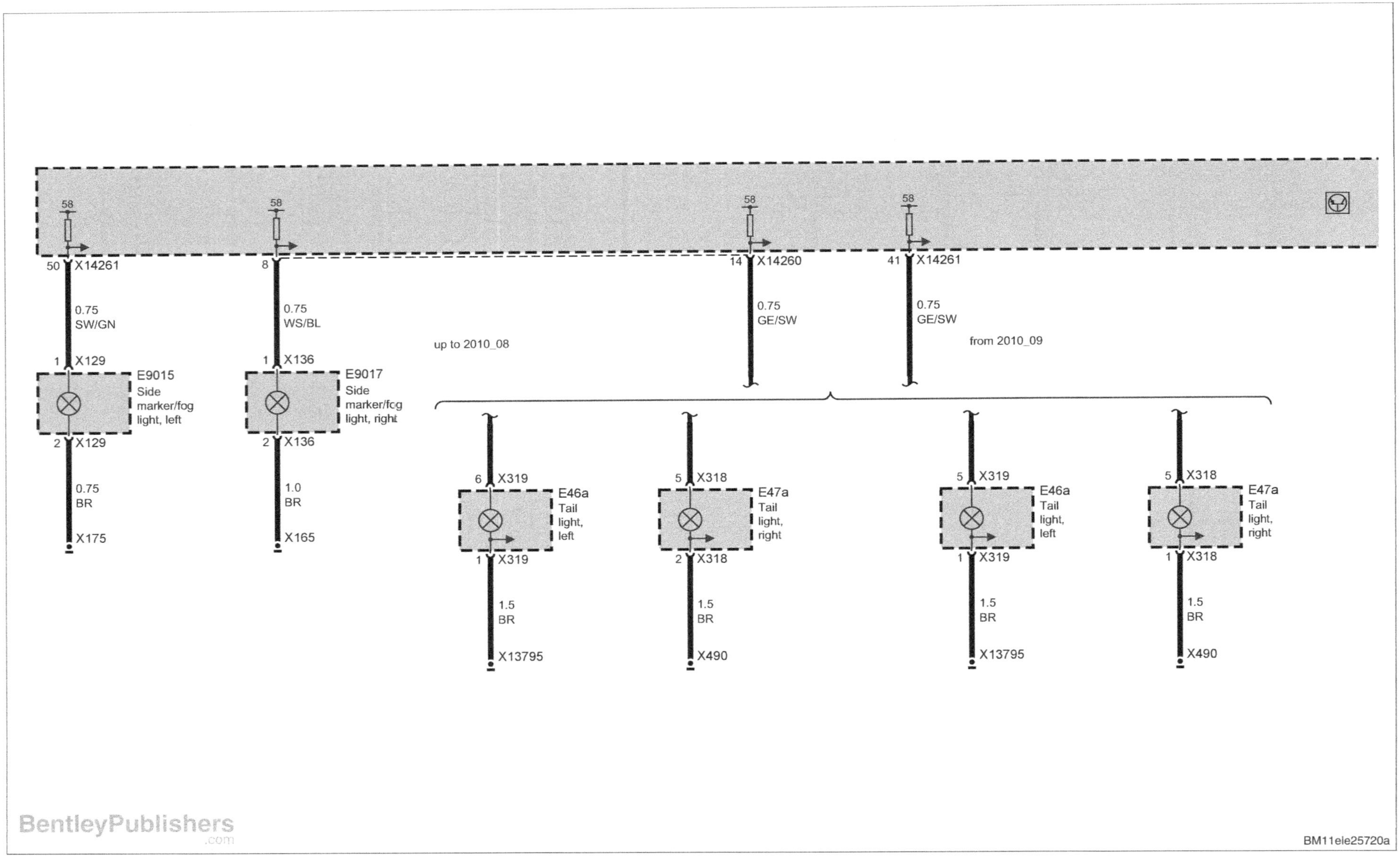

Side light, parking light, license plate light, side marker to 09/2010 page 2 of 2

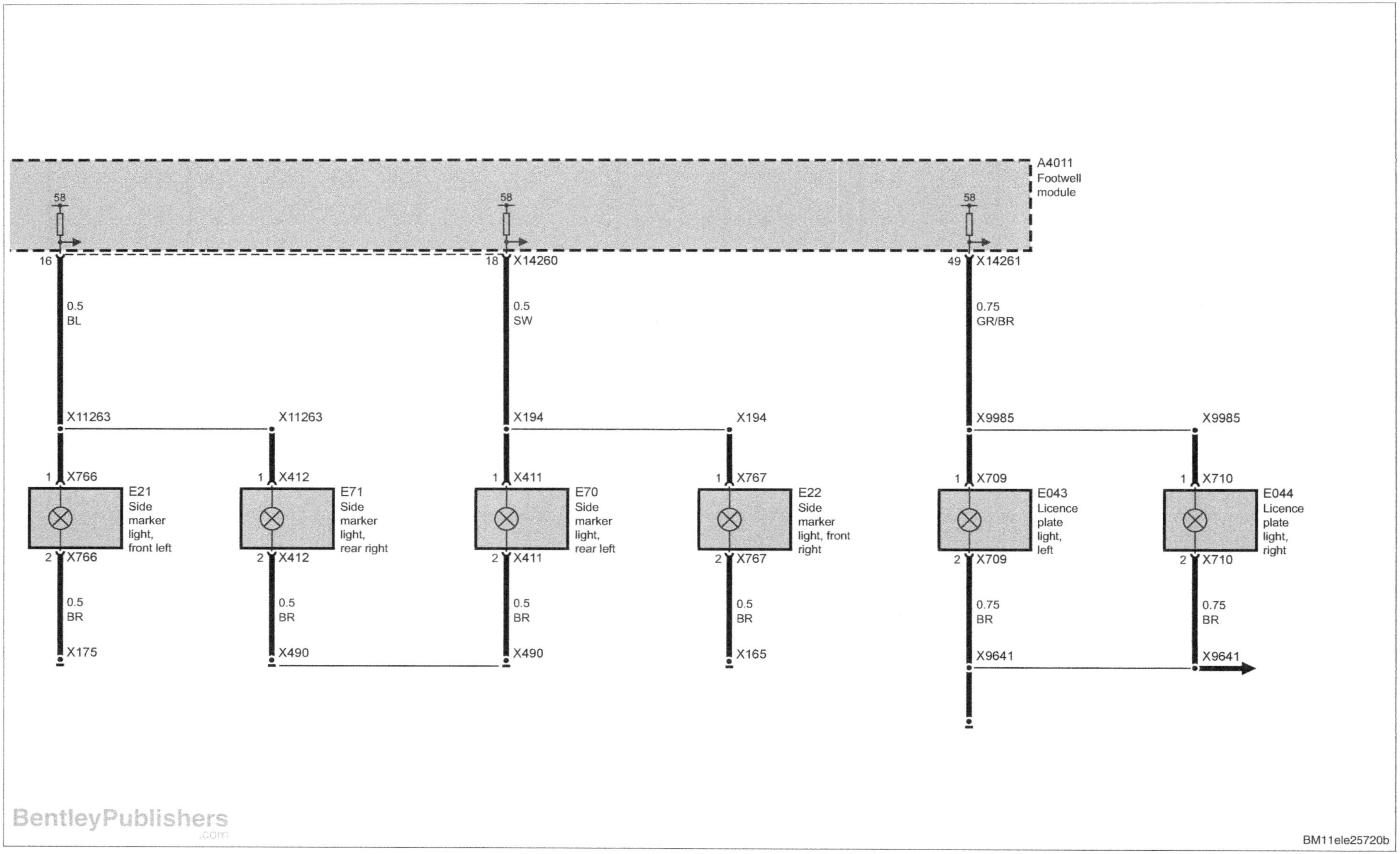

Side light, parking light, fog light, license plate light, side marker, trunk lid push button, tail light from 09/2010

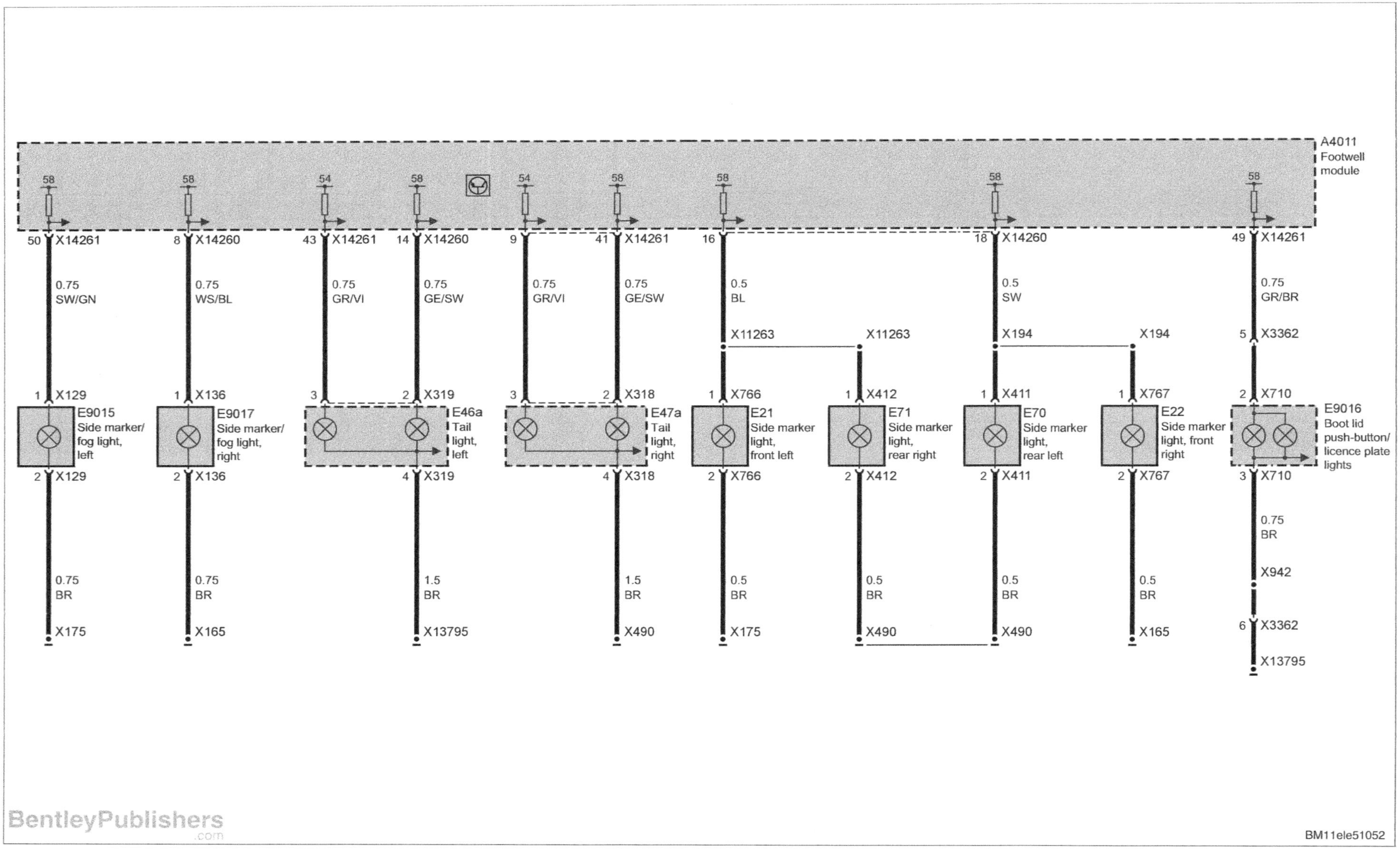

Central locking inputs (Convertible)

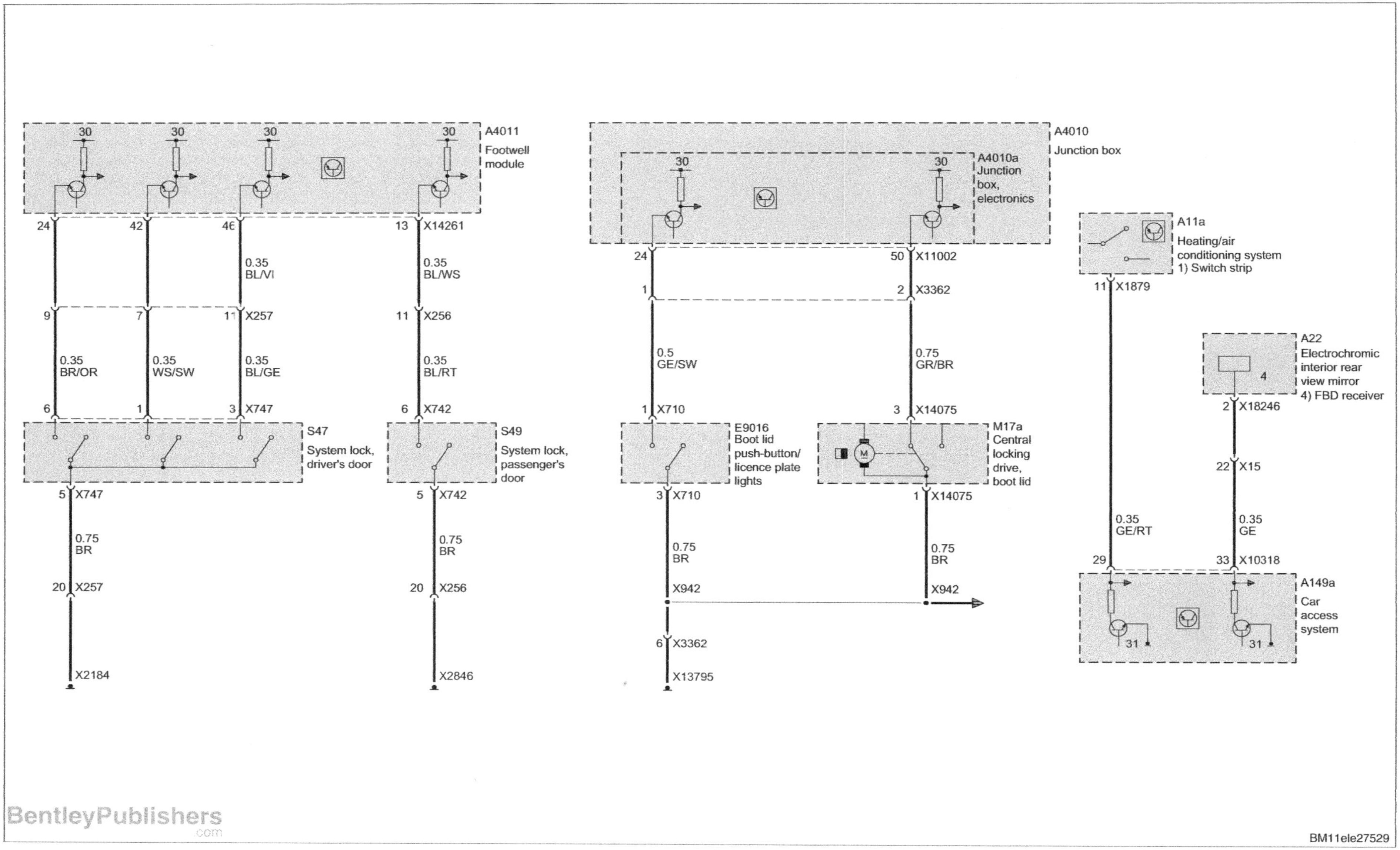

Central locking inputs (Hardtop)

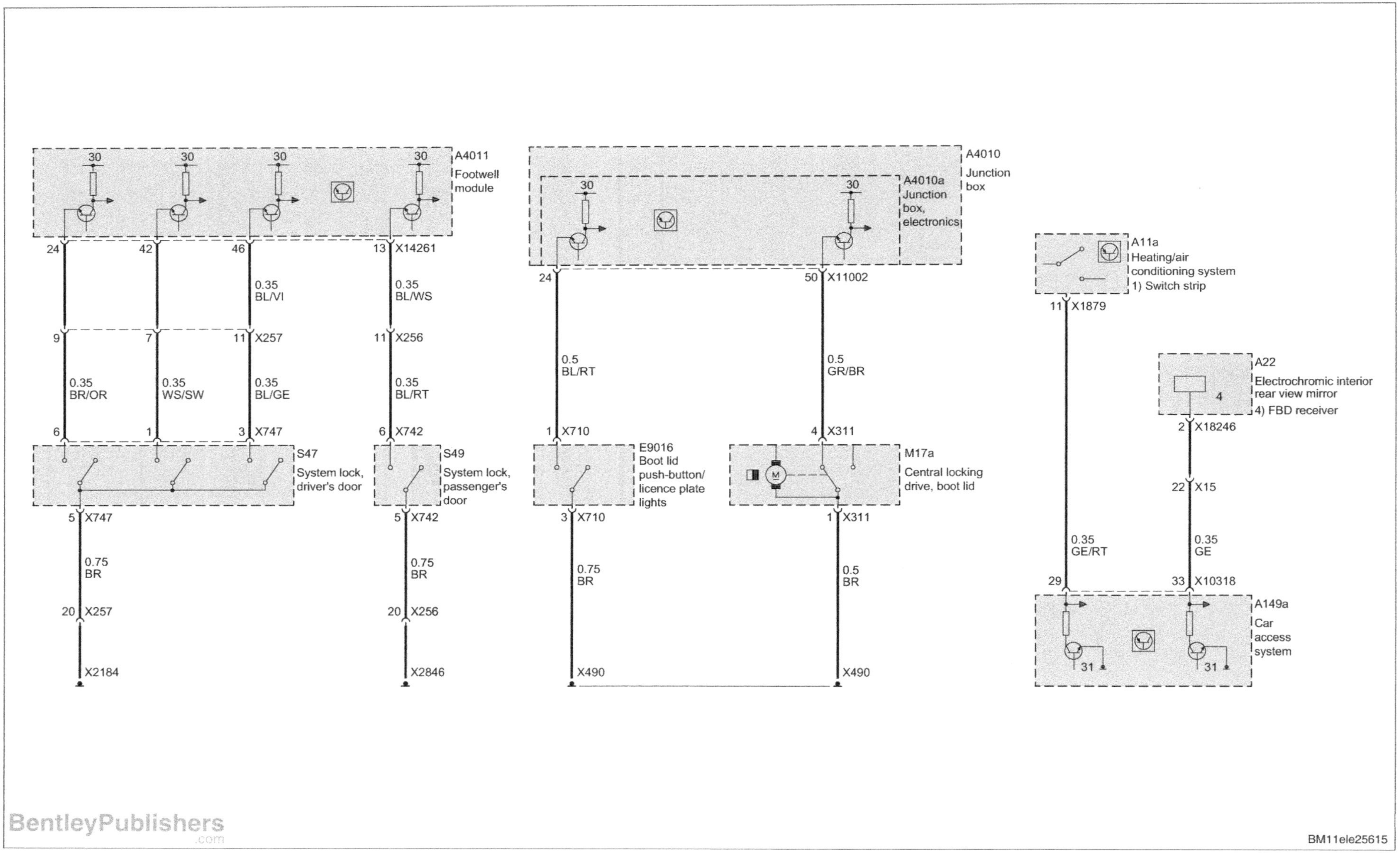

BM11ele25615

Cigarette lighter and charging socket

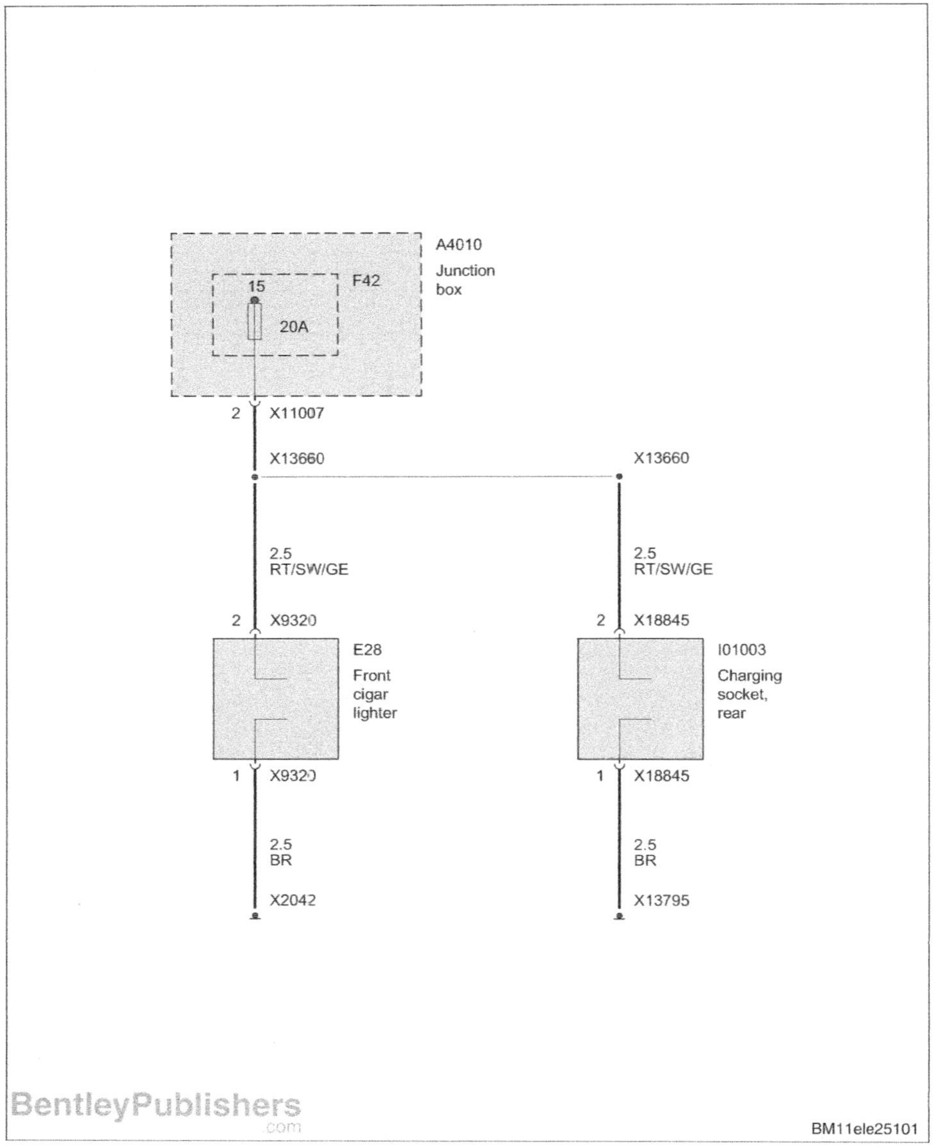

Wipe-wash system outputs, heated water jets, wiper module to 09/2010 page 1 of 2

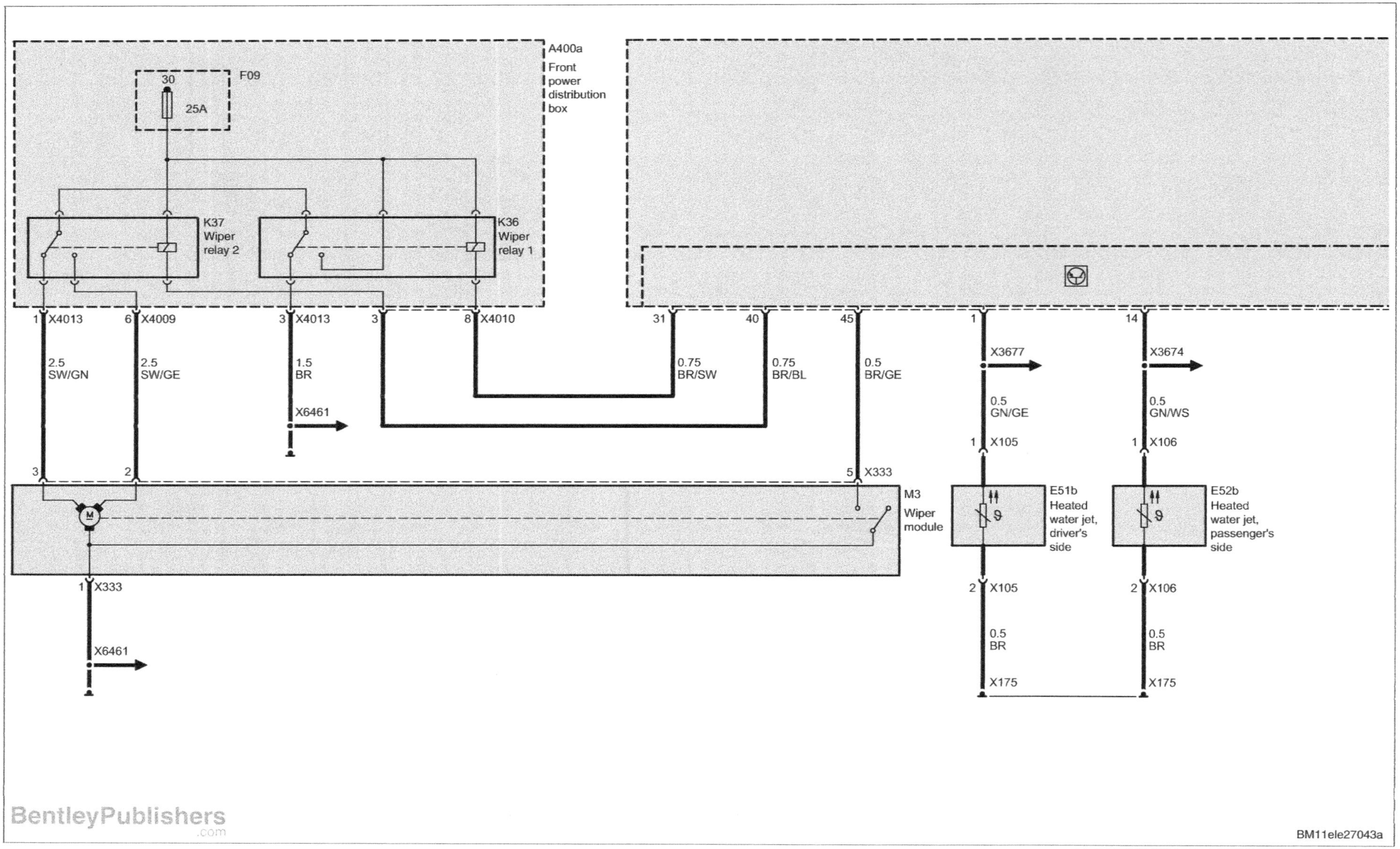

Wipe-wash system outputs, headlight washer pump, windshield washer pump, wiper drive units, wiper relay, to 09/2010, page 2 of 2

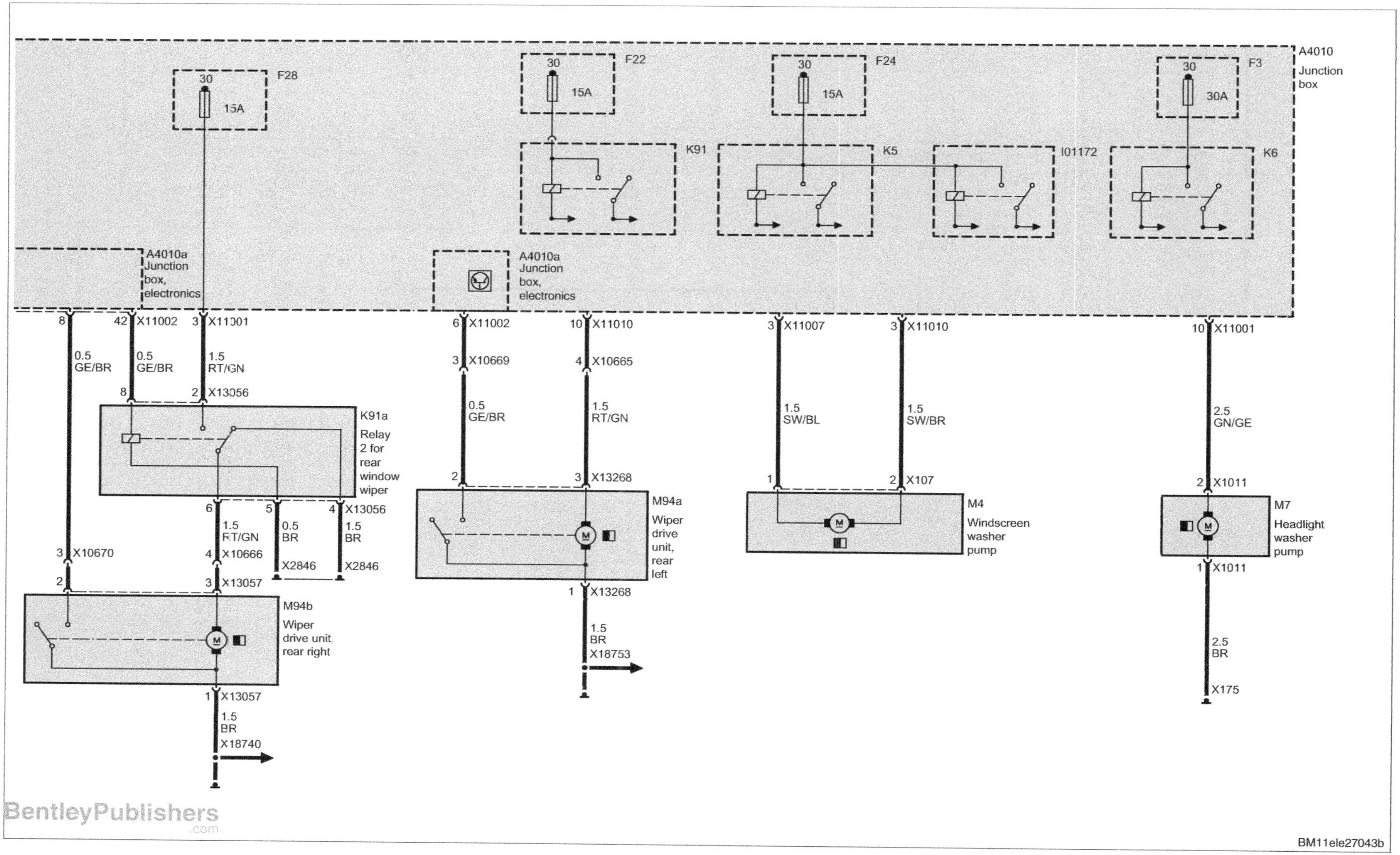

Wipe-wash system outputs, heated water jets
from 09/2010
page 1 of 2

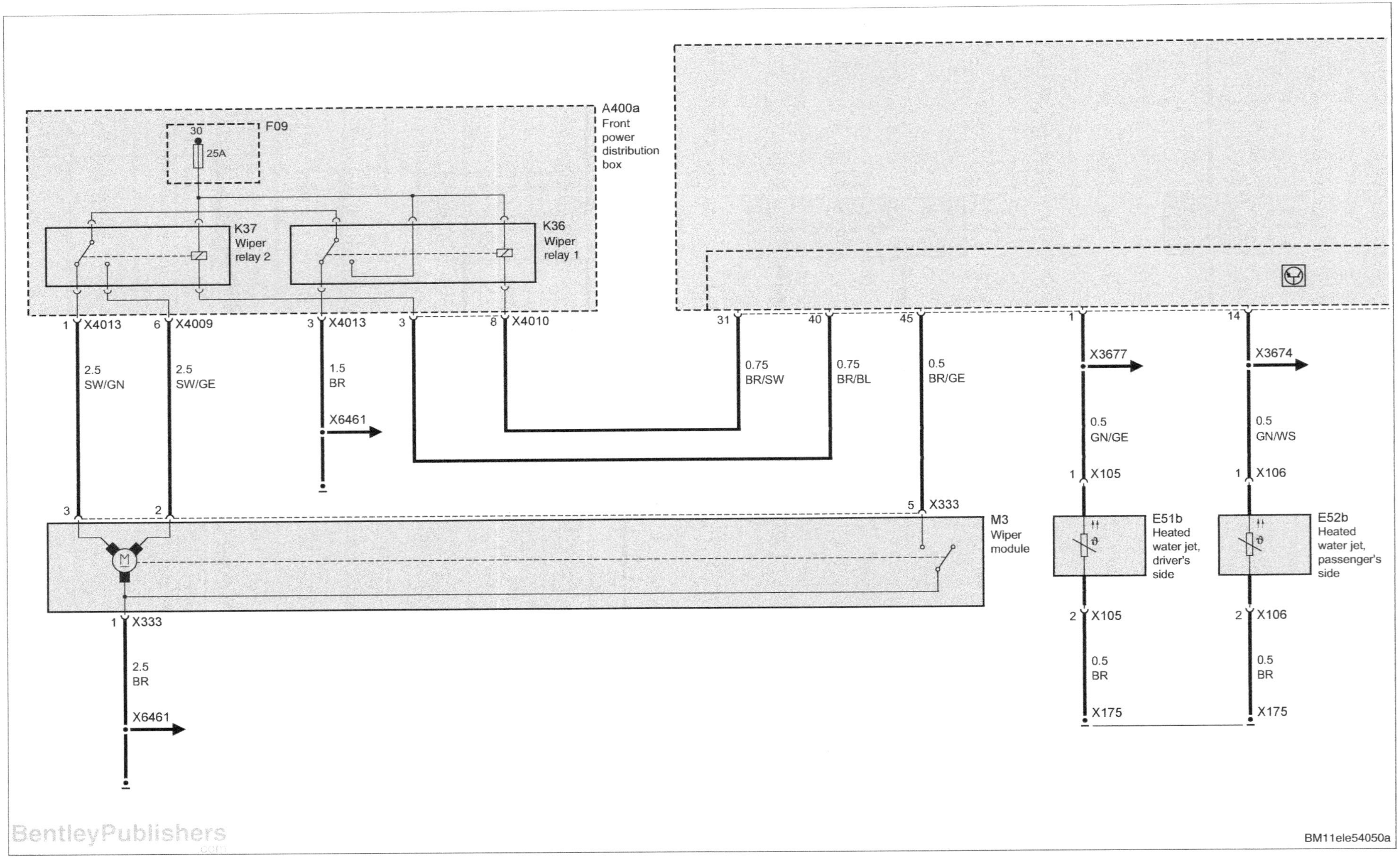

Wipe-wash system outputs, headlight washer pump, windshield washer pump, wiper drive unit from 09/2010, page 2 of 2

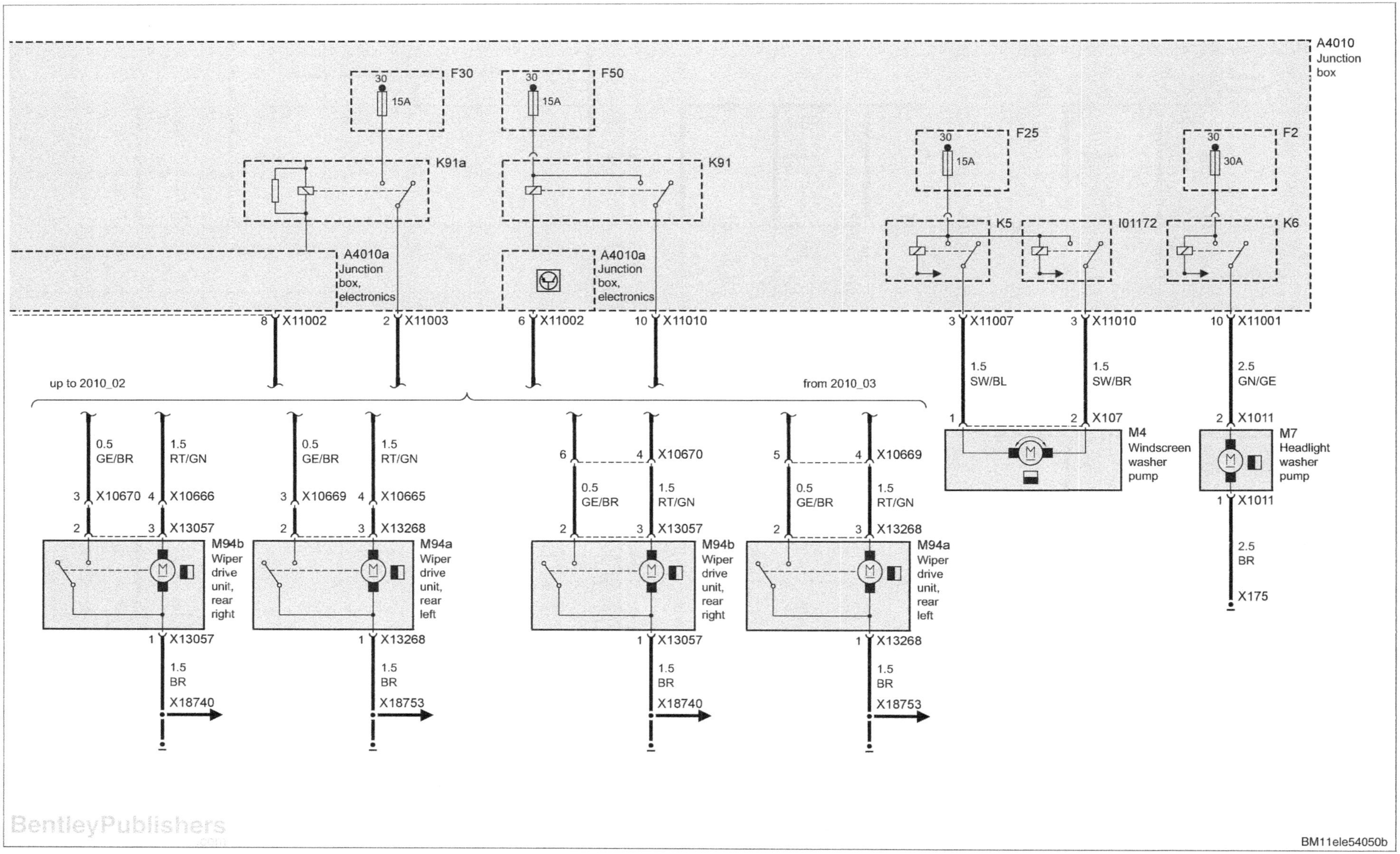

Junction box power supply

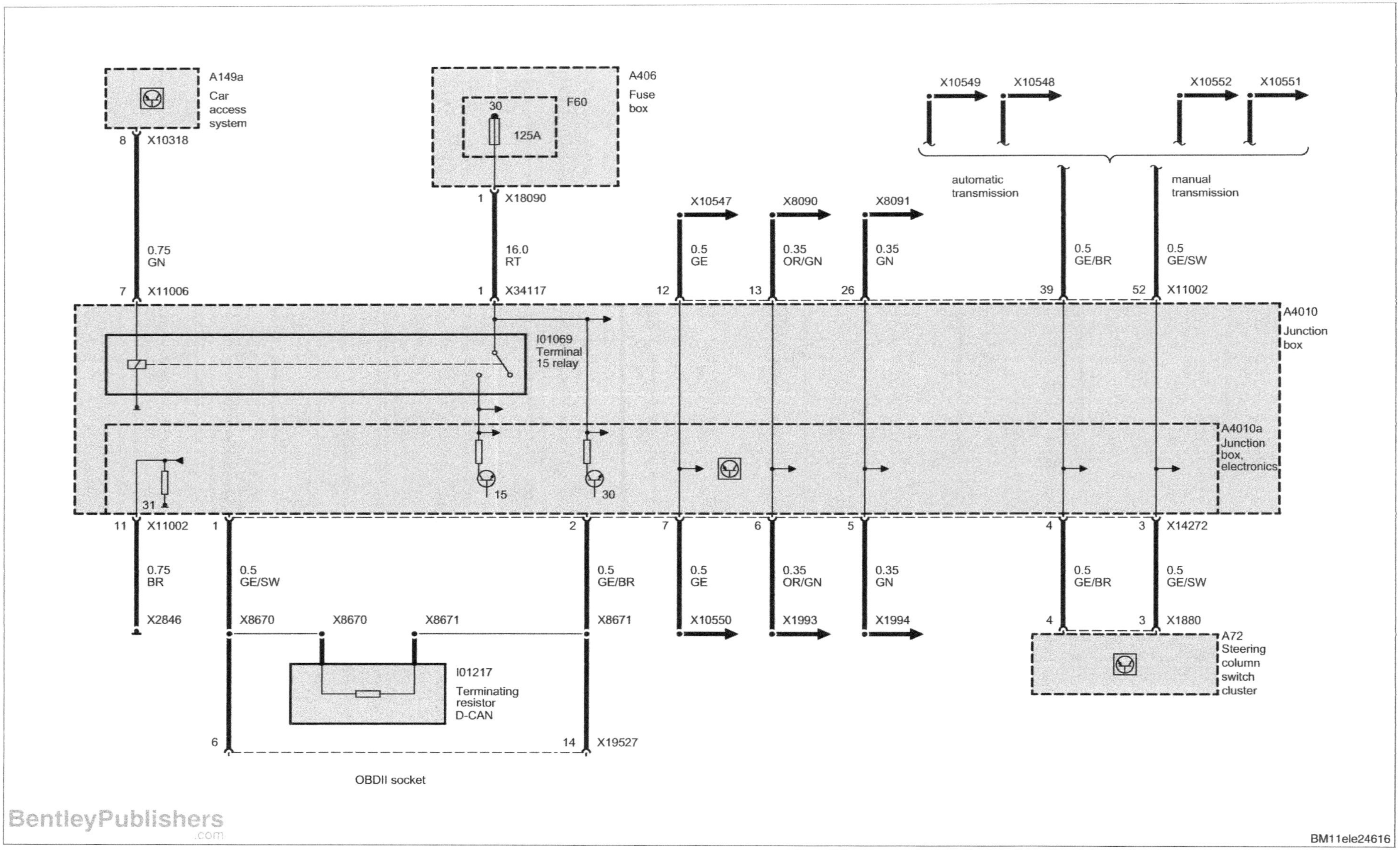

Horn

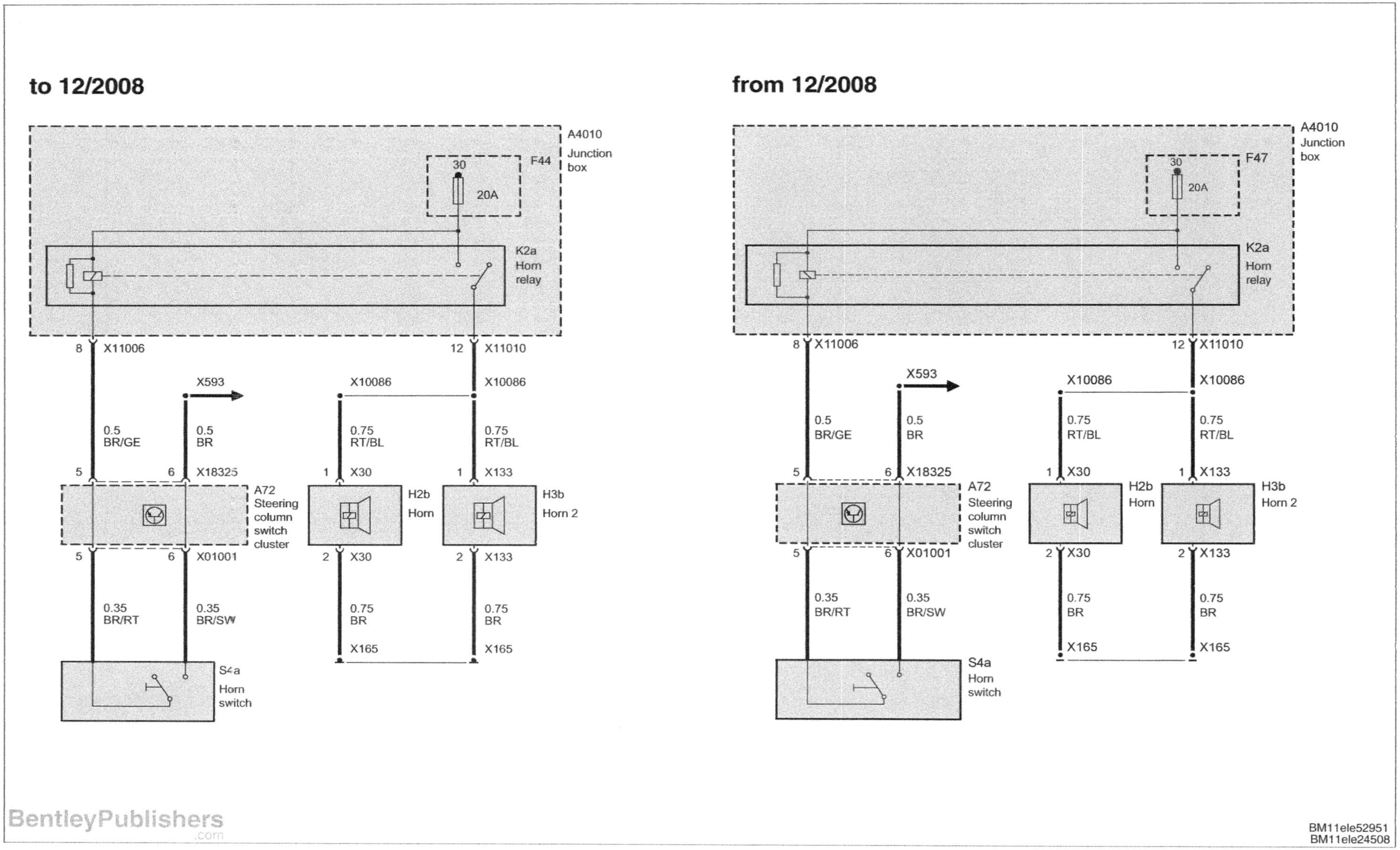
to 12/2008
A4010
Junction
box
F44
30
20A
K2a
Horn
relay
8
X11006
12
X11010
X593
X10086
X10086
0.5
BR/GE
0.5
BR
0.75
RT/BL
0.75
RT/BL
5
6
X18325
1
X30
1
X133
A72
Steering
column
switch
cluster
H2b
Horn
H3b
Horn 2
5
6
X01001
2
X30
2
X133
0.35
BR/RT
0.35
BR/SW
0.75
BR
0.75
BR
X165
X165
S4a
Horn
switch
from 12/2008
A4010
Junction
box
F47
30
20A
K2a
Horn
relay
8
X11006
12
X11010
X593
X10086
X10086
0.5
BR/GE
0.5
BR
0.75
RT/BL
0.75
RT/BL
5
6
X18325
1
X30
1
X133
A72
Steering
column
switch
cluster
H2b
Horn
H3b
Horn 2
5
6
X01001
2
X30
2
X133
0.35
BR/RT
0.35
BR/SW
0.75
BR
0.75
BR
X165
X165
S4a
Horn
switch
BM11ele52951
BM11ele24508

Electric outside mirror to 12/2008 page 1 of 2

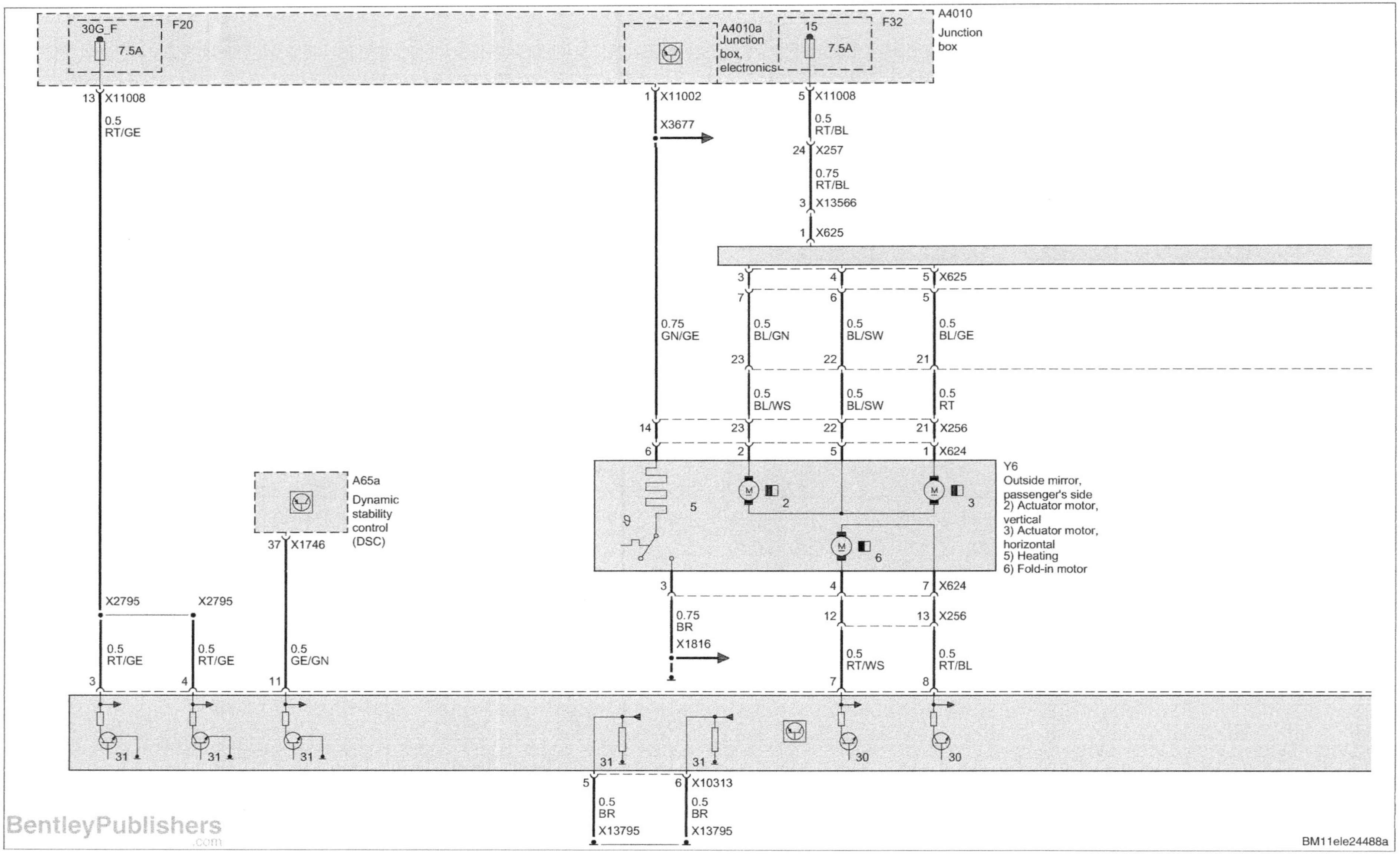

Electric outside mirror
to 12/2008
page 2 of 2

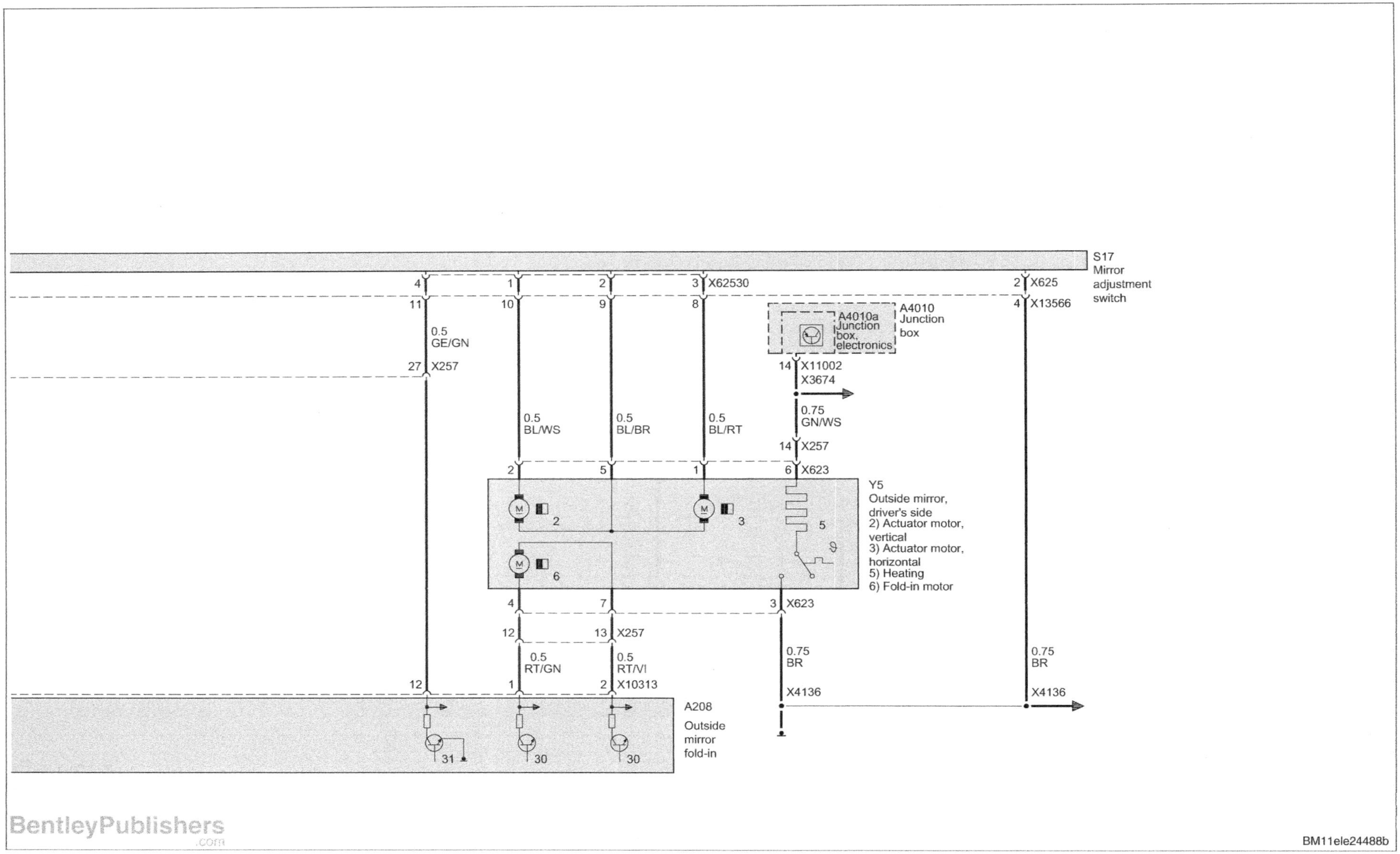

Electric outside mirror from 12/2008 page 1 of 2

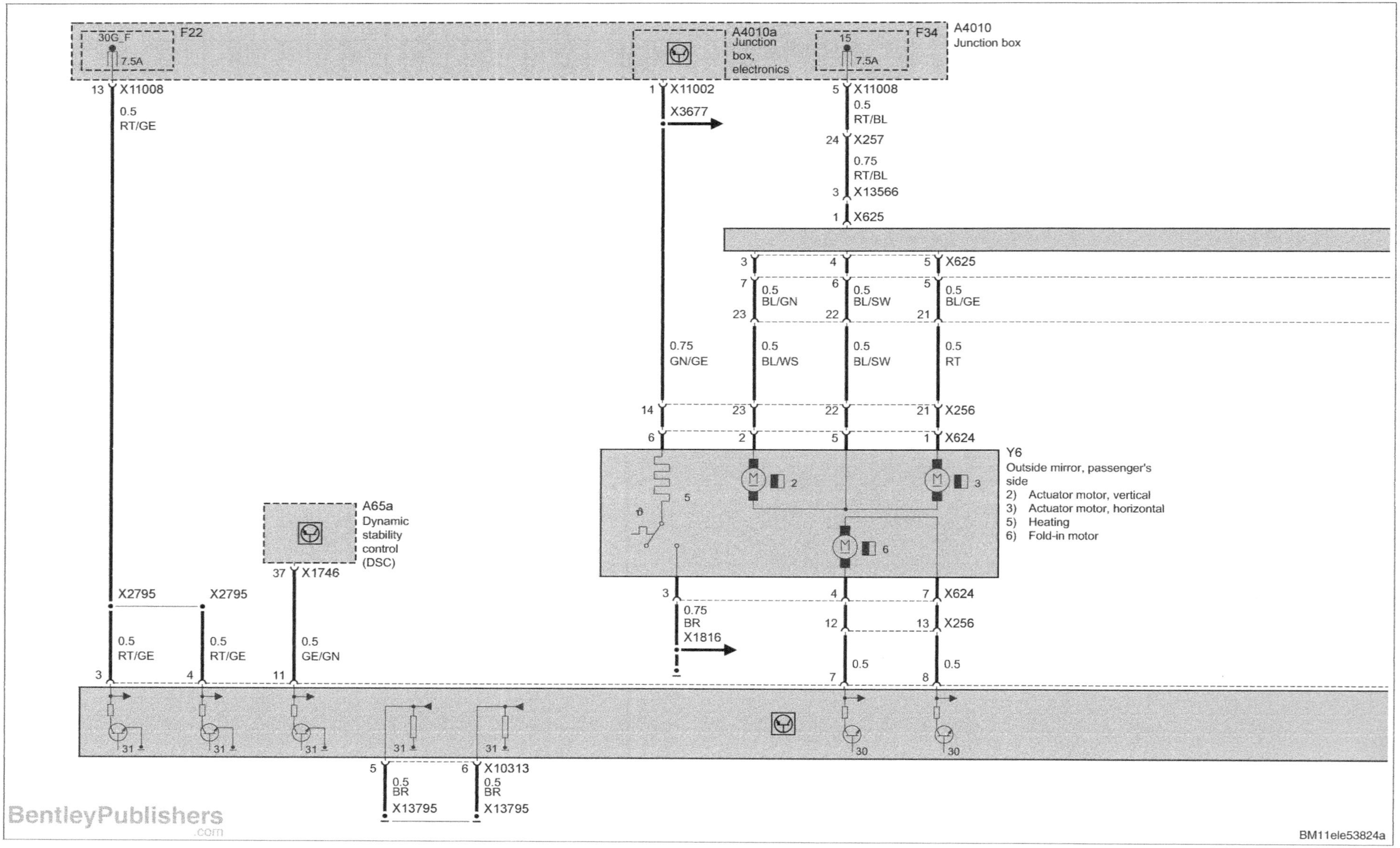

Electric outside mirror from 12/2008 page 2 of 2

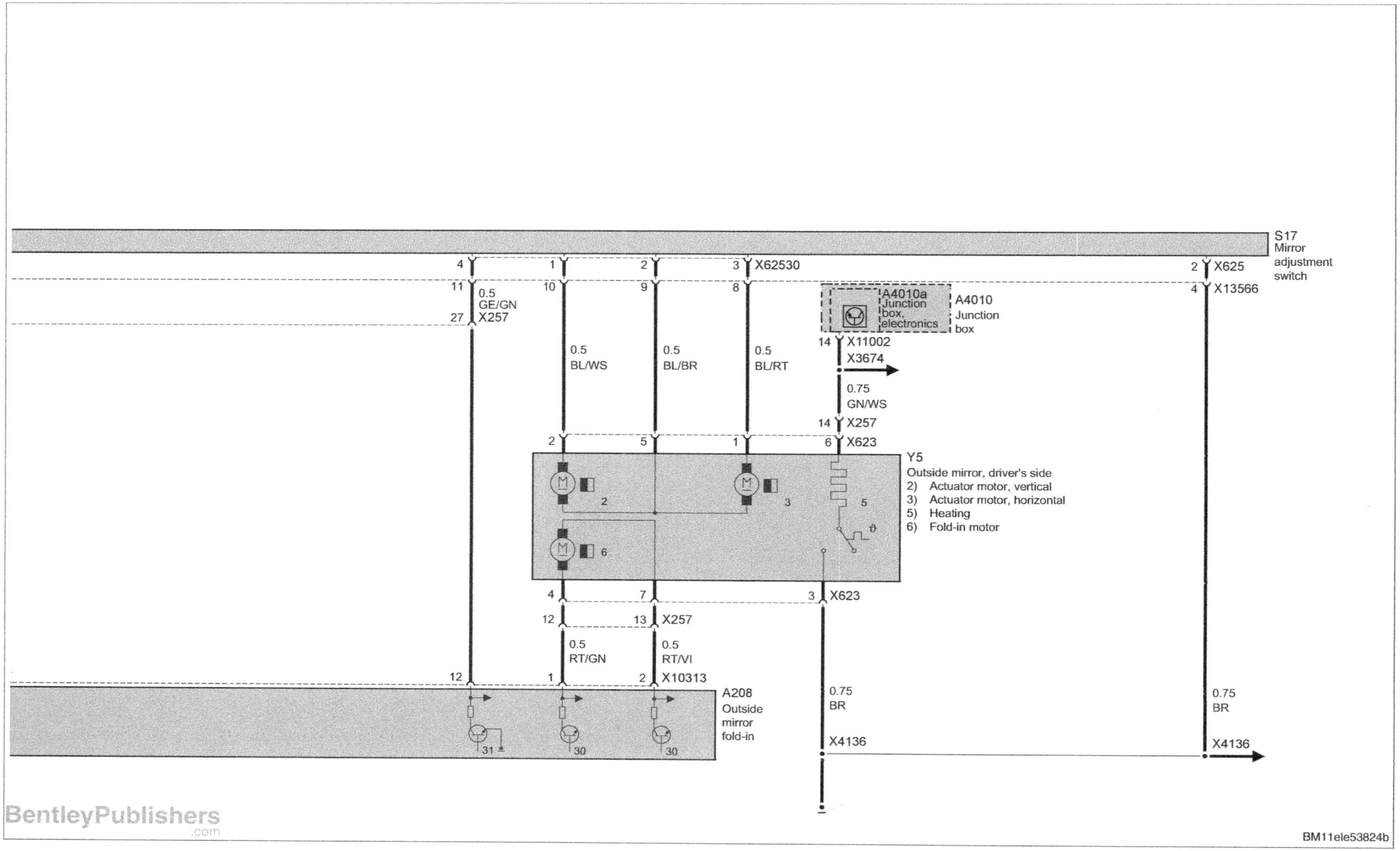

DSC switch functions, brake fluid level switch, brake light switch, center console switch cluster, handbrake warning switch

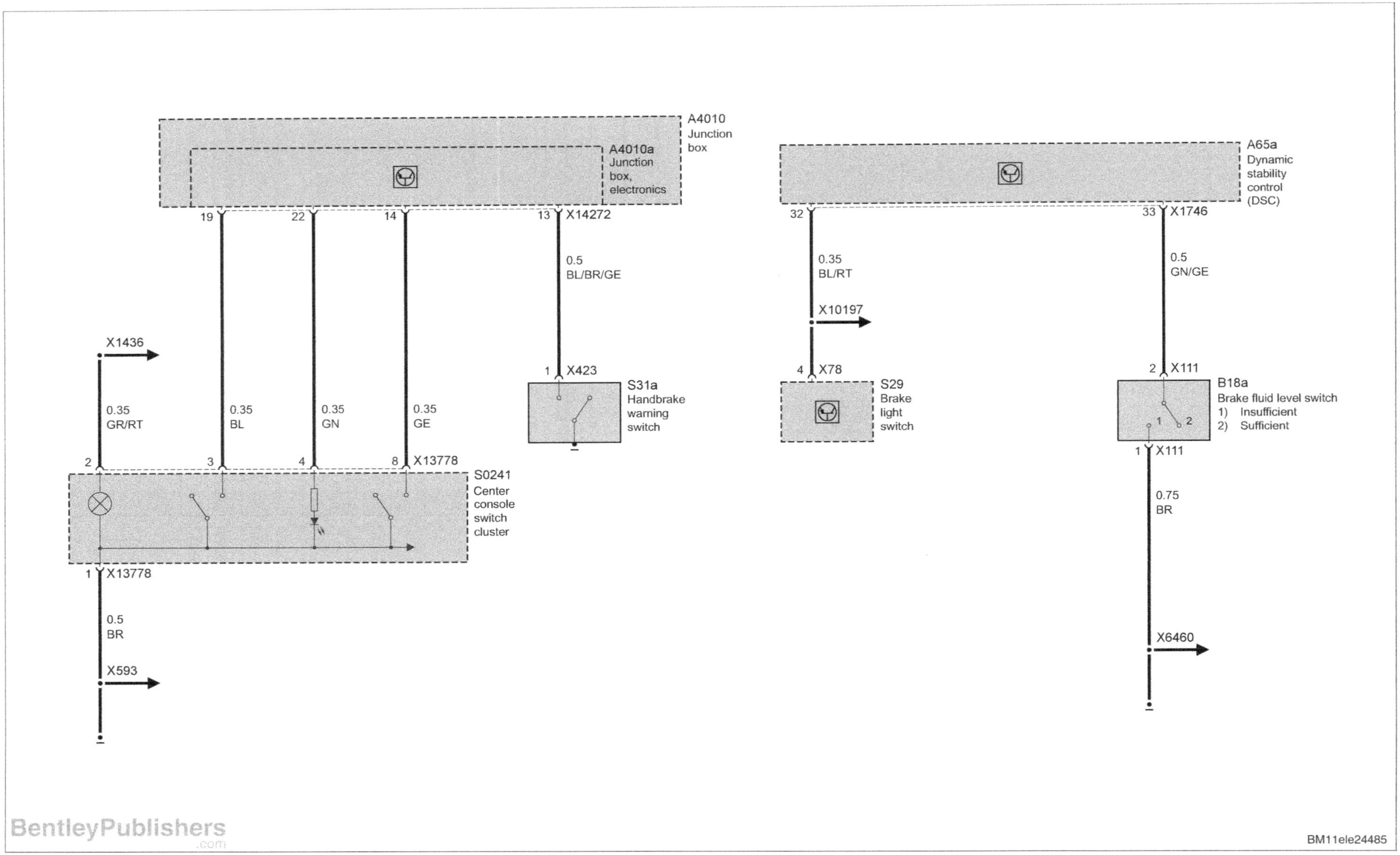

Dynamic stability control (DSC) power supply
page 1 of 2

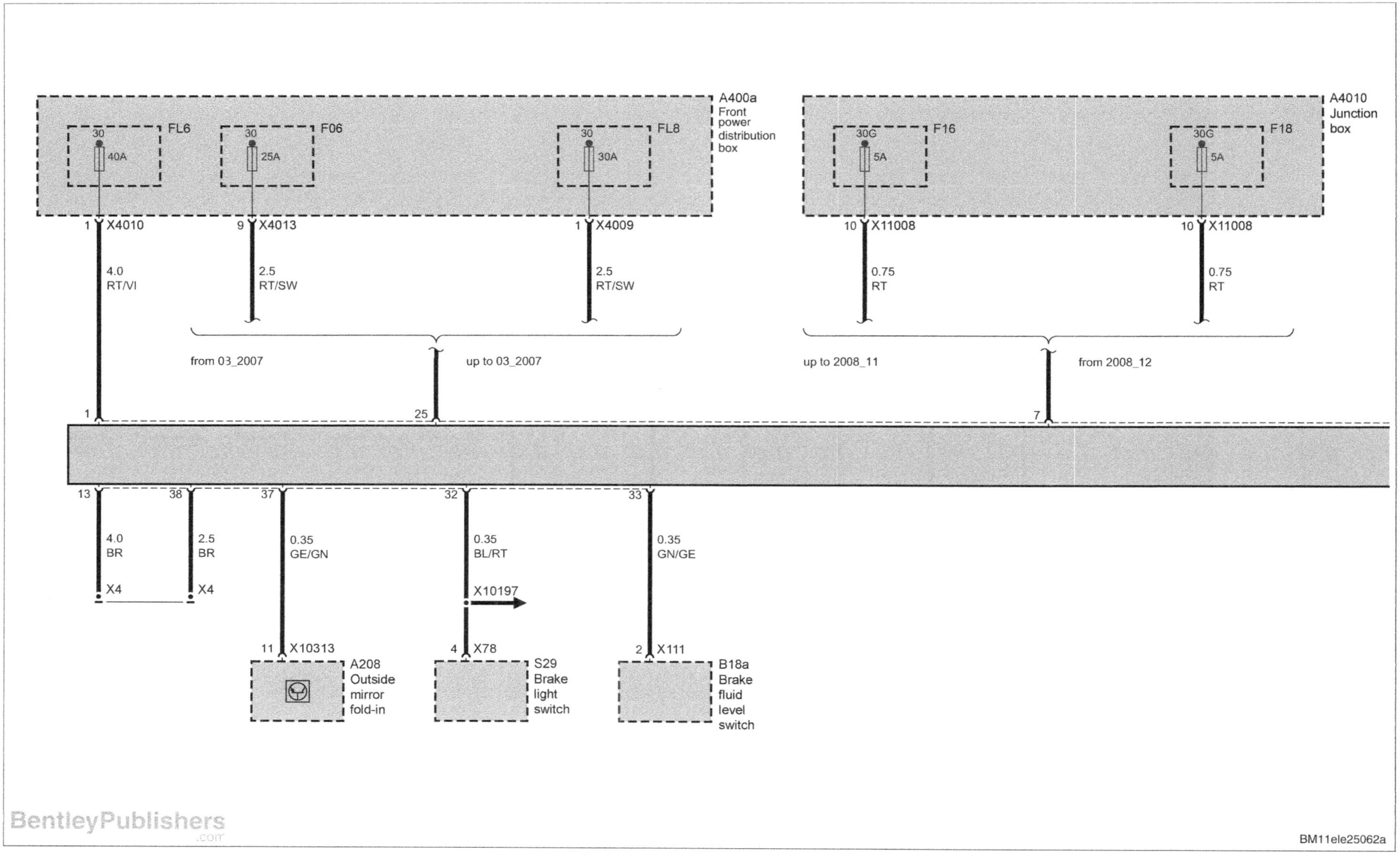

Dynamic stability control (DSC) power supply, wheel speed sensors page 2 of 2

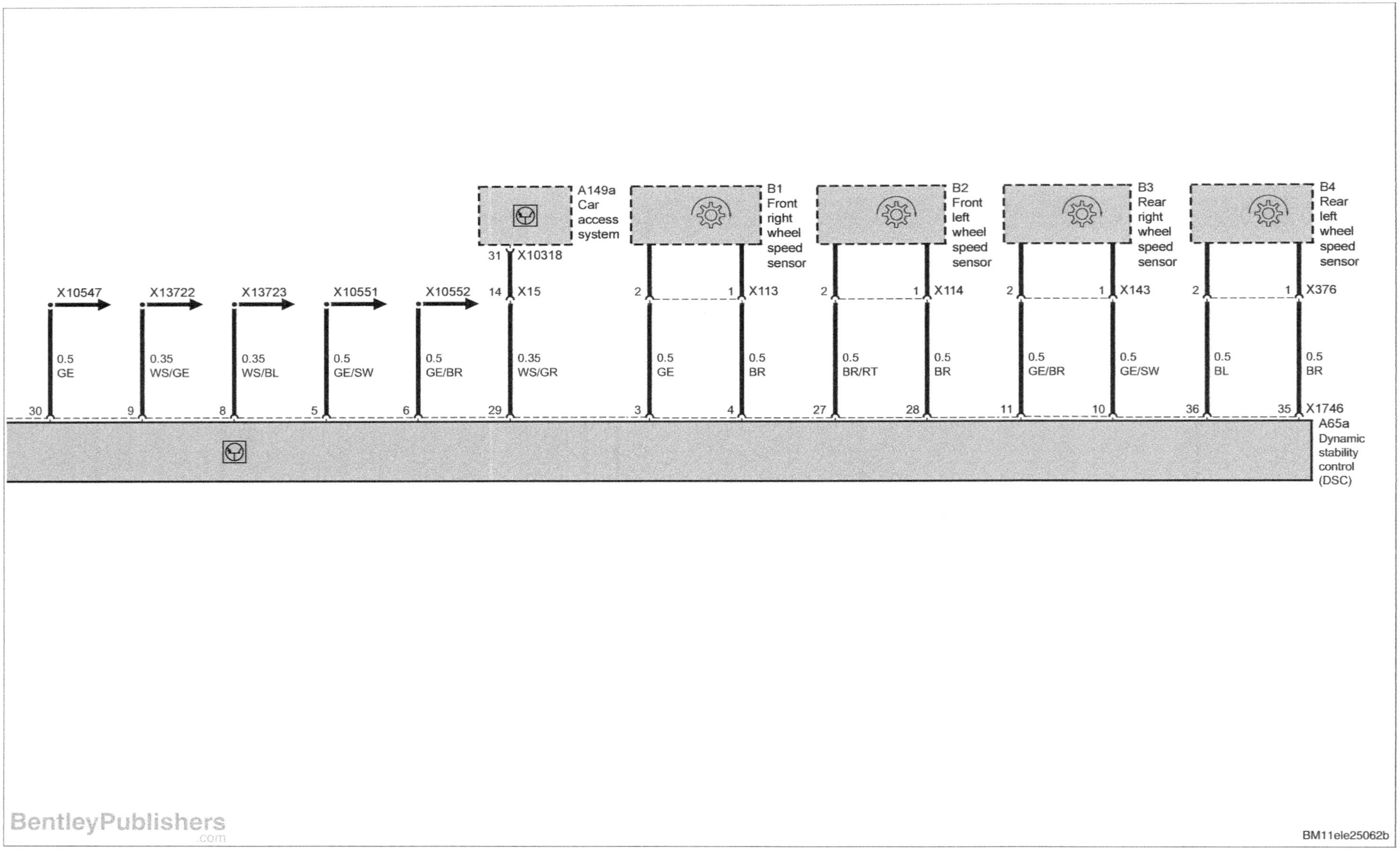

DSC sensor system
to 12/2008
page 1 of 2

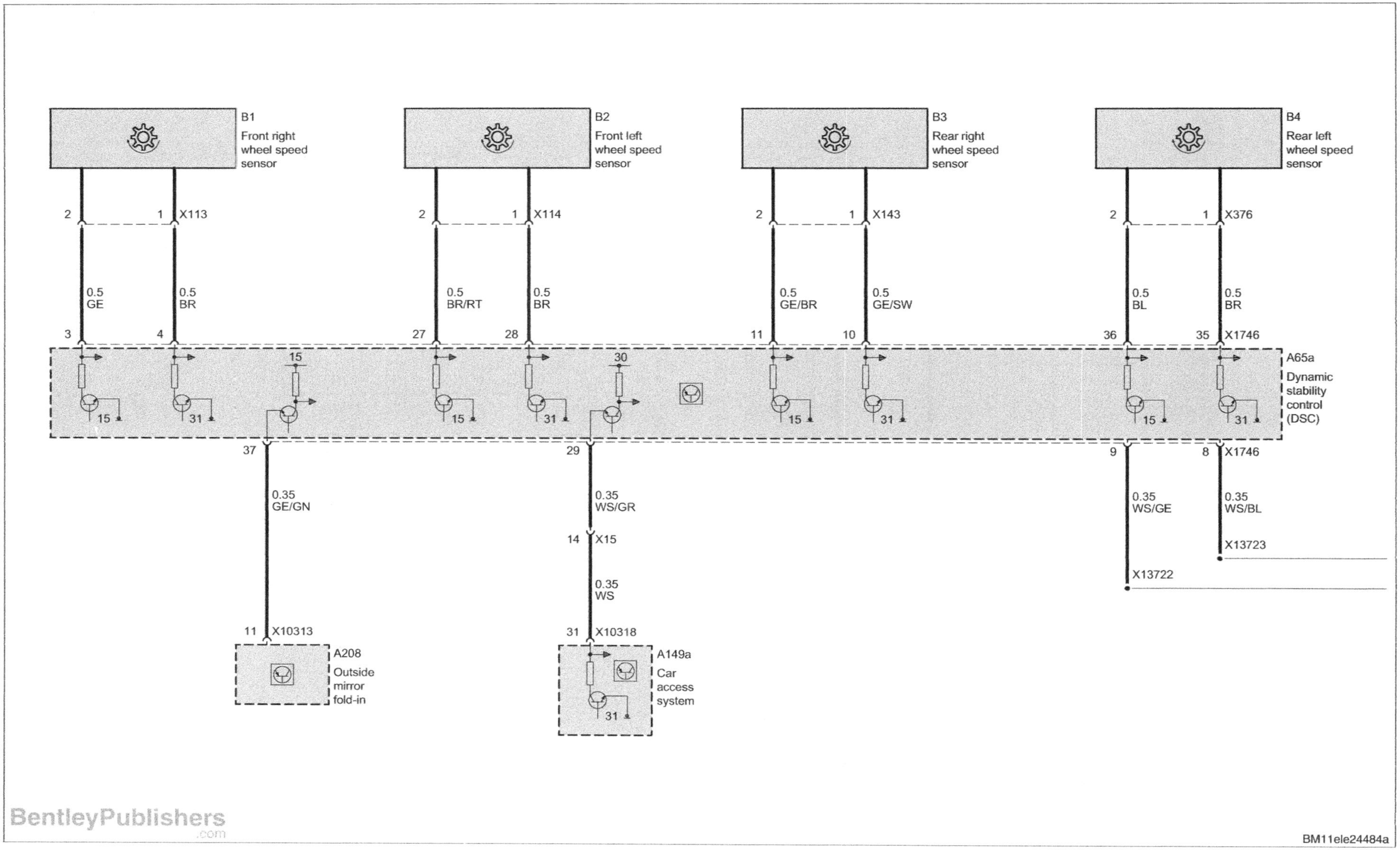

DSC sensor system
to 12/2008
page 2 of 2

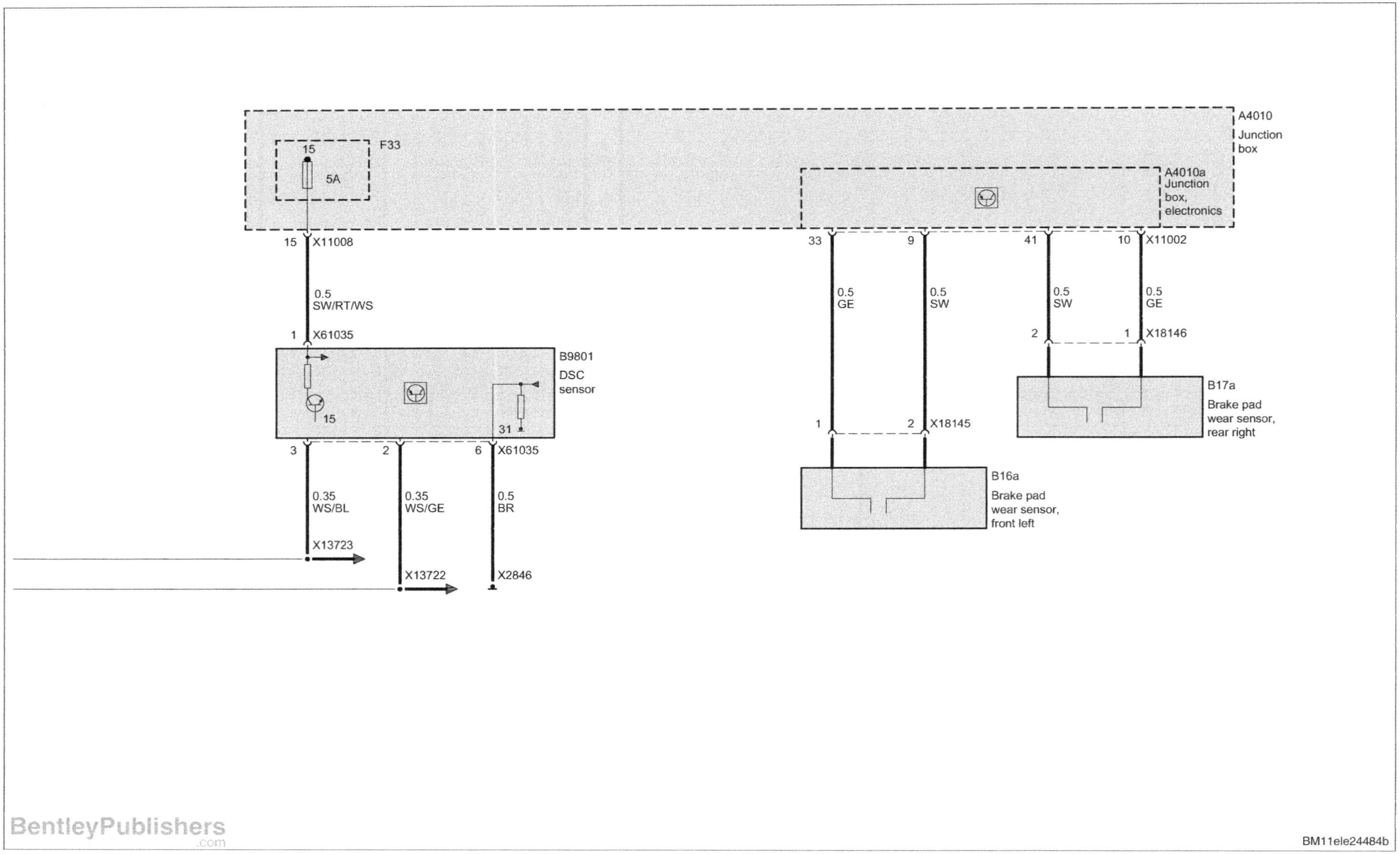

DSC sensors from 12/2008 page 1 of 2

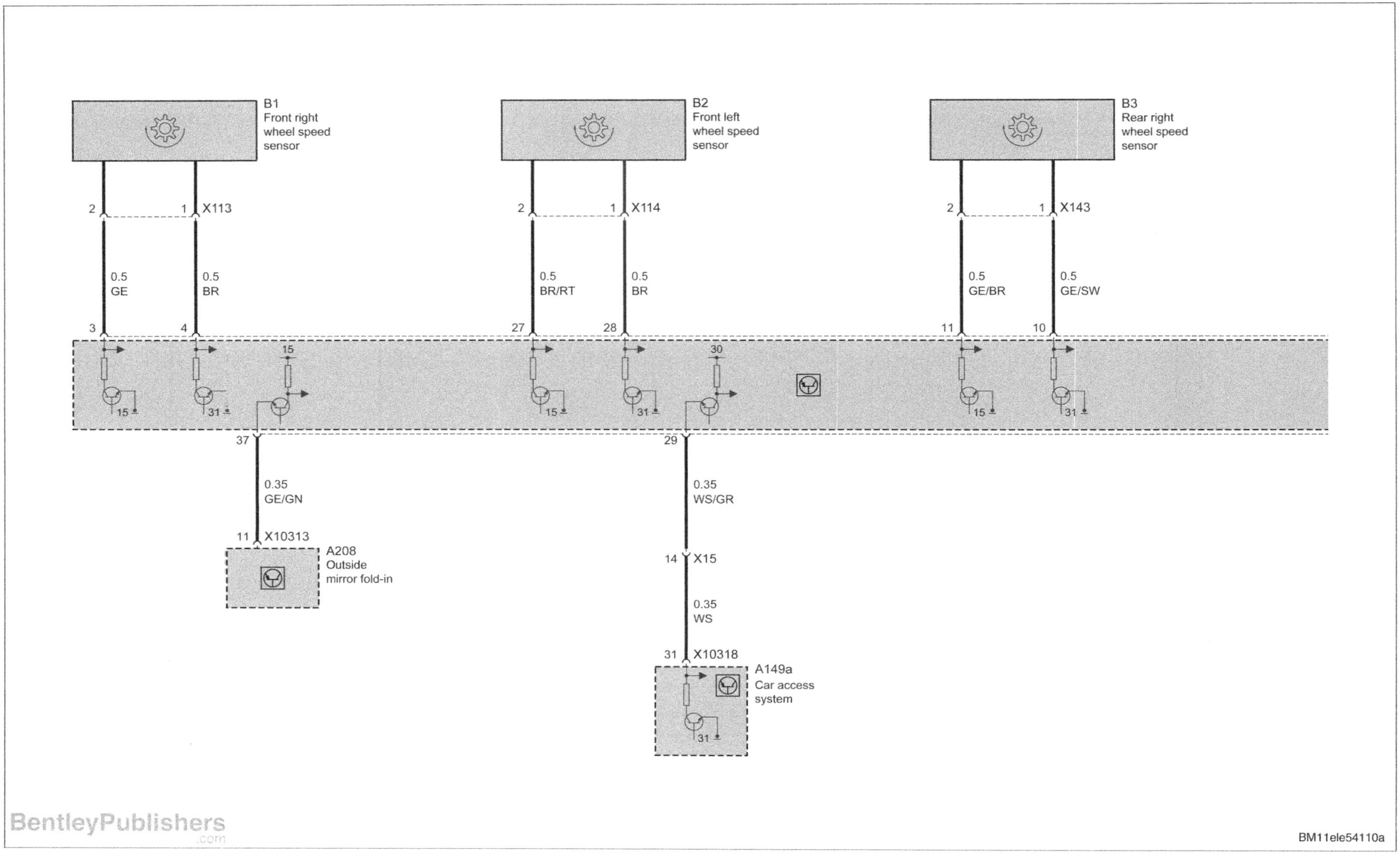

DSC sensors from 12/2008 page 2 of 2

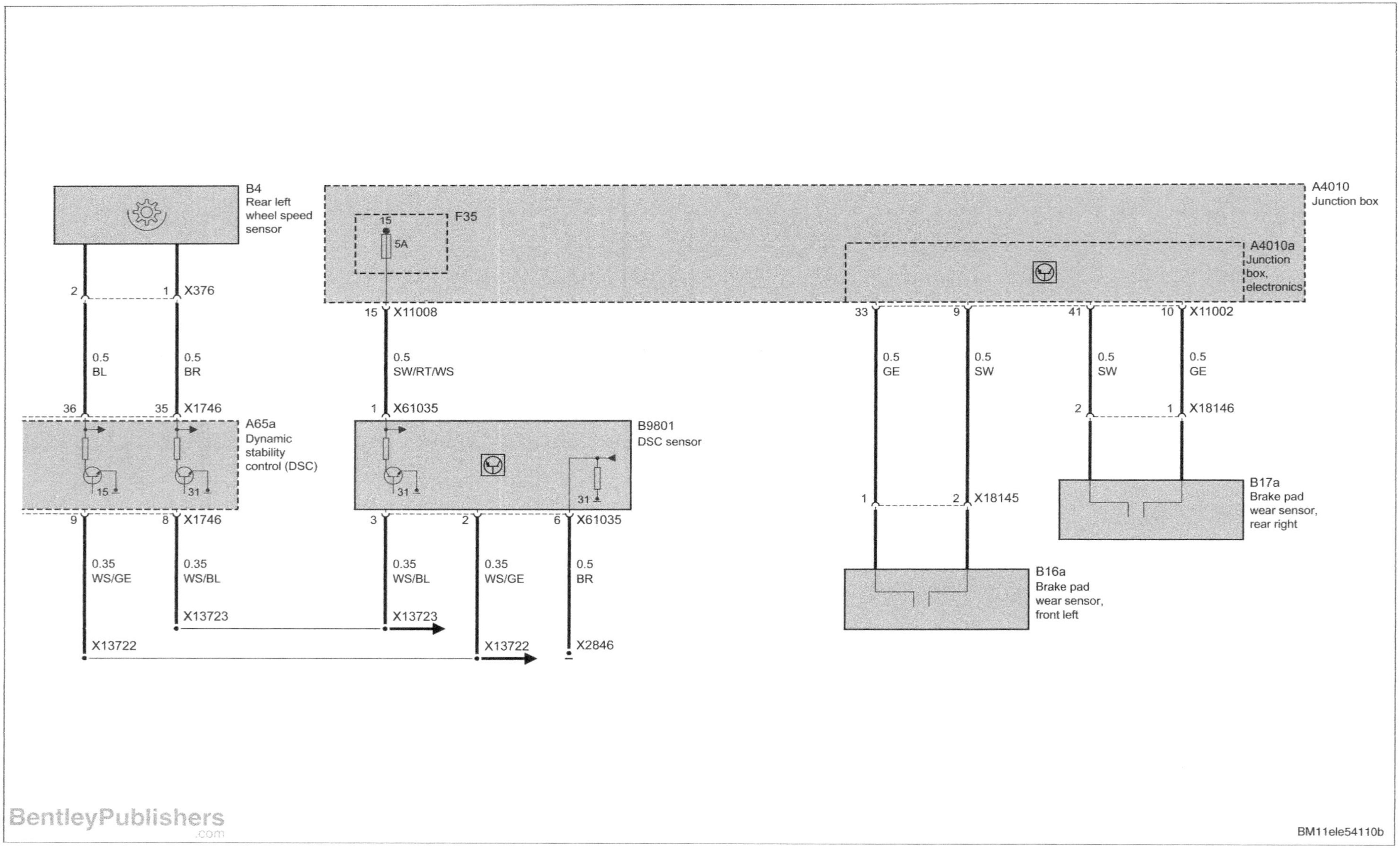

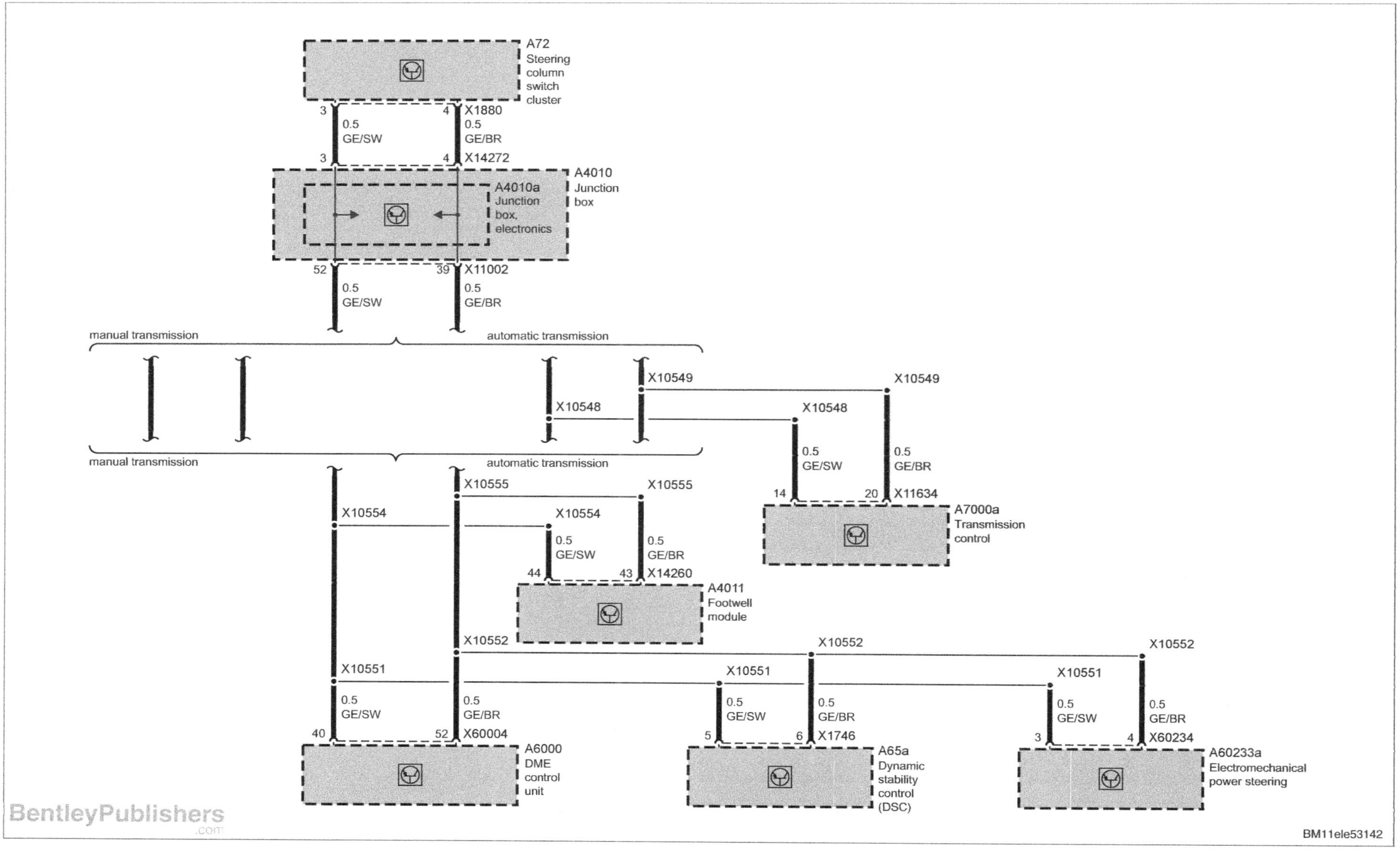
A72
Steering column switch cluster
3
4
X1880
0.5 GE/SW
0.5 GE/BR
3
4
X14272
A4010
Junction box
A4010a
Junction box, electronics
52
39
X11002
0.5 GE/SW
0.5 GE/BR
manual transmission
automatic transmission
X10549
X10548
X10549
X10548
0.5 GE/SW
0.5 GE/BR
14
20
X11634
A7000a
Transmission control
manual transmission
automatic transmission
X10555
X10554
X10555
X10554
0.5 GE/SW
0.5 GE/BR
44
43
X14260
A4011
Footwell module
X10552
X10552
X10552
X10551
X10551
X10551
0.5 GE/SW
0.5 GE/BR
40
52
X60004
A6000
DME control unit
0.5 GE/SW
0.5 GE/BR
5
6
X1746
A65a
Dynamic stability control (DSC)
0.5 GE/SW
0.5 GE/BR
3
4
X60234
A60233a
Electromechanical power steering
BM11ele53142

Driver's seat heating to 12/2008

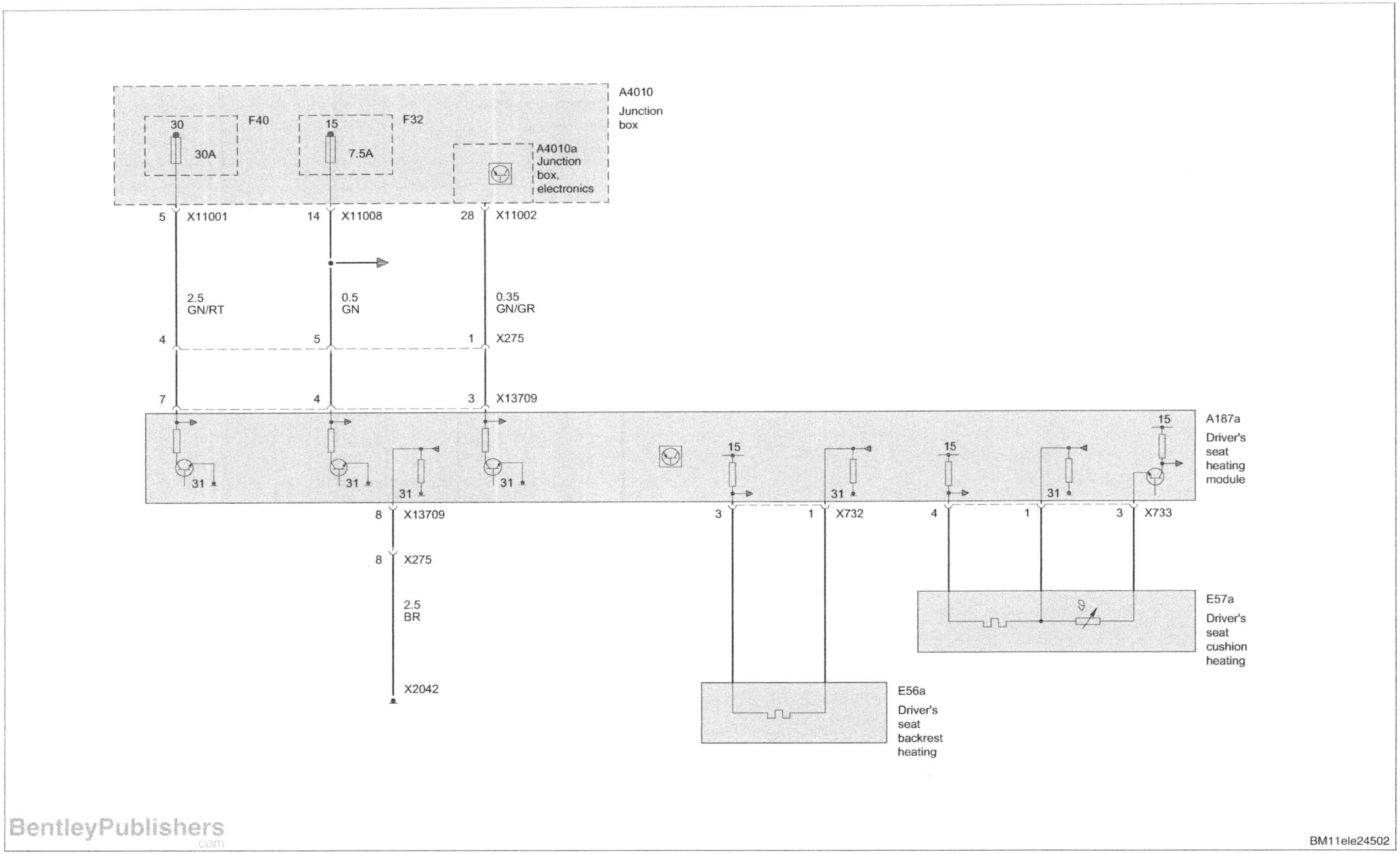

Driver's seat heating from 12/2008

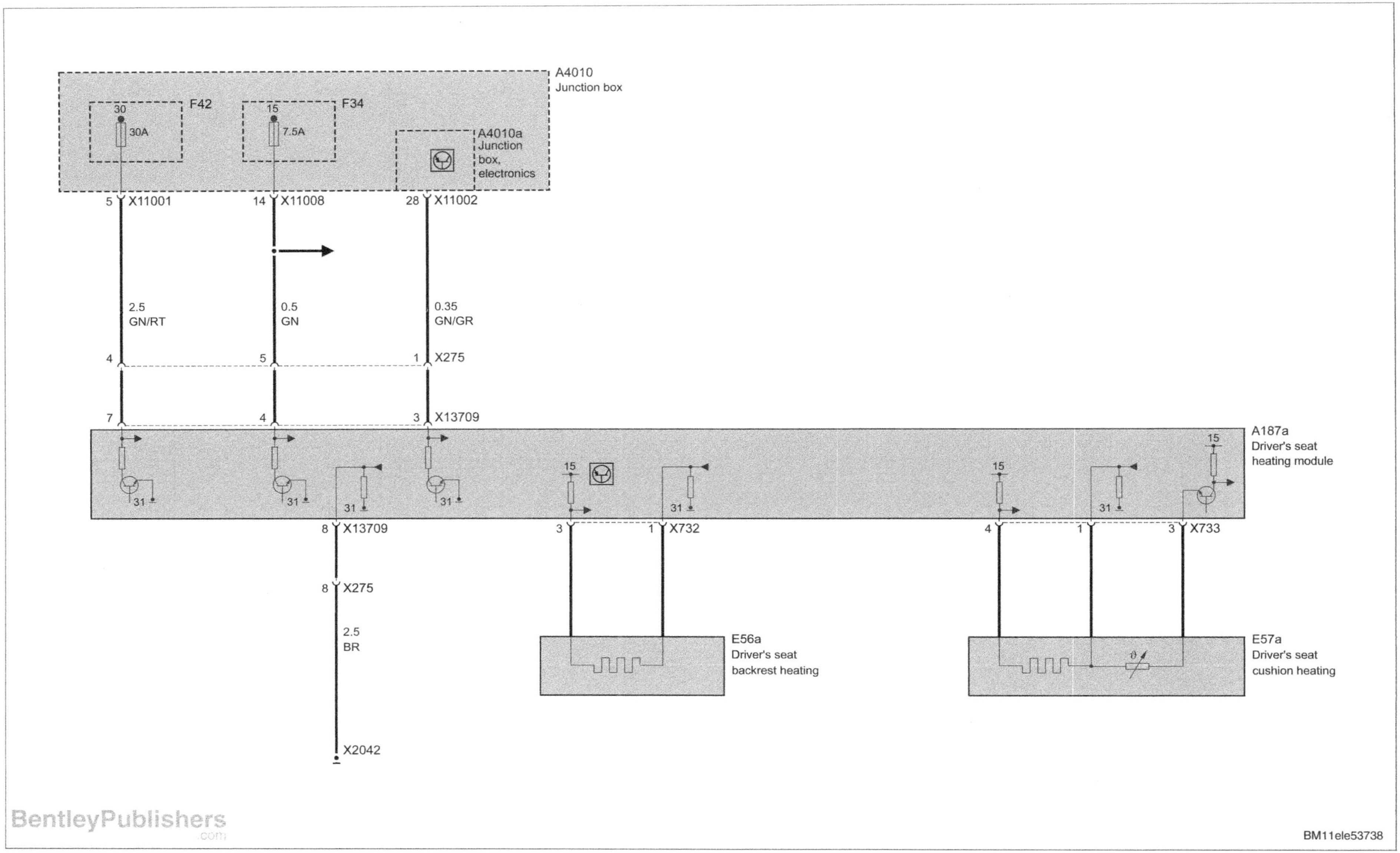

Passenger's seat heating to 12/2008

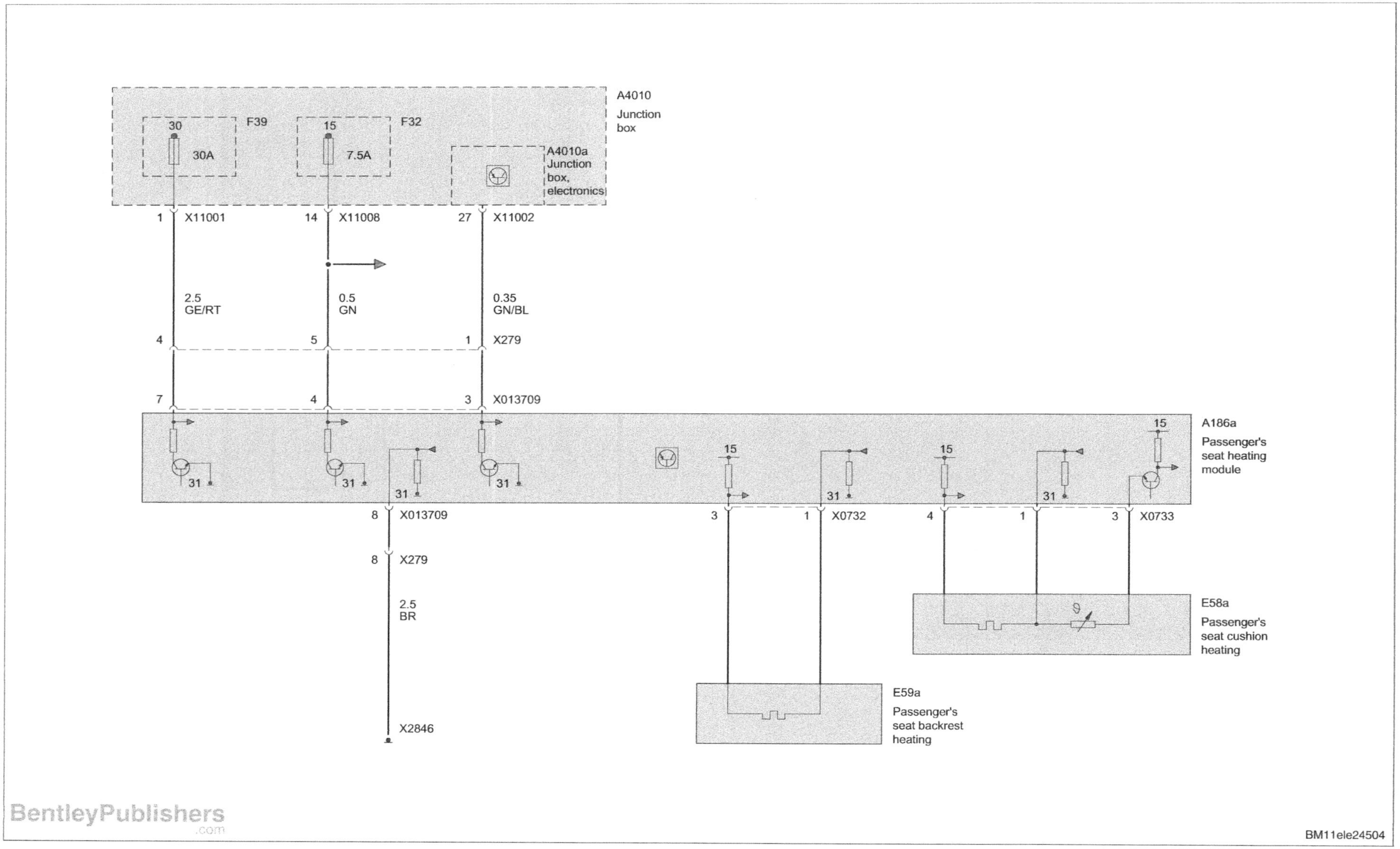

Passenger's seat heating from 12/2008

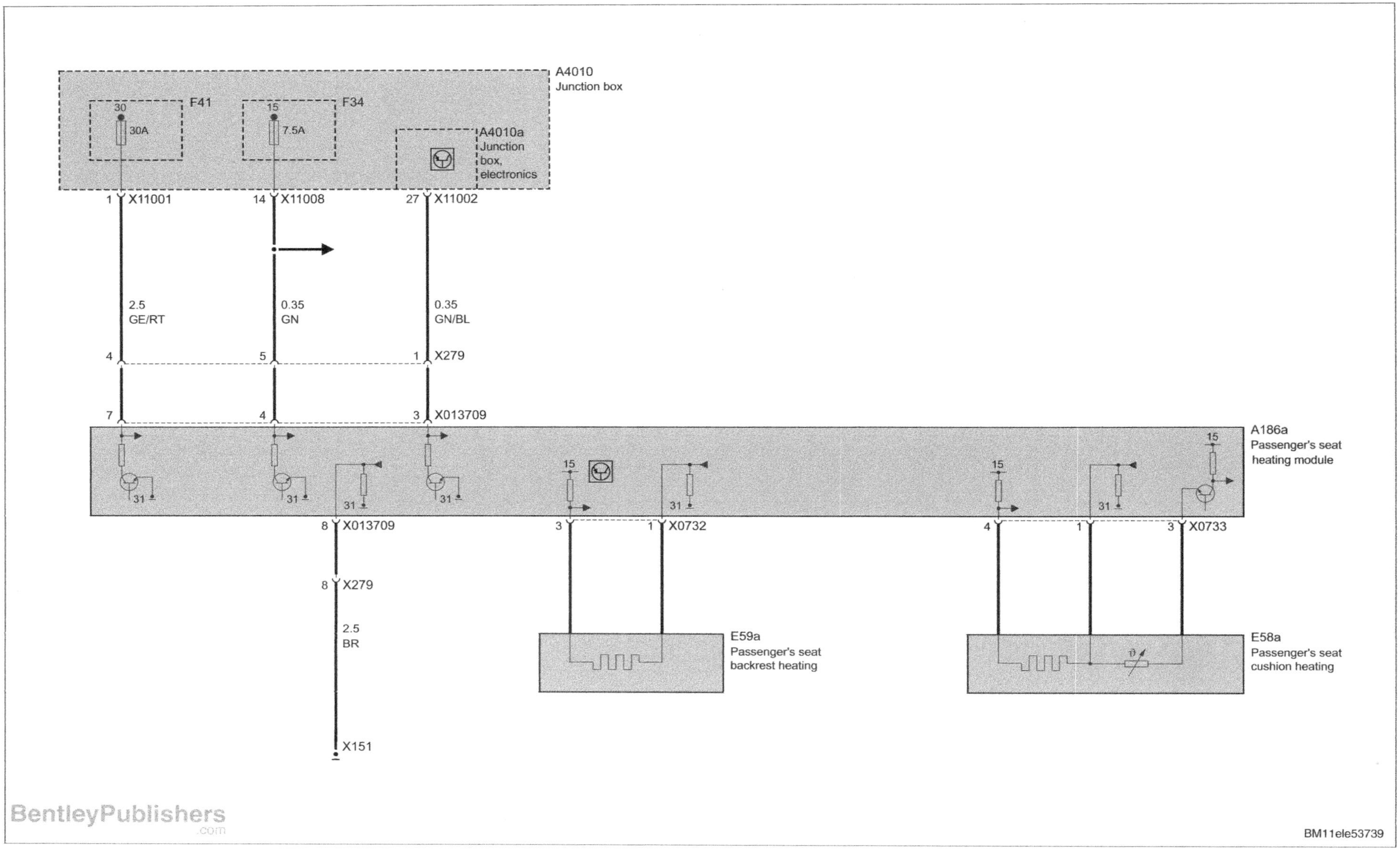

Automatic recirculated air control, AUC sensor, fresh air/recirculating flap motor

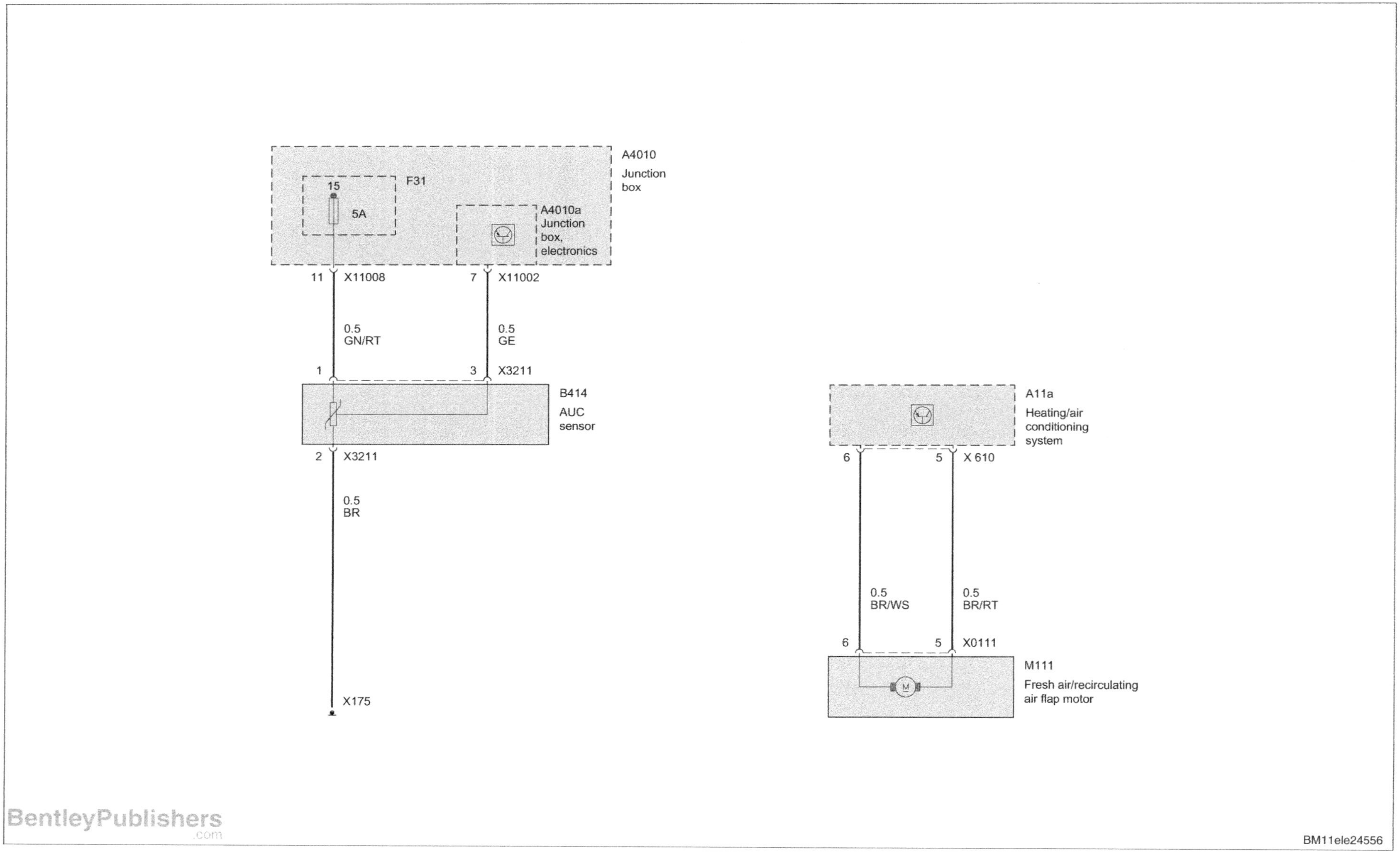

Blower motor ballast resistor

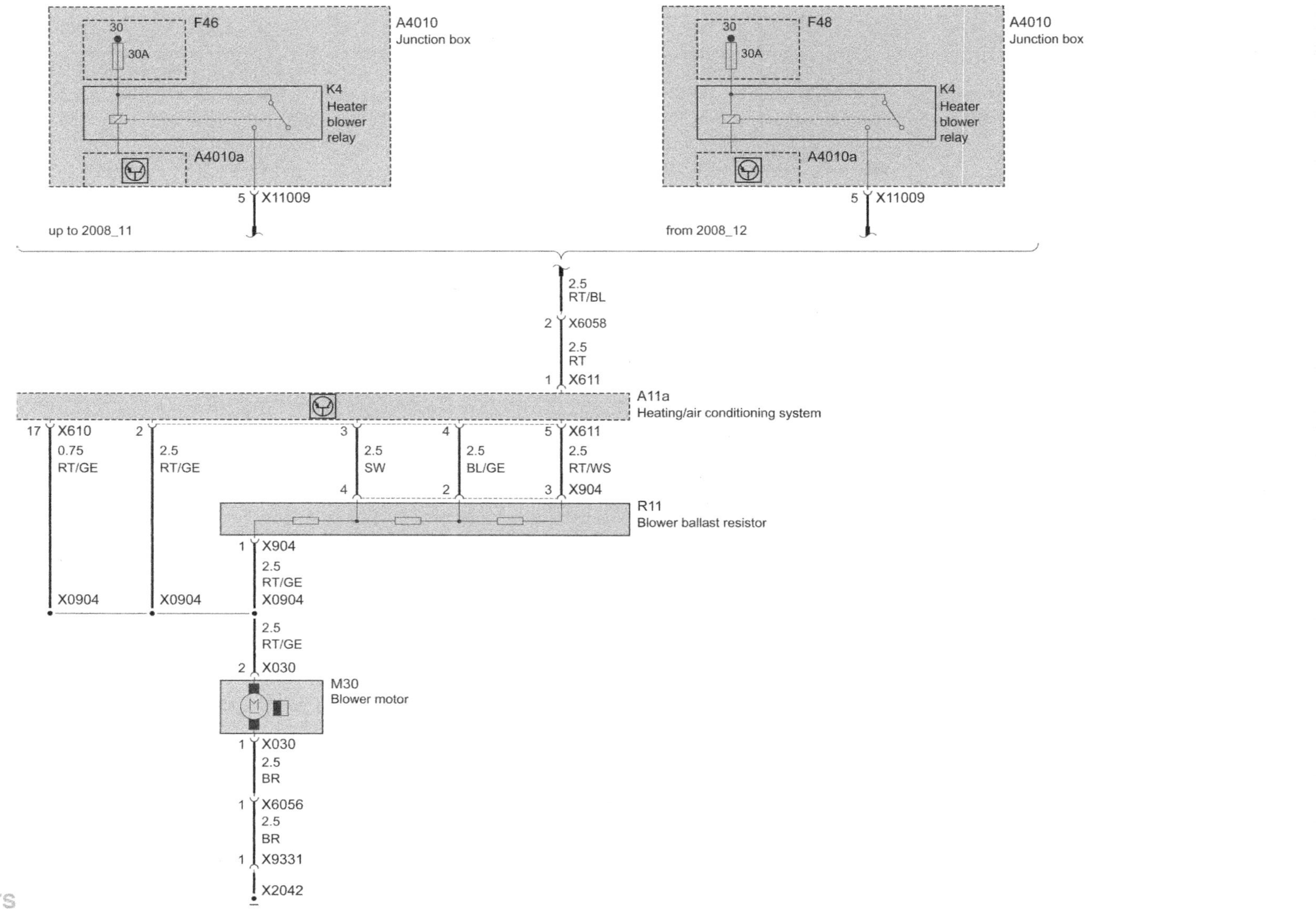

Heating and air conditioning system, A/C compressor clutch, temperature control

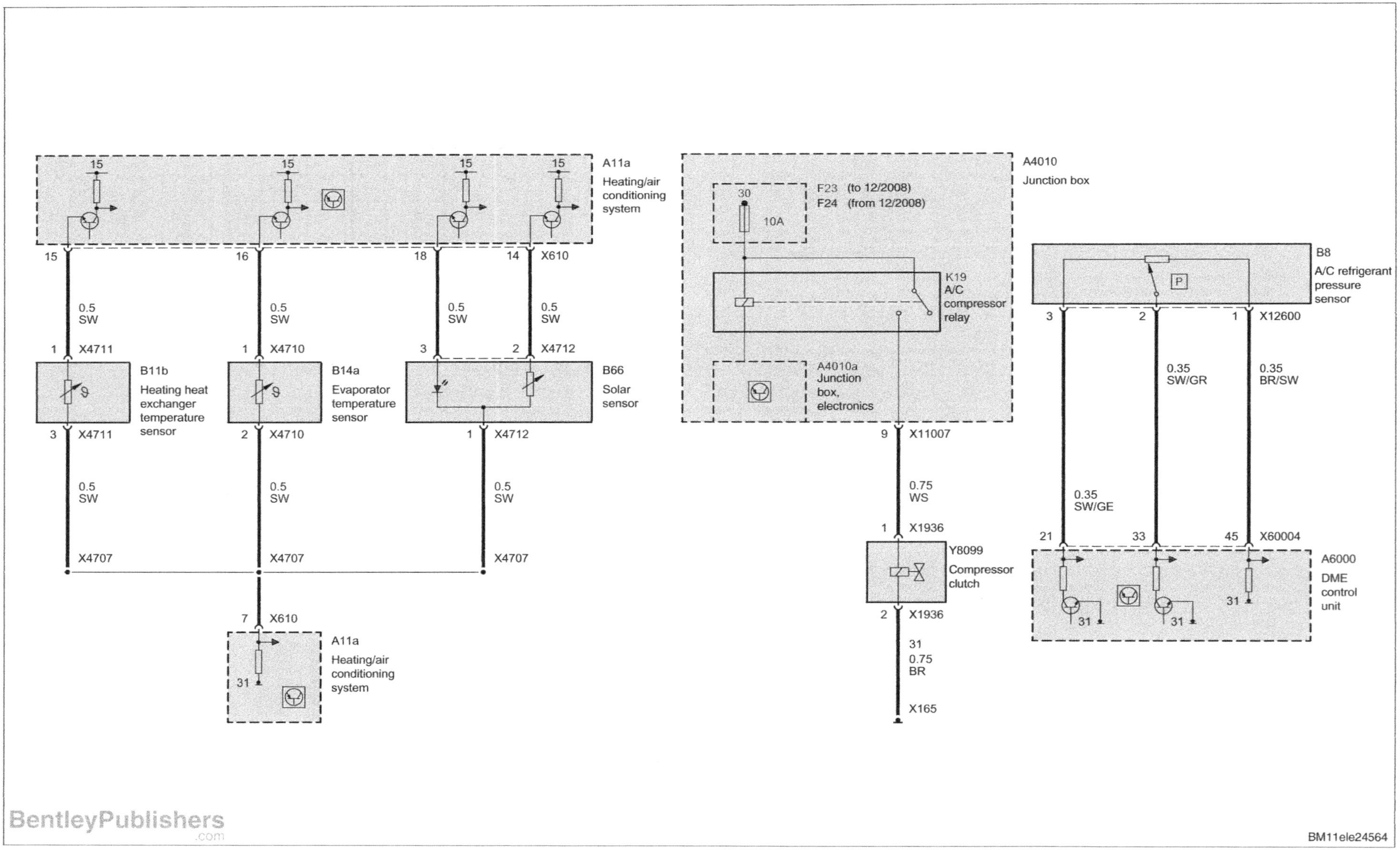

Heating and A/C system, air distribution (with blower resistors)

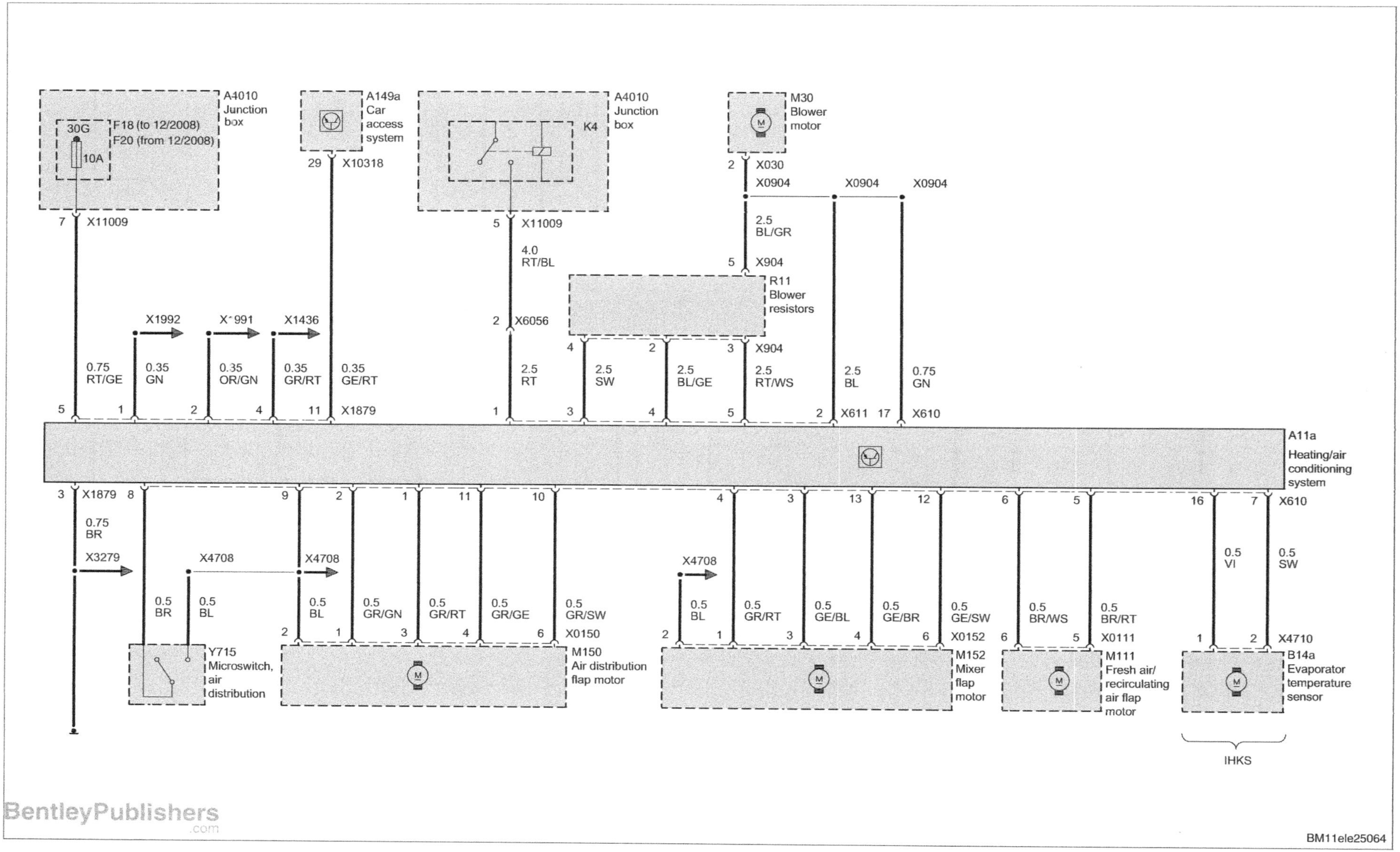

Heating and A/C system, air distribution (with blower output stage)

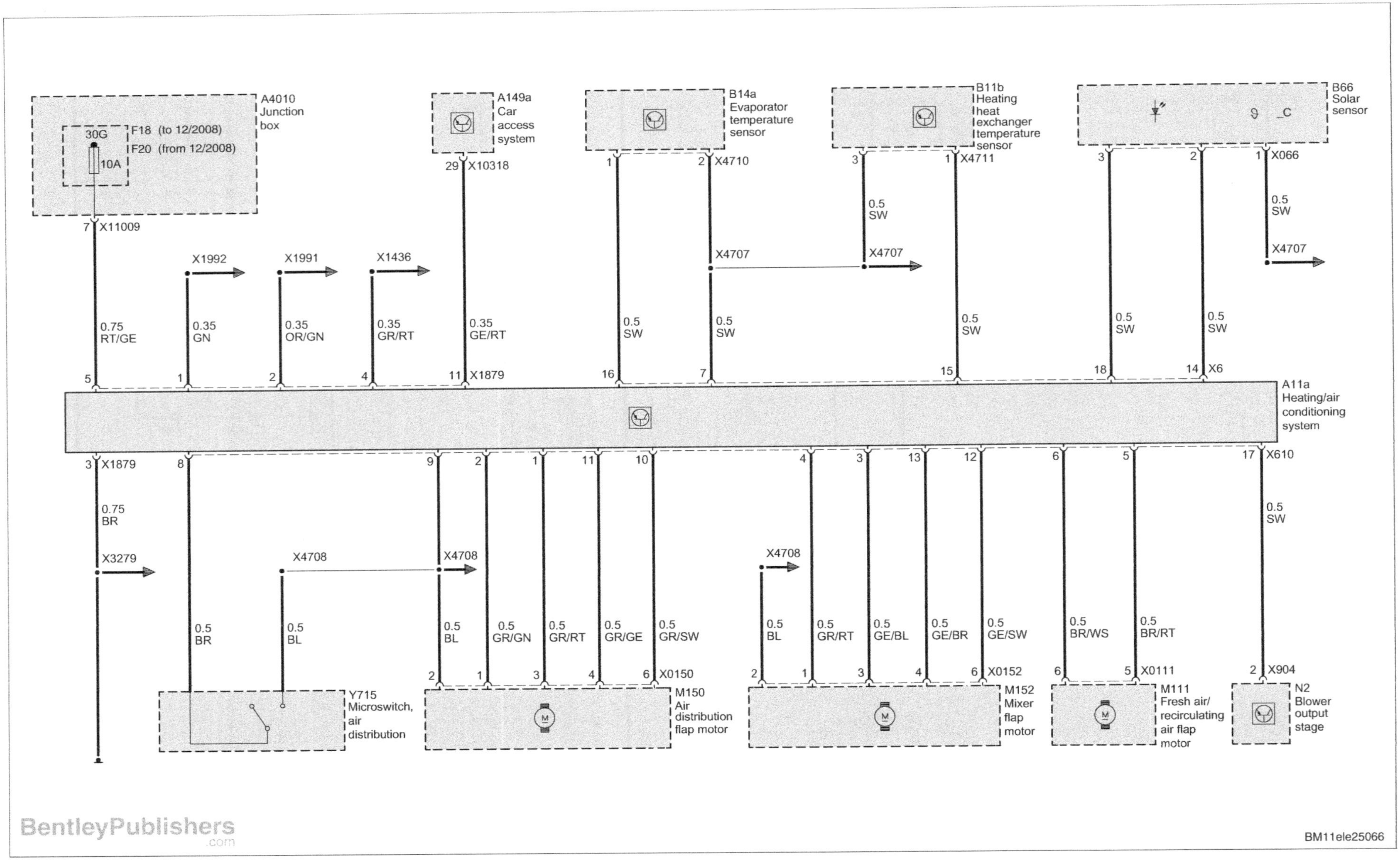

Power windows
page 1 of 2

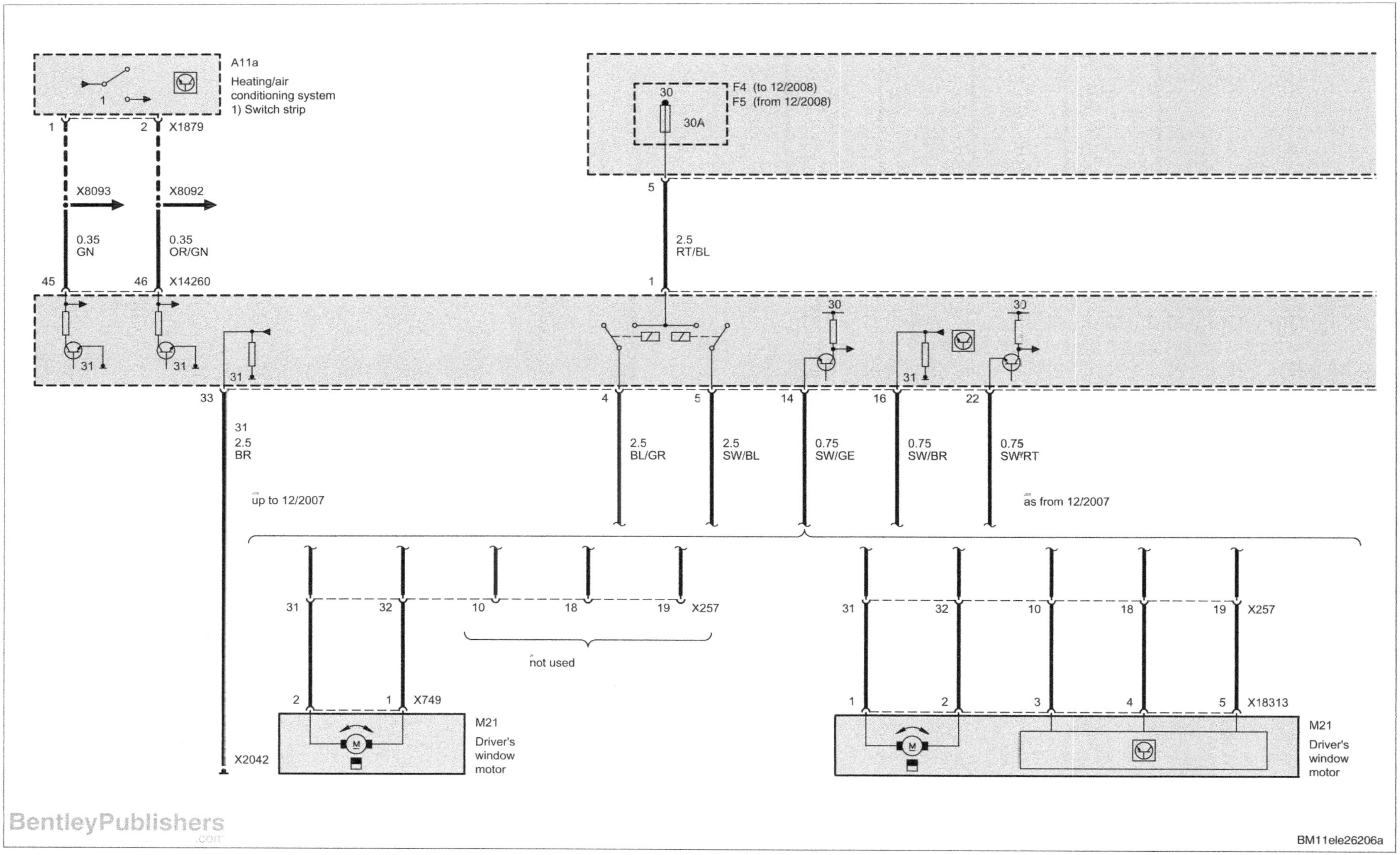

BM11ele26206a

Power windows
page 2 of 2

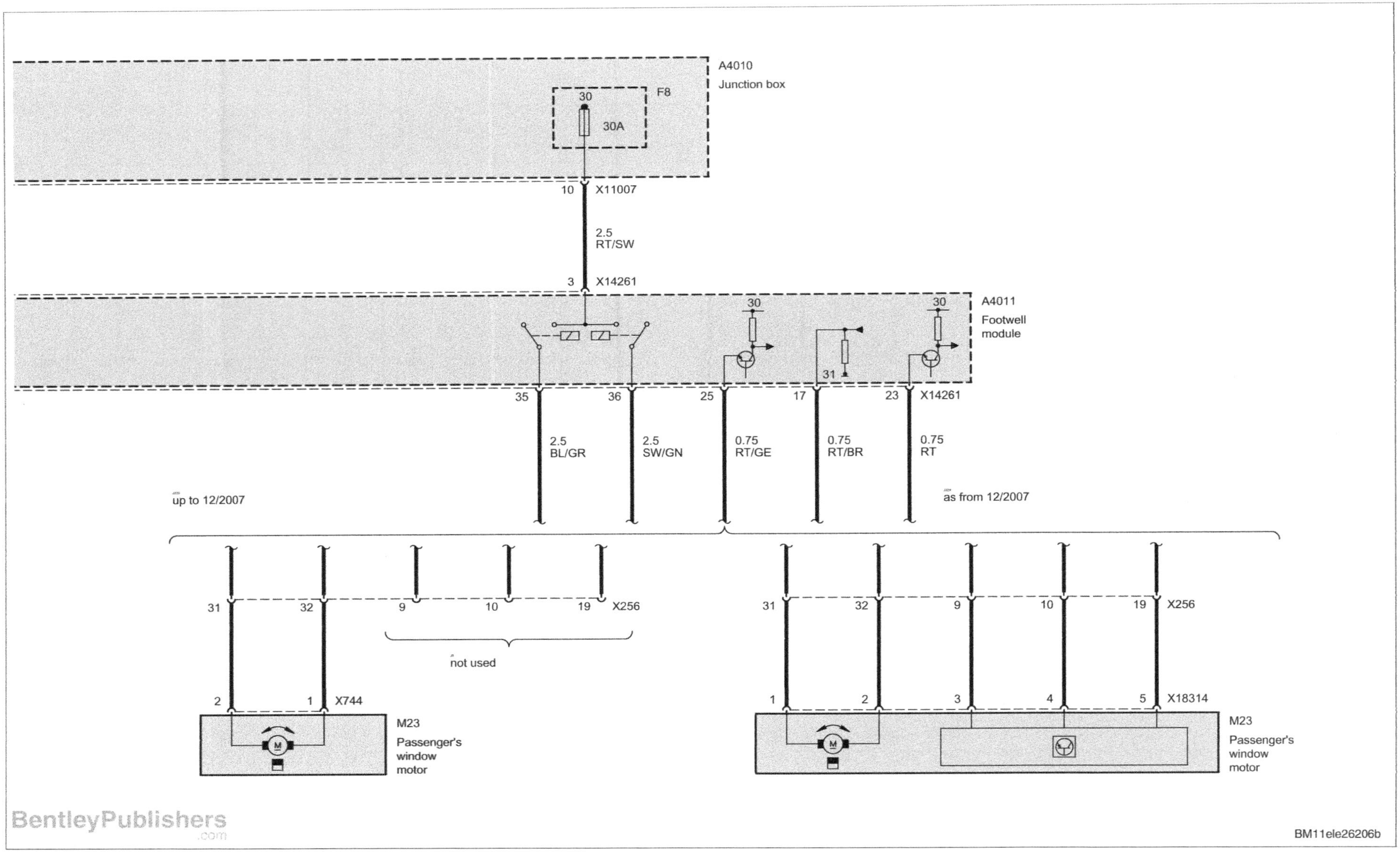

Multiple restraint system (MRS), passenger airbag deactivation switch

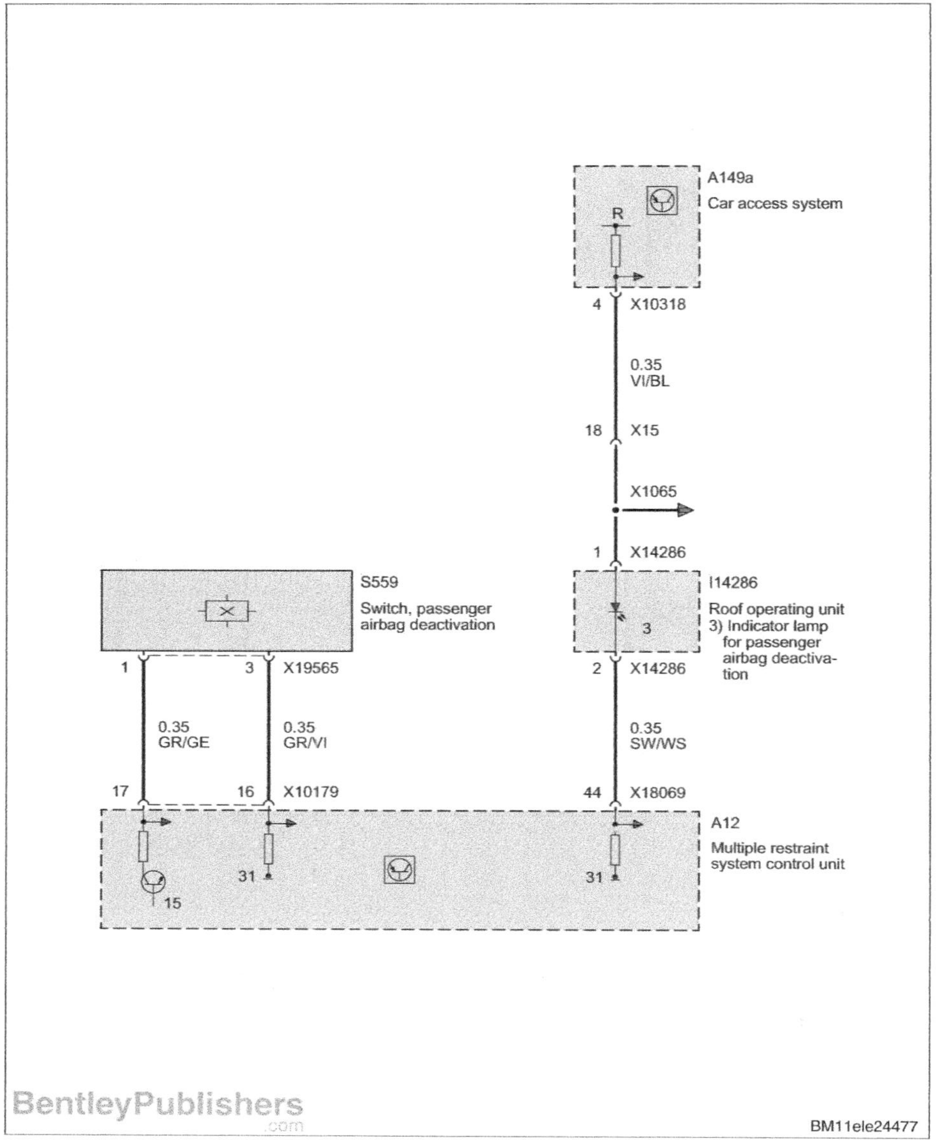

Central locking inputs (Clubman) page 1 of 2

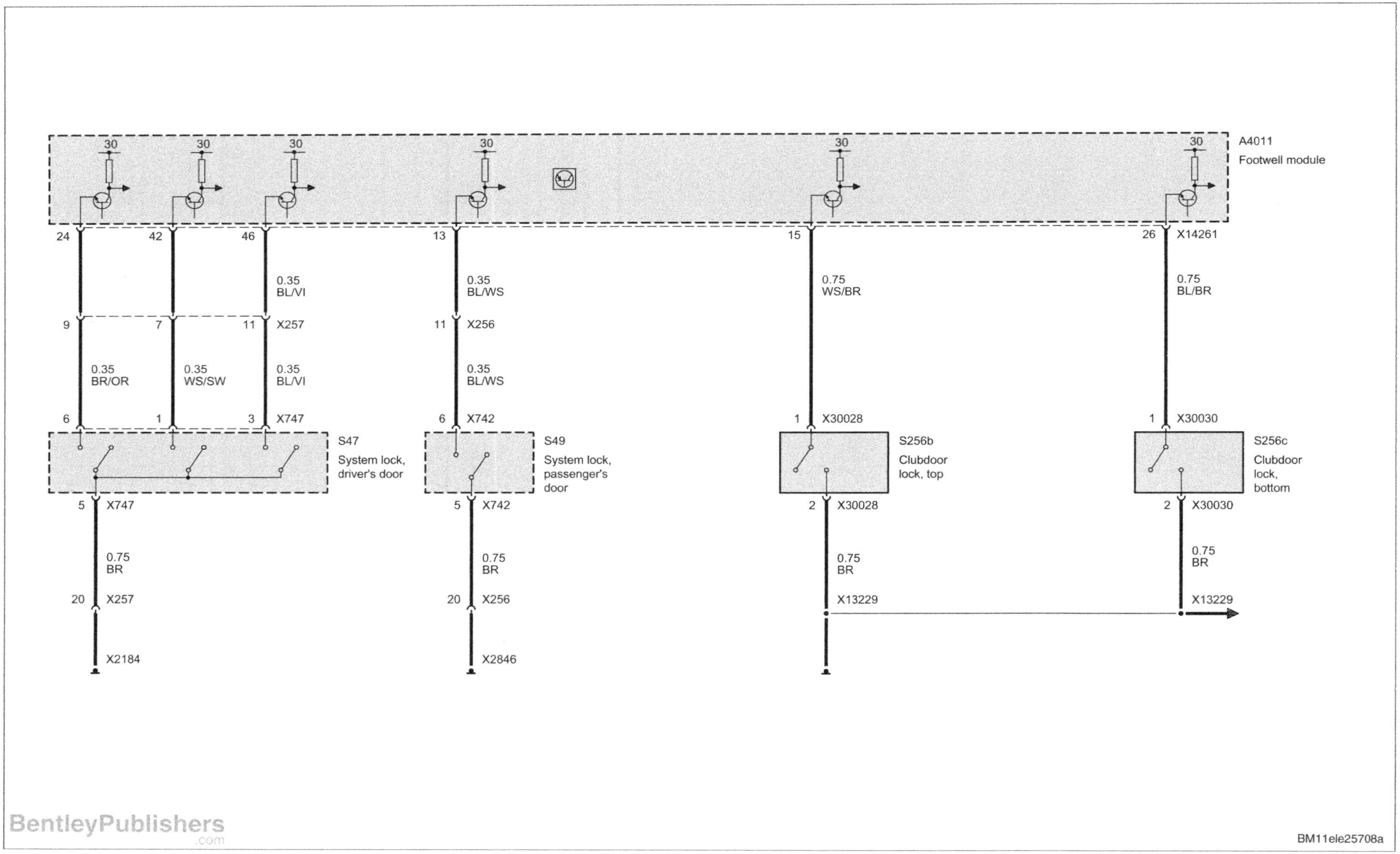

Central locking inputs (Clubman) page 2 of 2

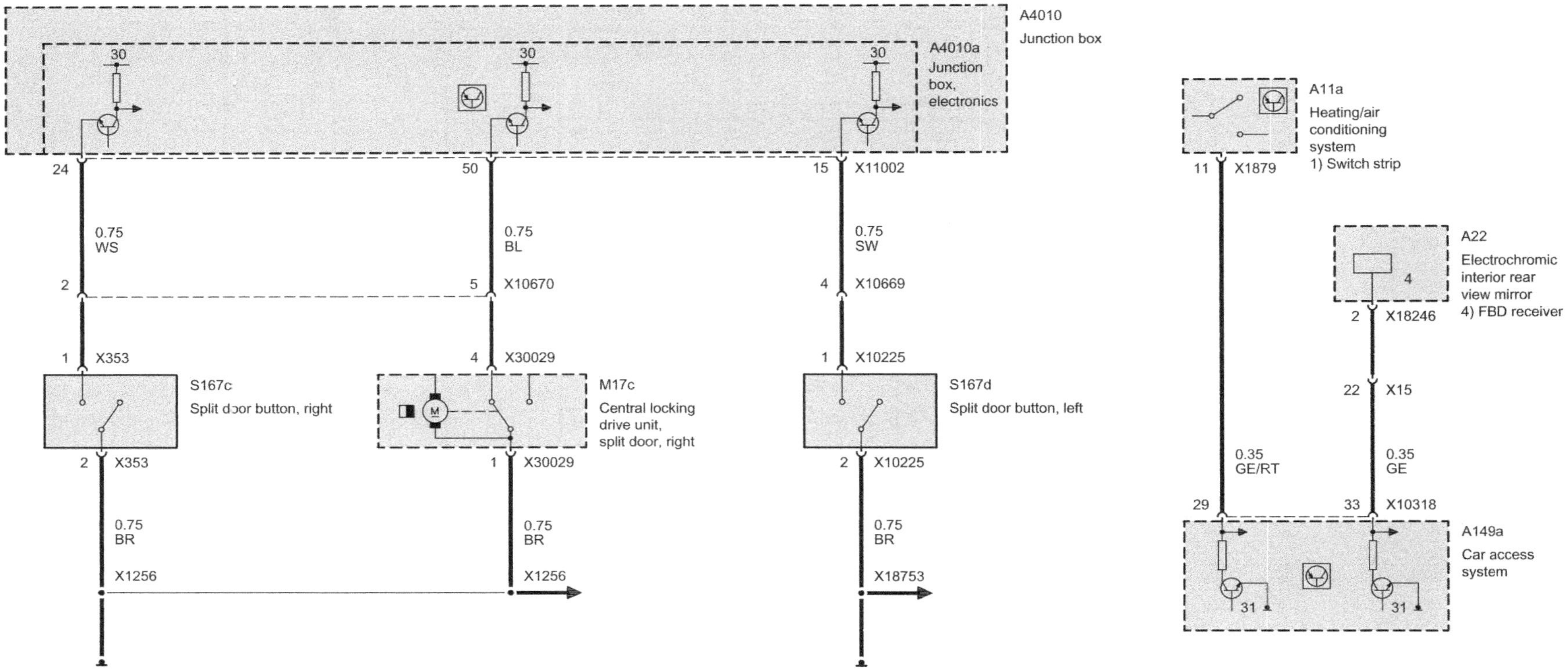

BM11ele25708b

Park distance control

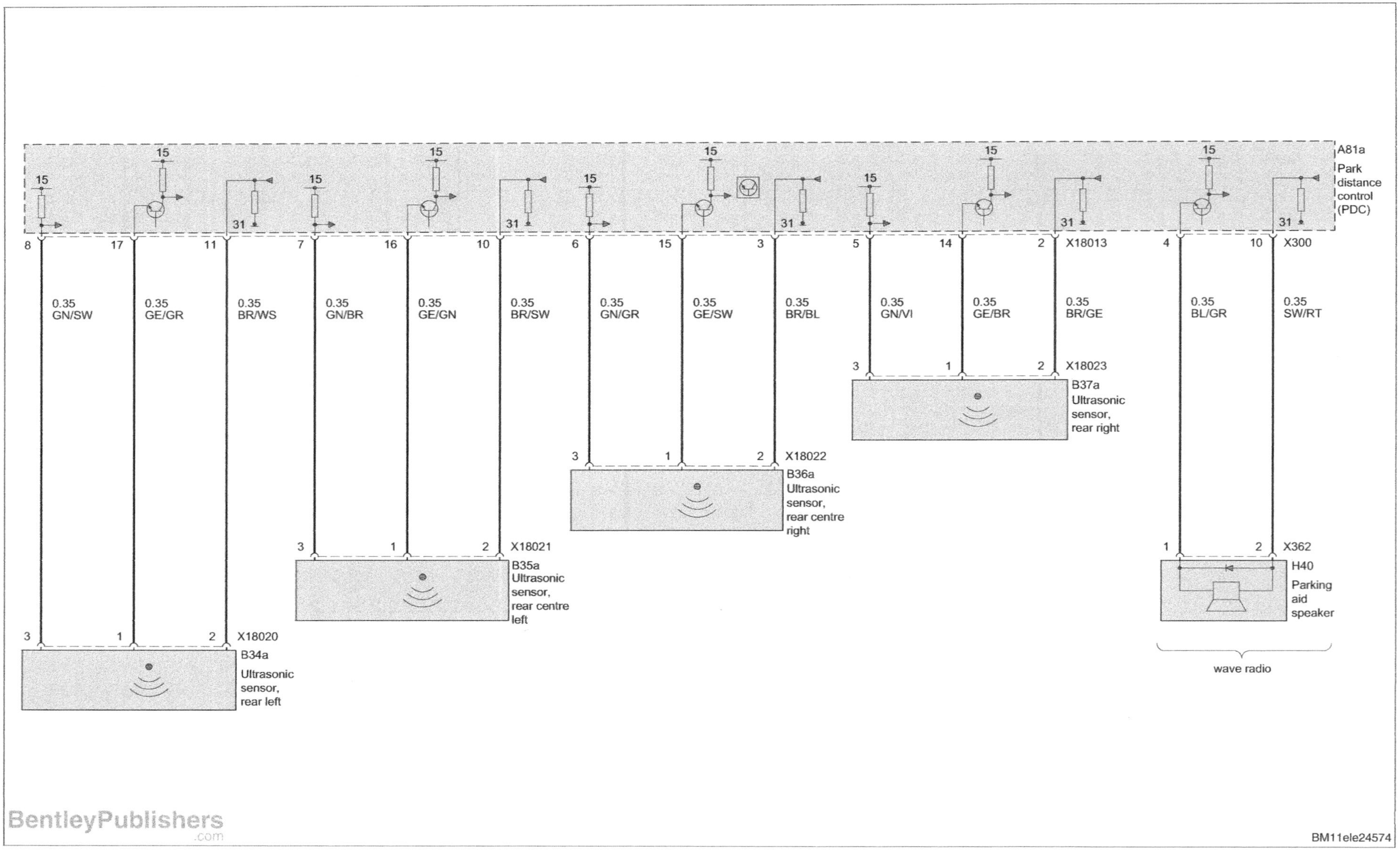

CD changer power supply

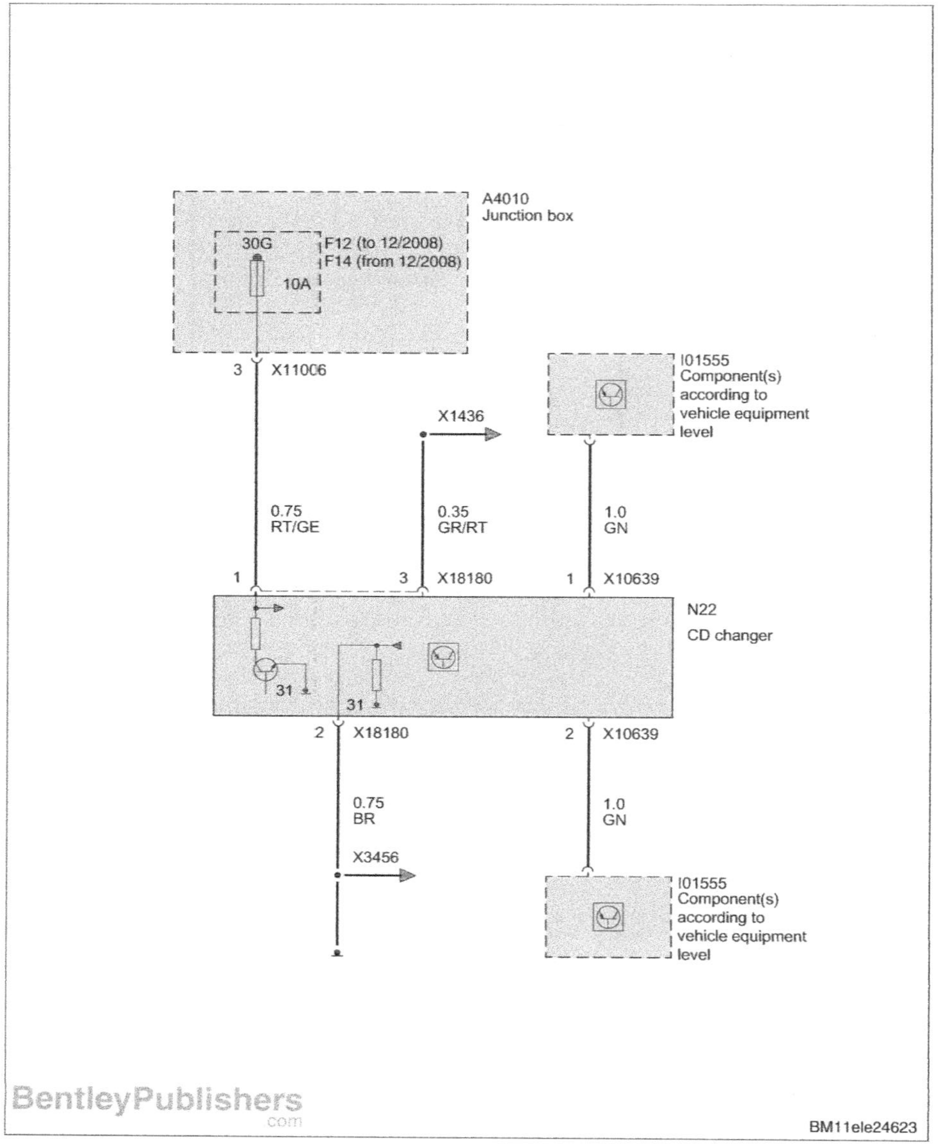

Satellite receiver, roof antenna

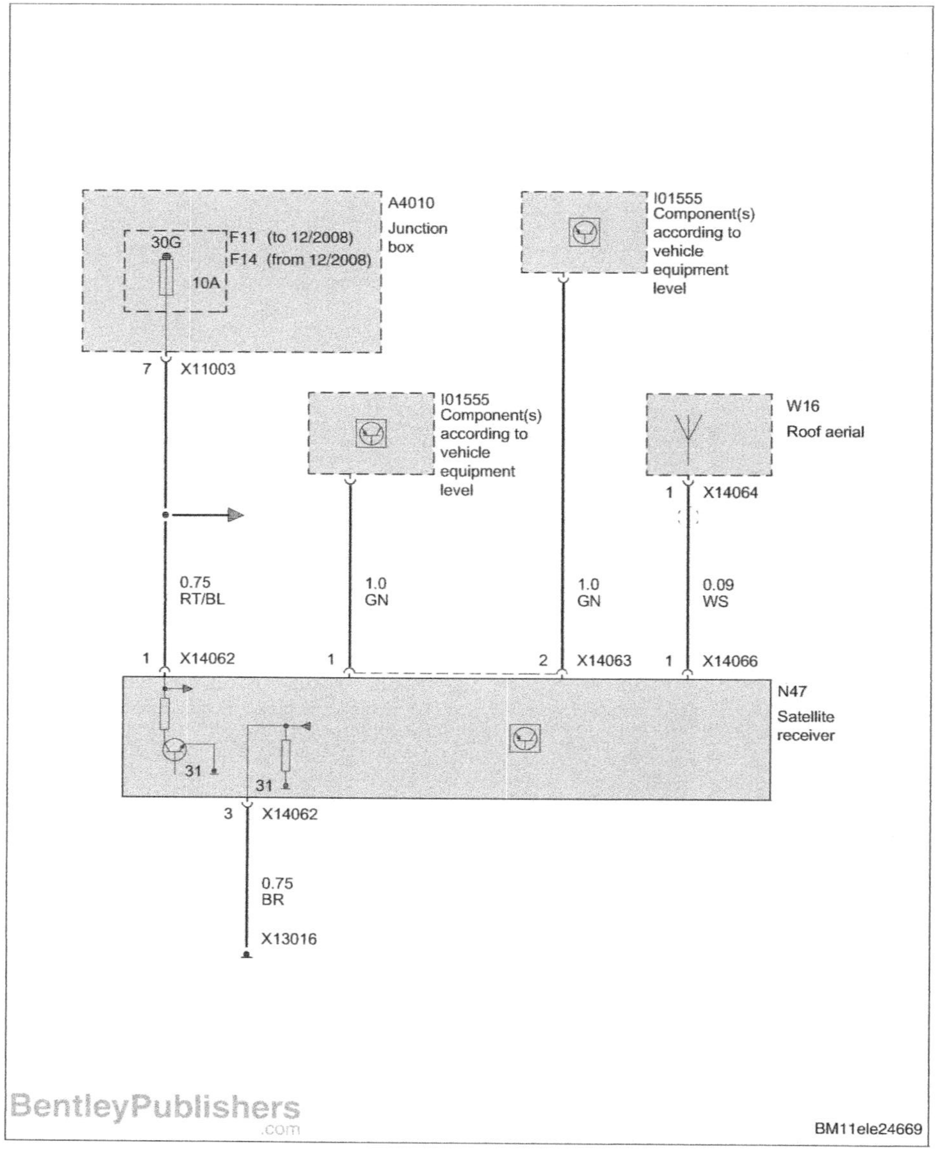

Audio output, mid-range speakers, radio, woofers

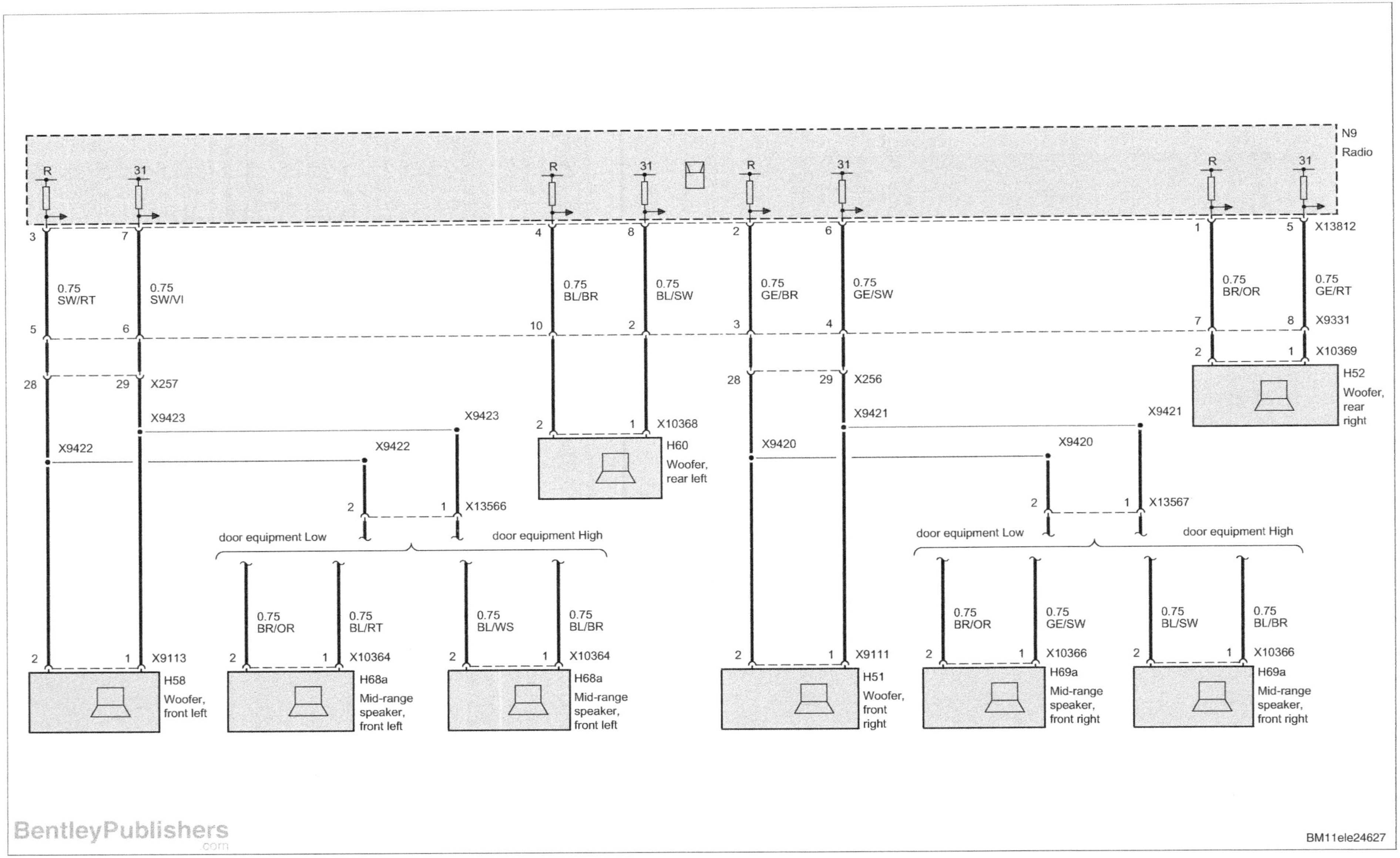

Audio output, CCC/M-ASK, mid-range speakers, radio, tweeters, woofers
page 1 of 2

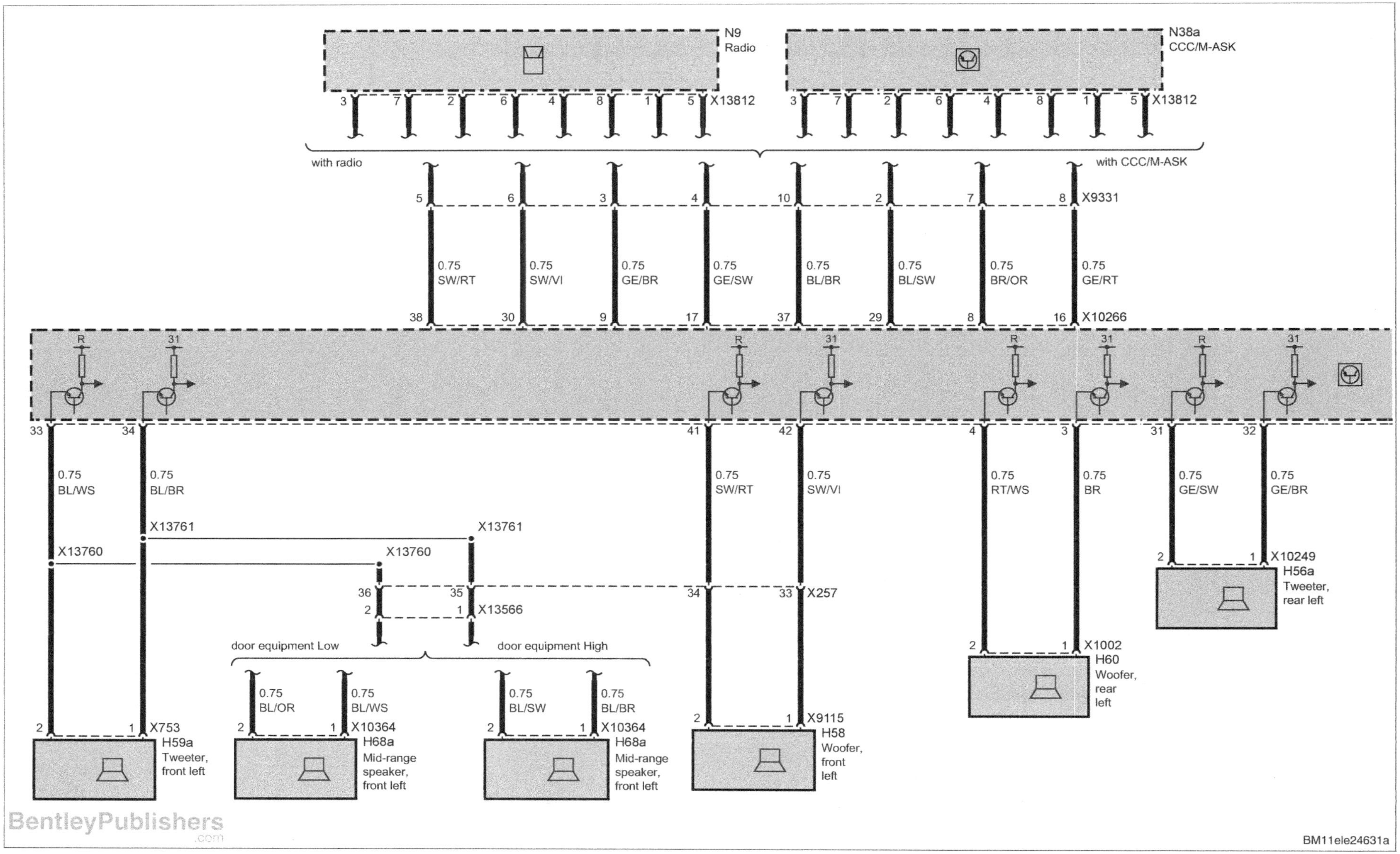

Audio output, amplifier, mid-range speakers, tweeters, woofers
page 2 of 2

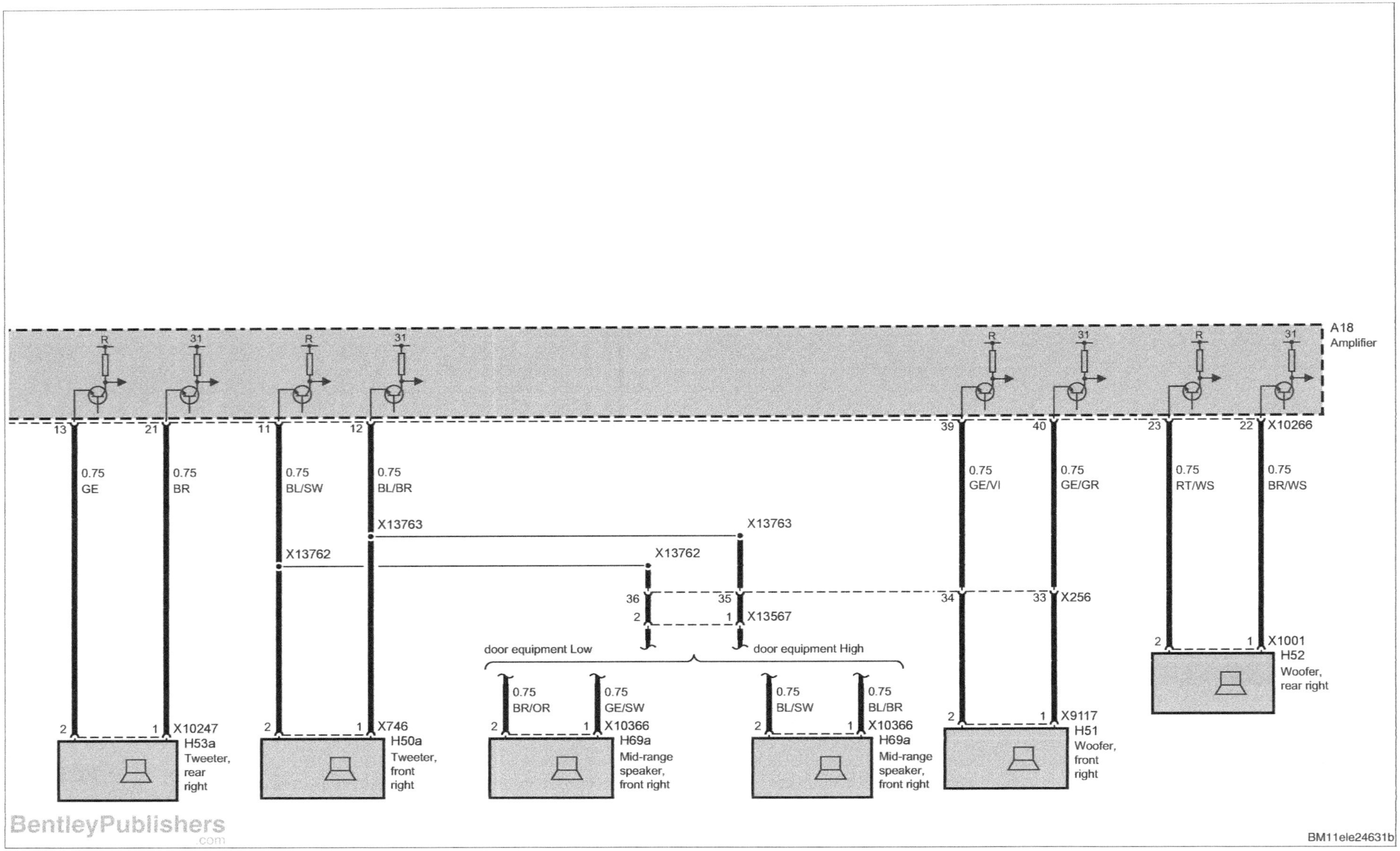

MOST-bus connector,

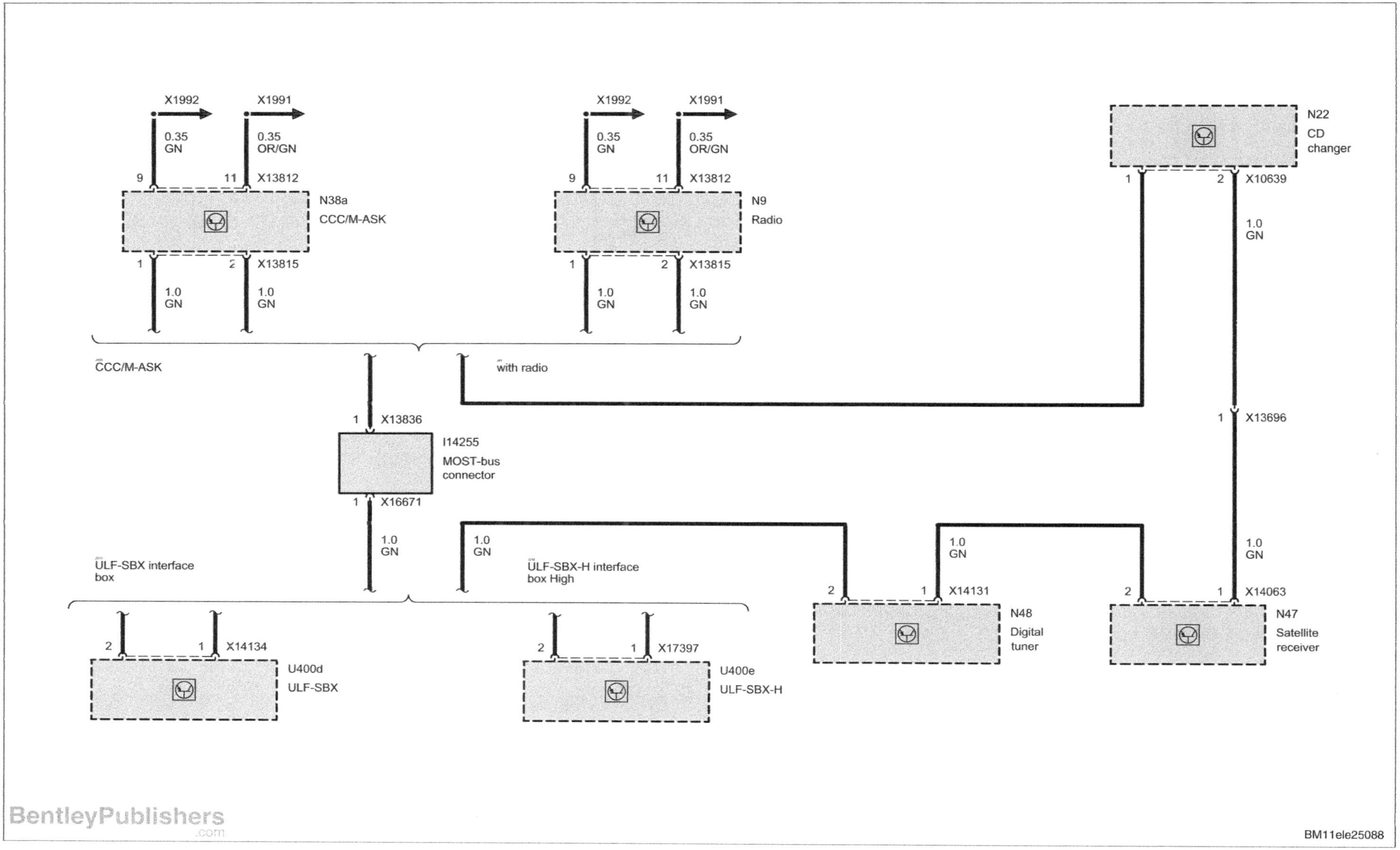

Radio reception, antenna diversity
page 1 of 2

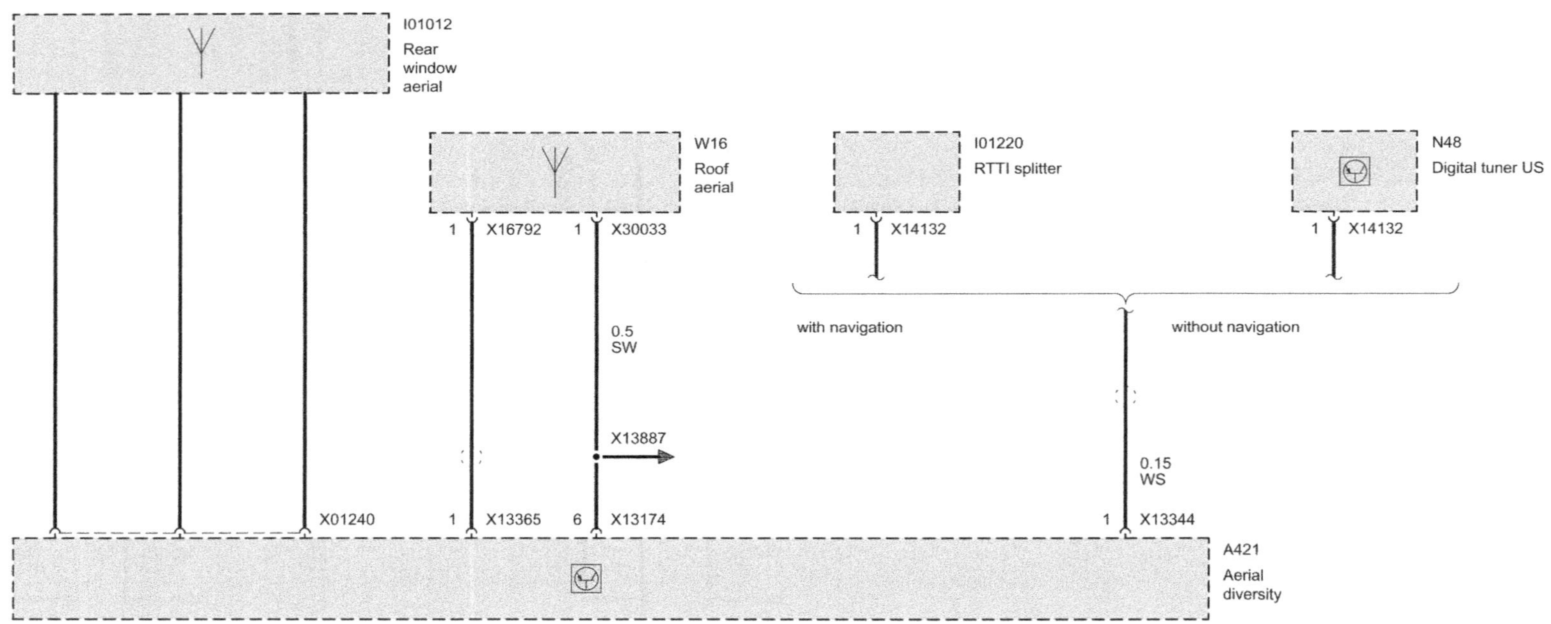

BM11ele25100

Radio reception, antenna diversity
page 2 of 2

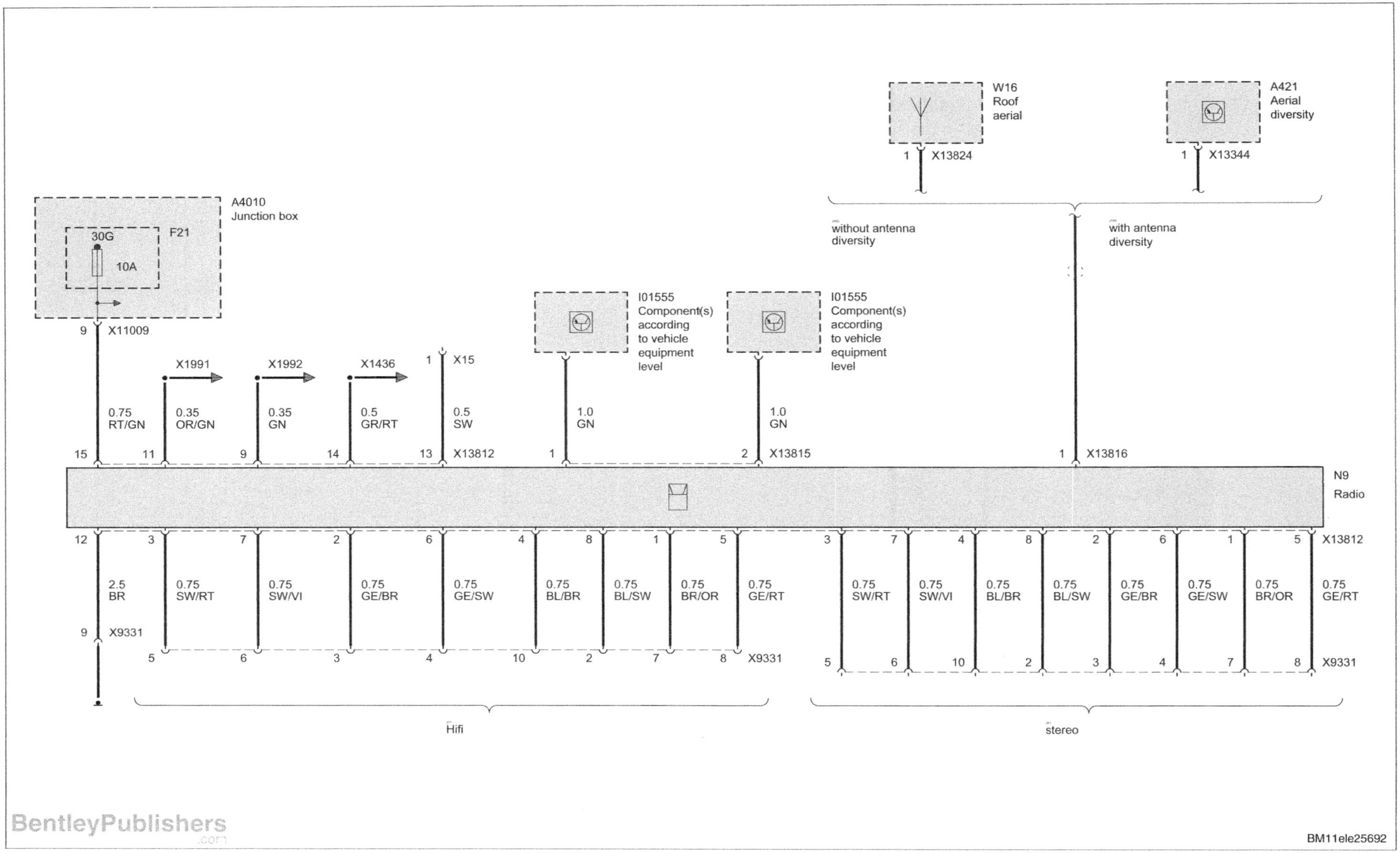

Antenna diversity, CCC/M-ASK, digital tuner, RTTI splitter from 12/2008

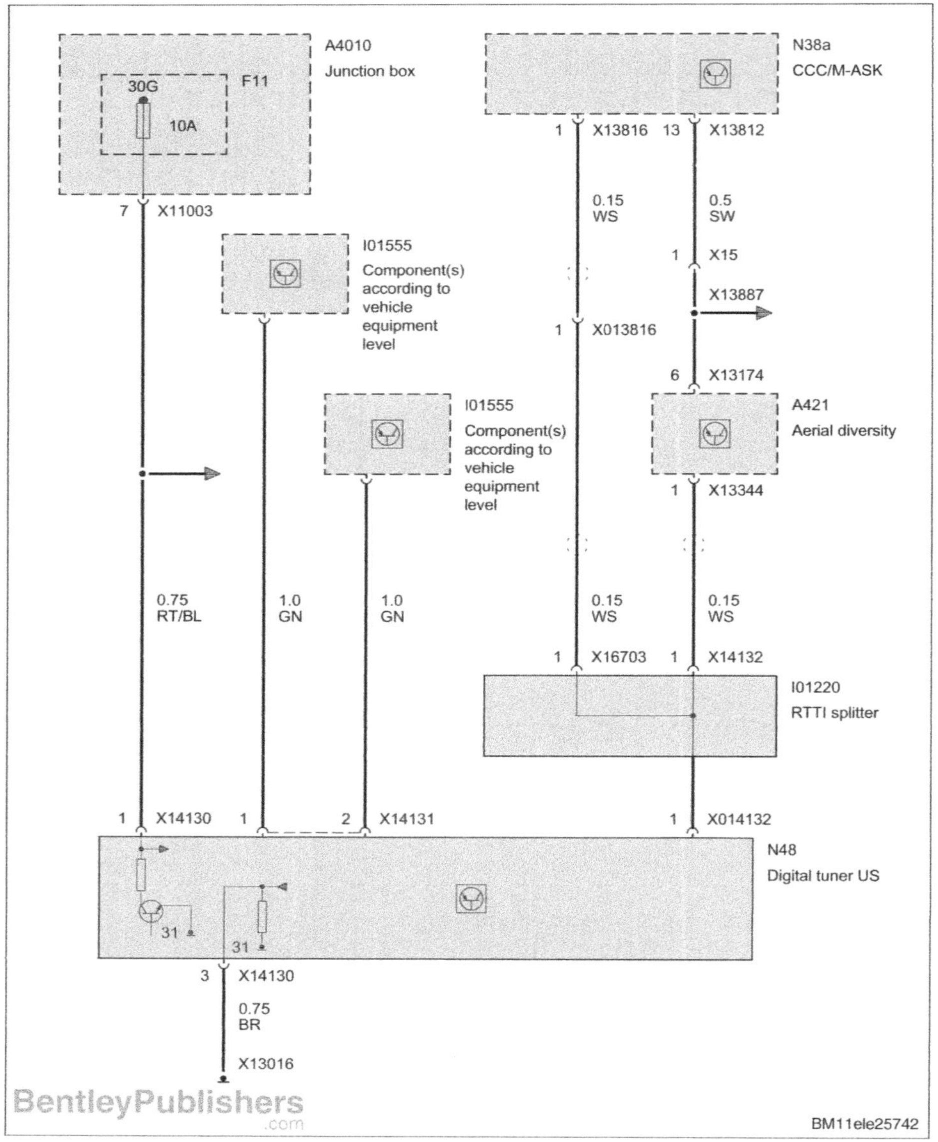

Car Information Computer (CIC), AUX-IN connection, base plate

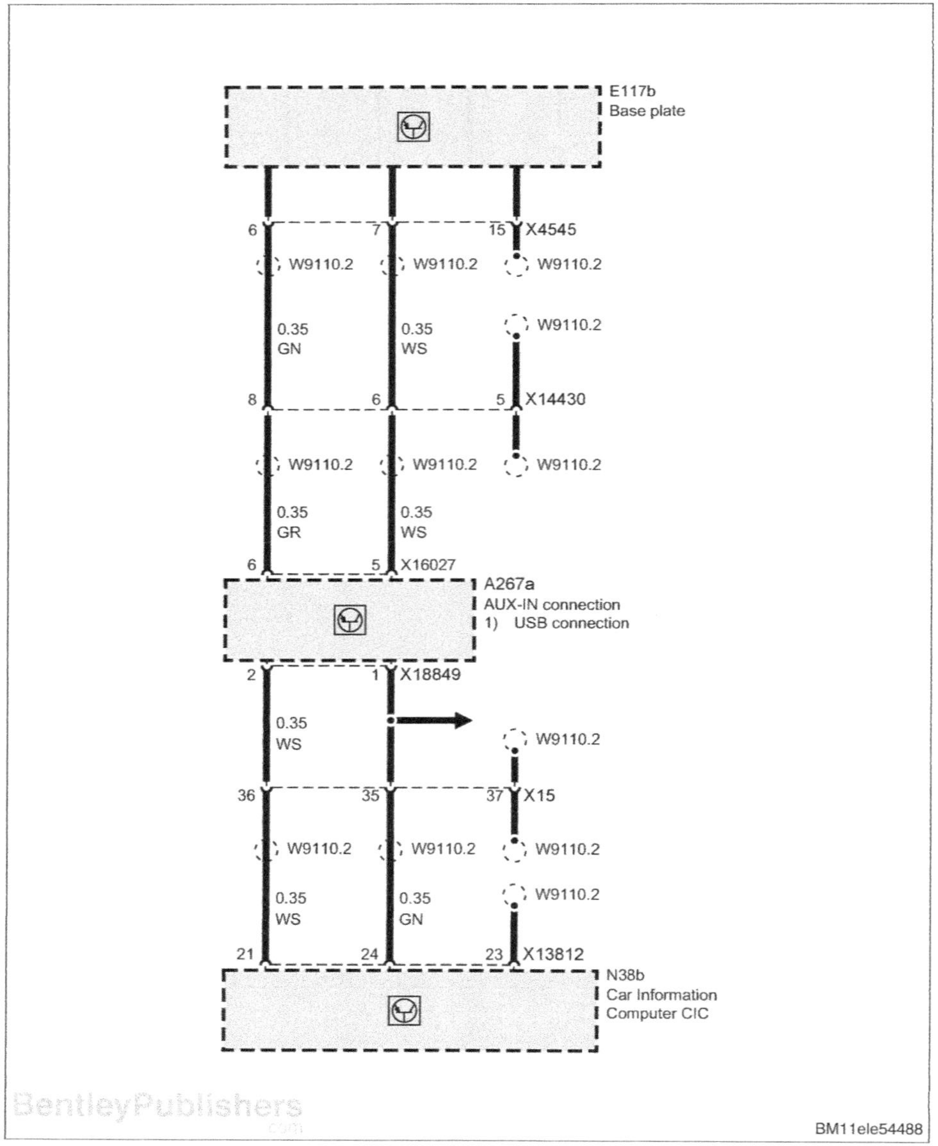

CCC/CHAMP, amplifier, antenna tuner with remote control receiver, roof antenna
page 1 of 2

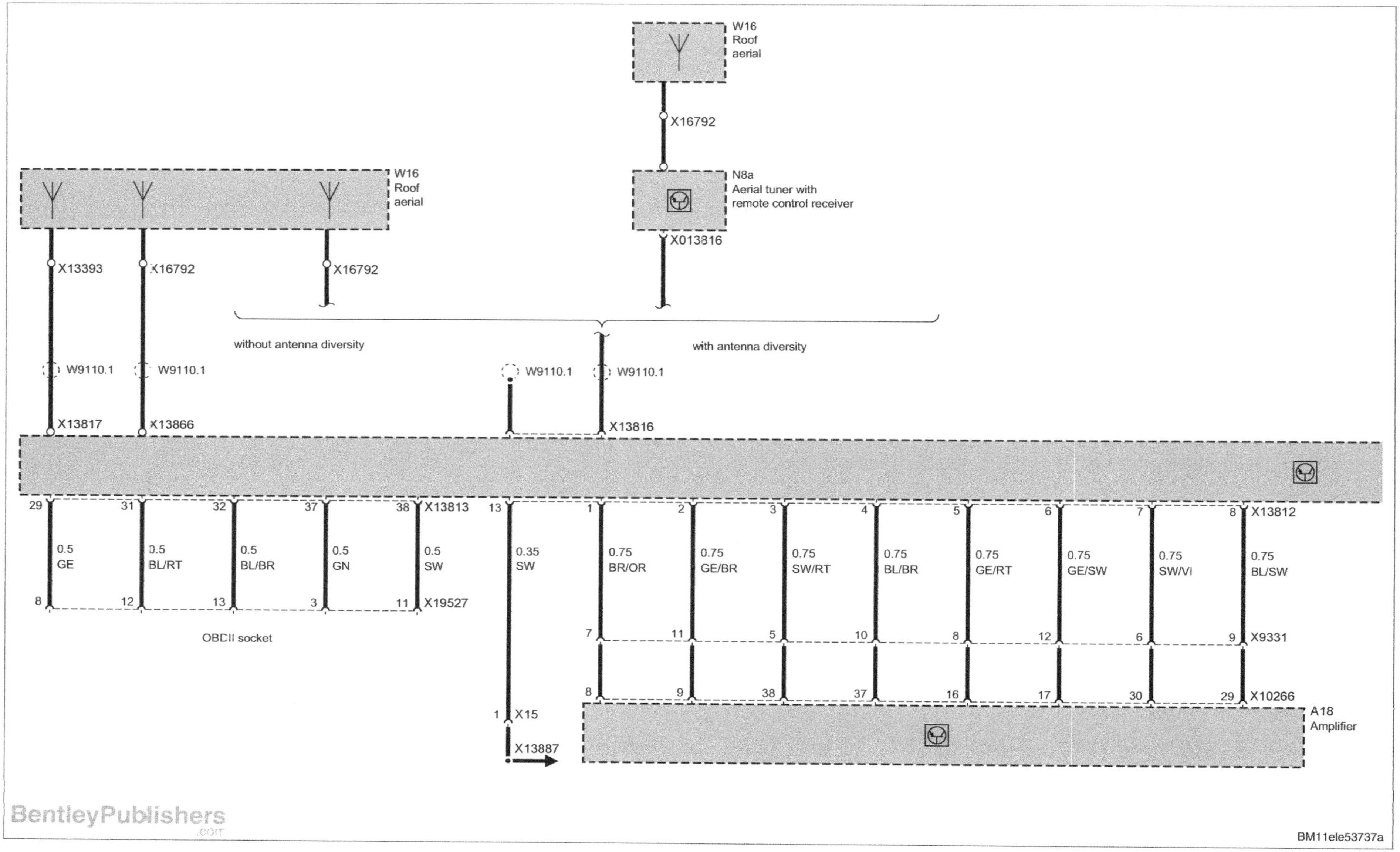

CCC/CHAMP, AUX-IN connection

page 2 of 2

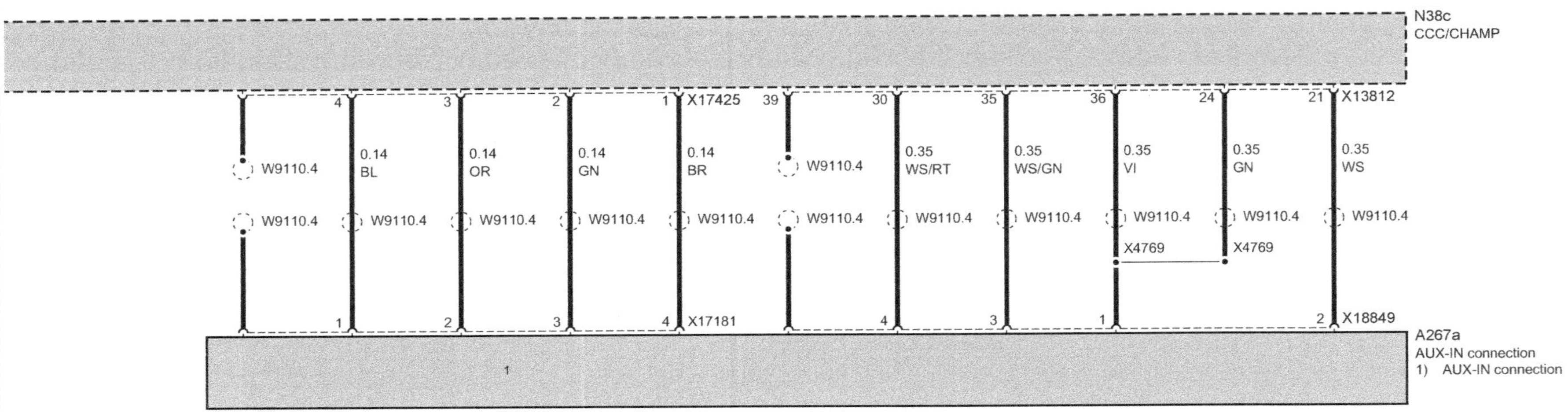

Multiple restraint system (MRS), control unit power supply

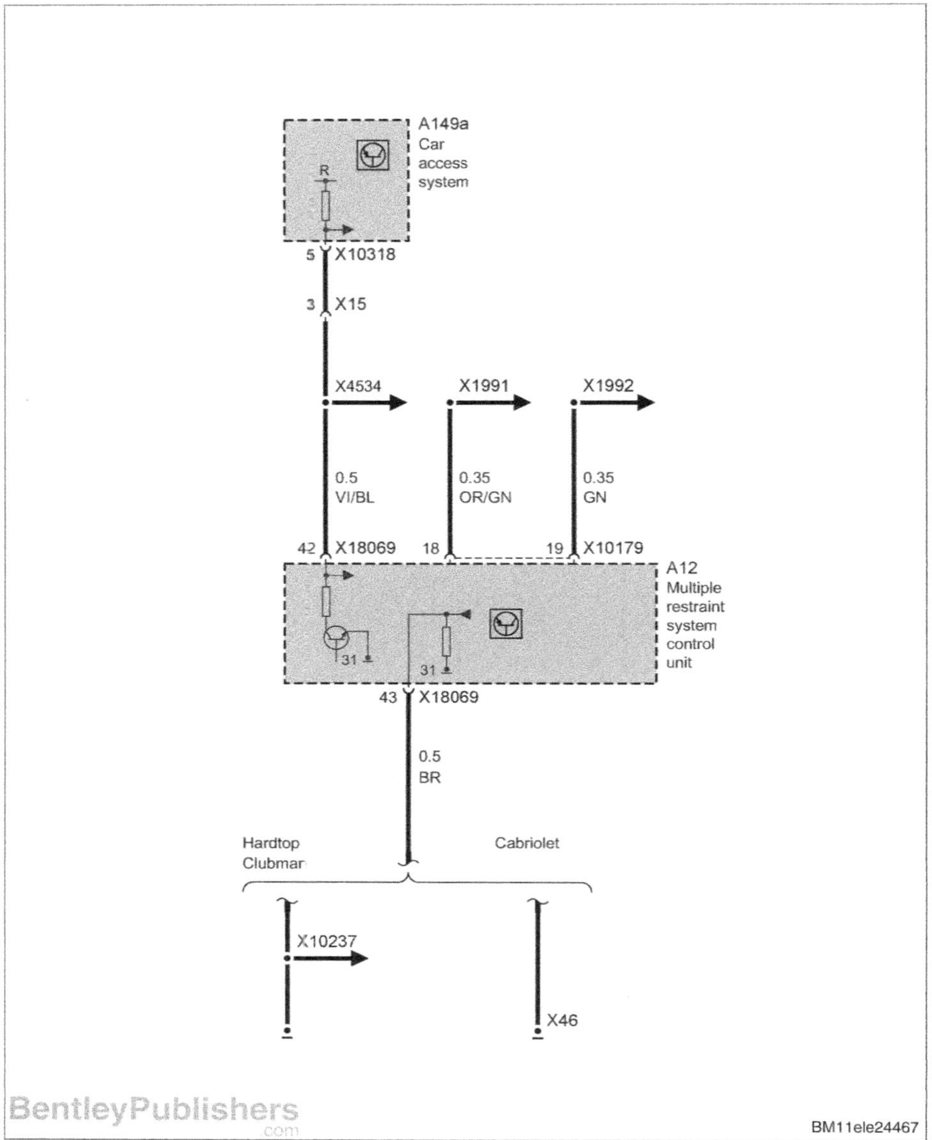

Multiple restraint system (MRS), passenger seat occupancy recognition

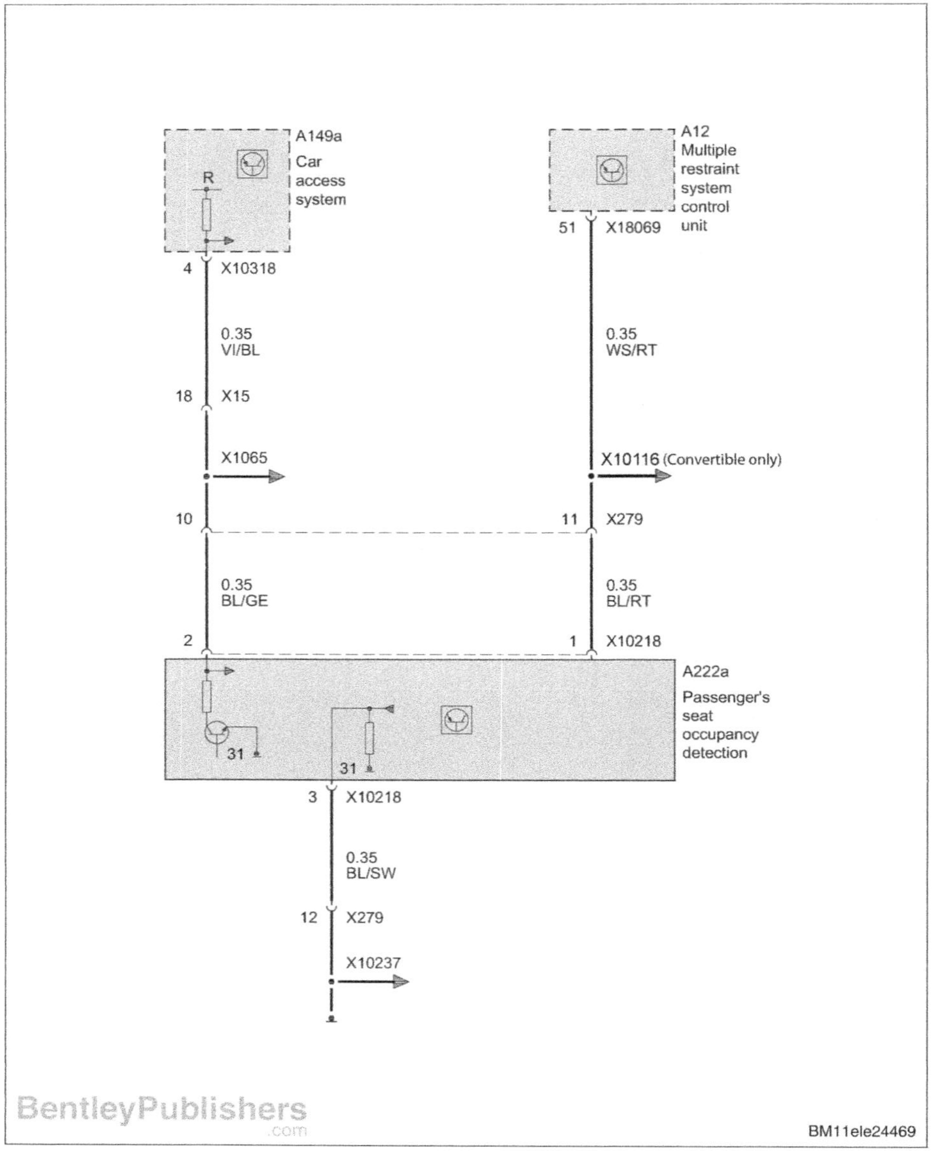

Multiple restraint system (MRS), airbag sensors, seat belt buckles, (4-wire B-pillar sensors) page 1 of 2

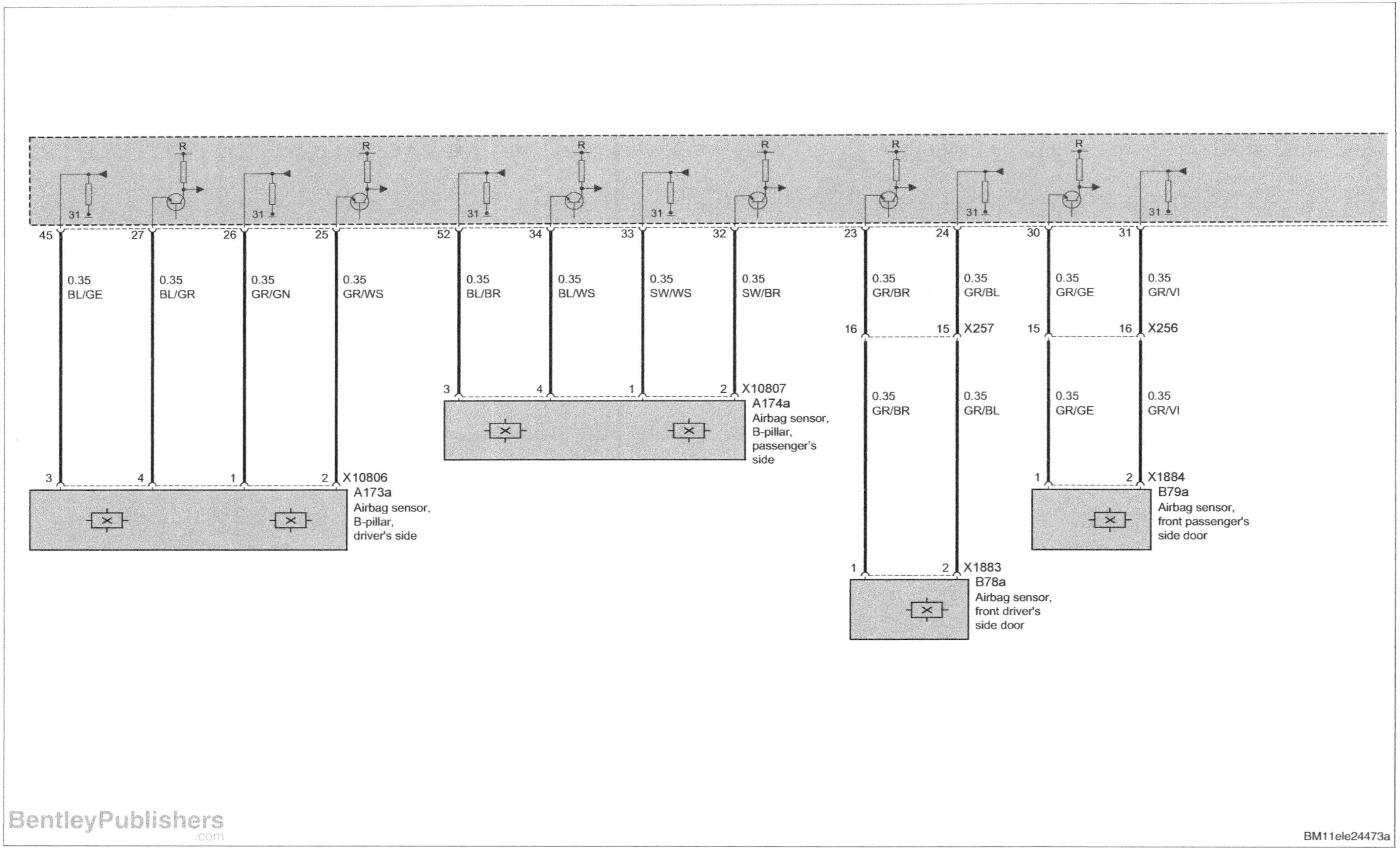

Multiple restraint system (MRS), airbag sensors, seat belt buckles, (4-wire B-pillar sensors) page 2 of 2

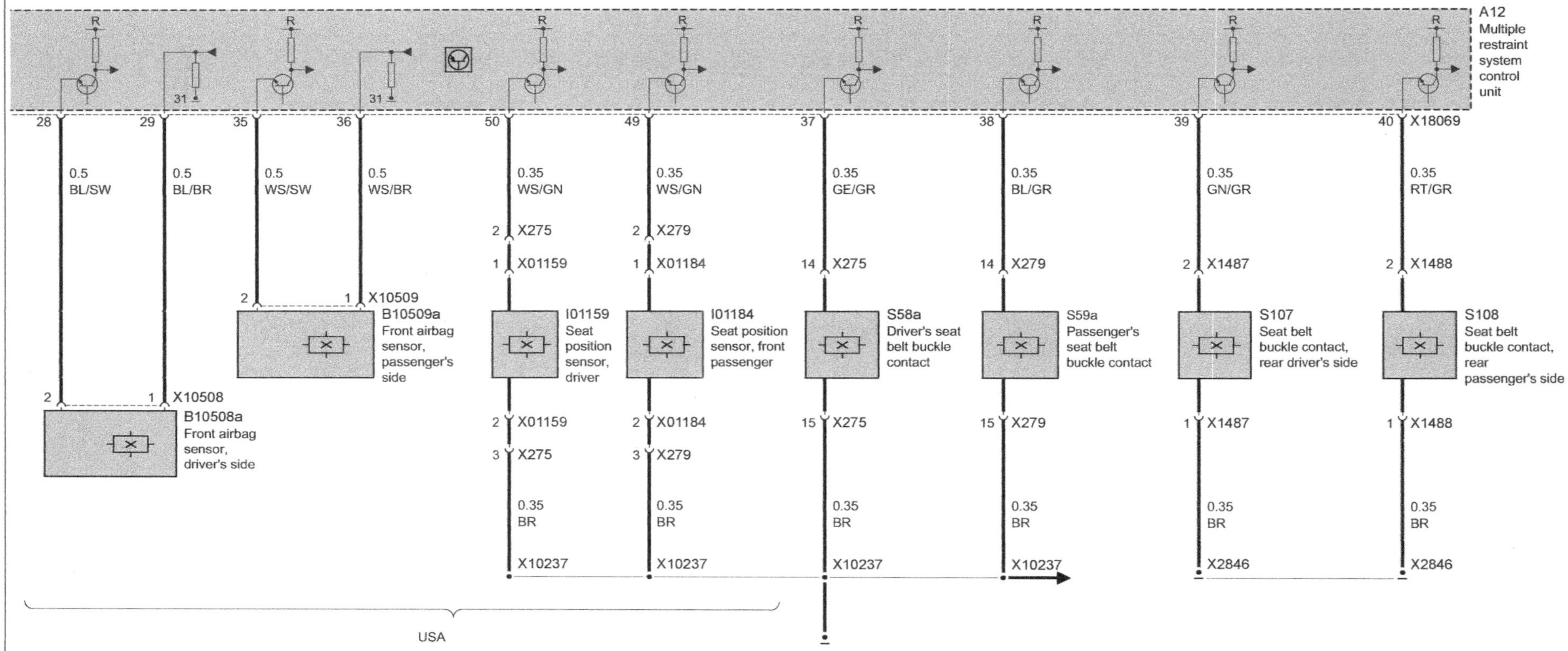

BM11ele24473b

Multiple restraint system, airbag sensors Convertible, (2-wire B-pillar sensors) page 1 of 2

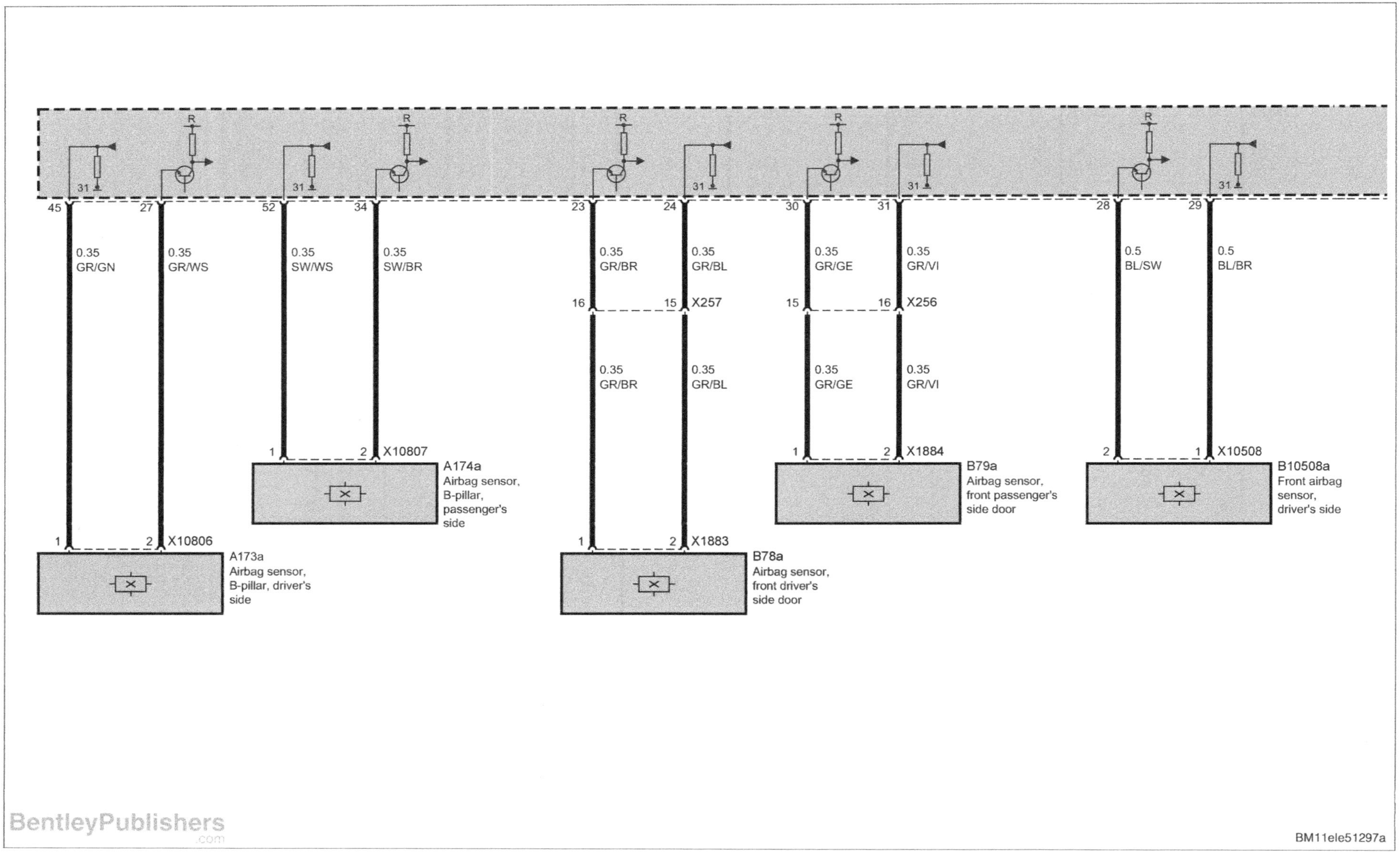

Multiple restraint system, airbag sensors Convertible, (2-wire B-pillar sensors) page 2 of 2

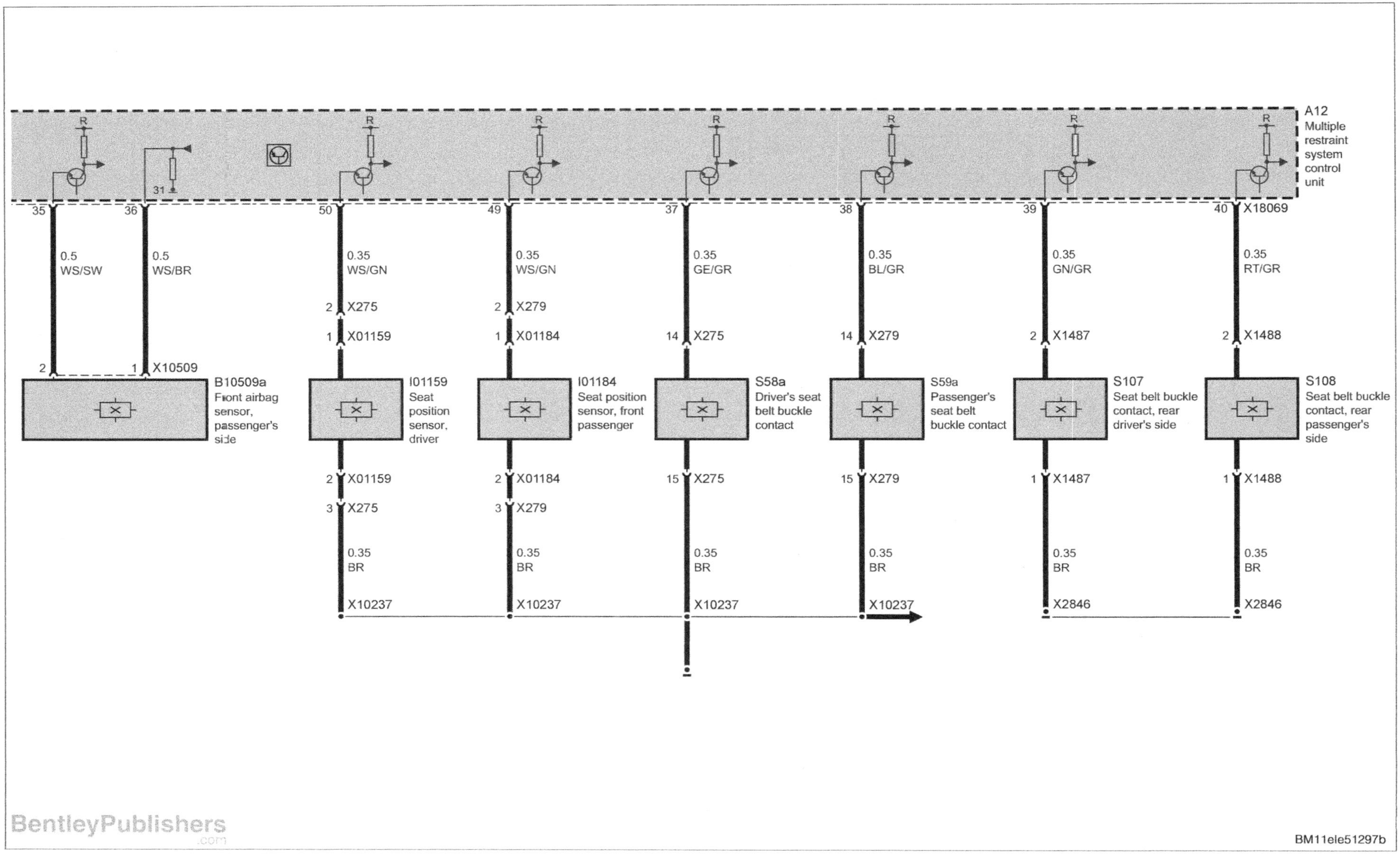

Multiple restraint system (MRS), airbag triggering circuits, airbag inflator assemblies, airbag sensors, seat belt buckles page 1 of 2

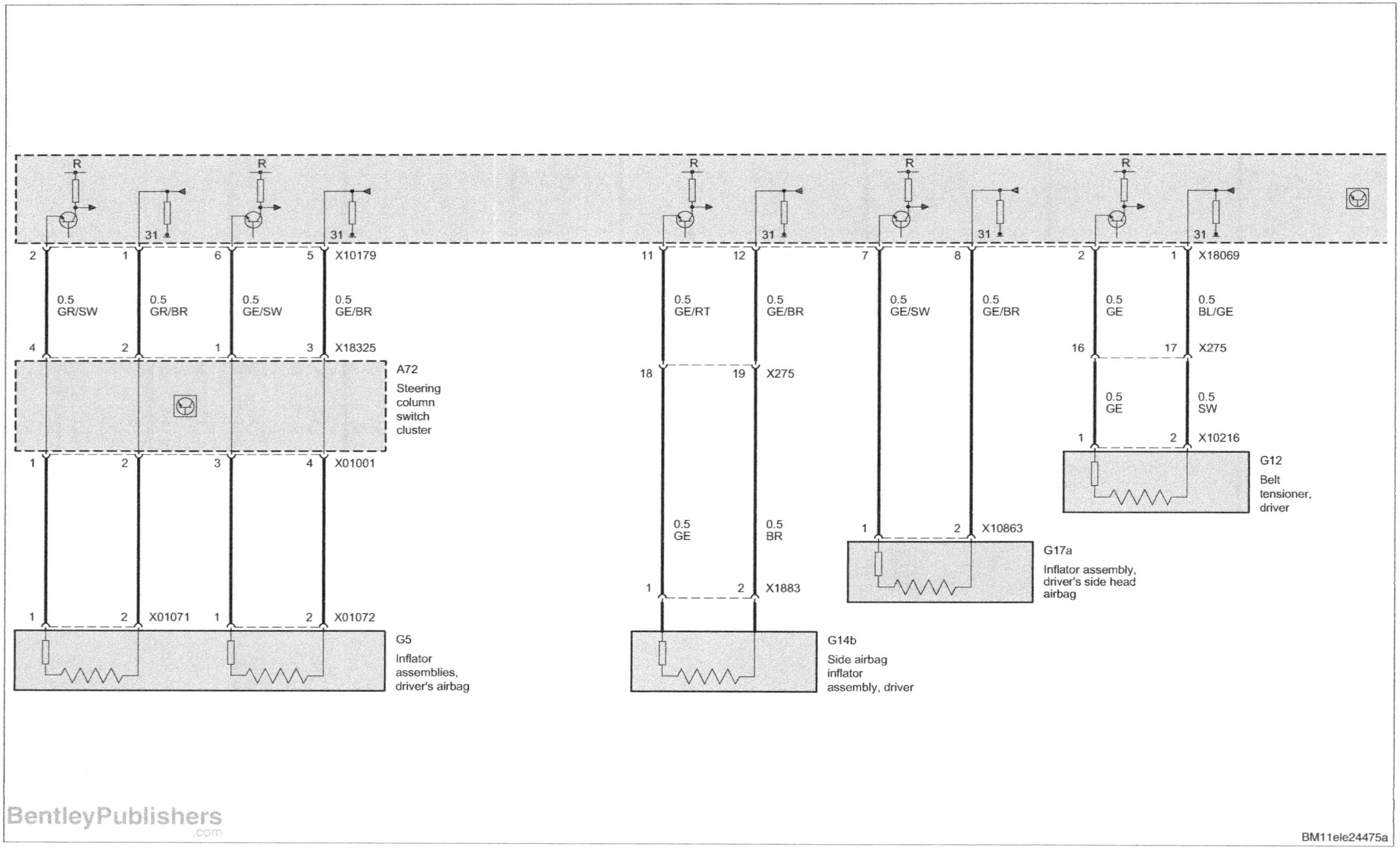

Multiple restraint system (MRS), airbag triggering circuits, airbag inflator assemblies, airbag sensors, seat belt buckles page 2 of 2

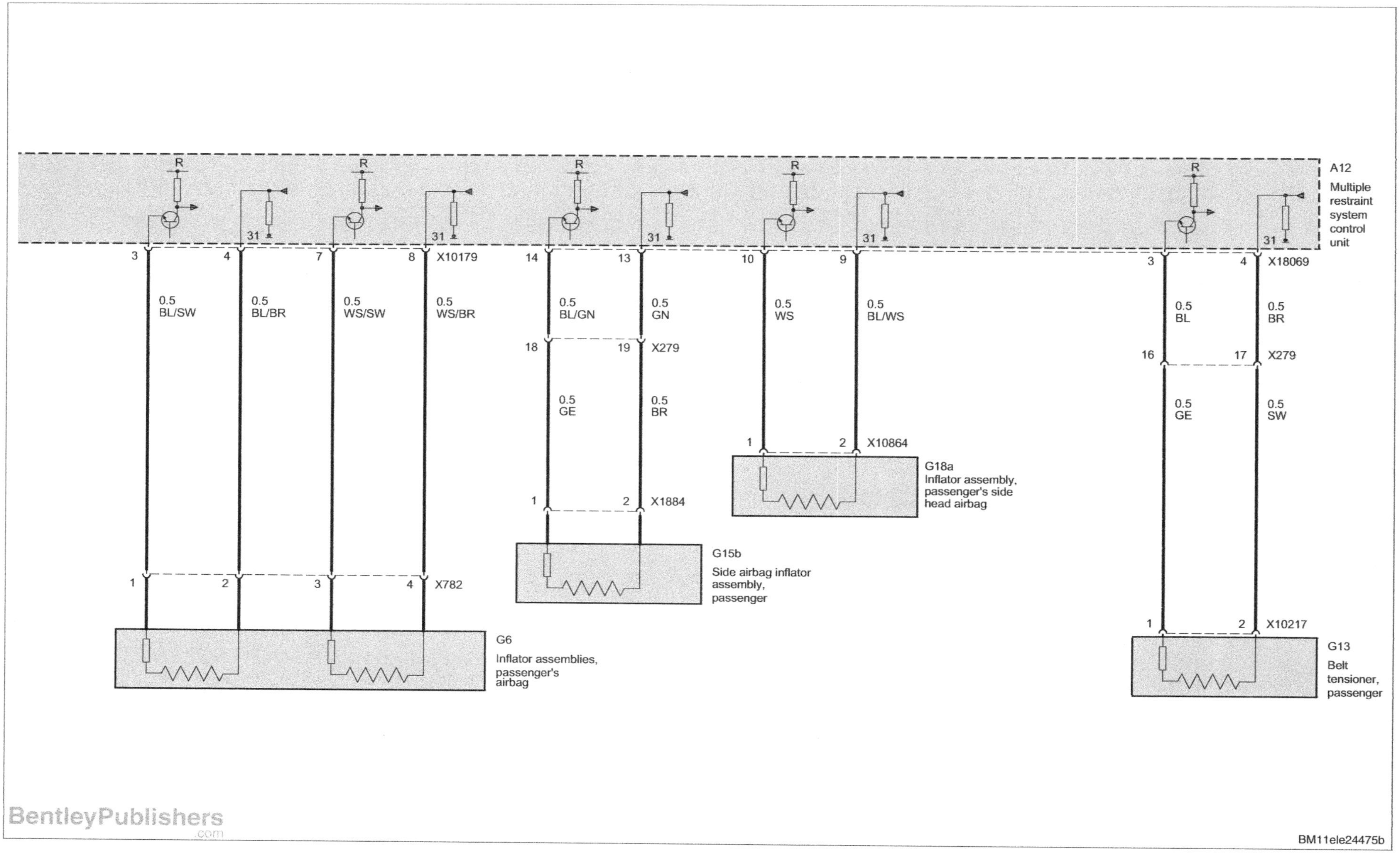

Electromechanical power steering, center console switch cluster, steering torque sensor

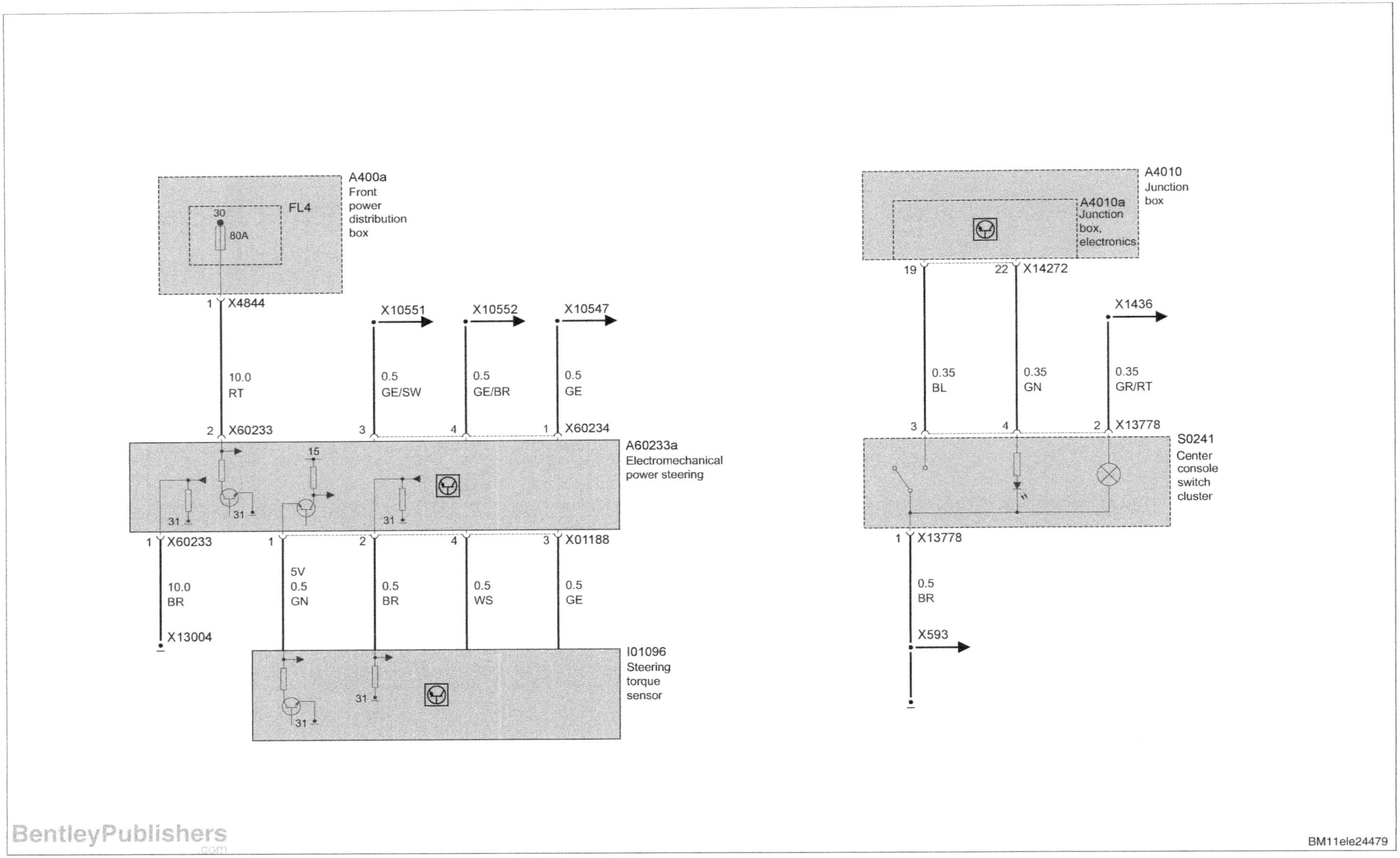

Rear window defogger, rear window defogger lockout circuit, windscreen heater, windscreen heater relay

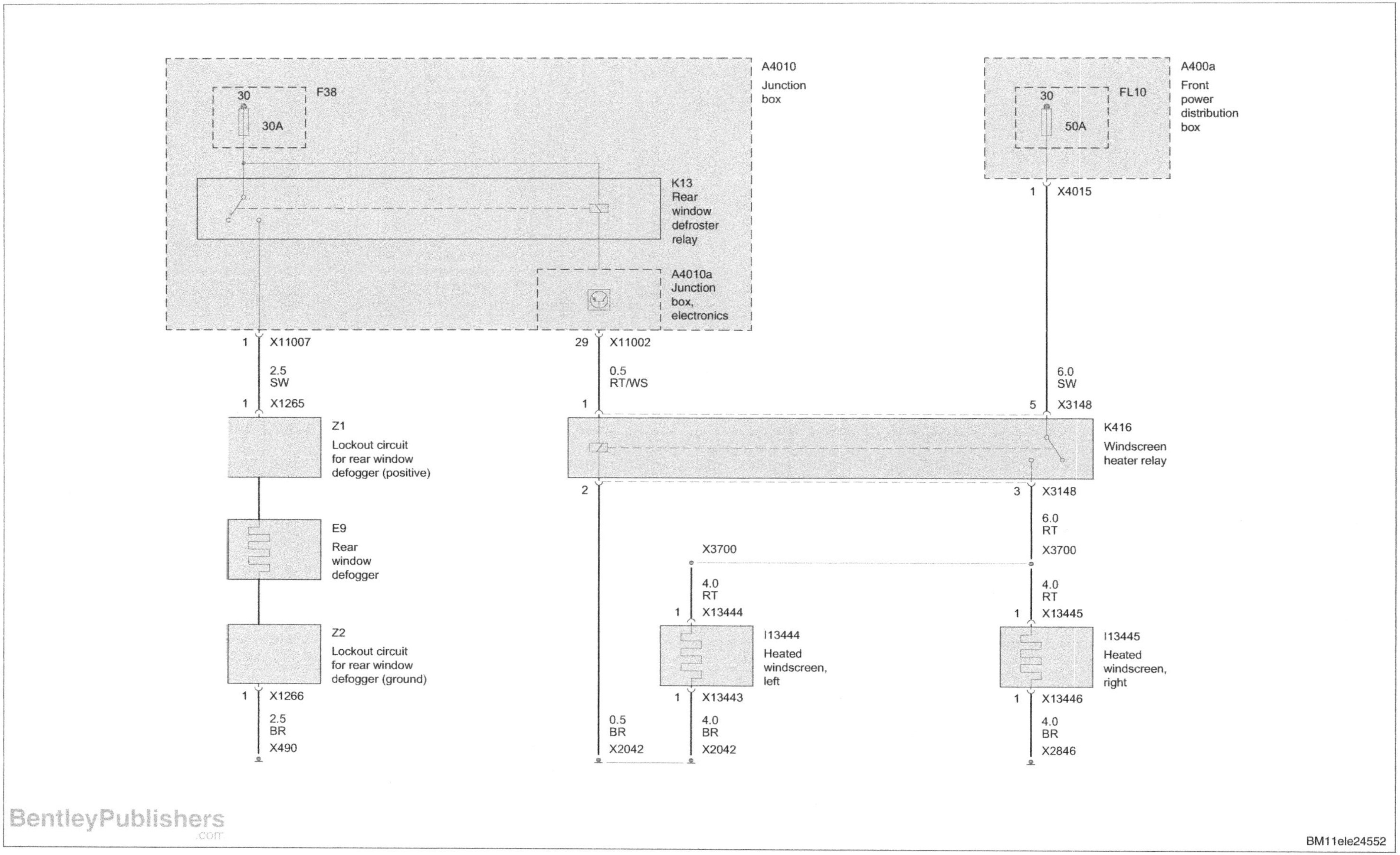

Multifunction steering wheel

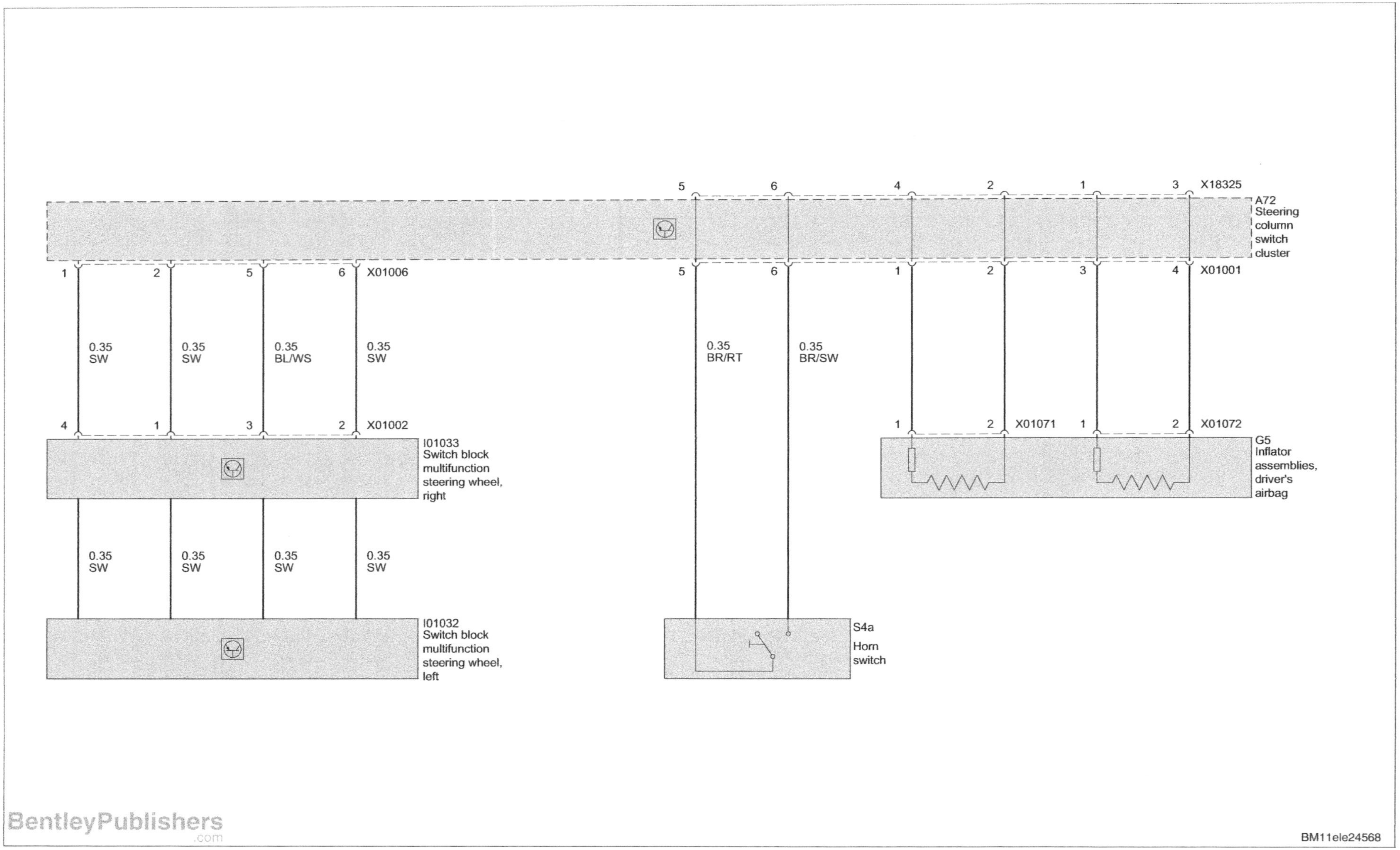

Antitheft alarm system (except Clubman)

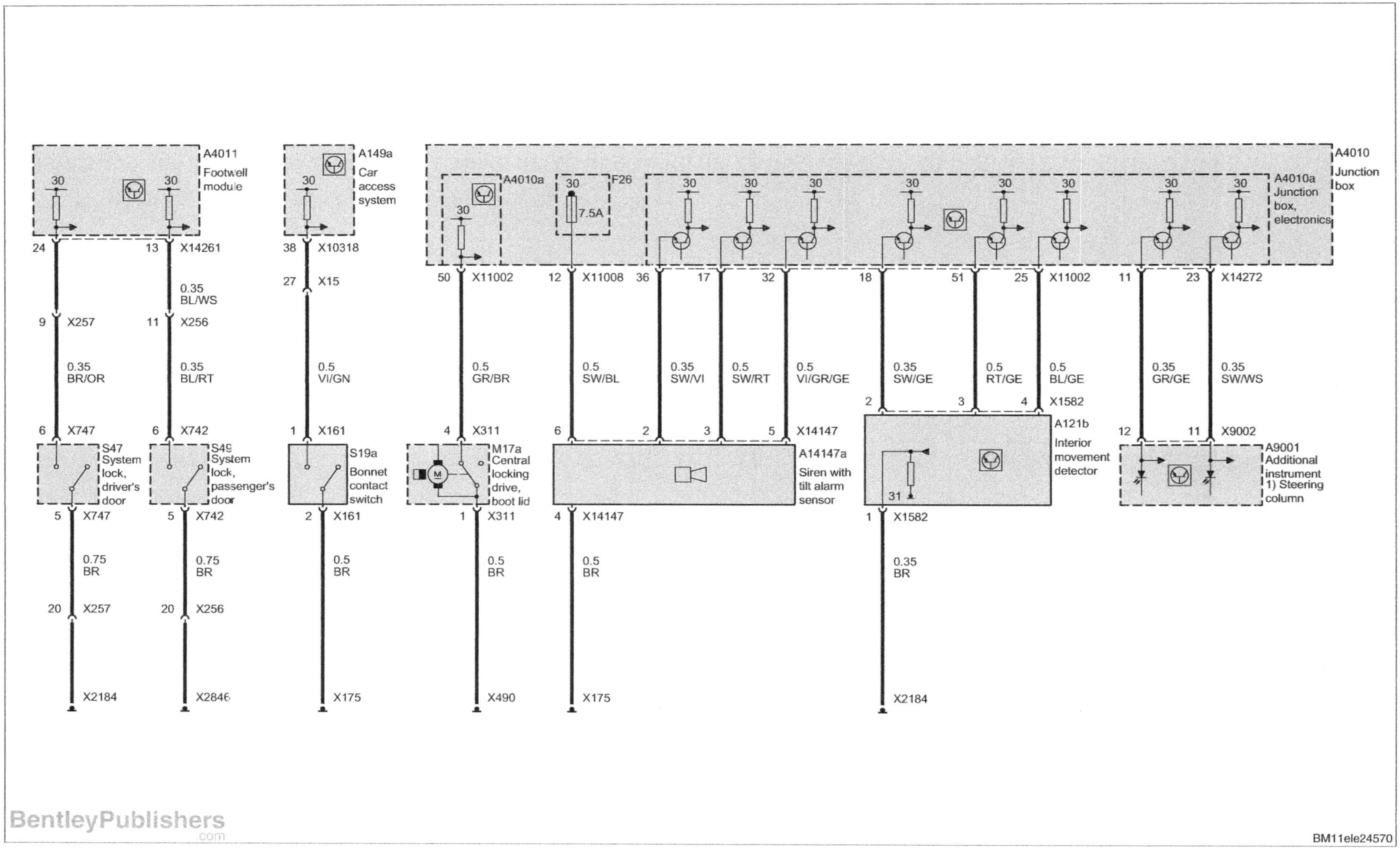

Park distance control power supply N12 engine, N14 engine, Clubman S

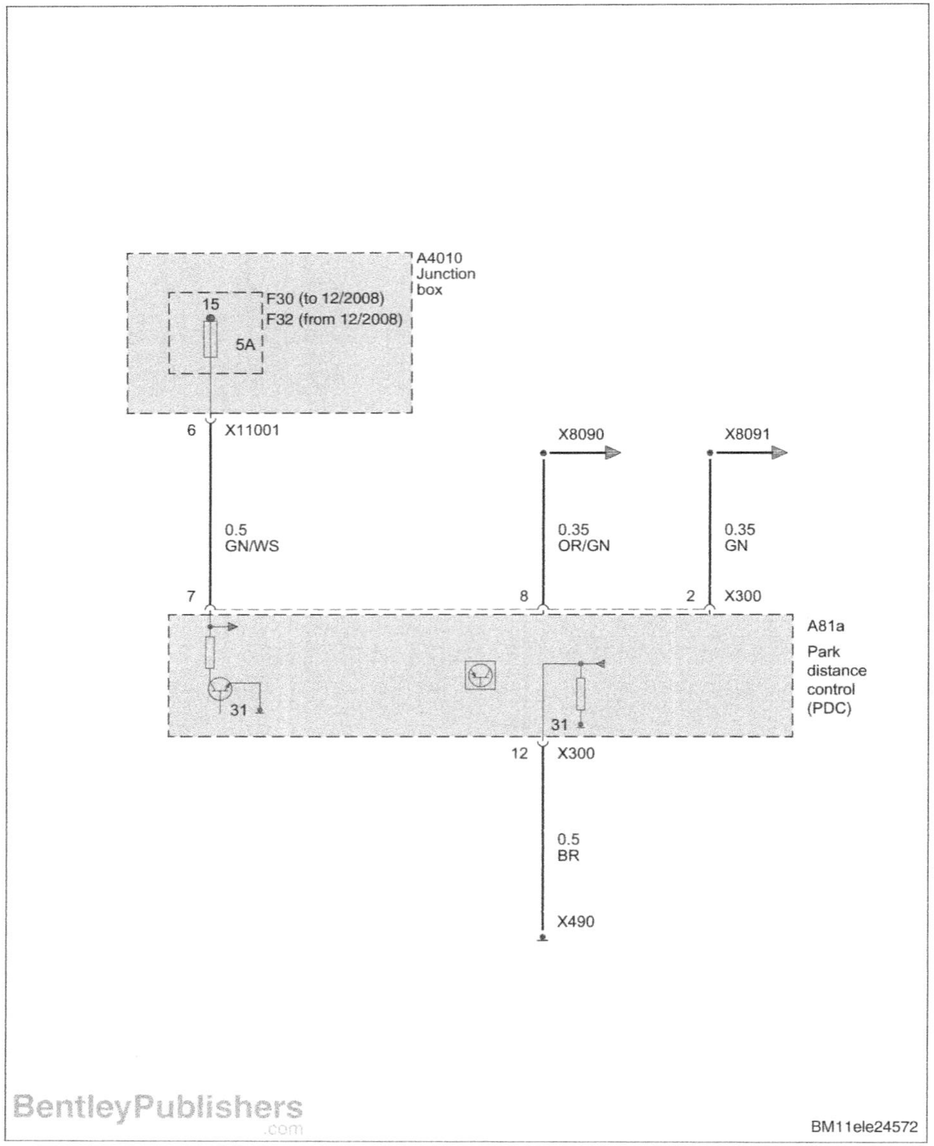

Instrument cluster control unit, outside temperature sensor

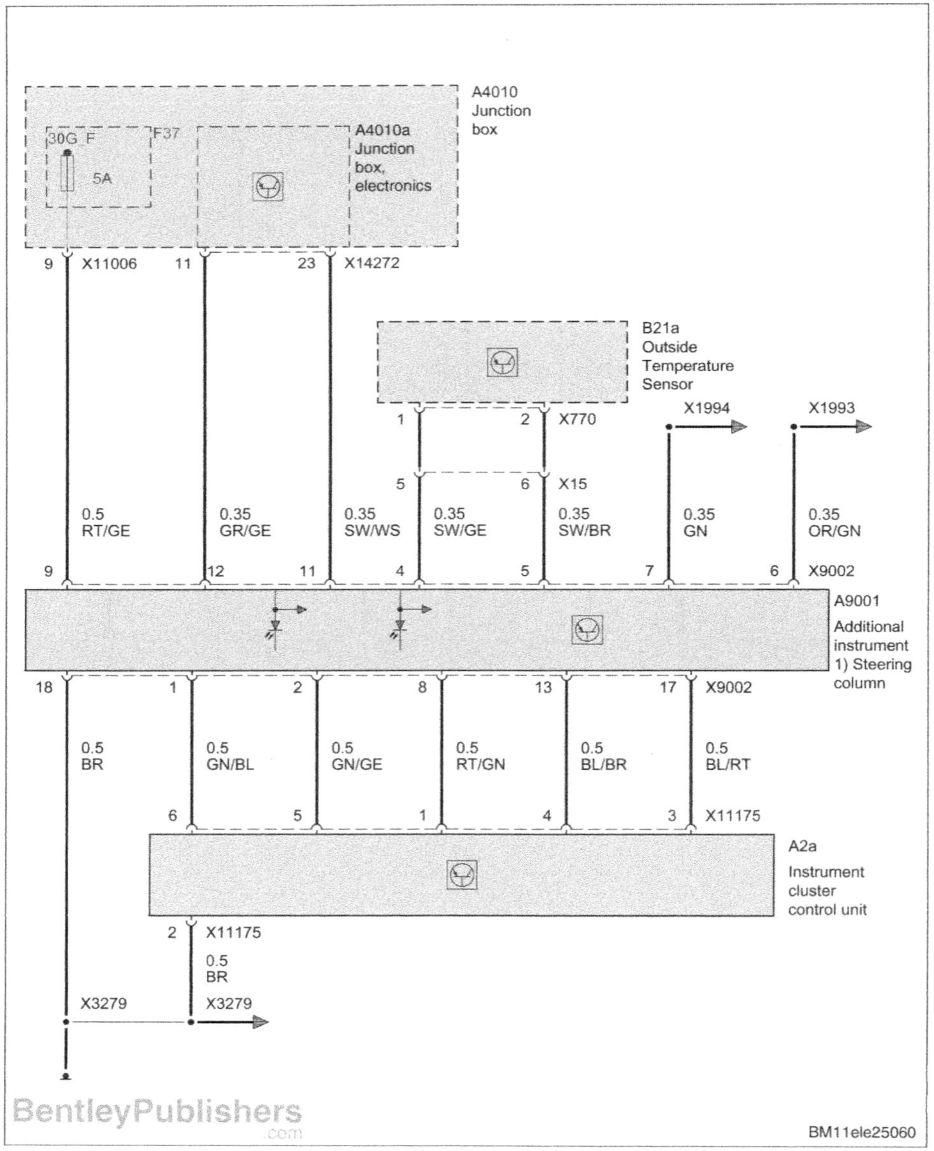

Tire pressure control, electronic wheel modules
2007 - 2009

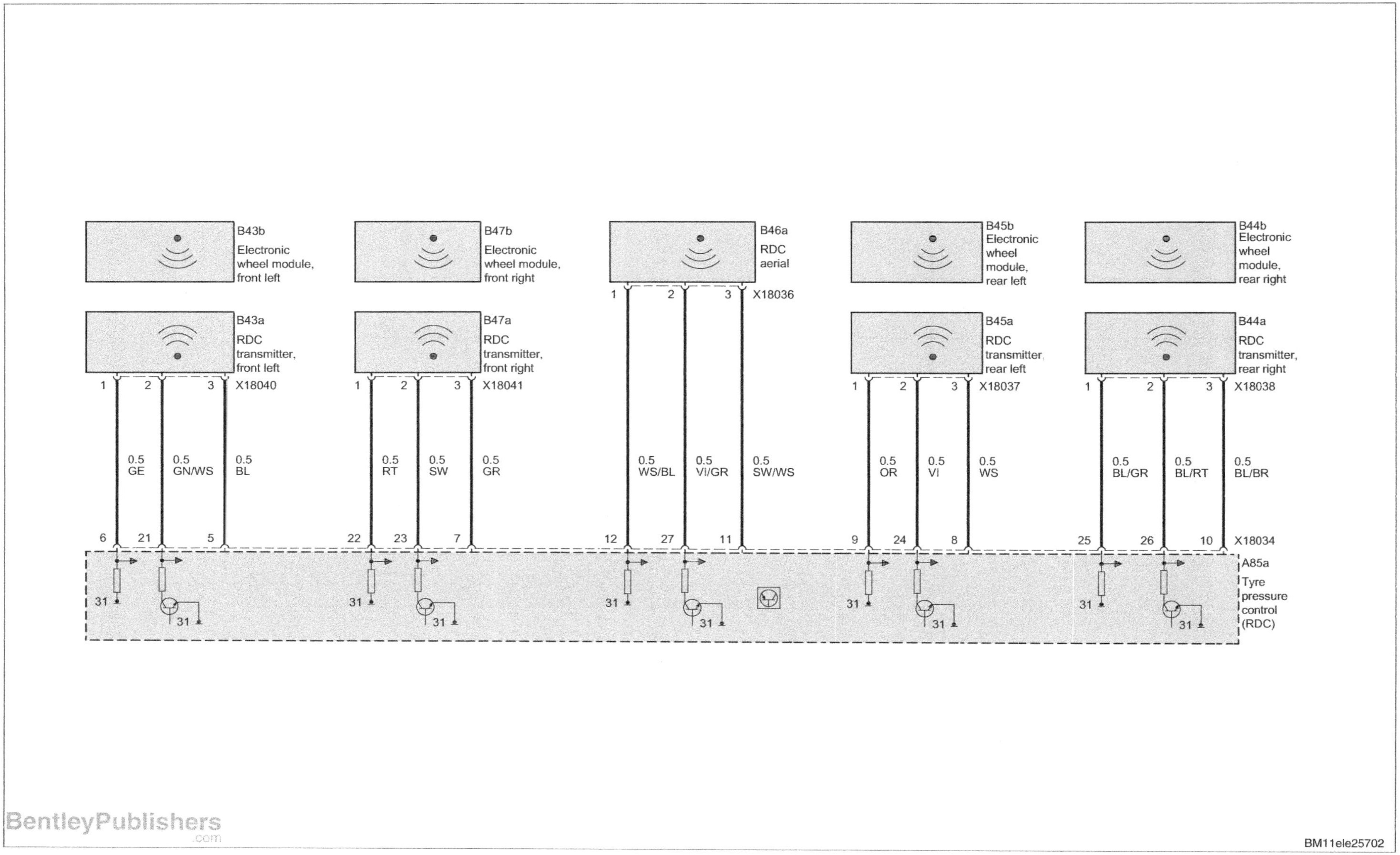

Steering column switch cluster
page 1 of 2

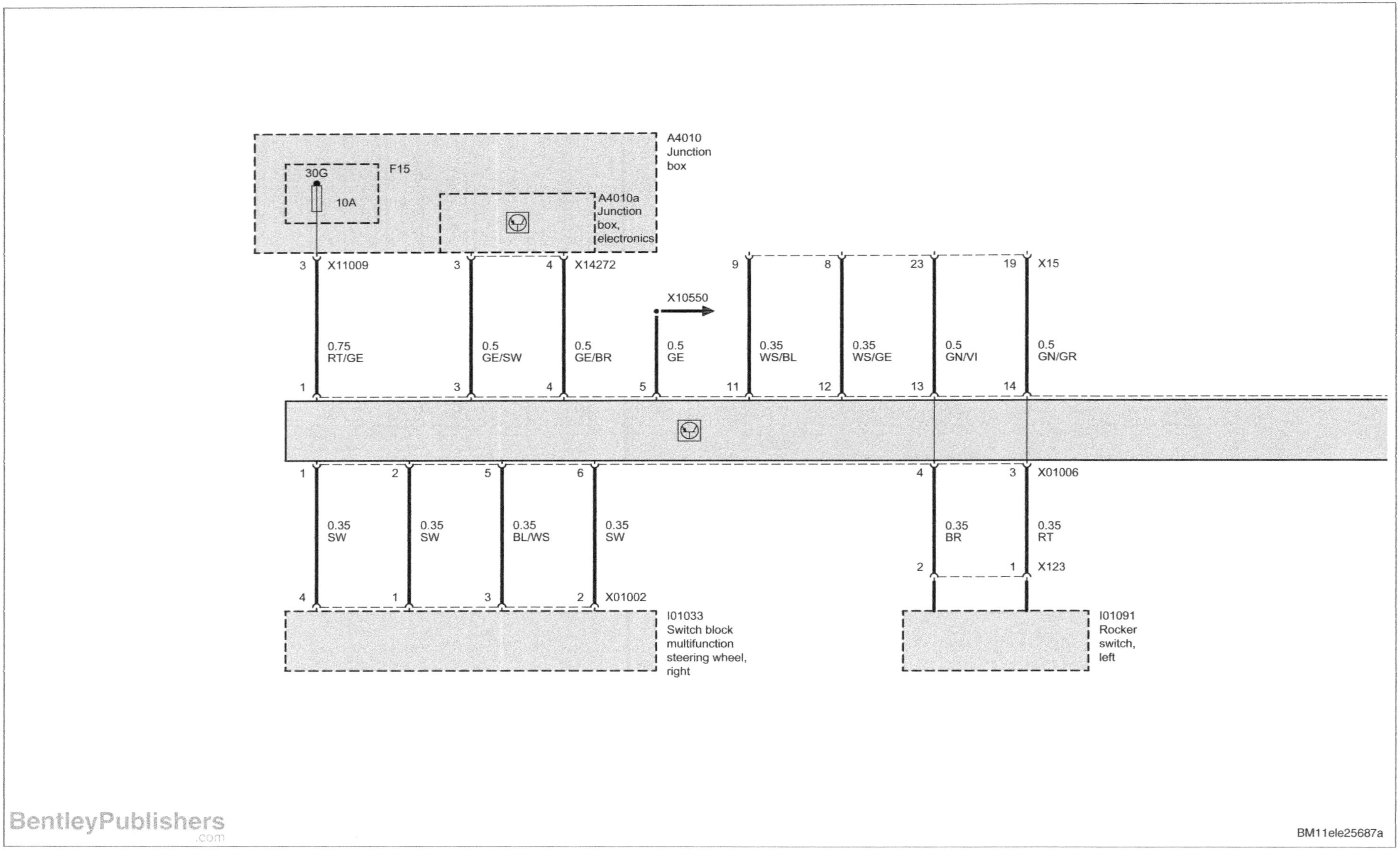

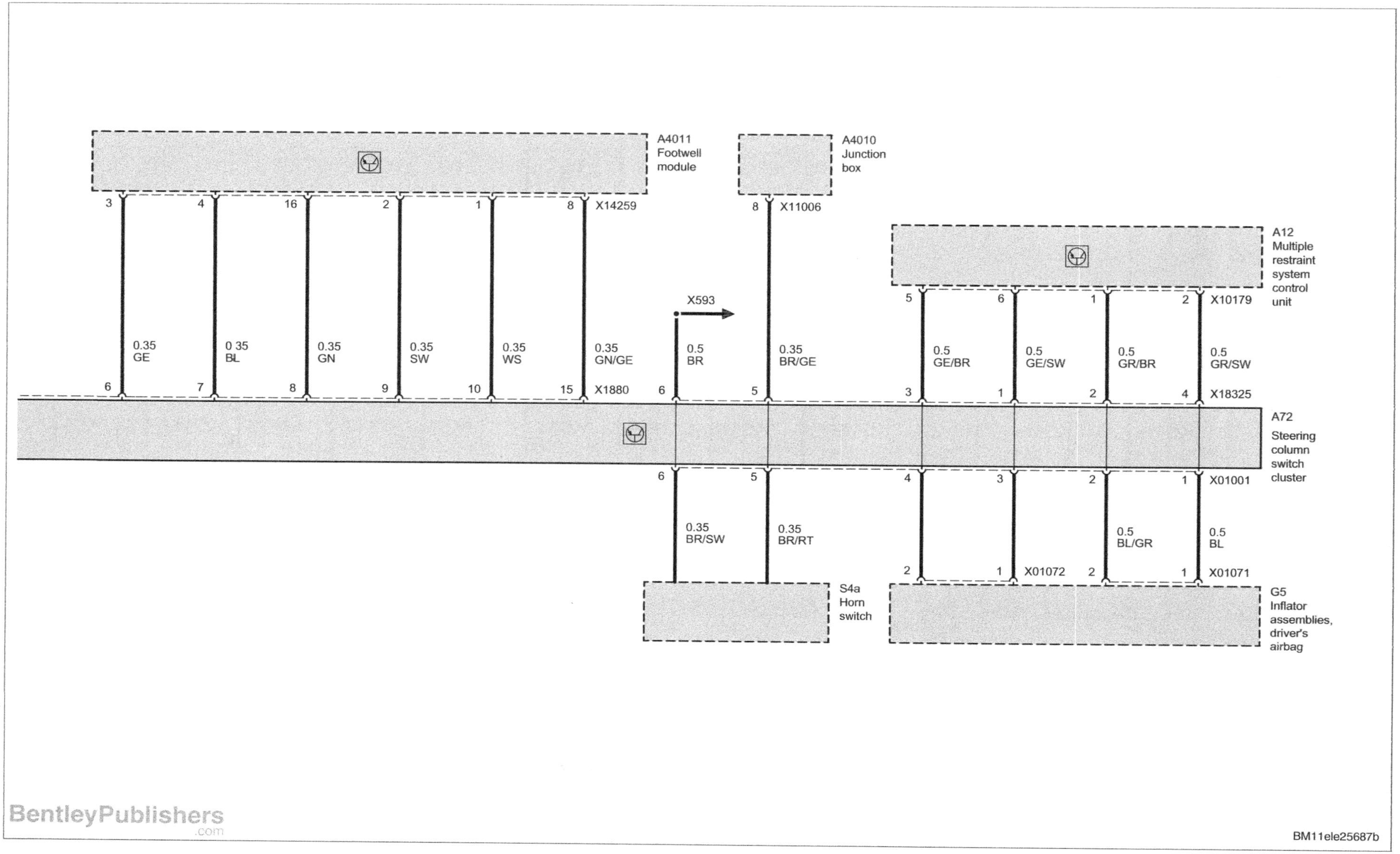
A4011
Footwell
module
3
4
16
2
1
8
X14259
0.35
GE
0 35
BL
0.35
GN
0.35
SW
0.35
WS
0.35
GN/GE
6
7
8
9
10
15
X1880
A4010
Junction
box
8
X11006
X593
0.5
BR
0.35
BR/GE
6
5
A12
Multiple
restraint
system
control
unit
5
6
1
2
X10179
0.5
GE/BR
0.5
GE/SW
0.5
GR/BR
0.5
GR/SW
3
1
2
4
X18325
A72
Steering
column
switch
cluster
6
5
4
3
2
1
X01001
0.35
BR/SW
0.35
BR/RT
0.5
BL/GR
0.5
BL
2
1
X01072
2
1
X01071
S4a
Horn
switch
G5
Inflator
assemblies,
driver's
airbag
BentleyPublishers
.com
BM11ele25687b

Instrument cluster control unit
page 1 of 2

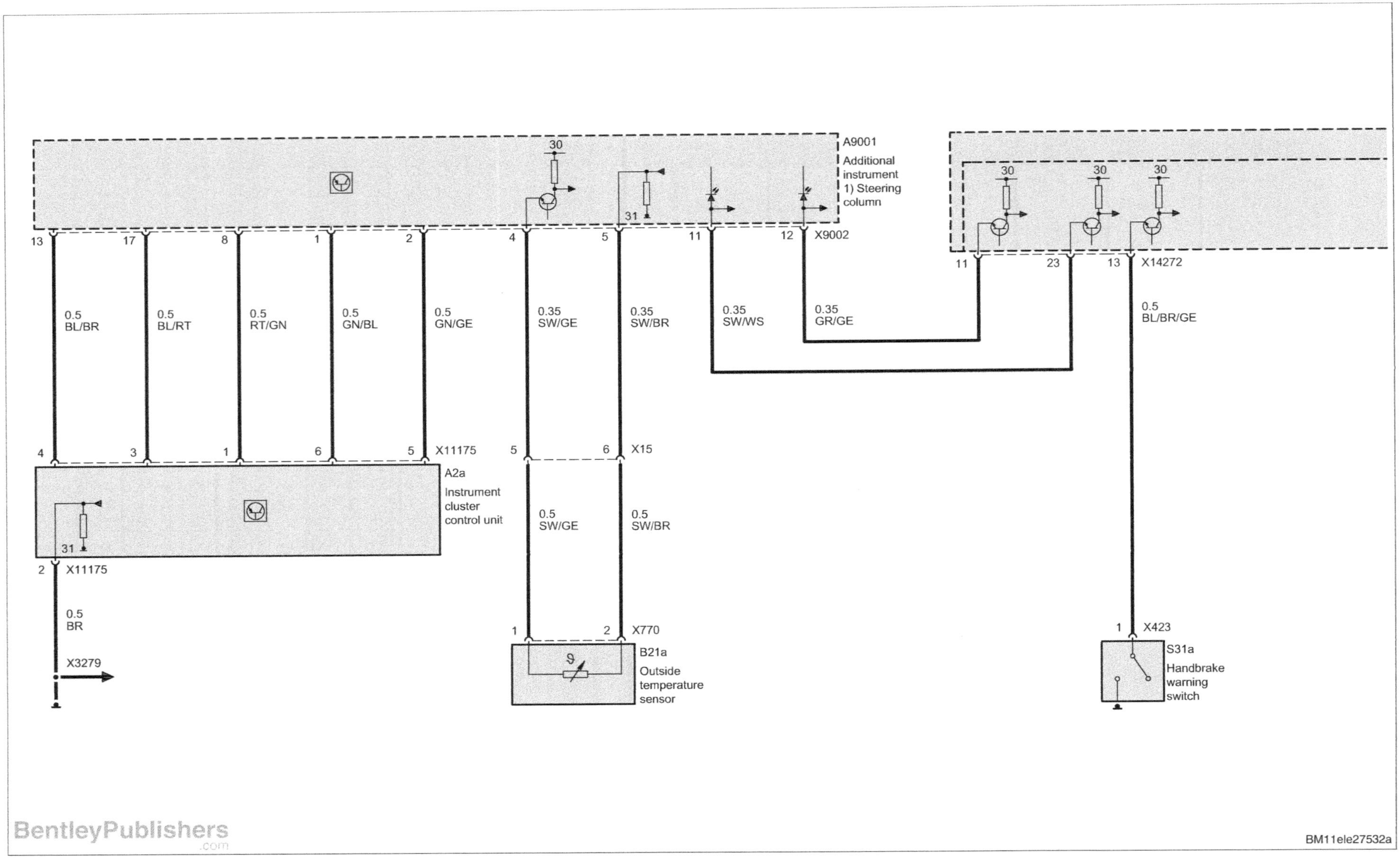

Instrument cluster control unit
page 2 of 2

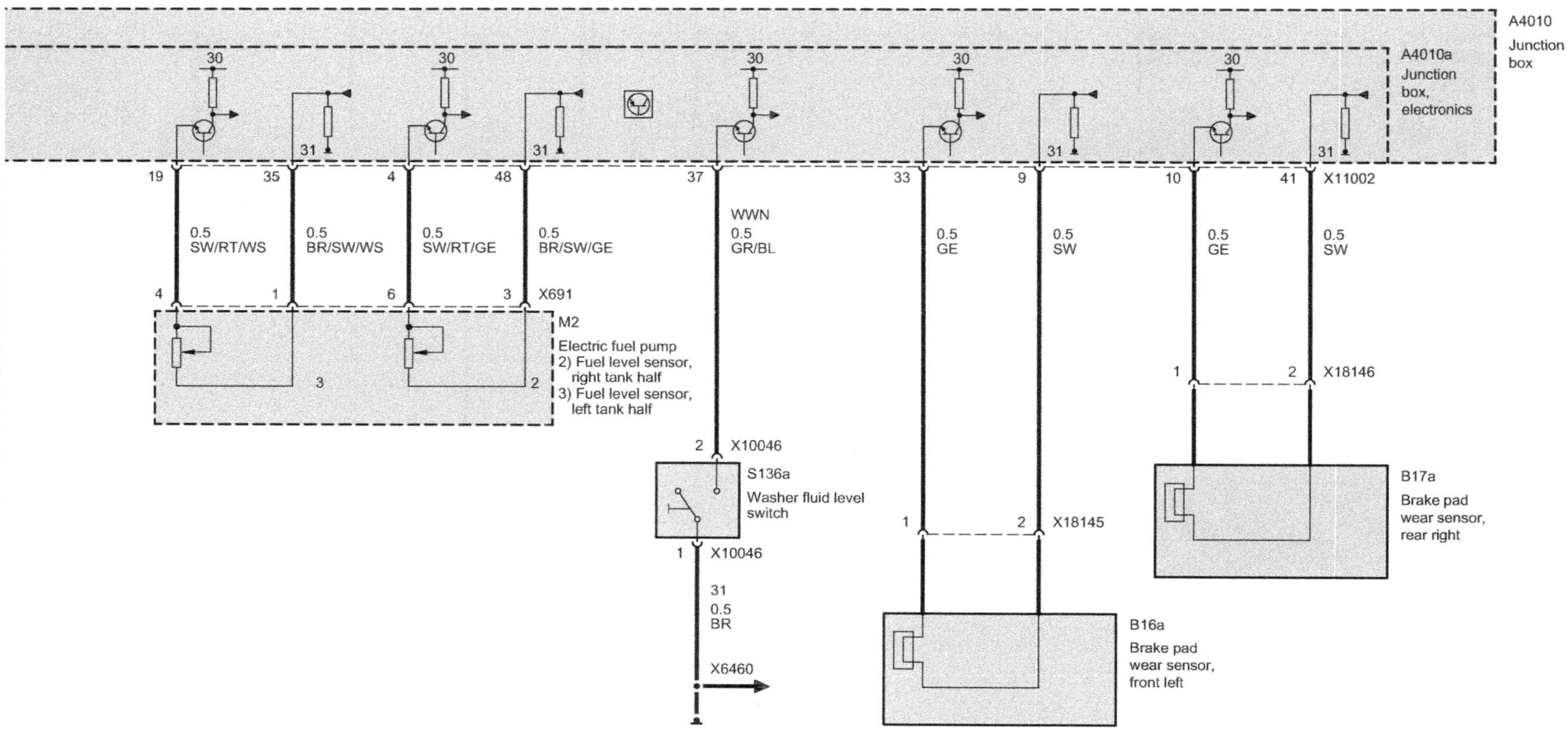

Electromechanical power steering

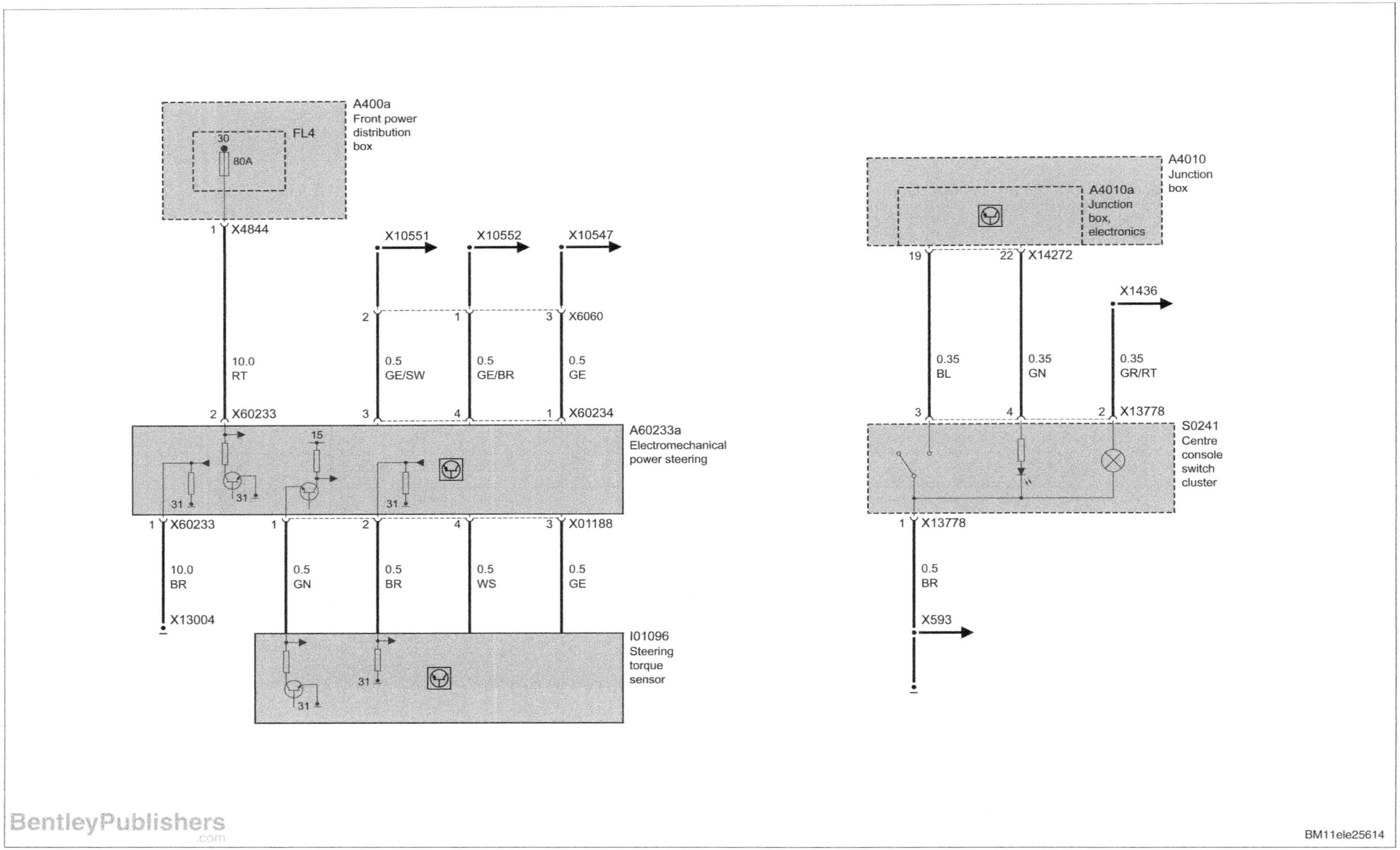

Antitheft alarm system, Clubman page 1 of 2

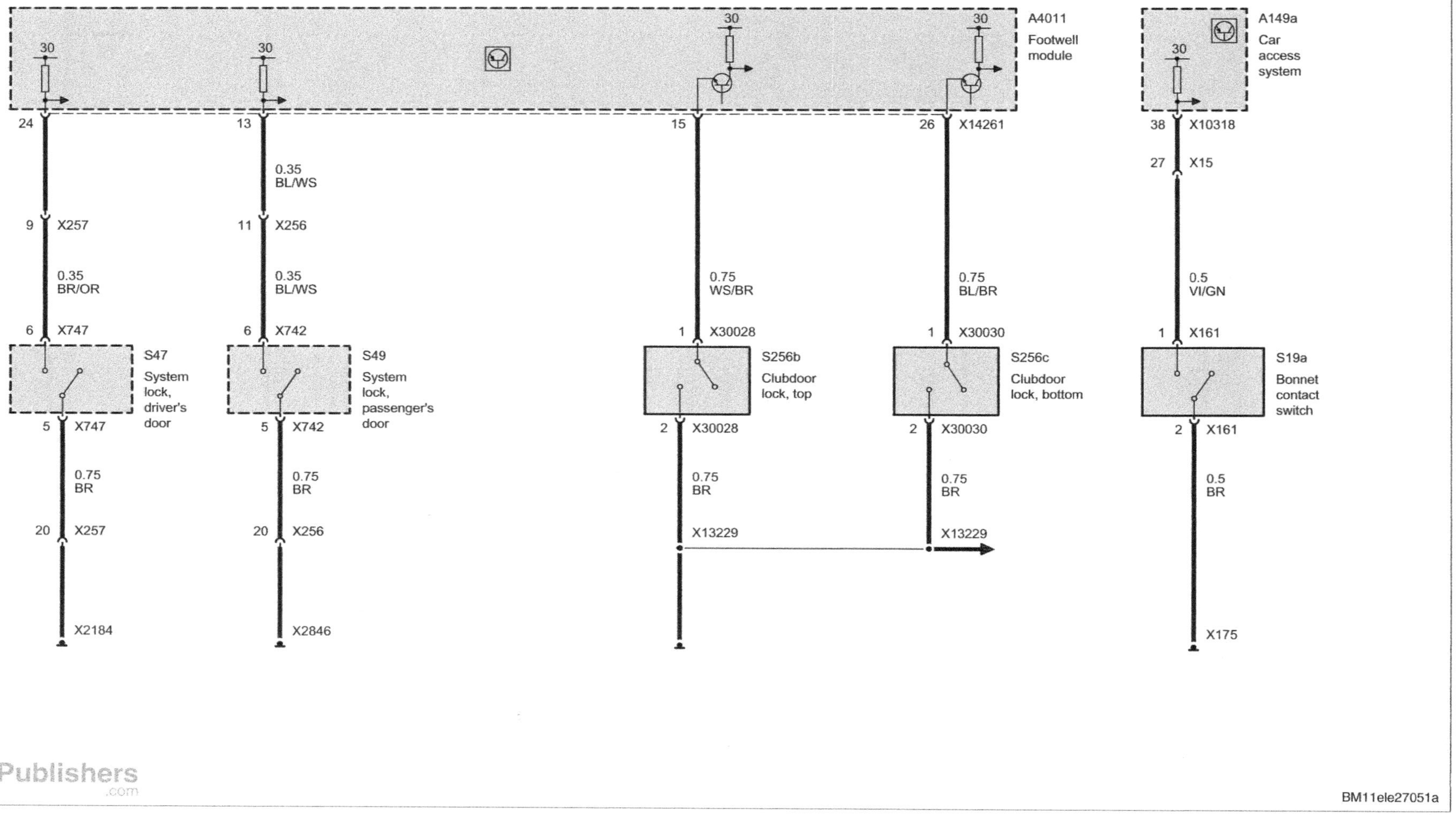

Antitheft alarm system, Clubman page 2 of 2

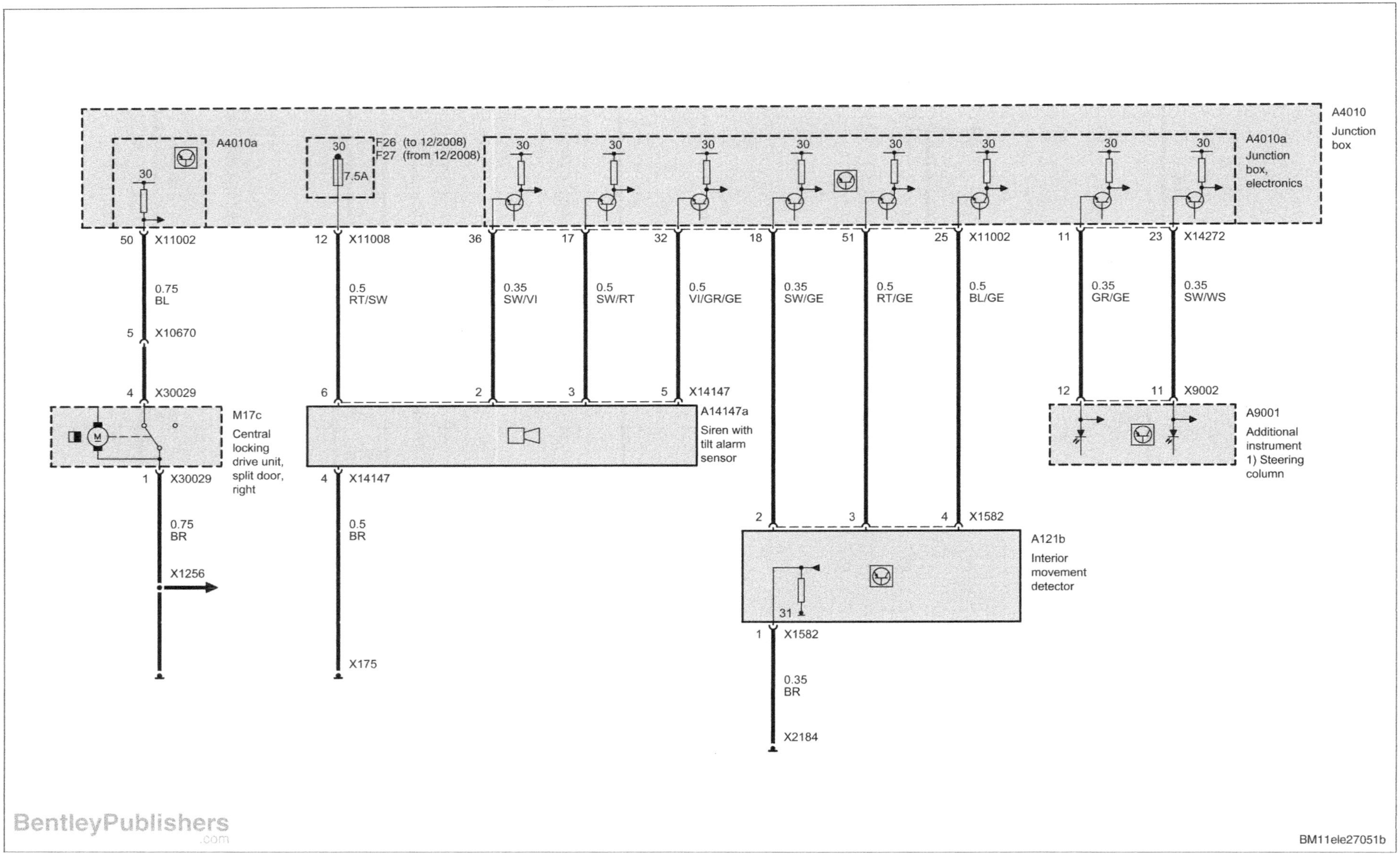

BM11ele27051b

Central locking drives (Clubman) to 12/2008

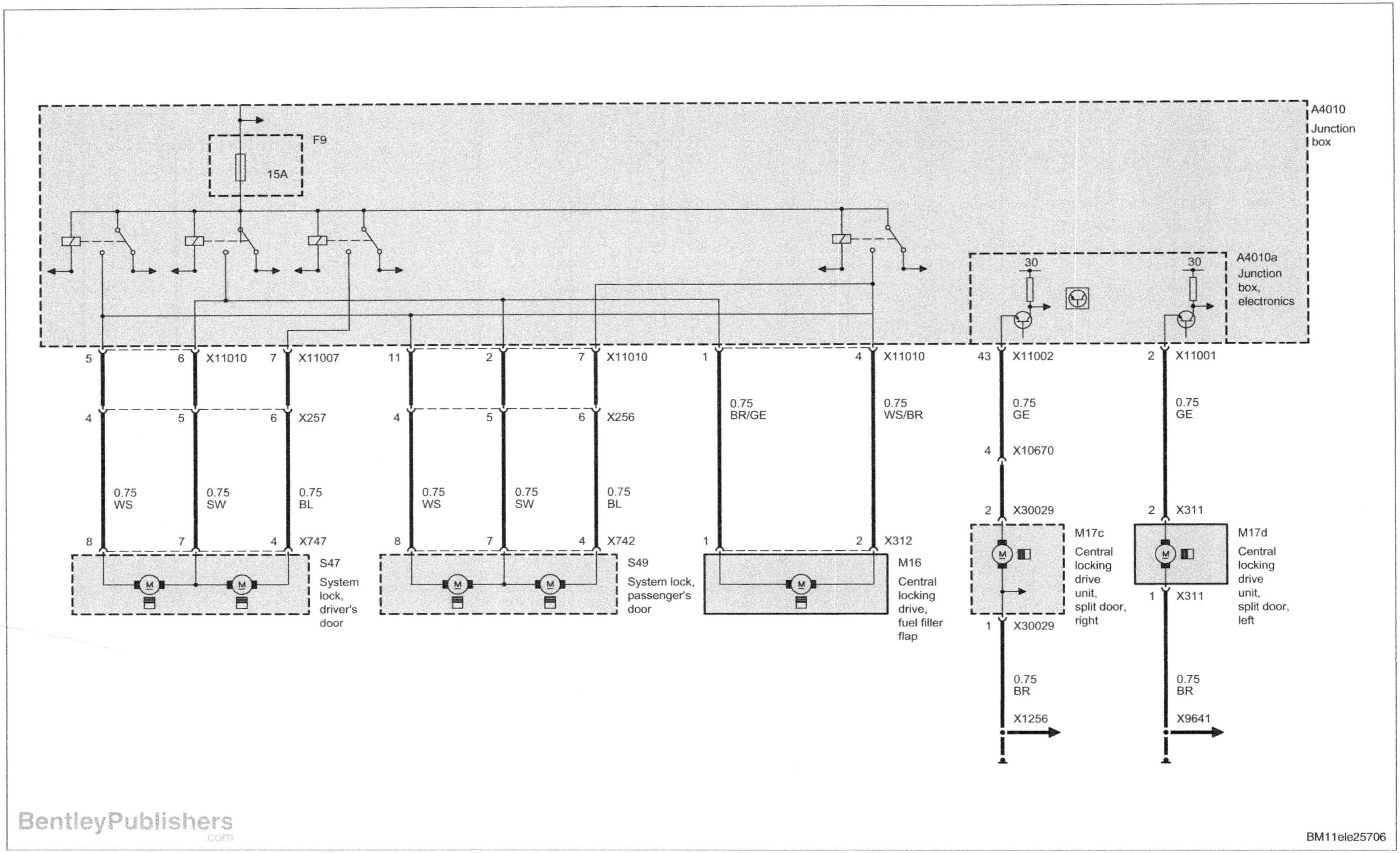

Central locking drives (Hardtop, Convertible) to 12/2008

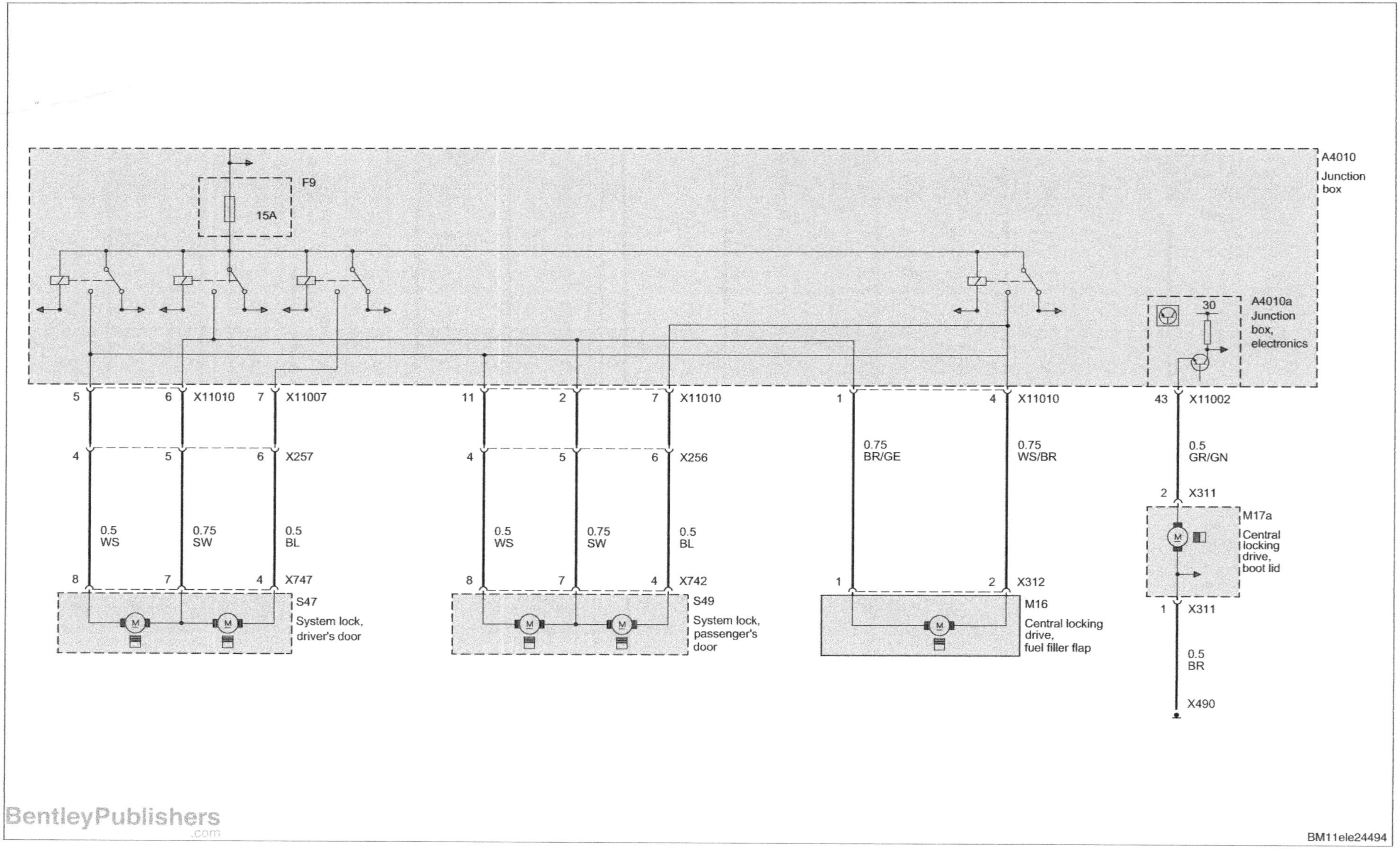

Window defogger
except Convertible

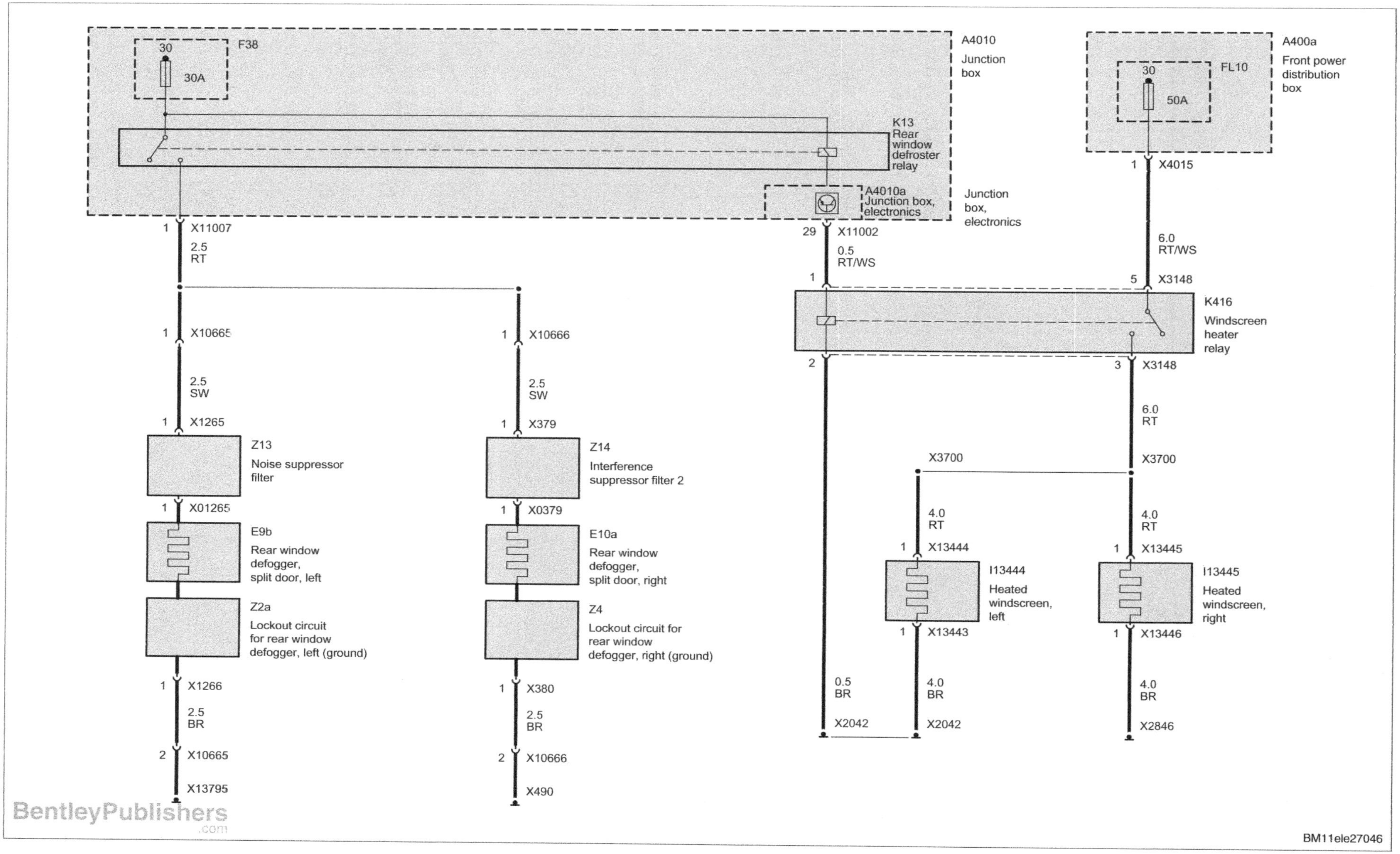

Central locking drives (Hardtop, Convertible) from 12/2008

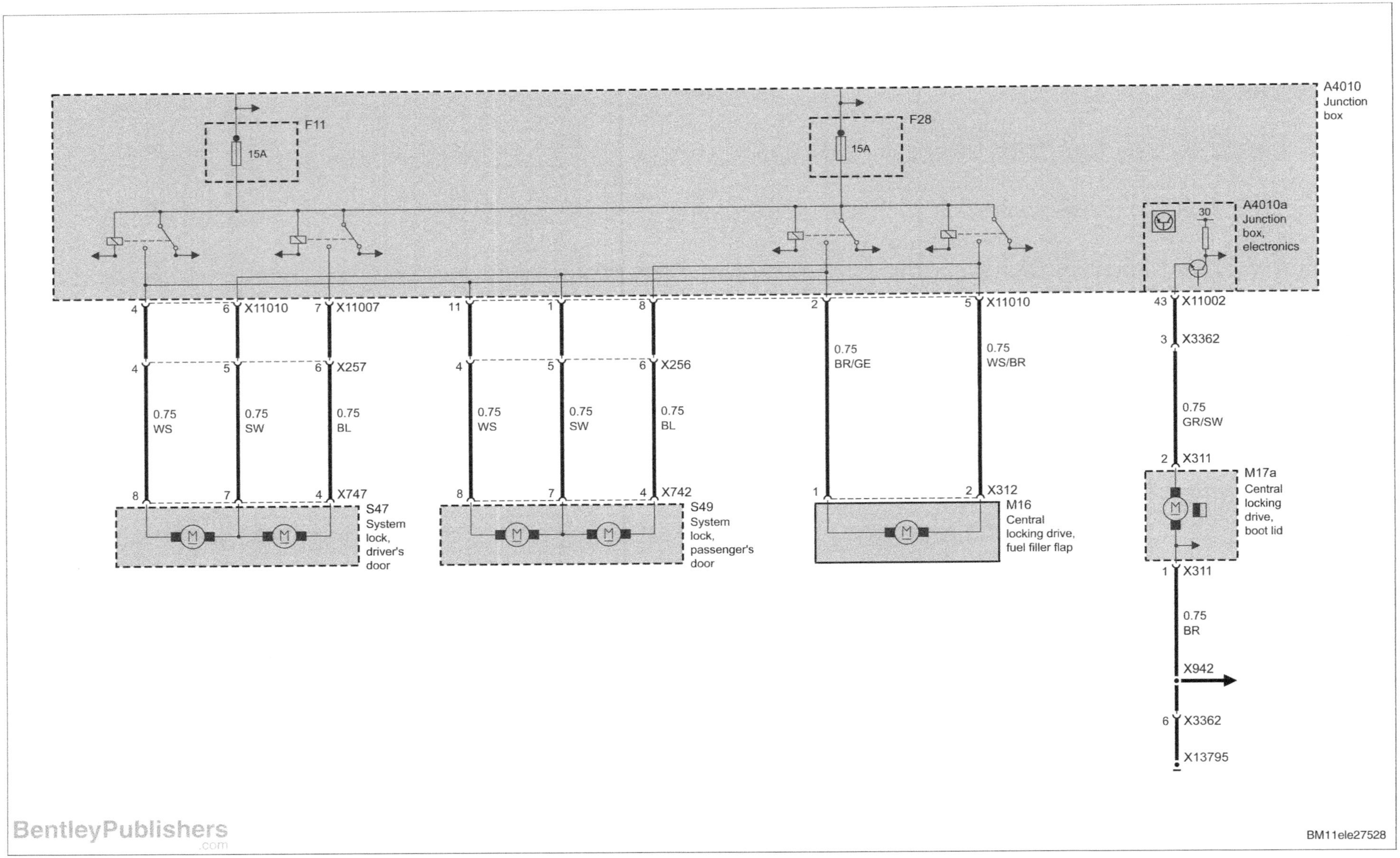

Convertible top
Convertible top power supply
page 1 of 4

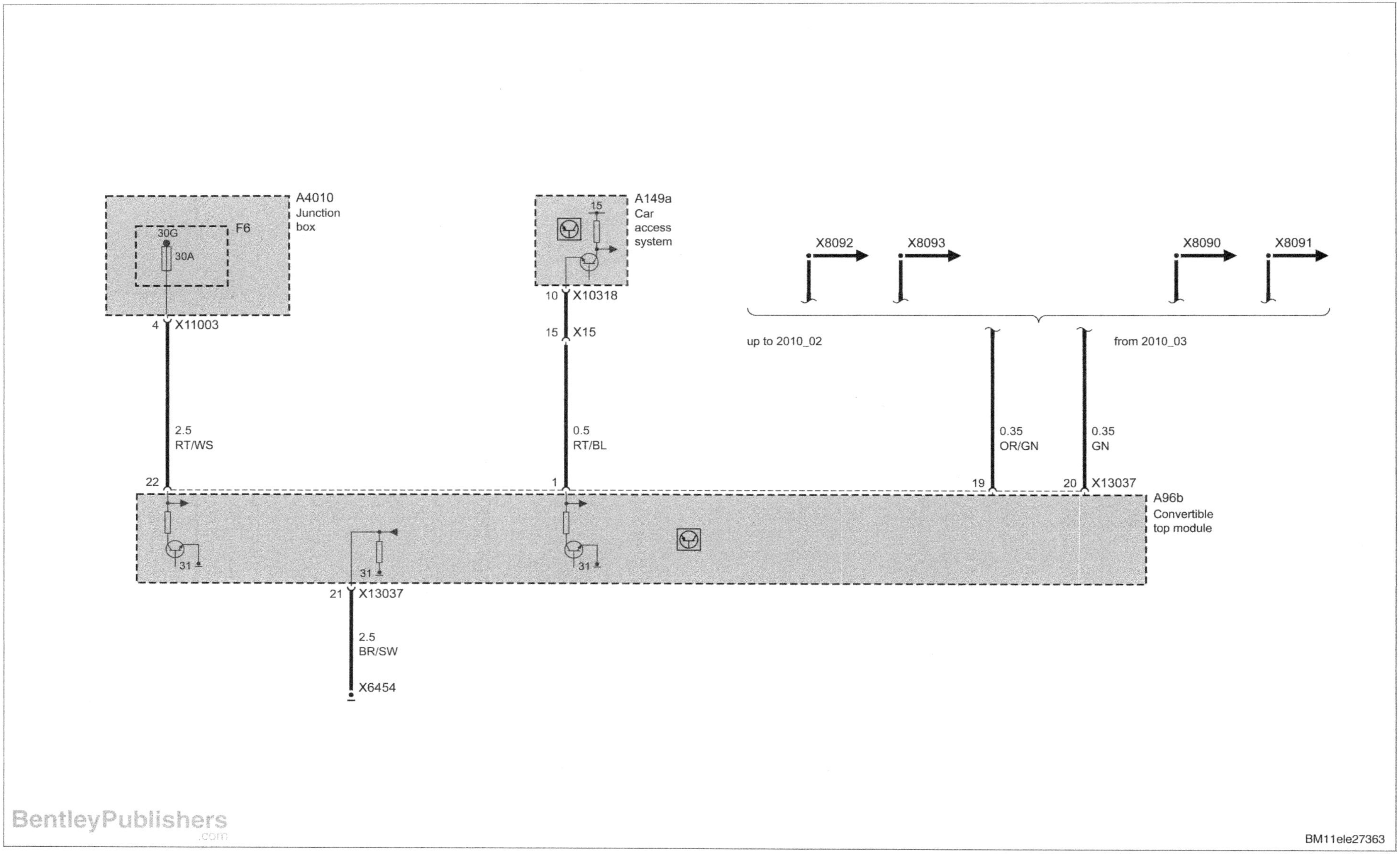

Convertible top
Convertible top drive
page 2 of 4

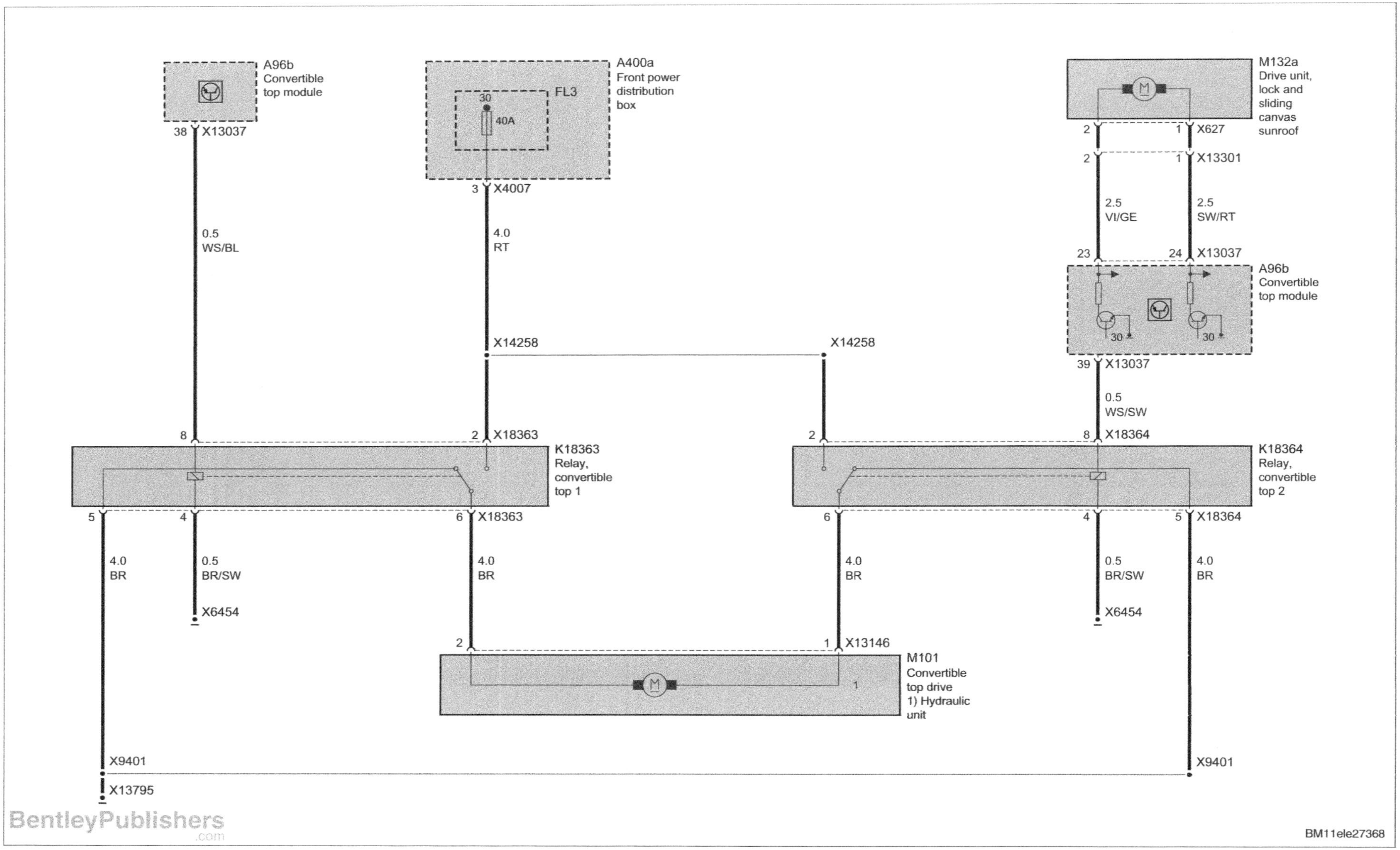

Convertible top
Convertible top module

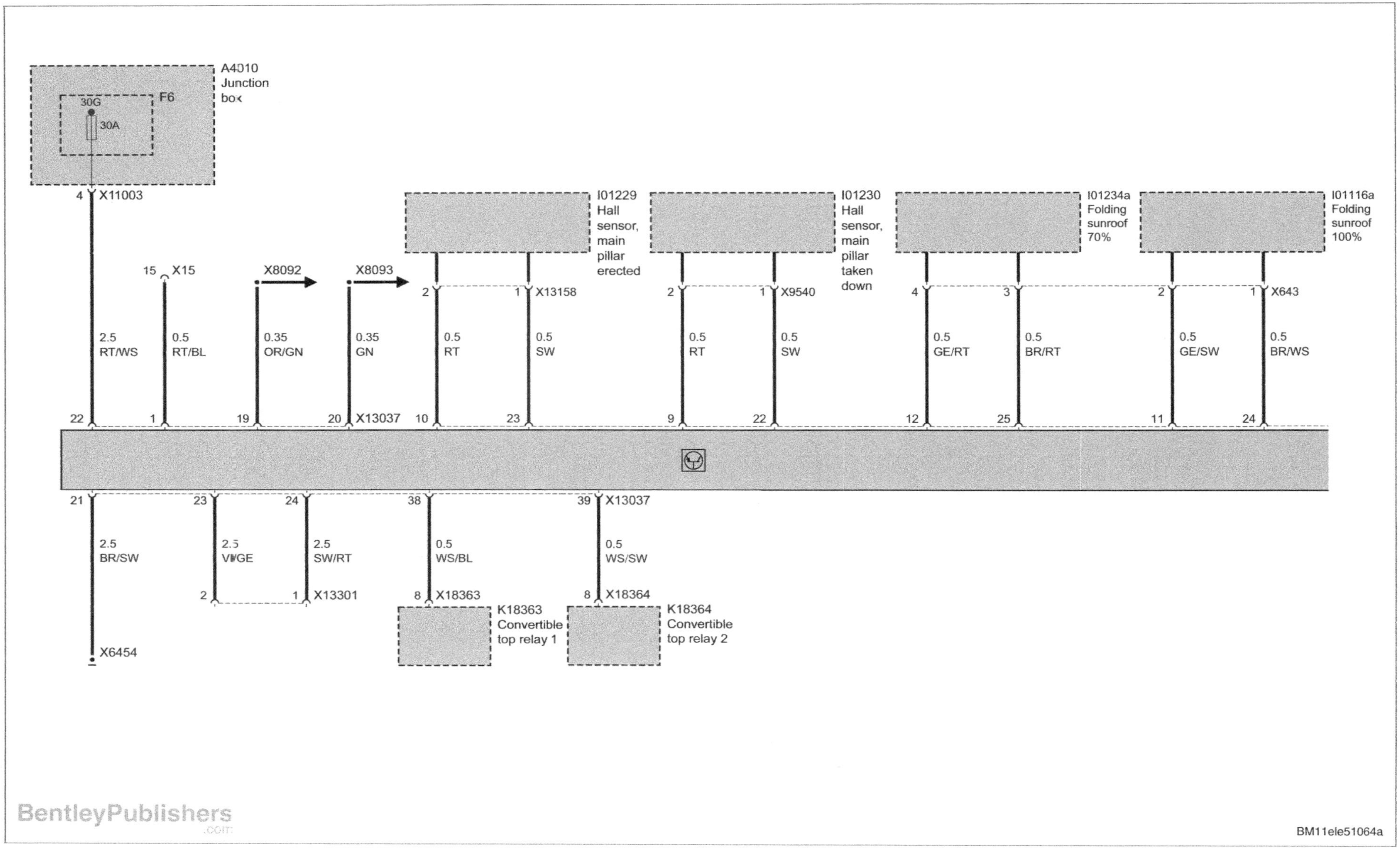

Convertible top
Convertible top module
page 4 of 4

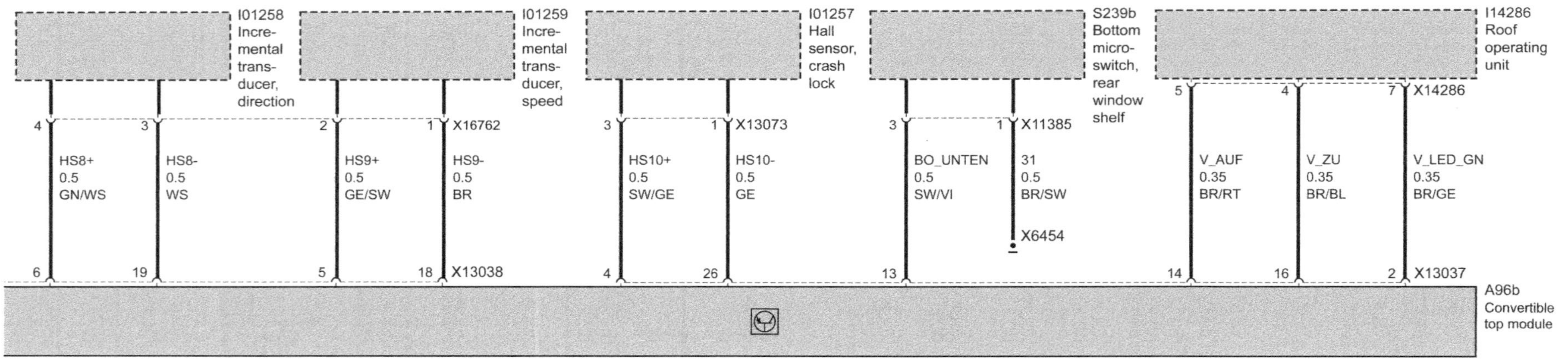

BM11ele51064b

Window defogger
Convertible

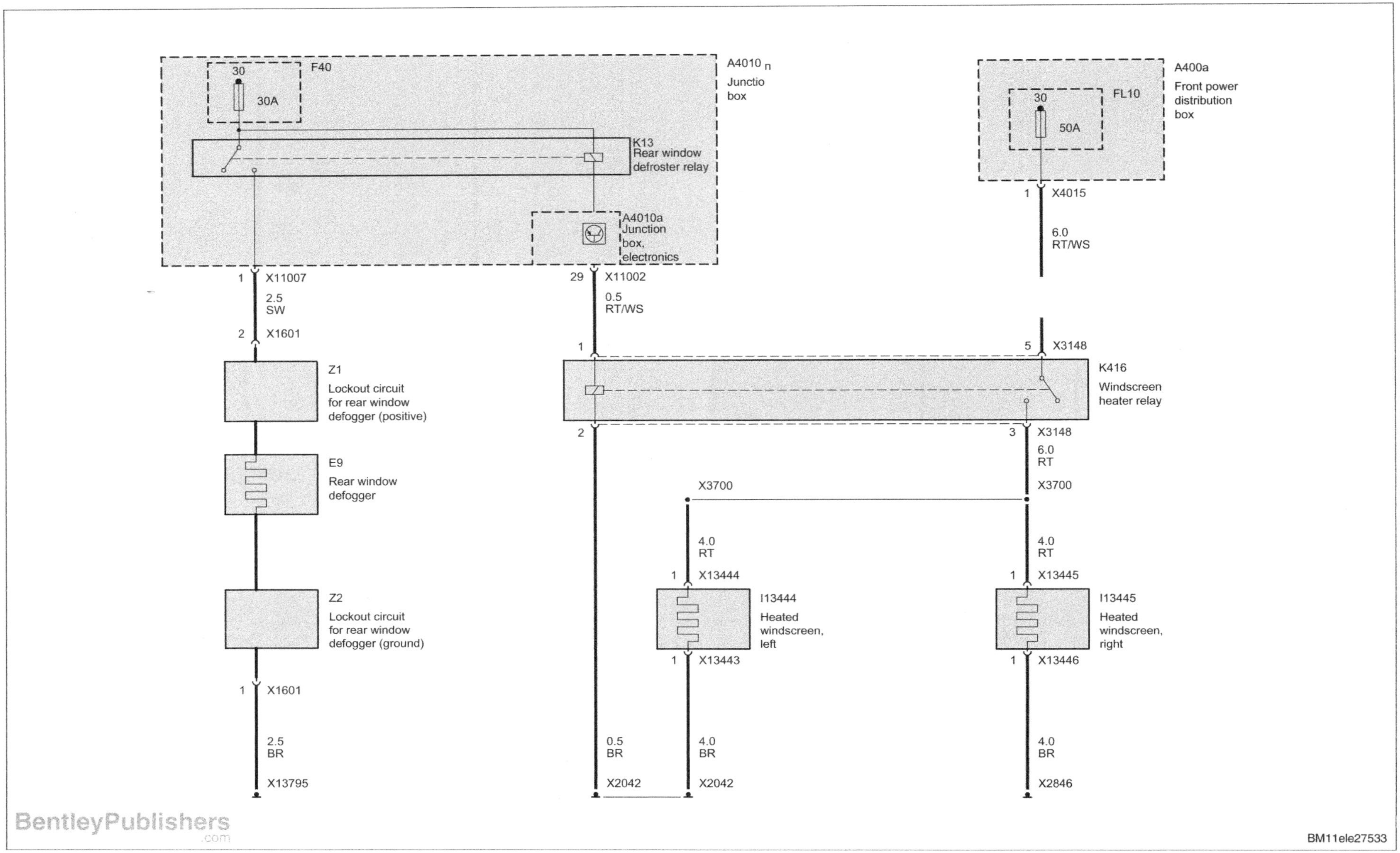

Antitheft alarm system
Convertible
page 1 of 2

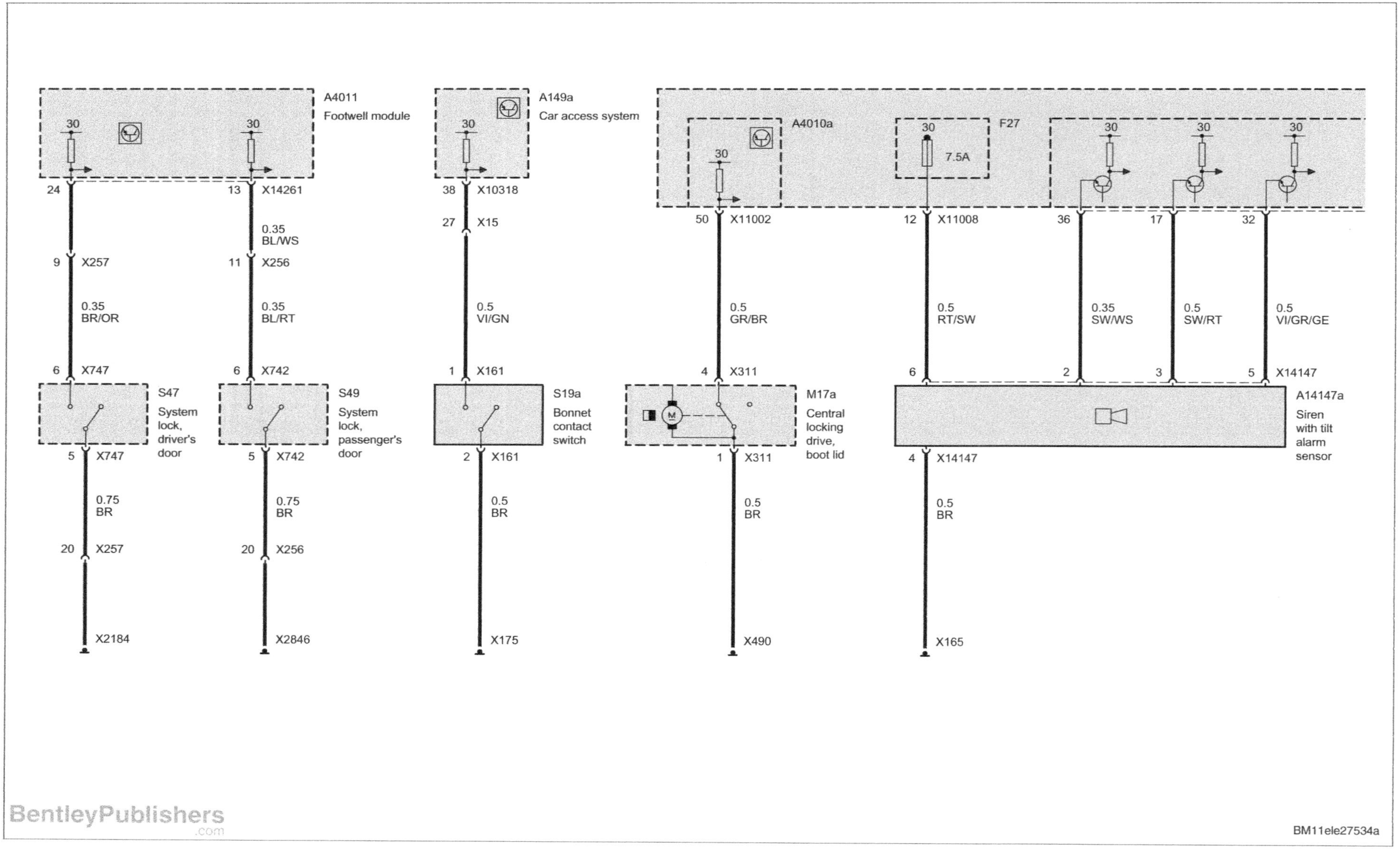

Antitheft alarm system
Convertible
page 2 of 2

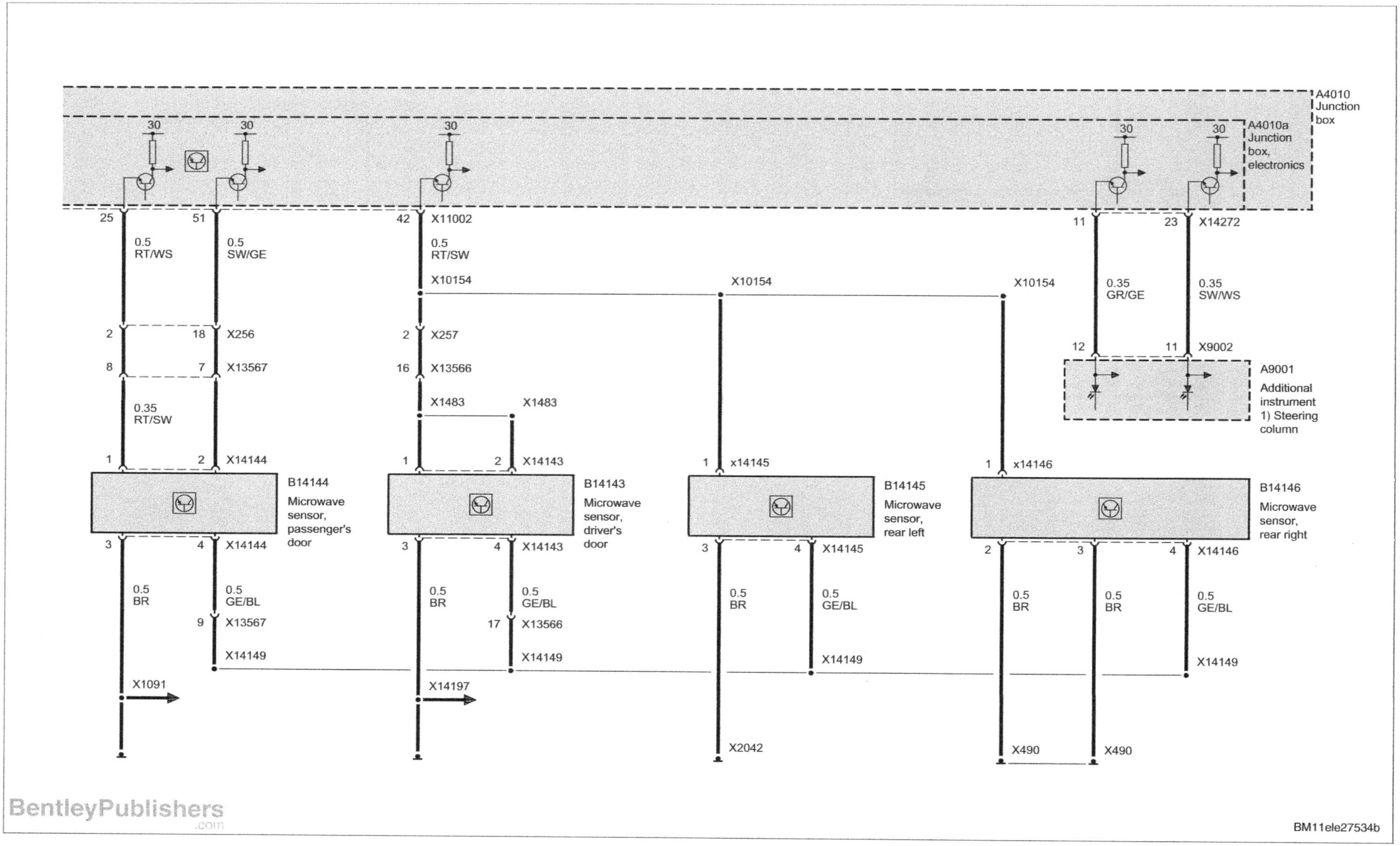

Tire pressure control, power supply 2007-2009

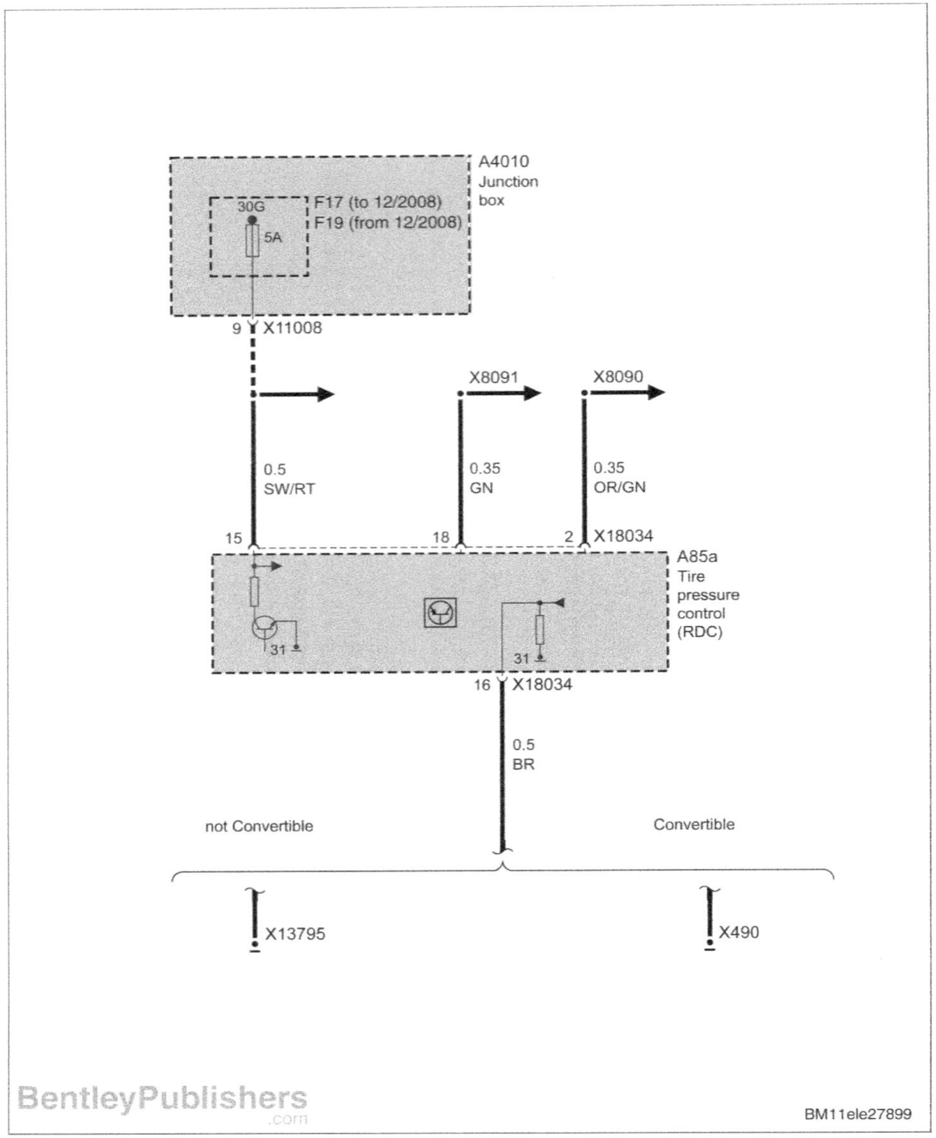

Tire pressure control, power supply 2010-2013

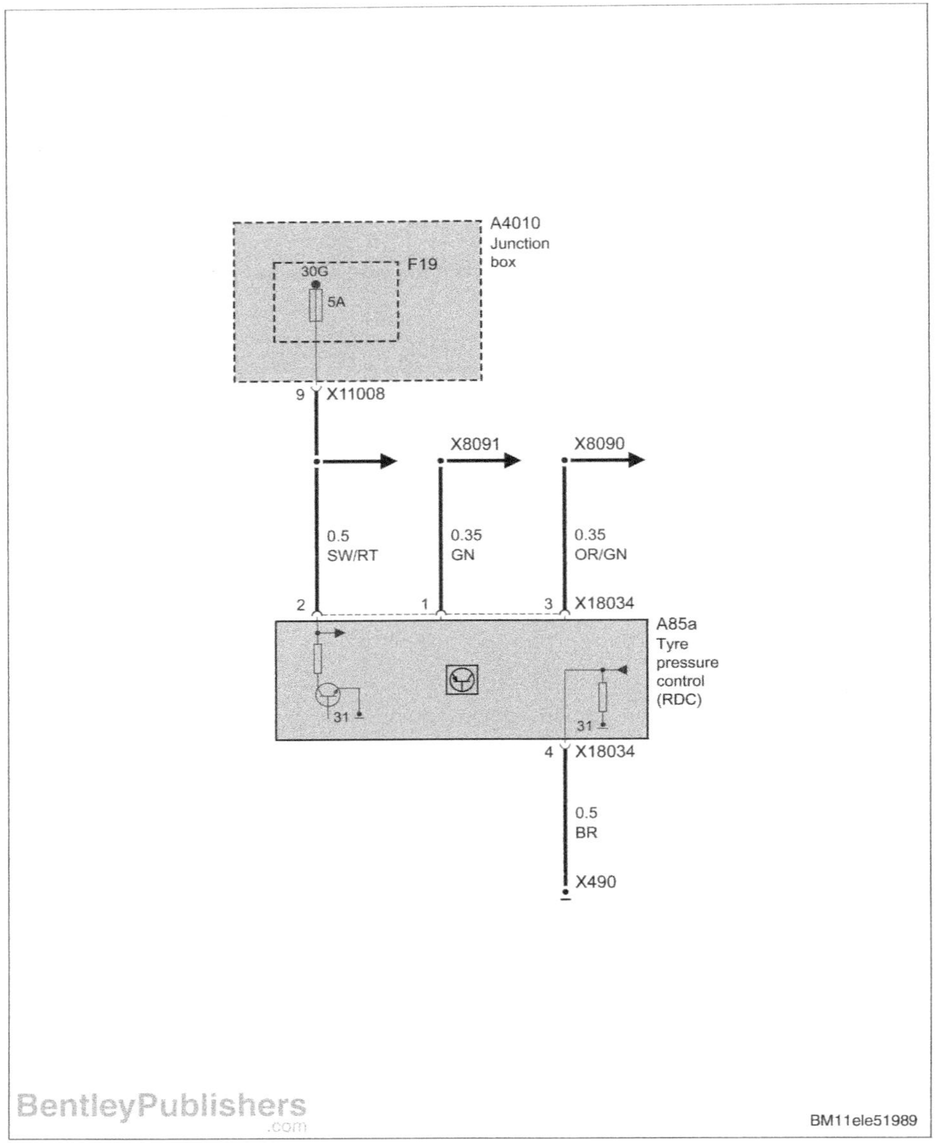

Roof operating unit
Convertible

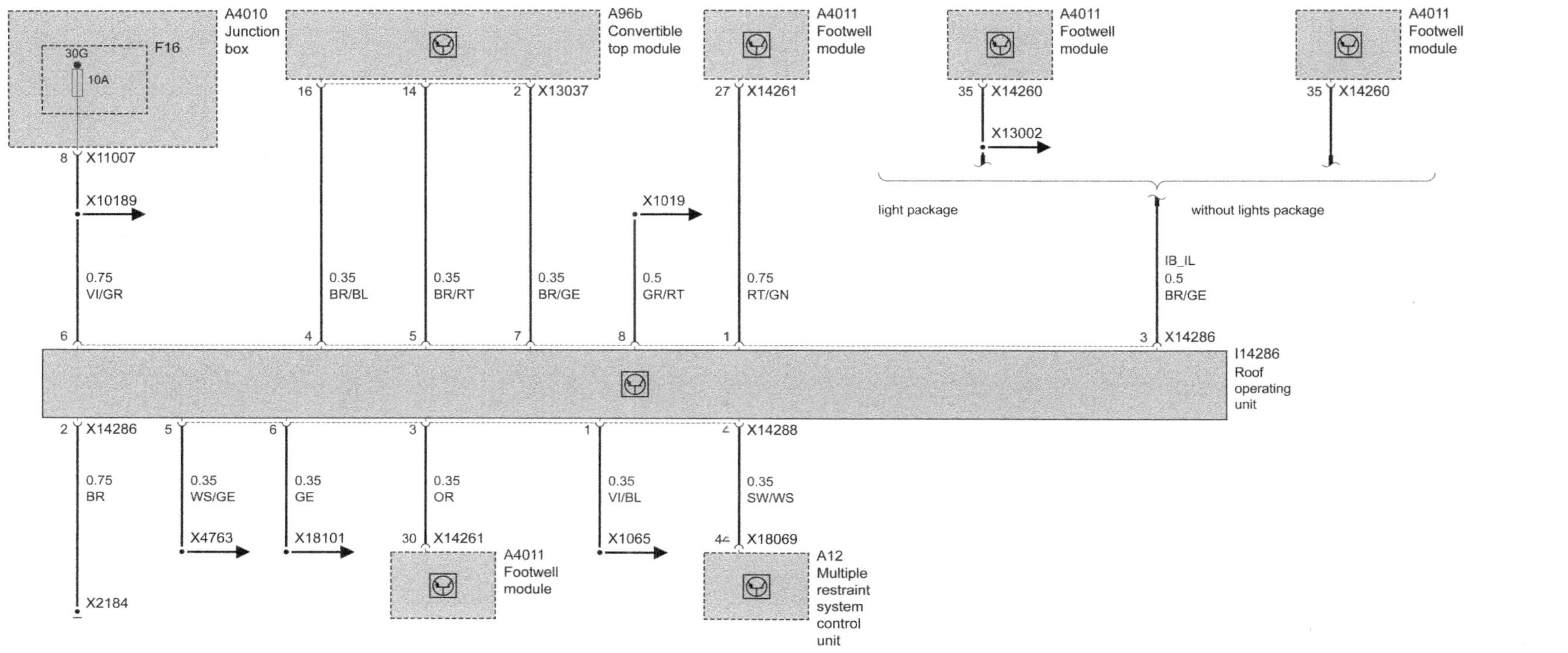

Central locking drives
Clubman from 12/2008

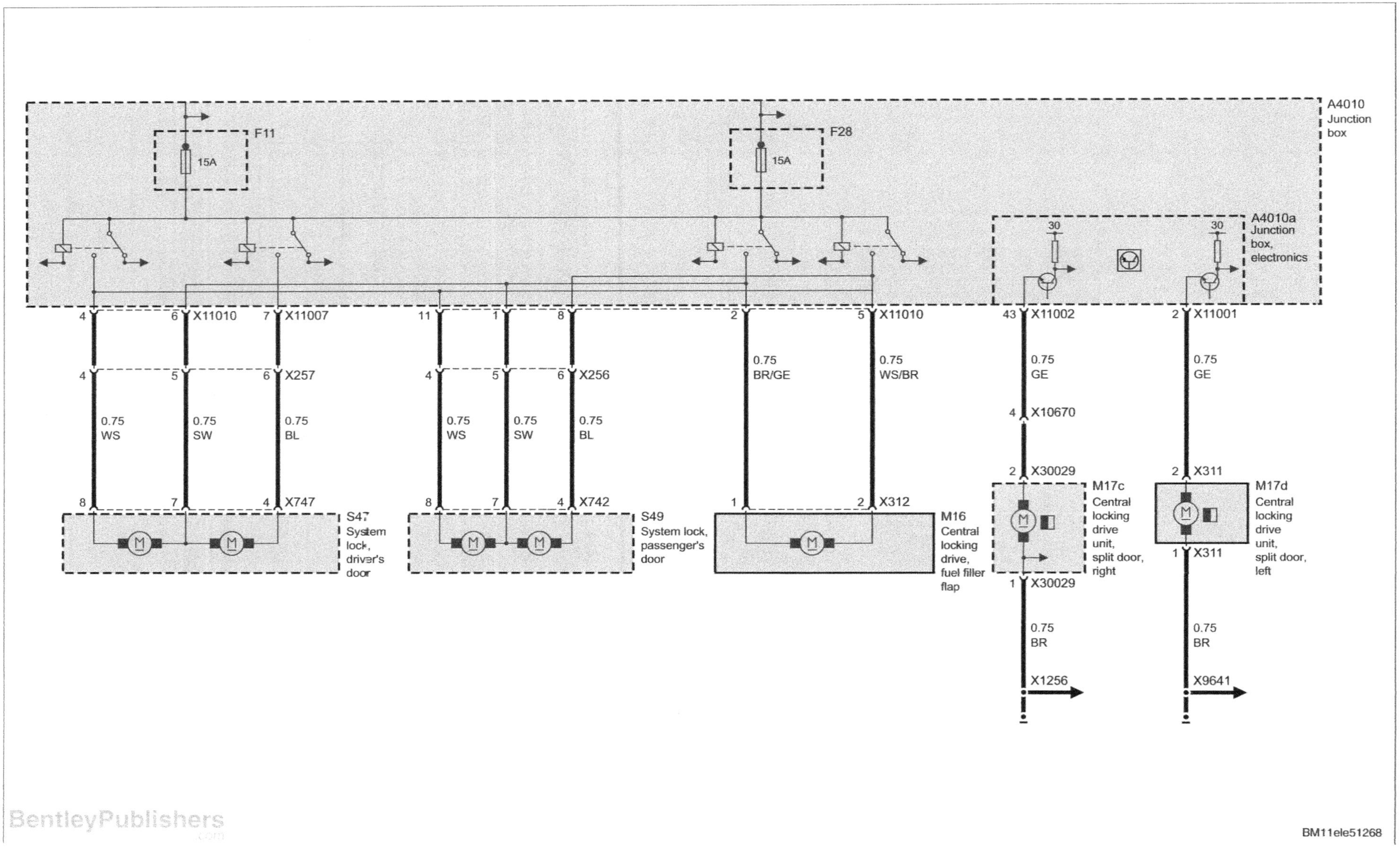

OBD On-Board Diagnostics

TABLE

GENERAL

This chapter outlines the fundamentals and equipment requirements of On-Board Diagnostics II (OBD II) standards as they apply to MINI vehicles. Also covered here is a listing of MINI and OBD II diagnostic trouble codes (DTCs).

ON-BOARD DIAGNOSTICS (OBD II)

OBD II standards were developed by the SAE (Society of Automotive Engineers) and CARB (California Air Resources Board). OBD II is the second generation of on-board self-diagnostic equipment requirements. These standards were originally mandated for California vehicles. Since 1996 they have been applied to all passenger vehicles sold in the United States.

On-board diagnostic capabilities are incorporated into the hardware and software of the engine control module (ECM) to monitor virtually every component that can affect vehicle emissions. The OBD II system works to ensure that emissions remain as clean as possible over the life of the vehicle.

Each emission-influencing component is checked by a diagnostic routine (called a monitor) to verify that it is functioning properly. If a problem or malfunction is detected, the **diagnostic executive** built into the OBD II system illuminates a malfunction indicator light (MIL) on the instrument panel.

The OBD II system also stores diagnostic trouble codes (DTCs) about the detected malfunction in the ECM so that a repair technician can accurately find and fix the problem. Specialized OBD II scan tool equipment is needed to access the fault memory and OBD II data.

The extra hardware needed to operate the OBD II system consists mainly of the following:

- Additional oxygen sensors downstream of the catalytic converters
- Fuel tank pressure sensor and device to pressurize fuel storage system
- Several engine and performance monitoring devices
- Standardized 16-pin OBD II connector under the dashboard
- Upgraded components for the federally required reliability mandate

Malfunction indicator light (MIL)

OBD II software cause the malfunction indicator light (MIL) to illuminate when emission levels exceed 1.5 times Federal standards.

For MINI vehicles covered by this manual, two MIL symbols are used.

1	2	Cause	What to do
SERVICE ENGINE SOON		Indication in US models Indicator lamp **1** flashes: Engine malfunction under high load. High engine load will result in damage to the catalytic converter	You can continue your journey, but moderate your speed and exercise due caution. Have the vehicle checked without delay.
		Indicator lamp **1** comes on: Engine malfunction with adverse effect on exhaust emissions	You can continue your journey, but moderate your speed and exercise due caution. Have the vehicle checked as soon as possible.
ABS BRAKE		Indication in US models Vehicle electronics failed	You cannot continue your journey. Contact your MINI Dealer.

BM13obd002

MIL illuminates under the following conditions:

- Engine management system fault detected for **two** consecutive OBD II drive cycles. See **Drive cycle** in this repair group.
- Catalyst damaging fault detected.
- Component malfunction causes emissions to exceed 1.5 times OBD II standards.
- Manufacturer-defined specifications exceeded.
- Implausible input signal.
- Misfire faults.
- Leak in evaporative system.
- Oxygen sensors observe no purge flow from purge valve / evaporative system.
- ECM fails to enter closed-loop operation within specified time.
- ECM or automatic transmission control module (TCM) in “limp home” mode.
- Ignition key ON before cranking (bulb check function).

OBD II fault memory (including the MIL) can only be reset using a special scan tool. Removing the connector from the ECM or disconnecting the battery does not erase the fault memory.

Additional MIL information:

- A fault code is stored within the ECM upon the first occurrence of a fault in the system being checked.
- Two **complete** consecutive drive cycles with the fault present illuminate the MIL. The exception to the two-fault requirement is a catalyst-damaging fault, which illuminates the MIL immediately.
- If the second drive cycle was not complete and the fault was not checked, the ECM counts the third drive cycle as the next consecutive drive cycle. The MIL illuminates if the system is checked and the fault is still present.
- Once the MIL is illuminated, it remains illuminated until the vehicle completes three consecutive drive cycles without detecting a fault.
- An existing fault code is cleared from memory automatically when the vehicle completes **40** consecutive drive cycles without the fault being detected.

In order to automatically clear a catalyst-damaging fault from memory, the condition under which the fault occurred must be evaluated for **80** consecutive drive cycles without the fault reoccurring.

A generic scan tool connected to the OBD II plug can display diagnostic trouble codes (DTCs), along with the conditions associated with the illumination of the MIL. Using a more advanced or MINI-dedicated scan tool, additional proprietary information is normally available.

Scan tool and scan tool display

The complexity of the OBD II system requires that all diagnostics begin by connecting a scan tool to the 16-pin OBD II plug inside the vehicle.

OBD II standards require that the 16-pin OBD II plug be located within three (3) feet of the driver and not require any tools to access.

Professional diagnostic scan tools available at the time of this printing include the MINI factory tools and a small number of aftermarket MINI-specific tools. See **020 Maintenance**.

In addition to the professional line of scan tools, inexpensive generic OBD II scan tool software programs and handheld units are readily available. Though limited, they are nonetheless powerful diagnostic tools. These tools read live data streams and freeze frame data as well as a host of other valuable diagnostic data.

Diagnostic monitors

Diagnostic monitors run tests and checks on specific emission control systems, components, and functions.

A complete drive cycle is required for the tests to be valid. See **Drive cycle** in this repair group. The diagnostic monitor signals the ECM of the loss or impairment of the signal or component and determines if a signal or sensor is faulty based on 3 conditions:

- Signal or component shorted to ground
- Signal or component shorted to B+
- Signal or component missing (open circuit)

The OBD II system monitors the emission control systems that are installed. Emission control systems vary by vehicle model and year. For example, engines in MINI models covered by this manual are not equipped with secondary air injection, so no secondary air readiness code are displayed.

OBD II software monitors the following:

- Oxygen sensors
- Catalysts
- Engine misfire
- Fuel tank evaporative control system
- Secondary air injection (not applicable to MINI)
- Fuel system

Oxygen sensor monitoring. When driving conditions allow, response rate and switching time of each oxygen sensor is monitored. The oxygen sensor heater function is also monitored. The OBD II system differentiates between precatalyst and post-catalyst oxygen sensors and reads each one individually. In order for the oxygen sensor to be effectively monitored, the system must be in closed loop operation.

Catalyst monitoring. This strategy monitors the two heated oxygen sensors. It compares the oxygen content going into the catalytic converter to the oxygen leaving the converter.

The diagnostic executive knows that most of the oxygen should be used up during the oxidation phase. If it sees higher than programmed values, a fault is set and the MIL illuminates.

Misfire detection. This strategy monitors crankshaft speed fluctuations and determines if an engine misfire occurs by monitoring variations in speed between each crankshaft sensor trigger point. This strategy is so finely tuned that it can determine the severity of the misfire.

The system determines if a misfire is occurring, as well as other pertinent misfire information such as:

- Specific cylinder(s)
- Severity of the misfire event
- Emissions relevant or catalyst damaging

Misfire detection is an on-going monitoring process that is only disabled under certain limited conditions.

Fuel system monitoring. This monitor looks at the fuel delivery needed (long / short term fuel trim) for proper engine operation based on programmed data. If too much or not enough fuel is delivered over a predetermined time, a DTC is set and the MIL is turned on.

Fuel trim refers to adjustments to base fuel schedule. Long-term fuel trim refers to gradual adjustments to the fuel calibration adjustment as compared to short term fuel trim. Long term fuel trim adjustments compensate for gradual changes that occur over time.

Fuel system monitoring monitors the calculated injection time (ti) in relation to engine speed, load, and precatalytic converter oxygen sensor(s) signals.

Using this data, the system optimizes fuel delivery for all engine operating conditions.

Evaporative system monitoring. This monitor checks the fuel storage system and related fuel lines for leaks. It can detect very small leaks anywhere in the system.

A leak detection unit (DMTL) is used to pressurize the EVAP system on a continuous basis (as the drive cycle allows) and to check system integrity.

Drive cycle

The OBD II drive cycle is an important concept in understanding OBD II requirements. The purpose of the drive cycle is to run all of the emission-related on-board diagnostics over a broad range of driving conditions.

A drive cycle is considered complete when all of the diagnostic monitors have run their tests without interruption. For a drive cycle to be initiated, the vehicle must be started cold and brought up to 160°F and at least 40°F above its original starting temperature.

Readiness codes

Inspection/maintenance (I/M) readiness codes are mandated as part of OBD II. The readiness code is stored after complete diagnostic monitoring of specified components and systems is carried out. The readiness code function was designed to prevent manipulating an I/M emission test procedure by clearing faults codes or disconnecting the ECM or battery.

Readiness codes indicate whether the OBD II system is actually ready to monitor the various emission control systems on the vehicle. The vehicle must complete a drive cycle to set readiness codes. The code is binary:

- 0 for ready
- 1 for not ready

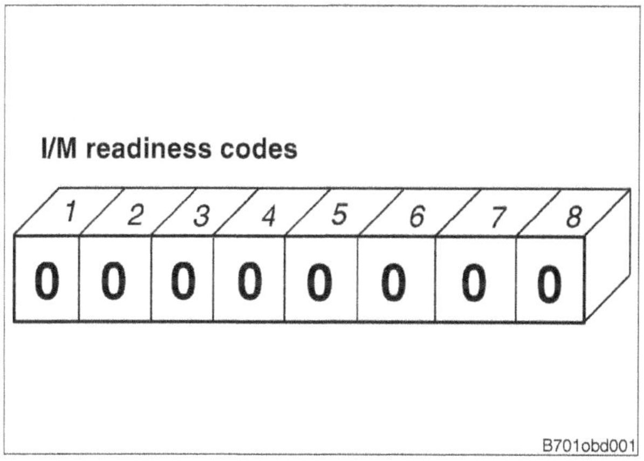

The parameters which are monitored for readiness are:

1. Catalyst efficiency
2. Catalyst heating
3. Fuel tank evaporative control
4. Secondary air injection
5. A/C refrigerant
6. Oxygen sensors
7. Oxygen sensor heaters
8. Exhaust gas recirculation

When all zeros are displayed, the system has established readiness. Readiness codes can be displayed using MINI and aftermarket scan tools.

Readiness codes are set to 1 (not ready) in the following cases:

- The battery or ECM is disconnected.
- DTCs are erased after completion of repairs and a drive cycle is not completed.

Diagnostic trouble codes (DTCs)

SAE standard J2012 mandates a 5-digit diagnostic trouble code (DTC) standard. Each digit represents a specific value. Emission related DTCs start with the letter P for power train. When the MIL illuminates it indicates that a DTC has been stored:

- DTCs are stored as soon as they occur, whether or not the MIL illuminates.
- DTCs store and display a time stamp.
- DTCs record the current fault status: Present, not currently present, or intermittent.

DTC digit interpretation	
1st digit P B C	 powertrain body chassis
2nd digit 0 1	 SAE MINI
3rd digit 0 1 2 3 4 5 6 7	 total system air / fuel induction fuel injection ignition system or misfire auxiliary emission control vehicle speed & idle control ECM inputs / outputs transmission
4th - 5th digits	individual circuits or components

DTC example: P 0 3 0 6

- **P:** A powertrain problem
- **0**: SAE sanctioned ('generic')
- **3:** Related to an ignition system / misfire
- **06** Misfire detected at cylinder #6

DTCs provide a freeze frame or snap-shot of a vehicle performance or emissions fault at the moment that the fault first occurs. This information is accessible through generic scan tools.

Freeze frame data contains, but isn't limited to, the following information:

- Engine load (calculated)
- Engine rpm
- Short and long-term fuel trim
- Vehicle speed
- Coolant temperature
- Intake manifold pressure
- Open / closed loop operation
- Fuel pressure (if available)

Diagnostic trouble codes (DTCs)

Table a. Diagnostic trouble codes (DTCs)		
P-code	**MINI FC**	**Definition**
P0001	0x28AE	Fuel Volume Regulator Control Circuit Open
P0001	0x2C3F	Fuel Volume Regulator Control Circuit Open
P0003	0x28AD	Fuel Volume Regulator Control Circuit Low
P0003	0x2C3E	Fuel Volume Regulator Control Circuit Low
P0004	0x287D	'A' Camshaft Position Slow Response (Bank 1)
P0004	0x28AC	Fuel Volume Regulator Control Circuit High
P0004	0x2C3D	Fuel Volume Regulator Control Circuit High
P000A	0x287D	'A' Camshaft Position Slow Response (Bank 1)
P0010	0x	VANOS intake-side solenoid valve, control-activation: Open circuit
P0010	0x287A	'A' Camshaft Position Actuator Circuit Open (Bank 1)
P0011	0x2870	'A' Camshaft Position Timing Over-Advanced or System Performance (Bank 1)
P0012	0x287D	'A' Camshaft Position Timing Over-Retarded (Bank 1)
P0012	0x287E	'A' Camshaft Position Timing Over-Retarded (Bank 1)
P0012	0x2D5A	VANOS intake: Closed-loop control fault, camshaft sticks
P0013	0x2842	VANOS exhaust-side solenoid valve, control-activation Open circuit
P0013	0x2D9D	'B' Camshaft Position Actuator Circuit Open (Bank 1)
P0014	0x283D	'B' Camshaft Position Timing Over-Advanced or System Performance (Bank 1)
P0015	0x2845	'B' Camshaft Position Timing Over-Retarded (Bank 1)
P0015	0x2D60	VANOS, exhaust: Closed-loop control fault, camshaft sticks
P0015	0x2D61	VANOS, exhaust: Closed-loop control fault, position not reached
P0016	0x2891	Camshaft Position Sensor Correlation (Bank 1 Sensor 'A')
P0017	0x288D	Camshaft Position Sensor Correlation (Bank 1 Sensor 'B')
P0030	0x2A63	HO2S Heater Control Circuit (Bank 1 Sensor 1)
P0030	0x2D0D	HO2S Heater Control Circuit (Bank 1 Sensor 1)
P0031	0x2A61	HO2S Heater Control Circuit Low (Bank 1 Sensor 1)
P0031	0x2D0C	HO2S Heater Control Circuit Low (Bank 1 Sensor 1)
P0032	0x2A60	HO2S Heater Control Circuit High (Bank 1 Sensor 1)
P0032	0x2D0B	HO2S Heater Control Circuit High (Bank 1 Sensor 1)
P0033	0x28B1	Turbocharger/Supercharger Bypass Valve Control Circuit
P0033	0x2C8A	Turbocharger/Supercharger Bypass Valve Control Circuit
P0034	0x28B0	Turbocharger/Supercharger Bypass Valve Control Circuit Low
P0034	0x2C89	Turbocharger/Supercharger Bypass Valve Control Circuit Low
P0035	0x28AF	Turbocharger/Supercharger Bypass Valve Control Circuit High
P0035	0x2C88	Turbocharger/Supercharger Bypass Valve Control Circuit High
P0036	0x2A53	HO2S Heater Control Circuit (Bank 1 Sensor 2)
P0036	0x2D11	HO2S Heater Control Circuit (Bank 1 Sensor 2)

Table a. Diagnostic trouble codes (DTCs) (continued)

P-code	MINI FC	Definition
P0037	0x2A51	HO2S Heater Control Circuit Low (Bank 1 Sensor 2)
P0037	0x2D10	HO2S Heater Control Circuit Low (Bank 1 Sensor 2)
P0038	0x2A50	HO2S Heater Control Circuit High (Bank 1 Sensor 2)
P0038	0x2D0F	HO2S Heater Control Circuit High (Bank 1 Sensor 2)
P0039	0x28AA	Turbocharger/Supercharger Bypass Valve Control Circuit Range/Performance
P0071	0x29A2	Ambient Air Temperature Sensor Circuit Range / Performance
P0071	0x29A3	Ambient Air Temperature Sensor Circuit Range / Performance
P0071	0x2F2F	Ambient Air Temperature Sensor Circuit Range / Performance
P0071	0x2F30	Ambient Air Temperature Sensor Circuit Range / Performance
P0072	0x299B	Ambient Air Temperature Sensor Circuit Low
P0072	0x2F26	Ambient Air Temperature Sensor Circuit Low
P0073	0x299A	Ambient Air Temperature Sensor Circuit High
P0073	0x2F25	Ambient Air Temperature Sensor Circuit High
P007B	0x29E2	Charge-air temperature sensor, voltage change: Too rapid
P007C	0x29E1	Charge-air temperature sensor, electrical: Short circuit to ground
P007D	0x29E0	Charge-air temperature sensor, electrical: Short circuit to positive
P0087	0x28BF	Fuel Rail/System Pressure - Too Low
P0087	0x2BF5	Fuel Rail/System Pressure - Too Low
P0088	0x28BE	Fuel Rail/System Pressure - Too High
P0088	0x2BF4	Fuel Rail/System Pressure - Too High
P00BC	0x2775	Mass Airflow System, Plausibility: Air mass too low
P00BC	0x2788	Mass Airflow System, Plausibility: Air mass too low
P00BC	0x278D	Mass Airflow System, Plausibility: Air mass too low
P00BD	0x2774	Mass Airflow System, Plausibility: Air mass too high
P00BD	0x2789	Mass Airflow System, Plausibility: Air mass too high
P00BD	0x278C	Mass Airflow System, Plausibility: Air mass too high
P0100	0x277A	Mass Air Flow or Volume 'A' Circuit
P0100	0x2B51	Mass Air Flow or Volume 'A' Circuit
P0102	0x2779	Mass or Volume Air Flow Circuit Low Input
P0102	0x2789	Mass or Volume Air Flow Circuit Low Input
P0102	0x2B50	Mass or Volume Air Flow Circuit Low Input
P0103	0x2778	Mass or Volume Air Flow Circuit High Input
P0107	0x281B	Manifold Absolute Pressure/Barometric Pressure Circuit Low Input
P0107	0x2B6D	Manifold Absolute Pressure/Barometric Pressure Circuit Low Input
P0108	0x281A	Manifold Absolute Pressure/Barometric Pressure Circuit High Input
P0108	0x2B6C	Manifold Absolute Pressure/Barometric Pressure Circuit High Input
P0111	0x2EFB	Intake Temperature Sensor Bank 1 Temperature Range / Performance

Table a. Diagnostic trouble codes (DTCs) (continued)

P-code	MINI FC	Definition
P0112	0x2EF2	Intake Air Temperature Sensor 1 Circuit Low
P0113	0x2EF1	Intake Air Temperature Sensor 1 Circuit High
P0116	0x2943	Coolant Temperature Sensor rationality Check
P0116	0x2F23	Coolant Temperature Sensor rationality Check
P0117	0x2936	Engine Coolant Temperature Circuit Low
P0117	0x2F1A	Engine Coolant Temperature Circuit Low
P0118	0x2937	Engine Coolant Temperature Circuit High
P0118	0x2F19	Engine Coolant Temperature Circuit High
P0120	0x2B0B	Throttle/Pedal Position Sensor/Switch 'A' Circuit
P0121	0x2B03	Throttle/Pedal Position Sensor/Switch 'A' Circuit Range/Performance
P0122	0x28A5	Throttle/Pedal Position Sensor/Switch 'A' Circuit Low
P0122	0x2B02	Throttle/Pedal Position Sensor/Switch 'A' Circuit Low
P0123	0x28A4	Throttle/Pedal Position Sensor/Switch 'A' Circuit High
P0123	0x2B01	Throttle/Pedal Position Sensor/Switch 'A' Circuit High
P0128	0x2F07	Coolant Thermostat (Coolant Temperature Below Thermostat Regulating Temperature)
P0128	0x348A	Program map thermostat: Sticking in open position
P0129	0x284D	Barometric pressure, operating range: Pressure too low
P0130	0x2AB4	O2 Sensor Circuit (Bank 1 Sensor 1)
P0130	0x2D2B	Before-catalyst oxygen sensor, electrical: Nernst-cell resistance or ceramic temperature implausible, wiring or heater fault
P0131	0x2A99	O2 Sensor Circuit Low Voltage (Bank 1 Sensor 1)
P0131	0x2D00	O2 Sensor Circuit Low Voltage (Bank 1 Sensor 1)
P0132	0x2A98	O2 Sensor Circuit High Voltage (Bank 1 Sensor 1)
P0132	0x2CFF	O2 Sensor Circuit High Voltage (Bank 1 Sensor 1)
P0133	0x2A35	O2 Sensor Circuit Slow Response (Bank 1 Sensor 1)
P0133	0x2CF4	O2 Sensor Circuit Slow Response (Bank 1 Sensor 1)
P0135	0x2A66	O2 Sensor Heater Circuit (Bank 1 Sensor 1)
P0135	0x2D18	Measured temperature of oxygen sensor before catalytic converter: Failure to achieve operating temperature in warmup phase
P0136	0x2A7F	O2 Sensor Circuit (Bank 1 Sensor 2)
P0136	0x2D22	Oxygen sensor behind catalytic converter, electrical: Open circuit
P0137	0x2A7D	O2 Sensor Circuit Low Voltage (Bank 1 Sensor 2)
P0137	0x2D20	O2 Sensor Circuit Low Voltage (Bank 1 Sensor 2)
P0138	0x2A7C	O2 Sensor Circuit High Voltage (Bank 1 Sensor 2)
P0139	0x2A77	O2 Sensor Circuit Slow Response (Bank 1 Sensor 2)
P0139	0x2D1E	Oxygen sensor behind catalytic converter, aging: Failure to reach overrun voltage threshold
P013A	0x2ACD	O2 Sensor Slow Response - Rich to Lean (Bank 1 Sensor 2)
P013A	0x2AD2	O2 Sensor Slow Response - Rich to Lean (Bank 1 Sensor 2)

Table a. Diagnostic trouble codes (DTCs) (continued)

P-code	MINI FC	Definition
P013A	0x2CED	Oxygen sensor behind catalytic converter, dynamic response, from lean to rich, Slow response
P013E	0x2ACD	O2 Sensor Delayed Response - Rich to Lean (Bank 1 Sensor 2)
P013E	0x2AD2	O2 Sensor Delayed Response - Rich to Lean (Bank 1 Sensor 2)
P013E	0x2D15	O2 Sensor Delayed Response - Rich to Lean (Bank 1 Sensor 2)
P0140	0x2A7E	O2 Sensor Circuit No Activity Detected (Bank 1 Sensor 2)
P0140	0x2D21	Oxygen sensor behind catalytic converter, electrical: Heating-cycle coupling to signal
P0141	0x2A56	O2 Sensor Heater Circuit (Bank 1 Sensor 2)
P0141	0x2D13	Heater for oxygen sensor behind catalytic converter, operation: Internal resistance too high
P0148	0x27CE	Fuel Delivery Error
P0171	0x2BC0	System Too Lean (Bank 1)
P0172	0x2BC1	System Too Rich (Bank 1)
P0190	0x2848	Fuel Rail Pressure Sensor ‘A’ Circuit
P0192	0x2849	Fuel Rail Pressure Sensor ‘A’ Circuit Low
P0192	0x2BDA	Fuel Rail Pressure Sensor ‘A’ Circuit Low
P0193	0x2BD9	Rail-pressure sensor, electrical: Short circuit to positive
P0201	0x2D57	Injector Circuit/Open - Cylinder 1
P0202	0x2D63	Injector Circuit/Open - Cylinder 2
P0203	0x2D5A	Injector Circuit/Open - Cylinder 3
P0204	0x2D5F	Injector Circuit/Open - Cylinder 4
P0221	0x2B07	Throttle/Pedal Position Sensor/Switch 'B' Circuit Range/Performance
P0222	0x28A9	Throttle/Pedal Position Sensor/Switch 'B' Circuit Low
P0222	0x2B06	Throttle/Pedal Position Sensor/Switch 'B' Circuit Low
P0223	0x28A8	Throttle/Pedal Position Sensor/Switch 'B' Circuit High
P0223	0x2B05	Throttle/Pedal Position Sensor/Switch 'B' Circuit High
P0234	0x2C56	Boost pressure control, upper value: Boost pressure too high
P0234	0x2C71	Boost pressure control, upper value: Boost pressure too high
P0234	0x2C83	Boost pressure control, upper value: Boost pressure too high
P0237	0x289E	Turbocharger Boost Sensor ‘A’ Circuit Low
P0237	0x2C70	Turbocharger Boost Sensor ‘A’ Circuit Low
P0238	0x289D	Turbocharger Boost Sensor ‘A’ Circuit High
P0238	0x2C6F	Turbocharger Boost Sensor ‘A’ Circuit High
P023A	0x2F37	Turbocharger ‘A’ Overboost Condition
P023B	0x2F36	Turbocharger ‘A’ Overboost Condition
P023C	0x2F35	Charge Air Cooler Coolant Pump Control Circuit High
P0243	0x2883	Turbocharger Wastegate Solenoid ‘A’
P0243	0x2CA3	Turbocharger Wastegate Solenoid ‘A’
P0245	0x2882	Turbocharger Wastegate Solenoid ‘A’ Low

Table a. Diagnostic trouble codes (DTCs) (continued)

P-code	MINI FC	Definition
P0245	0x2CA2	Turbocharger Wastegate Solenoid ‘A’ Low
P0246	0x2881	Turbocharger Wastegate Solenoid ‘A’ High
P0246	0x2CA1	Turbocharger Wastegate Solenoid ‘A’ High
P0261	0x2D55	Cylinder 1 Injector Circuit Low
P0262	0x2D54	Cylinder 1 Injector Circuit High
P0264	0x2D61	Cylinder 2 Injector Circuit Low
P0265	0x2D60	Cylinder 2 Injector Circuit High
P0267	0x2D59	Cylinder 3 Injector Circuit Low
P0268	0x2D58	Cylinder 3 Injector Circuit High
P0270	0x2D5D	Cylinder 4 Injector Circuit Low
P0271	0x2D5C	Cylinder 4 Injector Circuit High
P0299	0x2C57	Boost-pressure control, lower value: Boost pressure too low
P0299	0x2C72	Boost-pressure control, lower value: Boost pressure too low
P0299	0x2C84	Boost-pressure control, lower value: Boost pressure too low
P0300	0x2781	Cylinder Misfire, Several Cylinders
P0300	0x2782	Cylinder Misfire, Several Cylinders
P0300	0x2783	Cylinder Misfire, Several Cylinders
P0300	0x2EE0	Combustion miss, multiple cylinders: Fuel injection deactivation
P0300	0x2EE1	Combustion miss, multiple cylinders: Increased exhaust emissions
P0300	0x2EE2	Combustion miss, multiple cylinders: Increased exhaust emissions after start sequence
P0301	0x2771	Cylinder 1 Misfire Detected
P0301	0x2772	Cylinder 1 Misfire Detected
P0301	0x2773	Cylinder 1 Misfire Detected
P0301	0x2EE4	Combustion miss, cylinder 1: Fuel injection deactivation
P0301	0x2EE5	Combustion miss, cylinder 1: Increased exhaust emissions
P0301	0x2EE6	Combustion miss, cylinder 1: Deterioration in exhaust emissions after start sequence
P0302	0x277D	Cylinder 2 Misfire Detected
P0302	0x277E	Cylinder 2 Misfire Detected
P0302	0x277F	Cylinder 2 Misfire Detected
P0302	0x2EE7	Combustion miss, cylinder 2: Fuel injection deactivation
P0302	0x2EE8	Combustion miss, cylinder 2: Deterioration in exhaust emissions
P0302	0x2EE9	Combustion miss, cylinder 2: Deterioration in exhaust emissions after start sequence
P0303	0x2775	Cylinder 3 Misfire Detected
P0303	0x2776	Cylinder 3 Misfire Detected
P0303	0x2777	Cylinder 3 Misfire Detected
P0303	0x2EEA	Combustion miss, cylinder 3: Fuel injection deactivation
P0303	0x2EEC	Combustion miss, cylinder 3: Deterioration in exhaust emissions after start sequence

Table a. Diagnostic trouble codes (DTCs) (continued)		
P-code	**MINI FC**	**Definition**
P0304	0x2779	Cylinder 4 Misfire Detected
P0304	0x277A	Cylinder 4 Misfire Detected
P0304	0x277B	Cylinder 4 Misfire Detected
P0304	0x2EED	Combustion miss, cylinder 4: Fuel injection deactivation
P0304	0x2EEF	Combustion miss, cylinder 4: Deterioration in exhaust emissions
P0304	0x2EF0	Combustion miss, cylinder 4: Deterioration in exhaust emissions after start sequence
P0324	0x2D53	Knock Control System Error
P0324	0x303E	Knock control, fault check: Malfunction, system fault
P0327	0x2D8C	Knock Sensor 1 Circuit Low (Bank 1 or Single Sensor)
P0327	0x2D8E	Knock Sensor 1 Circuit Low (Bank 1 or Single Sensor)
P0327	0x2D9C	Knock Sensor 1 Circuit Low (Bank 1 or Single Sensor)
P0327	0x3049	Knock sensor signal: Electrical fault KS (loose contact) or KS loose
P0328	0x2D8B	Knock Sensor 1 Circuit High (Bank 1 or Single Sensor)
P0328	0x2D8D	Knock Sensor 1 Circuit High (Bank 1 or Single Sensor)
P0328	0x2D9B	Knock Sensor 1 Circuit High (Bank 1 or Single Sensor)
P0328	0x304B	Knock sensor signal: Engine produces excessive mechanical noise or crankshaft outside tolerance (sensitivity)
P0335	0x296F	Crankshaft Position Sensor 'A' Circuit
P0335	0x2FDA	Crankshaft sensor, signal: Missing
P0336	0x296E	Crankshaft Position Sensor 'A' Range
P0336	0x2FDB	Crankshaft sensor: Interference on crankshaft signal
P0340	0x2969	Camshaft Position Sensor 'A' Circuit (Bank 1 or Single Sensor)
P0340	0x2982	Camshaft Position Sensor 'A' Circuit (Bank 1 or Single Sensor)
P0341	0x2968	Camshaft Position Sensor 'A' Performance (Bank 1)
P0341	0x2D9F	Intake camshaft sensor: Signal implausible
P0342	0x2983	Camshaft Position Sensor 'A' Circuit Low (Bank 1)
P0342	0x300D	Intake camshaft position sensor: Signal too low
P0343	0x300C	Intake camshaft sensor: Signal is high
P0351	0x273D	Ignition Coil 'A' Primary / Secondary Circuit
P0352	0x2740	Ignition Coil 'B' Primary / Secondary Circuit
P0353	0x273E	Ignition Coil 'C' Primary / Secondary Circuit
P0354	0x273F	Ignition Coil 'D' Primary / Secondary Circuit
P0365	0x296C	Camshaft Position Sensor 'B' Circuit (Bank 1)
P0365	0x2986	Camshaft Position Sensor 'B' Circuit (Bank 1)
P0366	0x296B	Camshaft Position Sensor 'B' Performance (Bank 1)
P0366	0x2DA1	Exhaust camshaft sensor: Signal implausible
P0367	0x2987	Camshaft Position Sensor 'B' Circuit Low (Bank 1)

Table a. Diagnostic trouble codes (DTCs) (continued)		
P-code	**MINI FC**	**Definition**
P0367	0x300F	Exhaust camshaft sensor: Signal too low
P0368	0x300E	Exhaust camshaft sensor: Signal is high
P0420	0x276A	Catalyst System Efficiency Below Threshold (Bank 1)
P0420	0x3106	Catalytic converter: Operational efficiency below specified limit
P0441	0x27B0	Evaporative Emission System Incorrect Purge Flow
P0442	0x278C	Evaporative Emission System Leak Detected (small leak)
P0442	0x321C	Tank EVAP and purge-air system, minor leak: Leak larger than 1.0 mm
P0444	0x27B4	Evaporative Emission System Purge Control Valve Circuit Open
P0444	0x3157	Tank EVAP valve, control-activation: Open circuit
P0455	0x2710	Evaporative Emission System Leak Detected (large leak)
P0455	0x2711	Evaporative Emission System Leak Detected (large leak)
P0456	0x272F	Evaporative Emission System Leak Detected (very small leak)
P0456	0x321D	Tank EVAP and purge-air system, minute leak: Leak larger than 0.5 mm
P0458	0x27B2	Evaporative Emission System Purge Control Valve Circuit Low
P0458	0x3156	Tank EVAP valve, control-activation: Short circuit to ground
P0459	0x27B1	Evaporative Emission System Purge Control Valve Circuit High
P0459	0x3155	Tank EVAP valve, control-activation: Short circuit to positive
P0460	0x2755	Fuel Level Sensor 'A' Circuit
P0461	0x275D	Fuel Level Sensor A Performance
P0462	0x2756	Fuel Level Sensor 'A' Circuit Low
P0462	0x3188	Tank fuel-level sensor, left, signal: Short circuit to ground
P0463	0x2755	Fuel Level Sensor 'A' Circuit High
P0463	0x3187	Tank fuel-level sensor, left, signal: Short circuit to positive
P0480	0x2EE8	Fan 1 Control Circuit
P0480	0x31E9	Electric fan, control-activation: Open circuit
P0481	0x2F3F	Fan 2 Control Circuit
P0500	0x2FF7	Vehicle Speed Sensor 'A'
P0501	0x3000	Vehicle Speed Sensor 'A' Range/Performance
P0501	0x32DA	Vehicle speed, plausibility: Implausible road speed signal
P0503	0x2FF6	Vehicle Speed Sensor 'A' Intermittent / Erratic
P0503	0x32D0	Vehicle speed: Signal too high
P0506	0x2971	Idle Air Control System Lower Than Expected
P0506	0x2975	Idle Air Control System Lower Than Expected
P0506	0x3521	Idle control: Engine speed too low
P0507	0x2970	Idle Air Control System Higher Than Expected
P0507	0x2974	Idle Air Control System Higher Than Expected
P0507	0x3520	Idle control: Engine speed too high

Table a. Diagnostic trouble codes (DTCs) (continued)

P-code	MINI FC	Definition
P050B	0x2F83	Ignition angle adjustment at idle, in cold start: Ignition angle advanced too far
P0520	0x2FC4	Engine Oil Pressure Switch Circuit
P0521	0x342B	Engine oil-pressure sensor, plausibility: Signal hangs
P0521	0x342C	Engine oil-pressure sensor, plausibility: Open ground wire
P0522	0x3427	Engine-oil pressure sensor, electrical Short circuit to ground
P0523	0x3426	Engine-oil pressure sensor, electrical Short circuit to positive
P0524	0x3406	Oil Pump: Oil Pressure Too Low
P052B	0x2D55	VANOS, intake-side, cold start: No closed-loop control possible
P0532	0x2EE1	A/C Pressure Sensor 'A' Circuit Low
P0533	0x2EE0	A/C Pressure Sensor 'A' Circuit High
P053A	0x37DE	Engine PCV heating system, control-activation: Open circuit
P053B	0x37DD	Engine PCV heating, control-activation: Short circuit to ground
P053C	0x37DC	Engine PCV heating, control-activation: Short circuit to positive
P054B	0x2D54	VANOS, exhaust-side, cold start: No closed-loop control possible
P0557	0x2FC3	Brake Booster Pressure Sensor Circuit Low
P0558	0x2FC2	Brake Booster Pressure Sensor Circuit High
P0560	0x2C9A	System Voltage
P0560	0x3888	Onboard electrical system voltage: Analog-digital converter defective
P0562	0x2C99	System Voltage Low
P0562	0x3887	Onboard electrical system voltage: Voltage too low
P0562	0x5015	System Voltage Low
P0563	0x2C98	System Voltage High
P0563	0x5014	System Voltage High
P0571	0x2FAA	Brake Switch 'A' Circuit
P0571	0x3325	Brake light switch, plausibility: Signal implausible
P0597	0x2EE5	Thermostat Heater Control Circuit / Open
P0597	0x3490	Program map thermostat, control-activation: Open circuit
P0598	0x2EE3	Thermostat Heater Control Circuit Low
P0598	0x348F	Program map thermostat, control-activation: Short circuit to ground
P0599	0x2EE2	Thermostat Heater Control Circuit High
P0599	0x348E	Program map thermostat, control-activation: Short circuit to positive
P0604	0x4FB3	Internal Control Module Random Access Memory (RAM) Error
P0605	0x4FB0	Internal Control Module Read Only Memory (ROM) Error
P060A	0x36B8	DME, internal fault Monitoring-module fault
P060B	0x2BC0	Internal Control Module A/D Processing Performance
P060B	0x2BC1	Internal Control Module A/D Processing Performance
P060B	0x2BC2	Internal Control Module A/D Processing Performance

Table a. Diagnostic trouble codes (DTCs) (continued)		
P-code	**MINI FC**	**Definition**
P060B	0x2BC3	Internal Control Module A/D Processing Performance
P060D	0x36C3	DME, internal fault Accelerator pedal module or pedal-travel sensor signal plausibilizing monitor
P0615	0x2FE7	Starter Relay Circuit
P0615	0x36FC	Starter, control-activation: Open circuit
P0616	0x2FE6	Starter Relay Circuit Low
P0616	0x36FB	Starter, control-activation: Short circuit to ground
P0617	0x2FE5	Starter Relay Circuit High
P0617	0x36FA	Starter, control-activation: Short circuit to positive
P061F	0x28BA	Internal Control Module Throttle Actuator Performance
P061F	0x2B1F	Internal Control Module Throttle Actuator Performance
P0620	0x2E24	Generator Control Circuit
P0620	0x2E33	Generator Control Circuit
P0620	0x3840	Alternator, electrical: Malfunction
P0627	0x276F	Fuel Pump ‘A’ Control Circuit Open
P0627	0x2770	Fuel Pump ‘A’ Control Circuit Open
P0628	0x276E	Fuel Pump ‘A’ Control Circuit Low
P0629	0x276D	Fuel Pump ‘A’ Control Circuit High
P062F	0x36B5	DME, internal fault: Delete EEPROM faulty
P0634	0x2FF2	PCM/ECM/TCM Internal Temperature ‘A’ Too High
P0634	0x3584	DME, internal fault, inside temperature sensor: Value too high
P0641	0x2BD2	Sensor Reference Voltage ‘A’ Circuit/Open
P0651	0x2BD3	Sensor Reference Voltage ‘B’ Circuit/Open
P0686	0x2C9D	ECM Power Relay Control Circuit Low
P0687	0x2C9C	ECM Power Relay Control Circuit High
P0687	0x3714	Onboard electrical system voltage, DME main relay: Voltage too high
P0691	0x2EE7	Fan 1 Control Circuit Low
P0691	0x31E8	Electric fan, control-activation: Short circuit to ground
P0692	0x2EE6	Fan 1 Control Circuit High
P0692	0x2EE6	Fan 1 Control Circuit High
P0692	0x31E7	Electric fan, control-activation: Short circuit to positive
P0694	0x2F3C	Fan 2 Control Circuit High
P0697	0x2BD4	Sensor Reference Voltage ‘C’ Circuit/Open
P0704	0x2FC1	Clutch Switch Input Circuit
P0705	0x5089	Transmission Range Sensor ‘A’ Circuit (PRNDL Input)
P0706	0x5088	Transmission Range Sensor ‘A’ Circuit Range/Performance
P0711	0x4EF4	Transmission Fluid Temperature Sensor ‘A’ Circuit Range/Performance
P0712	0x4EF2	Transmission Fluid Temperature Sensor ‘A’ Circuit Low

Table a. Diagnostic trouble codes (DTCs) (continued)		
P-code	**MINI FC**	**Definition**
P0713	0x4EF3	Transmission Fluid Temperature Sensor 'A' Circuit High
P0715	0X4EE9	Input/Turbine Speed Sensor 'A' Circuit
P0716	0X4EEA	Input/Turbine Speed Sensor 'A' Circuit Range/Performance
P0717	0x4EE8	Input/Turbine Speed Sensor 'A' Circuit No Signal
P0720	0X4EEC	Output Speed Sensor Circuit
P0721	0x4EED	Output Speed Sensor Circuit Range/Performance
P0722	0x4EEB	Output Speed Sensor Circuit No Signal
P0729	0x4F97	Gear 6 Incorrect Ratio
P0732	0x4F93	Gear 2 Incorrect Ratio
P0733	0x4F94	Gear 3 Incorrect Ratio
P0734	0x4F95	Gear 4 Incorrect Ratio
P0735	0x4F96	Gear 5 Incorrect Ratio
P0741	0x4F53	Torque Converter Clutch Circuit Performance/Stuck Off
P0961	0x4E22	Pressure Control Solenoid 'A' Control Circuit Range/Performance
P0962	0x4E21	Pressure Control Solenoid 'A' Control Circuit Low
P0963	0x4E20	Pressure Control Solenoid 'A' Control Circuit High
P0969	0x4E36	Pressure Control Solenoid 'C' Control Circuit Range/Performance
P0970	0x4E35	Pressure Control Solenoid 'C' Control Circuit Low
P0971	0x4E34	Pressure Control Solenoid 'C' Control Circuit High
P0973	0x4E85	Shift Solenoid 'A' Control Circuit Low
P0974	0x4E84	Shift Solenoid 'A' Control Circuit High
P0976	0x4E88	Shift Solenoid 'B' Control Circuit Low
P0977	0x4E89	Shift Solenoid 'B' Control Circuit High
P0A3B	0x2E2C	Generator Over Temperature
P0A3B	0x2E30	Generator Over Temperature
P0A3B	0x3848	Alternator, temperature: Over temperature
P1000	0x2888	VVT-System Min. Stroke Adaptation Number of Stops Exceeded
P1004	0x2853	VVT Guiding Sensor Solenoid Loss (Bank 1)
P1005	0x2854	VVT Guiding Sensor Reset Error (Bank 1)
P1006	0x2856	VVT Eccentric Shaft Sensor Parity Error (Bank 1)
P1007	0x2855	VVT Eccentric Shaft Sensor Gradient Error (Bank 1)
P1012	0x2857	VVT Guiding Sensor Solenoid Loss (Bank 1)
P1013	0x2858	VVT Reference Sensor Reset Error (Bank 1)
P1014	0x285A	VVT Eccentric Shaft Sensor Parity Error (Bank 1)
P1015	0x2859	VVT Eccentric Shaft Sensor Gradient Error (Bank 1)
P1017	0x2861	VVT Guiding Sensor Plausibility (Bank 1)
P1017	0x2DD8	Valvetronic actuator motor, position sensors: Signal implausible

Table a. Diagnostic trouble codes (DTCs) (continued)

P-code	MINI FC	Definition
P1019	0x286D	VVT Supply Voltage Sensors High Input (Bank 1)
P101A	0x2D42	Valvetronic, adjustment range: Travel range not initialized
P101A	0x2D45	Valvetronic, adjustment range: Travel limit not learned owing to environmental conditions
P101B	0x285D	VVT Self-Learning Function Storage of Learning Values in EEPROM Not Possible
P101C	0x28B6	VVT Load Shedding Relay Control Circuit Input High
P101D	0x28B7	VVT Load Shedding Relay Control Circuit Input Low
P101E	0x28B9	VVT Load Shedding Relay Control Circuit
P1020	0x286E	VVT Sensor Supply Voltage Low Input (Bank 1)
P1023	0x285B	VVT Self Learning Function Faulty Adjustment Range (Bank 1)
P1023	0x2D43	Valvetronic, adjustment range: Range check fault
P1023	0x2D44	Valvetronic, adjustment range: Range check, deviation from base learning routine
P1024	0x285C	VVT Self Learning Function Faulty Lower Learning Range (Bank 1)
P1025	0x285E	VVT Self Learning Function No Positions Stored (Bank 1)
P1030	0x2869	VVT Control Motor Position Control Deviation (Bank1)
P1030	0x2DCE	Valvetronic system: Closed-loop control deviation too high
P1031	0x2868	VVT Actuator Monitoring Recognition of Direction of Rotation Plausibility (Bank 1)
P1047	0x284F	VVT Control Circuit High (Bank 1)
P1047	0x2DBF	Valvetronic actuator motor, control-activation Short circuit to positive
P1048	0x2850	VVT Control Circuit Low (Bank 1)
P1048	0x2DC0	Valvetronic actuator motor, control-activation: Short circuit to ground
P1049	0x2852	VVT Control Circuit Short (Bank 1)
P1050	0x2851	Valvetronic Module Power Stage Generic
P1050	0x2DC3	Valvetronic actuator motor, control-activation: Open circuit
P1056	0x286B	VVT Control Circuit High (Bank 1)
P1057	0x2847	VVT Motor Supply Voltage (Bank 1)
P1057	0x286C	VVT Motor Supply Voltage (Bank 1)
P105C	0x28B4	VVT Control Motor Tight
P105D	0x2EEE	Cold Start Intake Air Temperature - Too Low
P1062	0x2863	VVT Limp Home Request Full Stroke Position Reached (Bank 1)
P1063	0x2865	VVT Limp Home Request Air Mass Plausibility (Bank 1)
P1077	0x2860	VVT Overload Protection Control Motor Input Temperature (Bank 1)
P1078	0x285F	VVT Overload Protection Current Too High (Bank 1)
P10B0	0x29E4	Charge air temperature: Plausibility, temperature too high
P10B8	0x29E5	Charge air temperature: Signal, remains stationary
P10CE	0x2BF2	Rail-pressure sensor, operating range: Pressure too high
P10CF	0x2BF3	Rail-pressure sensor, operating range pressure too low
P10D0	0x29DC	Charge air temperature sensor, plausibility, cold start: Temperature too high

OBD

Table a. Diagnostic trouble codes (DTCs) (continued)		
P-code	**MINI FC**	**Definition**
P10D1	0x29DD	Charge air temperature sensor, plausibility, cold start: Temperature too low
P10D4	0x293B	Coolant temperature sensor, plausibility, cold start: Temperature too low
P10D5	0x293A	Coolant temperature sensor, plausibility, cold start: Temperature too high
P10D9	0x2BF8	Rail-pressure sensor, signal: No variation
P10DE	0x2D41	Valvetronic, adjustment range: Initial learning outside tolerances
P10DF	0x2DBA	Valvetronic, driver circuit component protection: System deactivation
P10E1	0x2DCF	Valvetronic system: No motion detected
P10E2	0x2DD6	Valvetronic actuator motor, position sensors: Short circuit or line break (open circuit)
P10E5	0x36B0	DME, internal fault, Valvetronic control: Malfunction
P10E6	0x2DD7	Valvetronic actuator motor, position sensors: Malfunction in voltage supply
P10E7	0x2DCA	Valvetronic, overload protection: Driver circuit overloaded
P10E8	0x2DCB	Valvetronic, overload protection: Actuator motor overloaded
P10F2	0x2C85	Boost pressure - barometric pressure, comparison: Boost air pressure too high
P10F3	0x2C86	Boost pressure - barometric pressure, comparison: Boost air pressure too low
P1104	0x2B16	Differential Pressure Sensor Intake Manifold Pressure Too Low Bank 1
P1104	0x2B6A	Differential Pressure Sensor Intake Manifold Pressure Too Low Bank 1
P1105	0x2B15	Differential Pressure Sensor Intake Manifold Pressure Too High Bank 1
P1105	0x2B69	Differential Pressure Sensor Intake Manifold Pressure Too High Bank 1
P110D	0x26AC	Throttle Position Sensor ‘A’ and ‘B’ Range / Performance
P110E	0x2752	Internal Code (Service/End-of-Line Test)
P110E	0x2753	Internal Code (Service/End-of-Line Test)
P110E	0x2754	Internal Code (Service/End-of-Line Test)
P110F	0x299C	Ambient Temperature Sensor Faulty CAN Signal
P110F	0x2F28	Ambient Temperature Sensor Faulty CAN Signal
P111A	0x2B67	Mass or Volume Air Flow va O2 Sensor Too High (Bank 1)
P111B	0x2B68	Mass or Volume Air Flow va O2 Sensor Too Low (Bank 1)
P111E	0x2EF9	Intake Temperature Sensor Bank 1 Maximum Temperature Implausible
P1124	0x2AFC	Differential Pressure Sensor Intake Manifold Pressure Offset Bank 1
P112B	0x2F22	Engine Coolant Temperature Sensor 1 Minimum Temperature Implausible
P112E	0x28A1	Throttle valve angle - Intake manifold pressure, comparison: Pressure too low
P112F	0x28A0	Manifold Absolute Pressure to Throttle Angle - Too High
P113A	0x2B5E	Mass or Volume Air Flow 1 Correction Signal Plausibility Too Long
P113A	0x30AE	Mass or Volume Air Flow 1 Correction Signal Plausibility Too Long
P113B	0x2B5F	Mass or Volume Air Flow 1 Correction Signal Plausibility Too Long
P115A	0x2B59	Mass or Volume Air Flow ‘A’ Maximum Exceeded
P115B	0x2B5A	Mass Air Volume Flow ‘A’ Minimum Fallen Below
P115C	0x2775	Mass Air Volume Flow ‘A’ Too Low

Table a. Diagnostic trouble codes (DTCs) (continued)		
P-code	**MINI FC**	**Definition**
P115C	0x2B56	Mass Air Volume Flow 'A' Too Low
P115C	0x2B58	Mass Air Volume Flow 'A' Too Low
P115C	0x2B5C	Mass Air Volume Flow 'A' Too Low
P115D	0x2774	Mass Air Volume Flow 'A' Too High
P115D	0x2B55	Mass Air Volume Flow 'A' Too High
P115D	0x2B57	Mass Air Volume Flow 'A' Too High
P115D	0x2B5B	Mass Air Volume Flow 'A' Too High
P115F	0x28B4	Throttle/Pedal Position Sensor/Switch 'A' / 'B' Synchronous Operation Correlation (Bank 1)
P115F	0x2B13	Throttle/Pedal Position Sensor/Switch 'A' / 'B' Synchronous Operation Correlation (Bank 1)
P116F	0x27AB	Air Mass Flow Sensor Signal Electrical
P118A	0x2F40	Engine Oil Separator Heating Circuit High
P118B	0x2F41	Engine Oil Separator Heating Circuit Low
P118C	0x2F42	Engine Oil Separator Heating Circuit Open
P1197	0x2AF9	Differential Pressure Sensor Intake Manifold High Input Bank 1
P1198	0x2AFA	Differential Pressure Sensor Intake Manifold Low Input Bank 1
P1199	0x2B6B	Differential Pressure Sensor Intake Manifold Low Input Bank 1
P119A	0x281A	Manifold Absolute Pressure Sensor High (Bank 1)
P119B	0x281B	Manifold Absolute Pressure Sensor Low (Bank 1)
P119D	0x2BC5	Fuel Trim, Injector Aging Long Term Adaptation Too High (Bank 1)
P11AA	0x2714	Throttle valve, operation: Resistance to motion, too slow
P11CB	0x284C	Barometric pressure, operating range: Pressure too high
P1216	0x27CF	Fuel Pump Emergency Operation
P1247	0x284E	Barometric pressure, operating range: Pressure implausible
P1247	0x284F	Barometric pressure, operating range: Pressure implausible
P1250	0x280E	Manifold Absolute Pressure Too High
P1250	0x2B71	Manifold Absolute Pressure Too High
P1255	0x280F	Manifold Absolute Pressure Too Low
P1255	0x2B72	Manifold Absolute Pressure Too Low
P1260	0x2C58	Boost-pressure control, deactivation: Boost pressure build-up blocked
P1296	0x2884	Boost Pressure Control Boost Pressure Too High
P1297	0x2885	Boost Pressure Control Boost Pressure Too Low
P129D	0x2B73	Manifold Absolute Pressure Maximum Pressure Implausible
P129E	0x2B74	Manifold Absolute Pressure Minimum Pressure Implausible
P12A0	0x28A2	Turbocharger/Supercharger Boost Pressure, Pressure Front of Throttle Valve Too High
P12A0	0x2C83	Turbocharger/Supercharger Boost Pressure, Pressure Front of Throttle Valve Too High
P12A1	0x28A3	Turbocharger/Supercharger Boost Pressure, Pressure Front of Throttle Valve Too Low
P12A1	0x2C84	Turbocharger/Supercharger Boost Pressure, Pressure Front of Throttle Valve Too Low

Table a. Diagnostic trouble codes (DTCs) (continued)		
P-code	**MINI FC**	**Definition**
P12A2	0x28A4	Turbocharger/Supercharger Boost Pressure, Maximum Pressure Front of Throttle Valve Implausible
P12A3	0x28A5	Turbocharger/Supercharger Boost Pressure, Minimum Pressure Front of Throttle Valve Implausible
P12A5	0x280E	Absolute pressure sensor, intake manifold, plausibility, shutdown phase: Pressure too high
P12A8	0x2C72	Boost-pressure sensor, plausibility, shutdown phase: Pressure too low
P12A9	0x2C71	Boost-pressure sensor, plausibility, shutdown phase: Pressure too high
P12B8	0x2842	Barometric pressure sensor, plausibility, shutdown phase: Pressure too low
P12B9	0x2841	Barometric pressure sensor, plausibility, shutdown phase: Pressure too high
P12DA	0x2C85	Boost pressure - barometric pressure, comparison: Boost pressure too high
P12DB	0x2C86	Boost pressure - barometric pressure, comparison: Boost pressure too low
P1338	0x296A	Camshaft Position Sensor Rationality Check (Inlet - Outlet)
P1338	0x2DA0	Intake camshaft: Angular offset to crankshaft is outside tolerance range
P1339	0x296D	Camshaft Position Sensor Rationality Check (Inlet - Outlet)
P1339	0x2DA2	Exhaust camshaft: Angular offset to crankshaft is outside tolerance range
P136C	0x2DC2	Knock Control Super Knocking Detected
P137A	0x2D51	Knock Control Super Knocking Caused by Defective Ignition Coil
P137A	0x2F82	Super-knock: Super-knock with ignition fault
P137B	0x2D50	Knock Control Super Knocking Caused by Defective Knock Sensor
P137B	0x2F81	Super-knock: Knock sensor fault, super-knock in limp-home mode
P137D	0x2DC0	Knock Control Torque Limitation Caused by Too High Number of Super Knocking
P137D	0x2F7F	Knock Control Torque Limitation Caused by Too High Number of Super Knocking
P137E	0x2DC1	Knock Control Permanent Torque Limitation Caused by Too High Number of Super Knocking
P137E	0x2F80	Knock Control Permanent Torque Limitation Caused by Too High Number of Super Knocking
P137F	0x2D52	Knock Control Fuel Cutoff Due to Super Knocking
P137F	0x2F7C	Knock Control Fuel Cutoff Due to Super Knocking
P13A0	0x2DCC	Knock Control Fuel Cut-Off Due to Super Knocking Cylinder 1
P13A0	0x2F76	Super-knock on Cylinder 1: Injection deactivation
P13A1	0x2DCB	Knock Control Fuel Cut-Off Due to Super Knocking Cylinder 2
P13A1	0x2F77	Super-knock on Cylinder 2: Injection deactivation
P13A2	0x2DCD	Knock Control Fuel Cut-Off Due to Super Knocking Cylinder 3
P13A2	0x2F78	Super-knock on Cylinder 3: Injection deactivation
P13A3	0x2DCA	Knock Control Fuel Cut-Off Due to Super Knocking Cylinder 4
P13A3	0x2F79	Super-knock on Cylinder 4: Injection deactivation
P13AE	0x3040	Knock sensor, electrical: Signal input A, short circuit to ground
P13AF	0x303F	Knock sensor, electrical: Signal input A, short circuit to positive
P13B8	0x3042	Knock sensor, electrical: Signal input B, short circuit to ground
P13B9	0x3041	Knock sensor, electrical: Signal input B, short circuit to positive
P13C0	0x2D5A	VANOS intake: Closed-loop control fault, camshaft sticks

Table a. Diagnostic trouble codes (DTCs) (continued)		
P-code	**MINI FC**	**Definition**
P13C9	0x2D5A	VANOS intake: Closed-loop control fault, camshaft sticks
P13CA	0x3011	Exhaust camshaft, mechanical: Assembly faulty
P13CB	0x3012	Exhaust camshaft, mechanical: Incorrect installation
P13EA	0x2F84	Ignition-angle adjustment at part-load, cold start: Ignition angle advanced too far
P1407	0x2757	Fuel Level Signal 1
P1407	0x318A	Tank fuel-level sensor, left, signal: CAN value implausible
P1408	0x3185	Tank fuel-level sensor, right, signal: Implausible CAN data
P1409	0x2758	Fuel Level 1 CAN Error
P1433	0x27C6	Fuel level 2 CAN error
P1434	0x2736	DMTL Tank Leakage
P1434	0x3220	DMTL, system fault: Abort due to current fluctuations during reference measurement
P1447	0x2735	Diagnostic Module Tank Leakage (DMTL) Pump Current Too High during Switching Solenoid Test
P1447	0x3221	DMTL, system fault: Pump current reaches limit during valve test
P1448	0x2734	Diagnostic Module Tank Leakage (DMTL) Pump Current Too Low
P1448	0x321F	DMTL, system fault: Pump current too low during reference measurement
P1449	0x2733	Diagnostic Module Tank Leakage (DMTL) Pump Current Too High
P1449	0x321E	DMTL, system fault: Pump current too high during reference measurement
P144A	0x275E	Fuel level / Tank Capacity Correlation
P144B	0x275F	Fuel level / Fuel Consumption Correlation
P144B	0x318D	Tank fuel-level sensor: Fuel consumption rate does not correlate with change fuel level
P1497	0x2B64	Downstream Throttle Air Leak
P150A	0x2E44	Battery Sensor Extended Communication Circuit
P150B	0x2E47	Battery Sensor Serial Data Interface
P150B	0x2E4F	Battery Sensor Serial Data Interface
P150C	0x2E46	Battery Sensor Firmware Implausible
P150C	0x38A5	Intelligent battery sensor, signal: Software fault
P150D	0x2E40	Battery Sensor Temperature Error
P150D	0x38A9	Intelligent battery sensor, function: Temperature fault
P150E	0x2E43	Battery Sensor Voltage Error
P150E	0x38AA	Intelligent battery sensor, function: Voltage fault
P150F	0x2E42	Battery Sensor Current Error
P1515	0x2FAC	Engine OFF Timer Plausibility
P151A	0x2E48	Battery Sensor Terminal 15 / 30 Wakeup Circuit
P151B	0x2E4A	Battery Sensor Wakeup Circuit
P151B	0x38B2	Intelligent battery sensor, signal transmission: Implausible wake-up wire signal level
P151C	0x2E4B	Battery Sensor System Error
P152A	0x32D8	Vehicle speed, plausibility: Minimum speed under load not reached

Table a. Diagnostic trouble codes (DTCs) (continued)

P-code	MINI FC	Definition
P152B	0x2FFF	Vehicle Speed Sensor Too Low During Coast Down Compared to Reference Under Load
P152B	0x32D9	Vehicle speed, plausibility: Minimum speed in coast mode not reached
P1551	0x2FAD	Engine OFF Timer Time-out
P1551	0x3390	Engine shutdown time: Timeout or invalidity value
P1561	0x2973	Cold Start Idle RPM Lower Than Expected
P1561	0x3525	Idle control, cold start: Engine speed too low
P1562	0x2972	Cold Start Idle RPM Higher Than Expected
P1562	0x3524	Idle control, cold start: Engine speed too high
P1577	0x2FF4	Speed Indication Instrument Pack
P159E	0x33FC	Engine oil-pressure control, dynamic: Pressure fluctuations
P159F	0x33FD	Engine oil-pressure control, static: Engine oil pressure too high, backup mode
P15A0	0x33FE	Engine oil-pressure control, static: Engine oil pressure too low, backup mode
P15A1	0x3408	Oil-pressure control valve: Sticks when under maximum current (minimum oil pressure)
P15A2	0x3409	Oil-pressure control valve: Sticks in no-voltage position (maximum oil pressure)
P15A3	0x3405	Oil pump: Oil pressure too high
P15A5	0x3404	Engine oil-pressure system: Control instability
P15A6	0x3429	Engine oil-pressure sensor, plausibility: Pressure too high before start
P15A7	0x342A	Engine oil-pressure sensor, plausibility: Pressure too low before start
P15B5	0x2FF5	Speed Indication Instrument Pack / DSC Signal Correlation
P15C7	0x2FF5	Oil-pressure control valve: Not plausible with component analysis
P15DA	0x32DD	Vehicle speed, wheelspeed sensor front/right, plausibility: Signal implausible
P15DB	0x32DF	Vehicle speed, wheelspeed sensor left/right, plausibility: Signal implausible
P15DC	0x32DC	Vehicle speed, wheelspeed sensor left/right, plausibility: Signal implausible
P15DD	0x32DE	Vehicle speed, wheelspeed sensor rear/right, plausibility: Implausible signal
P15DE	0x2BED	High-pressure fuel system, cold start: Pressure too high
P15DF	0x2BED	High-pressure fuel system, cold start: Pressure too low
P15E8	0x3392	Engine shutdown time, plausibility: Time too short relative to temperature drop in engine coolant
P15E9	0x3393	Engine shutdown time, plausibility: Time to long in correlation to engine-coolant cooling
P15EA	0x3401	Oil-pressure control valve, activation: Open circuit
P15EC	0x33FF	Oil-pressure control valve, activation: Short circuit to positive
P15FA	0x3394	Engine shutdown time: Too rapid with engine running
P15FB	0x3395	Engine shutdown time: Too slow with engine running
P15FC	0x3398	Engine shutdown time: Too rapid in shutdown phase
P15FD	0x3399	Engine shutdown time: Too slow in shutdown phase
P15FE	0x3396	Engine shutdown time, signal: Missing
P1603	0x2BD9	Control Module Self-Test, Torque Monitoring
P1603	0x2BD9	Control Module Self-Test, Torque Monitoring

Table a. Diagnostic trouble codes (DTCs) (continued)

P-code	MINI FC	Definition
P1605	0x2BCB	Safety Concept Torque Limitation Level 1
P160A	0x2C8D	Powermanagement Exhaustive Discharge
P160A	0x387C	Power management, battery monitoring: Deep discharge
P160B	0x2C8E	Powermanagement Defective
P160C	0x2C90	Powermanagement Overvoltage
P160C	0x3877	Power management, overvoltage: Overvoltage detected
P160D	0x2C91	Powermanagement Undervolatge
P160D	0x3878	Power management, undervoltage: Undervoltage detected
P160E	0x2C93	Powermanagement Operation Without Battery
P160E	0x3879	Power management: Operation without battery
P160F	0X2C96	Powermanagement No Load Current Error
P160F	0X387F	Power management, standby current monitor: Standby current violation
P1626	0x278A	DME, internal fault, mass air flow sensor: Fault in driver circuit
P1628	0x2B26	Throttle Valve Spring Does Not Open
P1629	0x2B25	Throttle Valve Spring Does Not Open
P1631	0x28BD	Throttle Valve Actuator Spring Test
P1631	0x2B22	Throttle Valve Actuator Spring Test
P1632	0x28CC	Throttle Valve Adaptation Conditions Not Met
P1633	0X2B2F	Throttle Valve Adaptation Limp-Home Position Unknown
P1633	0X2BC4	Throttle Valve Adaptation Limp-Home Position Unknown
P1634	0x28BC	Throttle Valve Adaptation Spring Test Failed
P1634	0x2B21	Throttle Valve Adaptation Spring Test Failed
P1635	0x2B3F	Throttle Valve Adaptation Lower Mechanical Stop not Adapted
P1637	0x2B2B	Throttle Valve Position Control, Control Deviation
P1638	0x2711	Throttle Valve Position Control Throttle Stuck Temporarily
P1638	0x2B32	Throttle Valve Position Control Throttle Stuck Temporarily
P1639	0x2710	Throttle Valve Position Control Throttle Stuck Permanently
P1639	0x2B31	Throttle Valve Position Control Throttle Stuck Permanently
P163A	0x2FF3	PCM/TECM/TCM Internal Temperature Too Low
P163A	0x3585	DME, internal fault, inside temperature sensor: Value too low
P163C	0x2BD1	Voltage Monitoring Control Module Overload
P163D	0x2BC4	Voltage Monitoring Control Module Cutoff Detected
P163D	0x2BCF	Voltage Monitoring Control Module Cut-OFF Detected
P163D	0x2BD1	Voltage Monitoring Control Module Overvoltage
P163E	0x2BC4	Voltage Monitoring Control Module Communication Error
P163E	0x2BC5	Voltage Monitoring Control Module Communication Error
P163E	0x2BCC	Voltage Monitoring Control Module Communication Error

Table a. Diagnostic trouble codes (DTCs) (continued)		
P-code	**MINI FC**	**Definition**
P163E	0x2BD0	Voltage Monitoring Control Module Communication Error
P1641	0x2B39	Throttle Valve Adaptation Stop Due to Environmental Conditions
P1642	0x2B3A	Throttle Valve Adaptation Stop Due to Environmental Values
P1643	0x28D8	Throttle Actuator Start Test Amplifier Balancing Faulty
P1643	0x2B47	Throttle Actuator Start Test Amplifier Balancing Faulty
P1644	0x28D4	Throttle Valve Adaptation Relearning Lower Mechanical Stop
P1644	0x2B43	Throttle Valve Adaptation Relearning Lower Mechanical Stop
P165A	0x2FD9	EWS Interface to ECM Error
P165A	0x32E3	Electronic immobilizer - DME interface: Hardware defect
P165B	0x2FDB	EWS Interface to ECM Checksum Error
P165C	0x2FDD	EWS Data, No Available Storage Possible
P165C	0x32E7	DME, internal fault, immobilizer data: No available memory
P165D	0x2FDE	EWS Data, Faulty Release Code
P165D	0x32E9	DME, internal fault, immobilizer data: Checksum fault
P165E	0x2FDF	EWS Data Checksum Error
P1660	0x2FDA	EWS Telegram Error
P1660	0x32E4	Electronic immobilizer - DME interface: Frame fault
P1661	0x2FDC	Time-out EWS Telegram
P1661	0x32E5	Electronic immobilizer - DME interface: Timeout
P1665	0x2FD7	EWS Rolling Code Tampering
P1667	0x2FD6	EWS Start Value not yet Programmed
P1667	0x32E1	Electronic immobilizer anti-tampering protection: No start value programmed
P1668	0x2FE0	EWS Start Value Destroyed
P167E	0x2BCD	Internal Control Module Error, 5 Volt Supply Voltage High
P167E	0x36B9	DME, internal fault, monitoring 5V supply: Overvoltage detected
P167F	0x2BCE	Internal Control Module Error, 5 Volt Supply Voltage Low
P167F	0x36BA	DME, internal fault, monitoring 5V supply: Undervoltage detected
P1680	0x2BDF	Throttle Valve Control Monitor Level 2/3 ADC Processor fault
P1681	0x2BDA	Throttle Valve Control Monitor Level 2/3 Engine Speed Error
P1686	0x2BE8	Throttle Valve Control Monitor Level 2/3 Pedal Position Sensor Error
P16A5	0x2A70	Time-out Control Module Multiple Output Stage SPI-Bus
P16A5	0x2A72	Time-out Control Module Multiple Output Stage SPI-Bus
P16A5	0x2A74	Time-out Control Module Multiple Output Stage SPI-Bus
P16A5	0x2D70	Time-out Control Module Multiple Output Stage SPI-Bus
P16A5	0x2D71	Time-out Control Module Multiple Output Stage SPI-Bus
P16A5	0x2D72	Time-out Control Module Multiple Output Stage SPI-Bus
P16BC	0x28D0	Throttle valve, adaptation: Initial adaptation, lower travel stop not learned

Table a. Diagnostic trouble codes (DTCs) (continued)

P-code	MINI FC	Definition
P16CF	0x32E2	Electronic immobilizer anti-tampering protection: Expected answer is implausible
P16E6	0x28CD	Throttle valve, adaptation: Compliance with prescribed environmental conditions not achieved; battery voltage too low
P16E7	0x36E2	DME, internal fault, monitoring 5V sensor supply: Voltage outside valid range
P16E8	0x36E3	DME, internal fault, monitoring of 5V sensor power supply 2: Voltage outside valid range
P16E9	0x36E4	DME, internal fault, monitoring of 5V sensor power supply 3: Voltage outside valid range
P16EA	0x36BB	DME, internal fault, watchdog output: Malfunction
P16EB	0x36BC	DME, internal fault, watchdog output: faulty question/answer communication
P16EC	0x36BD	DME, internal fault, watchdog output: Overvoltage detection
P1721	0x3096	CAN Engine Speed
P1727	0xCF2B	Engine Speed Signal Plausibility
P173B	0xCF2D	CAN Engine Coolant Temperature Invalid
P1772	0x4F81	Turbine Speed and Engine Speed in Transmission Range D Correlation
P1774	0xCF30	CAN Wheel Speed Rear Wheel Left No Signal
P1775	0xCF31	CAN Wheel Speed Rear Wheel Right No Signal
P1776	0xCF32	CAN Wheel Speed Front Wheel Left No Signal
P1777	0xCF33	CAN Wheel Speed Front Wheel Right No Signal
P1782	0xCF2C	CAN Brake Signal
P1791	0x4FB1	Internal Transmission Control Module Memory Check Sum/EEPROM Error
P1794	2DD3	TCM Checksum Error
P2065	0x27C4	Fuel Level Sensor 'B' Circuit
P2067	0x27C5	Fuel Level Sensor 'B' Circuit Low
P2067	0x3184	Tank fuel-level sensor, right, signal: Short circuit to ground
P2068	0x27C4	Fuel Level Sensor 'B' Circuit High
P2088	0x2878	'A' Camshaft Position Actuator Control Circuit Low (Bank 1)
P2088	0x2D52	VANOS intake-side solenoid valve, control-activation: Short circuit to ground
P2089	0x2877	'A' Camshaft Position Actuator Control Circuit High (Bank 1)
P2089	0x2D51	VANOS intake-side solenoid valve, control-activation: Short circuit to positive
P2090	0x2841	'B' Camshaft Position Actuator Control Circuit Low (Bank 1)
P2090	0x2D9C	VANOS exhaust-side solenoid valve, control-activation Short circuit to ground
P2091	0x2840	'B' Camshaft Position Actuator Control Circuit High (Bank 1)
P2091	0x2D9B	VANOS exhaust-side solenoid valve, control-activation Short circuit to positive
P2096	0x2AC2	Post Catalyst Fuel Trim System Too Lean (Bank 1)
P2096	0x2BCB	Pre-catalyst oxygen sensor, mixture fine-tuning: Exhaust gas behind catalytic converter too lean
P2096	0x2D34	Before-catalyst oxygen sensor, system check: Signal stuck at rich
P2097	0x2AC1	Post Catalyst Fuel Trim System Too Rich (Bank 1)
P2097	0x2BCA	Pre-catalyst oxygen sensor, mixture fine-tuning: Exhaust gas behind catalytic converter too rich

Table a. Diagnostic trouble codes (DTCs) (continued)

P-code	MINI FC	Definition
P2097	0x2D33	Before-catalyst oxygen sensor, system check: Signal stuck on lean
P2100	0x28BB	Throttle Actuator 'A' Control Motor Circuit Open
P2100	0x2B20	Throttle Actuator 'A' Control Motor Circuit Open
P2102	0x2B1E	Throttle Actuator 'A' Control Motor Circuit Low
P2103	0x2B1D	Throttle Actuator 'A' Control Motor Circuit High
P2103	0x2BB8	Throttle Actuator 'A' Control Motor Circuit High
P2118	0x28B9	DME, internal fault, throttle valve control: Overheated or current too high
P2120	0x2B4E	Throttle / Pedal Position Sensor 'D' Circuit
P2122	0x27DA	Throttle Pedal Position Sensor 'D' Circuit Low
P2122	0x2B4A	Throttle Pedal Position Sensor 'D' Circuit Low
P2123	0x27D9	Throttle Pedal Position Sensor 'D' Circuit High
P2123	0x2B49	Throttle Pedal Position Sensor 'D' Circuit High
P2127	0x27DC	Throttle Pedal Position Sensor 'E' Circuit Low
P2127	0x2B4D	Throttle Pedal Position Sensor 'E' Circuit Low
P2128	0x27DB	Throttle Pedal Position Sensor 'E' Circuit High
P2128	0x2B4C	Throttle Pedal Position Sensor 'E' Circuit High
P2138	0x27E8	Throttle / Pedal Position Sensor 'D' 'E' Voltage Correlation
P2138	0x2B4B	Throttle / Pedal Position Sensor 'D' 'E' Voltage Correlation
P213F	0x27D0	Fuel Pump Emergency Cut Off
P2146	0x2DBD	Fuel Injector Group 'A' Supply Voltage Circuit Open
P2147	0x2DBE	Fuel Injector Group 'A' Supply Voltage Circuit Low
P2148	0x2DBF	Fuel Injector Group 'A' Supply Voltage Circuit High
P2177	0x2745	Fuel Trim Limit Exceeded System Too Lean
P2177	0x274D	Fuel Trim Limit Exceeded System Too Lean
P2177	0x2BC0	Fuel Trim Limit Exceeded System Too Lean
P2178	0x2746	Fuel Trim Limit Exceeded System Too Rich
P2178	0x274E	Fuel Trim Limit Exceeded System Too Rich
P2178	0x2BC1	Fuel Trim Limit Exceeded System Too Rich
P2187	0x2785	System Too Lean at Idle (Bank 1)
P2187	0x2789	System Too Lean at Idle (Bank 1)
P2188	0x2786	System Too Rich at Idle (Bank 1)
P2195	0x2AC3	O2 Sensor Signal Stuck Lean (Bank 1 Sensor 1)
P2195	0x2D33	O2 Sensor Signal Stuck Lean (Bank 1 Sensor 1)
P2195	0x2D35	Pre-catalyst oxygen sensor, plausibility: Mixture behind catalytic converter too rich
P2196	0x2AC4	O2 Sensor Signal Stuck Rich (Bank 1 Sensor 1)
P2196	0x2D34	O2 Sensor Signal Stuck Rich (Bank 1 Sensor 1)
P2196	0x2D36	Pre-catalyst oxygen sensor, plausibility: Mixture gas behind catalytic converter too lean

Table a. Diagnostic trouble codes (DTCs) (continued)		
P-code	**MINI FC**	**Definition**
P2228	0x283D	Barometric Pressure Sensor 'A' Circuit Low
P2228	0x2FC6	Barometric Pressure Sensor 'A' Circuit Low
P2229	0x283C	Barometric Pressure Sensor 'A' Circuit High
P2229	0x2FC5	Barometric Pressure Sensor 'A' Circuit High
P2237	0x2A93	O2 Sensor Reference Circuit Continuity (Bank 1 Sensor 1)
P2243	0x2AA3	O2 Sensor Reference Voltage Circuit Open (Bank 1 Sensor 1)
P2243	0x2CFA	O2 Sensor Reference Voltage Circuit Open (Bank 1 Sensor 1)
P2251	0x2AAB	O2 Sensor Reference Circuit Continuity (Bank 1 Sensor 1)
P2251	0x2D27	Oxygen sensor before catalytic converter, electrical: Open circuit, virtual ground
P2270	0x2A74	O2 Sensor Signal Stuck Lean (Bank 1 Sensor 2)
P2270	0x2D1B	O2 Sensor Signal Stuck Lean (Bank 1 Sensor 2)
P2271	0x2A71	O2 Sensor Signal Stuck Rich (Bank 1 Sensor 2)
P2271	0x2A75	O2 Sensor Signal Stuck Rich (Bank 1 Sensor 2)
P2271	0x2D1C	O2 Sensor Signal Stuck Rich (Bank 1 Sensor 2)
P2400	0x272E	DMTL Activation Control Circuit Open
P2400	0x3227	DMTL leak diagnosis pump, control-activation: Open circuit
P2401	0x272C	DMTL Activation Control Circuit Low
P2401	0x3226	DMTL leak diagnosis pump, control-activation: Short circuit to ground
P2402	0x272B	DMTL Activation Control Circuit High
P2402	0x3225	DMTL leak diagnosis pump, control-activation: Short circuit to positive
P240A	0x271B	DMTL Pump Heater Circuit Open
P240A	0x3224	DMTL, heater, activation: Open circuit
P240B	0x2719	DMTL Pump Heater Circuit Low
P240B	0x3223	DMTL, heater, activation: Short circuit to ground
P240C	0x2718	DMTL Pump Heater Circuit High
P240C	0x3222	DMTL, heater, activation: Short circuit to positive
P2414	0x2ACB	O2 Sensor Oxygen Sample Error (Bank 1 Sensor 1)
P2414	0x2CF8	Oxygen sensor before catalytic converter, installation: Sensor is not installed correctly in exhaust system
P2418	0x272A	DMTL Solenoid Control Circuit Open
P2418	0x321B	DMTL solenoid valve, control-activation: Open circuit
P2419	0x2728	DMTL Solenoid Control Circuit Low
P2419	0x321A	DMTL solenoid valve, control-activation: Short circuit to ground
P2420	0x2727	Evaporative Emission System Switching Valve Control Circuit High
P2420	0x3219	DMTL solenoid valve, control-activation: Short circuit to positive
P2421	0x315A	EVAP evaporative emissions valve: Sticks in closed position
P2422	0x3159	EVAP evaporative emissions valve: jammed open
P2510	0x2C9E	ECM/PCM Power Relay Sense Circuit Range/Performance

Table a. Diagnostic trouble codes (DTCs) (continued)

P-code	MINI FC	Definition
P2600	0x2F3A	Coolant Pump ‘A’ Control Circuit Open
P2602	0x2F39	Coolant Pump ‘A’ Control Circuit Low
P2603	0x2F38	Coolant Pump ‘A’ Control Circuit High
P2610	0x3391	Engine shutdown time: Signal implausible
P2626	0x2A8B	O2 Sensor Pumping Current Circuit Open (Bank 1 Sensor 1)
P2637	0xCF16	Torque Management Feedback Signal ‘A’
P2719	0x4E40	Pressure Control Solenoid ‘D’ Control Circuit Range/Performance
P2720	0x4E3F	Pressure Control Solenoid ‘D’ Control Circuit Low
P2721	0x4E3E	Pressure Control Solenoid ‘D’ Control Circuit High
P2728	0x4E4A	Pressure Control Solenoid ‘E’ Control Circuit Range/Performance
P2729	0x4E49	Pressure Control Solenoid ‘E’ Control Circuit Low
P2730	0x4E48	Pressure Control Solenoid ‘E’ Control Circuit High
P2737	0x4E54	Pressure Control Solenoid ‘F’ Control Circuit Range/Performance
P2738	0x4E53	Pressure Control Solenoid ‘F’ Control Circuit Low
P2739	0x4E52	Pressure Control Solenoid ‘F’ Control Circuit High
P2762	0x4E2C	Torque Converter Clutch Pressure Control Solenoid Control Circuit Range/Performance
P2763	0x4E2A	Torque Converter Clutch Pressure Control Solenoid Control Circuit High
P2764	0X4E2B	Torque Converter Clutch Pressure Control Solenoid Control Circuit Low
P3007	0x287F	Fuel Rail Pressure Pressure-Rate-Controlled, Maximum Pressure Exceeded
P3012	0x2A6C	Adaptation value Too High (Bank 1 Sensor 1)
P3012	0x2D03	Adaptation value Too High (Bank 1 Sensor 1)
P3014	0x2A6D	O2 Sensor IC Supply Voltage Too Low (Bank 1 Sensor 1)
P3014	0x2D04	O2 Sensor IC Supply Voltage Too Low (Bank 1 Sensor 1)
P3016	0x2A67	O2 Sensor Ceramic Internal Resistance Adjusted Value Plausibility (Bank 1 Sensor 1)
P3016	0x2D19	O2 Sensor Ceramic Internal Resistance Adjusted Value Plausibility (Bank 1 Sensor 1)
P3018	0x2A90	O2 Sensor Lambda Controller Value Above Threshold Due to Open Pumping Circuit (Bank 1 Sensor 1)
P3018	0x2D23	Oxygen sensor before catalytic converter, pumping current wire: Oxygen sensor control value above threshold due to open pump wire
P301A	0x2DA5	Injector 1 Stuck Open
P301B	0x2DAE	Injector 2 Stuck Open
P301C	0x2DA8	Injector 3 Stuck Open
P301D	0x2DAB	Injector 4 Stuck Open
P301E	0x2DB1	Injector 1 or 3 Stuck Open
P301F	0x2DB4	Injector 2 or 4 Stuck Open
P3020	0x2A92	O2 Sensor Signal Too Low During Coast Down Fuel Cut-OFF (Bank 1 Sensor 1)
P3022	0x2A6F	O2 Sensor Disturbed SPI WRAF-IC (Bank 1 Sensor 1)
P3022	0x2D06	DME, internal fault, oxygen sensor before catalytic converter: Communication fault

Table a. Diagnostic trouble codes (DTCs) (continued)		
P-code	**MINI FC**	**Definition**
P3024	0x2A6E	O2 Sensor Initialization Error WRAF-IC (Bank 1 Sensor 1)
P3024	0x2D05	DME, internal fault, oxygen sensor before catalytic converter: Initialization fault
P3026	0x2A64	O2 Sensor Operating Temperature Not Reached (Bank 1 Sensor 1)
P3026	0x2D17	O2 Sensor Operating Temperature Not Reached (Bank 1 Sensor 1)
P302A	0x2CDD	High-pressure fuel system, fuel pressure: Maximum pressure exceeded
P302C	0x2C01	High-pressure fuel system, fuel pressure: Pressure falls below minimum threshold
P306A	2X2BDB	Fuel system: Pressure too high, emergency running mode with low pressure
P306C	2X2BDD	Fuel system: Pressure briefly too high, rpm and load factor limited
P306F	2X2BE9	Fuel pressure: Pressure drops below minimum, injection deactivation to protect catalyst
P3091	0x2880	Fuel Rail Pressure Pressure-Rate-Controlled, Fallen Below Minimum Pressure
P3100	0x2A4C	Cylinder 1 High Pressure Injector Low Side Circuit Open
P3100	0x2D7F	Cylinder 1 High Pressure Injector Low Side Circuit Open
P3101	0x2A01	Cylinder 1 High Pressure Injector Low Side Circuit Low
P3101	0x2D77	Cylinder 1 High Pressure Injector Low Side Circuit Low
P3102	0x29FF	Cylinder 1 High Pressure Injector Low Side Circuit High
P3102	0x2D73	Cylinder 1 High Pressure Injector Low Side Circuit High
P3103	0x2A40	Cylinder 1 High Pressure Injector Low Side Booster Time Error
P3103	0x2D7B	Cylinder 1 High Pressure Injector Low Side Booster Time Error
P3104	0x2A4D	Cylinder 2 High Pressure Injector Low Side Circuit Open
P3104	0x2D82	Cylinder 2 High Pressure Injector Low Side Circuit Open
P3105	0x2A05	Cylinder 2 High Pressure Injector Low Side Circuit Low
P3105	0x2D7A	Cylinder 2 High Pressure Injector Low Side Circuit Low
P3106	0x2A03	Cylinder 2 High Pressure Injector Low Side Circuit high
P3106	0x2D76	Cylinder 2 High Pressure Injector Low Side Circuit high
P3107	0x2A41	Cylinder 2 High Pressure Injector Low Side Booster Time Error
P3107	0x2D7E	Cylinder 2 High Pressure Injector Low Side Booster Time Error
P3108	0x2A43	Cylinder 3 High Pressure Injector Low Side Circuit Open
P3108	0x2D80	Cylinder 3 High Pressure Injector Low Side Circuit Open
P3109	0x2A09	Cylinder 3 High Pressure Injector Low Side Circuit Low
P3109	0x2D78	Cylinder 3 High Pressure Injector Low Side Circuit Low
P3110	0x2A07	Cylinder 3 High Pressure Injector Low Side Circuit High
P3110	0x2D74	Cylinder 3 High Pressure Injector Low Side Circuit High
P3111	0x2A42	Cylinder 3 High Pressure Injector Low Side Booster Time Error
P3111	0x2D7C	Cylinder 3 High Pressure Injector Low Side Booster Time Error
P3112	0x2A4F	Cylinder 4 High Pressure Injector Low Side Circuit Open
P3112	0x2D81	Cylinder 4 High Pressure Injector Low Side Circuit Open
P3113	0x2A0D	Cylinder 4 High Pressure Injector Low Side Circuit Low

Table a. Diagnostic trouble codes (DTCs) (continued)

P-code	MINI FC	Definition
P3113	0x2D79	Cylinder 4 High Pressure Injector Low Side Circuit Low
P3114	0x2D75	Cylinder 4 High Pressure Injector Low Side Circuit High
P3115	0x2A43	Cylinder 5 High Pressure Injector Low Side Booster Time Error
P3115	0x2D7D	Cylinder 5 High Pressure Injector Low Side Booster Time Error
P3148	0x2A30	Cylinder 1 High Pressure Injector High Side Shorted to Coil
P3148	0x2D6C	Cylinder 1 High Pressure Injector High Side Shorted to Coil
P3149	0x29FE	Cylinder 1 High Pressure Injector High Side Circuit Low
P3149	0x2D68	Cylinder 1 High Pressure Injector High Side Circuit Low
P3150	0x2A00	Cylinder 1 High Pressure Injector High Side Circuit High
P3150	0x2D64	Cylinder 1 High Pressure Injector High Side Circuit High
P3151	0x2A31	Cylinder 2 High Pressure Injector High Side Shorted to Coil
P3151	0x2D6F	Cylinder 2 High Pressure Injector High Side Shorted to Coil
P3152	0x2A02	Cylinder 2 High Pressure Injector High Side Circuit Low
P3152	0x2D6B	Cylinder 2 High Pressure Injector High Side Circuit Low
P3153	0x2A04	Cylinder 2 High Pressure Injector High Side Circuit High
P3153	0x2D67	Cylinder 2 High Pressure Injector High Side Circuit High
P3154	0x2A32	Cylinder 3 High Pressure Injector High Side Shorted to Coil
P3154	0x2D6D	Cylinder 3 High Pressure Injector High Side Shorted to Coil
P3155	0x2A06	Cylinder 3 High Pressure Injector High Side Circuit Low
P3155	0x2D69	Cylinder 3 High Pressure Injector High Side Circuit Low
P3156	0x2A08	Cylinder 3 High Pressure Injector High Side Circuit High
P3156	0x2D65	Cylinder 3 High Pressure Injector High Side Circuit High
P3157	0x2A33	Cylinder 4 High Pressure Injector High Side Shorted to Coil
P3157	0x2D6E	Cylinder 4 High Pressure Injector High Side Shorted to Coil
P3158	0x2A0A	Cylinder 4 High Pressure Injector High Side Circuit Low
P3158	0x2D6A	Cylinder 4 High Pressure Injector High Side Circuit Low
P3159	0x2A0C	Cylinder 4 High Pressure Injector High Side Circuit High
P3159	0x2D66	Cylinder 4 High Pressure Injector High Side Circuit High
P316B	0x2947	Engine Coolant Temperature Signal Stuck low
P3198	0x2943	Engine Coolant Temperature Gradient Too High
P3199	0x2948	Engine Coolant Temperature Signal Stuck
P3202	0x3091	Powertrain CAN Chip Cut-Off
P3202	0xCD87	PT CAN communications fault: CAN bus off or CAN bus defective
P3209	0x3094	CAN Message Monitoring ASC / DSC Alive Check
P3210	0x3095	CAN Monitoring ASC / DSC Plausibility
P3214	0x3098	CAN Message Monitoring ETC Plausibility
P3215	0x309D	CAN Message Monitoring IHKA No Signal

Table a. Diagnostic trouble codes (DTCs) (continued)

P-code	MINI FC	Definition
P3217	0x30A0	CAN Message Instrument Pack Plausibility
P3219	0x30A4	CAN Message SZL Alive Check
P321E	0x2FCD	Ambient Pressure Sensor Maximum Pressure Implausible
P321F	0x2FCE	Ambient Pressure Sensor Minimum Pressure Implausible
P3220	0x30A6	CAN Message SZL No Signal
P3221	0x30A5	CAN Message Monitoring SZL Plausibility
P3223	0x2E32	Generator Mechanical
P3223	0x2E34	Generator Mechanical
P3223	0x3850	Alternator, mechanical: Malfunction
P3225	0x2E31	Generator Communication Error
P322A	0x2FD0	Ambient Pressure Sensor Continuity
P3232	0x2BDE	Control Module Monitoring Ignition Timing Plausibility
P3233	0x2BDD	Control Module Monitoring Relative Charge Plausibility
P3235	0x236CC	DME, internal fault Version encoding monitor
P3235	0x2BE2	Control Module Monitoring Version Coding Plausibility
P3236	0x2BE1	Control Module Monitoring Injection Time Relative Fuel Quantity Plausibility
P3237	0x2BE0	Control Module Monitoring Fuel Correction Error
P3237	0x36C5	DME, internal fault: Mixture correction factor plausibilizing monitor
P3238	0x2BE3	Control Module Monitoring TPU Chip Defective
P3239	0x303E	Control Module - No Coding
P323C	0x2FCF	Ambient Pressure Sensor Continuity Error
P323D	0x2B7A	Control Module Monitoring Air Mass Flow Balancing, Control Range Monitoring
P323E	0x2BDB	Control Module Monitoring Fuel Pressure Sensor
P324A	0x3858	Alternator: Incorrect type
P324C	0x2E28	Generator Over Temperature Calculated
P324C	0x384C	Alternator, plausibility, temperature: Overtemperature calculated
P324E	0x3854	Alternator, regulator: Incorrect type
P325A	0x2E20	Generator Electrical Error Calculated
P325A	0x3844	Alternator, plausibility, electrical: Calculated
P325C	0x36C4	DME, internal fault, electric accelerator pedal monitor: RPM sensor
P325D	0x36C6	DME, internal fault Level 1 injection quantity limitation monitor
P325D	0x36C7	DME, internal fault Level 2 injection quantity limitation monitor
P325E	0x36CD	DME, internal fault, electric accelerator pedal monitor: Ignition-timing monitoring
P325F	0x36C9	DME, internal fault Relative fuel mass plausibilizing monitor
P325F	0x36CF	DME, internal fault Fuel mass plausibilizing monitor
P326A	0x36C1	DME, internal fault, electric accelerator pedal monitor: AD converter test voltage check
P326B	0x36C2	DME, internal fault, electric accelerator pedal monitor: Airflow calibration

Table a. Diagnostic trouble codes (DTCs) (continued)		
P-code	**MINI FC**	**Definition**
P326C	0x36CB	DME, internal fault, electric accelerator pedal monitor: Implausible drivetrain conversion ratio
P3337	0x2BDC	Function MOnitoring Lambda Plausibility
P3337	0x36C8	DME, internal fault: Monitoring of specified lambda value
U0001	0xCF07	High Speed CAN Communication Bus
U0101	0x3099	Lost Communication With TCM
U0121	0xCF15	Lost Communication With Anti-Lock Brake System (ABS) Control Module
U0167	0x2FE4	Lost Communication With Vehicle Immobilizer Control Module
U110F	0x3BDD	PT CAN, message (EGS torque request, 0xB5): omitted
U1115	0x30B0	Lost Communication With Vehicle Mode Status
U1116	0x30AF	Message Monitoring Vehicle Mode Status Check Sum Error
U1118	0x3BD7	PT CAN, message (speed of vehicle, 0x1A0): omitted
U111A	0x3BE1	PT CAN, message (transmission data, 0xBA): Missing
U111E	0x30A8	Lost Communication With Terminal Status
U111F	0x30A7	Message Monitoring Terminal Status Check Sum Error
U1126	0x3BD5	PT CAN, message (status of DSC, 0x19E): Missing
U1129	0x30B2	Lost Communication With Reverse Status
U112A	0x309F	Message Monitoring Instrument Pack Status Alive Check
U112B	0x30A1	Lost Communication With Instrument Pack Status
U112D	0x30A9	Lost Communication With Control Crash Cut-Off EKP
U112E	0xCF18	Lost Communication With Pedal Position Sensor
U1132	0x2E1C	No BSD message from generator
U1132	0x2E67	No BSD message from generator
U1134	0x30AB	Lost Communication With Lamp Status
U113C	0x30AD	Lost Communication With Time/Date
U114A	0xCF12	Message Monitoring Engine Control Torque 1 Alive Check
U114B	0xCF13	Lost Communication With Engine Control Torque 1
U114C	0xCF14	Message Monitoring Engine Control Torque 1 Check Sum Error
U115A	0xCF39	Message Monitoring Engine Control Torque 3 Alive Check
U115B	0xCF3A	Lost Communication With Engine Control Torque 3
U115C	0xCF3B	Message Monitoring Engine Control Torque 3 Check Sum Error
U115E	0xCF43	Lost Communication With Engine Data
U1166	0x2FE2	Message Monitoring EWS (Electronic Immobilizer) - Frame Error
U1166	0x2FE2	Message Monitoring EWS - Frame Error
U1169	0x30BA	Lost Communication With OBD-Sensor
U116D	0x3BD2	PT CAN, message (wheel speed, 0x0CE): Missing
U116F	0x30BB	Lost Communication With Sports Mode ETC
U1173	0x3BE0	PT CAN, message (transmission data, 0xBA): Checksum fault/Alive check

Table a. Diagnostic trouble codes (DTCs) (continued)		
P-code	**MINI FC**	**Definition**
U11C9	0x3BDC	PT CAN, message (EGS torque request, 0xB5): Checksum fault/Alive check
U11CA	0x3BD6	PT CAN, message (speed of vehicle, 0x1A0): Checksum fault/Alive check
U1202	0x3097	Message Monitoring TCM Alive Check

WARNING

Your common sense, good judgement and general alertness are crucial to safe and successful service work. Before attempting any work on your MINI, be sure to read **001 Warnings and Cautions** *and the copyright page at the front of the manual. Review these warnings and cautions each time you prepare to work on your car. Please also read any warnings and cautions that accompany the procedures in the manual.*

A

B

> ***WARNING***
>
> *Your common sense, good judgement and general alertness are crucial to safe and successful service work. Before attempting any work on your MINI, be sure to read* **001 Warnings and Cautions** *and the copyright page at the front of the manual. Review these warnings and cautions each time you prepare to work on your car. Please also read any warnings and cautions that accompany the procedures in the manual.*

> ***WARNING***
>
> *Your common sense, good judgement and general alertness are crucial to safe and successful service work. Before attempting any work on your MINI, be sure to read* **001 Warnings and Cautions** *and the copyright page at the front of the manual. Review these warnings and cautions each time you prepare to work on your car. Please also read any warnings and cautions that accompany the procedures in the manual.*

WARNING

Your common sense, good judgement and general alertness are crucial to safe and successful service work. Before attempting any work on your MINI, be sure to read **001 Warnings and Cautions** *and the copyright page at the front of the manual. Review these warnings and cautions each time you prepare to work on your car. Please also read any warnings and cautions that accompany the procedures in the manual.*

> ***WARNING***
>
> *Your common sense, good judgement and general alertness are crucial to safe and successful service work. Before attempting any work on your MINI, be sure to read* **001 Warnings and Cautions** *and the copyright page at the front of the manual. Review these warnings and cautions each time you prepare to work on your car. Please also read any warnings and cautions that accompany the procedures in the manual.*

T

U

V

Selected Books and Repair Information From Bentley Publishers

Motorsports

Alex Zanardi: My Sweetest Victory
Alex Zanardi and Gianluca Gasparini
ISBN 978-0-8376-1249-2

The Unfair Advantage *Mark Donohue and Paul van Valkenburgh*
ISBN 978-0-8376-0069-7

Driving Forces: The Grand Prix Racing World Caught in the Malestrom of the Third Reich
Peter Stevenson
ISBN 978-0-8376-0217-2

Equations of Motion – Adventure, Risk and Innovation
William F. Milliken
ISBN 978-0-8376-1570-7

The Racing Driver – The Theory and Practice of Fast Driving
Denis Jenkinson
ISBN 978-0-8376-0201-1

Engineering

The Hack Mechanic Guide to European Automotive Electrical Systems *Rob Siegel*
ISBN 978-0-8376-1751-0

Physics for Gearheads *Randy Beikmann*
ISBN 978-0-8376-1615-5

Bosch Automotive Handbook
Updated 9th Edition *Robert Bosch GmbH*
ISBN 978-0-8376-1732-9

Bosch Fuel Injection and Engine Management *Charles O. Probst, SAE*
ISBN 978-0-8376-0300-1

Maximum Boost: Designing, Testing, and Installing Turbocharger Systems
Corky Bell ISBN 978-0-8376-0160-1

Supercharged! Design, Testing and Installation of Supercharger Systems
Corky Bell ISBN 978-0-8376-0168-7

Race Car Aerodynamics *Joseph Katz*
ISBN 978-0-8376-0142-7

Audi Repair Manuals

Audi A4 Service Manual: 2002-2008, 1.8L Turbo, 2.0L Turbo, 3.0L, 3.2L
Bentley Publishers
ISBN 978-0-8376-1574-5

Audi A4 Service Manual: 1996-2001, 1.8L Turbo, 2.8L *Bentley Publishers*
ISBN 978-0-8376-1675-9

Audi TT Service Manual: 2000-2006, 1.8L turbo, 3.2 L *Bentley Publishers*
ISBN 978-0-8376-1625-4

Audi A6 (C5 platform) Service Manual: 1998-2004 *Bentley Publishers*
ISBN 978-0-8376-1670-4

BMW

BMW 3 Series (E90, E91, E92, E93) Service Manual: 2006-2011 *Bentley Publishers*
ISBN 978-0-8376-1723-7

BMW 3 Series (E36) Service Manual: 1992-1998 *Bentley Publishers*
ISBN 978-0-8376-1709-1

BMW X5 (E53) Service Manual: 2000-2006 *Bentley Publishers*
ISBN 978-0-8376-1643-8

BMW 5 Series (E39) Service Manual: 1997-2003 *Bentley Publishers*
ISBN 978-0-8376-1672-8

BMW Z3 Service Manual: 1996-2002
Bentley Publishers
ISBN 978-0-8376-1617-9

BMW 7 Series (E32) Service Manual: 1988-1994 *Bentley Publishers*
ISBN 978-0-8376-1619-3

BMW 7 Series (E38) Service Manual: 1995-2001 *Bentley Publishers*
ISBN 978-0-8376-1618-6

Memoirs of a Hack Mechanic
Rob Siegel ISBN 978-0-8376-1720-6

Porsche

Porsche 911 (Type 996) Service Manual: 1999-2005 *Bentley Publishers*
ISBN 978-0-8376-1710-7

Porsche 911 Carrera Service Manual: 1984-1989 *Bentley Publishers*
ISBN 978-0-8376-1696-4

Porsche 911 SC Service Manual: 1978-1983 *Bentley Publishers*
ISBN 978-0-8376-1705-3

Porsche 911 (964): Enthusiast's Companion *Bentley Publishers*
ISBN 978-0-8376-0293-6

Porsche Boxster. Boxster S (986) Service Manual: 1997-2004
Bentley Publishers
ISBN 978-0-8376-1645-2

Porsche: Excellence Was Expected
Karl Ludvigsen ISBN 978-0-8376-0235-6

Ferdinand Porsche – Genesis of Genius
Karl Ludvigsen ISBN 978-0-8376-1557-8

Porsche – Origin of the Species
Karl Ludvigsen ISBN 978-0-8376-1331-4

Volkswagen

Volkswagen Jetta, Golf, GTI Service Manual: 1999-2005 *Bentley Publishers*
ISBN 978-0-8376-1678-0

Volkswagen Jetta Service Manual: 2005-2010 *Bentley Publishers*
ISBN 978-0-8376-1616-2

Volkswagen Passat Service Manual: 1998-2005 *Bentley Publishers*
ISBN 978-0-8376-1669-8

Volkswagen Jetta, Golf, GTI: 1993-1999, Cabrio: 1995-2002 Service Manual *Bentley Publishers*
ISBN 978-0-8376-1660-5

Volkswagen GTI, Golf, Jetta Service Manual: 1985-1992 *Bentley Publishers*
ISBN 978-0-8376-1637-7

MINI Repair Manuals

MINI Cooper Service Manual: 2002-2006 *Bentley Publishers*
ISBN 978-0-8376-1639-1

MINI Cooper Service Manual: 2007-2013 *Bentley Publishers*
ISBN 978-0-8376-1730-5

Featured Repair Manuals

BMW 3 Series (F30, F31, F34) Service Manual: 2012-2015
Bentley Publishers
ISBN 978-0-8376-1752-7

BMW X3 (E83) Service Manual: 2004-2010
Bentley Publishers
ISBN 978-0-8376-1731-2

BMW 5 Series (E60, E61) Service Manual: 2004-2010
Bentley Publishers
ISBN 978-0-8376-1689-6

BMW 3 Series (E46) Service Manual: 1999-2005
Bentley Publishers
ISBN 978-0-8376-1657-5

Bentley Publishers®
Automotive Reference

Bentley Publishers has published service manuals and automobile books since 1950. For more information, please contact Bentley Publishers at 1734 Massachusetts Avenue, Cambridge, MA 02138 USA, or visit our web site at
`BentleyPublishers.com`